CARBON NANOMATERIALS SOURCEBOOK

Graphene, Fullerenes, Nanotubes, and Nanodiamonds

CARBON NANOMATERIALS SOURCEBOOK

VOLUME I
Graphene, Fullerenes, Nanotubes, and Nanodiamonds

Edited by

Klaus D. Sattler

University of Hawaii at Manoa, Honolulu, USA

CRC Press is an imprint of the
Taylor & Francis Group, an **informa** business

CRC Press
Taylor & Francis Group
6000 Broken Sound Parkway NW, Suite 300
Boca Raton, FL 33487-2742

First issued in paperback 2021

Version Date: 20160204

ISBN 13: 978-1-4822-5272-9 (set) (vol 1)
ISBN 13: 978-0-367-78308-2 (pbk)
ISBN 13: 978-1-4822-5268-2 (hbk)

Library of Congress Cataloging-in-Publication Data

Names: Sattler, Klaus D.
Title: Carbon nanomaterials sourcebook. Graphene, fullerenes, nanotubes, and nanodiamonds / [edited by] Klaus D. Sattler.
Description: Boca Raton : Taylor & Francis Group, 2016. | "A CRC title." | Includes bibliographical references and index.
Identifiers: LCCN 2016002144 | ISBN 9781482252682 (alk. paper)
Subjects: LCSH: Nanostructured materials. | Graphene. | Nanotubes. | Nanodiamonds.
Classification: LCC TA418.9.N35 C34224 2016 | DDC 620.1/15--dc23
LC record available at http://lccn.loc.gov/2016002144

Visit the Taylor & Francis Web site at
http://www.taylorandfrancis.com

and the CRC Press Web site at
http://www.crcpress.com

Contents

Preface

In the last three decades, zero-dimensional, one-dimensional, and two-dimensional carbon nanomaterials have attracted significant attention because of their unique electronic, optical, thermal, mechanical, and chemical properties. There is a need for understanding the science of carbon nanomaterials, the production methods, and the applications, to use these exceptional properties. In particular, in the last two decades, carbon nanotubes, fullerenes, nanodiamonds, and mesoporous carbon nanostructures became a new class of nanomaterials. Research on graphene, a single sheet of graphite, is one of the fastest growing fields today and holds the promise of someday replacing silicon in computers and electronic devices. Many scientists and engineers are redirecting their work toward carbon nanomaterials. It is in this context that the editor of this work determined that a new resource concentrating on this subject would be a unique and useful source of information and learning.

Carbon Nanomaterials Sourcebook is the most comprehensive reference that covers the field of carbon nanomaterials, reflecting its interdisciplinary nature that brings together physics, chemistry, materials science, molecular biology, engineering, and medicine. The two volumes describe fundamental properties, growth mechanisms, and processing of nanocarbons as well as their functionalization for electronic device, energy conversion and storage, and biomedical and environmental applications. It encompasses a wide range of areas from science to engineering. Moreover, in addition to addressing the latest advances, the *Sourcebook* presents core knowledge with basic mathematical equations, tables, and graphs to provide the reader with the tools necessary to understand current and future technology developments.

The contents are made up of 54 total chapters organized into nine subject areas, with each chapter covering one type of carbon nanomaterial. Materials have been selected to showcase exceptional properties, good synthesis and large-scale production methods, and strong current and future application prospects. Every chapter covers the three main areas: formation, properties, and applications. This setup makes the book a unique source where a reader can easily navigate to find the information about a particular material. The chapters will be written in tutorial style, where basic equations and fundamentals are included in an extended introduction.

Editor

Klaus D. Sattler pursued his undergraduate and master's courses at the University of Karlsruhe in Germany. He received his PhD under the guidance of Professors G. Busch and H.C. Siegmann at the Swiss Federal Institute of Technology (ETH) in Zurich, where he was among the first to study spin-polarized photoelectron emission. In 1976, he began a group for atomic cluster research at the University of Konstanz in Germany, where he built the first source for atomic clusters and led his team to pioneering discoveries such as "magic numbers" and "Coulomb explosion." He was at the University of California, Berkeley, for three years as a Heisenberg fellow, where he initiated the first studies of atomic clusters on surfaces with a scanning tunneling microscope.

Dr. Sattler accepted a position as professor of physics at the University of Hawaii, Honolulu, in 1988. There, he initiated a research group for nanophysics, which, using scanning probe microscopy, obtained the first atomic-scale images of carbon nanotubes directly confirming the graphene network. In 1994, his group produced the first carbon nanocones. He has also studied the formation of polycyclic aromatic hydrocarbons and nanoparticles in hydrocarbon flames in collaboration with ETH Zurich. Other research has involved the nanopatterning of nanoparticle films, charge density waves on rotated graphene sheets, bandgap studies of quantum dots, and graphene folds. His current work focuses on novel nanomaterials and solar photocatalysis with nanoparticles for the purification of water.

He is the editor of the seven-volume *Handbook of Nanophysics* (CRC Press, 2011) and *Fundamentals of Picoscience* (CRC Press, 2014). Among his many other accomplishments, Dr. Sattler was awarded the prestigious Walter Schottky Prize from the German Physical Society in 1983. At the University of Hawaii, he teaches courses in general physics, solid state physics, and quantum mechanics.

Contributors

Jandro L. Abot
Department of Mechanical Engineering
The Catholic University of America
Washington, District of Columbia

Mashkoor Ahmad
Nanomaterials Research Group (NRG)
Physics Division
PINSTECH
Nilore, Islamabad, Pakistan

Saniya Alwani
Drug Discovery and Development Research Group
College of Pharmacy and Nutrition
University of Saskatchewan
Saskatoon, Saskatchewan, Canada

Sambandam Anandan
Nanomaterials & Solar Energy Conversion Lab
Department of Chemistry
National Institute of Technology
Trichy, India

Muthupandian Ashokkumar
School of Chemistry
University of Melbourne
Victoria, Australia

Ayousha Ayaz
Nanomaterials Research Group (NRG)
Physics Division
PINSTECH
Nilore, Islamabad, Pakistan

Ildiko Badea
Drug Discovery and Development Research Group
College of Pharmacy and Nutrition
University of Saskatchewan
Saskatoon, Saskatchewan, Canada

Diana M. Brus
Institute of Chemistry
University of Bialystok
Bialystok, Poland

Luca Camilli
Department of Micro and Nanotechnology
Technical University of Denmark
Kongens Lyngby, Denmark

Huan-Cheng Chang
Institute of Atomic and Molecular Sciences
Academia Sinica
Taipei, Taiwan

Stuart J. Corr
Department of Surgery
Baylor College of Medicine
and
Department of Chemistry
and
Richard E. Smalley Institute for Nanoscale Science & Technology
Rice University
Houston, Texas

Steven A. Curley
Department of Surgery
Baylor College of Medicine
and
Department of Mechanical Engineering and Materials Science
Rice University
Houston, Texas

Panagiotis Dallas
Department of Materials
University of Oxford
Oxford, United Kingdom

Pablo A. Denis
Computational Nanotechnology
DETEMA
Facultad de Química
UDELAR
Montevideo, Uruguay

Valerie Dolmatov
Federal State Unitary Enterprise
Special Design and Technology Bureau
Russia

Luis Echegoyen
Department of Chemistry
University of Texas at El Paso
El Paso, Texas

Stephan Engels
JARA-FIT and II
Institute of Physics
RWTH Aachen University
Aachen, Germany

and

Peter Grünberg Institute (PGI-9)
Forschungszentrum Jülich
Jülich, Germany

Alexander Epping
JARA-FIT and II
Institute of Physics
RWTH Aachen University
Aachen, Germany

and

Peter Grünberg Institute (PGI-9)
Forschungszentrum Jülich
Jülich, Germany

Maria H. Fernandes
Laboratory for Bone Metabolism and Regeneration
Faculty of Dental Medicine
University of Porto
Porto, Portugal

Andrey S. Goryunov
Institute of Biology Karelian Research Center RAS
Petrozavodsk, Russia

Wesley Wei-Wen Hsiao
Institute of Atomic and Molecular Sciences
Academia Sinica
Taipei, Taiwan

Feng-Jen Hsieh
Institute of Atomic and Molecular Sciences
Academia Sinica
Taipei, Taiwan

Bolong Huang
Department of Physics and Materials Science
City University of Hong Kong

Muhammad Hussain
Nanomaterials Research Group (NRG)
Physics Division
PINSTECH
Nilore, Islamabad, Pakistan

Irum Khalid
Nanomaterials Research Group (NRG)
Physics Division
PINSTECH
Nilore, Islamabad, Pakistan

Niveen M. Khashab
Smart Hybrid Materials (SHMs)
Advanced Membranes and Porous Materials Center
King Abdullah University of Science and Technology
Thuwal, Kingdom of Saudi Arabia

Hong Koo Kim
Department of Electrical and Computer Engineering
and
Petersen Institute of NanoScience and Engineering
University of Pittsburgh
Pittsburgh, Pennsylvania

Myungji Kim
Department of Electrical and Computer Engineering
and
Petersen Institute of NanoScience and Engineering
University of Pittsburgh
Pittsburgh, Pennsylvania

Jianfu Li
Department of Physics and Materials Science
City University of Hong Kong
Kowloon Tong, Hong Kong

Hsin-Hung Lin
Institute of Atomic and Molecular Sciences
Academia Sinica
Taipei, Taiwan

and

Institute of Engineering in Medicine
University of California, San Diego
La Jolla, California

Maria A. Lopes
CEMUC
Metallurgical and Materials Engineering Department
Faculty of Engineering
University of Porto
Porto, Portugal

Vivian Machado de Menezes
Universidade Federal da Fronteira Sul–UFFS
Laranjeiras do Sul, PR, Brazil

Diogo Mata
CICECO
Materials and Ceramic Engineering Department
University of Aveiro
Aveiro, Portugal

Abha Misra
Department of Instrumentation and Applied Physics
Indian Institute of Science
Bangalore, Karnataka, India

Basem Moosa
Smart Hybrid Materials (SHMs)
Advanced Membranes and Porous Materials Center
King Abdullah University of Science and Technology
Thuwal, Kingdom of Saudi Arabia

Olena Mykhailiv
Institute of Chemistry
University of Bialystok
Bialystok, Poland

Christoph Neumann
JARA-FIT and II
Institute of Physics
RWTH Aachen University
Aachen, Germany

and

Peter Grünberg Institute (PGI-9)
Forschungszentrum Jülich
Jülich, Germany

Argyro T. Papastavrou
Laboratory of Organic Chemistry
Department of Chemistry
University of Athens
Panepistimiopolis, Athens, Greece

Eleftherios K. Pefkianakis
Laboratory of Organic Chemistry
Department of Chemistry
University of Athens
Panepistimiopolis, Athens, Greece

Marta E. Plonska-Brzezinska
Institute of Chemistry
University of Bialystok
Bialystok, Poland

Alexey Popov
Leibniz Institute for Solid State and Materials Research (IFW Dresden)
Dresden, Germany

Kyriakos Porfyrakis
Department of Materials
University of Oxford
Oxford, United Kingdom

Cijo Punnakattu Rajan
Department of Mechanical Engineering
The Catholic University of America
Washington, District of Columbia

Ilija Rašović
Department of Materials
University of Oxford
Oxford, United Kingdom

Gregory Rogers
Department of Materials
University of Oxford
Oxford, United Kingdom

Manolis M. Roubelakis
Laboratory of Organic Chemistry
Department of Chemistry
University of Athens
Panepistimiopolis, Athens, Greece

Sergey P. Rozhkov
Institute of Biology Karelian Research Center RAS
Petrozavodsk, Russia

Natalia N. Rozhkova
Institute of Geology Karelian Research Center RAS
Petrozavodsk, Russia

Maria Sarno
Department of Industrial Engineering
and
NANO_MATES Research Centre
University of Salerno
Salerno, Italy

Adolfo Senatore
Department of Industrial Engineering
and
NANO_MATES Research Centre
University of Salerno
Salerno, Italy

Hidetsugu Shiozawa
Faculty of Physics
University of Vienna
Vienna, Austria

Rui F. Silva
CICECO
Materials and Ceramic Engineering Department
University of Aveiro
Aveiro, Portugal

Christoph Stampfer
JARA-FIT and II
Institute of Physics
RWTH Aachen University
Aachen, Germany

and

Peter Grünberg Institute (PGI-9)
Forschungszentrum Jülich
Jülich, Germany

Christian Volk
JARA-FIT and II
Institute of Physics
RWTH Aachen University
Aachen, Germany

and

Peter Grünberg Institute (PGI-9)
Forschungszentrum Jülich
Jülich, Germany

Georgios C. Vougioukalakis
Laboratory of Organic Chemistry
Department of Chemistry
University of Athens
Panepistimiopolis, Greece

Lon J. Wilson
Department of Chemistry
and
Richard E. Smalley Institute for Nanoscale Science & Technology
Rice University
Houston, Texas

Rui-Qin Zhang
Department of Physics and Materials Science
City University of Hong Kong
Hong Kong SAR, China

I

Graphene

1

Suspended Graphene

Hong Koo Kim

Myungji Kim

1.1 Introduction

Graphene, a one-atom-thick carbon crystal in a honeycomb lattice, possesses many fascinating properties originating from the manifold potential for interactions at electronic, atomic, or molecular levels (Novoselov et al. 2004, 2005; Zhang et al. 2005; Geim & Novoselov 2007; Schedin et al. 2007; Bolotin et al. 2009; Geim 2009; Neto et al. 2009; Kotov et al. 2012). Graphene is remarkably strong for its atomic thinness and conducts heat and electricity with great efficiency. Electron motion in graphene is essentially governed by Dirac's relativistic equation. The charge carriers in graphene behave like relativistic particles with zero rest mass and have an effective "speed of light" of $\sim 10^6$ m s^{-1}. For a perfect graphene sheet free from impurities and disorder, the Fermi level lies at the so-called Dirac point, where the density of electronic states vanishes. Unlike the two-dimensional (2D) electron system in conventional semiconductors, where the charge carriers become immobile at low densities, the carrier mobility in graphene can remain very high, even with vanishing density of states at Dirac point (Du et al. 2008; Morozov et al. 2008).

Every carbon atom of graphene is on the surface and is therefore easily accessible from both sides. This surface-only nature renders graphene prone to interactions with (and disturbances from) surrounding atoms and molecules. Many of the predicted properties arising from the 2D nature of graphene can be significantly altered by perturbations from an underlying substrate or adsorbates (Hwang et al. 2007; Adam & Sarma 2008; Chen et al. 2008). Substrates, for example, introduce scattering centers, dopants, and corrugations that can obscure graphene's intrinsic properties. Suspended graphene structures have drawn increasing attention, driven by the motivations of exploring intrinsic/ultimate properties of graphene and also of exploring new applications that would not be possible with substrate-bound graphene, such as nanoelectromechanical resonators or transelectrode membranes (Bunch et al. 2008; Garaj et al. 2010; Hu et al. 2014).

From the perspective of thermodynamic stability, perfect crystals cannot exist in 2D space (Mermin 1968; Landau & Lifshitz 1980). Near-perfect crystals, however, can exist in 3D space through bending in the third dimension. A transmission electron microscopy study revealed that suspended graphene sheets are not perfectly flat but exhibit nanometer-scale random elastic deformations (out-of-plane deformation of ~1 nm and the surface normal varying by several degrees) (Figure 1.1) (Meyer et al. 2007). The interaction between bending and stretching long-wavelength phonons is believed to stabilize atomically thin membranes through their nanometer-scale deformation in the third dimension.

By isolating graphene from external sources of scattering, electron mobility exceeding 200,000 cm^2/Vs was demonstrated, which corresponds to a more than 10-fold enhancement compared with the conventional graphene on substrate (Figure 1.1) (Bolotin et al. 2008). It is interesting to note that this dramatic enhancement of carrier mobility required an annealing (current-induced Joule heating) of suspended

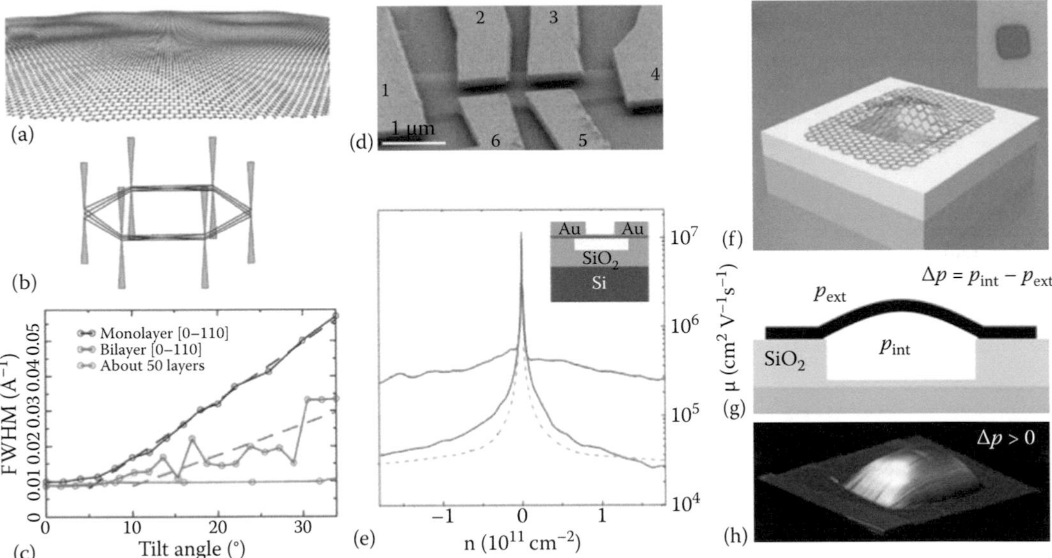

FIGURE 1.1 Morphology of suspended graphene (a–c). (a) A perspective view illustrating microscopic corrugation. The roughness shown imitates quantitatively the roughness found experimentally. (b) A superposition of the diffracting beams from microscopic flat areas effectively turns the rods (reciprocal space) into cone-shaped volumes so that diffraction spots become blurred at large angles. (c) Full-width half-maximum (FWHM) for the ($0\underline{1}10$) diffraction peak in monolayer and bilayer membranes and thin graphite (as a reference), as a function of tilt angle. Graphene roughness is measured from the diffraction patterns obtained at different tilt angles. In-plane carrier transport of suspended graphene (d, e). (d) SEM image of a suspended six-probe graphene device. (e) Mobility measured as a function of carrier density: before (middle, solid) and after (top, solid) current annealing; data from a traditional high-mobility device on substrate (bottom, dotted line) are shown for comparison. (Inset) device schematic, side view. Impermeable atomic membrane from graphene sheets (f–h). (f) Schematic of a graphene sealed microchamber. (Inset) optical image of a single atomic layer graphene drumhead on 440-nm-thick SiO_2. The dimensions of the microchamber are 4.75 μm × 4.75 μm × 380 nm. (g) Side view schematic. (h) AFM image of a ~9-nm-thick many-layer graphene drumhead with $\Delta p > 0$. The upward deflection at the center of the membrane is 90 nm. (a–c, Reprinted by permission from MacMillan Publishers Ltd. *Solid State Commun.* Meyer, J.C., Geim, A.K., Katsnelson, M.I., *Nature*, 446, 60–63, 2007. Copyright 2007. d, e, Reprinted from *Solid State Commun.*, 146, Bolotin, K.I., Sikes, K.J., Jiang, Z. et al., Ultrahigh electron mobility in suspended graphene, 351–355. Copyright 2008, with permission from Elsevier. f–h, Reprinted with permission from Bunch, J.S., Verbridge, S.S., Alden, J.S. et al., *Nanosci. Lett.*, 8, 2458–2462, 2008. Copyright 2008 American Chemical Society.)

graphene, which is designed to desorb adsorbates on either side of graphene. This suggests that impurities trapped between the substrate (SiO_2) and graphene are limiting the mobility of unsuspended graphene.

Graphene is shown to have high adhesion energy (~0.45 J m^{-2}), which is much larger than those measured in conventional micromechanical structures and is comparable with solid–liquid adhesion energies (Koenig et al. 2011). By pressing on suspended graphene with an atomic force microscope (AFM) tip with calibrated spring constant, atomic layers of graphene are shown to have large stiffness values, similar to bulk graphite (E ~ 1 TPa) (Frank et al. 2007; Lee et al. 2008). The impermeability of suspended graphene to gas molecules is also demonstrated (Figure 1.1) (Bunch et al. 2008). By applying a pressure difference across the membrane, both the elastic constants and the mass of a monolayer graphene were measured. The atomically thin sealed chambers can support pressures up to a few atmospheres. By adjusting the pressure difference, thereby the strain of suspended graphene, the mechanical resonance frequency was tuned by ~100 MHz.

Graphene is known to have an anomalously large, negative thermal expansion coefficient (Bao et al. 2009). This can cause a large amount of thermal stress/strain depending on the substrate material and the amount of temperature change. A flat suspended graphene, exfoliation-transferred to a trench on SiO_2/Si substrate, exhibits ripples (up to ~30 nm amplitude) after experiencing a thermal treatment (heating or cooling) by 100–200 K (Bao et al. 2009).

Graphene is transmissive to impinging electrons while being impermeable to atoms and molecules (Bunch et al. 2008). Harboring a 2D electron system (2DES) (Ando et al. 1982; Eisenstein et al. 1992; Ho et al. 2009; Li et al. 2011), graphene is expected to be interactive with out-of-plane incident electrons as well. A graphene electrode suspended on a nanoscale void channel (trenches or holes) provides an interesting configuration to investigate the interplay of in- and out-of-plane interactions of 2DESs mediated by electron transport in vacuum.

At electron energy <~10 eV, the de Broglie wavelength becomes greater than the lattice atomic spacing and crystalline diffraction is less likely to occur. Below 5–10 eV, the dominant scattering mechanism is expected to involve inelastic interactions such as electron excitations or electron–phonon interactions (Kuhr & Fitting 1999; Müllerová et al. 2010; Cazaux 2012). The damage threshold of graphene is known to be >15 eV, corresponding to incident electron energy >80 keV (Crespi et al. 1996; Krasheninnikov & Nordlund 2010; Börrnert et al. 2012). Considering the relatively large threshold, electrons of very low energy (<5 eV) are expected to induce no damage to graphene (Srisonphan et al. 2014).

In this chapter, we review the emission, capture, and transmission interactions with very-low-energy electrons and explore the potential of using graphene as an electron-transparent grid in low-voltage nanoscale vacuum electronic devices (Child 1911; Langmuir 1913; Spindt et al. 1976; Han et al. 2012; Srisonphan et al. 2012; Stoner & Glass 2012). The electron transparency of graphene has been the subject of debate in recent literature (Müllerová et al. 2010; Li et al. 2014). Most studies were performed in transmission electron microscopes at relatively high energies (>>100 eV) or in electron holography mode at low energies (~100 eV) demonstrating transparences >70% (Kreuzer et al. 1992; Morin et al. 1996; Mutus et al. 2011). In the case of very-low-energy electrons (<5 eV), however, reports are rare. Most work was performed in a triode (three-terminal) configuration, where the anode potential is designed to be sufficiently high to collect incoming electrons transmitted through a graphene grid; therefore, the field distribution around the graphene is inevitably altered by the anode field. In this study, we employ a diode (two-terminal) configuration and investigate the direct interplay of cathode and anode/grid (graphene) mediated by the ballistic transport of electrons through a nanoscale gap in air ambient (Srisonphan et al. 2014).

1.2 Fabrication

Suspended graphene can be fabricated with or without involving a transfer process. In the case of a non-transfer method, a graphene layer is initially grown on a chosen substrate, usually on a Cu or Ni foil by chemical vapor deposition (CVD) or on a SiC substrate by Si sublimation or by CVD (Jernigan et al. 2009;

Shivaraman et al. 2009; Mattevi et al. 2011; Suk et al. 2011). In the case of graphene on a metal substrate, the bottom side is lithographically patterned and etched away in a chemical solution. In the case of graphene on SiC, an undercut etching is performed through a lithographically defined etch mask. This substrate removal process leaves a graphene membrane supported by an unetched part of substrate.

In the case of a transfer method, a graphene flake is placed on a predefined microstructure (trench, hole, or grid) formed in a substrate (Meyer et al. 2007). Precision placement of graphene on a predetermined microstructure would be challenging, if it is a small size flake. In an alternative transfer method, a graphene flake is placed on a substrate and then metal electrodes are lithographically defined on top (Bolotin et al. 2008). An undercut etching is then performed to remove the substrate material underneath the graphene. Compared with the nontransfer method discussed previously, the transfer method generally allows a greater degree of freedom/flexibility in integrating with other functional substrates/devices.

In this study, we have employed the transfer method with CVD-grown graphene onto nanoscale trenches/holes formed in a SiO_2/Si substrate (Figure 1.2). CVD growth of graphene on metallic substrates is a well-established technique yielding large graphene flakes to a commercially viable scale (Kim et al. 2009; Li et al. 2009). Although CVD-grown graphene generally shows lower electron mobility and more defects than does graphene produced by an exfoliation method, the easy transfer from the metal substrate to other substrates is expected to enable many device applications at chip or wafer level.

First, a SiO_2 layer (~20 nm thickness) was grown by thermal oxidation on p-type Si (B-doped, 10 Ω-cm resistivity) or n-type Si (P-doped, 5 Ω-cm resistivity) wafers [(100)-oriented; 525 μm thickness]. A bottom-side electrode was prepared by depositing a 150-nm-thick Al layer (5 N purity) on Si by thermal evaporation, followed by Ohmic contact annealing at 350°C. Vertically etched trench or hole structures were then formed by employing a nanoscale patterning/lithography technique, such as focused ion beam (FIB) etching or electron-beam lithography (EBL), or photolithography in conjunction with use of plasma reactive ion etching (RIE).

FIB etching directly etches nanoscale patterns into substrate without involving an etch mask or a subsequent pattern transfer process. In this study, FIB etching was performed with the Seiko Dual Beam System (SMI-3050SE). A Ga ion beam (30 keV; 94 pA) was used with 0.5-μs dwell time in creating square wells, holes, or trenches with minimum lateral dimensions (trench width or hole diameter) down to ~70 nm and an etch depth up to 1–2 μm.

In the case of photolithography, a 50-nm-thick Cr layer was deposited on SiO_2/Si substrate by thermal evaporation (Figure 1.2a). A window of narrow stripe patterns (5–50 μm width; single or multiple channels of 8–10 mm length) was then opened in the Cr layer by performing photolithography and RIE. The Cr window etching was performed in Cl_2/O_2 ambient with an inductively coupled plasma RIE (ICP-RIE) system (Unaxis 790 ICP-RIE). Subsequently, a trench etching was performed to 500–1000 nm depth by RIE in CF_4/O_2 ambient with use of the Cr window as an etch mask. The remaining photoresist was removed in acetone. The Cr mask was removed in Cr etchant [$NaOH{:}K_3Fe(CN)_6{:}H_2O$ = 2 g:6 g:22 mL].

In the case of EBL, an e-beam resist, polymethylmethacrylate (PMMA, ~200 nm thickness) was spin-coated on the SiO_2/Si substrate (Figure 1.2b). EBL was then performed using the Raith e-Line system (10 keV, beam current, 220 pA) to define trench/hole patterns (with minimum lateral dimension of ~70 nm). The exposed PMMA was developed, and the EBL-defined pattern was subsequently transferred to SiO_2/Si substrate (to 100–200 nm etch depth) by performing RIE in CF_4/O_2 ambient with the PMMA pattern as an etch mask. A residual PMMA was removed in acetone.

Finally, a monolayer graphene was transferred to the trench/hole-etched SiO_2/Si substrate (Figure 1.2c) (Suk et al. 2011). To place a graphene membrane suspended on the etched SiO_2/Si substrate, we started with a monolayer graphene grown on Cu foil (purchased from ACS Material; CVD grown on 25-μm-thick Cu foil). A PMMA layer (MicroChem 950 PMMA A7, 4% in anisole) was spun-coated on the graphene-covered Cu foil ($0.3 \times 0.4\ cm^2$). The Cu foil was etched away in ferric chloride solution (Transene Cu Etchant CE-100). After ~30 min etching, the PMMA/graphene stacked film was floating in the etchant solution. A clear glass substrate was then submerged into the solution to lift the floating film. The film was then transferred to deionized (DI) water and was left floating for 10 min. This process

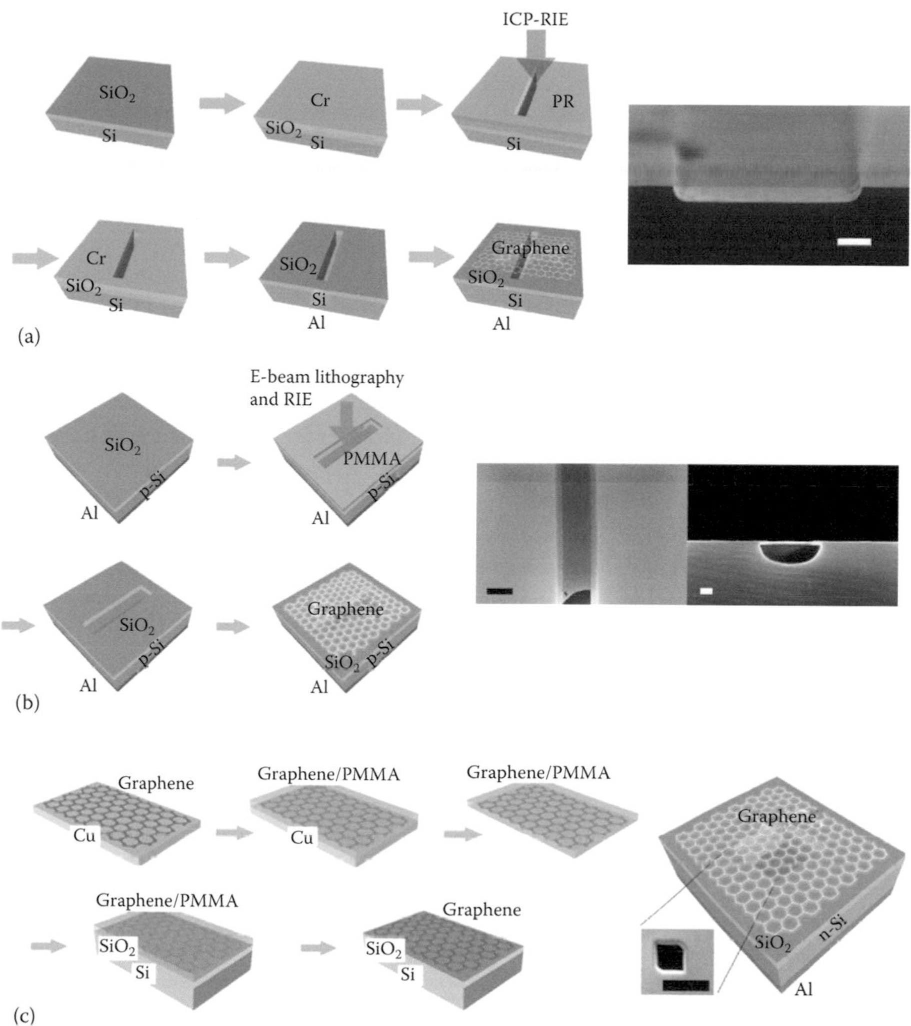

FIGURE 1.2 Fabrication of graphene suspended on a channel-etched SiO_2/Si substrate. (a) Process flow employing photolithography and RIE for channel etching. (Right) SEM image of suspended graphene on a trench-etched substrate. Scale bar, 1 μm. (b) Process flow employing EBL and RIE for channel etching. (Right) SEM images of suspended graphene: top view, left (scale bar, 300 nm); cross-section, right (scale bar, 100 nm). (c) Transfer process of CVD grown graphene onto a channel-etched substrate. (Right) SEM image of suspended graphene on a FIB-etched substrate. Scale bar, 1 μm.

(transfer to DI water) was repeated at least 3 times to remove residual etchant. A target substrate (trench/hole-etched SiO_2/Si wafer) was then immersed into the DI water, and the PMMA/graphene film was lifted up while being positioned by a needle. The PMMA/graphene stack, placed on the target substrate, was dried at ~70°C in air for 2 h to enhance adhesion of graphene to the substrate. The PMMA was removed in acetone, followed by rinse in methanol and DI water. Finally, the sample was dried at ~70°C for 2 h to remove moisture trapped in the void-channel. It should be mentioned that graphene is known to remain stable in humid oxygen ambient at up to 400°C (Liu et al. 2008). Therefore, thermal oxidation is not an issue to worry about during baking. Rather, a large thermal expansion mismatch with substrate is more the issue that may cause wrinkles in suspended graphene (Bao et al. 2009). Therefore, the post-transfer baking temperature needs to be in the moderate or mild range.

1.3 Low-Voltage Emission of 2D Electron Gas

In generating a constant flux of very-low-energy electrons, we exploit the phenomenon that a 2D electron gas (2DEG) induced at the SiO_2/Si interface of a metal-oxide-semiconductor (MOS) structure can easily emit into air (void channel) at low voltage (~1 V) (Srisonphan et al. 2012). This low-voltage emission, enabled by Coulombic repulsion of electrons in 2DEG, has the effect of negative electron affinity and demonstrates high current density emission (~10^5 A/cm^2). The emitted electrons ballistically travel in the nanovoid channel. The channel length (i.e., the thickness of oxide layer) is designed to be smaller than the mean free path of electrons in air (~60 nm). Therefore, emitted electrons should travel scattering-free in the ambient (air) channel, as if in a vacuum. The transit time is estimated to be 10–100 fs for 10–20-nm-thick SiO_2 at 1–10 V bias.

A MOS capacitor structure can harbor a quasi-2DES in a potential well (~2 nm width) developed in the semiconductor side (Ando et al. 1982; Torium et al. 1986). The metal side also develops band bending, accommodating charges of opposite polarity in a confined space (<1 nm) at the interface with the oxide layer (Mead 1961; Black & Welser 1999). Electrons residing inside the bulk Si of a MOS structure are basically confined by energy barriers at the surface (with air) and at the interface (with SiO_2), typically with a vacuum barrier greater than the SiO_2 barrier. Therefore, electrons cannot easily emit into air, unless the energy barrier is significantly lowered for thermionic or field emission. (Also, we note that in this work, the oxide layer thickness is designed to be ~20 nm; therefore, direct tunneling through the oxide barrier layer remains completely negligible.)

The situation can be very different for the electrons confined in a 2DEG layer. First, let us imagine an infinite extension of 2DEG formed at SiO_2/Si (Figure 1.3). The overall charge neutrality condition is maintained between the 2DEG (and also depletion charge) in Si and the positive charges induced in the metal side. Along the in-plane direction of 2DEG, Coulombic repulsion among electrons will cancel out because of the symmetry of electron distribution. Now, consider a MOS structure whose lateral extent is finite; i.e., the 2DEG layer is terminated by cleaved edge or a vertically etched trench structure. Electrons at the channel edge will then experience net repulsive force from neighboring electrons inside the 2DEG layer. In the case that the charge neutrality is maintained by relatively remote charges (e.g., opposite polarity charges induced across the oxide layer of MOS capacitor), strong in-plane Coulombic repulsion is expected in the local area around the edge of 2DEG, and this can significantly alter the electrostatic potential there (Han & Ihm 2000; Zheng et al. 2004; Mayer 2005). This in-plane Coulombic repulsion has the effect of lowering the vacuum barrier (Figure 1.3) for 2DEG at the edge. Similarly, the energy barrier in the metal side (metal/air interface) is lowered by the 2D positive charges at metal/SiO_2 interface,

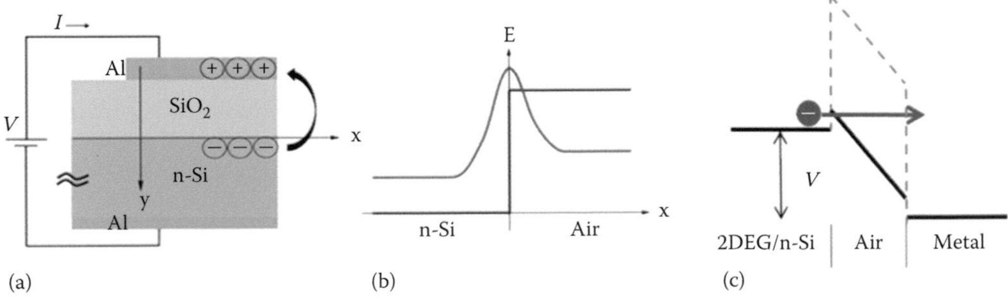

FIGURE 1.3 Energy band diagrams. (a) 2D electron or hole systems induced across the oxide layer of an Al/SiO_2/n-Si MOS capacitor structure. (b) Schematic illustration of electron potential (curved) and energy barrier (step profile) profiles on the plane of the 2DEG layer at the Si/SiO_2 interface. (c) Schematic energy band diagram illustrating the emission of 2DEG under accumulation bias. Coulombic interaction at the edge lowers the energy barriers at cathode (2DEG/n-Si) and anode (metal). 2DEG emits into air at low voltage, travels in the nanovoid channel (air), and is collected by anode.

and the applied capacitor voltage appears mostly across the air gap (i.e., the vertical void channel whose length is defined by the SiO_2 layer thickness).

1.3.1 Calculation of 2DEG Density in a MOS Capacitor Structure

First, we calculate the carrier density of 2DEG induced at SiO_2/Si under forward accumulation bias of a MOS capacitor structure having an infinite lateral extension. In a MOS structure, the space charge density in the semiconductor side can be determined by solving the Poisson equation and is expressed as follows for the case of p-MOS (i.e., on n-Si substrate) (Sze 1981):

$$Q_s = -\varepsilon_s E_s = \frac{-\sqrt{2}\varepsilon_s}{\beta L_D}\left[(e^{-\beta\varphi_s} + \beta\varphi_s - 1) + \frac{p_{no}}{n_{no}}(e^{\beta\varphi_s} - \beta\varphi_s - 1)\right]^{1/2} \tag{1.1}$$

where ε_s is the permittivity of the semiconductor and E_s is the electric field at the interface with the oxide layer. φ_s is the band bending at the semiconductor/oxide interface, called the surface potential. $\beta = q/kT$, and L_D is the extrinsic Debye length for electrons, given as $\sqrt{\frac{\varepsilon_s}{qn_{no}\beta}}$. n_{no} and p_{no} are the equilibrium densities of electrons and holes, respectively.

The applied capacitor voltage (V) appears across mainly three places (neglecting the band bending in the metal side): across the band bending region in semiconductor (φ_s), across the oxide layer (V_{ox}), and the flat band voltage (V_{FB}).

$$V = \varphi_s + V_{ox} + V_{FB} \tag{1.2}$$

The voltage drop across the oxide layer (V_{ox}) is related to the space charge (Q_s) and oxide capacitance ($C_{ox} = \varepsilon_{ox}/d$) as follows:

$$V_{ox} = Q_s/C_{ox}. \tag{1.3}$$

Solving Equations 1.1 to 1.3 simultaneously, the space charge density Q_s can be calculated as a function of applied voltage V.

Figure 1.4 shows the case of a Si p-MOS with $N_D = 1.0 \times 10^{15}$ cm^{-2}. At $V = 1$ V, the MOS is in the accumulation regime, and the space charge density is calculated to be 1×10^{12} cm^{-2}. The same amount of

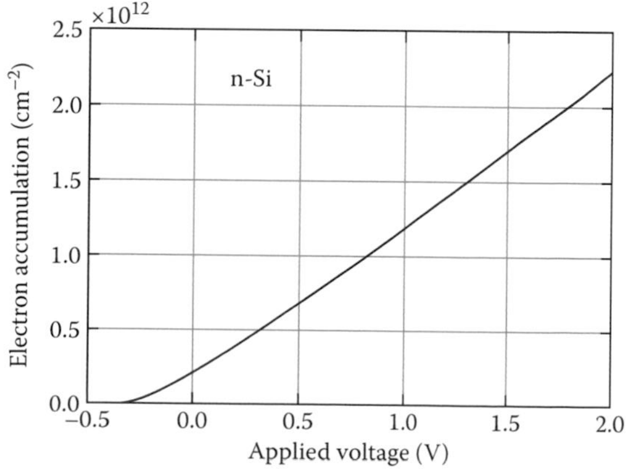

FIGURE 1.4 Accumulation electron density calculated as a function of applied voltage for an Al/SiO_2(23nm)/n-Si capacitor structure.

positive charges accumulates in the metal side. At this bias voltage, the surface potential φ_s is 0.21 V and the voltage drop across the oxide layer V_{ox} is 1.11 V.

In this calculation, the following numbers were assumed: the work function of Al, 4.1 eV; electron affinity of Si, 4.15 eV; electron affinity of SiO_2, 0.95 eV; dielectric constant of SiO_2, 3.9; dielectric constant of Si, 11.8 (Lide 2006).

This calculation confirms that the 2DEG density of ~10^{12} cm^{-2} is easily attainable at low bias voltage (~1 V). The average spacing between electrons in the 2DEG is ~10 nm, smaller than the oxide thickness (~20 nm). This will ensure that the in-plane interaction of 2DEG electrons becomes stronger than the dipole charge interaction across the oxide layer. When the 2DEG layer is terminated at one edge by cleaving the MOS wafer or by vertical etching into SiO_2/Si substrate, the Coulombic repulsion among electrons in 2DEG at the edge is expected to result in overcoming the potential barrier at the Si/air interface. This in-plane Coulombic repulsion will have the effect of lowering the energy barrier at the edge and will enable low-voltage emission of 2DEG into air (Figure 1.3).

1.3.2 Space-Charge-Limited Current Flow of 2DEG in a Nanoscale Void Channel

Nanoscale void channels were fabricated by performing FIB etching of a Si MOS structure (20-nm Al/23-nm SiO_2/n-Si substrate) (Figure 1.5). Square wells (0.5 × 0.5 μm², 1 × 1 μm², and 2 × 2 μm²) were etched to 1 or 2 μm depth. In this vertical structure, the channel length between anode and cathode was precisely determined by the oxide layer thickness (23 nm) and was designed to be smaller than the mean free path of air (~60 nm). The current-versus-voltage (*I*–*V*) characteristics of the Al/SiO_2/Si structure were measured in dark air ambient at room temperature with a semiconductor parameter analyzer (HP4145B) in conjunction with use of a probe station. Tungsten probes (tip radius of curvature, ~2 μm) were used in contacting the top and bottom electrodes. The voltage scan was performed with a step size of 0.02 V.

The two-terminal *I*–*V* characteristic shows a rectifying behavior with a forward slope of ~1.5 and a reverse slope of 0.5–1.0 in the log–log scale plots (Figure 1.5). The channel reveals a forward characteristic when the Al gate is positively biased. This implies that electron emission from the Si side is more efficient than from the metal side at the same bias voltage of opposite polarity. With a 0.5 × 0.5 × 1.0 μm³

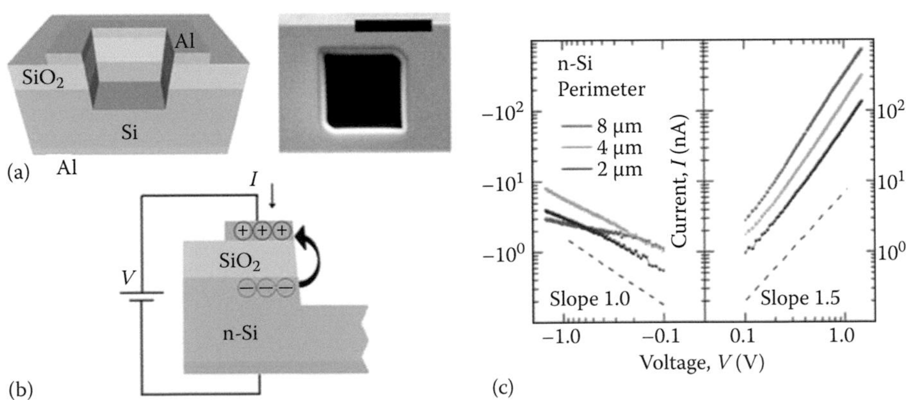

FIGURE 1.5 Ballistic transport of electrons in a nanovoid channel formed in an Al/SiO_2(23nm)/n-Si MOS capacitor structure. (a) Schematic drawing of a nanovoid channel fabricated by FIB etching (left). SEM image of a square well (1 × 1 μm²) etched to 1 μm depth (right). Scale bar, 1 μm. (b) Schematic of electron emission and transport in a nanovoid channel under forward, accumulation bias. (c) Measured *I*–*V* characteristic of square wells (with perimeter of 2, 4, or 8 μm) formed on n-Si. The dashed lines indicate the slope of 1.5 (forward) or 1.0 (reverse). The $V^{1.5}$ voltage dependence corresponds to the CL SCL current flow in vacuum. (Reprinted from Srisonphan, S., Jung, Y.S., Kim, H.K., *Nat. Nanotechnol.*, 7, 504–508, 2012.)

well formed on n-Si, for example, a channel current of 70 nA is observed at +1 V bias, whereas 3 nA is obtained at −1 V bias (Figure 1.5). Comparison of the three different well-size samples (perimeter of 2, 4, or 8 μm) reveals that the forward channel current is proportional to the perimeter of the well, not to the area of the well. This result suggests that electron emission occurs at the edge surface (periphery) on the vertical sidewalls of a well.

In the Al/SiO_2/n-Si MOS capacitor structure, the flat band voltage is −0.32 V, and the MOS at 1 V forward bias is accumulation biased by the amount of 1.32 V. The electron accumulation in Si is estimated to be 1×10^{12} cm^{-2} at this bias voltage (Figure 1.4). The accumulation electrons form a 2DES, and this layer serves as a reservoir of electrons that would be readily available for emission through the edge under forward bias (the top Al electrode positively biased with respect to n-Si substrate). Because of Coulombic repulsion of electrons around the aperture edge, the 2DEG in Si emits into air and travels toward the edge of 2DHS at Al/SiO_2.

The voltage dependence of electron injection is governed by the capacitor relationship, $Q_e \sim V$. A scattering-free transport of electrons in a void channel converts the potential energy (eV) to kinetic energy ($m^*v^2/2$). The terminal velocity is expressed as $\sqrt{\frac{2eV}{m^*}}$, and the average transit time across the channel has the following voltage dependence: $\tau_{av} \sim V^{-0.5}$. The voltage dependence of the channel current can then be expressed as $I = Q_e/\tau_{av} \sim V^{1.5}$, following the Child–Langmuir (CL) three-halves-power law. Here, it should be mentioned that the CL is governed by the space-charge-limited (SCL) emission at cathode. In our case, the electrons injected into air (i.e., nanovoid channel) form a space charge around the anode edge and this space charge field limit the emission of 2DEG at the cathode edge. In reverse bias, part of the bias voltage goes to depletion region formation in Si; therefore, the void-channel section receives less voltage than the accumulation case. This explains the reduced slope (0.5–1.0) in reverse bias (Figure 1.5).

The CL SCL current flow in vacuum is given as follows (Child 1911; Langmuir 1913; Grinberg et al. 1989; Sze 1990):

$$J = \frac{4}{9}\varepsilon\sqrt{\frac{2e}{m^*}}\frac{V^{3/2}}{d^2}, \tag{1.4}$$

where ε is the permittivity of gap medium, m^* is the effective mass of electron, d is the gap size, and V is the applied voltage.

The SCL current observed in this work demonstrates a scattering-free ballistic transport of electrons across the gap with a negligible barrier height for carrier injection.

To make sure that the observed $V^{3/2}$ dependence is from the electron transport through the air (nanoscale vacuum), not from a surface conduction that might be enabled by possible etch residue or deposit on oxide surface, the same vertical channel structure was fabricated by cleaving a MOS wafer [Al/SiO_2(23nm)/n-Si]. The cleaved samples clearly demonstrate the same rectifying I–V characteristic as the FIB samples (Figure 1.6). The leakage current through the oxide layer, measured before FIB etching, was ~20 pA at 2 V bias, far smaller than the channel current level described previously.

In conventional cold-cathode field-emission devices, the electron flow across a metal-nanogap structure usually involves a two-step process: field emission from metal surface, commonly described by the Fowler–Nordheim (FN) theory (Fowler & Nordheim 1928), and a subsequent transport through the gap, governed by the CL SCL current flow (Lau et al. 1994). The CL law assumes zero normal field at cathode surface, whereas the FN emission requires a surface normal field of significant strength (typically ~10 V/nm order for metals with work function of 4–5 eV) to enable tunneling emission through the potential barrier at cathode. This implies that the applied voltage initially goes to lowering the potential barrier at the cathode surface, being governed by the FN process. Once a significant amount of electrons are emitted into the gap, the injected electrons in transit form a space charge. As more electrons are injected, the space charge builds up, reducing the field at the cathode surface. At some point, the surface field is reduced to

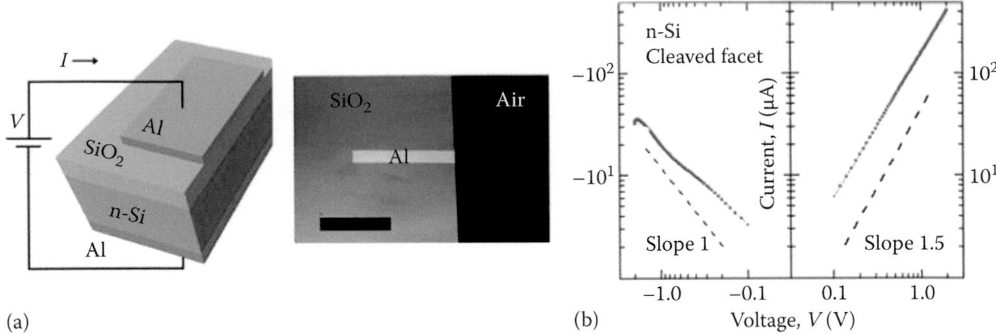

FIGURE 1.6 Nanovoid channel formed on the cleaved facet of an Al/SiO_2(23nm)/n-Si MOS wafer. (a) A schematic of a nanoscale void vertical channel prepared by cleaving a MOS capacitor structure. An Al electrode pad (stripe) was first deposited on an oxidized Si substrate. The MOS wafer was then mechanically cleaved into two pieces, with each cleaved facet comprising cross-sections of MOS layers (left). An optical micrograph of a top view of a cleaved sample with an Al electrode (right: scale bar, 300 μm). (b) Measured *I*–*V* characteristic. The cleaved sample clearly demonstrates the same rectifying *I*–*V* characteristic (forward slope of 1.5 and reverse slope of 0.5–1.0) as the FIB sample.

zero and the cathode emission, and therefore the channel current, becomes space-charge limited. The transitional relationship between FN and CL regimes in conventional cold-cathode structures is illustrated in Figure 1.7: The FN regime ($JD^2 \propto V^2e^{-D/V}$) evolves into the CL regime ($JD^2 \propto V^{1.5}$) at large bias. Here, *D* refers to the gap size. Note that in the conventional cold-cathode surface emission, the transition voltage (from FN to CL) decreases as the gap size is reduced (i.e., as the surface field is increased) (Figure 1.7a and c). In contrast, in the case of edge emission of 2DEG, the energy barriers at cathode and anode are reduced by Coulombic interactions, and this would result in low voltage emission (Figure 1.7b and c).

In the present work, the SCL regime begins to appear at very low voltage (~0.5 V), whereas the FN regime is absent in the voltage range tested (<2 V). The maximum surface field at this onset voltage is estimated to be ~0.02 V/nm (i.e., 0.5 V across 23-nm channel length), much smaller than the typical surface field required for FN emission (~10 V/nm). This observation is consistent with the earlier reports

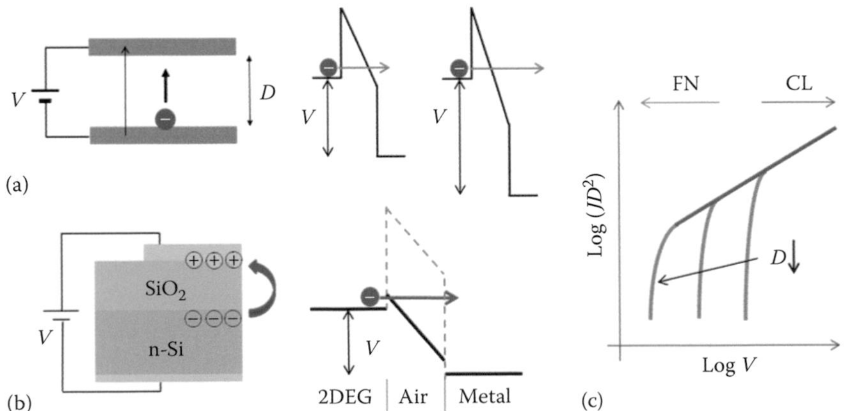

FIGURE 1.7 Evolution of FN emission to CL SCL emission current. (a) Schematic of conventional cold-cathode field emission. The FN tunneling emission requires sufficiently strong field on the cathode surface. The FN regime is governed by $JD^2 \propto V^2e^{-D/V}$. As the bias voltage is increased, space charges build up. The CL regime emerges at large bias, governed by $JD^2 \propto V^{1.5}$. (b) In the case of edge emission of 2DEG the Coulombic interaction lowers the energy barriers and the CL regime emerges at low voltage. (c) The transitional relationship between FN and CL regimes.

that the barrier height for electrons at cathode edges can be very low for SCL emission (Han & Ihm 2000; Zheng et al. 2004; Mayer 2005). In the 2DES with net accumulation charges, electron emission from cathode edges is virtually thresholdless, enabling very low voltage operation (similar to the negative electron affinity effect) of channel transport with high current density. Similarly, electrons approaching the anode edges will experience Coulombic attraction from the 2D positive charge system formed there, and this will help in capturing electrons into the anode. Unlike conventional cold cathodes (Yang et al. 1991; Mil'shtein et al. 1993; Yun et al. 1999), the nanovoid channel structure also demonstrates good stability and endurance in electron emission (Srisonphan et al. 2012).

1.4 Electron Capture and Transmission at Dark Forward Bias

A graphene membrane was placed on top of a void channel (500 nm × 500 nm cross-section; 1 μm depth) that was FIB etched or EBL/RIE etched into a SiO_2 (23 nm thickness)/n-Si (5 Ω-cm resistivity) substrate (Figure 1.8). A graphene/oxide(or air)/Si (GOS) structure, instead of MOS, was formed by introducing a monolayer graphene as a counterelectrode to the 2DEG layer at the SiO_2/Si interface.

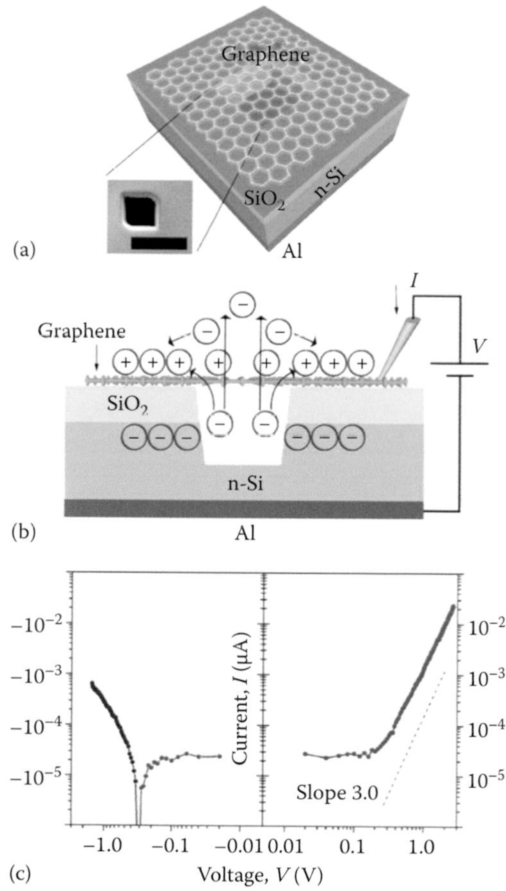

FIGURE 1.8 Transport of very-low-energy (<3 eV) electrons in a void channel covered with a suspended graphene. (a) Schematic of a graphene/SiO_2(23nm)/n-Si structure with a void channel and a top-view SEM image (inset) of a square well (500 × 500 nm²) etched into 1 μm depth by FIB. Scale bar, 1 μm. (b) Schematic of electron emission from the 2DEG at SiO_2/n-Si interface and capture/transmission at the graphene anode. (c) Measured I–V characteristic of a void channel (500-nm square well) covered with a monolayer graphene. Note the V^3 dependence ($V > 0.3$ V). (Reprinted from Srisonphan, S., Kim, M., Kim, H.K., *Sci. Rep.*, 4, 3764, 2014.)

Here, in the two-terminal mode of operation, the graphene serves as an anode while the n-Si substrate serves as a cathode. Under forward bias (i.e., graphene electrode positively biased with respect to n-Si substrate), a quasi-2DEG (accumulation) develops in the Si side while a 2D hole system (2DHS) forms in the graphene side (Figure 1.8). As a result of Coulombic repulsion of electrons around the aperture edge, the 2DEG in Si emits into air and travels up toward the edge of 2DHS at graphene/SiO_2 interface (Figure 1.8). Some of the incident electrons are captured at the graphene, while others transmit through, forming a space charge region outside the graphene.

It should be noted that in this work, no external collector is employed other than the graphene anode (Spindt 1968; Brodie 1989; Spindt et al. 1991). Therefore, the electrons transmitted through graphene are Coulombically attracted to and collected by the positively biased graphene anode on SiO_2 surface, satisfying the charge neutrality of the overall configuration. The closed-circuit nature (i.e., charge conservation) of this two-terminal operation is confirmed by performing measurements of anode current with the system ground connected to the bottom (cathode) or top (anode) electrode, which demonstrate the same amount of channel current for a given bias voltage (Srisonphan et al. 2014).

1.4.1 Emission and Transport of 2DEG

The dark *I–V* characteristic was analyzed to understand the emission and transport properties of the quasi-2DEG accommodated at the Si/SiO_2 interface. Figure 1.8 shows a measurement result of the void-channel *I–V* characteristic. The forward *I–V* characteristic reveals the V^3 dependence for $V > 0.3$ V. Note that the flat band voltage of this GOS structure is 0.25 V, and an electron accumulation layer begins to develop at around this voltage. At 1 V bias, the channel current is measured to be 1.3 nA. The V^3 regime is called the double injection or injected plasma regime (Lampert & Rose 1961). This corresponds to another type of SCL emission, differing from the CL $V^{1.5}$ dependence or the Mott-Gurney $V^{2.0}$ dependence (Child 1911; Langmuir 1913; Mott & Gurney 1940; Grinberg et al. 1989): The V^3 regime involves bipolar space charges (electrons and holes) injected into a void channel, whereas the latter ones are governed mostly by unipolar space charges (electrons).

In this bipolar space-charge regime (V^3), the cathode emission is governed by the availability of hole charges on graphene (Q_h). The carrier density in graphene has a quadratic dependence on Fermi energy: $n \sim E_F^2$, where E_F refers to the Dirac point (Novoselov et al. 2005; Zhang et al. 2005). In an accumulation-biased graphene/SiO_2/Si capacitor structure, the applied voltage (V) goes to primarily three places: across the oxide layer (V_{ox}), to compensate the flat band voltage (V_{FB}), and to shift the graphene's Fermi level (E_F) (see Section 4.3). As the bias voltage is increased, V_{ox} follows closely ($V_{ox} \sim \alpha V$: $\alpha < 1$). The Fermi level shift can then be expressed as $E_F \sim (V - \alpha V - V_{FB}) \propto (V - V_{offset})$. The hole concentration in graphene will then show a quadratic dependence on voltage ($Q_h \sim V^2$), and the amount of electrons being injected into a channel (Q_e) is expected to show the same voltage dependence ($Q_e \sim Q_h \sim V^2$). With enhanced injection of charge carriers in the channel region, electron transport is expected to show an average velocity that is proportional to the electric field and, therefore, bias voltage: $v_{av} \sim \mu\varepsilon \sim \mu V/L$. Here, μ is the electron mobility and L is the channel length. The average transit time of electron in the channel is then determined as $\tau_{av} = L/v_{av}$, and the channel current can be expressed as $I = Q_e/\tau_{av} = Q_e v_{av}/L \sim V^3$.

A graphene layer was present in a void channel, and therefore, the availability of holes in the aperture region is expected to affect the space charge field in the channel. In response to electron injection from cathode, for example, the graphene anode brings positive space charges into the void channel by inducing hole charges in the free-standing cover. This has the effect of neutralizing the electron space charges in transit in the channel region (Langmuir & Kingdon 1923; Wilson 1959). With a reduced space charge field on cathode surface, electron emission becomes easier, resulting in a higher channel current with stronger voltage dependence (i.e., V^3 instead of $V^{1.5}$ or $V^{2.0}$).

Besides altering the behavior of SCL emission under forward bias, a free-standing graphene appears to affect the reverse characteristic as well (Figure 1.8). At around −0.3 V, the current level drops to zero, switching the polarity from reverse to forward. Note that bias voltage was swept in the positive direction from −1.5 V to +1.5 V. The early reversal of current flow suggests a discharge of graphene anode around this bias. As the Fermi level is reduced toward the Dirac point, the carrier density of graphene monotonically decreases (Yu et al. 2009). Electrons then evacuate from the graphene at reduced bias, and this exiting (discharging) electron flow has an effect of compensating the reverse leakage (charging) current (Figure 1.8). At some bias point, the two current components cancel each other, causing a zero-current crossing (i.e., a dip in I–V).

To estimate the electron capture efficiency at the edge of graphene anode, the total electron emission from cathode needs to be measured. Here, we refer the electron capture to the directly captured component, not counting the electrons transmitted through graphene and then were collected back by graphene anode via Coulombic attraction. Otherwise, all emitted electrons would be eventually collected/captured by anode (graphene) in this two-terminal mode of operation. In an effort to measure cathode emission, the graphene anode was covered by placing a Ga droplet in the aperture area (Figure 1.9). Here, the Ga droplet size is designed to be much larger than the channel diameter (i.e., 500 μm vs. 500 nm) so that incident electrons are fully blocked by the Ga-covered graphene anode. The forward I–V characteristic clearly reveals the V^3 dependence for $V > 0.1$ V. By placing Ga on top, graphene's work function is expected to decrease slightly by ~0.2 eV (Giovannetti et al. 2008). This will then reduce the flat band voltage to ~0.12 V, as seen in the earlier onset of steeply rising channel current (Figure 1.9). The V^3 regime of the Ga-covered graphene sample shifted up almost parallel to that of the sample without Ga. This

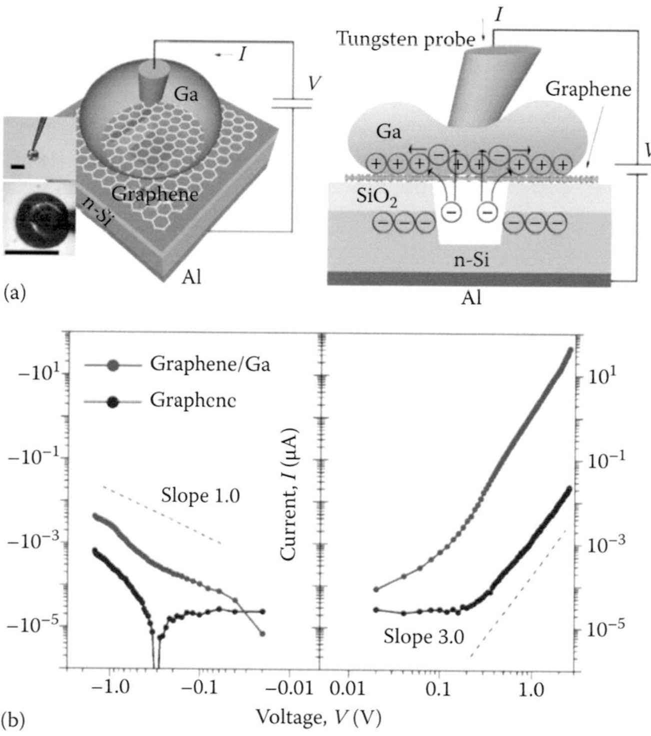

FIGURE 1.9 Electron capture efficiency of a suspended graphene anode. (a) Schematic of electron emission measurement. A Ga droplet is placed on top of the graphene anode covering the entire aperture area (inset: optical micrograph of a Ga droplet attached to a tungsten probe). Scale bars, 500 μm. (b) Measured I–V characteristics: with a Ga cover on graphene (top) and without Ga (i.e., graphene only) (bottom). The channel current increased by 1100 times (from 1.3 nA to 1.4 μA at 1 V bias) after placing a Ga cover, implying that the electron capture efficiency at a suspended graphene anode is estimated to be >~0.1%. (Reprinted from Srisonphan, S., Kim, M., Kim, H.K., *Sci. Rep.*, 4, 3764, 2014.)

indicates that the graphene layer underneath Ga still plays the same role of neutralizing electron space charge in the channel as in the case of the graphene-only sample via inducing holes in the suspended area. At 1 V bias, the channel current with graphene/Ga is now measured to be 1.4 μA, a 1.1×10^3 times increase from the current without Ga (1.3 nA).

When assuming all incident electrons are blocked and captured by the graphene/Ga anode, the total electron emission from the cathode equals the measured anode current (1.4 μA). (Here, we note that the reflectivity of very low energy electrons [~1 eV] at bulk metal surface is known to be ~10% [Herring & Nichols 1949; Cazaux 2012]. When the reflection effect at the Ga surface is taken into account, the total electron emission from the cathode is expected to be ~10% greater than the measured anode current.) If it is further assumed that the emission current of the graphene/Ga sample remains the same as that of the graphene-only sample, the electron capture efficiency of suspended graphene anode is estimated to be ~0.1% at 1 V bias. By placing a Ga cover on graphene, however, the space charges that might be present in air outside the graphene are expected to be eliminated, and this may further reduce the space charge field in the channel region, thereby enhancing cathode emission. Taking this possible effect into account, we note that the actual cathode emission without Ga might be less than the measured anode current with Ga. Based on this reasoning, the estimated capture efficiency (~0.1% at 1 V) should be considered as a lower limit.

The 1.4 μA channel current of the graphene/Ga sample at +1 V corresponds to an injection rate of $\sim 10^{13}$ electrons/s at cathode and the same rate of electron capture at Ga-covered graphene anode. The electron transit time in a nanovoid channel (channel length, 23 nm) is estimated to be ~100 fs at 1 V bias. This implies that, on average, one electron is in transit inside the void channel. In other words, an average amount of electron space charge is to be of single electron level. A similar amount of hole charges are expected to be induced on the suspended graphene area (500 nm × 500 nm). The resulting space-charge density in graphene is then estimated to be maximum ~4 holes/μm^2 or $\sim 4 \times 10^8$ holes/cm^2. The induction of holes at this level of density is expected to shift the graphene's Fermi level by no more than 0.1 eV at 1 V bias (Das et al. 2008; Xia et al. 2011; Yan et al. 2012). Overall, the result demonstrates graphene's enabling nature of enhancing cathode emission by inducing hole space charge at single electron level, thereby overcoming the CL space charge limit (Child 1911; Langmuir 1913).

Note that the zero-current crossing point in reverse bias (i.e., the dip at −0.3 V in the graphene-only sample) now shifted close to 0 V (~−0.01 V) with the graphene/Ga sample (Figure 1.9). This is explained by the fact that the Ga-covered graphene has a reduced work function (Fermi level), shifting the discharging of graphene to occur at lower bias. For a given bias voltage, the Fermi level shift at anode might have affected the band bending in the Si side, altering the density of 2DEG at SiO_2/Si and, therefore, cathode emission. To further investigate these possible effects of anode work function change (i.e., graphene Fermi level shift) on emission and capture at 2DES edges, an additional sample structure was prepared and characterized.

1.4.2 Space Charge Neutralization by Suspended Graphene

Without involving graphene, a Ga droplet was directly placed on top of a void-channel-etched SiO_2/n-Si substrate, and the resulting *I–V* characteristic was compared with that of the sample with graphene/Ga (Figure 1.10). Again, the Ga droplet size was designed to be significantly greater than the channel diameter (500 μm vs. 500 nm) so that incident electrons are fully captured.

The forward *I–V* characteristic reveals the V^2 dependence for $V > 0.1$ V (Figure 1.10). Without space charge neutralization by graphene in the void channel (i.e., without graphene), the voltage dependence of electron injection (Q_e vs. V) follows the capacitor relationship and is expressed as $Q_e \sim V$. With enhanced injection of electrons into the confined space, the electron transport can be expressed in terms of average velocity $v_{av} \sim \mu\varepsilon \sim \mu V/L$. The channel current is then determined as $I = Q_e/\tau_{av} = Q_e v_{av}/L \sim V^2$. At low bias ($V < 0.8$ V), the sample with Ga-only (top, with slope 2) shows larger current than does the sample with graphene/Ga (bottom). This is explained by the fact that the work function of Ga (4.3 eV)

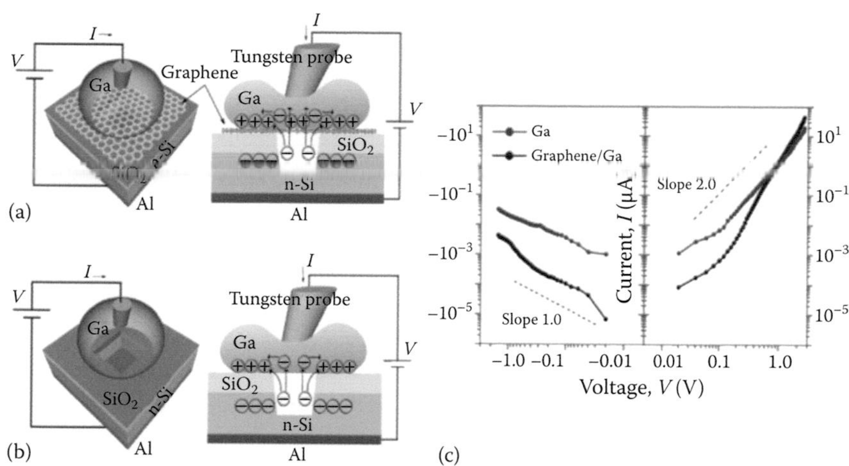

FIGURE 1.10 Enhancement of cathode electron emission by a suspended graphene anode. (a) Schematic of a void channel covered with a Ga droplet as a top cover: with a graphene layer placed underneath a Ga cover. (b) Schematic of a void channel covered with a Ga droplet as a top cover: without graphene underneath. (c) Measured *I–V* characteristics. The channel current of the sample covered with graphene/Ga (bottom) shows the V^3 voltage dependence, and the current level surpasses the CL SCL current of the sample without graphene (top, with slope 2) at $V > 0.8$ V. (Reprinted from Srisonphan, S., Kim, M., Kim, H.K., *Sci. Rep.*, 4, 3764, 2014.)

is smaller than that of the graphene under Ga (estimated to be 4.43 eV) (Giovannetti et al. 2008), and therefore, accumulation electrons build up more readily at low voltage for the Ga-only sample case. At 0.4 V, for example, the 2DEG density is calculated to be 3.0×10^{11} cm^{-2} or 1.4×10^{11} cm^{-2} for the Ga-only or the graphene/Ga sample, respectively (Figure 1.11). The ratio of the two electron densities (2.1) well corresponds to the ratio of channel currents at the same bias (148 nA vs. 53 nA). As bias voltage is increased over the flat band voltage, accumulation electrons build up fast, ensuing electron emission at cathode and space charge build-up in the void channel.

In the graphene/Ga sample case, hole space charges are induced in the suspended graphene area and the double injection regime emerges, as evidenced by a steep rise in channel current at $V > 0.2$ V (Figure 1.9). Note that the V^3 regime of the graphene/Ga sample surpasses the V^2 regime current of the Ga-only

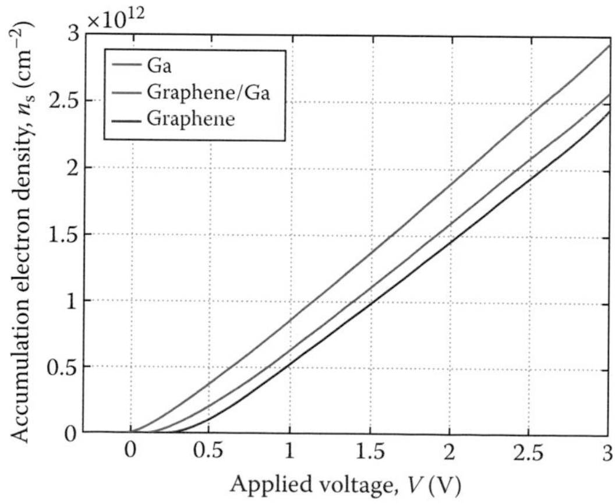

FIGURE 1.11 Accumulation electron density ($n_s = Q_s/q$) in a graphene/SiO_2/Si (GOS) (or MOS) capacitor structure calculated as a function of applied bias voltage V for 0–3 V. Three different cases are shown for top electrode: graphene (bottom); graphene/Ga (middle); Ga (top). SiO_2 thickness, 23 nm; $N_D = 1.0 \times 10^{15}$ cm^{-2}.

sample at 0.8 V. This hole-charge induction process in graphene has the effect of neutralizing the electron space charges in transit in the channel region (Langmuir & Kingdon 1923; Wilson 1959). With a reduced space charge field on cathode surface, electron emission becomes easier, resulting in higher channel current with stronger voltage dependence (i.e., V^3 instead of $V^{1.5}$ or $V^{2.0}$).

Overall, this comparison clarifies the roles played by graphene in different bias regimes: In low bias, the work function shift at anode alters the 2DEG density at cathode (and therefore, the channel current), whereas in large bias, the suspended graphene directly affects cathode emission by inducing hole space charge in the channel, thereby neutralizing electron space charge.

1.4.3 Calculation of 2DEG Density in Graphene/SiO_2/Si

Here, we calculate the 2DEG density induced at the Si/SiO_2 interface of a capacitor structure formed on a SiO_2/n-Si substrate for the cases of three different electrode materials/configurations: graphene only, graphene/Ga, or Ga only. This comparative study aims at developing a quantitative understanding of how the work function change of top electrode affects the 2DEG density induced in the Si side.

Similar to the MOS capacitor case discussed previously, the bias voltage (V) applied to a graphene/SiO_2/Si (GOS) capacitor structure is expended at mainly three places: the flat band voltage (V_{FB}), across the band bending region in semiconductor (φ_s), and across the oxide layer (V_{ox}).

$$V = V_{FB} + \varphi_s + V_{ox} \tag{1.5}$$

Here, the flat band voltage (V_{FB}) refers to the work function difference of graphene ($\phi_{graphene}$) and semiconductor (ϕ_{Si}), expressed as $V_{FB} = \phi_{graphene} - \phi_{Si}$.

The work function of intrinsic (undoped) graphene is ~4.56 eV (Yu et al. 2009; Yan et al. 2012). Unlike the metal or conventional semiconductor case, graphene offers relatively lower density of states (Luryi 1988). Therefore, the Fermi level of graphene (work function) is not fixed but can shift depending on the bias voltage and, thereby, the level of accommodation of carriers (electrons or holes), i.e., carrier concentration n_s. The Fermi level shift (referring to the Dirac point) can be characterized as

$$\Delta E_F = |v_F|\sqrt{\pi n_s}, \tag{1.6}$$

where v_F is the Fermi velocity, 1.1×10^8 cm/s.

Considering the dependence of flat band voltage on the Fermi level shift in graphene, Equation 1.5 is recast as follows.

$$V = [(\phi_{graphene} \pm \Delta E_F) - \phi_{si}] + \varphi_s + V_{ox} \tag{1.7}$$

Here, the positive sign is for p-type graphene and the negative sign is for the n-type case.

The space charge density (Q_s) in the semiconductor side (n-Si) can be determined by applying the same process as in the MOS case and is expressed as follows.

$$Q_s = -\varepsilon_s E_s = \frac{-\sqrt{2}\varepsilon_s}{\beta L_D}\left[(e^{-\beta\varphi_s} + \beta\varphi_s - 1) + \frac{p_{no}}{n_{no}}(e^{\beta\varphi_s} - \beta\varphi_s - 1)\right]^{1/2} \tag{1.8}$$

Similarly, the voltage drop across the oxide layer (V_{ox}) is related to the space charge in Si (Q_s) and oxide capacitance ($C_{ox} = \varepsilon_{ox}/d$) as follows:

$$V_{ox} = Q_s/C_{ox}. \tag{1.9}$$

Across the oxide layer, the same amount of charges n_s (= Q_s/q) (of opposite polarity) appear in the graphene side.

Solving Equations 1.6 to 1.9 simultaneously, the accumulation space charge density Q_s can be calculated as a function of applied voltage V.

Figure 1.11 shows Q_s for V in the range of 0 to 3 V for the structure on n-Si substrate with $N_D = 1.0 \times 10^{15}$ cm^{-3}. Three different cases were calculated for top electrode on SiO_2/n-Si substrate: graphene only, graphene/Ga, and Ga only.

In this calculation, the following numbers were assumed: the work function of Ga, 4.3 eV; electron affinity of Si, 4.15 eV; electron affinity of SiO_2, 0.95 eV; dielectric constant of SiO_2, 3.9; dielectric constant of Si, 11.8 (Lide 2006). The work function of Ga-covered graphene is expected to be similar to graphene's, but slightly reduced (to ~4.43 eV) because of the contact with Ga, which has a smaller work function than graphene (Giovannetti et al. 2008; Xia et al. 2011).

Now, consider the possible effect of graphene's Fermi level shift on cathode emission and, therefore, on the electron capture efficiency at suspended graphene anode. When a Ga droplet is placed on graphene, the graphene's Fermi level is expected to decrease slightly, from 4.56 eV to 4.43 eV. At low bias, this can make a significant increase in 2DEG density, e.g., at +0.4 V from 5.9×10^{10} cm^{-2} to 1.4×10^{11} cm^{-2} after placing Ga. At large bias, however, this effect becomes insignificant, e.g., at +1.0 V bias, the 2DEG density increases from 5.2×10^{11} cm^{-2} to 6.4×10^{11} cm^{-2}, only a 1.2 times increase (Figure 1.11). The cathode emission of the Ga-covered graphene sample is then estimated to have been affected by the same ratio. Overall, the result confirms good transparency of monolayer graphene to very low energy electrons that up to ~99.9% of incident electrons transmit through a suspended graphene electrode. This high level of electron transparency would be beneficial for low leakage current when a suspended graphene is utilized as a control gate (grid) in vacuum electronic devices.

1.5 Electron Capture and Transmission at Photo Reverse Bias

Under reverse bias, a MOS capacitor structure can support a 2DEG inversion layer at the Si/SiO_2 interface, which forms the basis of Si MOS field-effect-transistor technology (Sze 1981; King et al. 1998). This inversion 2DEG is of minority carriers, thermally generated in Si substrate. Under illumination of light, electron-hole pairs can be generated and separated by depletion field. Photogenerated minority carriers (electrons in p-Si substrate and holes in n-Si) drift to the Si/SiO_2 interface and become confined at the potential well, forming a 2DEG inversion layer. In this study, we investigate a graphene/SiO_2/p-Si (GOS) capacitor structure under reverse bias and optical illumination. The 2DEG at the SiO_2/Si interface is expected to emit into a nanovoid channel, being governed by the same principle as the accumulation 2DEG in a dark, forward-biased graphene/SiO_2/n-Si structure.

A graphene/SiO_2/n-Si structure with a void channel was fabricated by employing EBL, plasma etching, and graphene transfer processes (Figure 1.12). In brief, a vertical trench structure (120–440 nm width, 100–200 nm depth, 1-mm trench length) was formed by plasma RIE of a 23-nm-thick SiO_2-covered (100)-Si substrate (p-type doped with resistivity of 10 Ω-cm). A monolayer graphene (3 mm × 4 mm) was then transferred to the trench-etched substrate.

1.5.1 Photo-Induced 2DEG Inversion Layer

The photo I–V characteristic of graphene/SiO_2/p-Si structure was measured under illumination with a 633-nm laser light (1-mm beam diameter) (Figure 1.12). In reverse bias, the photocurrent saturates at 1.8–4.3 V for 0.1–1.0 mW input power. Under 0.25-mW illumination, the saturation photocurrent is read to be 0.25 mA at 2.7 V. This corresponds to a responsivity of 1.0 A/W and external quantum efficiency (EQE) of 200% (internal quantum efficiency [IQE] of 300%). The dark current is measured to be 1.6–3.6 μA at 2.2–7.0 V, resulting in an on/off current ratio of 225–270 at 1.0 mW input power (Figure 1.12).

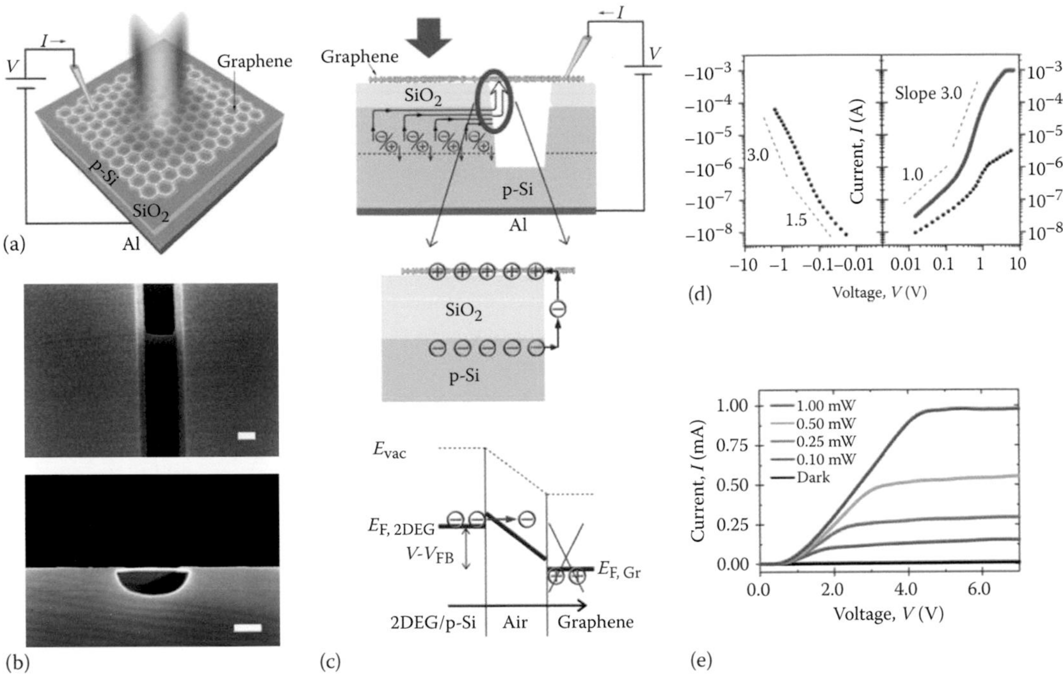

FIGURE 1.12 Emission and transport of photo-induced 2DEG in a graphene/SiO_2(23nm)/p-Si structure with a nanovoid channel under reverse bias. (a) Schematic of a GOS structure with a trench (340–440 nm width, 200 nm depth, 1 mm length) covered by a suspended graphene electrode (3 mm × 4 mm). A laser beam (1 mm diameter) illuminates the graphene electrode. (b) SEM images of monolayer graphene placed on top of a trench: top-view (top) and cross-section (bottom) images. Scale bars, 200 nm. (c) Schematic of photocurrent flow under reverse bias. Photocarriers generated in Si are separated by depletion field: photoelectrons form a 2DEG at Si/SiO_2, while photoholes flow to substrate. The 2DEG flows into/accumulate at the channel edge, while some are exiting into air and traveling toward graphene electrode (top and middle panels). Coulombic repulsion of 2DEG at the channel edge has the effect of lowering the energy barrier at the surface enabling low voltage emission into air. Similarly, the energy barrier in the graphene side is also lowered by the 2DHS in graphene (bottom panel). (d) Measured *I–V* characteristics: dark- (dotted) and photocurrent (solid) under 1-mW illumination (633 nm wavelength). (e) Measured photo *I–V* for different input power levels: (from bottom up) black, 0.10 mW, 0.25 mW, 0.50 mW, and 1.00 mW.

The responsivity measured at input power of 0.1 to 1.0 mW remains nearly constant at 0.9–1.1 A/W, demonstrating reasonably good linearity.

The absorption depth of Si is 3.0 μm at 633 nm wavelength. Since a monolayer graphene absorbs only 2.3% of incident light, most photons are absorbed in/near the depletion region (910–940 nm width at 2–5 V reverse bias) (Figure 1.12). The photogenerated carriers are separated by depletion field, and photoelectrons drift to the Si/SiO_2 interface, forming a 2DEG inversion layer. Similar to the dark forward-bias case (Srisonphan et al. 2012), Coulombic repulsion among electrons around the channel edge enables low-voltage emission of 2DEG into air. Emitted electrons travel ballistically in the nanovoid channel. Some of them are captured/collected at the edge of 2DHS induced in the graphene side, while the majority pass through the suspended graphene (Figure 1.12). The transmitted electrons form a space charge outside the suspended graphene. Once a space charge region is established, further transmitting electrons are collected by the graphene electrode on SiO_2. Photoholes separated in the depletion region drift down to the substrate side. Photoelectrons in 2DEG, traveling along the horizontal (longitudinal) direction, accumulate at the channel edge, while some exit through the edge, emitting into air. This local accumulation of electrons around the edge induces some of the photoholes to be held back near the depletion region boundary. The closed-circuit nature (i.e., charge conservation) of this two-terminal operation with a suspended graphene electrode was confirmed by performing *I–V* measurements in air or vacuum (~10^{-6} Torr) with the system ground connected to the bottom (cathode) or top (anode), which demonstrate the same amount of channel current for a given bias voltage.

Assuming 1-mW/mm^2 input power density at 633-nm wavelength and 2.3% absorption in graphene, the photocarrier generation rate in graphene is calculated to be 7.3×10^{15} s^{-1} cm^{-2}. Further assuming a minority carrier lifetime of ~1 ps (Rana et al. 2009), the photocarrier density in graphene is estimated to be ~7×10^3 cm^{-2}. This number is several orders of magnitude smaller than the 2DEG density in the Si side. This confirms that the photocurrent observed in this work originates from the photocarrier generation and separation occurring in the Si side.

1.5.2 Emission and Transport of Photocarrier 2DEG

Here, we note that the measured photo *I–V* of reverse-biased p-Si sample reveals the same voltage dependence as the dark forward-biased n-Si case: V^3 dependence of photocurrent at 0.3–1.0 V (Figure 1.12, solid), much faster than the $V^{1.5}$ dependence of CL space-charge-limited current. The 2DEG at SiO_2/Si is balanced by the 2DHS in graphene across the oxide layer, and therefore, the availability of electrons at the channel edge shows the same voltage dependence of hole concentration in graphene: $Q_e \sim Q_h \sim V^2$. Under high-level injection, the electron transport is scattering-limited, and the average velocity is proportional to the electric field, $v_{av} \sim \mu\varepsilon \sim \mu V/L$. This results in the V^3 dependence of channel current, $I = Q_e/\tau_{av} \sim V^{3.0}$.

The spectral dependence of photocurrent responsivity was characterized in the ultraviolet-to-near-infrared (UV-to-NIR) range (325–1064 nm) at input power of ~0.25 mW (Figure 1.13). The IQE shows a three-step cascade profile: an initial rise to 230% level at ~850 nm, followed by an increase to 300% level at ~650 nm, and a ramp-up to 380% at <400 nm. At steady state, the saturation photocurrent is balanced by the photocarrier generation in Si. An IQE greater than 100% indicates multiplication of photocarriers (Robbins 1980; Sano & Yoshii 1992; Kolodinski et al. 1993). The spectral dependence of IQE without a carrier multiplication effect is calculated and shown for comparison (Figure 1.13, dashed) (Sze 1981).

The saturation photocurrent (I_{ph}) was read at the knee point (V_{sat}) of photo *I–V* for ~0.25 mW input power (P_{in}). Photocurrent responsivity (I_{ph}/P_{in}) was calculated from the measured photocurrent and input power.

EQE is calculated from

$$\text{EQE} = \frac{I_{ph}/q}{P_{in}/h\nu}. \tag{1.10}$$

IQE is determined from

$$\text{IQE} = \frac{1}{1-R}\frac{I_{ph}/q}{P_{in}/h\nu} = \frac{\text{EQE}}{1-R}. \tag{1.11}$$

Here, *R* is the reflectance at the sample surface [graphene/SiO_2(23nm)/Si].

The IQE without a carrier multiplication effect (η_{in}) was calculated from (Gärtner 1959)

$$\eta_{in} = 1 - \frac{e^{-\alpha W_D}}{1+\alpha L_n}. \tag{1.12}$$

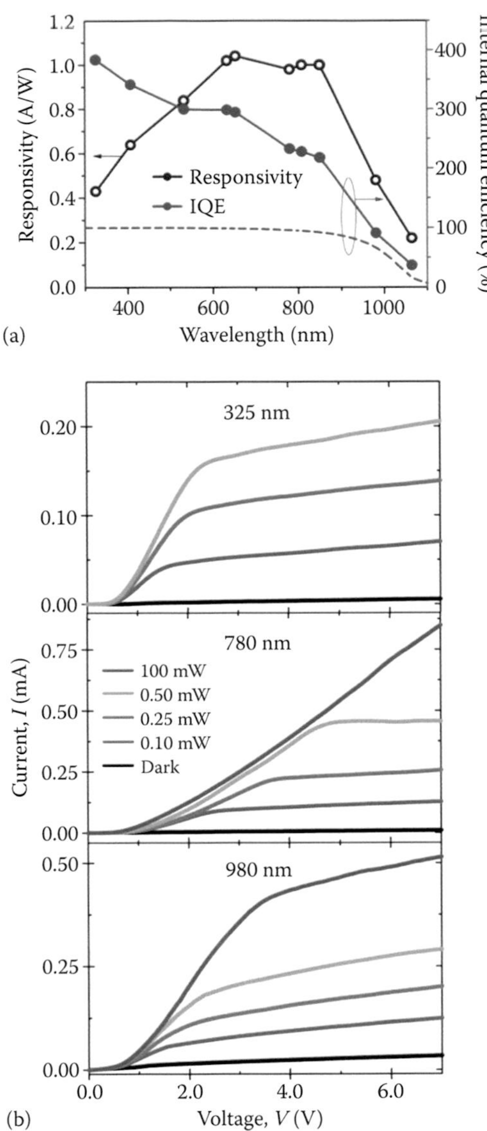

FIGURE 1.13 Spectral dependence of photocurrent responsivity and quantum efficiency. (a) The photo *I–V* characteristics of a graphene/SiO_2(23nm)/p-Si structure with a trench were measured at 325–1064 nm with input power of ~0.25 mW. The responsivity and IQE values were calculated from the knee points of saturation photocurrents. The spectral dependence reveals a three-step cascade profile with IQE of 220%–380% at <850 nm, implying broadband photocarrier multiplication. The IQE calculated without a multiplication effect is shown for comparison (dashed). (b) Photo *I–V* characteristics measured at three different wavelengths (325, 780, and 980 nm) with different input power.

Here, α is the absorption coefficient of silicon (Green & Keevers 1995; Lide 2006). L_n is the minority carrier diffusion length (Law et al. 1991). W_D is the dark depletion region width. The EQE without multiplication (η_{ex}) was calculated from $(1 - R)\eta_{in}$.

The three-step-cascade profile observed with the graphene/SiO_2/p-Si structure suggests that different mechanisms are involved in carrier multiplication depending on spectral range. The first regime that covers the NIR (>~800 nm) corresponds to near-band-edge absorption. Since the amount of above-bandgap excess energy is negligible in this regime, the underlying mechanism is believed to involve a field-assisted process.

The saturation photocurrent is determined by the photocarrier generation rate in Si and is expressed as $q\eta P_{in}/h\nu$, where q is the electron charge, η is the EQE, P_{in} is the power incident to graphene electrode, and $h\nu$ is the photon energy. For 1-mW illumination with 800-nm light (1-mm beam diameter), the photogeneration rate is calculated to be $2.7 \times 10^{17}\ s^{-1}\ cm^{-2}$. This corresponds to a photocarrier density of $8.2 \times 10^{12}\ cm^{-2}$ when assuming a minority carrier (electron) lifetime of 30 μs in Si (Law et al. 1991). The photogenerated electrons drift to the Si/SiO_2 interface and flow along the 2DEG channel. The channel edge at the trench becomes a bottleneck for continuous flow of electrons (i.e., exit flux into air), since the emission of 2DEG at the edge is limited by the space charge effect in the void channel. Photoelectrons then accumulate at the edge while some return. The local 2DEG density at the bottleneck is expected to stabilize at a certain level because of a negative feedback effect discussed further in the chapter.

Across the oxide layer of the capacitor structure, the graphene side will have an accumulation of positive charge (holes) to the same level as that of net negative charge in Si (Das et al. 2008). The peak hole concentration in graphene is expected to be limited to/stabilized at $\sim 1 \times 10^{13}\ cm^{-2}$. Beyond this level, the electric field in SiO_2 ($> \sim 5 \times 10^6$ V/cm) would reach the breakdown regime (Ponomarenko et al. 2013), and carriers (holes) generated by an avalanche process will neutralize the photoelectrons accumulated in Si, reducing the concentration to a stable level. At this level of hole concentration, the Fermi level of graphene is ~0.4 eV below the Dirac point (Das et al. 2008). A further rise in hole density would lower the graphene's Fermi level and thereby decrease the flat band voltage (V_{FB}) such that the band bending in Si becomes less, lowering the 2DEG density. Overall, this negative feedback effect is expected to limit/stabilize electron accumulation to $\sim 1 \times 10^{13}\ cm^{-2}$ level. In the Si side, the accumulated electrons will attract holes that are being generated in the depletion/neutral region. The photoholes induced by this Coulombic interaction will form a space charge region at the depletion/neutral boundary, altering the field and potential distributions.

1.6 Conclusion

We reviewed the recent progress in investigating suspended graphene's perpendicular interactions with very-low-energy (<5eV) impinging electrons. In this study, a graphene membrane is suspended on top of a nanoscale void channel formed in a SiO_2/Si substrate. In generating a constant flux of very-low-energy electrons, we exploit the phenomenon that a 2DEG induced at the SiO_2/Si interface of a MOS structure can easily emit into air (a void channel whose channel length is smaller than the mean free path) at low voltage (~1 V) and makes a ballistic transport toward the suspended graphene. Here, the 2DEG is induced by applying dark forward bias (accumulation) or photo reverse bias (inversion). We characterized the emission, capture, and transmission interactions of suspended graphene with the low-energy incident electrons. A small fraction (>~0.1%) of impinging electrons are captured at the edge of 2DHS in graphene, demonstrating good transparency (up to ~99.9%) to very-low-energy (<5 eV) electrons. While being transmissive, the suspended graphene is found to be highly responsive to impinging electrons. In response to electron injection, a graphene anode induces hole charges in the suspended area, thereby neutralizing electron space charge. This charge compensation dramatically enhances 2DEG emission at cathode to the level far surpassing the CL SCL emission. Besides electron transparency, graphene's ability to overcome the space charge limit in cathode emission offers promising potential for low-voltage, high-current-density nanoscale vacuum electronic devices.

Acknowledgments

This work was supported by the Office of Naval Research (grant no. N00014-1310465) and the National Science Foundation (grant no. ECCS-0925532).

References

Adam, S. & Sarma, D., "Transport in suspended graphene," *Solid State Commun.* 146 (2008): 356–360.

Ando, T., Fowler, A. B. & Stern, F., "Electronic properties of two-dimensional systems," *Rev. Mod. Phys.* 54 (1982): 437–625.

Bao, W., Miao, F., Chen, Z. et al., "Controlled ripple texturing of suspended graphene and ultrathin graphite membranes," *Nat. Nanotechnol.* 4 (2009): 562–566.

Black, C. T. & Welser, J. J., "Electric-field penetration into metals: Consequences for high-dielectric-constant capacitors," *IEEE Trans. Electron Devices* 46 (1999): 776–780.

Bolotin, K. I., Sikes, K. J., Jiang, Z. et al., "Ultrahigh electron mobility in suspended graphene," *Solid State Commun.* 146 (2008): 351–355.

Bolotin, K., Ghahari, F., Shulman, M. D. et al., "Observation of the fractional quantum Hall effect in graphene," *Nature* 462 (2009): 196–199.

Börrnert, F., Avdoshenko, S. M., Bachmatiuk, A. et al., "Amorphous carbon under 80 kV electron irradiation: A means to make or break graphene," *Adv. Mater.* 24 (2012): 5630–5635.

Brodie, I., "Physical considerations in vacuum microelectronics devices," *IEEE Trans. Electron Devices* 36 (1989): 2641–2644.

Bunch, J. S., Verbridge, S. S., Alden, J. S. et al., "Impermeable atomic membranes from graphene sheets," *Nano Lett.* 8 (2008): 2458–2462.

Cazaux, J., "Reflectivity of very low energy electrons (<10 eV) from solid surfaces: Physical and instrumental aspects," *J. Appl. Phys.* 111 (2012): 064903.

Chen, J. H., Jang, C., Xiao, S. et al., "Intrinsic and extrinsic performance limits of graphene devices on SiO_2," *Nat. Nanotechnol.* 3 (2008): 206–209.

Child, C. D., "Discharge from hot CaO," *Phys. Rev.* 32 (1911): 492–511.

Crespi, V. H., Chopra, N. G., Cohen, M. L. et al., "Anisotropic electron-beam damage and the collapse of carbon nanotubes," *Phys. Rev. B* 54 (1996): 5927–5931.

Das, A., Pisana, S., Chakraborty, B. et al., "Monitoring dopants by Raman scattering in an electrochemically top-gated graphene transistor," *Nat. Nanotechnol.* 3 (2008): 210–215.

Du, X., Skachko, I., Barker, A. et al., "Approaching ballistic transport in suspended graphene," *Nat. Nanotechnol.* 3 (2008): 491–495.

Eisenstein, J. P., Pfeiffer, L. N. & West, K. W., "Negative compressibility of interacting two-dimensional electron and quasiparticle gases," *Phys. Rev. Lett.* 68 (1992): 674–677.

Fowler, R. H. & Nordheim, L., "Electron emission in intense electric fields," *Proc. Roy. Soc. Lond.* 119 (1928): 173–181.

Frank, I. W., Tanenbaum, D. M., van der Zande, A. M. et al., "Mechanical properties of suspended graphene sheets," *J. Vac. Sci. Technol. B* 25 (2007): 2558.

Garaj, S., Hubbard, W., Reina, A. et al., "Graphene as a subnanometre trans-electrode membrane," *Nature* 467 (2010): 190–193.

Gärtner, W. W., "Depletion-layer photoeffects in semiconductors," *Phys. Rev.* 116 (1959): 84–87.

Geim, A., "Graphene: Status and prospects," *Science* 324 (2009): 1530–1534.

Geim, A. K. & Novoselov, K. S., "The rise of graphene," *Nat. Mater.* 6 (2007): 183–191.

Giovannetti, G., Khomyakov, P. A., Brocks, G. et al., "Doping graphene with metal contacts," *Phys. Rev. Lett.* 101 (2008): 026803.

Green, M. A. & Keevers, M., "Optical properties of intrinsic silicon at 300 K," *Prog. Photovoltaics* 3 (1995): 189–192.

Grinberg, A. A., Luryi, S., Pinto, M. R. et al., "Space-charge-limited current in a film," *IEEE Trans. Electron Devices* 36 (1989): 1162–1170.

Han, S. & Ihm, J., "Role of the localized states in field emission of carbon nanotubes," *Phys. Rev. B* 61 (2000): 9986–9989.

Han, J.-W., Oh, J. S. & Meyyappan, M., "Vacuum nanoelectronics: Back to the future?—Gate insulated nanoscale vacuum channel transistor," *Appl. Phys. Lett.* 100 (2012): 213505.

Herring, C. & Nichols, M. H., "Thermionic emission," *Rev. Mod. Phys.* 21 (1949): 185–270.

Ho, L. H., Micolich, A. P., Hamilton, A. R. et al., "Ground-plane screening of Coulomb interactions in two-dimensional systems: How effectively can one two-dimensional system screen interactions in another," *Phys. Rev. B* 80 (2009): 155412.

Hu, S., Lozada-Hidalgo, M., Wang, F. C. et al., "Proton transport through one-atom-thick crystals," *Nature* 516 (2014): 227–230.

Hwang, E. H., Adam, S. & Sarma, D., "Carrier transport in two-dimensional graphene layers," *Phys. Rev. Lett.* 98 (2007): 186806.

Jernigan, G. G., VanMil, B. L., Tedesco, J. L. et al., "Comparison of epitaxial graphene on Si-face and C-face 4H SiC formed by ultrahigh vacuum and RF furnace production," *Nano Lett.* 9 (2009): 2605–2609.

Kim, K. S., Jang, H., Lee, S. Y. et al., "Large-scale pattern growth of graphene films for stretchable transparent electrodes," *Nature* 457 (2009): 706–710.

King, Y.-C., Fujioka, H., Kamoharaet, S. et al., "DC electrical oxide thickness model for quantization of the inversion layer in MOSFETs," *Semicond. Sci. Technol.* 13 (1998): 963–966.

Koenig, S. P., Boddeti, N. G., Dunn, M. L. et al., "Ultrastrong adhesion of graphene membranes," *Nat. Nanotechnol.* 6 (2011): 543–546.

Kolodinski, S., Werner, J. H., Wittchen, T. et al., "Quantum efficiencies exceeding unity due to impact ionization in silicon solar cells," *Appl . Phys. Lett.* 63 (1993): 2405–2407.

Kotov, V. N., Uchoa, B., Vitor, M. et al., "Electron–electron interactions in graphene: Current status and perspectives," *Rev. Mod. Phys.* 84 (2012): 1067–1125.

Krasheninnikov, A. V. & Nordlund, K., "Ion and electron irradiation-induced effects in nanostructured materials. *J. Appl. Phys.* 107 (2010): 071301.

Kreuzer, H. J., Nakamura, K., Wierzbicki, A. et al., "Theory of the point source electron microscope," *Ultramicroscopy* 45 (1992): 381–403.

Kuhr, J.-C. & Fitting, H.-J., "Monte Carlo simulation of electron emission from solids," *J. Electr. Spectr. Rel. Phenom.* 105 (1999): 257–273.

Lampert, M. A. & Rose, A., "Volume-controlled, two-carrier currents in solids: The injected plasma case," *Phys. Rev.* 121 (1961): 26–37.

Landau, L. D. & Lifshitz, E. M., *Statistical Physics*, Part I, Sections 137 and 138 (Pergamon, Oxford, 1980).

Langmuir, I., "The effect of space charge and residual gases on thermionic currents in high vacuum," *Phys. Rev.* 2 (1913): 450–486.

Langmuir, I. & Kingdon, K. H., "Thermionic effects caused by alkali vapors in vacuum tubes," *Science* 57 (1923): 58–60.

Lau, Y. Y., Liu, Y. & Parker, R. K., "Electron emission: From the Fowler–Nordheim relation to the Child–Langmuir law," *Phys. Plasmas* 1 (1994): 2082–2085.

Law, M. E., Solley, E., Liyang, M. et al., "Self-consistent model of minority-carrier lifetime, diffusion length, and mobility," *IEEE Electron Device Lett.* 12 (1991): 401–403.

Lee, C., Wei, X., Kysar, J. W. & Hone, J., "Measurement of the elastic properties and intrinsic strength of monolayer graphene," *Science* 321 (2008): 385–388.

Li, X., An, J., Kim, S. et al., "Large-area synthesis of high-quality and uniform graphene films on copper foils," *Science* 324 (2009): 1312–1314.

Li, L., Richter, C., Paetel, S. et al., "Very large capacitance enhancement in a two-dimensional electron system," *Science* 332 (2011): 825–828.

Li, C., Cole, M. T., Lei, W. et al., "Highly electron transparent graphene for field emission triode gates," *Adv. Funct. Mater.* 24 (2014): 1218–1227.

Lide, D. R., ed., *CRC Handbook of Chemistry and Physics*, 87th ed. (CRC Press, Boca Raton, FL, 2006).

Liu, L., Ryu, S., Tomasik, M. R. et al., "Graphene oxidation: Thickness-dependent etching and strong chemical doping," *Nano Lett.* 8 (2008): 1965–1970.

Luryi, S., "Quantum capacitance devices," *Appl. Phys. Lett.* 52 (1988): 501–503.

Mattevi, C., Kim, H. & Chhowalla, M., "A review of chemical vapor deposition of graphene on copper," *J. Mater. Chem.* 21 (2011): 3324–3334.

Mayer, A., "Polarization of metallic carbon nanotubes from a model that includes both net charges and dipoles," *Phys. Rev. B* 71 (2005): 235333.

Mead, C. A., "Anomalous capacitance of thin dielectric structures," *Phys. Rev. Lett.* 6 (1961): 545–546.

Mermin, N. D., "Crystalline order in two dimensions," *Phys. Rev.* 176 (1968): 250–254.

Meyer, J. C., Geim, A. K. & Katsnelson, M. I., "The structure of suspended graphene sheets," *Nature* 446 (2007): 60–63.

Mil'shtein, S., Paludi, Jr., C. A., Chau, P. et al., "Perspectives and limitations of vacuum microtubes," *J. Vac. Sci. Technol. A* 11 (1993): 3126–3129.

Morin, P., Pitaval, M. & Vicario, E., "Low energy off-axis holography in electron microscopy," *Phys. Rev. Lett.* 76 (1996): 3979–3982.

Morozov, S. V., Novoselov, K. S., Katsnelson, M. I. et al., "Giant intrinsic carrier mobilities in graphene and its bilayer," *Phys. Rev. Lett.* 100 (2008): 016602.

Mott, N. F. & Gurney, R. W., *Electronic Processes in Ionic Crystals* (Oxford Univ Press, New York, 1940).

Müllerová, I., Hovorka, M., Hanzlíková, R. et al., "Very low energy scanning electron microscopy of free-standing ultrathin films," *Mater. Trans.* 51 (2010): 265–270.

Mutus, J. Y., Livadaru, L., Robinson, J. T. et al., "Low-energy electron point projection microscopy of suspended graphene, the ultimate 'microscope slide,'" *New J. Phys.* 13 (2011): 063011.

Neto, A. H. C., Guinea, F., Peres, N. M. R. et al., "The electronic properties of graphene," *Rev. Mod. Phys.* 81 (2009): 109–162.

Novoselov, K. S., Geim, A. K., Morozov, S. V. et al., "Electric field effect in atomically thin carbon films," *Science* 306 (2004): 666–669.

Novoselov, K. S., Geim, A. K., Morozov, S. V. et al., "Two-dimensional gas of massless Dirac fermions in graphene," *Nature* 438 (2005): 197–200.

Ponomarenko, L. A., Gorbachev, R. V., Yu, G. L. et al., "Cloning of Dirac fermions in graphene superlattices," *Nature* 497 (2013): 594–597.

Rana, F., George, P. A., Strait, J. H., Dawlaty, J., Shivaraman, S., Chandrashekhar, M. & Spencer, M. G., "Carrier recombination and generation rates for intravalley and intervalley phonon scattering in graphene," *Phys. Rev. B* 79 (2009): 115447.

Robbins, D. J., "Aspects of the theory of impact ionization in semiconductors (I)," *Phys. Stat. Sol. B* 97 (1980): 9–50.

Sano, N. & Yoshii, A., "Impact-ionization theory consistent with a realistic band structure of silicon," *Phys. Rev. B* 45 (1992): 4171–4180.

Schedin, F., Geim, A. K., Morozov, S. V. et al., "Detection of individual gas molecules adsorbed on graphene," *Nat. Mater.* 6 (2007): 652–655.

Shivaraman, S., Barton, R. A., Yu, X. et al., "Free-standing epitaxial graphene," *Nano Lett.* 9 (2009): 3100–3105.

Spindt, C. A., "A thin-film field-emission cathode," *J. Appl. Phys.* 39 (1968): 3504–3505.

Spindt, C. A., Brodie, I., Humphrey, L. et al., "Physical properties of thin-film field emission cathodes with molybdenum cones," *J. Appl. Phys.* 47 (1976): 5248–5263.

Spindt, C. A., Holland, C. E., Rosengreen, A. et al., "Field-emitter arrays for vacuum microelectronics," *IEEE Trans. Electron Devices* 38 (1991): 2355–2363.

Srisonphan, S., Jung, Y. S. & Kim, H. K., "Metal-oxide-semiconductor field-effect transistor with a vacuum channel," *Nat. Nanotechnol.* 7 (2012): 504–508.

Srisonphan, S., Kim, M. & Kim, H. K., "Space charge neutralization by electron-transparent suspended graphene," *Sci. Rep.* 4 (2014): 3764.

Stoner, B. R. & Glass, J. T., "Nanoelectronics: Nothing is like a vacuum," *Nat. Nanotechnol.* 7 (2012): 485–487.

Suk, J. W., Kitt, A., Magnuson, C. W. et al., "Transfer of CVD-grown monolayer graphene onto arbitrary substrates," *ACS Nano* 5 (2011): 6916–6924.

Sze, S. M., *Physics of Semiconductor Devices*, 2nd ed. (Wiley, New York, 1981).

Sze, S. M. ed., *High-Speed Semiconductor Devices* (Wiley, New York, 1990).

Toriumi, A., Yoshimi, M., Iwase, M. et al., "Experimental determination of finite inversion layer thickness in thin gate oxide MOSFETs," *Surf. Sci.* 170 (1986): 363–369.

Wilson, V. C., "Conversion of heat to electricity by thermionic emission," *J. Appl. Phys.* 30 (1959): 475–481.

Xia, F., Perebeinos, V., Lin, Y.-M. et al., "The origins and limits of metal-graphene junction resistance," *Nat. Nanotechnol.* 6 (2011): 179–183.

Yan, R., Zhang, Q., Li, W. et al., "Determination of graphene work function and graphene-insulator-semiconductor band alignment by internal photoemission spectroscopy," *Appl. Phys. Lett.* 101 (2012): 022105.

Yang, G., Chen, K. K. & Marcus, R. B., "Electron field emission through a very thin oxide layer," *IEEE Trans. Electron Devices* 38 (1991): 2373–2376.

Yu, Y.-J., Zhao, Y., Ryu, S. et al., "Tuning the graphene work function by electric field effect," *Nano Lett.* 9 (2009): 3430–3434.

Yun, M., Turner, A., Roedel, R. J. et al., "Novel lateral field emission device fabricated on silicon-on-insulator material," *J. Vac. Sci. Technol. B* 17 (1999): 1561–1566.

Zhang, Y., Tan, Y.-W., Stormer, H. L. et al., "Experimental observation of the quantum Hall effect and Berry's phase in graphene," *Nature* 438 (2005): 201–204.

Zheng, X., Chen, G., Li, Z. et al., "Quantum-mechanical investigation of field-emission mechanism of a micrometer-long single-walled carbon nanotube," *Phys. Rev. Lett.* 92 (2004): 106803.

2

Graphene Quantum Dots

Christian Volk

Christoph Neumann

Stephan Engels

Alexander Epping

Christoph Stampfer

Abstract

This chapter reviews experiments on graphene and bilayer graphene quantum dot devices. First, a brief theoretical background is given. The electronic and optical properties of graphene QDs are explained, and the differences from extended graphene sheets are emphasized.

The most common fabrication technique is based on micromechanical exfoliation of individual graphene flakes from bulk graphite followed by plasma etching the desired shape. Carving width modulated nanoribbons out of graphene flakes, it has been possible to realize single electron transistors, quantum dots, and double quantum dots. Following a bottom-up approach, graphene QDs can either be fabricated by ruthenium-catalyzed C60 transformation, by hydrothermal or electrochemical strategies from graphene oxide, or by wet chemical oxidization and cutting of micrometer-sized carbon fibers. These types of graphene QDs are used in optical and biological experiments.

The transport properties of graphene nanoribbons are summarized. They allow opening a band-gap in graphene by lateral confinement and thus serve as building blocks for many graphene quantum devices. Graphene quantum dots have intensively been studied in low-temperature DC measurements. Coulomb blockade and excited state spectra have been investigated in graphene quantum dots. Graphene nanoribbon-based charge sensors are commonly used to detect charging events in regimes where the current through the quantum dots is below the detection limit. Furthermore, time-resolved charge detection on a graphene quantum dots has been demonstrated. In magnetic field-dependent measurements, it was possible to determine spin-filing sequences and the g-factor in graphene quantum dots. The relaxation dynamics of excited states in quantum dots is addressed by pulsed gate spectroscopy. Measurements of transient currents through electronic excited states give an estimate for a lower bound for charge relaxation times on the order of 60–100 ns.

As the quality of graphene devices is regarded to be limited by disorder induced by surface potentials of the host substrate and edge states, alternative fabrication techniques have been investigated. Local anodic oxidation avoids the use of lithography and plasma etching and can thus potentially reduce the influence of edge disorder. Spin states have been investigated in devices built following this technique. To cope with the problem of substrate-induced disorder, graphene quantum dots on hexagonal boron nitride (hBN) have been investigated. A competitive study of devices on hBN and SiO_2 has shown a significant reduction of the influence of surface-induced disorder.

Bilayer graphene is of special interest as it allows to open a bandgap by applying perpendicular electric fields. This approach has been used to realize quantum dots by soft confinement. Bilayer graphene quantum dots in suspended flakes and in hBN/bilayer graphene/hBN heterostructures have been studied.

The device concepts known from quantum dots have been extended to realize double quantum dots. Excited state spectra, the gate control of the mutual capacitive coupling and the interdot tunnel coupling, have been studied. By pulsed gate control on a double quantum dot, a gigahertz charge pump has been demonstrated. Studies on a bilayer graphene double quantum dot have proven that the excited state level spacing in bilayer graphene is constant in contrast to single-layer graphene. A magnetic field dependency of excited states in agreement with Zeeman splitting has been observed.

2.1 Introduction

Quantum dots (QDs) allow a controlled investigation and manipulation of individual quantum systems. They are in particular interesting as promising hosts for spin qubits (Loss and DiVincenzo 1998). These are reasons why QDs have been intensively investigated in different material systems over the past years. So far, most progress has been made in QDs in two-dimensional (2D) electron gases, especially in GaAs-based heterostructures (Elzermann et al. 2003; Hanson 2005; Johnson et al. 2005; Petta et al. 2005; Koppens et al. 2006; Nowack et al. 2011), and elementary spin-qubit operations have been demonstrated.

However, these systems suffer from limited spin decoherence times originating from spin-orbit and hyperfine interactions (Khaetskii et al. 2002). Ways to cope with this issue have been explored, e.g., by polarizing the nuclear spin bath (Bluhm et al. 2010). To minimize the influence of nuclear spins, alternative materials are of great interest, especially group IV elements. Spin relaxation has been measured, e.g., in Ge/Si nanowire qubits (Hu et al. 2011) and in silicon QDs (Yang et al. 2013). Recently, electron spin resonance has been demonstrated in a Si/SiGe spin qubit, with decay timescales significantly larger compared with III/V QDs (Kawakami et al. 2014). One way to further reduce the influence of nuclear spins is the use of isotopically purified silicon.

Among the group IV elements, carbon materials are an interesting alternative. The spin-orbit interaction is small because of the low mass of the nucleus (Huertas-Hernando et al. 2006; Min et al. 2006). The hyperfine interaction is weak, as 99% of natural carbon is the isotope C^{12}, which has zero nuclear spin (Trauzettel et al. 2007). A valley-spin qubit in a carbon nanotube (CNT) has been studied by Hahn echo measurements (Laird et al. 2013). It is predicted that spin-qubits in graphene feature long coherence times (Trauzettel et al. 2007).

Despite the advantages concerning hyperfine and spin-orbit interaction, graphene has a significant drawback. Because of the absence of a bandgap and the pseudorelativistic Klein tunneling effect, it is challenging to confine electrons (Katsnelson et al. 2006; Castro Neto et al. 2009; Das Sarma et al. 2011).

A lot of effort has been spent so far to overcome this limitation. Most approaches are based on carving nanostructures out of graphene sheets. It has been shown that an energy gap will be opened in graphene nanoribbons (GNRs). Single-electron transistors (SETs), QDs, and double QDs (DQDs) have successfully been fabricated either by etching width-modulated nanostructures (Ponomarenko et al. 2008; Stampfer et al. 2008b; Güttinger et al. 2009; Moriyama et al. 2009; Molitor et al 2009b; Wang et al. 2010; Volk et al. 2011; Connolly et al. 2013; Volk et al. 2013; Engels et al. 2013b) or by electrostatic confinement of electrons in nanoribbons using gate electrodes (Liu et al. 2009, 2010). Another technique defines the nanostructures by local anodic oxidation (Neubeck et al. 2010; Puddy et al. 2013). Recently, the confinement of electrons by magnetic fields has been demonstrated (Moriyama et al. 2014). In addition, bilayer graphene offers the possibility for soft confinement of electrons by applying perpendicular electric fields (Allen et al. 2012; Goossens et al. 2012b).

Graphene QDs are not only investigated in electronic transport experiments but also their optical properties are explored, e.g., by photoluminescence (PL) and PL excitation (PLE) (Kim et al. 2012; Peng et al. 2012). Although no optical luminescence is observed in extended graphene sheets, the size-dependent confinement gap in graphene QDs results in size-dependent (and hence controllable) emission and absorption spectra, which makes graphene QDs promising for optoelectronic and bioimaging applications.

2.2 Band Structure of Graphene and Bilayer Graphene

Graphene is a 2D crystal where neighboring carbon atoms form strong covalent sp^2 σ-bonds with a binding angle of 120°, resulting in a hexagonal crystal structure that can be regarded as a trigonal lattice with two atoms per unit cell. The carbon–carbon bond length measures $a_0 = 1.42$ Å (see Figure 2.1a). The remaining p_z orbitals of the carbon atoms give rise to the so-called π-bands responsible for the electronic properties of graphene.

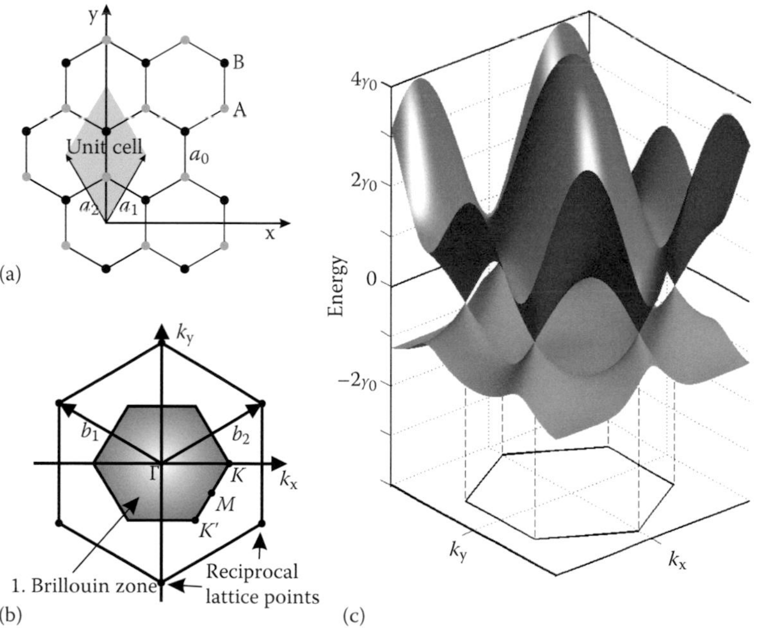

FIGURE 2.1 Crystal structure and band structure of single-layer graphene. (a) Crystal structure of graphene in real space. The colors distinguish between the A and B sublattice. The unit cell is shaded in gray. (b) Reciprocal lattice of graphene. The first Brillouin zone is highlighted, including the high symmetry points Γ, *M*, *K*, and *K′*. (c) Nearest-neighbor tight-binding band structure calculation (cf. Equation 2.1). The π- and π^*-bands touch at the *K*- and *K′*-points.

The first Brillouin zone shows a hexagonal symmetry reminding of the graphene crystal structure in real space (see Figure 2.1b). The two inequivalent corners of the Brillouin zone are $\mathbf{K} = (2\pi/3a_0)(1,\sqrt{3})$ and $\mathbf{K}' = (2\pi/3a_0)(1,-\sqrt{3})$.

The electronic band structure can be derived following a tight-binding approach (Wallace 1947). Taking into account both nearest-neighbor and next-nearest-neighbor hopping terms results in a band structure expressed by the relation (Castro Neto et al. 2009; Das Sarma et al. 2011; Katsnelson et al. 2012).

$$E_{\pm}(\mathbf{k}) = \pm\gamma_0\sqrt{3+f(\mathbf{k})} - \gamma_0' f(\mathbf{k}), \tag{2.1}$$

with

$$f(\mathbf{k}) = f(k_x,k_y) = 2\cos\left(\sqrt{3}k_y a_0\right) + 4\cos\left(\sqrt{3}/2k_y a_0\right)\cos(3/2k_x a_0). \tag{2.2}$$

The nearest-neighbor hopping energy measures $\gamma_0 \approx 2.8$ eV. According to ab initio calculations, the next-nearest-neighbor hopping energy γ_0' has been found to be on the order of $-0.02\gamma_0$ to $-0.2\gamma_0$ (Reich et al. 2002). The positive and negative branches of the dispersion relation correspond to the π- and π*-band, respectively. Finite γ_0' breaks the electron hole symmetry. A detailed calculation can be found, e.g., in review articles by Castro Neto et al. (2009) and Das Sarma et al. (2011). Figure 2.1c shows the band structure for $\gamma_0 = 2.8$ eV and $\gamma_0' = -0.2\gamma_0$.

At the center of the Brillouin zone, the Γ-point, the conduction and valence bands are separated by $6\gamma_0 \approx$ 17 eV, whereas the bands touch at the K-points of the Brillouin zone, making graphene a semimetal. In transport measurements, usually, only energies close to the Fermi energy are accessible; thus, only the dispersion relation close to the charge neutrality point is relevant. In this regime, the dispersion relation can be linearized around **K** as $E_{\pm}(\mathbf{k}) = \pm\hbar v_F|\mathbf{k}|$, where v_F is the Fermi velocity $3\gamma_0 a_0/2\hbar \approx 10^6$ m/s (Wallace 1947; Katsnelson et al. 2006; Neto et al. 2006; Katsnelson and Novoselov 2007). As the dispersion relation takes the form of the one valid for massless relativistic particles ($E = cp = \hbar c|k|$, where c is the speed of light), the K-points are called Dirac points. The density of states is given by $D(E) = 2|E|/\left(\pi\hbar^2 v_F^2\right)$ in contrast to conventional 2D electron gases (2DEGs), where it is independent of energy.

Bilayer graphene is formed by two graphene sheets Bernal stacked on top of each other. In close analogy to single-layer graphene, the band structure can be obtained by a tight-binding approach taking into account intralayer and interlayer hopping (γ_0, γ_1) (cf. Figure 2.2).

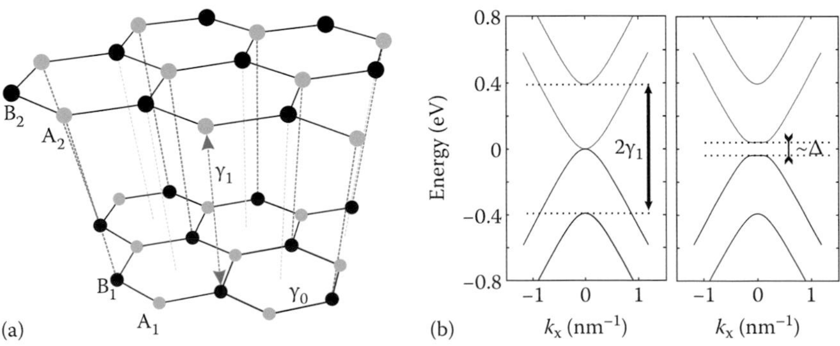

FIGURE 2.2 Crystal structure and band structure of bilayer graphene. (a) Crystal structure of bilayer graphene. The interlayer distance measures $c_0 = 3.35$ Å, the intralayer hopping energy $\gamma_0 \approx 2.8$ eV, and the interlayer hopping energy $\gamma_1 \approx 0.4$ eV. (From Du, X. et al., *Nature*, 462, 192–195, 2009.) The latter describes the coupling of B_1 atoms in the lower layer and A_2 atoms in the upper layer. (b) Band structure in the absence of an external electric field and the presence of a perpendicular external electric field leading to a potential difference of Δ = 80 mV, which breaks the lattice symmetry and thus leads to a bandgap opening.

The energy dispersion relation around the K-point is given by

$$E_{\pm}(k) = \pm\left(\hbar^2 v_F^2 k^2 + \gamma_1^2/2 \pm \sqrt{\gamma_1^4/4 + \gamma_1^2 \hbar^2 v_F^2 k^2}\right)^{1/2}. \tag{2.3}$$

The sign in front of the inner square root distinguishes between the first and second energy subbands, which are split by $2\gamma_1 \approx 0.78$ eV (Brandt et al. 1988; Dresselhaus and Dresselhaus 2002; McCann et al. 2006; Castro Neto et al. 2009; Das Sarma et al. 2011).

In the limit of low momenta ($k \ll \gamma_1/(2\hbar v_F)$), the dispersion relation simplifies to $E_{\pm}(k) = \pm\hbar^2 v_F^2 k^2/\gamma_1$, describing a parabolic dispersion relation. As a consequence, the quasiparticles have a finite effective mass $m^* = \gamma_1/2v_F^2$ and the density of states is constant: $D(E) = \gamma_1/\left(\pi\hbar^2 v_F^2\right)$. At large k, the dispersion relation converges toward the linear one known from single-layer graphene.

Interestingly, bilayer graphene allows the opening of a bandgap by applying a perpendicular electric field. Taking into account this effect, the energy dispersion relation reads as

$$E_{\pm}(k) = \pm\left(\Delta^2/4 + \hbar^2 v_F^2 k^2 + \gamma_1^2/2 \pm \sqrt{\gamma_1^4/4 + \left(\gamma_1^2 + \Delta^2\right)\hbar^2 v_F^2 k^2}\right)^{1/2}, \tag{2.4}$$

where Δ is a measure for the potential energy difference of the two graphene layers (Brandt et al. 1988; McCann et al. 2006; Das Sarma et al. 2011; Katsnelson 2012; Castro Neto et al. 2009). As long as Δ is small compared to γ_0, the dispersion relation remains parabolic around the K-point, while at larger Δ, the curve takes the typical "mexican hat"-like shape with a minimum bandgap of $\Delta - \Delta^3/2\gamma_1^2$. Figure 2.2b compares the band structure with and without an applied electric field.

2.3 Electronic and Optical Properties of Graphene and Bilayer Graphene Quantum Dots (QDs)

2.3.1 Graphene in Reduced Dimensions

Right after the experimental discovery of graphene, electrical transport experiments have been done on "bulk" graphene, typically micrometer-sized Hall bars. Interesting properties like the theoretically predicted temperature-independent mobility have been demonstrated (Novoselov et al. 2004). One of the famous results is the observation of the anomalous "half-integer" quantum Hall effect (Novoselov et al. 2005; Zhang et al. 2005). It is the quasi-relativistic analog of the integer quantum Hall effect in semiconductors with a parabolic dispersion relation. Furthermore, it has been observed that the minimum conductivity in graphene does not reach zero in the limit of zero carrier density but approaches a minimum of the order of e^2/h (Tworzydło et al. 2006; Tan et al. 2007). This is an intrinsic property of 2D Dirac fermions present in graphene crystals without impurities or lattice defects (Katsnelson et al. 2006): The experimental value depends on sample geometry, disorder, and overall doping of the graphene flake (Geim and Novoselov 2007; Tan et al. 2007). The presence of the residual conductivity has strong impact on the electrostatic tunability of graphene devices (Novoselov et al. 2004), making it difficult to fully pinch off currents in 2D graphene devices.

In GNRs (which typically have a width around 100 nm), an effective energy gap can be observed (see Figure 2.3b). Ideal GNRs (Brey and Fertig 2006; White et al. 2007) promise interesting quasi-1D physics in analogy to CNTs (Saito et al. 1999; Reich et al. 2004), which can be imagined as rolled-up GNRs. The overall semiconducting behavior of GNRs allows overcoming the limitations of the gapless graphene band structure. This makes them promising candidates for the fabrication of nanoscale graphene transistors (Wang et al. 2008), tunnel barriers, and QDs (Ponomarenko et al. 2008; Stampfer et al. 2008). Electronic transport through GNRs has been studied intensively (Chen et al. 2007; Han et al. 2007; Liu et al. 2009; Stampfer et al. 2009; Todd et al. 2009) and will be discussed in more detail in Section 2.5.1.

QDs (Sohn et al. 1997) are tiny objects, typically consisting of 10^3–10^9 atoms. The confinement of the electrons in all three spatial directions results in a quasi-0D system within a quantized energy spectrum

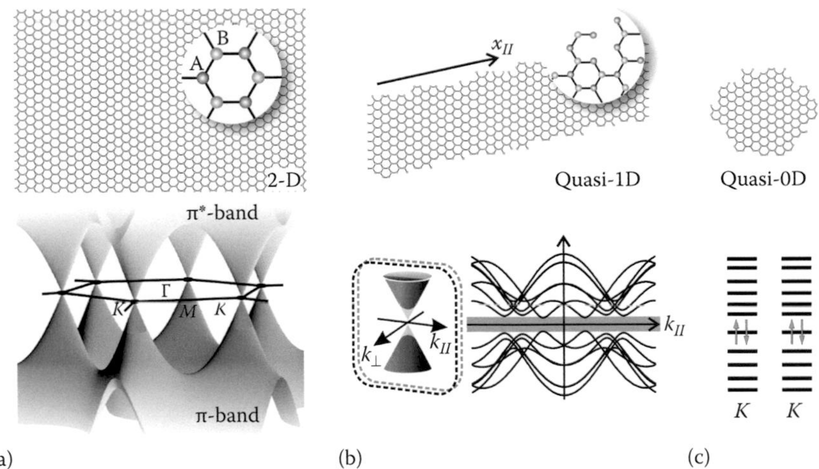

FIGURE 2.3 Graphene structures from 2D to 0D. (a) Illustration of the honeycomb-like lattice in extended graphene (2D) and the dispersion relation of graphene. (b) GNR (1D) with rough edges and schematics of the corresponding quasi-1D band structure. (c) QD (0D) in form of a nanometer-sized graphene flake and the expected discrete energy level spectrum (spin and valley degeneracy is assumed).

(see Figure 2.3c). QDs are therefore regarded as artificial atoms (Kastner 1993) where the single particle level spectrum can be set by the dimensions of the QD. The relevant energy scales in graphene QDs are discussed in Section 2.3.2. The designable level spectrum makes graphene QDs interesting for optical experiments (see Section 2.3.3). Furthermore, QDs und DQDs are regarded as promising hosts for spin qubits (Loss and DiVincenzo 1998). Because of the low spin-orbit and hyperfine interaction, graphene is of special interest. Over the past years, graphene QDs and DQDs have been studied intensively, which will be summarized in Section 2.3 and the following.

2.3.2 Relevant Energy Scales

There are several energy scales that have to be considered performing transport experiments in graphene QDs.

Coulomb energy: The Coulomb energy $E_C = E^2/C_\Sigma$ of a QD can be estimated by its self-capacitance. For QDs in planar 2DEGs, and in graphene, the QD is commonly approximated by a circular disc of radius r embedded in a material with dielectric constant ε (Sohn et al. 1997; Ihn 2010). Gates and contacts are neglected within this model. The self-capacitance then measures $C = 8\varepsilon\varepsilon_0 r$. Considering a graphene QD placed on a SiO_2 substrate, $\varepsilon \approx \left(\varepsilon_{SiO_2} + 1\right)/2 \approx 2.5$ can be estimated as the average dielectric constant of the substrate material and air. Experimentally observed addition energies, E_{add}, i.e., the sum of charging energy E_C and level spacing Δ in graphene QDs, are shown in Figure 2.4a.

Level spacing in single-layer graphene: The level spacing of electronic excited states (ESs) in graphene QDs can be derived from the density of states (Schnez et al. 2009; Schnez 2010). The density of states (DOS) in single-layer graphene is given by $D(E) = 2E/\left(\pi\hbar^2 v_F^2\right)$ (Castro Neto et al. 2009; Das Sarma et al. 2011). Approximating a QD by a circular island with an area $A = \pi r^2$ and number of charge carriers N, the total quantum mechanical energy is given by

$$E_{QM}(N) = A\int_0^{E_F} dE\, E\, D(E) = \frac{2A}{3\pi(\hbar v_F)^2} E_F^3. \tag{2.5}$$

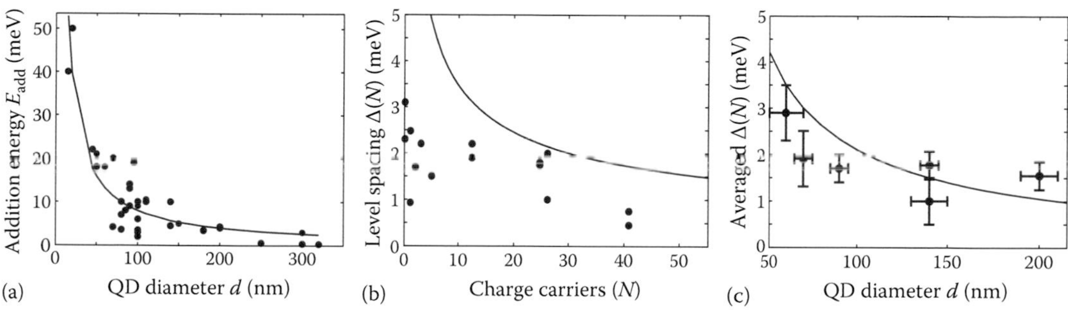

FIGURE 2.4 Addition energies and ES level spacings in graphene QDs. (a) Addition energies of single-layer and bilayer graphene QDs obtained by finite-bias spectroscopy measurements. The solid line is a fit $E_{add} \propto d^{-1}$. (b) Level spacing of the first ES as a function of the number of charge carriers on the QD estimated from finite-bias spectroscopy. The dashed line shows the theoretical energy dependence according to Equation 2.8. (c) Averaged level spacings observed in different samples as a function of effective dot diameter. The solid line shows the diameter dependence expected from Equation 2.8. (Experimental data are extracted from Ponomarenko, L.A. et al., *Science*, 320, 356, 2008; Stampfer, C. et al., *Nano Lett.*, 8, 2378, 2008; Güttinger, J. et al., *Phys. Rev. Lett.*, 103, 046810, 2009; Moriyama, S. et al., *Nano* 8, 2891, 2009; Volk, C. et al., *Nano Lett.*, 11, 3581, 2011; Volk, C. et al., *Nat. Commun.*, 4, 1753, 2013; Connolly, M.R. et al., *Nat. Nanotechnol.*, 8, 417–420, 2013; Liu, X. et al., *Phys. Rev. B*, 80, 121407, 2009; Liu, X.L. et al., *Nano Lett.*, 10, 1623–1627, 2010; Neubeck, S. et al., *Small*, 6, 14, 2010; Puddy, R.K. et al., *Appl. Phys. Lett.*, 103, 183117, 2013; Goossens, A.M. et al., *Nano Lett.*, 12, 4656–4660, 2012b; Allen, M.T. et al., *Nat. Commun.*, 3, 934, 2012; Schnez, S. et al., *Appl. Phys. Lett.*, 94, 012107, 2009; Güttinger, J. et al., *N. J. Phys.*, 10, 12029, 2008; Stampfer, C. et al., *Appl. Phys. Lett.*, 920, 012102, 2008; Molitor, F. et al., *Appl. Phys. Lett.*, 940, 222107, 2009; Güttinger, J. et al., *Phys. Rev. Lett.*, 1050, 116801, 2010; Güttinger, J. et al., *Phys. Rev. B*, 83, 165445, 2011; Güttinger, J. et al., *Nano. Res. Lett.*, 6, 253, 2011; Müller, T. et al., *Appl. Phys. Lett.*, 101, 012104, 2012; Dröscher, S. et al., *Appl. Phys. Lett.*, 101, 043107, 2012; Jacobsen, A. et al., *N. J. Phys.*, 14, 023052, 2012; Bischoff, B. et al., *N. J. Phys.*, 15, 083029, 2013; Wang, L.-J. et al., *Appl. Phys. Lett.*, 99, 112117, 2011; Wang, L. et al., *Science*, 342, 614, 2013; Zhou, C. et al., *Chin. Phys. Lett.*, 29, 117303, 2012; Wang, L.-J. et al., *Chin. Phys. Lett.*, 28, 067301, 2011; Kölbl, D., Zumbühl, D., 2013; Fringes et al., *Phys. Status Solidi B*, 248, 2684, 2011; Epping, A. et al., *Phys. Status Solidi B*, 25, 2682, 2013. Panels b and c adapted from Güttinger, J. et al., *Rep. Prog. Phys.*, 75, 126502, 2012. Copyright 2012 IOP Publishing.)

Applying the dispersion relation for single layer graphene $E = \hbar v_F |k_F|$ with $k_F = \sqrt{\pi n} = \sqrt{\pi N/A}$, Equation 3.1 can be rewritten as

$$E_{QM}(N) = \frac{2A\hbar^3 v_F^3}{3\pi(\hbar v_F)^2} k_F^3 = \frac{2\hbar v_F}{3}\sqrt{\pi/A}\,N^{3/2} = \frac{4\hbar v_F}{3}\frac{1}{d}N^{3/2}. \tag{2.6}$$

The chemical potential μ can be expressed as the following, where the approximation is valid in the regime of large N.

$$\mu(N) = E_{QM}(N+1) - E_{QM}(N) = \frac{4\hbar v_F}{3}\frac{1}{d}\left((N+1)^{3/2} - N^{3/2}\right) \approx 2\hbar v_F \frac{\sqrt{N}}{d}. \tag{2.7}$$

The ES level spacing Δ is thus given by the difference

$$\Delta(N) = \mu(N+1) - \mu(N) \approx \hbar v_F \frac{1}{d\sqrt{N}}. \tag{2.8}$$

The ES level spacing in graphene QDs obtained from finite-bias spectroscopy measurements is shown a function of the number of charge carriers in Figure 2.4b and as a function of the QD diameter in Figure 2.4c.

Level spacing in bilayer graphene: For bilayer graphene QDs, the level spacing can be calculated following the same approach taking into account the difference in density of states and dispersion relation. In the vicinity of the Dirac points, the dispersion relation of bilayer graphene is parabolic, and thus, the density of states equals the one of a 2DEG, $D(E) = \gamma_1/\left(\pi\hbar^2 v_F^2\right)$. This is in contrast to single-layer graphene and does not depend on the energy. Thus, the total energy is quadratic in E_F:

$$E_{QM}(N) = A\int_0^{E_F} dE\, E\, D(E) = \frac{A\gamma_1}{2\pi(\hbar v_F)^2} E_F^2. \tag{2.9}$$

As derived in Section 2.2, the low-energy dispersion relation takes the form of a 2DEG, $E = \frac{\hbar^2 k_F^2}{2m^*}$ with the effective mass defined by the interlayer hopping energy: $m^* = \gamma_1/2v_F^2$. Thus, the energy reads as

$$E_{QM}(N) = \frac{A\gamma_1}{2\pi(\hbar v_F)^2}\left(\frac{\pi\hbar^2}{2m^*}\right)^2\left(\frac{N}{A}\right)^2 = \frac{\pi\hbar^2 v_F^2}{2\gamma_1}\frac{N^2}{A} \tag{2.10}$$

and the chemical potential is given by

$$\mu(N) = \frac{\pi\hbar^2 v_F^2}{2\gamma_1}\frac{2N+1}{A}. \tag{2.11}$$

This results in an ES level spacing Δ, which is independent of the number of electrons on the QD (Volk et al. 2011):

$$\Delta(N) = \frac{\pi\hbar^2 v_F^2}{\gamma_1}\frac{1}{A}. \tag{2.12}$$

Tunnel coupling: The tunnel resistance R_t to the leads has to be sufficiently high such that an electron is either located in one of the leads or on the QD. The minimum R_t to fulfill this condition can be estimated according to Heisenberg's uncertainty relation $\Delta E \cdot \Delta t > h$, with the desired energy resolution ΔE and the time scale of a tunneling process $\Delta t = R_t C_\Sigma$. This yields the relation $R_t > h/C_\Sigma \Delta E$. Regarding the charging energy as the desired resolution, the condition $R_t > h/e^2$ has to be satisfied (Sohn et al. 1997; Ihn 2010).

Thermal energy: The thermal broadening of a Coulomb peak is proportional to $\cosh^{-2}\left(\frac{\alpha e \Delta V_G}{2k_B T}\right)$. This implies that the temperature has to be sufficiently low to resolve Coulomb charging effects ($k_B T \ll E_C$) and the ES spectrum ($k_B T \ll \Delta$).

2.3.3 Optical Properties

Because of the absence of a bandgap, no optical luminescence is observed in extended graphene sheets. It has been shown that a bandgap can be induced by shaping graphene into nanoribbons (GNRs) and dots (QDs) due to quantum confinement (Ponomarenko et al. 2008; Li and Yan 2010) and edge effects (Zhu et al. 2011), which will be discussed in Section 2.5.1. Theory predicts a $1/d$ dependence of the induced confinement gap on the QD diameter d. Graphene QDs have been studied with optical methods like absorption spectroscopy, PL, and PLE.

Figure 2.5a shows the ultra violet (UV)-visible absorption spectra of chemically synthesized graphene QDs of different diameters (Peng et al. 2012). A blue shift from 330 to 270 nm with decreasing size is

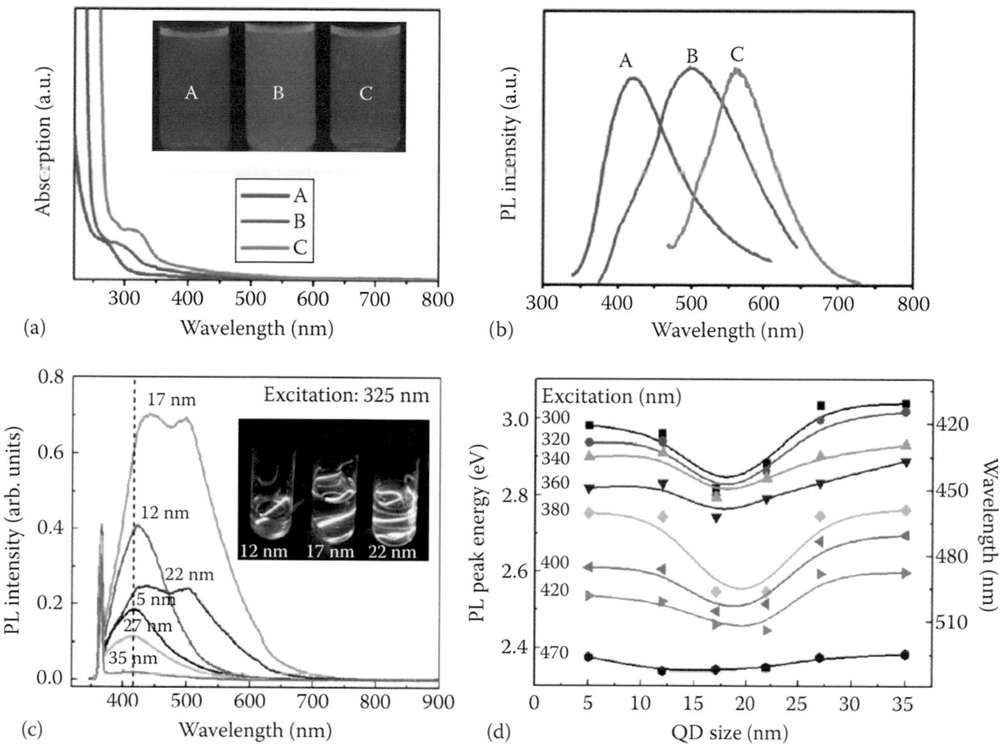

FIGURE 2.5 Optical properties of graphene QDs. (a) Absorption spectra in the UV to visible regime of solutions containing graphene QDs synthesized from carbon fibers under different conditions. The average diameter of the QDs measures 2–4 nm, 5–7 nm, and 8–10 nm respectively in samples A, B, and C. Inset: Photograph of the corresponding graphene QDs under UV light with 365 nm. (b) PL spectra of the QDs in panel a showing different emission colors. (c) PL spectra excited at 325 nm for graphene QDs of 5–35 nm average sizes in deionized (DI) water. Inset: different colors of luminescence from QDs depending on their average size (12, 17, and 22 nm). (d) PL peak energy as a function of the QD diameter measured with different excitation wavelengths from 300 to 470 nm. (a and b, Reprinted with permission from Peng, J. et al., *Nano Lett.*, 12, 844, 2012. Copyright 2012 American Chemical Society. c and d, Reprinted with permission from Kim, S. et al., *ACS Nano*, 6, 8203, 2012. Copyright 2012 American Chemical Society.)

observed. The inset shows the optical images of solutions containing three sizes of graphene QDs under UV light. Figure 2.5b shows the PL spectra of the same QDs.

Figure 2.5c shows PL spectra of graphene QDs synthesized by chemical cutting of graphene sheets obtained by thermal deoxidization of graphene oxide sheets (Kim et al. 2012). A clear size-dependent emission energy can be observed. Interestingly, the energy first decreases with increasing diameter, which is in agreement with the quantum confinement effect in graphene. Above $d = 17$ nm, the energy increases again (see Figure 2.5d).

These experiments show that the emission and absorption spectra of graphene QDs are size dependent, which is in agreement with the theory assuming a size-dependent confinement gap.

2.4 Fabrication of Graphene Nanostructures

2.4.1 Top–Down Approach

The most common technique to fabricate graphene nanostructures for electronic transport experiments is based on micromechanical exfoliation of graphene from natural bulk graphite followed by lithography and dry etching. This fabrication process dates back to the early ages of graphene and has been introduced by Novoselov et al. (2004, 2005) and Zhang et al. (2005). Alternative techniques as local anodic oxidation (Neubeck et al. 2010) or soft confinement of electrons in bilayer graphene (Allen et al. 2012; Goossens et al. 2012b) have been demonstrated more recently.

Although a lot of progress has been made in the growth of graphene over the past years, graphene flakes exfoliated from natural bulk graphite still provide the best crystal quality. Thus, research, especially transport experiments, is mainly carried out on such flakes. The technique of micromechanical exfoliation, often referred to as "Scotch tape technique," makes use of the fact that in graphite, the individual graphene layers are only weakly bond by van der Waals forces, in contrast to the strong covalent intralayer bonds. This allows overcoming the interlayer bonds using an adhesive tape and thus cleaving of individual graphene sheets from a graphite crystal. The graphene flakes are deposited on prepared silicon chips with a SiO_2 top layer (see Figure 2.7a and b). Graphene is highly transparent (absorption ≈ 2.3%), but thanks to interference effects, the visibility of graphene can be increased by tuning the oxide thickness and the wavelength of the incident light. It has been shown that the contrast has maxima at oxide thicknesses of 90 and 300 nm for green light ($\lambda \approx 550$ nm) (Novoselov et al. 2004; Blake et al. 2007; Tombros 2008). Raman spectroscopy is used to reliably identify single-layer and bilayer graphene (Ferrari et al. 2006; Gupta et al. 2006; Graf et al. 2007; Malard et al. 2009) among the flakes that have been deposited onto the substrate. Figure 2.6 shows the differences between the Raman spectra measured on a single-layer and a bilayer graphene flake. A detailed study of Raman spectra of graphitic flakes of different thicknesses can be found in Graf et al. (2007).

Plasma-based reactive ion etching (RIE) is commonly used to pattern graphene nanostructures (Novoselov et al. 2005; Zhang et al. 2005; Güttinger et al. 2012). The sample is coated with a resist, typically polymethylmethacrylate (PMMA). An etch mask is designed individually for each graphene flake, which is transferred to the resist by electron beam lithography (EBL) (see Figure 2.7c). With thin resist layers (≈ 50 nm) and optimized EBL parameters, structures as narrow as 20 nm can be routinely defined. After development, the graphene is etched by RIE. The advantage of dry etching is its high anisotropy and selectivity. It has been proven that this technique does not introduce bulk defects in the graphene sheet (Bischoff et al. 2011). For example, an argon/oxygen plasma (20% O_2) combines the physical impact of the argon with the chemical reactivity of the oxygen ions. Short etching times of typically below 10 s and low power (60 W) are sufficient to etch graphene. With increasing etching time or increasing power, the PMMA is cross-linked, making it challenging to remove it with organic solvents. Common treatments to remove hardened resist like oxidizing acids or plasma ashing cannot

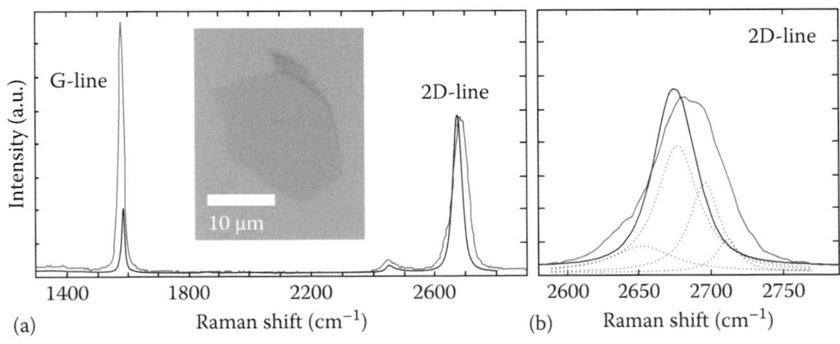

FIGURE 2.6 Raman spectroscopy on single-layer and bilayer graphene. (a) Comparison of the Raman spectra of single-layer graphene (black curve) and bilayer graphene (gray curve). In single-layer graphene, the 2D-line is often twice as high as the G-line. The inset shows an optical micrograph of a single-layer graphene flake on a Si/SiO_2 substrate. (b) Close-up on the 2D-lines shown in panel a. The bilayer 2D-Raman peak can be described by four Lorentzian line shapes (see dotted lines). (Data from Volk, C. et al., *Nano Lett.*, 11, 3581, 2011; Volk, C., Neumann, C., Kazarski, S. et al., *Nat. Commun.*, 4, 1753, 2013.)

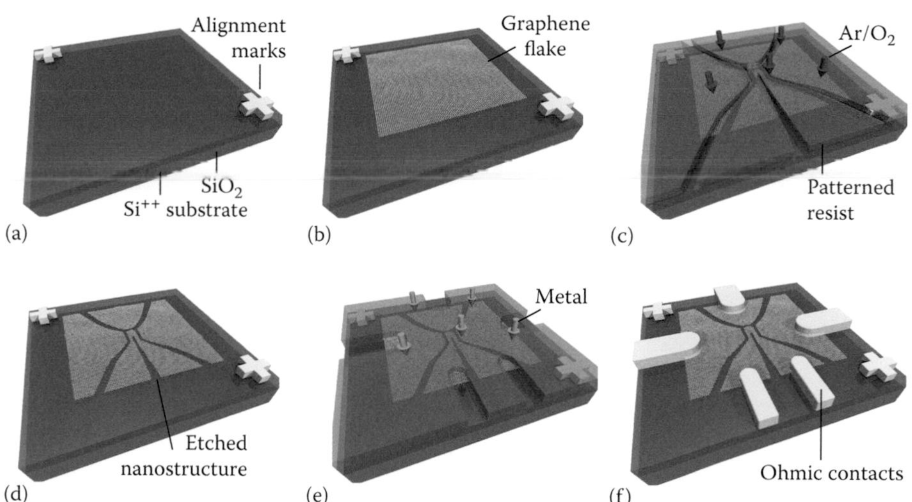

FIGURE 2.7 Typical process flow to fabricate graphene QD devices. (a) Highly p-doped Si substrate covered by 295 nm SiO_2 (see labels). Metal alignment marks have been deposited for further lithography steps. (b) Exfoliated graphene flake. (c) A layer of polymer resist has been patterned by EBL. The graphene is etched by an Ar/O_2-plasma (indicated by arrows). (d) Etched graphene nanostructure after removal of the resist. The example shows a graphene QD with three lateral gates. (e) Deposition of ohmic contacts by metal evaporation (typically Cr/Au) after a second EBL step. (f) Device after contacting which is now ready for measurements.

be used as these will remove graphene as well. The schematic in Figure 2.7d shows the etched graphene nanostructure.

Metal contacts connecting the graphene devices to bond pads are defined by an additional EBL, metallization, and lift-off step (see Figure 2.7e). A thin layer (typically 2 to 5 nm) of Cr or Ti is deposited as an adhesion layer, followed by 50 nm Au or Pt, leading to typical contact resistances in the range of kΩ. Figure 2.7f shows a contacted graphene QD device. Transmission line measurements have shown that graphene/metal contact resistances can be potentially reduced by using palladium (Song et al. 2012; Watanabe et al. 2012). However, in graphene nanostructures, the contact resistance is not very crucial since the resistance is limited by the nanostructure itself.

2.4.2 Bottom–Up Approach

Besides the fabrication technique based on EBL, graphene QDs have been fabricated, e.g., by ruthenium-catalyzed C60 transformation (Lu et al. 2011) suffering from extremely expensive raw materials and low yield. Graphene QDs prepared by multistep hydrothermal (Pan et al. 2010) or electrochemical strategies (Li et al. 2010) from graphene oxide have shown blue or green luminescence. As the bandgap in graphene QDs is size dependent, a controlled size of the QDs is especially important for PL emission. It has been shown that graphene QDs can be synthesized by wet chemical oxidization and cutting of micrometer-sized carbon fibers (see Figure 2.8a) (Peng et al. 2012). Commercially available carbon fibers are added into a mixture of concentrated H_2SO_4 and HNO_3. The solution is first sonicated and then stirred at different temperatures between 80°C and 120°C. The mixture is cooled and the pH is adjusted to 8 with Na_2CO_3. The final product solution is further dialyzed for 3 days (Peng et al. 2012). The as-synthesized graphene QDs are highly soluble in water and other polar organic solvents. Their lateral size ranges from 1 to 4 nm, and the QDs are typically one to three atomic layers thick. Figure 2.8b shows a high-resolution transmission electron microscopy (HRTEM) image of a graphene QD. The arrows indicate the zigzag direction of the lattice. The corresponding fast Fourier transform (FFT) pattern is shown in the inset.

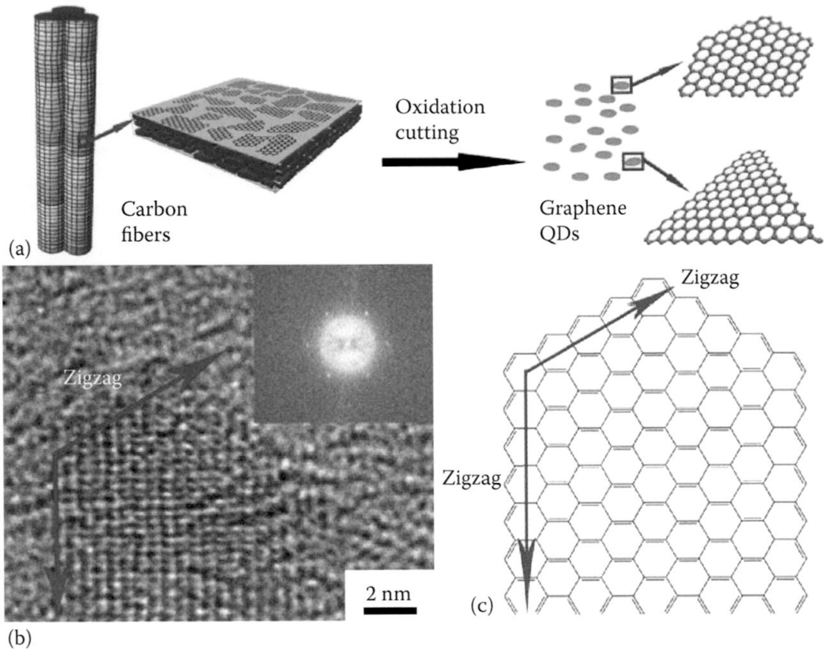

FIGURE 2.8 Chemical synthesis of graphene QDs. (a) Illustration of the wet chemical oxidation cutting of carbon fibers into graphene QDs. (b) High-resolution transmission electron micrograph of a graphene QD. The arrows indicate the zigzag edges of the QD. Inset: 2D FFT of the edge. (c) Schematic representation of the edge termination of the HRTEM image in panel b. (Reprinted with permission from Peng, J. et al., *Nano Lett.*, 12, 844, 2012. Copyright 2012 American Chemical Society.)

2.5 Graphene Single-Electron Transistors and QDs

2.5.1 GNRs and Nanostructures

As GNRs are building blocks of most graphene quantum devices, this section summarizes the relevant transport properties of these structures.

From CNTs, which can be imagined as rolled-up GNRs, it is well known that a bandgap opens depending on their orientation and their diameter (Saito et al. 1999). The orientation of CNTs is named after their circumference, while GNRs are classified according to their edges along the ribbon. N-aGNRs and N-zGNRs commonly denote armchair (a) and zigzag (z) GNRs with N dimers across the ribbon width. The zGNRs and aGNRs are well understood in theory. The band structure can be determined by applying vanishing boundary conditions to the Dirac Hamiltonian of graphene. In zGNRs, one edge is made up by A atoms, and the other one, by B atoms, and thus, the boundary condition can be applied to the sublattices separately. In aGNRs, the edges contain atoms of both sublattices, and thus, the boundary condition has to be fulfilled by both sublattices (Brey and Fertig 2006).

Alternatively, the band structure can be determined following a tight-binding approach depending on the orientation and the width of GNRs. Armchair GNRs have a metallic band structure if the condition $N = 3m - 1$ with integer m is fulfilled. Otherwise, a semiconducting band structure occurs (Kastner 1993; Nakada et al. 1996; Cresti et al. 2008). The bandgap E_g scales inversely with the ribbon width W. An estimate is given by Saito et al. (1999) and Güttinger (2011)

$$E_g = 2\hbar v_F \Delta k_F = 2\pi\hbar v_F / W, \tag{2.13}$$

where Δk_F is the minimum allowed wave number across the ribbon width.

Density fuctional theory (DFT) calculations taking into account next-nearest-neighbor hopping and a contraction of the bond length at the edges have shown that even metallic aGNRs have at least a small bandgap that depends on the ribbon width (Son et al. 2006). Zigzag GNRs have a gapless band structure independent on *N*. First-principle calculations have shown a high density of zero-energy states at the edges, which has been proved by scanning tunneling spectroscopy (Kobayashi et al. 2005). Magnetic ordering at the edges may lead to the opening of a small bandgap in zGNRs (Wakabayashi et al. 1999; Son et al. 2006).

Theory assumes either pure armchair or pure zigzag GNRs, where the edges are terminated by hydrogen atoms. Using common experimental techniques, it has not yet been possible to fulfill these conditions. Typically, arbitrary edges occur when GNRs are etched out of graphene sheets. When using an O_2-based plasma as the etchant, the edges will probably be oxygen terminated. In the following, electronic transport through GNRs will be described on a more phenomenological basis.

Field effect measurements have proven the presence of a transport gap (see, e.g., Figure 2.9a and b and Stampfer et al. [2009]; Terrés et al. [2011]), but still a number of sharp resonances can be observed within this gap. A common model to describe the electronic transport through etched GNRs (i.e., nanoribbons with rough edges) is based on stochastic Coulomb blockade (Sols et al. 2007; Stampfer et al. 2007). A disorder potential (e.g., substrate or edge disorder) can form electron and hole puddles close to the charge neutrality point. Because of the absence of a bandgap in bulk graphene and of the presence of the effect of Klein tunneling (Katsnelson et al. 2006), transport between the electron and hole regions is possible. The situation is different in nanoribbons, where a width-dependent confinement gap separates electrons and hole puddles. Thus, effectively, a large number of QDs or localized states formed and only tunneling transport is possible (Stampfer et al. 2009; Todd et al. 2009).

The presence of QDs has been demonstrated by finite bias spectroscopy measurements on GNRs (Molitor et al. 2009; Stampfer et al. 2009; Todd et al. 2009; Terrés et al. 2011). Most importantly, two characteristic energy scales can be extracted, the effective energy gap E_g and the transport gap ΔV_g (see Figure 2.9b).

The transport gap (Figure 2.9c) is correlated with the maximum amplitude of the disorder potential. An effective energy gap E_g can be defined by the largest observed charging energy corresponding to the smallest QD. It has been shown that this energy gap only weakly depends on the length of the nanoribbon (Terrés et al. 2011). The width dependence can be modeled by

$$E_g(W) = \alpha / W e^{-\beta W}, \tag{2.14}$$

(see Figure 2.9d and Han et al. [2007]; Molitor et al. [2010b]; Stampfer et al. [2011]).

2.5.2 Single-Electron Transistors

A SET can be imagined as a conductive island that is weakly coupled to lead electrodes, and its electrochemical potential can be controlled by at least one gate electrode. A simple model of such a configuration is illustrated in Figure 2.10a. Tunneling transport between the QD and the leads is allowed; the gate is coupled only electrostatically. Electronic transport through a SET is dominated by the Coulomb blockade effect, which is a consequence of the Coulomb interaction of electrons on the island leading to a repulsive force. Thus, a certain amount of energy—the so-called charging energy E_C—has to be supplied to add an additional electron to the island. Assuming the temperature and the bias voltage are small compared with the charging energy, electron transport is possible only if the electrochemical potential of the island is positioned between the electrochemical potentials of the source and the drain lead (the so-called transport window). An electron from the source lead can now enter the island and subsequently leave to the drain lead. The system is in the regime of sequential tunneling and a current can flow. If no level is in the transport window, the number of electrons occupying the island is fixed. The system is in Coulomb blockade (see Figure 2.10b). The potential of the island can be controlled by

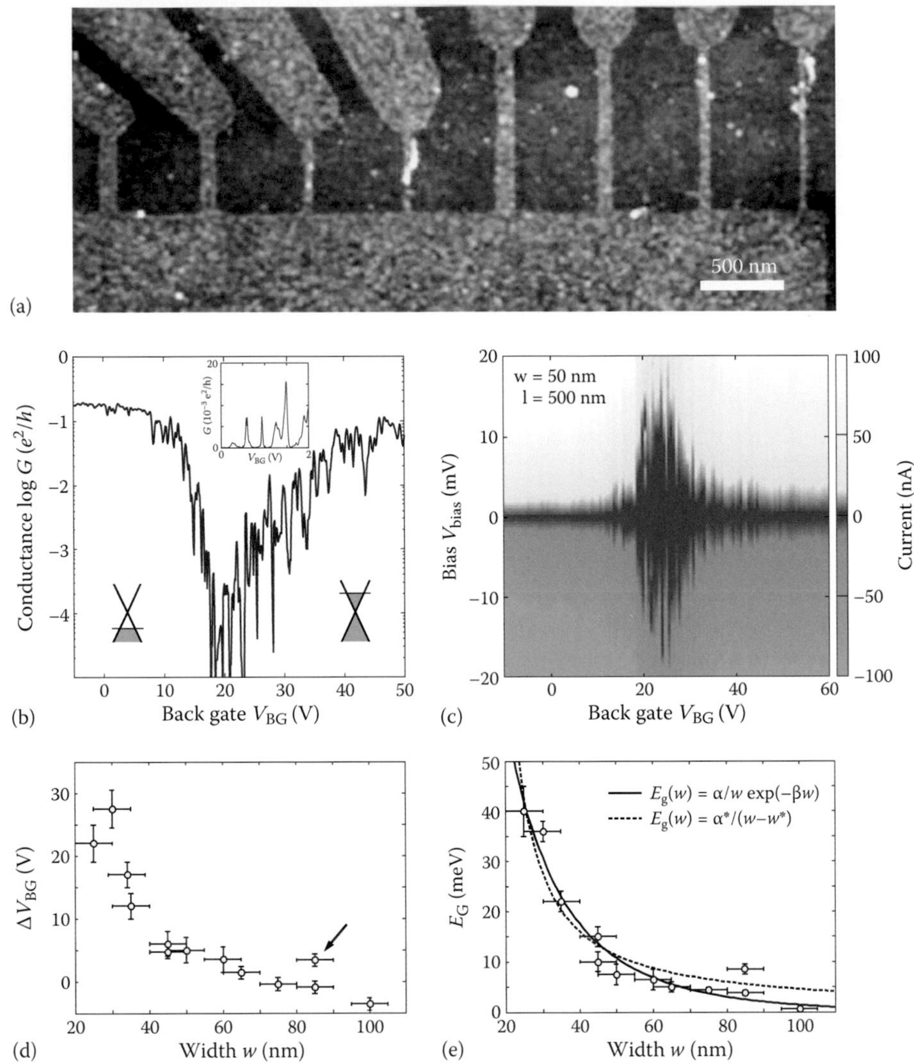

FIGURE 2.9 Characteristics of etched GNRs. (a) Scanning force micrographs of etched GNRs with different lengths and widths. (b) Conductance through a GNR (50 nm wide, 500 nm long) as a function of the back gate voltage and, thus, the Fermi level. Regions of electron and hole transport are separated by a transport gap. The inset shows a measurement within the transport gap of a 200-nm-wide nanoribbon. (c) Finite bias spectroscopy measurement allowing determination of the effective energy gap E_g and the transport gap correlated with ΔV_{BG}. (d) Transport gap as a function of the width of different nanoribbons. (e) Effective energy gap as a function of the width. Two models are fitted to the experimental data. (a–c, Reprinted with permission from Terrés, B. et al., *Appl. Phys. Lett.*, 98, 032109, 2011. Copyright 2011, American Institute of Physics. d and e, Adapted from Molitor, F., Stampfer, C., Güttinger, J. et al., *Semicond. Sci. Technol.*, 250, 034002, 2010. Copyright 2010 IOP Publishing.)

gates. Thus, measuring the current as a function of the gate voltage, regimes of conductance and of Coulomb blockade alternate, so-called Coulomb peaks, appear (see Figure 2.10d). The proportionality between the peak spacing Δ_{VG} and the addition energy is given by the lever arm $\alpha = E_{add}/e\Delta_{VG} = C_G/C_\Sigma$. It is a measure for the capacitive coupling of a gate to the QD.

By bias spectroscopy measurements, the current through a SET is recorded both as a function of the bias and the gate voltage. Diamond-shaped regions (so-called Coulomb diamonds) of suppressed

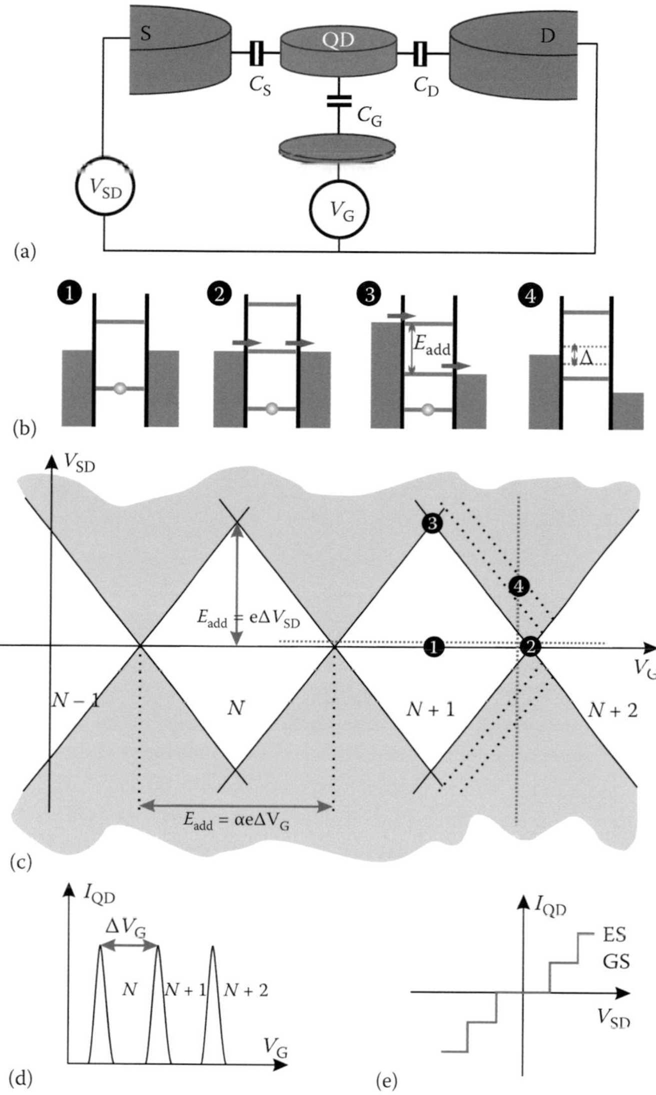

FIGURE 2.10 Transport through QDs. (a) Simple model of a QD capacitively coupled to the source and drain leads (C_S, C_D) and to one gate (C_G). Tunneling transport between the QD and the leads is allowed. These tunnel junction can be modeled by a capacitance and a resistance in parallel. (b) Schematics of four different configurations of a QD: (1) Zero bias and misalignment of the QD states with the lead potentials. (2) A QD state aligned with lead potentials. (3) The bias equals the addition energy. At least one state is within the transport window. (4) GS and first ES are aligned within the transport window; two possible transport channels are open. (c) Illustration of finite bias spectroscopy measurements on a QD. The current I_{QD} is plotted as a function of the bias V_{SD} and the gate voltage V_G. In the white regions, the device is in Coulomb blockade. The current and the number of electrons on the QD are fixed. In the gray shaded regime, transport occurs. The lever arm α is the proportionality factor between the addition energy E_{add} and the change in gate voltage ΔV_G necessary to add the next electron to the QD. (d) Cut along the gate axis at a small bias (horizontal line in panel c). Coulomb peaks with a spacing of ΔV_G appear. (e) Cut along the vertical line in panel c. The current increases each time another ES enters the transport window.

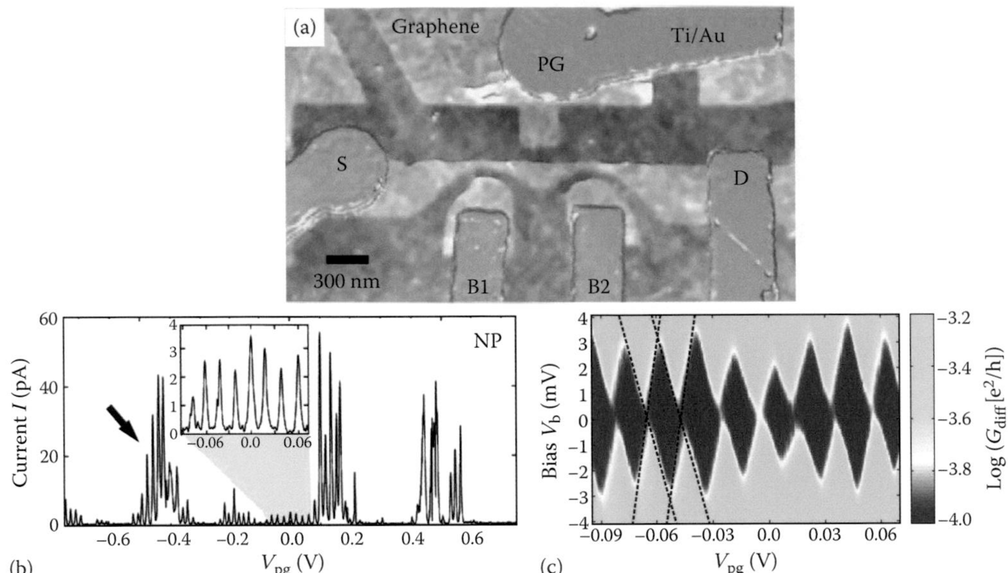

FIGURE 2.11 SET in a width-modulated graphene nanostructure. (a) False color scanning force micrograph of the device. Three lateral graphene gates are designed to locally tune the potential of the nanostructure. (b) Current through the SET as a function of the gate voltage. The inset shows a series of Coulomb peaks proving the operation as a SET. (c) Finite bias spectroscopy measurement in the same regime. (Reprinted with permission from Stampfer, C. et al., *Nano Lett.*, 8, 2378, 2008. Copyright 2008, American Chemical Society.)

conductance occur when the system is in Coulomb blockade (see schematics in Figure 2.10c). Within such a region, the number of charge carriers on the QD is constant. The extent of the diamonds in bias direction is a measure of the addition energy. Employing the gate lever arm α, the gate voltage axis can be converted into an energy scale ($E = \alpha e V_G$).

SETs have been fabricated by carving the desired shape out of graphene sheets using EBL followed by RIE (see Section 2.4). These devices consist of a graphene island connected to the source and drain electrodes via two GNRs. The devices make use of the fact that because of the narrow width of the GNRs, an effective transport gap is opened (see Section 2.5.1), and thus, they can be operated as tunable tunneling barriers. According to Equation 2.14, the gap scales approximately inversely with the ribbon width. Width modulation allows tailoring the transport gap along the ribbon axis. Close-by graphene gates and a global back gate tune the Fermi level in the nanostructure. A scanning force micrograph of a representative device is shown in Figure 2.11a. Carefully tuning the voltages on the two outer gates (B1 and B2), it is possible to bring the device into a regime where the transport gaps of both constrictions cross the Fermi level. The central island is electrically isolated, and only tunneling transport is possible between the island and the leads. The device can be operated as a SET. A series of distinct Coulomb peaks recorded as a function of the plunger gate voltage is shown in Figure 2.11b. A charging energy of ≈ 3.4 meV has been determined by finite bias spectroscopy measurements (see Figure 2.11c) on a graphene SET with 50 nm wide tunneling barriers and a central island measuring approximately 180 × 750 nm (Stampfer et al. 2008b).

2.5.3 QDs in Width-Modulated Nanostructures

The concept of width-modulated GNRs described in the previous section can be employed to design a graphene QD device. The smaller the central graphene island the more relevant quantum confinement effects become. Figure 2.12a shows a typical example of a graphene QD with lateral graphene gates.

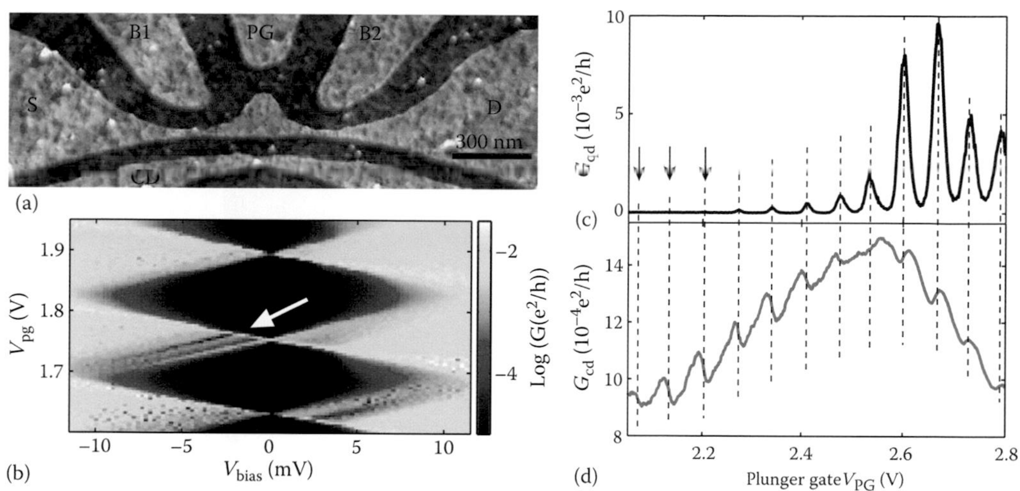

FIGURE 2.12 QD in a width-modulated graphene nanostructure and charge detection. (a) Scanning electron micrograph of a graphene QD with a close-by nanoribbon. (b) Finite bias spectroscopy measurement on an etched graphene QD resolving ESs (see white arrow). (c) Conductance through the QD shown in panels a and d through the nanoribbon as a function of the plunger gate voltage. The dashed lines are a guide to the eye, emphasizing the coincidence of the Coulomb peaks with the conductance steps in the nanoribbon. (a and c, Reprinted with permission from Güttinger, J. et al., *Appl. Phys. Lett.*, 930, 212102, 2008. Copyright 2008, American Institute of Physics. b, Reprinted with permission from Schnez, S. et al., *Appl. Phys. Lett.*, 94, 012107, 2009. Copyright 2009, American Institute of Physics.)

Finite-bias spectroscopy is a method used to investigate the excited level spectrum of a QD via direct transport experiments. If the bias exceeds the level spacing, it is possible that two QD levels, a ground state (GS) and an ES, are positioned within the transport window. Electrons can now tunnel through the QD via GS and ES transitions. As two channels contribute to transport, the current will increase (see Figure 2.10e). ES resonances appear as lines parallel to the Coulomb diamond edge in finite-bias spectroscopy (see schematics in Figure 2.10c).

A representative measurement is shown in Figure 2.12b, where the differential conductance through a 140-nm-wide QD is plotted (Schnez et al. 2009). From the size of the Coulomb diamonds, a charging energy of $E_C \approx 10$ meV can be determined. ES resonances are clearly visible in the form of lines of increased conductance, most prominent at negative bias voltage (see white arrow). A level spacing of $\Delta \approx 1.6$ meV has been extracted from the data set. QDs and DQDs based on width-modulated nanostructures have been fabricated and studied in transport experiments by a number of groups. Typical diameters of such devices measure between 50 and 150 nm. The addition energies follow roughly a trend inversely proportional to the diameter. The ES level spacing is typically on the order of 1 to 2 meV. On a 40-nm-wide QD, a level splitting of up to 10 meV has been observed (Ponomarenko et al. 2008).

2.5.4 Charge Detection

Charge sensors are commonly used in low-dimensional electronics to sensitively resolve individual localized charge states. Charge detectors (CDs) based on quantum point contacts (QPCs) (van Wees et al. 1988) have extensively been used in 2D electron systems (Field et al. 1993), especially in III/V heterostructures. In such devices, e.g., coherent spin and charge manipulation (Hayashi et al. 2003; Petta et al. 2005) and time resolved charge detection (Fujisawa et al. 2006; Gustavsson et al. 2006) have been demonstrated. Moreover, QPC-based CDs are regularly used to read out spin qubits realized in DQD systems in GaAs/AlGaAs heterostructures (Elzermann et al. 2003; Bluhm et al. 2010). This makes charge detection techniques interesting for read out of charge state of graphene QDs as well.

An all-graphene device where a nanoribbon has been used as a CD for a close-by QD is shown in Figure 2.12a (Güttinger et al. 2008a). The nanoribbon measures a width of 45 nm and is separated from the QD by 60 nm. The close distance allows a capacitive coupling between the two objects. As shown in Section 2.5.1, transport through GNRs is characterized by resonances originating from localized states. These resonances can be used to probe charging events on the coupled QD. Figure 2.12c shows a series of Coulomb peaks measured on the QD, and the simultaneously measured conductance of the detector is shown in Figure 2.12d. The overall peak shape is caused by a local resonance in the detector. Whenever an electron is added or removed from the QD, the potential of the detector is changed, resulting in a conductance step. The conductance steps are well aligned with the Coulomb peaks (see dashed lines as a guide to the eye). Using this method, charging events of the QD can even be detected in regimes where the direct current through the QD is below the detection limit (see arrows in the regime between 2 and 2.2 V).

The principle of charge detection can also be applied to finite bias spectroscopy measurements. Figure 2.13b shows the differential conductance through a 100-nm-wide QD (for a scanning force micrograph, see Figure 2.13a) (Neumann et al. 2013). A number of Coulomb diamonds and a rich spectrum of ESs (level spacing on the order of 1.5 to 2 meV) can be observed. The differential transconductance of the detector has been measured simultaneously (see Figure 2.13c). Prominent features aligned well with the Coulomb resonances in the QD conductance. The detector is most sensitive to the dominant tunneling barrier of the QD. For the two left Coulomb diamonds (marked by the black arrows), the dominant tunneling barrier changes with bias direction, indicating a rather strong capacitive coupling of the QD tunneling barriers to the source and the drain lead. Furthermore, transport via ES transitions can be identified in the differential transconductance of the CD, highlighted by white arrows.

Further studies have addressed the effect of back action of the CD on the QD. The peak current as well as the FWHM of the resonances increase with the applied voltage. It has even been shown that a further increase in the detector bias can fully lift the Coulomb blockade of the QD (Neumann et al. 2013).

So far, the time-averaged current through the CD has been investigated. Real-time detection measurements on graphene QDs give deeper insight into the tunneling processes of a graphene QD device (scanning force micrograph shown in Figure 2.14a) (Güttinger et al. 2011a; Müller et al. 2012). A schematic of an individual conductance step of the charge sensor signal is shown in Figure 2.14b. The step

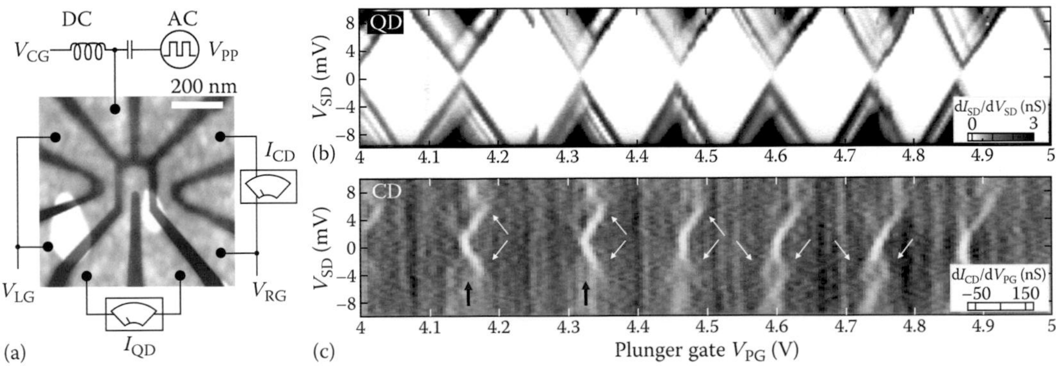

FIGURE 2.13 Charge sensing. (a) Scanning force micrograph of a 110 nm wide graphene QD with integrated charge sensor and gates. The CG controlling the QD potential is connected to a bias-tee mixing AC and DC signals. (b) Differential conductance dI_{SD}/dV_{SD} of a graphene QD. Coulomb diamonds as well as numerous ESs are visible. (c) The simultaneously recorded differential transconductance dI_{CD}/dV_{PG} of the CD exhibits features which can be associated with the Coulomb resonances and with ESs (see white arrows). (Adapted from Neumann, C. et al., *Nanotechnology*, 24, 444001, 2013. Copyright 2013, Institute of Physics.)

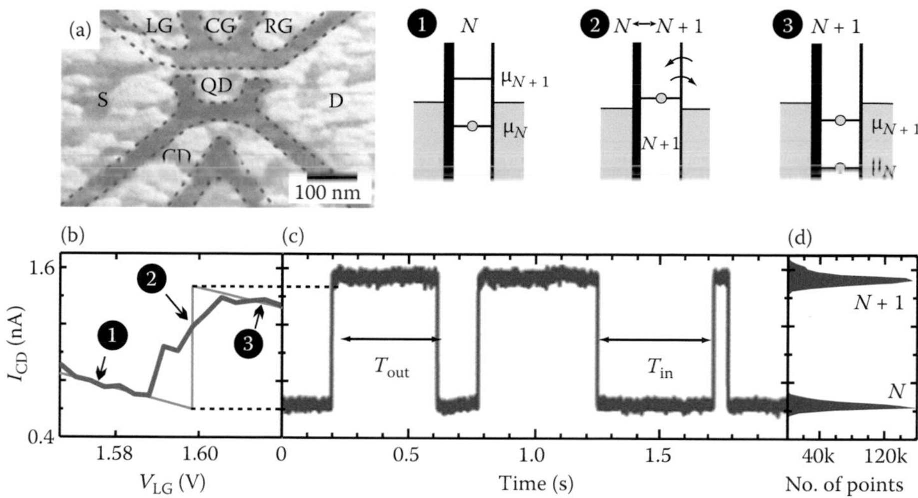

FIGURE 2.14 Time-resolved charge detection. (a) Scanning force micrograph of the investigated device. (b) Time-averaged current through the CD as a function of the gate voltage while scanning over Coulomb resonance. In regime (1) and (3), the QD is occupied by either N or $N + 1$ electrons. Position (2) is a metastable state where the occupation oscillates between N and $N + 1$. (c) Time-resolved current measured at position (2) marked in panel b. (d) Histogram of the current values for a time span of 60 s. (Reprinted with permission from Güttinger, J. et al., *Phys. Rev. B*, 83, 165445, 2011. Copyright 2011 by the American Physical Society.)

corresponds to one Coulomb peak in the QD conductance; when increasing the gate voltage, the number of electrons occupying the QD changes by one. The regions labeled 1 and 3 correspond to a fixed electron occupation of N and $N + 1$, respectively. Region 2—right on the conductance step—corresponds to a configuration where electrons can enter and leave the QD.

Experimental data of an individual conductance step is shown in Figure 2.14c. The gate voltage is then parked at the position marked by the arrow, and the time-resolved current through the CD is recorded (see Figure 2.14d). The two levels of current can be observed corresponding to an occupation of the QD with either N or $N + 1$ electrons. Each jump corresponds to an electron entering or leaving the QD. The histogram in Figure 2.14c shows the current distribution recorded in a time span of 60 s. The occupation probabilities of N and $N + 1$ can be altered by tuning the gate voltage.

2.5.5 Spin States in a Graphene QD

To address spin states in graphene QDs, the magnetic field dependency of a 70-nm-wide graphene QD device (see Figure 2.15a) has been studied (Güttinger et al. 2010). Sequences of neighboring Coulomb peaks have been recorded as a function of magnetic field at a bias voltage of 100 μV, which is significantly lower than the ES level spacing. An out-of-plane magnetic field allows determination of the spin filling of orbital states. At an in-plane magnetic field, the Zeeman splitting can be measured. Figure 2.15b shows the spacing of pairs of neighboring Coulomb peaks as a function of $B_{||}$ up to 12 T. The spacings depend linearly on the magnetic field with a slope of either zero or approximately $\pm 2\mu_B$, which is compatible with Zeeman spin splitting with a g factor of 2. Such slopes have been extracted from several pairs of neighboring Coulomb peaks. The results are plotted as a function of the corresponding gate voltage in Figure 2.15c. Orbital effects due to a misalignment of the magnetic field with respect to the sample orientation have been compensated. From the data, a spin-filling sequence of ↓↑↑↓↓↑↑↓ can be determined, in contrast to CNTs, where a sequence of ↑↓↑↓ has been observed (Güttinger et al. 2010).

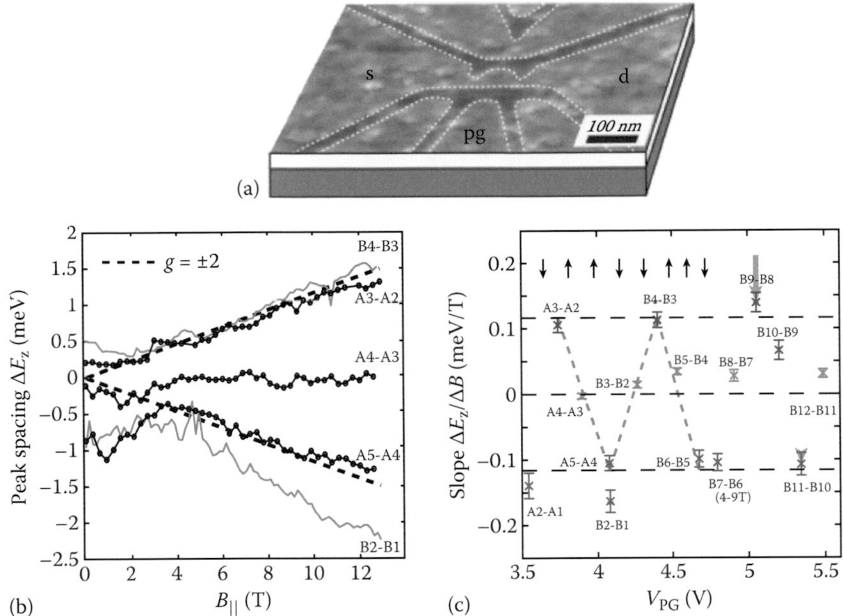

FIGURE 2.15 Spin states. (a) Scanning force microscopy (SFM) image of a 70-nm-wide graphene QD. (b) Spacing of neighboring Coulomb peaks as a function of $B_{||}$. The dashed lines correspond to slopes of $\pm g\mu_B$ for a g factor of 2. (c) Slopes of several Coulomb peak splittings. The obtained spin filling sequence is indicated on top of the plot. (Reprinted with permission from Güttinger, J. et al., *Phys. Rev. Lett.*, 1050, 116801, 2010. Copyright 2010 by the American Physical Society.)

2.5.6 Charge Relaxation in Graphene QDs

So far, graphene QDs have mainly been studied by DC transport. The relaxation dynamics of ESs can be studied in graphene QDs via a pulsed gate spectroscopy technique (Volk et al. 2013). For this purpose, a QD device with capacitively coupled charge sensors and gates has been etched out of a single-layer graphene sheet. The charge sensors also act as gates to tune the transparency of the barriers. The QD potential is controlled by a central gate (CG), which is connected to a bias-tee, to allow the application of AC and DC signals on the same gate (see Figure 2.13a).

The device has been characterized by DC finite-bias spectroscopy measurements. Figure 2.13b shows the differential conductance through the QD. ESs are clearly visible. As mentioned in Section 2.5.4, information on the asymmetry of the tunneling barriers can be obtained from the differential transconductance of the CD (see Figure 2.13c). Both edges of the Coulomb diamond can be resolved in the detector signal, indicating the QD is in a regime with comparable coupling to each of the leads. This helps to identify suitable regimes to investigate the relaxation dynamics by pulsed-gate spectroscopy. This technique requires the pulse rise time (τ_{rise}) to be the fastest time scale in the system. Thus, both tunnel rates (Γ_L, Γ_R) need to be significantly slower than τ_{rise}^{-1}. To fulfill this condition, the system is tuned into a low current regime with rather symmetric tunneling barriers.

The relaxation dynamics of ESs is studied by a pulsed-excitation scheme following the one introduced by Fujisawa et al. (2001a,b, 2002). The basic idea is to probe transient phenomena in a QD by measuring the average DC current. A rectangular pulse scheme is applied to the CG using a bias-tee (cf. Figure 2.13a). The current through the QD is recorded while the DC gate voltage scans over a Coulomb resonance. At low pulse frequency (e.g., 100 kHz in Figure 2.16a), the Coulomb resonance splits into two peaks (see also Dröscher et al. [2012]), shifting linearly with increasing amplitude. These peaks originate

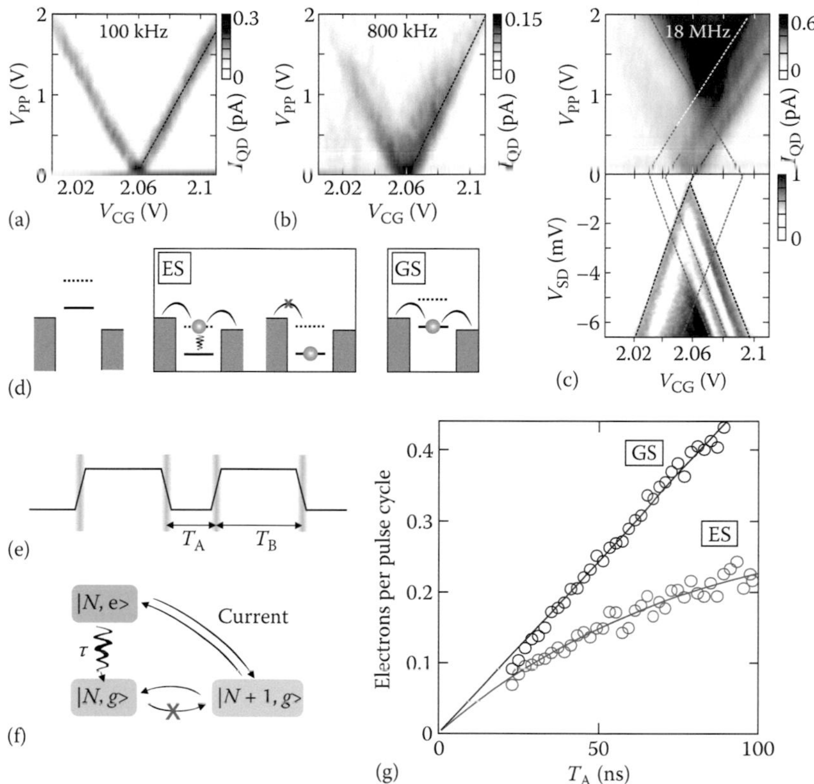

FIGURE 2.16 Pulsed-gate spectroscopy of ESs. (a) Current through the QD under the influence of a 100-kHz square pulse. With increasing pulse amplitude, the Coulomb peak splits linearly in two resonances. In panel b, the pulse frequency is 800 kHz, whereas in panel c, upper panel, the frequency measures 18 MHz. A number of additional resonances can be seen, corresponding to transient currents through ESs. The lower panel represents a DC finite-bias measurement at the same Coulomb resonance. The dashed lines are guides to the eyes. (d) Schematic of transport via GS and ES and of a possible initialization stage. (e) Pulse scheme used to study transient currents. The time T_A is varied keeping the pulse amplitude T_B constant. (f) Relevant transitions during T_A probing transient currents through the ES. $|N, g\rangle$ and $|N, e\rangle$ indicate the ground and ES of the dot with N excess electrons. (g) Average number of electrons transmitted per cycle via GS (black circles) and ES (gray circles) transitions as a function of T_A. The solid lines are a fit to the experimental data. (Reprinted by permission from MacMillan Publishers Ltd. *Nat. Commun.* Volk, C. et al., *Nat. Commun.*, 4, 1753, 2013. Copyright 2013.)

from the QD GS entering the bias window at two different values of the DC gate voltage, one for the lower pulse level and one for the upper one. At 800 kHz, the peaks broaden (Figure 2.16b).

At higher frequencies (e.g., 18 MHz in Figure 2.16c), a number of additional peaks appear, which are attributed to transient transport through ESs. The level spacing extracted from this measurement coincides with the one given by DC finite-bias measurements. Figure 2.16c compares the transient current spectroscopy with the DC measurement where the ES energies are clearly visible.

Each resonance corresponds to a situation in which the QD levels are pushed well outside the bias-window in the first half of the pulse and then brought into a position where transport can occur only through the ESs in the second one (see schematics in Figure 2.16d). In that case, electrons can tunnel from one lead to the other via the ES, as long as the GS remains unoccupied, which contributes to a current through the QD. Once the GS gets filled, transport is blocked until the system is initialized in the next pulse cycle again.

For a more quantitative analysis of the relaxation process, a different pulse scheme is used (see Figure 2.16e), where T_A is varied while keeping T_B and the amplitude fixed (Fujisawa et al. 2001a, 2002a,b; Volk et al. 2013).

At the beginning of each cycle, the dot is initialized in a GS with $N + 1$ electrons during T_B. The ES is pushed into the transport window during T_A, allowing transitions between $|N, e\rangle$ and $|N + 1, g\rangle$, a current can flow (see Figure 2.16f). If the system is in the ES $|N, e\rangle$, two competing processes (relaxation $|N, e\rangle \rightarrow |N, g\rangle$ and tunneling $|N, e\rangle \rightarrow |N, g + 1\rangle$) are possible. The transition $|N, g\rangle \rightarrow |N + 1, g\rangle$ is not energetically allowed at low bias voltage. Once in $|N, g\rangle$, the QD is blocked for further electrons until the initialization in $|N + 1, g\rangle$ in the next pulse cycle.

While applying the pulse sequence, a DC current through the QD averaged over a large number of pulse cycles is measured. This yields the average number of electrons tunneling through the device per cycle $\langle n\rangle = I(T_A + T_B)/e$. Figure 2.16g shows the number of electrons transmitted via the GS and via the ES as a function of the pulse length T_A. While the first one increases linearly with T_A, the number of electrons transmitted via the ES tends to saturate, indicating a transient effect. Fitting $n(T_A) = n_{sat}[1 - \exp(-\gamma T_A)]$ to the experimental data yields a characteristic blocking rate of $\gamma = 12.8$ MHz. As both tunneling and relaxation lead to this blocking, γ^{-1} gives a lower bound for the ES to GS relaxation time. Analysis of further states gives a lower bound of 60–100 ns.

2.5.7 Gate-Defined QDs in GNRs

GNRs are also the basis for another approach to fabricate graphene SETs and QDs. In this, technique graphene flakes are first patterned in GNRs (with a constant width along the GNR, typically below 100 nm) using EBL and RIE. In a subsequent lithography step, a narrow gate finger is positioned on top of the GNR, separated by thin dielectric. Ohmic contacts are deposited at the ends of the nanoribbon. Because of the narrow width of the GNR, a transport gap is opened, which enables electrostatic confinement by the top and back gate.

A series of such devices with different geometries (GNR width ranging from 40 to 60 nm, length from 520 to 2000 nm, and top gate width between 50 and 500 nm) has been studied (Liu et al. 2009). Figure 2.17a and b shows a scanning electron micrograph and a schematic of a typical device. The local top gate and the global back gate allow tailoring the potential landscape along the nanoribbon. The gates locally dope the nanoribbon and thus can create tunable pn-junctions. Choosing appropriate gate voltages, the device can be driven into a npn or pnp configuration (see Figure 2.17c). In such a regime, either holes or electrons are confined in between the two pn-junctions, effectively acting as tunneling barriers. Finite bias spectroscopy measurements prove the presence of Coulomb blockade that the device behaves as a SET with addition energies ranging from 0.5 to 3 meV. The addition energy scales inversely with the product of GNR width and top gate width. Even in nn′n and pp′p regimes (n′ and p′ denote strong n- and p-doping, respectively), Coulomb blockade effects can be observed (see Figure 2.17d). This is not expected because of the absence of tunneling barriers. Electron confinement is then attributed to disorder in a larger area than the gate geometry.

2.5.8 QDs Defined by Anodic Oxidation

All devices discussed so far are based on plasma etched graphene nanostructures. The quality of such devices is expected to be influenced by the quality of the edges. Therefore, the local anodic oxidation of graphene sheets has been introduced as an alternative technique (Neubeck et al. 2010). In a controlled atmosphere (humidity typically ≈ 70%), a biased conductive AFM tip carefully scans the graphene flake along desired lines. The AFM is operated in contact mode, and the tip current and the device resistance are monitored in situ. In the vicinity of the tip, water molecules dissociate and the resulting radicals react with the carbon atoms, thus creating locally nonconductive areas in the graphene sheet. Optimizing the humidity and the applied tip bias, this technique is capable writing line widths down to 15 nm. Furthermore, the choice of another etchant in this process allows functionalizing the edges of the graphene structure.

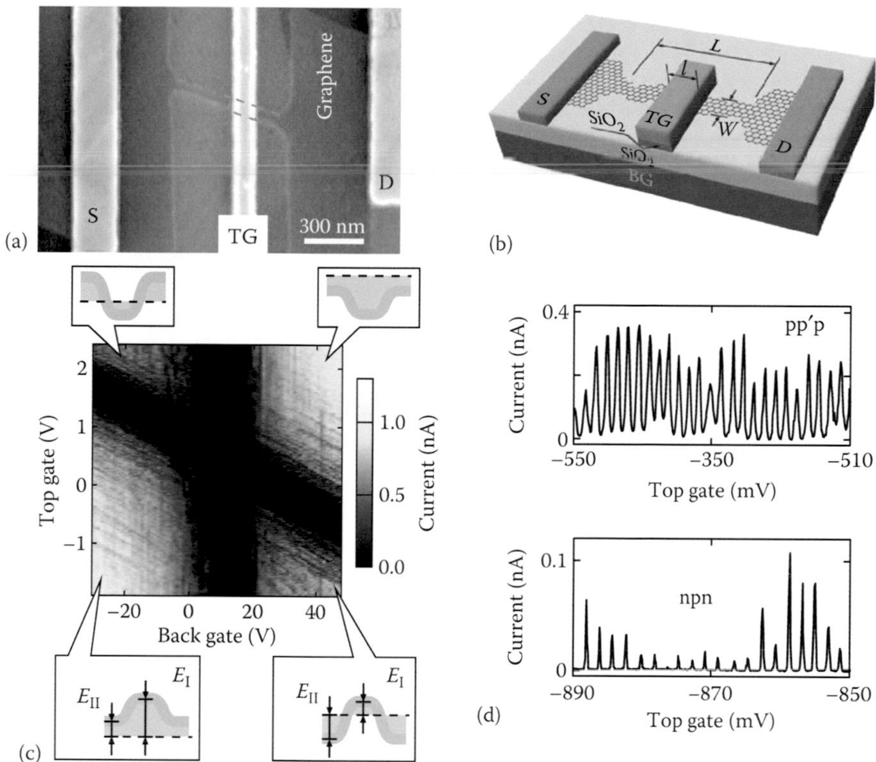

FIGURE 2.17 Gate-defined QD in a GNR. (a) Scanning electron micrograph of a top-gated GNR. The GNR is highlighted by dashed lines. (b) Schematic of the device. (c) Current as a function of back gate and top gate voltage. The insets illustrate the doping profile along the GNR in four different regimes. (d) Coulomb oscillations in the pp′p (V_{BG} = 0 V) and npn (V_{BG} = 81 V) configuration. (Reprinted with permission from Liu, X. et al., *Phys. Rev. B*, 80, 121407, 2009. Copyright 2009 by the American Physical Society.)

QDs with diameters as small as 20 nm—among the smallest graphene QDs reported so far—have been shaped. Figure 2.18a shows a scanning force micrograph of such a device. Addition energies of up to 50 meV have been observed as well as controlled tunability between electron and hole occupation of the QD. A finite bias spectroscopy measurement is shown in Figure 2.18b. On another device, the metal contacts have been evaporated using a shadow mask (Puddy et al. 2013). Thus, it has been possible to

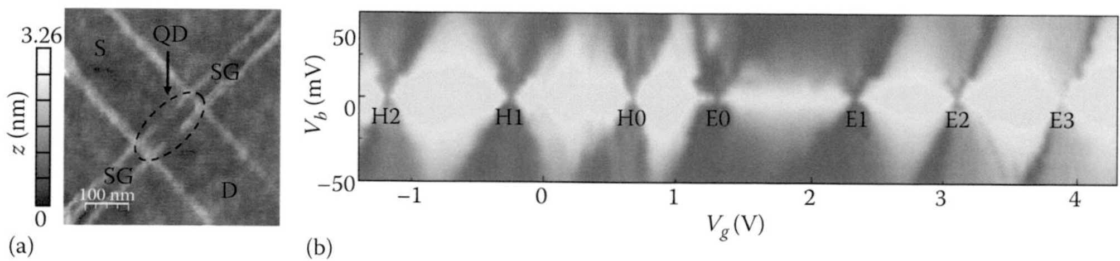

FIGURE 2.18 QD defined by anodic oxidation. (a) Scanning force micrograph of a graphene QD structure created by local anodic oxidation. The bright lines are the regions where the graphene has been oxidized. (b) Finite bias spectroscopy measurement on a QD measuring ≈ 20 nm in diameter. (Neubeck, S., Ponomarenko, L.A., Freitag, F. et al.: From one electron to one hole: Quasiparticle counting in graphene quantum dots determined by electrochemical and plasma etching. *Small*. 2010. 6. 14. Copyright Wiley-VCH Verlag GmbH & Co. KGaA. Reproduced with permission.)

avoid any EBL steps (and hence the use of any polymer resists) in the fabrication process. ES spectroscopy measurements have been performed on this device, and applying an in-plane magnetic field, a g-factor of ≈ 2 could be determined but no regular spin filling sequence.

2.5.9 QDs on Hexagonal Boron Nitride

Most graphene nanodevices including QDs are fabricated from graphene sheets resting on the host substrate SiO_2. This material is believed to cause substrate-induced disorder, limiting the device performance, e.g., making it hard to tune QDs into the few carrier regime. Placing graphene on hexagonal boron nitride (hBN) is one approach to reduce the substrate-induced disorder. Scanning tunneling microscopy studies have demonstrated that the size of charge puddles in graphene on hBN are one order of magnitude larger compared with graphene on SiO_2, where they measure a few tens of nm (Martin et al. 2008; Xue et al. 2011).

In a competitive study, graphene QDs of different diameters have been characterized on both SiO_2 and hBN (Engels et al. 2013). The graphene/hBN heterostructures have been fabricated by depositing exfoliated hBN flakes on a Si/SiO_2 substrate. Graphene flakes are deposited on individual hBN flakes in a controlled way following a transfer process introduced in Dean et al. (2010). An optical image of an hBN/graphene stack is shown in Figure 2.19a. The 20–30 nm thick hBN flake is visible in light gray, whereas the graphene is hardly visible (the dashed line is a guide to the eye). Scanning tunneling microscopy unveils a Moire pattern originating from the lattice mismatch of graphene and hBN (see Figure 2.19b), which indicates a high quality of the graphene. The unit cell vectors (a_1, a_2) measure approx. 3 nm, in agreement with a mismatch of less than 5°. Graphene nanostructures are then etched by RIE and contacted as described in Section 2.4. The device design is adapted from typical QDs on SiO_2. An example of a 300-nm-wide QD device is shown in Figure 2.19c.

Figure 2.19d shows a sequence of Coulomb peaks recorded on a graphene QD on hBN. For a quantitative comparison, the spacing between two subsequent Coulomb peaks is evaluated on several QDs on

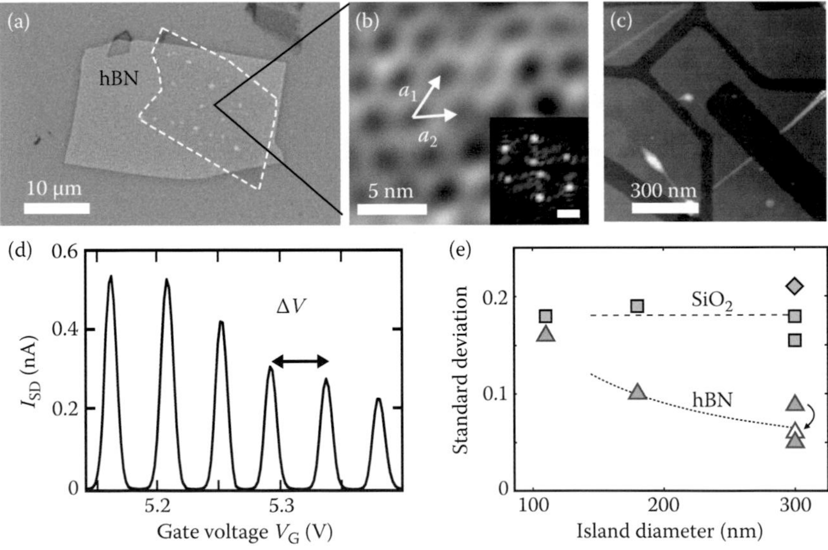

FIGURE 2.19 QDs on hBN. (a) Optical micrograph of a graphene flake on hBN. (b) Scanning tunneling micrograph of the stack (Fourier filtered) showing a Moire pattern. (c) Scanning force micrograph of an etched graphene QD on hBN (diameter: 300 nm). (d) Coulomb peaks measured on a 180-nm-wide QD on hBN. (e) Standard deviation of the distribution of Coulomb peak spacings as a function of the QD diameter for graphene on hBN and SiO_2. Over 600 peaks have been evaluated on each device. (Reprinted with permission from Engels, S. et al., *Appl. Phys. Lett.*, 103, 073113, 2013. Copyright 2013, American Institute of Physics.)

both substrates. Figure 2.19e shows a competitive analysis of the peak-spacing distribution of QDs of different sizes. The devices on hBN show a clear diameter dependence in contrast to the ones on SiO_2. The constant standard deviation of the SiO_2 devices hints that the substrate-induced disorder is dominant as the contributions due to edge roughness are expected to scale with QD diameter. Vice versa, the decrease in the standard deviation with the diameter indicates the edge roughness being dominant. The data allow estimation of a reduction in the substrate-induced disorder in graphene QDs on hBN by roughly one order of magnitude compared with SiO_2.

2.6 Bilayer Graphene QDs

2.6.1 Soft Confinement of QDs in Bilayer Graphene

The fact that a bandgap can be opened by applying a perpendicular electric field to a bilayer graphene sheet allows for a completely different method to confine electrons. Bilayer graphene QD devices with split gates have been fabricated based on a suspended flake and on an hBN/bilayer graphene/hBN sandwich (Allen et al. 2012; Goossens et al. 2012). What both devices have in common is that a voltage difference between the top gates and the global back gate is applied. The electric field perpendicular to the graphene plane induces a local bandgap under each of the gate fingers. The gap can be controlled by the voltage drop (cf. Equation 2.4). Adjusting the gate voltages while leaving the displacement field constant allows driving the Fermi level into the gap. Figure 2.20d shows a schematic of the band structure of a QD confined between the bandgaps opened under the split gates.

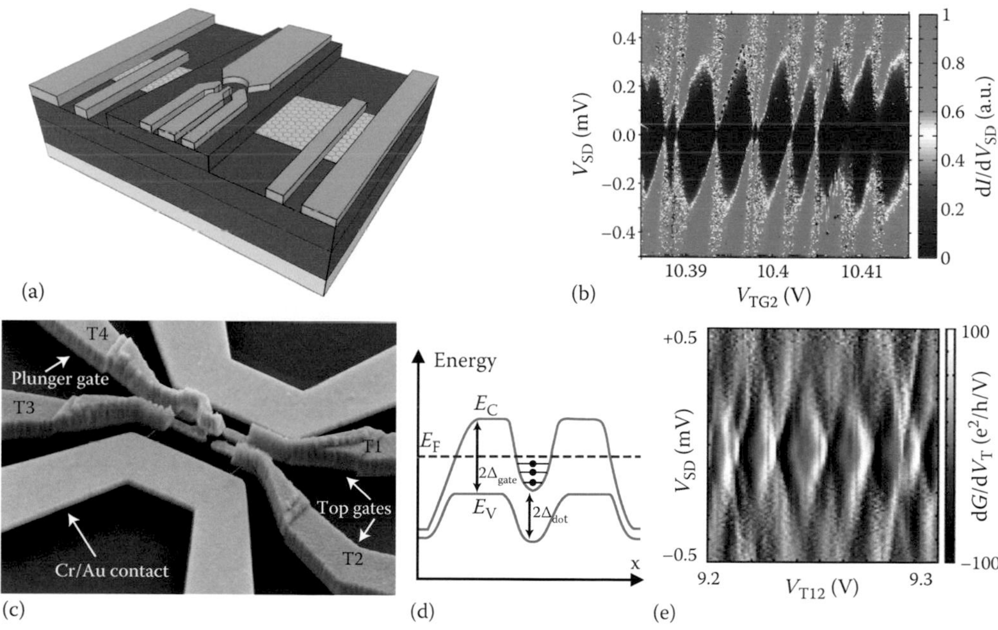

FIGURE 2.20 Soft confinement in bilayer graphene. (a) Schematic of an hBN/graphene/hBN heterostructure with top gates to confine a QD. (b) Finite bias spectroscopy on a device illustrated in panel a. (c) False color scanning electron micrograph of a suspended bilayer QD device with suspended top gates. (d) Band structure along a cross-section through a QD. Tunnel barriers are formed by inducing a bandgap tuned by an out-of-plane electric field. The back gate allows inducing charge accumulation in the dot and the leads. (e) Finite bias spectroscopy on the device shown in panel c. (a and b, Reprinted with permission from Goossens, A.M. et al., *Appl. Phys. Lett.*, 100, 073110, 2012a. Copyright 2012 by the American Physical Society. c–e, Reprinted by permission from MacMillan Publishers Ltd. Allen, M.T., Martin, J., Yacoby, A., *Nat. Commun.*, 3, 934, 2012. Copyright 2012.)

The soft confinement technique allows to electrostatically confine the electrons in the QD as it is commonly done in devices based on semiconductor heterostructures. The graphene flake does not need to be etched in the desired shape, which avoids the influence of edge disorder. Furthermore, using hBN as the host material for the graphene sheet or suspending the flake reduces or even eliminates the influence of the substrate induced disorder commonly observed in graphene devices resting on SiO_2.

The sandwich device has been fabricated by successive transfer steps. The split gates are arranged on top of the heterostructure in the geometry of a 320-nm-wide QD (see Figure 2.20a). The gate arrangement is reminiscent of layouts used to confine QDs in III/V heterostructures. A voltage of 60 V has been applied, leading to a displacement field of $\approx$ 0.6 V/nm, which should theoretically result in a bandgap of $\approx$ 50 meV. The experimentally observed transport gap was significantly smaller and conductance could not be tuned below $\approx 1e^2/h$. Nevertheless, Coulomb blockade has been achieved. An additional energy of 0.35 meV has been extracted from finite bias spectroscopy (see Figure 2.20b). The measurements did not show signatures of ESs, which is in agreement with the relatively large diameter of the island.

The suspended QD device has been fabricated by first contacting a bilayer graphene flake resting on an Si/SiO_2 chip and then covering it with a 150-nm-thick SiO_2 spacer layer. After depositing metal top gate electrodes, the oxide substrate as well as the sacrificial layer have been etched away, leaving the flake and the split gates suspended. Different lithographic device dimensions have been probed. Figure 2.20c shows a false color scanning electron microscopy (SEM) image of a representative device. Coulomb blockade measurements are in agreement with the geometric sizes (see Figure 2.20e). Conductance modulations only coupling to the back gate have been observed, which hint parasitic conductance channels under the gates limiting the device performance. A suspended QD offers the chance to study the coupling of quantized electronic and vibrational degrees of freedom.

2.7 DQDs in Graphene and Bilayer Graphene

2.7.1 DQDs in Graphene

After successful realization of graphene QDs in width-modulated nanoribbons, the device concept has been extended to DQDs: two graphene islands separated by a GNR acting as the interdot tunneling barrier. DQDs are of interest as they can host solid-state spin qubits.

Figure 2.21a shows an example of an etched graphene DQD device surrounded by five lateral gates. From a stability diagram, i.e., the current through the QD recorded as a function of two gate voltages, gate lever arms, charging energies, mutual capacitive coupling, and excited state spectra of the two QDs can be determined (Molitor et al. 2009, 2010; Moriyama et al. 2009, 2010). The CG voltage allows tuning the capacitive interdot coupling energy. As transport through GNRs is governed by a large number of sharp resonances, the coupling shows a highly nonmonotonic behavior. Figure 2.21b shows a detail of a finite bias stability diagram highlighting an individual pair of triple points. Triple points occur at gate configurations, where both QDs are in resonance and a current can flow through the device. Fine structure inside the triple points originates from ESs of the QDs, but some of the observed resonances are also attributed to modulations of the tunneling barriers. Another example of a graphene DQD in is shown in Figure 2.21c (Wei et al. 2013). Here, a different device geometry has been chosen and metal gate electrodes have been placed in close vicinity of the QDs. The corresponding charge stability diagram is shown in Figure 2.21d. The relative electron occupation of the two QDs is indicated in the plot. Not only the capacitive coupling can be influenced by a gate but also the quantum mechanical tunnel coupling. Gate-tunability of the tunnel coupling of a factor of 4 has been demonstrated (Wei et al. 2013). Furthermore, a significant dependency on the electron occupation has been observed. A DQD device can also be driven into a single QD regime in a controlled way. This effect is explained by a significant increase in the tunnel coupling as a response on the gate voltage (Wang et al. 2012; Zhou et al. 2012).

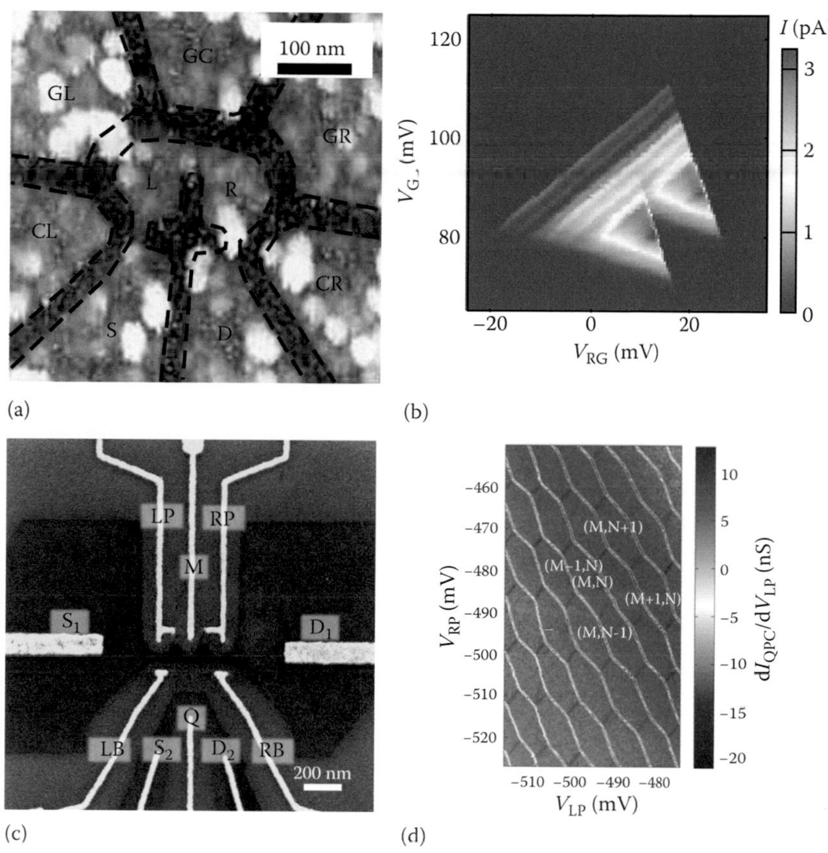

FIGURE 2.21 DQDs in single-layer graphene nanostructures. (a) Scanning force micrograph of a DQD structure with all-graphene lateral gates etched out of a graphene flake. (b) Stability diagram highlighting an individual pair of triple points (V_{SD} = 6 mV) measured on the device in panel a. (c) False color scanning electron micrograph of a DQD device with a capacitively coupled CD. The metal gate electrodes are placed in close vicinity to active structure. (d) Stability diagram recorded using the CD (dI_{QPC}/dV_{LP} plotted as a function of plunger gate voltages). (a and b, Adapted from Molitor, F. et al., *Europhys. Lett.*, 89, 67005, 2010. Copyright 2010 IOP Publishing. c and d, Reprinted by permission from MacMillan Publishers Ltd. Wei, D. et al., *Sci. Rep.*, 3, 3175, 2013. Copyright 2013.)

2.7.2 High-Frequency Gate Manipulation

Several experiments have addressed the mutual capacitive coupling, tunnel coupling, and the ES spectra in graphene DQDs. All experiments have been performed at DC. More recently, a graphene charge pump operating in the GHz regime has been demonstrated (Connolly et al. 2013). Such devices are of great interest among others for the realization of a current standard (Giblin et al. 2012), single-photon generation (Mueller et al. 2009) and read-out of spin-based graphene (Trauzettel et al. 2007).

A DQD (diameter ≈200 nm) surrounded by five graphene gates tuning the coupling energies and the QD potentials has been fabricated. Two gates, designed as plunger gates to control the QDs, are connected to bias-tees. A radio frequency (RF) voltage $V_{RF}(t)$ is applied to these gates, and a phase shifter controls the phase difference between them (see Figure 2.22a). A low bias charge stability diagram of the device is shown in Figure 2.22b. Because of the RF signal, the system describes a circle in the gate voltage plane, which is illustrated in the stability diagram. Whenever the DC gate voltages are tuned such that the system encircles a triple point, one electron is shuttled through the system per pulse cycle. When crossing position (1), the potential of first QD is lowered such that an electron from the source is loaded. At (2), the

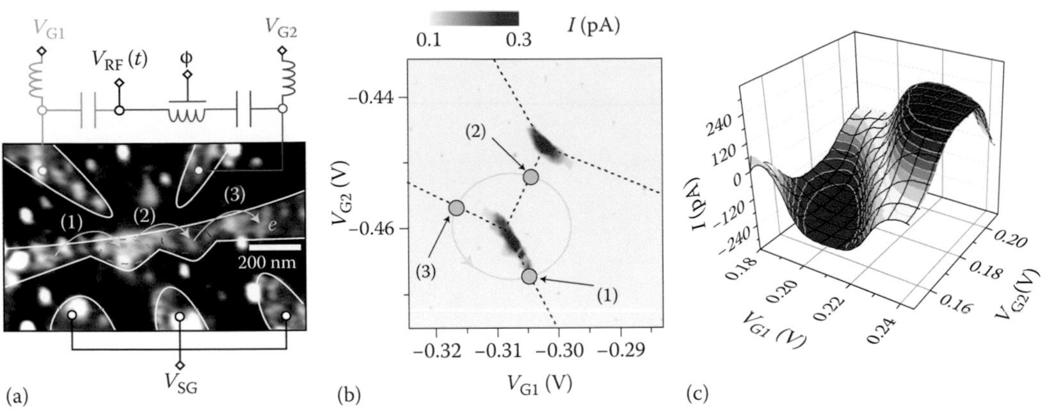

FIGURE 2.22 Charge pumping in a DQD. (a) Scanning force micrograph of the device. A RF voltage $V_{RF}(t)$ is added to the DC gates controlling the QD potentials. A controllable phase difference ϕ is added to the signal on gate 2. (b) Stability diagram at bias voltage below 1 μV. The circle illustrates a possible pumping cycle around a triple point. Transitions (1)–(3) correspond to changes of the electron occupation of the QDs (see text). (c) 3D representation of the pumped current as a function of the DC gate voltages at fixed frequency and power (f = 1.465 GHz, P = −15 dBm). (Reprinted by permission from MacMillan Publishers Ltd. Connolly, M.R. et al., *Nat. Nanotechnol.*, 8, 417–420, 2013. Copyright 2013.)

electron tunnels to the second dot. Crossing position (3), the electron leaves the second dot to the drain. The frequency determines the rate of the transferred charges and, thus, the current through the device, which is expected to be $I = e \cdot f$. Figure 2.22c shows a measurement of the pumped current as a function of the DC gate voltages at fixed frequency (f = 1.5 GHz) and power. Depending on which triple point is enclosed in the cycle, plateaus of different signs but same height are formed. The pumped current shows an almost linear frequency dependency with an error rate (i.e., a deviation of the measured current from the expected value $e \cdot f$) below 0.5% below 0.5 GHz. At higher frequencies, the error rate increases up to a few percentage as a result of an increase in failed pump cycles. It scales exponentially with $f \cdot RC$. In metal charge pumps, the same dependency is observed, but because of the approximately 10 times larger RC constant in such devices, the error rate of the graphene pump is smaller at the same frequency.

2.7.3 DQDs in Bilayer Graphene

A DQD device with a number of lateral gates has been etched out of a bilayer graphene sheet (see scanning force micrograph in Figure 2.23a) (Volk et al. 2011). A Raman spectrum showing the bilayer nature of the flake can be seen in Figure 2.6. Because of the narrow constrictions connecting the QDs, effective transport gaps are opened locally in the constrictions. The global back gate tunes the Fermi level such that it crosses these gaps (see Figure 2.23b). Five lateral gates control the electrostatic potential on each QD as well as the tunneling rates through the three barriers individually. By means of the CG, the device can be tuned between a single and a DQD configuration.

Figure 2.23c shows a stability diagram of the device. The extracted addition energies of E_a^L = 18 meV and E_a^R = 21 meV are in reasonable agreement with values from single-layer graphene QDs of a similar size (Güttinger et al. 2010). The stability diagram furthermore allows the quantitative extraction of the mutual capacitive coupling energy (E^m) between the two QDs from the separation of the two overlapping triangles (see arrows).

The coupling energy can be tuned by the CG, which is studied in Figure 2.24a on three individual nearby triple points (Fringes et al. 2011). The coupling energies of the neighboring triple points (M,N) and (M + 1,N) increase with V_{cg}, whereas it decreases for the triple point (M + 5,N − 1) which is in

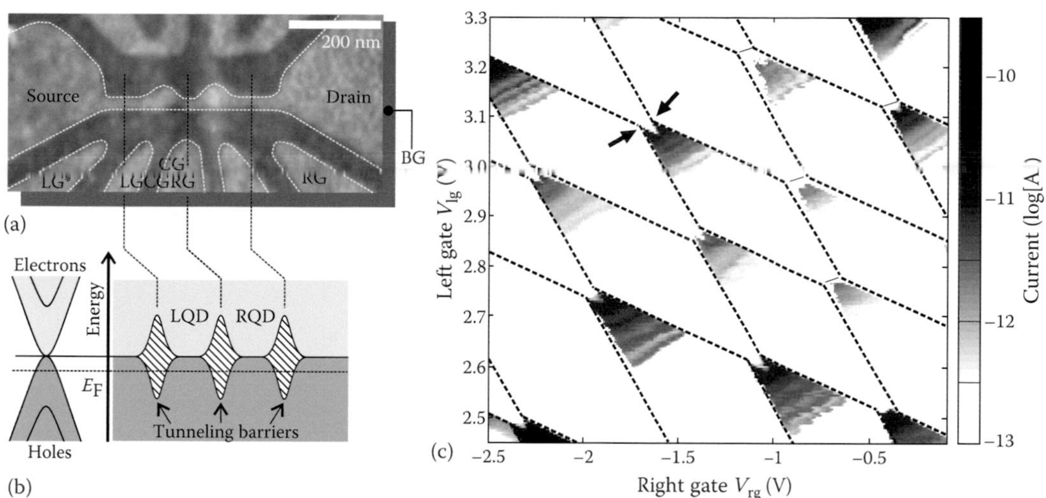

FIGURE 2.23 Bilayer graphene DQD. (a) Scanning force microscope image of a DQD device. The diameters of the etched QDs measures roughly 50 nm, while the width of the 100-nm-long constrictions leading to the dots measure 30 nm. (b) Schematic illustration of the effective band structure of the device highlighting the three tunneling barriers (hatched areas) induced by the local constrictions (see panel a). (c) Finite bias charge stability diagram recorded in the DQD regime (V_{SD} = 10 mV). (Reprinted with permission from Volk, C., Fringes, S., Terrés, B. et al., *Nano Lett.*, 11, 3581, 2011. Copyright 2011, American Chemical Society.)

agreement with studies on single-layer graphene QDs (Molitor et al. 2009; Liu et al. 2010), showing a strongly nonmonotonic dependence of the conductance on gate voltages due to the sharp resonances in the constrictions. Figure 2.24b and c compare close-ups of the very same triple point at two different interdot coupling energies E^m = 4.8 meV and E^m = 3.0 meV, highlighted by gray arrows in panel a.

The triple points in Figure 2.23c exhibit a number of clearly distinguishable ES resonances (visible as parallel lines to the base line). The energies of the five visible ESs of the highlighted triple point measure ε = 1.7, 3.3, 4.9, 6.5, and 8.1 meV, respectively. The energy spacing of two subsequent electronic ESs has been found to be constant over a large energy range with a value of $\Delta = 1.75 \pm 0.27$meV. This is in agreement with the single-particle confinement energy in disk-like bilayer graphene QDs, $\Delta_{BL} = \frac{4\hbar^2 v_F^2}{\gamma_1}\frac{1}{d^2}$ with the QD diameter d, the Fermi velocity v_F, and the interlayer hopping energy γ_1 = 0.39 meV. Please note that this result depends only on the QD diameter, especially that it is independent on the number of charge carriers N occupying the QD. Figure 2.24d compares the experimental data with the calculated single-level spacing Δ for 50-nm-diameter single-layer and bilayer graphene QDs. The level spacing in single-layer graphene is given by $\Delta_{SL}(N) = \hbar v_F / \left(d\sqrt{N}\right)$ and obviously does not suite the data.

The evolution of ESs as a function of the magnetic field oriented in parallel to the bilayer graphene plane proves the electronic nature of these states. Figure 2.24e and f compare a pair of triple points recorded at $B_{||}$ = 0 T and at $B_{||}$ = 2 T, respectively. When applying a finite magnetic field, the first ES splits into two separate peaks (see dotted lines). This splitting is linear in magnetic field up to roughly $B_{||}$ = 5 T, as shown in Figure 2.24g, where current line cuts along the detuning energy axis ε (see dashed line in Figure 2.24f) are plotted for different B-fields. The characteristic linear peak splitting measures 0.2 meV/T. Most likely, the observed peak splittings originate from Zeeman spin splitting, which is motivated by the linear $B_{||}$-field dependence of the peak splittings and their characteristic energy scale, which is in reasonable agreement with $g\mu_B$ = 0.116 meV/T. The position of the triangles with respect to the gate voltage plane almost does not change at all (see crosses and solid lines), proving that the orbital contributions of the wave functions are unaffected.

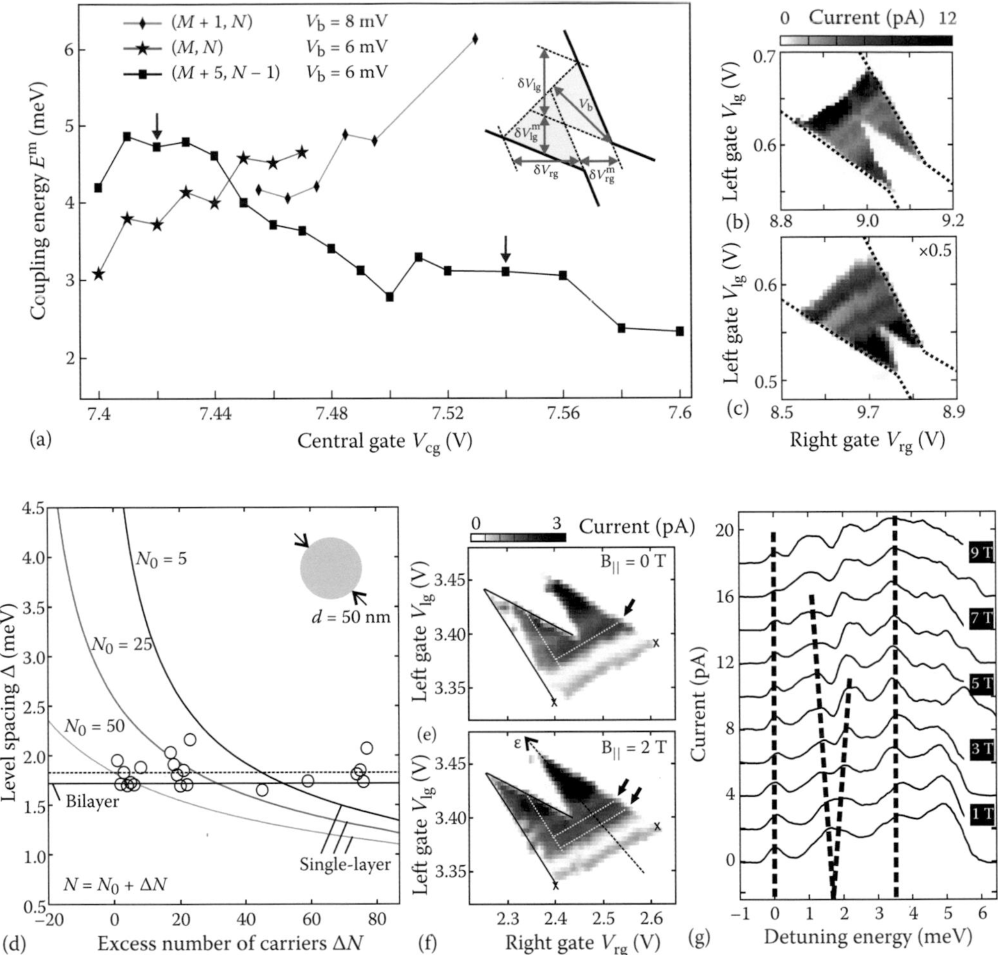

FIGURE 2.24 Capacitive interdot coupling and ES spectroscopy. (a) Mutual capacitive interdot coupling energy E^m as a function of CG voltage for three nearby triple points with the relative numbers of charge carriers (M,N), $(M+1,N)$ and $(M+5,N-1)$. The inset shows a schematic pair of triple points labeled with the quantities to deduce the coupling energy $E^m = \alpha\delta V^m$. (b, c) Same triple points in a stronger (V_{cg} = 7.42 V) and weaker (V_{cg} = 7.54 V) coupling regime, respectively. (d) Single-level spacing as a function of the excess number of charge carriers on the left dot. The circles represent the experimental data and the dashed horizontal line marks their mean value (Δ = 1.82 meV). The three curves represent the model of a single-layer QD for different absolute occupation numbers N_0. The solid horizontal line marks the constant level spacing estimated for a bilayer dot with a diameter of 50 nm. (e–g) DQD under the influence of a magnetic field. Pair of triple points measured at (e) $B_{||}$ = 0 T and (f) $B_{||}$ = 2 T. (g) Current as a function of the detuning energy measured along the dashed line indicated in panel f for different values of $B_{||}$. The traces are offset by 2 pA for clarity. (a–c, Fringes, S., Volk, C., Terrés, B., Dauber, J. et al.: Tunable capacitive inter-dot coupling in a bilayer graphene double quantum dot. *Phys. Status Solidi C*. 2011. 9. 169. Copyright Wiley-VCH Verlag GmbH & Co. KGaA. Adapted with permission. d–g, Reprinted with permission from Volk, C. et al., *Nano Lett.*, 11, 3581, 2011. Copyright 2011, American Chemical Society.)

2.8 Applications of Graphene QDs

Because of size-dependent absorption and emission properties, graphene QDs are of great interest for applications in solar cells and optoelectronics (Kim et al. 2012). Graphene QDs with absorption and emission spectra ranging from UV via visible to the infrared (IR) regime can be fabricated. Photovoltaic operation has been observed in a ZnO nanowire/graphene QD composite solar cell (Dutta et al. 2012).

Graphene QDs are interesting for biological applications like bioimaging, protein analysis, cell tracking, and gene technology. They offer the chance to obtain nanoparticles of low cytotoxicity and high biocompatibility. The use of graphene QD-based labels for imaging living cells has already been demonstrated (Peng et al. 2012; Zhu et al. 2011).

QDs in general are promising candidates for hosting spin qubits (Loss and DiVincenzo 1998). In this context, graphene is of special interest because of the low spin-orbit and hyperfine interaction. Thus, it is predicted that spin qubits in graphene feature long coherence times (Trauzettel et al. 2007).

2.9 Conclusions and Outlook

A variety of transport experiments on graphene and bilayer graphene QD devices have been presented. Coulomb blockade, ES spectra, and spin states have been studied in graphene QDs. GNR-based charge has proven to be a valuable tool to detect charging events on a nearby QD and even allows for time-resolved charge detection. By pulsed-gate transient current spectroscopy, charge relaxation times as long as 60–100 ns have been measured. The weak hyperfine and spin-orbit interaction promise long spin coherence times, which still has to be demonstrated.

On DQDs, ES spectra and interdot tunnel coupling have been investigated. On a bilayer device, gate control of the mutual capacitive coupling has been demonstrated and the ES level spacing turned out to be constant in contrast to single-layer graphene. By high-frequency gate manipulation, a highly efficient GHz charge pump has been realized. Pauli spin blockade, which is commonly used to read out semiconductor spin qubits, has not yet been observed in graphene DQDs.

Most of the graphene nanostructures studied so far have been fabricated based on plasma etched width modulated nanostructures on the substrate SiO_2. The quality of such devices is expected to be limited by edge roughness and substrate disorder. One idea to reduce the influence of edge roughness is the use of local anodic oxidation. ES spectra and spin states have been investigated. Comparing QDs on SiO_2 and hBN has shown that the influence of substrate-induced disorder can be reduced using hBN for QDs with diameters larger than 100 nm. For smaller devices, the influence from the edges seems to dominate. Devices based on soft confinement in bilayer graphene go even a step further. Using either an hBN/bilayer graphene/hBN sandwich or a suspended flake, they minimize the substrate disorder. Lateral confinement by gates instead of plasma etching reduces the edge disorder. Although the technique seems to be promising and first devices have been demonstrated, no ESs have been observed so far and parasitic conductance channels under the gates seem to limit the device performance.

For optical and biological applications where arrays of QDs rather than individual QD are investigated, graphene QDs obtained chemically from graphene oxide or by wet chemical oxidization and cutting of micrometer-sized carbon fibers have turned out to be suitable techniques. Moreover, absorption and emission spectra have been studied and first experiments in photovoltaics and bioimaging have been reported.

References

Allen, M. T., J. Martin, and A. Yacoby. Gate-defined quantum confinement in suspended bilayer graphene. *Nat. Commun.* **3**, 934 (2012).

Bischoff, D., J. Güttinger, S. Dröscher, T. Ihn, K. Ensslin, and C. Stampfer. Raman spectroscopy on etched graphene nanoribbons. *Appl. Phys. Lett.* **109**, 073710 (2011).

Bischoff, B., A. Varlet, P. Simonet, T. Ihn, and K. Ensslin. Electronic triple-dot transport through a bilayer graphene island with ultrasmall constrictions. *New J. Phys.* **15**, 083029 (2013).

Blake, P., E. W. Hill, A. H. C. Neto, K. S. Novoselov, D. Jiang, R. Yang, T. J. Booth, and A. K. Geim. Making graphene visible. *Appl. Phys. Lett.* **910** (6), 063124 (2007).

Bluhm, H., S. Foletti, I. Neder, M. Rudner, D. Mahalu, V. Umansky, and A. Yacoby. Dephasing time of GaAs electron-spin qubits coupled to a nuclear bath exceeding 200 μs. *Nat. Phys.* **7**, 109 (2010).

Brandt, N. B., Y. G. Chudinov, and S. M. Ponomarev. Modern Problems in Condensed Matter Sciences, vol. 20, 1. Amsterdam, The Netherlands: North-Holland (1988).

Brey, L. and H. Fertig. Electronic states of graphene nanoribbons studied with the Dirac equation. *Phys. Rev. B* **73**, 235411 (2006).

Castro Neto, A. H., F. Guinea, N. M. R. Peres, K. S. Novoselov, and A. K. Geim. The electronic properties of graphene. *Rev. Mod. Phys.* **810** (1), 109–162 (2009).

Chen, Z., Y.-M. Lin, M. J. Rooks, and P. Avouris. Graphene nano-ribbon electronics. *Physica E Low Dimens. Syst. Nanostruct.* **400** (2), 228–232 (2007).

Connolly, M. R., K. L. Chiu, S. P. Giblin, M. Kataoka, J. D. Fletcher, C. Chua, J. P. Griffiths et al. Gigahertz quantized charge pumping in graphene quantum dots. *Nat. Nanotechnol.* **8**, 417–420 (2013).

Cresti, A., B. Nemec, B. Biel, G. Niebler, F. Triozon, G. Cuniberti, and S. Roche. Charge transport in disordered graphene-based low dimensional materials. *Nano Res.* **1**, 361 (2008).

Das Sarma, S., S. Adam, E. H. Hwang, and E. Rossi. Electronic transport in two-dimensional graphene. *Rev. Mod. Phys.* **83**, 407 (2011).

Dean, C. R., A. F. Young, I. Meric, C. Lee, L. Wang, S. Sorgenfrei, K. Watanabe et al. Boron nitride substrates for high-quality graphene electronics. *Nat. Nanotechnol.* **5**, 722–726 (2010).

Dresselhaus, M. S. and G. Dresselhaus. Intercalation compounds of graphite. *Adv. Phys.* **51**, 1 (2002).

Dröscher, S., J. Güttinger, T. Mathis, B. Batlogg, T. Ihn, and K. Ensslin. High-frequency gate manipulation of a bilayer graphene quantum dot. *Appl. Phys. Lett.* **101**, 043107 (2012).

Du, X., I. Skachko, F. Duerr, A. Luican, and E. Y. Andrei. Fractional quantum Hall effect and insulating phase of Dirac electrons in graphene. *Nature* **462**, 192–195 (2009).

Dutta, M., S. Sarkar, T. Ghosh, and D. Basak. ZnO/Graphene quantum dot solid-state solar cell. *J. Phys. Chem. C* **116**, 20127 (2012).

Elzermann, J. M., R. Hanson, J. S. Greidanus, L. H. W. van Beveren, S. De Franceschi, L. M. K. Vandersypen, S. Tarucha, and L. P. Kouwenhoven. Few-electron quantum dot circuit with integrated charge read out. *Phys. Rev. B* **67**, 161308 (2003).

Engels, S., A. Epping, C. Volk, S. Korte, B. Voigtländer, K. Watanabe, T. Taniguchi, S. Trellenkamp, and C. Stampfer. Etched graphene quantum dots on hexagonal boron nitride. *Appl. Phys. Lett.* **103**, 073113 (2013a).

Engels, S., P. Weber, B. Terrés, J. Dauber, C. Meyer, C. Volk, S. Trellenkamp, U. Wichmann, and C. Stampfer. Fabrication of coupled graphene-nanotube quantum devices. *Nanotechnology* **24**, 035204 (2013b).

Epping, A., S. Engels, C. Volk, K. Watanabe, T. Taniguchi, S. Trellenkamp, and C. Stampfer. Etched graphene single electron transistors on hexagonal boron nitride in high magnetic fields. *Phys. Status Solidi B* **25**, 2682 (2013).

Ferrari, A. C., J. C. Meyer, V. Scardaci, C. Casiraghi, M. Lazzeri, F. Mauri, S. Piscanec et al. Raman spectrum of graphene and graphene layers. *Phys. Rev. Lett.* **97**, 187401 (2006).

Field, M., C. G. Smith, M. Pepper, D. A. Ritchie, J. E. F. Frost, G. A. C. Jones, and D. G. Hasko. Measurements of Coulomb blockade with a noninvasive voltage probe. *Phys. Rev. Lett.* **70**, 1311 (1993).

Fringes, S., C. Volk, C. Norda, B. Terrés, J. Dauber, S. Engels, S. Trellenkamp, and C. Stampfer. Charge detection in a bilayer graphene quantum dot. *Phys. Status Solidi B* **248**, 2684 (2011a).

Fringes, S., C. Volk, B. Terrés, J. Dauber, S. Engels, S. Trellenkamp, and C. Stampfer. Tunable capacitive inter-dot coupling in a bilayer graphene double quantum dot. *Phys. Status Solidi C* **9**, 169 (2011b).

Fujisawa, T., D. G. Austing, Y. Tokura, Y. Hirayama, and S. Tarucha. Allowed and forbidden transitions in artificial hydrogen and helium atoms. *Nature* **419**, 278–281 (2002a).

Fujisawa, T., D. G. Austing, Y. Tokura, Y. Hirayama, and S. Tarucha. Transport through a vertical quantum dot in the absence of spin-flip energy relaxation. *Phys. Rev. Lett.* **88**, 236802 (2002b).

Fujisawa, T., T. Hayashi, R. Tomita, and Y. Hirayama. Bidirectional counting of single electrons. *Science* **312**, 1634 (2006).

Fujisawa, T., Y. Tokura, and Y. Hirayama. Energy relaxation process in a quantum dot studied by DC current and pulse-excited current measurements. *Physica B* **298**, 573 (2001a).

Fujisawa, T., Y. Tokura, and Y. Hirayama. Transient current spectroscopy of a quantum dot in the Coulomb blockade regime. *Phys. Rev. B* **63**, 081304 (2001b).

Geim, A. K. and K. Novoselov. The rise of graphene. *Nat. Mater.* **6**, 183–191 (2007).

Giblin, S. P., M. Kataoka, J. D. Fletcher, P. See, T. Janssen, J. P. Griffiths, G. Jones, I. Farrer, and D. Ritchie. Towards a quantum representation of the ampere using single electron pumps. *Nat. Commun.* **3**, 290 (2012).

Goossens, A. M., V. E. Calado, A. Barreiro, K. Watanabe, T. Taniguchi, and L. M. K. Vandersypen. Mechanical cleaning of graphene. *Appl. Phys. Lett.* **100**, 073110 (2012a).

Goossens, A. M., S. C. M. Driessen, T. A. Baart, K. Watanabe, T. Taniguchi, and L. M. K. Vandersypen. Gate-defined confinement in bilayer graphene-hexagonal boron nitride hybrid devices. *Nano Lett.* **12**, 4656–4660 (2012b).

Graf, D., F. Molitor, K. Ensslin, C. Stampfer, C. H. A. Jungen, and L. Wirtz. Spatially resolved Raman spectroscopy of single- and few-layer graphene. *Nano Lett.* **70** (2), 238–242 (2007).

Gupta, A., G. Chen, P. Joshi, S. Tadigadapa, and P. Eklund. Raman scattering from high-frequency phonons in supported n-graphene layer films. *Nano Lett.* **6**, 2667 (2006).

Gustavsson, S., R. Leturcq, B. Simovic, R. Schleser, T. Ihn, P. Studerus, and K. Ensslin. Counting statistics of single electron transport in a quantum dot. *Phys. Rev. Lett.* **96**, 076605 (2006).

Güttinger, J. Graphene quantum dots. PhD thesis, Zurich, Switzerland: ETH Zürich (2011).

Güttinger, J., C. Stampfer, S. Hellmüller, F. Molitor, T. Ihn, and K. Ensslin. Charge detection in graphene quantum dots. *Appl. Phys. Lett.* **930** (21), 212102 (2008a).

Güttinger, J., C. Stampfer, F. Molitor, D. Graf, T. Ihn, and K. Ensslin. Coulomb oscillations in three-layer graphene nanostructures. *New J. Phys.* **10**, 12029 (2008b).

Güttinger, J., C. Stampfer, F. Libisch, T. Frey, J. Burgdörfer, T. Ihn, and K. Ensslin. Electron-hole crossover in graphene quantum dots. *Phys. Rev. Lett.* **103**, 046810 (2009).

Güttinger, J., T. Frey, C. Stampfer, T. Ihn, and K. Ensslin. Spin states in graphene quantum dots. *Phys. Rev. Lett.* **1050** (11), 116801 (2010).

Güttinger, J., J. Seif, C. Stampfer, A. Capelli, K. Ensslin, and T. Ihn. Time-resolved charge detection in graphene quantum dots. *Phys. Rev. B* **83**, 165445 (2011a).

Güttinger, J., C. Stampfer, T. Frey, T. Ihn, and K. Ensslin. Transport through a strongly coupled graphene quantum dot in perpendicular magnetic field. *Nano. Res. Lett.* **6**, 253 (2011b).

Güttinger, J., F. Molitor, C. Stampfer, S. Schnez, S. Jacobsen, S. Dröscher, T. Ihn, and K. Ensslin. Transport through graphene quantum dots. *Rep. Prog. Phys.* **75**, 126502 (2012).

Han, M. Y., B. Özyilmaz, Y. Zhang, and P. Kim. Energy band gap engineering of graphene nanoribbons. *Phys. Rev. Lett.* **98**, 206805 (2007).

Hanson, R. Electron spins in semiconductor quantum dots. PhD thesis, Delft, The Netherlands: Technische Universiteit Delft (2005).

Hayashi, T., T. Fujisawa, H. Cheong, Y. Jeong, and Y. Hirayama. Coherent manipulation of electronic states in a double quantum dot. *Phys. Rev. Lett.* **91**, 226804 (2003).

Hu, Y., F. Kuemmeth, C. M. Lieber, and C. M. Marcus. Spin relaxation in Ge/Si core-shell nanowire qubits. *Nat. Nanotechnol.* **7**, 47 (2011).

Huertas-Hernando, D., F. Guinea, and A. Brataas. Spin-orbit coupling in curved graphene, fullerenes, nanotubes, and nanotube caps. *Phys. Rev. B* **74**, 155426 (2006).

Ihn, T. *Semiconductor Nanostructures—Quantum States and Electronic Transport*. Oxford, United Kingdom: Oxford University Press (2010).

Jacobsen, A., P. Simonet, K. Ensslin, and T. Ihn. Transport in a three-terminal graphene quantum dot in the multi-level regime. *New J. Phys.* **14**, 023052 (2012).

Johnson, A. C., J. R. Petta, J. M. Taylor, A. Yacoby, M. D. Lukin, C. M. Marcus, M. P. Hanson, and A. C. Gossard. Triplet-singlet spin relaxation via nuclei in a double quantum dot. *Nature* **435**, 925 (2005).

Kastner, M. A. Artificial atoms. *Phys. Today* **46**, 24–31 (1993).

Katsnelson, M. Graphene—Carbon in Two Dimensions. Cambridge, United Kingdom: Cambridge University Press (2012).

Katsnelson, M. I. Zitterbewegung, chirality, and minimal conductivity in graphene. *Eur. Phys. J. B Condens. Matter Complex Syst.* **51**, 157–160 (2006).

Katsnelson, M. I. and K. S. Novoselov. Graphene: New bridge between condensed matter physics and quantum electrodynamics. *Solid State Comm.* **143**, 3 (2007).

Katsnelson, M., K. Novoselov, and A. Geim. Chiral tunnelling and the Klein paradox in graphene. *Nat. Phys.* **2**, 620 (2006).

Kawakami, E., P. Scarlino, D. R. Ward, F. R. Braakman, D. E. Savage, M. G. Lagally, M. Friesen, S. N. Coppersmith, M. A. Eriksson, and L. M. K. Vandersypen. Electrical control of a long-lived spin qubit in a Si/SiGe quantum dot. *Nat. Nanotechnol.* **9**, 666–670 (2014).

Khaetskii, A. V., D. Loss, and L. Glazman. Electron spin decoherence in quantum dots due to interaction with nuclei. *Phys. Rev. Lett.* **88**, 186802 (2002).

Kim, S., S. Hwang, M.-K. Kim, D. Y. Shin, D. H. Shin, C. Kim, B. Yang et al. Anomalous behaviors of visible luminescence from graphene quantum dots: Interplay between size and shape. *ACS Nano* **6**, 8203 (2012).

Kobayashi, Y., K.-I. Fukui, T. Enoki, K. Kusakabe, and Y. Kaburagi. Observation of zigzag and armchair edges of graphene using scanning tunneling microscopy and spectroscopy. *Phys. Rev. B* **710,** 193406 (2005).

Kölbl, D. and D. Zumbühl. Transport spectroscopy of disordered graphene quantum dots etched into a single graphene flake. arXiv:1307.8163v1 (2013).

Koppens, F., C. Buizert, K. Tielrooij, I. Vink, K. Nowack, T. Meunier, L. Kouwenhoven, and L. Vandersypen. Driven coherent oscillations of a single electron spin in a quantum dot. *Nature* **442**, 766–771 (2006).

Laird, E., F. Pei, and L. Kouwenhoven. A valley-spin qubit in a carbon nanotube. *Nat. Nanotechnol.* **8**, 565–568 (2013).

Li, L. and X. Yan. Colloidal graphene quantum dots. *J. Phys. Chem. Lett.* **1**, 2572 (2010).

Li, Y., Y. Hu, Y. Zhao, G. Shi, L. Deng, Y. Hou, and L. Qu. An electrochemical avenue to green-luminescent graphene quantum dots as potential electron-acceptors for photovoltaics. *Adv. Mater.* **23**, 776 (2010).

Liu, X. L., D. Hug, and L. M. K. Vandersypen. Gate-defined graphene double quantum dot and excited state spectroscopy. *Nano Lett.* **10**, 1623–1627 (2010).

Liu, X., J. B. Oostinga, A. F. Morpurgo, and L. M. K. Vandersypen. Electrostatic confinement of electrons in graphene nanoribbons. *Phys. Rev. B* **80**, 121407 (2009).

Loss D. and D. P. DiVincenzo. Quantum computation with quantum dots. *Phys. Rev. A* **570** (1), 120 (1998).

Lu, J., P. S. E. Yeo, C. K. Gan, P. Wu, and K. P. Loh. Transforming C60 molecules into graphene quantum dots. *Nat. Nanotechnol.* **5**, 247 (2011).

Malard, L., M. Pimenta, G. Dresselhaus, and M. Dresselhaus. Raman spectroscopy in graphene. *Phys. Rep.* **473**, 51–87 (2009).

Martin, J., N. Akerman, G. Ulbricht, T. Lohmann, J. H. Smet, K. von Klitzing, and A. Yacoby. Observation of electron-hole puddles in graphene using a scanning single-electron transistor. *Nat. Phys.* **4**, 144–148 (2008).

McCann, E., K. Kechedzhi, V. Fal'ko, H. Suzuura, T. Ando, and B. Altshuler. Weak localisation in graphene. *Phys. Rev. Lett.* **96**, 086805 (2006).

Min, H., J. E. Hill, N. A. Sinitsyn, B. R. Sahu, L. Kleinman, and A. H. MacDonald. Intrinsic and Rashba spin-orbit interactions in graphene sheets. *Phys. Rev. B* **74**, 165310 (2006).

Molitor, F., S. Dröscher, J. Güttinger, A. Jacobsen, C. Stampfer, T. Ihn, and K. Ensslin. Transport through graphene double dots. *Appl. Phys. Lett.* **940** (22), 222107 (2009a).

Molitor, F., A. Jacobsen, C. Stampfer, J. Güttinger, T. Ihn, and K. Ensslin. Transport gap in side-gated graphene constrictions. *Phys. Rev. B* **790** (7), 075426 (2009b).

Molitor, F., H. Knowles, S. Dröscher, U. Gasser, T. Choi, P. Roulleau, J. Güttinger et al. Observation of excited states in a graphene double quantum dot. *Europhys. Lett.* **89**, 67005 (2010a).

Molitor, F., C. Stampfer, J. Güttinger, A. Jacobsen, T. Ihn, and K. Ensslin. Energy and transport gaps in etched graphene nanoribbons. *Semicond. Sci. Technol.* **250** (3), 034002 (2010b).

Moriyama, S., Y. Morita, E. Watanabe, and D. Tsuya. Field-induced confined states in graphene. *Appl. Phys. Lett.* **104**, 053108 (2014).

Moriyama, S., Y. Morita, E. Watanabe, D. Tsuya, S. Uji, M. Shimizu, and K. Koji Ishibashi. Fabrication of quantum-dot devices in graphene. *Sci. Technol. Adv. Mater.* **11**, 054601 (2010).

Moriyama, S., D. Tsuya, E. Watanabe, S. Uji, M. Shimizu, T. Mori, T. Yamaguchi, and K. Ishibashi. Coupled quantum dots in a graphene-based two-dimensional semimetal. *Nano* **8**, 2891 (2009).

Müller, T., J. Güttinger, D. Bischoff, S. Hellmüller, K. Ensslin, and T. Ihn. Fast detection of single-charge tunneling to a graphene quantum dot in a multi-level regime. *Appl. Phys. Lett.* **101**, 012104 (2012).

Mueller, T., M. Kinoshita, M. Steiner, V. Perebeinos, A. A. Bol, D. B. Farmer, and P. Avouris. Efficient narrow-band light emission from a single carbon nanotube p–n diode. *Nat. Nanotechnol.* **5**, 27–31 (2009).

Nakada, N., M. Fujita, G. Dresselhaus, and M. Dresselhaus. Edge states in graphene nanoribbons: Nanometer size effects and edge shape dependence. *Phys. Rev. Lett.* **54**, 17954 (1996).

Neto, A. C., F. Guinea, and N. M. Peres. Drawing conclusion from graphene. *Phys. World* **19**, 33 (2006).

Neubeck, S., L. A. Ponomarenko, F. Freitag, A. J. M. Giesbers, U. Zeitler, S. V. Morozov, P. Blake, A. K. Geim, and K. S. Novoselov. From one electron to one hole: Quasiparticle counting in graphene quantum dots determined by electrochemical and plasma etching. *Small* **6**, 14 (2010).

Neumann, C., C. Volk, S. Engels, and C. Stampfer. Graphene-based charge sensors. *Nanotechnology* **24**, 444001 (2013).

Novoselov, K. S., A. K. Geim, S. V. Morozov, D. Jiang, M. I. Katsnelson, I. V. Grigorieva, S. V. Dubonos, and A. A. Firsov. Two-dimensional gas of massless Dirac fermions in graphene. *Nature* **438**, 197–200 (2005).

Novoselov, K. S., A. K. Geim, S. V. Morozov, D. Jiang, Y. Zhang, S. V. Dubonos, I. V. Grigorieva, and A. A. Firsov. Electric field effect in atomically thin carbon films. *Science* **306**, 666–669 (2004).

Nowack, K. C., M. Shafiei, M. Laforest, G. Prawiroatmodjo, L. Schreiber, C. Reichl, W. Wegscheider, and L. Vandersypen. Single-shot correlations and two-qubit gate of solid-state spin. *Science* **333**, 1269 (2011).

Pan, D. Y., J. C. Zhang, Z. Li, and M. H. Wu. Hydrothermal route for cutting graphene sheets into blue-luminescent graphene quantum dots. *Adv. Mater.* **22**, 734 (2010).

Peng, J., W. Gao, B. Gupta, Z. Liu, R. Romero-Aburto, L. Ge, L. Song et al. Graphene quantum dots derived from carbon fibers. *Nano Lett.* **12**, 844 (2012).

Petta, J. R., A. C. Johnson, J. M. Taylor, E. A. Laird, A. Yacoby, M. D. Lukin, C. M. Marcus, M. P. Hanson, and A. C. Gossard. Coherent manipulation of coupled electron spins in semiconductor quantum dots. *Science* **309**, 2180 (2005).

Ponomarenko, L. A., F. Schedin, M. I. Katsnelson, R. Yang, E. H. Hill, K. S. Novoselov, and A. K. Geim. Chaotic Dirac billiard in graphene quantum dots. *Science* **320**, 356 (2008).

Puddy, R. K., C. J. Chua, and M. R. Buitelaar. Transport spectroscopy of a graphene quantum dot fabricated by atomic force microscope nanolithography. *Appl. Phys. Lett.* **103**, 183117 (2013).

Reich, S., J. Maultzsch, C. Thomsen, and P. Ordejon. Tight-binding description of graphene. *Phys. Rev. B* **66**, 035412 (2002).

Reich, S., C. Thomsen, and J. Maultzsch. *Carbon Nanotubes: Basic Concepts and Physical Properties*. Wiley-VCH, Weinheim, Germany (2004).

Saito, R., G. Dresselhaus, and M. S. Dresselhaus. *Physical Properties of Carbon Nanotubes*. London, United Kingdom: Imperial College Press (1999).

Schnez, S. Transport properties and local imaging of graphene quantum dots. PhD thesis, Zurich, Switzerland: ETH Zurich (2010).

Schnez, S., F. Molitor, C. Stampfer, J. Güttinger, I. Shorubalko, T. Ihn, and K. Ensslin. Observation of excited states in a graphene quantum dot. *Appl. Phys. Lett.* **94**, 012107 (2009).

Sohn, L. L., L. P. Kouwenhoven, and G. Schön, editors. *Mesoscopic Electron Transport*, vol. 345. Alphen aan den Rijn, The Netherlands: Advanced Study Institute Kluwer (1997).

Sols, F., F. Guinea, and A. Castro Neto. Coulomb blockade in graphene nanoribbons. *Phys. Rev. Lett.* **99**, 166803 (2007).

Son, Y.-W., M. L. Cohen, and S. G. Louie. Energy gaps in graphene nanoribbons. *Phys. Rev. Lett.* **97**, 216803 (2006).

Song, S. M., J. K. Park, O. S. Sul, and B. J. Cho. Determination of work function of graphene under a metal electrode and its role in contact resistance. *Nano Lett.* **12**, 3887–3892 (2012).

Stampfer, C., J. Güttinger, F. Molitor, D. Graf, T. Ihn, and K. Ensslin. Tunable Coulomb blockade in nanostructured graphene. *Appl. Phys. Lett.* **920** (1), 012102 (2008a).

Stampfer, C., E. Schurtenberger, F. Molitor, J. Güttinger, T. Ihn, and K. Ensslin. Tunable graphene single electron transistor. *Nano Lett.* **8**, 2378 (2008b).

Stampfer, C., S. Fringes, J. Güttinger, F. Molitor, C. Volk, B. Terrés, J. Dauber et al. Transport in graphene nanostructures. *Front. Phys.* **6**, 271–293 (2011).

Stampfer, C., J. Güttinger, S. Hellmüller, F. Molitor, K. Ensslin, and T. Ihn. Energy gaps in etched graphene nanoribbons. *Phys. Rev. Lett.* **102**, 056403 (2009).

Tan, Y.-W., Y. Zhang, K. Bolotin, Y. Zhao, S. Adam, E. H. Hwang, S. D. Sarma, H. L. Stormer, and P. Kim. Measurement of scattering rate and minimum conductivity in graphene. *Phys. Rev. Lett.* **99**, 246803 (2007).

Terrés, B., J. Dauber, C. Volk, S. Trellenkamp, U. Wichmann, and C. Stampfer. Disorder induced Coulomb gaps in graphene constrictions with different aspect ratios. *Appl. Phys. Lett.* **98**, 032109 (2011).

Todd, K., H.-T. Chou, S. Amasha, and D. Goldhaber-Gordon. Quantum dot behavior in graphene nanoconstrictions. *Nano Lett.* **90** (1), 416–421 (2009).

Tombros, N. Electron spin transport in graphene and carbon nanotubes. PhD thesis, Groningen, The Netherlands: University of Groningen (2008).

Trauzettel, B., D. V. Bulaev, D. Loss, and G. Burkard. Spin qubits in graphene quantum dots. *Nat. Phys.* **30** (3), 192–196 (2007).

Tworzydło, J., B. Trauzettel, M. Titov, A. Rycerz, and C. W. J. Beenakker. Sub-Poissonian shot noise in graphene. *Phys. Rev. Lett.* **960** (24), 246802 (2006).

van Wees, B., H. van Houten, C. W. J. Beenakker, J. G. Williamson, L. P. Kouwenhoven, D. van der Marel, and C. T. Foxon. Quantized conductance of point contacts in a two-dimensional electron gas. *Phys. Rev. Lett.* **60**, 848 (1988).

Volk, C., S. Fringes, B. Terrés, J. Dauber, S. Engels, S. Trellenkamp, and C. Stampfer. Electronic excited states in bilayer graphene double quantum dots. *Nano Lett.* **11**, 3581 (2011).

Volk, C., C. Neumann, S. Kazarski, S. Fringes, S. Engels, F. Haupt, A. Müller, and C. Stampfer. Probing relaxation times in graphene quantum dots. *Nat. Commun.* **4**, 1753 (2013).

Wakabayashi, K., M. Fujita, H. Ajiki, and M. Sigrist. Electronic and magnetic properties of nanographite ribbons. *Phys. Rev. B* **590,** 8271 (1999).

Wallace, P. R. The band theory of graphite. *Phys. Rev. Lett.* **710** (9), 622 (1947).

Wang, L.-J., G. Cao, T. Tu, H.-O. Li, C. Zhou, X.-J. Hao, G.-C. Guo, and G.-P. Guo. Ground states and excited states in a tunable graphene quantum dot. *Chin. Phys. Lett.* **28**, 067301 (2011a).

Wang, L.-J., G.-P. Guo, D. Wei, G. Cao, T. Tu, M. Xiao, G.-C. Guo, and A. M. Chang. Gates controlled parallel-coupled double quantum dot on both single layer and bilayer graphene. *Appl. Phys. Lett.* **99**, 112117 (2011b).

Wang, L.-J., G. Cao, T. Tu, H.-O. Li, C. Zhou, X.-J. Hao, Z. Su, G.-C. Guo, H.-W. Jiang, and G.-P. Guo. A graphene quantum dot with a single electron transistor as an integrated charge sensor. *Appl. Phys. Lett.* **97**, 262113 (2010).

Wang, L.-J., L. Hai-Ou, T. Tu, G. Cao, C. Zhou, X.-J. Hao, Z. Su et al. Controllable tunnel coupling and molecular states in a graphene double quantum dot. *Appl. Phys. Lett.* **100**, 022106 (2012).

Wang, L., I. Meric, P. Huang, Q. Gao, Y. Gao, H. Tran, T. Taniguchi et al. One-dimensional electrical contact to a two-dimensional material. *Science* **342**, 614 (2013).

Wang, X., Y. Ouyang, X. Li, H. Wang, J. Guo, and H. Dai. Room-temperature all-semiconducting sub-10-nm graphene nanoribbon field-effect transistors. *Phys. Rev. Lett.* **1000** (20), 206803 (2008).

Watanabe, E., A. Conwill, D. Tsuya, and Y. Koide. Low contact resistance metals for graphene based devices. *Diam. Relat. Mater.* **24**, 171–174 (2012).

Wei, D., H.-O. Li, G. Cao, G. Luo, Z.-X. Zheng, T. Tu, M. Xiao, G.-C. Guo, H.-W. Jiang, and G.-P. Guo. Tuning inter-dot tunnel coupling of an etched graphene double quantum dot by adjacent metal gates. *Sci. Rep.* **3**, 3175 (2013).

White, C. T., J. Li, D. Gunlycke, and J. W. Mintmire. Hidden one-electron interactions in carbon nanotubes revealed in graphene nanostrips. *Nano Lett.* **70** (3), 825–830 (2007).

Xue, J., J. Sanchez-Yamagishi, D. Bulmash, P. Jacquod, A. Deshpande, K. Watanabe, T. Taniguchi, P. Jarillo-Herrero, and B. J. LeRoy. Scanning tunnelling microscopy and spectroscopy of ultra-flat graphene on hexagonal boron nitride. *Nat. Mater.* **100** (4), 282–285 (2011).

Yang, C. H., A. Rossi, R. Ruskov, N. S. Lai, F. A. Mohiyaddin, S. Lee, C. Tahan, G. Klimeck, A. Morello, and A. S. Dzurak. Spin-valley lifetimes in a silicon quantum dot with tunable valley splitting. *Nat. Commun.* **4**, 2069 (2013).

Zhang, Y., Y.-W. Tan, H. L. Stormer, and P. Kim. Experimental observation of the quantum Hall effect and Berry's phase in graphene. *Nature* **438**, 201–204 (2005).

Zhou, C., T. Tu, L. Wang, H.-O. Li, G. Cao, G.-C. Guo, and G.-P. Guo. Transport through a gate tunable graphene double quantum dot. *Chin. Phys. Lett.* **29**, 117303 (2012).

Zhu, S., J. Zhang, C. Qiao, S. Tang, Y. Li, W. Yuan, B. Li et al. Strongly green-photoluminescent graphene quantum dots for bioimaging applications. *Chem. Commun.* **47**, 6858 (2011).

3

Graphene Network

Mashkoor Ahmad

Irum Khalid

Ayousha Ayaz

Muhammad Hussain

Abstract

This chapter concerns graphene, the youngest member of the prominent carbon family. In recent decades, many allotropes and forms of carbon have been discovered and explored, including sheets such as graphene (http://en.wikipedia.org/wiki/Allotropes_of_carbon). Graphene, having the single layer of graphite, recently attracted much attention because of its potential applications with greatest similarity to carbon nanotubes (CNTs). It is recognized as promising building blocks for nanoelectronic and spintronic devices in the carbon family (Geim and Novoselov 2007).

3.1 History of Graphene

The fundamental breakthroughs toward the physical understanding of graphene and graphite were routed in the 1940s and 1950s. In recent history, the use of graphite as a neutron moderator to thermalize high-energy neutrons in nuclear reactors has been of great significance. Modern derivatives also include carbon nanofibers (with diameters less than 10 nm) prepared and studied extensively in the 1970s and 1980s (Saito et al. 1998; Dresselhaus et al. 2001). Graphene can also be conceptually thought of as a parent material for Bucky ball molecules and CNTs. Their discoveries in the 1980s and 1990s by R.F. Curl and coworkers and by S. Iijima, respectively, formed the basis of not only new fundamental research areas, but also an exciting new set of applications (Dresselhaus 1996, 2001; Jorio 2008).

3.2 Graphite to Graphene

We have mentioned previously that graphene is the building-block of graphite (http://www.graphene-info.com/introduction). Layers of graphene stacked on top of each other form graphite with an interplanar spacing of 0.335 nm. Its extended honeycomb network is the basic building block of other important allotropes; it can be stacked to form three-dimensional (3D) graphite, rolled to form 1D nanotubes, and wrapped to form 0D fullerenes, as shown in Figure 3.1 (Wan 2012). Long-range π-conjugation in graphene yields extraordinary thermal, mechanical, and electrical properties, which have long been the interest of many theoretical studies and more recently became an exciting area for experimentalists (Lu 1999).

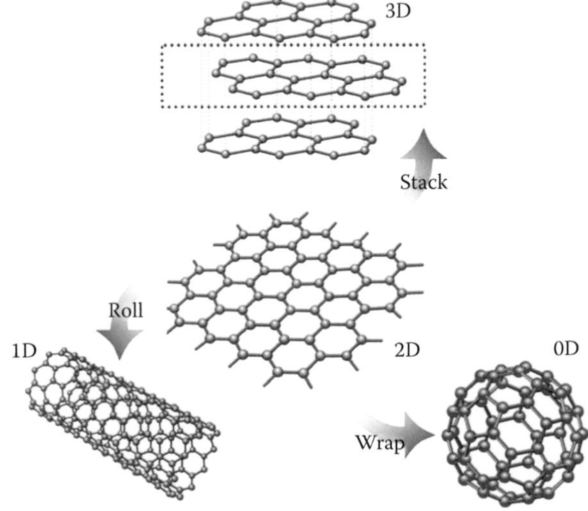

FIGURE 3.1 Graphene: the basic building block for other carbon allotropes, graphite (3D), fullerene (0D), and CNT (1D). (From Wan, X. et al., *Acc. Chem. Res.*, 45, 598, 2012.)

3.3 Introduction

Graphene (/ˈgræf.iːn/) (http://dictionary.cambridge.org/dictionary/british/graphene) is pure carbon in the form of a very thin, nearly transparent sheet, one atom thick. It is remarkably strong for its very low weight (100 times stronger than steel) (Andronico 2014).

Graphene is a 2D single atom-thick membrane of carbon atoms arranged in a honeycomb crystal (Wallace 1947; Saito et al. 1998; Geim and Novoselov 2007; Castro et al. 2009), as shown in Figure 3.2. It is a perfect example of a 2D electron system for a physicist, an elegant form of a 2D organic macromolecule consisting of benzene rings for a chemist, and a material with immense possibilities for an engineer due to its excellent electrical, magnetic, thermal, optical, and mechanical properties. Bilayer graphene is also an important material, as shown in Figure 3.3, and has very unique electronic structure and transport properties (McCann 2006; McCann and Fal'ko 2006; Novoselov et al. 2006).

The discovery of monolayer graphene in 2004 (Novoselov et al. 2004) has led to the demonstration of a host of novel physical properties in this most exciting of nanomaterials (Geim and Novoselov 2007). Graphene is generally made by micromechanical cleavage, a process whereby monolayers are peeled from graphite crystals. However, this process has significant disadvantages in terms of yield and throughput. As such, there has been significant interest in the development of a large-scale production method for graphene. In the long-term, for many research areas, the growth of graphene monolayers (Berger et al. 2006; Ohta et al. 2008; Kim et al. 2009) is by far the most desirable route. However, progress has been slow, and in any case, this technique will be unsuitable for certain applications. Thus, in the medium term, the most promising route is the exfoliation of graphite in the liquid phase to give graphene-like materials. The most common technique has been the oxidation and subsequent exfoliation of graphite to give graphene oxide (GO) (Niyogi et al. 2006; Stankovich et al. 2007; Becerril et al. 2008; Eda et al. 2008; Li et al. 2008b). Graphene has been attracting great interest because of its distinctive band structure and physical properties (Geim and Novoselov 2007). Due to some of these properties, such as very high carrier mobility (Novoselov et al. 2004) and ballistic transport at room temperature (Berger et al. 2006), it has been considered to be a promising candidate for future applications in semiconductor technology.

Of particular interest to materials scientists is the fact that graphene is the strongest and stiffest material known to man (Lee et al. 2008). This immediately suggests that graphene sheets would be an ideal reinforcing agent for polymer composites as the reinforcement potential scales linearly with filler stiffness and strength (Padawer and Beecher 1970). In fact, much work has already been done (Yang et al. 2008; Kuilla et al. 2010; Verdejo et al. 2011), showing that graphene and GO are potentially effective reinforcements (Stankovich et al. 2006; Liang et al. 2009; May et al. 2012).

FIGURE 3.2 Single graphene layer. (Available at http://en.wikipedia.org/wiki/Allotropes_of_carbon.)

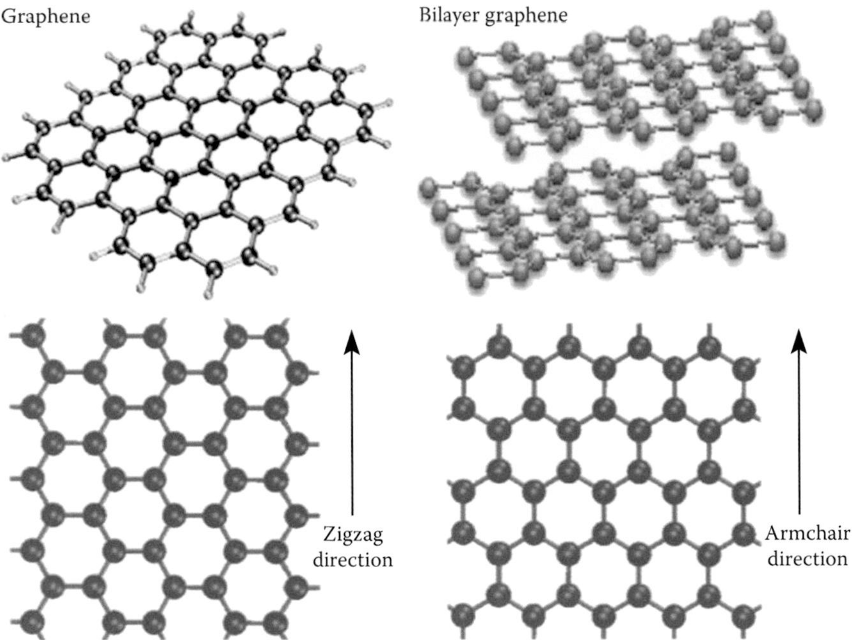

FIGURE 3.3 Graphene and its nanostructures. Two graphene membranes with Bernal stacking order form bilayer graphene. One-dimensional nanoribbons with armchair and zigzag edges conceptually extracted from the 2D graphene are shown. Atomic visualization was done using Hückel-NV. (From Oetting, L., Raza, T.Z., Raza, H., in preparation.)

3.3.1 2D Graphene

Although scientists knew that one-atom-thick, 2D crystal graphene existed, no one had worked out how to extract it from graphite. That was until it was isolated in 2004 by two Russian-born researchers at The University of Manchester, Andre Geim and Kostya Novoselov. In 2010, the Nobel committee awarded the Prize in Physics to Andre Geim and Konstantin Novoselov "for ground breaking experiments regarding the 2D material graphene." A model of a 2D graphene arranged in a honeycomb pattern is presented in Figure 3.4, in which each carbon atom is attached to its three neighboring atoms.

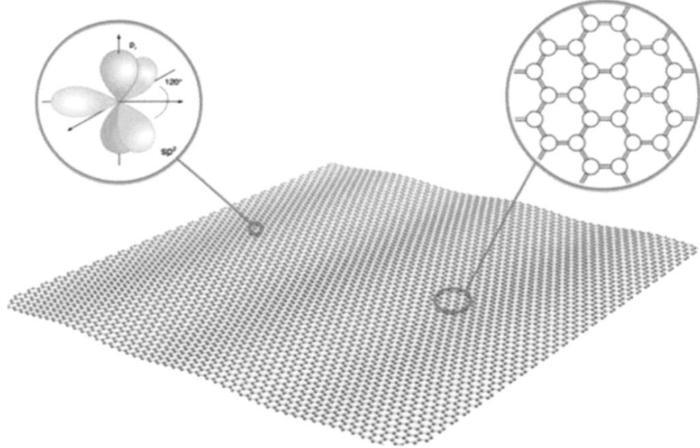

FIGURE 3.4 Graphene is a 2D crystal of carbon atoms, arranged in a honeycomb lattice. Each carbon atom is sp^2 hybridized and it is bound to its three neighbors. (Available at http://en.wikipedia.org/wiki/Allotropes_of_carbon.)

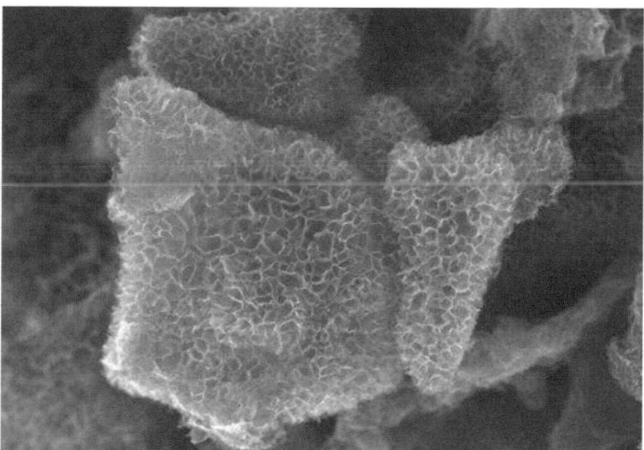

FIGURE 3.5 Field emission scanning electron microscopy (FESEM) image of 3D honeycomb-structured graphene. The novel material can replace platinum in dye-sensitized solar cells will virtually no loss of generating capacity. (From Wang, H. et al., *Angew. Chem. Int. Ed.*, 52, 9210, 2013.)

However, the debate about the real dimensionality of graphene is still open: Is it 2D or 3D? "Some of the debate surrounds the observation that the graphene sheets are not perfectly flat but can contain waves as distortions since the carbon rings are puckered," explains Professor Samantha Jenkins, College of Chemistry and Chemical Engineering, Hunan Normal University, China.

3.3.2 3D Graphene

From Euclidean geometry, a graphene sheet oriented in the 3D Cartesian x–y plane is defined as being 3D. This is because such a sheet has a finite extent along the x-, y-, and z-Cartesian axes. Because of their "thickness," the carbon atoms have a finite extent along the z-axis, and also the sheet may contain wave-like features creating displacements of the graphene sheet along the z axis. Within Euclidean geometry, there is the assumption that a 3D object will always remain 3D no matter how small it is (Wang et al. 2013). Regular graphene is a famously 2D form of carbon just a molecule or so thick. Hu and his team invented a novel approach to synthesize a unique 3D version with a honeycomb-like structure, as is depicted in the FESEM image in Figure 3.5 (Jenkins and Cavalleri 2013). To do so, they combined lithium oxide with carbon monoxide in a chemical reaction that forms lithium carbonate (Li_2CO_3) and the honeycomb graphene. The Li_2CO_3 helps shape the graphene sheets and isolates them from each other, preventing the formation of garden-variety graphite. Furthermore, the Li_2CO_3 particles can be easily removed from 3D honeycomb-structured graphene by an acid.

3.4 Synthesis of Graphene Network

3.4.1 Synthesis of Graphene

Graphene is a very special material, since it has the advantage of being both conducting and transparent. Accordingly, there are a lot of efforts to prepare graphene easily with the required properties. The methods described in this chapter are evaluated on the bases of different requirements: on the purity of the graphene, which is defined by the lack of intrinsic defects (Quality), as well as on the size of the obtained flakes or layers (Size). Another aspect is the amount of graphene, which can be produced simultaneously (Amount), or the complexity such as the requirement of labor or the need for specially designed machines (Complex). One last attribute is the controllability of the method to achieve reproducible results (Control).

3.4.2 Exfoliation

Basically, there are two different approaches to preparing graphene. On the one hand, graphene can be detached from an already existing graphite crystal, the so-called exfoliation methods; on the other hand, the graphene layer can be grown directly on a substrate surface. The first reported preparation of graphene was by Novoselov and Geim in 2004 (Novoselov et al. 2004) by exfoliation using a simple adhesive tape.

3.4.3 Dispersion of Graphite

Graphene can be prepared in liquid phase. This allows upscaling the production, to obtain a much higher amount of graphene. The easiest method would be to disperse the graphite in an organic solvent with nearly the same surface energy as graphite (Lotya et al. 2010). Thereby, the energy barrier is reduced, which has to be overcome to detach a graphene layer from the crystal. The solution is then sonicated in an ultrasound bath for several hundred hours or a voltage is applied (Su et al. 2011). After the dispersion, the solution has to be centrifuged to dispose of the thicker flakes. The quality of the obtained graphene flakes is very high in accordance with the micromechanical exfoliation. Its size, however, is still very small, neither is the controllability given. On the other hand, the complexity is very low, and as mentioned previously, this method allows preparing large amounts of graphene.

3.4.4 Graphite Oxide Exfoliation

The principle of liquid-phase exfoliation can also be used to exfoliate graphite oxide. Because of several functional groups like epoxide or hydroxyl, graphene oxide is hydrophilic and can be solved in water by sonication or stirring. Thereby, the layers become negatively charged, and thus, a recombination is inhibited by the electrical repulsion. After centrifugation, the GO has to be reduced to regular graphene by thermal or chemical methods. It is hardly possible to dispose of all the oxygen. In fact, an atomic carbon and oxygen ratio of about 10 still remains (Park and Ruoff 2009). The performance of this method is very similar to liquid-phase exfoliation of pristine graphene. The obtained GO has to be reduced afterward, using thermal treatments or chemicals again (Tkachev et al. 2011). The reduced graphene oxide (rGO) is of very bad quality compared with pristine graphene; nevertheless, GO could be the desired product. GO modified with Ca and Mg ions is capable of forming very tensile GO paper, as the ions are cross-linkers between the functional groups of the graphene flakes (Park et al. 2008).

3.5 Substrate Preparation

There are different methods for substrate preparation to use the dispersed graphene in a non-liquid-phase. By vacuum filtration, the solution is sucked through a membrane using a vacuum pump. As a result, the graphene flakes end up as filtration cake of graphene paper. The deposition of graphene on a surface can be done by simple drop-casting, where a drop of the solution is placed on top of the substrate. After the solvents have evaporated, the graphene flakes remain on the surface. To achieve a more homogeneous coating, the sample can be rotated using the spincoating method to disperse the solution with the help of the centrifugal force. With spray-coating, the solution is sprayed onto the sample, which allows the preparation of larger areas.

3.6 Growth on Surfaces

A totally different approach for obtaining graphene is to grow it directly on a surface. Consequently, the size of the obtained layers is not dependent on the initial graphite crystal. The growth can occur in two different ways. Either the carbon already exists in the substrate or it has to be added by chemical vapor deposition (CVD).

3.6.1 Epitaxial Growth

Graphene can be prepared by simply heating and cooling down a silicon carbide crystal (Forbeaux et al. 1998). Generally speaking, single-layer or bilayer graphene forms on the Si face of the crystal, whereas few-layer graphene grows on the C face (Cambaz et al. 2008). The results are highly dependent on the parameters used, like temperature, pressure, or heating rate. If temperatures and pressure are too high, the growth of nanotubes instead of graphene can occur. The graphitization of SiC was discovered in 1955, but it was regarded as an unwelcome side effect instead of a method of preparing graphene. The Ni(111) surface has a lattice structure very similar to the one of graphene, with a mismatch of the lattice constant at about 1.3%. Thus, by use of the nickel diffusion method, a thin Ni layer is evaporated onto an SiC crystal. Upon heating, the carbon diffuses through the Ni layer and forms a graphene or graphite layer on the surface, depending on the heating rate. The thus produced graphene is easier to detach from the surface than is the graphene produced by the growth on a simple SiC crystal without Ni (Enderlein 2012). The graphene is not perfectly homogeneous owing to defects or grain boundaries (GBs). Its quality therefore is not as good as that of exfoliated graphene, except the graphene would be grown on a perfect single crystal. However, the size of the homogeneous graphene layer is limited by the size of the crystal used. The possibility to produce large amounts of graphene by epitaxial growth is not as good as by liquid-phase exfoliation, although the controllability to gain reproducible results is given. Also, the complexity of these methods is comparatively low.

3.7 Useful and Most Simple Method

3.7.1 Chemical Vapor Deposition

The most promising, inexpensive, and readily accessible approach for the deposition of reasonably high-quality graphene is CVD onto transition-metal substrates such as Ni (Kim et al. 2009), Pd (Kwon et al. 2009), Ru (Sutter et al. 2008), Ir (Coraux et al. 2008), and Cu (Li et al. 2009). It is a well-known process in which a substrate is exposed to gaseous compounds. These compounds decompose on the surface to grow a thin film, whereas the by-products evaporate. There are a lot of different ways to achieve this, e.g., by heating the sample with a filament or with plasma. Graphene can be grown by exposing an Ni film to a gas mixture of H_2, CH_4, and Ar at about 1000°C (Kim et al. 2009). The methane decomposes on the surface, so that the hydrogene evaporates. The carbon diffuses into the Ni. After cooling down in an Ar atmosphere, a graphene layer grows on the surface, a process similar to the Ni diffusion method. Hence, the average number of layers depends on the Ni thickness and can be controlled in this way. Furthermore, the shape of the graphene can also be controlled by patterning of the Ni layer. These graphene layers can be transferred via polymer support, which will be attached onto the top of the graphene. After etching the Ni, the graphene can be stamped onto the required substrate and the polymer support gets peeled off or etched away. Using this method, several layers of graphene can be stamped onto each other to decrease the resistance. Because of rotation relatively to the other layers, the turbostratic graphite does not have the Bernal stacking, and consequently, the single graphene layers hardly change their electronic properties, since they interact marginally with the other layers (Casiraghi et al. 2007). Using copper instead of nickel as growing substrate results in single-layer graphene with less than 5% of few-layer graphene, which does not grow larger with time (Xuesong et al. 2009). This behavior is supposed to be caused by the low solubility of carbon in Cu. For this reason, Bae and coworkers developed a roll-to-roll production of a 30-inch graphene (Bae et al. 2010), as demonstrated in the schematic of Figure 3.6. Using CVD, a 30-inch graphene layer was grown on a copper foil and then transferred onto a PET film by a roll-to-roll process. CVD also allows doping of the graphene, e.g., with HNO_3, to decrease the resistance. Bae and colleagues stacked four doped layers of graphene onto a polyethylene terephthalate (PET) film and thus produced a fully functional touch-screen panel. It has about 90% optical transmission and about 30 Ω per square resistance, which is superior to indium tin oxide (ITO). The optical and electrical performance of graphene prepared by CVD is very high, but the purity, which would be necessary for laboratory research, is not given.

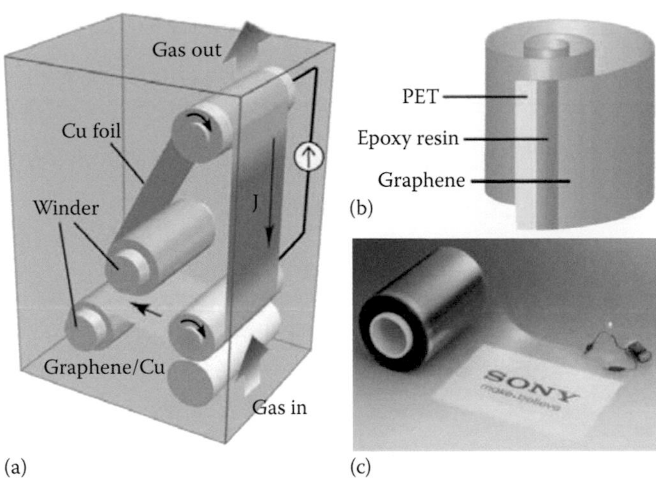

FIGURE 3.6 (a) Schematic of the Joule heating-induced roll-to-roll CVD system for continuous fabrication of graphene films on Cu. (b) Structure of graphene film transferred on PET with epoxy resin as an adhesive layer. (c) Photograph of the graphene/epoxy/PET roll. (From Casiraghi, C. et al., *Nano Lett.*, 7, 2711–2717, 2007.)

3.7.2 The "Scotch Tape Method"

The "Scotch Tape Method" is a simple and reproducible method for preparation of graphene. In this micromechanical exfoliation method, graphene is detached from a graphite crystal using adhesive tape. After peeling off the graphite, multiple-layer graphene remains on the tape. By repeated peeling, the multiple-layer graphene is cleaved into various flakes of few-layer graphene. Afterward, the tape is attached to the substrate and the glue solved, e.g., by acetone, to detach the tape. Finally, one last peeling with an unused tape is performed. The obtained flakes differ considerably in size and thickness, where the sizes range from nanometers to several tens of micrometers for single-layer graphene, depending upon the preparation of the used wafer. Single-layer graphene has an absorption rate of 2%; nevertheless, it is possible to see it under a light microscope on SiO_2/Si owing to interference effects (Bae et al. 2010). However, it is difficult to obtain larger amounts of graphene by this method, not even taking into account the lack of controllability. The complexity of this method is basically low; nevertheless, the graphene flakes need to be found on the substrate surface, which is labor intensive. The quality of the prepared graphene is very high, with almost no defects.

3.8 Preparation of 3D Graphene Networks

The synthesis of 3D graphene network was performed through a CVD approach. Commercial porous Ni was used as growth template. Before CVD growth, the porous Ni was immersed in a dilute solution of acetic acid for 30 min to remove oxide layer. The clean Ni template was heated to 1000°C in ~40 min under H_2 (200 sccm). After annealing for 40 min, a gas mixture flow of CH_4, H_2, and Ar was introduced to initiate graphene growth for 30 min, and the specific flow rate was 2/50/300 (CH_4/H_2/Ar) in standard cubic centimeter per minute. After growth, the sample was rapidly cooled to 500°C at a rate of ~200°C min^{-1} under Ar and H_2. The Ni template covered with graphene was drop-coated with a poly(methyl methacrylate) (PMMA) solution (4% in anisole) and then baked at 100°C for 2 h. The PMMA/graphene/Ni network was obtained after solidification. Then, the samples were put into a 1 M HCl solution at 60°C to completely dissolve the Ni template. Finally, 3D graphene network was obtained after removing PMMA in acetone.

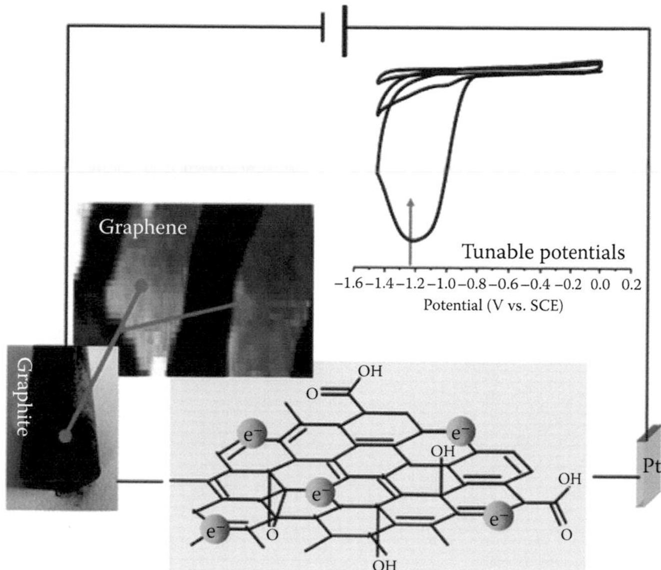

FIGURE 3.7 Experimental setup illustration of the graphite electrode and the GO suspension. Inset shows cyclic voltammograms of a GO-modified glassy carbon electrode. (Available at http://pubs.acs.org/doi/pdf/10.1021/nn900227d.)

3.8.1 Graphene Nanosheets

Graphene can be viewed as an individual atomic plane extracted from graphite, as unrolled single-walled CNT, or as an extended flat fullerene molecule. A facile approach to the synthesis of high-quality graphene nanosheets in large scale through electrochemical reduction of exfoliated graphite oxide precursor at cathodic potentials (completely reduced potential: −1.5 V) is reported. This method is green and fast and will not result in contamination of the reduced material. The electrochemically reduced graphene nanosheets have been carefully characterized by spectroscopic and electrochemical techniques in comparison with the chemically reduced graphene-based product. Particularly, a variety of the oxygen-containing functional groups have been thoroughly removed from the graphite oxide plane via electrochemical reduction. The chemically converted materials are not expected to exhibit graphene's electronic properties because of residual defects. Indeed, the high-quality graphene accelerates the electron transfer rate in dopamine electrochemistry (ΔE_p is as small as 44 mV, which is much smaller than that on a glassy carbon electrode), as illustrated in Figure 3.7. This approach opens up the possibility for assembling graphene biocomposites for electrocatalysis and the construction of biosensors (Guo et al. 2009).

3.9 Other Techniques

Zhuang-Jun Fan's group also synthesized graphene nanosheets from graphite oxide, which typically involves harmful chemical reductants that are undesirable for most practical applications of graphene. They demonstrate a green and facile approach to the synthesis of graphene nanosheets based on Fe reduction of exfoliated graphite oxide, resulting in a substantial removal of oxygen functionalities of the graphite oxide, as shown in Figure 3.8. More interestingly, the resulting graphene nanosheets with residual Fe show a high adsorption capacity of 111.62 mg/g for methylene blue at room temperature, as well as easy magnetic separation from the solution. This approach offers a potential for cost-effective, environmentally friendly, and large-scale production of graphene nanosheets (Fan et al. 2011).

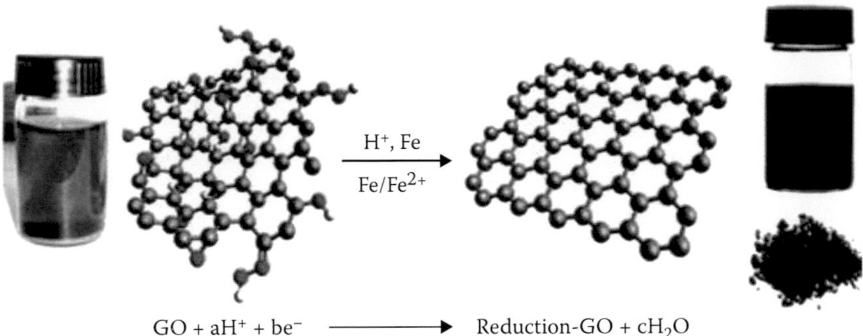

FIGURE 3.8 Preparation of graphene from Fe reduction of exfoliated graphite oxide. (Available at http://pubs.acs.org/doi/abs/10.1021/nn102339t.)

3.9.1 Graphene Fibers

Graphene fibers have been developed from cost-efficient aqueous graphite oxide (GO) suspensions by facile hydrothermal strategy. The fiber diameter and length can be controlled by either simply using the pipeline (made of glass and used as a rector) with predesigned length and inner diameter or adjusting the initial GO concentration.

3.9.2 Graphene Foams

Graphene foams have been grown by CVD (Cao et al. 2011b; Chen et al. 2011b; Yong et al. 2012), which promise to reduce the problems that result in compositing graphene with other nanostructure materials. Interconnected macrospores can be derived from 3D graphene aerogels (GAs) that can be prepared by hydrothermal assembly of GO, while the mesopores can be introduced by the silica networks uniformly grown on the graphene surface (Wu et al. 2012c).

3.9.3 The Metal-Graphene-CNT

The metal-graphene-CNT structure has been synthesized by directly growing graphene on porous nickel films, followed by the growth of controlled lengths of vertical CNT forests that emanate from the graphene surface.

For formation of polymerized polypyrrole-graphene PPy-G foam, 3D graphene had been prepared by hydrothermal treatment of homogeneous GO aqueous dispersion. Py monomer was introduced into the GO aqueous suspension to form a homogeneous solution before the hydrothermal process. 3D G (Py) was made to act as the working electrode in a three-electrode cell and a constant potential in 0.2 M $NaClO_4$ aqueous solution for polymerization of Py monomer, and the 3D PPy-G foam was obtained.

3.10 Properties

The first time graphene was artificially produced, scientists literally took a piece of graphite and dissected it layer by layer until only one single layer remained. This process is known as mechanical exfoliation. This resulting monolayer of graphite (known as graphene) is only one atom thick and is therefore the thinnest material possible to be created without becoming unstable when being open to the elements (temperature, air, etc.). Because graphene is only one atom thick, it is possible to create other materials by interjecting the graphene layers with other compounds, effectively using graphene as atomic scaffolding from which other materials are engineered. These newly created compounds could also be superlative materials, just like graphene, but with potentially even more applications.

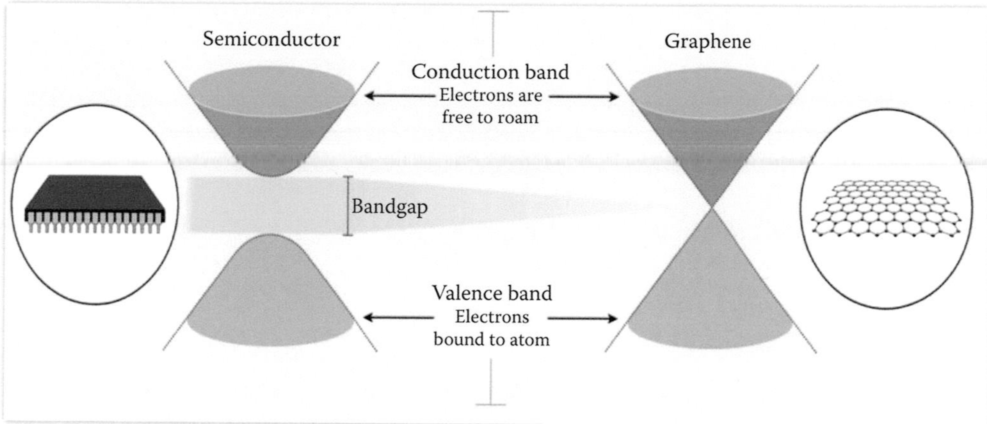

FIGURE 3.9 In an insulator or semiconductor, an electron bound to an atom can break free only if it gets enough energy from heat or passing photon to jump the "bandgap." But in graphene, the gap is infinitesimal. This is the main reason why graphene's electron can move easily and very fast.

3.10.1 Electronic Properties

Tests have shown that the electronic mobility of graphene is very high, with previously reported results above 15,000 $cm^2 V^{-1} s^{-1}$ and theoretically potential limits of 200,000 $cm^2 V^{-1} s^{-1}$ (limited by the scattering of graphene's acoustic photons). Compared with a semiconductor or insulator, the energy gap in graphene is infinitesimal enough for electrons to move freely and very fast, as is elaborated in Figure 3.9. It is said that graphene electrons act very much like photons in their mobility owing to their lack of mass. These charge carriers are able to travel submicrometer distances without scattering, a phenomenon known as ballistic transport. However, the quality of the graphene and that of the substrate that is used are the limiting factors. With silicon dioxide as the substrate, for example, mobility is potentially limited to 40,000 $cm^2 V^{-1} s^{-1}$.

3.10.2 Optical Properties

Graphene has the ability to absorb a rather large 2.3% of white. This is because its electrons act like massless charge carriers with very high mobility. Adding another layer of graphene increases the amount of white light absorbed by approximately the same value (2.3%). Because of these impressive characteristics, it has been observed that once optical intensity reaches a certain threshold (known as the saturation fluence), saturable absorption takes place (very-high-intensity light causes a reduction in absorption). This is an important characteristic with regard to the mode-locking of fiber lasers (http://www.graphenea.com/pages/graphene-uses-applications#.VJl99sA8).

3.10.3 Mechanical Properties

It was experimentally found that graphene shows both nonlinear elastic behavior and brittle fracture. According to experiments, graphene is characterized by Young modulus of $E = 1.0$ TPa, which is extremely large and close to that specifying CNTs. Brittle fracture of graphene occurs at a critical stress equal to its intrinsic strength of $\sigma_{int} = 130$ GPa. This value is the highest ever measured for real materials (Lee et al. 2008). These values of E and σ_{int} are extremely large and make graphene very attractive for structural and other applications. At the same time, graphene can be easily bent. Crystallographic characteristics of crack growth in monolayer graphene experimentally revealed that cracks or tears are generated and grow in suspended monolayer graphene membranes under unavoidable mechanically applied stress during their processing. In doing so, tears grow predominantly along straight lines—in either the armchair or the zigzag directions of the hexagonal crystal lattice of graphene, occasionally changing growth direction by 30° (Kim et al. 2012). The presence of defects in graphene is capable of significantly influencing its plastic deformation and fracture. Typical defects experimentally observed in graphene are vacancies, Stone-Wales defects (Meyer et al. 2008), dislocations (Warner et al. 2012), and GBs composed of dislocations (Kim et al. 2011; Yu et al. 2011b). In particular, dislocations can serve as carriers of plastic flow in graphene (Warner et al. 2012), whereas GBs dramatically decrease its strength characteristics (Huang et al. 2011a; Ruiz-Vargas et al. 2011).

3.10.4 Thermal Conductivity

The thermal conductivity of graphene is dominated by phonons and has been measured to be approximately 5000 W $m^{-1}K^{-1}$. Copper at room temperature has a thermal conductivity of 401 W $m^{-1}K^{-1}$. Thus, graphene conducts heat 10 times better than copper does.

3.10.5 Strength of Graphene

Graphene has a breaking strength of 42 N/m. Steel has a breaking strength in the range of 250–1200 MPa = 0.25–1.2 × 109 N/m^2. For a hypothetical steel film of the same thickness as graphene, this would give a 2D breaking strength of 0.084–0.40 N/m. Thus, graphene is more than 100 times stronger than the strongest steel (The Royal Swedish Academy of Sciences 2010).

3.11 2D Graphene

Two-dimensional graphene possesses electrical properties such as unusual mechanical strength and ultralarge specific surface area (Geim and Novoselov 2007; Eda and Chhowalla 2010; Zhu et al. 2010). In virtue of the intrinsic 2D structure, graphenes have been conformably assembled into 2D macroscopic configurations such as papers (Chen et al. 2008; Li et al. 2008b), transparent and conductive films (Eda et al. 2008; Li et al. 2008a), and even 3D frameworks (Li et al. 2008c; Xu et al. 2010a). However, it is of an extraordinary challenge to directly assemble 2D microcosmic graphene sheets into macroscopic fibers because the irregular size and shape of chemically derived graphenes and the movable layer-by-layer stacking of graphenes, in contrast to the highly tangled CNT assembly, could seriously obstruct the formation of graphene fibers and impair their macroscopic mechanical properties. As a consequence, up to date, few attempts have been successfully made to assemble graphenes into macroscopic fibers (Lee et al. 2010; Li et al. 2011).

3.12 Graphene Composite Properties

CNTs grown directly on bulk metal substrates results in inadequacy in the contacts and low surface area is utilized (Wang et al. 2003; Xu and Gao 2011). The use of graphene serves as an interfacial layer between metal and CNTs.

In current developments, chemically reduced graphene oxide is used as the substitute for pristine graphene because of the low-cost large-scale production enabled by the chemical exfoliation processes (Talapatra et al. 2006). However, the exceptional properties of graphene are severely impaired in rGO because of abundant defects and chemical moieties created in the synthesis procedures. In graphene composites, aggregation and stacking between individual graphene sheets driven by the strong π–π interaction greatly compromise the intrinsic high specific surface area of graphene. Furthermore, the high conductivity of graphene is also largely compromised because of intersheet contact resistance.

3.13 3D Graphene

Three-dimensional graphene-based frameworks (3DGFs), such as aerogels, foams, and sponges, are an important class of new-generation porous carbon materials, which exhibit continuously interconnected macroporous structures, low mass density, large surface area, and high electrical conductivity (Dong et al. 2010a; Tang et al. 2010; Worsley et al. 2010; Xu et al. 2010a,b). These materials can serve as a robust matrix for accommodating metal, metal oxide, and electrochemically active polymers.

3DGFs generally lack well-defined mesopores and/or micropores, which substantially limits the efficiency of mass transport and charge storage for electrochemical capacitors (ECs) through the small pores. Therefore, it is highly attractive to build up hierarchical porous architectures for 3DGFs by integrating small mesoporous channels within interconnected macroporous frameworks.

3D GA-based mesoporous carbon shows outstanding specific capacitance (226 F g^{-1}), high rate capability, and excellent cycling stability (no capacitance loss after 5000 cycles) when it is applied in ECs, demonstrating a synergistic effect of macropores and mesopores.

The 3D graphene/Co_3O_4 composite synthesized by CVD has shown to be capable of delivering high specific capacitance of 1100 F g^{-1} at a current density of 10 A g^{-1} with excellent cycling stability, and it can detect glucose with an ultrahigh sensitivity of 3.39 mA mM^{-1} cm^{-2} and a remarkable lower detection limit of <25 nM (S/N = 8.5).

The pristine 3D graphene structure tends to collapse under compression because of the relatively poor compressibility and springiness. As a result, although graphene has been integrated into the fabrication of capacitor electrodes (Chen et al. 2011c; Zhang et al. 2011; Zhao et al. 2011), there is still no report on the compressible supercapacitors based on the graphene electrodes. However, by formation of PPy-G foam, a highly compression-tolerant graphene-based supercapacitor has been demonstrated.

3.14 Graphene Fiber

Just like the daily used threads, the graphene fiber can be curved into coils and enlaced in bundles in wet and dry states and knotted, as shown in Figure 3.10. The fiber does not break as the knot is tightened,

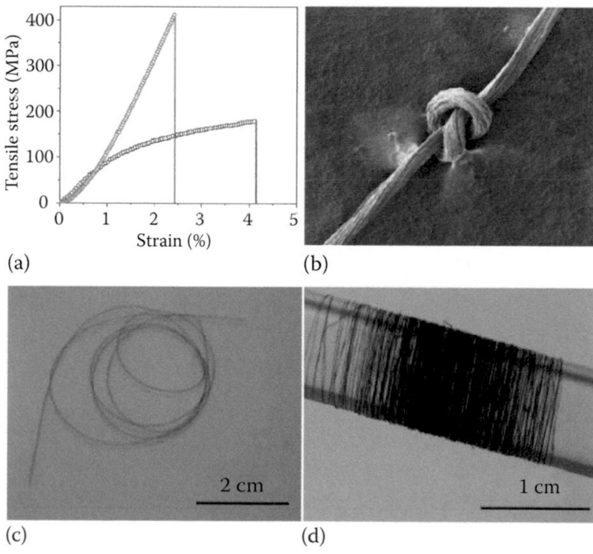

FIGURE 3.10 Typical stress-strain curves of graphene fiber and 800°C thermally treated graphene fiber (a); SEM image of the knotted graphene fiber (b); a photograph of wet graphene fiber coiled individually in water (c); a photograph of dry graphene fibers coiled in bundle around the glass rod (d).

and two-ply yarn can be obtained by twisting two fibers. These observations demonstrate the flexibility and resistance to torsion of graphene fibers, which is similar to CNT fibers (Basavaraja et al. 2011). The hydrothermally converted graphene fiber derived from micrometer-scale graphene oxide sheets has a measured tensile strength of up to 180 MPa. This value is close to that of the singles yarns of multiwalled CNTs (Vigolo et al. 2000; Li et al. 2004) but much better than those of single-walled CNT fibers fabricated by the wet spinning process (Ericson 2004; Zhang et al. 2004).

3.15 Graphene Foams

As a hydrogel, PPy-G has the capability of expansion/contraction upon absorption/desorption of solvent. As shown in Figure 3.11a–c, PPy-G foam can sustain large-strain deformations (e.g., $\sum = 50\%$) under manual compression and recover most of the material volume without structural fatigue within 10 s. During the compression process, it was found that the zigzag buckles without cracks were formed along the PPy-G body (Figure 3.11e), while the zigzag deformation entirely disappeared upon release of the load (Figure 3.11d and f).

Compression tests of several samples at differently set strains (40% to 80%), demonstrated in Figure 3.12, show reproducible results in which the unloading curves almost return to the origin, indicating complete volume recovery without plastic deformations.

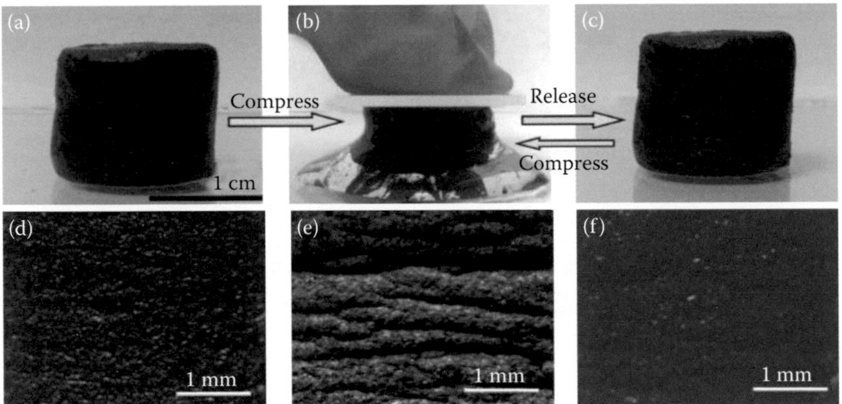

FIGURE 3.11 (a–c) The compression-recovery processes of PPy-G foam. (d–f) The surface views of PPy-G foam corresponding to the unloading-loading-unloading status in (a–c), respectively.

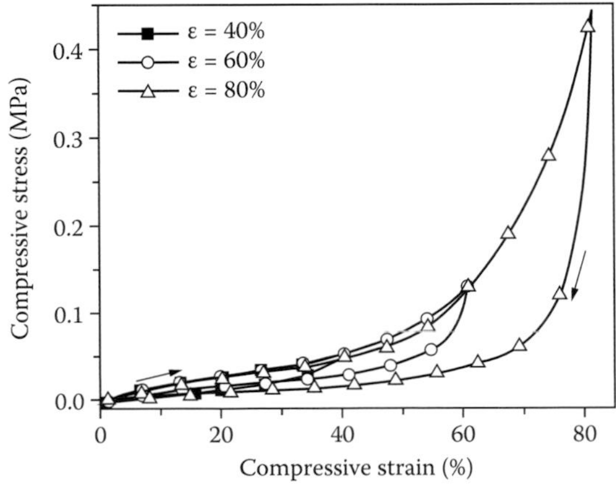

FIGURE 3.12 Stress–strain curves of PPy-G foam immersed in aqueous liquid at different set strains.

3.16 Applications

3.16.1 Display Screens

Researchers have found that graphene can replace indium-based electrodes in organic light emitting diodes (OLED). These diodes are used in electronic device display screens, which require low power consumption. The use of graphene instead of indium not only reduces the cost but also eliminates the use of metals in the OLED, which may make devices easier to recycle.

3.16.2 Lithium-Ion Batteries

These batteries use graphene on the anode surface. Defects in the graphene sheet (introduced using heat treatment) provide pathways for the lithium ions to attach to the anode substrate. Studies have shown that the time needed to recharge a battery using the graphene anode is much shorter than with conventional lithium-ion batteries.

3.16.3 Ultracapacitors

These ultracapacitiors store electrons on graphene sheets, taking advantage of the large surface of graphene to provide an increase in the electrical power that can be stored in the capacitor. Researchers are projecting that these ultracapacitors will have as much electrical storage capacity as lithium ion batteries but will be able to be recharged in minutes instead of hours.

3.16.4 Hydrogen Storage

Researchers have prepared graphene layers to increase the binding energy of hydrogen to the graphene surface in a fuel tank, resulting in a higher amount of hydrogen storage and, therefore, a lighter weight fuel tank. This could help in the development of practical hydrogen fueled cars.

3.16.5 Fuel Cells

Researchers at Ulsan National Institute of Science and Technology have demonstrated how to produce edge-halogenated graphene nanoplatelets that have good catalytic properties. The researchers prepared the nanoplatelets by ball-milling graphene flakes in the presence of chlorine, bromine, or iodine. They believe that these halogenated nanoplatelets could be used as a replacement for expensive platinum catalystic material in fuel cells.

3.16.6 Dye Sensitized Solar Cells

Researchers at Michigan Technological University have developed a honeycomb-like structure of graphene in which the graphene sheets are held apart by lithium carbonate. They have used this 3D graphene to replace the platinum in a dye sensitized solar cell and achieved 7.8% conversion of sunlight to electricity.

3.16.7 Electrodes

Researchers at Rice University have developed electrodes made from CNTs grown on graphene. The researchers first grew graphene on a metal substrate then grew CNTs on the graphene sheet. Because the base of each nanotube is bonded, atom to atom, to the graphene sheet, the nanotube–graphene structure is essentially one molecule with a huge surface area.

3.16.8 Transistors

The ability to build high-frequency transistors with graphene is possible because of the higher speed at which electrons in graphene move compared with electrons in silicon. Researchers are also developing lithography techniques that can be used to fabricate integrated circuits based on graphene.

3.16.9 Sensors

These sensors are based upon graphene's large surface area and the fact that molecules that are sensitive to particular diseases can attach to the carbon atoms in graphene. For example, researchers have found that graphene, strands of DNA, and fluorescent molecules can be combined to diagnose diseases. A sensor is formed by attaching fluorescent molecules to single-strand DNA and then attaching the DNA to graphene. When an identical single-strand DNA combines with the strand on the graphene, a double-strand DNA is formed that floats off from the graphene, increasing the fluorescence level. This method results in a sensor that can detect the same DNA for a particular disease in a sample. Fabrication of wireless graphene nanosensors on biomaterials via silk biosorption have also been reported. Graphene nanosensor tattoos on teeth have also been fabricated for the detection of very small amounts of chemical contaminants, virus, or bacteria in food systems, as shown in Figure 3.13.

3.16.10 Membranes

These membranes are made from sheets of graphene in which nanoscale pores have been created. Because graphene is only one atom thick, researchers believe that gas separation will require less energy than thicker membranes will.

3.16.11 Chemical Explosive Sensors

These sensors contain sheets of graphene in the form of a foam, which changes resistance when low levels of vapors from chemicals, such as ammonia, are present (Davis et al. 2009).

3.16.12 Coatings

Coating objects with graphene can serve different purposes. For instance, researchers have now shown that it is possible to use graphene sheets to create a superhydrophobic coating material that shows stable

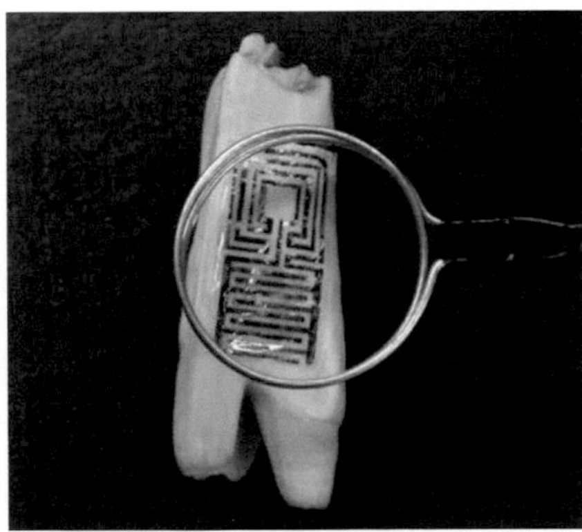

FIGURE 3.13 Optical image of the graphene wireless sensor biotransferred onto the surface of a tooth. (Available at http://www.nanowerk.com/spotlight/spotid=34184.php.)

superhydrophobicity under both static as well as dynamic (droplet impact) conditions, thereby forming extremely water repelling structures (http://www.nanowerk.com/spotlight/spotid=34184.php).

3.16.12.1 2D Graphene

2D graphene provides tremendous new advances in various fields, such as field-effect transistors (Novoselov et al. 2004; Xu et al. 2011; http://www.understandingnano.com/graphene-applications.html), biological/chemical sensors (Fowler et al. 2009; Dong et al. 2010b; Schwierz 2010; Huang et al. 2011b), energy storage (Wu et al. 2010; Zhu et al. 2011; Liu et al. 2012) and conversion devices (Yoo and Zhou 2011), and transparent conductors (Yu et al. 2011a). 3DGFs have various applications in ECs (Kasry et al. 2010; Cao 2011a; Choi et al. 2012; Wu et al. 2012a), batteries (Xiao et al. 2011; Dong et al. 2012), and catalysis (Chen et al. 2011a; Yong et al. 2012).

3.16.12.2 The Metal Graphene CNT

This type of structure is used directly to fabricate field-emitter devices and double-layer capacitors, demonstrating much improved performance over previously designed planar CNT-bulk metal structures (Wang et al. 2003; Lahiri et al. 2010; Wu et al. 2012b).

3.17 Future Prospects

The only problem with graphene is that high-quality graphene is a great conductor that does not have a bandgap (it cannot be switched off). Therefore, to use graphene in the creation of future nanoelectronic devices, a bandgap will need to be engineered into it, which will, in turn, reduce its electron mobility to that of levels currently seen in strained silicon films. This essentially means that future research and development need to be carried out for graphene to replace silicon in electrical systems in the future.

With graphene offering a large surface area, high electrical conductivity, thinness, and strength, it would make a good candidate for the development of fast and efficient bioelectric sensory devices, with the ability to monitor such things as glucose levels, hemoglobin levels, cholesterol, and even DNA sequencing. Eventually, we may even see engineered "toxic" graphene that is able to be used as an antibiotic or even anticancer treatment. Also, because of its molecular make-up and potential biocompatibility, it could be utilized in the process of tissue regeneration.

One particular area in which we will soon begin to see graphene used on a commercial scale is that in optoelectronics; specifically touch screens, liquid crystal displays, and OLEDs.

Currently, the most widely used material is indium tin oxide (ITO), and the development of manufacture of ITO over the last few decades time has resulted in a material that is able to perform very well in this application. However, recent tests have shown that graphene is potentially able to match the properties of ITO, even in current (relatively underdeveloped) states. Also, it has recently been shown that the optical absorption of graphene can be changed by adjusting the Fermi level. While this does not sound like much of an improvement over ITO, graphene displays additional properties that can enable very clever technology to be developed in optoelectronics by replacing the ITO with graphene.

Another standout property of graphene is that while it allows water to pass through it, it is almost completely impervious to liquids and gases (even relatively small helium molecules). This means that graphene could be used as an ultrafiltration medium to act as a barrier between two substances. The benefit of using graphene is that it is only one single atom thick and can also be developed as a barrier that electronically measures strain and pressures between the two substances (among many other variables). A team of researchers at Columbia University have managed to create monolayer graphene filters with pore sizes as small as 5 nm (currently, advanced nanoporous membranes have pore sizes of 30–40 nm). While these pore sizes are extremely small, as graphene is so thin, pressure during ultrafiltration is reduced.

Acknowledgments

The authors are grateful to Pakistan Science Foundation and TWAS for the financial support through projects PSF/Res/C-PINSTECH/Phys (172) and 13-319RG/MSN/AS-C-UNESCO FR:3240279202, respectively. The authors also acknowledge the contribution of NRG group members.

References

Andronico, M., 5 ways graphene will change gadgets forever. *Laptop*, April 14, 2014.

Available at http://dictionary.cambridge.org/dictionary/british/graphene.

Available at http://en.wikipedia.org/wiki/Allotropes_of_carbon.

Available at http://www.graphenea.com/pages/graphene-uses-applications#.VJl99sA8.

Available at http://www.nanowerk.com/spotlight/spotid=34184.php.

Available at http://www.understandingnano.com/graphene-applications.html.

Bae, S., H. Kim, Y. Lee, X. Xu, J.-S. Park, Y. Zheng, J. Balakrishnan et al., Roll-to-roll production of 30-inch graphene films for transparent electrodes. *Nature Nanotechnology* 5:574, 2010.

Basavaraja, C., W.J. Kim, P.X. Thinh, and D. Huh, Electrical conductivity studies on water-soluble polypyrrole–graphene oxide composites. *Polym. Comp.* 32:2076, 2011.

Becerril, H.A., J. Mao, Z. Liu, R.M. Stoltenberg, Z. Bao, and Y. Chen, Evaluation of solution-processed reduced graphene oxide films as transparent conductors. *ACS Nano* 2:463–470, 2008.

Berger, C., Z.M. Song, X.B. Li, X.S. Wu, N. Brown, C. Naud, D. Mayo et al., Electronic confinement and coherence in patterned epitaxial graphene. *Science* 312:1191–1196, 2006.

Cambaz, Z.G., G. Yushin, and S. Osswald, Non catalytic synthesis of carbon nanotubes, graphene and graphite on SiC. *Carbon* 46:841–849, 2008.

Cao, X.H., Y.M. Shi, W.H. Shi, G. Lu, X. Huang, Q.Y. Yan, Q. Zhang, and H. Zhang. *Small* 7:3163, 2011a.

Cao, X.H., Y.M. Shi, W.H. Shi, G. Lu, X. Huang, Q.Y. Yan, Q.C. Zhang, and H. Zhang, Preparation of novel 3D graphene networks for supercapacitor applications. *Small* 7:3163–3168, 2011b.

Casiraghi, C., A. Hartschuh, E. Lidorikis, H. Qian, H. Harutyunyan, T. Gokus, K.S. Novoselov, and A.C. Ferrari, Rayleigh imaging of graphene and graphene layers. *Nano Lett.* 7:2711–2717, 2007.

Castro Neto, A.H., F. Guinea, N.M.R. Peres, K.S. Novoselov, and A.K. Geim, The electronic properties of graphene. *Rev. Mod. Phys.* 81:109, 2009.

Chen, H., M.B. Müller, K.J. Gilmore, G.G. Wallace, and D. Li, Mechanically strong, electrically conductive, and biocompatible graphene paper. *Adv. Mater.* 20:3557, 2008.

Chen, W.F., S.R. Li, C.H. Chen, and L.F. Yan, *Adv. Mater.* 23:5679, 2011a.

Chen, Z., W. Ren, L. Gao, B. Liu, S. Pei, and H.M. Cheng, Three-dimensional flexible and conductive interconnected graphene networks grown by chemical vapor deposition. *Nat. Mater.* 10:424–428, 2011b.

Chen, Z.P., W.C. Ren, L.B. Gao, B.L. Liu, S.F. Pei, and H.M. Cheng, Three-dimensional flexible and conductive interconnected graphene networks grown by chemical vapour deposition. *Nat. Mater.* 10:424, 2011c.

Choi, B.G., M. Yang, W.H. Hong, J.W. Choi, and Y.S. Huh. *ACS Nano* 6:4020, 2012.

Coraux, J., A.T. N'Diaye, C. Busse, and T. Michely, *Nano Lett.* 8:565, 2008.

Davis, V.A., A.N.G. Parra-Vasquez, M.J. Green, P.K. Rai, N. Behabtu, V. Prieto, R.D. Booker et al., *Nat. Nanotechnol.* 4:830, 2009.

de La Fuente, J., *The Price of Graphene*. Available at http://www.graphenea.com/pages/graphene #.VIHeiDGUelc.

Dong, X.C., C.Y. Su, W.J. Zhang, J.W. Zhao, Q.D. Ling, W. Huang, P. Chen, and L.J. Li, Ultra-large single-layer graphene obtained from solution chemical reduction and its electrical properties. *Phys. Chem. Chem. Phys.* 12:2164–2169, 2010a.

Dong, X.C., Y.M. Shi, W. Huang, P. Chen, and L.J. Li, Electrical detection of DNA hybridization with single-base specific using transistors based on CVD-grown graphene sheets. *Adv. Mater.* 22:1649–1653, 2010b.

Dong, X.C., H. Xu, X.W. Wang, Y.X. Huang, M.B. Chan-Park, H. Zhang, L.H. Wang, W. Huang, and P. Chen. *ACS Nano* 6:3206, 2012.

Dresselhaus, M.S., G. Dresselhaus, and P. Avouris, *Carbon Nanotubes: Synthesis, Structure, Properties and Applications*, Springer, Berlin Heidelberg, New York, 2001.

Dresselhaus, M.S., G. Dresselhaus, and P.C. Eklund, *Science of Fullerenes and Carbon Nanotubes: Their Properties and Applications*, Academic Press, San Diego, 1996.

Eda, G. and M. Chhowalla, Chemically derived graphene oxide: Towards large-area thin-film electronics and optoelectronics. *Adv. Mater.* 22:2392–2415, 2010.

Eda, G., G. Fanchini, and M. Chhowalla, Large-area ultrathin films of reduced graphene oxide as a transparent and flexible electronic material. *Nat. Nanotechnol.* 3:270–274, 2008.

Enderlein, C., *Graphene and Its Interaction with Different Substrates Studied by Angular-Resolved Photoemission Spectroscopy.* Omniscriptum GmbH & Company Kg., 2012.

Ericson, L.M., H. Fan, H. Peng, V.A. Davis, W. Zhou, J. Sulpizio, Y. Wang et al., *Science* 305:1447, 2004.

Fan, Z.-J., W. Kai, J. Yan, T. Wei, L.-J. Zhi, J. Feng, Y.-m. Ren, L.-P. Song, and F. Wei, Facile synthesis of graphene nanosheets via Fe reduction of exfoliated graphite oxide. *ACS Nano*, 5(1):191–198, 2011.

Forbeaux, I., J.M. Themlin, and J.M. Debever, Heteroepitaxial graphite on 6H–SiC(0001): Interface formation through conduction-band electronic structure. *Phys. Rev. B* 58:16396–16406, 1998.

Fowler, J.D., M.J. Alle, V.C. Tung, Y. Yang, R.B. Kaner, and B.H. Weiller, Practical chemical sensors from chemically derived graphene. *ACS Nano* 3:301–306, 2009.

Geim, A.K. and K.S. Novoselov, The rise of graphene. *Nat. Mater.* 6:183, 2007.

Guo, H.-L., X.-F. Wang, Q.-Y. Qian, F.-B. Wang, and X.-H. Xia, A green approach to the synthesis of graphene nanosheets. *ACS Nano*, 3(9):2653–2659, 2009.

Huang, P.Y., C.S. Ruiz-Vargas, A.M. van der Zande, W.S. Whitney, M.P. Levendorf, J.W. Kevek, S. Garg et al., Grains and grain boundaries in single-layer graphene atomic patchwork quilts. *Nature* 469:389, 2011a.

Huang, Y.X., X.C. Dong, Y.X. Liu, L.J. Li, and P. Chen, Graphene-based biosensors for detection of bacteria and their metabolic activities. *J. Mater. Chem.* 21:12358–12362, 2011b.

Jenkins, S. and M. Cavalleri, Art work: Julio Maza. *International Journal of Quantum Chemistry*, John Wiley & Sons, July 2, 2013.

Jorio, A., M.S. Dresselhaus, and G. Dresselhaus, *Carbon Nanotubes: Advanced Topics in the Synthesis, Structure, Properties and Applications*, Springer, Berlin Heidelberg, New York, 2008.

Kasry, A., M.A. Kuroda, G.J. Martyna, G.S. Tulevski, and A.A. Bol, Chemical doping of large-area stacked graphene films for use as transparent, conducting electrodes. *ACS Nano* 4:3839–3844, 2010.

Kim, K., V.I. Artyukhov, W. Regan, Y. Liu, M.F. Crommie, B.I. Yakobson, and A. Zettl, Ripping graphene: Preferred directions. *Nano Lett.* 12:293, 2012.

Kim, K., Z. Lee, W. Regan, C. Kisielowski, M.F. Gommie, and A. Zettl, Grain boundary mapping in polycrystalline graphene. *ACS Nano* 5:2142, 2011.

Kim, K.S., Y. Zhao, H. Jang, S.Y. Lee, J.M. Kim, K.S. Kim, J.-H. Ahn, P. Kim, J.-Y. Choi, and B.H. Hong, Large-scale pattern growth of graphene films for stretchable transparent electrodes. *Nature* 457:706–710, 2009.

Kuilla, T., S. Bhadra, D. Yao, N.H. Kim, S. Bose, and J.H. Lee, Recent advances in graphene based polymer composites. *Prog. Polym. Sci.* 35(11):1350–1375, 2010.

Kwon, S.-Y., C.V. Ciobanu, V. Petrova, V.B. Shenoy, J. Bareno, V. Gambin, I. Petrov, and S. Kodambaka, *Nano Lett.* 9:3985, 2009.

Lahiri, I., R. Seelaboyina, J.Y. Hwang, R. Banerjee, and W. Choi, Enhanced field emission from multi-walled carbon nanotubes grown on pure copper substrate. *Carbon* 48:1531–1538, 2010.

Lee, C., X. Wei, J.W. Kysar, and J. Hone, Measurement of the elastic properties and intrinsic strength of monolayer graphene. *Science* 321:385–388, 2008.

Lee, S.H., H.W. Kim, J.O. Hwang, W.J. Lee, J. Kwon, C.W. Bielawski, R.S. Ruoff, and S.O. Kim, Three-dimensional self-assembly of graphene oxide platelets into mechanically flexible macroporous carbon films. *Angew. Chem. Int. Ed.* 49:10084, 2010.

Li, D., M.B. Müller, S. Gilje, R.B. Kaner, and G.G. Wallace, *Nat. Nanotechnol.* 3:101, 2008a.

Li, D., M.B. Muller, S. Gilje, R.B. Kaner, and G.G. Wallace, Processable aqueous dispersions of graphene nanosheets. *Nat. Nanotechnol.* 3:101–105, 2008b.

Li, X.L., G. Zhang, X. Bai, X. Sun, X. Wang, E. Wang, and H. Dai, Highly conducting graphene sheets and Langmuir–Blodgett films. *Nat. Nanotechnol.* 3:538, 2008c.

Li, X., W. Cai, J. An, S. Kim, J. Nah, D. Yang, R. Piner et al., *Science* 324:1312, 2009.

Li, X., T. Zhao, K. Wang, Y. Yang, J. Wei, F. Kang, D. Wu, and H. Zhu, Directly drawing self-assembled, porous, and monolithic graphene fiber from chemical vapor deposition grown graphene film and its electrochemical properties. *Langmuir* 27:12164, 2011.

Li, Y., I.A. Kinloch, and A.H. Windle, *Science* 304:276, 2004.

Liang, J., Y. Huang, L. Zhang, Y. Wang, Y. Ma, T. Guo, and Y. Chen, Molecular-level dispersion of graphene into poly(vinyl alcohol) and effective reinforcement of their nanocomposites. *Adv. Funct. Mater.* 19(14):2297–2302, 2009.

Liu, Y.X., X.C. Dong, and P. Chen, Biological and chemical sensors based on graphene materials. *Chem. Soc. Rev.* 41:2283–2307, 2012.

Lotya, M., P.J. King, U. Khan, S. De, and J.N. Coleman, High concentration, surfactant-stabilized graphene dispersions. *ACS Nano* 4:3155–3162, 2010.

Lu, X.K., M.F. Yu, H. Huang, and R.S. Ruoff, Personal perspectives on graphene: New graphene-related materials on the horizon. *Nanotechnology* 10:269, 1999.

May, P., U. Khan, A. O'Neill, and J.N. Coleman, Approaching the theoretical limit for reinforcing polymers with graphene. *J. Mater. Chem.* 22(4):1278–1282, 2012.

McCann E., Asymmetry gap in the electronic band structure of bilayer graphene. *Phys. Rev. B* 74:R161403, 2006.

McCann, E. and V.I. Fal'ko. Landau-level degeneracy and quantum Hall effect in a graphite bilayer. *Phys. Rev. Lett.* 96:086805, 2006.

Meyer, J.C., C. Kisielowski, R. Erni, M.D. Rossell, M.F. Gommie, and A. Zettl, Direct imaging of lattice atoms and topological defects in graphene membranes. *Nano Lett.* 8:3582, 2008.

Niyogi, S., E. Bekyarova, M.E. Itkis, J.L. McWilliams, M.A. Hamon, and R.C. Haddon, Solution properties of graphite and graphene. *J. Am. Chem. Soc.* 128:7720–7721, 2006.

Novoselov, K.S., A.K. Geim, S.V. Morozov, D. Jiang, Y. Zhang, S.V. Dubonos, I.V. Grigorieva, and A.A. Firsov, Electric field effect in atomically thin carbon films. *Science* 306:666–669, 2004.

Novoselov, K.S., E. McCann, S.V. Morozov, V.I. Fal'ko, M.I. Katsnelson, U. Zeitler, D. Jiang, F. Schedin, and A.K. Geim, Unconventional quantum Hall effect and Berry's phase of 2π in bilayer graphene. *Nat. Phys.* 2:177, 2006.

Ohta, T., F. El Gabaly, A. Bostwick, J.L. McChesney, K.V. Emtsev, A.K. Schmid, T. Seyller, K. Horn, and E. Rotenberg, Morphology of graphene thin film growth on SiC(0001). *New J. Phys.* 10:023034, 2008.

Padawer, G.E. and N. Beecher, On the strength and stiffness of planar reinforced plastic resins. *Polym. Eng. Sci.* 10(3):185–192, 1970.

Park, S. and R.S. Ruoff, Chemical methods for the production of graphenes. *Nat. Nanotechnol.* 4:217–224, 2009.

Park, S., K.S. Lee, G. Bozoklu, W. Cai, S.T. Nguyen, and R.S. Ruoff, Graphene oxide papers modified by divelent ions—Enhancing mechanical properties via chemical cross linking. *ACS Nano* 2:572–578, 2008.

Ruiz-Vargas, C.S., H.L. Zhuang, P.Y. Huang, A.M. van der Zande, S. Garg, P.L. McEuen, D.A. Miller, R.C. Hennig, and J. Park, Softened elastic response and unzipping in chemical vapor deposition graphene membranes. *Nano Lett.* 11:2259, 2011.

Saito, R., G. Dresselhaus, and M.S. Dresselhaus, *Physical Properties of Carbon Nanotubes*, Imperial College Press, London, 1998.

Schwierz, F., Graphene transistors. *Nat. Nanotechnol.* 5:487–496, 2010.

Stankovich, S., D.A. Dikin, G.H.B. Dommett, K.M. Kohlhaas, E.J. Zimney, E.A. Stach, R.D. Piner, S.T. Nguyen, and R.S. Ruoff, Graphene-based composite materials. *Nature* 442(7100):282–286, 2006.

Stankovich, S., D.A. Dikin, R.D. Piner, K.A. Kohlhaas, A. Kleinhammes, Y. Jia, Y. Wu, S.T. Nguyen, and R.S. Ruoff, Synthesis of graphene-based nanosheets via chemical reduction of exfoliated graphite oxide. *Carbon* 45:1558–1565, 2007.

Su, C.Y., A.Y. Lu, Y. Xu, F.R. Chen, A.N. Khlobystov, and L.J. Li, High-quality thin graphene films from fast electrochemical exfoliation. *ACS Nano* 5:2332–2339, 2011.

Sutter, P.W., J.-I. Flege, and E.A. Sutter, *Nat. Mater.* 7:406, 2008.

Talapatra, S., S. Kar, S.K. Pal, R. Vajtai, L. Ci, P. Victor, M.M. Shaijumon, S. Kaur, O. Nalamasu, and P.M. Ajayan, Direct growth of aligned carbon nanotubes on bulk metals. *Nat. Nanotechnol.* 1:112–116, 2006.

Tang, Z.H., S.L. Shen, J. Zhuang, and X. Wang, Noble-metal-promoted three-dimensional macroassembly of single-layered graphene oxide. *Angew. Chem. Int. Ed.* 49:4603, 2010.

The Royal Swedish Academy of Sciences, *Graphene*. 2010. Available at http://www.nobelprize.org/nobel_prizes/physics/laureates/2010/advanced-physicsprize2010.pdf.

Tkachev, S.V., E.U. Buslaeva, and S.P. Gubin, Graphene: A novel carbon nanomaterial. *Inorg. Mater.* 47:1–10, 2011.

Verdejo, R., M.M. Bernal, L.J. Romasanta, and M.A. Lopez-Manchado, Graphene filled polymer nanocomposites. *J. Mater. Chem.* 21(10):3301–3310, 2011.

Vigolo, B., A. Pénicaud, C. Coulon, C. Sauder, R. Pailler, C. Journet, P. Bernier, and P. Poulin, *Science* 290:1331, 2000.

Wallace, P.R., The band theory of graphite. *Phys. Rev.* 71:622, 1947.

Wan, X., Y. Huang, and Y. Chen, Focusing on energy and optoelectronic applications: A journey for graphene and graphene oxide at large scale. *Acc. Chem. Res.* 45:598–607, 2012.

Wang, B., X. Liu, H. Liu, D. Wu, H. Wang, J. Jiang, X. Wang, P. Hu, Y. Liu, and D. Zhu, Controllable preparation of patterns of aligned carbon nanotubes on metals and metal-coated silicon substrates. *J. Mater. Chem.* 13:1124–1126, 2003.

Wang, H., K. Sun, F. Tao, D. J. Stacchiola, and Y. H. Hu, 3D Honeycomb-like structured graphene and its high efficiency as a counter-electrode catalyst for dye-sensitized solar cells. *Angew. Chem. Int. Ed.*, 52:9210, 2013.

Warner, J.H., E.R. Margine, M. Mukai, A.W. Robertson, F. Guistino, and A.I. Kirkland, Dislocation-driven deformations in graphene. *Science* 337:209, 2012.

Worsley, M.A., P.J. Pauzauskie, T.Y. Olson, J. Biener, J.H. Satcher, and T.F. Baumann, Synthesis of graphene aerogel with high electrical conductivity. *J. Am. Chem. Soc.* 132:14067, 2010.

Wu, Q., Y.X. Xu, Z.Y. Yao, A.R. Liu, and G.Q. Shi, Supercapacitors based on flexible graphene/polyaniline nanofiber composite films. *ACS Nano* 4:1963–1970, 2010.

Wu, Z.S., A. Winter, L. Chen, Y. Sun, A. Turchanin, X. Feng, and K. Mullen, *Adv. Mater.* 24:5130, 2012a.

Wu, Z.S., S.B. Yang, Y. Sun, K. Parvez, X.L. Feng, and K. Müllen, *J. Am. Chem. Soc.* 134:9082, 2012b.

Wu, Z.-S., Y. Sun, Y.-Z. Tan, S. Yang, X. Feng, and K. Mullen, Three-dimensional graphene-based macro- and mesoporous frameworks for high-performance electrochemical capacitive energy storage. *J. Am. Chem. Soc.* 134:19532, 2012c.

Xiao, J., D.H. Mei, X.L. Li, W. Xu, D.Y. Wang, G.L. Graff, W.D. Bennett, Z.M. Nie, L.V. Saraf, I.A. Aksay, J. Liu, and J.G. Zhang, *Nano Lett.* 11:5071, 2011.

Xu, H., Z. Zhang, Z.X. Wang, S. Wang, X. Liang, and L.M. Peng, Quantum capacitance limited vertical scaling of graphene field-effect transistor. *ACS Nano* 5:2340–2347, 2011.

Xu, Y., K. Sheng, C. Li, and G. Shi, Self-assembled graphene hydrogel via a one-step hydrothermal process. *ACS Nano* 4:4324, 2010a.

Xu, Y.X., Q.O. Wu, Y.Q. Sun, H. Bai, and G.Q. Shi, Three-dimensional self-assembly of graphene oxide and DNA into multifunctional hydrogels. *ACS Nano* 4:7358, 2010b.

Xu, Z. and C. Gao, Graphene chiral liquid crystals and macroscopic assembled fibres. *Nature Commun.* 2:571, 2011.

Xuesong, L., W.W. Cai, J. An, Y.S. Kim, J. Nah, D.X. Yang, R. Piner et al., Large-area synthesis of high-quality and uniform graphene films on copper foils. *Science* 324:1312 1314, 2009.

Yang, Y.G., C.M. Chen, Y.F. Wen, Q.H. Yang, and M.Z. Wang, Oxidized graphene and graphene based polymer composites. *New Carbon Mater.* 23(3):193–200, 2008.

Yong, Y.C., X.C. Dong, M.B. Chan-Park, H. Song, and P. Chen, Macroporous and monolithic anode based on polyaniline hybridized three-dimensional graphene for high-performance microbial fuel cells. *ACS Nano*, 2012. doi:10.1021/nn204656d.

Yoo, E. and H.S. Zhou, Li-Air rechargeable battery based on metal-free graphene nanosheet catalysts. *ACS Nano* 5:3020–3026, 2011.

Yu, D., K. Park, M. Durstock, and L.M. Dai, Fullerence-grafted graphene for efficient bulk heterojunction polymer photovoltaic devices. *J. Phys. Chem. Lett.* 2:1113–1118, 2011a.

Yu, Q., L.A. Jauregui, W. Wu, R. Colby, J. Tian, Z. Su, H. Cao et al., Control and characterization of individual grains and grain boundaries in graphene grown by chemical vapour deposition. *Nat. Mater.* 10:443, 2011b.

Zhang, D.C., X. Zhang, Y. Chen, P. Yua, C.H. Wang, and Y.W. Ma, Enhanced capacitance and rate capability of graphene/polypyrrole composite as electrode material for supercapacitors. *J. Power Sources* 196:5990, 2011.

Zhang, M., K.R. Atkinson, and R.H. Baughman, *Science* 306:1358, 2004.

Zhao, Y., H. Bai, Y. Hu, Y. Li, L.Y. Qu, S.W. Zhang, and G.Q. Shi, Electrochemical deposition of polyaniline nanosheets mediated by sulfonated polyaniline functionalized graphenes. *J. Mater. Chem.* 21:13978, 2011.

Zhu, Y.W., S. Murali, M.D. Stoller, K.J. Ganesh, W. Cai, P. Ferreira, A. Pirkle et al., Carbon-based supercapacitors produced by activation of graphene. *Science* 332:1537–1541, 2011.

Zhu, Y.W., S. Murali, W.W. Cai, X.S. Li, J.W. Suk, J.R. Potts, and R.S. Ruoff, Graphene and graphene oxide: Synthesis, properties, and applications. *Adv. Mater.* 22:3906–3924, 2010.

4 Covalently Functionalized Graphene

Pablo A. Denis

4.1 Introduction

The exfoliation of graphene by the scotch tape method started a new era in current science. Scientists and engineers from all areas of knowledge use graphene to ameliorate their research and improve our lifestyle. The availability of graphene has eclipsed the studies in other carbon-related materials, such as fullerenes, nanotubes, and activated carbons. Graphene has unmatched properties like (Service 2009) high carrier mobility, exceptional mechanical properties due to its conjugated sp^2 framework, almost unlimited heat conduction, and room temperature quantum Hall effect, among many others. Notwithstanding the fact that these properties are quite impressive, before more graphene-based products reach the market, extensive work is required to fine tune the intrinsic characteristics of graphene. One method which facilitates the modification of the physicochemical properties of graphene is the chemical functionalization. That is the topic discussed in the present chapter.

The research on graphene is so vast that only 10 years after the exfoliation of graphene, the paper by Novoselov et al. (2004) has reached over 22,214 citations according to Google Scholar and more than 426,000 scientific documents about graphene are available in the literature. Also, the number of scientists working with graphene is extremely large. For this reason, it is a herculean task to be updated; it is an almost everyday job! Consequently, any attempt to make a complete description of just one area of research in graphene is condemned to failure. While we are writing these lines, groundbreaking research is being performed in laboratories across the world. The aim of this chapter is to introduce the reader to the rather complicated chemistry of graphene and help to visualize the capabilities of theoretical calculations, to shed light into the processes that occur at the nanoscale level. We describe how reactive graphene can be and understand the consequences of the chemical functionalization. In the first part of the chapter, we describe how reactive graphene is; the second one deals with the organic chemistry of graphene; then in the third, we briefly comment on the inorganic chemistry of graphene, while in the last one, we comment on the procedures that can increase the rather low reactivity of graphene.

4.2 How Reactive Can Graphene Be?

One of the most interesting aspects of graphene is that the sp^2 framework is not perfect. For this reason, we investigated (Denis 2013) the reactivity of the most popular defects: (a) single vacancies, (b) 585 double vacancies, (c) 555-777 reconstructed double vacancies, (d) Stone-Wales defects, and (e) hydrogenated armchair and zigzag edges. These are shown in Figure 4.1.

The removal of atoms leaves unbounded atoms, even in some cases where some reconstruction occurs, and thus, it is expected that the reactivity is enhanced. As a measure, we have taken the binding energy (BE) of hydrogen to perfect and defective graphene, which is 16.7 kcal/mol, at the M06-L/6-31G* level of theory (Denis 2013d). In Table 4.1, we present the BE determined for the addition of hydrogen to the atoms that belong to the defect.

Inspection of the results presented in Table 4.1 clearly indicates that the removal of atoms (single and double vacancies) and the translocations of CC bonds, such as the Stone-Wales defect, enhance reactivity. Interestingly, the removal of two atoms makes graphene less reactive than when only one atom is removed. In effect, the most reactive site assayed is the carbon atom number 7 of the single vacancy defective graphene. The second most reactive defect is the 585 one and then the 555-777 defect. The Stone-Wales defect is also very reactive despite not having atoms removed because electron density is concentrated in the C=C bond, which connects the two pentagons. A ranking of defect reactivity can be obtained when the most reactive site of each defect is considered. These values are gathered in Table 4.2. When edges are also taken into consideration, we find that the hydrogenated zigzag edge is the second most reactive site. Interestingly, the other type of edge, the armchair one, displays a much smaller reactivity, probably because the carbon atoms belonging to that present a stronger CC bonding.

It has been proposed in 2009 that the defect sites of graphene can be labeled employing mercapto radical (SH) radicals (Denis 2009). By means of first-principle calculations, it was found that a single SH radical does not bind to perfect graphene. However, for the Stone-Wales defect, the C-SH covalent BE is increased from 0 kcal/mol to 10.6 kcal/mol. Also, the thiol group can be dissociatively attached to the vacancy, as shown in Figure 4.2. This represents another method to produce sulfur-doped graphene. Finally, it is worth mentioning that the edges can also be labeled with SH radicals given that the BE of SH to a hydrogen saturated zigzag edge is 36.9 kcal/mol. In summary, graphene displays a rather small reactivity, which is significantly increased at defect sites.

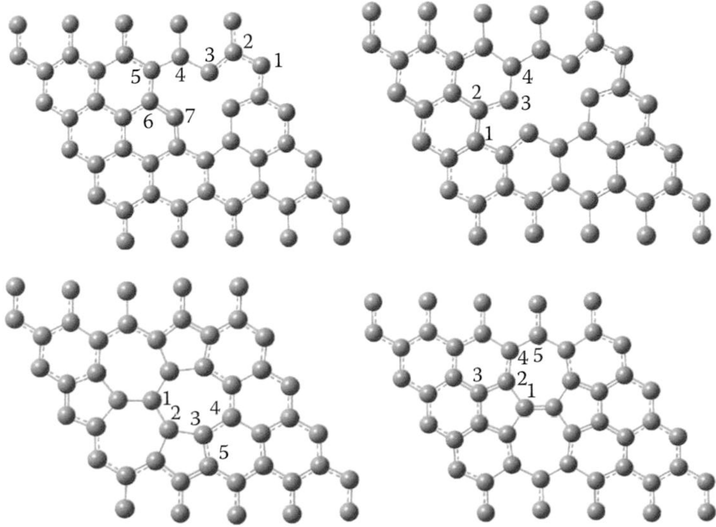

FIGURE 4.1 Most popular defect sites of graphene: (clockwise) single vacancies, 585 double vacancies, 555-777 reconstructed double vacancies, Stone-Wales defects.

TABLE 4.1 BEs Determined for the Addition of Hydrogen, Fluorine, and Phenyl Radicals onto Perfect and Defective Graphene

	H	F	Phenyl
	M06-L	M06-L	M06-L
G55 perfect	16.7	30.2	8.8
G55-VAC at 3 (956)	99.0	103.8	88.1
G55-VAC at 1 (665)	41.8	50.3	33.8
G55-VAC at 4 (966)	39.3	38.8	24.2
G55-VAC at 5 (966)	34.3	48.6	30.2
G55-VAC at 6 (966)	23.6	31.5	17.8
G55-VAC at 2 (966)	29.0	32.6	18.4
G55-SW-1 (775)	37.6	55.0	32.6
G55-SW-2 (765)	33.9	44.2	27.8
G55-SW-3 (665)	29.5	40.5	23.8
G55-SW-4 (766)	25.5	36.9	16.9
G55-SW-5 (766)	25.3	38.9	18.3
G55-555-777-2 (775)	49.6	55.2	40.0
G55-555-777-3 (765)	38.0	43.9	28.8
G55-555-777-5 (665)	30.2	36.8	24.5
G55-555-777-4 (766)	28.6	38.3	19.9
G55-555-777-1 (777)	21.7	37.0	13.9
G55-555-777-7 (665)	26.8	29.8	21.6
G55-585-3 (865)	51.4	56.1	41.5
G55-585-1 (566)	41.1	42.1	34.9
G55-585-2 (566)	34.6	42.2	29.6
G55-585-4 (866)	31.9	40.2	22.3

Source: Reproduced from Denis, P.A., *J. Phys. Chem. C*, 117, 19048–19055, 2013d.

TABLE 4.2 Binding Energies Determined for the Addition of Functional Groups to Perfect and Defective Graphene

	H	H	F	Aryl
	M06-L	VDW-DF + BSSE	M06-L	M06-L
G55 perfect	16.7	23.2	30.2	8.8
G55-VAC at 3 (956)	99.0	101.4	103.8	88.1
G55-SW-1 (775)	37.6	45.5	55.0	32.6
G55-555-777-2 (775)	49.6	55.9	55.2	40.0
G55-585-3 (865)	51.4	56.8	56.1	41.5
Zigzag edge		67.5		
Armchair edge		40.6		

Source: Reproduced from Denis, P.A., *J. Phys. Chem. C*, 117, 19048–19055, 2013d.

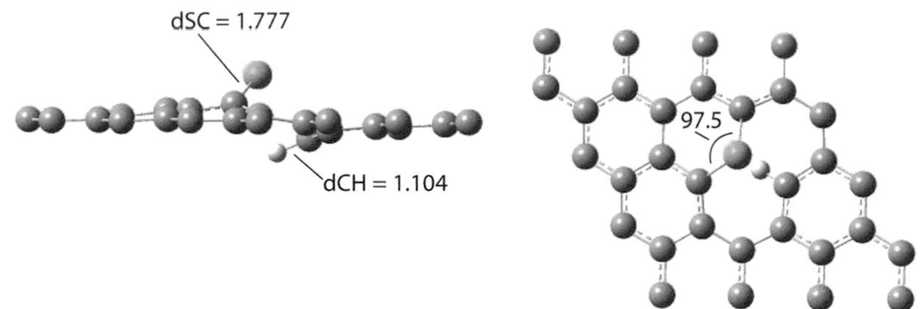

FIGURE 4.2 Optimized unit cell for the 4 × 4 graphene model containing a single vacancy defect and the SH group dissociatively added, side view (left) and top view (right). This structure is 2.6 eV more stable than that which has the SH group attached to the vacancy. Distances in Å, angles in degrees. (Reproduced from Denis, P.A., *J. Phys. Chem. C*, 113, 5612–5619, 2009.)

4.3 Organic Chemistry of Graphene

The conjugated sp^2 framework of graphene makes it a temptation for organic chemists to try all the panoply of reactions at their disposal. One of the first examples of an organic reaction performed onto graphene was the addition of aryldiazonium salts (Bekyarova et al. 2009; Sharma et al. 2010; Huang et al. 2011b; Denis 2013b). The product is a graphene sheet bearing an aryl group. This procedure has been shown to alter the electronic properties of graphene. Some research groups showed that this procedure opens a bang gap (Bekyarova et al. 2009), while others experimentally found that it increases conductivity (Huang et al. 2011b). In Table 4.3, we present the BE determined for the addition of one aryl group to perfect graphene. In general, there is a small variation in the BE with respect to the functional groups used, and in all cases, the size of the gap is similar (Denis 2013b).

Although the BE of one aryl group is small, it can be significantly increased if the aryl groups are agglomerated (Denis 2013b). In effect, if the addition is performed employing the pattern shown in Figure 4.3, the BE per aryl group can be as large as 36.1 kcal/mol, a value that represents an increment of more than 300%. We note that the addition must be performed on opposite sides of the sheet, to exploit at the highest level the agglomeration effect, which can be hampered by steric effects if the groups are added on the same side of the sheet; all the BEs are reported in Table 4.4.

One reaction that is more exothermic than the arylation is the [2 + 1] cicloaddition of nitrene (NH) groups (Choi et al. 2009; Denis and Iribarne 2011; Suggs et al. 2011). When a single nitrene group is attached, the BE is 23.0 kcal/mol. Although there is a gap at the Dirac point as shown in Figure 4.4, the system still maintains its semimetallic character because there is a crossing in the vicinity of the Dirac point. Interestingly, Suggs et al. (2011) showed that if perfluorophenylazide is employed instead of NH, a gap of 0.16 eV is opened.

TABLE 4.3 Electronic BEs (kcal/mol), Graphene-Aryl Bond Distances (Å), and Bang Gaps (eV) Determined for the Covalent Addition of Different Aryl Radicals C_6H_4R, onto 5 × 5 Graphene, at the M06-L/6-31G* Level of Theory

	BE	Dist Graphene-C_6H_4R	Gap
H	8.8	1.587	$\alpha = 0.70, \beta = 0.78$
CH_3	10.1	1.585	$\alpha = 0.71, \beta = 0.78$
NO_2	12.2	1.584	$\alpha = 0.70, \beta = 0.78$
Br	10.1	1.584	$\alpha = 0.70, \beta = 0.78$
Cl	9.9	1.584	$\alpha = 0.70, \beta = 0.78$
F	9.4	1.584	$\alpha = 0.70, \beta = 0.78$
HSO_3	8.3	1.585	$\alpha = 0.70, \beta = 0.78$

Source: Reproduced from Denis, P.A., *ChemPhysChem*, 14, 3271–3277, 2013b.

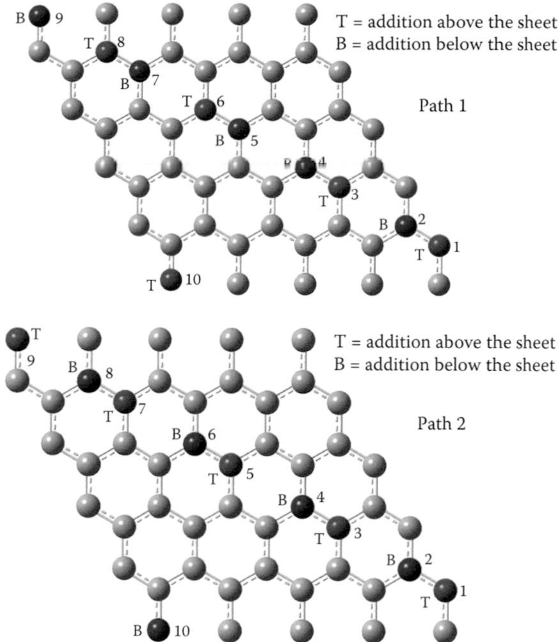

FIGURE 4.3 Functionalization paths employed for the addition of 10 aryl radicals in the ortho-position onto 5 × 5 graphene. (For BE, see Table 4.4.) (Reproduced from Denis, P.A., *ChemPhysChem*, 14, 3271–3277, 2013b.)

TABLE 4.4 Electronic BEs (kcal/mol) for the Covalent Addition of $n = 1$–10 Aryl Radicals C_6H_4R in the Ortho-Position, onto 5 × 5 Graphene (addition Paths 1 and 2 Are Presented in Figure 4.5)

		M06-L/6-31G*	
n	Configuration	BE (kcal/mol)	Gap (eV)
1		8.8	α = 0.7, β = 0.8
2	Ortho-line-1	28.1	0.0
3	Ortho-line-1	26.3	α = 1, β = 1
4	Ortho-line-1	29.8	0.0
5	Ortho-line-1	30.5	α = 1.1, β = 1.2
6	Ortho-line-1	31.5	0.0
7	Ortho-line-1	30.7	α = 1.1, β = 1.2
8	Ortho-line-1	33.8	1.5
9	Ortho-line-1	32.0	α = 1.3, β = 1.2
10	Ortho-line-1	36.1	α = 1.9
2	Ortho-line-2	28.1	0.0
3	Ortho-line-2	26.3	α = 1, β = 1
4	Ortho-line-2	29.8	0.0
5	Ortho-line-2	28.4	α = 1.2, β = 1.1
6	Ortho-line-2	30.6	0.0
7	Ortho-line-2	30.5	α = 1.1, β = 1.2
8	Ortho-line-2	33.5	1.5
9	Ortho-line-2	23.0	α = 1.4, β = 1.3
10	Ortho-line-2	35.5	1.9

Source: Reproduced from Denis, P.A., *ChemPhysChem*, 14, 3271–3277, 2013b.

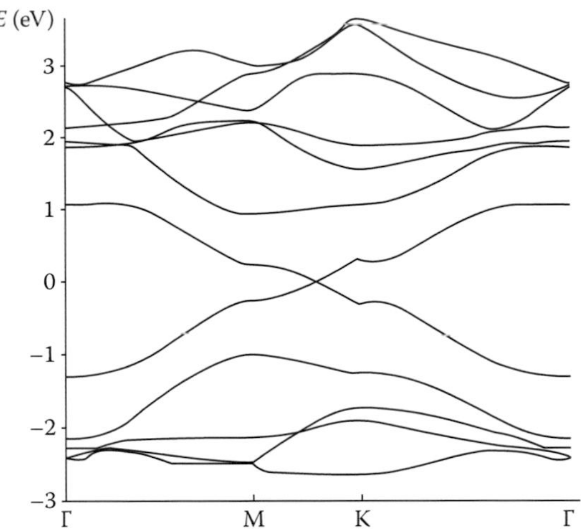

FIGURE 4.4 Density of states and band structure determined for a 4 × 4 graphene supercell functionalized with an NH group. (Reproduced from Denis, P.A., Iribarne, F., *J. Phys. Chem. C*, 115, 195–203, 2011.)

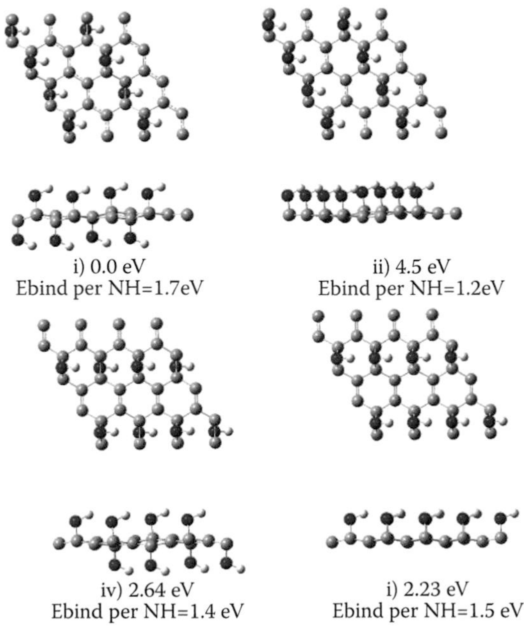

FIGURE 4.5 Graphene 4 × 4 functionalized with eight NH groups and the CC bonds unbroken (50% of functionalization), top and side views. Values indicate relative energies. (Reproduced from Denis, P.A., Iribarne, F., *J. Phys. Chem. C*, 115, 195–203, 2011.)

As observed for the aryl groups, we found that when 8 NH radicals are agglomerated as shown in Figure 4.5, the BEs are dramatically increased (Denis and Iribarne 2011). In Table 4.5, we present the BE for various levels of functionalization. When full coverage is reached, i.e., by adding 16 N groups as shown in Figure 4.6, the BE per NH unit is increased from 23.0 kcal/mol to 53.1 kcal/mol. The large change in BE is accompanied by the appearance of a bandgap of 1.63 eV.

TABLE 4.5 BEs (eV) and Bandgaps (eV) Determined for Different Graphene Models Functionalized with an NH Group

Graphene	BE[a]	Gap (eV)
4 × 4	0.97	Semimetallic
5 × 5	0.98	Semimetallic
6 × 6	1.00	Semimetallic
7 × 7	1.01	Semimetallic
8 × 8	1.00	Semimetallic
4 × 4 + 2NH	1.45	Semimetallic
4 × 4 + 8NH CC-not-broken (Figure 4.7)	1.73	0.30 (PBE)
4 × 4 + 8NH CC-not-broken (Figure 4.7)	1.50	2.19 (PBE)
4 × 4 + 16NH CC-not-broken (Figure 4.8)	1.93	4.56 (PBE)
4 × 4 + 16NH CC-not-broken (Figure 4.8)	2.30	0.64 (PBE) 1.63 (HSE06)

Source: Reproduced from Denis, P.A., Iribarne, F., *J. Phys. Chem. C*, 115, 195–203, 2011.
[a] Values determined at the Perdew-Burke-Ernzerhof (PBE) level.

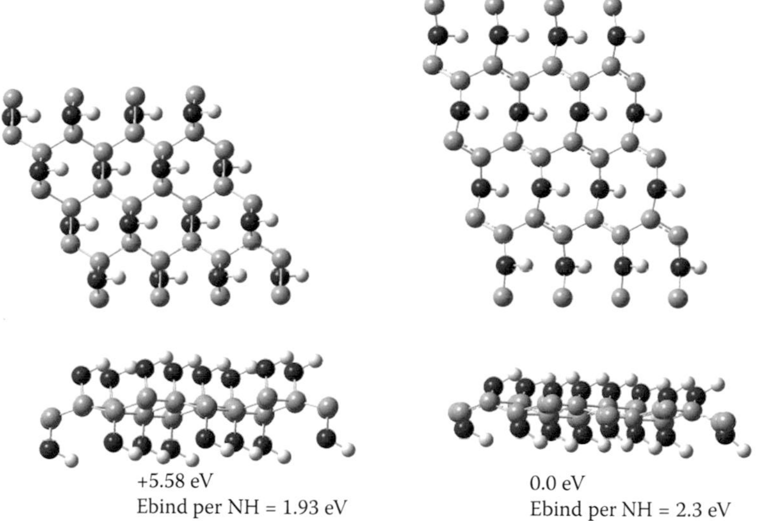

FIGURE 4.6 Graphene 4 × 4 functionalized with 16 NH groups (100% of functionalization), top and side views. Values indicate relative energies and BE per NH group. Left: unbroken CC bonds; right: broken CC bonds. (Reproduced from Denis, P.A., Iribarne, F., *J. Phys. Chem. C*, 115, 195–203, 2011.)

One of the most popular approaches to functionalize carbon nanotubes and fullerenes is through the use of azomethine ylides (Denis and Iribarne 2010, 2012b; Quintana et al. 2010; Cao and Houk 2011). Experimental work showed that this reaction can be performed onto graphene, and Quintana et al. (2010) employed these groups to homogeneously distribute gold nanoparticles onto graphene. In 2010, by means of first-principle calculations, we showed that the addition of this functional group onto perfect graphene is endothermic by 15.0 kcal/mol (Denis and Iribarne 2010). Yet, when Stone-Wales or edges are present, the reaction can easily proceed. Indeed, we found the enhanced electron density at the CC bond, which connects the two pentagons, makes the reaction exothermic by 19.1 kcal/mol. Our findings were supported by theoretical calculations performed by Cao and Houk (2011). The outcome of experimental and theoretical investigations was at odds because, on one hand, the reaction could be easily performed, but theoretical calculations highlighted that defects or edges are necessary for the reaction to proceed (Denis and Iribarne 2010; Cao and Houk 2011). In a second work on the topic, we investigated the topic in more detail (Denis and Iribarne 2012b). Bearing in mind that the BEs of aryl and nitrene groups increase when they are agglomerated onto graphene (Denis and Iribarne 2011), we reasoned that it might be possible to have a similar scenario for the azomethine ylide. In Table 4.6, we list the BE per functional group calculated for the addition of 1–10 azomethine groups to graphene, following the addition path showed in Figure 4.7.

The results presented here show that when 10 azomethine groups are added to a 5 × 5 graphene sheet, the reaction becomes favorable from an energetic standpoint. In effect, the BE reaches a value of 17.2 kcal/mol. This value is 28.6 kcal/mol larger than the BE computed for the addition of one azomethine group to graphene. For the latter experiment, we selected a 5 × 5 unit cell, but the question that obviously arises is what is the scenario when $2N$ functional groups are attached to an $N \times N$ graphene unit cell in terms of BE and bandgap opening. These results are gathered in Table 4.7. The convergence follows a wavelike pattern whose amplitude is reduced as we increase the size of the graphene model. In general, a periodicity of three is observed for the variation of the BE. The lowest BEs are obtained for $N = 4 + 3X$, $X = 1, 2, 3, 4$. The highest BEs can be found when $N = 3 + 3X$, $X = 2,3$. The only one value that

TABLE 4.6 BEs (kcal/mol) Determined for 1,3 Dipolar Cycloaddition of Azomethine Ylides onto a 5 × 5 Graphene Unit Cell, Determined at the M06-L/6-31G* Level of Theory

	X = azomethine
1X	−11.4
2X	4.9
3X	8.9
4X	12.1
5X	13.0
6X	14.2
7X	15.1
8X	15.9
9X	15.5
10X	17.2

Source: Reproduced from Denis, P.A., Iribarne, F., *Chem. Phys. Lett.*, 550, 111–117, 2012b.

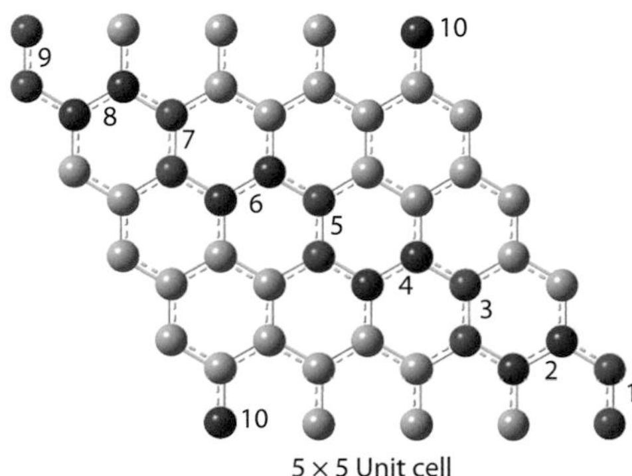

FIGURE 4.7 Functionalization path employed for the 4 × 4 and 5 × 5 graphene unit cells. Bonds with even numbers (odd numbers) denote functionalization above (below) the sheet. (Reproduced from Denis, P.A., Iribarne, F., *Chem. Phys. Lett.*, 2012, 550, 111–117, 2012b.)

TABLE 4.7 BE per Functional Group (kcal/mol) and Bandgaps (eV) Determined for 1,3 Dipolar Cycloaddition of Azomethine Ylides onto a $N \times N$ Graphene Unit Cell (With $2N$ Azomethine Groups), Determined at the M06-L/6-31G* Level of Theory

N	BE per Functional Group	Gap
4	15.6	1.8
5	18.3	1.5
6	16.8	1.9
7	16.1	0.7
8	17.0	0.9
9	17.1	1.4
10	16.5	0.5
11	17.0	0.3
12	17.1	1.1
13	16.7	0.3
14	16.9	0.5
16	16.8	0.3

Source: Reproduced from Denis, P.A., Iribarne, F., *Chem. Phys. Lett.*, 550, 111–117, 2012b.

seems out of place is $X = 1$ because the largest BE is observed for the 5×5 sheet. This behavior is observed because the distance between rows is optimal to exploit the H:::H contacts between the azomethine group. With regard to bandgaps, there is a huge variation with respect to the size of the unit cell. It can be as small as 0.3 eV for $N = 11$, 13, 16 or 1.9 eV for $N = 6$. Thus, the bandgap can be adjusted by varying the level of functionalization.

The third type of cicloaddition that we studied is the [2 + 2] addition of benzynes (Denis and Iribarne 2012c; Magedov 2013). This reaction is more likely to occur than the addition of azomethine ylides because the addition of one C_6H_4 group to graphene has an electronic BE of 13.9 kcal/mol. As observed for NH and the azomethine ylide, the BEs increase when benzynes are agglomerated following the path showed in Figure 4.8. The BEs are gathered in Table 4.8.

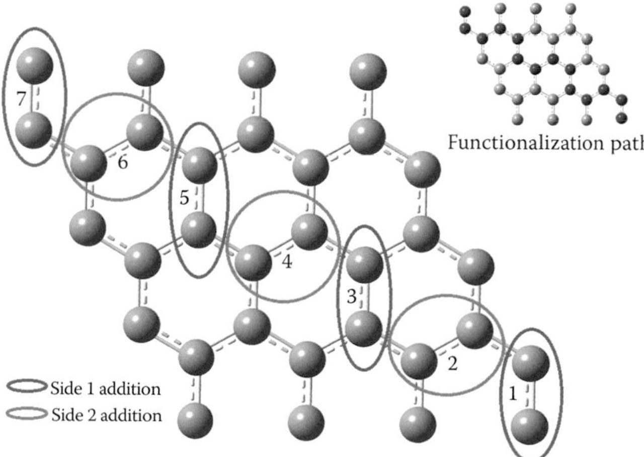

FIGURE 4.8 Functionalization path determined for the addition of benzyne groups. The C=C bonds marked with even numbers indicate functionalization on the top side of the sheet and those C=C bonds marked with odd numbers indicate functionalization on the opposite side. The sequence of addition is noted by the numbers near the functionalized C=C bond. (Reproduced from Denis, P.A., Iribarne, F., *J. Mater. Chem.*, 22, 5470–5477, 2012c.)

TABLE 4.8 Electronic BEs (kcal/mol per C_6H_4 unit) Determined for the Addition of Benzyne Groups Following the Functionalization Path Reported in Figure 4.8

	M06L/6-31G*
G4 × 4 + C_6H_4	13.9
G4 × 4 + $2C_6H_4$	29.4
G4 × 4 + $3C_6H_4$	34.7
G4 × 4 + $4C_6H_4$	35.4
G4 × 4 + $5C_6H_4$	37.8
G4 × 4 + $6C_6H_4$	40.0
G4 × 4 + $7C_6H_4$	42.0

Source: Reproduced from Denis, P.A., Iribarne, F., *J. Mater. Chem.*, 22, 5470–5477, 2012c.

Once we have studied the addition of NH groups (Denis and Iribarne 2011), azomethine ylides (Denis and Iribarne 2010, 2012b; Cao and Houk 2011), and benzynes (Denis and Iribarne 2012c), it is interesting to plot the variation of BE upon level of functionalization. This graph is shown in Figure 4.9. For comparative purposes, we also included the hydrogenation reaction. The four reactions considered clearly confirm that cooperative behavior can be expected for the organic reactions in a similar manner as found for the hydrogenation. There is a point at which the effect is almost negligible, for example, the fourth NH addition or the attachment of eight hydrogen atoms. Yet, when the infinite line of functional groups is attached, we observe in all cases the largest BE per functional group. Hydrogen displays the highest value (48.9 kcal/mol per H) when 16 H atoms are attached (16 H atoms correspond to 8 NH or C_6H_4 benzyne groups). It is closely followed by the [2 + 2] cycloaddition of benzynes, which bind to graphene with an energy of 44 kcal/mol per C_6H_4 group. The addition of imidogen radical possess a BE that is only 4 kcal/mol, smaller as compared with the [2 + 2] reaction. The 1,3 dipolar cycloaddition displays the lowest BE: 13.7 kcal/mol. This result was somewhat expected because the addition of a single azomethine ylide has the lowest BE among the four reactions considered. For comparison, it is important to recall that for 100% hydrogenated graphene (graphane), the BE per H atom is 57.8 kcal/mol per hydrogen atom, about 3.4 times larger than the BE determined for one H atom. The strongest cooperative behavior is observed for the addition of hydrogen atoms, while the weakest is verified for NH radicals. Surprisingly, the 1.3 dipolar cycloaddition is ranked third. We hypothesize that this ordering is related

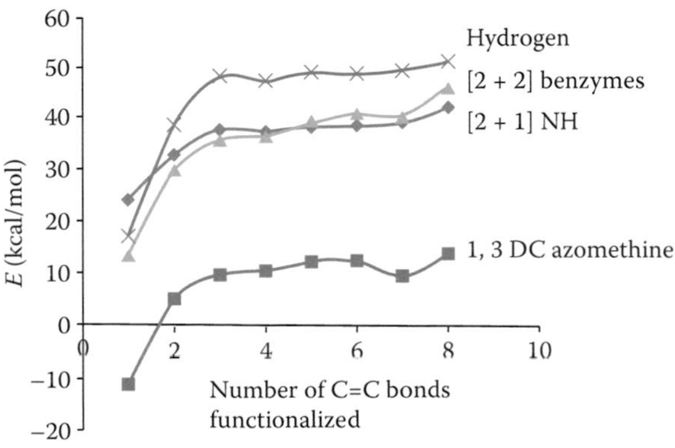

FIGURE 4.9 BE per functional group (kcal/mol) determined following the path presented in Figure 4.9. (Note: for hydrogen, twice as many atoms are needed to functionalize *X* bonds.) (Reproduced from Denis, P.A., Iribarne, F., *Chem. Phys. Lett.*, 550, 111–117, 2012b.)

to the way in which the functional groups interact between themselves once attached onto graphene. Benzynes interact via staking interactions, whereas the azomethine ylide establishes hydrogen bonding and H::H interactions. The addition of NH radicals is weakly favored by long-range NH:::H interactions.

In a following investigation, we considered the attachment of alkyl groups to graphene (Denis and Iribarne 2012a). To that end, we used methyl, ethyl, isopropyl, and tert butyl radicals. Isolated alkyl radicals are not likely to be attached onto perfect graphene. It was found that the covalent BEs are low, and because of the large entropic contribution, $\Delta G°_{298}$ is positive for methyl, ethyl, isopropyl, and tert butyl radicals. Although the alkylation may proceed by moderate heating, the desorption barriers are low. Methyl (Liao et al. 2013) and tert butyl radicals covalently bonded to graphene can be removed if 15.3 and 2.4 kcal/mol is given to surmount the energetic barrier. As observed for aryl groups, the BEs are increased, and alkyl radicals are agglomerated. For the addition in the ortho-position and opposite sides of the sheet, the BE of methyl is increased by 20 kcal/mol. The agglomeration turns the $\Delta G°_{298} < 0$. Interestingly, we found that for isolated alkyl radicals, physisorption is preferred to chemisorption, but only for the tert butyl radical, the free energy change is negative.

As discussed in Section 4.2, hydrogen terminated zigzag and armchair edges exhibit a considerably larger reactivity towards alkyl groups. The BEs are presented in Table 4.9.

Finally, the last organic reaction that we are going to discuss in this chapter is the Diels-Alder cycloaddition (Cao et al. 2013; Denis 2013a; Göstl 2014; Sarkar 2014; Seo and Baek 2014; Ji et al. 2015). Although experimental studies showed that this reaction can be performed for exfoliated and epitaxial graphene, theoretical calculations showed that the reaction may not proceed so easily. By means of periodic density functional calculations, we analyzed the diene and dienophile character of graphene. The dienophiles considered were tetracyanoethylene (TCNE) and maleic anhydride (MA), while the dienes were 2,3-dimethoxy-1,3 butadiene, 9-methylanthracene, and 9,10-dimethylanthracene (910DMA). When perfect graphene acted as dienophile, it was found that the cycloaddition products are 47–63 kcal/mol, less stable than reactants, making the reaction very difficult to proceed. The presence of Stone-Wales translocations, 585 double vacancies, or 555-777 reconstructed divacancies does not significantly improve reactivity since the cycloaddition products are still located above reactants. However, for the addition of 910DMA to single vacancies, the product is as stable as the separated reactants. With regard to the reactions with dienophiles, for TCNE, the cycloaddition product is metastable. In the case of MA, we were able to find a reaction product that is less stable than reactants by 50 kcal/mol. As observed for the reactions of graphene with dienes, it was found that the most promising defect in terms of reacting with dienophiles is the single vacancy since the other three defects were much less reactive. We conclude that the reactions with the previously mentioned dienes may proceed on the perfect or defective sheet with heating, despite being endergonic. The same statement applies to the dienophile MA. However, for TCNE, the reaction is likely to occur only onto single vacancies or unsaturated edges. We conclude that the dienophile character of graphene is slightly stronger than its behavior as diene.

TABLE 4.9 Electronic BEs (kcal/mol) and Bond Distances (Å) Determined at Different Levels of Theory for the Covalent Addition of a Single Alkyl Radical to 5 × 5 Graphene

	PBE/DZP	M06-L/6-31G*	PBE/DZP	M06-L/6-31G*
	E_{bind}		Distances	
CH_3	−7.6 (100%)	−7.3 (100%)	1.583	1.574
CH_2CH_3	−3.1 (41%)	−3.5 (48%)	1.612	1.601
CH_3CHCH_3	2.8	2.0	1.630	1.639
$C(CH_3)_3$	11.2	9.0	1.699	1.688

Source: Reproduced from Denis, P.A., Iribarne, F., *Chem. Eur. J.*, 18, 7568–7574, 2012a.
Note: DZP, double zeta plus polarization basis set.

4.4 Inorganic Chemistry of Graphene

The inorganic chemistry of graphene has been studied in less detail than its organic chemistry. The most intense research has been performed on the halogenation reaction. The fluorination of graphene is a well-known approach capable of opening a bandgap. As we move down in the periodic table, the BE of halogen decreases. As shown by Medeiros et al. (2010), the BEs are −2.98, −0.067, 1.23, and 2.5 eV for F, Cl, Br, and I, respectively. With regard to bandgaps, for fluorinated graphene, it is close to 3.5 eV at the PBE and 1.5 eV for chlorinated graphene. Thus, according to Medeiros et al. (2010), only F and Cl can be covalently added at full coverage to graphene. The other atomic reagent that has been extensively employed is hydrogen. Elias et al. (2009) confirmed the initial results by Sofo et al. (2007), which indicated that graphene can be fully hydrogenated. The material is called graphane and has a gap of 6.6 eV at the HSE06 level. As we observed for organic reagents, when free radicals are paired, the BEs increase dramatically. For example, if two hydrogen atoms are attached in the ortho-position on different sides of the sheet, the BE increases from 20.5 kcal/mol to 39.4 kcal/mol per H atom (Denis and Iribarne 2009). We note that if the addition is performed in the ortho-position but on the same side, the BE is slightly lower, 33.2 kcal/mol. Interestingly, although graphane is an insulator, the addition of two H atoms in the ortho-position on opposite sides of the sheet does not open a bandgap but maintains the semimetallic character. In addition to the halogens and H, we also studied the addition of oxygen and sulfur atoms to graphene (Denis 2009). At the PBE/DZP level, the BEs determined for the addition of O and S onto perfect graphene are 41.5 and 11.3 kcal/mol, lower. The BE determined for sulfur is almost four times lower than that corresponding to oxygen. In both cases, the electronic structure of graphene is retained (Denis 2009) and higher levels of functionalization are required to open a bandgap.

The last example that we consider in this subsection is the formation of organimetallic complexes by graphene, which is supposed to have an application in atomic spintronics. Avdoshenko et al. (2012) demonstrated that graphene|metal|ligand systems can open a new realm is surface magnetochemsitry. These authors studied complexes with Cr, and the ligands were CO or benzenes. When three COs were attached to a Cr that is coordinated to graphene, there is no magnetic moment. However, if the ligand is changed to benzene, a magnetic moment appears. Molecular dynamics showed that the systems are stable up to 2000 K.

4.5 How Can We Increase the Reactivity of Graphene?

As mentioned in Section 4.2, defects dramatically increase the reactivity of graphene, but are there any other methods? The answer to this question is yes. There are at least three methods that can increase the reactivity of graphene: (a) heteroatom doping, (b) surface roughness, and (c) charge puddles.

The substitutional doping of graphene with heteroatoms is a common approach to alter the properties of graphene (Ao et al. 2008; Denis et al. 2009; Dai and Yuan 2010; Denis 2010, 2011a, 2013c; Zou et al. 2011; Goyenola 2012; Zhou et al. 2012; Ramasse et al. 2013; Poh et al. 2014). In recent work (Denis and Huelmo 2015), we have investigated which is the effect of heteroatom doping on the reactivity of graphene. The heteroatoms employed to dope graphene were B, N, and O in the 2*p* period; Al, Si, P, and S in the 3*p* period; and Ga, Ge, As, and Se in the fourth period, and also, we considered the dual-doped graphene in which two C atoms of graphene are replaced with one 2*p* and one 3*p* element. To understand how much (or less) is increased reactivity, we present in Table 4.10 the BEs of hydrogen to 2*p* and 3*p* atom doped graphene.

To study the reactivity of monodoped graphene with B, N, or O, we added one hydrogen atom to the carbon atoms bonded to graphene or to the heteroatom itself. If we recall that the BE of H to perfect graphene is 16.7 kcal/mol, we can appreciate that in all cases, the substitutional doping increases the reactivity of the carbon atoms bonded to the dopant, with the one bonded to oxygen being the most reactive one, with a BE of 85.6 kcal/mol. When we turn our attention to the third period, we appreciate that reactivity is also increased. In all cases, the carbon atoms experience enhanced affinity toward hydrogen. When one 2*p* element and one 3*p* element are combined to attain the dual-doping, we also appreciate an

TABLE 4.10 BEs (kcal/mol) Determined for the Addition of One Hydrogen Atom to Different Dual-Doped Graphenes at the M06-L/6-31G* and VDW-DF/DZP Levels[a]

M06-L Dopants XY	X	C_X Up[b]	C_X Down[b]	Y	C_Y Up	C_Y Down	Average
AlB	72.1	–[c]	61.0	**74.7**	25.7	30.4	52.8
AlN	**71.0**	41.5	59.0	31.5	22.2	27.7	42.1
AlO	88.4	–	66.9	–	46.1	**95.5**	74.2
SiB	**73.5**	29.0	53.8	72.5	5.3	–	46.8
SiN	**85.4**	42.2	52.6	35.4	28.0	29.4	45.5
SiO	65.6	39.7	50.6	−6.9	43.5	**77.9**	45.1
PB	42.5	34.5	**43.6**	41.7	32.1	32.5	37.8
PN	33.4	35.3	**45.4**	20.9	28.1	29.8	32.2
PO	74.2	32.8	40.5	–	42.4	**74.7**	52.9
SB	11.1	46.7	59.0	**63.5**	32.5	30.2	40.5
SN	47.7	53.3	60.4	**71.8**	25.6	16.2	45.8
SO	11.5	37.7	**46.1**	−8.3	38.1	35.3	26.7
B	41.2	**47.1**					44.2
N	10.6	**42.9**					26.8
O	−7.5	**85.6**					46.6
Al	**69.5**	56.0	**69.5**				65.0
Si	**67.7**	35.0	40.0				47.6
P	**56.0**	45.4	54.3				51.9
S	−6.9	32.6	**60.9**				28.9

Source: Reproduced from Denis, P.A., Huelmo, C.P., *Carbon* 87, 106–115, 2015.

Note: Dashed lines indicate that when the H atoms were added to this site, the structure could not be optimized as the H atom migrated to the most reactive site.

[a] C_X denotes carbon atom bonded to dopant X, and C_Y denotes carbon atom bonded to dopant Y. See text for details. Most reactive sites are in bold.

[b] Up (Down) denotes addition on the same (opposite) side that the 3*p* element bulges.

enhancement of reactivity, but in this case, only for Al-O, S-N, P-O, and Si-B(para) codoped graphene, which presents reactivity higher than perfect, and monodoped graphenes. The most stables structures of the functionalized doped graphenes are presented in Figure 4.10.

Finally, we also studied the effect of 4*p* elements (Denis 2014). Taking the hydrogenation as the measure of reactivity, we present in Table 4.11 the BE of hydrogen to doped graphenes. The structure of the doped graphenes is presented in Figure 4.11.

Among the 4*p* doped graphene sheets, gallium doping is the most effective one, closely followed by As, probably because these two elements leave unpaired *e*- in graphene. In all cases except Ge, the C3 carbon is the most reactive site, thanks to the almost perfect sp^3 hybridization that it can adopt.

Another interesting aspect of graphene is that, in general, graphene is not flat but curved. In the regions in which graphene is curved, the reactivity is enhanced, as it has been shown recently (Fan et al. 2011). However, it is important to mention that this statement is true only for monolayer graphene. In effect, for bilayer graphene, the underneath layer reduces the curvature of the top layer, and thus, the sheet is less reactive, as it has been observed experimentally (Sharma et al. 2010).

Finally, the reactivity of graphene can be increased by the introduction of positive and negative charges (Tapia et al. 2011; Huang et al. 2011a; Denis 2011b,c; Logsdail et al. 2013; Li et al. 2015). The electron doping of graphene can be performed, for example, with lithium and potassium (Denis 2011c; Tapia et al. 2011). In effect, we showed that Li doping enhances the BEs of several functional groups onto graphene (Denis 2011c). Also, Logsdail et al. (2013) demonstrated that the adsorption of gold nanoparticles onto graphene is improved by Li doping. The most important results are presented in Table 4.12 and the structures are shown in Figure 4.12.

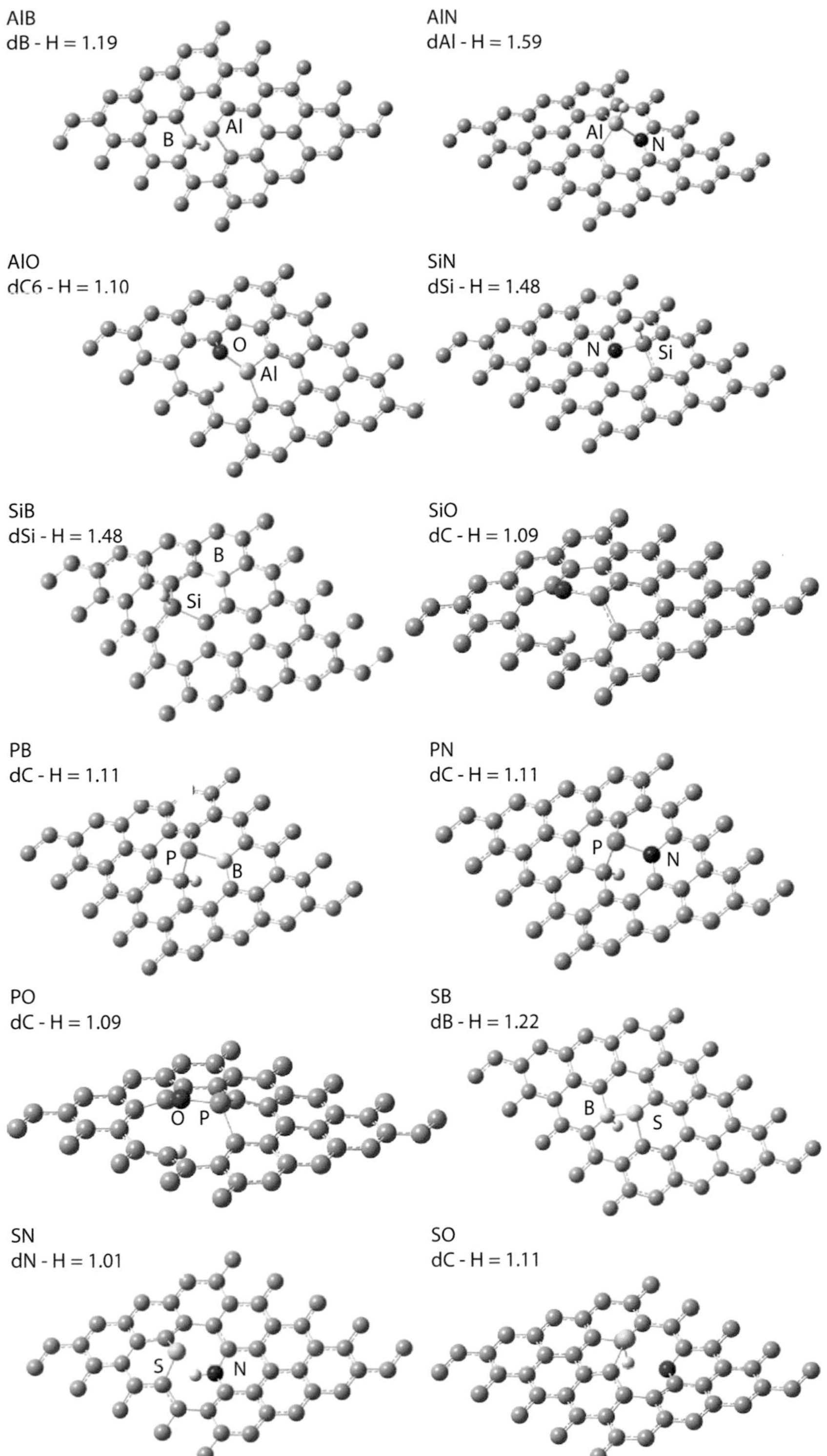

FIGURE 4.10 Optimized unit cells determined for the most stable hydrogenated dual doped graphenes at the VDW-DF/DZP level of theory. (Reproduced from Denis, P.A., Huelmo, C.P., *Carbon* 87, 106–115, 2015.)

TABLE 4.11 Reaction Energies (kcal/mol) for the Addition of Hydrogen to 5 × 5 Graphene Doped with 4*p* Elements: Ga, Ge, As, and Se, at the M06-L/6-31G* Level[a]

Addition Site	Ga	Ge	As	Se
C1-Up	43.2	31.5	45.5	28.6
C1-Down	38.2	22.0	35.1	20.2
C2-Up	32.7	19.6	15.6	16.2
C2-Down	31.2	18.4	14.9	16.3
C3-Up	56.6	14.6	48.8	32.8
C3-Down	**68.7**	29.0	**60.7**	**51.4**
4p Element	64.7	**58.2**	40.7	−3.0

[a] The reaction energy for the addition of hydrogen onto perfect graphene is 16.7 kcal/mol. Reproduced from Denis, P.A., *ChemPhysChem*, 15, 3994–4000, 2014.

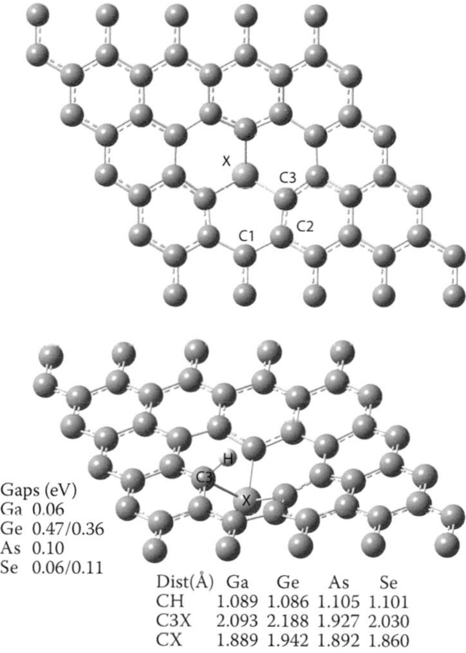

FIGURE 4.11 Addition sites studied (top) for the hydrogenation of graphene doped with Ga, Ge, As, and Se and most reactive site studied (bottom). (Reproduced from Denis, P.A., *ChemPhysChem*, 15, 3994–4000, 2014.)

As can be seen from Table 4.12, the presence of Li atoms has a profound effect on the BE. In the case of OH, the BE is increased by 23.5 kcal/mol, a finding that is in excellent agreement with the work of Nouchi. Positive charges can be introduced into graphene by the adsorption of molecules that can accept electrons from graphene (Chen et al. 2007; Denis 2011b; de Oliveira and Miwa 2015; Li et al. 2015). In recent work, we have observed that the adsorption of F4-TCNQ increases the BE of H, OH, and F onto graphene (Denis 2011b). This finding is also supported by theoretical studies which showed that artificial charges increased the reactivity of graphene (Huang et al. 2011a) and by the experimental work of Fan et al. (2011). Thus, in summary, the reactivity of graphene can be increased by substitutional doping, surface roughness, and through electron/hole doping.

TABLE 4.12 Electronic BEs (PBE/DZP, kcal/mol) for the Addition of Radicals and Azomethine to Bare and Lithium-Doped Monolayer and Bilayer Graphene

	Total BE	Extra Binding Energy	Dist C-X	Gap (eV)
Graphene 4 × 4 + Li	20.5			Metal
Graphene 5 × 5 + Li	22.6			Metal
Graphene 5 × 5 + F	34.1		1.502	Metal
Graphene 5 × 5 + OH	12.2		1.491	Metal
Graphene 5 × 5 + SH	0.0		2.949	Metal
Graphene 5 × 5 + CH_3	7.6		1.583	0.60
Graphene 5 × 5 + H	21.4		1.137	0.65
Graphene 5 × 5 + azomethine	15.9		1.600	Semimetal
Li-Graphene 5 × 5 + F	80.2	23.5	1.506	0.24
Li-Graphene 5 × 5 + OH	58.1	23.3	1.482	0.28
Li-Graphene 5 × 5 + SH	40.4	17.8	1.995	0.31
Li-Graphene 5 × 5 + CH_3	45.5	15.3	1.565	0.40
Li-Graphene 5 × 5 + H	57.6	13.6	1.136	0.40
Li-Graphene 5 × 5 + azomethine	44.5	6.0	1.586	Metal

Source: Reproduced from: Denis, P.A., *J. Phys. Chem. C*, 115, 13392–13398, 2011c.

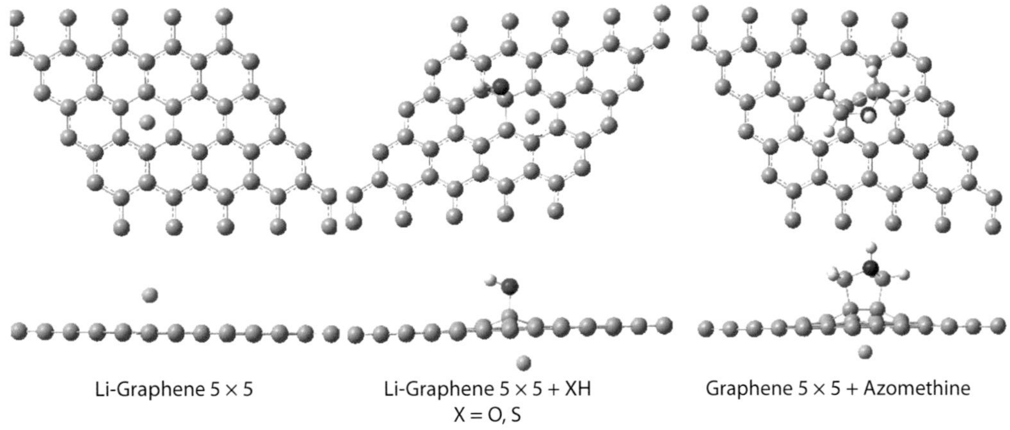

FIGURE 4.12 Optimized structures for lithium-doped graphene with and without functional groups. (Reproduced from Denis, P.A., *J. Phys. Chem. C*, 115, 13392–13398, 2011c.)

Acknowledgments

The author thanks PEDECIBA Quimica, CSIC, and ANII for sustained financial support.

References

Ao, Z.M., J. Yang, S. Li, Q. Jiang, *Chem. Phys. Lett.* 2008, 461, 276–279.

Avdoshenko, S.M., I.N. Ioffe, G. Cuniberti, L. Dunsch, A.A. Popov, *ACS Nano* 2012, 9, 9939.

Bekyarova, E., M.E. Itkis, P. Ramesh, C. Berger, M. Sprinkle, W.A. de Heer, R.C. Haddon, *J. Am. Chem. Soc.* 2009, 131, 1336.

Cao, Y., K.N. Houk, *J. Mater. Chem.* 2011, 21, 1503.

Cao, Y., S. Osuna, Y. Liang, R.C. Haddon, K.N. Houk, *J. Am. Chem. Soc.* 2013, 135, 17643.

Chen, W., S. Chen, D.C. Qi, X.Y. Gao, A.T.S. Wee, J. Am. Chem. Soc. 2007, 129, 10418–10422.
Choi, J., K.-J. Kim, B. Kim, H. Lee, S. Kim, *J. Phys. Chem. C* 2009, 113, 9433.
Dai, J., J. Yuan, *J. Phys. Condens. Matter* 2010, 22, 225501.
Denis, P.A., *J. Phys. Chem. C* 2009, 113, 5612–5619.
Denis, P.A., *Chem. Phys Lett.* 2010, 492, 251.
Denis, P.A., *Chem. Phys Lett.* 2011a, 508, 95.
Denis, P.A., *J. Phys. Chem. C* 2011b, 117, 3985.
Denis, P.A., *J. Phys. Chem. C* 2011c, 115, 13392–13398.
Denis, P.A., *Chem. Eur. J.* 2013a, 19, 15719.
Denis, P.A., *ChemPhysChem* 2013b, 14, 3271–3277.
Denis, P.A., *Comput. Mater. Sci.* 2013c, 67, 203–206.
Denis, P.A., *J. Phys. Chem. C* 2013d, 117, 19048–19055.
Denis, P.A., *ChemPhysChem*, 2014, 15, 3994–4000.
Denis, P.A., R. Faccio, A.W. Mombru, *ChemPhysChem* 2009, 10, 715.
Denis, P.A., C.P. Huelmo, *Carbon* 2015, 87, 106–115.
Denis, P.A., F. Iribarne, *J. Mol. Struct. Theochem* 2009, 907, 93.
Denis, P.A., F. Iribarne, *Int. J. Quantum Chem.* 2010, 110, 1764.
Denis, P.A., F. Iribarne, *J. Phys. Chem. C* 2011, 115, 195–203.
Denis, P.A., F. Iribarne, *Chem. Eur. J.* 2012a, 18, 7568–7574.
Denis, P.A., F. Iribarne, *Chem. Phys Lett.* 2012b, 550, 111–117.
Denis, P.A., F. Iribarne, *J. Mater. Chem.* 2012c, 22, 5470–5477.
de Oliveira, I.S.S., R.H. Miwa, *J. Chem. Phys.* 2015, 142, 044301.
Elias, D.C., R.R. Nair, T.M.G. Mohiuddin, S.V. Morozov, P. Blake, M.P. Halshall, A.C. Ferrari, D.W. Boukhvalov, M.I. Katsnelson, A.K. Geim, K.S. Novoselov, *Science* 2009, 323, 610.
Fan, X., R. Nouchi, K. Tanigaki, *J. Phys. Chem. C* 2011, 115, 12960.
Göstl, D.C.R., PhD. Thesis: Photocontrolling the Diels-Alder Reaction, 2014.
Goyenola, C., S. Stafstrom, L. Hultman, G.K. Gueorguiev, *J. Phys. Chem. C* 2012, 116, 21124–21131.
Huang, L.F., M.Y. Ni, G.R. Zhang, W.H. Zhou, Y.G. Li, X.H. Zheng, Z. Zeng, *J. Chem. Phys.* 2011a, 135, 064705.
Huang, P., H. Zhu, L. Jing, Y. Zhao, X. Gao, *ACS Nano* 2011b, 5, 7945–7949.
Ji, Z., J. Chen, L. Huang, G. Shi, *Chem. Commun.* 2015, 51, 2806.
Li, J.W., Y.Y. Liu, L.H. Xie, J.Z. Shang, Y. Qian, M.-D. Yi, T. Yu, W. Huang, *Phys. Chem. Chem. Phys.* 2015, 17, 4919.
Liao, L., Z. Song, Y. Zhou, H. Wang, Q. Xie, H. Peng, Z. Liu, *Small* 2013, 9, 1348.
Logsdail, A.J., R.L. Johnston, J. Akola, *J. Phys. Chem. C* 2013, 11, 22683.
Magedov, I.V., L.V. Frolova, M. Ovezmyradov, D. Bethke, E.A. Shaner, N.G. Kalugin, *Carbon* 2013, 54, 192.
Medeiros, P.V.C., A.J.S. Mascarenhas, F. de Brito Mota, C.M.C. de Castilho, *Nanotechnology* 2010, 21, 485701.
Novoselov, K.S., A.K. Geim, S.V. Morozov, D. Jiang, Y. Zhang, S.V. Dubonos, I.V. Grigorieva, A.A. Firsov, "Electric field effect in atomically thin carbon films," *Science* 2004, 306, 666.
Poh, H.L., Z. Sofer, M. Novacek, M. Pumera, *Chem. Eur. J.* 2014, 20, 4284–4291.
Quintana, M., K. Spyrous, M. Grzelczak, W.R. Browne, P. Rudolf, M. Prato, *ACS Nano* 2010, 4, 3527.
Ramasse, Q.M., C.R. Seabourne, D.-M. Kepaptsoglou, R. Zan, U. Bangert, A.J. Scott, *Nano Lett.* 2013, 13, 4989–4995.
Sarkar, S., PhD. Thesis: Chemistry at the Dirac point of graphene, 2014.
Seo, J.M., J.B. Baek, *Chem. Commun.* 2014, 50, 14651.
Service, R.F., "Carbon sheets an atom thick give rise to graphene dreams," *Science* 2009, 324, 875–877.
Sharma, R., J.H. Baik, C.J. Perera, M.S. Strano, *Nano Lett.* 2010, 10, 398.
Sofo, J.O., A.S. Chaudhari, G.D. Barber, *Phys. Rev. B* 2007, 75, 153401.

Suggs, K., D. Reuven, X.-Q. Wang, *J. Phys. Chem. C* 2011, 115, 3313.
Tapia, A., C. Acosta, R.A. Medina-Esquivel, G. Canto, *Comput. Mater. Sci.* 2011, 50, 2427.
Zhou, W., M.D. Kapetanakis, M.P. Prange, S.T. Pantelides, S.J. Pennycook, J.-C. Idrobo, *Phys. Rev. Lett.* 2012, 109, 206803.
Zou, Y., F. Li, Z.H. Zhu, M.W. Zhao, X.G. Xu, X.Y. Su, *Eur. Phys. J. B* 2011, 81, 475–479.

5

Few-Layer Graphene Oxide in Tribology

Adolfo Senatore

Maria Sarno

5.1 Introduction

In the research area of new promising materials in tribology, the nanoparticles as friction modifier additives in base oil or in solid media have been studied extensively in the last years. It has been found that nanoparticle inclusion in base oil enhances the extreme-pressure property and load-carrying capacity of the thin film in between the sliding surfaces with reduction in the friction coefficient. New liquid lubricants incorporating nanomaterials are expected to significantly contribute in the frictional processes of modern microscale devices as well as machines and industrial plants with considerable economical and environmental advantages.

Layered materials such as graphite or transition metal dichalcogenides such as molybdenum or tungsten disulfides (MoS_2, WS_2) are composed of vertically stacked, weakly interacting layers held together by Van der Waals interactions (Radisavljevic et al. 2011). Since their layered structure shears easily under sliding contact, these materials are widely applied as additives in oil (Black et al. 1969) or in grease (Gansheimer and Holinski 1972) or in solid lubrication (Bowden and Tabor 1964; Bhushan and Gupta 1991). MoS_2 and WS_2 are often used also in aerospace, tooling, drilling, medical, and semiconductor industries for better solid lubricants as antifriction coatings (Watanabe et al. 2004).

The development and use of materials at the nanometric scale, with different shape and morphology in the tribological environment such as inorganic fullerene-like (IF) materials (Zhao et al. 1996; Perfiliev et al. 2006; Brown et al. 2007; Joly-Pottuz et al. 2008; Yao et al. 2008; Rosentsveig et al. 2009), carbon nanotubes (CNTs) (Song et al. 2006; Qianming et al. 2008; Haiping et al. 2010; Hu et al. 2010; Bar-Sadan and Tenne 2011), and disulphide nanosheets (Hu et al. 2009; Altavilla et al. 2011a; Eswaraiah et al. 2011; Lin et al. 2011; Song and Li 2011; Zhang et al. 2011), have demonstrated marked improvement in tribological properties, both in friction reduction and wear resistance, in comparison with bulk microsized disulphides.

The outcomes of these researches are introduced in the scientific literature through the synthesis and preparation of nanoscale particles, the analysis of negligible or significant change in the rheological properties of the blending, and the tribological behavior of such novel lubricant formulations at the frictional interface.

Recently, nested spherical supramolecules of metal dichalcogenide have been synthesized by reaction of metal oxide nanoparticles with H_2S at elevated temperatures. Because of their nested fullerene-like structure, these species are known as IF nanoparticles. IF nanoparticles exhibited improved tribological behavior compared with microscale platelets for their robustness, unique shape, and enhanced flexibility (Yadgarov et al. 2013).

Fullerene-like (IF) nanoparticles and inorganic nanotubes (INT) of WS_2 and MoS_2 were discovered in 1992 (Tenne et al. 1992). Optical absorption experiments (Frey et al. 1998), Raman excitation measurements (Staiger et al. 2012), and first-principle theoretical calculations (Seifert et al. 2000) showed that the semiconductor behavior of MS_2 (M = W, Mo) is preserved in these structures, irrespective of their chirality and diameter. The semiconducting IF-MS_2 (M = Mo, W) nanoparticles exhibit also very good tribological behavior when used as additives for lubricating fluids (Chhowalla and Amaratunga 2000; Huang et al. 2005; Joly-Pottuz et al. 2005; Zou et al. 2006; Kalin et al. 2012), in self-lubricating coatings (Katz et al. 2006), and for improving the tribological and mechanical behavior of nanocomposites (Rapoport et al. 2004; Naffakh et al. 2007; Hou et al. 2008; Shneider et al. 2010). Recently, a new synthetic strategy was developed for obtaining rhenium-doped IF-MoS_2 (Re:IF-MoS_2) nanoparticles with outstanding tribological behavior in oil (Yadgarov et al. 2012).

In the past year, the metal nanosized particles have also attracted extensive research work, mainly because of the quantum effect, and their properties can be adjusted by particle size as well as by surface morphology. The main benefits of such metal nanoparticles as additive in oil come from the following issues: They are small and a good stability could be achieved through proper methods; they are liable to be trapped in the rubbing surfaces and to deposit on them. In Martin and Ohmae (2008), a review of the preparation and frictional properties of nanoparticles based on traditional soft metal lubricants such as Cu, Ag, Pb, Bi, and Sn is presented.

5.2 Nanocarbons in Tribology

Nanocarbon materials have received great attention by tribology researchers in the last three decades because of high load-bearing capacity, low surface energy, high chemical stability, and weak intermolecular and strong intramolecular bonding.

Well known as an effective and cheap material, graphite has been used as a solid lubricant for a long time. Diamond is known as the hardest material in nature. The friction coefficient for a dry diamond–diamond contact was found to be around 0.05, with excellent wear resistance (Bowden and Tabor 1950), and a much lower friction coefficient (0.001) was found for the same contact lubricated by water (Tzeng 1993).

The same results were obtained for a contact diamond–metal surface; thus, diamond is currently used for cutting tools. Furthermore, diamond films can be deposited by chemical vapor deposition, and diamond films are used to coat a broad family of tools and machinery parts. Diamond-like carbon coatings, introduced in 1973, are considered as amorphous or nanocrystalline diamond and are now widely studied for their excellent tribological properties (Kalin et al. 2004, 2014; Kalin and Vižintin 2005).

The discovery of fullerenes in 1985 (Kroto et al. 1985) opened a new field of research for carbon compounds in tribology. Indeed, carbon can also form nanometer-sized materials: CNTs, carbon onions, and carbon nanohorns, which present the advantage of easily entering the confined contact area. These nanolubricants are still in the early stage of tribological investigations, but some potential applications are already foreseen.

The purpose of the study of Lee et al. (2009) is the tribological testing of such nanoparticles dispersed in mineral oil for application as lubricant of the compressor in a refrigerator plant to decrease the friction coefficient and increase the load-carrying capacity of the bearings.

In Zin et al. (2013), the preparation of single-wall carbon nanohorns is discussed, along with their dispersion in engine oil classified as SAE40 and tribological outcomes in boundary, mixed, and elastohydrodynamic lubrication (EHL) regimes. In Chauveau et al. (2012), an original experimental approach allowed the measurement of the film thickness and the frictional properties of CNT dispersions in polyalphaolefin (PAO) base oil in EHL regime. The authors observed that the carbon-nanotube aggregates penetrate and travel through the EHL conjunction, whereas the contact inlet side behaves like a filter of aggregates. At lower sliding speed, the CNT aggregates propagate easily within the contact and the frictional behavior of the tribopair is dominated by these additives. At higher sliding velocity, the shear induces a reduction in the number of aggregates passing through the contact and the base oil rheology takes the control of the frictional response of the conjunction.

Moreover, to conjugate and to enhance, through a unique synergy, the performances of these unique materials, the synthesis of hybrid nanostructures, made by CNT and nanochalcogenides, has become an important goal not only in lubrication (Miura 2004; Zhang et al. 2009; Shu et al. 2010; Church et al. 2012) but also in other research fields such as energy conversion and storage (Soon and Loh 2007; Wang and Li 2007; Jeong-Hui et al. 2010; Ding et al. 2011), photoelectrocatalysis for the hydrogen evolution reaction (Laursen et al. 2012), reinforced nanocomposites (Li et al. 2011) and solar cells (Levy et al. 2010). For these reasons, binary systems such as MoS_2-coated CNTs have been recently synthesized by hydrothermal route (Ma et al. 2006; Koroteev et al. 2007), impregnation and annealing (Shang et al. 2007; Altavilla et al. 2011b), and electrodeposition (Shu et al. 2010). In Altavilla et al. (2013), an original synthetic strategy to obtain hybrid organic–inorganic oleylamine@MoS_2-CNT nanocomposites with different compositions is introduced along with their application as antifriction and antiwear additives for grease lubricants based on lithium and calcium soaps.

The lubrication mechanism of the carbon nanoparticles has not been yet completely understood, but their properties as frictional media are definitely based on structural modification. They present the same main advantage that nanoparticles made on metal dichalcogenides do: They are efficient even at room temperature. The comprehension of their action at the interfaces of frictional conjunction and the tailored functionalization will certainly aim at improving their excellent performance in this field.

5.3 Graphene and Graphene Oxide Nanosheets

Graphene has recently enjoyed extensive attention because of its excellent properties, such as high thermal conductivity, high Young's modulus, large specific surface area, electromagnetic interference shielding, and electrical conductivity (Geim 2009). These outstanding properties of graphene make it a bright prospect in various applications for advanced technologies, e.g., solar cells, field effect devices, practical sensors, transparent electrodes, nanocomposites, etc. (Jang and Zhamu 2008; Verdejo et al. 2011). Especially in the area of polymer nanocomposites, graphene is considered to be one of the most promising modifiers because of its excellent properties, and the research on polymer nanocomposites is addressed to the modification of neat polymers for newer and better properties, without sacrificing their processability or adding excessive weight (Tibbetts et al. 2007).

Focused on this material with extraordinary properties, the European Union's biggest ever research initiative has been launched in 2013: Graphene Flagship (Web reference to the project). With a budget of €1 billion, the project represents a new form of joint, coordinated research on an unprecedented scale. The Graphene Flagship is tasked with bringing together academic and industrial researchers to take graphene from the realm of academic laboratories into European society in the space of 10 years, with the goal of generating economic growth, new jobs, and new opportunities.

Although progress has been made in the use of graphene sheets as modifiers of polymer matrices, one main limit about its application is the poor compatibility between pristine graphene sheets and polymer matrix. In contrast to pristine graphene, there are plenty of oxygen-containing groups on the graphene oxide (GO) surface. These functional groups do not only allow the good dispersion of GO in aqueous solution but also facilitate the interaction between the host polymer and GO via covalent or noncovalent bonds (Wang et al. 2011). Thus, it would be more worthwhile to prepare GO/polymer nanocomposites and study their novel properties (Kim and Macosko 2009; Ajayan et al. 2011). In recent years, some studies have been reported with respect to the tensile strength, modulus, and electrical conductivity of GO/polymer nanocomposites. These studies have shown that the nanocomposites exhibited excellent electrical, mechanical, and thermal stability properties (Bai et al. 2011; Jung et al. 2011; Li and Bai 2011; Pan et al. 2011; Qiu and Wang 2011).

5.4 Graphene and GO in Tribology

As already stated previously, nanocarbon materials have received great attention by tribology researchers because of their high load-bearing capacity, low surface energy, high chemical stability, etc. However, few studies on the tribological applications of graphene and GO platelets have been reported so far. Graphene (Novoselov et al. 2004) has attracted considerable attention because of its unique mechanical, optical, and electronic properties (Bolotin et al. 2008; Lee et al. 2008; Nair et al. 2008). Reduced GO has been proven to deliver good friction reduction and antiwear ability on silicon wafers (Ou et al. 2010).

A number of researchers have reported that graphite (Wintterlin and Bocquet 2009), graphite derivatives (Bryant et al. 1964; Ramanathan et al. 2008), as well as other lubricant materials (Fusaro 1979; Hilton et al. 1992; Tian and Xue 1997), have together, the previously desirable properties. Lin et al. (2011) investigated the tribological properties of graphite nanosheets as an oil additive. As well known, this material is characterized by weak interatomic interactions between the layers (Van der Waals forces) and low-strength shearing (Bhushan and Gupta 1991). In Zhang et al. (2011), the tribological behaviors of graphene sheets modified by oleic acid and dispersed in lubricant oil were investigated using a four-ball tribometer.

Huang et al. (2006) investigated the tribological properties of graphite nanosheets as an oil additive. They found that the frictional behavior and antiwear ability of the lubricating oil were improved when graphite nanosheets were added to the paraffin oil at the optimal concentration.

In particular, a functionalization with long chain compounds (i.e., aliphatic amine to obtain the amide derivative) enhances the dispersion in nonpolar solvents (Lin et al. 2011) to ensure uniform dispersion without any agglomeration of the GO in the base oil, taking advantage of the surface –OH and –COOH introduced during the GO preparation.

On the other hand, additives such as carbon, and even more if functionalized with –OH and –COOH groups that increase their polarity, can be dispersed through the use of a dispersant (Zhang et al. 2011), avoiding further chemical reactions and using a consolidated methodology for the lubricant industry.

In D'Agostino et al. (2012) and Sarno et al. (2013), the tribological behavior of GO nanosheets in group I base mineral oil SN150 was investigated under a very wide spectrum of conditions, i.e., from boundary and mixed lubrication to the EHL regimes. A rotational tribometer with a ball on disc setup has been employed to explore the performances of the nanosheets in the lubricating fluid. Raman analysis on the steel ball worn surfaces was performed to investigate the presence of graphitic material on the mating surfaces after tribological tests, to verify the formation of a protective film on the rubbing surfaces due to the additive interactions during the sliding process.

5.5 GO Nanosheets Preparation

In Senatore et al. (2013), GO nanosheets were prepared by a modified Hummer method (Hummers and Offeman 1958). The oxidation of graphite particles was obtained from Lonza to graphitic oxide accomplished with a water-free mixture of concentrated sulfuric acid, sodium nitrate, and potassium permanganate. This process requires less than 2 h for completion at temperatures below 45°C. With the aid of further sonication step, the oxidized graphite layers were exfoliated from each other. Then

30% H_2O_2 was added to the suspension to eliminate the excess MnO_4^-. The desired products were rinsed with deionized water. The remaining salt impurities were eliminated with resinous anion and cation exchangers. The dry form of graphitic oxide was obtained by centrifugation, followed by dehydration at 40°C.

As stated previously about the use of dispersant agent (Zhang et al. 2011), a polyisobutyl succinic acid–polyamine ester was sonicated with the GO nanosheet in base oil to provide dispersion stability of the additive. The weight ratio GO/dispersant was 3.5. The dispersant has a polar head that attaches itself to the solid particles and a very long hydrocarbon tail that keeps it suspended in oil.

Once several dispersant polar heads have attached themselves to a solid particle, the dispersant can no longer combine with other particles to form large aggregates, enwrapping one nanoparticle to repel another and thereby form a uniform suspension.

Also in Li et al. (2012), GO was produced following the Hummers method. In brief, graphite flakes and $KMnO_4$ were gradually added into the mixture of concentrated H_2SO_4/H_3PO_4 and then heated to 50°C, followed by stirring 12 h. The mixture was cooled to room temperature and poured into cold (0°C) deionized water. Then, 30% H_2O_2 was added slowly into the mixture, until the solution turned into bright yellow. The resulting brilliant yellow mixture was centrifuged and washed with deionized water and 30% HCl for several times, until the pH of the mixture was neutral. Finally, the GO slurry was purified by dialysis for a week and then dried in a vacuum oven at 60°C. The composite GO-nitrile rubber (GO/nitrile rubber [NBR]) was first cut into fragments and dissolved in a flask with 400 mL acetone at 60°C by means of violent stirring. Simultaneously, the required amount of GO was dispersed in DMF by means of ultrasonication. Then, the GO/dimethyl formamide (DMF) suspension was gradually poured into the rubber solution and was stirred for 12 h. Finally, the deionized water was added to coagulate rubber and GO to produce the GO/NBR nanocomposites. The nanocomposites were dried in a vacuum oven at 80°C until its weight remained unchanged and mixed with the curing agents.

In Ou et al. (2012), with the aim of achieving a polydopamine/GO multilayer structure with potential application as a lubricant film in nanodevices such as micro- and nano-electromechanical systems, the authors provided the nanomaterials of hydrophobic outer surface, through fluoroalkylsilane molecules of 1H, 1H, 2H, 2H-perfluorodecyltrichlorosilane (PFDTS) onto the surface of the multilayer composites and the consequent film was labeled as polydopamine (PDA)/GO-PFDTS. The sample's preparation has been carried out by way of preparation of GO colloid solution and fabrication of PDA/GO-PFDTS.

In Ou et al. (2010), the preparation of the GO colloid solution started from expandable graphite, which was first heated at 1050°C in air for 15 s. The heat-treated expandable graphite powder (1 g) was then added to 98% H_2SO_4 (23 mL) in an ice bath with stirring, and $KMnO_4$ (3 g) was subsequently added slowly. The mixture was kept at 35°C in a water bath for 30 min. Ultrapure water (46 mL) was gradually added, and the mixture was immersed in ice water. After 15 min, the mixture was further treated with ultrapure water (140 mL) and 30% H_2O_2 solution (12.5 mL). The obtained mixture was first washed with ultrapure water until a pH level of 7 was reached and was then dialyzed with stirring until SO_4^{2-} anions could no longer be detected by the $BaCl_2$ solution (1 M). A diluted GO colloid solution with a concentration of 0.4 mg 3 mL^{-1} was employed in the succeeding process.

5.6 Graphene and GO Characterization

Scanning electron microscopy (SEM) pictures of graphite chips and GO nanosheets were obtained with a LEO 1525 microscope. The samples, without any pretreatment, were covered with a 250-Å-thick gold film using a sputter coater (Agar 108 A).

Graphite chips and GO nanosheets are shown in Figure 5.1a and b respectively. The images at higher magnification (inserts in Figure 5.1) evidence the loss of the chip's structure of original graphite to GO transparent and thin flakes.

Transmission electron microscopy (TEM) micrographs were obtained with a JEOL JEM 2010 electron microscope operating at 200 keV. The TEM images in Figure 5.2 reveal the thickness of the GO

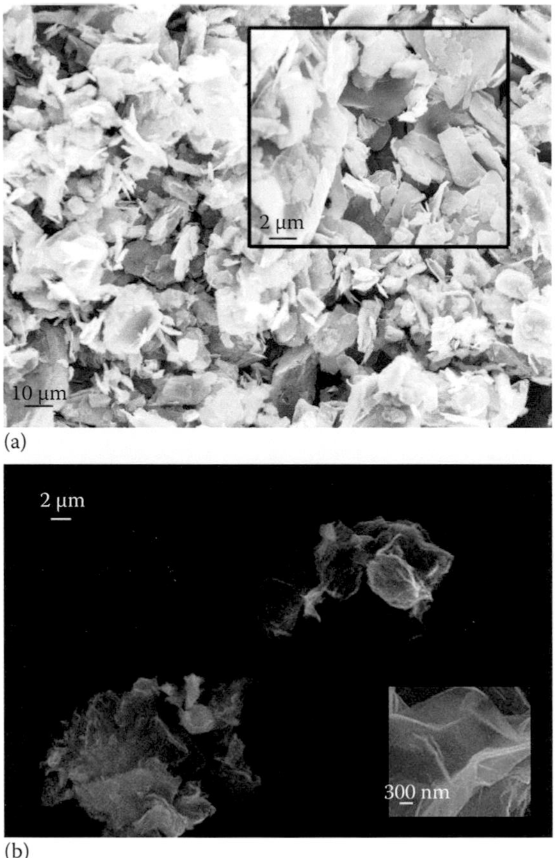

FIGURE 5.1 SEM images of graphite (a) and GO (b).

nanosheets that is in the range of 5–6 nm. The high-resolution transmission electron image quite permits measurement of the number of sheets, characterized by a not so high level of order.

X-ray diffraction (XRD) measurements were performed with a Bruker D8 X-ray diffractometer using CuKα radiation. X-ray diffraction spectra of original graphite and GO are shown in Figure 5.3, in the 2θ range of 20°–80°. The loss of the original graphite structure for GO is clearly evident.

In Figure 5.4, the thermogravimetric analysis (thermogravimetric and derivative thermogravimetric [TG-DTG]) at 20 K/min heating rate in flowing air performed with an SDTQ 500 Analyzer (TA

FIGURE 5.2 TEM image of GO at different magnifications.

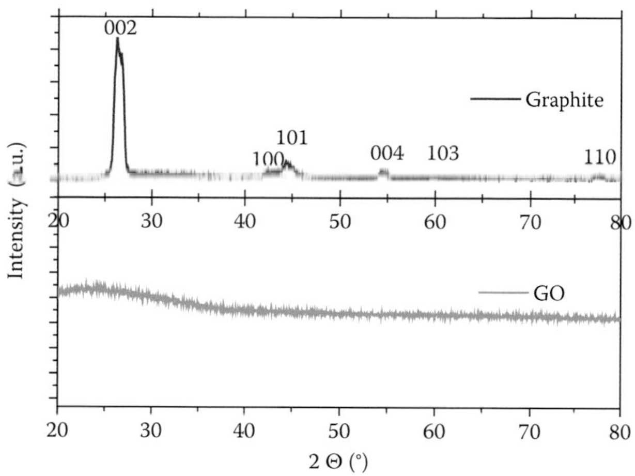

FIGURE 5.3 X-ray diffraction patterns of graphite and GO.

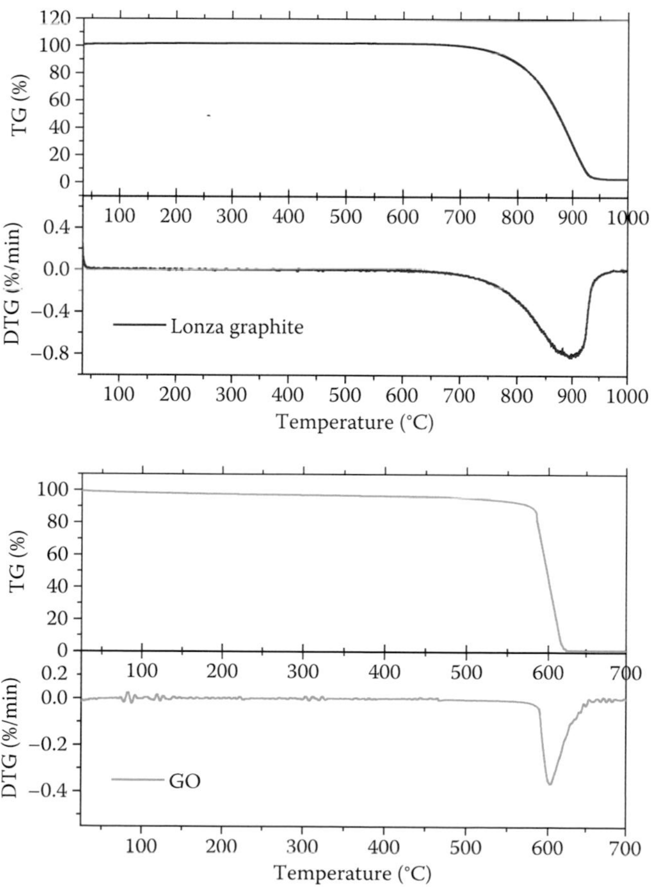

FIGURE 5.4 TG-DTG analysis of graphite (a) and GO (b).

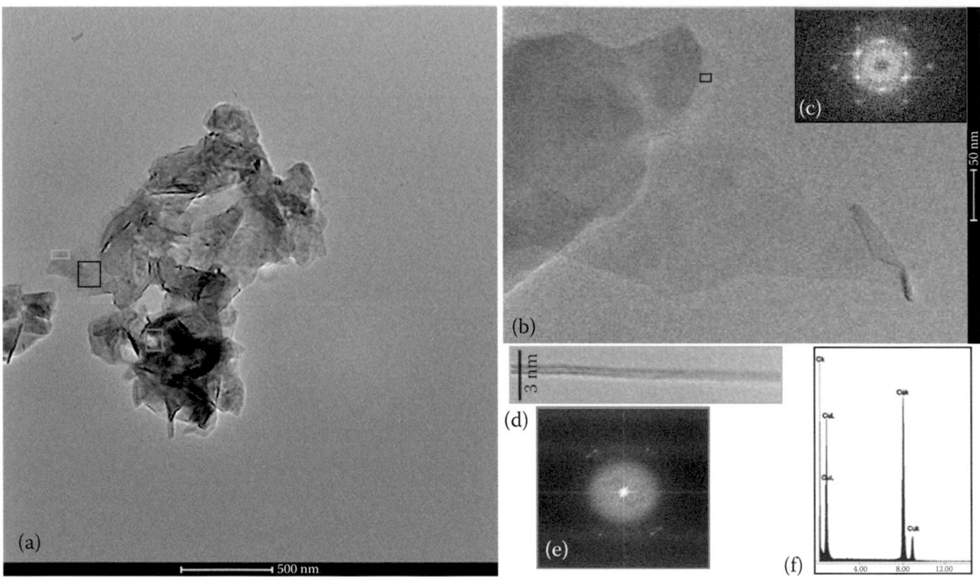

FIGURE 5.5 TEM images at two different magnifications of the graphene platelets (a, b). The inserts in panel a are high-resolution TEM image, in green area (d); FFT, in orange area (e); EDAX analysis, in the red area (f). The insert in panel b is electron diffraction pattern collected in the blue area (c).

Instruments) on graphite and GO samples is reported. The oxidation of the GO occurs as a one-step loss, centered at 605°C, well below the graphite oxidation temperature, likely because of the loss of the graphite original order.

Figure 5.5 show the TEM images of the graphene platelets used for the tribological tests, whose graphs are presented in Figures 5.13 and 5.14. The inserts in Figure 5.5a are high-resolution TEM image, in green area (d); fast fourier transform (FFT), in orange area (e); energy dispersive analysis of x-rays (EDAX) analysis, in the red area (f). The insert in Figure 5.5b is electron diffraction pattern collected in the blue area (c).

The suspension stability of two different samples based on SN350 mineral oil and modified graphene platelets has been checked in Lin et al. (2011) by ultraviolet–visible spectroscopy (UV-VIS) spectrophotometry, which measured the UV intensity of the lubricating oil solution by evaluating the rate of nanoparticle sedimentation. The relative concentration was calculated by the ratio of the particle concentration intensity of the supernatant fluid at each measurement time divided by the initial concentration intensity of the suspension. A value of the relative concentration close to 1.0 means excellent stability of the lubricating oil solution without particle sedimentation. Figure 5.6 shows that after a long period of centrifugation, rapid precipitation was observed for the suspension with the pristine graphene platelets, indicating that unmodified graphene platelets were heavily agglomerated in base oil (Lin et al. 2011). In contrast, little precipitation was observed in the suspension with the modified graphene platelets. This result indicates that the addition of modified graphene platelets to the lubricating oil produced stable suspension.

Such an improvement could be attributed to the effectiveness of the surface modification. According to Chen et al. (2005), when the platelets were dispersed in the base oil, the long hydrocarbon segments easily stretched into the base oil and therefore produced a typical steric hindrance effect, which effectively helped to separate the graphene platelets from each other.

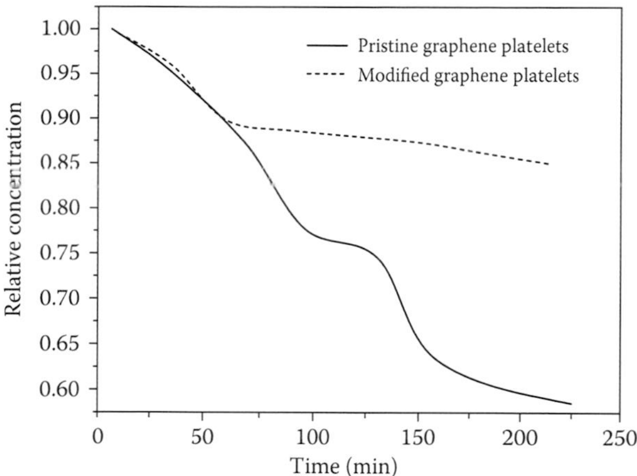

FIGURE 5.6 Suspension stability of the lubricating oils with pristine and modified graphene platelets. (Modified from Lin, J. et al., *Tribol. Lett.*, 41, 209–215, 2011.)

5.7 Tribological Testing of Graphene and GO in Oil

In Senatore et al. (2013), the sample formulated through dispersion of GO nanosheets in SN150 mineral oil, already characterized from the description in previous sections, was investigated from the tribological point of view in the range of high average contact pressure with a ball-on-disc setup on WAZAU TRM100 rotational tribometer. The main properties of SN150 base oil were as follows: kinematic viscosity, 29.7 cSt at 40°C, 5.1 cSt at 100°C; density at 20°C, 0.87 kg/dm^3. The tests were performed at different temperatures of the oil bath, with the following fully immersed specimens: The upper element of the tribopair was a X155CrVMo12-1 steel disc, 60 Rockwell Hardness value (HRC), roughness Ra = 0.50 μm, and 105 mm diameter; the lower one was a X45Cr13 steel ball, 52-54 HRC, 8 mm diameter.

The measured data are presented in Figure 5.7 according to the Stribeck curves representation, i.e., friction coefficient vs. sliding speed at different levels of average Hertzian contact pressure and lubricant temperature, along with the error bars denoting the standard deviation around the rolling mean.

These Stribeck graphs obtained from a test at average Hertzian pressures equal to 1.17 GPa, 1.47 GPa, and 1.68 GPa show a shape with a well-developed minimum, which is considered the transition from mixed lubrication regime to EHL regime for increasing speed (Kalin et al. 2009). This frontier divides the region with concurrent phenomena of solid-to-solid contacts, adhesion, and interaction between friction modifier additives and steel surface (mixed lubrication) from the other with predominant viscous stress and elastic deformation of the tribopair surfaces (EHL). For instance, the transition appears in the speed range 0.30–0.40 m/s for the test at 25°C and 0.50–0.60 m/s at the higher temperature, i.e., 80°C.

These results underline the decrease in the coefficient of friction (CoF) for increasing average Hertzian contact pressure for the formulated sample; the same behavior has been observed for the base oil according to a known point-contact effect as recalled in Yadgarov et al. (2013): the shear stress increases less in proportion to the contact pressure; this leads to a slight reduction in friction. As expected, for a given sample, the minimum of the Stribeck curve moves right for increasing temperature due to lower viscosity. Additionally, for each sample, the CoF increased with the temperature at a given level of speed and contact pressure. This observation could be addressed mainly by the effect of the lower lubricant viscosity and the ensuing GO precipitation at higher temperature.

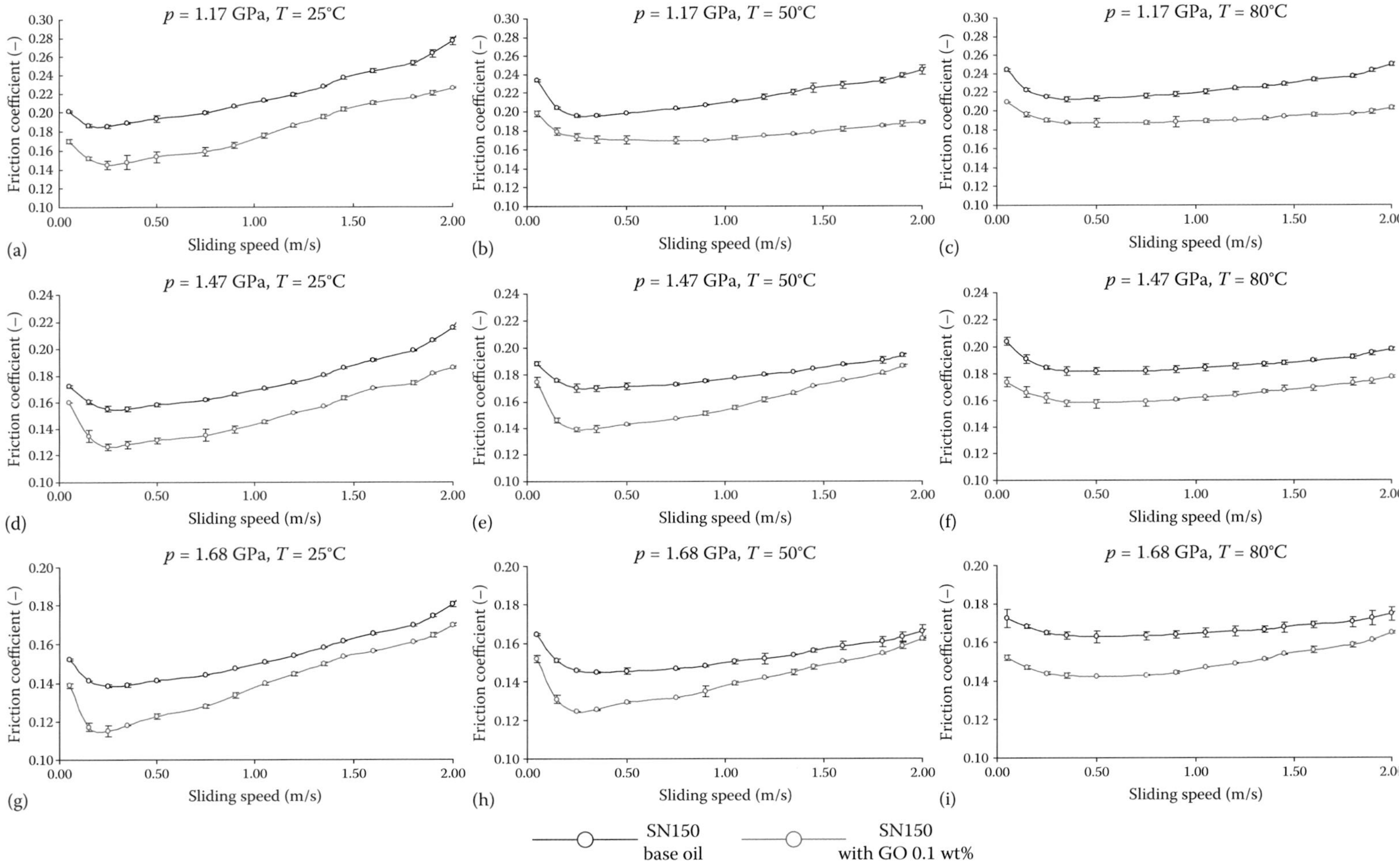

FIGURE 5.7 Stribeck curves from sweep-speed tests on SN150 base oil and SN150 with GO at 0.1 wt% concentration. (Modified from Senatore, A. et al. *ISRN Tribology* 425809, 2013.)

The average friction coefficient of GO–oil mixture decreased by 20% of the base lubricant value. Similar average reduction could be observed for all the combinations of the operating conditions. Even in the heaviest test conditions (high pressure and temperature) and in fully developed EHL regime where the viscous stresses are prevalent and the interface action of the solid friction modifiers is weak, the CoF reduction is still greater than 8% (Figure 5.7i).

Longer tests were carried out to analyze the influence of GO addition to the base oil on wear behavior of the steel ball/disc tribopair through steady-state pure sliding tests. For these tests, two levels of temperature (25°C and 80°C) and speed (5 and 500 mm/s) were selected. In this case, the load was maintained constant at 90 N, leading to 261 μm Hertzian contact diameter and a resulting average contact pressure of 1.68 GPa, assuming the elastic properties of the two previously mentioned materials used as tribopair parts. The worn surface of the steel ball has been measured ex situ by using an optical microscope to acquire the wear scar diameter (WSD). According to International Organization for Standardization (ISO)/International Electrotechnical Commission (IEC) Guide 98-3:2008, the WSD results are listed with expanded uncertainty equal to 20 μm, coverage factor $k = 2$, in Table 5.1. The corresponding lubrication regime is also stated according to the previous analysis based on the trend of the Stribeck graphs.

In Li et al. (2012), the friction and wear tests to explore the tribological behavior of the GO/NBR sample described in the preparation section were performed on a ring-block tribometer setup under dry sliding and water-lubricating condition. A stainless steel ring was used as a counterpart with an outer diameter of 49.22 mm and the rubber blocks with a size of 12.32 × 12.32 × 19.05 mm. Before each test, the surface of the counterpart was polished with metallographic abrasive paper. The test was conducted under ambient condition (temperature 20°C ± 2°C, humidity 30% ± 10%) and the sliding speed was 200 rpm under a normal load of 10 N for 1 h. At the end of each test, the width of the wear scar was measured by the optical microscopy.

The CoF of GO/NBR nanocomposites with different contents of GO under dry sliding conditions is shown in Figure 5.8. The graph underlines that under dry sliding condition, both the CoFs decreased dramatically at first, then increased with the increasing GO content. When the GO was incorporated, the

TABLE 5.1 Wear Scar Diameter After 1-h Sliding Test: Ball-on-Disc Contact, 1.68 GPa at 25°C and 80°C

Sample	Boundary Regime at 25°C	Boundary Regime at 80°C	EHL Regime at 25°C	Mixed Regime at 80°C
SN150–Base oil	600	620	900	910
SN150 with GO 0.1 wt%	530	540	630	660

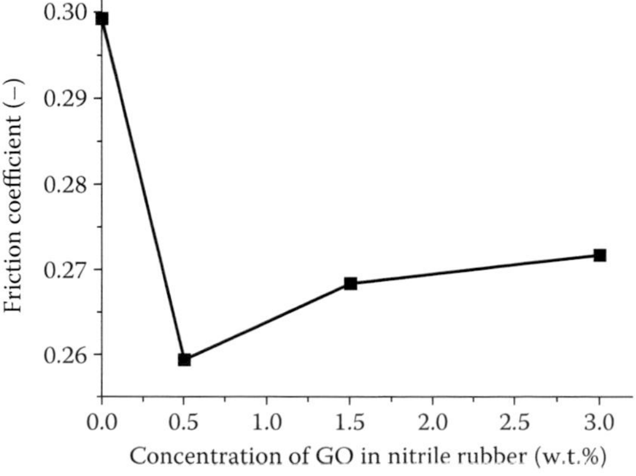

FIGURE 5.8 Friction coefficient of GO/NBR nanocomposites at different concentrations of GO under dry sliding condition. (Modified from Li, Y. et al., *J. Mater. Sci.*, 47, 730–738, 2012.)

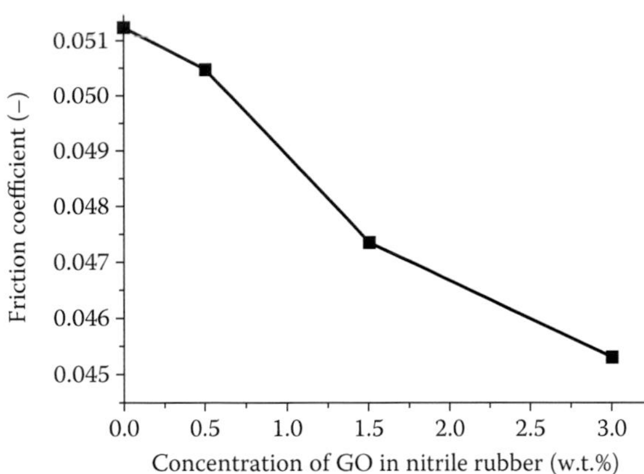

FIGURE 5.9 Friction coefficient of GO/NBR nanocomposites at different concentrations of GO under water-lubricated condition. (Modified from Li, Y. et al., *J. Mater. Sci.*, 47, 730–738, 2012.)

CoF of GO/NBR nanocomposites could be effectively reduced because the fundamental graphite lattices of GO consisted of loosely bound layers that intrinsically possess low shear strength (Lee et al. 2010).

As shown in Figure 5.9, under water-lubricated condition, the CoFs of the NBR and GO/NBR nanocomposites were obviously lower than that obtained from dry sliding tests. In this case, the CoF of the GO/NBR nanocomposites decreased with increasing GO content. Because the NBR matrix contained carbon-nitrogen (CN) polar group and the GO had hydrophilic groups, the surfaces of specimens could form strong hydrogen bonding with the water molecules, leading to the formation of continuous water film between the rubber block and countersurface, which could drastically reduce the frictional dissipation. The reasons that the water film could drastically improve the tribological properties of the nanocomposites could be addressed by the reduction in the direct contact area of metal slider and rubber block operated by the water-lubricated film and the debris-removing action due to the water flow so that it could effectively reduce the abrasive wear (Jia et al. 2004; Srinath and Gnanamoorthy 2007; Meng et al. 2009).

In Zhang et al. (2011), the tribological properties of oleic acid-modified graphene in oil were explored by using PAO9 base oil, with a density of about 0.83 g cm^{-3} and a viscosity of 9 cSt at 100°C. The oleic acid-modified graphene at concentration between 0.01 and 5 wt% was uniformly dispersed in PAO9 by ultrasonication for 15 min. The tribological properties in terms of the CoF and wear resistance were evaluated using a four-ball tribometer. The balls (12.7 mm in diameter) used in the experiments were made of GCr15 with a hardness of 64–66 HRC. The tribological test was conducted at room temperature under a load of 400 N and with a constant speed of 1450 rpm. The ex situ diameter of the wear scar on the ball was assumed as a key performance indicator of the antiwear properties of the formulated lubricants.

Figure 5.10 shows the CoF results for graphene additives at different concentrations in PAO9. The variation displays a "U" shape and the lowest CoF is obtained at a graphene concentration of 0.02 wt%. The CoF reduction compared with the pure PAO9 base oil is 17%. Beyond this point, the CoF gradually increases with the graphene concentration.

As shown in Figure 5.11, a similar trend in wear indicator is observed when different concentrations of graphene were added into PAO9. The result highlights that the addition of a small amount of graphene will improve the antiwear properties of the lubricants with reduced wear scar diameters. When 0.06 wt% of graphene was added, the reduction in the WSD is up to 14%.

The same authors proposed also an understanding path for the tribological mechanism at sliding interface in the presence of graphene. Graphene nanosheets are able to form a protective layer on the surface of each steel ball at lower concentrations; this behavior introduces enhanced antiwear protection. However, as the graphene concentration exceeds a critical value, the oil film becomes much more discontinuous, thus degrading the antiwear properties, finally leading to a nonnegligible share of dry friction (Figure 5.12).

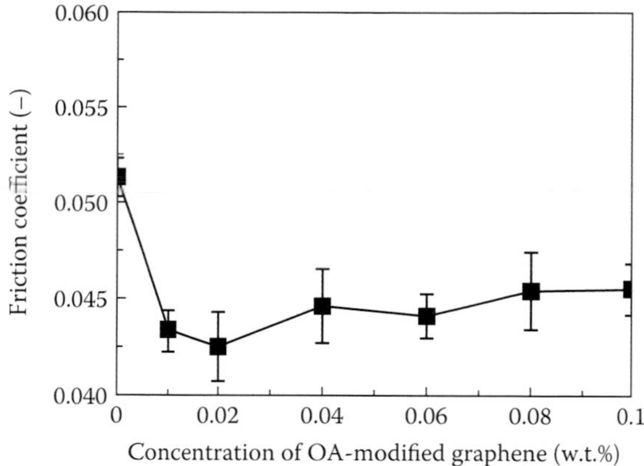

FIGURE 5.10 Friction coefficient from test of oleic acid (OA)-modified graphene in oil at different contents. (Modified from Zhang, W. et al., *J. Phys. D Appl. Phys.*, 44, 205303, 2011.)

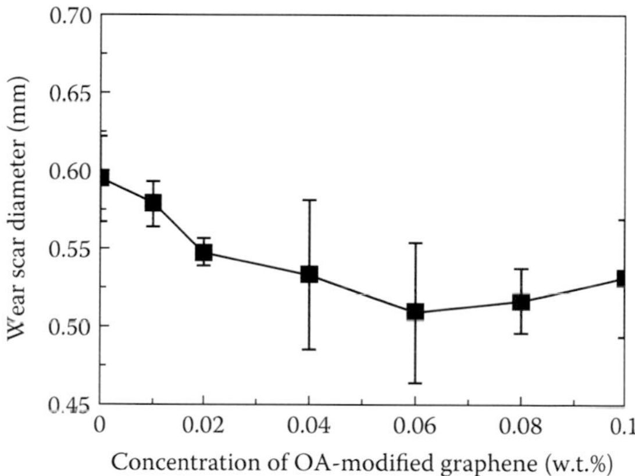

FIGURE 5.11 Wear scar diameter from test with oleic acid (OA)-modified graphene in oil at different concentrations. (Modified from Zhang, W. et al., *J. Phys. D Appl. Phys.*, 44, 205303, 2011.)

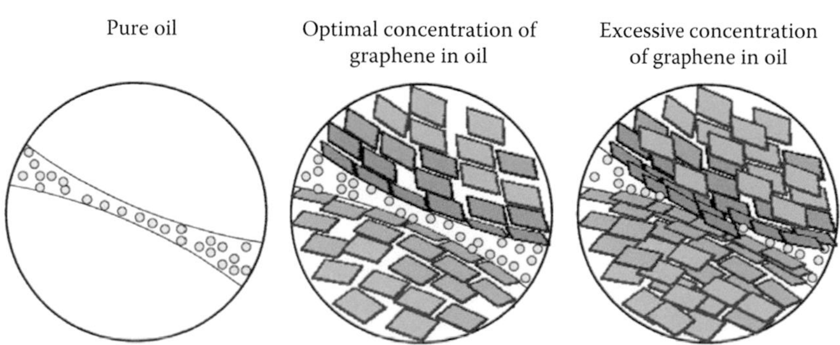

FIGURE 5.12 Schematic diagram of the tribological mechanism of graphene sheets as oil additives. (Modified from Zhang, W. et al., *J. Phys. D Appl. Phys.*, 44, 205303, 2011.)

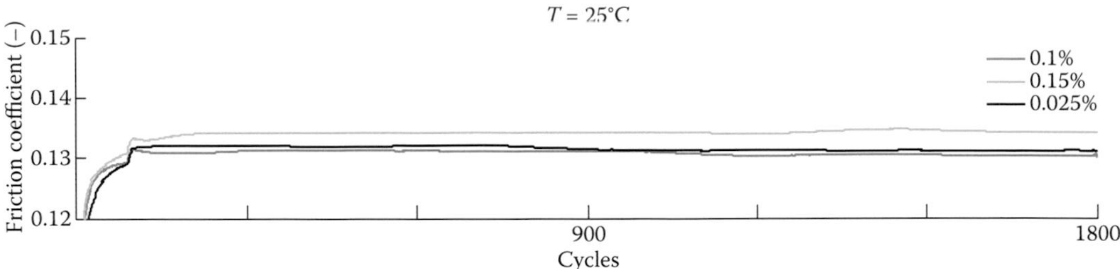

FIGURE 5.13 Friction coefficient vs. cycle in reciprocating sliding test on graphene in gearbox oil at different concentrations, $T = 25°C$.

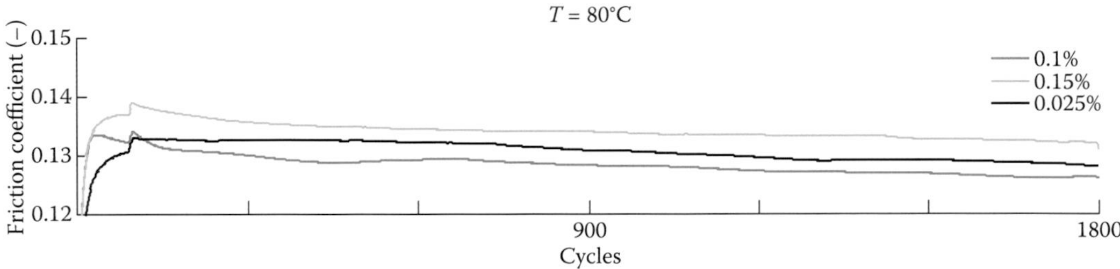

FIGURE 5.14 Friction coefficient vs. cycle in reciprocating sliding test on graphene in gearbox oil at different concentrations, $T = 80°C$.

The search for the optimal concentration of graphene nanosheets has also been carried out in SAE 75W85 commercial gearbox at three levels of loading: 0.025 wt%, 0.1 wt%, and 0.15 wt%. The experiments were carried out on a DUCOM TR-BIO 282 tribometer with linear alternative sliding motion between the counterparts. The selected materials were a X46Cr13 steel ball of 6 mm in diameter clamped on the upper holder and an EN-31 steel disc of 25 mm diameter. By applying 19 N as normal load, 141 μm Hertzian contact diameter was obtained with a resulting average Hertzian contact pressure of 1.22 GPa. The frequency chosen for the reciprocating movement was equal to 1 Hz, with a stroke length of 1 mm and a test duration of 30 min.

The frictional results at an oil bath mean temperature of 25°C and 80°C in Figures 5.13 and 5.14 prove that the optimal concentration is provided by the middle-loaded sample, with 0.1 wt% of graphene.

In Lin et al. (2011), it was stated that the wear resistance and load-carrying capacity of the base oil SN350 in other test conditions were greatly improved with the addition of modified graphene platelets at an optimal content of 0.075 wt%. Of course, also the friction coefficient of this sample was much lower than the base oil sample and the sample obtained with modified natural flake graphite.

5.8 Raman Spectrum and 3D Rubbed Surface Scan

In Figure 5.15, the Raman spectra of graphite and GO nanosheets are reported. These spectra were obtained at room temperature with a micro Raman spectrometer Renishaw inVia with 514 nm excitation wavelength (laser power 30 mW) in the range of 100–3000 cm^{-1}. The optical images were collected with an inline Leica DMLM optical microscope.

In the spectrum of graphite, the most prominent features (Senatore et al. 2013), the so-called G band appearing at 1582 cm^{-1} and the G' or 2D band at about 2700 cm^{-1}, using 514 nm excitation wavelength, are collected. The G' band at room temperature can be fitted with two Lorentzian lines. A broad D-band due to disorder or edge of a graphite sample can be also seen at about half of the frequency of the G' band (around 1350 cm^{-1} using 514 nm laser excitation). The oxide graphene shows, as expected, an improved

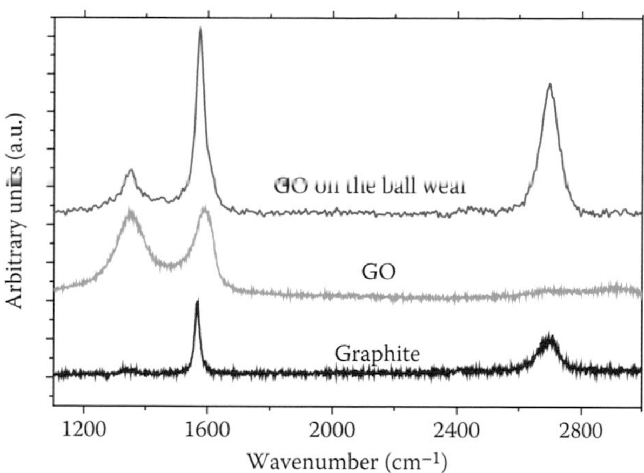

FIGURE 5.15 Raman spectra of graphite, GO, and further exfoliated GO on the ball wear after 1-h test at $p = 1.68$ GPa, $T = 25°C$, and $v = 0.5$ m/s.

D band intensity and flattening of the 2D line and displays a shift to higher frequencies (blue-shift) of a broader G band (Kudin et al. 2008).

In Figure 5.16, the Raman spectrum from more than 40 measurements collected on the worn surface of the steel ball specimen after 1-h test at $p = 1.68$ GPa, $T = 25°C$, $v = 0.5$ m/s is also reported. The same analyses were also carried out on the oil sample SN150, on the GO powder, and on their blending, GSN150_1, i.e., the oil sample collected from the tribometer pot after the 1-h frictional test. A notable fact is that the G band of this spectrum is located almost at the same frequency as that in graphite, while the G' band exhibits a single Lorentzian feature and D-band results reduced, evidencing a further exfoliation of the GO and the formation of a thin carbon film ("tribo-film") on the wear scar surface.

In the same figure, the Raman spectra of GO, SN150, GSN150_1h, and GO on the ball wear scar after the 1-h test at $p = 1.68$ GPa, $T = 25°C$, $v = 0.5$ m/s are also reported for comparison. For the Raman measurement of the oil and GSN150_1h, a drop was casted on a glass slide.

All the typical D, G, and 2D of carbon can be observed (grey arrows in Figure 5.16) in the spectrum of GSN150_1h, indicating that inside the oil at the end of the 1-h test, GO further exfoliated as that found on the ball wear scar (see in particular the presence of the 2D band as shown in the insert of Figure 5.16)

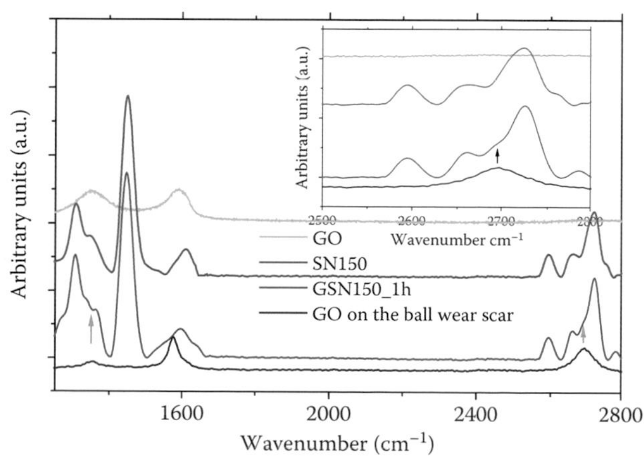

FIGURE 5.16 Raman spectra of GO, SN150, GSN150_1h, and GO on the ball wear scar after 1-h test at $p = 1.68$ GPa, $T = 25°C$, and $v = 0.5$ m/s.

is also present. According to the graphs of Figure 5.7, the prevalent lubrication regime was the EHL one. Thus, the Raman spectrum in Figure 5.16 is an ex situ proof that the GO interfacial behavior with rubbing surfaces of the steel specimens in the EHL regime results in further exfoliation of the packed layers of GO with adhesion on the progressive wearing surface. This evidence is in contrast with Chauveau et al. (2012), in which with in situ contact observation, the carbon-nanotube aggregates penetrate and travel through the frictional contact in the same lubrication regime but do not adsorb on the considered surfaces.

Figure 5.17 shows the 3D morphology of the rubbed surfaces after the frictional process with pure SN150 and SN150 with 0.1 wt% of GO samples. The image was obtained by means of a Sensofar PLu Neox 3D optical profiler based on confocal technology. In this case, the test length was 1 h in boundary lubrication regime: $p = 1.68$ GPa, $T = 25°C$, $v = 5.0$ mm/s. As shown, the rubbed surface after the test with the pure oil is rough with wide and deep furrows together with large and tall ridges. Addition of 0.1% of GO results in a reduced root mean squared (rms) roughness from 552 nm to 424 nm. On the other hand, not only the roughness is decreased, but also the morphology of the wear scar surface has changed, as it is possible to observe a succession of ridges and valleys of about 15 μm height. To better understand the lubrication mechanisms, the surface of the wear scar after the test on the blending SN150/GO was submitted to thermal oxidation in air at 200°C for 2 h to remove GO. The same roughness parameter acquired on the sample after the thermal oxidation was 500 nm (Figure 5.19): This result proves the GO filling of the deeper valleys and the formation of a thinner film on the asperities, since if the entire surface were covered with a uniform exfoliated GO thickness, the same roughness would be measured before and after the oxidation. On the other hand, the surface morphologies in Figures 5.18 and 5.19 are quite similar. Zhang et al. (2011) obtained similar results: addition of 0.06 wt% graphene significantly smoothed the surface, reducing the surface roughness from 464 to 220 nm. In contrast, 5 wt% of graphene loading resulted in dry friction, with many scratches from the ex situ specimen surface observation with a large roughness of 775 nm due to abrasive wear caused by graphene in excessive concentration.

In Song and Li (2011), the authors presented optical micrograph images of the worn surface of a steel ball after a 10-min sliding test against a steel plate at 0.431 m/s by using as lubricant media water, water added with 0.1 wt% of CNTs-COOH and water added with 0.1 wt% of GO nanosheets (Figure 5.20). They deduced that CNTs-COOH presented poor dispersion in water, since they form large CNTs-COOH bundles, resulting in the friction coefficient increasing, while GO nanosheets were found to be more easily dispersed in water.

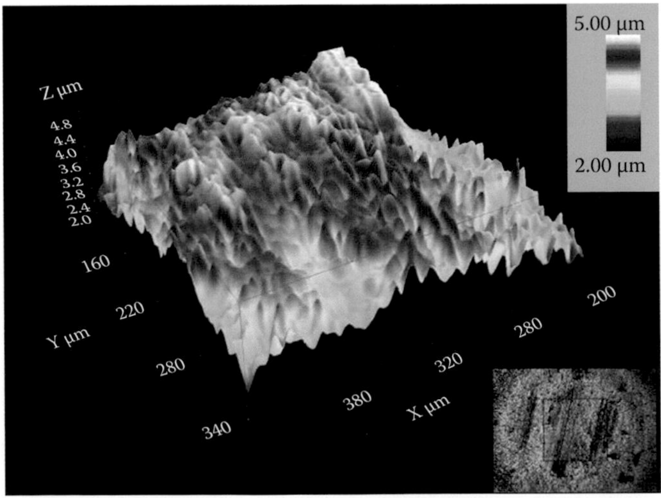

FIGURE 5.17 Surface morphologies of the wear scars after 1-h ball-on-disc test at $p = 1.68$ GPa, $T = 25°C$, and $v = 0.5$ m/s with SN150 base oil.

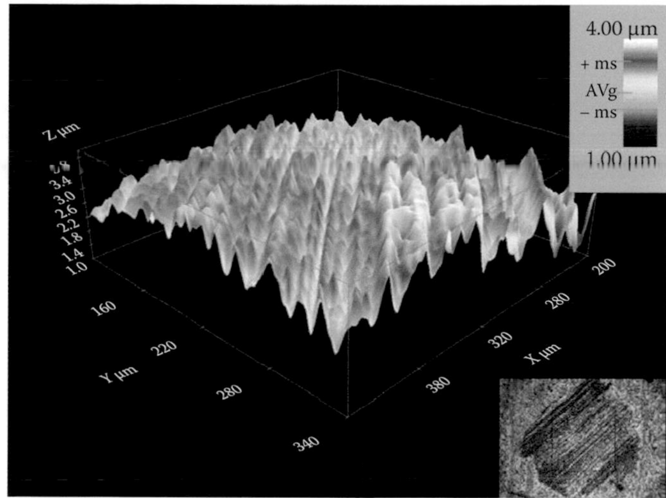

FIGURE 5.18 Surface morphologies of the wear scars after 1-h ball-on-disc test at p = 1.68 GPa, T = 25°C, and v = 0.5 m/s with base GO in SN150 oil at 0.1 wt%.

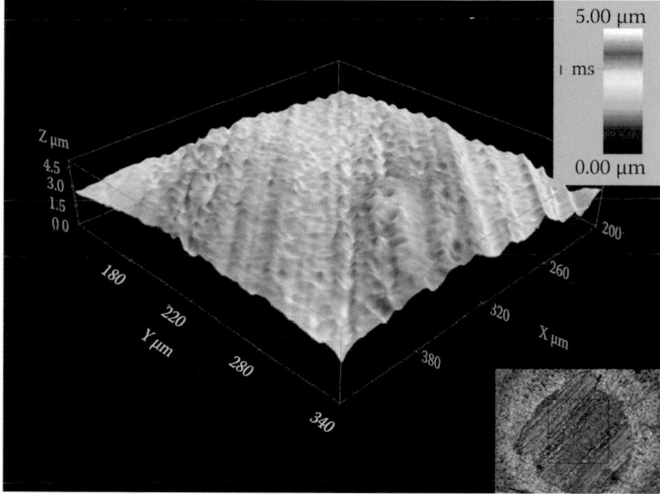

FIGURE 5.19 Surface morphologies of the wear scars after 1-h ball-on-disc test at p = 1.68 GPa, T = 25°C, and v = 0.5 m/s with base GO in SN150 oil at 0.1 wt%: wear scar analysis after thermal oxidation to remove GO.

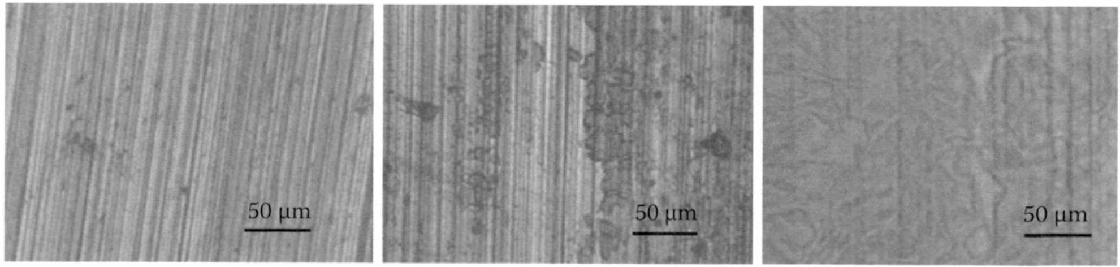

FIGURE 5.20 Optical micrographs of the transfer films on the ball surface. Left: pure water; center: water with 0.5 wt% of CNTs-COOH; right: water with 0.1 wt% of GO nanosheets. (Modified from Song, H.-J., Li, N., *Appl. Phys. A Mat. Sci. Proc.*, 105, 827, 2011.)

Therefore, two mating wear surfaces could be better filled by the dispersed GO nanosheets during the frictional test, and then GO nanosheets on wear surface could serve as spacers, preventing rough contact between the two mating wear surfaces. About the interface frictional mechanism, the authors proposed the following explanation: As a first step, the GO nanosheets enter the contact with the water and roll between the two rubbing surfaces; then, during the sliding, because of the high contact pressure creating stressed zones of traction/compression, a thin physical tribofilm is formed on the metal substrate. Since the physical tribofilm could not bear the whole normal load acting on the tribopair, it only partially prevents the solid-to-solid direct contact of the two mating metal surfaces. The increase in concentration could improve the role of GO as spacer, in the limits already stated previously about the search for the optimal concentration (Figures 5.10 through 5.14).

5.9 Discussion

The incessant literature dispute on the friction reduction mechanism introduced by nanoparticles as lubricant additives finds in the following list the more convincing physical explanations: rolling–sliding "rigid" motions together with flexibility properties (Tevet et al. 2011), nanoadditive exfoliation and material transfer to metal surface to form the so-called "tribofilm" or "tribolayer" (Chauveau et al. 2012; Yadgarov et al. 2013), electronic effects in tribological interfaces (Seifert et al. 2000), surface roughness improvement effect or "mending" (Liu et al. 2004); along with the more classical hypothesis of surface sliding on lower shear stress layers due to weak interatomic forces, valid also for microscale additives used for decades.

The observation of the surface morphologies reveals that the addition of GO considerably leads to a smoother surface, by reducing the surface rms roughness: The tribopair surface sliding on lower shear stress layers due to weak interatomic forces of GO sheets allows modified interaction between the asperities of the tribopair steel parts even from the qualitative point of view. Exfoliated GO fills the deeper scratches and covers with a thinner film the surface peaks.

The experimental evidences provided by several papers represent proof of the presence of a tribological film of reduced GO, which covers the whole worn surface of the steel ball, together with a mending effect (Wintterlin and Bocquet 2009; Lin et al. 2011; Sarno et al. 2013).

The since-start reduction of friction coefficient exhibited in the experiments outcomes in Senatore et al. (2013) can be addressed to the surface rubbing through the low shear stress GO layers.

The GO concentration equal to 0.1 wt% is quite lower than the usual one for inorganic nanoadditives (Huang et al. 2005; Zou et al. 2006; Kalin et al. 2012; Yadgarov et al. 2013) and in line with optimal loading of CNTs (Martin and Ohmae 2008). This feature is welcomed in the preparation of fully formulated engine or gearbox lubricants since the GO addition could only slightly modify the delicate equilibrium achieved by oil manufacturers between the essential and ubiquitous additives as antioxidant, viscosity modifier, pour-point depressant, and other minors.

The excellent properties as friction modifier and antiwear agent of graphene nanosheets are attributed to their small structure and extremely thin laminated structure, which allows easy entry in the contact area with the ability to deposit a more or less continuous protective film that prevents direct contact with the rough face. The microscale flakes of graphite are unable to equate the graphene result because of the larger size of their lamellae.

5.10 Conclusions

This chapter, after an overview of the results of last decades' research that deals with the inclusion of nanoparticles in oil lubricant as novel friction modifier materials, introduces the main benefits coming from nanocarbons in tribological applications. Among these materials, graphene platelets are today the focus of interest in studies about frictional interfacial mechanism owing to their weak interatomic interactions between the layers and ensuing low-strength shearing. The huge effort about the characterization of graphene and graphene nanosheets by way of advanced tools such as SEM, TEM, TG-DTG, Raman, along with successful synthesis routes, has been briefly summarized.

The excellent tribological behaviors of graphene and GO nanosheets have been proven in a broad range of operating conditions of rubbing surfaces and lubrication regimes, through a different principle of laboratory test, tribopair materials, liquid media, functionalization, and dispersion method, etc. Graphene and GO nanosheets are able to form a protective layer to lower frictional losses as well as material removing for progressive wear antifriction on the surfaces of the tribological conjunction at effective concentration lower than other nanocarbons and metal dichalcogenide-based nanoparticles. The two mating surfaces composing the sliding contact could be better filled by the dispersed nanosheets during the frictional test in comparison with microscale graphite additive. The nanosheets serve as spacers, preventing rough contact between the metal surfaces: They enter the contact with the liquid media, impact, and roll between the two rubbing surfaces. Because of the high contact pressure creating stressed zones of traction/compression in the central area of the contact, a thin physical tribofilm is formed on the metal substrate. This tribofilm could not account for the whole normal load acting on the tribopair and only partially prevents the direct contact of the two mating metal surfaces. Thus, the increase in concentration could improve the role of graphene and GO nanosheets, but it is upper-limited by the oil film discontinuities ensuing the excessive presence of sheets in the contact narrow leading to a nonnegligible share of solid-to-solid dry friction phenomenon.

Acknowledgment

The authors gratefully acknowledge Dr. Claudia Cirillo for helpful discussion and precious support.

References

Ajayan, P.M., Schadler, L.S., Giannaris, C. et al. Graphene oxide/polybenzimidazole composites fabricated by a solvent-exchange method. *Advanced Materials* 12(10):750, 2011.

Altavilla, C., Ciambelli, P., Sarno, M. et al. Tribological and rheological behaviour of lubricating greases with nanosized inorganic based additives. In: *Proceedings of the 3rd European Conference on Tribology Ecotrib 2011 and 4th Vienna International Conference on Nano-Technology VIENNANO 2011*, The Austrian Tribology Society, Vienna, 2011a, 903.

Altavilla, C., Sarno, M., and Ciambelli, P. A novel wet chemistry approach for the synthesis of hybrid 2D free-floating single or multilayer nanosheets of MS2@oleylamine (M = Mo, W). *Chemistry of Materials* 23:3879, 2011b.

Altavilla, C., Sarno, M., Ciambelli, P. et al. New "chimie douce" approach to the synthesis of hybrid nanosheets of MoS_2 on CNT and their anti-friction and anti-wear properties. *Nanotechnology* 24(12):125601, 2013.

Bai, X., Wan, C.Y., Zhang, Y. et al. Reinforcement of hydrogenated carboxylated nitrile–butadiene rubber with exfoliated graphene oxide. *Carbon* 49(5):1608–1613, 2011.

Bar-Sadan, M. and Tenne, R. Inorganic nanotubes and fullerene-like structures. In: *Inorganic Nanoparticles: Synthesis, Applications and Perspectives*, eds. C. Altavilla and E. Ciliberto, CRC Press, Boca Raton, FL, 2011, Ch 17.

Bhushan, B. and Gupta, B.K. *Handbook of Tribology*, McGraw-Hill, New York, 1991.

Black, A.L., Dunster, R.W., and Sanders, J.V. Comparative study of surface deposits and behaviour of MoS_2 particles and molybdenum dialkyl-dithio-phosphate. *Wear* 13(2):119–132, 1969.

Bolotin, K.I., Sikes, K.J., Jiang, Z. et al. Ultrahigh electron mobility in suspended grapheme. *Solid State Communications* 146:351–355, 2008.

Bowden, F.P. and Tabor, D. *The Friction and Lubrication of Solids*, Oxford University Press, New York, 1950.

Bowden, F.P. and Tabor, D. *The Friction and Lubrication of Solids*, Part II, Oxford University Press, London, 1964.

Brown, S., Musfeldt, J.L., Mihut, I. et al. Bulk vs nanoscale WS_2: Finite size effects and solid-state lubrication. *Nano Letters* 7:2365, 2007.

Bryant, P.J., Gutshall, P.L., and Taylor, L.H. A study of mechanisms of graphite friction and wear. *Wear* 7:118–126, 1964.

Chauveau, V., Mazuyer, D., Dassenoy, F. et al. In situ film-forming and friction-reduction mechanisms for carbon-nanotube dispersions in lubrication. *Tribology Letters* 47:467–480, 2012.

Chen, C.S., Chen, X.H., Xu, L.S. et al. Modification of multi-walled carbon nanotubes with fatty acid and their tribological properties as lubricant additive. *Carbon* 43:1660–1666, 2005.

Chhowalla, M. and Amaratunga, G.A.J. Thin films of fullerene-like MoS_2 nanoparticles with ultra-low friction and wear. *Nature* 407:164–167, 2000.

Church, A.H., Zhang, X.F., Sirota, B. et al. Carbon nanotube-based adaptive solid lubricant composites. *Advanced Science Letters* 5(1):188–191, 2012.

D'Agostino, V., Senatore, A., Petrone, V. et al. Tribological behaviour of graphene nanosheets as a lubricant additive. In: *Proceedings of the 15th International Conference on Experimental Mechanics ICEM15*, Porto, Portugal, July 22–27, 2012.

Ding, S.J., Chen, J.S., and Lou, X.W. Growth of MoS_2 nanosheets on CNTs for lithium storage. *Chemistry—A European Journal* 17:13142–13145, 2011.

Eswaraiah, V., Sankaranarayanan, V., and Ramaprabhu, S. Graphene-based engine oil nanofluids for tribological applications. *ACS Applied Materials & Interfaces* 3:4221, 2011.

Frey, G.L., Elani, S., Homyonfer, M. et al. Optical-absorption spectra of inorganic fullerene like MS_2 (M = Mo,W). *Physical Review B* 57:6666–6671, 1998.

Fusaro, R.L. Mechanisms of graphite fluoride [(CFx)n] lubrication. *Wear* 53:303–323, 1979.

Gansheimer, J. and Holinski, R. A study of solid lubricants in oils and greases under boundary conditions. *Wear* 19:439–449, 1972.

Geim, A.K. Graphene: Status and prospects. *Science* 324(5934):1530, 2009.

Haiping, H., Dustin, T., Waynick, A. et al. Carbon nanotube grease with enhanced thermal and electrical conductivities. *Journal of Nanoparticle Research* 12:529–535, 2010.

Hilton, M.R., Bauer, R., Didziulis, S.V. et al. Structural and tribological studies of MoS_2 solid lubricant films having tailored metal-multilayer nanostructures. *Surface and Coatings Technology* 53:13–23, 1992.

Hou, X., Shan, C., and Choy, K.L. Microstructures and tribological properties of PEEK-based nanocomposite coatings incorporating inorganic fullerene-like nanoparticles. *Surface and Coatings Technology* 202:2287–2291, 2008.

http://graphene-flagship.eu/. (official website of the EU's research initiative "Graphene Flagship").

Hu, K.H., Liu, M., Wang, Q.J. et al. Tribological properties of molybdenum disulfide nanosheets by monolayer restacking process as additive in liquid paraffin. *Tribology International* 42(1):33–39, 2009.

Hu, K.H., Hu, X.G., Xu, Y.F. et al. The effect of morphology on the tribological properties of MoS_2 in liquid paraffin. *Tribology Letters* 40:155–165, 2010.

Huang, H., Tu, J., Zou, T. et al. Friction and wear properties of IF-MoS_2 as additive in paraffin oil. *Tribology Letters* 20:247–250, 2005.

Huang, H.D., Tu, J.P., Gan, L.P. et al. An investigation on tribological properties of graphite nanosheets as oil additive. *Wear* 261:140–144, 2006.

Hummers, W. and Offeman, R. Preparation of graphitic oxide. *Journal of the American Chemical Society* 80:1339–1342, 1958.

Jang, B.Z. and Zhamu, A. Processing of nanographene platelets (NGPs) and NGP nanocomposites: A review. *Journal of Materials Science* 43(15):5092–5101, 2008.

Jeong-Hui, K., Hyo-Jun, A., Min-Sang, J. et al. The electrochemical properties of Li/TEGDME/MoS_2 cells using multi-wall carbon nanotubes as a conducting agent. *Research on Chemical Intermediates* 36:749, 2010.

Jia, J.H., Chen, J.M., Zhou, H.D. et al. Friction and wear properties of bronze–graphite composite under water lubrication. *Tribology International* 37(5):423–429, 2004.

Joly-Pottuz, L., Dassenoy, F., Belin, M. et al. Ultralow-friction and wear properties of IF-WS_2 under boundary lubrication. *Tribology Letters* 18:477–485, 2005.

Joly-Pottuz, L., Vacher, B., Ohmae, N. et al. Anti-wear and friction reducing mechanisms of carbon nano-onions as lubricant additives. *Tribology Letters* 30:69–80, 2008.

Jung, J.H., Jeon, J.H., Sridhar, V. et al. Electro-active graphene-Nafion actuators. *Carbon* 49(4):1279, 2011.

Kalin, M. and Vižintin, J. The tribological performance of DLC-coated gears lubricated with biodegradable oil in various pinion/gear material combinations. *Wear* 259(7–12):1270–1280, 2005.

Kalin, M., Vižintin, J., Barriga, J. et al. The effect of doping elements and oil additives on the tribological performance of boundary-lubricated DLC/DLC contacts. *Tribology Letters* 17(4):679–688, 2004.

Kalin, M., Velkavrh, I., and Vizintin, J. The Stribeck curve and lubrication design for non-fully wetted surfaces. *Wear* 267(5–8):1232–1240, 2009.

Kalin, M., Kogovsek, J., and Remskar, M. Mechanisms and improvements in the friction and wear behavior using MoS_2 nanotubes as potential oil additives. *Wear* 280–281:36–45, 2012.

Kalin, M., Kogovšek, J., Kovač, J. et al. The formation of tribofilms of MoS_2 nanotubes on steel and DLC-coated surfaces. *Tribology Letters* 55(3):381–391, 2014.

Katz, A., Redlich, M., Rapoport, L. et al. Self-lubricating coatings containing fullerene-like WS_2 nanoparticles for orthodontic wires and other possible medical applications. *Tribology Letters* 21:135–139, 2006.

Kim, H. and Macosko, C.W. Processing–property relationships of polycarbonate/graphene composites. *Polymer* 50(15):3797, 2009.

Koroteev, V.O., Okotrub, A., Mironov, Y.V. et al. Growth of MoS_2 layers on the surface of multiwalled carbon nanotubes. *Inorganic Materials* 43:236, 2007.

Kroto, H.W., Health, J.R., O'Brien, S.C. et al. C_{60}: Buckminsterfullerene. *Nature* 318:162–163, 1985.

Kudin, K.N., Ozbas, B., Schniepp, H.C. et al. Raman spectra of graphite oxide and functionalized graphene sheets. *Nano Letters* 8(1):36–41, 2008.

Laursen, A.B., Kegnæs, S., and Dahl, S. Molybdenum sulfides—efficient and viable materials for electro- and photoelectrocatalytic hydrogen evolution. *Energy & Environmental Science* 5:5577, 2012.

Lee, C., Wei, X.D., Kysar, J.W. et al. Measurement of the elastic properties and intrinsic strength of monolayer grapheme. *Science* 321:385–388, 2008.

Lee, K., Hwang, Y., Cheong, S. et al. Performance evaluation of nano-lubricants of fullerene nanoparticles in refrigeration mineral oil. *Current Applied Physics* 9:e128–e131, 2009.

Lee, C., Li, Q.Y., Kalb, W. et al. Frictional characteristics of atomically thin sheets. *Science* 328(5974):76, 2010.

Levy, M., Albu-Yaron, A., Tenne, R. et al. Synthesis of inorganic fullerene-like nanostructures by concentrated solar and artificial light. *Israel Journal of Chemistry* 50:417, 2010.

Li, J. and Bai, T. The effect of CNT modification on the mechanical properties of polyimide composites with and without MoS_2. *Mechanics of Composite Materials* 47:597, 2011.

Li, R., Liu, C.H., and Ma, J. Studies on the properties of graphene oxide-reinforced starch biocomposites. *Carbohydrate Polymers* 84(1):631, 2011.

Li, Y., Wang, Q., Wang, T. et al. Preparation and tribological properties of graphene oxide/nitrile rubber nanocomposites. *Journal of Materials Science* 47:730–738, 2012.

Lin, J., Wang, L., and Chen, G. Modification of graphene platelets and their tribological properties as a lubricant additive. *Tribology Letters* 41:209–215, 2011.

Liu, G., Li, X., Qin, B. et al. Investigation of the mending effect and mechanism of copper nano-particles on a tribologically stressed surface. *Tribology Letters* 17(4):961–966, 2004.

Ma, L., Chen, W.-X., Xu, Z.-D. et al. Carbon nanotubes coated with tubular MoS_2 layers prepared by hydrothermal reaction. *Nanotechnology* 17:571, 2006.

Martin, J.M. and Ohmae, N. *Nanolubricants*, Wiley & Sons, Chichester, England, 2008.

Meng, H., Sui, G.X., Xie, G.Y. et al. Friction and wear behavior of carbon nanotubes reinforced polyamide 6 composites under dry sliding and water lubricated condition. *Composites Science and Technology* 69(5):606, 2009.

Miura, K. *Encyclopedia of Nanoscience and Nanotechnology*, vol. 9, ed. H.S. Nalwa, American Scientific Publishers, Valencia, CA, 2004, p. 947.

Naffakh, M., Martin, Z., Fanegas, N. et al. Influence of inorganic fullerene-like WS_2 nanoparticles on the thermal behavior of isotactic polypropylene. *Journal of Polymer Science Part B: Polymer Physics* 45:2309–2321, 2007.

Nair, R.R., Blake, P., Grigorenko, A.N. et al. Fine structure constant defines visual transparency of graphene. *Science* 320:1308, 2008.

Novoselov, K.S., Geim, A.K., Morozov, S.V. et al. Electric field effect in atomically thin carbon films. *Science* 306(5696):666–669, 2004.

Ou, J.F., Wang, J.Q., Liu, S. et al. Tribology study of reduced graphene oxide sheets on silicon substrate synthesized via covalent assembly. *Langmuir* 26(20):15830–15836, 2010.

Ou, J., Liu, L., Wang, J. et al. Fabrication and tribological investigation of a novel hydrophobic polydopamine/graphene oxide multilayer film. *Tribology Letters* 48(3):407–415, 2012.

Pan, Y.Z., Wu, T.F., Bao, H.Q. et al. Green fabrication of chitosan films reinforced with parallel aligned graphene oxide. *Carbohydrate Polymers* 83(4):1908, 2011.

Perfiliev, V., Moshkovith, A., Verdyan, A. et al. A new way to feed nanoparticles to friction interfaces. *Tribology Letters* 21:89–93, 2006.

Qianming, G., Dan, L., Zhi, L. et al. Chapter 10. Tribology properties of carbon nanotube-reinforced composites. *Tribology of Polymeric Nanocomposites* 55:1–551, 2008.

Qiu, J.J. and Wang, S.R. Enhancing polymer performance through graphene sheets. *Journal of Applied Polymer Science* 119(6):3670, 2011.

Radisavljevic, B., Radenovic, A., Brivio, J. et al. Single-layer MoS_2 transistors. *Nature Nanotechnology* 6(3):147–150, 2011.

Ramanathan, T., Abdala, A.A., Stankovich, S. et al. Functionalized graphene sheets for polymer nanocomposites. *Nature Nanotechnology* 3:327–331, 2008.

Rapoport, L., Nepomnyashchy, O., Verdyan, A. et al. Polymer nanocomposites with fullerene-like solid lubricant. *Advanced Engineering Materials* 6:44–48, 2004.

Rosentsveig, R., Gorodnev, A., Feuerstein, N. et al. Fullerene-like MoS_2 nanoparticles and their tribological behavior. *Tribology Letters* 36:175, 2009.

Sarno, M., Senatore, A., Cirillo, C., Petrone, V., and Ciambelli, P. Oil lubricant tribological behaviour improvement through dispersion of few layer graphene oxide. *Journal of Nanoscience and Nanotechnology*, 14(7):4960–4968, 2014.

Seifert, G., Terrones, H., Terrones, M. et al. Structure and electronic properties of MoS_2 nanotubes. *Physical Review Letters* 85:146–149, 2000.

Senatore, A., D'Agostino, V., Petrone, V. et al. Graphene oxide nanosheets as effective friction modifier for oil lubricant: Materials, methods and tribological results. *ISRN Tribology* 2013:425809, 2013.

Shang, H., Liu, C., Xu, Y. et al. States of carbon nanotube supported Mo-based HDS catalysts. *Fuel Processing Technology* 88:117, 2007.

Shneider, M., Dodiuk, H., Kenig, S. et al. The effect of tungsten sulfide fullerene-like nanoparticles on the toughness of epoxy adhesives. *Journal of Adhesion Science and Technology* 24:1083–1095, 2010.

Shu, L., Feng, Y., and Yang, X. Influence of adding carbon nanotubes and graphite to Ag-MoS_2 composites on the electrical sliding wear properties. *Acta Metallurgica Sinica* 23(1):27–34, 2010.

Song, H.-J. and Li, N. Frictional behavior of oxide graphene nanosheets as water-base lubricant additive. *Applied Physics A: Materials Science & Processing* 105:827, 2011.

Song, X.C., Zheng, Y.F., Zhao, Y. et al. Hydrothermal synthesis and characterization of CNT@MoS_2 nanotubes. *Materials Letters* 60:2346–2348, 2006.

Soon, J.M. and Loh, K.P. Electrochemical double-layer capacitance of MoS_2 nanowall films. *Electrochemical and Solid-State Letters* 10:A250, 2007.

Srinath, G. and Gnanamoorthy, R. Sliding wear performance of polyamide 6–clay nanocomposites in water. *Composites Science and Technology* 67(3–4):399, 2007.

Staiger, M., Rafailov, P., Gartsman, K. et al. Excitonic resonances in WS_2 nanotubes. *Physical Review B* 86:165423, 2012.

Tenne, R., Margulis, L., Genut, M. et al. Polyhedral and cylindrical structures of WS_2. *Nature* 360:23, 1992.

Tevet, O., Von-Huth, P., Popovitz-Biro, R. et al. Friction mechanism of individual multilayered nanoparticles. *Proceedings of the National Academy of Sciences* 108(50):19901–19906, 2011.

Tian, J. and Xue, Q.J. The deintercalation effect of FeCl3-graphite intercalation compound in paraffin liquid lubrication. *Tribology International* 30:571–574, 1997.

Tibbetts, G.G., Lake, M.L., Strong, K.L. et al. A review of the fabrication and properties of vapor-grown carbon nanofiber/polymer composites. *Composites Science and Technology* 67(7–8):1709, 2007.

Tzeng, Y. Very low friction for diamond sliding on diamond in water. *Applied Physics Letters* 63:3586, 1993.

Verdejo, R., Bernal, M.M., Romasanta, L.J. et al. Graphene filled polymer nanocomposites. *Journal of Materials Chemistry* 21:3301–3310, 2011.

Wang, Q. and Li, J.H. Facilitated lithium storage in MoS_2 overlayers supported on coaxial carbon nanotubes. *Journal of Physical Chemistry C* 111:1675–1682, 2007.

Wang, Y., Shi, Z., Fang, J. et al. Graphene oxide/poly-benzimidazole composites fabricated by a solvent-exchange method. *Carbon* 49(4):1199–1207, 2011.

Watanabe, S., Noshiro, J., and Miyake, S. Friction properties of WS2/MoS_2 multilayer films under vacuum environment. *Surface and Coatings Technology* 188–189:644–648, 2004.

Wintterlin, J. and Bocquet, M.L. Graphene on metal surfaces. *Surface Science* 603:1841–1852, 2009.

Yadgarov, L., Rosentsveig, R., Leitus, G. et al. Controlled doping of MS2 (M = W,Mo) nanotubes and fullerene-like nanoparticles. *Angewandte Chemie International Edition* 51:1148–1151, 2012.

Yadgarov, L., Petrone, V., Rosentsveig, R. et al. Tribological studies of rhenium doped fullerene-like MoS_2 nanoparticles in boundary, mixed and elasto-hydrodynamic lubrication conditions. *Wear* 297(1–2):1103–1110, 2013.

Yao, Y., Wang, X., Guo, J. et al. Tribological property of onion-like fullerenes as lubricant additive. *Materials Letters* 62:2524, 2008.

Zhang, X., Luster, B., Church, A. et al. Carbon nanotube-MoS_2 composites as solid lubricants. *ACS Applied Materials & Interfaces* 1(3):735–739, 2009.

Zhang, W., Zhou, M., Zhu, H. et al. Tribological properties of oleic acid-modified graphene as lubricant oil additive. *Journal of Physics D: Applied Physics* 44:205303, 2011.

Zhao, W., Tang, J., Puri, A. et al. Tribological properties of fullerenes C60 and C70 microparticles. *Journal of Materials Research* 11:2749, 1996.

Zin, V., Agresti, F., Barison, S. et al. Investigation on tribological properties of nanolubricants with carbon nano-horns as additives at different temperatures. In: *Proceedings of World Tribology Congress 2013*, Turin, Italy, September 8–13, 2013.

Zou, T., Tu, J., Huang, H. et al. Preparation and tribological properties of inorganic fullerene-like MoS_2. *Advanced Engineering Materials* 8:289–293, 2006.

6

Graphene Oxide Nanodisks and Nanodots

Sambandam Anandan

Muthupandian Ashokkumar

Abstract

Graphene and graphene-derived materials such as graphene oxide and exfoliated graphene sheets have generated an extraordinary amount of interest among researchers across virtually in all scientific disciplines. This is mainly due to the presence of covalently bound oxygen functional groups in graphene materials as well as hexagonal lattice arrangement of carbon atoms with very strong sp^2 hybridized bonds in a two-dimensional matrix offering remarkable structural, mechanical, and electronic properties. This chapter will cover the fundamental principles involved in synthesizing graphene oxide nanomaterials of different morphology with a particular emphasis on graphene nanodisks and nanodots and the applications of such materials in various areas.

6.1 Introduction

After the quill pen was replaced by pencil in the 18th century, graphite had its strong impact on information technology. Graphene (an integral part of graphite) and graphene-derived materials such as graphene oxide (GO) and exfoliated graphene sheets have generated an extraordinary amount of interest among researchers across virtually in all scientific disciplines. This is mainly because of the presence of covalent bound oxygen functional groups in graphene materials as well as a hexagonal lattice arrangement of carbon atoms with very strong sp^2 hybridized bonds in a two-dimensional (2D) matrix offering remarkable structural, mechanical, and electronic properties. Even though graphene is the mother of all graphite forms including 0D fullerenes, 1D carbon nanotubes (CNTs) and 3D graphite (Figure 6.1) were given attention only after 2004, when Geim and Novoselov (2007) from Manchester University were awarded the 2010 Noble prize in physics for their groundbreaking experiments, i.e., isolation of graphene single-layer sheets from bulk graphite (Novoselov et al., 2004, 2007). However, looking back at graphene's history, the first observation was made by the British chemist Benjamin Brodie in (1859) by exposing graphite to strong acids and obtaining a new form of carbon with a molecular weight of 33 called carbonic acid, which he named "graphon" (Figure 6.2a). But it is a suspension of tiny crystals of GO that is a graphene sheet densely covered with hydroxyl and epoxide groups. Monolayers of the thinnest possible fragments

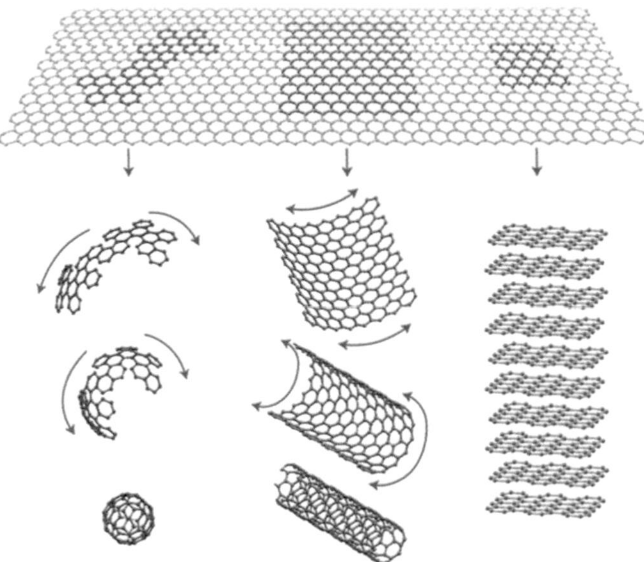

FIGURE 6.1 Graphene: mother of all graphitic forms. (Reprinted by permission from Macmillan Publishers Ltd. Geim, A.K., Novoselov, K.S., *Nat. Mater.*, 6, 183–191, 2007 Copyright 2007.)

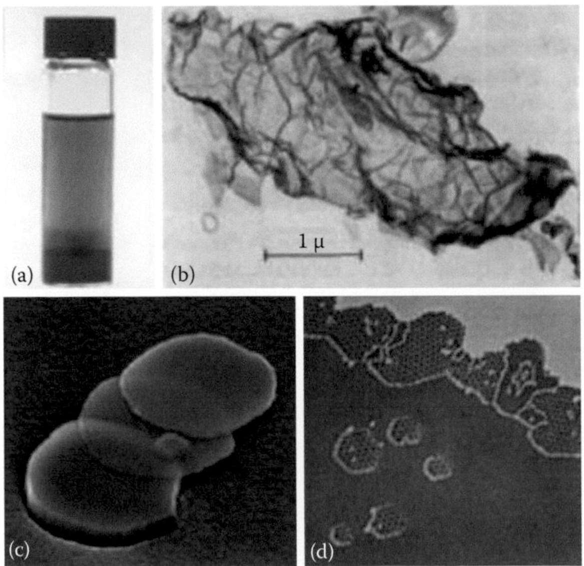

FIGURE 6.2 Prehistory of graphene. (a) Graphene as probably seen by Brodie 150 years ago. Graphite oxide at the bottom of the container dissolves in water, making the suspension of floating graphene flakes. (b) TEM image of ultrathin graphitic flakes from the early 1960s. (c) SEM image of thin graphite platelets produced by cleavage. (d) STM of graphene grown on Pt. The image is 100 × 100 nm^2 in size. The hexagonal superstructure has a period of ~22 Å and appears due to the interaction of graphene with the metal substrate. (Reprinted with permission from Geim, A.K., *Phys. Scr.* T146:014003, 2012. Copyright 2012, IOP Publishing.)

of reduced graphite oxide were identified under transmission electron microscopy (TEM) in 1962 by Boehm et al. (1962) (Figure 6.2b), and he introduced the term *graphene* by a combination of the word *graphite* and the suffix that refers to polycyclic aromatic hydrocarbons (Boehm et al., 1986); however, only after 40 years, researchers unambiguously identified the presence of graphene monolayers by counting the number of folding lines (Shioyama, 2001; Viculis et al., 2003; Horiuchi et al., 2004; Geim, 2012). Ruoff and his colleagues (Lu et al., 1999) observed thin graphite platelets (Figure 6.2c) under scanning electron microscopy (SEM), whereas Land et al. (1992) visualized thin graphitic films grown on metal substrates (Figure 6.2b) under scanning tunneling microscopy (STM). Likewise, graphite few-nanometer-thick "origami" was visualized under atomic force microscopy (AFM) on top of highly oriented pyrolytic graphite (HOPG) by Ebbesen and Hidefumi (1995). Furthermore, in 1990, peeling optically thin layers with transparent tape (scotch tape) were prepared from graphite by Kurz's group (Seibert et al., 1990). Novoselev et al. (2004) isolated graphene single-layer sheets from bulk graphite, which is large enough to do all sorts of measurements well beyond the scotch tape technique.

Single-layer graphene sheets have taken growing importance in solid-state physics (Sarrazin and Petit, 2014), and their remarkable properties make them a strategic material for future technologies probably known to solid-state realizations of a (2 + 1)-space times in which massless fermion live. This is because graphene is a one-atom-thick layer made up of sp^2 carbon atoms in a hexagonal lattice arrangement, and in the vicinity of the six corners (called Dirac points) of the 2D hexagonal Brillouin zone, the electric dispersion relation is linear for low energies (Figure 6.3). Electrons (and holes) can then be described by a Dirac equation for massless spin −1/2 particles in an effective (2 + 1)-space time. While massless Dirac fermions propagate at the speed of light in the (3 + 1) Minkowski space time, in graphene, the effective massless Dirac fermions propagate at the Fermi velocity ($v_f \approx 10^6$ ms^{-1}). Based on this theoretical point of view, exciton swapping is possible in graphene, which may be used to describe the concepts of low-dimensional electrodynamics and quantum dynamics, and in this context, the study of the specific features of graphene is of prime importance to develop new technological applications, which we are planning to demonstrate in this chapter.

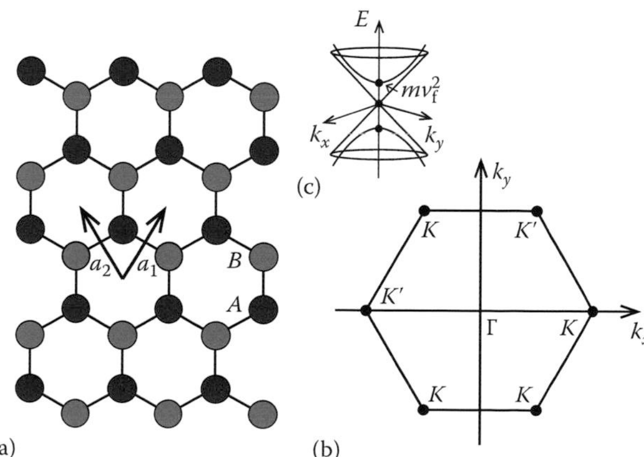

FIGURE 6.3 (a) Hexagonal lattice of graphene with the two sublattices *A* and *B*. a_1 and a_2 are the vectors of the unit cell. (b) Brillouin zone of the hexagonal lattice. (c) Energy behavior in the vicinity of the Dirac points *K* and *K*′. (With kind permission from Springer Science+Business Media: *Eur. Phys. J.*, Exciton swapping in a twisted graphene bilayer as a solid-state realization of a two-brane model, B87, 2014, 26.)

6.2 GO Nanoparticles of Different Morphologies

In recent years, the century-old GO synthesis (Bai and Shen, 2012) has taken a new turn after the emergence of both top–down and bottom–up approaches to isolate single graphene sheets. Planar graphene itself has been presumed not to exist in the free state, being unstable with respect to the formation of curved structures such as soot, fullerenes, and nanotubes. However, Novoselov et al. (2004) prepared few-layer graphene (FLG) sheets up to 10 μm in size by mechanical exfoliation of 1-μm-thick platelets of HOPG. That is, dry etching HOPG under oxygen plasma generates 5-μm-deep mesas on top of the platelets (mesas were squares of various sizes); upon baking, the mesas became attached to the photoresist layer and then were peeled using scotch tape flakes of graphite off the mesas. The films are found to be made of single-layer graphene sheets that exhibit electronic properties characteristic for a 2D semimetal upon developing multiterminal Hall bar devices (Figure 6.4).

For large-scale applications, chemical routes to fabricate graphene may offer significant advantages over the microcleaving of HOPG (Gan et al., 2003); however, such approaches require the use of a specific substrate material. Reina et al. (2009) followed a low-cost and scalable technique to fabricate large-area films of single- to few-layer graphene sheets, via ambient pressure chemical vapor deposition (CVD) on polycrystalline Ni films, that is, first by evaporating a 500-nm Ni film on a SiO_2/Si substrate followed by annealing at 900°C–1000°C under a flow rate of 600 sccm (mass flow unit) and 500 sccm of Ar and H_2, respectively, to generate Ni grains. Once again, CVD growth is carried out at 900°C or 1000°C under a flow rate of 5–25 sccm and 1500 sccm of CH_4 and H_2, respectively, for 5–10 min to generate graphene film. AFM, optical microscopy, and Raman spectroscopy (Figure 6.5) were used to confirm the relationship between the G to G′ intensity ratio ($I_G/I_{G'}$) and the number of graphene layers (one, two, and ~three layers) grown under CVD. Measurement of $I_G/I_{G'}$ ratio provides fast estimation of the number of layers under optical microscope. Height profiles are also shown (measured at the location indicated by the arrow in the AFM images), confirming the formation of one, two, and ~three layers of grapheme.

Ballistic conduction is one of the favorable electronic properties of CNTs, and unfortunately, incorporation of nanotubes in large-scale integrated electronic architectures proves to be so daunting that it

FIGURE 6.4 Graphene films. (a) Photograph (in normal white light) of a relatively large multilayer graphene flake with thickness ~3 nm on top of an oxidized Si wafer. (b) Atomic force microscope (AFM) image of 2 μm by 2 μm area of this flake near its edge. (c) AFM image of single-layer graphene. Notice the folded part of the film near the bottom, which exhibits a differential height of ~0.4 nm. (d) Scanning electron microscope image of one of our experimental devices prepared from FLG. (e) Schematic view of the device in (d). (From Novoselov, K.S., Geim, A.K., Morozov, S.V. et a l., *Science* 306: 666–669, 2004. Reprinted with permission from AAAS.)

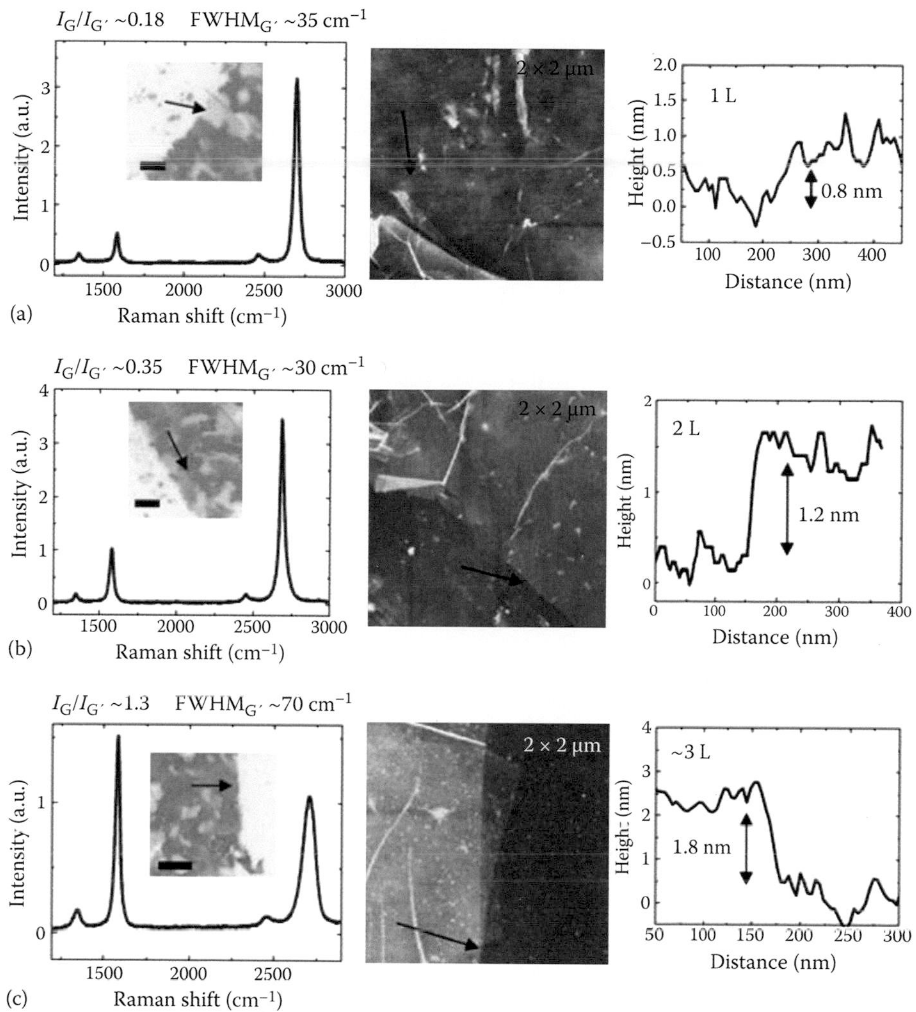

FIGURE 6.5 Raman spectra, optical, and AFM images of regions consisting of one (a), two (b), and ~three (c) graphene layers. Scale bars for optical images in the insets for the Raman spectra are 3 μm. (Reprinted with permission from Reina, A. et al. *Nano Lett.*, 9, 30–35, 2009. Copyright 2009 American Chemical Society.)

may never be realized. Harnessing these properties requires graphitic materials that are related to CNTs but that are more manageable. Precisely, these theoretical considerations led De Heer et al. (2007) to initiate experiments to grow graphene on single crystal silicon carbide specifically on the 0001 (silicon-terminated) and 0001 (carbon-terminated) faces of 4H- and 6H-SiC when crystals are heated to about 1300°C in ultrahigh vacuum, which is visible in STM (Figure 6.6).

Since CNTs are considered to be graphene nanosheets rolled up into seamless tubes, researchers think in other ways whether CNTs can be cleaved to form graphene nanosheets. In this aspect, Kosynkin et al. (2009) tried to isolate graphene nanosheets by suspending multiwalled CNTs (MWCNTs) in conc. H_2SO_4 followed by treatment with 500 wt% $KMnO_4$ at room temperature (22°C) for 1 h and for another 1 h by heating at 55°C–70°C. They proposed that the opening of the nanotubes appears to occur along a line, similar to the "unzipping" of graphite oxide. Also, they state that such cleavage occurs in a linear longitudinal cut or in a spiraling manner, depending upon the initial site

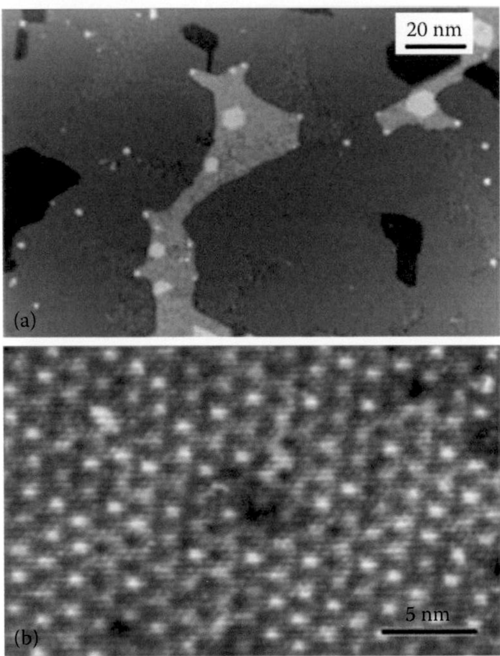

FIGURE 6.6 STM topographs (0.8 V sample bias, 100 pA) of nominally 1 ML epitaxial graphene on SiC(0001). (a) Image showing large flat regions of $6\sqrt{3} \times 6\sqrt{3}$ reconstruction and regions where the reconstruction has not fully formed. Next-layer islands are also seen. (b) A region of $6\sqrt{3} \times 6\sqrt{3}$ reconstruction, imaged through the overlying graphene layer. (Reprinted from *Solid. State. Comm.*, 143, De Heer, W.A., Berger, C., Wu, X. et al. Epitaxial graphene, 92–100, Copyright 2007, with permission from Elsevier.)

of attack and the chiral angle of the nanotube (Figure 6.7). Furthermore, such thin elongated strips of graphene possess straight edges, termed *graphene ribbons*, gradually transforming from semiconductors to semimetals as their width increases and representing a particularly versatile variety of graphene. The mechanism of opening is based on the oxidation of alkenes by permanganate in acid; i.e., the first step is the formation of manganate ester, which is the rate-determining step, followed by further oxidation to yield dione. The later occurs juxtaposition (an act or instance of placing close together or side by side) of the buttressing (a structure built against or protecting from a wall) ketones, which distorts the β-, γ-alkenes, and because of lessening of such buttressing-induced strain, there is more space for carbonyl projection, which initiates the tear of MWCNT from the end of the nanotube. Once the ketones turns into o-protonated form, the relief of bond angle strain takes place; hence, sequential bond cleavage occurs, which helps in opening of MWCNT nanotube to the graphene ribbon. Upon viewing under TEM, the prepared graphene ribbon size is found to be 40–80 nm in diameter, whereas the precursor MWCNT inner nanotube size is 15–20 nm. Single atomic layer is observed under AFM imaging, whereas 4-mm-long graphene nanoribbons are seen under SEM imaging (Figure 6.7).

A quantum dot is a nanometer-sized semiconductor particle where excitons are confined in all three spatial directions. Graphene consisting of a single atomic layer of graphite is a unique type of zero bandgap semiconductor with an infinite exciton Bohr radius because of a linear energy dispersion relation of the charge carriers. As a result, quantum confinement could take place in graphenes of any finite size and is expected to result in many interesting phenomena not obtainable in other semiconductor materials. Hence, Yan et al. (2010) were interested to synthesize uniform and size tunable large colloidal graphene quantum dots (GQDs) containing 168, 132, and 170 conjugated carbon atoms

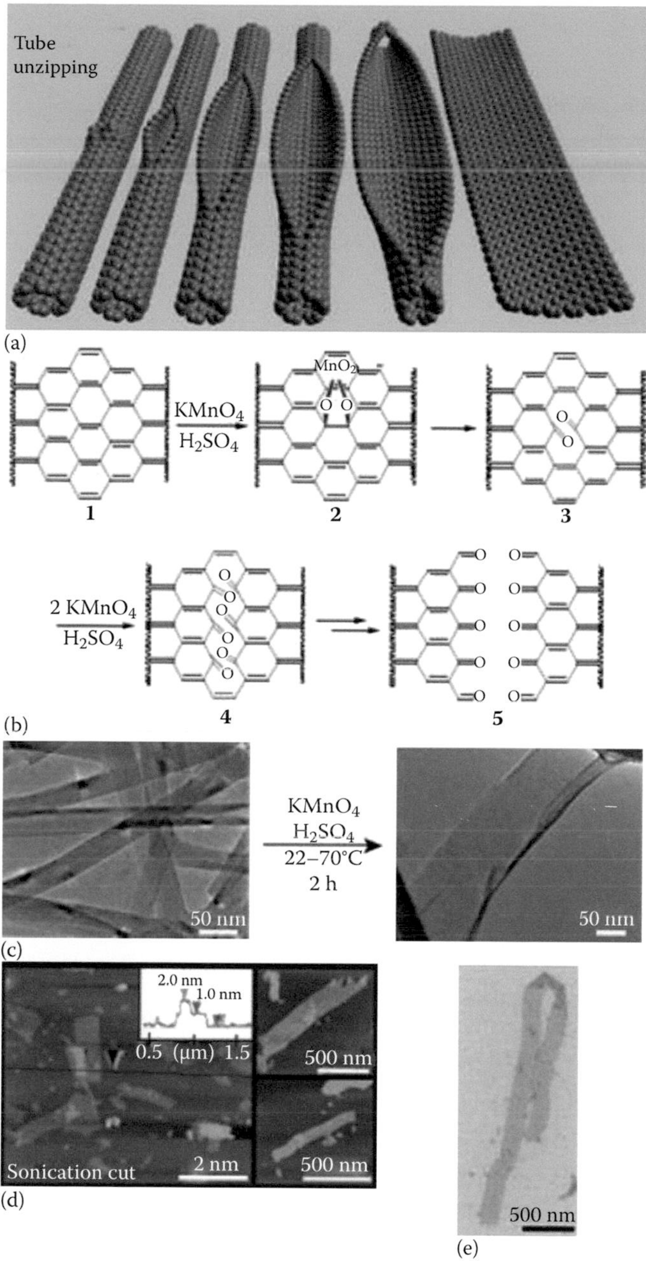

FIGURE 6.7 Nanoribbon formation and imaging. (a) Representation of the gradual unzipping of one wall of a carbon nanotube to form a nanoribbon. Oxygenated sites are not shown. (b) The proposed chemical mechanism of nanotube unzipping. The manganate ester in 2 could also be protonated. (c) TEM images depicting the transformation of MWCNTs (left) into oxidized nanoribbons (right). The right-hand side of the ribbon is partly folded onto itself. The dark structures are part of the carbon imaging grid. (d) AFM images of partly stacked multiple short fragments of nanoribbons that were horizontally cut by tip-ultrasonic treatment of the original oxidation product to facilitate spin-casting onto the mica surface. The height data (inset) indicates that the ribbons are generally single layered. The two small images on the right show some other characteristic nanoribbons. (e) SEM image of a folded, 4-mm-long single-layer nanoribbon on a silicon surface. (Reprinted by permission from Macmillan Publishers Ltd. Kosynkin, D.V. et al. *Nature* 458, 872–876, 2009. Copyright 2009.)

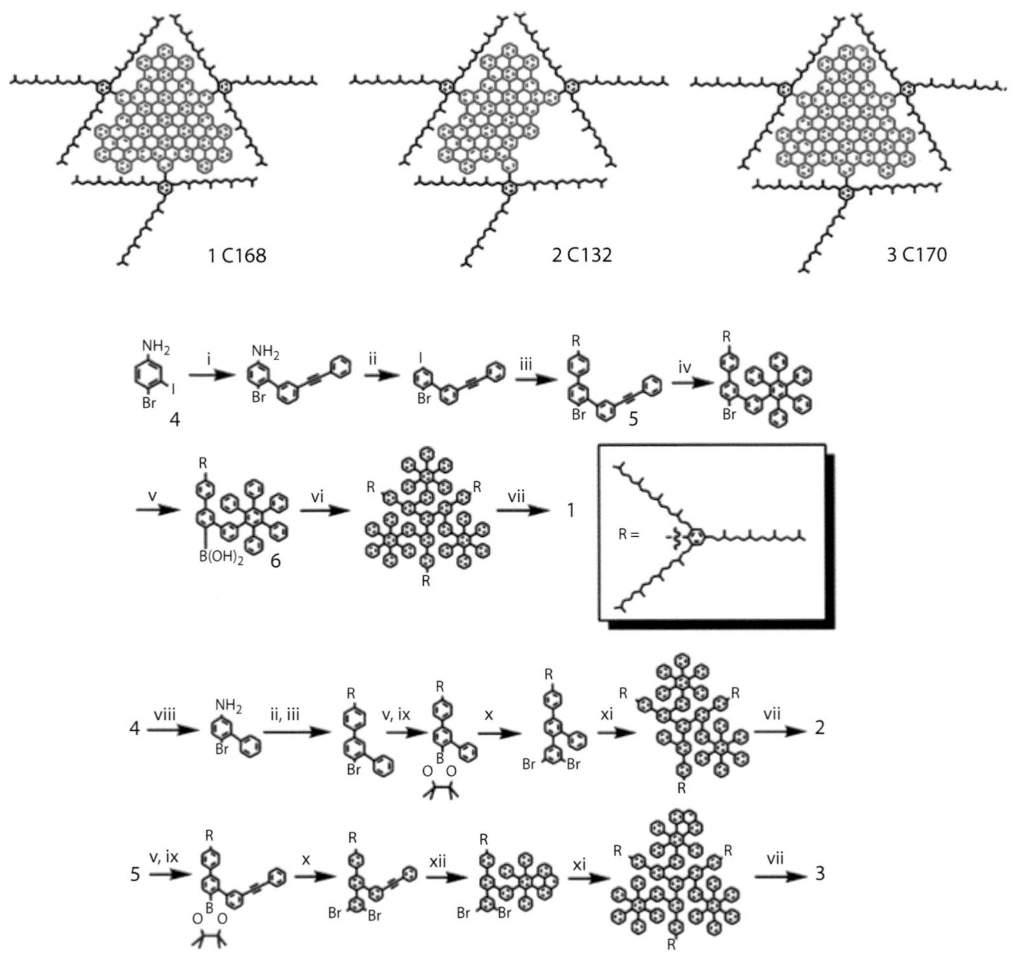

FIGURE 6.8 Scheme shows synthesis of graphene quantum dots 1–3. Conditions: (i) 3-(phenylethynyl)phenylboronic acid, $Pd(PPh_3)_4$, K_2CO_3, toluene, EtOH, H_2O, 80°C; (ii) I_2, *tert*-butyl nitrite, benzene, 5°C; (iii) 4-(2′,4′,6′-trialkylphenyl)phenylborate, $Pd(PPh_3)_4$, K_2CO_3, toluene, EtOH, H_2O, 80°C; (iv) tetraphenylcyclopentadienone, diphenyl ether, reflux; (v) (a) *n*-BuLi, THF, −78°C; (b) $B(i\text{-}PrO)_3$, (c) HCl, H_2O; (vi) 1,3,5-triiodobenzene, $Pd(PPh_3)_4$, K_2CO_3, H_2O, toluene, 80°C; (vii) $FeCl_3$, CH_2Cl_2, CH_3NO_2; (viii) phenylboronic acid, $Pd(PPh_3)_4$, K_2CO_3, toluene, EtOH, H_2O, 80°C; (ix) pinacol; (x) 1,3-dibromo-5-iodobenzene, $Pd(PPh_3)_4$, K_2CO_3, toluene, H_2O, EtOH, 60°C; (xi) 6, $Pd(PPh_3)_4$, K_2CO_3, toluene, H_2O, 80°C; (xii) 1,3-diphenylcyclopenta[*e*]pyren-2-one, diphenyl ether, reflux. (Reprinted with permission from Yan, X. et al., *J. Am. Chem. Soc.*, 132, 5944–5945, 2010. Copyright 2010 American Chemical Society.)

(Figure 6.8) through a solution chemistry approach, which is based on oxidative condensation reactions via a new solubilization strategy. The synthesis of such colloidal GQDs is outlined in Figure 6.8, starting from small molecule precursors such as 3-iodo-4-bromoaniline to generate polyphenylene dendritic precursors via Suzuki coupling reactions, which were then exposed to an excess of $FeCl_3$ in a dichloromethane/nitromethane mixture, yielding the GQDs.

In general, the colloidal suspension method stands out as the primary stratagem that can generate large amounts of graphene and, in addition, is best suited for chemical functionalization. Hence, Park et al. (2008) used KOH to produce aqueous homogeneous suspension containing chemically modified conducting graphene sheets from a precursor dispersion of GO in water (Figure 6.9). This is because KOH, being a strong base, confers a large negative charge by reacting with hydroxyl, epoxy, and carboxylic

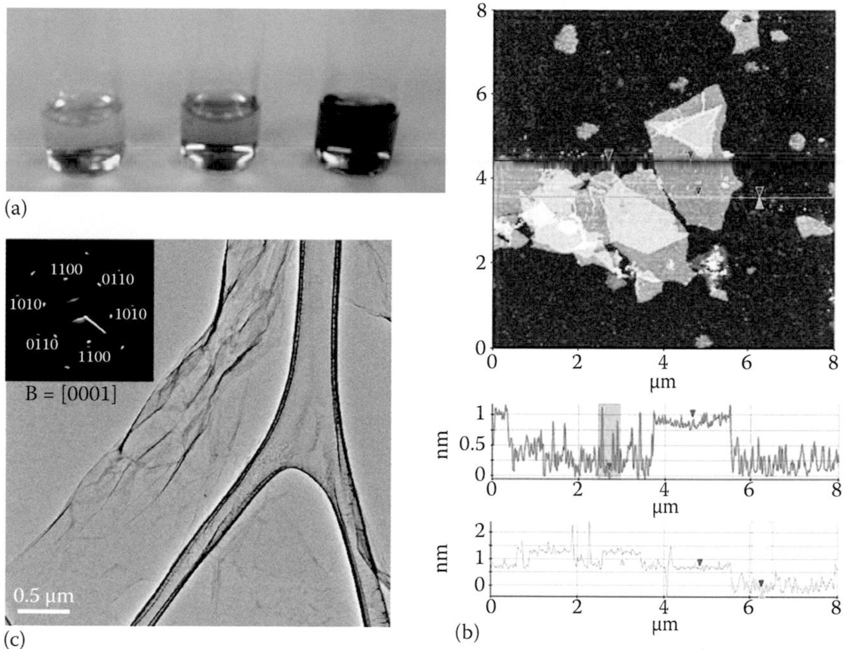

FIGURE 6.9 (a) Aqueous colloidal suspension from left: GO, K-modified GO, hKMG. (b) AFM image of hKMG sheets on a mica substrate. (c) BF TEM image of hKMG sheets; inset, selected area diffraction pattern of what were found to be two overlapping hKMG sheets. (Reprinted with permission from Park, S. et al., *Chem. Mater.*, 20, 6592–6594, 2008. Copyright 2008 American Chemical Society.)

acid groups present on the GO sheets, resulting in reduced GO (rGO) sheets, which appear slightly darker, and it is denoted as KMG (Figure 6.9). Further addition of hydrazine monohydrate to KMG and stirring at 35°C for 6 h generates hydrazine reduced KMG suspension (hKMG; darker suspension), which is stable for more than 4 months. From AFM and bright-field TEM (BF TEM) images, it is clear that hKMG sheets were in the range of several hundreds of nanometers to a few micrometers thickness, and from the selected area electron diffraction (SAED) pattern, it is clear that two overlapping hKMG sheets are found.

Likewise, similar chemical functionalization reactions of graphene layers have been attempted to prepare solubilized graphene sheets. Sometimes, such functionalization can also result in the loss of the intrinsic electronic properties upon the formation of sp^3 carbon centers, although in some cases, the electronic properties of the functionalized graphene could be recovered by simple reduction or annealing. Thus, covalent intercalation compounds of graphite such as graphite fluoride are appropriate starting materials for this approach, and recently, Worsley et al. (2007) produced soluble graphene layers starting from graphite fluoride by reaction with alkyl lithium reagents to form the product R-CFx, where x corresponds to the composition of the starting graphite fluoride, and R, to the alkyl group in the R-Li reagent used for functionalization (see Figure 6.10). Furthermore, they determined that the solubility and extinction coefficient of such products are found higher (0.54 gm/L and 7 L/gm.cm, respectively) than of functionalized graphene (0.5 gm/L and 3.3 L/gm.cm, respectively) prepared by other precursors (Niyogi et al., 2006). In addition, they restored the electronic properties by annealing the samples to do dealkylation.

Scaling up of graphene using an oven to heat large substrates becomes less energy efficient, and hence, a new approach has to be followed to yield graphene of quality comparable with or higher than that of current chemical vapor deposition technique. Piner et al. (2013) reported a route to synthesize graphene

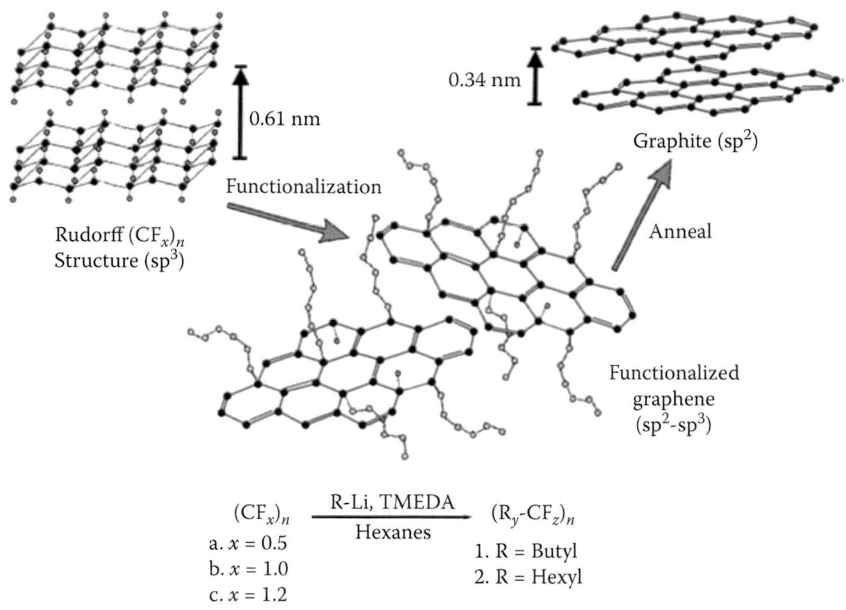

FIGURE 6.10 Scheme shows synthesis of soluble graphene from graphite fluoride. (Reprinted from *Chem. Phys. Lett.*, 445, Worsley, K.A., Ramesh, P., Mandal, S.K. et al., Soluble graphene derived from graphite fluoride, 51–56, Copyright 2007, with permission from Elsevier.)

by applying radio frequency (RF) magnetic fields to inductively heat copper foils. On applying RF induction, a free-standing conductor will have small eddy current induced on and near the metal surface (i.e., the metal may be heated directly without a hot environment) for large-scale and rapid manufacturing of graphene with much better energy efficiency. Figure 6.11 shows the heated copper foil in the presence of RF, and by this approach, copper foil can reach even 1035°C from room temperature within 2 min, which was monitored with an optical pyrometer. Formation of graphene film on copper substrate may be identified from SEM, Raman, and AFM. Further back-gated field effect transistors fabricated on a SiO_2/Si substrate showed carrier mobility up to ~14,000 $cm^2\,V^{-1}\,s^{-1}$ measured under ambient conditions.

Likewise, creation of 3D hierarchical superstructures via covalently interconnected nanoscale building blocks remains one of the fundamental challenges in nanotechnology. Sudeep et al. (2013) synthesized ordered, stacked macroscopic 3D solid scaffolds of polygraphene oxide via chemical cross-linking of 2D GO building blocks with glutaraldehyde (GAD) and resorcinol (Res) (Figure 6.12). That is, first, GAD interacts with GO via hydroxyl groups and then both GAD and Res interact to form a polycondensed product, which, upon controlled reduction using hydrazine, generates a reduced form of polygraphene oxide (poly-RGO), keeping the macroscopic 3D morphology, which is clearly visible in TEM. Furthermore, such 3D structures show superior properties toward gas storage applications, i.e., 2.7 mmol/g of CO_2 adsorbed at a pressure of 20 atm.

Similarly, a variety of graphene-based materials with different chemical structures and morphologies have been explored for various applications. For example, Kim T.Y. et al. (2013) prepared activated microwave-expanded graphite oxide as electrode materials for supercapacitor applications (Figure 6.13). The observed supercapacitor performance with these material is about 174 F g^{-1} at a current density of 8.4 A g^{-1}, attributed to their unique pore structure, which makes them potentially promising for diverse energy storage devices.

Unfortunately, prepared GO showed significant defects, which leads to degradation of its unique properties such as superior carrier mobility, mechanical strength, and chemical stability. Hence, other methods such as metal doping, organic compound hybridization, etc., have been performed to enhance

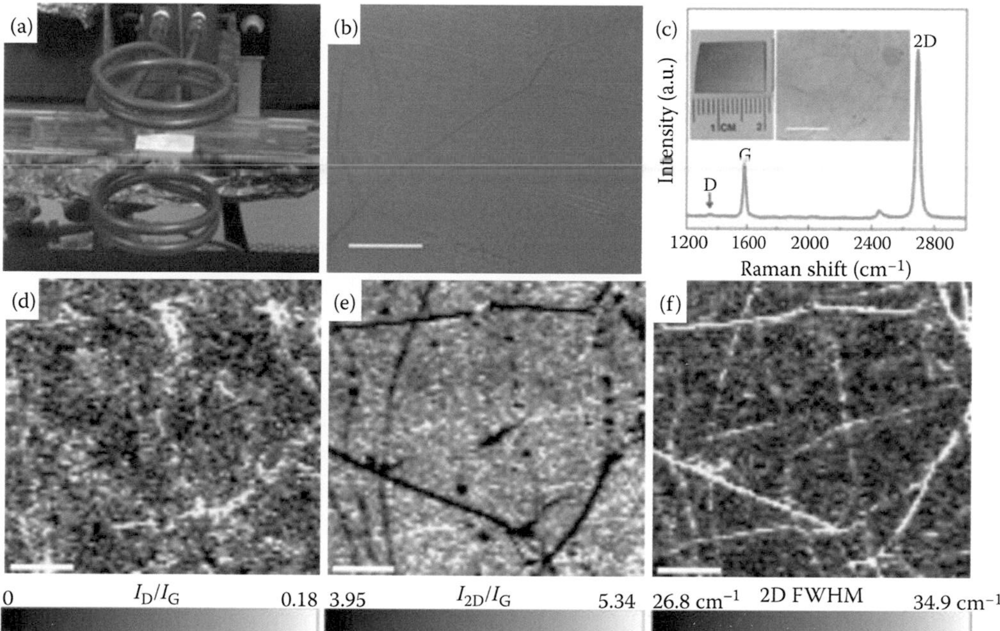

FIGURE 6.11 Photograph, SEM, and Raman characterization of RF graphene growth. (a) Photograph of a heated copper foil and RF coil. (b) SEM image of as-grown monolayer graphene on copper foil substrate, showing copper steps and graphene wrinkles. Scale bar is 3 μm. (c) Raman spectrum, G peak at 1582 cm^{-1}, 2D band at 2695 cm^{-1} (bandwidth ~30 cm^{-1}). Insets are photograph and optical microscopic image (scale bar is 20 μm) of transferred graphene on a SiO_2/Si substrate with 285-nm-thick thermal oxide layer. (d–f) Raman mapping of transferred graphene film. (d) Intensity ratio of D band (region from 1300 to 1400 cm^{-1}) to G band (1540 to 1640 cm^{-1}), I_D/I_G. (e) Intensity ratio of 2D band (2600 to 2800 cm^{-1}) to the G band, I_{2D}/I_G. (f) Full width at half-maximum of the 2D band. Dark lines in panel e and bright lines in panel f are from wrinkles. Scale bars are 6 μm. (Reprinted with permission from Piner, R. et al., *ACS Nano*, 7, 7495–7499, 2013. Copyright 2013 American Chemical Society.)

electrical properties. Kim K.Y. et al. (2013) followed a method to attach nickel nanoparticles to the oxygen functionalities of the GO surface, which acts as a catalyst for the repairing of defects in the GO sheets. It has been found that defects of the GO surfaces are significantly repaired by incorporating such nickel nanoparticles, which, in turn, results in high electrical conductivity of 18,620 S/m. That is, they first prepared GO–nickel nanoparticle composite by mixing the GO dispersion and $NiCl_2$ solution, followed by the addition of a reducing agent (hydrazine), which not only reduces the GO but also yields nickel hydrazine complex on the surface of the GO. After this, CVD was performed at 700°C in the presence of methane gas to generate multilayered graphene on nickel surface, which looks like graphene balls. Removal of nickel core by HCl solution was performed to yield repaired GO with multilayered graphene balls (repaired GO sheets/multilayered graphene ball [RGGB]) (Figure 6.14).

Most of the preparation methods involve reduction of GO and metal precursors using strong reducing agents (e.g., hydrazine). The disadvantage of such methods is the adsorption of reducing agents, leading to a lowering of their activity. Hence, the development of a new method for anchoring of metals on rGO under mild reduction conditions using minimal external reagents to get stable suspension is ideal. The sonochemical method offers such advantages for the coordinated reduction of metal precursors and GO to rGO to get a homogeneous dispersion of exfoliated graphene sheets with metal nanoparticles. In such a method, the shear forces generated by acoustic cavitation are enough to overcome the van der Waals forces between graphene sheets. Recently Vinodgopal et al. (2010) successfully prepared gold supported graphene sheets through sonochemical approach. Likewise, Anandan et al. (2012) prepared various nominal compositions of bimetallic supported graphene (Pt/Sn/rGO) nanoparticles, keeping

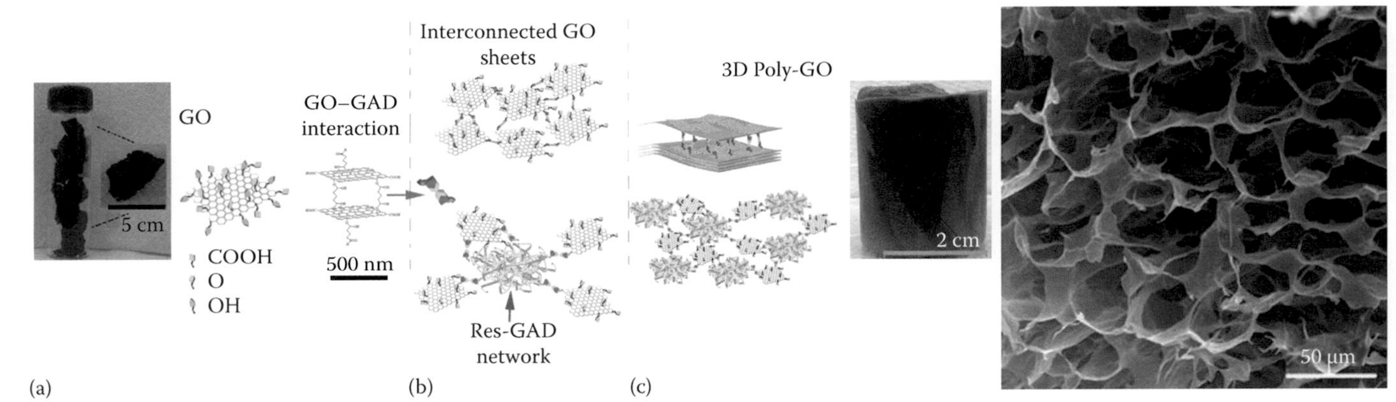

FIGURE 6.12 Schematic of the poly-GO synthesis process from initial GO powder. (a) Photograph showing the synthesized GO powders via improved technique (left); schematic of GO structure with main functional groups is shown (middle); a possible interaction of GAD with two neighboring GO sheets via the –OH groups in GO is also depicted. (b) Schematic of the linkage of GAD with two different –OH groups; schematic of two different mechanisms for large-scale GO cross-linking in water solution via pure GAD (top) and via GAD-Res (bottom). (c) Schematic of the 3D networked structure of poly-GO formed via two different mechanisms. Lyophilization (freeze-drying) of cross-linked solution leads to macroscopic solid structures with controllable shape and size. (Reprinted with permission from Sudeep, P.M. et al., *ACS Nano*, 7, 7034–7040, 2013. Copyright 2013, American Chemical Society).

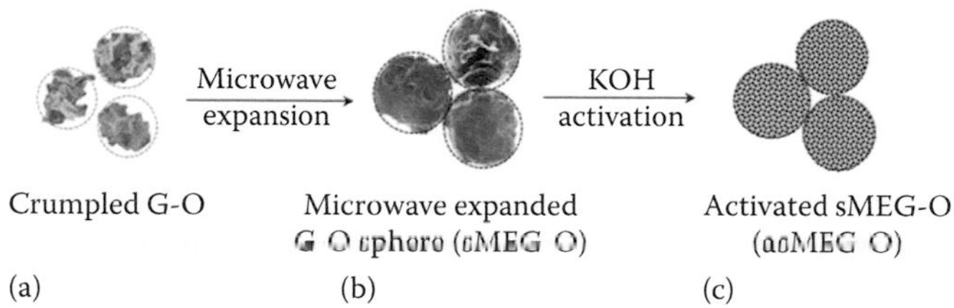

FIGURE 6.13 Scheme 1. Schematic of the fabrication of highly porous graphene-derived carbons with hierarchical pore structures: (a) GO sheets are transformed into crumpled ball-like G-O particles by aerosol spray drying technique, (b) under microwave irradiation, crumpled G-O particles form hollow graphene-based spheres, and (c) chemical activation with potassium hydroxide. (Reprinted with permission from Kim, T.Y. et al., *ACS Nano,* 7, 6899–6905, 2013. Copyright 2013, American Chemical Society.)

FIGURE 6.14 (a–c) Schematic illustration of the synthesis of repaired GO sheets/multilayered graphene ball 3D hybrids (RGGB). (Reprinted by permission from Macmillan Publishers Ltd. Kim, K.Y. et al., *Sci. Rep.*, 3, 3251, 2013. Copyright 2013.)

the graphene loading constant at 70%, to explore their electrochemical applications (direct methanol oxidation reaction). Figure 6.15 shows typical TEM images of simultaneously reduced hybrid nanoparticles with different atomic ratios of Pt/Sn dispersed on rGO sheets. Upon viewing the high-resolution TEM, the presence of both the Pt and Sn nanoparticles is seen heterogeneously on the rGO sheets. The SAED pattern showing polycrystalline ring pattern coupled with dotted pattern indicates the presence of both the metals, and in addition, energy-dispersive x-ray spectroscopy (EDX) also supports the same (Figure 6.15).

6.3 GO Nanodisks and Nanodots

Research is focused on tuning the bandgap of GO by modifying nanosheets into nanoribbons, hollow spheres, nanodisks, quantum dots, hybrid materials, etc. The high performance of modified GO sheets is a result of quantum confinement and edge effects (Ponomarenko et al., 2008; Guo et al., 2010; Zhu et al., 2011; Bai and Shen, 2012; Song et al., 2013).

Significant effort has been devoted to develop such self-assembled nanocomposites using various methods, including solution-based synthesis, solvothermal methods, electro-beam lithography, and ultrasonic spray pyrolysis; however, it is difficult to control size distribution by these methods (Novoselov et al., 2004; Ponomarenko et al., 2008). The development of new methods to generate uniform-size distribution under mild reduction conditions using minimal external reagents is a key challenge. In this regard, Guo et al. (2010) reported the synthesis of hollow GO spheres (HGOSs) using an emulsification process utilizing water-in-oil (W/O) without a surfactant. The formation process of HGOSs is illustrated in Figure 6.16, and it includes four steps: (1) delamination of graphite after intensive oxidation, (2) homogeneous mixture of graphene oxide nanosheet (GONs) and aqueous ammonia as well as the precipitation of large graphite oxide particles, (3) formation of a W/O emulsion containing GONs, and

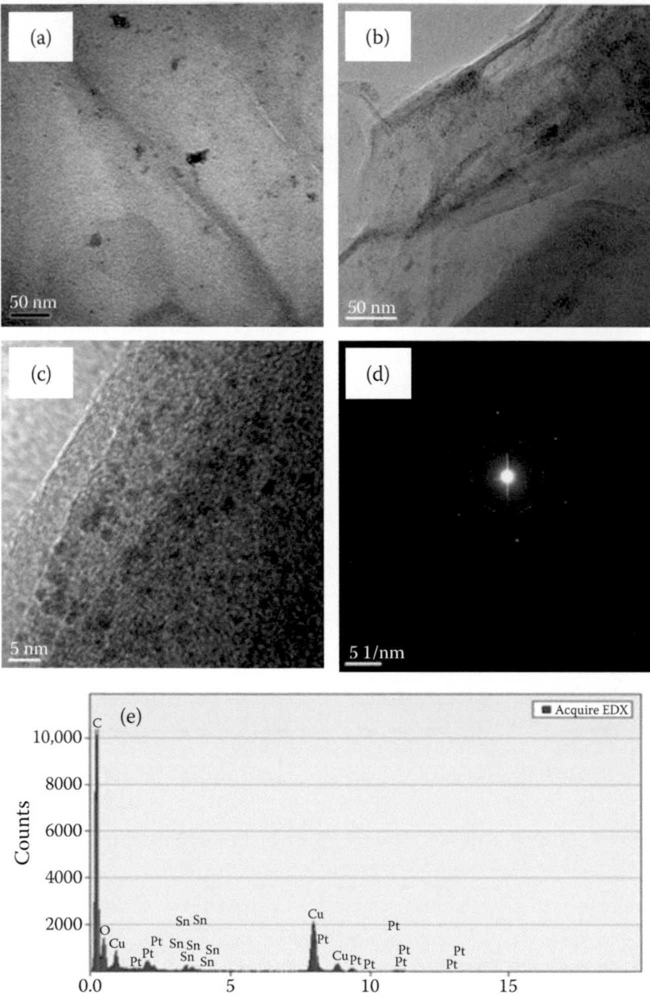

FIGURE 6.15 TEM images of sonochemically prepared hybrid samples: (a) Pt0.68/Sn0.32/rGO, (b) Pt0.46/Sn0.54/rGO, (c) HR-TEM of Pt0.68/Sn0.32/rGO, (d) corresponding SAED pattern of Pt0.68/Sn0.32/rGO O, (e) EDX spectrum of Pt0.68/Sn0.32/rGO. (Anandan, S., Manivel, A., Ashokkumar, M.: One-step sonochemical synthesis of reduced graphene oxide/Pt/Sn hybrid materials and their electrochemical properties. *Fuel Cells*. 2012. 12. 956–962. Copyright Wiley-VCH Verlag GmbH & Co. KGaA. Reproduced with permission.)

(4) removal of water and the separation of HGOSs from oil. Using HGOSs as anode material for lithium-ion batteries exhibited reversible capacities of about 485 mA hg^{-1} owing to hollow-structure, thin, and porous shells consisting of grapheme.

Likewise, Tian et al. (2013) reported synthesis of a novel erythrocyte-like graphene microspheres via electrospray-assisted self-assembly of GO suspension into a coagulation bath of cetyl trimethylammonium bromide (CTAB) solution at an applied voltage of 9 kV, followed by chemical reduction. Such prepared particles, upon viewing under optical and TEMs, showed a very interesting structural characteristic of perfect exterior doughnut shape and interior porous network (Figure 6.17). In addition, they exhibited 10 times higher capability for the removal of oil and toxic organic solvents from water compared with conventional sorbent materials.

In this regard, He et al. (2013) effectively prepared disk-shaped and nanosized graphene (DSNG) with higher edge density and fringe defects by metal ion–ion exchange resin framework, as shown in

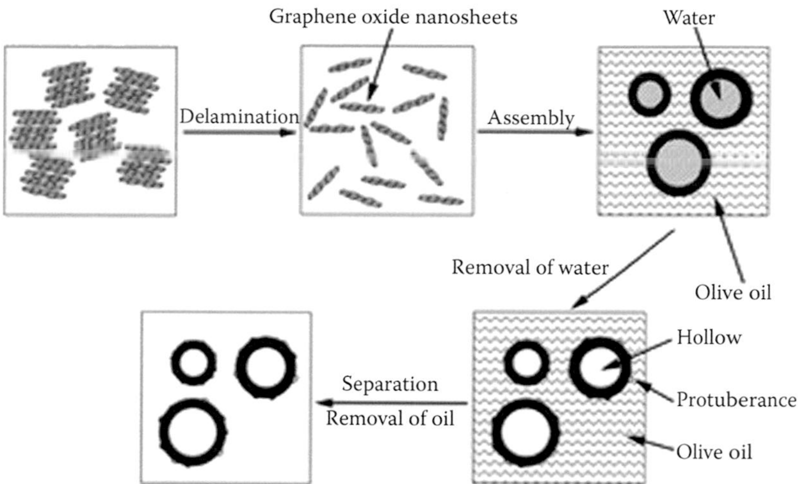

FIGURE 6.16 Schematic illustration of HGOS synthesis. (Guo, P. et al., *J. Mater. Chem.*, 20, 4867–4874, 2010. Reproduced by permission of The Royal Society of Chemistry.)

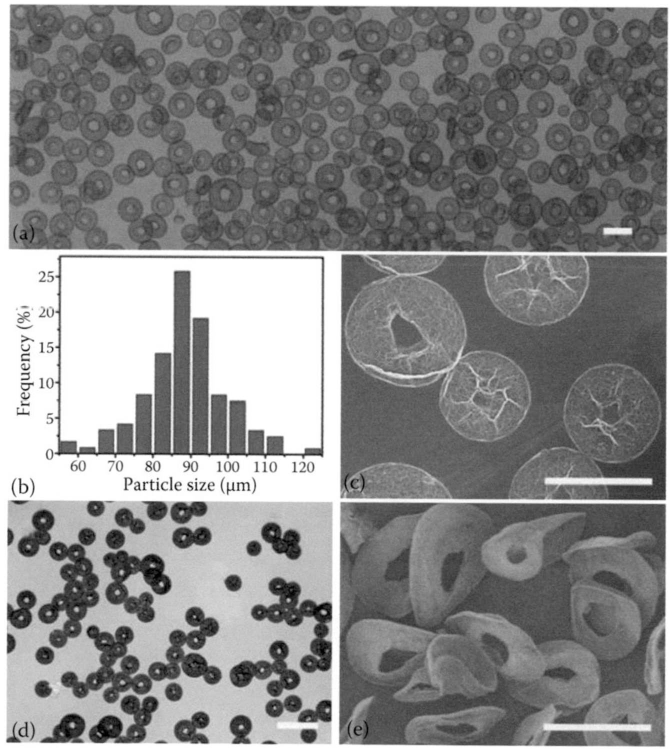

FIGURE 6.17 Microstructures of microspheres. (a) A general optical microscope view of the erythrocyte-like GO microspheres in CTAB solution by electrospray of the 12.5 mg/mL GO suspension in a coagulation bath of 0.75 mg/mL CTAB solution (scale bar, 200 mm); (b) size distribution of erythrocyte-like GO microspheres. (c) SEM image of the erythrocyte-like GO microspheres (scale bar, 100 mm). (d) Optical microscope image of erythrocyte-like GO microspheres after drying (scale bar, 200 mm). (e) SEM image of erythrocyte-like graphene microspheres reduced by hydrazine hydrate (scale bar, 100 mm). (Reprinted by permission from Macmillan Publishers Ltd. Tian, Y. et al., *Sci. Rep.*, 3, 3327, 2013. Copyright 2013.)

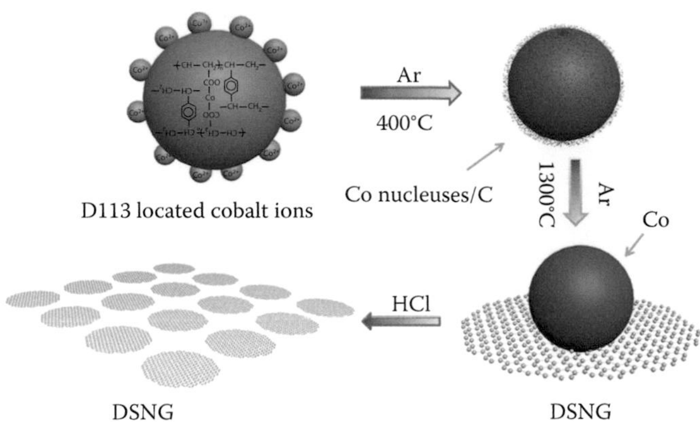

FIGURE 6.18 Schematic view of the approach for disk-shaped and nanosized graphene (DSNG) growth. (Reprinted by permission from Macmillan Publishers Ltd. He, C. et al., *Sci. Rep.*, 3, 2144, 2013. Copyright 2013.)

Figure 6.18, which was confirmed using various instrumental tools like AFM, SEM, X-ray powder diffraction (XRD), and X-ray photoelectron spectroscopy (XPS). The synthetic mechanism is as follows: DSNG was prepared using D113 resin and exchanged with cobalt ions by carbonizing in a tubular furnace at 400°C under Ar atmosphere for 2 h. In this process, cobalt atoms, encapsulated in the exchange sites of D113 resin, brought together the nuclei and at the same time D113 resin began to decompose. After this, the compound was heated at 1300°C under Ar atmosphere for 1 h, when the cobalt nucleus grew and formed polycrystalline morphology, during which carbon atom diffused out of cobalt upon cooling and formed nanographene. Such nanographenes expanded outward homogeneously in all directions, forming DSNG. In the case of CVD approach, a solid solution of cobalt and carbon was formed when the mixture of H_2 and amorphous carbon source was introduced to the hot cobalt film. Upon cooling, the high-energy carbon atoms in cobalt metal overcame the energy barrier and segregated from the interior of the cobalt film to form graphene on the surface, and such graphene grown on cobalt substrate was extensive, with low defects and homogeneous.

Fang et al. (2014) transferred such uniform atomic monolayer of graphene prepared by the CVD method onto a BaF_2 substrate coated with a thin layer of In-In_2O_3 via atomic layer deposition and patterned into closely packed nanodisk arrays using e-beam lithograpy (Figure 6.19). The formation of such graphene nanodisk arrays on a Si/SiO_2 substrate is identified under SEM (Figure 6.19). In addition, such nanopatterned graphene disk acts as an active medium for infrared electro-optic devices.

In this regard, we have recently synthesized RGO nanodisks stabilized by Sn nanoparticles by an integrated reduction method where GO sheets were used to generate RGO nanodisks with simultaneous deposition of Sn nanoparticles on the edges of the nanodisks (Anandan et al., 2014). The sonochemical procedure established in this study utilizes the physical and chemical effects generated during acoustic cavitation. The reason for the formation of disk-shaped RGO may be the probable conversion of acid (–COOH) functional groups to ester (–COOR), which occurs in the presence of $SnCl_2$, a Lewis acid catalyst (Da Silva et al., 2011). The shear forces that are generated by acoustic cavitation may break the nanosheets into nanodisks that may possess a lower surface energy (Zhang et al., 2003, 2011; Zhu et al., 2004; Guo et al., 2010; Xue et al., 2011). Such a conversion process may also be aided by the presence of Sn ions, which upon sonication produces Sn(0) nanoparticles. It is well known that the secondary reducing radicals generated during acoustic cavitation can reduce metal ions to form metal particles. The Sn nanoparticles seem to aggregate on the edges that provide stability to the disks or otherwise may lead to folded/scrolled edges. TEM images clearly show the presence of Sn particles predominantly on the edges of the nanodisks (Figure 6.20).

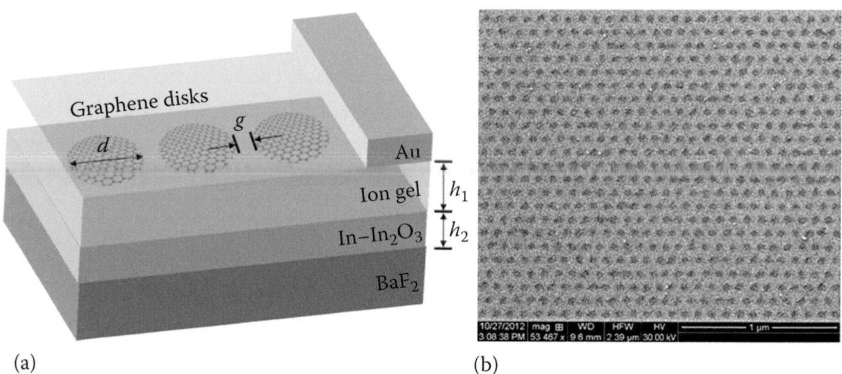

FIGURE 6.19 (a) Scheme of the measured devices. A graphene monolayer was transferred to a In–In_2O_3/BaF_2 substrate and subsequently patterned with e-beam lithography. The ion-gel was spin-coated on top of the graphene nanostructure, and the Au gate contact deposited on the top. The thickness of the ion-gel layer was h_1 = 100 nm and the thickness of In–In_2O_3 layer was h_2 = 20 nm. (b) SEM image showing a fabricated array on Si/SiO_2 of closely packed graphene nanodisks with diameter d = 60 nm and an edge-to-edge disk gap of ~30 nm. (Reprinted with permission from Fang, Z. et al., *Nano Lett.*, 14, 299–304, 2014. Copyright 2014 American Chemical Society.)

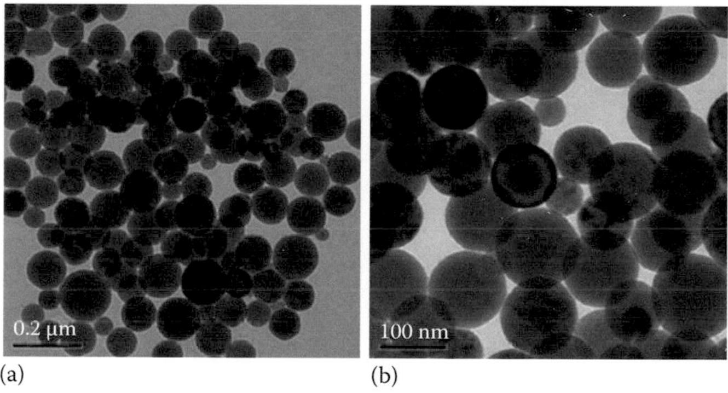

FIGURE 6.20 TEM image of sonochemically prepared Sn nanoparticles intersected reduced graphene spheres. (a) Is low magnification and (b) is high magnification. (Reprinted from *Ultrasonic Sonochem.*, 21, Anandan, S., Asiri, A.M., Ashokkumar, M., Ultrasound assisted synthesis of Sn nanoparticles-stabilized reduced grapheme oxide nanodiscs, 920–923, Copyright 2014, with permission from Elsevier.)

Theoretical and experimental evidences illustrate that specific morphology of graphene has excellent properties for broad application aspects (Bacon et al., 2014). In general, large-area graphene behaves as a conventional single-electron transistor, exhibiting periodic Coulomb blockade peaks. This is because large-area graphene flakes may not behave as a semiconductor because of the absence of an energy gap. When graphene is shaped up into quasi-1D structures with narrow widths (<~30 nm), it generates an energy gap up to 0.5 eV. When the size of graphene narrows down to 100 nm, or even sub-100 nm, which is referred to as GQDs, the peak becomes strongly nonperiodic because of quantum confinement and hence could be used as molecular scale electronics. Liu et al. (2013) produced homogeneous GQDs without oxygenous functional groups and GO quantum dots (GOQDs) with oxygenous functional groups from a much smaller and more uniform precursor graphite nanoparticle (GNP) than any other carbon source by simple exfoliation processes (Figure 6.21). Earlier, Hernandez et al. (2008) synthesized graphene through a liquid exfoliation approach using *N*-methylpyrrolidone, *N,N*-dimethylformamide, and dimethyl sulfoxide because such solvents supply the required energy to generate exfoliate graphene, and Liu et al. (2013) followed a similar rationale to produce GQDs. Furthermore, they followed modified

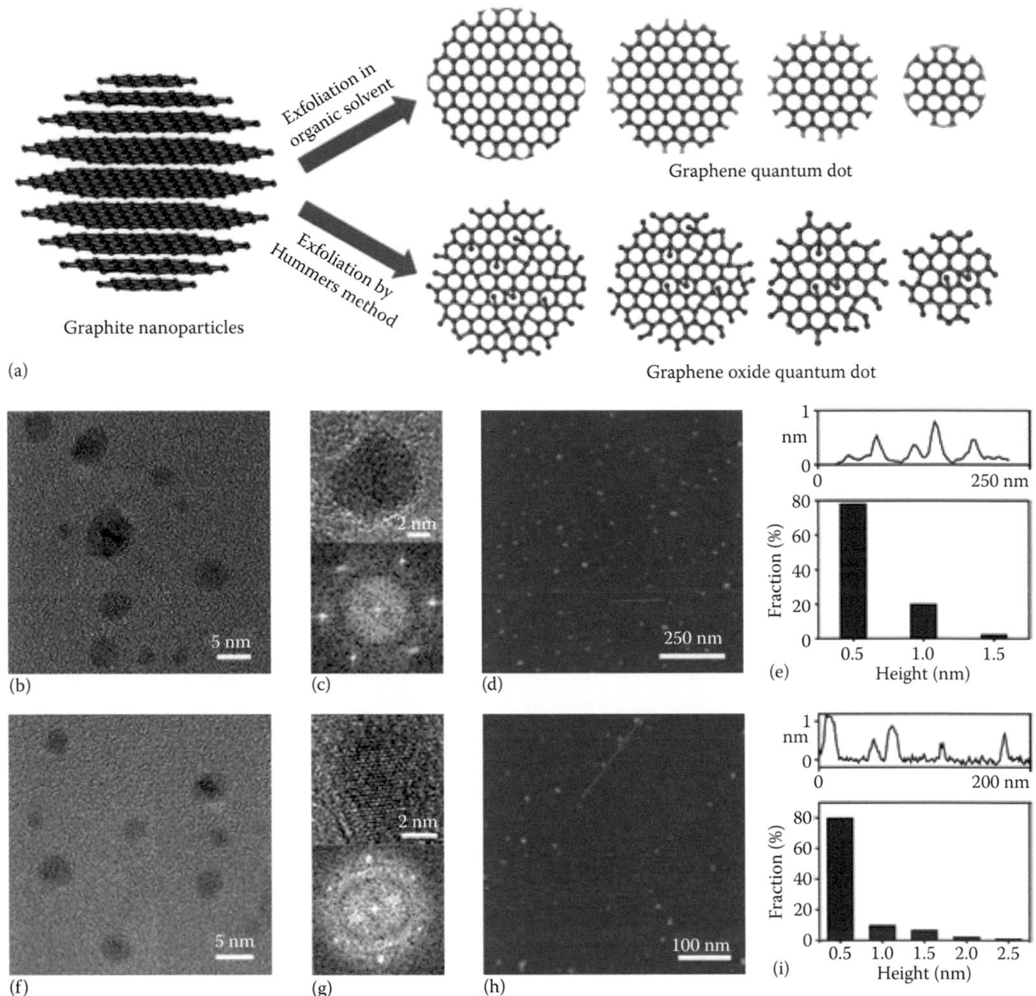

FIGURE 6.21 Synthesis and characterization of GQDs and GOQDs. (a) Synthetic scheme for GQDs and GOQDs using chemical exfoliation of GNPs (oxygenous sites are shown as red dots). (b) TEM image of GQDs. (c) High-resolution TEM image of GQDs and corresponding 2D FFT images. (d) AFM image of GQDs. (e) Height profile and height distribution of GQDs. (f) TEM image of GOQDs. (g) High-resolution TEM image of GOQDs and corresponding 2D FFT images. (h) AFM image of GOQDs. (i) Height profile and height distribution of GOQDs. (Liu, F., Jang, M.-H., Ha, H.D., Kim, J.-H., Cho, Y.-H., and Seo, T.S.: Graphene quantum dots: facile synthetic method for pristine graphene quantum dots and graphene oxide quantum dots: origin of blue and green luminescence. *Adv. Mater.* 2013. 25. 3657–3748. Copyright Wiley-VCH Verlag GmbH & Co. KGaA. Reproduced with permission.)

Hummers method to generate GOQDs from GNPs, which mainly exist as a single layer identified from the height profile and height distribution plots arrived from AFM data, and in addition identified that the observed crystal structure has the same lattice parameter as that of GQDs.

Likewise, GQDs have been fabricated by many methods such as electron-beam lithograpy, ruthenium catalyzed C_{60} transformation, etc., which are limited because of the requirement of special equipment, extremely rare raw materials, and low yield. Synthesis of GQDs through hydrothermal or electrochemical strategies required several days and, in addition, involves many chemical reagents. Peng et al. (2012) reported facile one-step, wet, chemically derived GQDs from acidic treatment of carbon fibers (CFs), which have a resin-rich surface and are highly soluble in water and other polar organic solvents (Figure 6.22).

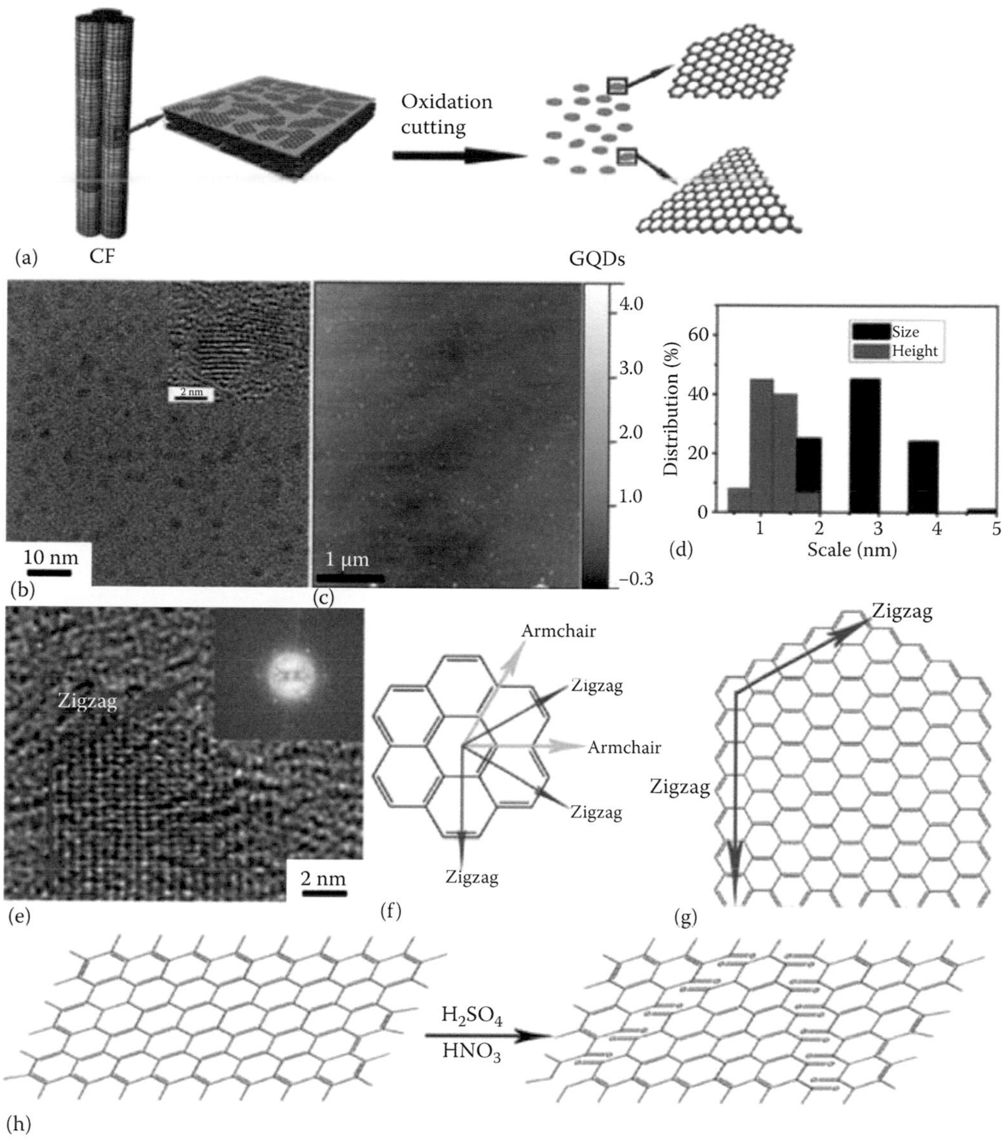

FIGURE 6.22 (a) Representation scheme of oxidation cutting of CF into GQDs. (b) TEM images of GQDs (synthesized reaction temperature at 120°C); inset of panel b is the high-resolution TEM of GQDs. (c) AFM image of GQDs. (d) Size and height distribution of GQDs. (e) HRTEM image of the edge of GQD; inset is the 2D Fast Fourier transform (FFT) of the edge in panel e. (f) Schematic illustration showing the orientation of the hexagonal graphene network and the relative zigzag and armchair directions. (g) Schematic representation of the edge termination of the HRTEM image in panel e. (h) Proposed mechanism for the chemical oxidation of CF into GQDs. (Reprinted with permission from Peng, J. et al., *Nano Lett.*, 12, 844–849, 2012. Copyright 2012 American Chemical Society.)

A TEM image of such fiber-derived GQDs shows a relatively narrow size distribution between 1 and 4 nm, while an AFM image illustrates a height profile between 0.4 and 2 nm corresponding to one to three graphene layers, which indicates that the prepared GQDs are in a 2D morphology with zigzag edge structure. Furthermore, such GQDs are found to be an excellent probe for high-contrast bioimaging and biosensing applications because of their luminescence stability, nanosecond lifetime, and high water solubility.

6.4 Summary

In this chapter, synthesis, properties, and application of graphene and graphene-derived materials such as GO nanodisks and nanodots have been highlighted. Most synthetic methods are not easy to use and are based on laboratory-scale "exploration." It is ideal if the concepts are tested under large-scale experimental conditions involving specific applications. While specific examples available in the literature were discussed, the authors sincerely hope that the information provided in this review would prompt such experimental investigation in a new dimension.

Acknowledgments

The research described herein was financially supported by the Department of Science and Technology, India and National Science Council (NSC), Australia, under the India-Australia collaborative research grant. In addition, SA thanks the Council of Scientific and Industrial Research (CSIR), New Delhi [CSIR Reference No. 02 (0021)/11/EMR-II] for financial support.

References

Anandan, S., Manivel, A., and Ashokkumar, M., "One-step sonochemical synthesis of reduced graphene oxide/Pt/Sn hybrid materials and their electrochemical properties," *Fuel Cells* 12 (2012): 956–962.

Anandan, S., Asiri, A.M., and Ashokkumar, M., "Ultrasound assisted synthesis of Sn nanoparticles-stabilized reduced grapheme oxide nanodiscs," *Ultrason. Sonochem.* 21 (2014):920–923.

Bacon, M., Bradley, S.J., and Nann, T., "Graphene quantum dots," *Part. Part. Syst. Charact.* 31 (2014):415–428.

Bai, S. and Shen, X., "Graphene-inorganic nanocomposites," *RSC Adv.* 2 (2012):64–98.

Boehm, H.P., Clauss, A., Fischer, G.O., and Hofmann, U., "The adsorption behavior of very thin carbon films," *Z. Anorg. Allg. Chem.* 316 (1962):119–127.

Boehm, H.P., Setton, R., and Stumpp, E., "Nomenclature and terminology of graphite intercalation compounds," *Carbon* 24 (1986):241–245.

Brodie, B.C. "On the atomic weight of graphite," *Phil. Trans. R. Soc. A* 149 (1859):249–259.

Da Silva, M.J., Goncalves, C.E., and Laier, L.O., "Novel esterification of glycerol catalysed by tin chloride (II): A recyclable and less corrosive process for production of bioadditives," *Catal. Lett.* 141 (2011):1111–1117.

De Heer, W.A., Berger, C., Wu, X., First, P.N., Conrad, E.H., Li, X., Li, T. et al., "Epitaxial grapheme," *Solid State Commun.* 143 (2007):92–100.

Ebbesen, T.W. and Hidefumi, H., "Graphene in 3-dimensions: Towards graphite origami," *Adv. Mater.* 7 (1995):582–586.

Fang, Z., Wang, Y., Schlather, A.E., Liu, Z., Ajayan, P.M., de Abajo, F.J.G., Nordlander, P., Zhu, X., and Halas, N.J., "Active tunable absorption enhancement with graphene nanodisk arrays," *Nano Lett.* 14 (2014):299–304.

Gan, Y., Chu, W., and Qiao, L., "STM investigation on interaction between superstructure and grain boundary in graphite," *Surf. Sci.* 539 (2003):120–128.

Geim, A.K. "Graphene prehistory," *Phys. Scripta* T146 (2012):014003.

Geim, A.K. and Novoselov, K.S., "The rise of graphene," *Nat. Mater.* 6 (2007):183–191.

Guo, P., Song, H., and Chen, X., "Hollow graphene oxide spheres self-assembled by W/O emulsion," *J. Mater. Chem.* 20 (2010):4867–4874.

He, C., Jiang, S.P., and Shen, P.K., "Large-scale and rapid synthesis of disk-shaped and nano-sized grapheme," *Sci. Rep.* 3 (2013):2144.

Hernandez, Y., Nicolosi, V., Lotya, M., Blighe, F.M., Sun, Z., and De, S., "High-yield production of graphene by liquid-phase exfoliation of graphite," *Nat. Nanotechnol.* 3 (2008):563–568.

Horiuchi, S., Gotou, T., Fujiwara, M., Asaka, T., Yokosawa, T., and Matsui, Y., "Single graphene sheet detected in a carbon nanofilm," *Appl. Phys. Lett.* 84 (2004):2403–2405.

Kim, K.Y., Yang, M., Cho, K.M., Jun, Y.-S., Lee, S.B., and Jung, H.-T., "High quality reduced graphene oxide through repairing with multi-layered graphene ball nanostructures," *Sci. Rep.* 3 (2013):3251.

Kim, T.Y., Jung, G., Yoo, S., Suh, D.S., and Ruoff, R.S., "Activated graphene-based carbons as supercapacitor electrodes with macro- and mesopores," *ACS Nano* 7 (2013):6899–6905.

Kosynkin, D.V., Higginbotham, A.L., Sinitskii, A., Lomeda, J.R., Dimiev, A., Price, B.K., and Tour, J.M., "Longitudinal unzipping of carbon nanotubes to form graphene nanoribbons," *Nature* 458 (2009):872–876.

Land, T.A., Michely, T., Behm, R.J., Hemminger, J.C., and Comsa, G., "STM investigation of single layer graphite structures produced on Pt(111) by hydrocarbon decomposition," *Surf. Sci.* 264 (1992):261–270.

Liu, F., Jang, M.-H., Ha, H.D., Kim, J.-H., Cho, Y.-H., and Seo, T.S., "Graphene quantum dots: Facile synthetic method for pristine graphene quantum dots and graphene oxide quantum dots: Origin of blue and green luminescence," *Adv. Mater.* 25 (2013):3657–3748.

Lu, X., Yu, M., Huang, H., and Ruoff, R.S., "Tailoring graphite with the goal of achieving single sheets," *Nanotechnology* 10 (1999):269–272.

Niyogi, S., Bekyarova, E., Itkis, M.E., McWilliams, J.L., Hamon, M.A., and Haddon, R.C. "Solution properties of graphite and grapheme," *J. Am. Chem. Soc.* 128 (2006):7720–7721.

Novoselov, K.S., Geim, A.K., Morozov, S.V., Jiang, D., Zhang, Y., Dubonos, S.V., Grigorieva, I.V., and Firsov, A.A., "Electric field effect in atomically thin carbon films," *Science* 306 (2004):666–669.

Novoselov, K.S., Jiang, Z., Zhang, Y., Morozov, S.V., Stormer, H.L., Zeitler, U., Maan, J.C., Boebinger, G.S., Kim. P., and Geim, A.K., "Room-temperature quantum hall effect in grapheme," *Science* 315 (2007):1379.

Park, S., An, J., Piner, R.D., Jung, I., Yang, D., Velamakanni, A., Nguyen, S.T., and Ruoff, R.S., "Aqueous suspension and characterization of chemically modified graphene sheets," *Chem. Mater.* 20 (2008):6592–6594.

Peng, J., Gao, W., Gupta, B.K., Liu, Z., Romero-Aburto, R., Ge, L., Song, L. et al., "Graphene quantum dots derived from carbon fibers," *Nano Lett.* 12 (2012):844–849.

Piner, R., Li, H., Kong, X., Tao, L., Kholmanov, I.N., Ji, H., Lee, W.H. et al., "Graphene synthesis *via* magnetic inductive heating of copper substrates," *ACS Nano* 7 (2013):7495–7499.

Ponomarenko, L.A., Schedin, F., Katsnelson, M.I., Yang, R., Hill, E.W., Novoselov, K.S., and Geim, A.K., "Chaotic Dirac billiard in graphene quantum dots," *Science* 320 (2008):356–358.

Reina, A., Jia, X., Ho, J., Nezich, D., Son, H., Bulovic, V., Dresselhaus, M.S., and Kong, J., "Large area, few-layer graphene films on arbitrary substrates by chemical vapor deposition," *Nano Lett.* 9 (2009):30–35.

Sarrazin, M. and Petit, F., "Exciton swapping in a twisted graphene bilayer as a solid-state realization of a two-brane model," *Eur. Phys. J. B* 87 (2014):26.

Seibert, K., Cho, G.C., Kütt, W., Kurz, H., Reitze, D.H., Dadap, J.I., Ahn, H., Downer, M.C., and Malvezzi, A.M., "Femtosecond carrier dynamics in graphite," *Phys. Rev. B* 42 (1990):2842–2851.

Shioyama, H. "Cleavage of graphite to graphene," *J. Mater. Sci. Lett.* 20 (2001):499–500.

Song, L.H., Lim, S.N., Kang, K.K., and Park, S.B., "Graphene-based mesoporous nanocomposites of spherical shape with a 2-D layered structure," *J. Mater. Chem. A* 1 (2013):6719–6722.

Sudeep, P.M., Narayanan, T.N., Ganesan, A., Shaijumon, M.M., Yang, H., Ozden, S., Patra, P.K. et al., "Covalently interconnected three-dimensional graphene oxide solids," *ACS Nano* 7 (2013):7034–7040.

Tian, Y., Wu, G., Tian, X., Tao, X., and Chen, W., "Novel erythrocyte-like graphene microspheres with high quality and mass production capability via electrospray assisted self-assembly," *Sci. Rep.* 3 (2013):3327.

Viculis, L.M., Mack, J.J., and Kaner, R.B., "A chemical route to carbon nanoscrolls," *Science* 299 (2003):1361.

Vinodgopal, K., Neppolian, B., Lightcap, I.V., Grieser, F., Ashokkumar, M., and Kamat, P.V., "Sonolytic design of graphene–Au nanocomposites. simultaneous and sequential reduction of graphene oxide and Au(III)," *J. Phys. Chem. Lett.* 1 (2010):1987–1993.

Worsley, K.A., Ramesh, P., Mandal, S.K., Niyogi, S., Itkis, M.E., and Haddon, R.C., "Soluble graphene derived from graphite fluoride," *Chem. Phys. Lett.* 445 (2007):51–56.

Xue, Y., Chen, H., Yu, D., Wang, S., Yardeni, M., Dai, Q., Guo, M. et al.,"Oxidizing metal ions with graphene oxide: The in situ formation of magnetic nanoparticles on self-reduced graphene sheets for multifunctional applications," *Chem. Commun.* 47 (2011):11689–11691.

Yan, X., Cui. X., and Li, L.S., "Synthesis of large, stable colloidal graphene quantum dots with tunable size," *J. Am. Chem. Soc.* 132 (2010):5944–5945.

Zhang, Z., Yu, L., Zhang, W., Zhu, Z., He, G., Chen, Y., and Hu, G., "Carbon spheres synthesized by ultrasonic treatment," *Phys. Lett. A* 307 (2003):249–252.

Zhang, Z., Zou, R., Song, G., Yu, L., Chen, Z., and Hu, J., "Highly aligned SnO_2 nanorods on graphene sheets for gas sensors," *J. Mater. Chem.* 21 (2011):17360–17365.

Zhu, Z., Su, D., Weinberg, G., and Schlogl, R., "Supermolecular self-assembly of graphene sheets: Formation of tube-in-tube nanostructures," *Nano Lett.* 4 (2004):2255–2261.

Zhu, S.J., Zhang, J., Qiao, C., Tang, S., Li, Y., Yuan, W., Li, B. et al., "Strongly green-photoluminescent graphene quantum dots for bioimaging applications," *Chem. Commun.* 47 (2011):6858–6860.

7 Natural Graphene-Based Shungite Nanocarbon

Natalia N. Rozhkova

Sergey P. Rozhkov

Andrey S. Goryunov

7.1 Shungite as a Natural Source of Nanocarbon Materials

Shungite carbon (ShC) has been attracting the attention of scientists and practical people for over two centuries. Now, the carbon from shungite rocks is of utmost interest as a natural nanomaterial source (Rozhkova 2011).

The geological event that led to the formation of tremendous quantities of carbonaceous rocks in the Shunga district, Russian Karelia, about 1.8–2.0 Ga ago was named the Shunga Event. Similar rocks have been described in Greenland, North America, Africa, and Fennoscandia. Shungite, together with anthraxolite, is a member of the pyrobitumen group (Parnell 1993). Atomic H/C and O/C ratios for shungite are less than 0.1 and 0.03, respectively, and are close to the ratios estimated for graphite (Cornelius 1987). However, volcanic-sedimentary rocks of various lithologo-petrological types contain only ShC, which occurs in almost all the rocks in the Onega Lake region of Russian Karelia over an area of more than 9000 km^2. ShC reserves are estimated at 25×10^{10} t (Buseck et al. 1997).

The origin of ShC remains a disputable problem. The hypotheses of its biogenic and abiogenic origin are most commonly discussed. The latter is associated with the concept of the deep source of carbon and the evolution of volcanism in the region assumed earlier to have been formed from sedimentary rocks. The mineral constituent of shungite rocks varies consistently, depending on the phase of volcanic activity. However, there is extensive geological evidence for the formation of shungite complex as a result of large-scale hydrothermal activity under relatively mild conditions at a temperature of no more than 450°C and a pressure of 700 MPa (Melezhik et al. 1999).

In accordance with the generally accepted classification, shungite rocks are subdivided into five types, depending on carbon content that varies from a percentage fraction to 98 wt%. Samples from the Shunga deposit contain almost pure vitreous carbon classified as shungite of type I (96–98 wt% carbon) (Figure 7.1). Rocks that contain 20–45 wt% carbon (shungite rocks of type III) are most common. They were produced as a fairly homogeneous ShC-silica (quartz) composite material whose constituents form interpenetrating three-dimensional nets. A similar homogeneous distribution is displayed by lydite—a cherty shungite rock that consists of 95% quartz and contains 2%–5% carbon (Sokolov et al. 1984, Zaidenberg et al. 1995).

Shungite in the form of stones and powder is now used dominantly to produce adsorbents and catalysts for water purification and water treatment, fillers for multiple-purpose polymers, and inorganic binders used in the production of electromagnetic radiation-shielding materials. As shungite displays antiseptic activity, it is also used for sanitary and therapeutic purposes. Large amounts of the rock are used as a coke and quartzite substitute in metallurgy (Rozhkova et al. 2004; Kalinin et al. 2008).

At the same time, nanostructurized carbonaceous materials are typically highly homogeneous in terms of the size and structure of primary carbon particles. These nanoparticles (NPs) are packed in a certain order to form a secondary particle (aggregate). Examples of such nanostructurized materials are ultradisperse diamonds, produced by detonation synthesis, and ShC. A detailed study of fullerene-like structures has shown that it is curved graphene layers that are typical of the unusual electronic states of new-generation carbon materials such as fullerenes, onion structures, and nanodiamonds (NDs). Therefore, attempts have been made to look for similarities in the structural and electronic properties of fullerenes and multilayered ShC globules (Kovalevski 1994; Berezkin et al. 1997).

Most of natural carbons are in a noncrystalline state. Their structure is commonly described as combinations of sp^2 or sp^3 bonds and can be identified as a graphite-like, diamond-like, or carbyne-like

FIGURE 7.1 The objects of the study: a sample of shungite type I Shunga deposit (a), powder of shungite type I (b), and its water dispersion (c).

form of noncrystalline carbon, depending on the dominant type of bond. The term *fullerene-like* has been coined recently to describe near-spherical carbon structures such as "onions" and hollow NPs that consist of concentric carbon layers inserted into each other (Kuznetsov et al. 2001).

Data obtained by the x-ray and selected area electron diffraction (SAED) methods show that multilayered globular or ellipsoid particles, from 6 to 10 nm in size, with an internal cavity form the structure of ShC. In the general case, domains of coherent scattering correspond to those of a single-type arrangement of layers of globular elements, and layers with considerable distortion that provide a link between the globules form an arbitrary net (Kovalevski et al. 2005).

Despite a great variety of shungite rocks, the structures of the carbon they contain display common features such as the high-resolution electron microscopy (HRTEM) images of carbon (Figure 7.2) and the parameters of SAED and x-ray diffraction. It has been shown (Kovalevski et al. 2001) that the structural patterns of carbon matter from the Ericson (Canada) and Sovetskoe (Russia) gold deposits and the Sudbury deposit, Ontario, Canada, are similar to that of ShC. Closed three-dimensional shells or, in a more general case, fragments of shells or regions that consist of curved packages of graphite-like, turbostratic oriented layers that envelope the pores, like the carbon of natural and artificial coke, have been described. In contrast to coke, all ShC samples were shown to be structurally homogeneous over the range of 1 μm to 1 mm and are ungraphitizable. Analysis of the characteristic features of fullerenes, such as the gentle curve of carbon layers that encase nanosized pores, compared with the trigonal symmetry of the graphite-like structural motif of ShC, has led the authors to conclude that similarity to fullerenes is the main structural feature of ShC. A globule, as a multilayered unit that consists of gently curved graphene layers that encase a nanosized pore, provided the basis for a model of the fullerene-like structure of ShC.

The development of fullerene synthesis has triggered another wave of interest in ShC as a possible natural source of fullerenes and contributed to the formulation of a scientific problem related to the conditions of formation and conservation of fullerenes and fullerene-like structures in nature (Lemanov et al. 1994; Osawa 1999). Although some indirect methods have indicated the presence of fullerenes in ShC, only traces of C_{60} were revealed in ShC using methods for colloid extraction by a mixture of organic solvents (Masterov et al. 1994).

It was hard to extract and stabilize the globular nanostructures of ShC to support their similarity to fullerenes. It has been shown that mild oxidative or electrochemical treatment of ShC reduces the bonds between a globule and a carbon matrix. Individual ShC NPs are restructured and modified upon dry grinding and thermal treatment, as indicated by HRTEM and by changes in their shape and size: Globular particles grow in size to tens to hundreds of nanometers and acquire a dominantly polyhedral shape (Zaidenberg et al. 1996).

Polyhedral particles were assumed to have been formed by the merging of globules. As ShC has a lot of elastic energy, local stresses and heating that arise upon the dispersal of a sample disturb the metastable state of the structure. As a result, the layers are disrupted and combined, internal globular pores

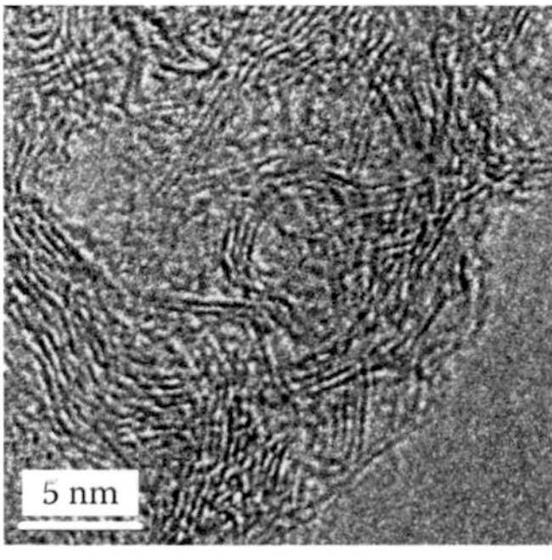

FIGURE 7.2 HRTEM image of ShC structure.

coalesce simultaneously, coarser particles are formed, and the surface layers are deformed (straightened), i.e., polyhedral particles are produced.

The restructuring of globules and the modification of the carbon structure upon thermal treatment provide evidence for the metastability of the ShC structure (Rozhkova 2002). The study of ShC structure-modifying mechanisms under mild conditions, i.e., metastability by which ShC activity can vary greatly, has identified a graphene fragment as the basic element of ShC and provided a better understanding of its activation and stabilization conditions.

Graphene as the two-dimensional nanofragment of a hexagonal graphite surface is a new nanocarbon material. It should be noted that all of the graphene properties predicted, such as high strength, transparency, anomalously high thermal conductivity, and excellent conductivity at room temperature, show that one-layered graphene is a promising material for nanoelectronics. However, no large-scale graphene production technology has yet been developed (An et al. 2010; Singh et al. 2011), and the presence of graphenes in a natural material provides good production opportunities.

The extraction and stabilization of the nanosized constituents of ShC is a key problem that restricts the use of shungite rocks in high-tech processes. Until now, conventional expensive technologies, similar to those known for synthetic nanocarbon materials, have been proposed. For example, carbon-rich shungite rock is treated at high temperatures over 1200°C. As a result, microfibrous silicon carbide is produced, and globular carbon clusters are shaped as hyperfullerene structures (Kovalevski et al. 1996). Autoclave treatment and strong chemical reagents, used to remove mineral constituents from carbon-rich shungite rocks, result in the microporosity and mesoporosity of shungite (Alekseev et al. 2006). The disadvantage of such a "rigid" ShC modification is the disappearance of amphiphilicity, a distinctive feature of ShC not displayed by conventional carbon materials such as graphite, carbon black, and a new generation of synthetic nanocarbon materials. The ability of ShC to be wetted by liquids, which differ in polarity, and water is indicated by homogeneity at a nanolevel of a three-dimensional heterophase structure formed from two interpenetrating carbon and silica nets. To retain this ability, new low-cost ecologically pure ShC treatment methods for the production of a new-generation nanocarbon material using low-temperature ozonation and the production of water dispersions are needed (Emel'yanova et al. 2004; Rozhkova et al. 2004).

7.2 Water in ShC

Water acts as a dispersion medium and a liquid that controls intermolecular interaction between carbon NPs. Water is expected to contribute greatly to the production of new composite materials using fullerene-like particles because the cohesive forces increase upon water evaporation, leading to the increased strength and chemical resistance of the net, which is formed from carbon NP. The goal of studies conducted these days is to better understand the microdynamic properties of water molecules near carbon NP and to estimate the effect of hydration on the structural and morphological characteristics and stability of nanocarbon complexes (Mchedlov-Petrossyan et al. 1997).

Available geological evidence shows that a shungite complex was produced by extensive hydrothermal activity. Consequently, polar solvents, primarily water, are expected to have affinity to ShC nanostructures. Water is encapsulated in nanosized shungite pores and is desorbed in two temperature intervals: 20°C–110°C (0.42 wt%) and 110°C–375°C (4.15 wt%), as shown by thermogravimetric analysis. When heated to 375°C, bulk shungite-I samples are disintegrated into plates, the specific surface and porosity change slightly, and conductivity increases (Zaidenberg et al. 1996).

Anomalies (jumps or gradual transitions) in the temperature dependence of the thermal conductivity and electrical conductivity (Figure 7.3), as well as thermal electromotive force and thermal capacity, of ShC, similar to those obtained on fullerenes, are associated with the sorption/desorption of water from nanosized pores in the ShC structure (Parfen'eva et al. 1994).

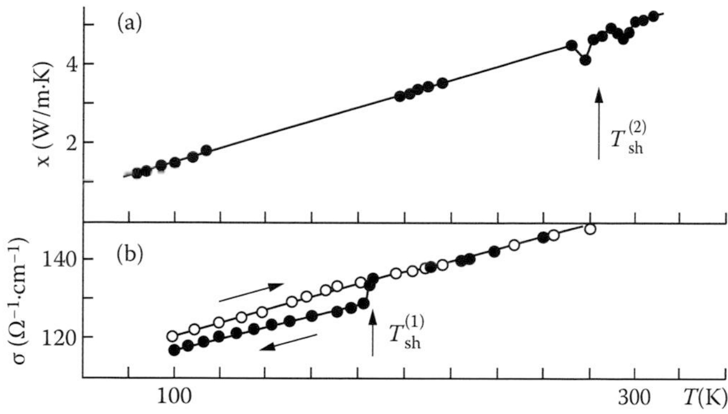

FIGURE 7.3 Temperature dependence of thermal conductivity (a) and electrical conductivity (b) of shungite type I Shunga deposit.

Water is also known to affect the electronic structure and electrophysical properties of nanographite. It has been shown that they can change reversibly upon water vapor sorption (Ziatdinov 2004). Deformed structures, used to distinguish between fullerenes and graphites, predetermine the high reactivity of fullerene structures.

Studies conducted by ^{13}C nuclear magnetic resonance (NMR) methods have supported the assumption that globular carbon clusters of ShC in water are split up into graphene fragments that interact with water. These results are in agreement with ^{1}H NMR data (Rozhkova et al. 2007a). The small-angle neutron scattering (SANS) and x-ray scattering (SAXS) methods corroborate the two-level structural pattern of both the pores and structural elements of ShC in the range of 1–100 nm (Avdeev et al. 2006).

Globular clusters, less than 6 nm in size, formed by nonplanar graphene (NGr) fragments less than 1 nm in size, were visualized by HRTEM (Figure 7.4). The size and curvature of the surface of a graphene as the basic structural element of ShC are responsible for its electron structure and dipole moment, providing the amphiphilicity of ShC and contributing to the stabilization of carbon NPs in water (Rozhkova et al. 2007b). The dipole moment of the structural elements of ShC was calculated by measuring the static dielectric polarization of NGrs diluted in benzene, toluene, and orthoxylene solutions at temperatures of 25°C to 60°C (Hairullin et al. 2012). The dipole moment of NGr in benzene is 6.5 D. It shows that NGrs are present as clusters, stabilized by the mobility of nonplanar fragments, even in highly diluted solutions.

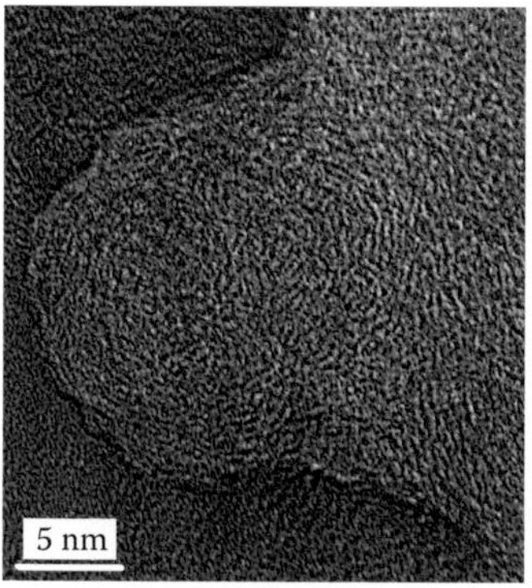

FIGURE 7.4 HRTEM image of a primary globular aggregate of graphene fragments ~0.5 nm in water dispersion.

7.2.1 Stable Water Dispersions of ShC NPs

Stable water dispersions of hydrophobic carbon NPs, synthesized in inert atmosphere (fullerenes, nanotubes, fullerene ash, NDs, and nanographite), can only be obtained after they are modified. The study of water dispersions of graphene oxide (GO), produced by modifying graphene in various ways, has aroused great interest (Zhu et al. 2010). Of special interest, however, are water-soluble forms of chemically unmodified carbon NP stabilized without oxidation and attempts to assess the conditions required for solvation and colloid dissolution of particles without salt impurities or surface-active substances (SASs), which is characteristic, for example, of water dispersions of hydrated fullerenes (C60FWS) (Andrievsky et al. 1995).

A comparative study of water dispersions of carbon NP, prepared from shungite, fullerene, and ND powders without using SAS, was conducted (Rozhkova et al. 2009; Rozhkova 2013). The removal of silica, a major impurity that passes into water from shungite I, from ShC water dispersion, was carried out by multiple treatment of the powder left on the filter. The composition of the ShC water dispersions obtained after such purification is comparable in the amount of impurities with the synthetic nanocarbon materials studied, as shown by mass spectrometry data (Rozhkova 2013). Attempts to obtain water dispersion from graphite powder under similar conditions have failed.

ShC water dispersions at a concentration of 0.1 mg/mL were grey-brown in color and opalescent. The average size of the clusters in water, estimated by dynamic light scattering (DLS), is ~50 nm at a dispersion pH of 6.3–6.8 units. Transmission electron microscopy (TEM) was conducted to identify particles 5 nm and their aggregates up to 100 nm in size (Figure 7.5).

The average sizes of nanoclusters in the dispersion medium, estimated with a Zetasizer Nano ZS NP size analyzer (Malvern Instruments) by the DLS method, are 50 nm, 36 nm, and 26 nm for ShC, fullerenes, and ND, respectively (Table 7.1).

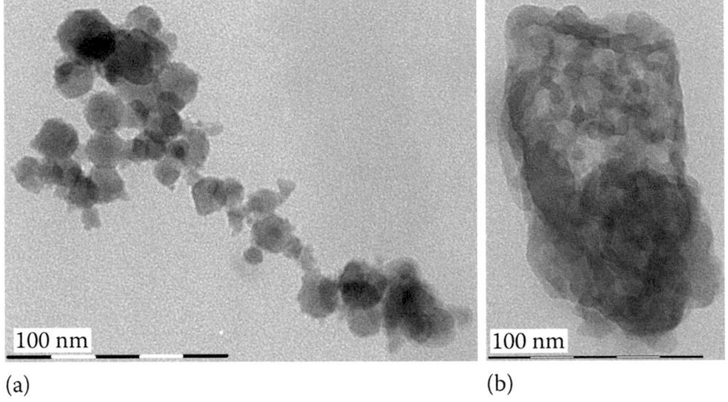

FIGURE 7.5 TEM of ShC NPs obtained upon drying of stable water dispersion: a chain of NPs (a) and a net of NPs (b).

TABLE 7.1 Average Size of Carbon Nanoparticles in Stable Water Dispersions Estimated from DLS Data

Samples	Average Size of a Structural Element (nm)	Nanoparticle Concentration (mg/mL)	Average Radius of Nanoparticles in Water Dispersion (nm)	Polydispersity
ShC	Globular cluster <6	0.024	50.0	0.41
ShC	Multilayered aggregate <6	0.048	52.0	0.28
C_{60}/C_{70}	Average diameter of a C_{60} molecule 0.7	0.027	36.8	0.42
ND	ND core diameter 4.5	0.24	26.4	0.30

Although structural elements aggregate in water medium and the average size of carbon clusters increases more than 5 times for all carbon NPs (Table 7.1), the dispersions retain sedimentation stability for several years.

7.2.2 Basic Structural Units and Graphene Nature of ShC

The basic structural element of ShC was extracted from larger nanosized carbon clusters in stable water dispersion and described by comparative analysis of ^{13}C NMR high-resolution spectra of ShC and the model NMR spectra of nonplanar carbon structures (Rozhkova et al. 2007a). A ^{1}H NMR spectrum in a solid body, obtained from a sample of ShC NPs dried from water dispersion, shows the presence of water molecules connected with water complexes on the basic ShC element and adsorbed bulk water.

Figure 7.6 shows the Raman spectra of ShC water dispersions at concentrations of 0.06 and 0.4 mg/mL, condensed dispersion, and original shungite powder obtained on a Nicolet Almega XR spectrometer (Thermo Scientific), laser 532 nm ND-YAG.

A close relation between original shungite and water dispersion and condensate of its NPs is clearly indicated by Raman spectra (Sheka and Rozhkova 2014). Figure 7.6 shows the spectra of two powders and two water dispersions. It is clear from Figure 7.6 (curves 1 and 2) that practically identical doublets, consisting of characteristic G- and D-bands, are the spectra of both solid bodies, undoubtedly suggesting their similarity. However, both condensate bands are slightly shifted toward higher frequencies of 1341 to 1348 cm^{-1} (D-band) and 1586 to 1596 cm^{-1} (G-band). This suggests that the stacking of the colloid particles of shungite water dispersions upon their conversion to condensate does not result in the complete recovery of the fractal structure of original shungite.

As valent O-H variations of ~3400 cm^{-1} dominate in the Raman spectra of water dispersions (curves 3 and 4), their low-frequency spectra, shown in Figure 7.6, are recorded under accumulation conditions. The Raman spectra of both dispersions are characterized by rather wide D-bands at 1350 cm^{-1} (curve 3) and 1353 cm^{-1} (curve 4). The corresponding G-bands of dispersions have to be much lower in intensity than D-bands and are expected to have a frequency of 1596 ± 4 cm^{-1}, which lies in the region of the low-frequency wings of intensive S-bands. Nevertheless, the maxima of the G-bands of the dispersions are higher than that of D-bands. The G-band maxima of the dispersion with a higher concentration of NP

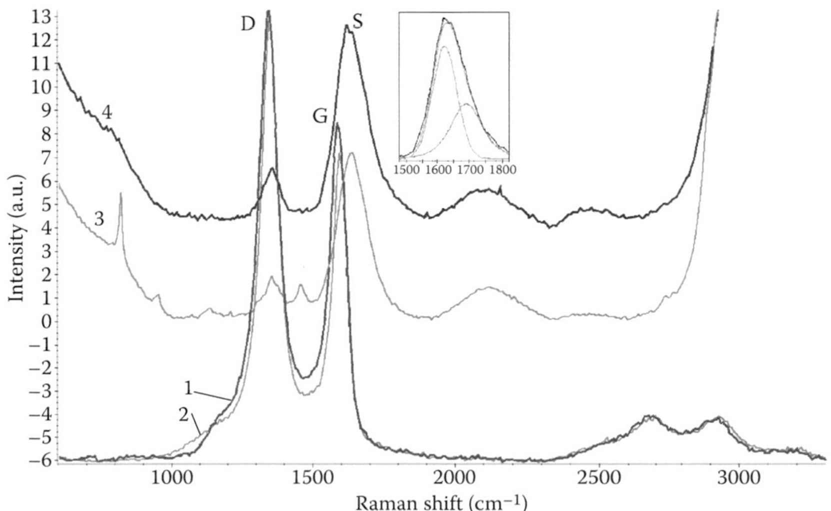

FIGURE 7.6 Raman spectra of (1) pristine shungite-I powder Shunga deposit (Shl-Sh) and (2) condensed water dispersion of ShC; water dispersion of ShC NPs with concentrations of (3) 0.06 mg/mL and (4) 0.4 mg/mL. Insert: deconvolution of the G line of spectrum 4.

(curve 4) is a superposition of two lines; common deconvolution shows that the basic lines are 1615 cm^{-1} and 1635 cm^{-1}. The first one could be attributed to carbon scatterer, and the second one, to bound water molecules. The shift in frequencies for the system of carbon NPs formed from graphene fragments could indicate the water molecules' intercalation between the fragments. Indeed, the appearance in the Raman spectrum of this variation with shifted frequency, prohibited for free water by symmetry, was observed in the spectra of water retained in clayey minerals (Amara 1997).

Experimental evidence for the similarity of a structural element to graphene has been supported by quantum-chemical calculations. It has been shown that a planar Gr can be converted to a curved one upon the unilateral adsorption of not only atomic hydrogen but also hydroxyl (Sheka and Popova 2012, 2013). Initially, the zero dipole moment of the carbon framework of the membrane, 0.8 nm in linear size, increases upon termination by hydrogen to 1.4 D and to 6.5 D when termination is done by oxygen. The latter dipole moment value practically coincides with the experimental value. This indicates the effect of oxygen on the formation of NGr ShC elements and attributes the stabilization of Gr in clusters to water molecules involved in intensive dipole–dipole interaction with a curved graphene membrane.

The graphene hypothesis of the ShC structure has been developed as a concept of graphene quantum dots (Razbirin et al. 2014). Thus, achievements in the molecular theory of graphene can be used to cast light on the origin of shungite and its unique properties (Sheka and Rozhkova 2014).

7.3 Multilevel Structural Organization of ShC

Shungite is presented as a multilevel fractal structure that is formed by successive aggregation of ~1 nm graphene fragments. Stacks of graphene fragments of 1–2 nm in thickness and globular composition of the stacks of 5–7 nm in size determine the secondary and tertiary levels of the structure. Aggregates of globules 20–100 nm complete the structure (Rozhkova 2013; Sheka and Rozhkova 2014).

The disaggregation and stabilization conditions of structural elements in various media, primarily water, a fundamental problem for all nanocarbon materials, can be assessed by studying the microstructure of the aggregates formed. To produce water dispersions of carbon NP, their aggregation resistance mechanism should be understood. As fullerene is aggregated, polycrystalline or amorphous fullerite, which interacts with water more actively than the original carbon material, is formed. Similar globular and acicular morphological structures have been obtained in ShC dispersion, fullerene, and ND condensates (Rozhkov et al. 2007; Osawa 2008; Rozhkova et al. 2009).

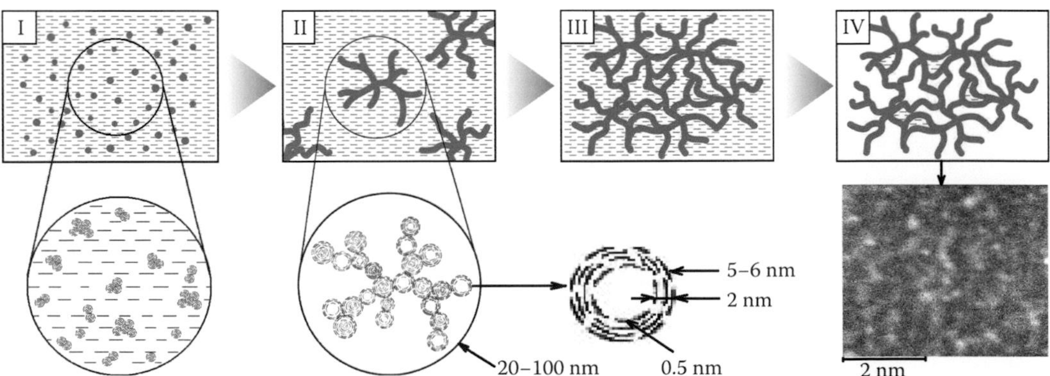

FIGURE 7.7 Scheme showing the clusterization of the graphene fragments of ShC upon the condensation of water dispersion. (I) Diluted dispersion, 0.1 mg/mL; (II) dendrite clusters upon the concentration of dispersion are formed of globular ones consisting of NGr fragments; (III) net formed upon the condensation of dispersion; (IV) as dispersion is dried, water remains in clusters and illustrates interaction between NGr fragments. SEM image of a film produced from ShC water dispersion on glass plate.

The structural changes induced by the condensation of ShC water dispersions and accompanied by the aggregation of primary nanoclusters and the formation of a three-dimensional net are illustrated by schemes in Figure 7.7 (II and III). Graphene fragments and their distinct interaction with water are an essential link in these transformations. Such particles (Figure 7.7, III and IV) can be spontaneously redispersed in water.

The structural pattern and properties of ShC vary at each dispersion concentration stage followed by condensation, as shown by a spectral study.

7.4 Modification of ShC Structure at Different Levels

The effect of chemical processes that activate carbon materials, e.g., vapor activation, ozonization, and transformation of nanosized constituents into water dispersions, on carbon structure is of utmost interest. We used shungite type I from Shunga deposit (Shl-Sh) (see Figure 7.1). Water dispersion of ShC, obtained by multiple consecutive ultrasonic treatments of shungite powder that remained on the filter, was condensed on a glass substrate until a film was formed. Diffraction patterns of the resulting film were obtained over a wide angle range at Co radiation. Two characteristic sizes of clusters on the diffraction pattern, 7.7 nm and >30.1 nm, were estimated (Table 7.2). As clusters, 7.7 nm in size, were revealed on thick and thin films obtained from the water dispersion of Shl-Sh and as the corresponding intensity increases in time, we assume that the average size of stable clusters, comparable in size with a globular aggregate, the structural element of shungite identified earlier by the x-ray and SAED methods, was obtained.

Changes in the porous structure, induced by the modification of Shl-Sh, were studied using SAXS, adsorption methods, TEM, and SAED (Rozhkova et al. 2009). The results of SAXS are presented in Table 7.2.

The diffraction patterns of all the ShC samples modified typically display two minimum sizes: a small peak in the region of 13.23° (0.39 nm) and 10.14° (0.51 nm). The first peak on the diffraction curve was found to be most intensive for the sample after vapor activation and less intensive on the ozonized sample, but there was practically no peak upon treatment of ShC by a mixture of water vapor and CO_2. This size coincides with the size of the micropores estimated on modified samples by adsorption methods (Tables 7.3 and 7.4). The second peak, 0.51 nm in size and more distinct than the one obtained after vapor activation, was estimated on ozonized ShC.

Ozonization results in the increased concentration of oxygen-containing groups on the ShC surface and the dominant growth of phenol and lactone groups and "nonfunctional" oxygen (Emel'yanova et al. 2004). Ozonization also exerts a substantial effect on the specific ShC surface value, increasing it by

TABLE 7.2 Sizes of the Structural Elements of Modified ShI-Sh Estimated by the SAXS Method

Samples and Modification Conditions	Fractal Dimension	Size of a Structural Element (nm)
Initial ShI-Sh	2.2	0.51, 6.0–4.5, 13.0; ≥30.2
Ozonized ShI-Sh	2.19	0.39, 0.51, 2.6, 4.5–5.8, 25.0–26.5
ShI-Sh after steam activation	1.88	0.39, 0.51, 2.2, 4.5–5.2, 30.5
ShI-Sh after activation by $H_2O + CO_2$	1.74	0.51, 2.2, 4.5–5.2, 25.0–26.5
The film of ShC water dispersion		0.51, 7.7; >30.1

TABLE 7.3 Adsorption Characteristics of Shungites

Shungite Samples and Source Sampling	S (m^2/g) BET	V_{pores} (cm^3/g)	r_{pores}, nm BET	r_{pores}, nm DR
Shunga deposit (ShI-Sh)	25.9	0.03	2.6	2.6
Nigozero deposit (ShI-N)	325.0	0.44	2.5	3.1
Maksovo deposit (ShI-M)	307.0	0.18	2.7	3.1
Chebolaksha deposit (ShI-Ch)	19.7	0.029	3.0	2.9

TABLE 7.4 Comparison of the Adsorption Parameters of Shungites and ShC Produced from Stable Water Dispersion

	In Terms of N_2 Adsorption <2 nm		In Terms of CO_2 Adsorption 0.33–0.7 nm	
Samples	S, m^2/g	V, cm^3/g	S, m^2/g	V, cm^3/g
Shl-Sh	25.9	0.03	87.7	0.033
Shl-N	325.0	0.41	103.5	0.040
ShC from stable water dispersion	325.4	0.45	144.2	0.055

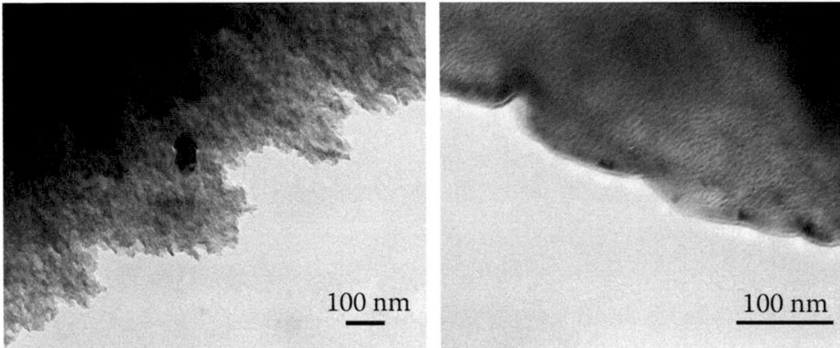

FIGURE 7.8 TEM images of two structures formed from ShI-Sh under treatment of ozone and steam: (left) porous structure with pores of 20–30 nm; (right) two-dimensional film 20–30 nm thick.

30%–50%, depending on ozonization time. The increase in specific surface is favored by the increasing internal porosity of "globular" aggregates and the oxidation of secondary aggregates, 20–50 nm in size, as shown by the TEM image of ozonized ShC (Figure 7.8). Adsorption data indicate the increased number of ~0.4 nm micropores and 10–50 nm mesopores. The increase in ultramicroporosity is supported by SAED data on variations in the half-width of diffraction maximum I (002) upon ozonization, which could be associated with the decrease in the coherent scatter region in the direction perpendicular to graphene layers (Rozhkova et al. 2009).

Ozonization can exert opposite effect on ShC surface values depending on the activity of ShC form: Shl-N or Shl-Sh. The total volume of the pores decreases, while the average pore radius of Shl-N remains unchanged, suggesting the oxidation of the basic structural elements of ShC (less than 1 nm), primarily on the surface. Conversely, the specific surface and volume increase within structural pores for Shl-Sh, as shown by x-ray and SAED data. The internal structural pattern of shungite mesopores has been determined earlier by the contrast variation method using SANS (Avdeev et al. 2006).

Ozonization, like thermal treatment of ShC, gives rise to two-dimensional structurized aggregates, 20–30 nm in size, impermeable for gas molecules. The presence of such aggregates is presumably responsible for a decrease in the porosity and specific surface of Shl-N and the structural transformations of ShC NPs obtained upon the precipitation of stable water dispersion. Aggregates in the film produced by drying ShC water dispersion are reproduced in size and structural pattern by globular aggregates on the cleaved surface of a natural sample Shl-Sh, from which ShC NPs were extracted (Rozhkova et al. 2011).

7.4.1 Modeling of Globular Nanocluster Formation

The condensation of stable ShC water dispersion is accompanied by the aggregation of NPs, followed by net formation. The various aggregation stages of ShC NPs in nature were modeled in the laboratory upon the concentration of stable water dispersion (Rozhkova et al. 2011).

Adsorption properties of original shungite powders produced from carbon-rich rocks that underwent different conditions were analyzed in comparison with the well-known Shunga deposit (sample Shl-Sh). According to geological evidences, Shl-N was formed as hydrothermally redeposited carbon; Shl-Ch subjected higher pressure and temperature, Shl-M was produced at elevated temperature. The samples under investigation are shown schematically in Figure 7.9. The adsorption characteristics of the shungite samples (Table 7.3) were compared with the powder produced from carbon NPs from Sh1-Sh water dispersion (Table 7.4). The average pore radius (*r*) of a Sh1-N sample by Brunauer, Emmett, and Teller (BET) is 2.5 nm at a maximum pore size of <21.2 нм (at $p/p° = 0.95$). Calculations made using the Dubinin-Radushkevich (DR) equation give a similar but higher value, 3.1 nm. The quantitative ratio of mesopores and micropores is ~2:1, and the total pore volume (V_{pores}) is 0.406 cm^3/g. Hence, Sh1-N is identified as a mesoporous–microporous adsorbent with a well-developed surface ($S = 325$ m^2/g).

ShC NPs, produced from water dispersion, coincide practically in adsorption parameters with Sh1-N (Table 7.4), suggesting the identity of the aggregate structure that forms the pores of the samples. The similarity of the data on the adsorption of CO_2 molecules indicates the identity of the microporous structure formed from the basic structural elements of ShC.

The adsorption isotherm of Sh1-Ch shows that it is a microporous adsorbent. The specific surface by BET is 19.7 m^2/g and the total volume of pores with a radius of <20.2 nm (up to $p/p° = 0.95$) is 0.029 cm^3/g. However, the average pore radius, 3.0 nm by BET, is practically the same as in all the Sh1-N, Sh1-M, and Sh1-Sh samples analyzed. It is a very important result that supports the integrity of the basic ShC structure (Table 7.4, Figure 7.10).

The pore size distribution observed for all shungites in which micropores and mesopores, <0.7–5.0 nm in size, predominate correlates with the size of basic structural elements (0.4–0.7 nm) released in water dispersion and with that of primary globular ShC aggregates, less than 6 nm in size. Secondary aggregates, measuring 10–100 nm, are formed later upon centrifugation of water dispersion and persist upon its condensation.

The results obtained have supported the similarity of the structural properties of shungites on the basic and primary level and, hence, the similarity of pore sizes and the dominance of small mesopores and micropores. The difference, which depends on the conditions of formation of shungites in the above deposits, is a result of the formation of the larger structural units of secondary aggregates indicated by such adsorption characteristics as total pore volume and the specific surface value.

Comparison of the porosity and specific surface values of ShC and ND clusters, obtained upon condensation of respective water dispersions, has shown that the specific surface values for N_2 (BET) are 283.8 and 325.0 m^2/g for ND and ShC, respectively. These values are sensitive to storage conditions because the samples are hygroscopic: The specific surfaces of ND and ShC based on the H_2O molecule

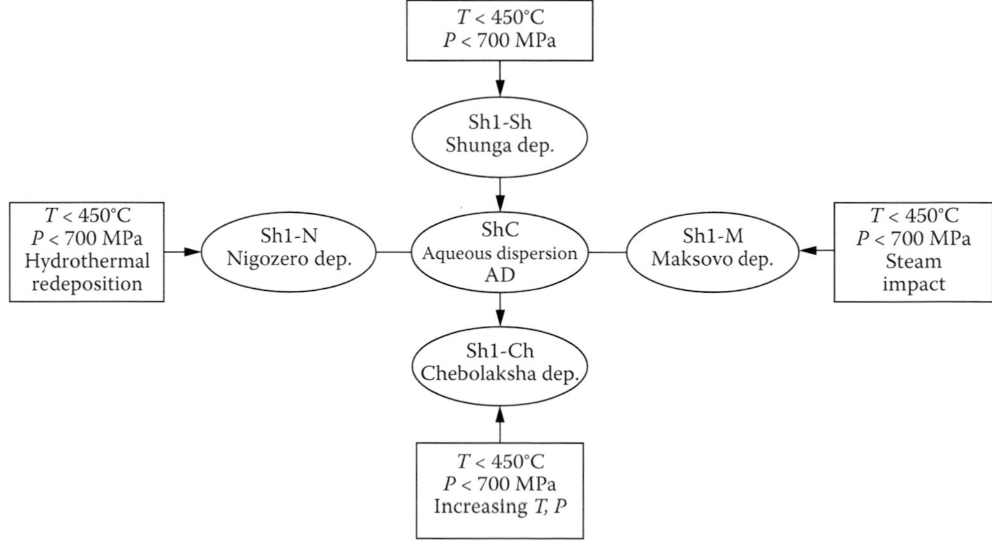

FIGURE 7.9 Scheme of the samples under study.

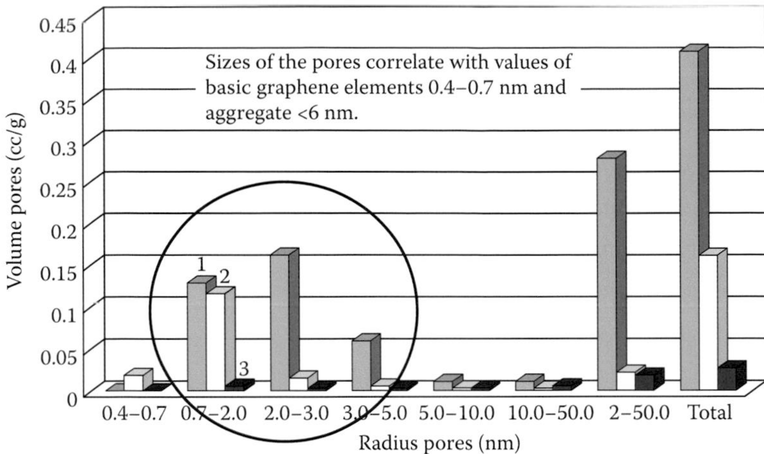

FIGURE 7.10 Pore size distribution for shungites from different deposits: (1) Nigozero (Shl-N), (2) Maksovo (Shl-M), and (3) Chebolaksha (Shl-Ch).

adsorption are 503.2 and 577.7 m^2/g, respectively. Both materials display a mesoporous structure and well-defined hysteresis on the desorption curves (Rozhkova et al. 2009).

If, however, ND can be described as a corpuscular adsorbent when obtained from stable water dispersion, then carbon, obtained from stable ShC water dispersion, has an adsorption isotherm similar to that of montmorillonite, the adsorption capacity of which varies owing to the mobility of the NPs that build up pore walls. In ShC, variations in adsorption capacity are induced by the pores formed from NGr fragments. The total pore volumes are 0.475 and 0.45 cm^3/g and the average pore radii are 3.4 and 3.1 nm for ND and ShC, respectively.

Nanoparticles react to ozonization in different ways: The specific surface of ND increases slightly and the concentration of oxygen-containing groups, primarily carboxyl groups, rises substantially, while the specific surface of ShC nanoclusters decreases, which could be a result of the high mobility of NGr fragments in ShC clusters and the formation of impermeable films.

7.5 Reduced GO in ShC

7.5.1 Graphene Molecular Chemistry in Defining the Origin of ShC

The study of the solubility of GOs, which, unlike graphite, are soluble in water and polar solvents, and less soluble reduced GOs (rGO) has been conducted intensively in the past few years (Stankovich et al. 2007). Their behavior is also characterized by the presence of interacting clusters that form gel-like liquid–crystalline systems of nematic type (Xu and Gao 2014; Zhao et al. 2014) with well-defined photoluminescent (PL) properties and the dependence of PL on solvent and excitation wave length (Cushing et al. 2014). The dipole moment and polar groups along the defective margins of rGO may impart hydrophilic properties to the surface of carbon NPs and contribute to the stability of clusters from NPs in water dispersion and their ability to intercalate large amounts of water within the structures (Xu et al. 2014). It seems most likely that rGO molecules themselves have a nonplanar structure with a dipole moment of about 2 D, as has been shown for corranulene–polyarene, which have a similar molecular mass, and their derivatives (Lovas et al. 2005). Curved polyarenes are capable of building up various crystal-packed structures (Filatov et al. 2010) and supramolecular compounds with interesting optical properties (Cocchi et al. 2013).

In accordance with the molecular theory of a graphene (Sheka 2012), any chemical reaction in which a graphene molecule is involved begins at its edge atoms because of the distinctive properties of the atoms. The termination of edge atoms by chemical addenda is the main reason for the cessation of the growth of a lamella, thereby restricting its linear size. This has been repeatedly observed empirically in the case of GO (Dreyer et al. 2010). The latter evidence has provided the basis for the large-scale production technology of technical graphene in the form of rGO (Dimiev et al. 2013). It has been shown under laboratory conditions that an oxidation reaction is accompanied by the destruction of original graphite and the transformation of macrosized graphene sheets into a combination of nanosized fragments (Pan and Aksay 2011). Thus, continuous 900-s oxidation results in the disintegration of a micron-sized graphene sheet into small fragments ~1 nm in size (Hui and Yun 2011). Further oxidation does not lead to a decrease in size, thus stabilizing the size of end fragments at the 1-nm level. This discovery suggests that shungite sheets, ~1 nm in size, were formed upon the geologically long oxidation of graphene sheets produced upon the graphenization of carbonaceous deposits.

Hydrothermal conditions that favored oxidation and accompanied shungite formation are regarded as a convincing argument in favor of the hypothesis of the origin of shungite in accordance with the GO scenario. However, analysis of the mass percentage of oxygen has shown that the oxygen content does not exceed 2 wt%. This discrepancy suggests either the complete or partial reduction of original GO during its geological formation.

Synthetic GO is commonly reduced in the laboratory by strong reducing agents, but it has been shown recently (Liao et al. 2011) that GO can also be reduced in water, but this process is much longer. Obviously, the geological time taken for ShC formation is long enough for the reduction of original GOs in water. Thus, water is crucial for the formation of the basic structural element of ShC.

It has been proven empirically that aggregation is also characteristic of synthetic GOs and rGO sheets. The data obtained by infrared absorption (Acik et al. 2010) and nonelastic neutron scatter (NNS) (Sheka et al. 2014) have shown that synthetic GO forms turbostratic structures that retain water. Similar structural characteristics have been proposed recently for synthetic rGO and ShC (Sheka et al. 2014). It has been proven by the neutron diffraction method that the average characteristic interfacial distance of graphite, d_{002}, is ~6.9 Å for GO and ~3.5 Å for rGO of both synthetic and natural origins because the planarity of rGO lamellar is restored upon chemical GO reduction.

Quantum-chemical calculations show that water molecules lie between the neighboring layers of the stacked GO structure, while water molecules in rGO can only be retained by edge atoms beyond the basal plane. Water molecules in the stacked rGO structure can only lie in the pores formed by stacks. NNS data (Sheka et al. 2014) show that 4 wt% of water is retained in shungite pores. In addition, the contribution of the nanofragment of rGO as the basic structural element of ShC has been proven, making it easier to imagine the elements at the next stages of ShC microstructure.

The characteristic diffraction d_{002} peaks in shungite are much wider than those in graphite. Hence, stack rGOs have a thickness of ~1.5 nm, and each stack consists of five to six layers. The same structural elements were reported in Jehlicka and Rouzaud (1993). Stacks build up the secondary structure of shungite. Stacks, in turn, are combined to form globules that represent a third level in the ShC structure (Sheka and Rozhkova 2014). Interaction between globules gives rise to larger aggregates, 20–100 nm across, that agglomerate to complete the formation of the fractal structure of shungite.

7.5.2 Hypothetic Phase Diagram of ShC Dispersions

It is known that the type and characteristics of the parameters of the pair potential of interaction between atoms and molecules in condensed matter are responsible for the basic thermodynamic characteristics of the system. An individual class of substances with common thermodynamic characteristics consists of systems of molecules with short-range pair interaction. The best-known examples are inert gas molecules (Smirnov 1993), fullerenes (Bezmel'nitsyn et al. 1998; Gripon et al. 1998), and globular proteins (Gripon et al. 1998), remarkable for the cluster organization of condensed phases. Consequently, a phase diagram (PD) describing the behavior of such molecules in various solvents can also be useful. Such diagrams have been proposed and described for the heterogeneous colloid systems (Zaccarelli 2007) and of dispersions of some globular proteins (Rozhkov and Goryunov 2010, 2014; Vekilov 2012), while no PDs for fullerenes have been constructed (Bezmel'nitsyn et al. 1998; Avdeev et al. 2010). ShC dispersions are even more poorly understood because the solubility of rGO, the basic structural element of ShC, has not been studied. At the same time, some experimental data show that the phase properties of C_{60} and ShC dispersions have many common features (Belousova et al. 2015).

Solid-phase samples of fullerenes and ShC, obtained by slow evaporation of water from their dispersions at temperatures of 1°C and 4°C, were studied by the SAED and TEM methods (Rozhkov et al. 2007). The study revealed at least two types of phases in the samples: fullerite (for fullerenes) with face-centered (fcc) cubic lattice and a new phase that displays an acicular morphology and the same interplanar distance (d_{002}) but a different diffraction peak intensity distribution. This crystalline structure has wide maxima with small intensity and diffuse halo consistent with the great contribution of noncrystalline constituents, which decrease with a drop in temperature to 1°C; i.e., structural ordering increases with decreasing temperature. A similar phase is also crystallized from ShC dispersion, but its composition is more complex.

Anomalies (jumps or gentle transitions) on the temperature dependences of thermal capacity, thermal conductivity, thermal electromotive force, and electrical conductivity, similar to those obtained on fullerenes, were revealed for solid ShC samples. They were interpreted as similar processes occurring in globular ShC and fullerene structures (Parfen'eva et al. 1994). The hysteresis of the temperature dependence of electrical conductivity was observed on the shungite samples upon heating and cooling. As the samples are completely dehydrated, hysteresis disappears.

The structural-dynamic properties of the water dispersions of ShC and fullerenes in the temperature range close to type 1 PT temperature for fullerenes (260K) were studied (Rozhkov et al. 2007) by the electron spin resonance (ESR) method using water-soluble 4-oxo-TEMPO spin probe. Variation in the rotational frequency of the spin probe in ShC dispersion is shown in Figure 7.11. A similar pattern is observed for C_{60} dispersion (Rozhkov et al. 2007).

An abrupt jump is caused by the freezing of overcooled water at 265K and a plateau in the 260K range for ShC (and C_{60}) with a phase transition. Obviously, correlation time varies substantially in nonlinear manner at temperatures close to 260K, which is characteristic of the spin probe method when recording the PT (Wasserman and Kovarsky 1986).

As variations in the crystalline structure of molecules affect the properties of solutions and vice versa (Bezmel'nitsyn et al. 1998), available data are indirectly indicative of a PD in ShC similar to that in

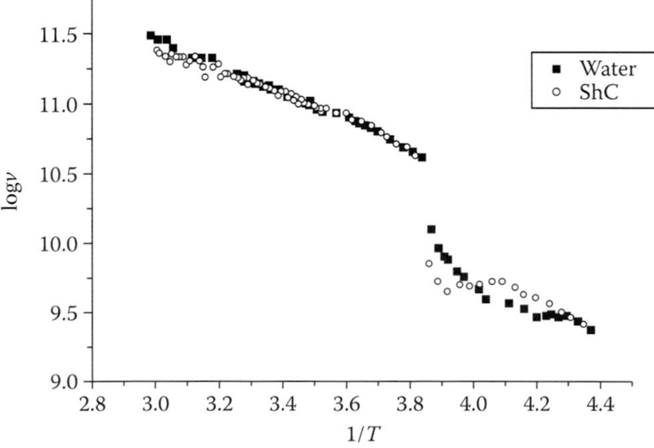

FIGURE 7.11 Arrhenius plots of the spin probe 4-oxo-TEMPO rotational frequency ν for water (solid squares) and 0.1 mg/mL ShC dispersion (open circles).

fullerite at 260K. Therefore, it is safe to assume that analysis of the phase properties of fullerene dispersions and the construction of a hypothetic PD for C_{60} will help to understand the PD of ShC dispersion.

There are models of fullerene cluster formation in solutions used to interpret the anomalous dependence of C_{60} solubility on temperature having maximum (Avdeev et al. 2010). Some attribute cluster formation to PT in fullerite at 260K (Ruoff et al. 1993; Schur et al. 2008), others to transition from crystals to crystallosolvates (Korobov et al. 2000). At the same time, the ability to form crystallosolvates could be related to PT of type 1 at 260K induced by variation in the parameters of the crystalline lattice of fullerite from simple cubic (sc) to fcc. Deformed body-centered (bcc) or body-centered tetragonal crystalline lattices of NPs may be formed as a result of interaction with solvent (Bian et al. 2011; Goodfellow and Korgel 2011; Wang et al. 2013). These crystalline polymorphs could be a result of the various electron states of the pentagon constituents of a fullerene molecule because of the thermally induced restructuring of the double bonds in the C_{60} frame (Schur et al. 2008).

Such a hypothetic PD is presented in temperature-packing density coordinates in Figure 7.12 by analogy with globular protein dispersions (Vekilov 2012). The left vertical line to the temperature 260K is consistent with a crystalline phase with sc-lattice formed by C_{60} (or rGO in ShC) molecules with a packing density of 0.52. PT of type 1 (horizontal line) occurs at 260K, and fullerene (or rGO in ShC) molecules form the densest (presumably amorphous, packing density 0.74) fcc-phase, shown by the right vertical line. However, as temperature rises, molecules with varying electron structures may appear in dispersion. The formation of structures with a distorted lattice, e.g., (bcc) with a packing density of 0.68, induced by interaction with solvent and crystallosolvate formation, would be more optimum for such molecules. There is a solubility curve for such molecules shown in Figure 7.12 by an ascending dotted line (curve 1). Solubility for molecular clusters is shown in Figure 7.12 by curve 2.

If packing density (which arises with increasing concentration of carbon) reaches the binodal, then a metastable, dense phase of such molecules forms (PT of liquid–liquid type, arrow in Figure 7.12) as microsized drops. As temperature rises above the critical point, dynamic clusters remain as phantoms of low-temperature phases instead of metastable phases (because the supercritical region is expected to be single phase) (Rozhkov and Goryunov 2014), and their "solubility" (curve 2) declines because crystallosolvates lose stability with rising temperature. In this case, electron bonds that no longer interact with the solvent are involved in intermolecular interaction between fullerene (or rGO in ShC) molecules, and solid-phase embryos with dense fcc lattice are formed.

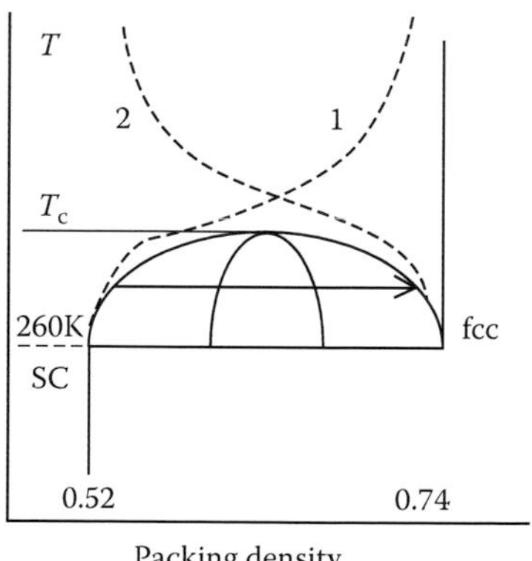

FIGURE 7.12 Hypothetic PD of C_{60} fullerene (or graphene) dispersion in temperature—packing density coordinates. The large semiellipse is a binodal. The small semiellipse is a spinodal. Curve 1 shows the solubility of molecules that create embryos with bcc lattice. The arrow indicates the metastable equilibrium of such solvated molecules and dense phase. Curve 2 shows the solubility of clusters in supercritical region. T_c is critical temperature. Phase transition of type 1 between sc and fcc at 260K is shown by a horizontal line.

At the critical point, the differences between the densities and electron structures of all fullerene (or rGO in ShC) molecules become indistinguishable, and solubility is expected to be maximum, considering the properties of a critical state.

If the concentration (packing density) of NPs reaches the spinodal boundary (small semiellipse in Figure 7.12), then the stability of the metastable phase pattern of the solution is disturbed, triggering the formation of gel-like nonequilibrium structures by aggregation.

Caused by evaporation or salting out, a solid phase may arise in the supercritical region as a multilevel structural pattern that consists of nanosized lattices and mesoscopic liquid–crystalline fractal-structured carbon clusters.

7.6 Interaction of ShC NPs with Blood and Membrane Proteins

Biomedical technologies have been among the most promising fields of nanocarbon materials applications. The known examples of carbon nanotube utilization include drug, gene material, and other biomacromolecule delivery as well as production of substrates for cell growth, implants, and low-impedance bioelectrodes (Chun et al. 2013; Wong et al. 2013). Considerable data on the biological activity of the fullerene NP and their derivatives and of pristine and modified C_{60} fullerenes including antiviral, neuromodulating, immunomodulating, and antioxidant effects (Nakamura and Mashino 2012; Yang et al. 2014) have been one of the main sources of understanding of the biological activity of nanocarbon. Fullerenes can be localized in the hydrophobic regions of membranes and protein structures owing to their lipophilicity. On the other hand, some results on the possible toxic, prooxidant, mutagenic, and carcinogenic effects of fullerenes (Sayes et al. 2005; Lee et al. 2010) have been reported with the development of water-soluble fullerene derivatives. A systematic investigation of graphene and its derivatives' cellular toxicity has demonstrated the reduction of cytotoxicity of GO on protein corona formation (Hu et al. 2011; Mao et al. 2013).

Of the carbonaceous materials that are the most prevalent form of nanomaterials found in the environment, shungite nanocarbon is one of the best examples as it presents a nanomaterial that is rather specific, widely spread, and intensively introduced into various cosmetic and bioactive products mostly without reliable substantiation. The mechanism of ShC biological activity is not well studied, with the role of nanocarbon released from shungite rocks upon its interaction with water being of particular interest. To this end, shungite nanocarbon-induced effects on vertebrate blood proteins and erythrocyte membranes have been studied and compared with those of fullerenes and NDs. In particular, the aim was to elucidate the effects of nanocarbon dispersions on the structural dynamic, thermodynamic, hydrodynamic, and redox properties of the biomacromolecules and cell membranes that are expected to underlie shungite nanocarbon biological activity.

A set of employed physicochemical methods includes spin probe and spin label ESR, differential scanning microcalorimetry (DSC), DLS, size exclusion chromatography (SEC), and scanning electron microscopy (SEM). The carbon nanomaterials were used in the form of water nanodispersions prepared via procedures described in previous chapters for ShC.

7.6.1 Redox Properties of ShC Nanoparticles

Oxygen and reactive oxygen species solubility is known to be enhanced in surface (hydration) water, and developed hydration shells of NPs, with their extremely high surface–volume ratio, can therefore contribute considerably to NP's biological activity in redox reactions, in particular. ESR studies of the effects of fullerenes and ShC suspensions on the reduction and oxidation of stable nitroxyl spin probe in the presence of iron Fe(II) (Rozhkov and Goryunov 2013) have shown that the spin probe in the presence of Fe(II) is most rapidly reduced to hydroxylamine, its paramagnetic properties being vanished in the nanocarbon dispersions (Figure 7.13a). The oxidation of hydroxylamine started immediately, most likely

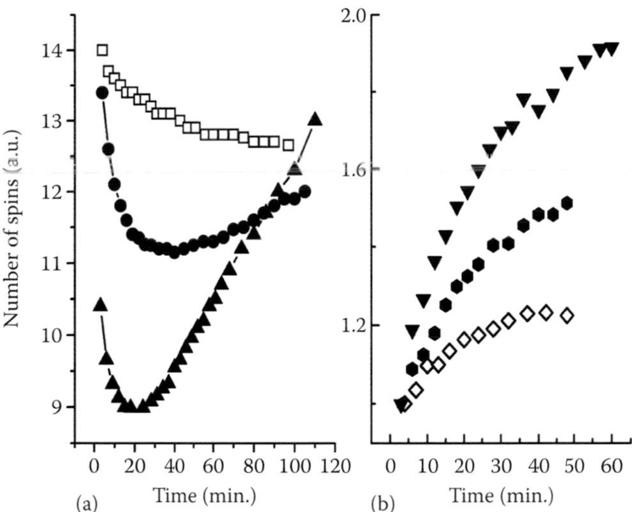

FIGURE 7.13 Effects of carbon NP on the $FeSO_4$-induced oxidation/reduction of spin probes: (a) TEMPO spin-probe in solution (squares), in C_{60} fullerene water dispersion (circles), and in ShC NPs' dispersion (triangles); (b) 5-DOXYL-stearic acid spin-probe in the erythrocyte membrane (diamonds), on addition of C_{60} fullerene dispersion (hexagons), and on addition of ShC NPs' dispersion (triangles).

because of the oxygen dissolved in the hydration water. The spin probe paramagnetism is thus restored, with both processes being accelerated with increasing nanocarbon concentration.

The spin probe paramagnetism has not been restored in the absence of nanocarbon as well as under the oxygen shortage. Analogous phenomena have been observed in suspensions of erythrocyte ghost membranes with the difference that, in this case, the spin probe reduction stage was not observed at all, probably because of the very high rate of reduction, so that only the second stage (oxidation of hydroxylamine) has been registered experimentally (Figure 7.13b).

A study (Goryunov and Borisova 2014) of the oxidative effect of nanocarbon dispersions on proteins, using the spontaneous oxidation (auto-oxidation) of hemoglobin (Hb), has shown that both shungite nanocarbon and hydrated C_{60} nanoclusters (C60FWS) induce a significant increase in the oxidized Hb concentration already upon mixing (Figure 7.14). No definite effect has been observed in the case of ND dispersions. Oxidation of Fe(II) of Hb molecule has been significantly accelerated in the presence of C60FWS and shungite nanocarbon even at high pH (above physiological values). Auto-oxidation is extremely slow at very high pH and can thus be neglected. The observed effect is an evidence of the prooxidant activity of shungite nanocarbon as well as C60FWS.

Hb auto-oxidation is generally considered to proceed as iron oxidation (oxyHb to metHb conversion). Oxygen reduction to superoxide anion is accompanied by nucleophilic displacement of the superoxide by water molecule and requires an imidazole group of distal histidine of Hb to be protonated; therefore, lower pH promotes such an auto-oxidation (Benesch et al. 1973). However, no effect of dispersion pH on the concentration-dependent ferrous heme oxidation in Hb caused by shungite nanocarbon and fullerene C_{60} has been revealed. Heme protein auto-oxidation pathway consideration suggests that the most probable mechanism of shungite nanocarbon and C_{60} effect on the Hb auto-oxidation process is determined by their electron-acceptor capacity and involves a pH-independent transfer of an electron from a deoxyheme to an O_2 molecule, but not the dissociation of superoxide-anion (Goryunov and Borisova 2014).

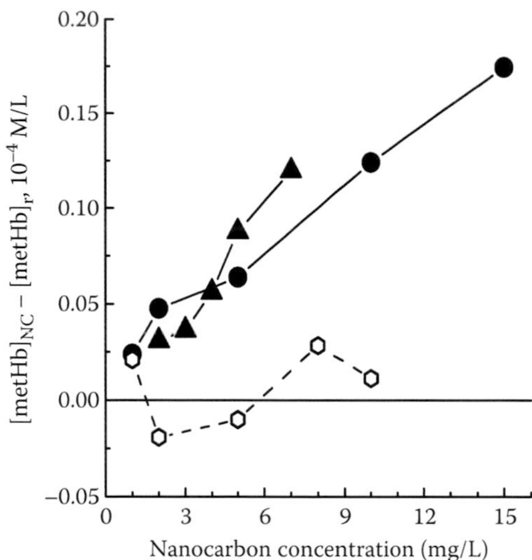

FIGURE 7.14 Magnitude of Hb (1.7 mg/mL) auto-oxidation at pH within 5.2–9.1 in the presence of varied nanocarbon concentrations: ShC NPs (triangles), fullerene C_{60} (C60FWS) (circles), and NDs (hexagons).

7.6.2 ShC Nanoparticles—Serum Albumin Complexation

Biological responses to the introduction of NPs into the environment of living systems is crucially influenced by the physicochemical surface properties of the NPs predominantly through the effects on the protein corona formation (protein adsorption), which is known as a general phenomenon for carbon NPs as well (Rahman et al. 2013; Saptarshi et al. 2013; Lee et al. 2014). This has been observed in the cases of NDs and fullerenes (Deguchi et al. 2007; Puzyr et al. 2007) and we have revealed a similar effect in the case of shungite nanocarbon.

Gel filtration and ultracentrifugation experiments have shown that under physiological conditions, human serum albumin (HSA)–nanocarbon dispersions are highly stable toward sedimentation and aggregation, whereas the stability of protein-free nanocarbon is minimal under these conditions. The complexes after exposure to ShC in HSA solution for periods of time sufficient to generate NP hard corona have also shown much higher chromatographic mobility as compared with the pure protein solution. On the other hand, dispersion of bare shungite nanocarbon has not passed through the column at all, likely because of efficient direct binding with the gel through nonspecific interactions caused by higher surface energy of the bare NPs (Lee et al. 2014). This indicates that ShC interacts effectively with HSA and penetrate the column as complex with the protein. Hard corona is retained on ShC when eluted, while soft corona is lost by the NP and eluted later.

DSC data demonstrate that HSA and bovine serum albumin molecules are expanded in water dispersion of nanocarbon particles as compared with the protein water solution. Deconvolution of DSC thermograms has shown changes in the intermolecular interactions of protein domains owing to the intermolecular interactions with ShC NPs that lead to divergent alterations in the domains' interaction. Similar data are available for NDs and C_{60} fullerenes (Rozhkov and Goryunov 2013). The thermograms' changes have shown that the ShC NP–protein interaction occurs not via the negatively charged sites of NP surface (Fleischer and Payne 2014) but via the fatty acid binding sites on serum albumins predominantly.

The protein structure destabilization upon the complex formation has been additionally confirmed by fatty acid spin-probe ESR. The spectral changes similar to those in the case of HSA denaturation with urea suggest partial unfolding of the protein upon interaction with the NP surface. However, at increasing HSA/NP concentration ratios, the protein structure has been somewhat stabilized, possibly because of protein–protein interactions in the course of layered adsorption (Rozhkov and Goryunov 2013). As follows from analogous DSC studies of Hb solutions, such an efficient complex formation with ShC is not typical of this heme protein as no structural differences of the protein have been revealed in presence of the ShC.

7.6.3 ShC Nanoparticles' Effects on Erythrocyte Membrane

The interaction with cell membrane is crucial in the view of both nanocarbon biocompatibility and cytotoxicity as well as the prospects of their applications for drug delivery. Using ESR spin-labeling, the proteins of the actin–spectrin complex of erythrocyte membrane were shown to increase their thermal stability in the presence of hydrated ShC, C60FWS, and ND (Rozhkov and Goryunov 2013). DSC data have also demonstrated the increased thermal stability of the proteins of the erythrocyte cytoskeleton complex, whereas their membrane proteins were not affected. Studies of the thermal and osmotic stability of erythrocytes in ShC dispersions have revealed that the major effect of the possible interaction is the decrease in hemolytic stability of the cells (Rozhkov and Goryunov 2013). From the analysis of SEM pictures, the formation of shapeless aggregates of erythrocytes in the presence of ShC dispersion has been much enhanced. Separation of ShC has led to reversion of the cells' aggregation (Goryunov et al. 2009). The molecular mechanism of action probably involves the interaction of ShC with the membrane surface proteins of erythrocytes. The effects depend on the NP concentration and temperature.

Thus, adsorption of serum albumins, and not Hbs, that reach its maximum values for the majority of carbon NP within 5 min (Zhu et al. 2010) leads to complexation with the formation of a layered ShC NP–protein corona. Such complexation is reported to reduce the intracellular uptake of carbon NPs and decrease their inherent cytotoxicity (Hu et al. 2011; Park and Khang 2012), including through the decrease in reactive oxygen species production (Mao et al. 2013). Expansion of serum albumin structure on ShC NP–protein interaction with occupation of some protein ligand binding sites can potentially lead to subsequent adverse biological responses.

7.7 Conclusion

The occurrence of shungite deposits in the Lake Onega area near graphite deposits in the vicinity of Lake Ladoga, characteristic of the Karelian region's geology, is noteworthy. This shows that the Karelian region is generally favorable for the formation of graphene layers and that graphite and shungite have originated from common sources. Potential chemical reactions in water at 300°C–400°C during the metamorphism of carbon substance are expected to result either in the growth of graphene layers or in the healing of pores (Sheka et al. 2014).

The water dispersions obtained display a variety of properties that, on the one hand, indicate their direct relation to the properties of ShC and, on the other hand, are similar to fullerenes and have properties typical of water dispersions of such quantum dots as gold and silver NPs, cadmium sulfide (CdS) and cadmium selenide (CdSe), and synthetic Gr quantum dots (Razbirin et al. 2014). Like the previously discussed dispersions, ShC NP dispersions react actively to an increase in nonlinear (Belousova et al. 2015) and spectral (Razbirin et al. 2014) optical properties. Shungite NP dispersions are similar in the great heterogeneity of morphological and optical properties to synthetic Gr quantum dots.

The hypothetic PD proposed and the existence of a mesoscopic supercritical region suggest that the formation of a heterogeneous multilevel cluster-aggregate system is not only a kinetic but also a thermodynamic phenomenon. Therefore, the conditions required can be selected to obtain unified ShC NP dispersions and condensation structures with certain morphology and ordering, consistent with the dispersions, which correspond to the varying types of nanocrystalline superlattices induced by interaction with solvents and ligands. A structural–dynamic study of the formation of ShC clusters and aggregates, affected by the varying concentrations and physicochemical properties of the dispersion medium, was conducted. The conditions required for obtaining the various structural types of ShC nanoclusters and their aggregates were determined for the first time. NPs in the films produced from water dispersions lie at net sites formed upon condensing dispersions. The size of the cells in the nets coincides with that of globular clusters (~100 nm). Ellipsoid clusters and chain as well as stacks and plate aggregates were obtained dominantly in films from different organic solvents.

The preliminary study of the effect of autoclaved ShC dispersions, injected intravenously into the animal organism, has shown that they are not toxic, but they activate the system of antioxidant protection from oxidative stress.

Acknowledgments

This work was supported by Basic Research Program, RAS, Earth Sciences Section-5, and grant RFBR 13-03-00422.

References

Acik, M., Mattevi, C., Gong, C. et al. "The role of intercalated water in multilayered graphene oxide." *ACS Nano* 4 (2010): 5861–5868.

Alekseev, N.I., Arapov, O.V., Bodyagin, B.O. et al. "Activation of carbonaceous component of shungite III and sorption activity of the material towards hydrogen." *Russ. J. Appl. Chem.* 79 (9) (2006): 1439–1443.

Amara, A.B.H. "X-ray diffraction, infrared and TGA/DTG analysis of hydrated nacrite." *Clay Minerals* 32 (1997): 463–470.

An, X., Simmons, T., Shah, R. et al. "Stable aqueous dispersions of noncovalently functionalized graphene from graphite and their multifunctional high-performance applications." *Nano Lett.* 10 (11) (2010): 4295–4301.

Andrievsky, G.V., Kosevich, M.V., Vovk, O.M., Shelkovsky, V.S., Vaschcenko, L.A. "On the production of an aqueous colloidal solution of fullerene." *J. Chem. Soc. Chem. Commun.* 12 (1995): 1281–1282.

Avdeev, M.V., Tropin, T.V., Aksenov, V.L., Rosta, L., Garamus, V.M., Rozhkova, N.N. "Pore structures in shungites as revealed by small-angle neutron scattering." *Carbon* 44 (2006): 954–961.

Avdeev, M.V., Aksenov, V.L., Tropin, T.V. "Models of cluster formation in solutions of fullerenes." *Russ. J. Phys. Chem. A* 84 (8) (2010): 1273–1283.

Belousova, I.M., Videnichev, D.A., Kislyakov, I.M., Krisko, T.K., Rozhkova, N.N., Rozhkov, S.S. "Comparative studies of optical limiting in fullerene and shungite nanocarbon aqueous dispersions." *Opt. Mater. Express* 5 (1) (2015): 169–175.

Benesch, R.E., Benesch, R., Yung, S. "Equations for the spectrophotometric analysis of hemoglobin mixtures." *Anal. Biochem.* 55 (1973): 245–248.

Berezkin, V.I., Konstantinov, P.P., Holodkevich, S.V. "Holl's effect in natural glass carbon of shungite." *Solid State Phys.* 39 (10) (1997): 1783–1786.

Bezmel'nitsyn, V.N., Eletskii, A.V., Okun', M.V. "Fullerenes in solutions." *Phys. Usp.* 41 (1998): 1091–1114.

Bian, K., Choi, J.J., Kaushik, A., Clancy, P., Smilgies, D.-M., Hanrath, T. "Shape-anisotropy driven symmetry transformations in nanocrystal superlattice polymorphs." *ACS Nano* 5 (4) (2011): 2815–2823.

Buseck, P.R., Galdobina, L.P., Kovalevski, V.V., Rozhkova, N.N., Valley, J.W., Zaidenberg, A.Z. "Shungites: The C-rich rocks of Karelia, Russia." *Can. Mineral.* 35 (6) (1997): 1363–1378.

Chun, Y.W., Crowder, S.W., Mehl, S.C., Wang, X., Bae, H., Sung, H.J. "Therapeutic application of nanotechnology in cardiovascular and pulmonary regeneration." *Comput. Struct. Biotechnol. J.* 7 (2013): 1–7.

Cocchi, C., Prezzi, D., Ruini, A., Caldas, M.J., Fasolino, A., Molinari, E. "Concavity effects on the optical properties of aromatic hydrocarbons." *J. Phys. Chem. C* 117 (2013): 12909–12915.

Cornelius, C.D. "Classification of natural bitumen: A physical and chemical approach," in *Exploration for heavy crude oil and natural bitumen*, ed. R.F. Meyer (AAPG Studies in Geology, 25. Tulsa, OK: Amer. Assoc. Petroleum Geologists, 1987), 165–174.

Cushing, S.K., Li, M., Huang, F., Wu, N. "Origin of strong excitation wavelength dependent fluorescence of graphene oxide." *ACS Nano* 8 (1) (2014): 1001–1013.

Deguchi, S., Yamazaki, T., Mukai, S.A., Usami, R., Horikoshi, K. "Stabilization of C_{60} nanoparticles by protein adsorption and its implications for toxicity studies." *Chem. Res. Toxicol.* 20 (2007): 854–858.

Dimiev, A.M., Alemany, L.B., Tour, J.M. "Graphene oxide. Origin of acidity, its instability in water, and a new dynamic structural model." *ACS Nano* 7 (2013): 576–584.

Dreyer, D.R., Park, S., Bielawski, C.W., Ruoff, R.S. "The chemistry of graphene oxide." *Chem. Soc. Rev.* 39 (2010): 228–240.

Emel'yanova, G.I., Gorlenko, L.E., Tikhonov, N.A., Rozhkova, N.N., Rozhkova, V.S., Lunin, V.V. "Oxidative modification of shungites." *Russ. J. Phys. Chem.* 78 (7) (2004): 1070–1076.

Filatov, A.S., Scott, L.T., Petrukhina, M.A. "π–π Interactions and solid state packing trends of polycyclic aromatic bowls in the indenocorranulene family: Predicting potentially useful bulk properties." *Cryst. Growth Des.* 10 (10) (2010): 4607–4621.

Fleischer, C.C., Payne, C.K. "Nanoparticle–cell interactions: Molecular structure of the protein corona and cellular outcomes." *Acc. Chem. Res.* 47 (2014): 2651–2659.

Goodfellow, B.W., Korgel, B.A. "Reversible solvent vapor-mediated phase changes in nanocrystal superlattices." *ACS Nano* 5 (4) (2011): 2419–2424.

Goryunov, A.S., Borisova, A.G. "Probable mechanism of haemoglobin autoxidation in water dispersions of carbon-based nanomaterials." *Trans. Karelian Res. Cntr. RAS* 5 (2014): 71–77.

Goryunov, A.S., Borisova, A.G., Rozhkov, S.P., Sukhanova, G.A., Rozhkova, N.N. "Morphology and aggregation of erythrocytes in carbon nanodispersions." *Trans. Karelian Res. Cntr. RAS* 3 (2009): 30–37.

Gripon, C., Legrand, L., Rozenmann, I., Boue, F., Regnaut, C. "Relation between the solubility and effective solute–solute interaction for C60 solutions and lysozyme solutions: A comparison using the stickyhard-sphere potential." *J. Crystal Growth* 183 (1998): 258–268.

Hairullin, A.R., Stepanova, T.P., Rozhkova, N.N., Gladchenko, S.V. "Static dialectical polarization of structural elements of shungite carbon in benzene types solvents." *St. Petersburg State Polytech. Univ. J. Phys. Math.* 4 (153) (2012): 111–114.

Hu, W., Peng, C., Lu, M. et al. "Protein corona-mediated mitigation of cytotoxicity of graphene oxide." *ACS Nano* 5 (2011): 3693–3700.

Hui, W., Yun, H.H. "Effect of oxygen content on structures of graphite oxides." *Ind. Eng. Chem. Res.* 50 (2011): 6132–6137.

Jehlicka, J., Rouzaud, J.N. "Transmission electron microscopy of carbonaceous matter in Precambrian shungite from Karelia," in *Bitumens in ore deposits*, eds. J. Parnell, H. Kucha, P. Landais (Berlin: Springer, 1993), 53–60.

Kalinin, Y.K., Kalinin, A.I., Skorobogatov, G.A. *Karelian shungites for new construction materials, chemical synthesis, gaze purification, water treatment and medicine* (St.-Petersburg: St.-Petersburg University, 2008) (in Russian).

Korobov, M.V., Smith, A.L. "Solubility of the fullerenes," in *Fullerenes. Chemistry, physics, and technology*, eds. K.M. Kadish, R.S. Ruoff (New York: J. Wiley & Sons, Inc.), 53–90.

Kovalevski, V.V. "Structure of shungite carbon." *Russ. J. Inorg. Chem.* 39 (1994): 28–32.

Kovalevski, V.V., Safronov, A.N., Markovski, Y.A. "Hollow carbon microspheres and fibres produced by catalytic pyrolysis and observed in shungite rocks." *Mol. Mater.* 8 (1996): 21–24.

Kovalevski, V.V., Buseck, P.R., Cowley, J.M. "Comparison of carbon in shungite rocks to other natural carbons: An X-ray and TEM study." *Carbon* 39 (2) (2001): 243–256.

Kovalevski, V.V., Prikhodko, A.V., Buseck, P.R. "Diamagnetism of natural fullerene-like carbon." *Carbon* 43 (2005): 401–405.

Kuznetsov, V.L., Butenko, Y.V., Chuvilin, A.L., Romanenko, A.I., Okotrub, A.V. "Electrical resistivity of graphitized ultra-disperse diamond and onion-like carbon." *Chem. Phys. Lett.* 336 (2001): 397–404.

Lee, J., Mahendra, S., Alvarez, P.J. "Nanomaterials in the construction industry: A review of their applications and environmental health and safety considerations." *ACS Nano* 4 (2010): 3580–3590.

Lee, Y.K., Choi, E.J., Webster, T.J., Kim, S.H., Khang, D. "Effect of the protein corona on nanoparticles for modulating cytotoxicity and immunotoxicity." *Int. J. Nanomedicine* 10 (2014): 97–113.

Lemanov, V.V., Balashova, E.V., Sherman, A.B., Zaidenberg, A.Z., Rozhkova, N.N. "Are there fullerenes in pre-Cambrian rock shungite?" *Mol. Mater.* 4 (1994): 205–208.

Liao, K.-H., Mittal, A., Bose, S., Leighton, C., Mkhoyan, K.A., Macosko, C.W. "Aqueous only route toward graphene from graphite oxide." *ACS Nano* 5 (2011): 1253–1258.

Lovas, F.J., McMahon, R.J., Grabow, J.-U., Schnell, M., Scott, L.T., Kuczkowski, R.L. "Interstellar chemistry: A strategy for detecting polycyclic aromatic hydrocarbon in space." *J. Am. Chem. Soc.* 127 (12) (2005): 4345–4349.

Mao, H., Chen, W., Laurent, S. et al. "Hard corona composition and cellular toxicities of the graphene sheets." *Colloids Surf. B Biointerfaces* 109 (2013): 212–218.

Masterov, V.F., Chudnovsky, F.A., Zaidenberg, A.Z. et al. "Microwave absorption in fullerene-containing shungites." *Mol. Mater.* 4 (1994): 213–216.

Mchedlov-Petrossyan, N.O., Klochkov, V.K., Andrievsky, G.V. "Colloidal dispersions of fullerene C60 in water: Some properties and regularities of coagulation by electrolytes." *Chem. Soc., Faraday Trans.* 93 (24) (1997): 4343–4346.

Melezhik, V.A., Fallick, A.E., Filippov, M.M., Larsen, O. "Karelian shungite—An indication of 2.0-Ga-old metamorphosed oil-shale and generation of petrolium: Geology, lithology and geochemistry." *Earth Sci. Rev.* 47 (1999): 1–40.

Nakamura, S., Mashino, T. "Water-soluble fullerene derivatives for drug discovery." *J. Nippon Med. Sch.* 79 (2012): 248–254.

Osawa, E. "Natural fullerenes—Will they offer a hint to the selective synthesis of fullerenes?" *Fullerene Sci. Technol.* 7 (4) (1999): 637–652.

Osawa, E. "Monodisperse single nanodiamond particulates." *Pure Appl. Chem.* 80 (2008): 1365–1379.

Pan, S., Aksay, I.A. "Factors controlling the size of graphene oxide sheets produced via the graphite oxide route." *ACS Nano* 5 (2011): 4073–4083.

Parfen'eva, L.S., Smirnov, I.A., Zaidenberg, A.Z., Rozhkova, N.N., Stefanovich, G.B. "Electrical conductivity of shungite carbon." *Fizika Tverdogo Tela* 36 (1) (1994): 234–236.

Park, S.J., Khang, D. "Conformational changes of fibrinogen in dispersed carbon nanotubes." *Int. J. Nanomedicine* 7 (2012): 4325–4333.

Parnell, J. "Introduction," in *Bitumens in ore deposits*, eds. J. Parnell, H. Kucha, P. Landais (Berlin: Springer, 1993), 3–10.

Puzyr, A.P., Purtov, K.V., Shenderova, O.A., Luo, M., Brenner, D.W., Bondar, V.S. "The adsorption of aflatoxin B1 by detonation-synthesis nanodiamonds." *Dokl. Biochem. Biophys.* 417 (2007): 299–301.

Rahman, M., Laurent, S., Tawil, N., Yahia, L., Mahmoudi, M. *Protein–nanoparticle interactions* (Berlin, Heidelberg: Springer-Verlag, 2013).

Razbirin, B.S., Rozhkova, N.N., Sheka, E.F., Nelson, D.K., Starukhin, A.N. "Fractals of graphene quantum dots in photoluminescence of shungite." *JETP* 145 (5) (2014): 838–850.

Rozhkov, S.P., Goryunov, A.S. "Thermodynamic study of protein phases formation and clustering in model water–protein–salt solutions." *Biophys. Chem.* 151(1–2) (2010): 22–28.

Rozhkov, S.P., Goryunov, A.S. "Interaction of shungite carbon nanoparticles with blood protein and cell components." *Russ. J. Gen. Chem.* 83 (2013): 2585–2595.

Rozhkov, S.P., Goryunov, A.S. "Dynamic protein clusterization in supercritical region of the phase diagram of water–protein–salt solutions." *J. Supercrit. Fluids* 95 (2014): 68–74.

Rozhkov, S.P., Kovalevskii, V.V., Rozhkova, N.N. "Fullerene-containing phases obtained from aqueous dispersions of carbon nanoparticles." *Russ. J. Phys. Chem. A* 81 (6) (2007): 952–958.

Rozhkova, N.N. "Role of fullerene-like structures in the reactivity of shungite carbon as used in new materials with advanced properties," in *Perspectives of fullerene nanotechnology*, ed. E. Osawa (Dordrecht, 2002), 237–251.

Rozhkova, N.N. *Shungite nanocarbon* (Petrozavodsk: Karelian Research Centre of RAS, 2011) (in Russian).

Rozhkova, N.N. "Aggregation and stabilization shungite carbon nanoparticles." *Russ. J. Gen. Chem.* 4 (2013): 240–251.

Rozhkova, N.N., Emel'yanova, G.I., Gorlenko, L.E., Lunin, V.V. "Shungite and its modification." *Russ. J. Gen. Chem.* XLVIII (5) (2004): 107–115.

Rozhkova, N.N., Gribanov, A.V., Khodorkovskii, M.A. "Water mediated modification of structure and physical chemical properties of nanocarbons." *Diamond Relat. Mater.* 16 (2007a): 2104–2108.

Rozhkova, N.N., Rozhkova, V.S., Emelianova, G.I., Gorlenko, L.E., Lunin, V.V. "Stabilization conditions of carbon nanoclusters in water." *Karbo LII* 4 (2007b): 207–211.

Rozhkova, N.N., Gorlenko, L.E., Emel'yanova, G.I., Korobov, M.V., Lunin, V.V., Osawa, E. "The effect of ozone on the structure and physico-chemical properties of ultradisperse diamond and shungite nanocarbon elements." *Pure Appl. Chem.* 81 (11) (2009): 2093–2105.

Rozhkova, N.N., Yemel'yanova, G.I., Gorlenko, L.E., Gribanov, A.V., Lunin, V.V. "From stable aqueous dispersion of carbon nanoparticles to the clusters of metastable shungite carbon." *Glass Phys. Chem.* 37 (6) (2011): 613–618.

Ruoff, R.S., Malhorta, R., Huestis, D.L., Tse, D.S., Lorents, D.C. "Anomalous solubility behaviour of C 60." *Nature* 362 (1993): 140–141.

Saptarshi, S.R., Duschl, A., Lopata, A.L. "Interaction of nanoparticles with proteins: Relation to bioreactivity of the nanoparticle." *J. Nanobiotechnol.* 11 (2013): 26.

Sayes, C.M., Gobin, A.M., Ausman, K.D., Mendez, J., West, J.L., Colvin, V.L. "Nano-C60 cytotoxicity is due to lipid peroxidation." *Biomaterials* 26 (2005): 7587–7595.

Schur, D.V., Zaginaichenko, S.Y., Zolotarenko, A.D., Veziroglu, T.N. "Solubility and transformation of fullerene C60 molecule," in *Carbon nanomaterials in clean energy hydrogen system*, eds. B. Baranovski, S. Zaginaichenko, D. Schur, V. Skorokhod, A. Veziroglu. (Dordrecht: Springer Sci. + Business Media B.V., 2008), 85–95.

Sheka, E.F. "Computational strategy for graphene: Insight from odd electrons correlation." *Int. J. Quant. Chem.* 112 (2012): 3076–3090.

Sheka, E.F., Popova, N.A. "Odd-electron molecular theory of the graphene hydrogenation." *J. Mol. Mod.* 18 (2012): 3751–3768.

Sheka, E.F., Popova, N.A. "Molecular theory of graphene oxide." *Phys. Chem. Chem. Phys.* 15 (2013): 13304–13322.

Sheka, E.F., Rozhkova, N.N. "Shungite as the natural pantry of nanoscale reduced graphene oxide." *Int. J. Smart Nano Mater.* 5 (2014): 1–16.

Sheka, E.F., Natkaniec, I., Rozhkova, N.N., Holderna-Natkaniec, K. "Neutron scattering study of reduced graphene oxide of natural origin." *JETP Lett.* 118 (2014): 735–746.

Singh, V., Joung, D., Zhai, L., Das, S., Khondaker, S.I., Seal, S. "Graphene based materials: Past, present and future." *Prog. Mater. Sci.* 56 (2011): 1178–1271.

Smirnov, B.M. "Clusters with close packing and filled shells." *Phys. Usp.* 36 (10) (1993): 933–955.

Sokolov, V.A., Kalinin, Y.K., Dyukkiev, E.F. (eds.). *Shungites: New carbonaceous raw materials* (Petrozavodsk: Karelian Research Centre, 1984) (in Russian).

Stankovich, S., Dikin, D.A., Piner, R.D. et al. "Synthesis of graphene-based nanosheets via chemical reduction of exfoliated graphite oxide." *Carbon* 45 (2007): 1558–1565.

Wasserman, A.M., Kovarskii, A.L. *Spin labels and probes in physical chemistry of polymers* (Moscow: Nauka, 1986) (in Russian).

Vekilov, P.G. "Phase diagrams and kinetics of phase transitions in protein solutions." *J. Phys. Condens. Matter* 24 (2012): 193101.

Wang, Z., Schliehe, C., Bian, K. et al. "Correlating superlattice polymorphs to internanoparticle distance, packing density, and surface lattice in assemblies of PbS nanoparticles." *Nano Lett.* 13 (3) (2013): 1303–1311.

Wong, B.S., Yoong, S.L., Jagusiak, A. et al. "Carbon nanotubes for delivery of small molecule drugs." *Adv. Drug Deliv. Rev.* 65 (2013): 1964–2015.

Xu, Z., Gao, C. "Graphene in macroscopic order: Liquid crystals and wet-spun fibers." *Acc. Chem. Res.* 47 (4) (2014): 1267–1276.

Xu, Y., Watermann, T., Limbach, H.-H., Gutmann, T., Sebastiani, D., Buntkowsky, G. "Water and small organic molecules as probes for geometric confinement in well-ordered mesoporous carbon materials." *Phys. Chem. Chem. Phys.* 16 (2014): 9327–9336.

Yang, X., Ebrahimi, A., Li, J., Cui, Q. "Fullerene-biomolecule conjugates and their biomedicinal applications." *Int. J. Nanomedicine.* 9 (2014): 77–92.

Zaccarelli, E. "Colloidal gels: Equilibrium and non-equilibrium routes." *J. Phys. Condens. Matter* 19 (2007): 323101.

Zaidenberg, A.Z., Kovalevski, V.V., Rozhkova, N.N. "Spheroidal fullerene-like carbon in shungite rock," in *The ECS Fullerene Symposium Proceedings* (Reno, NJ, 1995), 24–27.

Zaidenberg, A.Z., Rozhkova, N.N., Kovalevski, V.V., Lorents, D.C., Chevallier, J. "Physical chemical model of Fullerene-like Shungite carbon." *Mol. Mater.* 8 (1996): 107–110.

Zhao, X., Xu, Z., Xie, Y., Zheng, B., Kou, L., Gao, C. "Polyelectrolyte-stabilized graphene oxide liquid crystals against salt, pH and serum." *Langmuir* 30 (13) (2014): 3715–3722.

Zhu, Y., Li, W., Li, Q. et al. "Graphene and graphene oxide: Synthesis, properties, and applications." *Adv. Mater.* 22 (2010): 3906–3924.

Ziatdinov, A.M. "The structure and properties of nanographites and their compounds." *Russ. J. Gen. Chem.* XLVIII (5) (2004): 5–11.

II

Fullerenes

8

Solid Fullerenes under Compression

Jianfu Li

Bolong Huang

Rui-Qin Zhang

Abstract

This chapter reviews the properties and phase transitions of C_{60} and C_{20} solid fullerenes under high pressures through density functional tight binding simulations, augmented by an empirical van der Waals force. The geometric structures of C_{60} molecules situated in the solid fcc phase are considerably changed under compression due to variations in the type of hybridization of carbon atoms, and the HOMO–LUMO gap dramatically decreases in response to the overlap of π orbitals from neighboring C_{60} molecules. For C_{20} solid fullerene, three new carbon allotropes have been revealed by cold compression. The three predicted structures are shown to be energetically more favorable and stable than C_{20} molecule. The volume compression calculations suggest that phase III has a large bulk modulus (427 GPa) and a bandgap of 5.1 eV. The results indicate that C_{20} solid fullerene is a potential building block for transparent superhard carbon with more advantages than graphite. We believe that the high-pressure study of small fullerenes will continue to be a rewarding field of study in the future as the advance in synthesis technology of small fullerene.

8.1 Introduction

Dispersive force is a type of noncovalent force that widely exist between molecules. Examples include hydrogen bonding and van der Waals (vdW) interactions, which are crucial for the formation entropy, mechanical stability, and electro-optical properties of molecular materials and related devices. Applying compression can decrease the distance between molecules or atoms and induce a variety of interesting physical and chemical properties, offering an adjustable independent-variable in experiment to tune the properties of molecular materials (Wang et al. 1991, 2007, 2012; Sundqvist 1999; Chen et al. 2011).

Since the identification of the C_{60} molecule (Kroto et al. 1985) and the effective synthesis of its solid (Krätschmer et al. 1990; Zhou and Cox 1992), C_{60} has become an important research subject because of its potential application in functional materials and nanodevices (Wang et al. 2012). It is expected that the electronic and optical properties of the nanoscale fullerene may be modified under applied pressure, and thus, device performance could be altered or even deteriorated. Investigation of the mechanical stability of solid C_{60} under external stress is consequently to be a desirable mission to us before its successful application. Particular attention has been paid to the compressibility or bulk modulus of solid C_{60}, as a measure of its stiffness, by x-ray and neutron diffraction with a variety of pressure-transmitting media (Duclos et al. 1991; Fischer et al. 1991; Wang et al. 1991; Haines and Léger 1994; Schirber et al. 1995; Soifer et al. 1999; Sundqvist 1999; Horikawa et al. 2000; Rhee et al. 2003). However, the available experimental data from different groups are often quite different, probably because of nonhydrostaticity of the pressure generated, the different methods of pressurization and measurements, as well as amorphization, polymerization, or phase transition of the sample (Horikawa et al. 2000).

To overcome the well-known difficulties in the experimental measurements of mechanical properties, theoretical calculations could be a very good choice to establish a reference platform. Attempts in the past include the theoretical investigation of compression properties of solid C_{60} using simple empirical intermolecular potentials with Lennard-Jones or empirical interatomic interactions (Li et al. 1992; Burgos et al. 1993; Yu et al. 1994; Prilutski and Shapovalov 1997; Venkatesh and Gopala Rao 1997), with the focuses on testing models of the potentials and also giving a large variation of the predicted bulk modulus. For studying C_{60}'s derivatives and other complexes, more transferable methods are required.

A weak vdW-type force is the primary intermolecular interaction among C_{60} molecules in the solid phase, with long-range dispersion interactions being the major source of attraction. At present, the omnipresent vdW interactions can only be accounted for properly by extreme high-level quantum-chemical wavefunction calculations or Quantum Monte Carlo simulations, but with high demands of computational cost. Quantitative information about weakly bound solid C_{60} can be obtained by theoretically studying the effect of weak intermolecular interactions. The most popularly used method pointing to this issue is the second-order Møller–Plesset (MP2) perturbation theory. But this leads to huge computational costs. The density functional theory (DFT) is a good choice of accurate solutions for large clusters or periodic structures, but conventional DFT methods are unable to describe the dispersion interaction owing to its limitations on approximating the exact wavefunction of each atom. This drawback can be overcome by introducing an empirical correction to calculate the additional attraction energy (Venkatesh and Gopala Rao 1997; Elstner et al. 1998; van Duijneveldt et al. 1999). To date, there exists a variety of hybrid semiempirical solutions that introduce damped pairwise interatomic dispersion corrections of the form C_6R^{-6} in the DFT formalism, also aiming to simulate the correct long-range interaction tail. Such a correction has been implemented also in the recently developed self-consistent charge density functional tight binding (SCC-DFTB) method, augmented with the empirical London dispersion correction (acronym SCC-DFTB-D) (Wu and Yang 2002).

Another physics essence we address is the subtle relation between the bonding and physical properties of solids. As is well known, carbon has a large variety of structures existing in nature, with either crystalline or disordered forms, because of its flexibility of three hybridized bondings, sp^3, sp^2, and sp^1. Thus, there are numerous allotropes in the carbon family, such as diamond, graphite, hexagonal diamond (lonsdaleite), amorphous carbon, carbon nanotubes, fullerenes, and graphene.

As in diamond, carbon atom's four valence electrons are assigned to a tetrahedrally directed sp^3 orbital, which makes a strong σ bond to an adjacent atom. As in graphite or graphene, with threefold coordinated sp^2 bonding configuration, three of the four valence electrons enter trigonally directed sp^2, forming σ bonds in a plane, while the fourth electron of the atom sits in a $p\pi$ orbital normal to the σ bonding plane. Such a π orbital forms a weaker π bond with a π orbital of one or more neighboring carbon atoms.

Based on the view of bonding, it is possible to explain why carbons have various extreme physical properties, such as the largest bulk modulus of any solids to date, the highest atom density, the largest room temperature thermal conductivity, the smallest thermal expansion coefficient, and even the largest electron-hole velocities among semiconductors. Therefore, it is a promising road in front of us for carrying this concept over to the exploring bonding and properties under compression.

To date, searching for new carbon allotropes has become a hot topic since different polymorphs are scientifically important and can offer different electronic, optical, and mechanical properties. According to experiments, graphite, amorphous carbon, nanotube, and fullerene are promising starting materials for the synthesis of new carbon phase. Graphite (honeycomb) can be converted to lonsdaleite (hexagonal) and diamond (tetrahedral) without catalysts at above 15 GPa, but high temperature (1600–2500 K) is needed (Aust and Drickamer 1963; Bundy et al. 1996; Irifune et al. 2003; Sumiya and Irifune 2007). On the contrary, compression of graphite at room temperature (called cold compression) produces transparent (in optical) and hard (in mechanical) phases, different from lonsdaleite and diamond (Hanfland et al. 1989; Yagi 1991; Yagi et al. 1992; Mao 2003). Recently, first-principles calculations have predicted several superhard phases from, for example, monoclinic *M*-carbon with *C2/m* structure (8 atoms = cell) (Li et al. 2009), cubic body center bct C_4 with the space group *I4/mmm* (Umemoto et al. 2010), orthorhombic W-carbon with *Pnma* symmetry (Wang et al. 2011), and C-C turned to be orthorhombic C_8 (Zhao et al. 2011). Developments based on cold compression have indicated that carbon atom in graphite needs large lattice distortions to interact with other carbons of adjacent graphite layers, which have a large energy barrier to overcome between different phases.

In the process of high-temperature compression, pure C_{60} undergoes a subtle bonding evolution and forms various interesting polymerized structures, whose cages transform into graphitic carbon or a mixture of sp^2 and sp^3 superhard phases (Iwasa et al. 1994; Marques 1999; Sundqvist 1999). As the smallest member of the fullerene family, the C_{20} molecule, with a dodecahedral cage structure, unlike C_{60}, consists of 12 pentagons and no hexagons. Its extreme cage surface curvature makes the bond angles (108°) closer to sp^3 (109.5°) than to sp^2 hybridization, with certain sp^2 hybridization characteristics. Thus, the hybridization type of C_{20} can be changed especially easily. On the other hand, the C_{20} molecule has the largest carbon mass density (3.24 g/cm³) among the fullerene family. The properties introduced previously are considered to be the prerequisites for searching novel superhard carbon allotropes.

In this chapter, we review the high-pressure behaviors of C_{60} and C_{20} fullerene solids. A retrospective investigation on the change in the geometry of C_{60} molecules caused by external pressure is given, as well as the variation in the energy gap with the compression. We also describe the prediction of three stable carbon phases with C_{20} molecules as starting materials employing the SCC-DFTB-D method.

8.2 Method and Models

The SCC-DFTB approach is derived from a second order expansion of the Kohn–Sham equation of the DFT formalism, in terms of charge density fluctuations. The Hamiltonian matrix elements are calculated with a two-center approximation, which is considered together with the overlap matrix elements as a function of the interatomic distance. A detailed description of this method is available in the literature (Elstner et al. 1998; Frauenheim et al. 2002). It has been reported to be computationally efficient and rather reliable in the simulation of large systems with hundreds of atoms or periodic materials.

To describe the vdW force, an empirical dispersion term is introduced into the SCC-DFTB total energy for large distances, together with a damping of this term with the onset of overlap of the charge density. The vdW interaction energy is defined in terms of damping function:

$$E_{\mathrm{vdW}} = -\sum_{\alpha\beta} f(R_{\alpha\beta}) C_6^{\alpha\beta} (R_{\alpha\beta})^{-6},$$

where $f(R)$ is the damping function as defined by

$$f(R) = \left[1 - \exp\left(-d * \left(\frac{R}{R_0}\right)^N\right)\right]^M,$$

where $d = 3.0$, $N = 7$, and $M = 4$ for all types of atoms. R_0 is the range of the overlap of two atoms and is 4.8 Å for the second row elements. The C_6 coefficient for a given atom α is parameterized as

$$C_6^{\alpha} = 0.75\sqrt{N_{\alpha} p_{\alpha}^3},$$

where N_α is the Slater–Kirkwood effective number of electrons and p_α is the polarizability of atom α. Also,

$$C_6^{\alpha\beta} = \frac{2C_6^{\alpha} C_6^{\beta} p_{\alpha} p_{\beta}}{p_{\alpha}^2 C_6^{\alpha} + p_{\beta}^2 C_6^{\beta}}.$$

In our study, we used the same damping function $f(R)$ and C_6 coefficient as used by Elstner et al. (2001). The method we introduced previously was found to be appropriate for predicting the geometrical structure and binding energy of weakly interacting systems (Elstner et al. 2001; Lin et al. 2005; Feng et al. 2007, 2008).

At ambient condition, crystalline C_{60} forms a face-centered-cubic (fcc) phase, with the C_{60} molecules rotating almost freely and thus are disordered in orientation (Heiney et al. 1991; Girifalco 1992; Zubov et al. 1996; Bohnen and Heid 1999). According to x-ray and neutron diffraction studies, the molecular rotation is not completely free, but with an intermolecular orientation correlation (Chow et al. 1992; Pintschovius 1996; Pintschovius et al. 1995, 1999; David et al. 1993). However, to calculate compressibility, the C_{60} molecule can be treated as a freely rotating ball with a negligible orientation correlation. Consequently, in the model of solid fcc C_{60}, all C_{60} molecules are oriented parallel to the supercell vectors. Thus, their mirror planes are all aligned with the cell faces, as shown in Figure 8.1a–c, with each C_{60} molecule being oriented such that the centers of 8 of its 20 hexagons are aligned along the cubic 111 directions (Rosseinsky 1995). As such, a unique orientation of each molecule is assumed with respect

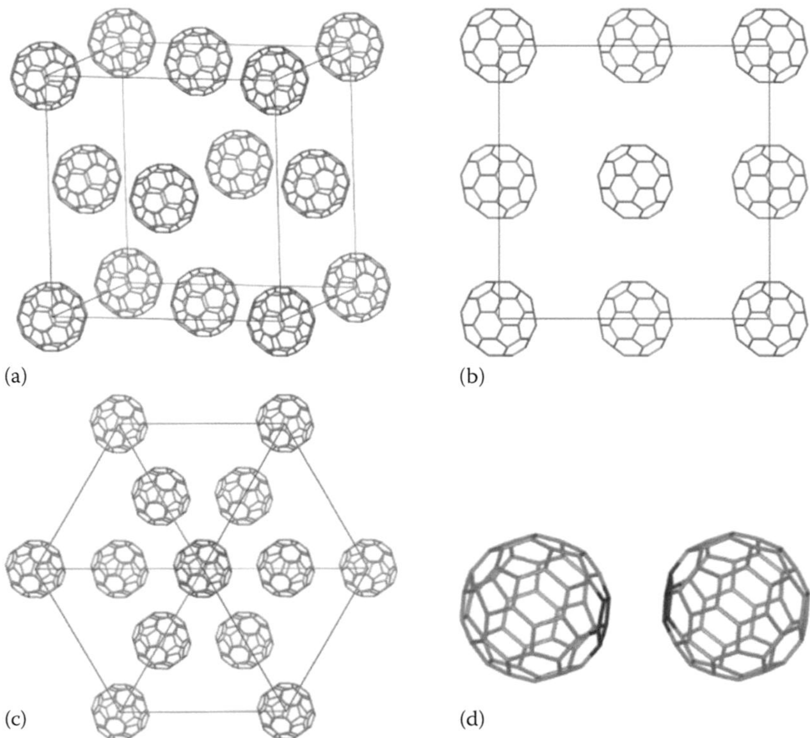

FIGURE 8.1 (a) The fcc structure of C_{60} in a supercell and the view along the (b) [001] and (c) [111] directions, respectively. (d) The relative orientation of two neighboring C_{60} molecules with two face-to-face contact pentagons colored in dark gray. (Reprinted with permission from Feng, C. et al., *Phys. Condens. Matter*, 20, 275240, 2008. Copyright 2008, IOP Publishing Ltd.)

to their neighboring molecules. As depicted in Figure 8.1d, one pentagon of a molecule faces another pentagon of its neighbor with a parallel displacement.

For the simulation, we took external pressure as a constraint to study the high-pressure behaviors of C_{20}. Both internal positions of carbon atoms and lattice parameters are allowed to be optimized during the structural relaxation. At ambient condition, crystalline C_{20} forms a hexagonal structure (Wang et al. 2001; Li and Zhang 2013), with C_{20} acting as building blocks. C_{20} solid under compressing takes a hexagonal symmetry as an initial structure with one C_{20} molecule in a unit cell.

8.3 C_{60} Fullerene Solid under Compression

The bulk modulus B of a given material can be calculated from

$$B = V_0 \frac{\partial^2 E}{\partial V^2},$$

where E is the total energy and V is the unit cell volume. We first performed the geometry optimization of an internal C_{60} molecule and yielded an optimized geometry and total energy $E(d)$ of the C_{60} molecule as functions of d and a. Next, we repeated the following procedure with elongating the d by a small deviation of δd, starting the optimization calculations, obtaining the corresponding total energies of the local minima, and determining the global minimum $E_0 = \min_d E(d)$. As a result, we obtained the diagram of the function $E(d)$ around the global minimum and the minimal total energy $E_0 = E(d_0)$ vs. the optimized lattice constant d_0 for the relaxed structure, as shown in Figure 8.2. The optimal lattice constant d_0 by calculation is 9.9 Å, in good agreement with the measured closest contact distance of 10.0 Å by center-to-center C_{60}–C_{60} (Rosseinsky 1995). The total energy curve as a function of the unit cell volume $E(V)$ is shown on the inset of Figure 8.2. The bulk modulus B was calculated to be 9.1 GPa, close to the experimental value of 9.6 GPa (Pintschovius et al. 1999). The result can be considered as a proper estimation of the bulk modulus of solid fcc C_{60} in the low-pressure range and a reference for theoretical calculations (Sundqvist 1999). Note that the previous theoretical study gave a rather high value of 18.1 GPa (Duclos et al. 1991).

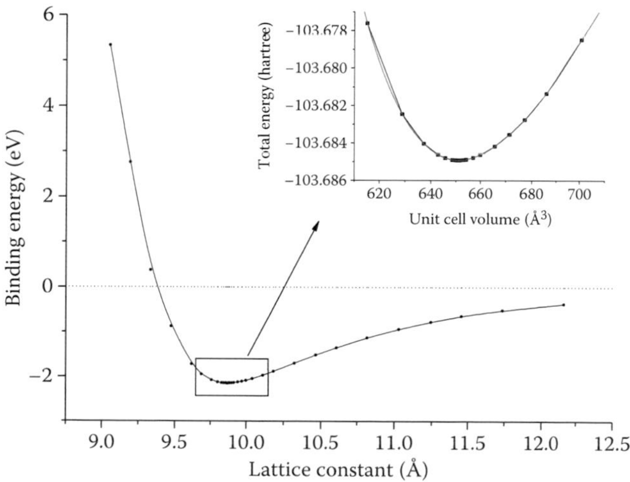

FIGURE 8.2 Binding energy vs. lattice constant. The inset is the total energy curve as a function of the unit cell volume near the as-calculated global system energy minimum. (Reprinted with permission from Feng, C. et al., *Phys. Condens. Matter*, 20, 275240, 2008. Copyright 2008, IOP Publishing Ltd.)

In the C_{60} molecule, each carbon atom is still sp^2 hybridized with threefold coordinated bonding by σ bonds, although the smallest motif is a distorted pentagon. The remaining valence electron enters into delocalized π orbital. However, the pentagonal rings make electron delocalization poorer. Accordingly, there are two different bonds in the C_{60} molecule, short bonds (SBs) between atoms located between pairs of hexagons and long bonds (LBs) between a hexagon and a pentagon. Our calculations show that the lengths of SBs and LBs of C_{60} within the stable fcc phase are in the ranges of 1.406–1.408 Å and 1.450–1.456 Å, respectively, consistent with the 1.407 Å and 1.453 Å measured in the isolated C_{60} molecule. Our predicted lengths of SBs are close to that of the C–C bond within the benzene molecule (1.40 Å), indicating a similar sp^2 configuration. In addition, each carbon atom is threefold coordinated, with an interior angle of the pentagon (α) and two interior angles of the hexagon (β). In solid fcc C_{60}, the predicted bond angle α is about 107.9°, and β ranges from 119.9° to 120.1°. These values show close correspondence to values of 108.0° and 120.0° in an isolated C_{60} cage. Meanwhile, the diameter of the C_{60} molecule was calculated to be 6.842 Å, only 0.029 Å smaller than that of an isolated C_{60} molecule. Therefore, we find that the C_{60} molecules have less sensitive response to minor external pressure compression, as they shrink slightly.

However, with strong compression induced by external pressure, the geometries of C_{60} molecules change considerably, despite the strong intramolecular bonds. Figure 8.3 shows that the geometries of C_{60} molecules are evidently "wrinkled" when the lattice constant is compressed to 8.8 Å, considerably reduced from that value at zero pressure. As shown in Figure 8.3, the geometry of the C_{60} cage changes anisotropically along three lattice vectors because of the anisotropic intermolecular interactions, regardless of the isotropic compression. The distances between the two closest carbon atoms from adjacent C_{60} molecules decrease to 1.39–1.71 Å, indicating weak π orbital coupling between neighboring C_{60} molecules. The LBs vary from 1.406 to 1.497 Å, while SBs change from 1.373 to 1.456 Å, both about 3% different from those in the most stable structure at zero pressure. From the bond length distribution, the C–C bonds possibly have both single and double bonds (1.34 Å). The angles α and β change to 100.8°–115.2° and 103.2°–128.7°, respectively, close to the bond angles range (107°–111°) of sp^3 configuration. Therefore, in the C_{60} molecule, the sp^2-hybridized carbon atoms must be bent to form the closed sphere, resulting in an angular strain, which,

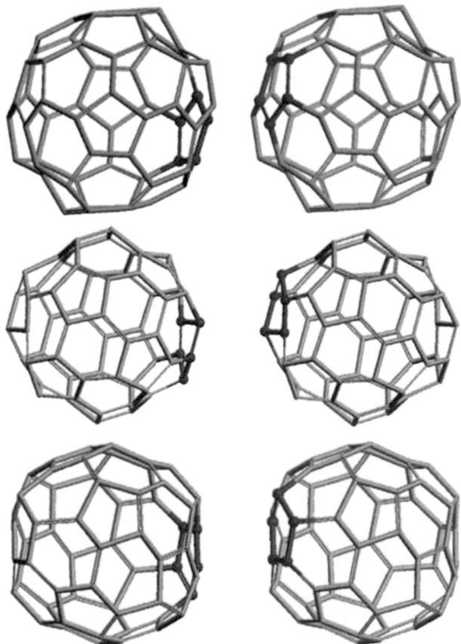

FIGURE 8.3 Relaxed "wrinkling" of C_{60} molecular local structures in an fcc lattice under strong external compression. Pairs of neighboring C_{60} molecules are viewed from three directions, perpendicular to the a, b, and c unit cell vectors in the top, middle, and lower panels, respectively. The two face-to-face contact pentagons are colored in dark gray to highlight the orientational correlation of two neighboring C_{60} molecules by weak vdW interactions. (Reprinted with permission from Feng, C. et al., *Phys. Condens. Matter*, 20, 275240, 2008. Copyright 2008, IOP Publishing Ltd.)

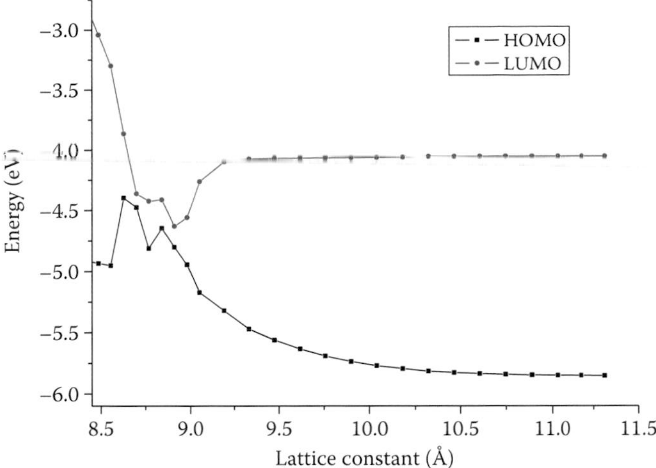

FIGURE 8.4 Evolution of energies of HOMO and LUMO with change in the lattice constant of fcc C_{60} under compression. (Reprinted with permission from Feng, C. et al., *Phys. Condens. Matter*, 20, 275240, 2008. Copyright 2008, IOP Publishing Ltd.)

however, can be reduced by changing partial sp^2-hybridized carbons into sp^3-hybridized ones under compression by external stress. The change in hybridized orbitals causes the bond angles to transform to about 109.5° of sp^3 orbitals, which allows the bonds to bend less when closing to a spherical configuration. The π–π orbital coupling becomes more localized, deteriorating aromaticity within the hexagonal rings. Thus, the intramolecular bond lengths and bond angles vary, resulting in the distortion of C_{60} molecules.

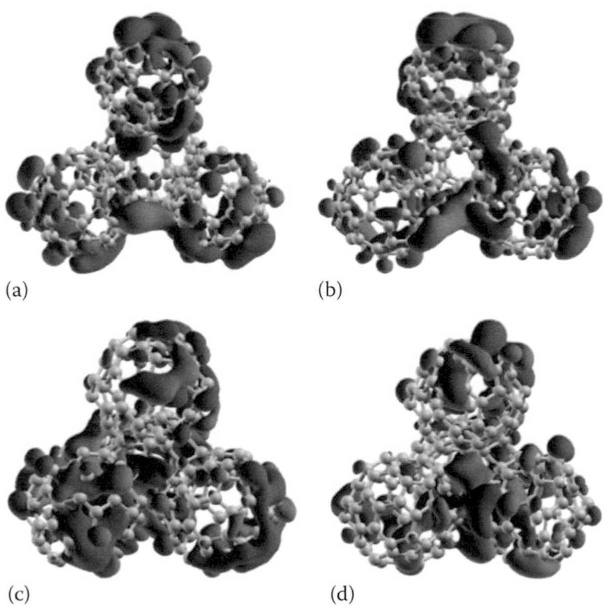

FIGURE 8.5 The HOMOs (a, c) and LUMOs (b, d) derived from bands at the Γ point for the most stable fcc structures of C_{60} in a supercell at zero pressure (d = 9.9 Å, upper) and under severe compression (d = 8.9 Å, lower), respectively. The isovalue is ±0.02 au. (Reprinted with permission from Feng, C. et al., *Phys. Condens. Matter*, 20, 275240, 2008. Copyright 2008, IOP Publishing Ltd.)

As previously, the external compression decreases the intermolecular distance and thus increases the intermolecular interactions between neighboring C_{60} molecules, which further alters their physical and chemical properties, such as their electronic properties. Figure 8.4 shows that the energy of the highest occupied molecular orbital (HOMO) gradually increases when the lattice constant is reduced from 11.3–8.9 Å as more valence electrons entered into higher π orbitals of carbon atoms, while the energy of the lowest unoccupied molecular orbital (LUMO) remains nearly unchanged, causing drastic reduction in the HOMO–LUMO gap from 1.80 to 0.17 eV induced by the external compression. This is primarily a result of the increase in overlap between π orbitals from neighboring C_{60} molecules. From Figure 8.5, the compression of lattice constant to 8.9 Å leads the overlap between intermolecular π orbitals to transform to be much more delocalized. With such severe compression, covalent bonds may be formed between neighboring C_{60} molecules, with the closest intermolecular C–C distance of only about 1.49 Å and thus the enhanced delocalization of valence electrons and reduced HOMO–LUMO gap. However, compressing the lattice constant to below 8.8 Å can change quite dramatically the energies of HOMO and LUMO. Therefore, solid C_{60} is shown to acquire special distinctive tunable electronic properties under severe external pressure.

8.4 C_{20} Solid under High Pressure

The C_{20} solid unit cell volume as a function of pressure and the molecule structures in different compressions are shown in Figure 8.6. Two discontinuities in volume appear in the pressure range of 0 to 400 GPa. The C_{20} molecule changes not only in volume but also in geometric structure. The discontinuity of unit cell volume versus the external pressuring compression indicates the appearance of phase transitions. And the corresponding three phases are named phase I, II, and III, respectively. Phase I exists in the 0 to 100 GPa pressure range and the structure of the C_{20} molecule almost remains the same compared with that in vacuum conditions. Phase II appears in 105 GPa, and two pentagons of the C_{20} molecule collapse, resulting in transformation into triangles and quadrilaterals under compression. Phase III appears when external pressure reaches 290 GPa. The C–C bonds in phase III in the top and bottom pentagons of the C_{20} molecule are elongated from 1.51 Å to 2.35 Å. The simulated compressing and decompressing processes for testing the mechanical stabilities of the three obtained phases are shown in Figure 8.7, with phase I for comparison. Elastic deformation is seen under 100 GPa or less. With their shapes with 12 pentagons maintained except volume, the C_{20} molecules present deformed crystal structures, with their initial shape restored when decompressed back to ambient pressure. Every C_{20} molecule is connected to six neighboring molecules through eight intermolecular bonds,

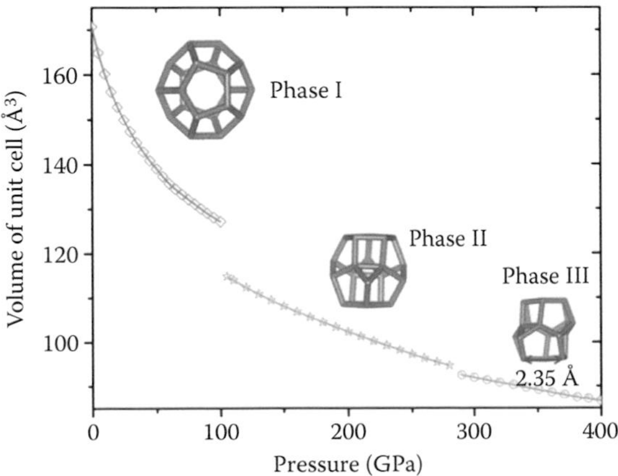

FIGURE 8.6 Volume of unit cell versus external pressure and the geometric structures of C_{20} molecule in different phases. Two discontinuities between diagrams denoting two different phase transitions. (Reprinted from New superhard carbon allotropes based on C_{20} fullerene. *Carbon N. Y.*, 63, Li, J., Zhang, R. Q., 571, Copyright 2013, with permission from Elsevier.)

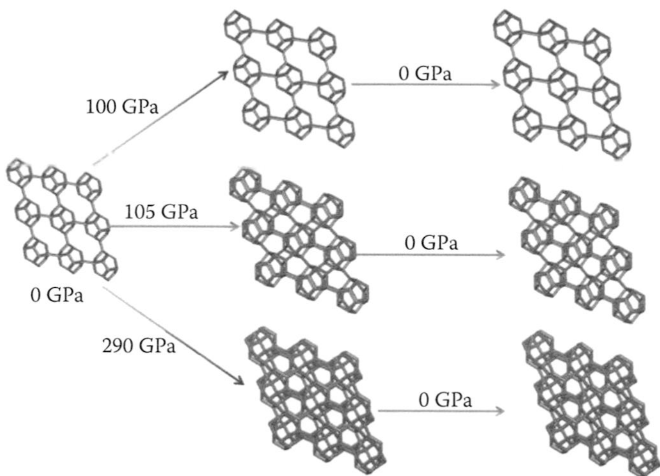

FIGURE 8.7 The calculated crystalline structure of C_{20} solid phases under different compression and decompression conditions. The bond length is smaller than 1.8 Å. (Reprinted from *Carbon N. Y.*, 63, Li, J., Zhang, R. Q., 571, Copyright 2013, with permission from Elsevier.)

which indicates that the C_{20} molecule starts to break the sp^2 hybridization and forms sp^3 hybrid C–C bonds. However, when the pressure is increased to 105 GPa, inelastic deformations are found and another phase transition is triggered. The C_{20} solid transforms to phase II. Meanwhile, the sp^3 hybridization is enhanced, with each C_{20} molecule being connected to eight neighbor molecules through 14 intermolecular bonds. When the pressure is increased to 290 GPa, phase III appears. The structure of C_{20} solid presents a slight volume change, with the sp^2 hybrid C–C double bonds transforming to sp^3 hybrid.

The compressibility of C_{20} solid in different phases under pressure is compared in Figure 8.8, with data of *c*-BN and diamond also included for reference. The largest incompressibility among the three phases is found for phase III.

The bulk modulus (B_0) of phase I, II, and III by fitting the calculated pressure as a function of volume to the third-order Birch-Murnaghan equation of state is given in Table 8.1. Phase III presents a bulk modulus larger than that of *c*-BN (391 GPa) and smaller than that of diamond (452 GPa). Thus, a large

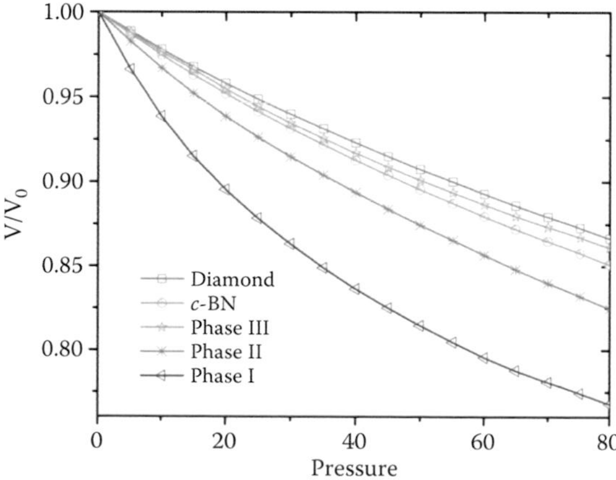

FIGURE 8.8 Volume compressions versus external pressure for different phases of solid C_{20} compared with *c*-BN and diamond. (Reprinted from *Carbon N. Y.*, 63, Li, J., Zhang, R. Q., 571, Copyright 2013, with permission from Elsevier.)

TABLE 8.1 Calculated Energy Difference ΔE (eV) per Atom between Different Carbon Materials and C_{20} Molecule, Carbon Density D (g/cm^3), Bandgap E_g (eV) and Bulk Modulus B_0 (GPa)

	ΔE	D	E_g	B_0
Phase I	−0.052	2.34	2.7	186
Phase II	−0.11	2.79	4.0	316
Phase III	−0.331	3.16	5.1	427
Graphite	2.106	2.23	–	–
c-BN	–	–	–	391,387[a]
Diamond	−1.235	3.63	5.5[b]	452,446[b]

Source: Reprinted from *Carbon N. Y.*, 63, Li, J., Zhang, R. Q., 571, Copyright 2013, with permission from Elsevier.

Note: Some experimental data are also listed for comparison.

[a] Goncharov, A., Crowhurst, J., Dewhurst, J. et al., *Phys. Rev. B* 75, 224114, 2007.

[b] He, C., Sun, L., Zhang, C. et al. *Solid State Commun.* 152, 1560, 2012.

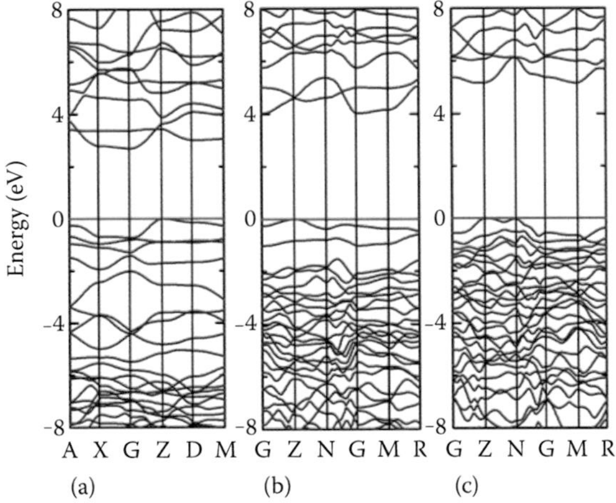

FIGURE 8.9 Electronic band structures for phases I (a), II (b), and III (c) of solid C_{20} at 0 GPa, showing different bandgaps and paths along high symmetric points of reciprocal space. (Reprinted from *Carbon N. Y.*, 63, Li, J., Zhang, R. Q., 571, Copyright 2013, with permission from Elsevier.)

mass density of carbon and a high degree of sp^3 hybridization are important prerequisites for a superhard carbon material. Phase III has a carbon density of 3.16 g/cm^3, which is close to 3.33–3.45 g/cm^3 of recently predicted carbon phases (Goncharov et al. 2007; He et al. 2012). Therefore, phase III of C_{20} solid is predicted to be a superhard material.

Figure 8.9 gives the calculated band structures for phase I, II, and III at 0 GPa. It is seen that the valence band tops of the three phases are all at the Z point, while the conduction band bottoms are at G, G, and Z, respectively. The first two are insulating with indirect bandgap 2.7 and 4.0 eV, while phase III is a direct bandgap of 5.1 eV at Z point of reciprocal space, which is larger than that of reported M carbon (3.6 eV) (Li et al. 2009) and bct C_4 (2.5 eV) (Umemoto et al. 2010) but closer to 5.5 eV of diamond (Occelli et al. 2003).

8.5 Summary and Outlook

We reviewed the study of the geometric structures and electronic properties of C_{60} and C_{20} fullerene solids under external compression predicted by the density functional tight binding method, augmented with an empirical dispersion correction.

The calculated bulk modulus of solid fcc C_{60} is 9.1 GPa, consistent with experimental values and much superior to the results from empirical molecular mechanics calculations. With external compression, the internal molecular structures are severely "wrinkled," with intramolecular bond length and bond angle changes, owing to modifications in the hybridization of the carbon atoms. Changes caused include HOMO–LUMO gap decrease from 1.80 to 0.17 eV when compressing the lattice constant from 11.3 to 8.9 Å, mainly as a result of the overlap of π orbitals and the formation of covalent bonds between neighboring C_{60} molecules.

For C_{20} fullerene solid, compression can result in three new carbon phases that are stable at ambient pressure conditions with a lower energy than the C_{20} molecule. In particular, phase III possesses the highest bulk modulus (427 GPa) and is energetically more favorable than the other two phases. It is an insulator with a wide energy gap over 5 eV. Our results demonstrate that C_{20} fullerene is a potential building block for superhard materials.

References

Aust, R. B. and H. G. Drickamer, Carbon: A new crystalline phase, *Science* **140**, 817 (1963).

Bohnen, K.-P. and R. Heid, Ab initio intermolecular potential of solid C_{60} in the low-temperature phase *Phys. Rev. Lett.* **83**, 1167 (1999).

Bundy, F. P., W. A. Bassett, M. S. Weathers, R. J. Hemley, H. U. Mao, and A. F. Goncharov, The pressure-temperature phase and transformation diagram for carbon; updated through 1994, *Carbon* **34**, 141 (1996).

Burgos, E., E. Halac, and H. Bonadeo, Calculation of static, dynamic, and thermodynamic properties of solid C_{60}, *Phys. Rev. B* **47**, 13903 (1993).

Chen, X. J., C. Zhang, Y. Meng, R. Q. Zhang, H. Q. Lin, V. V. Struzhkin, and H. K. Mao, β–tin→*Imma*→*sh* Phase transitions of germanium, *Phys. Rev. Lett.* **106**, 135502 (2011).

Chow, P. C., X. Jiang, G. Reiter, P. Wochner, S. C. Moss, J. D. Axe, J. C. Hanson et al., Synchrotron x-ray study of orientational order in single crystal C_{60} at room temperature, *Phys. Rev. Lett.* **69**, 2943 (1992).

David, W. I. F., R. M. Ibberson, and T. Matsuo, High Resolution Neutron Powder Diffraction: A case study of the structure of C_{60}, *Proc. R. Soc. A Math. Phys. Eng. Sci.* **442**, 129 (1993).

Duclos, S. J., K. Brister, R. C. Haddon, A. R. Kortan, and F. A. Thiel, Effects of pressure and stress on C_{60} fullerite to 20 GPa, *Nature* **351**, 380 (1991).

Elstner, M., D. Porezag, G. Jungnickel, J. Elsner, M. Haugk, T. Frauenheim, S. Suhai, and G. Seifert, Self-consistent-charge density-functional tight-binding method for simulations of complex materials properties, *Phys. Rev. B* **58**, 7260 (1998).

Elstner, M., P. Hobza, T. Frauenheim, S. Suhai, and E. Kaxiras, Hydrogen bonding and stacking interactions of nucleic acid base pairs: A density-functional-theory based treatment, *J. Chem. Phys.* **114**, 5149 (2001).

Feng, C., R. Q. Zhang, S. L. Dong, T. A. Niehaus, and T. Frauenheim, Signatures in vibrational spectra of ice nanotubes revealed by a density functional tight binding method, *J. Phys. Chem. C* **111**, 14131 (2007).

Feng, C., C. Zhang, R. Zhang, T. Frauenheim, and M. Van Hove, Mechanical properties of solid C_{60} studied with density functional tight binding method augmented by an empirical dispersion term, *J. Phys. Condens. Matter* **20**, 275240 (2008).

Fischer, J. E., P. A. Heiney, A. R. Mcghie, W. J. Romanow, A. M. Denenstein, J. P. Mccauley, and A. B. Smith, Compressibility of solid C_{60}, *Science* **252**, 1288 (1991).

Frauenheim, T., G. Seifert, M. Elstner, T. Niehaus, C. Köhler, M. Amkreutz, M. Sternberg et al., Atomistic simulations of complex materials: Ground-state and excited-state properties, *J. Phys. Condens. Matter* **14**, 3015 (2002).

Girifalco, L. A., Molecular properties of fullerene in the gas and solid phases, *J. Phys. Chem.* **96**, 858 (1992).

Goncharov, A. F., J. C. Crowhurst, J. K. Dewhurst, S. Sharma, C. Sanloup, E. Gregoryanz et al., Thermal equation of state of cubic boron nitride: Implications for a high-temperature pressure scale, *Phys. Rev. B* **75**, 224114 (2007).

Haines, J. and J. M. Léger, An x-ray diffraction study of C_{60} up to 28 GPa, *Solid State Commun.* **90**, 361 (1994).

Hanfland, M., H. Beister, and K. Syassen, Graphite under pressure: Equation of state and first-order Raman modes, *Phys. Rev. B* **39**, 12598 (1989).

He, C., L. Sun, C. Zhang, X. Peng, K. Zhang, and J. Zhong, New superhard carbon phases between graphite and diamond, *Solid State Commun.* **152**, 1560 (2012).

Heiney, P. A., J. E. Fischer, A. R. McGhie, W. J. Romanow, A. M. Denenstein, J. P. McCauley et al., Orientational ordering transition in solid C_{60}, *Phys. Rev. Lett.* **66**, 2911 (1991).

Horikawa, T., T. Kinoshita, K. Suito, and A. Onodera, Compressibility measurement of C_{60} using synchrotron radiation, *Solid State Commun.* **114**, 121 (2000).

Irifune, T., A. Kurio, S. Sakamoto, T. Inoue, and H. Sumiya, Correction: Ultrahard polycrystalline diamond from graphite, *Nature* **421**, 806 (2003).

Iwasa, Y., T. Arima, R. M. Fleming, T. Siegrist, O. Zhou, R. C. Haddon et al., New phases of C_{60} synthesized at high pressure, *Science* **264**, 1570 (1994).

Krätschmer, W., L. D. Lamb, K. Fostiropoulos, and D. R. Huffman, Solid C_{60}: A new form of carbon, *Nature* **347**, 354 (1990).

Kroto, H. W., J. R. Heath, S. C. O'Brien, R. F. Curl, and R. E. Smalley, C_{60}: Buckminsterfullerene, *Nature* **318**, 162 (1985).

Li, J. and R. Q. Zhang, New superhard carbon allotropes based on C_{20} fullerene, *Carbon* **63**, 571 (2013).

Li, Q., Y. Ma, A. R. Oganov, H. Wang, H. Wang, Y. Xu, T. Cui, H.-K. Mao, and G. Zou, Superhard monoclinic polymorph of carbon, *Phys. Rev. Lett.* **102**, 175506 (2009).

Li, X., J. P. Lu, and R. M. Martin, Ground-state structural and dynamical properties of solid C_{60} from an empirical intermolecular potential, *Phys. Rev. B* **46**, 4301 (1992).

Lin, C. S., R. Q. Zhang, S. T. Lee, M. Elstner, T. Frauenheim, and L. J. Wan, Simulation of Water Cluster Assembly on a Graphite Surface, *J. Phys. Chem. B* **109**, 14183 (2005).

Mao, W. L., Bonding changes in compressed superhard graphite, *Science* **302**, 425 (2003).

Marques, L., 'Debye-Scherrer Ellipses' from 3D fullerene polymers: An anisotropic pressure memory signature, *Science* **283**, 1720 (1999).

Occelli, F., P. Loubeyre, and R. LeToullec, Properties of diamond under hydrostatic pressures up to 140 GPa, *Nat. Mater.* **2**, 151 (2003).

Pintschovius, L., Neutron studies of vibrations in fullerenes, *Reports Prog. Phys.* **59**, 473 (1996).

Pintschovius, L., O. Blaschko, G. Krexner, and N. Pyka, Bulk modulus of C_{60} studied by single-crystal neutron diffraction, *Phys. Rev. B* **59**, 11020 (1999).

Pintschovius, L., S. L. Chaplot, G. Roth, and G. Heger, Evidence for a pronounced local orientational order in the high temperature phase of C_{60}, *Phys. Rev. Lett.* **75**, 2843 (1995).

Prilutski, Y. I. and G. Shapovalov, Effect of pressure on the structure and lattice dynamics of fullerene crystal C_{60}, *Phys. Status Solidi (b)* **201**, 361 (1997).

Rhee, J. H., D. Singh, Y. Li, and S. C. Sharma, Crystal structure of C_{60} following compression under 31.1 GPa in diamond anvil cell at room temperature, *Solid State Commun.* **127**, 295 (2003).

Rosseinsky, M. J., Fullerene intercalation chemistry, *J. Mater. Chem.* **5**, 1497 (1995).

Schirber, J. E., G. H. Kwei, J. D. Jorgensen, R. L. Hitterman, and B. Morosin, Room-temperature compressibility of C_{60}: Intercalation effects with He, Ne, and Ar, *Phys. Rev. B* **51**, 12014 (1995).

Soifer, Y. M., N. Kobelev, R. K. Nikolaev, and V. M. Levin, Integral and local elastic properties of C_{60} single crystals, *Phys. Status Solidi (b)* **214**, 303 (1999).

Sumiya, H. and T. Irifune, Hardness and deformation microstructures of nano-polycrystalline diamonds synthesized from various carbons under high pressure and high temperature, *J. Mater. Res.* **22**, 2345 (2007).

Sundqvist, B., Fullerenes under high pressures, *Adv. Phys.* **48**, 1 (1999).

Umemoto, K., R. M. Wentzcovitch, S. Saito, and T. Miyake, Body-centered tetragonal C_4: A viable sp^3 carbon allotrope, *Phys. Rev. Lett.* **104**, 125504 (2010).

van Duijneveldt, F. B., J. G. C. M. D. de Rijdt, and B. P. van Eijck, Transferable ab initio intermolecular potentials. 1. Derivation from methanol dimer and trimer calculations, *J. Phys. Chem. A* **103**, 9872 (1999).

Venkatesh, R. and R. V. Gopala Rao, Elastic and other associated properties of C_{60}, *Phys. Rev. B* **55**, 15 (1997).

Wang, J.-T., C. Chen, and Y. Kawazoe, Low-temperature phase transformation from graphite to sp^3 orthorhombic carbon, *Phys. Rev. Lett.* **106**, 075501 (2011).

Wang, L., B. Liu, H. Li, W. Yang, Y. Ding, S. V Sinogeikin et al., Long-Range Ordered Carbon Clusters: A crystalline material with amorphous building blocks, *Science.* **337**, 825 (2012).

Wang, L., B. Liu, D. Liu, M. Yao, S. Yu, Y. Hou, B. Zou et al., Synthesis and high pressure induced amorphization of C_{60} nanosheets, *Appl. Phys. Lett.* **91**, 103112 (2007).

Wang, Y., D. Tománek, and G. F. Bertsch, Stiffness of a solid composed of C_{60} clusters, *Phys. Rev. B* **44**, 6562 (1991).

Wang, Z., X. Ke, Z. Zhu, F. Zhu, M. Ruan, H. Chen, R. Huang, and L. Zheng, A new carbon solid made of the world's smallest caged fullerene C_{20}, *Phys. Lett. A* 280, **351** (2001).

Wu, Q. and W. Yang, Empirical correction to density functional theory for van der Waals interactions, *J. Chem. Phys.* **116**, 515 (2002).

Yagi, T., W. Utsumi, M. Yamakata, T. Kikegawa, and O. Shimomura, High-pressure in situ x-ray-diffraction study of the phase transformation from graphite to hexagonal diamond at room temperature, *Phys. Rev. B* **46**, 6031 (1992).

Yagi, W. U. A. T., Light-transparent phase formed by room-temperature compression of graphite, *Science* **252**, 1542 (1991).

Yu, J., L. Bi, R. K. Kalia, and P. Vashishta, Intermolecular and intramolecular phonons in solid C_{60}: Effects of orientational disorder and pressure, *Phys. Rev. B* **49**, 5008 (1994).

Zhao, Z., B. Xu, X.-F. Zhou, L.-M. Wang, B. Wen, J. He, Z. Liu, H.-T. Wang, and Y. Tian, Novel superhard carbon: C-centered orthorhombic C_8, *Phys. Rev. Lett.* **107**, 215502 (2011).

Zhou, O. and D. E. Cox, Structures of C_{60} intercalation compounds, *J. Phys. Chem. Solids* **53**, 1373 (1992).

Zubov, V. I., N. P. Tretiakov, J. F. Sanchez, and A. A. Caparica, Thermodynamic properties of the C_{60} fullerite at high temperatures: Calculations taking into account the intramolecular degrees of freedom and strong anharmonicity of the lattice vibrations, *Phys. Rev. B* **53**, 12080 (1996).

9

Open-Cage Fullerenes

Contents

Argyro T. Papastavrou

Eleftherios K. Pefkianakis

Manolis M. Roubelakis

Georgios C. Vougioukalakis

9.1 Introduction

Almost simultaneously with the discovery of fullerenes (Kroto et al. 1985; Krätschmer et al. 1990; Hirsch and Brettreich 2005), it was found that these hollow structures can act as host molecules encapsulating guests such as small atoms, molecules, or ions within their cavity, thereby forming endohedral fullerenes (represented as $M@C_{2n}$) (Kikuchi et al. 1993, 1994; Saunders et al. 1993, 1994, 1996; Shinohara et al. 1993, 1994; Jimenez-Vasquez et al. 1994; Akasaka et al. 1995; Suzuki et al. 1995; Ding and Yang 1996; Kirbach and Dunsch 1996; Kubozono et al. 1996a,b; Murphy et al. 1996; Nagase et al. 1996; Tellgmann et al. 1996; Shimshi et al. 1997; Rubin 1999; Stevenson et al. 1999; Khong et al. 2000; Liu and Sun 2000; Shinohara 2000; Hirsch 2001; Akasaka and Nagase 2002; Syamala et al. 2002; Kato et al. 2003; Komatsu et al. 2005a,b; Murata et al. 2006). Interestingly, the inner surface of the fullerene cage, in contrast to its exterior, is chemically inert (Murphy et al. 1996). Fullerenes enclosing metal atoms (endohedral metallofullerenes) bear significantly altered physical properties compared with those of empty fullerenes. This is because of electron transfer from the encaged metal atom to the fullerene core. A transition metal atom inside the fullerene cage (diameter ~3.5 Å) may, for example, result in higher temperature superconductors or lead to less cytotoxic contrast agents for magnetic resonance imaging (Kato et al. 2003).

Currently, endohedral fullerenes are prepared by one of the following physicochemical methods: (a) covaporization of carbon and metal atoms in an arc by using graphite rods doped with the metal oxides or carbides (Kikuchi et al. 1993, 1994; Shinohara et al. 1993; Suzuki et al. 1995; Ding and Yang 1996;

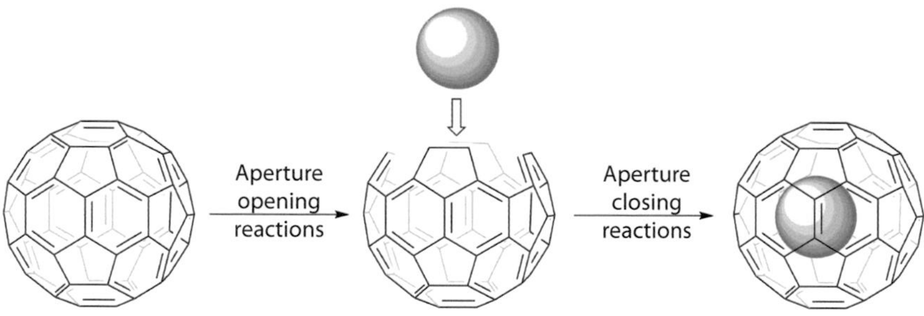

SCHEME 9.1 The "molecular surgery" approach.

Shinohara 2000); (b) high-pressure/high-temperature treatment of the pure fullerenes with noble gases or nitrogen (Saunders et al. 1993, 1994, 1996; Syamala et al. 2002); (c) by using beams of fast ions or atoms (Murphy et al. 1996; Tellgmann et al. 1996; Shimshi et al. 1997); or (d) by using the recoil energy from nuclear processes to shoot atoms inside the fullerene cage (Jimenez-Vasquez et al. 1994; Khong et al. 2000). However, all of these techniques lack efficiency as isolation and purification of the endohedral complexes from the produced soot demand difficult and time-consuming procedures, while the isolated yields for pure endohedral fullerenes are up to only 1%.

An alternative route to endohedral fullerenes, which uses only organic transformations, is called the "molecular surgery" approach (Scheme 9.1) (Rubin 1997). According to this strategy, a hole is temporarily opened and then closed back onto the original fullerene carbon framework, after the introduction of a small atom, molecule, or ion in the cage. This technique has been materialized by the isolation of endohedral [60]fullerenes incorporating either a hydrogen molecule (Komatsu and Murata 2005; Komatsu et al. 2005b; Murata et al. 2006) or a helium atom (Morinaka et al. 2010), i.e., $H_2@C_{60}$ or He@ C_{60}, as well as by the isolation of endohedral [70]fullerenes incorporating either one or two hydrogen molecules (Murata et al. 2008a) or a helium atom (Morinaka et al. 2010), i.e., $H_2@C_{70}$, $(H_2)_2@C_{70}$, or He@ C_{70}, respectively. The chemical synthesis of an endohedral C_{60} encapsulating a water molecule has also been achieved (Kurotobi and Murata 2011).

The term *open-cage fullerenes* usually refers to fullerene derivatives that have undergone more than one σ-bond scissions, and their π-system is significantly disturbed. These fullerene adducts are prepared by the self-photooxygenation of C_{60} (Taliani et al. 1993) and a number of exohedral fullerenes (Hummelen et al. 1995c; Inoue et al. 2001; Murata et al. 2001a,c; Weisman et al. 2001), as well as through a combination of rearrangements (mainly [4 + 4] intramolecular cycloadditions of cyclohexadiene fullerene derivatives, followed by retro [2 + 2 + 2] ring opening of the fullerene framework) (Arce et al. 1996; Hsiao et al. 1998; Schick et al. 1999; Qian et al. 2000, 2003; Nierengarten 2001; Iwamatsu et al. 2002; Qian and Rubin 2002; Suresh et al. 2003). Besides self-photooxygenation, ring expansion reactions of cage-opened fullerene adducts include the addition of aromatic hydrazines (Iwamatsu et al. 2003a,b; Vougioukalakis et al. 2004b; Iwamatsu and Murata 2005; Roubelakis et al. 2007a), hydrazones (Iwamatsu et al. 2004b; Iwamatsu and Murata 2005), and 1,2-phenylenediamines (Iwamatsu et al. 2004a,c; Iwamatsu and Murata 2005) to open-cage diketone fullerene derivatives, as well as sulfur or selenium atom insertion into the rim of the orifice of open-cage fullerenes bearing an α,β-unsaturated carbonyl moiety (Murata et al. 2003b; Vougioukalakis et al. 2004a; Chuang et al. 2007a; Roubelakis et al. 2007b). A different approach to open-cage fullerenes through peroxide-mediated chemical transformations has been also developed (Xiao et al. 2007).

This chapter discusses all chemical transformations aiming at the creation, expansion, or reduction and complete closure of an opening on the fullerene cage and covers the literature until December 2014. The related encapsulation studies are outside the scope of the present chapter (Murata et al. 2008b; Gan et al. 2010; Vougioukalakis et al. 2010a).

9.2 Fulleroids and Azafulleroids

The first fullerene derivative resulting from the cleavage of one σ-bond on the fullerene core, [5,6]-dihydrofulleroid H_2C_{61}, was isolated in 1992 via the reaction of C_{60} with diazomethane (Suzuki et al. 1992a). The electronic properties and number of π electrons in this adduct remain unaltered with regard to the parent fullerene, although it has an open transannular bond (methanoannulene structure). Fullerene adducts with such structure are called fulleroids (Wudl 1992). Addition of organic azides to C_{60} leads to the isolation of azafulleroids, the nitrogen substituted analogs of fulleroids (Prato et al. 1993a). In this section, we will briefly discuss the synthesis of fulleroids and azafulleroids. Some of these molecules were prepared in the early stages of the development of the "molecular surgery" approach and many of the subsequently synthesized open-cage adducts originate from these compounds.

The corresponding pyrazolines (**1**, Scheme 9.2) are formed through the addition of diazomethanes (Suzuki et al. 1991, 1992a,b; Shi et al. 1992; Wudl 1992; Isaacs et al. 1993a; Prato et al. 1993b; Smith et al. 1993; Diederich et al. 1994b), diazoacetates (Wudl 1992; Isaacs et al. 1993b), and diazoamides (Skiebe and Hirsch 1994) at the [6,6]-ring junctions of fullerenes via [3 + 2] cycloadditions. Thermal extrusion of N_2 leads to the closed [6,6]-bridged (**2**, Scheme 9.2) and open [5,6]-bridged fullerene adducts (**3a** and **3b**, Scheme 9.2) in different ratios (Diederich et al. 1994a). The reason for the exclusive formation of closed [6,6]-bridged and open [5,6]-bridged with the simultaneous absence of open [6,6]-bridged (**4**, Figure 9.1) and closed [5,6]-bridged (**5**, Figure 9.1) valence isomers is the preservation of the favorable [5]radialene-type bonding pattern. This effect is basically the principle for the minimization of [5,6]-double bonds (highlighted bonds in **4** and **5**, Figure 9.1) in the fullerene framework (Taylor 1992; Taylor et al. 1993). According to International Union of Pure and Applied Chemistry (IUPAC), the terms *fulleroids* and *methanofullerenes* are used to indicate the open [5,6]-bridged (methanoannulene-type bonding) and closed [6,6]-bridged (bisnorcaradiene- or cyclopropane-type bonding) fullerene isomers.

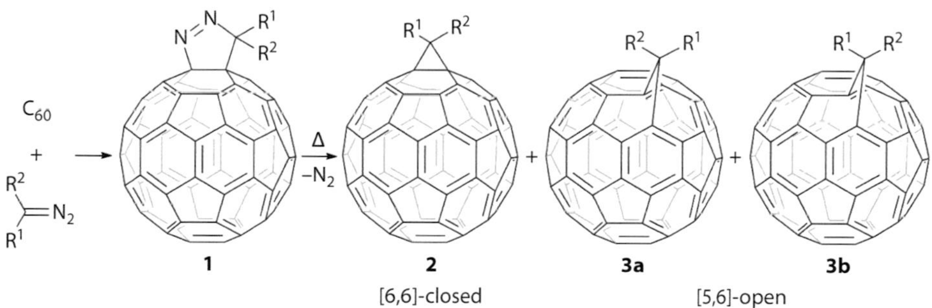

SCHEME 9.2 The reaction of diazo compounds with C_{60}.

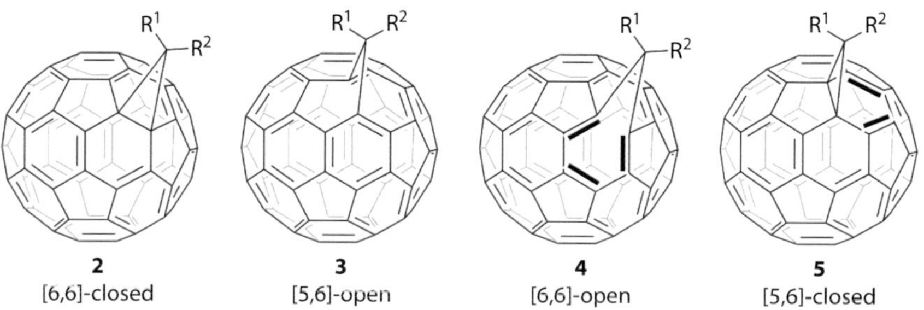

FIGURE 9.1 The four different isomers formed from a single diazomethane addition to C_{60}.

The preference for closed [6,6]- or open [5,6]-bridged fullerenes was proposed to be independent from substituent effects, in contrast to the chemistry of methanoannulenes (Isaacs et al. 1993a). This preference was explained on the basis of the [5,6]-double bonds minimization principle (Taylor 1992; Taylor et al. 1993), that is, the energetically favorable preservation of the greatest possible number of doubly bonded [6,6]-junctions, which was confirmed theoretically (Prato et al. 1993b; Raghavachari and Sosa 1993; Diederich et al. 1994b). The closed [6,6]-bridged fullerene adducts are more stable than their open [5,6]-bridged isomers, as a result of the violation of Bredt's rule as well. A series of theoretical calculations in combination with experimental work suggested that the thermal transformation of the open [5,6]- to the closed [6,6]-bridged valence isomers can occur only in cases where the corresponding closed [5,6]-bridged structure is located in a local energy minimum (Diederich et al. 1994b).

Fulleroids are generally unstable (thermally, photochemically, and under acidic conditions), usually providing the corresponding methanofullerenes. The methodology of using prolonged heating of the reaction mixture in the diazo-addition reactions to C_{60} to get rid of the fulleroid isomers, leading to a single, closed [6,6]-bridged fullerene adduct, is quite usual. Nevertheless, some notable examples of [5,6]-fulleroid adducts are presented in Figure 9.2. For example, adducts **6–8** (Figure 9.2) have been synthesized during the synthesis of amino acid and amido derivatives of C_{60} (Skiebe and Hirsch 1994), whereas **9** (Figure 9.2) has been synthesized as a fullerene derivative bearing a versatile functional group that facilitates anchoring of C_{60} on various substrates (Hummelen et al. 1995b). The reaction of C_{60} with tosylhydrazone lithium salts in refluxing toluene afforded fulleroids **10–14** (Figure 9.2) in low to moderate yields (Li et al. 1996).

The first [6,6]-open structure was reported in 2007 (Pimenova et al. 2007). According to this work, a mixture of $C_{60}(CF_2)_n$ adducts (n = 1–3) and unreacted [60]fullerene was obtained by refluxing C_{60} with $CF_2ClCOONa$ and 18-crown-6 in *o*-dichlorobenzene (ODCB). Monoadduct $C_{60}(CF_2)$, isolated by means of semipreparative high performance liquid chromatography (HPLC), was found to consist of [6,6]- (major product) and [5,6]- (minor product) isomers, both having an open structure. The [6,6]-open

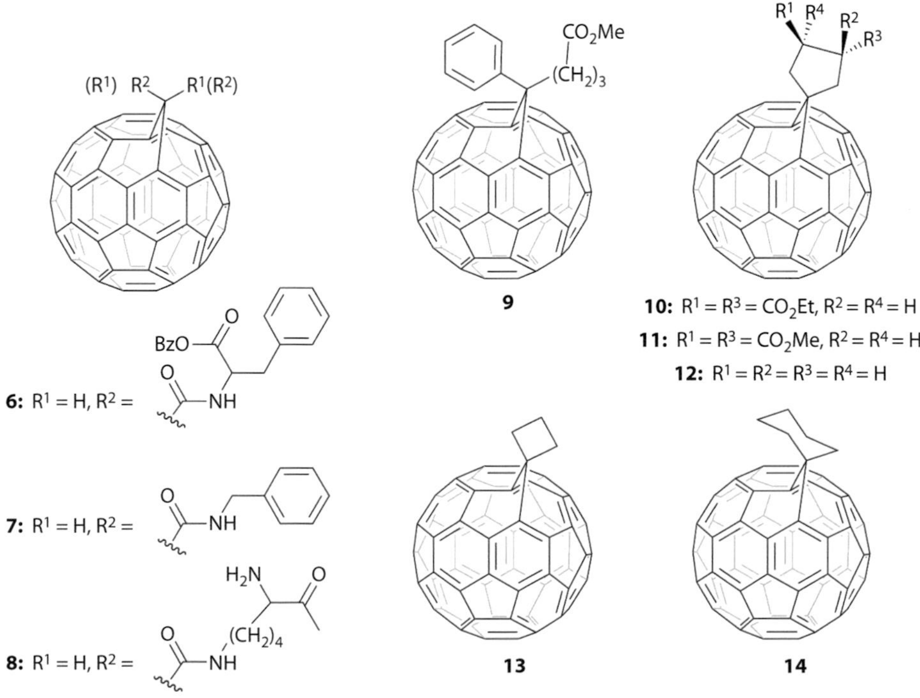

FIGURE 9.2 Fulleroids **6–14**.

structure was deduced from the ^{13}C nuclear magnetic resonance (NMR) spectrum of the adduct, as no chemical shifts were present in the typical region for cyclopropane sp^3 carbon atoms. Ultraviolet-visible spectroscopy (UV-Vis) and infrared (IR) spectroscopy, as well as theoretical calculations, confirmed this result. It was suggested that this unprecedented structure gains stabilization via a redistribution of the π-electron density of the cleaved [6,6]-bond to the adjacent pentagonal rings.

The isolation and characterization of C_{70} fulleroids are rather difficult tasks because of the existence of eight different bond types in C_{70} (four of which are [6,6]-carbon–carbon bonds), compared with the two types contained in C_{60}. In other words, the low symmetry of C_{70} usually leads to a large number of isomers. However, two contributions concerning the diazomethane addition to C_{70} have been published by Smith et al. (1994, 1995). Specifically, the addition of ethereal diazomethane to a toluene solution of C_{70} afforded a 12/1/2 mixture of three pyrazolines. Photolysis of this mixture afforded two isomeric methano[70]fullerenes, whereas thermolysis gave two isomeric [70]fulleroids, the first to be reported for C_{70}. In a subsequent study, the first crystal structure of a fulleroid, obtained by treating C_{70} with $PhHgCCl_2Br$ in refluxing toluene, was acquired (Kiely et al. 1999). By the use of preparative HPLC, three isomeric monoadducts, $C_{71}Cl_2$, were isolated in 28% combined yield. Two of these [70]fullerene adducts were proved to be methanofullerenes, and the third one, a C7–C8 fulleroid. This is actually the strongest evidence that, indeed, the frameworks obtained by additions to the [5,6]-bonds are open, homoconjugated annulenes.

Azides add to C_{60} thermally, affording azafulleroids **15–18** (Figure 9.3) as the final addition products (Prato et al. 1993a). The assignment of the open [5,6]-azabridged annulene structure (azafulleroid), bearing an intact 60 π-electron system and an open transannular bond, was basically based on ^{13}C NMR and UV-vis data. Although the expected structure from a [3 + 2] cycloaddition of an organic azide to C_{60} is a triazoline, the direct observation of such a moiety had to wait until 1996, when the [6,6]-bridged fullerotriazoline **19** (Figure 9.3) was characterized through single-crystal x-ray analysis (Nuber et al. 1996).

Since then, several research groups have managed to isolate various other azafulleroid adducts. For instance, to prepare stable, well-defined fullerene monolayers, Hawker et al. (1994) synthesized the amphiphilic [5,6]-azafulleroid **20a** (Figure 9.4), in which a short hydrophilic linear chain is covalently attached to the hydrophobic C_{60} nucleus. Soon thereafter, the isolation of *N*-methoxy-ethoxymethyl (MEM)-substituted [5,6]-azafulleroid **20b** (Figure 9.4) was reported via the thermal addition of MEM-N_3 to C_{60} (Nuber et al. 1996).

The same addition reaction has been also carried out with C_{70} (Bellavia-Lund 1997c). Interestingly, in this case, only three of the possible six triazoline isomers are formed, chemo- and regio-selectively. Moreover, it was observed that the major product arises from azide addition to the double bond of C_{70} possessing the greatest local curvature and that thermolysis of the triazoline isomers produced mixtures of C_{70}, aza[70]fulleroids, and [70]fulleroaziridines (the closed [6,6]-bridged aza-analogs of methanofullerenes).

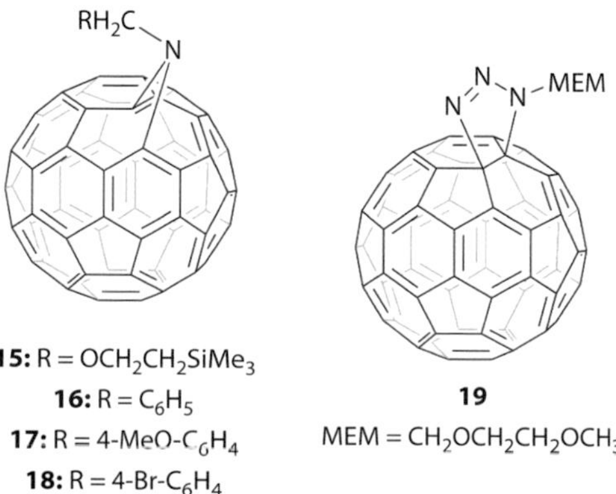

FIGURE 9.3 Azafulleroids **15–18** and fullerotriazoline **19**.

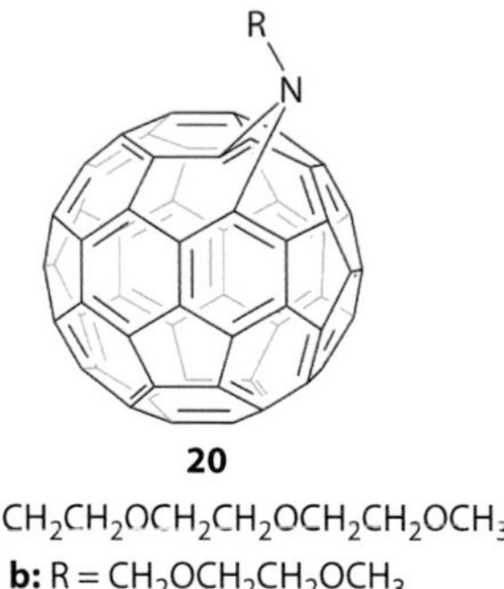

FIGURE 9.4 Amphiphilic azafulleroid **20a** and *N*-methoxyethoxymethyl (MEM)-substituted [5,6]-azafulleroid **20b**.

The photochemically induced rearrangement of azafulleroids to aziridinofullerenes involves isomerization of azafulleroids **21a** and **21b** to aziridinofullerenes **22a** and **22b** respectively (Scheme 9.3) (Averdung and Mattay 1996). Furthermore, thermolysis of the triazolinofullerene derivatives **23a** and **23b** affords mainly azafulleroids **21a** and **21b**, whereas aziridinofullerenes **22a** and **22b** are selectively isolated from the corresponding photolysis (Scheme 9.3). The exclusive formation of **22** upon photolysis of **23** supports a reaction mechanism via nitrene intermediates.

The reaction of azafulleroid **24a** with a second alkylazide moiety (Scheme 9.4) leads to only one mixed [6,6]-triazoline-[5,6]-iminofullerene isomer (**25a**) (Grösser et al. 1995). This highly regioselective pathway was rationalized on the basis of the fact that **24a** behaves as a strained electron-poor vinylamine. Given that the highest positive Mulliken charges (austin model 1—AM1—calculations) are located at C-1 and C-6 and the lowest at C-2 and C-5 (Scheme 9.4), as well as that the most negatively polarized nitrogen atom of the azide (AM1) is the one bearing R, a kinetically controlled attack of the azide should predominantly lead to **25a**. Upon heating, **25a** eliminates nitrogen, leading mainly to diazafulleroid **26a** (Scheme 9.4). The analogous bisadduct **26b** was more recently obtained as a by-product during the preparation of **24b** in a one-pot reaction (Yang et al. 2009). Treatment of C_{60} with two equivalents of methyl 5-azido-5-phenylpentanoate leads to a [3 + 2] cycloaddition, followed by the in situ thermal elimination of nitrogen, thus affording bisadduct **26b** (Kim et al. 2012).

The vinylamine moiety of azafulleroids affords another regioselective ring expansion reaction. More specifically, azafulleroid **20b** is self-photooxygenated to the corresponding *N*-MEM-ketolactam **28** (Scheme 9.5), apparently through the dioxetane intermediate **27** (Hummelen et al. 1995c;

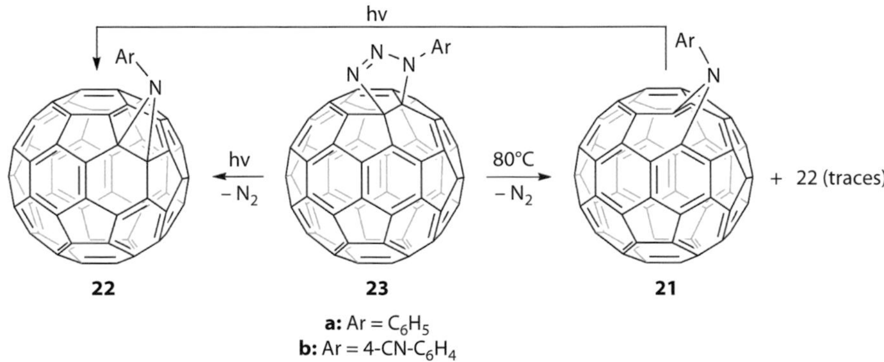

SCHEME 9.3 Thermolysis or photolysis of triazolinofullerene derivative **23** affords azafulleroid **21** or aziridinofullerene **22**, respectively.

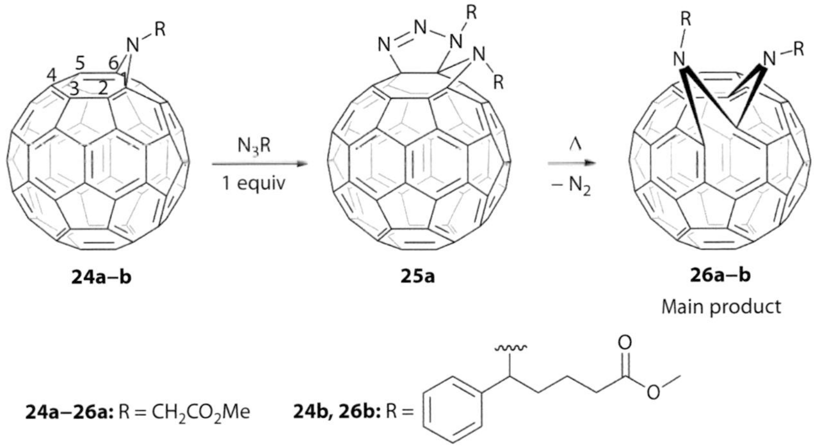

SCHEME 9.4 Regioselective formation of bisazafulleroids **26a** and **26b**.

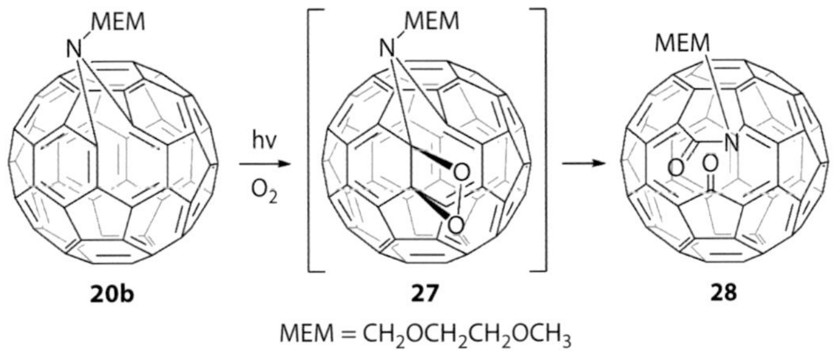

SCHEME 9.5 Regioselective self-photooxygenation of azafulleroid **20b**.

Vassilikogiannakis et al. 1998). This compound is the precursor of azafullerene $(C_{59}N)_2$ (Hummelen et al. 1995a), the only heterofullerene isolated in preparative scale to date.

The first example of open [6,6]-bridged azafullerenes was reported in 1996 (Banks et al. 1996). Upon treatment with zinc and glacial acetic acid, closed [6,6]-bridged *N*-oxycarbonylaziridinofullerenes **29a–29d** (Figure 9.5) were selectively reduced at the bridgehead C–C bond to give the dihydro derivatives **30a–30d** (Figure 9.5), bearing a bridged 10-membered-ring opening. The *N*-*t*-butoxycarbonyl protecting group in **30b** was next removed under acidic conditions to afford $C_{60}H_2NH$ (**31**, Figure 9.5). Adduct **31**

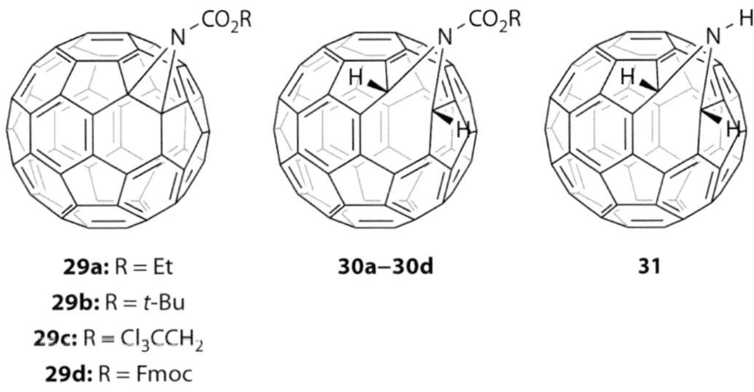

FIGURE 9.5 Fullerene adducts **29a–29d** and open [6,6]-bridged azahomofullerenes **30a–30d** and **31**.

is remarkably air-stable; nevertheless, under basic conditions, it undergoes oxidative ring closure to the closed [6,6]-bridged parent fulleroaziridine $C_{60}NH$.

In addition to azides, bisazides add to C_{60}, affording the corresponding 1:1 bisadducts **32–33** (Figure 9.6) in moderate to good yields (Dong et al. 1995; Shiu et al. 1995). The addition occurs regioselectively at two [5,6]-ring junctions of the same five-membered ring, providing bisazafulleroids with C_s symmetry. Similarly, the addition of C_2 chiral 1,4-*t*-alkoxy-2,3-bisazidobutanes **34a–34b** to C_{60} gives racemic mixtures of bisazafulleroids **35a–35b** (Figure 9.6), whose enantiomeric pairs exhibit mirror image circular dichroitic curves (Shen et al. 1996). It should be also noted that the addition of bisazidobutadiene system **36** (Figure 9.6) to C_{60} promoted one of the first effective openings on the fullerene framework concerning its ability to encapsulate gas atoms or molecules (vide infra) (Schick et al. 1999; Nierengarten 2001).

The preparation of bisazafulleroids **41a–41b** via the synthetic approach shown in Scheme 9.6 was reported in 1995 (Lamparth et al. 1995; Schick et al. 1996). Besides being the first dicyclo adducts with addends-bound in *cis*-1 positions, **41a–41b** are the first examples of fullerene bisadducts bearing open transannular [6,6]-bonds. Although three energetically unfavorable [5,6]-double bonds have to be introduced in **41**, this structure is 3 kcal/mol lower in energy (AM1 calculations) than **40** (Lamparth et al. 1995). Treatment of **41b** with trifluoroacetic acid (TFA) at room temperature quantitatively led to bisadduct **42** (Scheme 9.6), in which the transannular [6,6]-bonds are closed, by an intraring 2π to 2σ isomerization on the fullerene shell (Schick et al. 1996). To date, this is the only reported chemical modification of fullerenes that allows the synthesis of both open and closed valence isomers with the same addition pattern. In the same work (Lamparth et al. 1995), the hydroamination of **26** (Figure 9.7) with one equivalent of butylamine in the presence of 1,8-diazabicyclo[5.4.0]undec-7-ene was also reported. The isolated products, namely, a mixture of the monohydroaminated regioisomers (**43**, Figure 9.7), as well

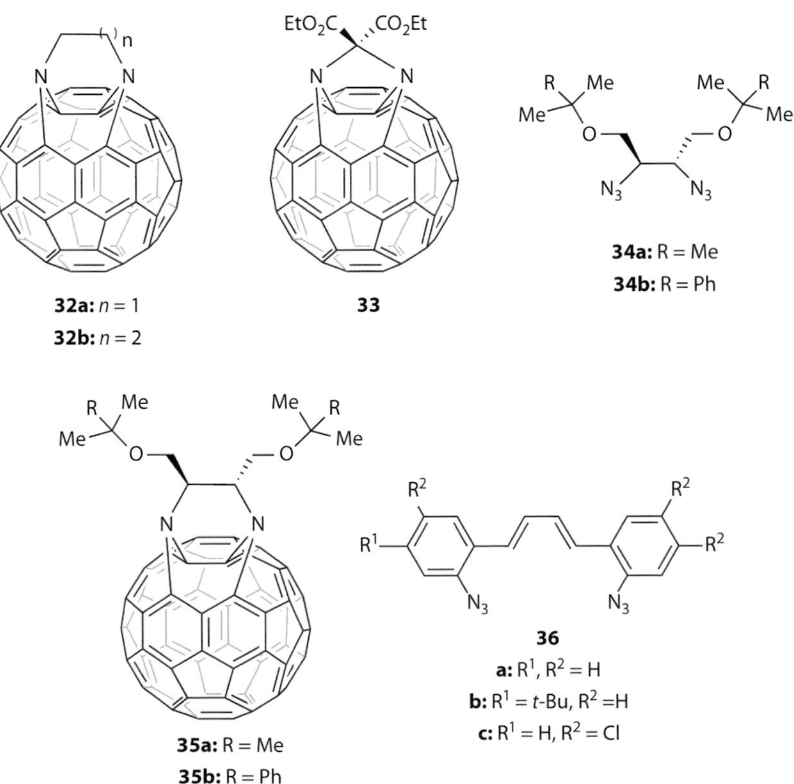

FIGURE 9.6 Bisazafulleroid adducts **32**, **33**, and **35**, **1**,**4**-t-alkoxy-**2**,**3**-bisazidobutanes **34a** and **34b**, and bisazidobutadienes **36a–36c**.

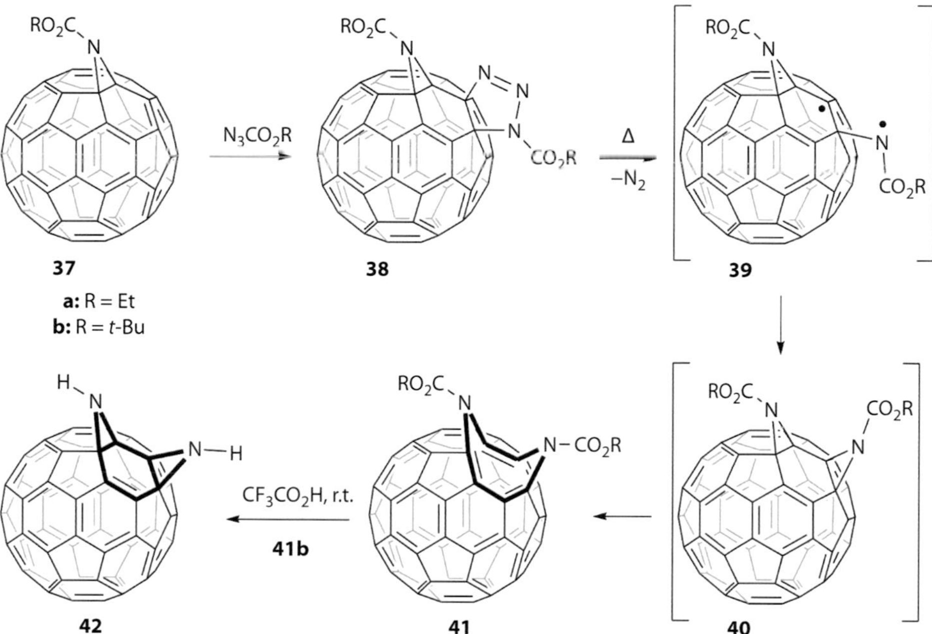

SCHEME 9.6 Proposed mechanism for the production of **41** and the conversion of **41b** to **42**.

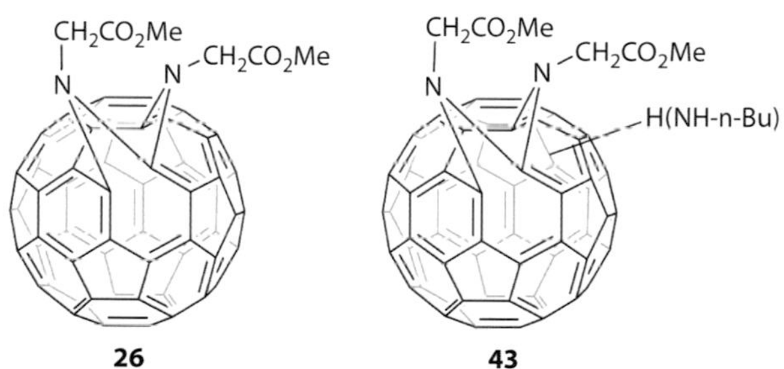

FIGURE 9.7 Bisazafulleroid adduct **26** and its monohydroaminated analog **43**.

as their C_{70} analogs, were also used as precursors in the synthesis of azafullerenes $(C_{59}N)_2$ and $(C_{69}N)_2$ (Nuber and Hirsch 1996).

In 2013, Gan and coworkers published a work on the skeletal modification of azafullerenes (Huang et al. 2013). Starting from azafullerene derivative **44**, the bromine atom is substituted leading to derivatives **45a–d** (Scheme 9.7). Treatment of the hydroxylamine adduct **45d** with PCl_5 affords azahomofullerene derivative **46** in a step involving a Beckman-type rearrangement. The chlorine atom in **46** is slightly more reactive toward alcohol substitution than the bromine atom in **44**, as the extra nitrogen atom increases the stability of the cationic intermediate in the S_N1 reaction performed. Compounds **48a–f** were prepared in a manner analogous to the one followed for the formation of **45**. Adduct **50** was isolated as a by-product of the reaction performed to form trifluoroethyl derivative **48d**. The reaction between silver hexafluoroantimonate and **46** results in the replacement of the *tert*-butylperoxo group next to the enamine and the formation of compound **49**. Treating trifluoroethoxy adduct **48d** with BBr_3 and PPh_3, in turn, leads to the formation of **51** (Scheme 9.8). To grow crystals, compound **51** was kept in a benzene–methanol mixture for 2 wk when the methanol-addition product **52** was isolated. Further treatment of **51** with 4-methylphenol in the presence of TsOH leads to the formation of **53** after a double

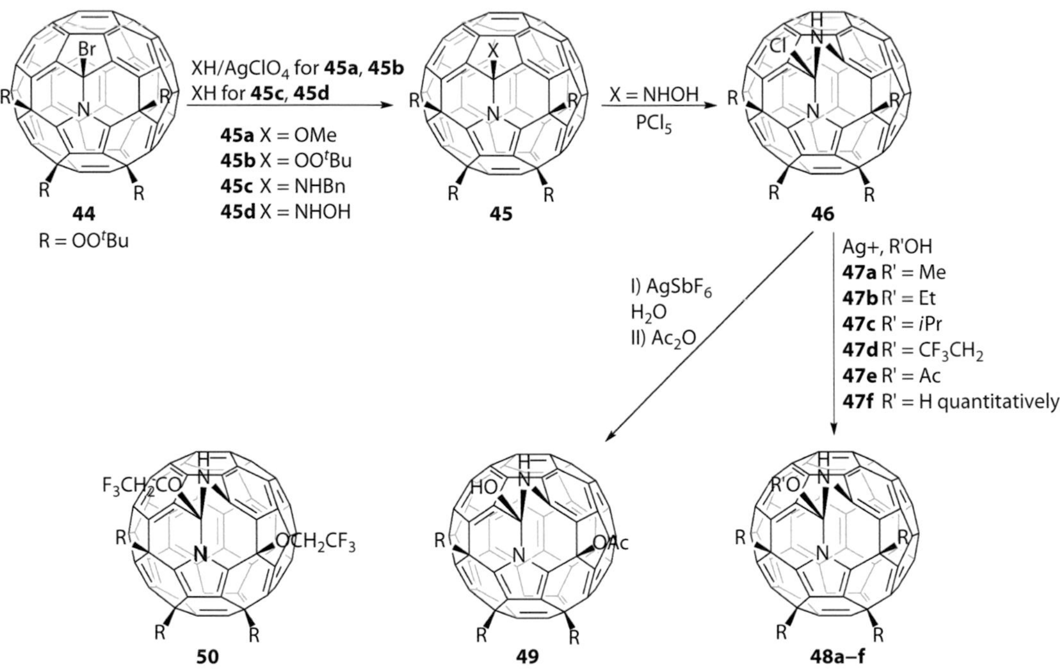

SCHEME 9.7 Fullerene adducts **44–50**.

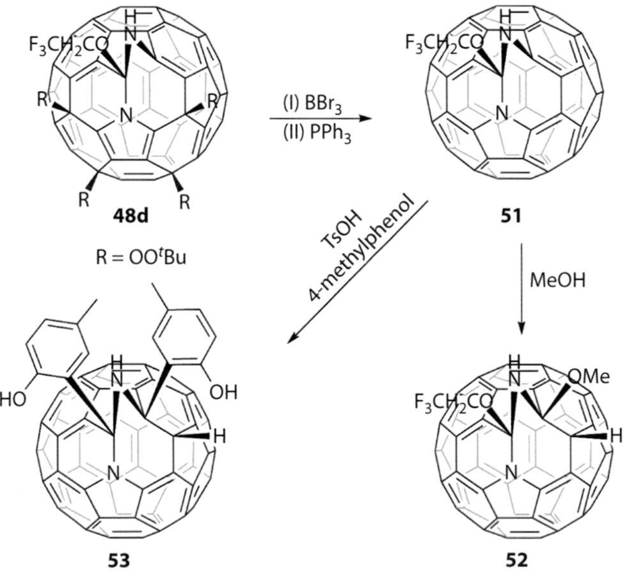

SCHEME 9.8 Fullerene adducts **52–53**.

Friedel-Crafts reaction. Finally, irradiation of **51** with a blue-diode light in the presence of molecular oxygen affords compound **54** (Scheme 9.9). To further explore the hydrolysis of adduct **51**, Gan and coworkers treated a chlorobenzene solution of **51** with TsOH obtaining **55**. **55** was converted back to **54** under an acidic treatment similar to that followed for the transformation of **51** to **54**. According to the authors, this fact suggests that the main route toward **54** under photochemical conditions involves a first, rate-determining oxygenation step to form **A** (Scheme 9.9).

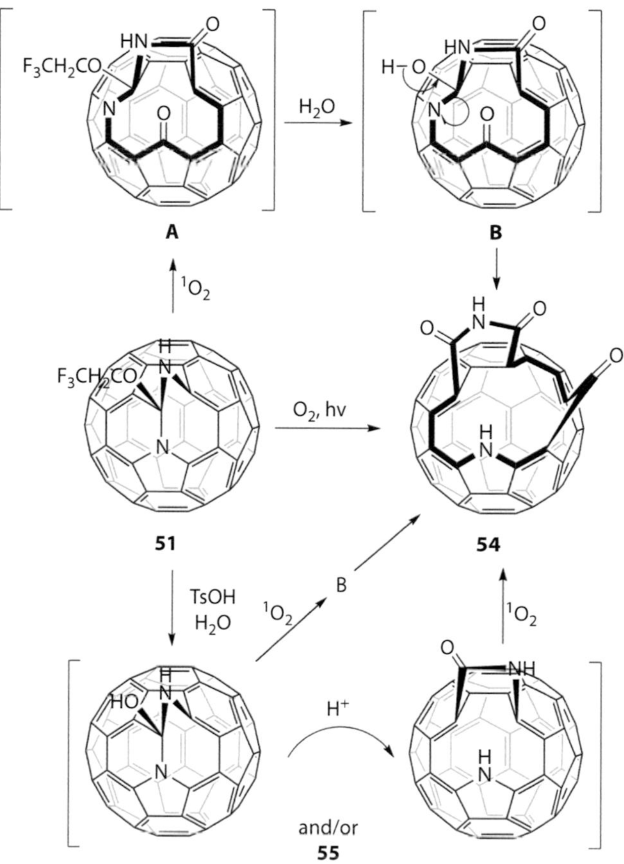

SCHEME 9.9 Photooxidation of azafulleroid **51** to **54**.

9.3 Intramolecular [4 + 4] Cycloadditions Followed by [2 + 2 + 2] Cycloreversions and Benzyne Cycloadditions

This strategy is highly efficient for the preparation of open-cage fullerene derivatives having relatively large openings. The first step toward this direction was reported in 1996 (Arce et al. 1996), when Rubin and coworkers managed to isolate and characterize the cobalt(III) complex **56** (Scheme 9.10). Open-cage bisfulleroid **57**, which is the precursor of complex **56**, was prepared (Arce et al. 1996) by the photochemically promoted rearrangement of cyclohexadiene-fused derivative **58** (An et al. 1995). This transformation was proposed to occur via a [4 + 4] photocycloaddition (Masamune et al. 1968), followed by a thermally allowed [2 + 2 + 2] cycloreversion (Scheme 9.10). Alternatively, photolysis of alcohol **59**, under reflux and acidic conditions (Scheme 9.10), afforded the same bismethano[12]annulene **57**, bypassing the isolation of oxygen-sensitive diene **58** (Arce et al. 1996). Later on, the isolation of bisadducts **60** and **61** (Figure 9.8) was also reported (Qian and Rubin 2002). Moreover, according to a pressure-tuning vibrational spectroscopic study of **56** (Edwards et al. 1998), no evidence for encapsulation of the cobalt atom was found upon application of high external pressures (up to ~45 kbar).

Likewise, open-cage bisfulleroids **62a–62e** (Hsiao et al. 1998) and **64a–64c** (Inoue et al. 2000) were synthesized from the corresponding cyclohexadiene-fused derivatives **63a–63e** and **65a–65c** in high yields (85%–98%) upon irradiation (Figure 9.9). The mechanistic rationalization offered for these transformations was again based on the previously mentioned [4 + 4] intramolecular photocycloaddition with a concomitant [2 + 2 + 2] cycloreversion.

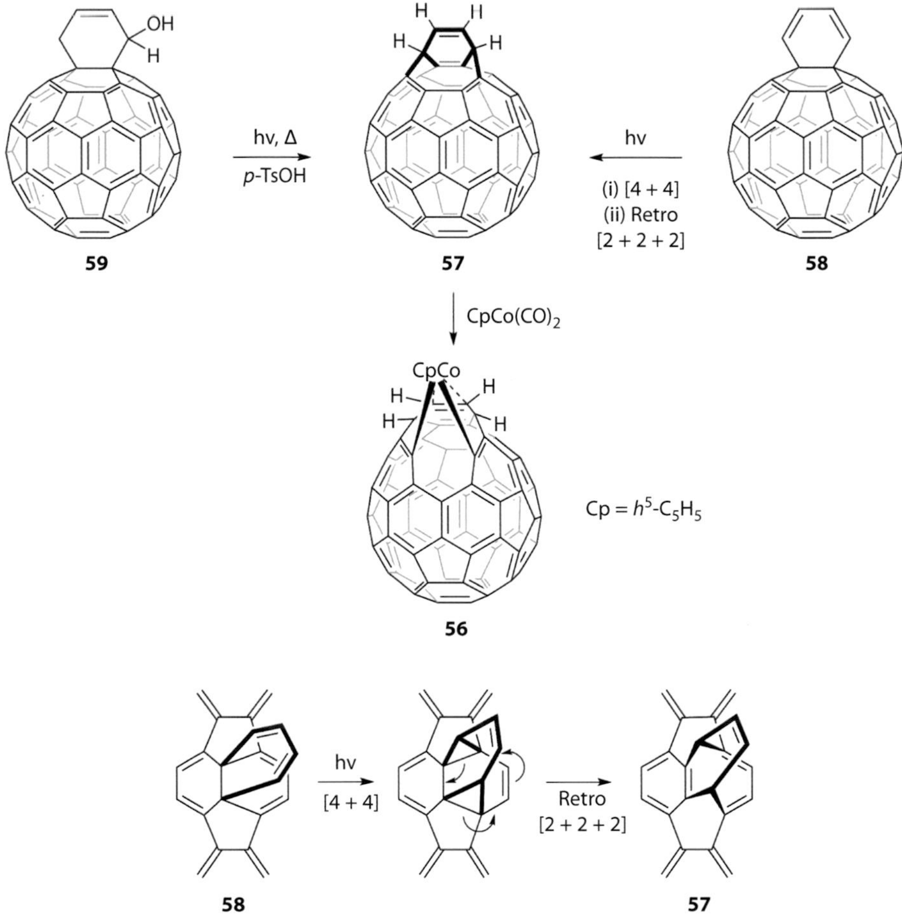

SCHEME 9.10 Preparation of open-cage fullerene adduct **57**, from diene **58** or alcohol **59**, followed by oxidative cobalt insertion.

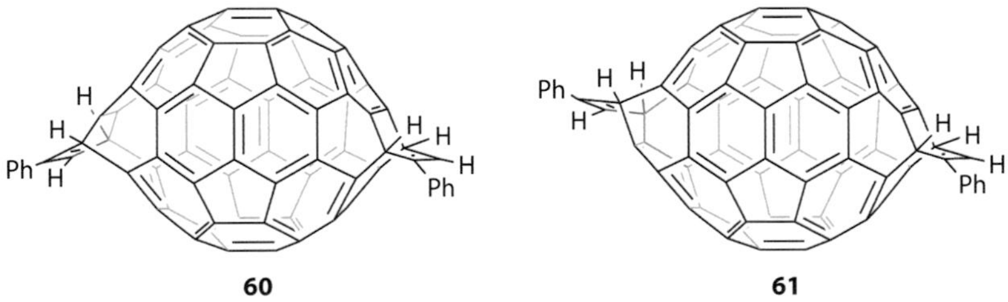

FIGURE 9.8 Open-cage fullerene adducts **60** and **61**.

A remarkable study on the photochemical reactivity of cyclohexadiene fullerene derivatives was published in 2002 (Iwamatsu et al. 2002). Once the photorearrangement reaction of **65b** (Figure 9.9) was carried out under mild photolytic conditions (20 W fluorescent lamp [Iwamatsu et al. 2002] instead of 500 W tungsten halogen lamp [Inoue et al. 2000]), bismethanofullerene **66** and unsymmetrical fullerene adduct **67** (Figure 9.10) were formed, apart from **64b** (Figure 9.9). Moreover, in addition to bisfulleroid **64a** (Inoue et al. 2000), bismethanofullerene **68** (Figure 9.10) was isolated from the benign photochemical rearrangement of **65a** (Figure 9.9), although in very low yield (<5%). Since neither **64a** nor **64b** gives

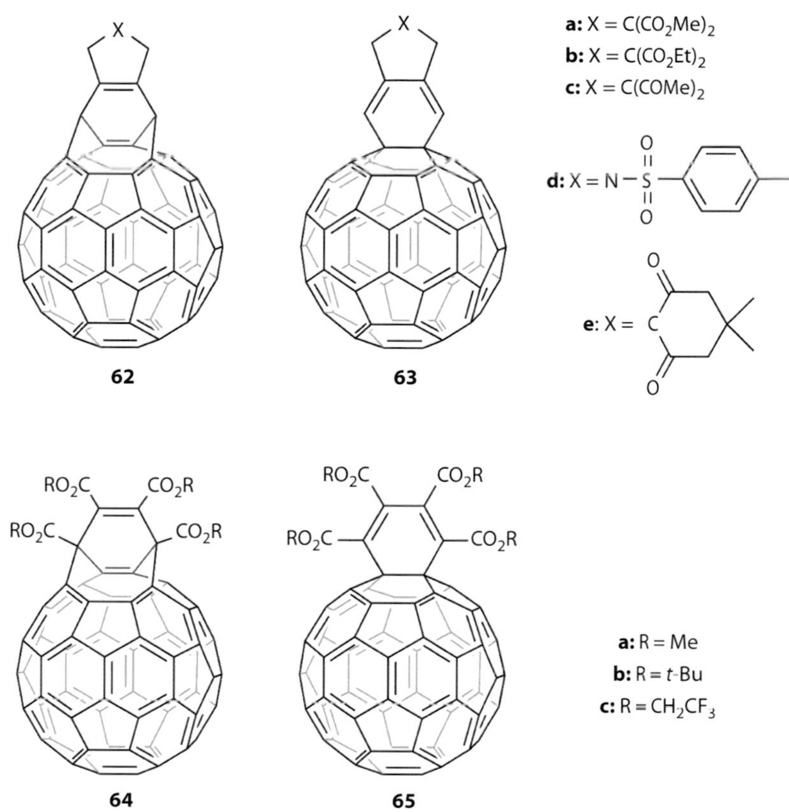

FIGURE 9.9 Open-cage bisfulleroids **62a–62e** and **64a–64c** and cyclohexadiene derivatives **63a–63e** and **65a–65c**.

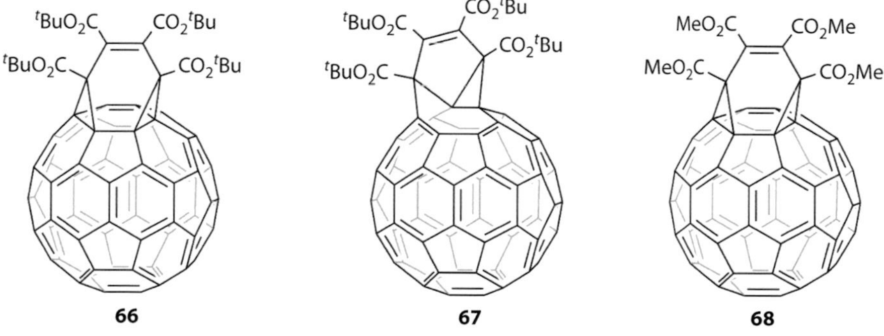

FIGURE 9.10 Bismethanofullerenes **66** and **68** and the unsymmetrical fullerene adduct **67**.

68 or **66**, respectively, and **67** affords a 1:1 equimolecular mixture of **64b** and **66** upon irradiation, it was proposed that in the reaction coordinate for the photorearrangement of **65b**, the unsymmetrical adduct **67** is the initial, common intermediate, concomitantly transformed either to bisfulleroid **64b** or to bismethanofullerene **66**. As also inferred from that work, the mechanism leading to bismethanofullerenes is significant only for substances with bulky substituents (Iwamatsu et al. 2002).

While trying to prepare azacyclohexadiene or diazacyclohexadiene fullerene derivatives (**69**, Figure 9.11) that would lead to the corresponding aza-open-cage fullerene derivatives (**70**), via the previously mentioned [4 + 4]/retro [2 + 2 + 2] route, Komatsu and coworkers studied the thermal reaction of C_{60}

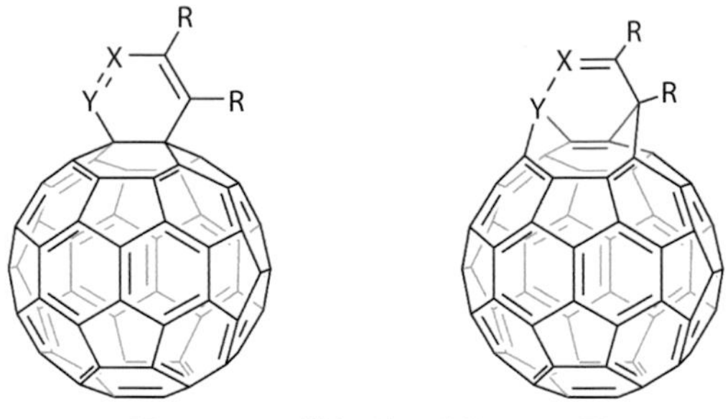

FIGURE 9.11 Conceptual fullerene derivatives **69** and **70**.

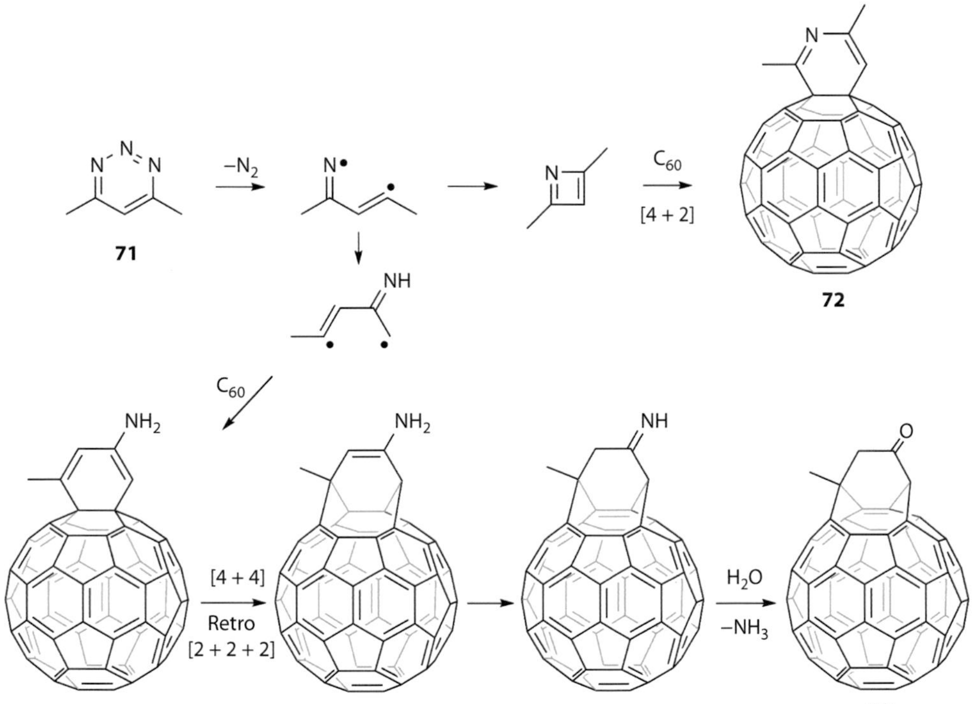

SCHEME 9.11 Proposed mechanism for the formation of azacyclohexadiene-fused derivative **72** and open-cage fullerene derivative **73**.

with 4,6-dimethyl-1,2,3-triazine **71** (Scheme 9.11) (Murata et al. 2001a). In contrast to their expectations, they only managed to isolate the azacyclohexadiene-fused derivative **72** (Scheme 9.11) together with open-cage fullerene **73** in low yield (<10% each). No transformation of **72** to **73** was observed. The mechanistic rationalization they offered to account for their observations is shown in Scheme 9.11.

The first "stable" higher fullerene according to the isolated pentagon rule (IPR) is C_{70} (Fowler and Manolopoulos 1995). IPR predicts fullerene structures with all pentagons isolated by hexagons to be stabilized against structures with adjacent pentagons. All intermediate fullerenes C_{62}–C_{68} include at least two energetically unfavorable fusions of two 5-membered rings within their structures. Moreover, the high calculated strain energy for fullerenes containing four-membered rings (Gao and Herndon 1993;

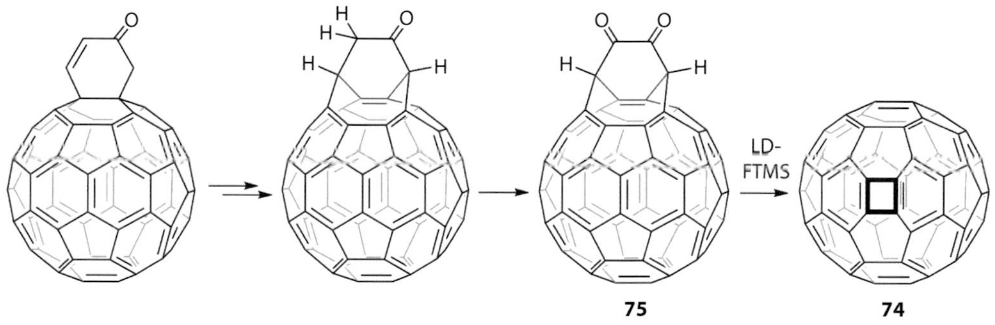

SCHEME 9.12 Preparation of open-cage fullerene derivative **75** and the LD-FTMS formation of the four-membered-ring isomer of C_{62}, **74**.

Aihara 1995; Fowler et al. 1996) renders them unlikely to be isolated from the fullerene soot produced at elevated temperatures.

Nevertheless, Houk, Wilkins, Rubin, and coworkers (Qian et al. 2000) proposed a rational synthetic approach to a four-membered-ring isomer of C_{62} (**74**, Scheme 9.12). Compound **75**, synthesized from C_{60} by the stepwise strategy shown in Scheme 9.12, was chosen as the key precursor. Indeed, although they did not manage to prepare C_{62} in solution, laser desorption Fourier transform mass spectrometry (LD-FTMS, CO_2 laser, negative ion mode) on **75** produced a strong signal corresponding to the C_{62} radical anion of **74** (m/z = 744). Three years later, they reported the one-pot synthesis and isolation of the four-membered C_{62} adducts **76a–76c** (Scheme 9.13) in low to moderate yields (Qian et al. 2003) and provided the x-ray structure of **76a**. As shown in Scheme 9.10, **76a–76c** have C_s symmetry. The two cyclobutene sp^3 carbons of **76a–76c** have characteristic ^{13}C NMR signals at 72.38, 73.57, and 72.50 ppm, respectively, while their UV-vis spectra are very similar to those of bisfulleroids.

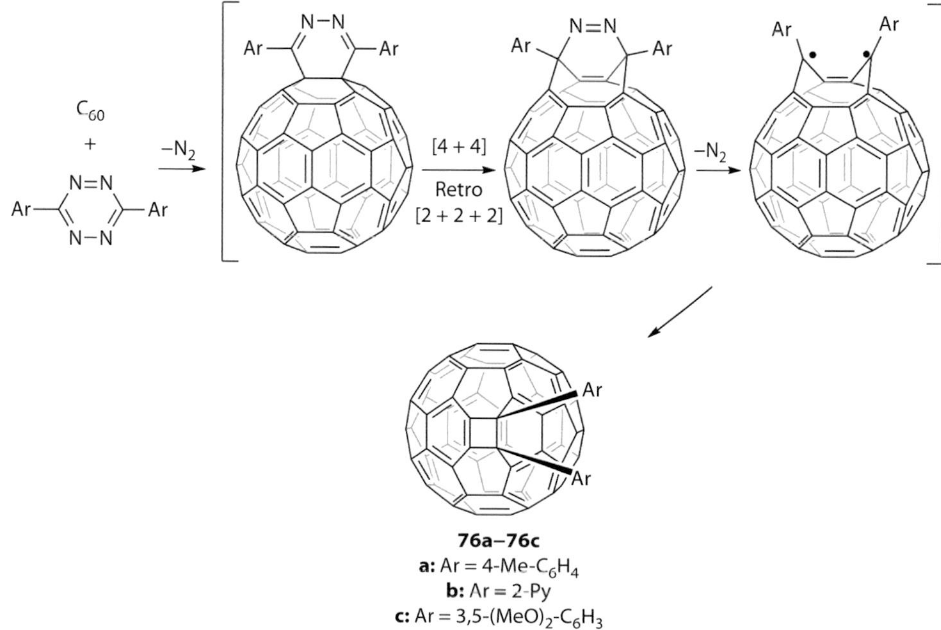

SCHEME 9.13 Proposed mechanism for the formation of the C_{62} adducts **76a–76c**.

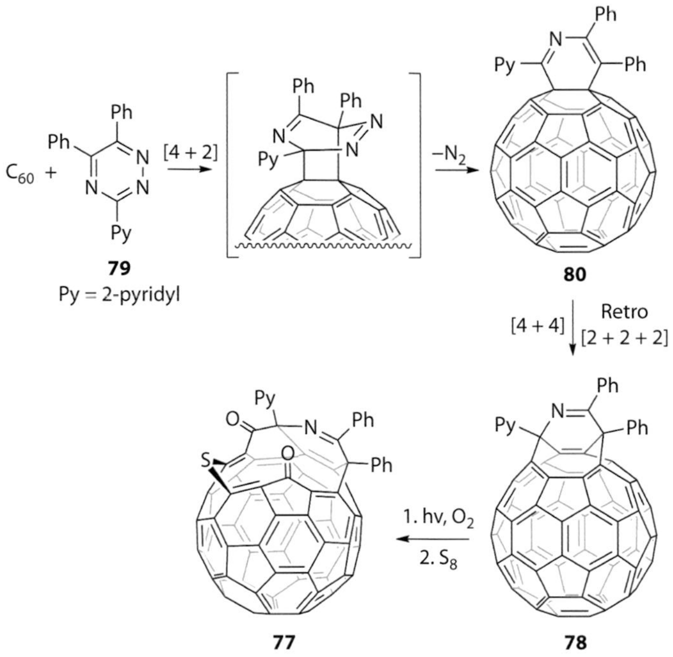

SCHEME 9.14 Formation of open-cage fullerene adducts **77** and **78**.

One of the most effective openings on the fullerene cage in encapsulating helium and molecular hydrogen has been created by Komatsu and coworkers (Murata et al. 2003b). Specifically, adduct **77** (Scheme 9.14), which has a 13-membered-ring orifice on the C_{60} shell, is obtained from the open-cage adduct **78** (Scheme 9.14) via self-photooxygenation and concomitant sulfur insertion. Open-cage derivative **78** is prepared in one step from C_{60} and 1,2,4-triazine derivative **79**. Its isolated yield is 50% (85% based on consumed C_{60}) and no bisfunctionalized product is obtained. Moreover, its structure was unambiguously determined by x-ray crystallography. The mechanism proposed for the formation of **78** is shown in Scheme 9.14 and includes a thermal [4 + 2] cycloaddition reaction between C_{60} and **79**, followed by nitrogen extrusion to give cyclohexadiene derivative **80**, and subsequent [4 + 4]/retro [2 + 2 + 2] intraconversion (also see Scheme 9.10).

The synthesis of a similar cage-opened derivative (**81**, Scheme 9.15) was later on realized by the same research group (Murata et al. 2010). The solubility of this new compound in common organic solvents

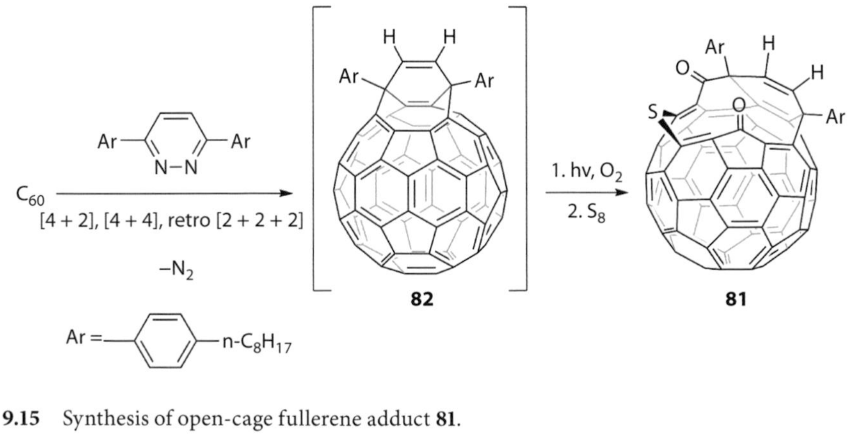

SCHEME 9.15 Synthesis of open-cage fullerene adduct **81**.

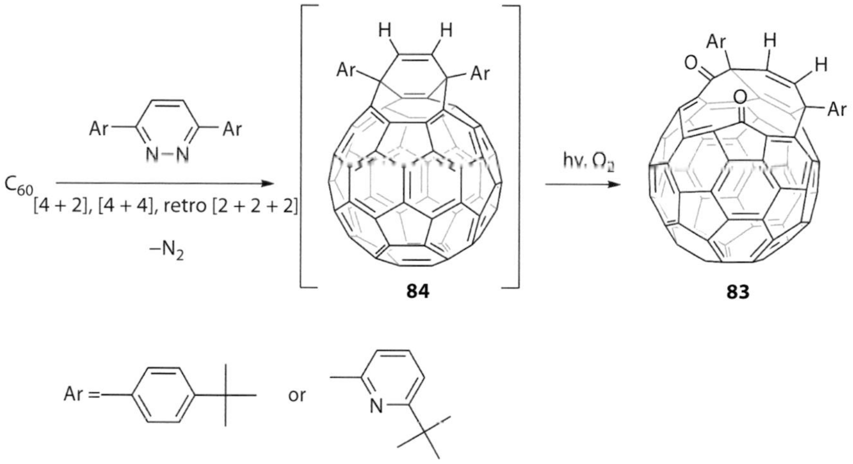

SCHEME 9.16 Preparation of open-cage fullerene **83**.

is improved as compared with adduct **77** owing to the presence of the two *n*-octyl chains. In another publication by Kurotobi et al., compound **83** (Scheme 9.16) was also obtained from C_{60} via a synthetic route involving a thermal [4 + 2] cycloaddition, in the initial step, followed by intramolecular [4 + 4] and retro [2 + 2 + 2] cycloadditions (Kurotobi and Murata 2011).

In 2011, Chuang and coworkers published the synthesis of **85**, an analog of **84** (Scheme 9.16), starting from C_{60} and a diazine derivative (Scheme 9.17) (Chen et al. 2011). Treating **85** photochemically in CS_2 in the presence of O_2 afforded, within 3 h, two different diastereoisomers: **86a** and **86b**. Thermal treatment of a solution of these diastereoisomers and powdered sulfur with tetrakis (dimethylamino)ethylene (TDAE) in ODCB led to fullerene adducts bearing 13-membered-ring orifices **87a** and **87b** (analogs of adduct **81**) in 10 min. Similarly, thermal treatment of **86** with *o*-phenylendiamine expands the orifice to a 20-membered ring, while treating **86** with phenylhydrazine expands the orifice into a 16 membered ring (Scheme 9.17).

Two years later, Yang and coworkers reported the formation of the [5,6]-open annulene derivative **90** from the reaction of C_{60} with the in situ generated benzyne adduct obtained from 2-amino-4,5-dibutoxybenzoic acid (Scheme 9.18) (Kim et al. 2013). The benzyne intermediate was suggested to first form the [2 + 2] adduct, which facilitates the subsequent rearrangement. The NMR spectra used to determine the structure of **90** suggest that the product has two geometric isomers with different conformations of the dibutoxybenzene moiety that proves the formation of the open [5,6]-structure.

9.4 Oxidation and Self-Photooxygenation Reactions

The photochemical degradation of C_{60} in the solid state was reported in 1993 (Taliani et al. 1993). This was the first report dealing with the oxidative cleavage of the fullerene framework. Based principally on differential IR spectroscopy studies of C_{60} thin films, the authors came to the conclusion that irradiation of C_{60} gives rise to open-cage $C_{60}O_2$ via the reaction of a [6,6]-double bond with electronically excited molecular oxygen, $^1O_2$ (singlet oxygen, $^1\Delta_g$). Note that $^1O_2$ production by triplet energy transfer from $^3C_{60}$ to ground-state triplet oxygen ($^3\Sigma_g$) is well documented (Zhang et al. 1993; Orfanopoulos and Kambourakis 1994, 1995; Tokuyama and Nakamura 1994; Vassilikogiannakis et al. 1998; Chronakis et al. 2002). The existence of carbonyl groups on the rim of the orifice was proven by the observation of the characteristic C = O stretching and bending IR bands. Although the cage opening reaction was found to be far from completion, and the calculated distance between the previously bound carbon atoms is only 2.68 Å, the authors essentially suggested that "surgery" at the molecular level may be viable: "The slit

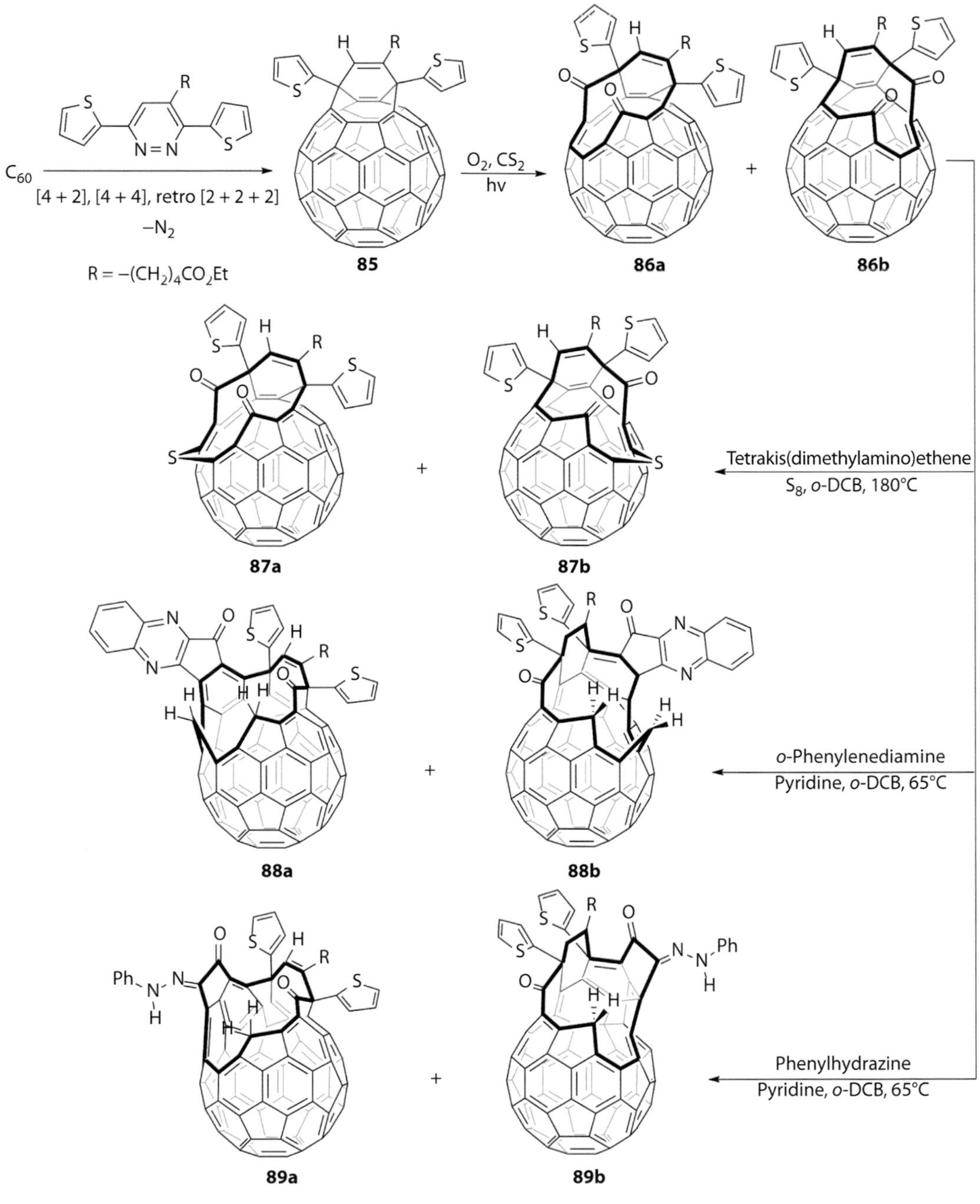

SCHEME 9.17 Synthesis of open-cage fullerenes **86**, **87**, **88**, and **89**.

in the C_{60} framework could also be used to make small atoms or ions to slip into the cage. The incision could then be sealed…restoring the original carbon-carbon double bond."

Nevertheless, the breakthrough in the controlled chemical opening of the fullerene cage commenced from a contribution by Wudl and coworkers (Hummelen et al. 1995c), showing that self-photooxygenation of azafulleroid **20b** affords *N*-MEM-ketolactam **28** in 65% yield, most probably through the intermediate dioxetane **27** (Scheme 9.5). Ketolactam **28** has been fully characterized (Hummelen et al. 1995c;

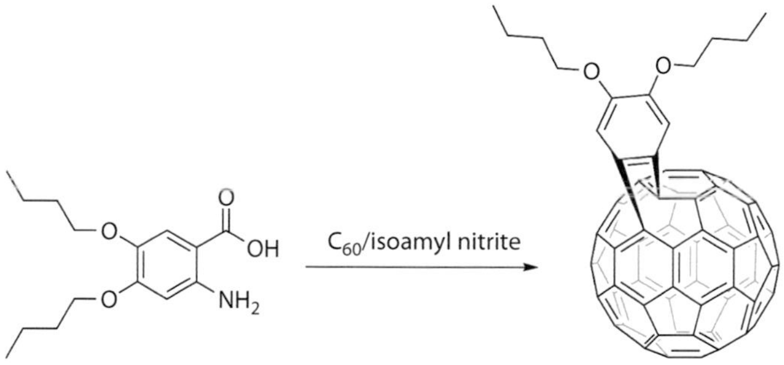

SCHEME 9.18 Preparation of open-cage fullerene **90**.

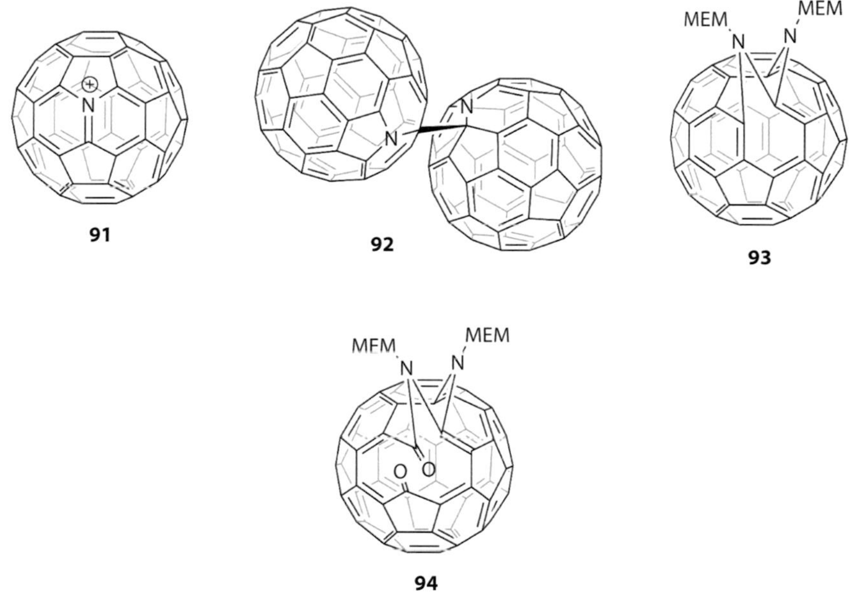

FIGURE 9.12 Aza[60]fullerene carbocation **91**, aza[60]fullerene dimer **92**, and open-cage fullerene adducts **93** and **94**.

Bellavia-Lund et al. 1997a) and is isolated as a racemic mixture of two enantiomers (Hummelen et al. 1998), which bear an 11-membered-ring chiral orifice. Enantiomers (+)-**28** and (−)-**28** were separated, on an analytical chiral stationery phase HPLC column, in 80% and 92% enantiomeric excess, respectively, as determined by their specific rotations at 589 nm (Hummelen et al. 1998). The gas phase generation of azafullerenium carbocation **91** (Figure 9.12) during mass spectrometry (MS) studies on **28** led to the first synthetic route toward aza[60]fullerene **92** (Figure 9.12) (Hummelen et al. 1995a; Keshavarz et al. 1996; Nuber and Hirsch 1998; Vougioukalakis et al. 2003b, 2004c, 2010b; Roubelakis et al. 2010). Later on, a second route to $(C_{59}N)_2$ was developed by Nuber and Hirsch (1996). $(C_{69}N)_2$ has also been synthesized and characterized (Nuber and Hirsch 1996; Bellavia-Lund et al. 1997b,c; Hasharoni et al. 1997; Hummelen et al. 1999; Tagmatarchis et al. 2000). Moreover, quite similarly to **20b** (Scheme 9.5), self-photooxygenation of bisazafulleroid **93** (Figure 9.12) afforded open-cage ketolactam **94**, albeit in low yield (Hummelen et al. 1995a, 1999).

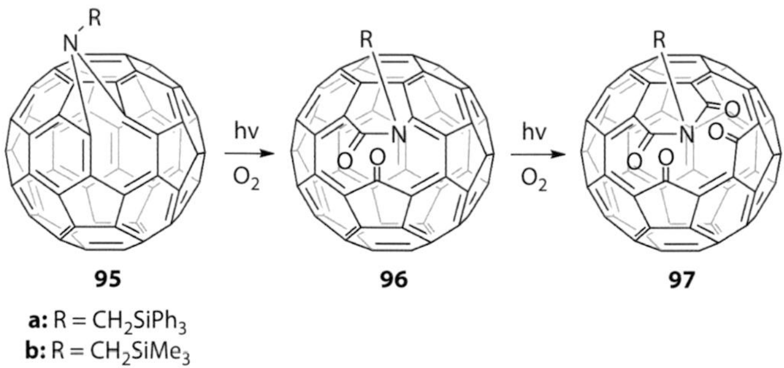

SCHEME 9.19 Photooxygenation of azafulleroids **95a** and **95b**.

The double photooxygenation of azafulleroids was reported for the first time in 2009 (Hachiya and Kabe 2009). Specifically, silyl-modified azafulleroids **95a–95b** were oxidized to diketo imides **97a–97b** upon prolonged irradiation in the presence of oxygen (Scheme 9.19). A stepwise addition of $^1O_2$ to azafulleroid **95** took place, affording keto-lactam **96** bearing an 11-membered-ring orifice as the major product, along with diketo imide **97** having a 15-membered-ring opening. The structure of **97** was assigned on the basis of its ^1H and ^{13}C NMR spectra, as well as via matrix-assisted laser desorption ionization time-of-flight (MALDI-TOF) mass spectroscopy.

In 1999, Rubin and colleagues reported the cleavage of four bonds on the fullerene cage by the reaction of the bisazidobutadienes **36a–36c** (Figure 9.6) with C_{60} (Schick et al. 1999). They proposed that the Diels-Alder cycloaddition of the 1,4-diarylbutadiene moiety, favored by the intramolecular effect (Brieger and Bennett 1980; Craig 1987), follows azide cycloaddition, affording bisiminofullerene derivative **98** (Scheme 9.20), which is impossible to be isolated because of its high reactivity toward $^1O_2$ (Schick et al. 1999). Thus, it is subsequently self-photooxygenated (Zhang et al. 1993; Vassilikogiannakis et al. 1998; Chronakis et al. 2002) and dehydrogenated to afford open-cage fullerene adduct **99** (Scheme 9.20). Derivative **98** encompasses an electron-rich 1,4-diaminobutadiene moiety having very large highest occupied molecular orbital (HOMO) coefficients at the 1- and 4-carbon atoms attached to nitrogen. Singlet oxygen, which is photochemically produced by **98** (Zhang et al. 1993; Orfanopoulos and Kambourakis 1994, 1995; Tokuyama and Nakamura 1994; Vassilikogiannakis et al. 1998; Chronakis et al. 2002), attacks at these positions via a [4 + 2] cycloaddition reaction giving *endo*- and *exo*-endoperoxides **100**; these concurrently rearrange to **101** through a [2 + 2 + 2] spontaneous ring opening reaction (Scheme 9.20). The addition of $^1O_2$ to **98** is the first example of a [4 + 2] cycloaddition reaction in which a diene moiety within the fullerene core participates as a 4π-electron system. Open-cage fullerene **99** is finally obtained by dehydrogenation of **101**. Dehydrogenation occurs much more easily for *endo*-**101** than for its corresponding *exo* isomer, which takes 2 h at 100°C in ODCB to be completed. The photophysical properties of **99** and **101** were also extensively studied (Stackow et al. 2000). Both open-cage fullerene derivatives remain good photosensitizers for the formation of $^1O_2$ with high quantum yields (Φ_Δ), and their photophysics still resemble pristine C_{60}. Open-cage bislactam derivative **99** was also subjected to a high accuracy ONIOM(G2MS) (our own N-layered integrated molecular orbital and molecular mechanics [G2-type model system]) molecular orbital study (Dapprich et al. 1999; Froese et al. 1997), in which the Li^+ insertion barrier to it was found to be less than 20 kcal/mol (Irle et al. 2002).

Open-cage bisfulleroids **63a–63c** (Inoue et al. 2000) and **102a–102c** (Inoue et al. 2001) were self-photooxygenated to the corresponding open-cage diketone derivatives **103a–103c** and **104a–104c** (Scheme 9.21), bearing 12-membered-ring orifices, in high yields (Inoue et al. 2001). This oxidative cleavage is regioselective and occurs exclusively on the C=C bond with the largest HOMO density (AM1 calculations), which is the most reactive bond toward electrophilic attack of $^1O_2$, via the intermediate dioxetane. In the same way, Murata and Komatsu (2001) carried out the self-photooxygenation of

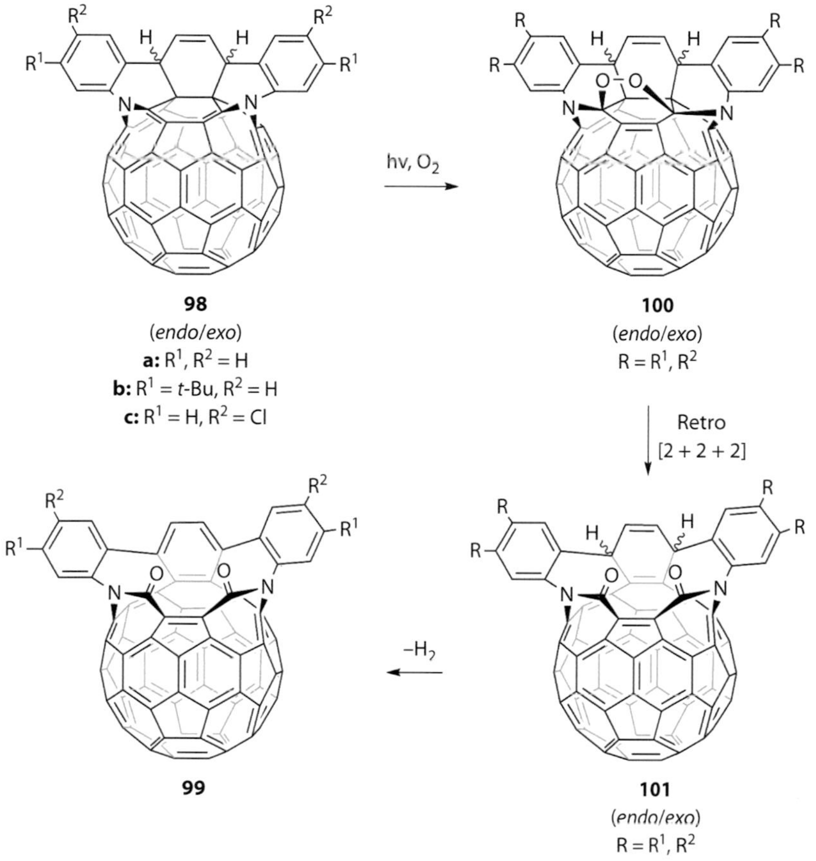

SCHEME 9.20 Proposed mechanism for the formation of **99**.

bisfulleroid **105** (Murata et al. 2001b) to open-cage fullerene **106** (Scheme 9.22). The latter was shown to have a 12-membered-ring orifice bearing an enol and a carbonyl group on the rim. The enol group was proposed to be stabilized by intramolecular hydrogen bonding with the adjacent carbonyl group.

In 2003, Komatsu and colleagues synthesized one of the most effective in encapsulating helium and molecular hydrogen open-cage fullerene derivative (**77**, Scheme 9.14) (Murata et al. 2003b). In brief, self-photooxygenation of the open-cage fullerene adduct **78** (Scheme 9.23) afforded its oxidation products **107**, **108**, and **109**, bearing significantly enlarged orifices in 60%, 31%, and 2% yields, respectively (Murata et al. 2003b). Once again, $^1O_2$ electrophilic attack occurs at the C=C bonds with the higher HOMO coefficients. Although the formation of dioxetane **110** (Scheme 9.23) should be slightly unfavorable, compared with the formation of **111** (according to theoretical calculations at the B3LYP/6-311G** level of theory), the close proximity of two bulky phenyl groups leads mainly to the kinetic formation of **110**. More recently, self-photooxygenation of compounds **82** (Murata et al. 2010) (Scheme 9.15) and **84** (Kurotobi and Murata 2011) (Scheme 9.16) was reported to give the corresponding open-cage diketone derivatives with a 12-membered-ring orifice. Quite similarly, the cage opening in fullerene adduct **73** (Scheme 9.24) was enlarged by regioselective $^1O_2$ attack at the double bond with the higher HOMO coefficients, affording the triketone open-cage fullerene **112** that has a 12-membered-ring orifice (Murata et al. 2001a).

In 2014, Yeh and coworkers functionalized adducts **107** and **108** further with ruthenium-bearing fragments (Chen et al. 2014). Thus, treating **107** with equimolar [$Ru_3(CO)_{12}$] in chlorobenzene at reflux affords the trinuclear cluster complex **113**, in which the fullerene bears an eight-membered ring orifice after

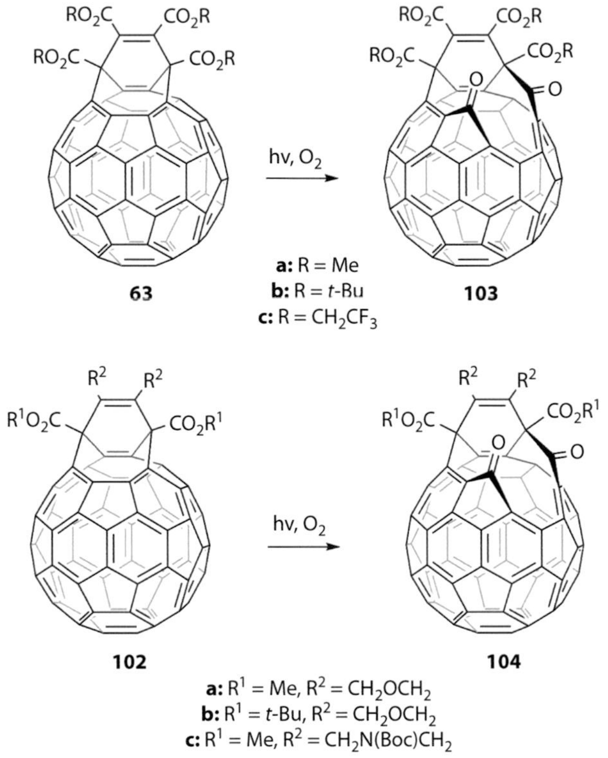

RO2C
CO2R
RO2C
CO2R
hv, O2
RO2C
CO2R
RO2C
CO2R
O
O
a: R = Me
b: R = t-Bu
c: R = CH2CF3
63
103
R2
R2
R1O2C
CO2R1
hv, O2
R2
R2
R1O2C
CO2R1
O
O
102
104
a: R1 = Me, R2 = CH2OCH2
b: R1 = t-Bu, R2 = CH2OCH2
c: R1 = Me, R2 = CH2N(Boc)CH2

SCHEME 9.21 Self photooxygenation of open-cage bisfulleroids **63a–63c** and **102a–102c** to open-cage diketone derivatives **103a–103c** and **104a–104c**, respectively.

H
H
hv, O2
H
O
OH
105
106

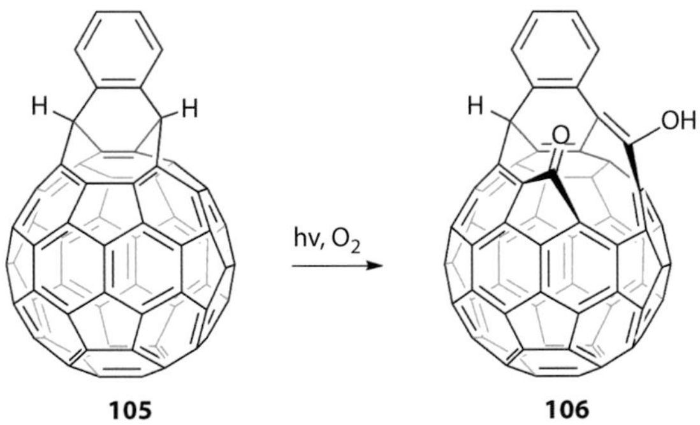

SCHEME 9.22 Self-photooxygenation of bisfulleroid **105** to the open-cage derivative **106**.

the former ketone carbons are coupled to generate a vicinal dioxy moiety (Scheme 9.25). Furthermore, pyrolysis of solid **113** at 350°C under vacuum for 30 min generates the corresponding dinuclear complex (**114**) and expands the orifice into an 11-membered ring. During this rearrangement, the two C–O bonds of **113** are cleaved. The phenyl group connected to the imine carbon undergoes an *ortho*-metalation reaction on a Ru atom, while another Ru atom is inserted into the pentagon facing the imine group. Treatment of adduct **108** with the same Ru complex shows a very different reactivity, in comparison with the one exerted by **107**. In this case, the Ru_3 cluster is fragmented to become a mononuclear

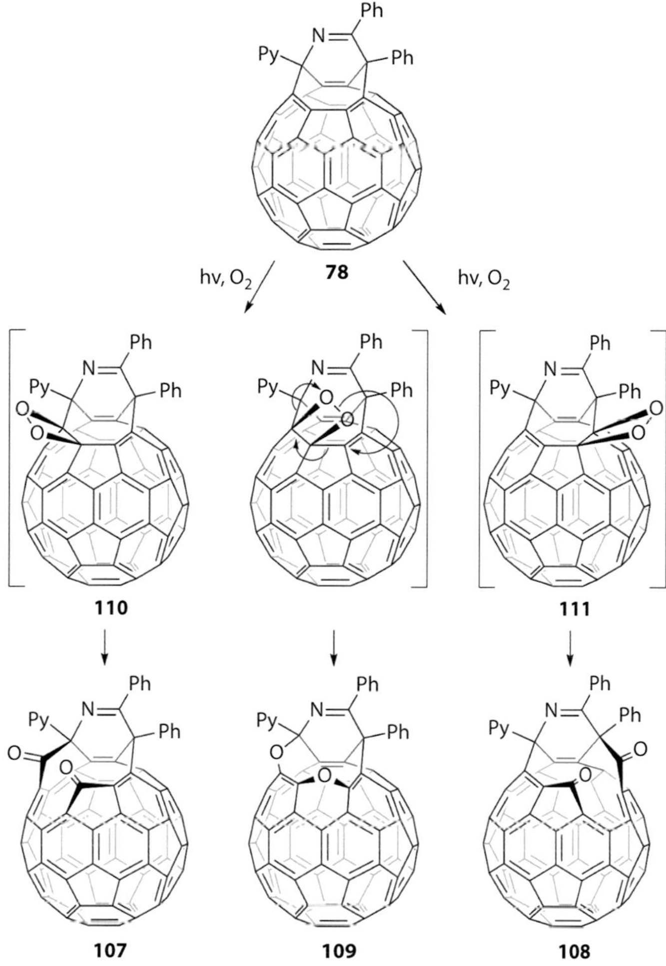

SCHEME 9.23 Possible mechanism for the formation of open-cage fullerene adducts **107–109**.

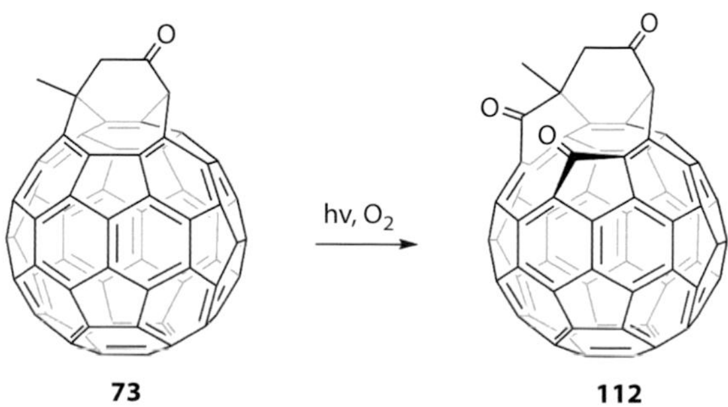

SCHEME 9.24 Formation of open-cage triketone fullerene adduct **112**.

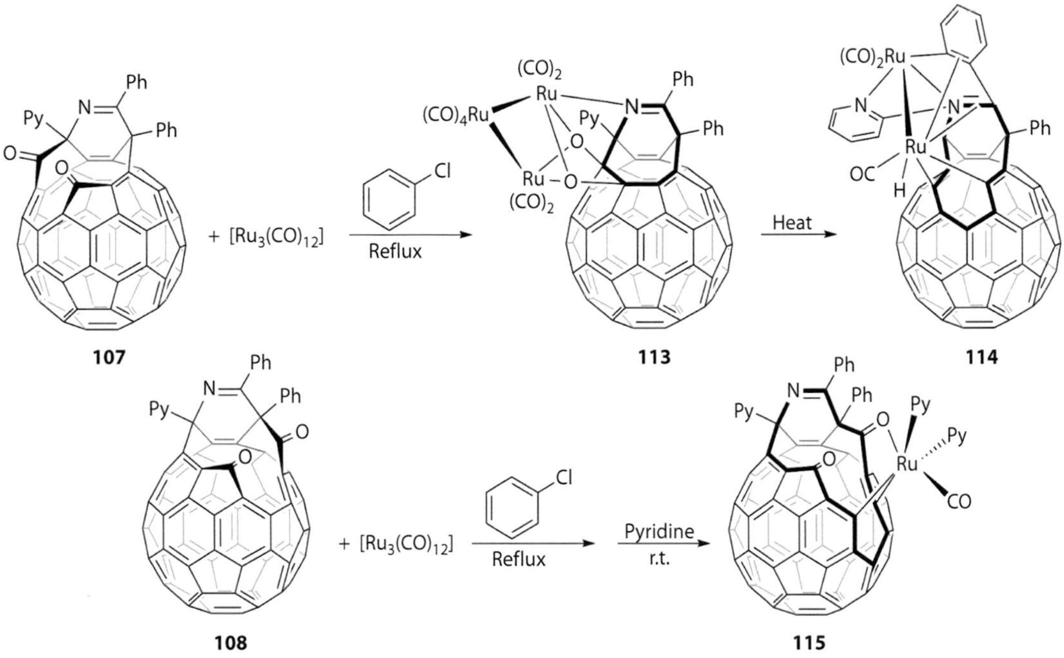

SCHEME 9.25 Formation of the open-cage fullerene derivatives **114** and **115**.

complex, and the Ru atom is inserted into the pentagon located far away from the imine group, forming a 15-membered ring orifice on the fullerene skeleton.

The first study dealing with the self-photooxygenation of silyl fulleroids was reported in 2009 (Kabe et al. 2009). In specific, the addition of silyl substituted diazomethanes **116a–116d** to C_{60} proceeds in a diastereoselective fashion to give silyl fulleroids **117a–117d**, with the silyl groups located above a five-membered ring (Scheme 9.26). Self-photooxygenation of compounds **96a** and **117b** was carried out in CS_2 at −60°C, affording the silyl enol ether derivatives **118a** and **118b**, after 1,3-silyl migration of the intermediate diketone derivatives **119a–119b** (Scheme 9.26). The later is formed after a [2 + 2] cycloaddition of **117a** and **117b** with $^1O_2$, followed by ring opening of the resulting dioxetane. The reactivity of the bridgehead carbon–carbon double bond of these silyl substituted fulleroids toward $^1O_2$ is also predicted by π-orbital axis vector analyses.

Despite the fact that the closed [6,6]-epoxide $C_{60}O$ (**120**, Figure 9.13) has been known since 1991 (Wood et al. 1991; Creegan et al. 1992; Elemes et al. 1992) and, according to theoretical calculations, the

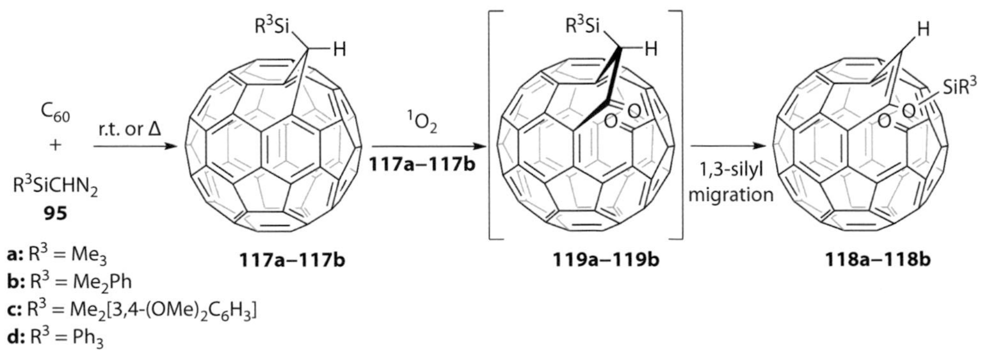

SCHEME 9.26 Formation and photooxygenation of silyl fulleroids **117**.

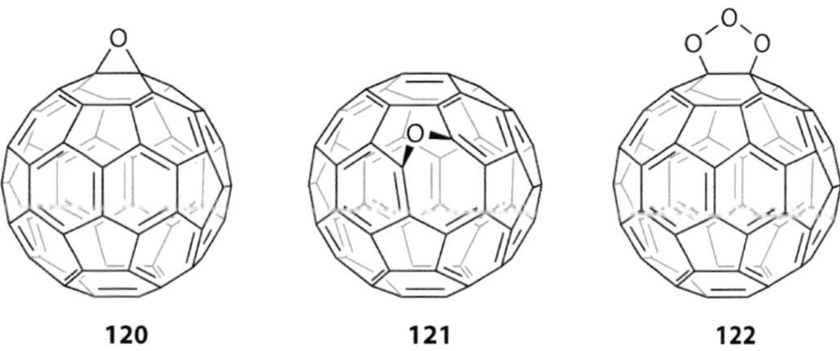

FIGURE 9.13 The structures of closed [6,6]-epoxide **120**, open [5,6]-oxidoannulene **121**, and closed [6,6]-ozone adduct **122**.

stability of a hypothetical [5,6]-open oxidoannulene $C_{60}O$ structure (**121**, Figure 9.13) is comparable with the epoxide one (Raghavachari 1992; Raghavachari and Sosa 1993; Wang et al. 1998), its isolation and characterization were first reported in 2001 (Weisman et al. 2001). [60]Fullerene oxide **121**, which has an ether structure, was obtained upon photolysis of the closed [6,6]-ozone adduct **122** (Figure 9.13) almost quantitatively. The oxidoannulene isomer of $C_{60}O$ (**121**) is stable enough to persist for several days at room temperature, under laboratory light, and its exponential thermolysis lifetime in toluene was estimated at 100 h. Prompted by this finding, Strongin and coworkers reinvestigated the photooxidation of C_{60} to ascertain whether oxidoannulene **121** forms during a major preparative method used for the preparation of the [60]fullerene epoxide (Escobedo et al. 2002). In contrast to **120**, no trace of **121** was detected in the singlet oxygen sensitized photolysis reaction of C_{60}. Oxidoannulene **121** is actually the second oxahomofullerene reported since Boltalina et al. (2000) published the isolation and characterization of $C_{60}F_{18}O$ in 2000. However, $C_{60}F_{18}O$ significantly differs from typical fullerene derivatives in the way that its π-system is exhausted by fluorination.

Besides self photooxygenation, ring openings on the fullerene shell have been also created by oxidative cleavage of one or more fullerene carbon–carbon bonds (Chiang et al. 1993; Al-Matar et al. 2001). In 1993, Chiang et al. reported the characterization of hemiketal moieties (**123**, Figure 9.14) incorporated in the [60]fullerene framework. The [60]fullerols encompassing these hemiketal moieties were prepared by the aqueous reaction of C_{60} with strong sulfuric and nitric acids. Upon treatment of these [60]fullerols with a dilute aqueous solution of HCl, the formation of several different hydroxyl ketone products, encompassing the partial structure **124** (Figure 9.14), took place. Later on, Taylor and coworkers reported the isolation and characterization of bisepoxide ketone $C_{60}Me_5O_3H$ (**125**, Figure 9.14) from the reaction of $C_{60}Cl_6$ with methyllithium, followed by hydrolysis (Al-Matar et al. 2001). Bisepoxide **125** has a nine-membered-ring orifice on the fullerene cage and its carbon–carbon bond scission is proposed to be driven by cage strain.

In 2011, the thermal reaction of endohedral metallofullerene I_h-$Sc_3N@C_{80}$ with 2-amino-4,5-diisopropoxybenzoic acid and isoamyl nitrite in ODCB at 60°C in the presence of air was reported to produce

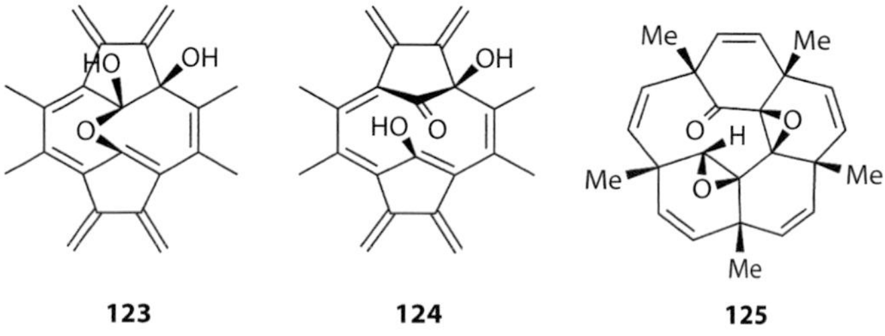

FIGURE 9.14 The structures of cage-opened fullerenes **123–125**.

the corresponding open-cage metallofullerene bearing a 13-membered-ring orifice, the largest ever reported for a metallofulerene derivative (Wang et al. 2011). The structure of this compound was determined by single-crystal x-ray analysis, which showed that both a benzyne unit and an oxygen atom have added at [5,6] ring junctions in the same pentagon. Two adjacent carbon–carbon bonds are cleaved on the C_{80} skeleton, but the resulting opening is spanned by the oxygen atom and the benzyne addend.

9.5 Ring Expansions with Hydrazines, Hydrazones, and 1,2-Phenylenediamines

According to density functional theory (DFT) calculations dealing with compound **107** (Scheme 9.23), its lowest unoccupied molecular orbital is located mainly on the conjugated butadiene moiety (Murata et al. 2003b). Thus, adduct **107** and other open-cage fullerene derivatives with such an opening motif were found reactive toward nucleophilic reagents such as hydrazines (Iwamatsu et al. 2003a,b; Vougioukalakis et al. 2004b; Iwamatsu and Murata 2005; Roubelakis et al. 2007a), hydrazones (Iwamatsu et al. 2004b; Iwamatsu and Murata 2005; Roubelakis et al. 2007a), 1,2-phenylenediamines (Iwamatsu et al. 2004a,c Iwamatsu and Murata 2005), sulfur (Murata et al. 2003b, 2010; Vougioukalakis et al. 2004a; Chuang et al. 2007a; Roubelakis et al. 2007b), selenium (Chuang et al. 2007a), sodium borohydride (Chuang et al. 2007b), Grignard reagents (Murata et al. 2010), and *N*-methylmorpholine *N*-oxide (NMMO) (Kurotobi and Murata 2011). In the following paragraphs, we will present these nucleophilic addition reactions.

A related regioselective addition reaction between the α,β-unsaturated carbonyl structure of the open-cage fullerene adduct **103a** and aromatic hydrazines **126a–126e** (Scheme 9.27) was reported in 2003 (Iwamatsu et al. 2003a; Iwamatsu and Murata 2005). The reaction proceeds with migration of two hydrogen atoms from the hydrazine to the fullerene framework, affording open-cage adducts **127a–127e** in 50%–89% yield. The enlarged open-cage fullerenes **127a–127e**, having a methylene carbon along their orifice, contain a 16-membered-ring opening in contrast to the 12-membered-ring orifice in precursor 103a.

An analogous cage scission via another hydroamination reaction, that is, the migrative addition reaction between *N*-MEM-ketolactam **28** and phenylhydrazines **128a–128b** (Scheme 9.28), was also reported (Iwamatsu et al. 2003b). In that work, Iwamatsu and coworkers proposed open-cage fullerenes **129a–129b**, bearing a 15-membered-ring orifice on the fullerene framework, as the cage enlargement products but did not offer any rationalization or experimental support for such a choice.

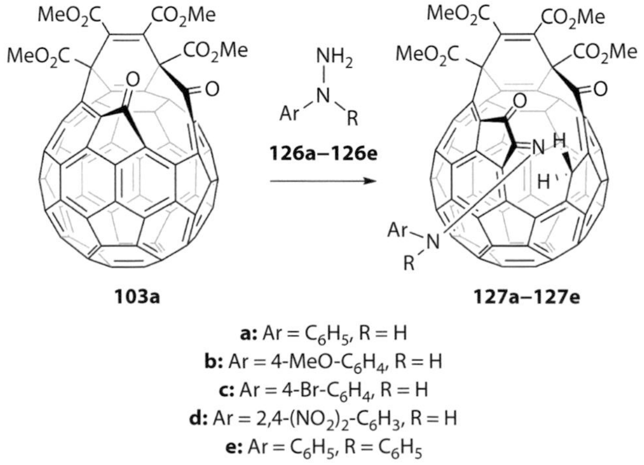

103a → (**126a–126e**) → **127a–127e**

a: Ar = C_6H_5, R = H
b: Ar = 4-MeO-C_6H_4, R = H
c: Ar = 4-Br-C_6H_4, R = H
d: Ar = 2,4-$(NO_2)_2$-C_6H_3, R = H
e: Ar = C_6H_5, R = C_6H_5

SCHEME 9.27 Formation of the open-cage fullerene derivatives **127a–127e**.

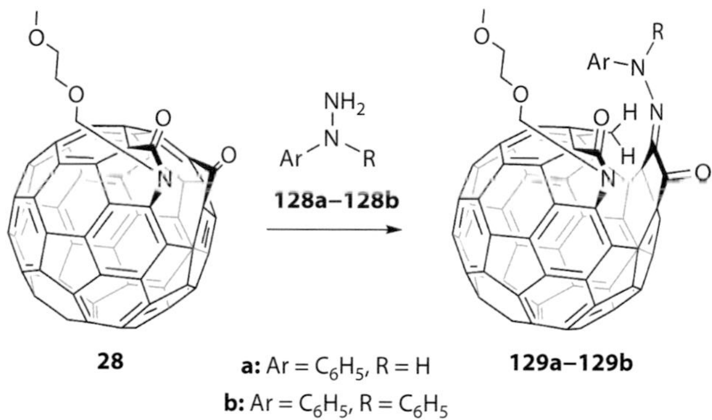

SCHEME 9.28 Proposed structures for open-cage fullerene derivatives **129a** and **129b**.

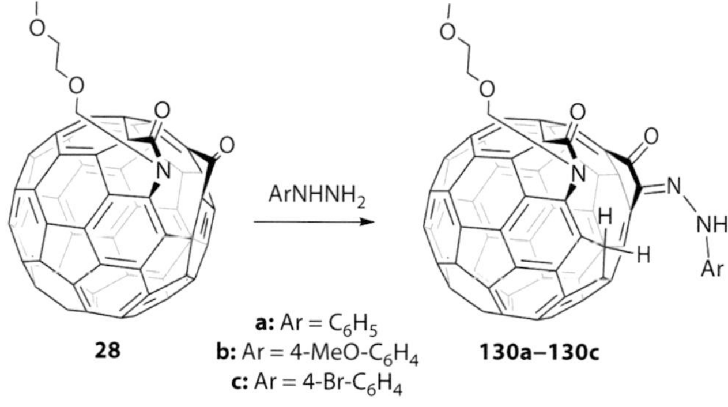

SCHEME 9.29 Formation of open-cage fullerene adducts **130a–130c**.

In a more detailed study concerning the same hydroamination reaction (Vougioukalakis et al. 2004b), Orfanopoulos and coworkers synthesized and characterized open-cage adducts **130a–130c** (Scheme 9.29), pointing out that the assumed structural assignment by Iwamatsu and coworkers is incorrect. Their structural assignment was based on the results of ^{1}H NMR, ^{13}C NMR, and nuclear Overhauser effect (NOE) difference experiments as well as MS, Fourier transform infrared spectroscopy (FT-IR), and UV-vis spectroscopy.

It was thus found that the ketone ^{13}C NMR absorption is upfield shifted from 198.5 ppm in starting material **28** to around δ 188 ppm in adducts **130a–130c**, while the lactam ^{13}C chemical shift change is insignificant. This change in the ketone carbonyl environment was also confirmed by IR spectroscopy, with the C=O absorption shifting from 1727 cm^{-1} to near 1670 cm^{-1}. The shift of the lactam carbonyl position is again insignificant. As inferred from these observations, the bond scission to provide the products occurs at one of the two double bonds (i) or (ii) next to the ketone group, shown in Figure 9.15, in a similar fashion to the reaction between **85a** and phenyl hydrazines **126a–126e** (Scheme 9.27) (Iwamatsu et al. 2003a).

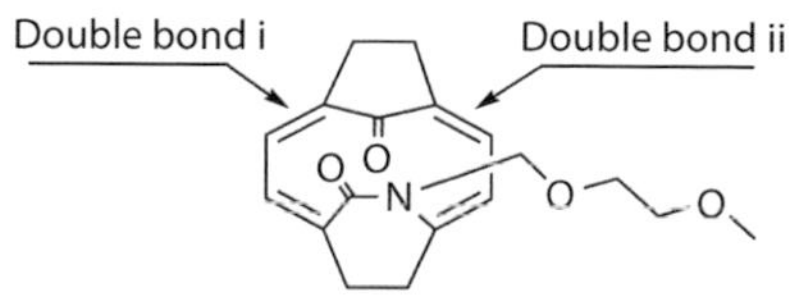

FIGURE 9.15 Double bonds (i) and (ii) next to the ketone group of open-cage fullerene adduct **28**.

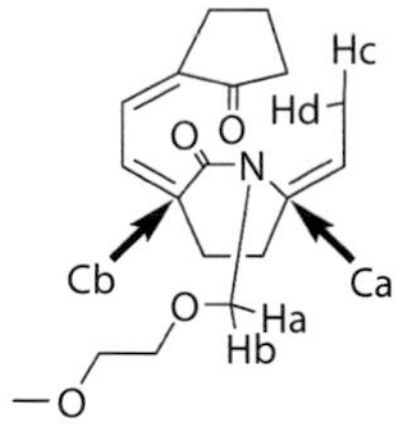

FIGURE 9.16 The cleaved α,β-unsaturated carbonyl structure of open-cage fullerene **28**.

By combining NOE enhancement measurements with theoretical calculations, it was concluded that double bond (ii) (Figure 9.15) was the one that had been cleaved, and therefore, the structures **130a–130c** presented in Scheme 9.23 were proposed (Vougioukalakis et al. 2004b). Although the assignment of both carbon atoms C_a and C_b (Figure 9.16) is quite difficult even in starting material **28** (Bellavia-Lund et al. 1997a), comparison between ^{13}C-^{1}H-coupled and selectively decoupled ^{13}C spectra (by irradiating one proton at a time) of **130a–130c** also leads to the conclusion that double bond (ii) is the one that reacts (Vougioukalakis et al. 2004b). In particular, the carbon atom that absorbs at 138.51 ppm (C_a) is coupled both with protons H_a and H_b and with H_c and H_d (Figure 9.16). Taking into account that coupling between a carbon atom and a proton can be observed only for carbon atoms that are no more than three bonds away from a specific proton, the proposed structures **130a–130c** are the only ones possible. If this were not the case and reaction had occurred at double bond (i), the H_a and H_b protons would have been four bonds away from carbon C_b and no coupling between them would have been observed. Although indisputable structural assignment would arise from the crystallographic analysis of a single crystal, which had not been possible to obtain, another confirmation of the structures **130a–130c** comes from the observation of the same reactivity trend in the regioselective sulfur-atom insertion at the central C–C bond of the butadiene unit of **28** that contains double bond (ii), shown in Figure 9.15 (Vougioukalakis et al. 2004a).

Among others, the orifice on the fullerene framework of open-cage diketone derivative **103a** has been regioselectively enlarged by its reaction with aromatic hydrazones **131** and **132** (Scheme 9.30) (Iwamatsu et al. 2004b). In this fashion, fullerene adducts **133** and **134** were obtained in 44% and 57% yield, respectively. The formation of **134** was proposed to proceed upon hydrolysis of the intermediate **135**.

Isomeric diketone open-cage derivatives **107** and **108** (Scheme 9.23) were also found to be reactive toward aromatic hydrazines and hydrazones (Roubelakis et al. 2007a). Both derivatives possess the characteristic α,β-unsaturated carbonyl structure along their rim; therefore, expansion of the 12-membered-ring orifice of **107** and **108** to a 16-membered one was realized in a same manner as discussed previously for **103a** (Scheme 9.31). Cage-opened fullerene adducts such as **136** (Scheme 9.31) were isolated in moderate to excellent yields (Roubelakis et al. 2007a).

In 2004, Iwamatsu et al. reported another reaction sequence that led to fullerene adducts bearing 19- and 20-membered-ring orifices, the largest openings constructed thus far on the fullerene framework (Iwamatsu et al. 2004a,c; Iwamatsu and Murata 2005). Specifically, the bowl-shaped fullerene adducts **137a–137b** (Scheme 9.32) were prepared by regioselective, sequential cage-scission reactions of the open-cage fullerene adduct **103a** with 1,2-phenylenediamine (**138a**) and 4,5-dimethyl-1,2-phenylenediamine (**138b**) in the presence of pyridine (Iwamatsu et al. 2004a). Fullerene adducts **137a–137b** have a 20-membered-ring orifice on the fullerene shell and were prepared in 52% and 60% yield, respectively. When the same reactions were carried out at room temperature in the absence of pyridine, open-cage adducts **139a–139b** (Scheme 9.32), having a 16-membered-ring orifice on the fullerene shell, were obtained in high yield (85% and 74%, respectively). Quite similarly, open-cage adducts **140** and **141a** (Scheme 9.33) were prepared by regioselective, sequential carbon–carbon bond scissions on **28** via its reaction with **138a** (Iwamatsu et al. 2004c). The MEM group in **141a** was subsequently removed by treatment with TFA, to afford **141b** and **141c**. Note that open-cage fullerene adduct **137b**

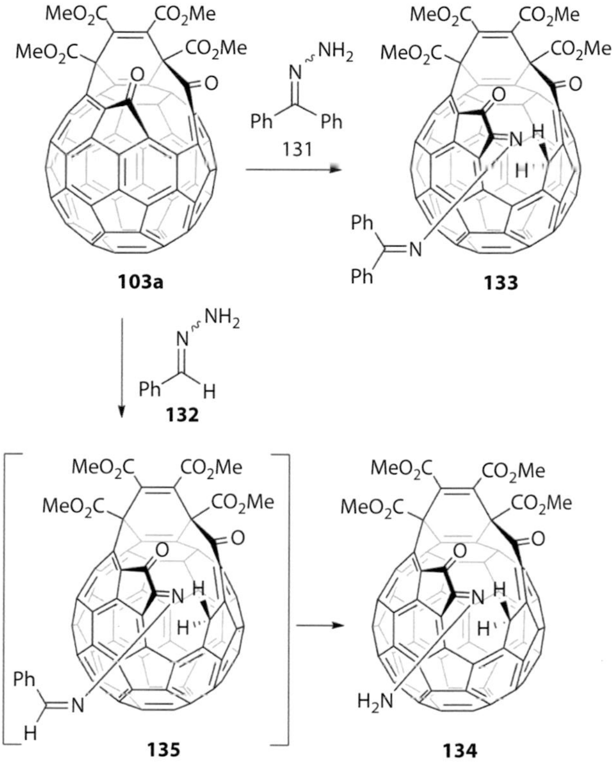

SCHEME 9.30 Formation of open-cage fullerenes **133** and **134**.

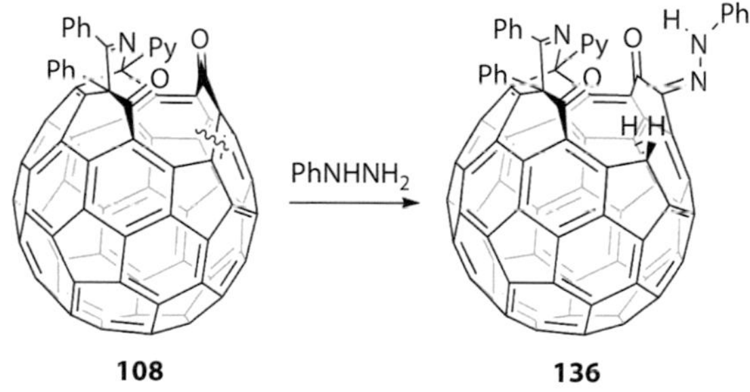

SCHEME 9.31 Reaction of open-cage derivative **108** with phenyl hydrazine.

forms an endohedral carbon monoxide complex (i.e., CO@**137b**), as well as an endohedral ammonia complex (i.e., NH_3@**137b**), and an endohedral methane complex (i.e., CH_4@**137b**), whereas **137a–137b** and **141a–141c** encapsulate a water molecule into the fullerene cage (Murata et al. 2008b; Gan et al. 2010; Vougioukalakis et al. 2010a). In related, more recent work, Ding and coworkers published the synthesis of fullerene derivatives **143a** and **143b** (Scheme 9.34) (Xiao et al. 2012). They used different diamine adducts (**142a–b**) and open-cage fullerene **103a**, prepared according to Murata's cage-opening process, and thus expanded the orifice to a 20-membered ring. These fullerene derivatives were shown to absorb both in the visible and the near-IR regions of the electromagnetic spectrum.

In 2014, Gan and colleagues started from open-cage derivatives **144a–b** and reduced them with cuprous bromide to remove the two *tert*-butyl groups, thus obtaining **145a–b** (Scheme 9.35) (Yu et al.

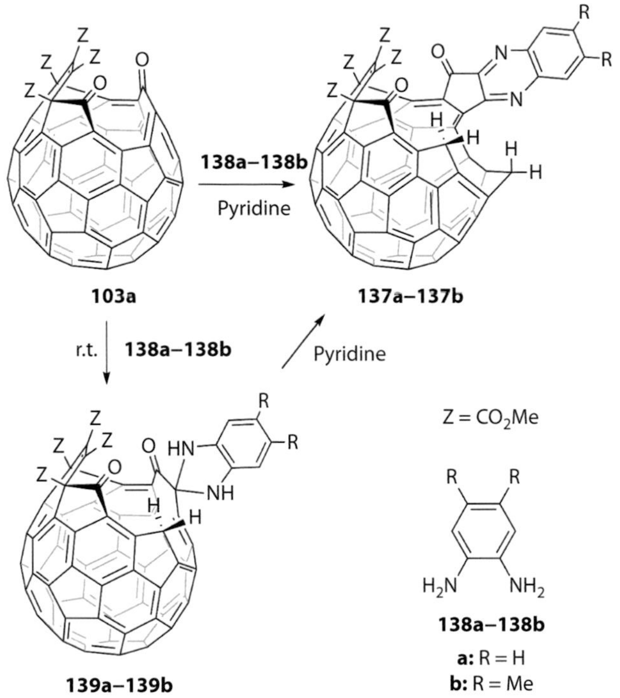

SCHEME 9.32 Formation of the bowl-shaped fullerene adducts **137a**, **137b** and **139a**, **139b**.

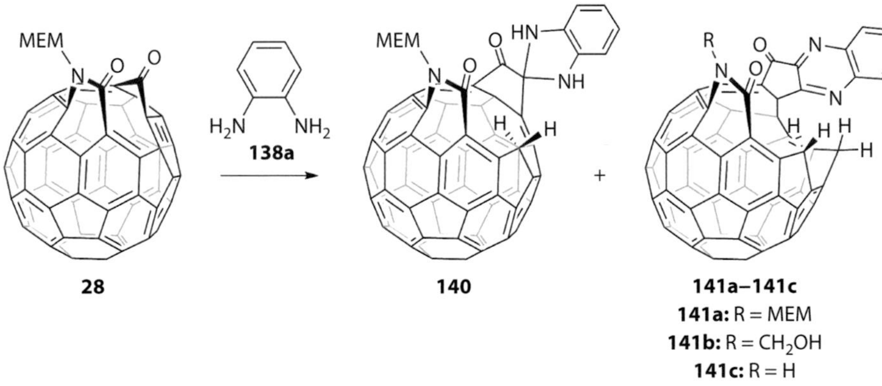

SCHEME 9.33 Formation of the bowl-shaped fullerene adducts **140** and **141a–141c**.

2014a). Further treatment of adducts **145** with PCl_5 and $SnCl_2$ resulted in the formation of derivative **146**, through the reductive aromatization and opening of the iodo acetal moiety. **146** was further reacted with *o*-diaminoaromatic compounds, affording products **147a–b** and **148a**. Treatment of **145** with trifluoromethanesulfonic acid led to the formation of a mixture of **149** and **150** (Scheme 9.36). The obtained mixture of adducts **149** and **150** was found difficult to separate, through column chromatography, because of the transformation of **149** to **150** on silica gel. Thus, the crude mixture of **149** and **150** was treated with *o*-diaminobenzene, resulting in compound **151** in the same way applied to intermediate **146**.

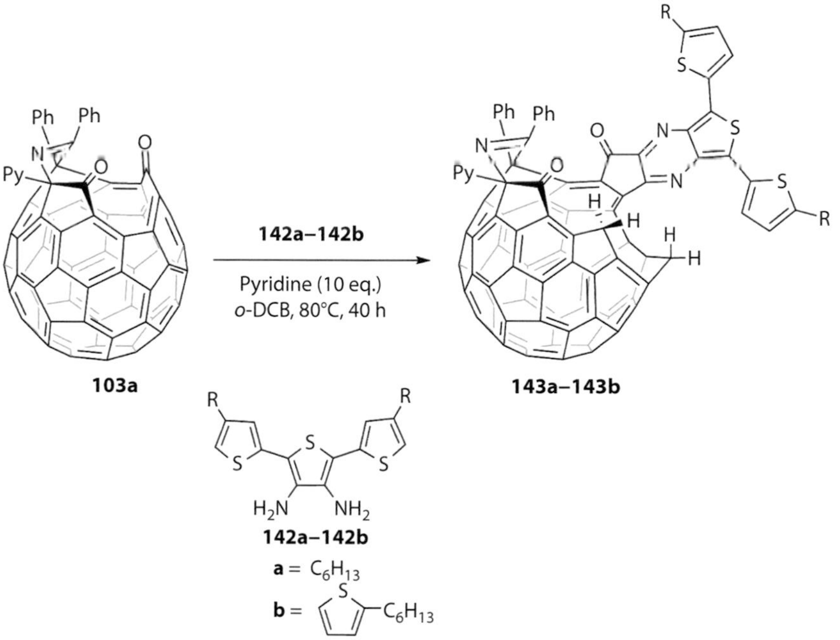

SCHEME 9.34 Formation of open-cage fullerene adducts **143a** and **143b**.

9.6 Sulfur and Selenium Atom Insertion Reactions

Komatsu and coworkers reported the first sulfur atom insertion at the rim of the orifice of an open-cage fullerene derivative in 2003 (Murata et al. 2003b). This approach is based on the fact that sulfur atoms can be inserted into activated carbon–carbon single bonds (Toda and Tanaka 1979). Thus, heating **105**, which has a 12-membered-ring orifice on the fullerene framework, together with S_8 in the presence of TDAE afforded **77** (Scheme 9.37), bearing an enlarged 13-membered-ring orifice, in 77% yield. TDAE was proposed either to activate **107** by an electron transfer or to lead to an intermediate complex that renders the electrophilic addition of elemental sulfur to **107** more feasible. The sulfur atom insertion at the rim of the orifice of **77** makes the cage opening rather circular, as shown by its single-crystal x-ray structure, increasing the possibility for the encapsulation of small molecules or atoms in the interior of the fullerene cage. The same experimental conditions were applied for the synthesis of compound **81** (Scheme 9.15) from its corresponding 12-membered-ring precursor (**Murata** et al. 2010).

In another approach, when compound **107** was stirred with elemental sulfur at ambient temperature, in the presence of 2.2 equivalents of sodium methylthiolate as a reducing agent, **77** was readily formed in 80% yield (Scheme 9.38) (Chuang et al. 2007a). The later method was also successfully applied for the insertion of a selenium atom at the rim of the aperture. Thus, heating a mixture of **107** and elemental selenium in the presence of sodium methylthiolate in ODCB resulted in the formation of open-cage fullerene **152** (Scheme 9.38) in 46% yield (Chuang et al. 2007a). The structure of **152** was unambiguously determined by x-ray crystallography. The orifice of **152** is slightly larger than the orifice of **77**, in accordance with the longer C-Se bond as compared with the C-S bond. This slightly enlarged orifice on the fullerene framework has an impact on the encapsulation efficiency of the cage.

Later on, Orfanopoulos and coworkers managed to synthesize and characterize open-cage fullerene derivatives **153**–**156**, shown in Scheme 9.39, containing 12- and 13-membered-ring openings, respectively, on the surface of the [60]fullerene (Vougioukalakis et al. 2004a; Roubelakis et al. 2007b). Interestingly, adducts **155** and **156** are the first open-cage [60]fullerene derivatives without any organic addend on the rim of their orifice.

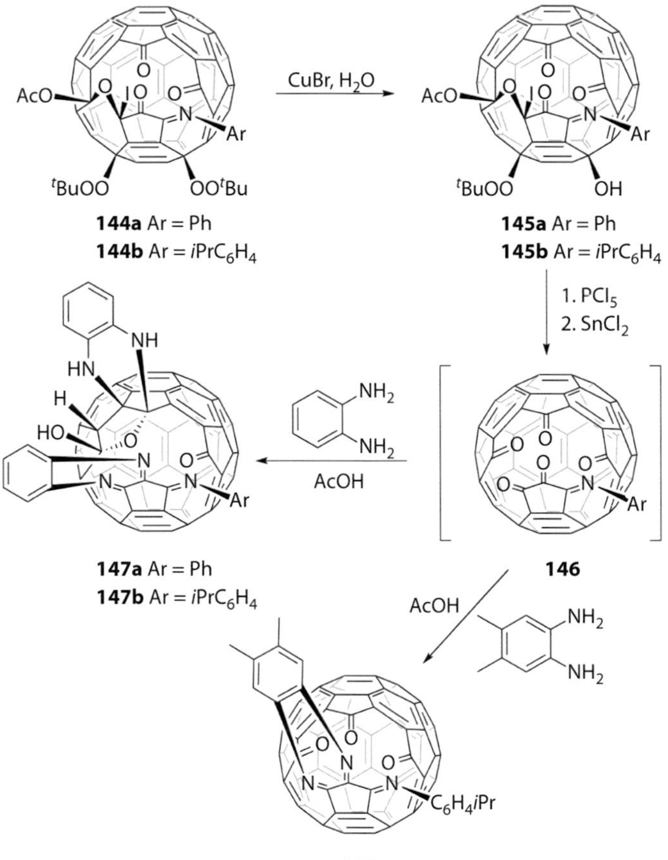

SCHEME 9.35 Synthesis of open-cage fullerenes **147** and **148**.

Cage-opened adduct **153** was prepared in 72% yield by the enlargement of ketolactam's **28** orifice via a sulfur atom insertion (Vougioukalakis et al. 2004a). Given that the lactam's carbon and carbonyl absorption shift in the corresponding ^{13}C NMR and IR spectra of the new adduct **153** is insignificant compared with that of the ketone's group, the possibility that the sulfur atom is inserted into the central carbon–carbon bond α of the butadiene unit of the starting material **28** (Figure 9.17), as one might expect in analogy to the reaction of **107** with elemental sulfur (Figure 9.17), was ruled out. Thus, sulfur insertion occurs at C–C single bond β in **28** (Figure 9.17) affording adduct **153** (Scheme 9.39). Another fullerene adduct was also isolated, in 11% yield. Based on NMR and MALDI-TOF mass spectroscopy data, it was suggested that this is the result of a double sulfur-insertion reaction (open-cage adduct **154**), according to which one sulfur atom is inserted into each of the central C–C single bonds of the two butadiene units in starting material **28** (i.e., both α and β bonds are attacked, Figure 9.17) (Roubelakis et al. 2007b).

Next, the removal of the MEM protecting group, aiming at the preparation of the thio-aza-[60]fullerene **157**, via the hypothetical intermediate radical **158** (Figure 9.18), by utilizing the azafullerene production method (Hummelen et al. 1995a) to adduct **153** was attempted. However, adduct **155** was isolated instead of **157** (Scheme 9.39) in 58% yield (Vougioukalakis et al. 2004a); **155** is the first open-cage fullerene adduct without organic addends on the rim of the orifice. Similarly, thermal treatment of compound **154** under acidic conditions afforded fully deprotected derivative **156** (Scheme 9.39) in

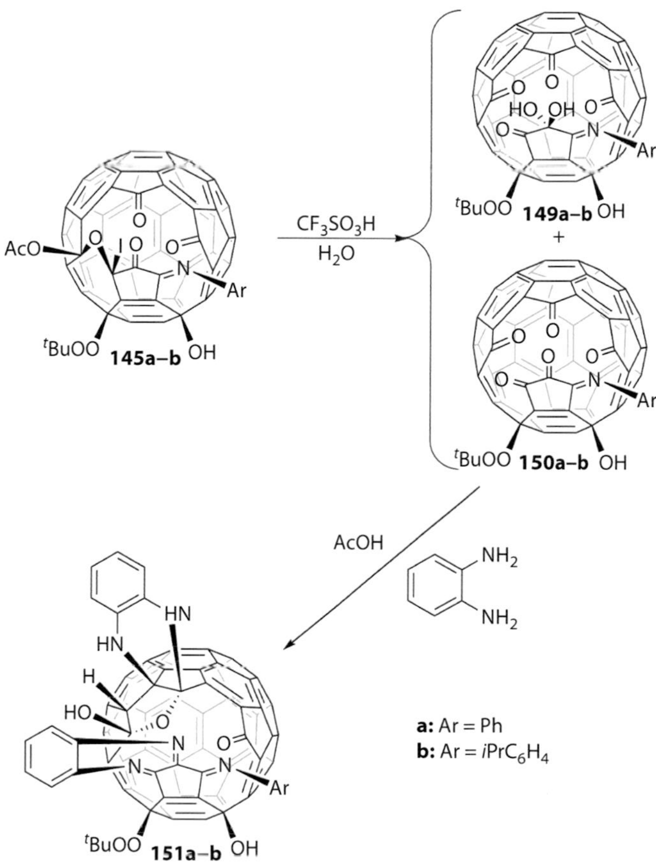

SCHEME 9.36 Formation of fullerene adducts **150** and **151**.

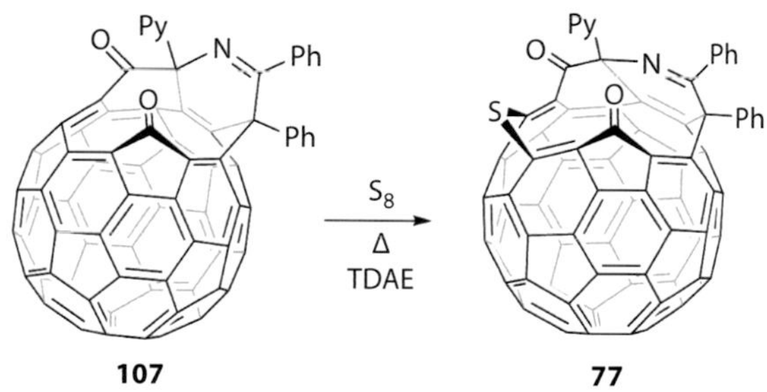

SCHEME 9.37 Formation of the open-cage fullerene 77.

78% isolated yield (Roubelakis et al. 2007b). Several attempts toward the partial removal of the MEM protecting group from **153** to obtain dimer **157**, utilizing milder deprotection conditions, exclusively resulted in the formation of new open-cage [60]fullerene adducts (Roubelakis et al. 2007b). Apparently, the sulfur atom insertion at the rim of the orifice of **28** disturbs the proper orientation required for the groups that participate in the cage-restoration reactions (namely, the cyclopentanone carbonyl group and the *N*-methyl carbonium ion; Hummelen et al. 1995a). This orientation seems to be appropriate for **28** but not for **153** or for **154**.

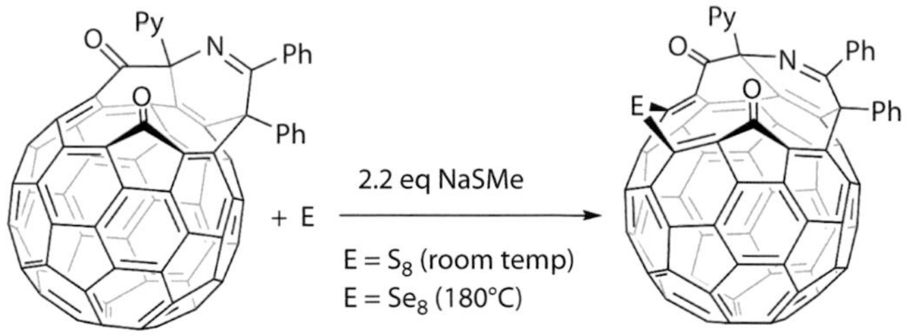

Py
O
N
Ph
Ph
O
+ E
2.2 eq NaSMe
E = S_8 (room temp)
E = Se_8 (180°C)
107
77 E = S (80%)
152 E = Se (46%)

SCHEME 9.38 Synthesis of compounds **77** and **152**.

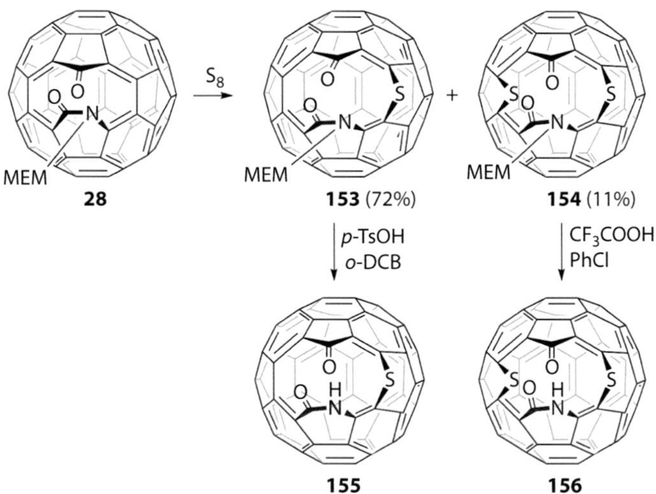

SCHEME 9.39 Formation of the open-cage fullerene derivatives **153**, **154**, **155**, and **156**.

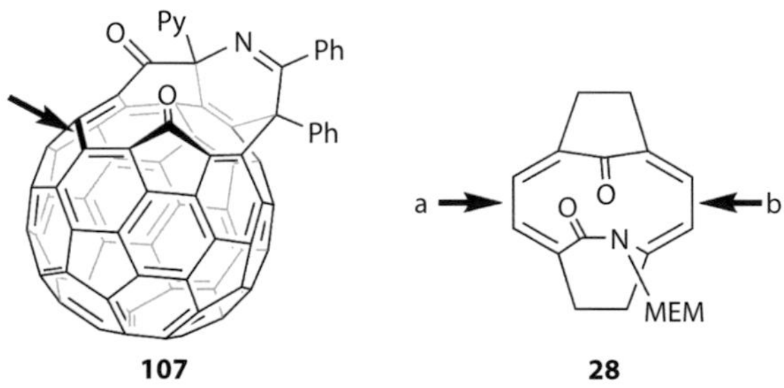

FIGURE 9.17 The central carbon–carbon bond of the butadiene unit of **107** cleaved by sulfur insertion and the corresponding carbon–carbon bonds in open-cage fullerene adduct **28**.

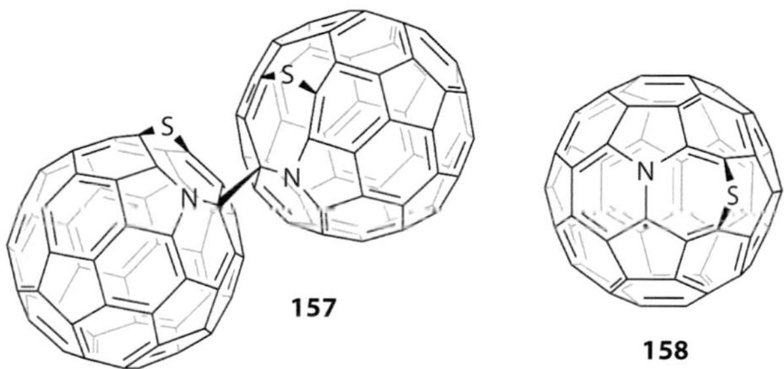

FIGURE 9.18 Hypothetical thio-aza-[60]fullerenes **157** and **158**.

9.7 Grignard Reagents, Sodium Borohydride, and NMMO Additions

Komatsu and coworkers developed a method for the reduction of the orifice size of cage-opened compound **77** in 2007 (Chuang et al. 2007b). More specifically, when **77** reacts with 15 equivalents of sodium borohydride in a mixture of ODCB and tetrahydrofuran (THF) at room temperature, the new open-cage adduct **159**, with an aperture size narrowed from a 13-membered to an 11-membered ring, is isolated in 86% yield (Scheme 9.40). This technique is based on the nucleophilic attack of the hydride to the carbonyl carbon of the five-membered ring of the fullerene structure of **77**, followed by the addition of the resulting alkoxide anion to the imino carbon located near the negatively charged oxygen.

In 2010, the same research group reported the nucleophilic addition of methyl and allyl Grignard reagents to compound **81** (Murata et al. 2010). Thus, treatment of **81** with R′MgCl in THF at −20°C, followed by quenching with acid, affords alcohol **160**, that is, the result of a selective 1,2-addition of the Grignard reagent to one of the two carbonyl groups (Scheme 9.41). Alcohol **160a** is stable under ambient conditions; however, upon refluxing a toluene solution of **160** for 12 h in the presence of 2.9 equivalents of *p*-TsOH, a transannular cyclization takes place, giving compound **161a** in 75% yield.

In a related work, Murata and coworkers reported the reaction of derivative **83** (Scheme 9.16) with 2.3 equivalents of NMMO, a nucleophilic oxidant, in wet THF at room temperature (Scheme 9.42) (Kurotobi and Murata 2011). Single-crystal x-ray analysis showed that the final product **163** has a 13-membered-ring opening in which two hemiacetal and two carbonyl carbons are present. It was reported that this molecule derives from tetraketone **162** after the attack of a water molecule, present in the solvent, on one of the four carbonyl carbons.

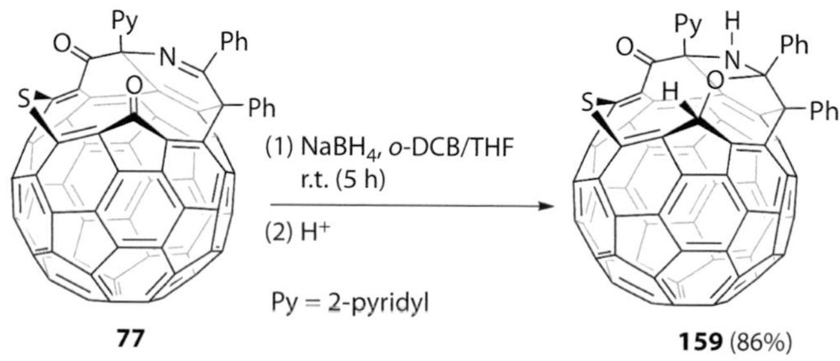

SCHEME 9.40 Orifice size reduction reaction of 77.

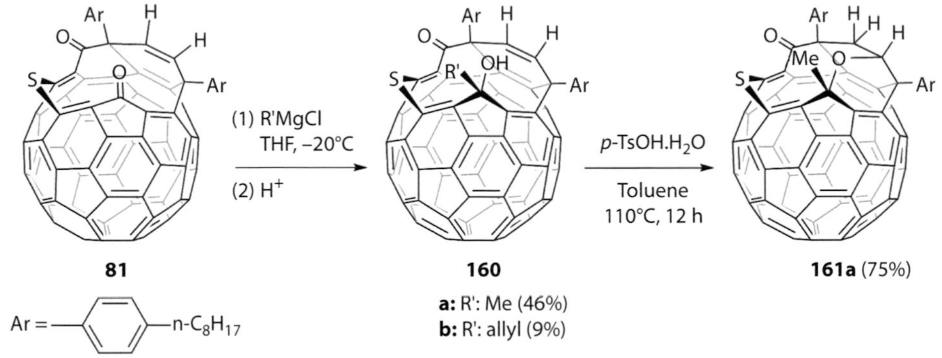

SCHEME 9.41 Addition of organometallic nucleophiles to **81** and subsequent transannular cyclization of the obtained alcohol.

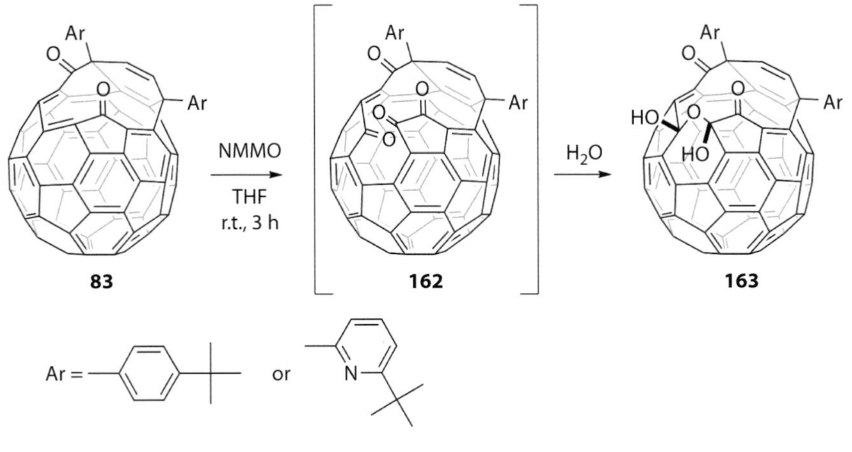

SCHEME 9.42 Addition of NMMO to **83** in wet THF.

9.8 Beginning from Fullerene-Mixed Peroxides

Gan, Zhang, and coworkers have reported the synthesis of fullerene-mixed peroxides, derived from the addition of *t*-butyl peroxy radicals to C_{60} (Gan et al. 2002; Huang et al. 2004). Subsequent studies on the reactivity of these oxygen-rich compounds revealed that they can be transformed into open-cage derivatives. The *t*-butyl peroxy groups directly attached to the fullerene skeleton serve as excellent oxygen sources for the formation of hydroxyl and carbonyl groups.

Open-cage compound **164** was isolated as the major product after visible light irradiation of an iodine-containing solution of the hexa-adduct **165** (Scheme 9.43) (Huang et al. 2004). The proposed mechanism for this reaction is shown in Scheme 9.43. Homolytic cleavage of the O–O peroxo bond on the central pentagon with sequential loss of a *t*-butoxy radical from **165** forms allyl radical **166** bearing an epoxide on the same pentagon. Next, homolytic cleavage of a C–O bond gives dioxetane intermediate **167**. Ring opening of the dioxetane moiety of intermediate **167** results in the final open-cage product **164**. This rearrangement is analogous to that for the formation of the *N*-MEM-ketolactam **28** from intermediate **27** (Scheme 9.5) reported by Wudl (Hummelen et al. 1995c). However, 3 years later, the proposed structure of derivative **164** was proved incorrect. On the basis of single-crystal x-ray analysis of a closely related compound, structure **168** (Scheme 9.43) instead of **164** was deduced to be valid (Xiao

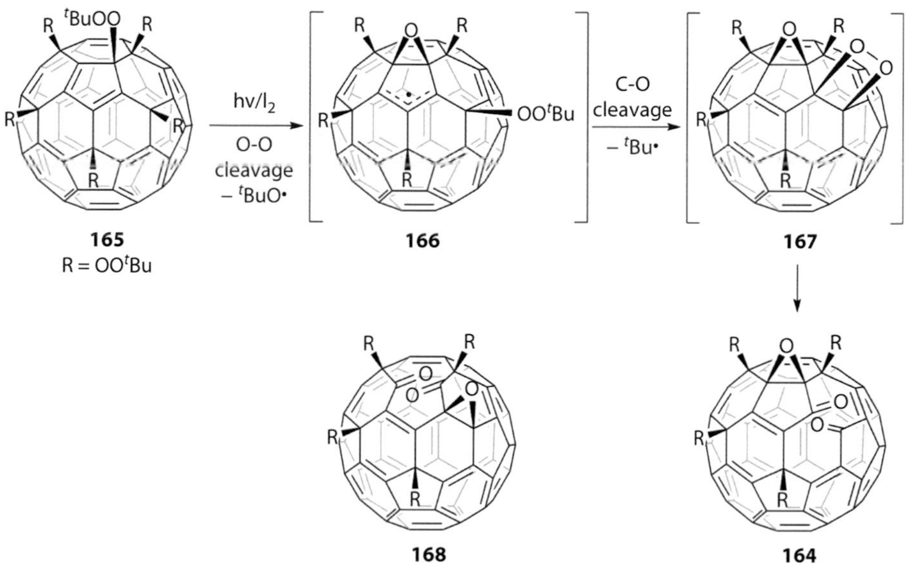

SCHEME 9.43 Proposed mechanism for the formation of **164** and revised structure **168**.

et al. 2007). A reaction mechanism similar to that shown in Scheme 9.43, involving homolytic cleavage of C–O and O–O bonds followed by rearrangement of a 1,2-dioxetane intermediate, was again proposed to explain the formation of **168** from **165** (Xiao et al. 2007).

The reactivity of hexa-adduct **165** was examined under thermal conditions as well (Yao et al. 2009). Heating a solution of **165** in chlorobenzene at 100°C promotes homolysis of the O–O bond on the central pentagon, eventually providing the two isomeric products **169** and **170**, bearing an epoxide and an ether group, and two epoxide moieties, respectively (Scheme 9.44). No interconversion between the two isomers was detected. Both the epoxide ring and the ether group of **169** were opened after the addition

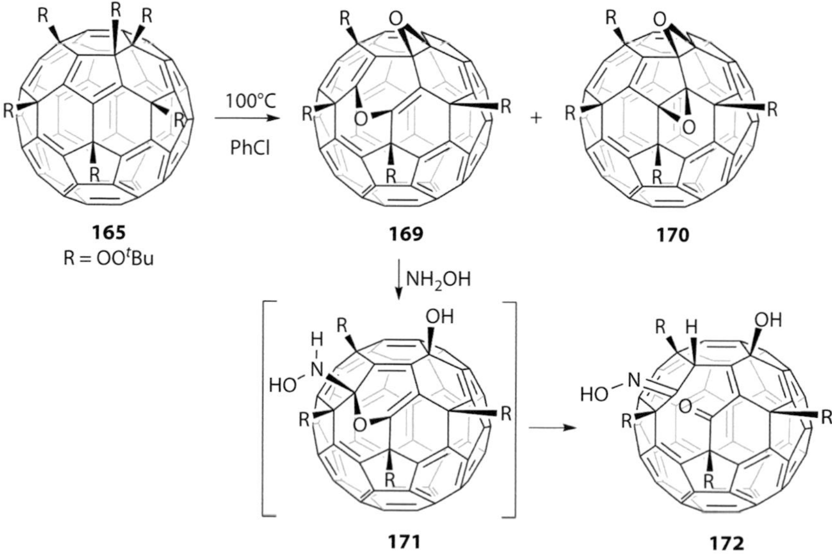

SCHEME 9.44 Thermolysis of the fullerene-mixed peroxide **165** and formation of the keto-ketoxime derivative **172**.

of hydroxylamine, affording open-cage derivative **172** through the unstable intermediate **171** (Scheme 9.44). The structure of compounds **169** and **172** was clarified through both spectroscopic and single-crystal x-ray analysis (Yao et al. 2009). It was found that the ether oxygen atom of the pyran ring in **169** is tilted toward the six-membered ring; furthermore, the existence of a carbonyl and a ketoxime group in **172** was confirmed. On the basis of spectroscopic data, the hydrogen atom in **172** could be either next to the ketoxime group (structure depicted in Scheme 9.44) or next to the carbonyl group; the exact location of the hydrogen was not determined because of the disorder in the x-ray structure of **172**.

In 2010, fullerene peroxide **165** was once again used as a precursor for the synthesis of other cage-opened structures with nine-membered holes (Yang et al. 2010). Specifically, **165** was converted into compound **173** encompassing a 2*H*-pyran moiety (Scheme 9.45). This moiety was hydrolyzed in the presence of potassium carbonate to give the already known compound **168**, along with a new one possessing two epoxy rings on the fullerene skeleton (**174**). Several Lewis acids promoted the hydrolysis of the *tert*-butoxyl group in **174** to afford alcohol **175**, while the epoxy groups remained intact under these conditions (Scheme 9.46). Next, the hydroxyl group of **175** was transformed into acetoxy (OAc) and octadecyltrimethoxysilane (OTMS) under mild conditions, forming compound **176** (Scheme 9.46).

Open-cage diketone derivative **177** was synthesized from fullerene-mixed peroxide **178** following the two-step reaction shown in Scheme 9.47 (Huang et al. 2006). The epoxy ring in **178** is hydrolyzed in the presence of trispentafluorophenyl boron to the vicinal fullerendiol **179**, and subsequently, **179** is

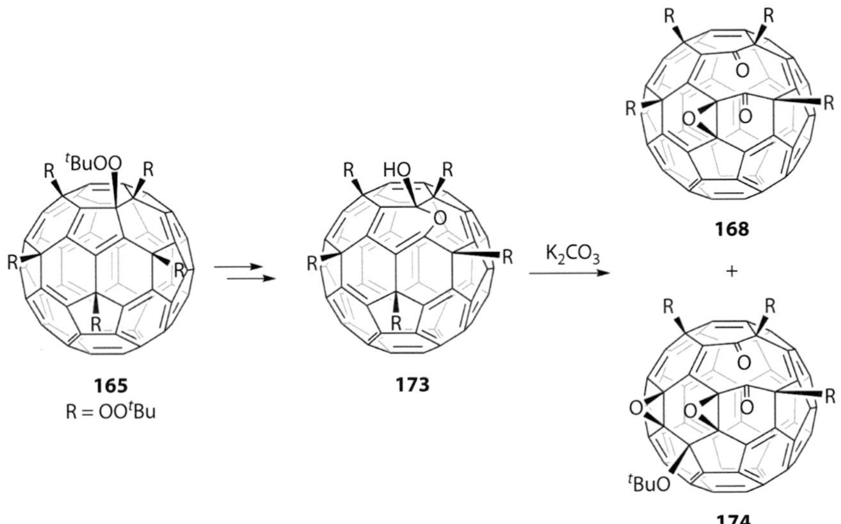

SCHEME 9.45 Formation of cage-ruptured fullerenes **168** and **174**.

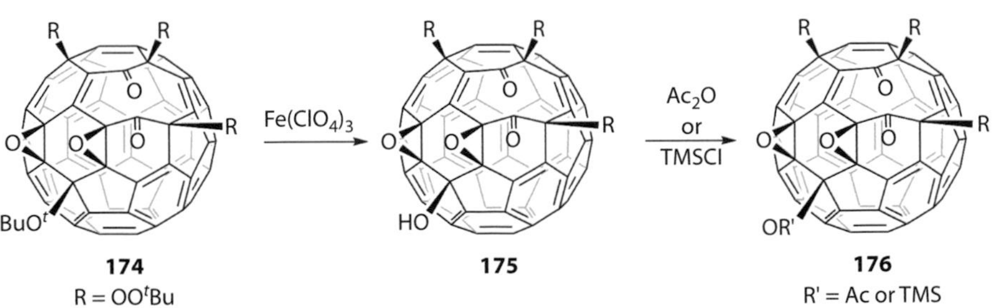

SCHEME 9.46 Reactivity of a *tert*-butoxyl group on an open-cage fullerene skeleton.

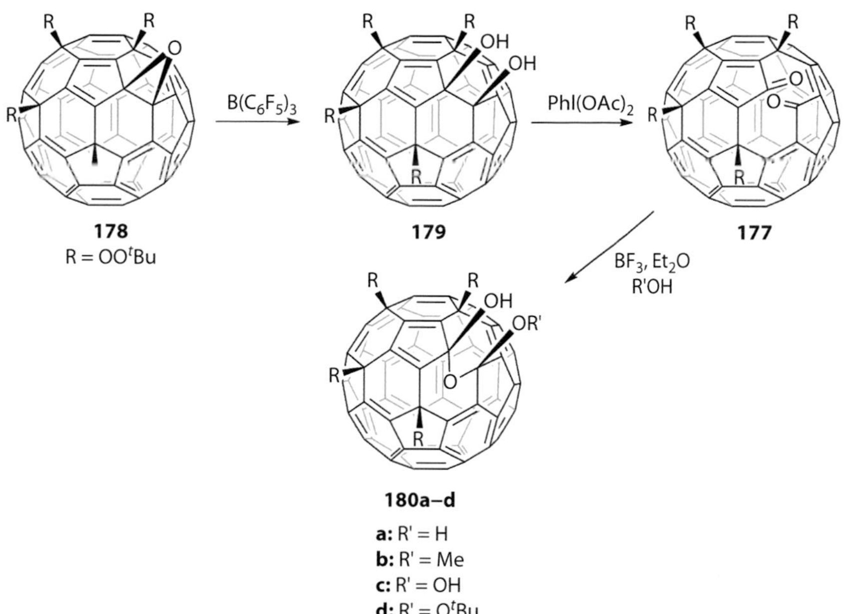

SCHEME 9.47 Synthetic route to open-cage fullerene derivatives **180a–180d**.

oxidized by (diacetoxy)iodobenzene $PhI(OAc)_2$ (DIB) to form the expected fullerendione **177** in good yield. Further treatment of **177** with BF_3 in diethyl ether, in the presence of water, results in coupling of the carbonyl groups and formation of open-cage adducts **180a–d**. These new derivatives bear two hemiketal moieties bridged by an oxygen atom (Huang et al. 2006). Later on, a series of similar DIB-induced oxidation reactions of alike fullerendiols resulted in the formation of a family of open-cage diketone derivatives (Jiang et al. 2008). Interestingly, oxidation of compound **181** in chloroform with DIB furnished the bisketal fullerendione **182** with a 12-membered-ring opening (Scheme 9.48) (Zhang et al. 2011a). An orifice-closing reaction took place on **182** after further treatment with BF_3•Et_2O in the presence of methanol. The two carbonyl groups were coupled into ketal moieties, giving **183** that contains three oxygen atoms bridging six carbon atoms of former [6,6] junctions of the fullerene cage.

A different perspective in creating holes on the fullerene cage, namely, the formation of more than one orifice on the same C_{60} molecule, was published by Gan and coworkers in 2010 (Xiao et al. 2010). In brief, homofullerene derivative **184** reacts with trispentafluorophenyl boron to give triol **185**, which is subsequently oxidized with DIB, affording diketone **186** with a 10-membered-ring opening within its structure (Scheme 9.49). Epoxidation with *meta*-chloroperoxybenzoic acid (*m*-CPBA) converts two double bonds on the C_{60} skeleton into two epoxy rings in compound **187**; then, treatment of the later

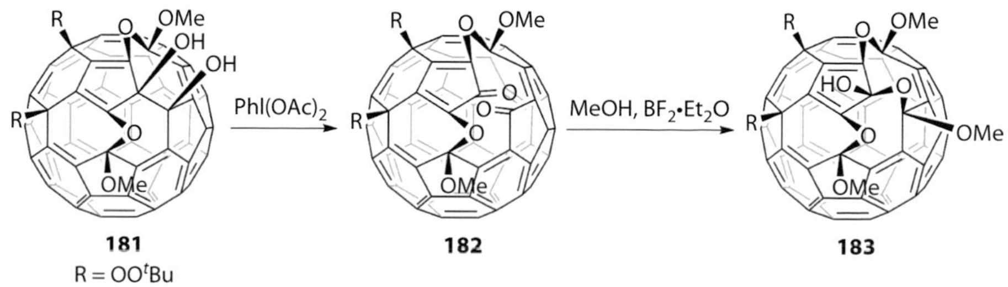

SCHEME 9.48 Orifice opening and closing reactions on a fullerene-mixed peroxide platform.

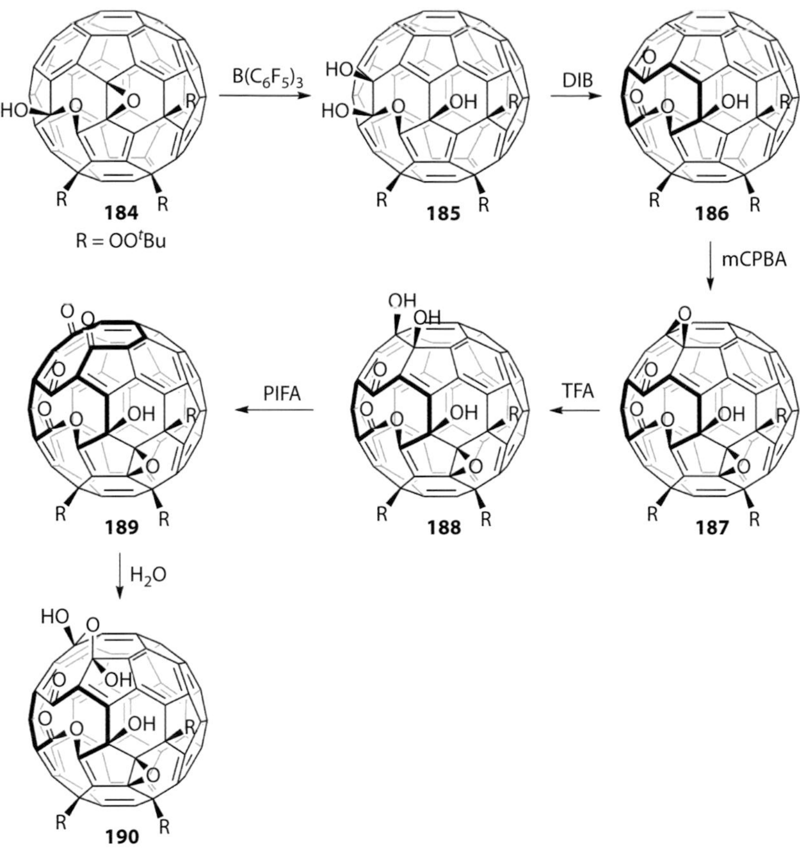

SCHEME 9.49 Preparation of an open-cage fullerene **189** having two 10-membered-ring orifices positioned next to each other.

with TFA results in cleavage of the less crowded epoxy group furnishing the vicinal diol moiety in **188**. Finally, oxidation of adduct **188** with either DIB or $PhI(OOCCF_3)_2$ forms a second 10-membered-ring opening on the same fullerene molecule (**189**, Scheme 9.49). This second orifice easily undergoes a cage-closing reaction to provide **190**, most probably as a result of enhanced ring strain energy (Xiao et al. 2010).

Removal of one carbon atom from the fullerene skeleton of fullerene-mixed peroxides was shown to give open-cage adduct **193** [$C_{59}(O)_3(OH)_2(OO^tBu)_2$, Scheme 9.50] embedded with furan and lactone moieties (Wang et al. 2007). Lewis acid treatment of the fullerene-mixed peroxide **191** (Wang et al. 2006) with BF_3 caused heterolytic scission of O–O, C–C, and C–O bonds to form compound **192**, which may possess either of the two structures depicted in Scheme 9.41 (it was not possible to distinguish which one). Thermolysis of ketolactone **192** at 75°C afforded oxafulleroid **193**, following elimination of a fullerene carbon atom as CO and conversion of two ^{t}BuOO moieties into OH groups (Wang et al. 2007). The carbon atom was then replaced by an oxygen atom to furnish a furan ring. In the final step, one of the two hydroxyl groups in **193** was selectively converted to an alkoxyl group in the presence of hydrochloric acid and a nucleophile (R′OH). All products in Scheme 9.41 are racemates and were characterized by NMR, MS, IR, and UV-vis spectroscopy. Notably, structural assignment of these compounds was facilitated by single-crystal x-ray analysis of **194a**. The crystal structure of **194a** verified the existence of the furan moiety within the fullerene framework and revealed that the lactone segment, as well as the furan oxygen, is lifted up from the cage surface because of steric reasons.

In 2013, Gan and coworkers reported on the synthesis of an open-cage fullerene bearing a ketolactone moiety embedded into the fullerene skeleton through a three-step procedure (**199**, Scheme 9.51) (Xin et

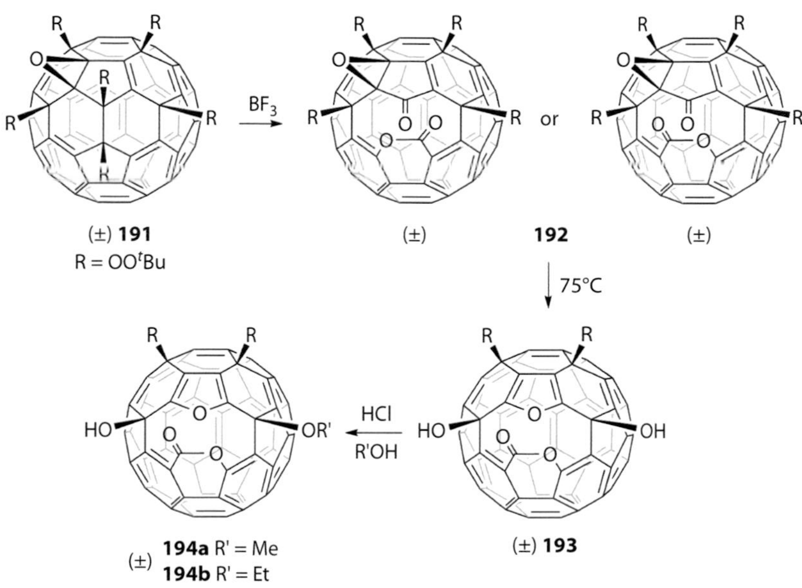

SCHEME 9.50 Synthesis of cage-opened compounds **192–194**.

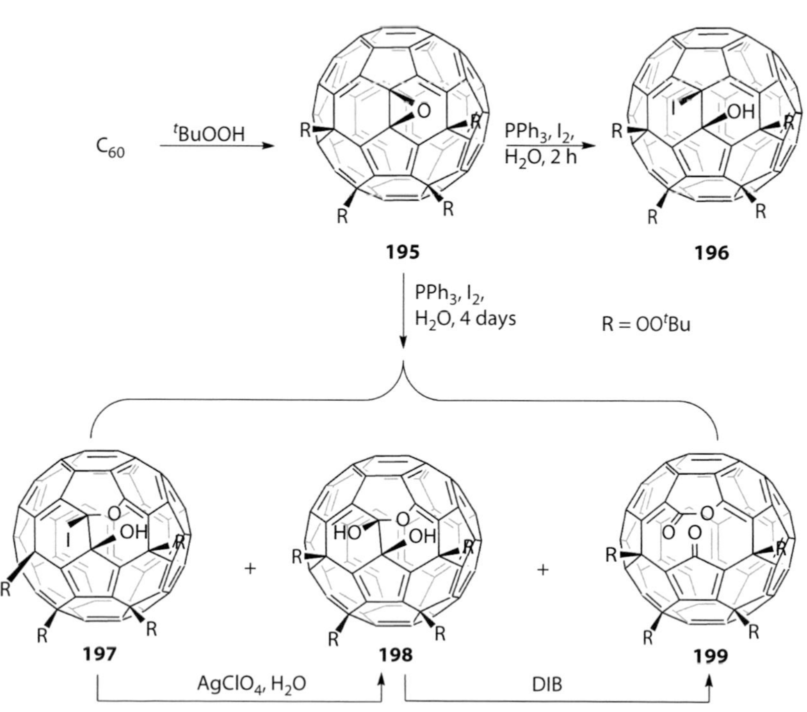

SCHEME 9.51 Formation of fullerenes **196–199**.

al. 2013). Treatment of C_{60} with *tert*-butyl hydroperoxide (*t*-BuOOH) results in the epoxy derivative **195**, whose further reaction with triphenylphosphine and iodine produces the iodo derivative **196**. When compound **196** is stored under ambient conditions, it slowly decomposes into several other products. These products include compounds **197**, **198**, and **199**, which can also be formed in moderate yields directly from the reaction solution that produces **196** if the reaction time is extended from a few hours to a few days. The formation of compounds **197**, **198**, and **199** involves fullerene skeleton bond cleavage. Upon treatment with silver chloride in water, fullerene **197** leads to **198**. Moreover, $PhI(OAc)_2$ was shown to efficiently oxidize compound **198** to **199**, while this family of open-cage derivatives can be converted back to C_{60} through deoxygenation with triphenylphosphine.

Treatment of fullerene derivative **191** with ferric chloride afforded, besides the already known compound **192**, a new open-cage adduct (**200**, Scheme 9.52) (Zhang et al. 2010b). Next, the reaction of benzylamine as well as hydroxylamine with compound **200** was carried out (Scheme 9.53). Both nucleophilic additions occurred at the less hindered carbonyl carbon of **200** to give new products **201** and **202** with a hemiacetal moiety. Treatment of **202** with DIB eliminated the hydroxyl amine group, giving back **200** (Scheme 9.53).

In 2011, Gan and coworkers showed that photochemical treatment of **191** in the presence of iodine produces **192** in low yield (Scheme 9.54) (Zhang et al. 2011b). Upon thermal treatment in benzene, adduct **192** leads to bislactone **203**. Thermolysis of **191** also leads to bislactone **203**, along with its isomer **204**. Possible mechanistic pathways for the formation of **203** and **204** involve radical intermediates when **192** is treated either thermally or photochemically.

Gan, Wang, and coworkers reported the synthesis of open-cage fullerene derivatives with 18- and 19-membered-ring orifices in 2007 (Xiao et al. 2007). Initially, aminoketal/hemiketal derivatives **205a–205c**, derived from adduct **168** after an epoxide ring opening reaction and coupling of the adjacent carbonyl groups through treatment with an arylamine, were oxidized by DIB to afford open-cage adducts **206a–206c** (Scheme 9.55). The three hydroxyl groups in **205** were converted into carbonyls, whereas the amino group was also transformed into an imine. Compounds **206a–206c** bear an anhydride bridge, the removal of which could afford a large opening on the fullerene cage. Indeed, primary

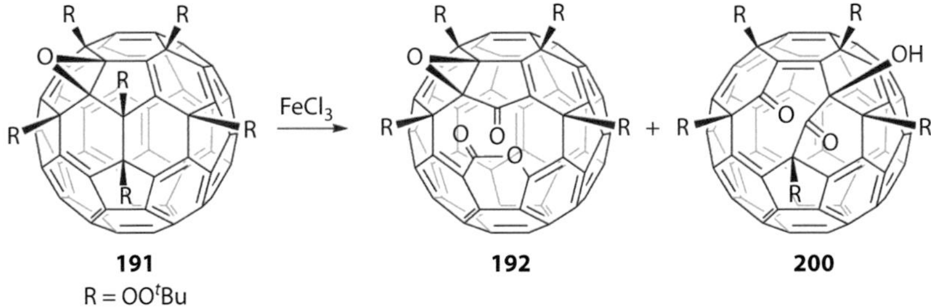

SCHEME 9.52 Cage-opening reaction catalyzed by ferric chloride.

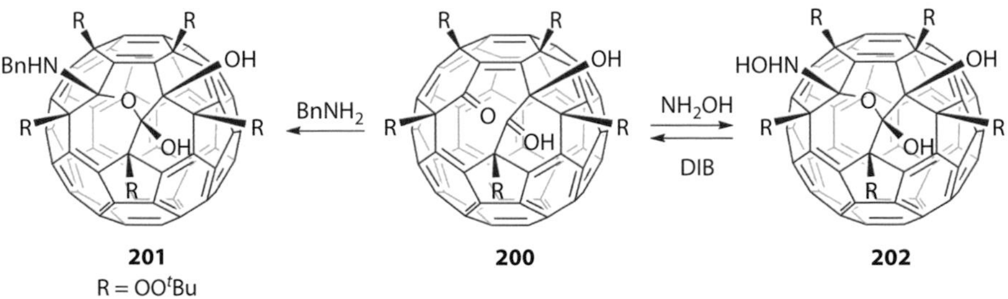

SCHEME 9.53 Addition of primary amines to adduct **200** and interconversion between **200** and **202**.

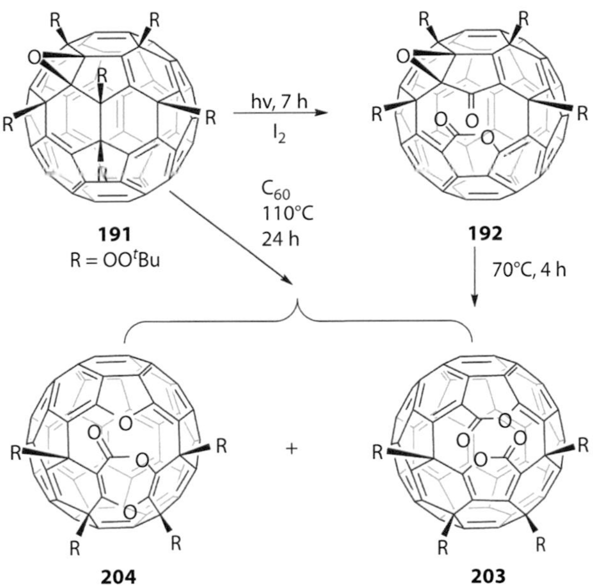

SCHEME 9.54 Synthesis of open-cage fullerene derivatives **203** and **204**.

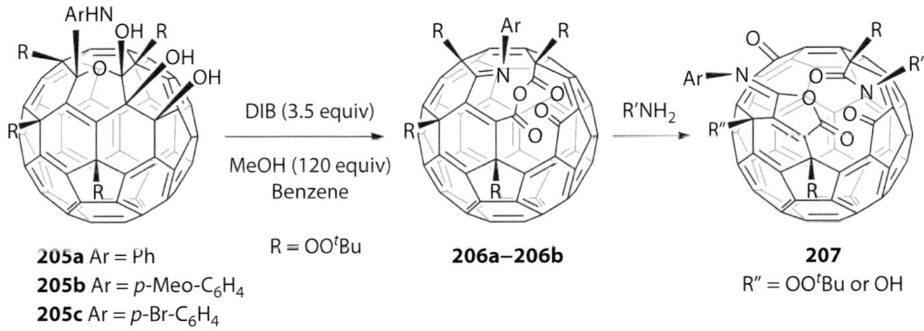

SCHEME 9.55 Oxidation of **205a–205c** by DIB and aminolysis of the anhydride bridge in **206a–206c**.

amines (R′NH$_2$) reacted with this anhydride to give 18-membered-ring orifice product **207** after multiple bond cleavage in one step. The structure of **207**, i.e., the exact position of all functional groups, was elucidated by means of single-crystal x-ray analysis on one of these compounds. The x-ray structure revealed that one of the carbon atoms at the initial pentagon is separated, forming an amide moiety, and at the same time, the initial central pentagon is converted into an isomaleimide moiety, which is completely lifted up from the cage surface. An intramolecular domino rearrangement mechanism was proposed for this unusual transformation (Xiao et al. 2007).

Despite the fact that compound **207** has a pretty large opening, the presence of the amide group blocks the entrance, as evinced in its x-ray structure. Hence, to make the aperture more accessible, the authors attempted to remove the amide moiety. At first, adduct **208** was treated with Br_2 to provide bromoazo product **209** as a mixture of *E* and *Z* isomers (Scheme 9.56). Then, treatment of **209** with $AgClO_4$ removed the acyl group, affording **211** through the unstable carboxylic acid intermediate **210**. In this step, decarboxylation takes place and one carbon atom is expelled as carbon dioxide. Conversion of **211** (R″ = OH) into the *p*-bromo substituted adduct **212**, in the presence Br_2, has been proven useful for the

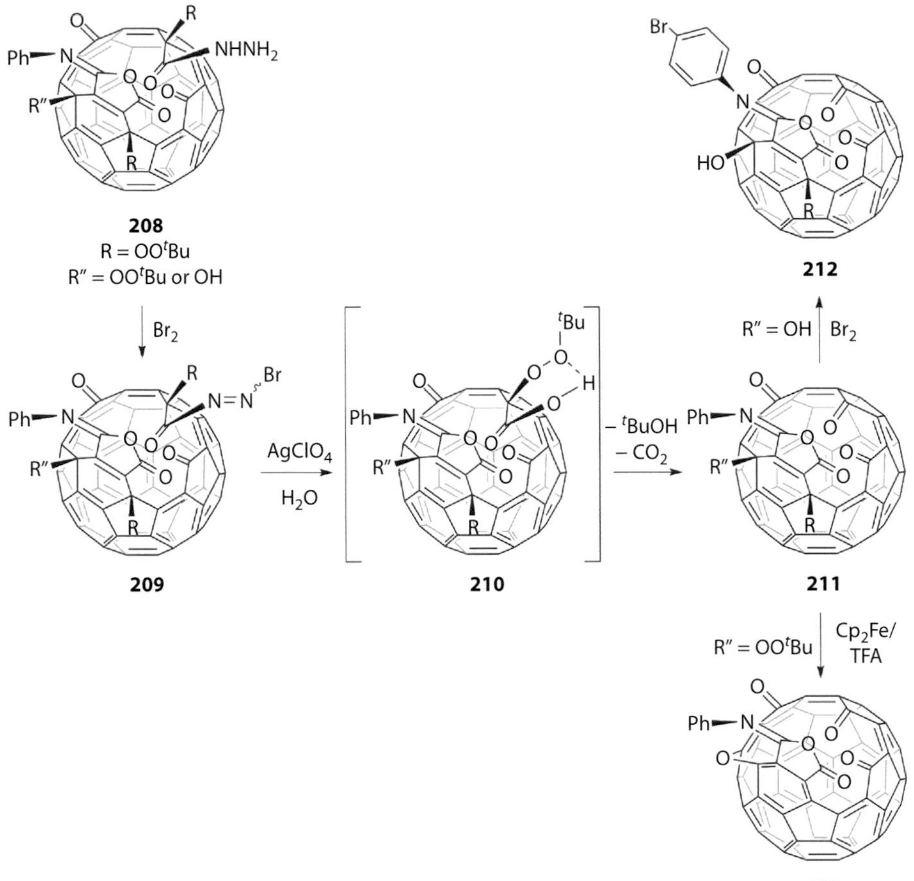

SCHEME 9.56 Preparation of open-cage fullerene derivatives **211**, **212**, and **213**.

preparation of crystals suitable for x-ray analysis. The x-ray structure of **212** clearly shows that the amide moiety is absent, in contrast to the crystal structure of **207** obtained previously, and that access to the 18-membered-ring opening is now improved (Xiao et al. 2007).

The orifice of compound 211 (R″ = OO^{t}Bu) was slightly enlarged by the insertion of an oxygen atom into the rim. This was accomplished after treatment with ferrocene in the presence of TFA, resulting in the formation of open-cage fullerene **213** bearing a 19-membered-ring orifice (Scheme 9.56) (Xiao et al. 2007). Compounds **211–213** turned out to be suitable for water encapsulation (Murata et al. 2008b; Gan et al. 2010; Vougioukalakis et al. 2010a).

Gan, Wang, and coworkers have also prepared open-cage fullerenes resulting from elimination of one carbon atom from the fullerene core, the so-called norfullerenes (Yao et al. 2008). Initially, nucleophilic addition of hydroxylamine to one carbonyl carbon of derivative **168** gave oxahomofullerene derivative **214** with a hydroxylaminoketal moiety (Scheme 9.57). Then, treatment of **214** with PCl_5 caused one carbon atom to be extruded out of the cage network forming an amide group, inducing the migration of one *t*-butoxyl group onto the nitrogen atom to form product **215**. The structure of **215** was determined on the basis of spectroscopic data and x-ray diffraction analysis of related structures (Yao et al. 2008). Also, removal of the amide moiety of **215** was achieved: A four-step reaction procedure furnished the C_{59} norfullerene derivative **216** after the elimination of one *t*-butyl group, one amide group, two chlorine atoms, and one *t*-butyl peroxy group (Scheme 9.57). Reaction of **216** with ICl and Br_2 formed adducts

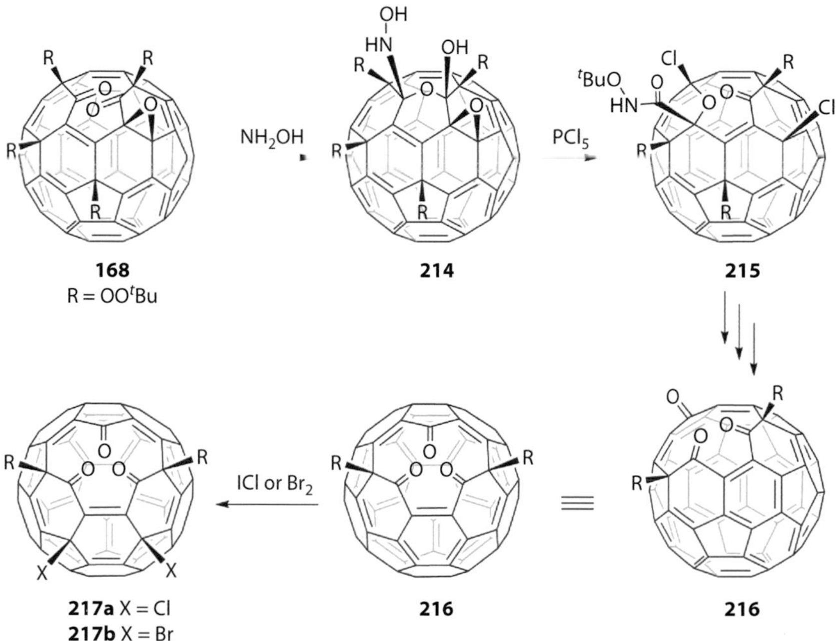

SCHEME 9.57 Formation of C_{59} norfullerene derivatives **216** and **217a**, **217b**.

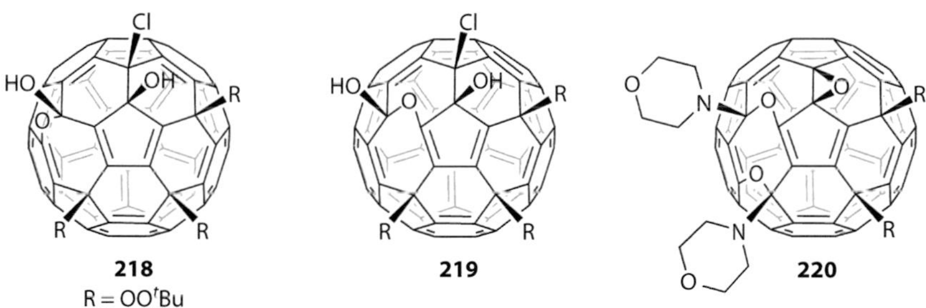

FIGURE 9.19 [5,6]- and [6,6]-oxahomofullerene derivatives **218**, **219**, and **220**.

217a and **217b**, respectively. Compounds **217a** and **217b** gave crystals suitable for x-ray diffraction analysis that permitted the structure assignment of **217a**, **217b**, and **216**, as well as of their precursors (not shown here).

[5,6]- and [6,6]-oxahomofullerene derivatives like **218** and **219** (Figure 9.19), as well as bisoxahomoaminoketal-fullerenes like **220** (Figure 9.19), have been also reported by Zhang, Chen, Gan, and coworkers, as a result of fullerene-mixed peroxides transformations (Huang et al. 2005; Hu et al. 2007).

In 2009, oxahomofullerene **219** was converted into diketone derivative **221** (Scheme 9.58) through a three-step procedure (Zhang et al. 2009). Next, **221** reacted with anilines ($ArNH_2$) to form open-cage adducts **222a–222c** and **223a–223c** (Scheme 9.58) (Zhang et al. 2009). In the later, all three *t*-butoxyl groups of **221** are removed, whereas in the former, one of the *t*-butylperoxo groups in **221** remains. This group can be efficiently reduced to the hydroxyl group in **223a–223c** with CuBr. The structure of compound **223a** was undoubtedly elucidated by single-crystal x-ray analysis. It possesses an 18-membered-ring orifice where the central pentagon is completely lifted up from the cage surface. Compound **223a** was later on found to bind reversibly a phosphate moiety that floats over the orifice as a stopper and

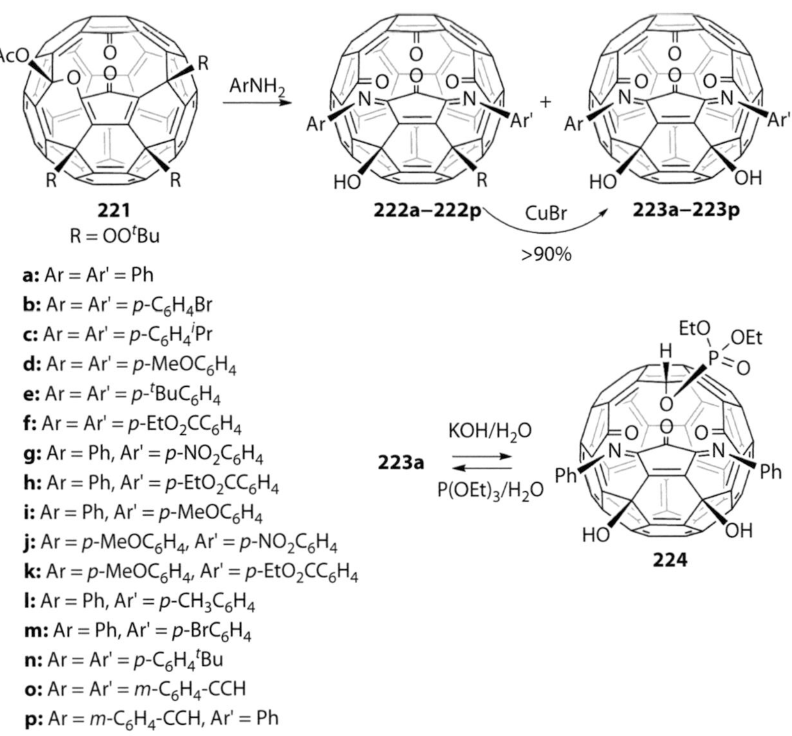

SCHEME 9.58 Formation of **222** and **223** and interconversion between **223a** and **224**.

blocks the opening of the cage (Scheme 9.58) (Zhang et al. 2010a). As shown in Scheme 9.58, treatment of **223a** with nucleophile triethylphosphite affords product **224**, where the phosphate group is located above the rim. This derivative can be converted back to **223a** after hydrolysis of the phosphate group under basic conditions. Hence, in this way, one can block and unblock the passage to and from the interior of a fullerene molecule in a single-step reaction.

Along the same lines, Li and coworkers published the synthesis of a whole family of **223** derivatives in 2011 and 2014 (Scheme 9.58) (Yu et al. 2011). One of these open-cage derivatives was further utilized in a click reaction with benzyl azide under standard click conditions (Scheme 9.59) (Yu et al. 2014b).

In 2012, the same research group, starting from **226**, opened a 19-member-ring orifice on the fullerene skeleton, in a two-step reaction procedure (Scheme 9.60) (Liu et al. 2012). **226** reacts with iodine

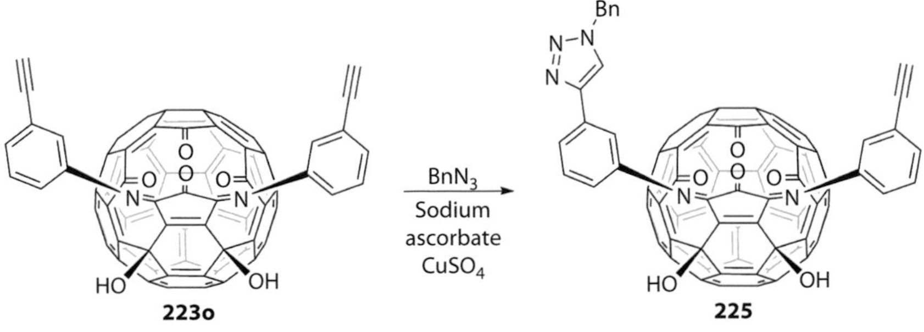

SCHEME 9.59 Functionalization of adduct **223o** via a click reaction.

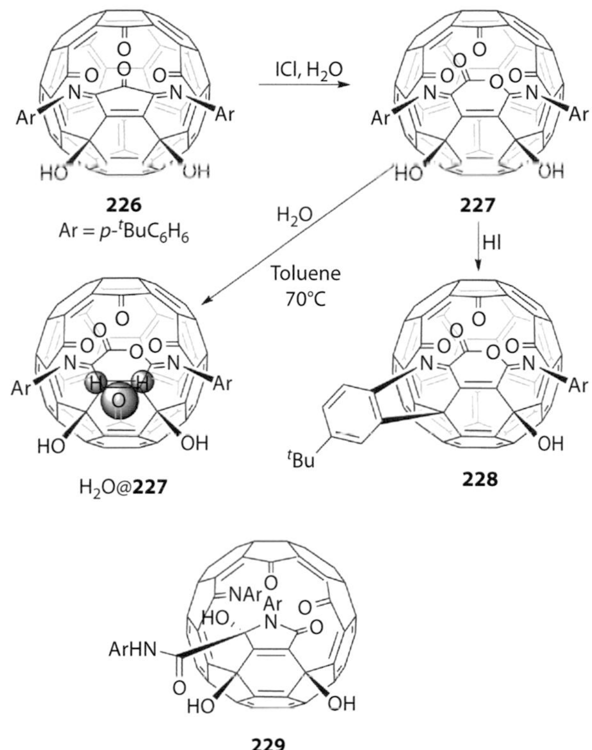

SCHEME 9.60 Synthesis of open-cage fullerene **228**, endohedral fullerene H_2O@**227**, and open-cage derivative **229**.

chloride to form a Baeyer-Villiger type product **227**, while product **228** is formed upon hydroiodic acid addition. It is also worth mentioning that upon heating of **227** with water, the corresponding water-encapsulated complex with an encapsulation ratio of 15% is obtained. Furthermore, **227**-type adducts react with aniline under acidic catalysis to afford **229**-type adducts through a domino reaction sequence involving lactone-opening, imino hydrolysis, aminal formation, and eventually imine formation (Liu and Gan 2014). On a different note, in 2013, a number of dienes were attached to the rim of the orifice of **223** analogs, leading to mono- and bis-adducts in high yields (Xu et al. 2013). As the double bond of the cyclopentenone moiety of adducts **223** behave as a good dienophiile, it can easily afford a variety of Diels-Alder products. These [4 + 2] addition reactions can be easily controlled by varying the reaction time and the ratio of the respective reactants.

In a parallel effort, the same research group treated derivative **229** with iodine using acetic acid as solvent to cleave the C–C bond in the epoxy moiety and in this way added iodo and aceto functionalities, obtaining **230** (Scheme 9.61) (Yu et al. 2013). Further treatment of **230** with silver perchlorate expands the orifice to give derivative **231**. Selective reduction of one of the peroxides in **231**, followed by an S_N1 substitution of the resulting hydroxyl group from a chloride atom, leads to **232**. Treatment of **232** with tin dichloride leads to the formation of **233**, which, upon heating in the presence of trifluoroacetic anhydride and acetic acid, finally affords product **234**. The orifice of adduct **234** has the ability to alter its size and hence either expand quite enough to be able to host H_2O molecules at room temperature, or can be small enough to retain H_2 and prevent it from escaping the cavity. This ability of adduct **234** to alter the effective size of its orifice lies on the rotation of the N-aryl bond at the rim of the orifice. As a final note, fullerene-mixed peroxides have been successfully used as starting materials for the preparation of

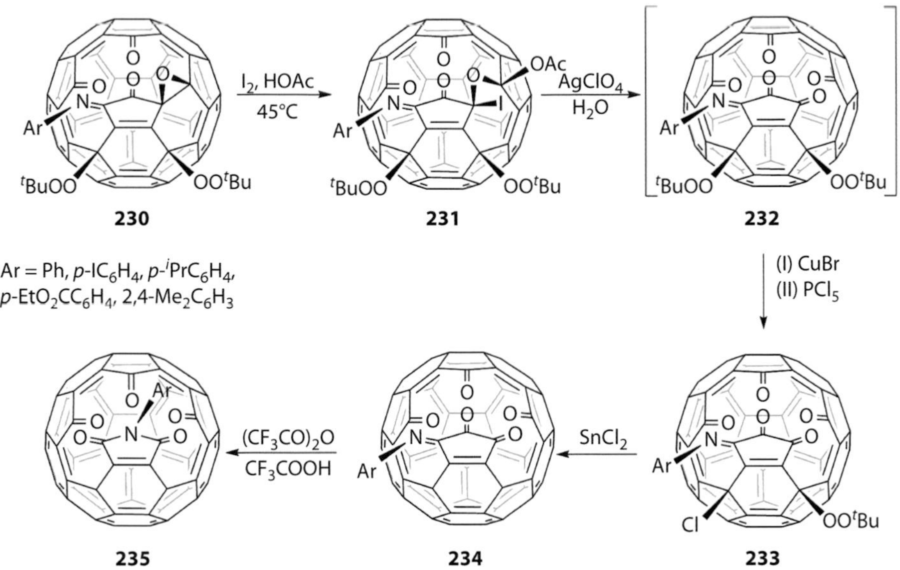

SCHEME 9.61 Synthesis of open-cage derivative **235**.

azafullerene derivatives (Keshavarz et al. 1996; Nuber and Hirsch 1998; Vougioukalakis et al. 2003a,b, 2004c, 2010b; Zhang et al. 2008; Roubelakis et al. 2010).

9.9 Open-Cage C_{70} and Higher Fullerene Derivatives

Cage-opened C_{70} derivatives are relatively rare. As discussed previously, the basic reason is that the lower symmetry of C_{70} (D_{5h}) can give rise to a large number of isomers.

The first open-cage [70]fullerene derivative was reported in 1995 (Birkett et al. 1995). Spontaneous oxidation of $C_{70}Ph_8$ afforded a bislactone derivative of [70]fullerene, namely, $C_{70}Ph_8O_4$ (**236**, shown in Scheme 9.62), bearing an 11-membered-ring orifice (Birkett et al. 1995; Avent et al. 1998). The formation of $C_{70}Ph_8O_4$ was proposed to proceed initially via oxygen insertion into two [5,6]-single bonds of $C_{70}Ph_8$, giving a bis(vinyl ether) intermediate, followed by oxidative cleavage of the adjacent [5,6]-double bond, eventually producing two carbonyl groups.

Wudl and coworkers were the first to report the synthesis of aza[60]fullerene **92** (Figure 9.12) in bulk quantities (Hummelen et al. 1995a). Open-cage [60]-*N*-MEM-ketolactam **28** (Scheme 9.5) was the key intermediate in that synthesis. The same experimental procedure was followed in the case of C_{70} to isolate the corresponding aza[70]fullerene dimer, i.e., $(C_{69}N)_2$. Thus, thermal reaction of C_{70} with

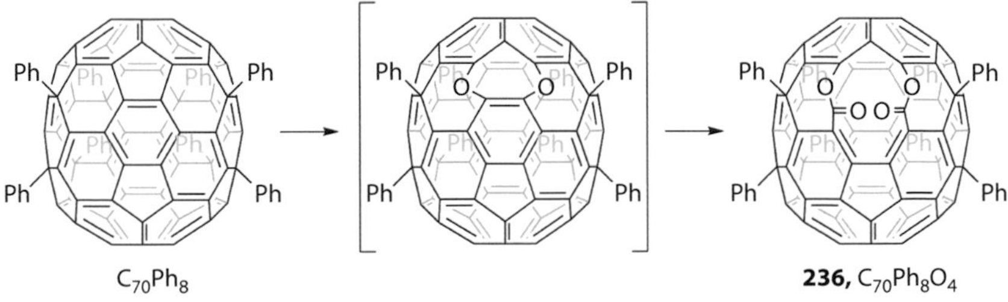

SCHEME 9.62 Suggested path for the formation of $C_{70}Ph_8O_4$ from $C_{70}Ph_8$.

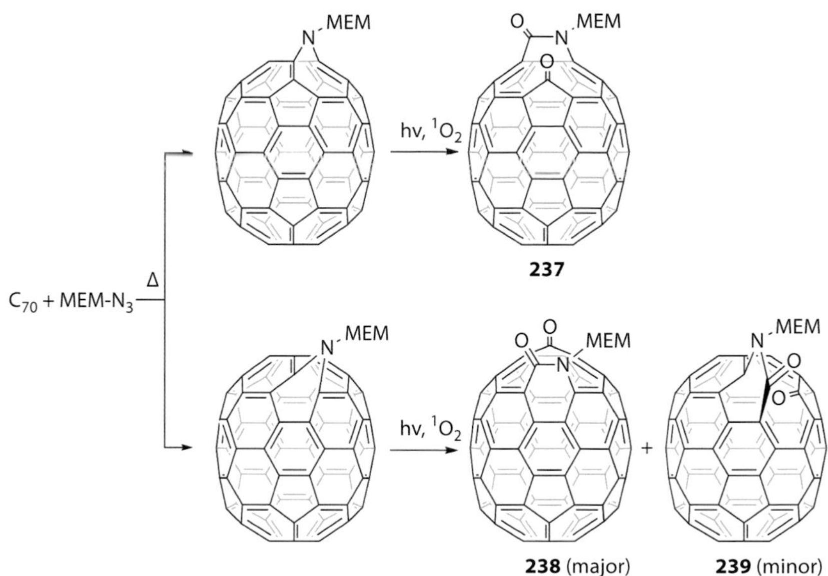

SCHEME 9.63 Formation of [70]-*N*-MEM-ketolactams **237**, **238**, and **239**.

MEM-azide and subsequent self-photooxygenation of the resulting azafulleroids afforded open-cage [70]-*N*-MEM-ketolactam isomers **237–239** (Scheme 9.63) (Bellavia-Lund et al. 1997b,c; Hasharoni et al. 1997; Hummelen et al. 1999; Tagmatarchis et al. 2000). Eleven-membered-ring cage-opened C_{70} derivatives **237–239** were fully separated and readily led to three discrete aza[70]fullerene dimers (Hasharoni et al. 1997; Hummelen et al. 1999; Tagmatarchis et al. 2000).

In 2008, Komatsu and coworkers published another work on open-cage [70]fullerene derivatives (Murata et al. 2008c) The method employed earlier for the production of adduct **77** from C_{60} (Murata et al. 2003b) was successfully applied to C_{70}, leading to the formation of a novel open-cage C_{70} derivative with a 13-membered-ring hole (**240** in Scheme 9.64), in a one-pot procedure. Thus, thermal reaction of C_{70} with 3,6-di(2-pyridyl)pyridazine **241** afforded adduct **242**, bearing an eight-membered-ring orifice. Photooxidation of one of the C–C double bonds in **242** gave cage-opened derivative **243**, having a 12-membered-ring hole. Finally, a sulfur atom insertion at the rim of the orifice in **243** was achieved, according to the procedures discussed previously. Single-crystal x-ray analysis of **240** allowed the determination of its exact structure.

More recently, the same research group prepared another family of open-cage C_{70} derivatives (Scheme 9.65) (Zhang et al. 2014). In that work, they showed that open-cage C_{70} diketones exert

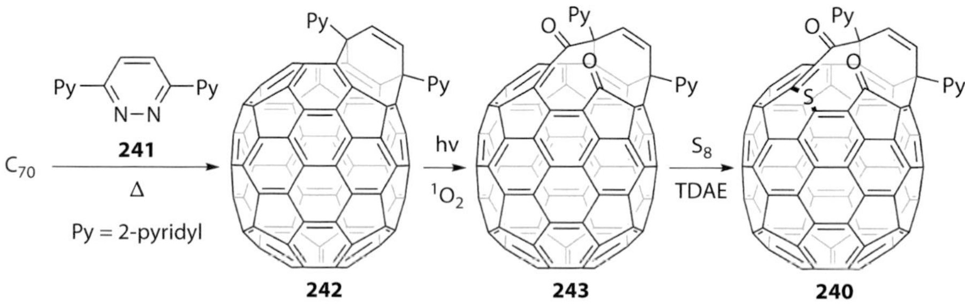

SCHEME 9.64 Synthesis of cage-opened C_{70} derivative **240**.

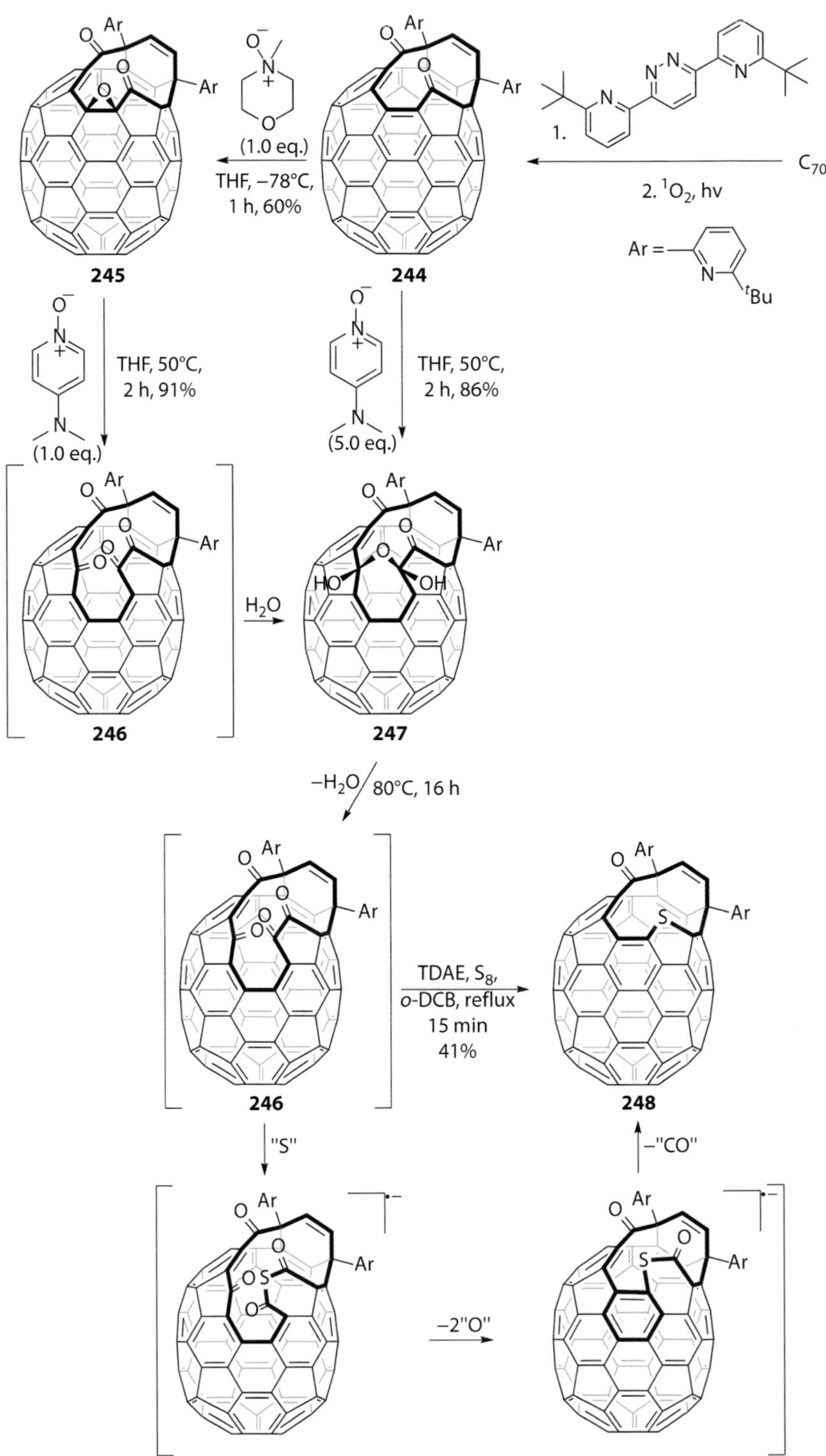

SCHEME 9.65 Cage-opened C_{70} derivatives **244–248**.

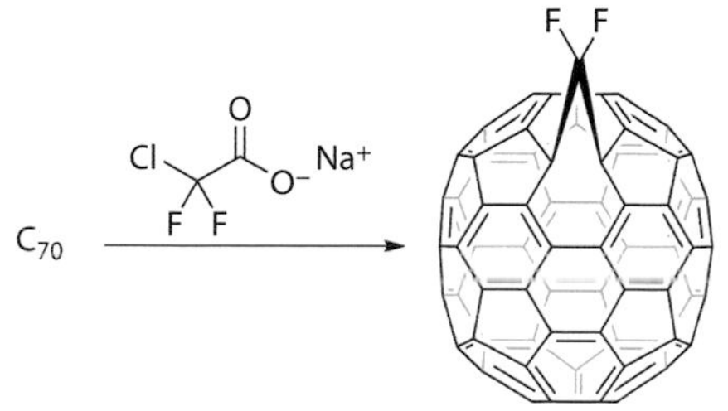

SCHEME 9.66 Synthesis of [6,6] open-cage derivative of C_{70}.

remarkably different reactivity when compared with their C_{60} analogs. Specifically, C_{70}, upon treatment with 3,6-bis(6-(t-butyl) pyridin-2-yl)pyridazine, followed by photochemical oxidation, affords open-cage C_{70} diketone **244** on which the *t*-butyl groups are introduced to increase its solubility. The reaction of **244** with NMMO at room temperature results in the formation of complex product mixtures. This is indicative of different reactivity patterns between the corresponding C_{70} and C_{60} derivatives. This reaction proceeds via the initial formation of isolated epoxide **245**, which affords tetraketone **246** after reacting with a second equivalent of the oxidant. This is followed by the addition of one molecule of water to one of the carbonyl groups, affording **247**. Heating of bis(hemiacetal)-diketone **247** at 80°C for 16 h expands its orifice by the elimination of a water molecule. The opening of **247** leads to a diketone and bis(hemiacetal) moiety, which adopt the same structural motif as in the C_{60} analogs. Further heating at 180°C in ODCB for 15 min in the presence of 1 equivalent of elemental sulfur and 0.2 equivalent of TDAE leads to the unexpected open-cage $C_{69}S$ thiafullerene **248** with a 12-member opened ring.

In 2013, Goryunkov and coworkers reported mono-, bis-, and tris-adducts of the general type $C_{70}(CF_2)_n$ (Samoylova et al. 2013). The major products are the isomeric mono-adducts $C_{70}(CF_2)$ with [6,6]-open and [6,6]-closed configurations. [6,6]-open $C_{70}(CF_2)$ **249** (Scheme 9.66) constitutes the first example of a non-hetero [70]fullerene derivative in which bond opening occurs at the most reactive bond in the polar region of the cage. The [6,6]-open isomer exhibits higher electron affinity than the closed one, whose affinity is close to that of pristine C_{70}. In situ electron spin resonance (ESR) spectro-electrochemical studies and DFT calculations of the hyperfine coupling constants of the $C_{70}(CF_2)$ radical anions suggest an interconversion between the [6,6]-closed and [6,6]-open forms of the respective cage-opened fullerene. This procedure was proposed to be driven by an electrochemical one-electron transfer, and therefore, the [6,6]-closed $C_{70}(CF_2)$ constitutes an interesting example of a redox-switchable fullerene derivative.

Many research groups have investigated the chemical reactivity of endohedral metallofullerenes, the properties of which have been found to differ greatly from those of empty fullerenes. The encaged metal atoms and the carbon cages have cationic and anionic character, respectively, as a result of electron transfer. As anticipated, exohedral chemical functionalization is an effective method for modifying the physical and chemical properties of endohedral metallofullerenes. Introduction of heteroatoms, such as electropositive silicon, directly onto the fullerene surface has inducted remarkable changes in the electronic characteristics of fullerenes. Along these lines, Nagase and coworkers have showed that silylated I_h-C_{80} fullerene derivatives have more negative charge on the cage than the parent I_h-C_{80} (Sato et al. 2012). Moreover, they proved that the number of silicon atoms on the fullerene surface plays an important role in the electronic properties and the movement of the encapsulated metal atoms. Functionalization of endohedral metallofulerenes has been shown to differ depending on the type of treatment (photochemical or thermal). In addition, silylation alters the electrostatic potentials and, thereby, the dynamic behavior of the cationic metals.

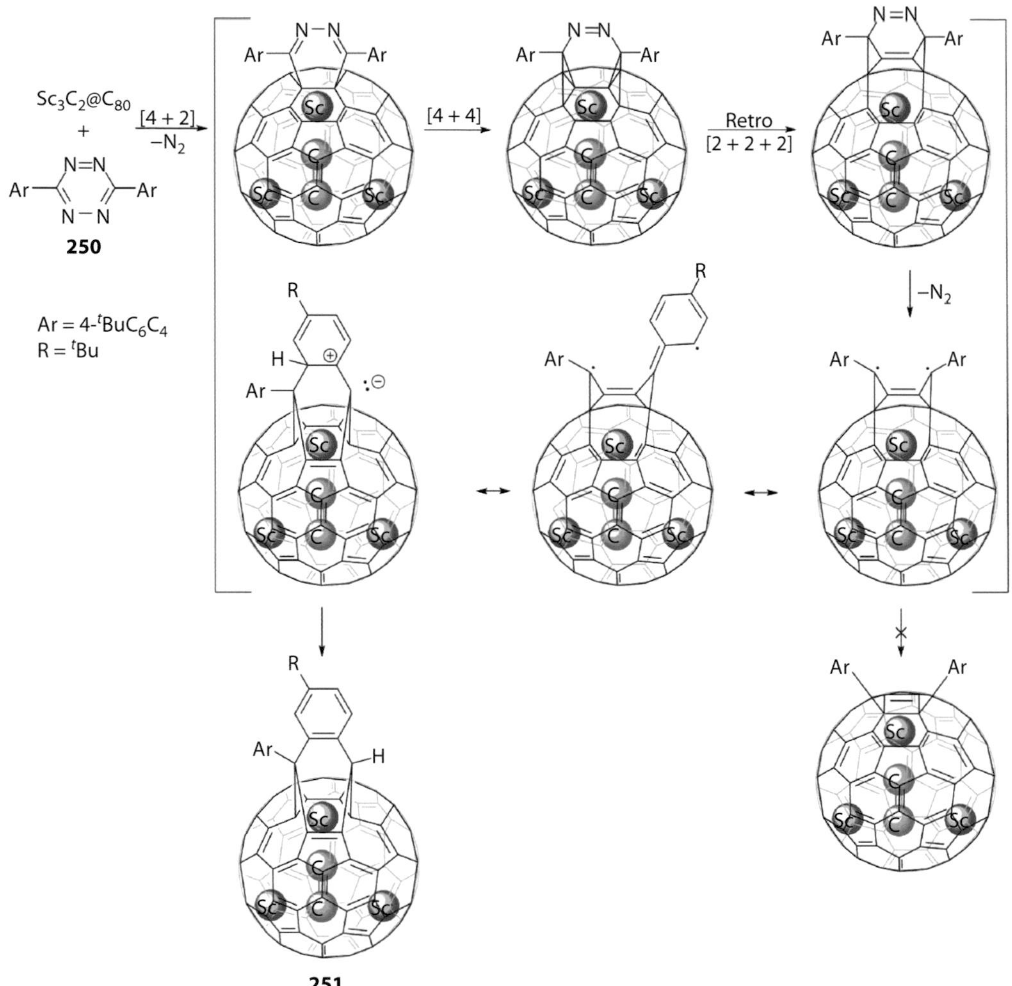

SCHEME 9.67 Formation of open-cage [80]fullerene adduct **251**.

In another recent work concerning I_h-C_{80} open-cage fullerenes, the reaction of endohedral fullerene $Sc_3C_2@C_{80}$ with tetrazine **250** exclusively affords open-cage derivative **251** (Scheme 9.67) instead of the expected analog of the C_2-inserted derivative of C_{60} **76** (Scheme 9.14) (Kurihara et al. 2012). The mechanism proposed for the formation of **251** includes a thermal [4 + 2] cycloaddition reaction between $Sc_3C_2@C_{80}$ and 1,2,4,5-tetrazine **250**, followed by a nitrogen molecule extrusion and a subsequent [4 + 4]/retro [2 + 2 + 2] transformation followed by a second dinitrogen extrusion to give a biradical intermediate. The free radicals of the biradical intermediate are delocalized over the p-orbital system of the phenyl group and eventually neutralized to afford **251**. The structure of **251** was unequivocally established, and it is considered to be the first step toward the release of the internal metal atoms in endohedral fullerenes.

In 2013, Echegoyen, Balch, and coworkers reported the synthesis and characterization of six diphenylmethano derivatives of $Sc_3N@I_h$-C_{80} in good yields under mild conditions (Izquierdo et al. 2013). These adducts were obtained through photoirradiation of the appropriate carbene precursors in a solution of $Sc_3N@I_h$-C_{80} in ODCB at 0°C for 10 min. The structures of these methanofullerenes were confirmed by x-ray crystallography. In a related work, an even higher endohedral metallofullerene, namely,

Yb@C_{84}, was chemically functionalized by Lu and coworkers (Zhang et al. 2013). In brief, diazine was photochemically added to Yb@C_{84}, forming the respective [5,6]-open cage adduct Yb@C_2(13)-C_{84}, the structure of which was studied, among others, by x-ray analysis. This was the first x-ray study of an endohedral metallofullerene derivative having a cage larger than C_{82}.

9.10 Aperture-Closing Strategies

Komatsu and coworkers were the first to introduce H_2 molecules into an open-cage fullerene derivative at a 100% encapsulation ratio (Murata et al. 2003a). Cage-opened fullerene **77** bearing a 13-membered-ring orifice on the fullerene framework, being 5.64 and 3.75 Å along the long and the short axis, respectively, was used (Murata et al. 2003b). According to theoretical calculations at the B3LYP/3-21G level of theory using the 6-31G** basis set, the energies required for He and H_2 insertion into **99b** were predicted to be 24.5 and 41.4 kcal/mol, respectively (Rubin et al. 2001), whereas the corresponding calculated encapsulation energies for **77** are 18.9 and 30.1 kcal/mol, respectively (Murata et al. 2003a). The encapsulation of H_2 gas into **77** was accomplished in an autoclave at an applied H_2 pressure of 800 atm, upon 8 h of heating at 200°C, while hydrogen encapsulation, affording H_2@**77** (Scheme 9.68), was confirmed by its ^{1}H NMR spectrum, where encapsulated hydrogen appears at δ −7.25 ppm as a single peak. Moreover, the encapsulation rate was found to be highly dependent on the applied H_2 pressure, affording 90% and 51% incorporation yields under 560 and 180 atm, respectively. More significantly, the gas phase generation of H_2@C_{60} during MALDI-TOF mass spectrometry experiments of H_2@**77** demonstrated that even highly modified fullerene derivatives, such as **77**, can regenerate the pristine C_{60} cage by self-restoration (Murata et al. 2003a). Moreover, according to a solid-state NMR spectroscopic study of H_2@**77**, published in 2004, the rotational motion of the endohedral H_2 is only slightly perturbed, while its rotational correlation time (τ_c) was found to be 2.3 ps at 295 K and 15.3 ps at 119 K, with a linear dependence in between (Carravetta et al. 2004). A single crystal of H_2@**77** was also studied by x-ray diffraction with synchrotron radiation (Sawa et al. 2005). As inferred from that study, the environment of

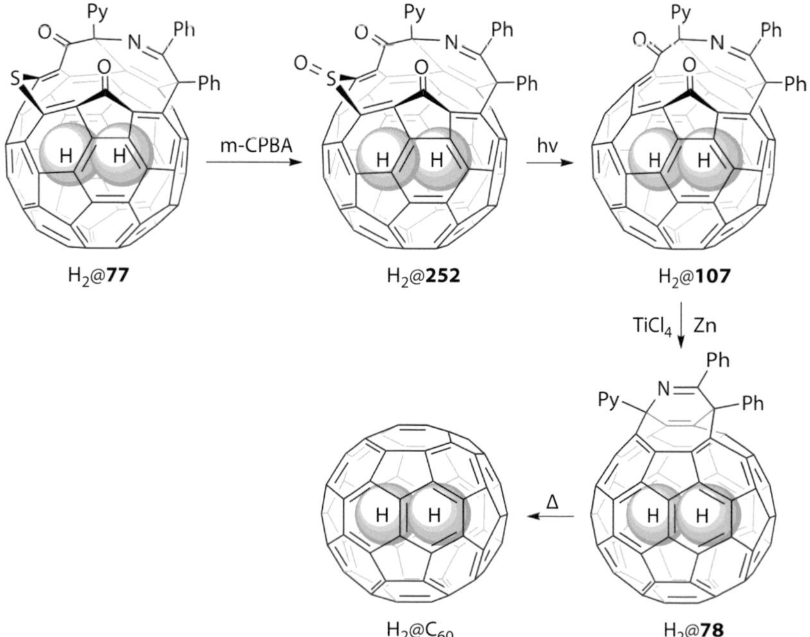

SCHEME 9.68 The four-step process for the closure of the 13-membered-ring orifice encompassed in H_2@77.

the encapsulated H_2, which is found to be located at the center of the fullerene cage, is similar to that of the H_2 molecule incorporated in $H_2@C_{60}$, and furthermore, there is no appreciable charge transfer between the encapsulated H_2 and the cage.

In another very interesting work published in 2005 (Komatsu et al. 2005b), Komatsu and colleagues reported a four-step synthetic route toward the complete closure of the 13-membered-ring orifice encompassed in $H_2@$**77** (Scheme 9.68). In that study, which is essentially the first practical confirmation of the "molecular surgery" approach and complements the total synthesis of C_{60} published by Scott et al. (2002), $H_2@C_{60}$ was prepared for the first time in vitro by chemical procedures. The total yield on going from empty C_{60} to hydrogen encapsulating $H_2@C_{60}$ was 9%. The ^{13}C NMR spectrum of the isolated $H_2@C_{60}$ exhibits a singlet at δ 142.844 ppm (the carbon atoms of empty C_{60} absorb at δ 142.766 ppm), while the ^1H NMR signal for the encapsulated hydrogen is observed at δ –1.44 ppm. A gradual downfield shift of the encapsulated hydrogen signal was observed at the second and the third steps of the procedure. This shift reflects the restoration of a fully π-conjugated pentagon, which induces a strong deshielding effect. On the other hand, the very small downfield shift observed for the ^{13}C NMR signal of $H_2@C_{60}$ (0.078 ppm), together with the coincidence of the UV-vis and the IR spectra of C_{60} and $H_2@C_{60}$, indicates that the electronic properties of the fullerene cage are essentially unaffected by the encapsulation of H_2, contrary to $Kr@C_{60}$ (Yamamoto et al. 1999) and $Xe@C_{60}$ (Syamala et al. 2002). The same trend is also confirmed by cyclic voltammetry, since the four reversible and one irreversible oxidation peaks of $H_2@C_{60}$ were observed at the same potentials as that of C_{60}.

Complete closure of the orifice of **240** (Scheme 9.64) to afford pristine C_{70} impregnated with one and two hydrogen molecules was realized in 2008 (Murata et al. 2008a). The same four-step procedure used in the case of $H_2@$**77** (Scheme 9.68) was successfully applied to a mixture of $H_2@$**240** and $(H_2)_2@$**240** (97:3), to provide endohedral fullerenes $H_2@C_{70}$ and $(H_2)_2@C_{70}$, shown in Figure 9.20, contaminated with 10% empty C_{70}. In the proton NMR of the product in ODCB-d_4, there exist two peaks at high fields: one sharp signal at δ –23.97 ppm, corresponding to encapsulated H_2 of $H_2@C_{70}$, and a weaker one at δ –23.80 ppm, corresponding to the two hydrogen molecules of $(H_2)_2@C_{70}$. The initial 97:3 molar ratio remained intact in the products as was concluded by the integrated peak areas of the ^1H NMR spectrum. In the ^{13}C NMR spectrum of $H_2@C_{70}$ in ODCB-d_4, five signals are present slightly downfield shifted, as compared with those of empty C_{70}. Finally, although incorporated hydrogen interacts poorly with the C_{70} cage, a difference in reactivity between $H_2@C_{70}$ and $(H_2)_2@C_{70}$ was observed concerning the Diels-Alder reaction with 9,10-dimethylanthracene (DMA). In particular, $(H_2)_2@C_{70}$ was found to be less reactive toward DMA because of the greater interaction between two molecules of H_2 and the C_{70} cage (Murata et al. 2008a).

Two years later, following an orifice-closing reaction sequence similar to the one depicted in Scheme 9.51, C_{60} and C_{70} molecules encapsulating a helium atom were also prepared by the same research group (Morinaka et al. 2010). Nevertheless, because of the smaller size of the helium atom, instead of using the orifice of compounds **77** and **240** to perform the encapsulation reaction, the authors used sulfoxide

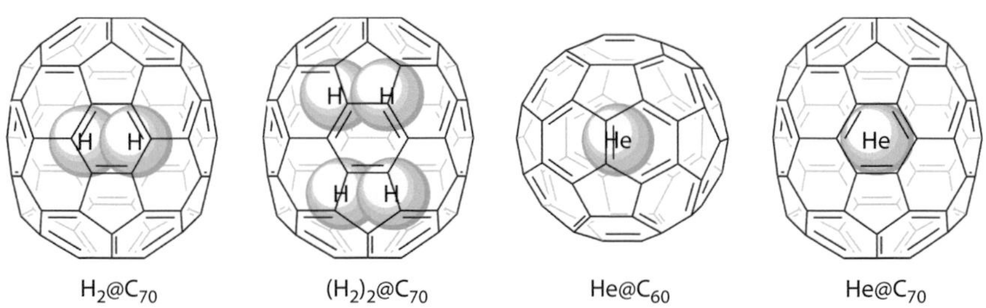

FIGURE 9.20 Endohedral fullerenes C_{60} and C_{70} encapsulating hydrogen molecules and helium atoms.

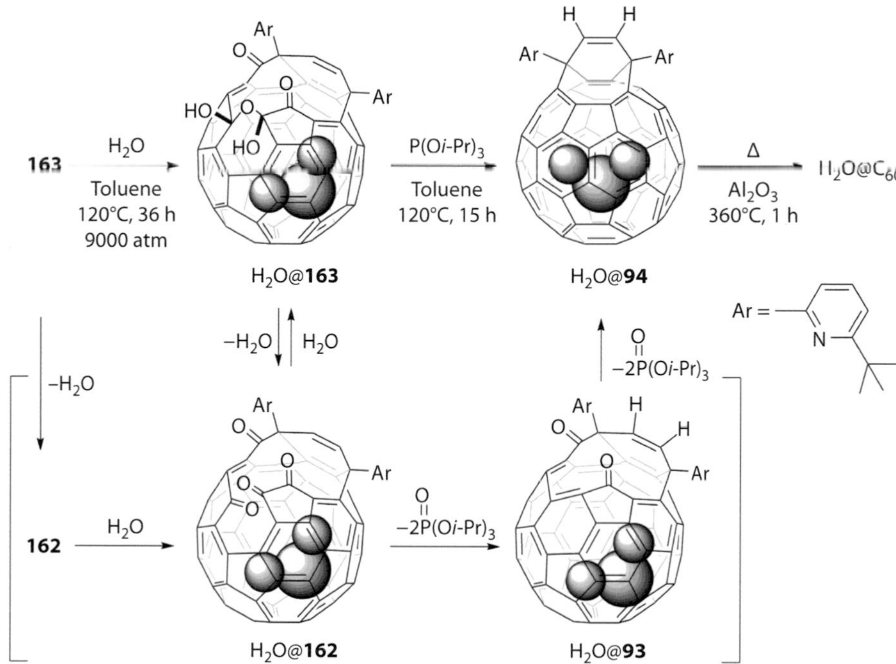

SCHEME 9.69 Dynamic control of the opening size of **163** for the encapsulation of a water molecule inside an open-cage fullerene and the chemical synthesis of $H_2O@C_{60}$.

derivative **252** (Scheme 9.68) and the corresponding C_{70} sulfoxide derivative (Murata et al. 2008a), whose orifice size can be immediately reduced after He insertion by removing the SO group. Thus, in this way, endohedral fullerenes $He@C_{60}$ and $He@C_{70}$ (Figure 9.20) were synthesized by means of organic synthesis at occupation levels of 30%. In the case of C_{70}, formation of $He_2@C_{70}$ was not detected in contrast to the synthesis of $(H_2)_2@C_{70}$ reported before (Murata et al. 2008a).

In 2011, the synthesis of a cage-closed C_{60} sphere with a water molecule inside its cavity, that is, $H_2O@C_{60}$, was reported by Murata and coworkers (Scheme 9.69) (Kurotobi and Murata 2011). As mentioned previously, derivative **163** is the result of a nucleophilic addition of a water molecule to tetraketone **162** (Scheme 9.42). The insertion of a water molecule into this tetraketone seems more probable since the size of the opening on **162** is bigger than that in **163**. Thus, when hydrate **163** was refluxed in wet toluene under a high pressure of 9000 atm for 36 h, quantitative encapsulation of a water molecule into **163** was confirmed by ^1H NMR analysis (Scheme 9.52). The formation of the endohedral H_2O@**163** was explained on the basis of the dynamic equilibrium established between compound **163** and tetraketone derivative **162**. Elimination of a water molecule from **163** gives **162**. The 16-membered-ring orifice of the later generated in situ permits the penetration of a H_2O molecule into its cavity, and a subsequent nucleophilic addition of a water molecule regenerates **163** encapsulating the H_2O molecule (Scheme 9.69). The proton NMR spectrum of this compound confirms the encapsulation of a H_2O molecule as it displays a sharp signal at −9.87 ppm corresponding to the protons of the encapsulated H_2O. Next, Murata and coworkers succeeded in restoring the ruptured cage of **163** to prepare pristine C_{60} with a H_2O molecule inside its cavity (Scheme 9.69). First, the two carbonyl groups on H_2O@**163** were coupled after treatment with a phosphite ester ($P(Oi\text{-}Pr)_3$) to afford H_2O@**94**. As shown in Scheme 9.69, this step proceeded via the in situ generated H_2O@**162** featuring two successive carbonyl couplings. Finally, the organic addends on the rim of H_2O@**94** were removed after thermal treatment with neutral Al_2O_3 under vacuum for 1 h.

The formation of the water encapsulating C_{60} molecule, i.e., $H_2O@C_{60}$, was undoubtedly confirmed by single-crystal x-ray analysis on single crystals obtained from a 1:2 mixture of the prepared $H_2O@C_{60}$ and nickel(II) octaethylporphyrin in *o*-xylene solution.

9.11 Conclusions

The cage opening and then closing of fullerenes via "molecular surgery" is a highly attractive alternative strategy toward the synthesis of endohedral fullerenes, in comparison with the conventional physical methods of production that are characterized by substantial shortcomings. Significant progress has been made in this direction, with a great number of cage-opened derivatives reported to date. On the basis of the progress realized in this field, we see many prospects for further open-cage fullerene chemistry development. This progress could, among others, lead to exciting applications of cage-opened fullerene derivatives.

Acknowledgment

M.M.R. and G.C.V. are grateful to Professor Michael Orfanopoulos (University of Crete) for introducing them to the field of fullerene chemistry.

References

Aihara, J., "Kinetic stability of fullerenes with four-membered rings," *J. Chem. Soc., Faraday Trans.* 91 (1995): 4349–53.

Akasaka, T., Nagase, S., *Endofullerenes: A new family of carbon clusters* (Dordrecht: Kluwer Academic Publishers, 2002).

Akasaka, T., Nagase, S., Kobayashi, K. et al., "Exohedral derivatization of an andohedral metallofullerene $Gd@C_{82}$," *J. Chem. Soc., Chem. Commun.* 13 (1995): 1343–4.

Al-Matar, H. A., Abdul-Sada, K., Avent, A. G., Taylor, R., "Isolation and characterization of unsymmetrical $C_{60}Me_5O_3H$, a cage-opened bisepoxide ketone: Tautomerism involving a fullerene cage bond," *Org. Lett.* 3 (2001): 1669–71.

An, Y.-Z., Ellis, G. A., Viado, A. L., Rubin, Y., "A methodology for the reversible solubilization of fullerenes," *J. Org. Chem.* 60 (1995): 6353–61.

Arce, M. J., Viado, A. L., An, Y. Z., Khan, S. I., Rubin, Y., "Triple scission of a six-membered ring on the surface of C_{60} via consecutive pericyclic reactions and oxidative cobalt insertion," *J. Am. Chem. Soc.* 118 (1996): 3775–6.

Avent, A. G., Benito, A. M., Birkett, P. R., "The structure of fullerene compounds," *J. Mol. Struct.* 436–437 (1998): 1–9.

Averdung, J., Mattay, J., "Exohedral functionalization of [60]fullerene by [3+2] cycloadditions: Syntheses and chemical properties of triazolino-[60]fullerenes and 1,2-(3,4-dihydro-2H-pyrrolo)-[60]fullerenes," *Tetrahedron* 52 (1996): 5407–20.

Banks, M. R., Cadogan, J. I. G., Gosney, I. et al., "Unprecedented ring expansion of [60]fullerene: Incorporation of nitrogen at an open 6,6-ring juncture by regiospecific reduction of oxycarbonylaziridino-[2′,3′:1,2][60]fullerenes. Synthesis of 1a-aza-1(6a)-homo[60]fullerene, $C_{60}H_2NH$," *Chem. Commun.* 4 (1996): 507–8.

Bellavia-Lund, C., Keshavarz-K, M., Collins, T., Wudl, F., "Fullerene carbon resonance assignments through ^{15}N-^{13}C coupling constants and location of the sp^3 carbon atoms of $(C_{59}N)_2$," *J. Am. Chem. Soc.* 119 (1997a): 8101–2.

Bellavia-Lund, C., Hummelen, J. C., Keshavarz-K, M., González, R., Wudl, F., "New developments in the organic chemistry of fullerenes," *J. Phys. Chem. Solids* 58 (1997b): 1983–90.

Bellavia-Lund, C., Wudl, F., "Synthesis of [70]azafulleroids: Investigations of azide addition to C_{70}," *J. Am. Chem. Soc.* 119 (1997c): 943–6.

Birkett, P. R., Avent, A. G., Darwish, A. D., Kroto, H. W., Taylor, R., Walton, D. R. M., "Holey fullerenes! A bis-lactone derivative of [70]fullerene with an eleven-atom orifice," *J. Chem. Soc., Chem. Commun.* 18 (1995): 1869–70.

Boltalina, O. V., de La Vaissiere, B., Fowler, P. W. et al., "$C_{60}F_{18}O$, the first characterised intramolecular fullerene ether," *Chem. Commun.* 14 (2000): 1325–6.

Brieger, G., Bennett, J. N., "The intramolecular Diels-Alder reaction," *Chem. Rev.* 80 (1980): 63–97.

Carravetta, M., Murata, Y., Murata, M. et al., "Solid-state NMR spectroscopy of molecular hydrogen trapped inside an open-cage fullerene," *J. Am. Chem. Soc.* 126 (2004): 4092–3.

Chen, C. P., Lin, Y. W., Horg, J. C., Chuang, S. C., "Open-cage fullerenes as n-type materials in organic photovoltaics: Relevance of frontier energy levels, carrier mobility and morphology of different sizable open-cage fullerenes with power conversion efficiency in devices," *Adv. Energy Mater.* 1 (2011): 776–80.

Chen, C. S., Lin, Y. F., Yeh, W. Y., "Activation of open-cage fullerenes with ruthenium carbonyl clusters," *Chem. Eur. J.* 20 (2014): 936–40.

Chiang, L. Y., Upasani, R. B., Swirczewski, J. W., Soled, S., "Evidence of hemiketals incorporated in the structure of fullerols derived from aqueous acid chemistry," *J. Am. Chem. Soc.* 115 (1993): 5453–7.

Chronakis, N., Vougioukalakis, G. C., Orfanopoulos, M., "Synthesis and self-photooxygenation of alkenyl-linked [60]fullerene derivatives. A regioselective ene reaction," *Org. Lett.* 4 (2002): 945–8.

Chuang, S.-C., Murata, Y., Murata, M. et al., "Fine tuning of the orifice size of an open-cage fullerene by placing selenium in the rim: Insertion/release of molecular hydrogen," *Chem. Commun.* 12 (2007a): 1278–80.

Chuang, S.-C., Murata, Y., Murata, M., Komatsu, K., "The outside knows the difference inside: Trapping helium by immediate reduction of the orifice size of an open-cage fullerene and the effect of encapsulated helium and hydrogen upon the NMR of a proton directly attached to the outside," *Chem. Commun.* 17 (2007b): 1751–3.

Craig, D., "Stereochemical aspects of the intramolecular Diels–Alder reaction," *Chem. Soc. Rev.* 16 (1987): 187–238.

Creegan, K. M., Robbins, J. L., Win, K. et al., "Synthesis and characterization of $C_{60}O$, the first fullerene epoxide," *J. Am. Chem. Soc.* 114 (1992): 1103–5.

Dapprich, S., Komaromi, I., Byun, K. S., Morokuma, K., Frisch, M. J., "A new ONIOM implementation in Gaussian98. Part I. The calculation of energies, gradients, vibrational frequencies and electric field derivatives," *J. Mol. Struct. (THEOCHEM.)* 461 (1999): 1–21.

Diederich, F., Isaacs, L., Philp, D., "Syntheses, structures, and properties of methanofullerenes," *Chem. Soc. Rev.* 23 (1994a): 243–55.

Diederich, F., Isaacs, L., Philp, D., "Valence isomerism and rearrangements in methanofullerenes," *J. Chem. Soc., Perkin Trans. 2* 3 (1994b): 391–4.

Ding, J., Yang, S., "Isolation and characterization of the dimetallofullerene $Ce_2@C_{80}$," *Angew. Chem., Int. Ed. Engl.* 35 (1996): 2234–5.

Dong, G. X., Li, J. S., Chan, T. H., "Reaction of [60]fullerene with diethyl diazidomalonate: A doubly bridged fulleroid," *J. Chem. Soc., Chem. Commun.* 17 (1995): 1725–6.

Edwards, M. C., Butler, I. S., Qian, W., Rubin, Y., "Pressure-tuning vibrational spectroscopic study of $(\eta^5\text{-}C_5H_5)Co(C_{64}H_4)$ Can endohedral fullerenes be formed under pressure?," *J. Mol. Struct.* 442 (1998): 169–74.

Elemes, Y., Silverman, S. K., Sheu, C. et al., "Reaction of C_{60} with dimethyldioxirane—Formation of an epoxide and a 1,3-dioxolane derivative," *Angew. Chem., Int. Ed. Engl.* 31 (1992): 351–3.

Escobedo, J. O., Frey, A. E., Strongin, R. M., "Investigation of the photooxidation of [60]fullerene for the presence of the [5,6]-open oxidoannulene $C_{60}O$ isomer," *Tetrahedron Lett.* 43 (2002): 6117–9.

Fowler, P. W. and Manolopoulos, D. E., *An atlas of fullerenes* (New York: Oxford University Press, 1995).

Fowler, P. W., Heine, T., Manolopoulos, D. E. et al., "Energetics of fullerenes with four-membered rings," *J. Phys. Chem.* 100 (1996): 6984–91.

Froese, R. D. J., Humbel, S., Svensson, M., Morokuma, K., "IMOMO(G2MS): A new high-level G2-like method for large molecules and its applications to Diels–Alder reactions," *J. Phys. Chem. A* 101 (1997): 227–33.

Gan, L. B., Huang, S. H., Zhang, X. et al., "Fullerenes as a tert-butylperoxy radical trap, metal catalyzed reaction of tert-butyl hydroperoxide with fullerenes, and formation of the first fullerene mixed peroxides $C_{60}(O)(OO^tBu)_4$ and $C_{70}(OO^tBu)_{10}$," *J. Am. Chem. Soc.* 124 (2002): 13384–5.

Gan, L. B., Yang, D. Z., Zhang, Q. Y., Huang, H., "Preparation of open-cage fullerenes and incorporation of small molecules through their orifices," *Adv. Mater.* 22 (2010): 1498–507.

Gao, Y. D., Herndon, W. C., "Fullerenes with four-membered rings," *J. Am. Chem. Soc.* 115 (1993): 8459–60.

Grösser, T., Prato, M., Lucchini, V., Hirsch, A., Wudl, F., "Ring expansion of the fullerene core by highly regioselective formation of diazafulleroids," *Angew. Chem., Int. Ed. Engl.* 34 (1995): 1343–5.

Hachiya, H., Kabe, Y., "Production of a 15-membered ring orifice in open-cage fullerenes by double photooxygenation of azafulleroid," *Chem. Lett.* 38 (2009): 372–3.

Hasharoni, K., Bellavia-Lund, C., Keshavarz-K, M., Srdanov, G., Wudl, F., "Light-induced ESR studies of the heterofullerene dimers," *J. Am. Chem. Soc.* 119 (1997): 11128–9.

Hawker, C. J., Saville, P. M., White, J. W., "The synthesis and characterization of a self-assembling amphiphilic fullerene," *J. Org. Chem.* 59 (1994): 3503–5.

Hirsch, A., "New cages and unusual guests: Fullerene chemistry continues to excite," *Angew. Chem., Int. Ed.* 40 (2001): 1195–7.

Hirsch, A. and Brettreich, M., *Fullerenes: Chemistry and reactions* (Weinheim: Wiley-VCH Verlag GmbH & Co. KGaA, 2005).

Hsiao, T. Y., Santhosh, K. C., Liou, K. F., Cheng, C. H., "Nickel-promoted first ene-diyne cycloaddition reaction on C_{60}: Synthesis and photochemistry of the fullerene derivatives," *J. Am. Chem. Soc.* 120 (1998): 12232–6.

Hu, X. Q., Jiang, Z. P., Jia, Z., "Amination of [60]fullerene by ammonia and by primary and secondary aliphatic amines—Preparation of amino[60]fullerene peroxides," *Chem.-Eur. J.* 13 (2007): 1129–41.

Huang, S. H., Xiao, Z., Wang, F. D. et al., "Selective preparation of oxygen-rich [60]fullerene derivatives by stepwise addition of tert-butylperoxy radical and further functionalization of the fullerene mixed peroxides," *J. Org. Chem.* 69 (2004): 2442–53.

Huang, S. H., Xiao, Z., Wang, F. D. et al., "Preparation of [5,6]- and [6,6]-oxahomofullerene derivatives and their interconversion by Lewis Acid assisted reactions of fullerene mixed peroxides," *Chem.-Eur. J.* 11 (2005): 5449–56.

Huang, S. H., Wang, F. D., Gan, L. B., Yuan, G., Zhou, J., Zhang, S. W., "Preparation of fullerendione through oxidation of vicinal fullerendiol and intramolecular coupling of the dione to form hemiketal/ketal moieties," *Org. Lett.* 8 (2006): 277–9.

Huang, H., Zhang, G., Wang, D. et al., "Synthesis of an azahomoazafullerene $C_{59}N(NH)R$ and gas-phase formation of the diazafullerene $C_{58}N_2$," *Angew. Chem. Int. Ed.* 52 (2013): 5037–40.

Hummelen, J. C., Knight, B., Pavlovich, J., Gonzalez, R., Wudl, F., "Isolation of the heterofullerene $C_{59}N$ as its dimer $(C_{59}N)_2$," *Science* 269 (1995a): 1554–6.

Hummelen, J. C., Knight, B. W., LePeq, F., Wudl, F., Yao, J., Wilkins, C. L., "Preparation and characterization of fulleroid and methanofullerene derivatives," *J. Org. Chem.* 60 (1995b): 532–8.

Hummelen, J. C., Prato, M., Wudl, F., "There is a hole in my bucky," *J. Am. Chem. Soc.* 117 (1995c): 7003–4.

Hummelen, J. C., Keshavarz-K, M., van Dongen, J. L. J., Janssen, R. A. J., Meijer, E. W., Wudl, F., "Resolution and circular dichroism of an asymmetrically cage-opened [60]fullerene derivative," *Chem. Commun.* 2 (1998): 281–2.

Hummelen, J. C., Bellavia-Lund, C., Wudl, F., "Heterofullerenes," in *Topics in current chemistry: Fullerenes and related structures*, ed. A. Hirsch (Berlin Heidelberg: Springer-Verlag 1999), 93–134.

Inoue, H., Yamaguchi, H., Suzuki, T., Akasaka, T., Murata, S., "A novel and practical synthesis of alkoxycarbonyl-substituted bis(fulleroid)," *Synlett* 8 (2000): 1178–80.

Inoue, H., Yamaguchi, H., Iwamatsu, S.-I. et al., "Photooxygenative partial ring cleavage of bis(fulleroid): Synthesis of a novel fullerene derivative with a 12-membered ring," *Tetrahedron Lett.* 42 (2001): 895–7.

Irle, S., Rubin, Y., Morokuma, K., "ONIOM study of ring opening and metal insertion reactions with derivatives of C_{60}: Role of aromaticity in the opening process," *J. Phys. Chem. A* 106 (2002): 680–8.

Isaacs, L., Diederich, F., "Structures and chemistry of methanofullerenes: A versatile route into N-[(methanofullerene)carbonyl]-substituted amino acids," *Helv. Chim. Acta* 76 (1993a): 2454–64.

Isaacs, L., Wehrsig, A., Diederich, F., "Improved purification of C_{60} and formation of σ- and π-homoaromatic methano-bridged fullerenes by reaction with alkyl diazoacetates," *Helv. Chim. Acta* 76 (1993b): 1231–50.

Iwamatsu, S.-I., Murata, S., "Open-cage fullerenes: Synthesis, structure, and molecular encapsulation," *Synlett* 14 (2005): 2117–29.

Iwamatsu, S.-I., Vijayalakshmi, P. S., Hamajima, M. et al., "A novel photorearrangement of a cyclohexadiene derivative of C_{60}," *Org. Lett.* 4 (2002): 1217–20.

Iwamatsu, S.-I., Ono, F., Murata, S., "A novel migrative addition reaction of hydrazines to the diketone derivative of C_{60}," *Chem. Commun.* 11 (2003a): 1268–9.

Iwamatsu, S.-I., Ono, F., Murata, S., "A novel ring expansion of the holey ketolactam derivative of C_{60}," *Chem. Lett.* 32 (2003b): 614–5.

Iwamatsu, S.-I., Uozaki, T., Kobayashi, K., Suyong, R., Nagase, S., Murata, S., "A bowl-shaped fullerene encapsulates a water into the cage," *J. Am. Chem. Soc.* 126 (2004a): 2668–9.

Iwamatsu, S.-I., Kuwayama, T., Kobayashi, K., Nagase, S., Murata, S., "Regioselective carbon-carbon bond cleavage of an open-cage diketone derivative of [60]fullerene by reaction with aromatic hydrazones," *Synthesis* 18 (2004b): 2962–4.

Iwamatsu, S.-I., Murata, S., "H_2O@open-cage fullerene C_{60}: Control of the encapsulation property and the first mass spectroscopic identification," *Tetrahedron Lett.* 45 (2004c): 6391–4.

Izquierdo, M., Ceron, R. M., Olmstead, M. M., Balch, L. A., Echegoyen, L., "[5,6]-Open methanofullerene derivatives of Ih-C_{80}," *Angew. Chem. Int. Ed.* 52 (2013): 11826–30.

Jiang, Z. P., Zhang, Y., Gan, L. B., Wang, Z. M., "Preparation of fullerenol, fullerenone, and aminofullerene derivatives through selective cleavage of fullerene C–H, C–C, C–N, and C–O bonds in fullerene-mixed peroxide derivatives," *Tetrahedron* 64 (2008): 11394–403.

Jimenez-Vasquez, H. A., Cross, R. J., Saunders, M., Poreda, R. J., "Hot-atom incorporation of tritium atoms into fullerenes," *Chem. Phys. Lett.* 299 (1994): 111–4.

Kabe, Y., Hachiya, H., Saito, T. et al., "Diastereoselective syntheses and oxygenation of silyl fulleroids," *J. Organomet. Chem.* 694 (2009): 1988–97.

Kato, H., Kanazawa, Y., Okumura, M., Taninaka, A., Yokawa, T., Shinohara, H., "Lanthanoid endohedral metallofullerenols for MRI contrast agents," *J. Am. Chem. Soc.* 125 (2003): 4391–7.

Keshavarz, K. M., Gonzalez, R., Hicks, R. G. et al., "Synthesis of hydroazafullerene $C_{59}HN$, the parent hydroheterofullerene," *Nature* 383 (1996): 147–50.

Khong, A., Cross, R. J., Saunders, M., "From $^3He@C_{60}$ to $^3H@C_{60}$: Hot-atom incorporation of tritium in C_{60}," *J. Phys. Chem. A* 104 (2000): 3940–3.

Kiely, A. F., Haddon, R. C., Meier, M. S. et al., "The first structurally characterized homofullerene (Fulleroid)," *J. Am. Chem. Soc.* 121 (1999): 7971–2.

Kikuchi, K., Suzuki, S., Nakao, Y. et al., "Isolation and characterization of metallofullerene LaC_{82}," *Chem. Phys. Lett.* 216 (1993): 67–71.

Kikuchi, K., Nakao, Y., Suzuki, S., Achiba, Y., Suzuki, T., Maruyama, Y., "Characterization of the isolated $Y@C_{82}$," *J. Am. Chem. Soc.* 116 (1994): 9367–8.

Kim, B., Lee, J., Seo, J. H., Wudl, F., Park, S. H., Yang, C., "Regioselective 1,2,3-bisazfulleroid: Doubly N-bridged bisimino-PCBMs for polymer solar cells," *J. Mater. Chem.* 22 (2012): 22958–63.

Kim, G., Lee, K. C., Kim, J. et al., "An unprecedented [5,6]-open adduct via a direct benzyne-C60 cycloaddition," *Tetrahedron* 69 (2013): 7354–9.

Kirbach, U., Dunsch, L., "The existence of stable Tm@C_{82} isomers," *Angew. Chem., Int. Ed. Engl.* 35 (1996): 2380–3.

Komatsu, K., Murata, Y., "A new route to an endohedral fullerene by way of σ-framework transformations," *Chem. Lett.* 34 (2005a): 886–91.

Komatsu, K., Murata, M., Murata, Y., "Encapsulation of molecular hydrogen in fullerene C_{60} by organic synthesis," *Science* 307 (2005b): 238–40.

Krätschmer, W., Lamb, L. D., Fostiropoulos, K., Huffman, D. R., "Solid C_{60}: A new form of carbon," *Nature* 347 (1990): 354–8.

Kroto, H. W., Heath, J. R., O'Brien, S. C., Curl, R. F., Smalley, R. E., "C_{60}: Buckminsterfullerene," *Nature* 318 (1985): 162–3.

Kubozono, Y., Hiraoka, K., Takabayashi, Y. et al., "Enrichment of Ce@C_{60} by HPLC technique," *Chem. Lett.* 25 (1996a): 1061–2.

Kubozono, Y., Maeda, H., Takabayashi, Y. et al., "Extractions of Y@C_{60}, Ba@C_{60}, La@C_{60}, Ce@C_{60}, Pr@C_{60}, Nd@C_{60} and Gd@C_{60} with aniline," *J. Am. Chem. Soc.* 118 (1996b): 6998–9.

Kurihara, H., Iiduka, Y., Rubin, Y. et al., "Unexpected formation of a Sc_3C_2@C_{80} bisfulleroid derivative," *J. Am. Chem. Soc.* 134 (2012): 4092–5.

Kurotobi, K., Murata, Y., "A single molecule of water encapsulated in fullerene C_{60}," *Science* 333 (2011): 613–6.

Lamparth, I., Nuber, B., Schick, G., Skiebe, A., Grösser, T., Hirsch, A., "$C_{59}N^+$ and $C_{69}N^+$: Isoelectronic heteroanalogues of C_{60} and C_{70}," *Angew. Chem., Int. Ed. Engl.* 34 (1995): 2257–9.

Li, Z., Bouhadir, K. H., Shevlin, P. B., "Convenient synthesis of 6,5 open and 6,6 closed cycloalkylidenefullerenes," *Tetrahedron Lett.* 37 (1996): 4651–4.

Liu, S., Sun, S., "Recant progress in the studies of endohedral metallofullerenes," *J. Organomet. Chem.* 599 (2000): 74–86.

Liu, S., Gan, L., "Aniline induced domino ring contraction process on the rim of an open-cage fullerene with carbonyl, imino and lactone moieties," *Chin. J. Chem.* 32 (2014): 819–21.

Liu, S., Zhang, C., Xie, X. et al., "Synthesis of a green [60]fullerene derivative through cage-opening reactions," *Chem. Commun.* 48 (2012): 2531–3.

Masamune, S., Seidner, R. T., Zenda, H., Wiesel, M., Nakatsuka, N., Bigam, G., "Low-temperature photolysis of bicyclo[6.2.0]deca-2,4,6,9-tetraene and trans- and cis-9,10-dihydronaphthalenes. Tetracyclo[4.4.0.02,10.05,7]deca-3,8-diene," *J. Am. Chem. Soc.* 90 (1968): 5286–8.

Morinaka, Y., Tanabe, F., Murata, M., Murata, Y., Komatsu, K., "Rational synthesis, enrichment, and ^{13}C NMR spectra of endohedral C_{60} and C_{70} encapsulating a helium atom," *Chem. Commun.* 46 (2010): 4532–4.

Murata, Y., Murata, M., Komatsu, K., "The reaction of fullerene C_{60} with 4,6-dimethyl-1,2,3-triazine: Formation of an open-cage fullerene derivative," *J. Org. Chem.* 66 (2001a): 8187–91.

Murata, Y., Kato, N., Komatsu, K., "The reaction of fullerene C_{60} with phthalazine: The mechanochemical solid-state reaction yielding a new C_{60} dimer versus the liquid-phase reaction affording an open-cage fullerene," *J. Org. Chem.* 66 (2001b): 7235–9.

Murata, Y., Komatsu, K., "Photochemical reaction of the open-cage fullerene derivative with singlet oxygen," *Chem. Lett.* 30 (2001c): 896–7.

Murata, Y., Murata, M., Komatsu, K., "100% Encapsulation of a hydrogen molecule into an open-cage fullerene derivative and gas-phase generation of H_2@C_{60}," *J. Am. Chem. Soc.* 125 (2003a): 7152–3.

Murata, Y., Murata, M., Komatsu, K., "Synthesis, structure, and properties of novel open-cage fullerenes having heteroatom(s) on the rim of the orifice," *Chem.-Eur. J.* 9 (2003b): 1600–9.

Murata, M., Murata, Y., Komatsu, K., "Synthesis and properties of endohedral C_{60} encapsulating molecular hydrogen," *J. Am. Chem. Soc.* 128 (2006): 8024–33.

Murata, M., Maeda, S., Morinaka, Y., Murata, Y., Komatsu, K., "Synthesis and reaction of fullerene C_{70} encapsulating two molecules of H_2," *J. Am. Chem. Soc.* 130 (2008a): 15800–1.

Murata, M., Murata, Y., Komatsu, K., "Surgery of fullerenes," *Chem. Commun.* 46 (2008b): 6083–94.

Murata, Y., Maeda, S., Murata, M., Komatsu, K., "Encapsulation and dynamic behavior of two H_2 molecules in an open-cage C_{70}," *J. Am. Chem. Soc.* 130 (2008c): 6702–3.

Murata, M., Morinaka, Y., Kurotobi, K., Komatsu, K., Murata, Y., "Reaction of cage-opened fullerene derivative with grignard reagents and subsequent transannular cyclization," *Chem. Lett.* 39 (2010): 298–9.

Murphy, T. A., Pawlik, T., Weidinger, A., Höhne, M., Alcala, R., Spaeth, J. M., "Observation of atomlike nitrogen in nitrogen-implanted solid C_{60}," *Phys. Rev. Lett.* 77 (1996): 1075–8.

Nagase, S., Kobayashi, K., Akasaka, T., "Endohedral metallofullerenes: New spherical cage molecules with interesting properties," *Bull. Chem. Soc. Jpn.* 69 (1996): 2131–42.

Nierengarten, J. F., "Ring-opened fullerenes: An unprecedented class of ligands for supramolecular chemistry," *Angew. Chem., Int. Ed.* 40 (2001): 2973–4.

Nuber, B., Hirsch, A., "A new route to nitrogen heterofullerenes and the first synthesis of $(C_{69}N)_2$," *Chem. Commun.* 12 (1996): 1421–2.

Nuber, B., Hirsch, A., "Facile synthesis of arylated heterofullerenes $ArC_{59}N$," *Chem. Commun.* 3 (1998): 405–6.

Nuber, B., Hampel, F., Hirsch, A., "X-ray structure of 1′-(2-methoxyethoxymethyl)triazolinyl-[4′,5′:1,2]-1,2-dihydro[60]fullerene," *Chem. Commun.* 15 (1996): 1799–800.

Orfanopoulos, M., Kambourakis, S., "Fullerene C_{60} and C_{70} photosensitized oxygenation of olefins," *Tetrahedron Lett.* 35 (1994): 1945–8.

Orfanopoulos, M., Kambourakis, S., "Chemical evidence of singlet oxygen production from C_{60} and C_{70} in aqueous and other polar media," *Tetrahedron Lett.* 36 (1995): 435–8.

Pimenova, A. S., Kozlov, A. A., Goryunkov, A. A. et al., "Synthesis and characterization of difluoromethylene-homo[60]fullerene, $C_{60}(CF_2)$," *Chem. Commun.* 4 (2007): 374–6.

Prato, M., Chan Li, Q., Wudl, F., Lucchini, V., "Addition of azides to fullerene C_{60}: Synthesis of azafulleroids," *J. Am. Chem. Soc.* 115 (1993a): 1148–50.

Prato, M., Lucchini, V., Maggini, M. et al., "Energetic preference in 5,6 and 6,6 ring junction adducts of C_{60}: Fulleroids and methanofullerenes," *J. Am. Chem. Soc.* 115 (1993b): 8479–80.

Qian, W., Rubin, Y., "Convergent, regioselective synthesis of tetrakisfulleroids from C_{60}," *J. Org. Chem.* 67 (2002): 7683–7.

Qian, W., Bartberger, M. D., Pastor, S. J., Houk, K. N., Wilkins, C. L., Rubin, Y., "C_{62}, a non-classical fullerene incorporating a four-membered ring," *J. Am. Chem. Soc.* 122 (2000): 8333–4.

Qian, W., Chuang, S.-C., Amador, R. B. et al., "Synthesis of stable derivatives of C_{62}: The first nonclassical fullerene incorporating a four-membered ring," *J. Am. Chem. Soc.* 125 (2003): 2066–7.

Raghavachari, K., "Structure of $C_{60}O$: Unexpected ground state geometry," *Chem. Phys. Lett.* 195 (1992): 221–4.

Raghavachari, K., Sosa, C., "Fullerene derivatives. Comparative theoretical study of $C_{60}O$ and $C_{60}CH_2$," *Chem. Phys. Lett.* 209 (1993): 223–8.

Roubelakis, M. M., Murata, Y., Komatsu, K., Orfanopoulos, M., "Efficient synthesis of open-cage fullerene derivatives having 16-membered-ring orifices," *J. Org. Chem.* 72 (2007a): 7042–5.

Roubelakis, M. M., Vougioukalakis, G. C., Orfanopoulos, M., "Open-cage fullerene derivatives having 11-, 12-, and 13-membered-ring orifices: Chemical transformations of the organic addends on the rim of the orifice," *J. Org. Chem.* 72 (2007b): 6526–33.

Roubelakis, M. M., Vougioukalakis, G. C., Nye, L. C., Drewello, T., Orfanopoulos, M., "Exploring the photoinduced electron transfer reactivity of aza[60]fullerene iminium cation," *Tetrahedron* 66 (2010): 9363–9.

Rubin, Y., "Organic approaches to endohedral metallofullerenes: Cracking open or zipping up carbon shells," *Chem.-Eur. J.* 3 (1997): 1009–16.

Rubin, Y., "Ring opening reactions of fullerenes: Designed approaches to endohedral metal complexes," in *Fullerenes and related structures*, ed. A. Hirsch (Berlin Heidelberg: Springer-Verlag, 1999), 67–91.

Rubin, Y., Jarrosson, T., Wang, G. W. et al., "Insertion of helium and molecular hydrogen through the orifice of an open fullerene," *Angew. Chem., Int. Ed.* 40 (2001): 1543–6.

Samoylova, N. A., Bemolv, N. M., Brotsman, V. A. et al., "[6,6]-Open and [6,6]-closed isomers of $C_{70}(CF_2)$: Synthesis, electrochemical and quantum chemical investigation," *Chem. Eur. J.* 19 (2013): 17969–79.

Sato, K., Kako, M., Suzuki, M. et al., "Synthesis of silylene-bridged endohedral metallofullerene Lu_3N@ Ih-C_{80}," *J. Am. Chem. Soc.* 134 (2012): 16033–9.

Saunders, M., Jimenez-Vasquez, H. A., Cross, R. J., Poreda, R. J., "Stable compounds of helium and neon: He@C60 and Ne@C60," *Science* 259 (1993): 1428–30.

Saunders, M., Jimenez-Vasquez, H. A., Cross, R. J. et al., "Incorporation of helium, neon, argon, krypton, and xenon into fullerenes using high pressure," *J. Am. Chem. Soc.* 116 (1994): 2193–4.

Saunders, M., Cross, R. J., Jimenez-Vasquez, H. A., Shimshi, R., Khong, A., "Noble gas atoms inside fullerenes," *Science* 271 (1996): 1693–7.

Sawa, H., Wakabayashi, Y., Murata, Y., Murata, M., Komatsu, K., "Floating single hydrogen molecule in an open-cage fullerene," *Angew. Chem., Int. Ed.* 44 (2005): 1981–3.

Schick, G., Hirsch, A., Mauser, H., Clark, T., "$C_{59}N^+$ and $C_{69}N^+$: Isoelectronic heteroanalogues of C_{60} and C_{70} opening and closure of the fullerene cage in cis-bisimino adducts of C_{60}: The influence of the addition pattern and the addend," *Chem.-Eur. J.* 2 (1996): 935–43.

Schick, G., Jarrosson, T., Rubin, Y., "Formation of an effective opening within the fullerene core of C_{60} by an unusual reaction sequence," *Angew. Chem., Int. Ed.* 38 (1999): 2360–3.

Scott, L. T., Boorum, M. M., McMahon, B. J., "A rational chemical synthesis of C_{60}," *Science* 295 (2002): 1500–3.

Shen, C. K. F., Chien, K. M., Juo, C. G., Her, G. R., Luh, T. Y., "Chiral bisazafulleroids," *J. Org. Chem.* 61 (1996): 9242–4.

Shi, S., Khemani, K. C., Li, Q. C., Wudl, F., "A polyester and polyurethane of diphenyl C_{61}: Retention of fulleroid properties in a polymer," *J. Am. Chem. Soc.* 114 (1992): 10656–7.

Shimshi, R., Cross, R. J., Saunders, M., "Beam implantation: A new method for preparing cage molecules containing atoms at high incorporation levels," *J. Am. Chem. Soc.* 119 (1997): 1163–4.

Shinohara, H., "Endohedral metallofullerenes," *Rep. Prog. Phys.* 63 (2000): 843–92.

Shinohara, H., Yamaguchi, H., Hayashi, N. et al., "Isolation and spectroscopic properties of scandium fullerenes (Sc_2@C_{74}, Sc_2@C_{82} and Sc_2@C_{84})," *J. Phys. Chem.* 97 (1993): 4259–61.

Shinohara, H., Inakuma, M., Hayashi, N. et al., "Spectroscopic properties of isolated SC_3@C_{82} mettalofullerene," *J. Phys. Chem.* 98 (1994): 8597–9.

Shiu, L. L., Chien, K. M., Liu, T. Y., Lin, T. I., Her, G. R., Luh, T. Y., "Bisazafulleroids," *J. Chem. Soc., Chem. Commun.* 11 (1995): 1159–60.

Skiebe, A., Hirsch, A., "A facile method for the synthesis of amino acid and amido derivatives of C_{60}," *J. Chem. Soc., Chem. Commun.* 3 (1994): 335–6.

Smith III, A. B., Strongin, R. M., Brard, L. et al., "1,2-Methanobuckminsterfullerene ($C_{61}H_2$), the parent fullerene cyclopropane: Synthesis and structure," *J. Am. Chem. Soc.* 115 (1993): 5829–30.

Smith III, A. B., Strongin, R. M., Brard, L. et al., "$C_{71}H_2$ cyclopropanes and annulenes: Synthesis and characterization," *J. Chem. Soc., Chem. Commun.* 18 (1994): 2187–8.

Smith III, A. B., Strongin, R. M., Brard, L. et al., "Synthesis of prototypical fullerene cyclopropanes and annulenes. Isomer differentiation via NMR and UV spectroscopy," *J. Am. Chem. Soc.* 117 (1995): 5492–502.

Stackow, R., Schick, G., Jarrosson, T., Rubin, Y., Foote, C. S., "Photophysics of open C_{60} derivatives," *J. Phys. Chem. B* 104 (2000): 7914–8.

Stevenson, S., Rice, G., Glass, T. et al., "Small-bandgap endohedral metallofullerenes in high yield and purity," *Nature* 401 (1999): 55–7.

Suresh, C. H., Vijayalakshmi, P. S., Iwamatsu, S.-I., Murata, S., Koga, N., "Rearrangement of the cyclohexadiene derivatives of C_{60} to bis(fulleroid) and bis(methano)fullerene: Structure, stability, and mechanism," *J. Org. Chem.* 68 (2003): 3522–31.

Suzuki, T., Li, Q., Khemani, K. C., Wudl, F., Almarsson, Ö., "Systematic inflation of bckminsterfullerene C_{60}: Synthesis of diphenyl fulleroids C_{61} to C_{66}," *Science* 254 (1991): 1186–8.

Suzuki, T., Li, Q. C., Khemani, K. C., Wudl, F., "Dihydrofulleroid H_3C_{61}: Synthesis and properties of the parent fulleroid," *J. Am. Chem. Soc.* 114 (1992a): 7301–2.

Suzuki, T., Li, Q., Khemani, K. C., Wudl, F., Almarsson, Ö., "Synthesis of m-phenylene- and p-pheny lenebis(phenylfulleroids): Two-pearl sections of pearl necklace polymers," *J. Am. Chem. Soc.* 114 (1992b): 7300–1.

Suzuki, T., Maruyama, Y., Kato, T. et al., "Electrochemistry and ab initio study of the dimetallofullerene $La_2@C_{80}$," *Angew. Chem., Int. Ed. Engl.* 34 (1995): 1094–6.

Syamala, M. S., Cross, R. J., Saunders, M., "^{129}Xe NMR spectrum of xenon inside C_{60}," *J. Am. Chem. Soc.* 124 (2002): 6216–9.

Tagmatarchis, N., Okada, K., Tomiyama, T., Shinohara, H., "Synthesis and spectroscopic characterization of the second isomer of $(C_{69}N)_2$ (II) heterofullerene," *Synlett* 12 (2000): 1761–4.

Taliani, C., Ruani, G., Zamboni, R. et al., "Light-induced oxygen incision of C60," *J. Chem. Soc., Chem. Commun.* 3 (1993): 220–2.

Taylor, R., "Rationalisation of the most stable isomer of a fullerene C_n," *J. Chem. Soc., Perkin Trans.* 2 1 (1992): 3–4.

Taylor, R., Wasserman, E., Haddon, R. C., Kroto, H. W., "The pattern of additions to fullerenes [and discussion]," *Phil. Trans., R. Soc. Lond. Ser. A* 343 (1993): 87–101.

Tellgmann, R., Krawez, N., Lin, S. H., Hertel, I. V., Campbell, E. E. B., "Endohedral fullerene production," *Nature* 382 (1996): 407–8.

Toda, F., Tanaka, K., "Isolation of 5-oxa, 5-thia, and 5-tosylaza-2,8-di-t-butyl-4,6,10,11-tetraphenyl tricyclo[7.2.0.0^{3,7}]undeca-1,3,6,8,10-pentaenes, their $HClO_4$ salts, and 4-selena-2,8-di-t-butyl-5,6,10,11-tetramethyl tricycle[7.2.0.0^{3,7}]undeca-1,3(7),5,8,10-pentaene: New antiaromatic systems," *Chem. Lett.* 8 (1979): 1451–4.

Tokuyama, H., Nakamura, E., "Synthetic chemistry with fullerenes. Photooxygenation of olefins," *J. Org. Chem.* 59 (1994): 1135–8.

Vassilikogiannakis, G., Chronakis, N., Orfanopoulos, M., "A new [2 + 2] functionalization of C_{60} with alkyl-substituted 1,3-butadienes: A mechanistic approach. Stereochemistry and isotope effects," *J. Am. Chem. Soc.* 120 (1998): 9911–20.

Vougioukalakis, G. C., Orfanopoulos, M., "Functionalization of azafullerene $C_{59}N$. Radical reactions with 9-substituted fluorenes," *Tetrahedron Lett.* 44 (2003a): 8649–52.

Vougioukalakis, G. C., Chronakis, N., Orfanopoulos, M., "Addition of electron-rich aromatics to azafullerenium carbocation. A stepwise electrophilic substitution mechanism," *Org. Lett.* 3 (2003b): 4603–6.

Vougioukalakis, G. C., Prassides, K., Orfanopoulos, M., "Novel open-cage fullerenes having a 12-membered-ring orifice: Removal of the organic addends from the rim of the orifice," *Org. Lett.* 6 (2004a): 1245–7.

Vougioukalakis, G. C., Prassides, K., Campanera, J. M., Heggie, M. I., Orfanopoulos, M., "Open-cage fullerene derivatives with 15-membered-ring orifices," *J. Org. Chem.* 69 (2004b): 4524–6.

Vougioukalakis, G. C., Orfanopoulos, M., "Photoinduced electron transfer reactivity of aza[60]fullerene: Three discrete functionalization pathways with a single substrate," *J. Am. Chem. Soc.* 126 (2004c): 15956–7.

Vougioukalakis, G. C., Roubelakis, M. M., Orfanopoulos, M., "Open-cage fullerenes: Towards the construction of nanosized molecular containers," *Chem. Soc. Rev.* 2 (2010a): 817–44.

Vougioukalakis, G. C., Roubelakis, M. M., Orfanopoulos, M., "Radical reactivity of aza[60]fullerene: Preparation of monoadducts and limitations," *J. Org. Chem.* 75 (2010b): 4124–30.

Wang, B.-C., Chen, M., Chou, Y., "Binding between the CD4 receptor and polysulfonated azo-dyes. An exploratory theoretical study on action-mechanism," *J. Mol. Struct. (THEOCHEM.)* 422 (1998): 153–9.

Wang, F. D., Xiao, Z., Yao, Z. P. et al., "Lewis acid promoted preparation of isomerically pure fullerenols from fullerene peroxides C_{60}(OOt-Bu)$_6$ and C_{60}(O)(OOt-Bu)$_6$," *J. Org. Chem.* 71 (2006): 4374–82.

Wang, F. D., Xiao, Z., Gan, L. B. et al., "From fullerene-mixed peroxide to open-cage oxafulleroid $C_{59}(O)_3(OH)_2(OO^tBu)_2$ embedded with furan and lactone motifs," *Org. Lett.* 9 (2007): 1741–3.

Wang, G.-W., Liu, T.-X., Jiao, M. et al., "The cycloaddition reaction of Ih-Sc_3N@C_{80} with 2-amino-4,5-diisopropoxybenzoic acid and isoamyl nitrite to produce an open-cage metallofullerene," *Angew. Chem. Int. Ed.* 50 (2011): 4658–62.

Weisman, R. B., Heymann, D., Bachilo, S. M., "Synthesis and characterization of the "missing" oxide of C_{60}: [5,6]-open $C_{60}O$," *J. Am. Chem. Soc.* 123 (2001): 9720–1.

Wood, J. M., Kahr, B., Hoke II, S. H., Dejarme, L., Cooks, R. G., Ben-Amotz, D., "Oxygen and methylene adducts of C_{60} and C_{70}," *J. Am. Chem. Soc.* 113 (1991): 5907–8.

Wudl, F., "The chemical properties of buckminsterfullerene (C_{60}) and the birth and infancy of fulleroids," *Acc. Chem. Res.* 25 (1992): 157–61.

Xiao, Z., Yao, J. Y., Yang, D. Z. et al., "Synthesis of [59]fullerenones through peroxide-mediated stepwise cleavage of fullerene skeleton bonds and X-ray structures of their water-encapsulated open-cage complexes," *J. Am. Chem. Soc.* 129 (2007): 16149–62.

Xiao, Z., Yao, J., Yu, Y., Jia, Z., Gan, L., "Carving two adjacent holes on [60]fullerene through two consecutive epoxide to diol to dione transformations," *Chem. Commun.* 46 (2010): 8365–7.

Xiao, Z., Ye, G., Liu, Y. et al., "Pushing fullerene absorption into the near-IR region by conjugately fusing oligothiophenes," *Angew. Chem. Int. Ed.* 51 (2012): 9038–41.

Xin, N., Yang, X., Zhou, Z., Zhang, J., Zhang, S., Gan, L., "Synthesis of C60(O)3: An open-cage fullerene with a ketolactone moiety on the orifice," *J. Org. Chem.* 78 (2013): 1157–62.

Xu, L., Zhang, Q., Zhang, G., Liang, S., Yu, Y., Gan, L., "Regioselective Diels–Alder reactions directed by carbonyl groups on the rim of open-cage fullerene derivatives," *Eur. J. Org. Chem.* 32 (2013): 7272–26.

Yamamoto, K., Saunders, M., Khong, A. et al., "Isolation and spectral properties of Kr@C_{60}, a stable van der Waals molecule," *J. Am. Chem. Soc.* 121 (1999): 1591–6.

Yang, C., Cho, S., Heeger, A. J., Wudl, F., "Heteroanalogues of PCBM: N-bridged imino-PCBMs for organic field-effect transistors," *Angew. Chem., Int. Ed.* 48 (2009): 1592–5.

Yang, D., Shi, L., Huang, H., Zhang, J., Gan, L., "Synthesis and reactivity of 2H-pyran moiety in [60] fullerene cage skeleton," *J. Org. Chem.* 75 (2010): 4567–73.

Yao, J. Y., Xiao, Z., Gan, L. B., Yang, D. Z., Wang, Z. M., "Towards the rational synthesis of norfullerenes. Controlled deletion of one carbon atom from C_{60} and preparation of 2,5,9-trioxo-1-nor(C_{60}-Ih) [5,6]fullerene $C_{59}(O)_3$ derivatives," *Org. Lett.* 10 (2008): 2003–6.

Yao, J. Y., Yang, D. Z., Xiao, Z., Gan, L. B., Wang, Z. M., "Synthesis of fullerene oxides containing both 6,6-closed epoxide and 5,6-open ether moieties through thermolysis of fullerene peroxides," *J. Org. Chem.* 74 (2009): 3528–31.

Yu, Y., Xie, X., Zhang, T. et al., "Synthesis of 18-membered open-cage fullerenes through controlled stepwise fullerene skeleton bond cleavage processes and substituent-mediated tuning of the redox potential of open-cage fullerenes," *J. Org. Chem.* 76 (2011): 10148–53.

Yu, Y., Shi, L., Yang, D., Gan, L., "Molecular containers with a dynamic orifice: Open-cage fullerenes capable of encapsulating either H_2O or H_2 under mild conditions," *Chem. Sci.* 4 (2013): 814–8.

Yu, Y., Xu, L., Huang, X., Gan, L. J., "Near-infrared absorbing compounds based on π-extended tetrathiafulvalene open-cage fullerenes," *Org. Chem.* 79 (2014a): 2156–62.

Yu, Y., Zhang, T., Gan, L., "Synthesis of open-cage fullerenes with 4-alkynylphenyl groups on the rim of the orifice," *Fuller. Nanotubes Carbon Nanostruct.* 22 (2014b): 54–60.

Zhang, X., Romero, A., Foote, C. S., "Photochemical [2 + 2] cycloaddition of N,N-diethylpropynylamine to C_{60}," *J. Am. Chem. Soc.* 115 (1993): 11024–5.

Zhang, G. H., Huang, S. H., Xiao, Z., Chen, Q., Gan, L. B., Wang, Z. M., "Preparation of azafullerene derivatives from fullerene-mixed peroxides and single crystal x-ray structures of azafulleroid and azafullerene," *J. Am. Chem. Soc.* 130 (2008): 12614–5.

Zhang, Q. Y., Jia, Z. S., Liu, S. M. et al., "Efficient cage-opening cascade process for the preparation of water-encapsulated [60]fullerene derivatives," *Org. Lett.* 11 (2009): 2772–4.

Zhang, Q., Pankewitz, T., Liu, S., Klopper, W., Gan, L., "Switchable open-cage fullerene for water encapsulation," *Angew. Chem. Int. Ed.* 49 (2010a): 9935–8.

Zhang, J., Yang, D., Xiao, Z., Gan, L., "The chemistry of open-cage fullerene, hydroxylamine mediated hole-closing and -opening reactions," *Chin. J. Chem.* 28 (2010b): 1673–7.

Zhang, G., Zhang, Q., Jia, Z., Liang, S., Gan, L., Li, Y., "Preparation of a 12-membered open-cage fullerendione through silane/borane-promoted formation of ketal moieties and oxidation of a vicinal fullerendiol," *J. Org. Chem.* 76 (2011a): 6743–8.

Zhang, J., Wang, F., Xin, N., Yang, D., Gan, L., "Preparation of ketolactone and bislactone [60]fullerene derivatives and their conversion into open-cage fullerenes with a 12- or 15-membered orifice," *Eur. J. Org. Chem.* 27 (2011b): 5366–73.

Zhang, W., Suzuki, M., Xie, Y. et al., "Molecular structure and chemical property of a divalent metallofullerene $Yb@C_2(13)$-C_{84}," *J. Am. Chem. Soc.* 135 (2013): 12730–5.

Zhang, R., Futagoishi, T., Murata, M., Wakamiya, A., Murata, Y., "Synthesis and structure of an open-cage thiafullerene $C_{69}S$: Reactivity differences of an open-cage C_{70} tetraketone relative to its C_{60} analogue," *J. Am. Chem. Soc.* 136 (2014): 8193–6.

10

Endohedral Fullerenes: Optical Properties and Biomedical Applications

Panagiotis Dallas

Ilija Rašović

Gregory Rogers

Kyriakos Porfyrakis

10.1 Introduction

Endohedral metallofullerenes (EMFs) are fascinating nanomaterials composed of a metal or a metal cluster encapsulated inside a carbon cage. Their unusual structures are stabilized through charge transfer between the metal and the cage, and exciting electronic, electrochemical, magnetic, and optical properties arise from this combination. These materials have been proposed for applications in photovoltaics, owing to their suitable bandgap with polymeric electron donors; in medicine, owing to their paramagnetic and photosensitizing properties; and in quantum information processing, owing to their exceptionally long spin lifetimes. In this chapter, we present insights on three topical areas: the optical properties of endofullerenes, their surface functionalization, and their potential for biomedical applications. We also critically discuss how they stand against other carbon and inorganic nanomaterials.

10.1.1 Photoluminescence of Fullerenes and EMFs

Fullerenes are widely used as excellent electron acceptors. Besides this feature, the optical and electrochemical properties of fullerenes have been of significant importance for a range of applications in biomedicine, green energy (Page et al. 2014), and sensors (Sherigara et al. 2003). Their extensive conjugated system provides them with absorbance in the visible range and emission in the far-red/near-infrared

(NIR) region (Gareis et al. 1998). Their tunable fluorescent properties from the visible to the NIR region make them suitable candidates for a wealth of biomedical applications, such as cell labeling and photoacoustic imaging (Krishna et al. 2010; Chen et al. 2012). The most important class of molecules and materials for these applications are highly fluorescent NIR dyes (Bandi et al. 2014). However, the organic dyes suffer from low biocompatibility and high toxicity, in contrast with fullerenes. Pristine fullerenes are not sufficiently fluorescent to be used in confocal microscopy to examine their intracellular uptake; however, upon suitable functionalization, this picture can change (Kwagza et al. 2013).

This weak photoluminescence (PL) with particularly low quantum yields (because of low-lying electronic transitions being only weakly allowed) is a major drawback for cell labeling and imaging applications of fullerenes. In nonpolar solvents like toluene and CS_2, the empty cage fullerenes like C_{60} and C_{70} exhibit emissions in the 570 nm and 660–700 nm regions, with their decay lifetimes depending on the solvent and the presence, or lack, of groups covalently bonded to the surface. To that end, the functionalization of fullerenes has been proven to be an effective way of increasing the quantum yields of emission by lowering the symmetry of the species. This emission is governed by p–p transitions and an intersystem crossing (ISC) to a triplet state with an energy transfer to oxygen accompanying this transition.

Figure 10.1 shows the transition representing the ISC for C_{60} and the energy transfer to oxygen leading to the formation of the highly reactive oxygen singlet state. Because of this transition, fullerenes are the most efficient photosensitizers known for the generation of the excited singlet oxygen state, $^1O_2$ (Yamakoshi et al. 1998)—this makes them excellent candidates for application in photodynamic therapy. The lifetime of the singlet oxygen sensitized by fullerenes has been found to be dependent on the aggregation state of the fullerenes in suspension and the localization of the oxygen inside these aggregates (Wang et al. 2013).

Another advantage for their use as photosensitizers in photodynamic therapy is their remarkably low toxicity and higher photostability compared with small organic molecules. An ideal material for biomedical applications should be able to be delivered to the specific tissue required. Indeed, fullerenes have demonstrated this ability with a tissue-vectored bisphosphonate fullerene designed to target bone tissue (Gonzalez et al. 2002).

EMFs may present strong emission at the NIR range because of the presence of transition or rare earth metals inside the cage. Erbium-encapsulating fullerenes are the most important class of these materials, and their optical properties have been extensively studied (Ito et al. 2007). These erbium endohedral fullerenes have attracted significant attention because of their strong PL, arising from the Er^{3+} ions, which is centered at 1520 nm. These rare earth erbium ions display this characteristic NIR PL

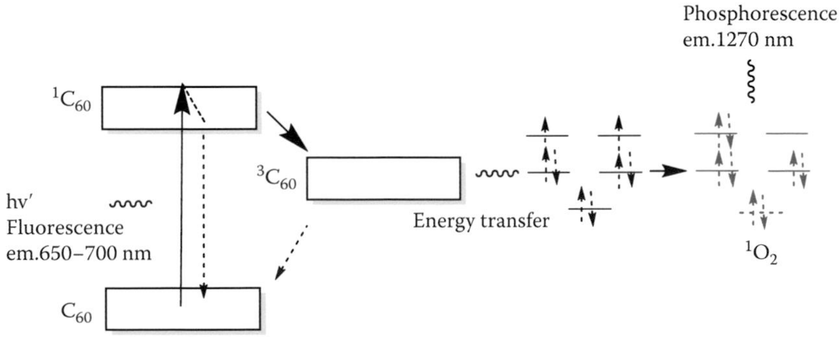

FIGURE 10.1 Schematic representation of the ISC of the fullerenes and the subsequent energy transfer to triplet oxygen ground state for the formation of the excited state of the highly reactive singlet oxygen state. Fullerenes are among the strongest photosensitizers for the formation of singlet oxygen, resulting in efficient photodynamic therapy.

owing to their f–f transitions. However, from the wealth of different EMFs, only erbium-encapsulating fullerenes so far have exhibited sufficiently strong NIR PL.

The PL observed from $Er_2@C_{82}$ and $Er_3N@C_{80}$ in various solvents (e.g., CS_2 and decalin) and even as thin films at low temperature has been assigned to the $^4I_{13/2} \rightarrow {}^4I_{15/2}$ transition of the Er^{3+} ion (Morton et al. 2008). This transition is shown in Figure 10.2 for $Er_2@C_{82}$ and the corresponding carbide $Er_2C_2@C_{82}$. The Shinohara group proposed an initial excitation from the singlet S_0 ground state (highest occupied molecular orbital) to a singlet S_n excited state followed by a fast relaxation to the singlet S_1 (lowest unoccupied molecular orbital) state and then an ISC to the triplet T state. From this triplet state, an energy transfer to the $^4I_{13/2}$ first excited state of the erbium ions Er^{3+} takes place. The $^4I_{13/2}$ state decays to the $^4I_{15/2}$ with an emission at 1520 nm.

Tiwari et al. (2007) observed a temperature-dependent PL in trimetallic nitride template (TNT) metallofullerenes, erbium-scandium $Er_2ScN@C_{80}$, and $ErSc_2N@C_{80}$. In these TNTs, the lifetime of Er^{3+} emission at 1520 nm is reported to be 1500 ns, an exceptionally long lifetime compared with small organic molecules, albeit with a low quantum efficiency of 10^{-4}. Furthermore, Jones et al. (2006) demonstrated that the PL of the TNT fullerene $Er_3N@C_{80}$ can be magnetically controlled. The authors performed detailed studies at high magnetic fields, up to 19.5 T, which caused the PL spectrum to split, corresponding to transitions from the lowest field-split doublet of the $^4I_{13/2}$ to the four lowest field-split levels of the $^4I_{15/2}$ state.

The interactions of light with Er^{3+} are assigned as non-cage-mediated optical interactions. These optical and spin properties make the erbium-based metallofullerenes excellent candidates not only for biomedical applications but also as white light emitters (Hutchison et al. 1999) and quantum information processing devices (Filidou et al. 2012) since it is possible to map the energy level structure of the encapsulated ion and to control its quantum state coherently. In these metallofullerenes, a Zeeman splitting of the observed spectrum was observed, which was linear with field, with the principal peak exhibiting a splitting in transition energy of 1.8 cm^{-1} T^{-1}. Under a magnetic field, the peaks became better defined.

Since the symmetry of the electrostatic crystal field of the Er^{3+} ion is C_{2v}, the splitting of the energy levels in a magnetic field depends on its direction with respect to the symmetry axes. Lower temperature below the melting point of the solvent preserves this orientation imbalance and leads to a modified and sharpened field-split spectrum. Concerning other EMFs, their quantum yields and emission intensities are very low. However, their PL is enhanced by two orders of magnitude upon coupling with noble metal nanoparticles—this provides promise for a substantial increase in their quantum yields upon the design of suitable dyads (Bharadwaj & Novotny 2010). As a comparison, coupling of the same gold

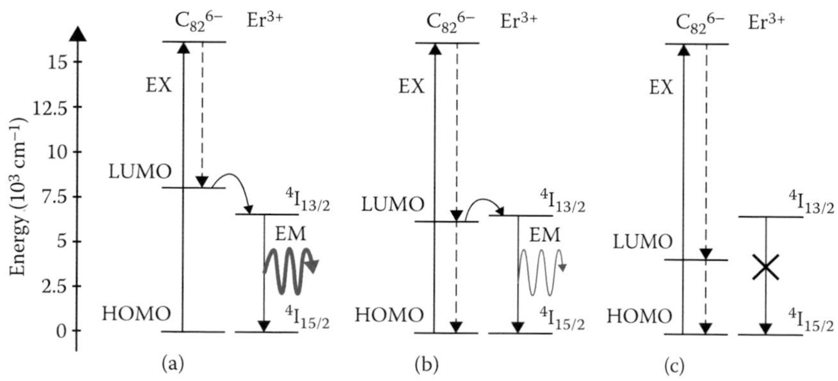

FIGURE 10.2 The proposed electronic transitions from the energy levels of the erbium carbide $Er_2C_2@C_{80}$ metallofullerene that results in their strong PL at 1520 nm. (Reprinted with permission from Ito, Y., Okazaki, T., Okubo, S. et al. *ACS Nano* 1(2007), 456. Copyright American Chemical Society.)

nanoparticles with a single Nile blue molecule resulted in an 8–10-fold increase in PL intensity while the increase was 100-fold for $Y_3N@C_{80}$.

10.1.2 Donor–Acceptor Dyads

Fullerenes and their derivatives are known to be exceptional electron acceptors that form unique donor–acceptor complexes (Imahori et al. 2004) and dyads (Imahori et al. 2001). To that end, incorporation of donor moieties on the surface of fullerenes can lead to the formation of unique donor–acceptor charge-separated states. As an example of an advanced bioapplication of these dyads, the charge-separated states of fullerene donor–acceptor dyads have been utilized by the Imahori group in cell membrane potentials and ion transport control (Numata et al. 2012).

Among the various available side groups, porphyrin macrocycles are popular candidates for light-emitting dyads with fullerenes because of the versatility of their structures, high quantum yields, and easy functionalization (Oswald et al. 2014). Previous work on porphyrin–C_{60} dyads has shown that there is a blue shift in the emission if electron transfer is from the macrocycle to the fullerene; red if from the fullerene to the macrocycle. This shift was found to depend upon the orientation of the porphyrin ligands with respect to the fullerene cage (Accorsi and Armaroli 2010).

The Imahori group has synthesized a ferrocene–porphyrin–fullerene triad (Fc-P-C_{60}), a charge-separated molecule with an exceptionally long lifetime of 0.01 ms. In this triad, photoexcitation results in a photoinduced electron transfer from the porphyrin excited state to the C_{60}, followed by a second electron transfer from the ferrocene to the porphyrin radical cation. This photoinduced charge separated state has a remarkable electric field of 10^6 V cm^{-1}, which could affect the electrophysiological activity of cell membranes. Indeed, they proved that this triad achieves photoinhibition of ion transport and subsequent depolarization of cells.

Further advances on the system could be the use of EMFs that possess a different bandgap and even stronger electron acceptor affinity compared with empty-cage C_{60}. Longer-lived charge-separated states and higher electric fields could possibly be achieved (Figure 10.3).

Other systems that have been extensively studied as models for optical properties are the methanofullerenes. Comparison between mono- and bis-methanofullerenes suggests that the excited singlet states of mono-methanofullerenes are more prone to undergo electron transfer than those of bis-methanofullerenes, because of a smaller internal reorganization energy. It has also been demonstrated that the monoadducts are usually slightly more easily reduced than bisadducts (Holler et al. 2006). This has been attributed to a more thermodynamically favorable process in the case of monofunctionalized

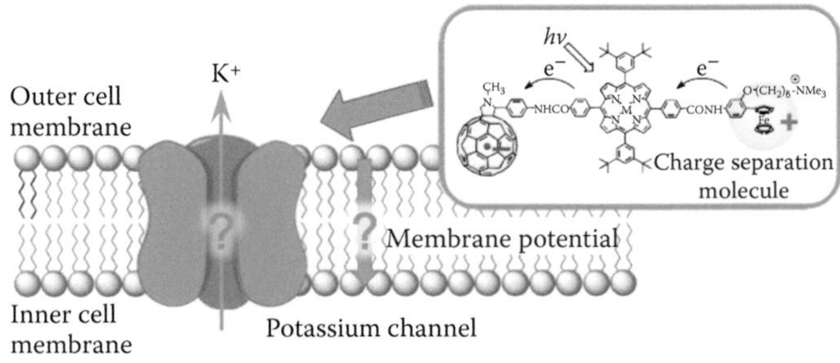

FIGURE 10.3 Fullerene derivatives have been utilized for the control of membrane potential and ion transport as charge-separation molecules upon light irradiation. The charge-separation molecule was delivered to the plasma membrane of PC12 cells by a nanocarrier, and light irradiation led to depolarization of the membrane potential as well as inhibition of the potassium ion flow across the membrane. (Reprinted with permission from Numata, T., Murakami, T., Kawashima, F., *Journal of the American Chemical Society* 134(2012), 6092. Copyright American Chemical Society.)

derivatives (Armaroli et al. 1998). Larger fullerene cages, such as C_{84}, and endohedral fullerenes can further expand these concepts owing to their more complex structure, the presence of a large number of isomers, and spin–spin interactions between incarcerated species (Plant et al. 2013).

10.1.3 Fullerenes vs. Inorganic and Carbon Nanomaterials for Biomedical Applications

Compared with other nanomaterials like carbon nanotubes (CNTs) (Liu et al. 2013), carbon dots (Baker & Baker 2010), and inorganic nanoparticles such as CdSe/CdS quantum dots (Fahmi et al. 2010), fullerene derivatives present a number of advantages and disadvantages with respect to their optical properties and their ability to be decorated with functional groups suitable for energy or biomedical applications. In Figure 10.4, we classify the most important families of nanomaterials and their PL properties.

CdSe and CdS quantum dots have been proposed as the most efficient light-emitting nanomaterials because of their size- and shape-dependent bandgap and high quantum yields. The bandgap ranges from 4.5 to 2.5 eV in CdS and 2.4 to 1.7 eV in CdSe, and their PL can also be tuned upon functionalization with aromatic amines and polymers, as demonstrated by Šišková et al. (2011). However, serious concerns have been raised regarding their biocompatibility because of the presence of highly toxic cadmium.

In the field of biomedicine, cell labeling applications are in need of sufficiently small and fluorescent carriers that can penetrate the cell and enable visualization of where they have been attached through light irradiation. Carbon dots have emerged as the most promising carbon material because of their biocompatibility and very bright PL, with quantum yields in the range of 10%–40%. Their main disadvantage is that their PL largely remains in the blue and the mechanism is still widely debated (Krysmann et al. 2012).

CNTs, on the other hand, exhibit emission in the NIR range (Choi et al. 2007), which has been found to be enhanced in the presence of sufficiently high magnetic fields of 30 T (Alexander-Webber et al. 2014). Potentially, Gd@C_{82}-CNT peapods can find applications in theranostics, where the nanotubes act as NIR emitters and the Gd metallofullerenes as magnetic resonance imaging (MRI) contrast agents (Ohashi et al. 2011). CNTs have already been utilized as carriers in drug delivery experiments; however, their large size and length limit their ability to penetrate biological barriers compared with the ultrasmall fullerenes and carbon dots.

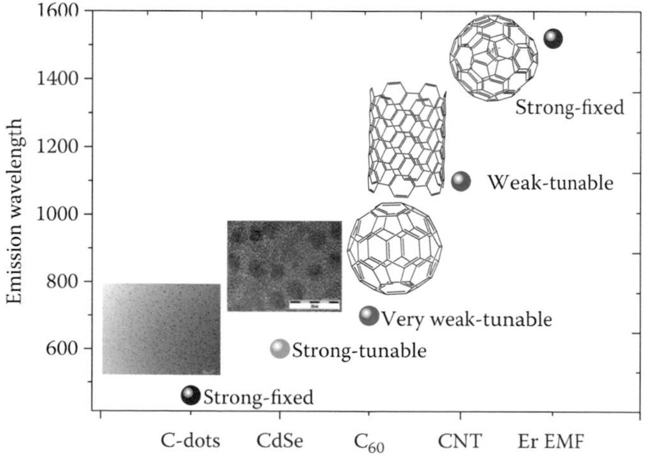

FIGURE 10.4 The PL emission maximum for a series of inorganic and organic nanomaterials and fullerenes. From left to right: Carbon dots, CdSe quantum dots, C_{60}, CNTs, and erbium endohedral metalofullerenes. (From Šišková, K., Kubala, M., Dallas, P. et al. *Journal of Materials Chemistry* 21(2011), 1086–1093.)

10.2 Exohedral Functionalization of Endohedral Fullerenes

A range of chemical reactions have been performed on endohedral fullerenes. These span almost the full range of reactions, which are possible with empty fullerene cages. However, not all reactions are possible with all endohedral fullerenes—for example, the fullerene La@C_{82} does not undergo the desilylation reaction under any conditions (Akasaka et al. 1995a). Both the reactivity of the fullerene cage and the regioisomeric product of the reaction also depend on the type of endohedral fullerene. The exact nature of the relationship between the encapsulated species and the reactivity of the molecule is not yet fully understood. However, what trends exist will be highlighted throughout the text that follows.

10.2.1 Photochemical Reactions

Akasaka et al. (1995a) reported the first functionalization of an EMF in 1995. The EMF La@C_{82} was reacted with 1,1,2,2-tetrakis(2,4,6-trimethylphenyl)-1,2-disilirane (**1**) in a de-aerated toluene solution at room temperature under illumination from a tungsten-halogen lamp to yield a 1:1 mixture of monoadduct isomers, which was characterized by fast atom bombardment mass spectrometry (MS) and electron paramagnetic resonance (EPR) spectroscopy (Akasaka et al. 1995a) (Figure 10.5). The reaction was also found to proceed in the absence of the lamp when heated to 80°C; this was unexpected as frontier orbital considerations showed that the reaction was forbidden for C_{82} as well as C_{70} and C_{60} (Akasaka et al. 1995b).

La@C_{82} functionalization with 1,1,2,2-tetrakis(2,6-diethylphenyl)-1,2-digermirane was found to proceed both under photochemical excitation and on heating to yield the monoadduct, which was characterized by MS and EPR spectroscopy (Akasaka et al. 1996). This reactivity is explained by the greater electron affinity and lower ionization potential of the EMF La@C_{82} relative to empty cage fullerenes (see Table 10.1). It is therefore possible to control the exohedral reactivity of the fullerene cage by changing the endohedral species.

The regioselective addition of 2-adamantane-2,3-[3H]-diazirine (**2**) to La@C_{82}—using a high-pressure mercury lamp to irradiate the sample with an excess of **2** in a solution of 1,2,5-trichlorobenzene (TCB) and toluene has been reported (Maeda et al. 2004) (Figure 10.6). This reaction was carried out in a sealed tube at room temperature, yielding a single monoadduct, as verified by matrix-assisted laser desorption/ionization time-of-flight MS, high-performance liquid chromatography (HPLC), EPR spectroscopy, and x-ray crystallography. The x-ray crystallography was able to show the La molecule sitting very close to the addition site. This again shows how the endohedral species can dictate the reactivity of the cage. The actual site of addition is affected by the endohedral species as well as the timescale of the reaction (Matsunaga et al. 2006).

This reaction has been used by Nagase and coworkers to study a range of EMF crystal structures (Iiduka et al. 2007; Akasaka et al. 2008), the most unusual of which was the reaction of **2** with the

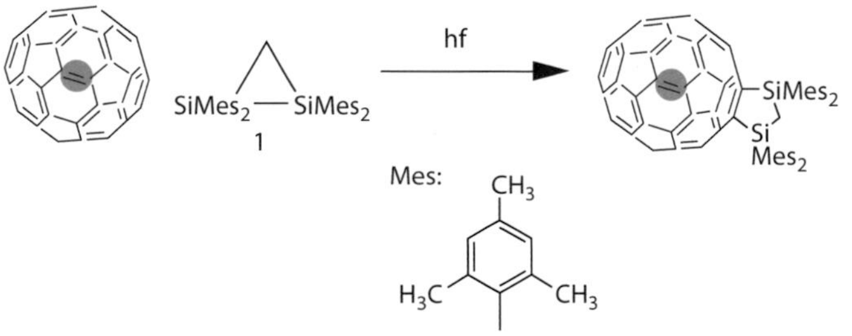

hf
SiMes2 SiMes2
1
Mes:
CH3
H3C CH3
SiMes2
Si
Mes2

FIGURE 10.5 Photochemical reaction of an endohedral fullerene with compound 1. Please note in all cases, the example of N@C_{60} is used for simplicity.

TABLE 10.1 Redox Potentials (V vs. Fc^+/Fc) and Reactivity with Compound 1 for Various Fullerenes and Endohedral Fullerenes

Compound	$^{Ox}E_1$/ev	$^{Red}E_1$/ev	Thermal Reactivity with **1**	Photochemical Reactivity with **1**
C_{60}	+1.21	−1.12	N	Y
C_{70}	+1.19	−1.09	N	Y
$N@C_{60}$	+1.21	−1.12	N	Y
$Y@C_{2v}$-C_{82}	+0.07	−0.42	Y	Y
$La@C_{2v}$-C_{82}	+0.08	−0.41	Y	Y
$Ce@C_{2v}$-C_{82}	+0.07	−0.39	Y	Y
$Gd@C_{2v}$-C_{82}	+0.09	−0.39	Y	Y
$La@C_s$-C_{82}	−0.07	−0.47	N	Y
$La_2@I_h$-C_{80}	+0.22	−0.36	N	N
$Sc_3N@I_h$-C_{80}	+0.62	−1.22	N	Y
$Sc_2C_2@C_{3v}$-C_{80}	+0.16	−0.95	N	Y

Source: Maeda, Y.J. et al., *J. Am. Chem. Soc.*, 127, 12190–12191, 2005.

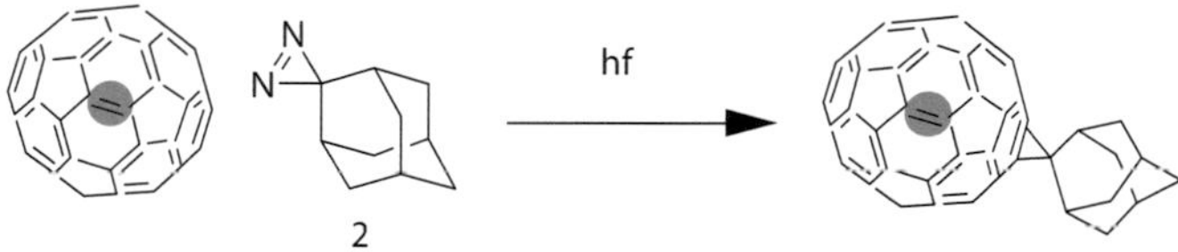

FIGURE 10.6 Photochemical reaction of an endohedral fullerene with compound 2.

isolated pentagon rule-breaking (non-IPR) molecule $La_2@C_{72}$ (D_2:10611), which contains two double pentagons located at each pole. For a shorter reaction time, six isomers of the monoadduct were generated (Lu et al. 2008b); longer reaction times yielded 15 isomers of the bisadduct (Lu et al. 2008b). In the bisadducts, both additions occurred at the poles. However, the reactions occurred at the [5,6] sites adjacent to the [5,5] sites, even though these [5,5] sites have the greatest strain. This is explained by the La atoms stabilizing the normally highly reactive [5,5] sites.

In 2002, Tagmatarchis et al. reported the perfluoroalkylation of $La@C_{82}$ by placing a degassed and cooled solution of $La@C_{82}$ with an excess of perfluoroctyl iodine in toluene under a UV lamp for 15 h. The product of this reaction was seven different bisadducts of $La@C_{82}$-$(C_8F_{17})_2$, as confirmed by HPLC and MS (Tagmatarchis et al. 2002). Using a similar method, Shu et al. were able to benzylate $Sc_3N@C_{80}$ by irradiating a deoxygenated solution of $Sc_3N@C_{80}$ with an excess of benzyl bromide at 355 nm for 1 h. Based on both x-ray crystal structure evidence and density functional theory calculations, it was concluded that the product was $Sc_3N@C_{80}$-$(CH_2C_6H_5)_2$ with a 1,4 structure and a [5,6,6] junction (Shu et al. 2008b).

10.2.2 Diels-Alder Reactions

In 2002, Iezzi et al. were able to report the first organically functionalized TNT EMF. $Sc_3N@C_{80}$ was refluxed in TCB for 24 h with an excess of 6,7-dimethoxyisochroman-3-one (**3**) (99%, ^{13}C labelled) to produce a single monoadduct. Carbon nuclear magnetic resonance spectroscopy (^{13}C NMR) showed that the methylene carbons were equivalent, while ^1H NMR showed that the protons were not, indicating a symmetry plane that passes through the pyrrolidine ring; thus, it was deduced that the reaction occurred on a [5,6] ring of the I_h isomer (Iezzi et al. 2002). X-ray crystallography was then used to show

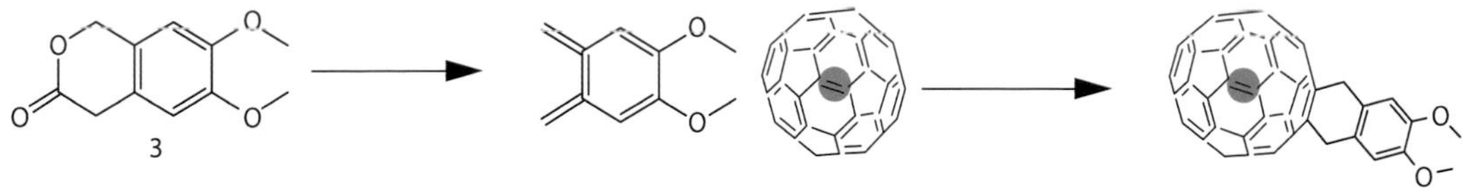

FIGURE 10.7 Diels-Alder reaction between compound 3 and an endohedral fullerene.

FIGURE 10.8 Diels-Alder reaction of 4 with an endohedral fullerene.

that the Sc_3N cluster was shifted away from the addition site, in contrast to the reaction between La@C_{82} and **2** (Lee et al. 2002) (Figure 10.7).

A reversible and regioselective reaction was reported by Maeda et al. in 2005 for the reaction between La@C_{82} and cyclopentadiene (**4**). An excess of **4** was added to a degassed solution of La@C_{82} in toluene and the reaction was monitored by EPR. The single monoadduct was isolated by HPLC in 44% yield—the octet signal of the EPR spectrum verified that there was a single product. The retro-Diels-Alder reaction occurred at 298 K in toluene, showing that it has a lower activation energy than the retro reaction of C_{60} with **4** (Maeda et al. 2005) (Figure 10.8).

As seen in other reactions, purely metallic EMFs are more reactive in Diels-Alder reactions than TNT EMFs. This fact allowed Ge et al. (2005) to employ the Diels-Alder reaction in the purification of TNT EMFs from mixed fullerene soot.

10.2.3 1,3-Dipolar Cycloaddition (Prato Reaction)

The first 1,3-dipolar cycloaddition reaction (well known in organic chemistry and first applied to fullerenes by the group of Prof. M. Prato) with an endohedral fullerene was performed by Cao et al. in 2004 with La@C_{82}. A toluene solution of the EMF, *N*-methyl-glycine, and paraformaldehyde was heated in a sealed tube at 100°C for 30 min, producing three bisadducts and two monoadducts. HPLC was used to isolate isotopically pure forms of one of the monoadducts and one of the bisadducts, before EPR was used to show that the hyperfine interactions had not changed during the reaction (Figure 10.9). EPR measurements showed that the monoadduct was electronically similar to the unfunctionalized EMF, while the bisadduct showed clear deviations in electronic structure (Cao et al. 2006).

The Prato reaction has been used extensively for the functionalization of the endohedral fullerene N@C_{60}, as, until recently, it was one of the few reactions that did not lead to the significant loss of nitrogen atoms from within the cage (see Bingel-Hirsch reaction). The first such reaction was performed by Franco et al. in

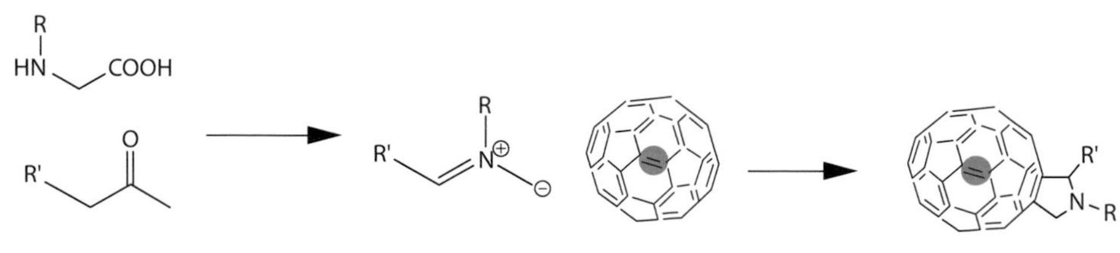

FIGURE 10.9 A 1,3-dipolar cycloaddition reaction performed with an amino acid, aldehyde, and an endohedral fullerene.

2006, attaching a range of species to the outside of the cage. EPR studies were also used in this case to determine the change in zero-field splitting of the nitrogen atom, showing that, in general, splitting increases with increasing size of ring bound to the surface of the fullerene. This is explained by the ability of the addend to withdraw electron density from the endohedral fullerene and so disturb the symmetry (Franco et al. 2006).

10.2.4 Bingel-Hirsch Reactions

The first Bingel-Hirsch reaction on an endohedral fullerene was performed by Dietel et al. in 1999. Solutions of the fullerene and diethyl bromomalonate (**5**) were reacted at room temperature. When using sodium hydride and allowing the reaction to proceed for 2 h, a monoadduct was produced, whereas a hexadduct was produced when using 1,8-diazobicyclo[5.4.0]undec-7-ene (DBU) and allowing the reaction to proceed for 45 h (Dietel et al. 1999) (Figure 10.10).

The first water-soluble Gd@C_{60} was produced by Bolskar et al., who used the Bingel-Hirsch reaction to add malonate groups, which were later converted to the solubilizing carboxylic acid group. Gd@C_{60} was reacted with an excess of **5** and an alkali metal hydride in THF, and a maximum of 10 malonate groups were added, as confirmed by MS. These groups were then hydrolyzed to create the water-soluble molecule Gd@$C_{60}(C(CO_2H)_2)_{10}$ (Bolskar et al. 2003).

Cardona et al. were the first to report a mono-methanofullerene from the I_h isomer of Y_3N@C_{80} by reaction with **5** and DBU. This reaction occurred only at a [6,6] double bond to yield a single product, as characterized by x-ray crystallography. Further study of this product showed that the [6,6] position has an extremely high stability since no retro-reaction took place, unlike with C_{60} (Cardona et al. 2005). In 2008 the same group studied this reaction with several Gd_3N EMFs, using the cages C_{80}, C_{84}, and C_{88}. With C_{80}, a monoadduct and bisadduct could be formed in a very fast reaction at room temperature; for C_{84}, it took 20 min for a single monoadduct to form at room temperature; and for C_{88}, no reaction occurred even on heating to 60°C. This trend in the reactivity was explained by considering the degree of pyramidalization of the carbon–carbon bonds, which is reduced when moving to a larger cage (Cardona et al. 2006). The strain is also reduced, which also explains the decreasing reactivity.

Previously, the Bingel-Hirsch reaction has not been suitable for N@C_{60} because of the high degree of spin loss, indicating that the nitrogen atom had escaped the fullerene cage. Zhou et al. have recently reported successful Bingel-Hirsch reactions on N@C_{60}, which are retained in excess of 90% of the spin signal, as measured by EPR spectroscopy. This result was achieved through use of a stoichiometric ratio of diluted DBU. A mechanism is also proposed for the escape of the nitrogen atom from the fullerene cage (Zhou et al. 2015).

10.2.5 Free Radical Reactions

The first reaction of this kind was performed in 1995 by Suzuki et al., who were able to show the addition of up to three groups on La@C_{82} in a reaction with diphenyldiazomethane in toluene at 60°C for 3.7 h (Suzuki et al. 1995). Many further reactions have since been performed, with no uniform trends in the nature of the reactions (Kareev et al. 2005; Shustova et al. 2007).

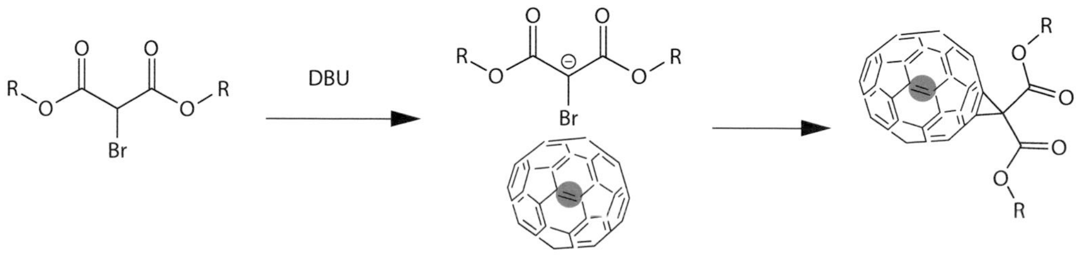

FIGURE 10.10 The Bingel-Hirsch reaction between a suitable reactant and an endohedral fullerene.

10.2.6 Hydroxylation

The first water-solubilization reaction of an EMF was performed by Zhang et al. in 1997—a mixture of gadolinium-containing EMFs were refluxed in toluene with potassium metal under nitrogen for 2 h. An average of 20 hydroxyl groups were added to each cage (Zhang et al. 1997). Wilson et al. (1999) were later able to demonstrate the hydroxylation of $^{165}Ho@C_{82}$ using a phase-transfer reaction to produce molecules with between 24 and 26 hydroxyl groups. Both methods have since been used to expand the range of water-soluble endohedral fullerenes (Kato et al. 2000; Shu et al. 2006a).

10.3 Medical Applications of EMFs

The size, stability, and electrochemical and photophysical properties of fullerenes make them attractive prospects for future deployment in medical contexts. The numerous methods for exohedral functionalization of the fullerene cage arising from its unique electronic structure provide myriad pathways to render it useful in the human body—"the hallmark of rational drug design" (Jain 2009). Water-soluble derivatives are the primary goal of exohedral functionalization, given the high water content of blood and inherent hydrophobicity of pristine fullerenes. The imposition of hydrophilicity also leads to a longer retention time in the bloodstream because of decreased recognition as a foreign body by the immune system (Rieger et al. 2007). Active targeting of certain sites in the body can be achieved by functionalizing the fullerene cage with appropriate ligands, e.g., Fan et al. (2013).

There are several excellent reviews and secondary sources in the literature on the chemical functionalization of EMFs (Bakry et al. 2007; Cataldo & Da Ros 2008; Partha & Conyers 2009; Carini et al. 2012), which show that the primary focus of endohedral fullerene research in the medical realm has been on those containing gadolinium for use as MRI contrast agents.

10.3.1 Therapeutics vs. Diagnostics

The properties of empty-cage fullerenes have been exploited for a number of potential therapeutic applications (Bakry et al. 2007; Cataldo & Da Ros 2008; Partha & Conyers 2009; Ananta & Wilson 2010; Carini et al. 2012; Lu et al. 2012; Popov et al. 2013)—in diagnostics, C_{60} has been proposed for incorporation into biosensors (Afreen et al. 2014). The incarceration of atomic species to give endohedral fullerenes transforms these molecules into powerful standalone diagnostic tools. Excitingly, they are also emerging as strong candidates for multimodal imaging as well as in the nascent field of theranostics (Kelkar & Reineke 2011; Lim et al. 2015).

10.3.2 Gadolinium-Containing EMFs

The paramagnetic gadolinium(III) complexes are used in MRI contrast agents because the large magnetic moment and seven unpaired, isotropically distributed 4*f* electrons (the highest number for any element) of gadolinium dramatically reduce the T_1 relaxation times of any surrounding water protons, thus giving increased contrast in images (Hagberg & Scheffler 2013). Gadolinium-containing EMFs offer several advantages over the most common commercial Gd(III)-chelates (Caravan et al. 1999). The major drawback in the use of these complexes is that gadolinium is toxic and is particularly dangerous to patients with kidney problems (Kuo et al. 2007; Penfield & Reilly 2007)—encapsulating the gadolinium ion in a fullerene negates this toxicity issue as it cannot escape. Despite the fullerene cage stopping water molecules directly coordinating with the gadolinium ion, the relaxivity effects can be over an order of magnitude greater than clinically available contrast agents (Hartman & Wilson 2007)—this is likely a result of the "spin leakage" exhibited by EMFs (Koltover 2004). Thus, lower doses can be administered for the same image contrast. A novel method to increase image contrast and decrease dosage used empty-cage C_{60} as a central tethering moiety to which were covalently attached 4-5 Gd-DOTA (1,4,7,10-tetraazacyclododecane-1,4,7,10-tetraacetic acid) chelates (Wang et al. 2015a). However, unlike EMFs, this still suffers from the possibility of toxic Gd breaking free from its chelated state.

Gadolinium-EMFs for medical applications are generally based on three templates:

1. $Gd@C_{82}$ is the most easily retrieved gadolinium-EMF owing to its good solubility in common organic solvents (Bolskar 2008).

   Hydroxylation of the cage ($Gd@C_{82}(OH)_x$) is most commonly employed to impart water solubility, with the majority of investigations looking at this derivative's viability as an MRI contrast agent (Zhang et al. 1997, 2007; Cagle et al. 1999; Mikawa et al. 2001; Shu et al. 2008a). Other potential MRI contrast agents include amino acid derivatives of $Gd@C_{82}$ ($Gd@C_{82}O_m(OH)_n$-$(NHCH_2CH_2COOH)_l$) which have shown increased water proton relaxivities compared with commercially available agents (Shu et al. 2006b). A recent study on a hybrid carbon nanomaterial—unmodified $Gd@C_{82}$ on graphene oxide—has actually shown better contrast in magnetic resonance images compared with $Gd@C_{82}(OH)_x$ derivatives (Cui et al. 2014).

   Away from MRI applications, $Gd@C_{82}(OH)_{22}$ has been shown to be a good cytoprotector by scavenging reactive oxygen species (Yin et al. 2009) as well as displaying promising antineoplastic properties (Chen et al. 2005; Meng et al. 2013; Liu et al. 2015). Aggregation studies have been carried out on $Gd@C_{82}$ functionalized by the Bingel reaction to yield $Gd@C_{82}[C(COOCH_2CH_3)_2]$ and its hydrolyzed salt, $Gd@C_{82}[C(COONa)_2]$ (He et al. 2013).
2. $Gd@C_{60}$ suffers from being less soluble in the organic solvents used to separate $Gd@C_{82}$; thus, it and its derivatives have been studied to a smaller extent.

   This issue was resolved by making $Gd@C_{60}$ soluble through covalent functionalization of the cage (Bolskar et al. 2003). The ensuing $Gd@C_{60}[C(COOCH_2CH_3)_2]_{10}$ derivative was hydrolyzed to the water-soluble $Gd@C_{60}[C(COOH)_2]_{10}$, which was the first water-soluble EMF species not to localize in the mononuclear phagocyte system. Subsequent studies on water-soluble $Gd@C_{60}$ derivatives have focused on $Gd@C_{60}[C(COOH)_2]_{10}$ and the hydroxylated $Gd@C_{60}(OH)_x$ and their potential use as MRI contrast agents (Sitharaman et al. 2004; Laus et al. 2005; Toth et al. 2005).
3. $Gd_xLn_{3-x}N@C_{80}$ are a family of so-called TNT EMFs that share a similar solubility in common organic solvents as $Gd@C_{82}$ but which were discovered at a later date (Stevenson et al. 1999; Zhang et al. 2013). They are synthesized using standard arc discharge in the presence of a nitrogen source.

   $Gd_3N@C_{80}$ has been variously functionalized for water solubility and subsequent use as an MRI contrast agent. The first such derivative was polyethylene glycol (PEG)ylated and hydroxylated to afford $Gd_3N@C_{80}[DiPEG5000(OH)_x]$—its imaging capabilities were tested in vivo on healthy and tumor-bearing rat brains (Fatouros et al. 2006). Different PEG chain lengths were introduced (350–5000 Daltons), and the relaxivities of these derivatives were measured to pinpoint the optimum molecular structure (Zhang et al. 2010). Higher relaxivities were achieved with the highly water-soluble $Gd_3N@C_{80}(OH)_{\sim 26}(CH_2CH_2COOM)_{\sim 16}$ where M = Na or H (Shu et al. 2009), while another $Gd_3N@C_{80}$ derivative has been used for the detection of atherosclerosis (Dellinger et al. 2013).

   Investigations of $Sc_xGd_{3-x}N@C_{80}O_m(OH)_n$ ($x = 1$ or 2) as an MRI contrast agent also showed good increases in relaxivities compared with commercially available agents (Zhang et al. 2007). This is a promising result as $Sc_xGd_{3-x}N@C_{80}$ was synthesized in greater yields than both $Gd@C_{82}$ and $Gd_3N@C_{80}$ and was actually the third most abundant fraction after C_{60} and C_{70}.

10.3.3 Other Gadolinium-EMFs

A new addition to the TNT family has been introduced and hydroxylated for use as an MRI contrast agent—the lower-symmetry $Gd_3N@C_{84}(OH)_x$ species was found to have a higher proton relaxivity than the related $Gd_3N@C_{80}$ derivative (Zhang et al. 2014). A comprehensive review of different gadolinium-EMF structures specifically for use as MRI contrast agents has recently been published, with additional coverage of the rarer larger-cage gadolinium-EMFs (Ghiassi et al. 2014). Looking at more exotic species, an exceptionally unreactive endohedral heterofullerene (EHF), $Gd_2@C_{79}N$, was synthesized and the first exohedral functionalization of an EHF was carried out using a Bingel reaction (Fu et al. 2011).

10.3.4 A Note on Solubility

The hydroxylated fullerenes (fullerenols) mentioned in the preceding paragraphs have often had ill-defined numbers of hydroxyl groups attached to the surface; a recent theoretical study has been carried out on C_{60}, $Gd@C_{60}$, and $Gd@C_{82}$ to better understand the mechanisms by which this functionalization occurs (Wang et al. 2015b) and, thus, how one can control the structures synthesized, which is crucial to the reproducible methods required in pharmaceutical production.

10.3.5 Other Lanthanide EMFs

Despite holmium having the highest magnetic moment of any natural element, holmium-containing EMFs have not found use as MRI contrast agents because interactions with the fullerene cage dramatically reduce the magnetic moment (Huang et al. 1999). A study on (nonfullerene) MRI contrast agents based on dysprosium and holmium showed that while their T_1 relaxation times are not of much use, they could be promising for T_2-weighted images (Norek & Peters 2011). Another study, this time on hydroxylated lanthanide-EMFs, produced results in agreement with this (Kato et al. 2003). The species synthesized and investigated were $M@C_{82}(OH)_x$ where M = La, Ce, Gd, Dy, or Er.

Away from MRI contrast agents, similar hydroxylated EMFs have been posited as potential x-ray contrast agents (Miyamoto et al. 2006). However, the derivatives—$M@C_{82}(OH)_{40}$, where M = Dy, Er, Gd, Eu, Lu_2—did not perform better than the iodinated materials already used in clinics. An iodinated empty-cage C_{60} derivative had previously been synthesized for the same purpose, with more promise (Wharton & Wilson 2002).

10.3.6 Theranostics and Multimodal Imaging

The surface of $Gd_3N@C_{80}$ has been functionalized with β-emitting ^{177}Lu, thus providing a theranostic platform for simultaneous MRI and delivery of brachytherapy (Shultz et al. 2011). A dual-modal imaging platform (positron emission tomography and MRI) has also been synthesized by first hydroxylating and carboxylating $Gd_3N@C_{80}$ and then labeling this with radioactive ^{124}I (Luo et al. 2012).

The intrinsic ability of TNTs to incarcerate three metallic species leads to the possibility of multimodal imaging by picking which atoms to include within the fullerene cage. ^{177}Lu has been incorporated in a TNT EMF to give $^{177}Lu_xLu_{3-x}N@C_{80}$, which was functionalised with a targeting peptide moiety (Shultz et al. 2010). Other novel lutetium-containing TNT EMFs have been synthesized without exohedral functionalization of the cages—$Lu_2GdN@C_{80}$, $HoLu_2N@C_{80}$, and $Ho_2LuN@C_{80}$—with the aim of providing the basis for dual-modal imaging agents, contrast being achieved in both MRI and x-ray imaging (Iezzi et al. 2002).

10.3.7 Stumbling Blocks to the Clinical Use of Endohedral Fullerenes

10.3.7.1 Scalability

Achieving the synthesis of EMFs in high yields has been a major goal since the discovery of scaled-up fullerene production using the arc discharge method (Kratschmer et al. 1990). Significant advances have been made, particularly in the synthesis of TNT EMFs (Dunsch et al. 2004) and $Gd@C_{82}$ (Kozlov et al. 2014). A recent adaptation of the arc discharge method has shown a significant increase in yield (over

400% for $Lu_3N@C_{80}$) at a reduced cost—this was achieved using a three-phase reactor without helium and with a metal oxide powder feed and solid graphite electrodes, cf., the drilled and packed electrodes normally used (Bezmelnitsyn et al. 2014).

The synthesis of endohedral fullerenes on the pharmaceutical market scale is not yet possible; this must change if any of the diverse research into their medical applications is to be realized in a day-to-day clinical setting.

10.3.7.2 Toxicity

The toxicity, or lack thereof, of fullerenes and their derivatives is a hotly debated topic. As discussed in this chapter, irradiation of pristine C_{60} and C_{70} leads to the formation of their respective singlet excited states, which then decay to triplets via ISC—these triplet states are devastatingly efficient at sensitizing the formation of singlet oxygen, which is highly reactive and damaging to many biomolecular species (Jensen et al. 1996). It has been shown that the cytotoxicity of empty-cage fullerenes decreases with increased water solubility achieved by surface functionalization (Sayes et al. 2004); similar functionalizations of the larger EMF cages also show promise, as demonstrated by the number of successful in vivo tests carried out with such derivatives.

As with many nanomaterials, there is still a lack of comprehensive and unambiguous health and safety protocols established to regulate this developing research area. Progress is being made (Rauscher et al. 2012; Sharma et al. 2012; Kyzyma et al. 2015) and an excellent European Union-funded review is available (Stone et al. 2009).

10.4 Perspectives

Significant progress has been made in the various techniques for functionalizing endohedral fullerenes since the first reactions were reported less than 20 years ago. Many of these surface functionalizations are vital to the utilization of endohedral fullerenes for a wealth of applications. It is clear that in some cases, the nature of the endohedral species dominates the reactivity, while in others, the size and shape of the cage have the greatest effect. The incorporation of paramagnetic, spin active, or photoluminescent metal clusters or ions inside the fullerene cage makes this unique class of materials a versatile tool for applications in biomedicine, green energy, quantum information processing, and spintronics. In this chapter, we presented a compact and accessible review of the most recent advances in the optical properties and medical applications of covalently functionalized EMFs.

References

Accorsi, G., Armaroli, N. *Journal of Physical Chemistry C.* 114(2010), 1385.

Afreen, S., Muthoosamy, K., Manickam, S. et al. *Biosensors and Bioelectronics* 63(2014), 354–364.

Akasaka, T., Kato, T., Kobayashi, K. et al. *Nature* 374(1995a), 600–601.

Akasaka, T., Mitsuhida, E., Ando, W. et al. *Journal Chemical Society, Chemical Communication* (1995b), 1529–1530.

Akasaka, T., Kato, T., Nagase, S. et al. *Tetrahedron* 52(1996), 5015–5020.

Akasaka, T., Kono, T., Matsunaga, Y. et al. *Journal of Physical Chemistry A* 112(2008), 1294–1297.

Alexander-Webber, J.A., Faugeras, C., Kossacki, P. et al. *Nano Letters* 14(2014), 5194–5200.

Ananta, J.S., Wilson, L.J. Gadonanostructures as magnetic resonance imaging contrast agents. In T. Akasaka, F. Wudl, and S. Nagase (Eds.), *Chemistry of Nanocarbons* (pp. 287–300). John Wiley & Sons Ltd., 2010.

Armaroli, N., Diederich, F., Dietrich-Buchecker, C.O. et al. *Chemistry—A European Journal* 4(1998), 406.

Baker, S.N., Baker, G.A. *Angewandte Chemie* 49(2010), 6726–6744.

Bakry, R., Vallant, R.M., Najam-ul-Haq, M. et al. *International Journal of Nanomedicine* 2(2007), 639–649.

Bandi, V., El-Khouly, M.E., Ohkubo, K. et al. *Journal of Physical Chemistry C* 118(2014), 2321–2332.

Bezmelnitsyn, V., Davis, S., Zhou, Z. Efficient synthesis of endohedral metallofullerenes in 3-phase arc discharge. *Fullerenes, Nanotubes and Carbon Nanostructures* 23(2014), 612–617.

Bharadwaj, P., Novotny, L. *Journal of Physical Chemistry C* 114(2010), 7444–7447.

Bolskar, R. Gadolinium endohedral metallofullerene-based MRI contrast agents. In F. Cataldo, T. Da Ros (Eds.), *Medicinal Chemistry and Pharmacological Potential of Fullerenes and Carbon Nanotubes SE-8* (Vol. 1, pp. 157–180). Springer Netherlands, 2008.

Bolskar, R.D., Benedetto, A.F., Husebo, L.O. et al. *Journal of the American Chemical Society* 125(2003), 5471–5478.

Cagle, D., Kennel, S., Mirzadeh, S. et al. *Proceedings of the National Academy of Sciences of the United States of America* 96(1999), 5182–5187.

Cao, B., Wakahara, T., Maeda, Y. et al. *Chemistry—A European Journal* 10(2006), 716–720.

Caravan, P., Ellison, J., McMurry, T. et al. *Chemical Reviews* 99(1999), 2293–2352.

Cardona, C.M., Kitaygorodskiy, A., Echegoyen, L. *Journal of the American Chemical Society* 127(2005), 10448–10453.

Cardona, C.M., Elliott, B., Echegoyen, L. *Journal of the American Chemical Society* 128(2006), 6480–6485.

Carini, M., Đorđević, L., Da Ros, T. Fullerenes in biology and medicine. In F. D'Souza, K. Kadish (Eds.), *Handbook of Carbon Nano Materials: Medicinal and Bio-Related Applications* (Vol. 3). World Scientific, 2012.

Cataldo, F., Da Ros, T. *Medicinal Chemistry and Pharmacological Potential of Fullerenes and Carbon Nanotubes*. Springer, Netherlands, 2008.

Chen, C., Xing, G., Wang, J. et al. Multihydroxylated $[Gd@C_{82}(OH)_{22}]_n$ nanoparticles: Antineoplastic activity of high efficiency and low toxicity. *Nano Letters* 5(2005), 2050–2057.

Chen, Z., Ma, L., Liu, Y. et al. *Theranostics* 2(2012), 238–250.

Choi, J.H., Nguyen, F.T., Barone, P.W. et al. *Nano Letters* 7(2007), 861.

Cui, R., Li, J., Huang, H. et al. Novel carbon nanohybrids as highly efficient magnetic resonance imaging contrast agents. *Nano Research* 8(2014), 1259–1268.

Dellinger, A., Olson, J., Link, K. et al. *Journal of Cardiovascular Magnetic Resonance: Official Journal of the Society for Cardiovascular Magnetic Resonance* 15(2013), 7.

Dietel, E., Hirsch, A., Pietzak, B. et al. *Journal of the American Chemical Society* 121(1999), 2432.

Dunsch, L., Krause, M., Noack, J. et al. *Journal of Physics and Chemistry of Solids* 65(2004), 309–315.

Fahmi, A., Pietsch, T., Bryszewska, M. et al. *Advanced Functional Materials* 20(2010), 1011.

Fan, J., Fang, G., Zeng, F. et al. *Small* 9(2013), 613–621.

Fatouros, P.P., Corwin, F.D., Chen, Z.-J. et al. *Radiology* 240(2006), 756–764.

Filidou, V., Simmons, S., Karlen, S.D. et al. *Nature Physics* 8(2012), 596.

Franco, L., Ceola, S., Corvaja, C. et al. *Chemical Physics Letters* 422(2006), 100–105.

Fu, W., Zhang, J., Fuhrer, T., Champion, H., Furukawa, K., Kato, T., Mahaney, J.E. et al. Gd2@C79N: Isolation, characterization, and monadduct formation of a very stable heterofullerene with a magnetic spin state of S = 15/2. *Journal of the American Chemical Society* 133(2011), 25, 9741–9750.

Gareis, T., Kothe, O., Daub, J. *European Journal of Organic Chemistry* 8(1998), 1549.

Ge, Z., Duchamp, J.C., Cai, T. et al. *Journal of the American Chemical Society* 127(2005), 16292–16298.

Ghiassi, K.B., Olmstead, M.M., Balch, A. *Dalton Transactions* 43(2014), 7346–7358.

Gonzalez, K.A., Wilson, L.J., Wu, W. et al. *Bioorganic & Medicinal Chemistry* 10(2002), 1991–1997.

Hagberg, G., Scheffler, K. Effect of r1 and r2 relaxivity of gadolinium-based contrast agents on the T1-weighted MR signal at increasing magnetic field strengths. *Contrast Media & Molecular Imaging* 8(2013), 456–465.

Hartman, K.B., Wilson, L.J. Carbon nanostructures as a new high-performance platform for MR molecular imaging. In W.C.W. Chan (Ed.), *Bio-Applications of Nanoparticles* (Vol. 620, pp. 74–84). Springer-Verlag, New York, 2007.

He, R., Zhao, H., Liu, J. et al. *Fullerenes, Nanotubes and Carbon Nanostructures* 21(2013), 549–559.

Holler, M., Cardinali, F., Mamlouk, H. et al. *Tetrahedron* 62(2006), 2060.

Huang, H.J., Yang, S.H., Zhang, X.X. *Journal of Physical Chemistry* 103(1999), 5928–5932.
Hutchison, K., Gao, J., Schick, G. et al. *Journal of the American Chemical Society* 121(1999), 5611.
Iezzi, E.B., Duchamp, J.C., Fletcher, K.R. et al. Lutetium-based trimetallic nitride endohedral metallofullerenes: New contrast agents. *Nano Letters* 11(2002), 1187–1190.
Iiduka, Y., Wakahara, T., Nakajima, K. et al. *Angewandte Chemie* 119(2007), 5658.
Imahori, H., Tamaki, K., Guldi, D.M. et al. *Journal of the American Chemical Society* 123(2001), 2607–2617.
Imahori, H., Yamada, H., Nishimura, Y. et al. *Journal of Physical Chemistry B* 104(2004), 2099–2108.
Ito, Y., Okazaki, T., Okubo, S. et al. *ACS Nano* 1(2007), 456.
Jain, K.K. The role of nanobiotechnology in drug discovery. In C.A. Guzman, G.Z. Feuerstein (Eds.), *Pharmaceutical Biotechnology* (pp. 37–43). Springer, New York, 2009.
Jensen, A., Wilson, S., Schuster, D. Biological applications of fullerenes. *Bioorganic & Medicinal Chemistry* 6(1996), 767–779.
Jones, M.A.G., Morton, J.J.L., Taylor, R.A. et al. *Physical Status Solidi* 243(2006), 3037–3041.
Kareev, I.E., Lebedkin, S.F., Bubnov, V.P. et al. *Angewandte Chemie International Edition* 117(2005), 1880–1883.
Kato, H., Suenaga, K., Mikawa, M. et al. *Chemical Physics Letters* 324(2000), 255–259.
Kato, H., Kanazawa, Y., Okumura, M. et al. Lanthanoid endohedral metallofullerenols for MRI contrast agents. *Journal of the American Chemical Society* 125(2003), 4391–4397.
Kelkar, S.S., Reineke, T.M. Theranostics: Combining imaging and therapy. *Bioconjugate Chemistry* 22(2011), 1879–1903.
Koltover, V.K. Spin-leakage of the fullerene shell of endometallofullerenes: EPR, ENDOR and NMR evidences. *Carbon* 42(2004), 1179–1183.
Kozlov, V.S., Suyasova, M.V., Lebedev, V.T. *Russian Journal of Applied Chemistry* 87(2014), 121–127.
Kratschmer, W., Lamb, L.D., Fostiropoulos, K. et al. *Nature* 347(1990), 354–358.
Krishna, V., Singh, A., Sharma, P. et al. *Small* 6(2010), 2236–2241.
Krysmann, M.J., Kelarakis, A., Dallas, P. et al. *Journal of the American Chemical Society* 134(2012), 747.
Kuo, P.H., Kanal, E., Abu-Alfa, A.K. et al. *Radiology* 242(2007), 647–649.
Kwagza, D.S., Park, K., Oh, K.T. et al. *Chemical Communications* 49(2013), 282–284.
Kyzyma, E., Tomchuk, A., Bulavin, L. et al. Structure and toxicity of aqueous fullerene C60 solutions. *Journal of Surface Investigation. X-Ray, Synchrotron and Neutron Techniques* 9(2015), 1–5.
Laus, S., Sitharaman, B., Tóth, É. et al. *Journal of the Chemical Society* 127(2005), 10–13.
Lee, H.M., Olmstead, M.M., Iezzi, E.B. et al. *Journal of the American Chemical Society* 124(2002), 3494–3495.
Lim, E., Kim, T., Paik, S. et al. Nanomaterials for theranostics: Recent advances and future challenges. *Chemical Reviews* 115(2015), 327–394.
Liu, A., Zhai, S., Zhang, B. et al. *Trends in Analytical Chemistry* 48(2013), 1.
Liu, Y., Chen, C., Qian, P. et al. *Nature Communications* 6(2015), 5988.
Lu, X., Nikawa, H., Nakahodo, T. et al. *Journal of the American Chemical Society* 130(2008a), 9129–9136.
Lu, X., Nikawa, H., Tsuchiya, T. et al. *Angewandte Chemie International Edition* 120(2008b), 8770–8773.
Lu, X., Akasaka, T., Nagase, S. *Chemical Communications* 47(2011), 5942–5957.
Lu, X., Feng, L., Akasaka, T. et al. Current status and future developments of endohedral metallofullerenes. *Chemical Society Reviews* 41(2012), 7723–7760.
Luo, J., Wilson, J.D., Zhang, J. et al. *Applied Sciences* 2(2012), 465–478.
Maeda, Y., Matsunaga, Y., Wakahara, T. *Journal of the American Chemical Society* 126(2004), 6858–6859.
Maeda, Y.J., Miyashita, T., Hasegawa, T. *Journal of the American Chemical Society* 127(2005), 12190–12191.
Matsunaga, Y., Maeda, Y., Wakahara, T. et al. *ITE Letters* 7(2006), C1.
Meng, J., Liang, X., Chen, X. et al. Biological characterizations of $[Gd@C_{82}(OH)_{22}]_n$ nanoparticles as fullerene derivatives for cancer therapy. *Integrative Biology* 5(2013), 43–47.
Mikawa, M., Kato, H., Okumura, M. et al. *Bioconjugate Chemistry* 12(2001), 510–514.
Miyamoto, A., Okimoto, H., Shinohara, H. et al. *European Radiology* 16(2006), 1050.
Morton, J.J.L., Tiwari, A., Dantelle, G. et al. *Physical Review Letters* 101(2008), 013002.

Norek, M., Peters, J.A. *Progress in Nuclear Magnetic Resonance Spectroscopy* 59(2011), 64–82.
Numata, T., Murakami, T., Kawashima, F. *Journal of the American Chemical Society* 134(2012), 6092.
Ohashi, K., Imazu, N., Kitaura, R. et al. *Journal of Physical Chemistry C* 115(2011), 3968.
Oswald, F., Islam, D.M.S., El-Khouly, M.L. *Physical Chemistry Chemical Physical* 16(2014), 1443.
Page, Z.A., Liu, Y., Duzhko, V.V. et al. *Science* 346(2014), 441.
Partha, R., Conyers, J.L. *International Journal of Nanomedicine* 4(2009), 261–275.
Penfield, J.G., Reilly, R.F. *Nature Clinical Practice Nephrology* 3(2007), 654–668.
Plant, S.R., Jevric, M., Morton, J.J.L. et al. *Chemical Science* 4(2013), 2971–2975.
Popov, A., Yang, S., Dunsch, L. Endohedral fullerenes. *Chemical Reviews* 113(2013), 5989–6113.
Rauscher, H., Sokull-Klüttgen, B., Stamm, H. *Nanotoxicology* 7(2012), 1–3.
Rieger, J., Jérôme, C., Jérôme, R. et al. *Nanotechnologies for the Life Sciences*. John Wiley & Sons, 2007.
Sayes, C.M., Fortner, J.D., Guo, W. et al. *Nano Letters* 4(2004), 1881–1887.
Sharma, A., Madhunapantula, S.V., Robertson, G.P. *Expert Opinion on Drug Metabolism & Toxicology* 8(2012), 47–69.
Sherigara, B.S., Kutner, W., D'Souza, F. *Electroanalysis* 15(2003), 753–772.
Shu, C.Y., Gan, L.H., Wang, C.R. et al. *Carbon* 44(2006a), 496–500.
Shu, C., Zhang, E., Xiang, J. et al. *Journal of Physical Chemistry B* 110(2006b), 15597–15601.
Shu, C.Y., Wang, C.R., Zhang, J.F. et al. *Chemistry of Materials* 20(2008a), 2106.
Shu, C., Slebodnick, L., Xu, H. et al. *Journal of the American Chemical Society* 130(2008b), 17755.
Shu, C., Corwin, F.D., Zhang, J. et al. *Bioconjugated Chemistry* 20(2009), 1186–1193.
Shultz, M.D., Duchamp, J.C., Wilson, J.D. et al. *Journal of the American Chemical Society* 132(2010), 4980–4981.
Shultz, M.D., Wilson, J.D., Fuller, C.E. et al. *Radiology* 261(2011), 136–143.
Shustova, N.B., Popov, A.A., Mackey, M.A. et al. *Journal of the American Chemical Society* 129(2007), 11676–11677.
Šišková, K., Kubala, M., Dallas, P. et al. *Journal of Materials Chemistry* 21(2011), 1086–1093.
Sitharaman, B., Bolskar, R.D., Rusakova, I. et al. *Nano Letters* 4(2004), 2373–2378.
Stevenson, S., Rice, G., Glass, T. et al. *Nature* 80(1999), 80–82.
Suzuki, T., Maruyama, Y., Kato, T. et al. *Journal of the American Chemical Society* 117(1995), 9606–9607.
Stone V., Hankin S., Aitken R. et al. ENRHES-Engineered Nanoparticles: Review of Health and Environmental Safety, European Commission, 2009.
Tagmatarchis, N.A., Taninaka, H., Shinohara, H. *Chemical Physics Letters* 355(2002), 226–232.
Toth, E., Bolskar, R., Borel, A. et al. *Journal of the American Chemical Society* 127(2005), 799–805.
Tiwari, A., Dantelle, G., Porfyrakis, K. et al. *Journal of Chemical Physics* 127(2007), 194504.
Wang, J., Leng, J., Yang, H.P. et al. *Langmuir* 29(2013), 9051–9056.
Wang, L., Zhu, X., Tang, X. et al. *Chemical Communications* 51(2015a), 4390–4393.
Wang, Z., Lu, Z., Zhao, Y. et al. *Nanoscale* 7(2015b), 2914–2925.
Wharton, T., Wilson, L.J. *Tetrahedron Letters* 43(2002), 561–564.
Wilson, L.J., Cagle, D.W, Thrash, T.P. et al. *Coordination Chemistry Review* 190–192(1999), 199–207.
Yamakoshi, Y.S., Sueyoshi, K., Fukuhara, N. et al. *Journal of the American Chemical Society* 120(1998), 12363–12364.
Yin, J.J., Lao, F., Fu, P.P. et al. *Biomaterials* 30(2009), 611–621.
Zhang, S., Sun, D., Li, X. et al. *Fullerene Science and Technology* 5(1997), 1635–1643.
Zhang, E.Y., Shu, C.Y., Feng, L. et al. *The Journal of Physical Chemistry B* 111(2007), 14223–14226.
Zhang, J., Shu, C., Reid, J. et al. *Bioconjugate Chemistry* 21(2010), 610–615.
Zhang, J., Stevenson, S., Dorn, H.C. *Accounts of Chemical Research* 46(2013), 1548–1557.
Zhang, J., Ye, Y., Chen, Y. et al. *Journal of the American Chemical Society* 136(2014), 2630–2636.
Zhou, S., Rašović, I., Briggs, G.A.D. et al. *Chemical Communications* 51(2015), 7096–7099.

11

Carbon Nano-Onions

Diana M. Brus

Olena Mykhailiv

Luis Echegoyen

Marta E. Plonska-Brzezinska

11.1 Introduction

In recent years, the ability of carbon to exist in different allotropic forms has provided, besides C_{60}, new varieties of nanoscale structures with fascinating physicochemical properties, such as the "higher" fullerenes, endohedral fullerenes, carbon nano-onions (CNOs), graphene, and single-walled and multiwalled carbon nanotubes (CNTs), among others. A significant amount of research has been done in the field of fullerenes after their discovery in 1985 by Smalley, Curl, and Kroto (Kroto et al. 1985). Fullerenes exhibit interesting properties such as high thermal stability, large surface area, and broad absorption spectra (Wang et al. 2004; Janssen et al. 2005; Shin et al. 2006). One of the most attractive and potentially useful properties of fullerenes is their ability to reversibly accept multiple electrons (Xie et al. 1992). Recently, larger morphological variations of fullerene-like all carbon structures such as CNTs and graphene have received enormous attention. The remarkable physical and chemical properties of these structures are of considerable interest. These properties include high conductivity and strong mechanical, thermal, and environmental resistance (Novoselov et al. 2004; Zhang et al. 2005; Han et al. 2007; Bekyarova et al. 2009; Rao et al. 2009). Within this large group of allotropic carbon nanomaterials, we focus on the onion-like layered structures. Sumio Iijima discovered CNOs in 1980 while looking at a sample of carbon black using a transmission electron microscope (TEM) (Iijima 1980). CNOs are spherical structures that consist of a hollow spherical fullerene core surrounded by concentric graphene layers (larger fullerenes) with increasing diameter (Iijima 1980). That is why they are referred to as "nano-onions," "onion-like carbon," multilayer fullerenes, multilayered round carbon particles, or "buckyonions" (Kuznetsov et al. 1994; Bates et al. 1998; Santiago et al. 2012; Costa et al. 2014). In this chapter, we discuss the preparation and characterization of CNOs; their structural, physical, and chemical properties as well as of their derivatives; and their potential applications and we offer some future perspectives.

11.2 CNOs: Structure, Production, and Formation Mechanisms

CNOs consist of a hollow spherical fullerene core surrounded by concentric and curved graphene layers with progressively increasing diameter and an interlayer distance of 0.335 nm. The distance is very close to that in bulk graphite (0.334 nm) (Bacon et al. 1960; Al-Jishi et al. 1982), owing to the van der Waals interactions between neighboring shells. CNOs have a cage-within-cage structure, with smaller fullerenes nested inside larger ones like in Russian dolls. Spherical CNOs contain the ideal fullerene series "$C_{60}@C_{240}@C_{540}@\ldots@C_{60}n^2$," where n is the ordinal number of the corresponding layer, and these are separated by a constant inter shell distance (Bates et al. 1998).

Most of the methods used for the preparation of CNOs result in spherical structures. The surface of CNOs is composed mainly of hexagonal rings like graphite layers, but they also possess some pentagonal and heptagonal rings (Figure 11.1a) as well as occasional quadrangular and octagonal ones (Bates et al. 1998). The carbon nanoparticles contain defects and holes. The structural defects present, such as edges, dangling bonds, vacancies, dislocations (Bom et al. 2002), and strain, induced through high curvature (Saxby et al. 1992), lead to relatively high chemical reactivities for CNOs. For molecules with more than a dozen or so graphene layers, deformation from spherical symmetry is observed and highly faceted shapes are seen (Bates et al. 1998).

CNOs can be divided according to the following:

1. Size: big or small (Figure 11.2a and b) (Tomita et al. 2001; Sano et al. 2002);
2. Shape: spherical (Figure 11.2a, b, e, and f), (Banhart et al. 1998; Tomita et al. 2001; Sano et al. 2002; Gan et al. 2008) and polyhedral (Figure 11.2c and d) (Tomita et al. 2001; Street et al. 2004);
3. Type of core: dense (Figure 11.2a, b, and c) (Tomita et al. 2001; Sano et al. 2002) and hollow (Figure 11.2d and e) (Street et al. 2004; Gan et al. 2008). The hollow cores may be empty or filled with different metals (Figure 11.2f) (Lamber et al. 1990; Konno et al. 1995; Cabioc'h et al. 1997, 1998; Banhart et al. 1998; Thune et al. 2002).

As already mentioned briefly, the first observation of CNOs was made during an in situ TEM analysis of amorphous carbon films prepared by vacuum evaporation at 10^{-6} torr and about 3700°C, which exhibited diameters between 8 and 15 nm (Iijima 1980). This process promotes the progressive reorganization of carbon atoms as a result of elastic collisions that lead to the formation of spherical carbon structures with several layers (Iijima 1980). In 1992, Ugarte (1992, 1995) observed that carbon soot particles and tubular graphitic structures are transformed into quasi-spherical CNOs by intense electron-beam

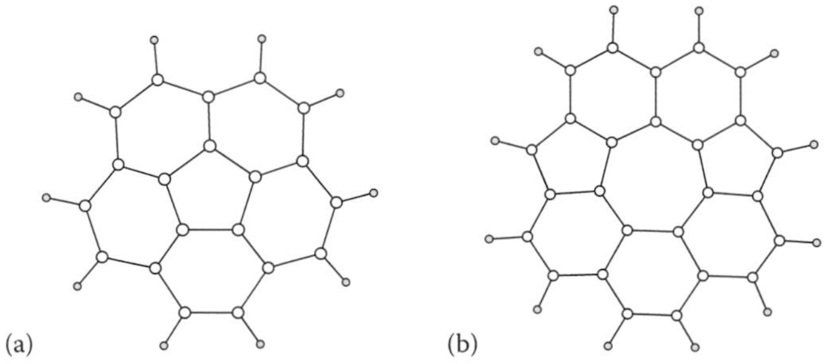

(a) (b)

FIGURE 11.1 Structures of (a) corannulene and (b) a graphene-surrounded 5-7-5 cluster, both fully optimized using 3-21G/HF. (Reproduced from Bates et al., 1998. Copyright 1998, with permission of Springer.)

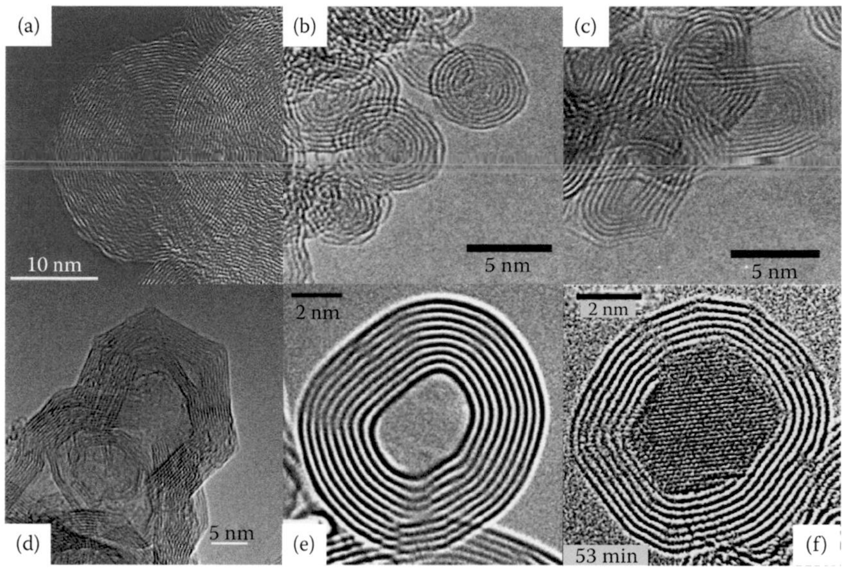

FIGURE 11.2 High-resolution TEM images of CNOs: (a) big (Reprinted with permission from Sano, N. et al., *J. Appl. Phys.*, 92, 2783–2788, 2002, American Institute of Physics.); (b) small (Reprinted with permission from Tomita et al. 2001, American Institute of Physics.); (c, d) polyhedral (c, Reprinted with permission from Streeta et al. 2004, American Institute of Physics; d, Reprinted with permission from Tomita et al. 2001, American Institute of Physics.); (e) hollow (Reproduced from Gan et al., 2008. Copyright 2008 John Wiley & Sons.); (f) with a metal core (Reprinted from Banhart et al. Copyright 1998, with permission from Elsevier.).

irradiation in TEM. With this method, it was possible to produce CNOs with diameters ranging from 6 to 47 nm, but this method is not suitable for the preparation of bulk quantities.

Another preparation method proposed by Kuznetsov et al. (1994) was the graphitization of nanodiamond (ND) particles (5 nm) at high temperatures under high vacuum. Under these conditions, they obtained mainly quasi-spherical particles with enclosed concentric graphite shells. The preparation of CNOs from NDs was performed at different temperatures (Figure 11.3) (Kuznetsov et al. 2008). On the basis of TEM analysis, Kuznetsov et al. (1999) proposed a mechanism for the formation of the spherical structures based on what was described as a zipper-type mechanism. Annealing ND particles at temperatures between 1500°C and 1800°C leads to the formation of particles that are called "small" CNOs. The nanostructures have six to eight shells and diameters of 5–6 nm and are obtained in gram quantities with high purity and narrow size dispersity. Increasing the annealing temperature results in an increase in the density of states at the Fermi level of the conductive electrons of the nanoparticles. High CNO conductivity correlates well with the formation of more perfect graphitic surfaces. Further annealing up to 1900°C leads to the formation of polyhedral onion-like structures (Tomita et al. 2001).

Other methods used for CNO preparation are presented in Table 11.1. Carbon ion implantation into different substrates: Cu (Cabioc'h et al. 1997), Ag (Cabioc'h et al. 1998; Thune et al. 2002), Ni (Lamber et al. 1990), and Co (Konno et al. 1995), is often used for the preparation of CNOs because of its flexibility, which provides an opportunity to obtain a wide variety of nanostructures. The process is conducted at high temperatures (500°C–1000°C) (Cabioc'h et al. 1997). In this process, the concentric spherical particles were obtained with diameters between 10 and 200 nm in large quantities. Using this method, empty spherical CNOs (hollow-core CNOs) are formed with metal-encapsulated cores. Obtaining macroscopic quantities of these materials with high purity was also possible using chemical vapor deposition methods (Chen et al. 2001; He et al. 2006, 2009).

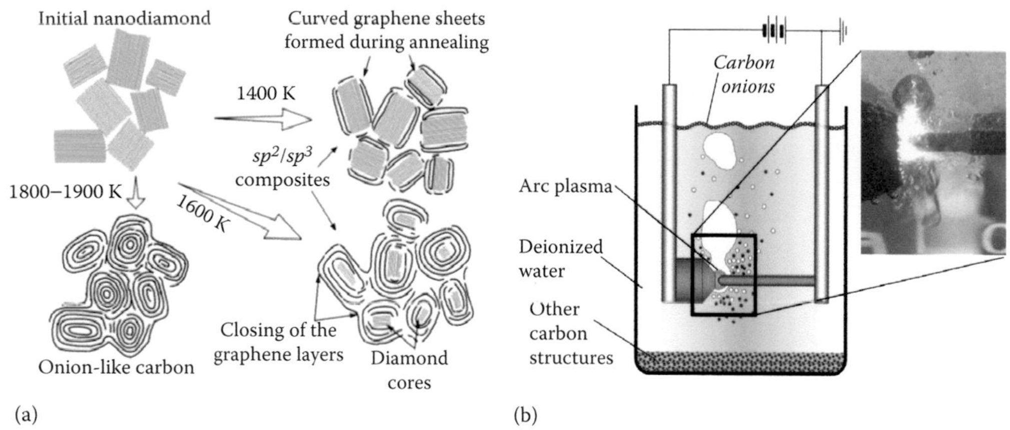

FIGURE 11.3 (a) Scheme of the formation of CNOs. (Reproduced with permission from Kuznetsov et al., 2008. Copyright 2008 John Wiley & Sons.) (b) Schematic of the apparatus used for arc discharge under water with a digital image of the discharge. (Reprinted with permission from Sano, N. et al., *J. Appl. Phys.*, 92, 2783–2788, 2002. Copyright 2002, American Institute of Physics.)

The preparation of spherical CNOs in large quantities was possible without the need of vacuum equipment (Sano et al. 2001). Carbon nanoparticles were obtained as a floating powder on a water surface using a direct current arc-discharge process (30 A, 17 V) between two graphite electrodes submerged in water (Figure 11.3b) (Sano et al. 2002). When using the arc-discharge method, the formation of graphene layers occurs as a result of the presence of large amounts of ions and carbon radicals transported by gases created as a result of water electrolysis (O_2 and H_2). The resulting structures have diameters between 15 and 25 nm, which correspond to approximately 20–30 graphene layers (Palkar et al. 2007). These compounds are referred to as "large" CNOs. Although the arc-discharge method is economical, the efficiency of product formation is quite low and the synthesized material contains impurities such as CNTs and amorphous carbon. Using this method, the synthesis of CNOs was performed at low temperatures (450°C–650°C) in the presence of different catalysts Ni/Al (Chen et al. 2001; He et al. 2009) and Co (He et al. 2006). These lead to quasi-spherical and onion-shaped graphite layers with hollow cores and diameters from 5 to 50 nm.

11.3 Physical Properties of CNOs

As mentioned previously, the physical and chemical properties of CNOs depend on their structure, the number of layers, and the distance between these. The properties are also dependent on the preparation method and conditions. Further in this chapter, we describe the physical properties of CNOs prepared by two different methods: by annealing ND particles (N-CNOs) and via arc-discharge under water (A-CNOs).

11.3.1 Physical Properties of CNOs Prepared by Annealing Nanodiamonds (N-CNOs)

To characterize the surface and structural properties of CNOs derived from NDs, Raman analysis (Tomita et al. 2001; Portet et al. 2007), nuclear magnetic resonance (NMR; Costa et al. 2014), ultraviolet-visible absorption spectroscopy (Tomita et al. 2002), x-ray emission studies (Okotrub et al. 2001), energy-loss near-edge structure (Tomita et al. 2000), x-ray diffraction (XRD) (Lee et al. 2007), and differential thermogravimetric analyses have been utilized (Palkar et al. 2007; Plonska-Brzezinska et al. 2014).

TABLE 11.1 Methods Used for CNO Synthesis

Method	Principle of Operation	Condition	Size and Shape	Advantages	Disadvantages	Yield	Reference
Electron beam irradiation	Annealing of polyvinyl alcohol and amorphous carbon	400°C in air, 100 A/cm^2, 1250 kV	Dense-core tetrahedral CNOs, 0.8–1.4 nm	–	Small quantities	Low yield	Iijima 1980; Narita et al. 2001; Oku et al. 2004
	Irradiation of carbon soot	300 kV TEM microscope	Pristine dense-core spherical CNOs, 6–47 nm	–	Small quantities	Low yield	Ugarte 1992
	Irradiation of NDs	18–40 A/cm^2, 120–200 kV TEM	Pristine dense-core spherical CNOs, 5–20 nm	Low current density	–	High yield	Jin et al. 2009; Roddatis et al. 2002
Annealing	Annealing of ultradispersed NDs (2–6 nm)	High-vacuum, 1000°C–2100°C	Pristine dense-core, spherical, polyhedral, and elongated CNOs, 5–10 nm	Very high purity and narrow size dispersity	Expensive	Gram quantities	Kuznetsov et al. 1994
	Copper dichloride hydrate ($CuCl_2x2H_2O$) and calcium carbide (CaC_2)	600°C, drying at 60°C for 10 h	Hollow- and irregular-core, spherical CNOs, 5–30 nm	Simple, efficient, and economical method, high purity	–	Large quantities	Han et al. 2011
	Acetylene carbon black	1000°C in the presence of ferric nitrate	Hollow-core, spherical CNOs, 40–100 nm	The ability to regulate yield of carbonized product by change in ferric nitrate (FN) amount	By-products	Low yield	Lian et al. 2008
Arc-discharge	Contacting a pure grounded graphite electrode in water	16–17 V, 30 A, ~3700°C	Dense-core, spherical CNOs, 20–40 nm	Large quantities, high purity product, low cost	–	Low yield	Sano et al. 2001
	Heating a carbon filament in liquid ethanol alcohol by passing an electric current	~2300°C in the presence of gas (CO, H_2, CO_2, CH_4, C_2H_6, C_2H_2, C_2H_4)	Spherical with large inner defect space CNOs, 20–50 nm	Simple method, high concentration, low cost	Imperfect layers	Low yield	Fan et al. 2012
	Reduction of Pt nanoparticles in an aqueous solution	Pt nanoparticles (5 nm), 30 A, 22–28 V, ~3700°C	Dense-core, spherical CNOs, 10–60 nm	Simple method	High cost	Low yield	Guo et al. 2009

(*Continued*)

TABLE 11.1 (CONTINUED) Methods Used for CNO Synthesis

Method	Principle of Operation	Condition	Size and Shape	Advantages	Disadvantages	Yield	Reference
Carbon–ion implantation	Silver substrates	~600°C, 120 keV carbon ions C_{12}^{+}	Dense-core, spherical CNOs, 10–20 nm	Large quantities	Difficulties with separation, by-product	High yield	Cabioc'h et al. 1998
		850°C, 10 h under vacuum (2 × 10^{-5} Pa)	Dense-core, spherical CNOs, 20–30 nm				
	Copper substrates	580°C–1000°C, 120 keV carbon ions C_{12}^{+}	Dense-core, spherical CNOs, 10–20 nm				Cabioc'h et al. 1997
	Cobalt substrates	2 × 10^{-7} torr, 300 kV, 500°C–600°C	Dense-core, spherical CNOs, 10–50 nm				Konno et al. 1995
Chemical vapor deposition (CVD)	Catalytic decomposition of methane over nano-sized catalyst	Ni/Al catalyst, Ni content >60 wt%, 80 wt%, 450°C–600°C	Hollow-core, spherical CNOs, 5–50 nm	Control of catalyst characteristics and reaction, 90% purity	Small quantities	Low yield	He et al. 2006, 2009
	Radiofrequency plasma-enhanced CVD process	Co catalyst, 1 torr, 35–50 V, 500°C	Pristine dense-core, spherical CNOs, 50 nm	Large quantities, high purity	-	High yield	Chen et al. 2001
	Catalysts	Annealing at 1000°C, Ni/Fe alloy catalyst, 850°C	Pristine hollow-core, spherical CNOs, 10–15 nm	-	By-products	Low yield	Zhang et al. 2012
	Mo_2C	300°C, 40 MPa, C/Mo	Dense-core, spherical CNOs, 4–6 nm	60%–80% yield, high purity	-	High yield	Du et al. 2011
High throughput combustion method	Flash pyrolysis of a polyaromatic hydrocarbon (naphthalene)	79°C–87°C in air	Pristine dense-core, spherical CNOs, 50 nm	Free of impurities and catalyst-free	-	Gram-scale yield	Choucair 2012
Laser irradiation	Carbon black suspension using millisecond pulse in deionized water	Room temperature, 4-h repetition rate of 20 Hz, pulse length of 0.4 ms	Hollow-core and the incomplete graphitic shells, spherical CNOs	-	Small quantities	Low yield	Hu et al. 2009

(Continued)

TABLE 11.1 (CONTINUED) Methods Used for CNO Synthesis

Method	Principle of Operation	Condition	Size and Shape	Advantages	Disadvantages	Yield	Reference
	Decomposition of polycyclic aromatic hydrocarbons into C_2 species	C_2H_4 and O_2	Pristine dense-core, spherical CNOs	Large quantities, a catalyst-free and highly efficient method	–	Gram-scale yield	Gao et al. 2011
Laser ablation	High-temperature annealing of concentric fullerene-like structures within foam-like carbon	~1700°C–5700°C, Ar pressure (2, 50, 200 torr)	Hollow-core, spherical CNOs	Controlling of microstructure	High number of defects	Low yield	Lau et al. 2007
	Mixed graphite–metal targets	~4700°C, 20 µs, Ni- and Ni-Co-doped graphite targets in O_2 and air	Pristine dense-core spherical CNOs, 100–200 nm	–	Small quantities, high temperature	Low yield	Radhakrishnan et al. 2007
Pyrolysis	A laminar premixed propane/oxygen flame (1.8 stoichiometric coefficient)	~200°C, Al surface	Hollow-core, spherical and polyhedral CNOs, 9–210 nm	Primary effective and low cost system, high concentration	Difficulties with separation	High yield	Garcia-Martin et al. 2013
	Pyrolysis of plastic wastes (poly-ethylene, acrylate, vinyl chloride, styrene, ethylene terephthalate) under static atmosphere	500°C ± 10°C	Dense-core, spherical CNOs, 60 ± 10 nm	Simple method, low cost	By-products	High yield	Sawat et al. 2013
Plate-impact heavy shock process	SiC polycrystalline powder (5 wt%) and copper powder (95 wt%)	~5200°C–6500°C, 1.2–1.4 MBar	Amorphous carbon core, spherical CNOs, 10–30 nm	–	By-product, high cost	Low yield	Kobayashi et al. 2003
Halogen gas treatment	Carbide derived carbon produced by etching SiC in halogens	600°C–1000°C, 200–300 kV, 3.5% Cl_2, 0%–2% H_2	Pristine hollow-core spherical and polyhedral CNOs	–	Difficulties with separation, by-products	Low yield	Welz et al. 2006
Low-temperature solution-phase reactions	Dissolved block copolymer P123 in toluene solution	180–200°C	Pristine dense-core spherical CNOs, 30–80 mm	Low-temperature	Difficulties with separation, by-products	High yield	Yan et al. 2007

(Continued)

TABLE 11.1 (CONTINUED) Methods Used for CNO Synthesis

Method	Principle of Operation	Condition	Size and Shape	Advantages	Disadvantages	Yield	Reference
Solid-state carbonization	Phenolic resin (PH), FN	1000°C for 10 h	Pristine dense-core spherical CNOs, 30–50 nm	Large quantities	–	High yield	Zhao et al. 2007
Catalyst-free by thermolysis	Redox reaction: sodium azide (NaN_3) and hexachlorobenzene (C_6Cl_6)	Ar, air atmosphere, 1 MPa	Pristine dense-core spherical CNOs, 30–100 nm	Simple and low-cost reagents, without sophisticated equipment	Difficulties with separation, by-products	Low yield	Bystrzejewski 2010
Metal-organic precursor vapor decomposition	Decomposition reaction of $Cu(acac)_2$ and coagulation processes	705°C–1216°C, 0.27 Pa, inert nitrogen atmosphere	Pristine hollow-core spherical CNOs, 5–30 nm	Large quantities	–	High yield	Nasibulin et al. 2003
Low-temperature synthesis	Black carbon from the flame of burning ghee (clarified butter)	800°C under an inert atmosphere	Dense-core, spherical CNOs, 40–45 nm	Cheap and green approach	Small quantities	Low yield	Azhagan et al. 2014

Using these methods, the densities (Butenko et al. 2000), conductivities (Bushueva et al. 2008), specific surface areas (SSAs) (Butenko et al. 2000), and average pore sizes (Portet et al. 2007) of CNOs have been determined.

Upon high-temperature annealing, the graphitization process of NDs starts on the surface and progressively extends into the bulk of the particle, initially resulting in the formation of graphene shells surrounding an ND core (Butenko et al. 2000; Portet et al. 2007; Kuznetsov et al. 2011). Increasing the annealing temperature results in an increase in the weight fractions of sp^2-carbons in the CNOs (Table 11.2) (Kuznetsov et al. 2011). Formation of the perfect graphitic shells occurs only after the complete transformation of the diamond core, which decreases the density (from 3.07 ± 0.02 to 2.03 ± 0.02 g/cm^3), while the density of the starting NDs is 3.49 g/cm^3 (Butenko et al. 2000). The final values are close to that for graphitic carbon (2.265 g/cm^3) (Butenko et al. 2000).

Table 11.2 shows the textural characteristics of the resulting CNOs, such as the SSAs, densities, and average pore sizes as a function of the annealing temperature utilized (Butenko et al. 2000; Portet et al. 2007). CNOs obtained by annealing NDs at 900°C–1900°C display high values of the SSA. The SSAs determined by nitrogen gas adsorption are between ca. 390 and ca. 580 m^2/g (Butenko et al. 2000). The materials obtained at lower temperatures have lower SSA values because of the presence of diamond in the structures. The differences in the values of the SSAs for pristine NDs and for the annealed particles at different temperatures arise as a consequence of the decrease in particle density and volume expansion during the diamond-to-graphite transformation (Portet et al. 2007). At 1500°C, NDs are totally transformed into spherical, defect-free CNO structures, which exhibit the maximum SSA values. Annealing NDs in the 1100°C–1900°C temperature range leads to the creation of spherical CNOs with a 6-nm diameter. The nanoparticles exhibit only a small fraction (~5 vol%) of micropores ($d < 2$ nm) and most of the volume is composed of 0.50–36-nm pores (Portet et al. 2007). The electrical conductivity of the resulting materials increases as the annealing temperature increases up to about 1500°C (Bushueva et al. 2008). When the NDs are annealed at temperatures higher than 1700°C, the electrical conductivities are very close to those of perfect graphite structures and exhibit saturation of the conductivity curves.

Raman spectroscopy is the method most often used to characterize the structural properties of carbon materials. Figure 11.4a shows the Raman spectra for CNOs obtained at annealing temperatures between 1500°C and 2000°C. All the spectra show two broad Raman bands around 1350 cm^{-1} ("D," the disorder-induced band) and 1580 cm^{-1} ("G," the graphitic band) (Portet et al. 2007). For single crystals of highly oriented pyrolitic graphite, the G-band is assigned to the E_{2g} states of the infinite crystal (Tuinstra et al. 1970). This band can be used to characterize the crystal size and can provide quantitative information about sample composition and about the degree of disorder as well as of the sp^2-to-sp^3 ratio of carbon atoms in the sample (Knight et al. 1989; Dresselhaus et al. 2002). The D-band is more pronounced for materials that consist of small crystallites, usually less than 1000 Å (Wang et al. 1990). The band is attributed to the A_{1g} mode of the small crystallites or to the boundaries between larger crystallites of graphite. For diamond, a single D line is observed at 1332 cm^{-1} which is related to the zone-center optical

TABLE 11.2 Textural Characteristics of CNOs Formed During Annealing of NDs

	Temperature Annealing (°C)						
	~300	~900	~1100	~1300	~1500	~1900	Reference
Weight fractions of sp^2-carbon	–	–	0.39	0.68	0.98	~1	Kuznetsov et al. 2011
True density (g/cm^3)	3.11 ± 0.02	3.07 ± 0.02	2.90 ± 0.02	2.53 ± 0.02	2.14 ± 0.02	2.03 ± 0.02	Butenko et al. 2000
Average pore size (nm)	–	–	6	6	6	5	Portet et al. 2007
SSA (m^2/g)	342 ± 5	394 ± 5	425 ± 5	490 ± 5	577 ± 5	572 ± 5	Butenko et al. 2000

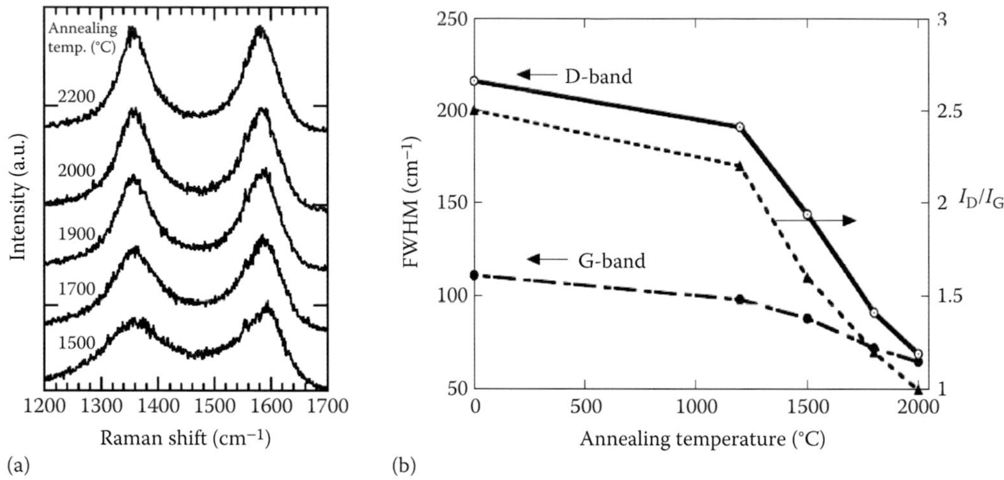

FIGURE 11.4 (a) Raman spectra obtained for different annealing temperatures. (Reprinted with permission from Tomita et al., 2001. Copyright 2001, American Institute of Physics.) (b) Variation of the I_D/I_G ratio and FWHM of ND (ND annealed at 1200°C) with the annealing temperature. (Reprinted from Portet et al., Copyright 2007, with permission from Elsevier.)

phonon with F_{2g} symmetry (Solin et al. 1970). Figure 11.4b shows the dependence of the full width at half maximum (FWHM) of the G-band with the annealing temperature. At the highest annealing temperature, the material is transformed into a three-dimensional ordered hexagonal graphite. The I_D/I_G ratios from the Raman spectra measure the extent of defective sites relative to those in a graphitic structure. The intensity ratio I_D/I_G decreases with increasing annealing temperatures, confirming an increase in the texture ordering and in the degree of graphitization.

Solid-state ^{13}C NMR spectroscopy also enables the observation of sp^2 hybridized carbons (Figure 11.5a) (Costa et al. 2014). A clear isotropic peak, positioned at 110 or 120 ppm, was observed for CNOs synthesized at 1500°C and 1800°C for 3 h, respectively. This peak is typical for graphite and CNOs (Crestani et al. 2005; Shames et al. 2007; Panich et al. 2013). The difference in peak position may be caused by the curvature of the CNO layers that can lead to changes in the angle of the C–C bonds. The optical extinction properties of spherical and polyhedral CNOs have also been studied using ultraviolet-visible spectroscopy (Figure 11.5b) (Tomita et al. 2002). As the annealing temperature is increased, the extinction coefficients at higher wave numbers decreases. The UV-Vis spectra exhibit a broad peak at about 3.7 μm^{-1} at 1100°C (Figure 11.5b). The transformation of the diamond nanoparticles into CNOs starts around 900°C. The particles with diamond cores result in a broad peak around 3.9 μm^{-1} (above 1700°C) or 4.6 μm^{-1} (2100°C) for polyhedral CNOs (Figure 11.5b).

The electronic structures of onion-like carbon particles were investigated using x-ray emission studies (Figure 11.5c) (Okotrub et al. 2001; Lee et al. 2007). X-ray emission arises as a result of electron transitions from occupied valence states to previously created core holes and is governed by the dipole selection rules (Okotrub et al. 2001). For homogeneous materials, like diamond or graphite, the C-Kα spectrum reflects the density of 2p electronic states of sp^3- or sp^2-hybridized carbon (Okotrub et al. 2001). Carbon atoms in diamond only have σ-type interactions, while the electronic structure of graphite includes σ- and π-states (Skytt et al. 1994; Okotrub et al. 2001). The C-Kα spectra have a maximum around 277.5 eV (σ-state interaction), a high energy shoulder around 279.2 eV, and a short-wave maximum at 281.6 eV for π-state interactions (Tomita et al. 2000; Okotrub et al. 2001).

The XRD patterns of the NDs and CNOs are shown in Figure 11.5c (Lee et al. 2007). The CNOs were prepared by annealing of the NDs at 1600°C under N_2 for 1 h. These main peaks were observed at 2θ equal to 43.5° (111), 75° (220), and 90.5° (311), which correspond to diamond (Satoshi et al. 2002). The two

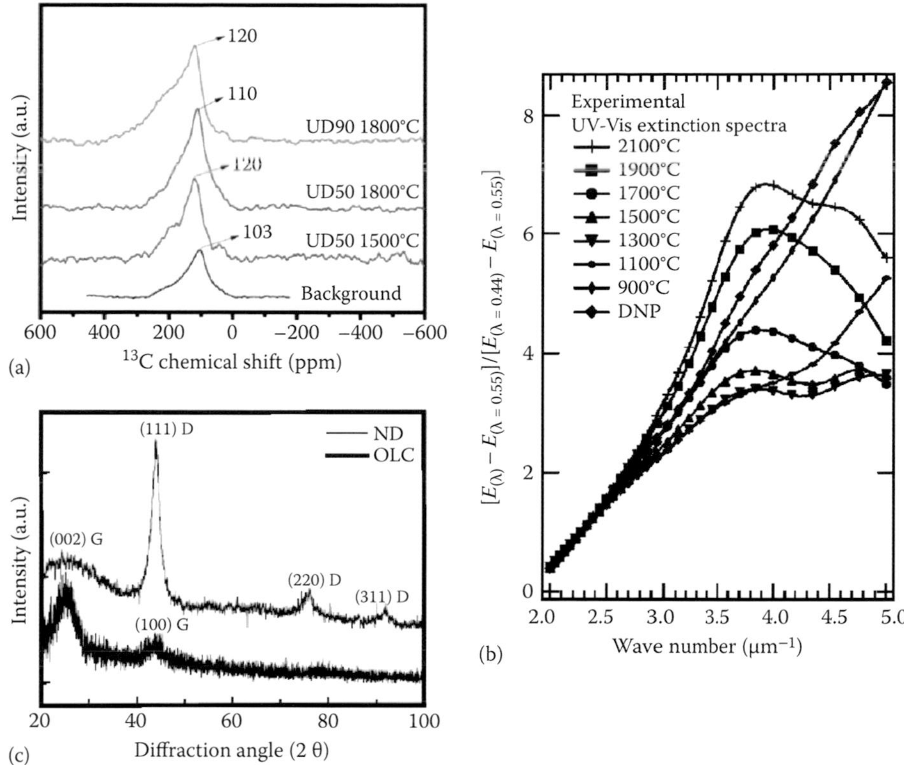

FIGURE 11.5 (a) ^{13}C NMR spectra of onion-like carbon structures synthesized at different temperatures for 3 h. (Reprinted from Costa et al., Copyright 2014, with permission from Elsevier.) (b) Experimental UV-Vis extinction spectra at different annealing temperatures. (Reprinted with permission from Tomita et al., 2002. Copyright 2002, American Institute of Physics.) (c) XRD patterns for ND particles and CNOs synthesized by heat treatment of the ND at 1600°C under N_2 for 1 h. (Reproduced from Lee et al., 2007. Copyright 2007, with permission from Springer.)

peaks at 2θ equal 25.8° (002), and 43.5° (100) are characteristic of hexagonal graphite. No diamond phase was found in the XRD profile for CNOs.

The crystallite size of carbon materials can be defined by L_c (thickness of crystallite perpendicular to the graphene sheets along the c-direction) and L_a (longitudinal crystallite size) (Guillaume et al. 1999). L_c can be calculated from the peaks corresponding to the (002) and (100) crystal planes using the Scherrer equation (Baldan et al. 2007). L_a can be calculated from the ratio of the integrated intensities of the D- and G-bands (I_D/I_G ratio) from the Raman spectra (Cancado et al. 2006). The L_a and L_c values for CNOs obtained from NDs are 1.06 nm and 2.23 nm, respectively (Cancado et al. 2006; Baldan et al. 2007).

The thermal stability of CNOs has been assessed by differential-thermogravimetric analyses (TGA-DTG-DTA) (Palkar et al. 2007; Plonska-Brzezinska et al. 2014). The measurements show that CNOs are stable up to around 520°C and start to decompose around 650°C, with an inflection temperature of 620°C (in an air atmosphere at 10°C/min). The lower thermal stability of N-CNOs compared with those of other carbon materials such as graphite and CNTs (~650°C) is probably caused by the high curvature and the presence of structural defects.

The spherical structure of CNOs can be gradually damaged by various physical and chemical factors. One of the most interesting properties of these structures is self-compression when the nanoparticles are exposed to electron or ion irradiation at high temperatures (Wesolowski et al. 1997; Banhart et al. 2002), annealing at 500°C in an air atmosphere (Tomita et al. 2000), using MeV Ne^+ (Wesolowski et al. 1997) and to CO_2 ions or laser irradiation (Wei et al. 1998). The distance between the compressed graphitic shells can reach values as low as 0.22 nm (Wesolowski et al. 1997). HRTEM observations and electron energy loss spectroscopy measurements clearly indicate that the transformation of CNOs into NDs is possible because of the presence of carbon atoms with sp^3 hybridization in the CNO structures in the presence of an oxygen atmosphere.

11.3.2 Physical Properties of CNOs Obtained from Arc-Discharge (A-CNOs)

"Big" CNOs (A-CNOs) obtained by arcing graphite underwater have been studied much less than N-CNOs. The size distribution of A-CNOs is around 15–25 nm (Imasaka et al. 2006), and their density is equal to 1.640 g/cm^3 (Sano et al. 2002), lower than for graphite (2.265 g/cm^3). The SSA of the floating A-CNO powder was determined by nitrogen gas adsorption based on the Brunauer–Emmett–Teller adsorption isotherm (Sano et al. 2002). The result was found to be 984.3 m^2/g. Raman spectroscopy of A-CNOs shows well-defined D- and G-bands at around 1313 and 1581 cm^{-1}, respectively (Palkar et al. 2007). The I_D/I_G ratio indicates a low number of defects on the graphitic layers (~0.8). The relatively low curvature of these structures (compared with those of N-CNOs) and the lack of defects result in low solubilities in polar solvents and in pronounced chemical inertness.

Finally, it should be noted that the Raman and TGA studies showed that N-CNOs are more defective than A-CNOs. The graphitic layers created during the formation of spherical CNOs have defects that may occur at specific points, in lines or as small volumes, and they can be described as (1) holey shells, (2) spiral-like, (3) "Y"-junction, and (4) protuberance structures (Roddatis et al. 2002). N-CNOs are more reactive and show better dispersion in organic and inorganic solvents than A-CNOs do because of their smaller size, high curvature, and surface defects (Palkar et al. 2007).

11.4 Chemical Reactivity of CNOs

It is well known that the reactivity of fullerene-like structures, including CNOs, decreases with increasing size as a result of a decrease in the curvature of the surface because of decreased strain (Hirsch 1999). The ability to functionalize CNO surfaces depends on the presence of defects on the carbon surface as well as on the presence of carbon atoms with sp^2 hybridization (Hamon et al. 2001). Covalent reactions can be performed on CNOs to attach multiple groups to their surfaces to improve their solubility. According to theoretical as well as experimental studies, two types of C–C bonds exist in the graphene layers of CNOs (Klein et al. 1986; David et al. 1991; Thilgen et al. 1997). These bonds range between 1.455 and 1.460 Å for the C–C bond between a hexagon and a pentagon and between 1.400 and 1.391 Å for the C–C bond between two hexagons, like in fullerenes. The lengths of these bonds correspond to the lengths observed for polyenic bonds in polyolefins, which are susceptible to addition reactions (Cataldo 2002).

11.4.1 Covalent Functionalization of CNOs

The main reason for functionalizing CNOs is to improve their solubility. Pristine CNOs are almost totally insoluble in protic and aprotic solvents (Bartelmess and Giordani 2014). One of the most often used reactions to increase the solubility of carbon nanoparticles in water solutions is strong oxidation with acids. This reaction also serves as the first stage for further functionalization. Oxidative treatments are very useful to introduce oxygen-containing groups on the surface of carbon nanostructures. Chemical oxidation on the surface of CNOs leads to the formation of aldehydes, ketones, esters, alcohols, and carboxylic acid moieties and creates defects on their graphitic structures (Hu et al. 2003; Zhang et al. 2003). Although oxidations lead to better dispersibility of CNOs in many protic solvents and provides the basis for further functionalization, the processes also cause structural changes on the carbon surfaces that disrupt the graphene sheet structures (Plonska-Brzezinska et al. 2011). After chemical oxidation, especially with acids, some of the inorganic ions (e.g., sulfide or nitride) are adsorbed on the carbon surface, which affects the physicochemical properties of these materials. CNOs can be oxidized with nitric acid (3M) by heating at reflux for 48 h (Palkar et al. 2007) or by using a mixture of nitric acid and sulfuric acid in a 1:3 volume ratio (Palkar et al. 2008).

Another oxidative method used to improve the dispersion of CNOs is ozonolysis, which is a milder method than treatment with strong acids (Plonska-Brzezinska et al. 2011). The final product, ozonated CNOs (oz-CNOs), is easily dispersed in aqueous solvents owing to their increased wettability and hydrophilicity. The process does not result in any major structural damage to the external graphene layer. oz-CNOs exhibit higher specific capacitances than pristine CNOs do. In 2013, the activation of the outer shells of CNOs using potassium hydroxide in concentrations from 4 to 7 mol/L was reported (Gao et al. 2013). Figure 11.6 shows the progressive destruction of the outer layer, which enhanced the porosity of the CNOs and improved their electrochemical properties. The specific capacitance for "activated" CNOs typically increases several-fold compared with their pristine precursors.

In 2008, Butenko et al. opened the carbon shells of CNOs and filled them up with potassium (Butenko et al. 2008). The opening process relied on the treatment of the pristine CNOs with a flow of carbon dioxide by partial oxidation to result in the "opened" surface CNOs. This stop allows the creation of core-shell structures or nanocapsules.

In 2014, nitrogen-doped onion-like carbon was synthesized (Lin et al. 2014). The CNOs were pretreated with concentrated nitric acid at 120°C for 4 h. The product was annealed at four different temperatures under an ammonia atmosphere for 4 h. The N-doped CNOs contained from 2.06% to 3.95% of nitrogen.

Various approaches using covalent or supramolecular strategies can be employed to solubilize CNOs. The 1,3-dipolar cycloaddition reaction with "big" CNOs was reported by Georgakilas et al. (2003), which led to increased dispersibility in many organic solvents (Scheme 11.1). A few years later and using the same 1,3-dipolar cycloaddition approach, pyrrolidine-derivatized CNOs were obtained (Scheme 11.1b), using α-amino acid and aldehydes with different alkyl chain lengths (Rettenbacher et al. 2006). "Small" CNOs were also functionalized using the 1,3-dipolar cycloaddition reaction in 2009 (Scheme 11.1c) (Cioffi et al. 2009). The pristine CNOs were functionalized by reacting with the amino acid and paraformaldehyde under reflux for 5 days. CNOs with covalently attached ferrocenes were thus obtained, representing the first reported donor–acceptor derivatives of CNOs.

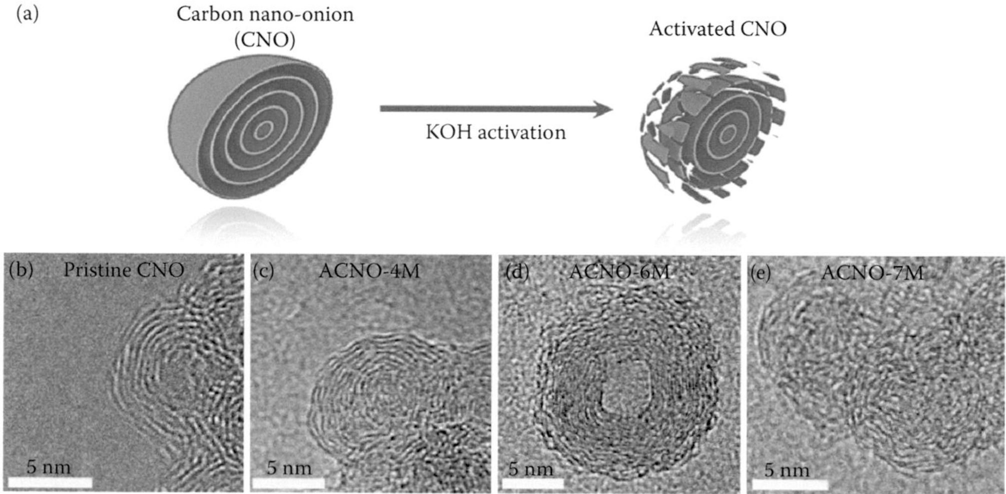

FIGURE 11.6 (a) A schematic showing the chemical activation of CNOs in KOH. TEM images of (b) pristine CNO, (c) ACNO-4M, (d) ACNO-6M, and (e) ACNO-7M. ACNO-nM denotes the activated CNO prepared using n mol/L KOH solution. (Reprinted from *Carbon*, 51, Gao, Y., Zhou, Y.S., Qian et al., Chemical activation of carbon nano-onions for high-rate supercapacitor electrodes, 52–58, Copyright 2013, with permission from Elsevier.)

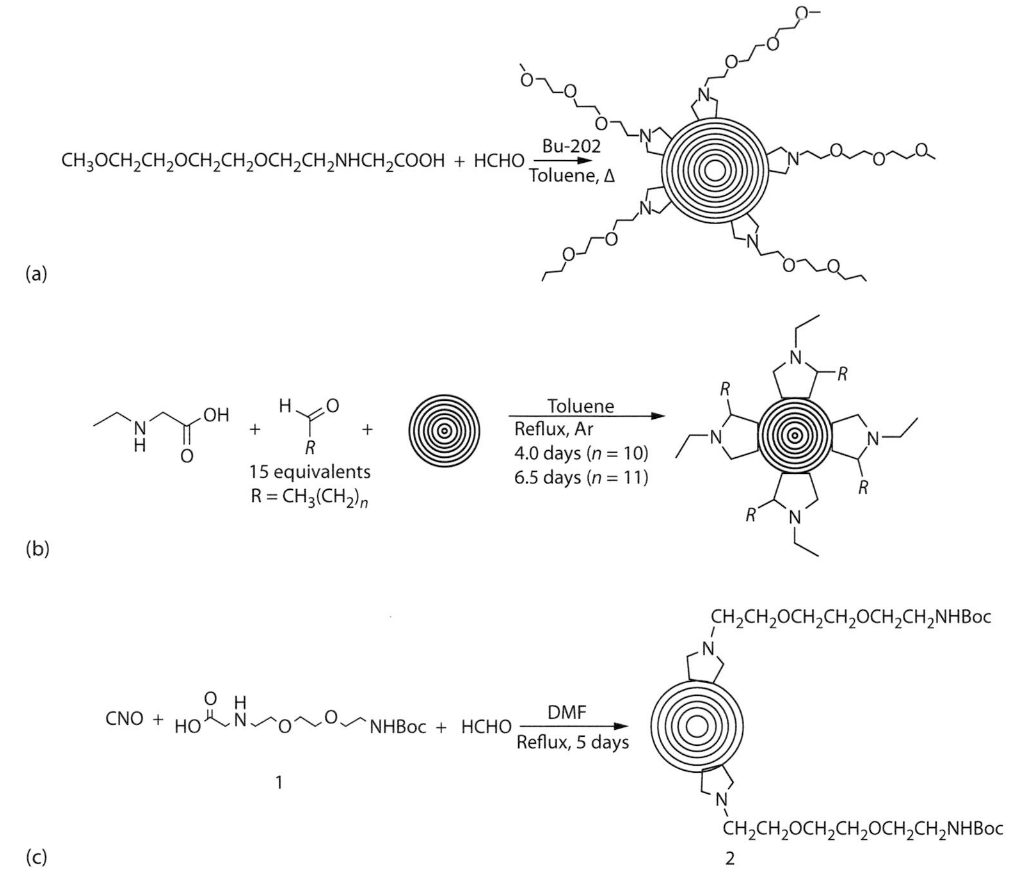

$CH_3OCH_2CH_2OCH_2CH_2OCH_2CH_2NHCH_2COOH$ + HCHO $\xrightarrow[\text{Toluene, } \Delta]{\text{Bu-202}}$

(a)

15 equivalents
$R = CH_3(CH_2)_n$

Toluene
Reflux, Ar
4.0 days ($n = 10$)
6.5 days ($n = 11$)

(b)

$CH_2CH_2OCH_2CH_2OCH_2CH_2NHBoc$

CNO + HO... NHBoc + HCHO $\xrightarrow[\text{Reflux, 5 days}]{\text{DMF}}$

1

$CH_2CH_2OCH_2CH_2OCH_2CH_2NHBoc$

(c) 2

SCHEME 11.1 1,3-dipolar cycloadditions to CNOs using (a) ((2-(2-methoxyethoxy)-ethoxy)ethylamino)acetic acid and paraformaldehyde. (Reprinted with permission from Georgakilas, V., Guldi, D.M., Signorini, R. et al., *J. Am. Chem. Soc.*, 125, 14268–14269, 2003. Copyright 2003, American Chemical Society.) (b) An α-amino acid and an aldehyde with different alkyl chains lengths. (Reproduced with permission from Rettenbacher, A.S., Elliott, B., Hudson, J.S., *Chem. Eur. J.*, 12, 376–387, 2006. Copyright 2006 John Wiley & Sons.) (c) An amino acid and paraformaldehyde. (Reproduced with permission from Cioffi, C.T., Palkar, A., Melin, F. et al., *Chem. Eur. J.*, 15, 4419–4427, 2009. Copyright 2009, John Wiley & Sons.)

The [2 + 1] cycloaddition reaction, known as the [2 + 1] Bingel-Hirsh cyclopropanation reaction, was used to functionalize CNOs. In 2007, Palkar et al. (2007) attempted the functionalization of "big" and "small" CNOs using the [2 + 1] cycloaddition reaction in the presence of dodecyl malonate ester, carbon tetrabromide, and 1,8-diazabicyclo[5.4.0]undec-7-ene in *o*-dichlorobenzene. This reaction was only successful to functionalize the "small" CNOs. This is likely because of their smaller size, high curvature, and surface defects (Hirsch 1999; Hamon et al. 2001).

The functionalization of "big" CNOs was also accomplished using a [2 + 1] cycloaddition of nitrenes to introduce hydroxyl and bromine groups on the surface (Zhou et al. 2009). These reactions were based on thermal decomposition of 2-azidoethanol and azidoethyl 2-bromo-2-methyl propanate to yield CNO-OH and CNO-Br, respectively. The products from these reactions were used as matrix materials for biodegradable poly(ε-carpolactone)- and polystyrene-grafted CNOs.

Amidation, a nondirect functionalization, occurs as a consequent modification of the oxidized CNOs. In 2006, functionalization of "big" CNOs by amidation and the preparation of water-soluble derivatives were reported for the first time (Rettenbacher et al. 2006). Amidation involving polyethylene

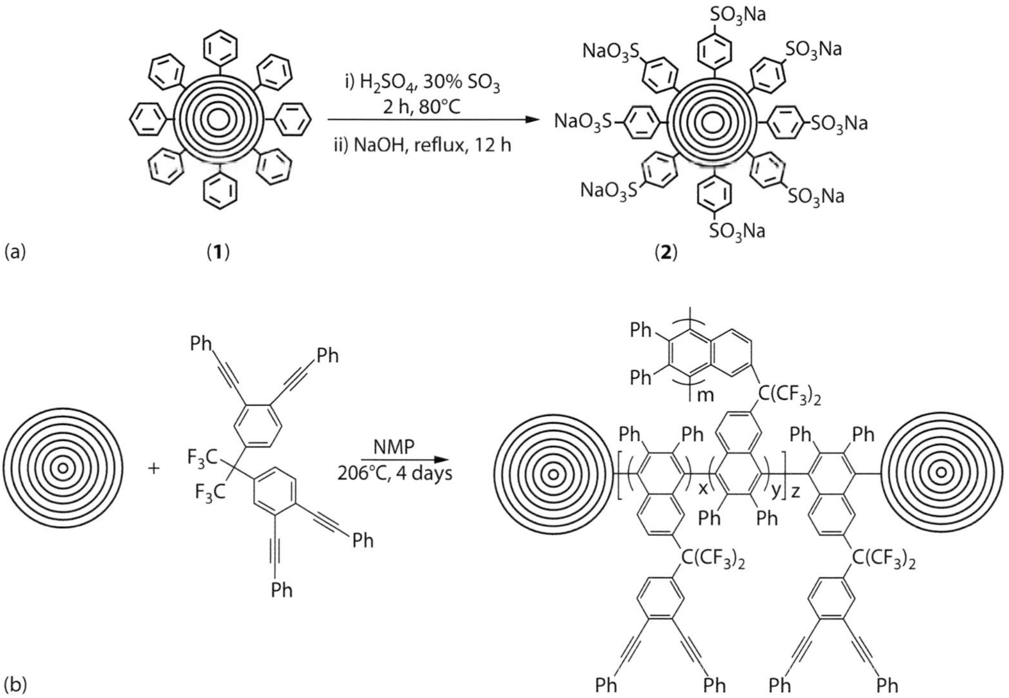

SCHEME 11.2 Examples of free-radical reaction: (a) sulfonation of CNOs. (Reproduced with permission from Palkar, A., Melin, F., Cardona, C.M. et al., *Chem. Asian J.*, 2, 625–633, 2007. Copyright 2007 John Wiley & Sons.) (b) Synthesis of BODA-CNO copolymers. (Reprinted with permission from Rettenbacher, A.S., Perpall, M.W., Echegoyen, L. et al., *Chem. Mater.*, 19, 1411 1417, 2007. Copyright 2007, American Chemical Society.)

glycol (PEG_{1500N}) or 1-octadecylamine (ODA) groups on the oxidized "big" CNOs was also successfully realized. The yield of this reaction was strongly dependent with time. The solid-state reaction between CNOs and ODA led to the functionalization of CNOs with the oligomeric alkyl amide chains. A few years later, the first water-soluble "small" CNOs prepared by amidation using 4-aminopyridine was obtained (Palkar et al. 2008), which were dispersible in water.

Free-radical reactions have been conducted with CNTs and similar procedures have been explored with CNOs. A free-radical reaction was conducted with "small" CNOs and benzoyl peroxide, which was the precursor of phenyl radicals, but only insoluble products (**1**) were obtained (Scheme 11.2a). Thus, the phenylated CNOs (Scheme 11.2a, product **1**) were further functionalized with 30% SO_3 in H_2SO_4 (Scheme 11.2a, product **2**) (Palkar et al. 2007). Another free-radical addition reaction was also employed with arcing CNOs using bis-*o*-diynylarene (BODA) as a monomer (Rettenbacher et al. 2007) directly in the presence of fluorine and hydrogen gases under different heating conditions (Liu et al. 2007). The fluorinated CNOs were used for further organic functionalization.

Electrophilic addition on the outer CNO shells was reported for the first time in 2013 (Molina-Ontoria et al. 2013). This reaction was conducted in two steps, using an initial strong reductive process with Na-K alloy in 1,2-dimethoxyethane solution, followed by the addition of 1-bromohexadecane as electrophile. As a result, truly soluble CNOs were obtained.

11.4.2 Composites Including CNOs

Composite materials based on the integration of carbon structures with other substances can lead to interesting materials possessing properties of the individual components. Many parameters, such as the degree of hydration and crystallinity of the materials, the potential degradation of the polymers used for the composites, and the decrease in the surface area, have to be taken into consideration for the successful preparation of composite materials (Furtado et al. 2004). For this material, both covalent as well as noncovalent functionalizations of CNOs have been utilized.

11.4.2.1 Composites with Polymers and Solid Polyelectrolytes

The combination of carbon materials with polymers yields new materials with interesting properties and promising application for electrochemical or photovoltaic purposes (Kuillaa et al. 2010; Pan et al. 2010; Spitalsky et al. 2010; Sun et al. 2011). A lot of attention has been paid to the preparation of composites of carbon nanostructure with polyaniline (Wu et al. 2006; Baranauskas et al. 2007; Lapinski et al. 2012; Plonska-Brzezinska et al. 2012b, 2013a; Cong et al. 2013; Wang et al. 2014). CNOs were also covalently functionalized with polyaniline (Lapinski et al. 2012; Plonska-Brzezinska et al. 2012b, 2013a). Using *p*-phenylenediamine (*p*-PDA) or 4-aminobenzoic acid (4-ABAC) in the first step of the reaction led to increased dispersibility of CNOs in protic solvents. This functionalization protocol enabled the homogenous polymerization of aniline on the CNO surfaces (Plonska-Brzezinska et al. 2012b).

One of the most promising modifications of CNOs, based on noncovalent functionalization by polymers, is functionalization with poly(ethylene glycol)/polysorbate (PEG/P20) or poly(4-vinylpyridyne-co-styrene) (PVPS) (Plonska-Brzezinska et al. 2013b). This approach provides a matrix that can be further functionalized by thiol derivatives, such as 3-mercaptopropionic and 2-mercapto-4-methyl-5-thiazoleacetic acids.

Poly(3,4-ethylenedioxythiophene):poly(styrenesulfonate) (PEDOT:PSS) is an important and common electrode material that possesses a conjugated backbone, which provides efficient transport of delocalized electrons (Kim et al. 2011; Alemu et al. 2012). CNOs have been modified with PEDOT:PSS using different mass ratios of polymer to CNOs (Plonska-Brzezinska et al. 2012a). Combining a highly conductive carbon material with an optically transparent polymer provides a potentially attractive composite for photovoltaic or electronic application (Antiohos et al. 2011; Sriprachuabwong et al. 2012).

The incorporation of CNOs into solid polyelectrolytes was achieved by ultrasonication of chitosan (Chit) or poly(diallyldimethylammonium chloride) (PDDA) solutions with dispersed carbon materials (Breczko et al. 2010). The composite electrodes showed a higher specific capacitance and a higher conductivity than the corresponding pristine components (CNOs or PDDA/Chit).

11.4.2.2 Integration CNOs with Metal Oxides or Hydroxides

One of the most popular modifications of carbon materials is their integration with inorganic components, such as metal oxides or hydroxides (Flahauta et al. 2000; Mazloumi et al. 2013). As a consequence, there is a significant increase in the capacitance, which allows these composites to be utilized as electrode material (Rakhi et al. 2011; Zhi et al. 2013). Metal oxides or hydroxides possess very low conductivity, low SSAs, poor power density, and low charge exchange rates (Wang et al. 2012a). Combining CNOs with these redox-active materials led to higher SSAs, energy densities, and specific capacitances (Borgohain et al. 2012; Wang et al. 2012; Plonska-Brzezinska et al. 2013a). The first functionalization of "small" CNOs by metal oxides was published in 2012 (Borgohain et al. 2012). The oxidized CNOs were added to a mixture of $RuCl_2$, glycolic acid, and NaOH. By chemical precipitation, CNOs oxidized in 3M HNO_3 (ox-CNOs) decorated with $RuO_2{\cdot}xH_2O$ and various RuO_2 loadings were obtained. In 2013, Plonska-Brzezinska et al. reported noncovalent functionalization of "small" CNOs using nickel hydroxide or oxide (Plonska-Brzezinska et al. 2013). These composites were obtained by chemical precipitation in the presence of a pyridyl derivative (4-dimethylaminopirydine). In 2012, Wang et al. published the first composites of "big" CNOs (20–30 nm) with manganese oxide (MnO_2) (Wang et al. 2012). The CNO/MnO_2 composites, obtained by a hydrothermal reaction, showed high electrochemical stability. All synthesized composites were used as electrode materials and showed higher specific capacitances than the corresponding pristine components.

11.4.2.3 CNOs and Metal Nanoparticles

Another frequently used functionalization method involves the decoration of carbon materials with metal nanoparticles (Georgakilas et al. 2007; Santiago et al. 2012; Yasin et al. 2012). These composites, especially with platinum, hold great promise as catalyst materials in new-generation fuel cells, e.g., in direct methanol fuel cells (Xu et al. 2006; Santiago et al. 2012). CNOs containing platinum nanoparticles can be synthesized using different methods. Xu et al. (2006) obtained a CNO/Pt catalyst using a reductive method. Immobilization of the metal particles on the CNO's surface was achieved by reduction of H_2PtCl_6 in formaldehyde for 8 h, and the final product contained a 20 wt% of the metal. Doping with Pt nanoparticles was also achieved via an electrochemical method using a rotating disk–slurry electrode with K_2PtCl_6 used as the platinum precursor (Santiago et al. 2012). Metal loading on carbon surface was calculated using TGA analysis and amounted to 11.5 wt%. The decorated CNOs showed high catalytic activity. For immobilization of Pt on the CNO surface, Goh et al. (2013) used a vortex fluidic device (VFD). The reaction was accomplished in VFD using an H_2PtCl_6 solution and ascorbic acid as the reducing agent. The VFD system was also used to attach palladium nanoparticles to the CNO surfaces (Yasin et al. 2012). The pristine CNOs were treated with *p*-phosphonic acid calix[8]arene to obtain a stable suspension of the carbon materials in water. Plasma etching resulted in a stable colloidal suspension without using surfactants. An analogous reaction using an H_2PdCl_4 solution and hydrogen gas as the reducing agent resulted in CNO/Pd composites (Yasin et al. 2012).

Composites of "small" ox-CNOs and C_{60}-Pd were investigated by Gradzka et al. (2013). The CNO films were deposited on a gold disk electrode via drop coating and C_{60}-Pd layers were then electrochemically deposited on them. The electrochemical polymerization of C_{60}-Pd was performed in an acetonitrile and toluene (4:1, v/v) mixture with palladium (II) acetate ($Pd(ac)_2$) and 0.1 M tetra(*n*-butyl)ammonium perchlorate as the supporting electrolyte. This led to a higher specific capacitance than that observed for the pristine CNOs, suggesting a potentially promising application as supercapacitors.

11.5 Potential Applications of CNOs

Do CNOs offer any real advantages over other carbon-based nanostructures? The main interest in CNOs arises from their unusual physicochemical properties as well as from promising applications in electronics, in optics, as hyperlubricants, and in energy conversion and storage. Although many areas in which CNOs may potentially be applied have already been established, most studies are still at the fundamental research level. Practical applications of these carbon nanostructures still await further advances in their synthesis, purification, and characterization.

11.5.1 Electrochemical, Electromagnetic, and Electro-Optical Applications of CNOs

Electrochemical energy storage devices such as batteries or capacitors, ultracapacitors, etc., have received considerable attention in recent years. An electrochemical capacitor is a device able to store charge at the electrode/electrolyte interface (Feng et al. 2013). This accumulation of charge is based mainly on two types of processes: electrostatic interactions (electrical double layer capacitors) or faradaic processes (pseudocapacitors, also called faradaic supercapacitors), which are based on redox-reactions (Simon and Gogotsi 2008). Supercapacitors possess high power densities, high energy densities, and long cycling stabilities (Miller and Simon 2008; Bose et al. 2012; Li and Wei 2013). The ideal material for supercapacitors should possess good conductivity, high SSA, high porosity, and high chemical stability (Bushueva et al. 2008). The use of carbon nanostructures in composites results in increases in the specific capacitances, decreases in electrode resistances, and increases in their mechanical stabilities (Bushueva et al. 2008). The CNO surfaces are easily accessible to electrolyte ions, thus making them attractive for high-rate supercapacitor electrodes (Gao et al. 2013). Recently, a lot of attention has been paid to the capacitive

TABLE 11.3 Specific Capacitance of CNOs and Their Composites

Material	Specific Capacitance (F/g)	Electrolyte	Reference
CNOs	~7	0.1 M NaCl	Plonska-Brzezinska et al. 2012a
CNO/KOH activation	122	2 M KNO_3	Gao et al. 2013
CNO/HNO_3 activation	~3	0.1 M NaCl	Plonska-Brzezinska et al. 2012a
ox-CNOs	45	1.0 M H_2SO_4	Borgohain et al. 2012
oz-CNOs	~4–43	Different electrolytes	Plonska-Brzezinska et al. 2011
CNOs	27	1.0 M H_2SO_4	Borgohain et al. 2012
ox-CNOs	45		
ox-CNOs/$RuO_2{\cdot}H_2O$ (14.9–67.5 wt% RuO_2)	96–334		
CNOs	31–39[a]	1 M KOH	Plonska-Brzezinska et al. 2013c
CNOs/$Ni(OH)_2$	1225–727[a]		
CNOs/NiO	291–273[a]		
CNOs/MnO_2	178	1 M Na_2SO_4	Wang et al. 2012
CNOs/MnO_2	1207–216[b]	0.5 M H_2SO_4	Azhagan et al. 2014
CNOs/PEDOT:PSS (1:0.1)	~36	0.1 M NaCl	Plonska-Brzezinska et al. 2012a
CNOs/PEDOT:PSS (1:0.25)	~40		
CNOs/PEDOT:PSS (1:0.5)	~75		
CNOs/PEDOT:PSS (1:1)	~95		
Ox-CNOs/PEDOT:PSS (1:1)	~47		
CNOs	30[c]	1 M H_2SO_4	Anjos et al. 2013
CNOs/1,4-naphthoquinone	91[c]		
CNOs/9,10-phenthrenequinone	124–267[c]		
CNOs/4,5-pyrenedione	130[c]		
ox-CNOs/C_{60}-Pd (2:1)	20–62[d]	0.1 M $TBAClO_4$ in acetonitryle	Gradzka et al. 2013
ox-CNOs/C_{60}-Pd (5:1)	14–86[d]		
ox-CNOs/C_{60}-Pd (10:1)	17–189[d]		
ox-CNOs/C_{60}-Pd (15:1)	17–284[d]		
ox-CNOs/*p*-PDA	~9[e]	0.1 M H_2SO_4	Plonska-Brzezinska et al. 2012b
CNOs/4-ABAC	34[e]		
ox-CNOs/*p*-PDA/polyaniline	155[e]		
CNOs/4-ABAC/polyaniline	~207[e]		
CNOs/Chitosan (1:1.5)	~9–23[f]	0.1 M H_2SO_4	Breczko et al. 2010
CNOs/Chitosan (1:6.5)	4–32[f]		
CNOs/PDDA (1:1)	9–19[f]	0.1 M H_2SO_4	Breczko et al. 2010
CNOs/PDDA (1:4.5)	6–33[f]		

Note: Given in parentheses are the mass ratios of CNO: material (m/m).
[a] Capacitance calculated for v = 5–30 mV/s.
[b] Capacitance calculated for scan rate from 1 to 100 mV/s.
[c] Capacitance calculated for v > 5 mV/s.
[d] Capacitance was calculated per gram of composites or per gram of C_{60}-Pd.
[e] Capacitance calculated for v = 5 mV/s.
[f] Capacitance was calculated per gram of composites or per gram of CNOs.

behavior of pristine CNOs and their composites (Table 11.3). Composites including CNOs show excellent properties such as long cycle life and high reversible capacity (Han et al. 2011), so they can be used as electrode materials in rechargeable lithium ion batteries (Gu et al. 2013). Liu et al. (2013) investigated Ni/C nanocapsules with CNO shells as an anode material for lithium ion batteries and they exhibited excellent properties, such as high charge–discharge rates and cycling stabilities.

CNTs are potentially attractive as point sources, but the main disadvantage is their length (Okano et al. 1994; Bonard et al. 2002). The longer CNTs show lower image resolutions (Treacy et al. 1996), and a possible improvement could result by using shorter multiwalled CNTs that somewhat resemble CNOs. One report described the use of CNOs for the construction of a point source field-emitter (see Figure 11.7) (Wang et al. 2010). The emitter showed a low threshold voltage (Mamezaki et al. 2000), a large emission current, and good emission endurance (Wang et al. 2010). Sek et al. (2013) have studied the electronic conductance of a single CNO and showed that the conductance exhibited metallic behavior. This result is promising for the use of CNOs in molecular electronic applications.

Electromagnetic wave attenuation by CNOs over a broad frequency range (12–230 THz; Shenderova et al. 2008) showed that CNOs are potentially useful for electromagnetic shielding (Maksimenko et al. 2007). The electromagnetic shielding properties of CNOs result from defects on their outer shell, their small sizes, and their high reactivities (Shenderova et al. 2008). Research on terahertz probing was also conducted using CNO composites, and these showed excellent spectral features up to 3 THz, which depend significantly on the films' preparation and concentration of CNO inclusions (Macutkevic et al. 2008). These results show a promising high potential of CNO-polymethyl methacrylate composites for the design of electromagnetic shielding materials over a broad microwave (2–38 GHz) frequency range.

The nonlinear optical response of CNOs has also been studied (Koudoumas et al. 2002). From these studies, it was found that the optical limiting action of CNOs is very strong, higher than that observed for C_{60} and ND (Koudoumas et al. 2002). In 2003, Georgakilas et al. studied the nonlinear optical properties of modified CNOs with ((2-(2-methoxyethoxy)-ethoxy)ethylamino)acetic acid and paraformaldehyde (Georgakilas et al. 2003). CNOs are very attractive candidates as optical limiters to protect delicate optical devices and the human eye (Gao et al. 2011).

11.5.2 Biomedical Applications of CNOs

The potential application of nanomaterials in biosystems necessarily requires that they exhibit no toxicity. Ding et al. (2005) investigated the cytotoxicity of "big" CNOs on human skin fibroblasts. They

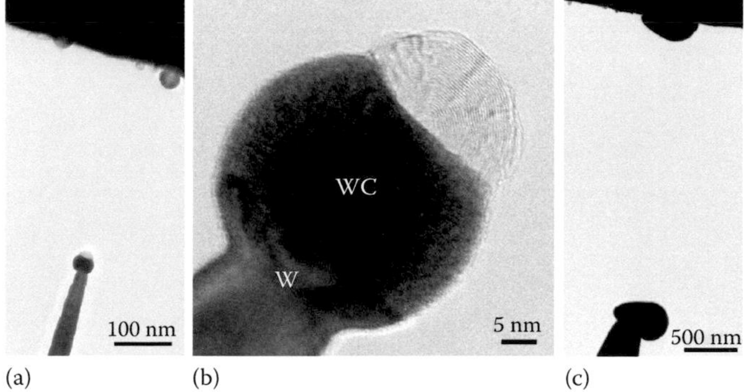

FIGURE 11.7 (a) A CNO emitter prepared for an FE measurement and (b) the corresponding TEM image of the emitter end. (c) The final structure of the emitter destroyed because of an arcing event at a high FE current. (Reprinted with permission from Wang, M.-S. et al., *ACS Nano*, 8, 4396–4402, 2010. Copyright 2010, American Chemical Society.)

reported that these nanostructures can seriously impact cellular functions during growth and differentiation. The cytotoxicity of "small" CNOs on human fibroblasts was also investigated (Luszczyn et al. 2010). It was found that pristine CNOs and their derivatives (oxidized and pegylated) are nontoxic and can be safely used in biological applications. As mentioned previously, oxidized CNOs can be used as a matrix for fabricating novel supramolecular architectures. Luszczyn et al. (2010) reported the covalent functionalization of "small" CNOs with biomolecules, making use of the well-established noncovalent but strong interaction between avidin and vitamin H (biotin). In these studies, surface plasmon resonance was employed to detect self-assembled organic monolayers and to monitor biomolecular interactions. The oxidized CNOs were incorporated into self-assembled monolayers of the optical biosensor and they changed the structural properties of the surface, leading to an enhanced signal for avidin. It was shown that protein molecules attached to the CNOs retain their biological activity. These studies suggest that CNOs can be potentially used as a platform for immobilizing biomolecules since they retain their bioactivity and increase the sensitivity of the method.

The high conductivity of CNO composites and their ion-exchange properties make them useful for the determination of some biologically active substances. Composites containing CNOs in combination with cationic polyelectrolytes (PDDA) can be used for the electrochemical determination of dopamine (DA, 2-(3,4-dihydroxyphenyl)ethylamine) over a broad concentration range in the presence of ascorbic and uric acids (Breczko et al. 2012). These two acids exist together with DA in high concentrations in biological samples. A CNO/PDDA electrode was also employed to quantitate drugs such as methyldopa. The CNO/PDDA composite exhibited reversible redox peaks for the active substances. The CNO/PVPS and CNO/PEG/P20 composites were applied as matrices for the immobilization of flavonoids (Plonska-Brzezinska et al. 2013b). These systems provide many opportunities for further functionalization and can be used as biosensors or as carriers of active substances.

Another potential application of CNOs is to incorporate fluorescent reagents in fruit flies (Figure 11.8) (Sonkar et al. 2012). This is a noninvasive and nontoxic technique that can be conducted in vivo (Ghosh et al. 2011). In 2014, CNOs were covalently functionalized with a high fluorescent dye-boron

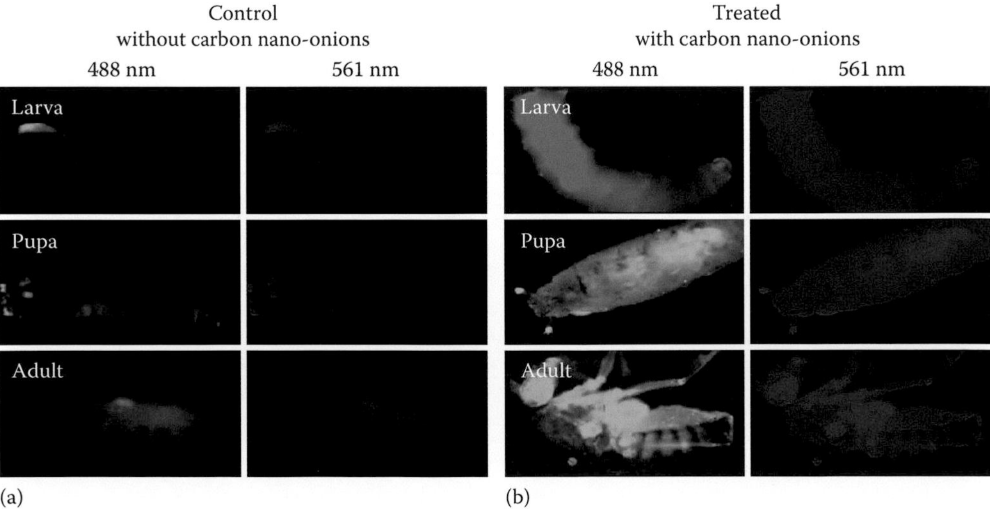

FIGURE 11.8 (a) Autofluorescence images of different development stages of *Drosophila melanogaster* from larva to adult. (b) Water-soluble CNO-fed *Drosophila melanogaster* under 488- and 561-nm filters. For imaging the larval to adult stages, movements of the living objects were controlled by anaesthetization using diethyl ether vapor. (Ghosh, M., Sonkar, S. K., Saxena, M., Sarkar, S.: Carbon nano-onions for imaging the life cycle of *Drosophila melanogaster*. *Small*. 2011. 22. 3170–7. Copyright Wiley-VCH Verlag GmbH & Co. KGaA. Reproduced with permission.)

dipyrromethene (BODIPY) (Bartelmess et al. 2014). This covalent construct was used for the high-resolution imaging of MCF-7 human breast cancer cells. CNOs could play a crucial role as nanoplatforms for a variety of applications and could be used for different biological and diagnostic applications (Bartelmess et al. 2014).

11.5.3 Other Applications of CNOs

CNOs have shown excellent tribological properties (Sano et al. 2001; Street et al. 2004). They exhibit better antiwear and friction-reducing properties than graphite powder does (Joly-Pottuz et al. 2008a,b) and have exhibited better tribological properties than bigger CNOs did (Hirata et al. 2004; Echegoyen et al. 2010). CNOs have large specific areas and chemical stabilities (Butenko et al. 2000; Borgohain et al. 2012), which are excellent properties for new catalytic nanomaterials. The combination of CNOs with Pt results in high thermal stability and makes it possible to use them for the efficient electrochemical oxidation of methanol in fuel cells (Xu et al. 2006; Santiago et al. 2012). CNOs have also been studied as catalysts for the oxidative dehydrogenation of ethylbenzene to styrene (Keller et al. 2002; Su et al. 2007). In 2014, Lin et al. studied nitrogen-doped CNOs as novel metal-free catalysts for the epoxidation of styrene (Lin et al. 2014). CNOs are also promising materials for gas storage because they possess high SSAs (around 984.3 m^2/g) (Sano et al. 2002). Zhang et al. showed that CNOs could be used for hydrogen storage in fuel cells. After electrochemical measurements, a 1.76 wt% of hydrogen storage capacity was found (Zhang et al. 2012).

11.6 Conclusions

CNOs represent one of the most interesting forms of carbon, mainly because of their unique 0D structure, small diameter, high electrical conductivity, and relatively easy dispersion in several solvents compared with other larger structures such as nanotubes and graphene. The unique spherical structure, which can be easily modified to other forms, offers a lot of opportunities. The high CNO conductivity is potentially important for energy storage devices, mainly as active materials for supercapacitor electrodes and for low-temperature devices using ionic liquid electrolytes. CNOs have large specific areas and chemical stabilities, which are promising for new catalytic nanomaterials and for gas storage devices. Another potential application is as biosensors and in bioelectrochemistry, which requires excellent biocompatibility and negligible cytotoxicities. Many reports indicate that CNOs are nontoxic over a wide range of concentrations, both as nonmodified and as functionalized nanostructures. Additionally, the size (diameter of 5–6 nm) of CNOs makes them compatible with in vitro and in vivo studies. CNOs could play a crucial role as nanoplatforms for a variety of applications, including biological and diagnostic.

Acknowledgments

We gratefully acknowledge the financial support of the National Science Centre, Poland (grants 2011/01/B/ST5/06051 and 2012/05/E/ST5/0380)0 to M.E.P.-B. L.E. thanks the Robert A. Welch Foundation for an endowed chair (grant H-0033) and the US NSF (grants DMR-1205302 and CHE-1408865).

References

Alemu, D., Wei, H.-Y., Ho, K.-C., Chu, C.-W., "Highly conductive PEDOT:PSS electrode by simple film treatment with methanol for ITO-free polymer solar cells," *Energy Environ. Sci.* 5 (2012): 9662–71.

Al-Jishi, R., Dresselhaus, G., "Lattice-dynamical model for graphite," *Phys. Rev. B* 26 (1982): 4514–22.

Anjos, D. M., McDonough, J. K., Perre, E. et al., "Pseudocapacitance and performance stability of quinone-coated carbon onions," *Nano Energy* 2 (2013): 702–12.

Antiohos, D., Folkes, G., Sherrell, P. et al., "Compositional effects of PEDOT-PSS/single walled carbon nanotube films on supercapacitor device performance," *J. Mater. Chem.* 21 (2011): 15987–94.

Azhagan, M. V. K., Vaishampayan, M. V., Shalke, M. V., "Synthesis and electrochemistry of pseudo-capacitive multilayer fullerenes and MnO_2 nanocomposites," *J. Mater. Chem. A* 2 (2014): 2152–9.

Bacon, R., "Growth, structure, and properties of graphite whiskers," *J. Appl. Phys.* 31 (1960): 283–90.

Banhart, F., "Structural transformations in carbon nanoparticles induced by electron irradiation," *Phys. Solid State* 44 (2002): 399–404.

Banhart, F., Redlich, Ph., Ajayan, P. M., "The migration of metal atoms through carbon onions," *Chem. Phys. Lett.* 292 (1998): 554–60.

Baranauskas, V., Ceragioli, H. J., Peterlevitz, A. C., Quispe, J. C. R., "Properties of carbon nanostructures prepared by polyaniline carbonization," *J. Phys. Conf. Ser.* 61 (2007): 71–4.

Bartelmess, J., Giordani, S., "Carbon nano-onions (multi-layer fullerenes): Chemistry and applications," *Beilstein J. Nanotechnol.* 5 (2014): 1980–98.

Bartelmess, J., De Luca, E., Signorelli, A. et al., "Boron dipyrromethene (BODIPY) functionalized carbon nano-onions for high resolution cellular imaging," *Nanoscale* 6 (2014): 13761–9.

Bates, K. R., Scuseria, G. E., "Why are buckyonions round?," *Theor. Chem. Acc.* 99 (1998): 29–33.

Bekyarova, E., Itkis, M. E., Ramesh, P. et al., "Chemical modification of epitaxial graphene: Spontaneous grafting of aryl groups," *J. Am. Chem. Soc.* 131 (2009): 1336–7.

Bom, D., Andrews, R., Jacques, D. et al., "Thermogravimetric analysis of the oxidation of multiwalled carbon nanotubes: Evidence for the role of defect sites in carbon nanotube," *Chem. Nano. Lett.* 2 (2002): 615–9.

Bonard, J.-M., Croci, M., Klinke, C., Kurt, R., Noury, O., Weiss, N., "Carbon nanotube films as electron field emitters," *Carbon* 40 (2002): 1715–28.

Borgohain, R., Li, J., Selegue, J. P., Cheng, Y.-T., "Electrochemical study of functionalized carbon nano-onions for high-performance supercapacitor electrodes," *J. Phys. Chem. C* 116 (2012): 15068–75.

Bose, S., Kuila, T., Mishra, A. K., Rajasekar, R., Kim, N. H., Lee, J. H., "Carbon-based nanostructured materials and their composites as supercapacitor electrodes," *J. Mater. Chem.* 22 (2012): 767–84.

Breczko, J., Winkler, K., Plonska-Brzezinska, M. E., Villalta-Cerdas, A. V., Echegoyen, L., "Electrochemical properties of composites containing small carbon nano-onions and solid polyeletrolytes," *J. Mater. Chem.* 20 (2010): 7761–8.

Breczko, J., Plonska-Brzezinska, M. E., Echegoyen, L., "Electrochemical oxidation and determination of dopamine in the presence of uric and ascorbic acids using a carbon nano-onion and poly(diallyldimethylammonium chloride) composite," *Electrochim. Acta* 72 (2012): 61–7.

Bushueva, E. G., Galkin, P. S., Okotrub, A. V. et al., "Double layer supercapacitor properties of onion-like carbon materials," *Phys. Status Solidi B* 245 (2008): 2296–9.

Butenko, Y. V., Kuznetsov, V. L., Chuvilin, A. L. et al., "Kinetics of the graphitization of dispersed diamonds at 'low' temperatures," *J. Appl. Phys.* 88 (2000): 4380–8.

Butenko, Y. V., Chakraborty, A. K., Peltekis, N. et al., "Potassium intercalation of carbon onions 'opened' by carbon dioxide treatment," *Carbon* 46 (2008): 1133–40.

Bystrzejewski, M., Rummeli, M. H., Gemming, T., Lange, H., Huczko, A., "Catalyst-free synthesis of onion-like carbon nanoparticles," *New Carbon Mater.* 25 (2010): 1–8.

Cabioc'h, T., Kharbach, A., Le Roy, A., Riviere J. P., "Fourier transform infra-red characterization of carbon onions produced by carbon-ion implantation," *Chem. Phys. Lett.* 285 (1998): 216–20.

Cabioc'h, T., Jaouen, M., Riviere, J. P., Delafond, J., Hug, G., "Characterization and growth of carbon phases synthesized by high temperature carbon ion implantation into copper," *Diam. Relat. Mater.* 6 (1997): 261–5.

Cataldo, F., "Polymeric fullerene oxide (fullerene ozopolymers) produced by prolonged ozonation of C_{60} and C_{70} fullerenes," *Carbon* 40 (2002): 1457–67.

Chen, X. H., Deng, F. M., Wang, J. X. et al., "New method of carbon onion growth by radio-frequency plasma-enhanced chemical vapor deposition," *Chem. Phys. Lett.* 336 (2001): 201–4.

Choucair, M., Stride, J. A., "The gram-scale synthesis of carbon onions," *Carbon* 50 (2012): 1109–15.

Cioffi, C. T., Palkar, A., Melin, F. et al., "A carbon nano-onion–ferrocene donor–acceptor system: Synthesis, characterization and properties," *Chem. Eur. J.* 15 (2009): 4419–27.

Cong, H.-P., Ren, X.-C., Wang, P., Yu, S.-H., "Flexible graphene–polyaniline composite paper for high-performance supercapacitor," *Energy Environ. Sci.* 6 (2013): 1185–91.

Crestani, M. G., Puente-Lee, I., Rendon-Vazquez, L. et al., "The catalytic reduction of carbon dioxide to carbon onion particles by platinum catalysts," *Carbon* 43 (2005): 2621–4.

David, W. I. F., Ibberson, R. M., Matthewman, J. C. et al., "Crystal structure and bonding of ordered C_{60}," *Nature* 353 (1991): 147–9.

Ding, L., Stilwell, J., Zhang, T. et al., "Molecular characterization of the cytotoxic mechanism of multiwall carbon nanotubes and nano-onions on human skin fibroblast," *Nano Lett.* 12 (2005): 2448–64.

Dresselhaus, M. S., Dresselhaus, G., Jorio, A., Souza Filho, A. G., Saito, R. "Raman spectroscopy of isolated single wall carbon nanotubes," *Carbon* 40 (2002): 2043–61.

Du, J., Zhao, R., Zhu, Z., "A facile approach for synthesis and in situ modification of onion-like carbon with molybdenum carbide," *Phys. Status Solidi A* 208 (2011): 878–81.

Echegoyen, L., Ortiz, A., Chaur, M. N., Palkar, A. J., "Carbon nano onions," in *Chemistry of Nanocarbons*, eds. T. Akasaka, F. Wudl and S. Nagase (Chichester, John Wiley & Sons, 2010), 463–83.

Fan, J.-C., Sung, H.-H., Lin, C.-R., Lai, M.-H., "The production of onion-like carbon nanoparticles by heating carbon in a liquid alcohol," *J. Mater. Chem.* 22 (2012): 9794–7.

Feng, G., Li, S., Presser, V., Cummings, P. T., "Molecular insights into carbon supercapacitors based on room-temperature ionic liquids," *J. Phys. Chem. Lett.* 4 (2013): 3367–76.

Flahauta, E., Peigneya, A., Laurenta, C., Marlièreb, C., Chastela, F., Rousseta, A., "Carbon nanotube–metal–oxide nanocomposites: Microstructure, electrical conductivity and mechanical properties," *Acta Mater.* 48 (2000): 3803–12.

Furtado, C. A., Kim, U. J., Gutierezz, H. R., Pan, L., Dickey, E. C., Eklund, P. C., "Debundling and dissolution of single-walled carbon nanotubes in amide solvents," *J. Am. Chem. Soc.* 126 (2004): 6095–105.

Gan, Y., Banhart, F., "The mobility of carbon atoms in graphitic nanoparticles studied by the relaxation of strain in carbon onions," *Adv. Mater.* 20 (2008): 4751–4.

Gao, Y., Zhou, S. Z., Park, J. B. et al., "Resonant excitation of precursor molecules in improving the particle crystallinity, growth rate and optical limiting performance of carbon nano-onions," *Nanotechnology* 22 (2011): 165604–10.

Gao, Y., Zhou, Y. S., Qian et al., "Chemical activation of carbon nano-onions for high-rate supercapacitor electrodes," *Carbon* 51 (2013): 52–8.

Garcia-Martin, T., Rincon-Arevalo, P., Campos-Martin, G., "Method to obtain carbon nano-onions by pyrolisys of propane," *Cent. Eur. J. Phys.* 11 (2013): 1548–58.

Georgakilas, V., Guldi, D. M., Signorini, R., Bozio, R., Prato, M., "Organic functionalization and optical properties of carbon onions," *J. Am. Chem. Soc.* 125 (2003): 14268–9.

Georgakilas, V., Gournis, D., Tzitzios, V., Pasquato, L., Guldi, D. M., Prato, M., "Decorating carbon nanotubes with metal and semiconductor nanoparticles," *J. Mater. Chem.* 17 (2007): 2679–94.

Ghosh, M., Sonkar, S. K., Saxena, M., Sarkar, S., "Carbon nano-onions for imaging the life cycle of *Drosophila melanogaster*," *Small* 22 (2011): 3170–7.

Goh, Y. A., Chen, X., Yasin, F. M. et al., "Shear flow assisted decoration of carbon nano-onions with platinum nanoparticles," *Chem. Commun.* 49 (2013): 5171–3.

Gradzka, E., Winkler, K., Borowska, M., Plonska-Brzezinska, M. E., Echegoyen, L., "Comparison of the electrochemical properties of thin films of MWCNTs/C_{60}-Pd, SWCNTs/C_{60}-Pd and ox-CNOs/C_{60}-Pd," *Electrochim. Acta* 96 (2013): 274–84.

Gu, W., Peters, N., Yushin, G., "Functionalized carbon onions, detonation nanodiamond and mesoporous carbon as cathodes in Li-ion electrochemical energy storage devices," *Carbon* 53 (2013): 292–301.

Guo, J., Wang, X., Xu, B., "One-step synthesis supported platinum nanoparticles by discharge in an aqueous solution," *Mater. Chem. Phys.* 113 (2009): 179–82.

Hamon, M. A., Hu, H., Bhowmik, P. et al., "End-group and defect analysis of soluble single-walled carbon nanotubes," *Chem. Phys. Lett.* 347 (2001): 8–12.

Han, M. Y., Oezyilmaz, B., Zhang, Y., Kim, P., "Energy bandgap engineering of graphene nanoribbons," *Phys. Rev. Lett.* 98 (2007): 206805-1–4.

Han, F.-D., Yao, B., Bai, Y.-J., "Preparation of carbon nano-onions and their application as anode materials for rechargeable lithium-ion batteries," *J. Phys. Chem. C* 115 (2011): 8923–7.

He, C., Zhao, N., Shi, C., Du, X., Li, J., "Carbon nanotubes and onions from methane decomposition using Ni/Al catalysts," *Mater. Chem. Phys.* 97 (2006): 109–15.

He, C. N., Zhao, N. Q., Shi, C. S., Song, S. Z., "Fabrication of carbon nanomaterials by chemical vapor deposition," *J. Alloy Comp.* 484 (2009): 6–11.

Hirata, A., Igarashi, M., Kaito, T., "Study on solid lubricant properties of carbon onions produced by heat treatment of diamond clusters or particles," *Tribol. Int.* 37 (2004): 899–905.

Hirsch, A., "Principles of fullerene reactivity," in *Fullerenes and Related Structures*, ed. A. Hirsch (Berlin: Springer, 1999), 1–65.

Hu, H., Zhao, B., Itkis, M. E., Haddon, R. C., "Nitric acid purification of single-walled carbon nanotubes," *J. Phys. Chem. B* 107 (2003): 13838–42.

Hu, S., Bai, P., Tian, F., Cao, S., Sun, J., "Hydrophilic carbon onions synthesized by millisecond pulsed laser irradiation," *Carbon* 47 (2009): 876–83.

Imasaka, K., Kanatake, Y., Ohshiro, Y., Suehiro, J., Hara, M., "Production of carbon nanoonions and nanotubes using an intermittent arc discharge in water," *Thin Solid Films* 506–507 (2006): 250–4.

Janssen, R. A. J., Hummelen, J. C., Saricifti, N. S., "Polymer–fullerene bulk heterojunction solar cells," *MRS Bull.* 30 (2005): 33–6.

Jin, C., Suenaga, K., Iijima, S., "In situ formation and structure tailoring of carbon onions by high-resolution transmission electron microscopy," *Chem. Phys. Lett.* 113 (2009): 5043–6.

Joly-Pottuz, L., Matsumoto, N., Kinoshita, H. et al., "Diamond-derived carbon onions as lubricant additives," *Tribol. Int.* 41 (2008a): 69–78.

Joly-Pottuz, L., Vacher, B., Ohmae, N., Martin, J. M., Epicier, T., "Anti-wear and friction reducing mechanisms of carbon nano-onions as lubricant additives," *Tribol. Lett.* 30 (2008b): 69–80.

Keller, N., Maksimova, N. I., Roddatis, V. V. et al., "The catalytic use of onion-like carbon materials for styrene synthesis by oxidative dehydrogenation of ethylbenzene," *Angew. Chem. Int. Ed.* 41 (2002): 1885–8.

Kim, Y. H., Sachse, C., Machala, M. L., May, C., Müller-Meskamp, L., Leo, K., "Highly conductive PEDOT:PSS electrode with optimized solvent and thermal post-treatment for ITO-free organic solar cells," *Adv. Funct. Mater.* 21 (2011): 1076–81.

Klein, D. J., Seitz, W. A., Schmalz, T. G., "Icosahedral symmetry carbon cage molecules," *Nature* 323 (1986): 703–6.

Knight, D. S., White, W. B., "Characterization of diamond films by Raman spectroscopy," *J. Mater. Res.* 4 (1989): 385–93.

Kobayashi, T., Sekine, T., He, H., "Formation of Carbon Onion from Heavily Snocked SiC," *Chem. Mat.* 15 (2003): 2681–3.

Konno, T. J., Sinclairt, R., "Crystallization of amorphous carbon in carbon-cobalt layered thin films," *Acta Met. Mater.* 43 (1995): 471–84.

Koudoumas, E., Kokkinaki, O., Konstantaki, M. et al., "Onion-like carbon and diamond nanoparticles for optical limiting," *Chem. Phys. Lett.* 357 (2002): 336–40.

Kroto, H. W., Heath, J. R., Obrien, S. C., Curl, R. F., Smalley, R. E., "C_{60}: Buckminsterfullerene," *Nature* 318 (1985): 162–3.

Kuillaa, T., Bhadrab, S., Yaoa, D., Kimc, N. H., Bosed, S., Lee, J. H., "Recent advances in graphene based polymer composites," *Prog. Polym. Sci.* 35 (2010): 1350–75.

Kuznetsov, V. L., Moseenkov, S. I., Elumeeva, K. V. et al., "Comparative study of reflectance properties of nanodiamonds, onion-like carbon and multiwalled carbon nanotubes," *Phys. Status Solidi B* 248 (2011): 2572–6.

Kuznetsov, V. L., Zilberberg, I. L., Butenko, Y. V., Chuvilin, A. L., Segall, B., "Theoretical study of the formation of closed curved graphite-like structures during annealing of diamond surface," *J. Appl. Phys.* 86 (1999): 863–70.

Kuznetsov, V., Moseenkov, S., Ischenko, A. et al., "Controllable electromagnetic response of onion-like carbon based materials," *Phys. Status Solidi* 245 (2008): 2051–4.

Lamber, R., Jaeger, N., Schulz-Ekloff, G., "Electron microscopy study of the interaction of Ni, Pd and Pt with carbon: Interaction of palladium with amorphous carbon," *Surface Science* 227 (1990): 15–23.

Lapinski, A., Dubis, A. T., Plonska-Brzezinska, M. E., Mazurczyk, J., Breczko, J., Echegoyen, L., "Vibrational spectroscopic study of carbon nano-onions coated with polyaniline," *Phys. Status Solidi C* 9 (2012): 1210–2.

Lau, D. W. M., McCulloch, D. G., Marks, N. A., Madsen, N. D., Rode, A. V., "High-temperature formation of concentric fullerene-like structures within foam-like carbon: Experiment and molecular dynamics simulation," *Phys. Rev. B* 75 (2007): 233408-1–4.

Li, X., Wei, B., "Supercapacitors based on nanostructured carbon," *Nano Energy* 2 (2013): 159–73.

Lian, W., Song, H., Chen, X., Li, L. et al., "The transformation of acetylene black into onion-like hollow carbon nanoparticles at 1000°C using an iron catalyst," *Carbon* 46 (2008): 525–30.

Lin, Y., Pan, X., Qi, W., Zhang, B., Su, D. S., "Nitrogen-doped onion-like carbon: A novel and efficient metal-free catalyst for epoxidation reaction," *J. Mater. Chem. A* 2 (2014): 12475–83.

Liu, Y., Vander Wal, R. L., Khabashesku, V. N., "Functionalization of carbon nano-onions by direct fluoriantion," *Chem. Mater.* 19 (2007): 778–86.

Liu, X., Or, S. W., Jin, C., Lv, Y., Feng, C., Sun, Y., "NiO/C nanocapsules with onion-like carbon shell as anode material for lithium ion batteries," *Carbon* 60 (2013): 215–20.

Luszczyn, J., Plonska-Brzezinska, M. E., Palkar, A. et al., "Small noncytotoxic carbon nano-onions: First covalent functionalization with biomolecules," *Chem. Eur. J.* 16 (2010): 4870–80.

Macutkevic, J., Adomavicius, R., Krotkus, A. et al., "Terahertz probing of onion-like carbon-PMMA composite films," *Diamond Relat. Mater.* 17 (2008): 1608–12.

Maksimenko, S. A., Rodionova, V. N., Slepyan, G. Y. et al., "Attenuation of electromagnetic waves in onion-like carbon composites," *Diamond Relat. Mater.* 16 (2007): 1231–5.

Mamezaki, O., Adachi, H., Tomita, S., Fujii, M., Hayashi, S., "Thin films of carbon nanocapsules and onion-like graphitic particles prepared by the cosputtering method," *Jpn. J. Appl. Phys.* 39 (2000): 6680–3.

Mazloumi, M., Shadmehr, S., Rangom, Y., Nazar, L. F., Tang, X. S., "Fabrication of three-dimensional carbon nanotube and metal oxide hybrid mesoporous architectures," *ACS Nano* 7 (2013): 4281–8.

Miller, J. R., Simon, P., "Electrochemical capacitors for energy management," *Science* 321 (2008): 651–2.

Molina-Ontoria, A., Chaur, M. N., Plonska-Brzezinska, M. E., Echegoyen, L., "Preparation and characterization of soluble carbon nano-onions by covalent functionalization, employing a Na-K alloy," *Chem. Commun.* 49 (2013): 2406–8.

Narita, I., Oku, T., Suganuma, K., Hiraga, K. Aoyagi, E., "Formation and atomic structure of tetrahedral carbon onion," *J. Mater. Chem.* 11 (2001): 1761–2.

Nasibulin, A. G., Moisala, A., Brown, D. P., Kauppinen, E. I., "Carbon nanotubes and onions from carbon monoxide using Ni(acac)2 and Cu(acac)2 as catalyst precursors," *Carbon* 41 (2003): 2711–24.

Novoselov, K. S., Geim, A. K., Morozov, S. V. et al., "Electric field effect in atomically thin carbon films," *Science* 306 (2004): 666–9.

Okano, K., Hoshina, K., Iida, M., Koizumi, S., Inuzuka, T., "Fabrication of a diamond field emitter array," *Appl. Phys. Lett.* 64 (1994): 2742–4.

Okotrub, A. V., Bulusheva, L. G., Kuznetsov, V. L., Butenko, Yu. V., Chuvilin, A. L., Heggie, M. I., "X-ray Emission Studies of the Valence Band of Nanodiamonds Annealed at Different Temperatures," *J. Phys. Chem. A* 105 (2001): 9781–7.

Oku, T., Narita, I., Nishiwaki, A., "Formation, atomic structural optimization and electronic structures of tetrahedral carbon onion," *Diam. Relat. Mater.* 13 (2004): 1337–41.

Palkar, A., Melin, F., Cardona, C. M. et al., "Reactivity differences between carbon nano onions (CNOs) prepared by different methods," *Chem. Asian J.* 2 (2007): 625–33.

Palkar, A., Kumbhar, A., Athans, A. J., Echegoyen, L., "Pyridyl-functionalized and water-soluble carbon nano onions: First supramolecular complexes of carbon nano onions," *Chem. Mater.* 20 (2008): 1685–7.

Pan, H., Li, J., Feng, Y. P., "Carbon nanotubes for supercapacitor," *Nanoscale Res. Lett.* 5 (2010): 654–68.

Panich, A. M., Shames, A. I., Sergeev N. A. et al., "Nanodiamond graphitization—A magnetic resonance study," *J. Phys. Condens. Matter.* 25 (2013): 245303-1-3.

Pech, D., Brunet, M., Durou, H. et al., "Ultrahigh-power micrometer-sized supercapacitors based on onion-like carbon," *Nat. Nanotechnol.* 5 (2010): 651–4.

Plonska-Brzezinska, M., Lapinski, A., Wilczewska, A. Z. et al., "The synthesis and characterization of carbon nano-onions produced by solution ozonolysis," *Carbon* 49 (2011): 5079–89.

Plonska-Brzezinska, M. E., Lewandowski, M., Blaszczyk, M., Molina-Ontoria, A., Lucinski, T., Echegoyen, L., "Preparation and characterization of carbon nano-onion/PEDOT:PSS composites," *Chem. Phys. Chem.* 13 (2012a): 4134–41.

Plonska-Brzezinska, M. E., Mazurczyk, J., Palys, B. et al., "Preparation and characterization of composites that contain small carbon nano-onions and conducting polyaniline," *Chem. Eur. J.* 18 (2012b): 2600–8.

Plonska-Brzezinska, M. E., Breczko, J., Palys, B., Echegoyen, L., "The electrochemical properties of nanocomposite films obtained by chemical in situ polymerization of aniline and carbon nanostructures," *Chem. Phys. Chem.* 14 (2013a): 116–24.

Plonska-Brzezinska, M. E., Brus, D. M., Breczko, J., Echegoyen, L., "Carbon nano-onions and biocompatible polymers for flavonoid incorporation," *Chem. Eur. J.* 19 (2013b): 5019–24.

Plonska-Brzezinska, M. E., Brus, D. M., Molina-Ontoria, A., Echegoyen, L., "Synthesis of carbon nano-onion and nickel hydroxide/oxide composites as supercapacitor electrodes," *RSC Adv.* 3 (2013c): 25891–901.

Plonska-Brzezinska, M. E., Echegoyen, L., "Carbon nano-onions for supercapacitor electrodes: Recent developments and applications," *J. Mater. Chem. A* 1 (2013d): 13703–14.

Plonska-Brzezinska, M. E., Molina-Ontoria, A., Echegoyen, L., "Post-modification by low-temperature annealing of carbon nano-onions in the presence of carbohydrates," *Carbon* 67 (2014): 304–17.

Portet, G. Yushin, Gogotsi, Y., "Electrochemical performance of carbon onions, nanodiamonds, carbon black and multiwalled nanotubes in electrical double layer capacitors," *Carbon* 45 (2007): 2511–18.

Radhakrishnan, G., Adams, P. M., Bernstein, L. S., "Plasma characterization and room temperature growth of carbon nanotubes and nano-onions by excimer laser ablation," *Appl. Surf. Sci.* 253 (2007): 7651–5.

Rakhi, R. B., Chen, W., Chaa, D., Alshareef, H. N., "High performance supercapacitors using metal oxide anchored graphene nanosheet electrodes," *J. Mater. Chem.* 21 (2011): 16197–204.

Rao, C. N. R., Sood, A. K., Subrahmanyam, K. S., Govindaraj, A., "Graphene: The new two-dimensional nanomaterial," *Angew. Chem. Int. Ed.* 48 (2009): 7752–77.

Rettenbacher, A. S., Elliott, B., Hudson, J. S., Amirkhanian, A., Echegoyen, L., "Preparation and functionalization of multilayer fullerenes (carbon nano-onions)," *Chem. Eur. J.* 12 (2006): 376–87.

Rettenbacher, A. S., Perpall, M. W., Echegoyen, L., Hudson, J., Smith Jr., D. W., "Radical addition of a conjugated polymer to multilayer fullerenes (carbon nano-onions)," *Chem. Mater.* 19 (2007): 1411–7.

Roddatis, V. V., Kuznetsov, V. L., Butenko, Y. V., Su, D. S., Schlögl, R., "Transformation of diamond nanoparticles into carbon onions under electron irradiation," *Phys. Chem. Chem. Phys.* 4 (2002): 1964–7.

Sano, N., Wang, H., Chhowalla, M., Alexandrou, I., Amaratunga, G. A. J., "Synthesis of carbon 'onions' in water," *Nature* 414 (2001): 506–7.

Sano, N., Wang, H., Alexandrou, I. et al., "Properties of carbon onions produced by an arc discharge in water," *J. Appl. Phys.* 92 (2002): 2783–8.

Santiago, D., Rodríguez-Calero, G. G., Palkar, A., "Platinum electrodeposition on unsupported carbon nano-onions," *Langmuir* 49 (2012): 17202–10.

Satoshi, T., Burian, A., Dore, J. C. et al., "Diamond nanoparticles to carbon onions transformation: X-ray diffraction studies," *Carbon* 30 (2002): 1469–74.

Sawat, S. Y., Somani, R. S., Panda, A. B., Bajaj, H. C., "Formation and characterization of onions shaped carbon soot from plastic wastes," *Mater. Lett.* 94 (2013): 132–5.

Saxby, J. D., Chatfield, S. P., Palmisano, A. J. et al.,"Thermogravimetric analysis of buckminsterfullerene and related materials in air," *J. Phys. Chem.* 96 (1992): 17–8.

Sek, S., Breczko, J., Plonska-Brzezinska, M. E., Wilczewska, A. Z., Echegoyen, L., "STM-based molecular junction of carbon nano-onion," *Chem. Phys. Chem.* 14 (2013): 96–100.

Shames, A. I., Panich, A. M., Mogilko, E., Grinblat, J., Prilutskiy, E. W., Katz, E. A., "Magnetic resonance study of fullerene-like glassy carbon," *Diam. Relat. Mater.* 16 (2007): 2039–43.

Shenderova, O., Grishko, V., Cunningham, G., Moseenkov, S., McGuire, G., Kuznetsov, V., "Onion-like carbon for terahertz electromagnetic shielding," *Diamond Relat. Mater.* 17 (2008): 462–6.

Shin, W. H., Yang, S. H., Goddard, W. A., Kang, J. K., "Ni-dispersed fullerenes: Hydrogen storage and desorption properties," *App. Phys. Lett.* 88 (2006): 053111–053111-3.

Simon, P., Gogotsi, Y., "Materials for electrochemical capacitors," *Nat. Mater.* 7 (2008): 845–54.

Skytt, P., Glans, P., Mancini, D. C. et al., "Angle-resolved soft-x-ray fluorescence and absorption study of graphite," *Phys. Rev. B* 50 (1994): 10457–61.

Solin, S. A., Ramdas, A. K. "Characterization of diamond films by Raman spectroscopy," *Phys. Rev. B* 1 (1970): 1687–98.

Sonkar, S. K., Roy, M., Babar, D. G., Sarkar, S., "Water soluble carbon nano-onions from wood wool as growth promoters for gram plants," *Nanoscale* 4 (2012): 7670–5.

Spitalsky, Z., Tasis, D., Papagelis, K., Galiotis, C., "Carbon nanotube–polymer composites: Chemistry, processing, mechanical and electrical properties," *Prog. Polym. Sci.* 35 (2010): 357–401.

Sriprachuabwong, C., Karuwan, C., Wisitsorrat, A. et al., "Inkjet-printed graphene-PEDOT:PSS modified screen printed carbon electrode for biochemical sensing," *J. Mater. Chem.* 22 (2012): 5478–85.

Street, K. W., Marchetti, M., Vander Wal, R. L., Tomasek, A. J., "Evaluation of the tribological behavior of nano-onions in Krytox 143AB," *Tribol. Lett.* 16 (2004): 143–9.

Su, D., Maksimowa, N. I., Mestl, G. et al., "Oxidative dehydrogenation of ethylbenzene to styrene over ultra-dispersed diamond and onion-like carbon," *Carbon* 45 (2007): 2145–215.

Sun, Y., Wu, Q., Shi, G., "Graphene based new energy materials," *Energy Environ. Sci.* 4 (2011): 1113–32.

Thilgen, C., Herrmann, T., Diederich, F., "The covalent chemistry of higher fullerenes: C_{70} and beyond," *Angew. Chem. Int. Ed.* 36 (1997): 2268–80.

Thune, E., Cabioc'h, T., Guérin, Ph., Denanot, M.-F., Jaouen, M., "Nucleation and growth of carbon onions synthesized by ion-implantation: A transmission electron microscopy study," *Mater. Lett.* 54 (2002): 222–8.

Tomita, S., Fujii, M., Hayashi, S., "Optical extinction properties of carbon onions prepared from diamond nanoparticles," *Phys. Rev. B* 66 (2002): 245424-1-7.

Tomita, S., Fujii, M., Hayashi, S., Yamamoto, K., "Transformation of carbon onions to diamond by low-temperature heat treatment in air," *Diam. Relat. Mater.* 9 (2000): 856–60.

Tomita, S., Sakurai, T., Ohta, H., Fujijii M., Hayashi, S., "Structure and electronic properties of carbon onions," *J. Chem. Phys.* 114 (2001): 7477–82.

Treacy, M. M. J., Ebbesen, T. W., Gibson, J. M., "Exceptionally high Young's modulus observed for individual carbon nanotubes," *Nature* 381 (1996): 678–80.

Tuinstra, F., Koenig, J. L., "Raman Spectrum of Graphite," *J. Chem. Phys.* 53 (1970): 1126–30.

Ugarte, D., "Curling and closure of graphitic networks under electron-beam irradiation," *Nature* 359 (1992): 707–9.

Ugarte, D., "Onion-like graphitic particles," *Carbon* 33 (1995): 989–93.

Wang, Y., Alsmeyer, D. C., McCreery, R. L., "Raman spectroscopy of carbon materials: Structural basis of observed spectra," *Chem. Mater.* 2 (1990): 557–63.

Wang, S. Z., Gao, R. M., Zhou, F. M., Selke, M., "Nanomaterials and singlet oxygen photosensitizers: Potential applications in photodynamic therapy," *J. Mater. Chem.* 14 (2004): 487–93.

Wang, M.-S., Golberg, D., Bando, Y., "Carbon 'onions' as point electron sources," *ACS Nano* 8 (2010): 4396–402.

Wang, Y., Yu, S. F., Sun, C. Y., Zhu, T. J., Yang, H. Y., "MnO_2/onion-like carbon nanocomposites for pseudocapacitors," *J. Mater. Chem.* 22 (2012): 17584–8.

Wang, L., Lu, X., Lei, S., Song, Y., "Graphene-based polyaniline nanocomposites: Preparation, properties and applications," *J. Mater. Chem. A* 2 (2014): 4491–509.

Wei, B., Zhang, J., Liang, J., Wu, D., "The mechanism of phase transformation from carbon nanotube to diamond," *Carbon* 36 (1998): 997–1001.

Welz, S., McNallan, M. J., Gogotsi, Y., "Carbon structures in silicon carbide derived carbon," *J. Mater. Process. Tech.* 179 (2006): 11–22.

Wesolowski, P., Lyutovich, Y., Banhart, F., Carstanjen, H. D., Kronmüller, H., "Formation of diamond in carbon onions under MeV ion irradiation," *Appl. Phys. Lett.* 71 (1997): 1948–50.

Wu, T.-M., Lin, Y.-W., "Doped polyaniline/multi-walled carbon nanotube composites: Preparation, characterization and properties," *Polymer* 47 (2006): 3576–82.

Xie, Q., Perez-Cordero, E., Echegoyen, L., "Electrochemical detection of C_{60}^{6-} and C_{70}^{6-}: Enhanced stability of fullerides in solution," *J. Am. Chem. Soc.* 114 (1992): 3978–80.

Xu, B., Yang, X., Wang, X., Guo, J., Liu, X., "A novel catalyst support for DMFC: Onion-like fullerenes," *J. Power Sources* 162 (2006): 160–4.

Yan, Y., Yang, H., Zhang, F., Tu, B., Zhao, D., "Low-temperature solution synthesis of carbon nanoparticles, onions and nanoropes by the assembly of aromatic molecules," *Carbon* 45 (2007): 2209–16.

Yasin, F. M., Boulos, R. A., Hong, B. Y. et al., "Microfluidic size selective growth of palladium nanoparticles on carbon nano-onions," *Chem. Commun.* 48 (2012): 10102–4.

Zhang, J., Zou, H., Qing, Q. et al., "Effect of chemical oxidation on the structure of single-walled carbon nanotubes," *J. Phys. Chem. B* 107 (2003): 3712–8.

Zhang, Y., Tan, J. W., Stormer, H. L., Kim, P., "Experimental observation of the quantum Hall effect and Berry's phase in graphene," *Nature* 438 (2005): 201–4.

Zhang, C., Li, J., Liu, E. et al., "Synthesis of hollow carbon nano-onions and their use for electrochemical hydrogen storage," *Carbon* 50 (2012): 3513–21.

Zhao, M., Song, H., Chen, X., Lian, W., "Large-scale synthesis of onion-like carbon nanoparticles by carbonization of phenolic resin," *Acta Materialia* 55 (2007): 6144–50.

Zhi, M., Xiang, C., Li, J., Lia, M., Wu, N., "Nanostructured carbon–metal oxide composite electrodes for supercapacitors: A review," *Nanoscale* 5 (2013): 72–88.

Zhou, L., Gao, C., Zhu, D. et al., "Facile functionalization of multilayer fullerenes (carbon nano-onions) by nitrene chemistry and 'grafting from' strategy," *Chem. Eur. J.* 15 (2009): 1389–96.

12

Endohedral Clusterfullerenes

Contents

Alexey Popov

12.1 Synthesis of Endohedral Metallofullerenes

One of the attractive properties of fullerenes is their ability to encapsulate atoms, ions, cluster, and even small molecules. The term *endohedral* for application to fullerenes with other species in their inner space was first introduced in 1991 (Cioslowski and Fleischmann 1991; Weiske et al. 1991) and originates from a combination of Greek ἔνδον (*endon*—within) and ἕδρα (*hedra*—face of a geometrical figure). Endohedral metallofullerenes (EMFs) are then fullerenes with metal atoms within the carbon cage. Encapsulation can be achieved in two pathways: either species are "implanted" into empty fullerene molecules, which are synthesized beforehand, or encapsulation happens during the fullerene formation (Figure 12.1). EMFs are usually obtained by the second procedure.

EMFs were first anticipated in 1985. Soon after the experiments on the synthesis of C_{60} and C_{70} by laser ablation of graphite (Kroto et al. 1985), the same group also studied the laser ablation of graphite impregnated with $LaCl_3$ and discovered the species corresponding to La@C_{60} (Heath et al. 1985). High stability of such ions was the first indication of the endohedral encapsulation of La within the C_{60} cage. The next milestone in the fullerene field was the discovery of the macroscopic fullerene production by arc-discharge synthesis in 1990 (Kratschmer et al. 1990). The use of metal-filled graphite targets or electrodes in the laser evaporation or arc-discharge synthesis resulted in the formation of appreciable amounts of EMFs along with much larger amounts of empty fullerenes

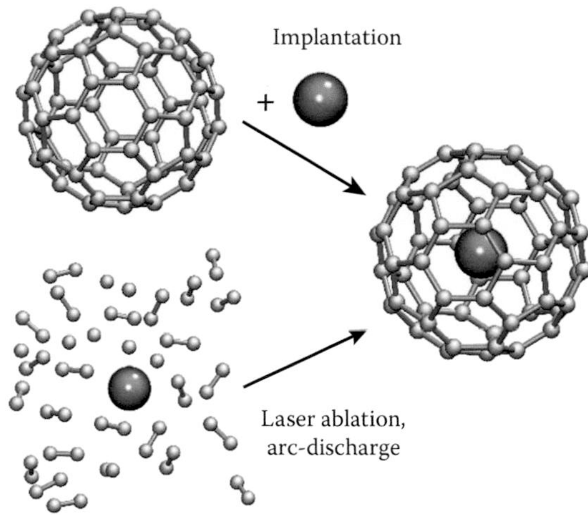

FIGURE 12.1 Two conceptually different pathways to endohedral fullerenes: (i) implantation, which here denotes all methods starting with the empty fullerenes cage and includes ion-beam implantation (From Tellgmann, R. et al., *Nature*, 382, 407–408, 1996.), molecular surgery (From Komatsu, K. et al., *Science*, 307, 238–240, 2005.), and high-pressure–high-temperature treatment (From Saunders, M. et al., *Science*, 259, 1428–1430, 1993.); (ii) encapsulation during the fullerene formation like in laser-ablation or arc-discharge methods.

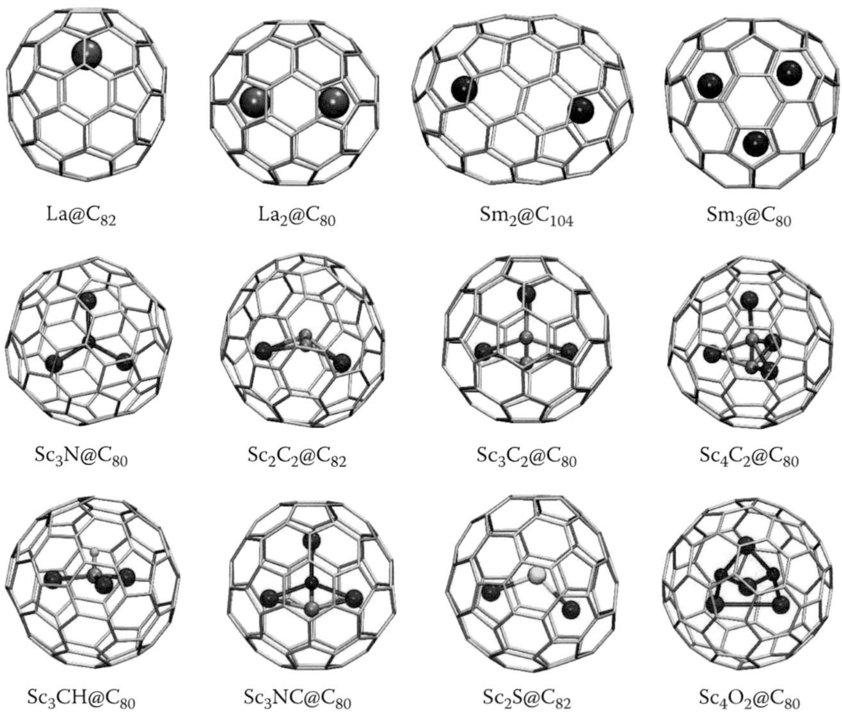

FIGURE 12.2 Representative examples of different classes of EMFs: monometallofullerenes $La@C_{82}$-C_{2v}(9); dimetallofullerenes $La_2@C_{80}$-I_h(7), and $Sm_2@C_{104}$-D_{3d}(822); trimetallofullerene $Sm_3@C_{80}$-I_h(7); nitride clusterfullerene $Sc_3N@C_{80}$-I_h(7); carbide clusterfullerenes $Sc_2C_2@C_{82}$-C_{3v}(8), $Sc_3C_2@C_{80}$-I_h(7), $Sc_4C_2@C_{80}$-I_h(8); methano clusterfullerene $Sc_3CH@C_{80}$-I_h(7); cyano clusterfullerene $Sc_3NC@C_{80}$-I_h(7); sulfide clusterfullerene $Sc_2S@C_{82}$-C_{3v}(8); and oxide clusterfullerene $Sc_4O_2@C_{80}$-I_h(7). Carbon cages are shown in gray; La atoms, brown; Sm, violet; Sc, magenta; N, blue; internal carbon, orange; H, lime; S, yellow; O, red.

(Alvarez et al. 1991; Chai et al. 1991; Johnson et al. 1992). Since then, the arc-discharge remains the main method for the EMF synthesis. Laser ablation coupled to mass-spectrometry detection of the formed species is now used mainly to study the mechanism of the fullerene formation (Dunk et al. 2012a, 2014).

When metal-filled electrodes are evaporated in the inert atmosphere (usually He), mainly monometallofullerenes and dimetallofullerenes are formed. Besides, acetylide C_2 units (which are the main species in the hot carbon plasma) can be also trapped inside the fullerene cages during the EMF formation, resulting in the so-called carbide clusterfullerenes. The first molecular structure of such EMF, $Sc_2C_2@C_{84}$, was elucidated in 2001 (Wang et al. 2001), when detailed structural characterization of EMF molecule became possible. In 1999, it was discovered that the presence of a small amount of nitrogen gas in the arc-discharge generator afforded $Sc_3N@C_{80}$, a new type of EMF with trimetal nitride clusters (Stevenson et al. 1999). The use of NH_3 reactive gas instead of molecular nitrogen allowed much higher selectivity in the synthesis of nitride clusterfullerenes (NCFs) as the yield of empty fullerenes in such conditions was reduced dramatically (Dunsch et al. 2004). Discovery of NCFs triggered exhaustive studies of other clusterfullerenes. Modification of the EMF synthesis by using either reactive gas (NH_3, CH_4, SO_2, CO_2) or solid chemicals added to the graphite electrodes allowed the synthesis of new types of EMFs with endohedral sulfur, oxygen, CH, CN, and other units as shown in Figure 12.2 (S. Yang et al. 2011; Lu et al. 2012; Popov, Yang, and Dunsch 2013).

Usually, the arc-discharge synthesis simultaneously produces many different EMF structures, and extended chromatographic procedures, such as high-performance liquid chromatography (HPLC) and recycling HPLC, are required to separate EMFs from empty fullerenes and from each other to obtain them in compositionally and isomerically pure forms (Figure 12.3). The need for chromatographic separation is therefore one of the main bottlenecks (along with the low yield in the arc-discharge synthesis) on the way to the broader availability of EMFs. To circumvent this problem, several nonchromatographic

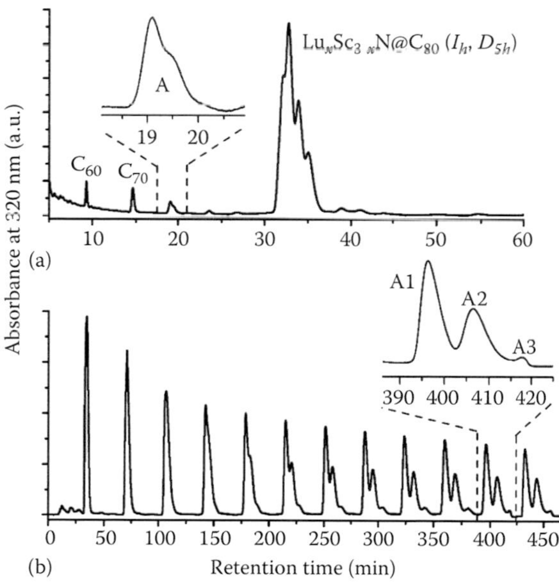

FIGURE 12.3 An example of chromatographic separation of the EMF mixture: (a) HPLC curve of the raw $Lu_xSc_{3-x}N@C_{2n}$ mixture ($x = 1–3$, $2n = 68–88$) extracted from the soot, which was obtained by arc-discharge synthesis with NH_3 as reactive gas (note very small amount of empty fullerenes in such conditions). Fraction A collected at 18.5–20.5 min is then subjected to recycling HPLC separation shown in (b). After 11 cycles, three components of the fraction can be separated as shown in the inset. Mass spectrometry showed that A1 is $Sc_3N@C_{68}$, A2 is $LuSc_2N@C_{68}$, and A3 is $Lu_2ScN@C_{68}$. (Based on the data from Yang, S.F. et al., *Chem. Commun.* 2885–2887, 2008.)

procedures based on the variation of redox and chemical properties of different EMFs were proposed (Bolskar and Alford 2003; Tsuchiya et al. 2004; Ge et al. 2005; Stevenson et al. 2006, 2009; Akiyama et al. 2012; Cerón, Li, and Echegoyen 2013).

12.2 Metals Forming EMFs

Fullerenes are good electron acceptors, so metal atoms inside fullerenes transfer their valence electrons to the carbon cage, thus forming nondissociative "salts" such as $La^{3+}@C_{82}^{3-}$. Certainly, the ease of ionization of metal ions and stability of negatively charged carbon cages play important roles in EMF formation. Figure 12.4 shows the periodic table of elements and highlights the elements that are able to form EMFs.

Encapsulation of alkali metals within the carbon cage can be achieved either by laser ablation (Dunk et al. 2014) or by exposing thin films of fullerenes to the high-energy M^+-ion beams (Tellgmann et al. 1996; Campbell et al. 1997). Both methods produce mainly $M@C_{60}$, but the yield is rather low. Bulk amounts are obtained so far only for $Li@C_{60}$ (Gromov et al. 1997), which can be stabilized in the form of cationic salts such as $[Li^+@C_{60}]SbCl_6^-$ (Aoyagi et al. 2010). Among the alkali earth metals, Be and Mg are not forming EMFs, whereas Ca, Sr, and Ba afford monometallofullerenes $M@C_{2n}$ in the arc-discharge synthesis with a broad range of carbon cage sizes (Kubozono et al. 1995; Xu, Nakane, and Shinohara 1996; John, Dennis, and Shinohara 1998). Divalent lanthanides (Sm, Eu, Tm, Yb) behave similarly to alkali earth in terms of the broad variety of the monometallofullerenes they form in the arc-discharge synthesis (Kirbach and Dunsch 1996; Kuran et al. 1998; Okazaki et al. 2000; Xu et al. 2004). For Sm-EMFs, a series of dimetallofullerenes $Sm_2@C_{2n}$ (Mercado et al. 2009; H. Yang et al. 2011a) and even a trimetallofullerene $Sm_3@C_{80}$ have been also characterized (Xu et al. 2013a). Neither alkali nor alkali earth or divalent rare earth metals are known to form clusterfullerenes.

The group III elements Sc, Y, and La, as well as trivalent rare earths, are the most versatile metals in the EMF field as they afford a broad range of different types of EMFs, from monometallofullerenes and dimetallofullerenes to a variety of clusterfullerenes. Their structures will be discussed in more detail further in the chapter.

In group IV, formation of EMFs was achieved for Ti, Zr, and Hf. Laser ablation gives monometallofullerenes $M@C_{2n}$ ($2n$ = 26–46), with the most abundant species at $2n$ = 28 and 44 (Guo et al.

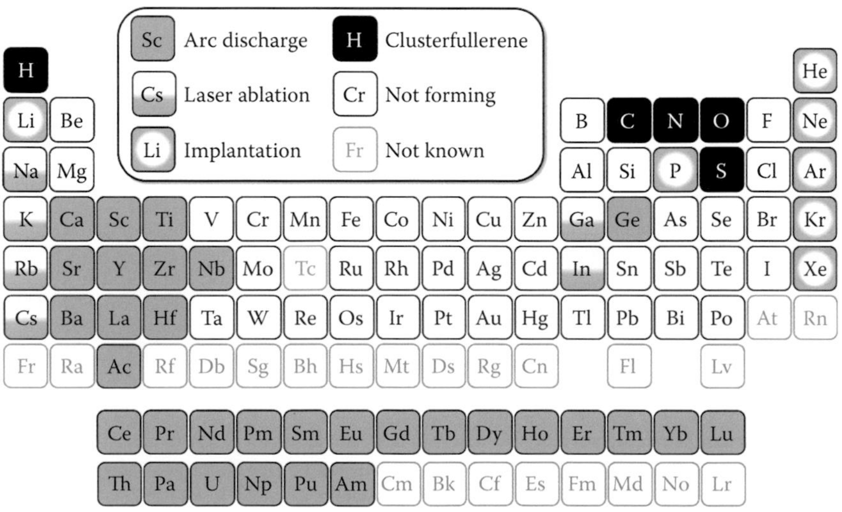

FIGURE 12.4 Periodic table of the elements highlighting the elements that were successfully encapsulated within the fullerene cages.

1992; Dunk et al. 2012b). Unfortunately, bulk amounts of these small-cage monometallofullerenes are not available. Arc-discharge synthesis of Ti-EMFs yields several structures with one and two Ti atoms within one cage, mainly Ti@C_{88}, Ti@C_{90}, Ti_2C_{80}, and Ti_2C_{84} (Cao et al. 2001, 2002). Ti_2C_{80} was found to be a carbide clusterfullerene Ti_2C_2@C_{78} (Tan and Lu 2005; Sato et al. 2006), while molecular structures of other Ti-EMFs are not known yet. Ti has been also encapsulated in EMFs in the form of several types of clusterfullerenes, including aforementioned Ti_2C_2@C_{78}, mixed-metal NCFs TiM_2N@C_{80} (M = Sc and Y; Ti alone cannot form NCFs) (Yang et al. 2009; C. Chen et al. 2012), sulfide Ti_2S@C_{78} (Li et al. 2013), and carbide $TiLu_2C$@C_{80} with double bond between Ti and the central carbon atom (Svitova et al. 2014a). The information on the arc-discharge synthesis of Zr and Hf-EMFs is rather scarce. Both metals were found to form small amounts of EMFs (Sueki et al. 1999). In a later work, Hf@C_{84} and Hf_2C_{80} were isolated and studied spectroscopically, but their molecular structures were not elucidated (Akiyama et al. 2000).

Very limited availability of actinides precludes detailed studies of their EMFs. Formation of EMFs with Ac, Th, Pa, U, Np, and Am in the arc-discharge process was detected by radio-chromatographic method (Akiyama et al. 2001, 2002, 2003, 2009). By their retention behavior, the actinide-EMFs were classified into two groups: Ac, U, Pu, Np, and Am are similar to the trivalent lanthanide analogs, whereas Th and Pa are substantially different. The amounts sufficient for spectroscopic characterization were obtained for U@C_{82} and Th@C_{84} (Akiyama et al. 2002). In laser ablation studies, uranium produces series of monometallofullerenes with the largest abundance of U@C_{28} and U_2@C_{60} (Guo et al. 1992), whereas Th gives a series of monometallofullerenes with the maxima at Th@C_{36} and Th@C_{44} (Dunk et al. 2014).

There is a consensus in the literature that most of transition metals are not forming EMFs. Inconclusive data on the possible formation of EMFs in the arc-discharge process are obtained on Nb (Sueki et al. 1999) and Ge (Roy et al. 2009). Laser-ablation studies showed that Ga and In form monometallofullerenes with a structure distribution similar to that of alkali metals (Dunk et al. 2014).

12.3 Molecular Structures of EMFs

During the first decade of the EMF research, the progress in the structural studies was rather modest (Shinohara 2000). In addition to the low availability of the structurally and isomerically pure samples, the main structural tools such as single-crystal x-ray diffraction and nuclear magnetic resonance (NMR) (particularly ^{13}C NMR) had severe difficulties. Single-crystal x-ray diffraction studies of fullerenes are hampered by the rotational disorder of the fullerene molecules in their crystals, which usually precludes direct determination of the carbon cage structures. Furthermore, endohedral species often have several possible positions inside the carbon cage, which makes the disorder problem even more complicated for EMFs. Low sensitivity of ^{13}C NMR spectroscopy and paramagnetism of many EMF molecules were also serious drawbacks in the structure elucidation. Besides, ^{13}C NMR determines only the symmetry of the carbon cage, which remains ambiguous for low-symmetry fullerenes when many cage isomers are possible. Fortuitously, the situation changed dramatically in the 2000s. For x-ray studies, the strategies were developed to circumvent the disorder problem, which are either based on cocrystallization (particularly successful and commonly used cocrystallizing agents are Ni- or Co-octaethylporphyrines; Olmstead et al. 1999; Rodriguez-Fortea, Balch, and Poblet 2011) or include exohedral chemical derivatization of the EMF, usually via cycloaddition (Yamada, Akasaka, and Nagase 2010; Lu, Akasaka, and Nagase 2011). Both approaches hinder rotation of EMF molecules in their crystals and often (but not always) reduce disorder in the positions of endohedral species, thus enabling determination of atomic coordinates. With the use of these approaches, the molecular structures of dozens of EMFs were elucidated by single crystal x-ray diffraction. X-ray structure elucidation is especially important for clusterfullerenes, since alternative approaches such as NMR spectroscopy are not able to determine the geometry of the endohedral cluster.

NMR spectroscopic studies of EMFs became also more accessible in the last decade. ^{13}C NMR studies of those paramagnetic EMFs, whose paramagnetism is caused by the odd number of electrons transferred to the carbon cage, can be studied in their ionic forms obtained by electrolysis. For example, C_{2v} symmetry of the carbon cage in paramagnetic La@C_{82} was determined by the study of its diamagnetic anion (Akasaka et al. 2001; Tsuchiya et al. 2005). Paramagnetic ^{13}C NMR studies of lanthanide-based EMFs were found to be useful for the determination of the position of metal atoms and their dynamics (Yamada et al. 2008a,b, 2009). Likewise, dynamics of endohedral species is conveniently studied by multinuclear NMR techniques, such as ^{89}Y, ^{14}N, ^{139}La, and especially ^{45}Sc (Fu et al. 2009, 2011a; Kurihara et al. 2011; Popov et al. 2011; Suzuki et al. 2013; Zhang and Dorn 2014; Feng et al. 2015).

Table 12.1 lists different classes of EMFs and the metals that form them, formal charges of endohedral species, and the most abundant fullerene cages for each class of EMFs; examples of EMF structures are shown in Figure 12.2. It is common practice to label the cage isomers of fullerenes by their symmetry and the number of the isomer in accordance with the spiral algorithm (Fowler and Manolopoulos 1995). Usually, a short form of numbering system is used, in which only the isomers following the isolated pentagon rule (IPR) are numbered. However, a large number of the non-IPR isomers found for EMFs require the use of the extended notation, which includes all possible isomers for a given number of carbon atoms. In this chapter, we use a twofold numbering system which is de facto employed in many publications: The short notation is retained for all IPR isomers (i.e., their numbers are usually below 1000), while the full notation is adopted for the non-IPR isomers (Popov 2009; Popov, Yang, and Dunsch 2013).

TABLE 12.1 Families of EMFs with Their Cluster Composition, Formal Charge, Most Abundant Carbon Cages, and Typical Metals

Cluster	Formal Charge	Main Cage(s)	Other Cages	Metals
M^{I}	+1	C_{60}-I_h(1)	C_{70}	Li
M^{II}	+2	C_{82}-C_2(5), C_{82}-C_s(6), C_{82}-C_{2v}(9)	C_{60}, C_{72}, C_{74}, C_{76}, C_{78}, C_{80}, C_{84}, C_{86}, C_{88}, C_{90}, C_{92}, C_{94}	Ca, Sr, Ba, Sm, Eu, Tm, Yb
MCN	+2	C_{82}-C_2(5), C_{82}-C_s(6), C_{82}-C_{2v}(9)		Y, Tb
M^{III}	+3	C_{82}-C_{2v}(9)	C_{82}, C_{60}, C_{72}, C_{74}, C_{80}, C_{82}	Sc, Y, La, Ce, Pr, Nd, Gd, Tb, Dy, Ho, Er, Lu
M_2	+4	C_{82}-C_{3v}(8), C_{82}-C_s(6)	C_{66}, C_{76}, C_{82}, C_{88}, C_{90}, C_{92}, C_{104}	Sc, Y, Er, Lu, Tm, Sm
M_2C_2	+4	C_{82}-C_{3v}(8)	C_{82}, C_{80}, C_{84}, C_{84}, C_{92}	Sc, Y, Gd, Tb, Dy, Er
M_2S	+4	C_{82}-C_{3v}(8), C_{82}-C_s(6)	C_{70}-C_{94}	Sc, Y, Dy
M_2O	+4	C_{82}-C_s(6)	C_{70}	Sc
M_2	+5	C_{79}N-[I_h]		Y, Gd, Tb
M_2	+6	C_{80}-I_h(7)	C_{80}, C_{72}, C_{76}, C_{78}, C_{80}, C_{100}	La, Ce, Pr
M_3	+6	C_{80}-I_h(7)		Sm, Y, Tb
M_3N	+6	C_{80}-I_h(7)	C_{68}, C_{70}, C_{78}, C_{82}	Sc
M_3N	+6	C_{80}-I_h(7)	C_{78}, C_{82}, C_{84}, C_{86}, C_{88}	Y, Gd, Tb, Dy, Ho, Er, Tm, Lu
M_3N	+6	C_{88}-D_2(35)	C_{84}, C_{86}, C_{92}, C_{96}, C_{104}	La, Ce, Pr, Nd,
M_2TiN	+6	C_{80}-I_h(7)		Sc, Y
M_2TiC	+6	C_{80}-I_h(7), C_{80}-D_{5h}(6)		Sc, Y, Nd, Gd, Dy, Er, Lu
M_3CH	+6	C_{80}-I_h(7)		Sc
M_3C_2	+6	C_{80}-I_h(7)	C_{88}	Sc, Lu
M_4C_2	+6	C_{80}-I_h(7)		Sc
M_4O_2	+6	C_{80}-I_h(7)		Sc
M_4O_3	+6	C_{80}-I_h(7)		Sc
M_3CN	+6	C_{80}-I_h(7)	C_{78}	Sc
M_3C_2CN	+6	C_{80}-I_h(7)		Sc

12.3.1 Monometallofullerenes, Dimetallofullerenes, and Trimetallofullerenes

Although La@C_{60} was the first EMF ever detected, the isolation of EMFs with C_{60} cage was proved to be difficult because of the low kinetic stability. The first attempts of the bulk synthesis showed that for trivalent metals (Sc, Y, many lanthanides), the most abundant isolable monometallofullerenes have a C_{82} cage. Its isomeric structure, $C_{2v}(9)$, was proposed in the synchrotron x-ray diffraction studies of powder samples (Takata et al. 2003) and confirmed by ^{13}C NMR spectroscopic study of the La@C_{82} anion (Akasaka et al. 2000). The position of the metal atom inside the carbon cage was a matter of numerous studies in the 1990s when the endohedral structure was still questioned by some researchers. The first unambiguous proof of the endohedral position of metal atoms was obtained in a synchrotron x-ray diffraction study of the Y@C_{82} powder samples (Takata et al. 1995). Electron density distribution of Y@C_{82} determined from the diffraction data showed the maximum of the density at the interior of the C_{82} cage. Importantly, Y atom was found close to the wall of the fullerene cage rather than in its center. The noncentral position of metal atoms is typical for all EMFs and is caused by strong bonding interaction between the metal ion and the carbon cage. The more precise position of metal atom inside the C_{82}-$C_{2v}(9)$ cage shown in Figure 12.2 was determined by single-crystal x-ray diffraction studies of M@C_{82} EMFs (Sato et al. 2012; Suzuki et al. 2012). M^{III}@C_{82} EMFs also have another, less abundant isomer with $C_s(6)$ cage (Akasaka et al. 2001). Akasaka and coworkers also found that many La@C_{2n} EMFs remain in the carbon soot during the standard extraction procedure. However, the use of 1,2,4-trichlorobenzene as extraction solvent promoted reaction with insoluble EMFs with the formation of soluble La@C_{2n}($C_6H_3Cl_2$) monoadducts. Structural characterization of such adducts by single crystal x-ray diffraction proved the formation of La@C_{72}-$C_2(10612)$, La@C_{74}-$D_{3h}(1)$, La@C_{80}-$C_{2v}(3)$, and La@C_{82}-$C_{3v}(7)$ (Akasaka and Lu 2012).

Mono-EMFs with divalent metals are also dominated by M^{II}@C_{82} species, but here, four isomers with $C_2(5)$, $C_s(6)$, C3v(7), and $C_{2v}(9)$ are usually obtained in comparable yield (Kirbach and Dunsch 1996; Kodama et al. 2002, 2003). Detailed studies were also reported for M^{II}@C_{74} structures isolated for Ca, Ba, Eu, Sm, and Yb (Kuran et al. 1998; Reich et al. 2004; Xu et al. 2013b). ^{13}C NMR showed the fluxional motion of metal atoms between several equivalent positions within the symmetric C_{74}-$D_{3h}(1)$ carbon cage (Kodama et al. 2004; Xu et al. 2006). Many Yb@C_{2n} and especially Sm@C_{2n} mono-EMFs with $2n = 74$, 80, 82 (four isomers), 84 (three isomers), 90 (four isomers), 92 (two isomers), and 94 (four isomers) were isolated and partially structurally characterized (Xu et al. 2004, 2012a,b, 2013b; Lu et al. 2010; H. Yang et al. 2011b, 2012a,b, 2013; Jin et al. 2012). The largest characterized mono-EMF is M@C_{94}-$C_{3v}(134)$ with M = Ca, Tm, or Sm (Che et al. 2009; Jin et al. 2012). The position of metal atom inside the fullerene cage in M^{II}-mono-EMFs is often not well defined, resulting in a large disorder in x-ray determined structures.

Formation of dimetallofullerenes (di-EMFs) was discovered along with mono-EMFs in the early 1990s, when La_2@C_{80} and Y_2@C_{82} were obtained (Alvarez et al. 1991; Shinohara et al. 1992). La_2@C_{80} was shown to have icosahedral $I_h(7)$ cage with freely rotating La_2 cluster (Akasaka et al. 1997; Nishibori et al. 2001). Analogous di-EMF structures were later described for Ce and Pr. In addition to M_2@C_{80}-$I_h(7)$, the family of di-EMFs based on early lanthanides (M = La, Ce, Pr) includes a number of other cages, including C_{72}-$D_2(10611)$, C_{76}-$C_s(17490)$, C_{78}-$D_{3h}(5)$, and C_{80}-$D_{5h}(6)$ (Akasaka et al. 1997; Kato et al. 2003; Cao et al. 2004; Yamada et al. 2005, 2008a,b, 2009). The largest structurally characterized M^{III}-di-EMF known so far is La_2@C_{100}-$D_5(450)$, while mass spectrometry showed that di-EMFs up to La_2@C_{138} were formed (Beavers et al. 2011).

Sc, Y, and "late" lanthanides (Er, Lu) also form di-EMF, but with a rather peculiar bonding situation. Although these metals prefer a trivalent state in mono-EMFs, in di-EMFs, they adopt a formal 2+ charge state with the metal–metal bond in the M_2 cluster (Popov et al. 2012a). The most abundant di-EMFs of this type are M_2@C_{82}-$C_{3v}(8)$ and M_2@C_{82}-$C_s(6)$ (Olmstead et al. 2002a,b; Inoue et al. 2004; Ito et al. 2007; Kurihara et al. 2012a). Also known are Er_2@C_{82}-$C_{2v}(9)$ (Plant et al. 2009) and Lu_2@C_{76}-$T_d(1)$ (Umemoto et al. 2010). At the same time, in Sc_2@C_{66}-$C_{2v}(4059)$, the smallest isolated di-EMF, Sc, is trivalent (Yamada et al. 2014). Among the truly divalent metals, di-EMFs are described so far for only Tm ($Tm_2@C_{82}$) and Sm [a series of di-EMFs, including $Sm_2@C_{88}$-D_2(35), $Sm_2@C_{90}$-C_2(21), $Sm_2@C_{92}$-D_3(85), and $Sm_2@C_{104}$-D_{3d}(822)] (Kikuchi et al. 2000; Mercado et al. 2009; H. Yang et al. 2011a).

A special group of di-EMFs to be mentioned are paramagnetic azafullerenes $M_2@C_{79}N$ (M = Y, Tb, Gd) described by Dorn and coworkers (Zuo et al. 2008; Fu et al. 2011c). The carbon cage in these EMFs is quasi C_{80}-I_h(7), with one of the carbon atoms substituted by nitrogen. Two metal atoms form a single-electron bond and both have large spin population.

Genuine trimetallofullerenes are very rare. Possible formation of $M_3@C_{80}$ species for Er, Tb, and Y was shown by mass spectrometry without further structure elucidation. The only unambiguously characterized trimetallofullerene is $Sm_3@C_{80}$-I_h(7), whose structure was elucidated by single-crystal x-ray diffraction (Xu et al. 2013a).

Although EMFs can comprise up to four metal atoms, the large number of metal atoms can be achieved only in the form of the clusters with nonmetals described below in Sections 12.3.2 through 12.3.5. Endohedral metal atoms bear significant positive charges, which results in strong Coulomb repulsion when two or more metals are encapsulated inside one carbon cage. For instance, repulsion energy between two La atoms in $La_2@C_{80}$ is estimated as 9 eV. As a result, metal atoms in polymetallic EMFs tend to be as far from each other as possible. The large number of metal atoms in EMFs can be easier realized in clusters with electronegative nonmetals: The latter bear a negative charge (the formal charge of nitrogen in NCFs is −3; oxygen, sulfur, and acetylide unit C_2 all have a formal charge of −2) and hence partially compensate Coulomb repulsion between the metal atoms.

12.3.2 Nitride Clusterfullerenes

In 1999, the molecular structure of the first isolated NCF, $Sc_3N@C_{80}$-I_h(7), was determined by single-crystal x-ray diffraction and ^{13}C NMR spectroscopy (Stevenson et al. 1999). The Sc atoms were found to form an equilateral triangle with nitrogen atom in the center of the planar cluster (Figure 12.2). Eventually, NCFs turned out to be the most abundantly produced EMFs (Dunsch and Yang 2007; Zhang, Stevenson, and Dorn 2013), and $Sc_3N@C_{80}$-I_h(7) is the third most abundantly produced fullerene (after C_{60} and C_{70}). $Sc_3N@C_{80}$ also has another, less abundant isomer with D_{5h}(6) cage, whose structure was determined by ^{13}C NMR spectroscopy (Duchamp et al. 2003) and single-crystal x-ray study (Cai et al. 2006). In addition to $Sc_3N@C_{80}$, the family of Sc-based NCFs includes less abundant $Sc_3N@C_{68}$, $Sc_3N@C_{70}$, $Sc_3N@C_{78}$, and $Sc_3N@C_{82}$. The non-IPR cage structure with D_3(6140) symmetry and three pairs of adjacent pentagons was first proposed for $Sc_3N@C_{68}$ in 2000 based on the ^{13}C NMR and computational data (Stevenson et al. 2000) and then confirmed in 2003 by single-crystal x-ray diffraction study (Olmstead et al. 2003). $Sc_3N@C_{78}$-D_{3h}(5) was first isolated in 2001 and characterized by ^{13}C NMR and single-crystal x-ray diffraction (Olmstead et al. 2001). The yield of $Sc_3N@C_{70}$ is almost 50 times lower than that of $Sc_3N@C_{68}$, and this NCF with the non-IPR C_{70}-C_{2v}(7854) cage was first isolated in 2007 and characterized by vibrational spectroscopy and density functional theory computations (Yang, Popov, and Dunsch 2007). Because of its low kinetic stability, structural characterization of elusive $Sc_3N@C_{82}$-C_{2v}(9) was accomplished only in 2015 (Wei et al. 2015).

With the increase in the ionic radii of metals, the cage size distribution of $M_3N@C_{2n}$ clusterfullerenes is shifting to larger cages (Figure 12.5). Yttrium and elements from the second half of the lanthanide row (Gd–Lu, except for Yb) form isolable $M_3N@C_{2n}$ NCFs with $2n$ = 78–88 (Krause and Dunsch 2005; Krause, Wong, and Dunsch 2005; Yang and Dunsch 2005; Beavers et al. 2006; Chaur et al. 2007; Zuo et al. 2007; Fu et al. 2009, 2011b; Xu et al. 2011). $M_3N@C_{80}$-I_h(7) is still the most abundant NCF for these metals. Other structurally characterized cage isomers are C_{78}-C_2(22010) (Popov et al. 2007; Beavers et al. 2009), C_{82}-C_s(39663) (Mercado et al. 2008), C_{84}-C_s(51365) (Beavers et al. 2006), C_{86}-D_3(19) (Zuo et al. 2007), and C_{88}-D_2(35) (Zuo et al. 2007).

For the first half of the lanthanide row (La, Ce, Pr, Nd), the distribution of the fullerene size shifts further to larger cages with the highest yield for $M_3N@C_{88}$ (Melin et al. 2007; Chaur et al. 2008b,c). The relative yield of $M_3N@C_{96}$ is also gradually increasing with the increase in the ionic radius from Nd to

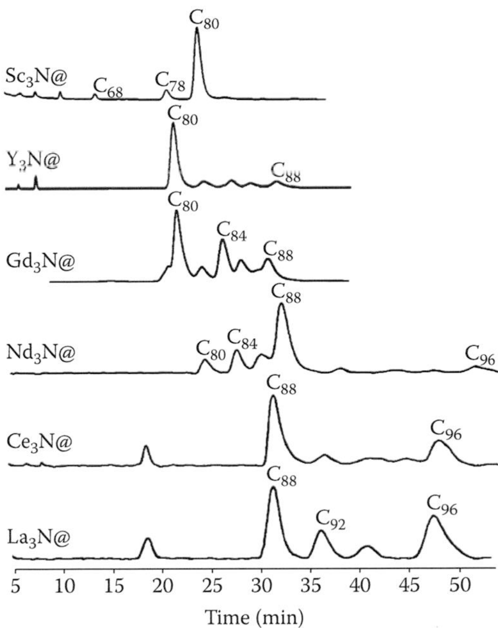

FIGURE 12.5 HPLC curves of the $M_3N@C_{2n}$ extracts for Sc, Y, and different lanthanides obtained with NH_3 as reactive gas (Buckyprep-M column, detection at 372 nm). Each curve is normalized to the highest intensity component of the mixture. The peaks eluting before 10 min correspond to empty fullerenes. The data are compiled from several works by Echegoyen's group. (From Chaur, M.N. et al., *Tetrahedron*, 64, 11387–11393, 2008a; Chaur, M.N. et al., *Chem. Eur. J.*, 14, 8213–8219, 2008b; Chaur, M.N. et al., *Chem. Eur. J.*, 14, 4594–4599, 2008c.)

La, so that the yield of $La_3N@C_{96}$ equals that of $La_3N@C_{88}$ (Figure 12.5). Neither ^{13}C NMR spectroscopic nor crystallographic studies have been performed so far for any $M_3N@C_{2n}$ with $2n > 88$. C_{92}-$T(86)$ and C_{96}-$D_2(186)$ cage isomers were tentatively assigned based on computational studies (Popov and Dunsch 2007; Chaur et al. 2009; Valencia et al. 2009a).

The evolution of the fullerene cage distribution with the increase in the metal size from Sc (Shannon ionic radius $R^{3+} = 0.745$ Å) to Y, Gd–Lu ($R^{3+} = 0.86$–0.94 Å), and further to La–Nd (0.98–1.03 Å) demonstrates that the size of the cluster is a very important parameter determining the preferable fullerene cage. The increase in the M_3N cluster size increases the strain in $M_3N@C_{2n}$ molecules when larger and larger clusters are encapsulated in the given cage and hence shifts the distribution toward the larger cage sizes. The strain experienced by the M_3N cluster can lead to the significant changes in its geometry: Whereas nitride clusters are usually planar, the Gd_3N cluster is pyramidalized in the $Gd_3N@C_{80}$ molecule because of the limited space (Stevenson et al. 2004). In larger cages (such as $Gd_3N@C_{86}$), the strain is reduced and the Gd_3N cluster is also planar (Chaur et al. 2010). Another illustrative example of the cluster size factor is isomerism of $M_3N@C_{78}$: Whereas Sc_3N cluster is encapsulated in the C_{78}-$D_{3h}(5)$ cage isomer, lanthanides form $M_3N@C_{78}$ molecule with the non-IR C_{78}-$C_2(22010)$ cage (Popov et al. 2007). The reason is that the latter cage has a flattened shape and can easier accommodate large M_3N clusters, whereas the C_{78}-$D_{3h}(5)$ isomer is more suitable for clusters of smaller size such as Sc_3N.

Mixing "small" Sc and "larger" lanthanide metals in the M_3N cluster enables the tuning of the cluster size (from small Sc_3N via MSc_2N and M_2ScN to larger M_3N) and opens a possibility for carbon cages to adopt the cluster of the most suitable size (Svitova, Popov, and Dunsch 2013). Whereas the carbon cages smaller than C_{78} are not accessible for non-Sc NCFs, mixing Sc with lanthanide makes it possible: For instance, $LuSc_2N@C_{68}$ and $Lu_2ScN@C_{68}$ can be formed in the mixed Lu-Sc system (Yang, Popov, and Dunsch 2008). A drawback of the mixed-metal NCFs is that the separation of individual compounds often becomes more complicated, as illustrated in Figure 12.3 for a $Lu_xSc_{3-x}N@C_{68}$ mixture.

12.3.3 Clusterfullerenes with Endohedral Carbon

Whereas NCFs remain the most abundantly produced clusterfullerenes, the clusterfullerenes with carbon in the endohedral clusters is the most versatile family in terms of possible cluster compositions (Lu, Akasaka, and Nagase 2013; Jin, Tang, and Chen 2014). In addition to its obvious role of forming the fullerene cage, carbon is also found in the core of at least five types of endohedral clusters.

Carbon vapor in conditions of the fullerene synthesis consists mainly of C_2 species, and formation of carbide clusterfullerenes with acetylide unit is believed to be a result of the C_2 trapping. The inability of conventional mass spectrometry to distinguish if the C_2 unit is endohedral or is a part of the carbon cage bluffed the EMF researchers for two decades and still remains a problem when new EMF compositions are detected in mass spectra. The most common endohedral carbide clusters have M_2C_2 composition, where M is a trivalent metal (Sc, Y, lanthanides Gd, Tb, Dy, Ho, Er, Lu), and the acetylide unit has a formal charge of −2, so that the total charge of the cluster is +4. A special case is $Ti_2C_2@C_{78}$, in which Ti atoms are in the four-valent state and the charge of the cluster is +6 (Tan and Lu 2005). The shape of the M_2C_2 cluster depends on the size of the carbon cage: In smaller cages, the C_2 unit is perpendicular to the M···M axis, whereas in large cages, more linear arrangement is possible (Zhang et al. 2012; Kurihara et al. 2012b; Deng and Popov 2014). Fullerene cages in structurally characterized $M_2C_2@C_{2n}$ molecules range from C_{72}-C_s(10528) (Feng et al. 2013) to C_{92}-D_3(85) (Yang et al. 2008); the fullerenes of larger size are also formed, but not structurally characterized yet (Yang et al. 2008; Zhang et al. 2012). The most abundant carbon cages for $M_2C_2@C_{2n}$ clusterfullerenes are C_{82}-C_{3v}(8), followed by C_{82}-C_s(6).

The number of metal atoms in carbide clusterfullerenes with endohedral acetylide unit can be also three or four. The former group has two examples so far, $Sc_3C_2@C_{80}$ (Iiduka et al. 2005) and $Lu_3C_2@C_{88}$ (Xu et al. 2011). The only acetylide clusterfullerene with four metal atoms known so far is $Sc_4C_2@C_{80}$ (Wang et al. 2009). In Sc_4C_2, Sc atoms form a tetrahedron with one carbon atom in the center and one above the Sc_3 face; the C_2 unit in $Sc_4C_2@C_{80}$ adopts a formal charge of −6 (Tan, Lu, and Wang 2006). In a recent report, Wang et al. showed the endohedral C_2 unit can also bear a bonded hydrogen atom forming a paramagnetic $Sc_4C_2H@C_{80}$ (Feng et al. 2014).

A single carbon atom can be a central atom in two types of endohedral clusters. In $Sc_3CH@C_{80}$, the central carbon atom is bonded by a hydrogen (Krause et al. 2007), whereas in the recently discovered $TiLu_2C@C_{80}$, the μ_3-carbido ligand forms a double bond with Ti atom (Svitova et al. 2014a). Interestingly, both EMF structures with single carbon atoms exhibit a certain analogy to NCFs. The difference in the valence of carbon and nitrogen is compensated here either by an additional hydrogen atom (C–H vs. N), or via encapsulation of the tetravalent metal and double bond formation (C = Ti vs. N–Sc). Recently, an EMF with presumable composition $Sc_4C@C_{80}$ was detected by mass spectrometry and studied computationally (Deng, Junghans, and Popov 2015), but it is still awaiting detailed structural characterization.

Carbon was also encapsulated in the form of cyanide fragment coordinated either to three metal atoms ($Sc_3NC@C_{80}$ [Wang et al. 2010] and $Sc_3NC@C_{78}$ [Wu et al. 2011]) or to only one metal atom ($YNC@C_{82}$ [S. Yang et al. 2013] or $TbNC@C_{82}$ [Liu et al. 2014]). The formal charge of the Sc_3NC cluster is +6, whereas MNC (M = Y, Tb) is only twofold charged. A combination of both acetylide and cyanide fragments in one EMF molecule, $Sc_3C_2CN@C_{80}$, was described recently (Wang, Wu, and Feng 2014).

12.3.4 Oxide Clusterfullerenes

Oxygen-containing endohedral clusterfullerenes were discovered in 2008. Arc-discharge synthesis in the presence of air and $Cu(NO_3)_2$ afforded the formation of $Sc_4O_2@C_{80}$-I_h(7) and $Sc_4O_3@C_{80}$-I_h(7)

(Stevenson et al. 2008). Both molecules have Sc_4 tetrahedron, with two or three oxygen atoms located above the Sc_3 faces in the μ_3-fashion (Stevenson et al. 2008; Mercado et al. 2010a). In $Sc_4O_2@C_{80}$, two Sc atoms are in the formal divalent state with the Sc–Sc bond, whereas in $Sc_4O_3@C_{80}$, all Sc atoms are in the +3 oxidation state (Valencia et al. 2009b; Popov et al. 2012b). The formal charge of both Sc_4O_2 and Sc_4O_3 clusters is +6. Stevenson et al. also discovered $Sc_2O@C_{82}$-$C_s(6)$ with bended Sc–O–Sc cluster (Mercado et al. 2010b). Recently, a family of $Sc_2O@C_{2n}$ oxide clusterfullerenes ($2n = 70$–94) was synthesized using CO_2 as a reactive gas, and $Sc_2O@C_{70}$-$C_2(7892)$ was isolated and characterized (Zhang et al. 2014). Stevenson also reported that the use of Y or Lu instead of Sc in the oxide clusterfullerene synthesis resulted in formation of $Y_2O@C_{2n}$ and $Lu_2O@C_{2n}$ structures ($2n = 80$–92) (Stevenson 2014). The formation of $M_4O_2@C_{2n}$ or $M_4O_3@C_{2n}$ clusterfullerenes was not detected.

12.3.5 Sulfide Clusterfullerenes

Sulfide clusterfullerenes were discovered in 2010 in the form of $M_2S@C_{82}$ molecules (M = Sc, Y, Dy, Lu) when guanidinium thiocyanate $C(NH_2)_3{\cdot}SCN$ was used for the synthesis of NCFs. Their absorption spectra closely resembled those of $M_2C_2@C_{82}$-$C_{3v}(8)$, which indicated that isolated $M_2S@C_{82}$ molecules also had $C_{3v}(8)$ cage isomer. With the use of SO_2 as a reactive gas, Echegoyen et al. synthesized a family of $Sc_2S@C_{2n}$ clusterfullerenes ($2n = 70$–100) (Chen et al. 2010) and $Ti_2S@C_{78}$-$D_{3h}(5)$ (Li et al. 2013). The structures of two isomers of $Sc_2S@C_{82}$ with $C_s(6)$ and $C_{3v}(8)$ cage isomers as well as that of $Sc_2S@C_{72}$-$C_s(10528)$ were elucidated by single-crystal x-ray diffraction (Mercado et al. 2011; Chen et al. 2013). The Sc_2S cluster in these molecules has bent structure with the Sc–S–Sc angles in the 97°–114°. Isolation and spectroscopic and electrochemical (EC) characterization was also reported for $Sc_2S@C_{70}$-$C_2(7892)$ (Chen et al. 2013).

12.3.6 Carbon Cage Isomerism of EMFs

An overview of the most abundant fullerene cages in EMFs (Table 12.1) shows that few cages are repeatedly preferred for different types of clusterfullerenes. For instance, C_{82}-$C_{2v}(9)$ is preferred for M^{III} monometallofullerenes, and some dimetallofullerenes and EMFs with M_2C_2, M_2S, and M_2O clusters are usually formed with the C_{82}-$C_{3v}(8)$/C_{82}-$C_s(6)$ cages, whereas many other clusterfullerenes are predominantly obtained with the C_{80}-$I_h(7)$ cage. Computational studies showed that the cage isomerism of EMFs is largely determined by the formal charge of the fullerene and that the lowest energy isomers for the same fullerene size are usually different in different charge states. For example, whereas the most stable isomer of C_{80} is $D_2(2)$, the most stable isomer of C_{80}^{6-} is $I_h(7)$ (Nakao, Kurita, and Fujita 1994); furthermore, C_{80}^{6-}–$I_h(7)$ has enhanced stability in comparison with many other cage sizes because of the very uniform spatial distribution of the pentagons (Popov and Dunsch 2007; Rodriguez-Fortea et al. 2010). Thus, this cage is preferred for sixfold charged clusters such as $(M^{3+})_2$, $(M^{3+})_3N^{3-}$, $(Sc^{3+})_4(C_2)^{6-}$, etc. Likewise, C_{82}-$C_{3v}(8)$ is the most suitable cage for the fourfold charged clusters such as $(M^{2+})_2$, $(M^{3+})_2S^{2-}$, or $(M^{3+})_2(C_2)^{2-}$ (Valencia, Rodríguez-Fortea, and Poblet 2008). Thus, carbon cage isomers of EMFs can be reliably predicted by a broad search of empty fullerene isomers in a proper charge state (Popov and Dunsch 2007). However, in some cases, the cluster's form factor (e.g., unsuitable size or shape) can cause deviations from expected cage isomers as discussed previously for $M_3N@C_{78}$.

A large formal charge and a strong metal-cage interaction lead to a specific structural feature of EMFs, a violation of the IPR. This rule imposes a strict limitation on the possible isomers of *empty* fullerenes but is often violated in EMFs (Figure 12.6). The possibility of the non-IPR EMFs was predicted by Kobayashi et al. in computational studies of $Ca@C_{72}$ isomers (Kobayashi et al. 1997) and the first experimental evidence was provided in 2000 by the synthesis of $Sc_2@C_{66}$ (Wang et al. 2000) and $Sc_3N@C_{68}$ (Stevenson et al. 2000). Neither C_{66} nor C_{68} has IPR cage isomers, and therefore, both EMFs are inevitably non-IPR ones. The most common non-IPR motif in EMFs is the adjacent pentagon pair (APP; also known as pentalene unit). Most of the characterized non-IPR EMFs have from one to three

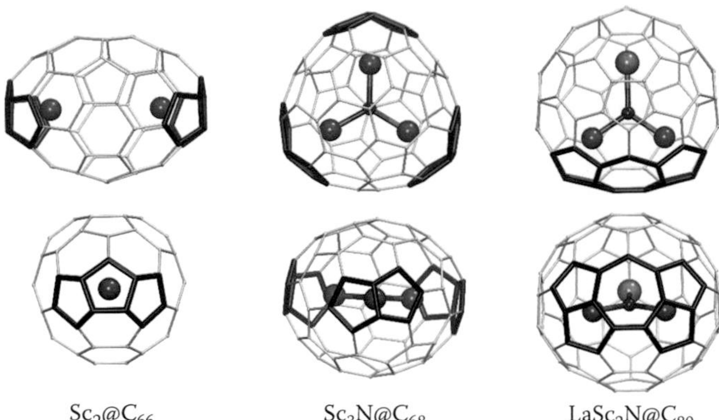

FIGURE 12.6 Examples of non-IPR EMFs with different structural motifs. $Sc_2@C_{66}$-C_{2v}(4059) with two unsaturated linear triquinanes, $Sc_3N@C_{68}$-D_3(6140) with three adjacent pentagon pairs, $LaSc_2N@C_{80}$-*Cs*(hept) with heptagon and two adjacent pentagon pairs coordinated by Sc atoms. Nonclassic structural fragments of the carbon cages are highlighted in black, and each structure is shown in two projections (top and bottom rows).

APPs. Since pentalene units are stabilized by coordinating metal atoms, the number of APPs in a given EMF molecule cannot exceed the number of metal atoms. Furthermore, the number of APPs usually decreases with the increase in the cage size. For instance, among the non-IPR NCFs, $Sc_3N@C_{68}$-D_3(6140) and $Sc_3N@C_{70}$-C_{2v}(7854) have three APPs, $DySc_2N@C_{76}$-C_s(17490) and $M_3N@C_{78}$-C_2(22010) have two APPs, and $M_3N@C_{82}$-C_s(39663) and $M_3N@C_{84}$-C_s(51365) have one APP.

Two other violations of the IPR in EMFs are presented by one example each. The structure of $Sc_2@C_{66}$, one of the first non-IPR EMFs, was unambiguously elucidated only in 2014 and was found to have a C_{66}-C_{2v}(4059) cage with two unsaturated linear triquinanes (Figure 12.6), i.e., two groups of three fused pentagons (Yamada et al. 2014). Very recently, $LaSc_2N@C_{80}$-C_s(hept), the first EMF with one heptagon and 13 pentagons, was characterized (Zhang et al. 2015a). Its carbon cage has two pentalene units, which are fused to the heptagon ring (Figure 12.6). Note that this structure has no number in the spiral numbering system since the latter was designed for fullerene isomers with only pentagons and hexagons.

12.4 Electrochemical Properties of EMFs

Fullerenes are good electron acceptors and undergo multiple reversible single-electron redox processes in solution. For instance, EC studies of C_{60} showed that it undergoes four reversible single-electron reductions at room temperature, whereas optimized conditions and reduced temperature allow observation of up to six reduction steps (Xie, Perezcordero, and Echegoyen 1992). Oxidation of the most abundant C_{60} and C_{70} fullerenes occurs at relatively high potentials, and it is not straightforward; however, many higher fullerenes exhibit one to two reversible oxidation steps in normal conditions. Encapsulation of metal atoms and clusters in EMFs can result in more complex redox behavior than that of empty fullerenes because both the carbon cage and the endohedral species can exhibit redox activity (Figure 12.7). Fullerene-based redox activity is typical for all mono-EMFs. The redox behavior of $M^{III}@C_{2n}$ mono-EMFs (M^{III} is Y, La, Ce, Pr, Gd, Er, etc.) is nevertheless substantially different from empty fullerenes because $M^{III}@C_{2n}$ molecules are radicals with an unpaired electron delocalized over the fullerene cage. As a result, their EC gap is relatively small (less than 1 V), and they are normally easier to reduce or oxidize than empty fullerenes (Suzuki et al. 1993, 1996; Popov, Yang, and Dunsch 2013). Redox potentials of $M^{II}@C_{2n}$ mono-EMFs (M^{II} = Sm, Tm, Yb), which have a diamagnetic carbon cage, are more similar to those of empty fullerenes with several reversible reduction steps, one to two oxidation steps, and EC gaps more than 1 V (Lu et al. 2010; Popov, Yang, and Dunsch 2013). Cage-based redox activity is also typical for many clusterfullerenes (e.g., oxidation of NCFs and reduction of non-Sc NCFs).

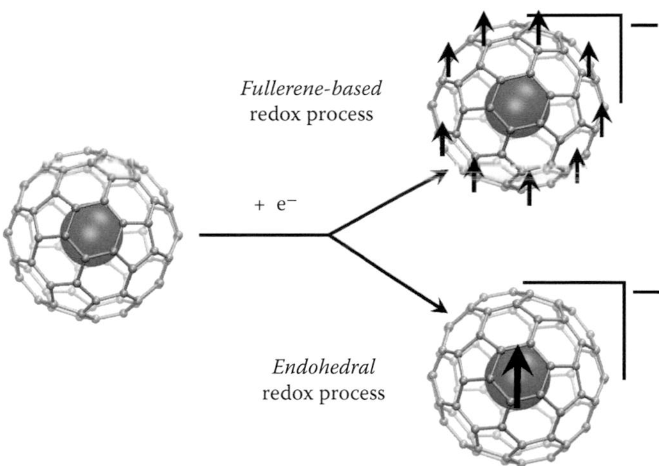

FIGURE 12.7 Schematic description of the fullerene-based and endohedral redox processes in endohedral fullerenes. Thick arrows indicate localization of spin and extraneous charge endohedral cluster (lower part) or delocalization over the fullerene cage (upper part).

When the endohedral cluster is redox-active, the carbon cage merely acts as an inert container, transparent to electrons. This type of electron transfer is known as an endohedral electron transfer process and is the main subject of endohedral electrochemistry, also known as *electrochemistry in cavea* (Popov and Dunsch 2011; Zhang and Popov 2014). An obvious, but not always necessary, prerequisite for endohedral redox activity is a suitable energy of the metal-based molecular orbitals, which should be the frontier molecular orbitals (the highest occupied molecular orbital [HOMO] or the lowest unoccupied molecular orbital [LUMO]) of the EMF molecule. In principle, this condition can be fulfilled for all EMFs whose endohedral species are more complex than a single metal atom. Experimentally, the endohedral redox processes can be revealed via unexpected redox behavior (e.g., shifted potential when compared with analogous molecules) and/or with the use of ex situ or in situ spectroelectrochemical methods (such as electron spin resonance or NMR spectroelectrochemistry).

Five groups of EMFs with well-established endohedral redox activity are described: (1) dimetallofullerenes, (2) $Sc_3N@C_{80}$ and its derivatives, (3) titanium-based EMFs, (4) cerium-based NCFs, and (5) oxide clusterfullerene $Sc_4O_2@C_{80}$. The LUMO of M^{III}-di-EMFs or the HOMO of M^{II}-di-EMFs usually has pronounced metal–metal bonding character, and therefore, reduction or oxidation of these molecules affects the metal–metal bonding character. These di-EMFs thus have either high reduction potentials (e.g., $La_2@C_{80}$) or low oxidation potentials (e.g., in $Sc_2@C_{82}$). The LUMO of $Sc_3N@C_{80}$ is equally delocalized over three Sc atoms, resulting in a Sc_3N-based reduction, whose mechanism can be modified by exohedral derivatization (Popov et al. 2014). Titanium is a rare example of a transition metal that can be encapsulated within fullerenes, and its valence state in Ti-EMFs can be tuned via EC reactions. For instance, $TiSc_2N@C_{80}$ and $TiY_2N@C_{80}$ both have Ti-based reduction and oxidation, whereas in $TiLu_2C@C_{80}$ and $Ti_2S@C_{78}$ only, reduction is a Ti-based process. Cerium exhibits endohedral redox activity in many NCFs, allowing for the redox potential of the Ce^{IV}/Ce^{III} redox couple to be tuned by varying a composition of the endohedral cluster and the size of the carbon cage (Zhang, Popov, and Dunsch 2014). The reason for such flexibility is an inherent strain, experienced by the Ce-containing clusters when they are encapsulated in the carbon cages of insufficient size. Oxidation of Ce^{III} to Ce^{IV} reduces its ionic radius from 1.02 Å to 0.87 Å and the strain is partially released. Therefore, the molecules with large strain have more negative Ce-based oxidation potentials. For instance, oxidation potentials of $CeY_2N@C_{80}$ and $CeSc_2N@C_{80}$ are −0.07 and +0.33 V, respectively, and the difference of 0.40 V between these EMFs is caused by a larger size (and hence larger strain) of the CeY_2N cluster (Zhang et al. 2013). Finally, $Sc_4O_2@C_{80}$ has cluster-based HOMO and LUMO, and both

TABLE 12.2 Redox Potentials of Representative EMFs

EMF	Ox-I	Ox-II	Red-I	Red-II	gap_{EC}
C_{60}-I_h(1)	1.21		−1.12	−1.50	2.33
C_{82}-C_2(3)	0.72		−0.69	−1.04	1.41
La@C_{82}-C_{2v}(9)	0.07	1.07	−0.42	−1.37	0.49
Yb@C_{82}-C_2(5)	0.38	0.90	−0.86	−0.98	1.24
Yb@C_{82}-C_{2v}(9)	0.61		−0.46	−0.78	1.07
La_2@C_{80}-I_h(7)	0.56	0.95	−0.31	−1.72	0.87
Sm_3@C_{80}-I_h(7)	0.30	0.78	−0.83	−1.88	1.13
Sc_3N@C_{80}-I_h(7)	0.59	1.09	−1.26	−1.62	1.85
$TiSc_2N$@C_{80}-I_h(7)	0.16		−0.94	−1.58	1.10
Y_3N@C_{80}-I_h(7)	0.64		−1.41	−1.83	2.05
CeY_2N@C_{80}-I_h(7)	−0.07		−1.36	−1.88	1.30
$CeSc_2N$@C_{80}-I_h(7)	0.33		−1.31	−1.83	1.64
$ScLu_2N$@C_{80}-I_h(7)	0.64	1.10	−1.38	−1.69	2.02
$TiLu_2C$@C_{80}-I_h(7)	0.63	1.10	−0.87	−1.53	1.50
Sc_3C_2@C_{80}-I_h(7)	−0.03		−0.50	−1.64	0.47
Sc_4C_2@C_{80}-I_h(7)	0.40	1.10	−1.16	−1.65	1.56
Sc_4O_2@C_{80}-I_h(7)	0.00	0.79	−1.10	−1.73	1.10
Sc_2@C_{82}-C_{3v}(8)	0.05		−1.10	–	1.15
Sc_2C_2@C_{82}-C_{3v}(8)	0.47	0.93	−0.94	−1.15	1.41
Sc_2S@C_{82}-C_{3v}(8)	0.47		−1.03	−1.16	1.50

Note: All values are measured in o-dichlorobenzene solution and are given vs. potential of the $Fe(Cp)_2^{+/0}$ pair.

its reduction and oxidation are based on the Sc_4O_2 cluster, as demonstrated by electron spin resonance spectroscopy (Popov et al. 2012b).

Importantly, when the redox process is determined by the carbon cage, redox potential is almost independent on the encapsulated species and is only dependent on the carbon cage isomer. Table 12.2 lists the redox potentials of a series of clusterfullerenes with C_{80}-I_h(7) isomer. In Y_3N@C_{80}, both reduction at −1.41 V and oxidation +0.64 V are cage-based processes (potentials are vs. $Fe(Cp)_2^{+/0}$ pair). Therefore, when other EMFs with C_{80}-I_h(7) cage have their redox potentials close to these values, cage-based redox activity can be expected, whereas significant deviations from these potentials are an indication of the endohedral redox process. For instance, La_2@C_{80} has cage-based oxidation (+0.56 V) and metal-based reduction (−0.31 V), and CeY_2N@C_{80} has cluster-based oxidation (−0.07 V) and cage-based reduction (−1.36 V), whereas oxidation and reduction of Sc_4O_2@C_{80} at 0.00 and −1.10 V are both cluster-based electron transfer. Similarly, for EMFs with the C_{82}-C_{3v}(8) cage, oxidation at +0.47 V observed for both Sc_2S@C_{82} and Sc_2C_2@C_{82} is attributed to the carbon cage, whereas a shift to +0.05 V found for the first oxidation step of Sc_2@C_{82} is an indication of the metal-based process.

12.5 EMF-Based Photovoltaic and Donor–Acceptor Dyads

The high electron affinity and low reorganization energy of fullerenes make them attractive as electron acceptors in photovoltaic application (Kirner, Sekita, and Guldi 2014). So far, fullerene C_{60} and its derivatives (such as phenyl-C61-butyric acid methyl ester [PCBM]) is the most popular electron acceptor used in organic photovoltaics. For a long time, EMFs were not considered in this role, in part because of the low amounts that precluded detailed studies of the charge and energy transfer phenomena, and a systematic exploration of their utility for photovoltaic applications was started only during the last years. PCBM-like derivatives of $Lu_3N@C_{80}$ were used as electron acceptors to prepare bulk heterojunction solar cells with conjugated polymer poly(3-hexylthiophene) (P3HT) as a donor component (Ross et al. 2009a,b). Transient absorption studies proved formation of the long-lived (>1 ms) charge-separated state in such systems upon their photoexcitation. Photocurrent efficiency of the collar cell with the $Lu_3N@C_{80}$-containing active layer exhibited noticeably higher power conversion efficiency, 4.2%, than analogous C_{60}-PCBM-based device, 3.4% (Figure 12.8). The reason for the improved performance of the EMF-based cell is the negative shift of the first reduction potential of the $Lu_3N@C_{80}$ derivative by 0.28 V compared with the C_{60}-PCBM value, which results in a higher open circuit voltage (V_{OC} = 0.81–0.89 V) than in C_{60}-based device (V_{OC} = 0.63 V) (Ross et al. 2009a). However, another group showed that the short circuit current, J_{SC}, in $Lu_3N@C_{80}$-based cells is typically smaller than in C_{60}-based analogs, which is rooted in a more efficient formation of the triplet excited state of the donor P3HT in EMF-based cells (Liedtke et al. 2011).

Different types of EMFs were also used to prepare molecular donor–acceptor dyads (i.e., donor and acceptor fragments are linked covalently within one molecule) and to study the mechanism of the intramolecular photoinduced electron transfer (Rudolf et al. 2012). The photophysical studies showed that the EMF-based dyads can have longer lifetimes of the photoinduced charge-separated state than C_{60} analogs (Pinzon et al. 2008, 2009), and the lifetimes exceeding 1 μs could be achieved in $Sc_3N@C_{80}$/zinc tetraphenylporphyrin dyads when the donor and acceptor parts were connected by linkers longer than 30 Å (Feng et al. 2011).

Importantly, EMFs are not only good electron acceptors but also can be good electron donors (see Table 12.2 for redox potentials) and hence can exhibit ambivalent behavior in the donor–acceptor dyads (Schubert et al. 2013). For example, $La_2@C_{80}$ behaves as an electron acceptor in the dyad with π-extended

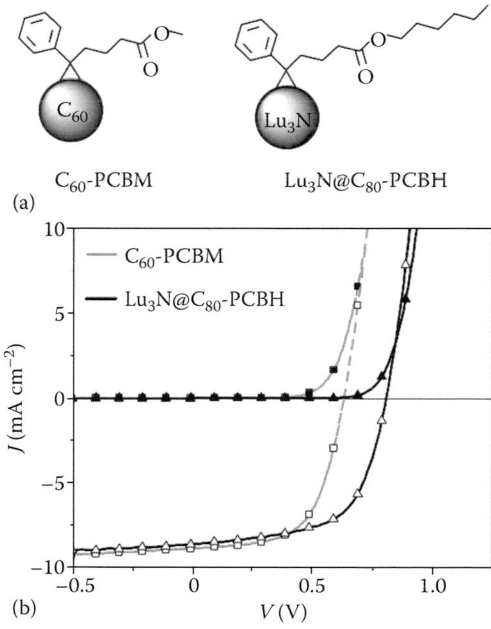

FIGURE 12.8 (a) Molecular structures of C_{60}-PCBM and $Lu_3N@C_{80}$-PCBH acceptors. (b) Current–voltage characteristics of bulk heterojunction solar cells prepared from blends of fullerene acceptors with P3HT donor. Filled symbols show the dark curves and open symbols show devices under light. For P3HT/$Lu_3N@C_{80}$-PCBH (triangles), power conversion efficiency (PCE) is 4.2%, open circuit voltage V_{oc} is 810 mV, short circuit current J_{sc} is 8.64 mA cm^{-2}, and filling factor (FF) is 0.61. Analog parameters of P3HT/C_{60}-PCBM (squares and dashed lines) are PCE = 3.4%, V_{oc} = 630 mV, J_{sc} = 8.9 mA cm^{-2}, and FF = 0.61. (Reproduced with permission from Ross, R.B. et al., *Nature Materials* 8, 208–212, 2009a.)

tetrathiafulvalene (Takano et al. 2010) but is an electron donor when linked to 11,11,12,12-tetracyano-9,10-anthra-*p*-quinodimethane (Takano et al. 2012).

12.6 Magnetic Properties of EMFs

The magnetic properties of EMFs are determined by a combination of several factors. First, if metal atoms transfer an odd number of electrons to the carbon cage (e.g., three electrons in $La@C_{82}$), the paramagnetic state of the EMF molecule is achieved with an unpaired electron delocalized over the carbon cage. Such EMFs are stable radicals and were studied by electron spin resonance techniques as soon as the very first sample of EMFs became available (Johnson et al. 1992). Coupling of the electron spin to the nuclear spin I of the metal atoms ($I = 7/2$ in Sc and La, or 1/2 in Y) gives characteristic hyperfine coupling patterns.

On the other hand, encapsulated metal ions can have intrinsic magnetic properties such as all lanthanides with partially filled 4f shell. If the carbon cage also bears an unpaired electron as in $M^{III}@C_{82}$ EMFs, the resulting magnetic moment is then determined by the interaction of the cage and lanthanide spins. Magnetic properties of lanthanide $M^{III}@C_{82}$ EMFs have been intensely studied since 1995 by electron spin resonance, superconducting quantum interference device (SQUID) magnetometry, and x-ray magnetic circular dichroism (XMCD) (Funasaka et al. 1995; Huang, Yang, and Zhang 2000; Furukawa et al. 2003; De Nadai et al. 2004). An EPR study of $Gd@C_{82}$ showed that the antiferromagnetic coupling of the Gd and fullerene spins is more stable than the state with ferromagnetic coupling by 14.4 cm^{-1} (Furukawa et al. 2003). SQUID and XMCD studies showed that $M@C_{82}$ EMFs exhibit paramagnetic behavior without hysteresis in magnetization curves. Effective magnetic moments are often reduced in comparison with those of free lanthanide ions by 20%–40%, which is usually explained by crystal-field-induced anisotropy and hybridization with the π-electrons of the carbon cage.

Clusterfullerenes usually have a closed-shell carbon cage, and hence, their magnetic properties are determined solely by encapsulated lanthanide ions and intramolecular exchange and dipolar magnetic interactions between the ions within one cluster. Numerous magnetization studies have been performed for $M_3N@C_{80}$ and $M_xSc_{3-x}N@C_{80}$ ($x = 1, 2$) NCFs with M = Gd, Tb, Dy, Ho, Er, and Tm (see Popov, Yang, and Dunsch 2013 and references therein). Most of these EMFs exhibited paramagnetic behavior and did not show measurable hysteresis down to 2 K (except for Dy, discussed further in the chapter). In Gd-NCFs, coupling of magnetic moments of Gd ions is ferromagnetic (Náfrádi et al. 2012; Svitova et al. 2014b). In Tb-, Dy-, and Ho-based NCFs, the magnetic moment of lanthanides is aligned along the M–N bonds. Ferromagnetic coupling of such moments in $M_3N@C_{80}$ results in frustrated magnetic states because simultaneous ferromagnetic coupling of all three moments in a triangular cluster is not possible (Wolf et al. 2005; Westerström et al. 2014). In Er- and Tm-based NCFs, magnetic anisotropy of lanthanide ions is different from other lanthanides, and magnetic moments are aligned perpendicular to the M–N bonds (Zhang et al. 2015b).

$DySc_2N@C_{80}$ and $Dy_2ScN@C_{80}$ were found to show single molecular magnetic (SMM) behavior (Figure 12.9). In $DySc_2N@C_{80}$, characteristic butterfly-shaped hysteresis curve with tunneling reversal of magnetization near zero field was observed at 5 K and below. The effective barrier to the thermally driven magnetization relaxation was estimated to be near 24 K (Westerström et al. 2012). In $Dy_2ScN@C_{80}$, ferromagnetic exchange and dipolar coupling of two Dy^{3+} moments resulted in an additional energy barrier, which prevented zero-field magnetization relaxation and led to the remanence in the magnetization curve (Westerström et al. 2014). The hysteresis was observed up to 6 K, and 100 s blocking temperature of about 5.5 K was determined, being among the highest temperatures for single molecule magnets. Relaxation of magnetization was described as a combination of two temperature-dependent processes with the barriers of 8.5 and 50 K, the former being ascribed to the exchange/dipolar interactions. Slow magnetic relaxation in $Dy_2ScN@C_{80}$ allowed observation of the alignment of magnetic moments and hysteresis at 4 K in the EMF monolayer sublimed onto the Rh(111) substrate (Westerström et al. 2015). Thus, Dy-based clusterfullerenes are a new class of SMM materials that are very promising for the spintronic applications because of the high thermal and chemical stability and protecting role of the carbon cage, which preserves endohedral spin states.

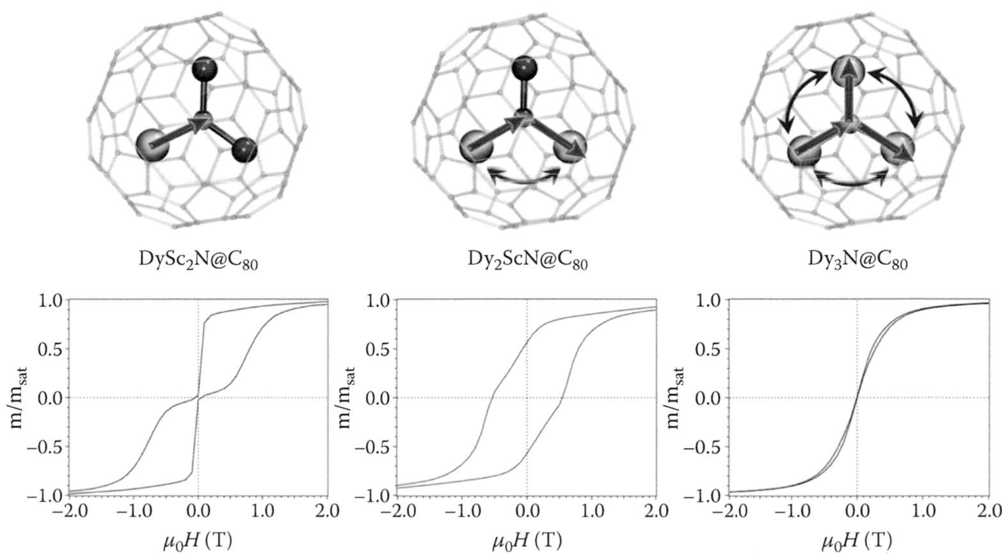

FIGURE 12.9 Molecular structures (top row) and hysteresis magnetization curves measured at 2 K with a field sweep rate of 0.8 mT s^{-1} (bottom row) of $DySc_2N@C_{80}$, $Dy_2ScN@C_{80}$, and $Dy_3N@C_{80}$. Straight arrows show alignment of Dy magnetic moments along the Dy–N bonds (with ferromagnetic coupling of two moments in $Dy_2ScN@C_{80}$), curved arrows denote exchange and dipolar interaction between Dy ions within of EMF molecule. Magnetization curves. (Reproduced with permission from Westerström, R. et al., *Physical Review B* 89, 060406, 2014.)

12.7 Biomedical Application of EMFs

Effective isolation of endohedral metals from the surrounding environment by the carbon cage makes EMF promising for different types of biomedical applications. Particularly, extended studies were performed for Gd-EMFs considered as a new type of contrast agents in magnetic resonance imaging (MRI). Nowadays, the most commonly used commercially available MRI contrast agents are gadolinium chelate complexes such as Gd-DTPA (DTPA: diethylenetriamino-pentaacetic acid), which uses the Gd^{3+} ion to enhance the relaxation rate of water protons. Unavoidable release of toxic Gd^{3+} in such complexes, although very small, may lead to unpleasant consequences. The use of Gd-EMFs (the most studied are $Gd@C_{82}$ or $Gd_3N@C_{80}$) as contrast agents solves the problem of the Gd^{3+} release. Besides, Gd-EMFs were shown to induce a much higher relaxation rate of water protons than Gd-DTPA and similar complexes (Mikawa et al. 2001; Fatouros et al. 2006; Shu et al. 2009; Zhang et al. 2010). It should be noted, however, that EMFs are not soluble in physiological liquids and hence have to be chemically modified first. Typical derivatization routes are hydroxylation, addition of poly(ethylene glycol) units, or carboxylation. Water solubility of such derivatives is enhanced dramatically; however, derivatized EMFs tend to aggregate in solution, forming nanoparticles. More sophisticated procedures include conjugation of EMFs with the biological targeting modules (e.g., special proteins), which would deliver EMF-based agents to only special kinds of cells (Figure 12.10). This approach can drastically improve the specificity of EMF-based contrast agents and at the same time reduce required amounts (Braun et al. 2010; Fillmore et al. 2011; Svitova et al. 2012).

Gd-EMFs can be used not only to enhance MRI contrast but also as antitumor drugs. Particularly intense were the studies of anticancer activity of $[Gd@C_{82}(OH)_{22}]_n$ nanoparticles (Z. Chen et al. 2012; Chen, Mao, and Liu 2012; Meng et al. 2013). Another promising route for applications of EMFs in medicine is their use as radiotracers or radiopharmaceuticals (Kobayashi et al. 1995; Cagle et al. 1999; Shultz et al. 2010, 2011; Horiguchi, Kudo, and Nagasaki 2011). Such approaches include labeling of lanthanide-EMFs with ^{140}La, ^{166}Ho, ^{177}Lu, or neutron activation of Gd-EMFs.

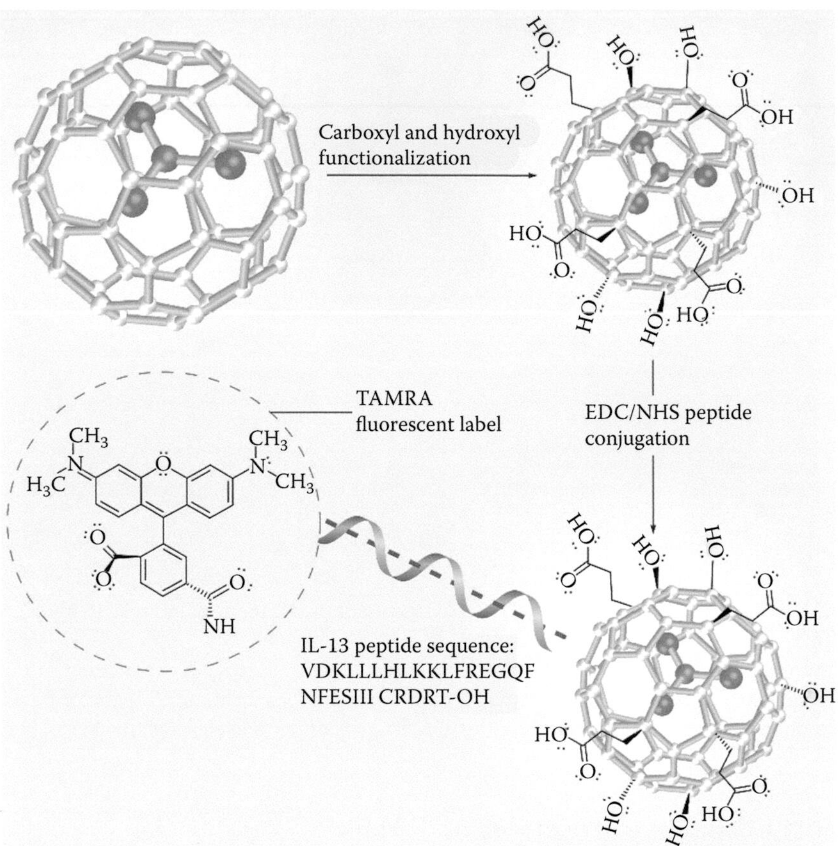

FIGURE 12.10 $Gd_3N@C_{80}$ functionalization and conjugation with IL-13 peptides for targeting and imaging glial tumors. The first step involves the addition of carboxyl groups for conjugation and hydroxyl groups for MRI contrast; both groups provide aqueous colloidal stability. The second step links the $Gd_3N@C_{80}$ molecules to the fluorescent-tagged glioma-targeting peptide through amide bond formation. EDC: *N*-(3-dimethylaminopropyl)-*N'*-ethylcarbodiimide; NHS: *N*-hydroxysuccinimide. (Reproduced from Fillmore, H.L. et al., *Nanomedicine*, 6, 449–458, 2011. With permission.)

References

Akasaka, T., and Lu, X. "Structural and electronic properties of endohedral metallofullerenes." *Chemical Record* 12 (2012): 256–269.

Akasaka, T., Nagase, S., Kobayashi, K., Walchli, M., Yamamoto, K., Funasaka, H., Kako, M., Hoshino, T., and Erata, T. "^{13}C and ^{139}La NMR studies of $La_2@C_{80}$: First evidence for circular motion of metal atoms in endohedral dimetallofullerenes." *Angewandte Chemie-International Edition in English* 36 (1997): 1643–1645.

Akasaka, T., Wakahara, T., Nagase, S., Kobayashi, K., Waelchli, M., Yamamoto, K., Kondo, M. et al. "$La@C_{82}$ anion. An unusually stable metallofullerene." *Journal of the American Chemical Society* 122 (2000): 9316–9317.

Akasaka, T., Wakahara, T., Nagase, S., Kobayashi, K., Waelchli, M., Yamamoto, K., Kondo, M. et al. "Structural determination of the $La@C_{82}$ isomer." *Journal of Physical Chemistry B* 105 (2001): 2971–2974.

Akiyama, K., Sueki, K., Kodama, T., Kikuchi, K., Takigawa, Y., Nakahara, H., Ikemoto, I., and Katada, M. "New fullerenes of a group IV element: Hf metallofullerenes." *Chemical Physics Letters* 317 (2000): 490–496.

Akiyama, K., Zhao, Y. L., Sueki, K., Tsukada, K., Haba, H., Nagame, Y., Kodama, T. et al. "Isolation and characterization of light actinide metallofullerenes." *Journal of the American Chemical Society* 123 (2001): 181–182.

Akiyama, K., Sueki, K., Tsukada, K., Yaita, T., Miyake, Y., Haba, H., Asai, M. et al. "Study of metallofullerenes encapsulating actinides." *Journal of Nuclear and Radiochemical Sciences* 3 (2002): 151–154.

Akiyama, K., Sueki, K., Haba, H., Tsukada, K., Asai, M., Yaita, T., Nagame, Y., Kikuchi, K., Katada, M., and Nakahara, H. "Production and characterization of actinide metallofullerenes." *Journal of Radioanalytical and Nuclear Chemistry* 255 (2003): 155–158.

Akiyama, K., Haba, H., Tsukada, K., Asai, M., Toyoshima, A., Sueki, K., Nagame, Y., and Katada, M. "A metallofullerene that encapsulates ^{225}Ac." *Journal of Radioanalytical and Nuclear Chemistry* 280 (2009): 329–331.

Akiyama, K., Hamano, T., Nakanishi, Y., Takeuchi, E., Noda, S., Wang, Z., Kubuki, S., and Shinohara, H. "Non-HPLC rapid separation of metallofullerenes and empty cages with $TiCl_4$ Lewis acid." *Journal of the American Chemical Society* 134 (2012): 9762–9767.

Alvarez, M. M., Gillan, E. G., Holczer, K., Kaner, R. B., Min, K. S., and Whetten, R. L. "La_2C_{80}—A soluble dimetallofullerene." *Journal of Physical Chemistry* 95 (1991): 10561–10563.

Aoyagi, S., Nishibori, E., Sawa, H., Sugimoto, K., Takata, M., Miyata, Y., Kitaura, R. et al. "A layered ionic crystal of polar Li@C_{60} superatoms." *Nature Chemistry* 2 (2010): 678–683.

Beavers, C. M., Zuo, T. M., Duchamp, J. C., Harich, K., Dorn, H. C., Olmstead, M. M., and Balch, A. L. "$Tb_3N@C_{84}$: An improbable, egg-shaped endohedral fullerene that violates the isolated pentagon rule." *Journal of the American Chemical Society* 128 (2006): 11352–11353.

Beavers, C. M., Chaur, M. N., Olmstead, M. M., Echegoyen, L., and Balch, A. L. "Large metal ions in a relatively small fullerene cage: The structure of $Gd_3N@C_2(22010)$-C_{78} departs from the isolated pentagon rule." *Journal of the American Chemical Society* 131 (2009): 11519–11524.

Beavers, C. M., Jin, H., Yang, H., Wang, Z., Wang, X., Ge, H., Liu, Z., Mercado, B. Q., Olmstead, M. M., and Balch, A. L. "Very large, soluble endohedral fullerenes in the series La_2C_{90} to La_2C_{138}: Isolation and crystallographic characterization of $La_2@D_5(450)$-C_{100}." *Journal of the American Chemical Society* 133 (2011): 15338–15341.

Bolskar, R. D., and Alford, J. M. "Chemical oxidation of endohedral metallofullerenes: Identification and separation of distinct classes." *Chemical Communications* (2003): 1292–1293.

Braun, K., Dunsch, L., Pipkorn, R., Bock, M., Baeuerle, T., Yang, S. F., Waldeck, W., and Wiessler, M. "Gain of a 500-fold sensitivity on an intravital MR contrast agent based on an endohedral gadolinium-cluster-fullerene-conjugate: A new chance in cancer diagnostics." *International Journal of Medical Sciences* 7 (2010): 136–146.

Cagle, D. W., Kennel, S. J., Mirzadeh, S., Alford, J. M., and Wilson, L. J. "In vivo studies of fullerene-based materials using endohedral metallofullerene radiotracers." *Proceedings of the National Academy of Sciences of the United States of America* 96 (1999): 5182–5187.

Cai, T., Xu, L. S., Anderson, M. R., Ge, Z. X., Zuo, T. M., Wang, X. L., Olmstead, M. M., Balch, A. L., Gibson, H. W., and Dorn, H. C. "Structure and enhanced reactivity rates of the D_{5h} $Sc_3N@C_{80}$ and $Lu_3N@C_{80}$ metallofullerene isomers: The importance of the pyracylene motif." *Journal of the American Chemical Society* 128 (2006): 8581–8589.

Campbell, E. E. B., Tellgmann, R., Krawez, N., and Hertel, I. V. "Production and LDMS characterisation of endohedral alkali-fullerene films." *Journal of Physics and Chemistry of Solids* 58 (1997): 1763–1769.

Cao, B. P., Hasegawa, M., Okada, K., Tomiyama, T., Okazaki, T., Suenaga, K., and Shinohara, H. "EELS and ^{13}C NMR characterization of pure $Ti_2@C_{80}$ metallofullerene." *Journal of the American Chemical Society* 123 (2001): 9679–9680.

Cao, B. P., Suenaga, K., Okazaki, T., and Shinohara, H. "Production, isolation, and EELS characterization of $Ti_2@C_{84}$ dititanium metallofullerenes." *Journal of Physical Chemistry B* 106 (2002): 9295–9298.

Cao, B. P., Wakahara, T., Tsuchiya, T., Kondo, M., Maeda, Y., Rahman, G. M. A., Akasaka, T., Kobayashi, K., Nagase, S., and Yamamoto, K. "Isolation, characterization, and theoretical study of $La_2@C_{78}$." *Journal of the American Chemical Society* 126 (2004): 9164–9165.

Cerón, M. R., Li, F.-F., and Echegoyen, L. "An efficient method to separate $Sc_3N@C_{80}$ I_h and D_{5h} isomers and $Sc_3N@C_{78}$ by selective oxidation with acetylferrocenium $[Fe(COCH_3C_5H_4)Cp]^+$." *Chemistry—A European Journal* 19 (2013): 7410–7415.

Chai, Y., Guo, T., Jin, C. M., Haufler, R. E., Chibante, L. P. F., Fure, J., Wang, L. H., Alford, J. M., and Smalley, R. E. "Fullerenes with metals inside." *Journal of Physical Chemistry* 95 (1991): 7564–7568.

Chaur, M. N., Melin, F., Elliott, B., Athans, A. J., Walker, K., Holloway, B. C., and Echegoyen, L. "$Gd_3N@C_{2n}$ (n = 40, 42, and 44): Remarkably low HOMO-LUMO gap and unusual electrochemical reversibility of $Gd_3N@C_{88}$." *Journal of the American Chemical Society* 129 (2007): 14826–14829.

Chaur, M. N., Athans, A. J., and Echegoyen, L. "Metallic nitride endohedral fullerenes: synthesis and electrochemical properties." *Tetrahedron* 64 (2008a): 11387–11393.

Chaur, M. N., Melin, F., Ashby, J., Kumbhar, A., Rao, A. M., and Echegoyen, L. "Lanthanum nitride endohedral fullerenes $La_3N@C_{2n}$ ($43<n<55$): Preferential formation of $La_3N@C_{96}$." *Chemistry—A European Journal* 14 (2008b): 8213–8219.

Chaur, M. N., Melin, F., Elliott, B., Kumbhar, A., Athans, A. J., and Echegoyen, L. "New $M_3N@C_{2n}$ endohedral metallofullerene families (M = Nd, Pr, Ce; n = 40–53): Expanding the preferential templating of the C_{88} cage and approaching the C_{96} cage." *Chemistry—A European Journal* 14 (2008c): 4594–4599.

Chaur, M. N., Valencia, R., Rodriguez-Fortea, A., Poblet, J. M., and Echegoyen, L. "Trimetallic nitride endohedral fullerenes: Experimental and theoretical evidence for the $M_3N^{6+}@C_{2n}^{6-}$ model." *Angewandte Chemie-International Edition* 48 (2009): 1425–1428.

Chaur, M. N., Aparicio-Angles, X., Mercado, B. Q., Elliott, B., Rodriguez-Fortea, A., Clotet, A., Olmstead, M. M., Balch, A. L., Poblet, J. M., and Echegoyen, L. "Structural and electrochemical property correlations of metallic nitride endohedral metallofullerenes." *Journal of Physical Chemistry C* 114 (2010): 13003–13009.

Che, Y., Yang, H., Wang, Z., Jin, H., Liu, Z., Lu, C., Zuo, T., Dorn, H. C. et al. "Isolation and structural characterization of two very large, and largely empty, endohedral fullerenes: $Tm@C_{3v}$-C_{94} and $Ca@C_{3v}$-C_{94}." *Inorganic Chemistry* 48 (2009): 6004–6010.

Chen, N., Chaur, M. N., Moore, C., Pinzon, J. R., Valencia, R., Rodriguez-Fortea, A., Poblet, J. M., and Echegoyen, L. "Synthesis of a new endohedral fullerene family, $Sc_2S@C_{2n}$ (n = 40–50) by the introduction of SO_2." *Chemical Communications* 46 (2010): 4818–4820.

Chen, C., Liu, F., Li, S., Wang, N., Popov, A. A., Jiao, M., Wei, T., Li, Q., Dunsch, L., and Yang, S. "Titanium/yttrium mixed metal nitride clusterfullerene $TiY_2N@C_{80}$: Synthesis, isolation, and effect of the group-III metal." *Inorganic Chemistry* 51 (2012): 3039–3045.

Chen, Z., Ma, L., Liu, Y., and Chen, C. "Applications of functionalized fullerenes in tumor theranostics." *Theranostics* 2 (2012): 238–250.

Chen, Z., Mao, R., and Liu, Y. "Fullerenes for cancer diagnosis and therapy: Preparation, biological and clinical perspectives." *Current Drug Metabolism* 13 (2012): 1035–1045.

Chen, N., Mulet-Gas, M., Li, Y.-Y., Stene, R. E., Atherton, C. W., Rodriguez-Fortea, A., Poblet, J. M., and Echegoyen, L. "$Sc_2S@C_2(7892)$-C_{70}: A metallic sulfide cluster inside a non-IPR C_{70} cage." *Chemical Science* 4 (2013): 180–186.

Cioslowski, J., and Fleischmann, E. D. "Endohedral complexes—Atoms and ions inside the C_{60} cage." *Journal of Chemical Physics* 94 (1991): 3730–3734.

De Nadai, C., Mirone, A., Dhesi, S. S., Bencok, P., Brookes, N. B., Marenne, I., Rudolf, P., Tagmatarchis, N., Shinohara, H., and Dennis, T. J. S. "Local magnetism in rare-earth metals encapsulated in fullerenes." *Physical Review B* 69 (2004): 184421.

Deng, Q., and Popov, A. A. "Clusters encapsulated in endohedral metallofullerenes: How strained are they?" *Journal of the American Chemical Society* 136 (2014): 4257–4264.

Deng, Q., Junghans, K., and Popov, A. A. "Carbide clusterfullerenes with odd number of carbon atoms: Molecular and electronic structures of $Sc_4C@C_{80}$, $Sc_4C@C_{82}$, and $Sc_4C_3@C_{80}$." *Theoretical Chemistry Accounts* 134 (2015): 10, DOI:10.1007/s00214-014-1610-6.

Duchamp, J. C., Demortier, A., Fletcher, K. R., Dorn, D., Iezzi, E. B., Glass, T., and Dorn, H. C. "An isomer of the endohedral metallofullerene $Sc_3N@C_{80}$ with D_{5h} symmetry." *Chemical Physics Letters* 375 (2003): 655–659.

Dunk, P. W., Kaiser, N. K., Hendrickson, C. L., Quinn, J. P., Ewels, C. P., Nakanishi, Y., Sasaki, Y., Shinohara, H., Marshall, A. G., and Kroto, H. W. "Closed network growth of fullerenes." *Nature Communications* 3 (2012a): 855.

Dunk, P. W., Kaiser, N. K., Mulet-Gas, M., Rodríguez-Fortea, A., Poblet, J. M., Shinohara, H., Hendrickson, C. L., Marshall, A. G., and Kroto, H. W. "The smallest stable fullerene, $M@C_{28}$ (M = Ti, Zr, U): Stabilization and growth from carbon vapor." *Journal of the American Chemical Society* 134 (2012b): 9380–9389.

Dunk, P. W., Mulet-Gas, M., Nakanishi, Y., Kaiser, N. K., Rodríguez-Fortea, A., Shinohara, H., Poblet, J. M., Marshall, A. G., and Kroto, H. W. "Bottom-up formation of endohedral monometallofullerenes is directed by charge transfer." *Nature Communications* 5 (2014): 5844.

Dunsch, L., and Yang, S. "Metal nitride cluster fullerenes: Their current state and future prospects." *Small* 3 (2007): 1298–1320.

Dunsch, L., Krause, M., Noack, J., and Georgi, P. "Endohedral nitride cluster fullerenes—Formation and spectroscopic analysis of $L_{3-x}M_xN@C_{2n}$ ($0 <= x <= 3$; n = 39, 40)." *Journal of Physics and Chemistry of Solids* 65 (2004): 309–315.

Fatouros, P. P., Corwin, F. D., Chen, Z. J., Broaddus, W. C., Tatum, J. L., Kettenmann, B., Ge, Z., Gibson, H. W. et al. "In vitro and in vivo imaging studies of a new endohedral metallofullerene nanoparticle." *Radiology* 240 (2006): 756–764.

Feng, L., Gayathri Radhakrishnan, S., Mizorogi, N., Slanina, Z., Nikawa, H., Tsuchiya, T., Akasaka, T., Nagase, S., Martín, N., and Guldi, D. M. "Synthesis and charge-transfer chemistry of $La_2@I_h$-C_{80}/$Sc_3N@I_h$-C_{80}-zinc porphyrin conjugates: Impact of endohedral cluster." *Journal of the American Chemical Society* 133 (2011): 7608–7618.

Feng, Y., Wang, T., Wu, J., Feng, L., Xiang, J., Ma, Y., Zhang, Z., Jiang, L., Shu, C., and Wang, C. "Structural and electronic studies of metal carbide clusterfullerene $Sc_2C_2@C_s$–C_{72}." *Nanoscale* 5 (2013): 6704–6707.

Feng, Y., Wang, T., Wu, J., Zhang, Z., Jiang, L., Han, H., and Wang, C. "Electron-spin excitation by implanting hydrogen into metallofullerene: The synthesis and spectroscopic characterization of $Sc_4C_2H@I_h$-C_{80}." *Chemical Communications* 50 (2014): 12166–12168.

Feng, Y., Wang, T., Xiang, J., Gan, L.-H., Wu, B., Jiang, L., and Wang, C.-R. "Tuneable dynamics of a scandium nitride cluster inside an I_h–C_{80} Cage." *Dalton Transactions* 44 (2015): 2057–2061.

Fillmore, H. L., Shultz, M. D., Henderson, S. C., Cooper, P., Broaddus, W. C., Chen, Z. J., Shu, C.-Y. et al. "Conjugation of functionalized gadolinium metallofullerenes with IL-13 peptides for targeting and imaging glial tumors." *Nanomedicine* 6 (2011): 449–458.

Fowler, P., and Manolopoulos, D. E. *An Atlas of Fullerenes* (Oxford, UK: Clarendon Press, 1995), Original edition.

Fu, W., Xu, L., Azurmendi, H., Ge, J., Fuhrer, T., Zuo, T., Reid, J., Shu, C., Harich, K., and Dorn, H. C. "^{89}Y and ^{13}C NMR cluster and carbon cage studies of an yttrium metallofullerene family, $Y_3N@C_{2n}$ (n = 40–43)." *Journal of the American Chemical Society* 131 (2009): 11762–11769.

Fu, W., Wang, X., Azuremendi, H., Zhang, J., and Dorn, H. C. "^{14}N and ^{45}Sc NMR study of trimetallic nitride cluster $(M_3N)^{6+}$ dynamics inside a icosahedral C_{80} cage." *Chemical Communications* 47 (2011a): 3858–3860.

Fu, W., Zhang, J., Champion, H., Fuhrer, T., Azuremendi, H., Zuo, T., Zhang, J., Harich, K., and Dorn, H. C. "Electronic properties and ^{13}C NMR structural study of $Y_3N@C_{88}$." *Inorganic Chemistry* 50 (2011b): 4256–4259.

Fu, W., Zhang, J., Fuhrer, T., Champion, H., Furukawa, K., Kato, T., Mahaney, J. E. et al. "$Gd_2@C_{79}N$: Isolation, characterization, and monoadduct formation of a very stable heterofullerene with a magnetic spin state of S = 15/2." *Journal of the American Chemical Society* 133 (2011c): 9741–9750.

Funasaka, H., Sakurai, K., Oda, Y., Yamamoto, K., and Takahashi, T. "Magnetic-properties of $Gd@C_{82}$ metallofullerene." *Chemical Physics Letters* 232 (1995): 273–277.

Furukawa, K., Okubo, S., Kato, H., Shinohara, H., and Kato, T. "High-field/high-frequency ESR study of $Gd@C_{82}$-I." *Journal of Physical Chemistry A* 107 (2003): 10933–10937.

Ge, Z. X., Duchamp, J. C., Cai, T., Gibson, H. W., and Dorn, H. C. "Purification of endohedral trimetallic nitride fullerenes in a single, facile step." *Journal of the American Chemical Society* 127 (2005): 16292–16298.

Gromov, A., Kratschmer, W., Krawez, N., Tellgmann, R., and Campbell, E. E. B. "Extraction and HPLC purification of $Li@C_{60/70}$." *Chemical Communications* (1997): 2003–2004.

Guo, T., Diener, M. D., Chai, Y., Alford, M. J., Haufler, R. E., McClure, S. M., Ohno, T., Weaver, J. H., Scuseria, G. E., and Smalley, R. E. "Uranium stabilization of C_{28}—A tetravalent fullerene." *Science* 257 (1992): 1661–1664.

Heath, J. R., O'Brien, S. C., Zhang, Q., Liu, Y., Curl, R. F., Tittel, F. K., and Smalley, R. E. "Lanthanum complexes of spheroidal carbon shells." *Journal of the American Chemical Society* 107 (1985): 7779–7780.

Horiguchi, Y., Kudo, S., and Nagasaki, Y. "$Gd@C_{82}$ metallofullerenes for neutron capture therapy—Fullerene solubilization by poly(ethylene glycol)-block-poly(2-(N, N-diethylamino)ethyl methacrylate) and resultant efficacy in vitro." *Science and Technology of Advanced Materials* 12 (2011): 044607.

Huang, H. J., Yang, S. H., and Zhang, X. X. "Magnetic properties of heavy rare-earth metallofullerenes $M@C_{82}$ (M = Gd, Tb, Dy, Ho, and Er)." *Journal of Physical Chemistry B* 104 (2000): 1473–1482.

Iiduka, Y., Wakahara, T., Nakahodo, T., Tsuchiya, T., Sakuraba, A., Maeda, Y., Akasaka, T. et al. "Structural determination of metallofullerene Sc_3C_{82} revisited: A surprising finding." *Journal of the American Chemical Society* 127 (2005): 12500–12501.

Inoue, T., Tomiyama, T., Sugai, T., Okazaki, T., Suematsu, T., Fujii, N., Utsumi, H., Nojima, K., and Shinohara, H. "Trapping a C_2 radical in endohedral metallofullerenes: Synthesis and structures of $(Y_2C_2)@C_{82}$ (Isomers I, II, and III)." *Journal of Physical Chemistry B* 108 (2004): 7573–7579.

Ito, Y., Okazaki, T., Okubo, S., Akachi, M., Ohno, Y., Mizutani, T., Nakamura, T., Kitaura, R., Sugai, T., and Shinohara, H. "Enhanced 1520 nm photoluminescence from Er^{3+} ions in di-erbium-carbide metallofullerenes $(Er_2C_2)@C_{82}$ (isomers I, II, and III)." *ACS Nano* 1 (2007): 456–462.

Jin, H., Yang, H., Yu, M., Liu, Z., Beavers, C. M., Olmstead, M. M., and Balch, A. L. "Single samarium atoms in large fullerene cages. Characterization of two isomers of $Sm@C_{92}$ and four isomers of $Sm@C_{94}$ with the x-ray crystallographic identification of $Sm@C_1(42)$-C_{92}, $Sm@C_s(24)$-C_{92}, and $Sm@C_{3v}(134)$-C_{94}." *Journal of the American Chemical Society* 134 (2012): 10933–10941.

Jin, P., Tang, C., and Chen, Z. "Carbon atoms trapped in cages: Metal carbide clusterfullerenes." *Coordination Chemistry Reviews* 270–271 (2014): 89–111.

John, T., Dennis, S., and Shinohara, H. "Production, isolation, and characterization of group-2 metal-containing endohedral metallofullerenes." *Applied Physics A-Materials Science & Processing* 66 (1998): 243–247.

Johnson, R. D., Devries, M. S., Salem, J., Bethune, D. S., and Yannoni, C. S. "Electron-paramagnetic resonance studies of lanthanum-containing C_{82}." *Nature* 355 (1992): 239–240.

Kato, H., Taninaka, A., Sugai, T., and Shinohara, H. "Structure of a missing-caged metallofullerene: $La_2@C_{72}$." *Journal of the American Chemical Society* 125 (2003): 7782–7783.

Kikuchi, K., Akiyama, K., Sakaguchi, K., Kodama, T., Nishikawa, H., Ikemoto, I., Ishigaki, T., Achiba, Y., Sueki, K., and Nakahara, H. "Production and isolation of the isomers of dimetallofulerenes, HoTm@C_{82} and Tm_2@C_{82}." *Chemical Physics Letters* 319 (2000): 472–476.

Kirbach, U., and Dunsch, L. "The existence of stable Tm@C_{82} isomers." *Angewandte Chemie-International Edition in English* 35 (1996): 2380–2383.

Kirner, S., Sekita, M., and Guldi, D. M. "25th anniversary article: 25 years of fullerene research in electron transfer chemistry." *Advanced Materials* 26 (2014): 1482–1493.

Kobayashi, K., Kuwano, M., Sueki, K., Kikuchi, K., Achiba, Y., Nakahara, H., Kananishi, N., Watanabe, M., and Tomura, K. "Activation and tracer techniques for study of metallofullerenes." *Journal of Radioanalytical and Nuclear Chemistry-Articles* 192 (1995): 81–89.

Kobayashi, K., Nagase, S., Yoshida, M., and Osawa, E. "Endohedral metallofullerenes. Are the isolated pentagon rule and fullerene structures always satisfied?" *Journal of the American Chemical Society* 119 (1997): 12693–12694.

Kodama, T., Ozawa, N., Miyake, Y., Sakaguchi, K., Nishikawa, H., Ikemoto, I., Kikuchi, K., and Achiba, Y. "Structural study of three isomers of Tm@C_{82} by ^{13}C NMR spectroscopy." *Journal of the American Chemical Society* 124 (2002): 1452–1455.

Kodama, T., Fujii, R., Miyake, Y., Sakaguchi, K., Nishikawa, H., Ikemoto, I., Kikuchi, K., and Achiba, Y. "Structural study of four Ca@C_{82} isomers by ^{13}C NMR spectroscopy." *Chemical Physics Letters* 377 (2003): 197–200.

Kodama, T., Fujii, R., Miyake, Y., Suzuki, S., Nishikawa, H., Ikemoto, I., Kikuchi, K., and Achiba, Y. "^{13}C NMR study of Ca@C_{74}: The cage structure and the site-hopping motion of a Ca atom inside the cage." *Chemical Physics Letters* 399 (2004): 94–97.

Kratschmer, W., Lamb, L. D., Fostiropoulos, K., and Huffman, D. R. "Solid C_{60}—A new form of carbon." *Nature* 347 (1990): 354–358.

Krause, M., and Dunsch, L. "Gadolinium nitride Gd_3N in carbon cages: The influence of cluster size and bond strength." *Angewandte Chemie-International Edition* 44 (2005): 1557–1560.

Krause, M., Wong, J., and Dunsch, L. "Expanding the world of endohedral fullerenes—The Tm_3N@C_{2n} (39 < = n < = 43) clusterfullerene family." *Chemistry—A European Journal* 11 (2005): 706–711.

Krause, M., Ziegs, F., Popov, A. A., and Dunsch, L. "Entrapped bonded hydrogen in a fullerene: The five-atom cluster Sc_3CH in C_{80}." *Chemphyschem* 8 (2007): 537–540.

Kroto, H. W., Heath, J. R., O'Brien, S. C., Curl, R. F., and Smalley, R. E. "C_{60}—buckminsterfullerene." *Nature* 318 (1985): 162–163.

Kubozono, Y., Ohta, T., Hayashibara, T., Maeda, H., Ishida, H., Kashino, S., Oshima, K., Yamazaki, H., Ukita, S., and Sogabe, T. "Preparation and extraction of Ca@C_{60}." *Chemistry Letters* 24 (1995): 457–458.

Komatsu, K., Murata, M., and Murata, Y. "Encapsulation of molecular hydrogen in fullerene C 60 by organic synthesis." *Science* 307 (2005): 238–240.

Kuran, P., Krause, M., Bartl, A., and Dunsch, L. "Preparation, isolation and characterisation of Eu@C_{74}: The first isolated europium endohedral fullerene." *Chemical Physics Letters* 292 (1998): 580–586.

Kurihara, H., Lu, X., Iiduka, Y., Mizorogi, N., Slanina, Z., Tsuchiya, T., Akasaka, T., and Nagase, S. "Sc_2C_2@C_{80} rather than Sc_2@C_{82}: Templated formation of unexpected $C_{2v}(5)$-C_{80} and temperature-dependent dynamic motion of internal Sc_2C_2 cluster." *Journal of the American Chemical Society* 133 (2011): 2382–2385.

Kurihara, H., Lu, X., Iiduka, Y., Mizorogi, N., Slanina, Z., Tsuchiya, T., Nagase, S., and Akasaka, T. "Sc_2@$C_{3v}(8)$-C_{82} *vs.* Sc_2C_2@$C_{3v}(8)$-C_{82}: Drastic effect of C_2 capture on the redox properties of scandium metallofullerenes." *Chemical Communications* 48 (2012a): 1290–1292.

Kurihara, H., Lu, X., Iiduka, Y., Nikawa, H., Hachiya, M., Mizorogi, N., Slanina, Z., Tsuchiya, T., Nagase, S., and Akasaka, T. "X-ray structures of Sc_2C_2@C_{2n} (n = 40–42): In-depth understanding of the core-shell interplay in carbide cluster metallofullerenes." *Inorganic Chemistry* 51 (2012b): 746–750.

Li, F.-F., Chen, N., Mulet-Gas, M., Triana, V., Murillo, J., Rodriguez-Fortea, A., Poblet, J. M., and Echegoyen, L. "$Ti_2S@D_{3h}(24109)$-C_{78}: A sulfide cluster metallofullerene containing only transition metals inside the cage." *Chemical Science* 4 (2013): 3404–3410.

Liedtke, M., Sperlich, A., Kraus, H., Baumann, A., Deibel, C., Wirix, M. J. M., Loos, J., Cardona, C. M., and Dyakonov, V. "Triplet exciton generation in bulk-heterojunction solar cells based on endohedral fullerenes." *Journal of the American Chemical Society* 133 (2011): 9088–9094.

Liu, F., Wang, S., Guan, J., Wei, T., Zeng, M., and Yang, S. "Putting a terbium-monometallic cyanide cluster into the C_{82} Fullerene cage: $TbCN@C_2(5)$-C_{82}." *Inorganic Chemistry* 53 (2014): 5201–5205.

Lu, X., Slanina, Z., Akasaka, T., Tsuchiya, T., Mizorogi, N., and Nagase, S. "$Yb@C_{2n}$ (n = 40, 41, 42): New fullerene allotropes with unexplored electrochemical properties." *Journal of the American Chemical Society* 132 (2010): 5896–5905.

Lu, X., Akasaka, T., and Nagase, S. "Chemistry of endohedral metallofullerenes: The role of metals." *Chemical Communications* 47 (2011): 5942–5957.

Lu, X., Feng, L., Akasaka, T., and Nagase, S. "Current status and future developments of endohedral metallofullerenes." *Chemical Society Reviews* 41 (2012): 7723–7760.

Lu, X., Akasaka, T., and Nagase, S. "Carbide cluster metallofullerenes: Structure, properties, and possible origin." *Accounts of Chemical Research* 46 (2013): 1627–1635.

Melin, F., Chaur, M. N., Engmann, S., Elliott, B., Kumbhar, A., Athans, A. J., and Echegoyen, L. "The large $Nd_3N@C_{2n}$ (40< = 2n< = 49) cluster fullerene family: Preferential templating of a C_{88} cage by a trimetallic nitride cluster." *Angewandte Chemie-International Edition* 46 (2007): 9032–9035.

Meng, J., Liang, X., Chen, X., and Zhao, Y. "Biological characterizations of $[Gd@C_{82}(OH)_{22}]_n$ nanoparticles as fullerene derivatives for cancer therapy." *Integrative Biology* 5 (2013): 43–47.

Mercado, B. Q., Beavers, C. M., Olmstead, M. M., Chaur, M. N., Walker, K., Holloway, B. C., Echegoyen, L., and Balch, A. L. "Is the isolated pentagon rule merely a suggestion for endohedral fullerenes? The structure of a second egg-shaped endohedral fullerene—$Gd_3N@C_s(39663)$-C_{82}." *Journal of the American Chemical Society* 130 (2008): 7854–7855.

Mercado, B. Q., Jiang, A., Yang, H., Wang, Z., Jin, H., Liu, Z., Olmstead, M. M., and Balch, A. L. "Isolation and structural characterization of the molecular nanocapsule $Sm_2@D_{3d}(822)$-C_{104}." *Angewandte Chemie International Edition* 48 (2009): 9114–9116.

Mercado, B. Q., Olmstead, M. M., Beavers, C. M., Easterling, M. L., Stevenson, S., Mackey, M. A., Coumbe, C. E. et al. "A seven atom cluster in a carbon cage, the crystallographically determined structure of $Sc_4(\mu_3\text{-}O)_3@I_h$-$C_{80}$." *Chemical Communications* 46 (2010a): 279–281.

Mercado, B. Q., Sruart, M. A., Mackey, M. A., Pickens, J. E., Confait, B. S., Stevenson, S., Easterling, M. L. et al. "$Sc_2(\mu_2\text{-}O)$ Trapped in a fullerene cage: The isolation and structural characterization of $Sc_2(\mu_2\text{-}O)@C_s(6)$-$C_{82}$ and the relevance of the thermal and entropic effects in fullerene isomer selection." *Journal of the American Chemical Society* 132 (2010b): 12098–12105.

Mercado, B. Q., Chen, N., Rodriguez-Fortea, A., Mackey, M. A., Stevenson, S., Echegoyen, L., Poblet, J. M., Olmstead, M. M., and Balch, A. L. "The shape of the $Sc_2(\mu_2\text{-}S)$ unit trapped in C_{82}: Crystallographic, computational, and electrochemical studies of the isomers, $Sc_2(\mu_2\text{-}S)@C_s(6)$-$C_{82}$ and $Sc_2(\mu_2\text{-}S)@C_{3v}(8)$-$C_{82}$." *Journal of the American Chemical Society* 133 (2011): 6752–6760.

Mikawa, M., Kato, H., Okumura, M., Narazaki, M., Kanazawa, Y., Miwa, N., and Shinohara, H. "Paramagnetic water-soluble metallofullerenes having the highest relaxivity for MRI contrast agents." *Bioconjugate Chemistry* 12 (2001): 510–514.

Náfrádi, B., Antal, Á., Pásztor, Á., Forró, L., Kiss, L. F., Fehér, T., Kováts, É., Pekker, S., and Jánossy, A. "Molecular and spin dynamics in the paramagnetic endohedral fullerene $Gd_3N@C_{80}$." *The Journal of Physical Chemistry Letters* 3 (2012): 3291–3296.

Nakao, K., Kurita, N., and Fujita, M. "Ab-initio molecular-orbital calculation for C_{70} and seven isomers of C_{80}." *Physical Review B* 49 (1994): 11415–11420.

Nishibori, E., Takata, M., Sakata, M., Taninaka, A., and Shinohara, H. "Pentagonal–dodecahedral La_2 charge density in [80-I_h]fullerene: $La_2@C_{80}$." *Angewandte Chemie-International Edition* 40 (2001): 2998–2999.

Okazaki, T., Lian, Y. F., Gu, Z. N., Suenaga, K., and Shinohara, H. "Isolation and spectroscopic characterization of Sm-containing metallofullerenes." *Chemical Physics Letters* 320 (2000): 435–440.

Olmstead, M. M., Costa, D. A., Maitra, K., Noll, B. C., Phillips, S. L., Van Calcar, P. M., and Balch, A. L. "Interaction of curved and flat molecular surfaces. The structures of crystalline compounds composed of fullerene (C_{60}, $C_{60}O$, C_{70}, and $C_{120}O$) and metal octaethylporphyrin units." *Journal of the American Chemical Society* 121 (1999): 7090–7097.

Olmstead, M. M., de Bettencourt-Dias, A., Duchamp, J. C., Stevenson, S., Marciu, D., Dorn, H. C., and Balch, A. L. "Isolation and structural characterization of the endohedral fullerene $Sc_3N@C_{78}$." *Angewandte Chemie-International Edition* 40 (2001): 1223–1225.

Olmstead, M. M., de Bettencourt-Dias, A., Stevenson, S., Dorn, H. C., and Balch, A. L. "Crystallographic characterization of the structure of the endohedral fullerene {$Er_2@C_{82}$ Isomer I} with C_s cage symmetry and multiple sites for erbium along a band of ten contiguous hexagons." *Journal of the American Chemical Society* 124 (2002a): 4172–4173.

Olmstead, M. M., Lee, H. M., Stevenson, S., Dorn, H. C., and Balch, A. L. "Crystallographic characterization of Isomer 2 of $Er_2@C_{82}$ and comparison with Isomer 1 of $Er_2@C_{82}$." *Chemical Communications* (2002b): 2688–2689.

Olmstead, M. M., Lee, H. M., Duchamp, J. C., Stevenson, S., Marciu, D., Dorn, H. C., and Balch, A. L. "$Sc_3N@C_{68}$: Folded pentalene coordination in an endohedral fullerene that does not obey the isolated pentagon rule." *Angewandte Chemie-International Edition* 42 (2003): 900–903.

Pinzon, J. R., Plonska-Brzezinska, M. E., Cardona, C. M., Athans, A. J., Gayathri, S. S., Guldi, D. M., Herranz, M. A., Martin, N., Torres, T., and Echegoyen, L. "$Sc_3N@C_{80}$-ferrocene electron-donor/acceptor conjugates as promising materials for photovoltaic applications." *Angewandte Chemie International Edition* 47 (2008): 4173–4176.

Pinzon, J. R., Gasca, D. C., Sankaranarayanan, S. G., Bottari, G., Torres, T., Guldi, D. M., and Echegoyen, L. "Photoinduced charge transfer and electrochemical properties of triphenylamine I_h-$Sc_3N@C_{80}$ donor–acceptor conjugates." *Journal of the American Chemical Society* 131 (2009): 7727–7734.

Plant, S. R., Dantelle, G., Ito, Y., Ng, T. C., Ardavan, A., Shinohara, H., Taylor, R. A., Briggs, A. D., and Porfyrakis, K. "Acuminated fluorescence of Er^{3+} centres in endohedral fullerenes through the incarceration of a carbide cluster." *Chemical Physics Letters* 476 (2009): 41–45.

Popov, A. A. "Metal-cage bonding, molecular structures and vibrational spectra of endohedral fullerenes: Bridging experiment and theory." *Journal of Computational and Theoretical Nanoscience* 6 (2009): 292–317.

Popov, A. A., and Dunsch, L. "Structure, stability, and cluster-cage interactions in nitride clusterfullerenes $M_3N@C_{2n}$ (M = Sc, Y; 2n = 68–98): A density functional theory study." *Journal of the American Chemical Society* 129 (2007): 11835–11849.

Popov, A. A., and Dunsch, L. "Electrochemistry in cavea: Endohedral redox reactions of encaged species in fullerenes." *Journal of Physical Chemistry Letters* 2 (2011): 786–794.

Popov, A. A., Krause, M., Yang, S. F., Wong, J., and Dunsch, L. "C_{78} cage isomerism defined by trimetallic nitride cluster size: A computational and vibrational spectroscopic study." *Journal of Physical Chemistry B* 111 (2007): 3363–3369.

Popov, A. A., Schiemenz, S., Avdoshenko, S. M., Yang, S., Cuniberti, G., and Dunsch, L. "The state of asymmetric nitride clusters in endohedral fullerenes as studied by ^{14}N NMR spectroscopy: Experiment and theory." *The Journal of Physical Chemistry C* 115 (2011): 15257–15265.

Popov, A. A., Avdoshenko, S. M., Pendás, A. M., and Dunsch, L. "Bonding between strongly repulsive metal atoms: An oxymoron made real in a confined space of endohedral metallofullerenes." *Chemical Communications* 48 (2012a): 8031–8050.

Popov, A. A., Chen, N., Pinzón, J. R., Stevenson, S., Echegoyen, L. A., and Dunsch, L. "Redox-active scandium oxide cluster inside a fullerene cage: Spectroscopic, voltammetric, electron spin resonance spectroelectrochemical, and extended density functional theory study of $Sc_4O_2@C_{80}$ and its ion radicals." *Journal of the American Chemical Society* 134 (2012b): 19607–19618.

Popov, A. A., Yang, S., and Dunsch, L. "Endohedral fullerenes." *Chemical Reviews* 113 (2013): 5989–6113.

Popov, A. A., Pykhova, A. D., Ioffe, I. N., Li, F.-F., and Echegoyen, L. "Anion radicals of isomeric [5,6] and [6,6] benzoadducts of $Sc_3N@C_{80}$: Remarkable differences in endohedral cluster spin density and dynamics." *Journal of the American Chemical Society* 136 (2014): 13436–13441.

Reich, A., Panthofer, M., Modrow, H., Wedig, U., and Jansen, M. "The structure of $Ba@C_{74}$." *Journal of the American Chemical Society* 126 (2004): 14428–14434.

Rodriguez-Fortea, A., Alegret, N., Balch, A. L., and Poblet, J. M. "The maximum pentagon separation rule provides a guideline for the structures of endohedral metallofullerenes." *Nature Chemistry* 2 (2010): 955–961.

Rodriguez-Fortea, A., Balch, A. L., and Poblet, J. M. "Endohedral metallofullerenes: A unique host-guest association." *Chemical Society Reviews* 40 (2011): 3551–3563.

Ross, R. B., Cardona, C. M., Guldi, D. M., Sankaranarayanan, S. G., Reese, M. O., Kopidakis, N., Peet, J. et al. "Endohedral fullerenes for organic photovoltaic devices." *Nature Materials* 8 (2009a): 208–212.

Ross, R. B., Cardona, C. M., Swain, F. B., Guldi, D. M., Sankaranarayanan, S. G., Van Keuren, E., Holloway, B. C., and Drees, M. "Tuning conversion efficiency in metallo endohedral fullerene-based organic photovoltaic devices." *Advanced Functional Materials* 19 (2009b): 2332–2337.

Roy, D., Tripathi, N. K., Ram, K., and Sathyamurthy, N. "Synthesis of germanium encapsulated fullerene." *Solid State Communications* 149 (2009): 1244–1247.

Rudolf, M., Wolfrum, S., Guldi, D. M., Feng, L., Tsuchiya, T., Akasaka, T., and Echegoyen, L. "Endohedral metallofullerenes—Filled fullerene derivatives towards multifunctional reaction center mimics." *Chemistry—A European Journal* 18 (2012): 5136–5148.

Sato, Y., Yumura, T., Suenaga, K., Moribe, H., Nishide, D., Ishida, M., Shinohara, H., and Iijima, S. "Direct imaging of intracage structure in titanium-carbide endohedral metallofullerene." *Physical Review B* 73 (2006): 193401.

Sato, S., Nikawa, H., Seki, S., Wang, L., Luo, G., Lu, J., Haranaka, M., Tsuchiya, T., Nagase, S., and Akasaka, T. "A co-crystal composed of the paramagnetic endohedral metallofullerene $La@C_{82}$ and a nickel porphyrin with high electron mobility." *Angewandte Chemie International Edition* 51 (2012): 1589–1591.

Saunders, M., Jiménez-Vázquez, H. A., Cross, R. J., and Poreda, R. J. "Stable Compounds of Helium and Neon - $He@C_{60}$ and $Ne@C_{60}$." *Science* 259 (1993): 1428–1430.

Schubert, C., Rudolf, M., Guldi, D. M., Takano, Y., Mizorogi, N., Herranz, M. Á., Martín, N., Nagase, S., and Akasaka, T. "Rates and energetics of intramolecular electron transfer processes in conjugated metallofullerenes." *Philosophical Transactions of the Royal Society of Chemistry A* 371 (2013): 20120490.

Shinohara, H. "Endohedral metallofullerenes." *Reports on Progress in Physics* 63 (2000): 843–892.

Shinohara, H., Sato, H., Saito, Y., Ohkohchi, M., and Ando, Y. "Mass spectroscopic and ESR characterization of soluble yttrium-containing metallofullerenes YC_{82} and Y_2C_{82}." *The Journal of Physical Chemistry* 96 (1992): 3571–3573.

Shu, C., Corwin, F. D., Zhang, J., Chen, Z., Reid, J. E., Sun, M., Xu, W. et al. "Facile preparation of a new gadofullerene-based magnetic resonance imaging contrast agent with high 1H relaxivity." *Bioconjugate Chemistry* 20 (2009): 1186–1193.

Shultz, M. D., Duchamp, J. C., Wilson, J. D., Shu, C. Y., Ge, J. C., Zhang, J. Y., Gibson, H. W. et al. "Encapsulation of a radiolabeled cluster inside a fullerene cage, $(Lu_xLu_{3-x})N\text{-}Lu^{177}@C_{80}$: An interleukin-13-conjugated radiolabeled metallofullerene platform." *Journal of the American Chemical Society* 132 (2010): 4980–4981.

Shultz, M. D., Wilson, J. D., Fuller, C. E., Zhang, J., Dorn, H. C., and Fatouros, P. P. "Metallofullerene-based nanoplatform for brain tumor brachytherapy and longitudinal imaging in a murine orthotopic xenograft model." *Radiology* 266 (2011): 136–143.

Stevenson, S., "Metal oxide clusterfullerenes." In *Endohedral Fullerenes. From Fundamentals to Applications*, edited by Shangfeng Yang and Chun-Ru Wang (Singapore: World Scientific, 2014), Original edition, 179–210.

Stevenson, S., Rice, G., Glass, T., Harich, K., Cromer, F., Jordan, M. R., Craft, J. et al. "Small-bandgap endohedral metallofullerenes in high yield and purity." *Nature* 401 (1999): 55–57.

Stevenson, S., Fowler, P. W., Heine, T., Duchamp, J. C., Rice, G., Glass, T., Harich, K., Hajdu, E., Bible, R., and Dorn, H. C. "Materials science—A stable non-classical metallofullerene family." *Nature* 408 (2000): 427–428.

Stevenson, S., Phillips, J. P., Reid, J. E., Olmstead, M. M., Rath, S. P., and Balch, A. L. "Pyramidalization of Gd_3N inside a C_{80} cage. The synthesis and structure of $Gd_3N@C_{80}$." *Chemical Communications* (2004): 2814–2815.

Stevenson, S., Harich, K., Yu, H., Stephen, R. R., Heaps, D., Coumbe, C., and Phillips, J. P. "Nonchromatographic 'stir and filter approach' (SAFA) for isolating $Sc_3N@C_{80}$ metallofullerenes." *Journal of the American Chemical Society* 128 (2006): 8829–8835.

Stevenson, S., Mackey, M. A., Stuart, M. A., Phillips, J. P., Easterling, M. L., Chancellor, C. J., Olmstead, M. M., and Balch, A. L. "A distorted tetrahedral metal oxide cluster inside an icosahedral carbon cage. Synthesis, isolation, and structural characterization of $Sc_4(\mu_3\text{-}O)_2@I_h\text{-}C_{80}$." *Journal of the American Chemical Society* 130 (2008): 11844–11845.

Stevenson, S., Mackey, M. A., Pickens, J. E., Stuart, M. A., Confait, B. S., and Phillips, J. P. "Selective complexation and reactivity of metallic nitride and oxometallic fullerenes with Lewis acids and use as an effective purification method." *Inorganic Chemistry* 48 (2009): 11685–11690.

Sueki, K., Kikuchi, K., Akiyama, K., Sawa, T., Katada, M., Ambe, S., Ambe, F., and Nakahara, H. "Formation of metallofullerenes with higher group elements." *Chemical Physics Letters* 300 (1999): 140–144.

Suzuki, T., Maruyama, Y., Kato, T., Kikuchi, K., and Achiba, Y. "Electrochemical properties of $La@C_{82}$." *Journal of the American Chemical Society* 115 (1993): 11006–11007.

Suzuki, T., Kikuchi, K., Oguri, F., Nakao, Y., Suzuki, S., Achiba, Y., Yamamoto, K., Funasaka, H., and Takahashi, T. "Electrochemical properties of fullerenolanthanides." *Tetrahedron* 52 (1996): 4973–4982.

Suzuki, M., Lu, X., Sato, S., Nikawa, H., Mizorogi, N., Slanina, Z., Tsuchiya, T., Nagase, S., and Akasaka, T. "Where does the metal cation stay in $Gd@C_{2v}(9)\text{-}C_{82}$? A single-crystal x-ray diffraction study." *Inorganic Chemistry* 51 (2012): 5270–5273.

Suzuki, M., Mizorogi, N., Yang, T., Uhlik, F., Slanina, Z., Zhao, X., Yamada, M. et al. "$La_2@C_s(17\,490)\text{-}C_{76}$: A new non-IPR dimetallic metallofullerene featuring unexpectedly weak metal–pentalene interactions." *Chemistry—A European Journal* 19 (2013): 17125–17130.

Svitova, A., Braun, K., Popov, A. A., and Dunsch, L. "A platform for specific delivery of lanthanide–scandium mixed-metal cluster fullerenes into target cells." *Chemistry Open* 1 (2012): 207–210.

Svitova, A. L., Popov, A. A., and Dunsch, L. "Gd/Sc-based mixed metal nitride cluster fullerenes: The mutual influence of the cage and cluster size and the role of Sc in the electronic structure." *Inorganic Chemistry* 52 (2013): 3368–3380.

Svitova, A. L., Ghiassi, K., Schlesier, C., Junghans, K., Zhang, Y., Olmstead, M., Balch, A., Dunsch, L., and Popov, A. A. "Endohedral fullerene with μ_3-carbido ligand and titanium-carbon double bond stabilized inside a carbon cage." *Nature Communications* 5 (2014a): 3568, DOI:10.1038/ncomms4568.

Svitova, A. L., Krupskaya, Y., Samoylova, N., Kraus, R., Geck, J., Dunsch, L., and Popov, A. A. "Magnetic moments and exchange coupling in nitride clusterfullerenes $Gd_xSc_{3-x}N@C_{80}$ (x = 1–3)." *Dalton Transactions* 43 (2014b): 7387–7390.

Takano, Y., Herranz, M. A., Martin, N., Radhakrishnan, S. G., Guldi, D. M., Tsuchiya, T., Nagase, S., and Akasaka, T. "Donor–acceptor conjugates of lanthanum endohedral metallofullerene and pi-extended tetrathiafulvalene." *Journal of the American Chemical Society* 132 (2010): 8048–8055.

Takano, Y., Obuchi, S., Mizorogi, N., García, R., Herranz, M. Á., Rudolf, M., Guldi, D. M., Martín, N., Nagase, S., and Akasaka, T. "An endohedral metallofullerene as a pure electron donor: Intramolecular electron transfer in donor–acceptor conjugates of $La_2@C_{80}$ and 11,11,12,12-tetracyano-9,10-anthra-*p*-quinodimethane (TCAQ)." *Journal of the American Chemical Society* 134 (2012): 19401–19408.

Takata, M., Umeda, B., Nishibori, E., Sakata, M., Saito, Y., Ohno, M., and Shinohara, H. "Confirmation by x-ray-diffraction of the endohedral nature of the metallofullerene $Y@C_{82}$." *Nature* 377 (1995): 46–49.

Takata, M., Nishibori, E., Sakata, M., and Shinohara, H. "Synchrotron radiation for structural chemistry—Endohedral natures of metallofullerenes found by synchrotron radiation powder method." *Structural Chemistry* 14 (2003): 23–38.

Tan, K., and Lu, X. "Ti_2C_{80} is more likely a titanium carbide endohedral metallofullerene $(Ti_2C_2)@C_{78}$." *Chemical Communications* (2005): 4444–4446.

Tan, K., Lu, X., and Wang, C. R. "Unprecedented μ_4 - C_2^{6-} anion in $Sc_4C_2@C_{80}$." *Journal of Physical Chemistry B* 110 (2006): 11098–11102.

Tellgmann, R., Krawez, N., Lin, S. H., Hertel, I. V., and Campbell, E. E. B. "Endohedral fullerene production." *Nature* 382 (1996): 407–408.

Tsuchiya, T., Wakahara, T., Shirakura, S., Maeda, Y., Akasaka, T., Kobayashi, K., Nagase, S., Kato, T., and Kadish, K. M. "Reduction of endohedral metallofullerenes: A convenient method for isolation." *Chemistry of Materials* 16 (2004): 4343–4346.

Tsuchiya, T., Wakahara, T., Maeda, Y., Akasaka, T., Waelchli, M., Kato, T., Okubo, H., Mizorogi, N., Kobayashi, K., and Nagase, S. "2D NMR characterization of the $La@C_{82}$ anion." *Angewandte Chemie-International Edition* 44 (2005): 3282–3285.

Umemoto, H., Ohashi, K., Inoue, T., Fukui, N., Sugai, T., and Shinohara, H. "Synthesis and UHV-STM observation of the Td-symmetric Lu metallofullerene: $Lu_2@C_{76}(T_d)$." *Chemical Communications* 46 (2010): 5653–5655.

Valencia, R., Rodríguez-Fortea, A., and Poblet, J. M. "Understanding the stabilization of metal carbide endohedral fullerenes $M_2C_2@C_{82}$ and related systems." *Journal of Physical Chemistry A* 112 (2008): 4550–4555.

Valencia, R., Rodriguez-Fortea, A., Clotet, A., de Graaf, C., Chaur, M. N., Echegoyen, L., and Poblet, J. M. "Electronic structure and redox properties of metal nitride endohedral fullerenes $M_3N@C_{2n}$ (M = Sc, Y, La, and Gd; 2n = 80, 84, 88, 92, 96)." *Chemistry—A European Journal* 15 (2009a): 10997–11009.

Valencia, R., Rodriguez-Fortea, A., Stevenson, S., Balch, A. L., and Poblet, J. M. "Electronic structures of scandium oxide endohedral metallofullerenes, $Sc_4(\mu_3\text{-}O)_n@I_h\text{-}C_{80}$ (n = 2, 3)." *Inorganic Chemistry* 48 (2009b): 5957–5961.

Wang, C. R., Kai, T., Tomiyama, T., Yoshida, T., Kobayashi, Y., Nishibori, E., Takata, M., Sakata, M., and Shinohara, H. "C_{66} fullerene encaging a scandium dimer." *Nature* 408 (2000): 426–427.

Wang, C. R., Kai, T., Tomiyama, T., Yoshida, T., Kobayashi, Y., Nishibori, E., Takata, M., Sakata, M., and Shinohara, H. "A scandium carbide endohedral metallofullerene: $(Sc_2C_2)@C_{84}$." *Angewandte Chemie-International Edition* 40 (2001): 397–399.

Wang, T.-S., Chen, N., Xiang, J.-F., Li, B., Wu, J.-Y., Xu, W., Jiang, L. et al. "Russian-doll-type metal carbide endofullerene: Synthesis, isolation, and characterization of $Sc_4C_2@C_{80}$." *Journal of the American Chemical Society* 131 (2009): 16646–16647.

Wang, T.-S., Feng, L., Wu, J.-Y., Xu, W., Xiang, J.-F., Tan, K., Ma, Y.-H. et al. "Planar quinary cluster inside a fullerene cage: Synthesis and structural characterizations of $Sc_3NC@C_{80}\text{-}I_h$." *Journal of the American Chemical Society* 132 (2010): 16362–16364.

Wang, T., Wu, J., and Feng, Y. "Scandium carbide/cyanide alloyed cluster inside fullerene cage: Synthesis and structural studies of $Sc_3(\mu_3\text{-}C_2)(\mu_3\text{-}CN)@I_h\text{-}C_{80}$." *Dalton Transactions* 43 (2014): 16270–16274.

Wei, T., Wang, S., Liu, F., Tan, Y., Zhu, X., Xie, S., and Yang, S. "Capturing the long-sought small-band-gap endohedral fullerene $Sc_3N@C_{82}$ with low kinetic stability." *Journal of the American Chemical Society* 137 (2015): 3119–3123, DOI:10.1021/jacs.5b00199.

Weiske, T., Bohme, D. K., Hrusak, J., Kratschmer, W., and Schwarz, H. "Endohedral cluster compounds—Inclusion of helium within C_{60}^{+} and C_{70}^{+} through collision experiments." *Angewandte Chemie-International Edition in English* 30 (1991): 884–886.

Westerström, R., Dreiser, J., Piamonteze, C., Muntwiler, M., Weyeneth, S., Brune, H., Rusponi, S., Nolting, F., Popov, A., Yang, S., Dunsch, L., and Greber, T. "An endohedral single-molecule magnet with long relaxation times: $DySc_2N@C_{80}$." *Journal of the American Chemical Society* 134 (2012): 9840–9843.

Westerström, R., Dreiser, J., Piamonteze, C., Muntwiler, M., Weyeneth, S., Krämer, K., Liu, S.-X. et al. "Tunneling, remanence, and frustration in dysprosium-based endohedral single-molecule magnets." *Physical Review B* 89 (2014): 060406.

Westerström, R., Uldry, A.-C., Stania, R., Dreiser, J., Piamonteze, C., Muntwiler, M., Matsui, F. et al. "Surface aligned magnetic moments and hysteresis of an endohedral single-molecule magnet on a metal." *Physical Review Letters* 114 (2015): 087201.

Wolf, M., Muller, K. H., Eckert, D., Skourski, Y., Georgi, P., Marczak, R., Krause, M., and Dunsch, L. "Magnetic moments in $Ho_3N@C_{80}$ and $Tb_3N@C_{80}$." *Journal of Magnetism and Magnetic Materials* 290 (2005): 290–293.

Wu, J., Wang, T., Ma, Y., Jiang, L., Shu, C., and Wang, C. "Synthesis, isolation, characterization, and theoretical studies of $Sc_3NC@C_{78}$–C_2." *The Journal of Physical Chemistry C* 115 (2011): 23755–23759.

Xie, Q. S., Perezcordero, E., and Echegoyen, L. "Electrochemical detection of C_{60}^{6-} and C_{70}^{6-}-enhanced stability of fullerides in solution." *Journal of the American Chemical Society* 114 (1992): 3978–3980.

Xu, Z. D., Nakane, T., and Shinohara, H. "Production and isolation of $Ca@C_{82}$ (I–IV) and $Ca@C_{84}$ (I,II) metallofullerenes." *Journal of the American Chemical Society* 118 (1996): 11309–11310.

Xu, J. X., Lu, X., Zhou, X. H., He, X. R., Shi, Z. J., and Gu, Z. N. "Synthesis, isolation, and spectroscopic characterization of ytterbium-containing metallofullerenes." *Chemistry of Materials* 16 (2004): 2959–2964.

Xu, J. X., Tsuchiya, T., Hao, C., Shi, Z. J., Wakahara, T., Mi, W. H., Gu, Z. N., and Akasaka, T. "Structure determination of a missing-caged metallofullerene: $Yb@C_{74}$ (II) and the dynamic motion of the encaged ytterbium ion." *Chemical Physics Letters* 419 (2006): 44–47.

Xu, W., Wang, T.-S., Wu, J.-Y., Ma, Y.-H., Zheng, J.-P., Li, H., Wang, B., Jiang, L., Shu, C.-Y., and Wang, C.-R. "Entrapped planar trimetallic carbide in a fullerene cage: Synthesis, isolation, and spectroscopic studies of $Lu_3C_2@C_{88}$." *Journal of Physical Chemistry C* 115 (2011): 402–405.

Xu, W., Niu, B., Feng, L., Shi, Z., and Lian, Y. "Access to an unexplored chiral C_{82} cage by encaging a divalent metal: Structural elucidation and electrochemical studies of $Sm@C_2(5)$-C_{82}." *Chemistry—A European Journal* 18 (2012a): 14246–14249.

Xu, W., Niu, B., Shi, Z., Lian, Y., and Feng, L. "$Sm@C_{2v}(3)$-C_{80}: Site-hopping motion of endohedral Sm atom and metal-induced effect on redox profile." *Nanoscale* 4 (2012b): 6876–6879.

Xu, W., Feng, L., Calvaresi, M., Liu, J., Liu, Y., Niu, B., Shi, Z., Lian, Y., and Zerbetto, F. "An experimentally observed trimetallofullerene $Sm_3@I_h$-C_{80}: Encapsulation of three metal atoms in a cage without a nonmetallic mediator." *Journal of the American Chemical Society* 135 (2013a): 4187–4190.

Xu, W., Hao, Y., Uhlik, F., Shi, Z., Slanina, Z., and Feng, L. "Structural and electrochemical studies of $Sm@D_{3h}$-C_{74} reveal a weak metal-cage interaction and a small band gap species." *Nanoscale* 5 (2013b): 10409–10413.

Yamada, M., Nakahodo, T., Wakahara, T., Tsuchiya, T., Maeda, Y., Akasaka, T., Kako, M. et al. "Positional control of encapsulated atoms inside a fullerene cage by exohedral addition." *Journal of the American Chemical Society* 127 (2005): 14570–14571.

Yamada, M., Wakahara, T., Tsuchiya, T., Maeda, Y., Akasaka, T., Mizorogi, N., and Nagase, S. "Spectroscopic and theoretical study of endohedral dimetallofullerene having a non-IPR fullerene cage: $Ce_2@C_{72}$." *Journal of Physical Chemistry A* 112 (2008a): 7627–7631.

Yamada, M., Wakahara, T., Tsuchiya, T., Maeda, Y., Kako, M., Akasaka, T., Yoza, K., Horn, E., Mizorogi, N., and Nagase, S. "Location of the metal atoms in $Ce_2@C_{78}$ and its bis-silylated derivative." *Chemical Communications* (2008b): 558–560.

Yamada, M., Mizorogi, N., Tsuchiya, T., Akasaka, T., and Nagase, S. "Synthesis and characterization of the D_{5h} isomer of the endohedral dimetallofullerene $Ce_2@C_{80}$: Two-dimensional circulation of encapsulated metal atoms inside a fullerene cage." *Chemistry—A European Journal* 15 (2009): 9486–9493.

Yamada, M., Akasaka, T., and Nagase, S. "Endohedral metal atoms in pristine and functionalized fullerene cages." *Accounts of Chemical Research* 43 (2010): 92–102.

Yamada, M., Kurihara, H., Suzuki, M., Guo, J. D., Waelchli, M., Olmstead, M. M., Balch, A. L. et al. "$Sc_2@C_{66}$ Revisited: An endohedral fullerene with scandium ions nestled within two unsaturated linear triquinanes." *Journal of the American Chemical Society* 136 (2014): 7611–7614.

Yang, S. F.; and Dunsch, L. "A large family of dysprosium-based trimetallic nitride endohedral fullerenes: $Dy_3N@C_{2n}$ (39 < = n < = 44)." *Journal of Physical Chemistry B* 109 (2005): 12320–12328.

Yang, S. F., Popov, A. A., and Dunsch, L. "Violating the Isolated Pentagon Rule (IPR): The endohedral non-IPR cage of $Sc_3N@C_{70}$." *Angewandte Chemie-International Edition* 46 (2007): 1256–1259.

Yang, H., Lu, C., Liu, Z., Jin, H., Che, Y., Olmstead, M. M., and Balch, A. L. "Detection of a family of gadolinium-containing endohedral fullerenes and the isolation and crystallographic characterization of one member as a metal-carbide encapsulated inside a large fullerene cage." *Journal of the American Chemical Society* 130 (2008): 17296–17300.

Yang, S. F., Popov, A. A., and Dunsch, L. "Large mixed metal nitride clusters encapsulated in a small cage: The confinement of the C_{68}-based clusterfullerenes." *Chemical Communications* (2008): 2885–2887.

Yang, S., Chen, C., Popov, A., Zhang, W., Liu, F., and Dunsch, L. "An endohedral titanium(III) in a clusterfullerene: Putting a non-group-III metal nitride into the C_{80}-I_h fullerene cage." *Chemical Communications* (2009): 6391–6393.

Yang, H., Jin, H., Hong, B., Liu, Z., Beavers, C. M., Zhen, H., Wang, Z., Mercado, B. Q., Olmstead, M. M., and Balch, A. L. "Large endohedral fullerenes containing two metal ions, $Sm_2@D_2(35)$-C_{88}, $Sm_2@C_1(21)$-C_{90}, and $Sm_2@D_3(85)$-C_{92}, and their relationship to endohedral fullerenes containing two gadolinium ions." *Journal of the American Chemical Society* 133 (2011a): 16911–16919.

Yang, H., Jin, H., Zhen, H., Wang, Z., Liu, Z., Beavers, C. M., Mercado, B. Q., Olmstead, M. M., and Balch, A. L. "Isolation and crystallographic identification of four isomers of $Sm@C_{90}$." *Journal of the American Chemical Society* 133 (2011b): 6299–6306.

Yang, S., Liu, F., Chen, C., Jiao, M., and Wei, T. "Fullerenes encaging metal clusters—Clusterfullerenes." *Chemical Communications* 47 (2011): 11822–11839.

Yang, H., Jin, H., Wang, X., Liu, Z., Yu, M., Zhao, F., Mercado, B. Q., Olmstead, M. M., and Balch, A. L. "X-ray crystallographic characterization of new soluble endohedral fullerenes utilizing the popular C_{82} Bucky cage. Isolation and structural characterization of $Sm@C_{3v}(7)$-C_{82}, $Sm@C_s(6)$-C_{82}, and $Sm@C_2(5)$-C_{82}." *Journal of the American Chemical Society* 134 (2012a): 14127–14136.

Yang, H., Yu, M., Jin, H., Liu, Z., Yao, M., Liu, B., Olmstead, M. M., and Balch, A. L. "The isolation of three isomers of $Sm@C_{84}$ and the x-ray crystallographic characterization of $Sm@D_{3d}(19)$-C_{84} and $Sm@C_2(13)$-C_{84}." *Journal of the American Chemical Society* 134 (2012b): 5331–5338.

Yang, H., Wang, Z., Jin, H., Hong, B., Liu, Z., Beavers, C. M., Olmstead, M. M., and Balch, A. L. "Isolation and crystallographic characterization of $Sm@C_{2v}(3)$-C_{80} through cocrystal formation with Ni^{II}(octaethylporphyrin) or Bis(ethylenedithio)tetrathiafulvalene." *Inorganic Chemistry* 52 (2013): 1275–1284.

Yang, S., Chen, C., Liu, F., Xie, Y., Li, F., Jiao, M., Suzuki, M. et al. "An improbable monometallic cluster entrapped in a popular fullerene cage: $YCN@C_s(6)$-C_{82}." *Scientific Reports* 3 (2013): 1487.

Zhang, J., and Dorn, H. C. "NMR studies of the dynamic motion of encapsulated ions and clusters in fullerene cages: A wheel within a wheel." *Fullerenes, Nanotubes and Carbon Nanostructures* 22 (2014): 35–46.

Zhang, Y., and Popov, A. A. "Transition-metal and rare-earth-metal redox couples inside carbon cages: Fullerenes acting as innocent ligands." *Organometallics* 33 (2014): 4537–4549.

Zhang, J. F., Fatouros, P. P., Shu, C. Y., Reid, J., Owens, L. S., Cai, T., Gibson, H. W. et al. "High relaxivity trimetallic nitride (Gd_3N) metallofullerene MRI contrast agents with optimized functionality." *Bioconjugate Chemistry* 21 (2010): 610–615.

Zhang, J., Fuhrer, T., Fu, W., Ge, J., Bearden, D. W., Dallas, J. L., Duchamp, J. C. et al. "Nanoscale fullerene compression of a yttrium carbide cluster." *Journal of the American Chemical Society* 134 (2012): 8487–8493.

Zhang, J., Stevenson, S., and Dorn, H. C. "Trimetallic nitride template endohedral metallofullerenes: Discovery, structural characterization, reactivity, and applications." *Accounts of Chemical Research* 46 (2013): 1548–1557.

Zhang, Y., Schiemenz, S., Popov, A. A., and Dunsch, L. "Strain-driven endohedral redox couple Ce^{IV}/Ce^{III} in nitride clusterfullerenes $CeM_2N@C_{80}$ (M = Sc, Y, Lu)." *The Journal of Physical Chemistry Letters* 4 (2013): 2404–2409.

Zhang, M., Hao, Y., Li, X., Feng, L., Yang, T., Wan, Y., Chen, N., Slanina, Z., Uhlik, F., and Cong, H. "Facile synthesis of an extensive family of $Sc_2O@C_{2n}$ (n = 35–47) and chemical insight into the smallest member of $Sc_2O@C_2(7892)$-C_{70}." *The Journal of Physical Chemistry C* 118 (2014): 28883–28889.

Zhang, Y., Popov, A. A., and Dunsch, L. "Endohedral metal or a fullerene cage based oxidation? Redox duality of nitride clusterfullerenes $Ce_xM_{3-x}N@C_{78-88}$ (x = 1, 2; M = Sc and Y) dictated by the encaged metals and the carbon cage size." *Nanoscale* 6 (2014): 1038–1048.

Zhang, Y., Ghiassi, K. B., Deng, Q., Samoylova, N. A., Olmstead, M. M., Balch, A. L., and Popov, A. A. "Synthesis and structure of $LaSc_2N@C_s$(hept)-C_{80} with one heptagon and thirteen pentagons." *Angewandte Chemie International Edition* 52 (2015a): 495–499.

Zhang, Y., Krylov, D., Rosenkranz, M., Schiemenz, S., and Popov, A. A. "Magnetic anisotropy of endohedral lanthanide ions: Paramagnetic NMR study of $MSc_2N@C_{80}$-I_h with M running through the whole 4f row." *Chemical Science* 6 (2015b): 2328–2341, DOI:10.1039/C5SC00154D.

Zuo, T. M., Beavers, C. M., Duchamp, J. C., Campbell, A., Dorn, H. C., Olmstead, M. M., and Balch, A. L. "Isolation and structural characterization of a family of endohedral fullerenes including the large, chiral cage fullerenes $Tb_3N@C_{88}$ and $Tb_3N@C_{86}$ as well as the I_h and D_{5h} isomers of $Tb_3N@C_{80}$." *Journal of the American Chemical Society* 129 (2007): 2035–2043.

Zuo, T., Xu, L., Beavers, C. M., Olmstead, M. M., Fu, W., Crawford, T. D., Balch, A. L., and Dorn, H. C. "$M_2@C_{79}N$ (M = Y, Tb): Isolation and characterization of stable endohedral metallofullerenes exhibiting M-M bonding interactions inside aza[80]fullerene cages." *Journal of the American Chemical Society* 130 (2008): 12992–12997.

III

Nanotubes

13

CVD-Synthesized Carbon Nanotubes

Maria Sarno

Adolfo Senatore

13.1 Introduction

Carbon nanotubes (CNTs) were discovered in 1991, soon after the successful laboratory synthesis of fullerenes. Since that time, there has been an intensive research activity, not only because of their fascinating structural features and electronic, mechanical, and thermal properties but also because of their potential technological applications.

The story of CNT discovery is closely connected with that of the disclosure and investigation of fullerenes. In 1985, a confluence of events led to an unexpected and unplanned experiment with a new kind of microscope, resulting in the discovery of a new molecule made purely of carbon—the very element chemists felt there was nothing more to learn about. Later, Iijima (1991) observed in 1991 that nanotubules of graphite were deposited on the negative electrode during the direct-current arcing of graphite for the preparation of fullerenes. These nanotubes are concentric graphitic cylinders closed at either end because of the presence of five-membered rings. Nanotubes can be multiwalled (MWNT), with a central tube of nanometric diameter surrounded by graphitic layers separated by about 0.34 nm, while single-walled nanotubes (SWNTs) are constituted of only one graphitic layer. An SWNT can be visualized by cutting C_{60} along the center and spacing apart the hemispherical corannulene end-caps by a cylinder of graphite of the same diameter. CNTs are the only form of carbon with extended bonding and yet with no dangling bonds. Since CNTs are derived from fullerenes, they have been referred to as tubular fullerenes or buckytubes.

The method for producing nanotubes described by Iijima in 1991 gave relatively poor yields, making further research into their structure and properties difficult. A significant advance came in July 1992, when Thomas Ebbesen and Pulickel Ajayan, working at the same Japanese laboratory as Iijima, described a method for making gram quantities of nanotubes (Ebbesen and Ajayan 1992). The syntheses were conducted in helium flow in the range of 2–3 atm, considerably higher than the pressure of gas used in the production of fullerene. A current of 60–100 A across a potential drop of about 25 V gives high yields of CNTs. The synthesized MWNTs have lengths on the order of 10 μm and diameters in the range of 5–30 nm. The nanotubes are typically bound together by strong van der Waals interactions and form tight bundles. MWNTs produced by arc discharge are very straight, indicative of their high crystallinity. For as-grown materials, there are few defects such as pentagons or heptagons existing on the sidewalls of nanotubes. Again, this was a serendipitous discovery: Ebbesen and Ajayan had been trying to make fullerene derivatives when they found that increasing the pressure of helium in the arc-evaporation chamber dramatically improved the yield of nanotubes in the cathodic soot. In the arc discharge, carbon atoms are evaporated by plasma of helium gas ignited by high currents passed through opposing carbon anode and cathode (Figure 13.1). CNTs can be obtained by controlling the growth conditions such as the pressure of inert gas in the discharge chamber and the arcing current. CNTs have been produced also by plasma arc jets (Hatta and Murata 1994), and in large quantities (Colbert et al. 1994), by optimizing the quenching process in an arc between a graphite anode and a cooled copper electrode. The product of these processes is a mixture of CNTs with different percentages of amorphous carbon and fullerene. The tubes thus produced are straight and show a small amount of structural defects. Laser ablation technique (Figure 13.2) is able to produce CNTs with the highest degree of purity. The method utilized intense laser pulses to ablate a carbon target containing a percent of nickel and/or cobalt. The target was placed in a tube furnace heated to 1200°C. During laser ablation, a flow of inert gas was passed through the growth chamber to carry the grown nanotubes downstream to be collected on a cold finger. In contrast to the arc method, direct vaporization allows far greater control over growth conditions, permits continuous operation, and produces nanotubes in higher yield and better quality.

After this, bulk quantities of CNTs have been produced by the chemical vapor deposition (CVD) method (Chesnokov et al. 1994; Ivanov et al. 1994; Dai et al. 1996; Li et al. 1996). The availability of

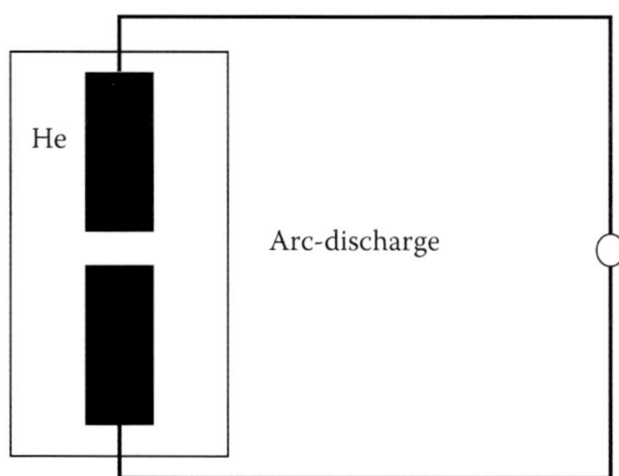

FIGURE 13.1 Arc-discharge schematic experimental setup.

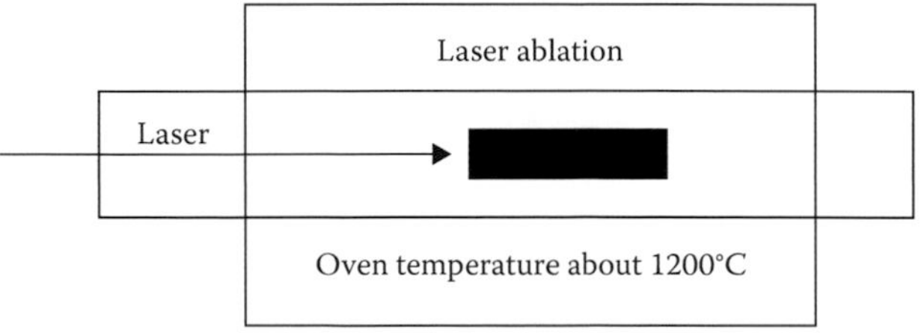

FIGURE 13.2 Laser ablation schematic experimental setup.

nanotubes in bulk gave an enormous boost to the pace of research worldwide. CVD is now the standard method for CNT production.

One idea that attracted early interest was that of using CNTs and nanoparticles as "molecular containers." A landmark in this field was the demonstration by Ajayan and Iijima that nanotubes could be filled with molten lead and thus be used as moulds for nanowires. Subsequently, more controlled methods of opening and filling nanotubes were developed, enabling a wide range of materials, including biological ones, to be placed inside. The resulting opened or filled tubes might have fascinating properties, with possible applications in catalysis or as biological sensors. Filled carbon nanoparticles may also have important applications in areas as diverse as magnetic recording and nuclear medicine.

Perhaps the largest volume of research into nanotubes has been devoted to their electronic properties. The theoretical work that predated Iijima's discovery has already been mentioned. A short time after the publication of Iijima's 1991 letter in *Nature*, two other papers appeared on the electronic structure of CNTs. The MIT group and Noriaki Hamada and colleagues from Iijima's laboratory in Tsukuba carried out band structure calculations on narrow properties using a tight-binding model and demonstrated that electronic properties were a function of both tube structure and diameter. These remarkable predictions stimulated a great deal of interest, but attempting to determine the electronic properties of nanotubes experimentally presented great difficulties. Since 1996, however, experimental measurements have been carried out on individual nanotubes, which appear to confirm the theoretical predictions. These results have prompted speculation that nanotubes might become components of future nanoelectronic devices.

Determining the mechanical properties of CNTs also presented formidable difficulties, but once again, experimentalists have proved equal to the colleagues. Measurements carried out using transmission electron microscopy (TEM) and atomic force microscopy (AFM) have demonstrated that the mechanical characteristics of CNTs may be just as exceptional as their electronic properties. As a result, there is growing interest in using nanotubes in composite materials.

A variety of other possible applications of nanotubes are currently exciting interest. For example, a number of groups are exploring the idea of using nanotubes as tips for scanning probe microscopy. With their elongated shapes, pointed caps, and high stiffness, nanotubes would appear to be ideally suited for this purpose, and initial experiments in this area have produced some extremely impressive results. Nanotubes have also been shown to have useful field emission properties, which might lead to their being used in flat-panel displays. Overall, the volume of nanotubes research is growing, at an astonishing rate, together with commercial applications.

13.2 CVD Synthesis of CNTs

An effective method of generating CNTs is based on the CVD of hydrocarbons over supported transition metal particles that constitute the active species. This is a successful way to synthesize CNTs because of the lower growth temperature (below 1000°C) compared with arc-discharge or laser vaporization methods and of the possibility to control the inner diameter and length of the tubes. A schematic experimental setup for CVD growth is depicted in Figure 13.3. The growth process involves heating a catalyst material to high temperatures in a tube furnace and flowing a hydrocarbon vapor through the

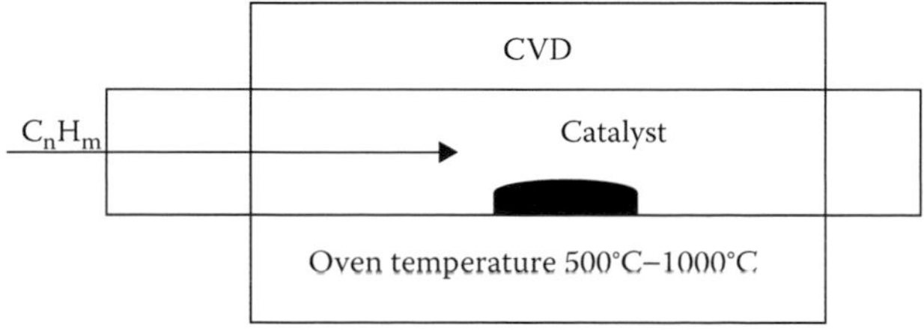

FIGURE 13.3 CVD schematic experimental setup.

tube reactor for a period of time. Materials grown over the catalyst are collected upon cooling the system to room temperature.

The general nanotube growth mechanism in a CVD process involves the dissociation of hydrocarbon molecules catalyzed by a transition metal and the dissolution and saturation of carbon atoms in the metal nanoparticles. The precipitation of carbon from the saturated metal leads to the formation of tubular carbon solids in sp^2 structure. Tubule formation is favored over other forms of carbon such as graphitic sheets with open edges. This is because a tube contains no dangling bonds and therefore is in a low energy form.

The key parameters in nanotube CVD growth are the nature of hydrocarbon and catalyst and the reaction temperature.

For MWNT growth, most of the CVD methods employ ethylene or acetylene as carbon feedstock and the growth temperature is typically in the range of 550°C–750°C (Ivanov et al. 1994; Ciambelli et al. 2005, 2007; Altavilla et al. 2009, 2013; Di Bartolomeo et al. 2009; Ciambelli et al. 2011; Giubileo et al. 2012; Sarno et al. 2012b; Funaro et al. 2013).

The catalytic system consists of transition metals, such as Fe, Ni, Cu, and Co, of several nanometers in size, used as catalysts, dispersed on a support. Concerning the carbon source, the most preferred in CVD are hydrocarbons such as methane (Palizdar et al. 2011), ethane (Tomie et al. 2010), ethylene (Narkiewicz et al. 2010), acetylene (He et al. 2011), xylene (Atthipalli et al. 2011; Shirazi et al. 2011), eventually their mixture (Li et al. 2010), isobutene (Santangelo et al. 2010), or ethanol (Hou et al. 2011; Yong et al. 2011). In the case of gaseous carbon source, the CNT's growth efficiency strongly depends on the reactivity and concentration of gas phase intermediates produced together with reactive species and free radicals as a result of hydrocarbon decomposition. Thus, it can be expected that the most efficient intermediates, which have the potential of chemisorption or physisorption on the catalyst surface to initiate CNT growth, should be produced in the gas phase (Skukla et al. 2010). Commonly used substrates are Ni, Si, SiO_2, Cu, Cu/Ti/Si, stainless steel, or glass, rarely $CaCO_3$; alumina, graphite, and tungsten foil or other substrates were also tested (Afolabi et al. 2011; Dumpala et al. 2011).

Considerable variations in the carbon yield of the reaction are observed varying the support used, alumina has emerged as really promising (Ciambelli et al. 2005, 2007). The effectiveness of the method mostly depends on the surface properties of the support through interaction with the active metal phases.

A similar effect is observed to vary the hydrocarbon source, passing from a less reactive to a more reactive hydrocarbon (e.g., ethylene-acetylene), even if the changes induced are definitely less consistent than varying the support. In particular, BEA zeolite supporting cobalt particles permits the growth of multiwall CNTs (MWCNTs) by ethylene catalytic chemical vapor deposition (CCVD) (Ciambelli et al. 2005). The nanotubes are sinuous and entangled, with an inner diameter determined by metal particle dimensions. Carbon deposit yield increases with reaction time and temperature, while reaction yield increases with reaction time. By wet impregnation of gibbsite ($Al(OH)_3$) with Co and Fe 2.5 wt.% of each metals were obtained catalysts capable of ensuring, in appropriate conditions identified thanks to an intense experimentation, a selectivity to CNTs equal to 100%, with a yield of carbon equal if not superior to the best results reported in the literature for similar synthesis and with a reaction yield of approximately 100% (value for the first time reported), with an average life of the catalyst of a few hours (Ciambelli et al. 2007, 2011; Altavilla et al. 2009; Di Bartolomeo et al. 2009; Giubileo et al. 2012; Sarno et al. 2012b; Funaro et al. 2013). The choice of catalyst is one of the most important parameters affecting the CNT's growth. Therefore, its preparation is also a crucial step in CNT synthesis. Recently, Fotopoulos and Xanthakis discussed the traditionally accepted models, which are base growth and tip growth. In addition, they mentioned a hypothesis that SWNTs are produced by base growth only; i.e., the cap is formed first, and then by a liftoff process, the CNT is created by addition of carbon atoms at the base. They refer to recent in situ video rate TEM studies that have revealed that the base growth of SWNT in thermal CVD is accompanied by a considerable deformation of the Ni catalyst nanoparticle and the creation of a subsurface carbon layer. These effects may be produced by the adsorption on the catalyst nanoparticle during pyrolysis (Fotopoulos and Xanthakis 2010). To produce SWNTs, the size of the nanoparticle catalyst must be smaller than about 3 nm. It has been suggested that the nanotube diameter is determined by the size of the metal particles (Rodriguez 1993). The length of nanotubes varies from several micrometers to several tens of micrometers. The rationale for choosing these metals as a catalyst for CVD growth of nanotubes lies in the phase diagram of the system metal–carbon. At high temperature, carbon has finite solubility in these metals, which leads to the formation of metal–carbon solutions. Iron, cobalt, and nickel are the favored catalytic metals in laser ablation and arc discharge too. This simple fact may hint that laser, arc, and CVD growth methods may share a common nanotube growth mechanism, although very different approaches are used to provide carbon feedstock. The yield of nanotubes in a sample depends on the type of catalyst: In the presence of iron, for example, it is much higher than for cobalt. For copper catalyst, a great part of carbon precipitates in an amorphous form (Chesnokov et al. 1994). The typical appearance is a tangle of tubes that after sonication exhibits the aspect shown in Figure 13.4. In particular, SWNTs, also after sonication, due to van der Waals forces between the tubes, are typically arranged in ropes (Figure 13.5). The nanotubes, in general, can present open and closed tips, in Figure 13.6 a high-resolution TEM image of a closed double-walled CNT end is visible. Two scanning electron microscopy (SEM) images of as produced CVD carbon nanotubes bundles are shown in Figure 13.7. Finally, the TEM image of Figure 13.8 shows double-walled CNTs before purification, the catalyst is visible as black spots in the figure. Recent interest in CVD nanotube is also due to the idea that aligned and ordered nanotube structures can be grown on a surface with a control that is not possible with arc

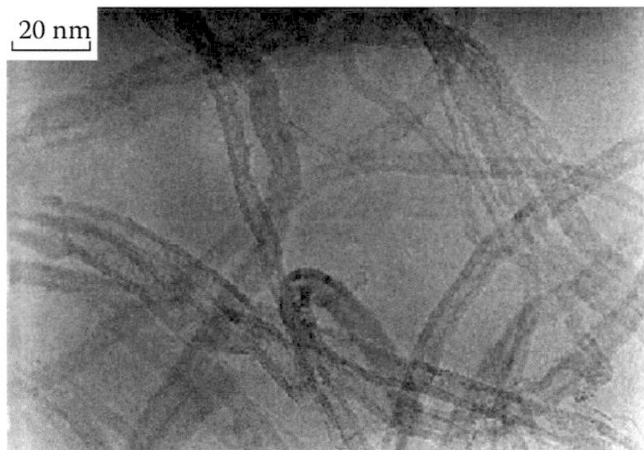

FIGURE 13.4 TEM images of CNTs produced by the CVD method.

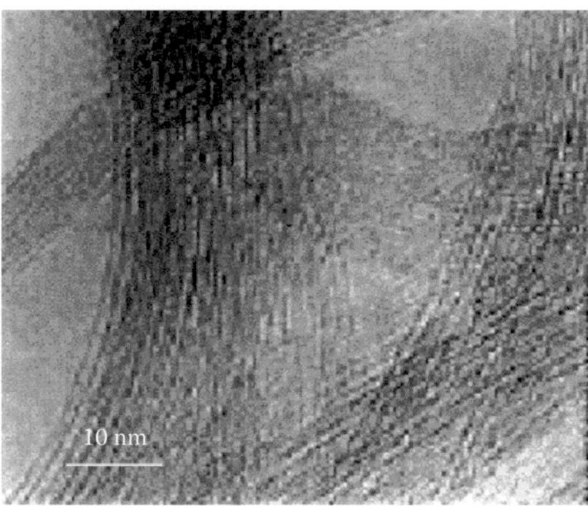

FIGURE 13.5 High-resolution TEM image of SWNTs ropes produced by the CVD method.

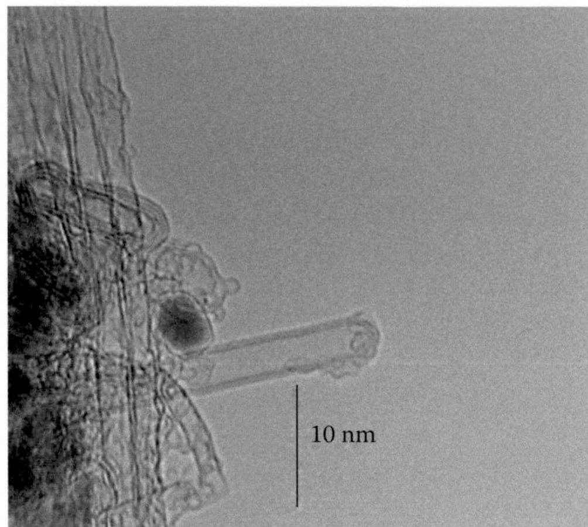

FIGURE 13.6 High-resolution TEM image of a double-walled CNT end.

FIGURE 13.7 Scanning electron microscopy images of CNT bundles at lower (a) and higher (b) magnifications.

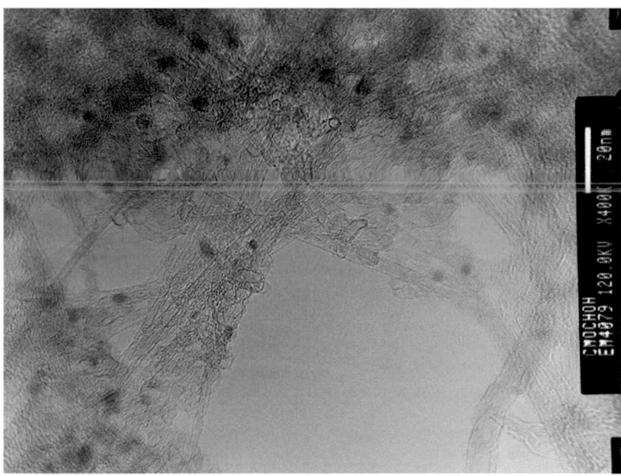

FIGURE 13.8 TEM image of unpurified double-walled CNTs.

discharge or laser ablation techniques (Dai et al. 1996). Methods that have been developed to obtain aligned MWNTs include CVD growth of nanotubes in the pores of mesoporous silica, an approach developed by Xie's group at the Chinese Academy of Science. The catalyst used in this case is iron oxide particles formed inside the pores of silica. The carbon feedstock is 9% acetylene in nitrogen at 180 torr pressure, and the growth temperature is 600°C. Remarkably, nanotubes with lengths up to millimeters are obtained (Li et al. 1996). Ren has grown relatively large-diameter, well-aligned MWNTs, for field emission display application, forming oriented "forest" on glass substrate using a plasma-assisted CVD method with nickel as catalyst and acetylene as carbon source around 660°C (Ren et al. 1998). An extension of the method reported by Li et al. (1996) has produced self-oriented patterned arrays of CNTs on iron-coated porous silicon (Fan et al. 1999). Dai's group has been devising growth strategies for ordered MWNT architectures by CVD on catalytically patterned substrate. They have found that MWNTs can self-assemble into aligned structures as they grow, and the driving force for self-aligned is the van der Waals interaction between nanotubes (Dai et al. 1996). The growth approach involves catalyst patterning and rational design of the substrate to enhance catalyst–substrate interactions and control the catalyst particle size. Porous silicon is found to be an ideal substrate for this approach and can be obtained by electrochemical etching of n-type silicon wafers in hydrofluoric acid–methanol solution. Ren et al. employed plasma-enhanced CVD on nickel-coated glass with acetylene and ammonia mixtures for this purpose (Ren et al. 1998). The mechanism of growth of nanotubes by this method and the exact role of the metal particles were investigated; a nucleation process involving the metal particles is considered to be important. Fan et al. have obtained aligned nanotubes by employing catalytic decomposition on porous silicon and plain silicon substrates patterned with iron films. A typical image of a CNT forest on a silicon substrate is shown in Figure 13.9. Also in this case, the role of the metal particles is not completely known (Fan et al. 1999). The role of the transition metal particles assumes importance in view of the report of Pan et al. that aligned nanotubes can be obtained by pyrolysis of acetylene over iron/silica catalytic surfaces (Pan et al. 1999). Flahaut et al. reported the influence of catalyst preparation conditions for the synthesis of CNTs by CCVD. In their work, the catalysts were prepared by the combustion route using either urea or citric acid as the fuel. They found that the milder combustion conditions obtained in the case of citric acid can either limit the formation of carbon nanofibers or increase the selectivity of the CCVD synthesis toward CNTs with fewer walls, depending on the catalyst composition (Flahaut et al. 2005). Xiang et al. prepared CNTs via CCVD of acetylene on a series of catalysts derived from Co/Fe/Al layered double hydroxides. They observed that the content of Co in the precursors had a distinct effect on the growth of CNTs. Increasing Co content enhanced the carbon yield because of good dispersion of a large number of active Co species. Higher

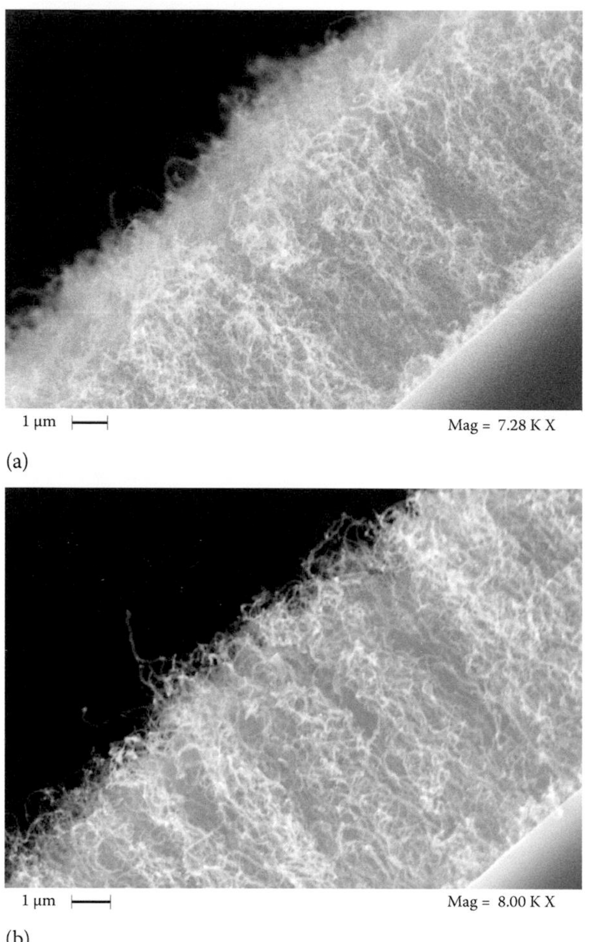

FIGURE 13.9 Scanning electron microscopy images of CNT forests on a silicon wafer at lower (a) and higher (b) magnifications.

Co content led to the formation of CNTs with smaller diameters and less structural disorder (Xiang et al. 2009). Lyu et al. produced high-quality and high-purity double-walled carbon nanotubes (DWNTs) by catalytic decomposition of benzene as an ideal carbon source and Fe-Mo/Al_2O_3 as a catalyst at 900°C. They obtained DWNT bundles free of amorphous carbon covering on the surface and of a low defect level in the atomic carbon structure (Lyu et al. 2003). Zhang et al. prepared MWNTs by the catalytic decomposition of methane with diameters of 40–60 nm at 680°C for 120 min using nickel oxide/silica binary aerogels as the catalyst (Zhang et al. 2005, 2006). Sano and colleagues evaluated two systems of metallic catalyst/carbon sources for CNT growth: ethanol/Co and benzene/Fe (Sano et al. 2010). Jiang et al. studied the growth of CNTs in situ on a pretreated graphite electrode via CCVD using $Ni(NO_3)_2$ as the catalyst (Jiang et al. 2008). The prepared CNTs had 80 and 20 nm in outer and inner diameter, respectively. Scheibe et al. tested Fe and Co for MWNTs preparation. Additionally, the authors were interested in concentrations of the carboxyl and hydroxyl groups on the CNT surface, which are essential features for applications in many science branches such as nanomedicine, biosensors, or polymer nanocomposites (Scheibe et al. 2010). Feng et al. used acetone as a carbon source, ferrocene for the Fe catalyst, and thiophene as promoter to synthesize high-quality double-walled CNT thin films in a one-step CVD reaction process in an argon flow (Feng et al. 2010). Liu and colleagues recently published the comprehensive review dealing with direct CVD growth of aligned, ultralong SWNTs on substrate surfaces, which are attractive building blocks for nanoelectronics. They discussed the key technical points, mechanisms, advantages, and limitations of this method (Liu et al. 2010). Kim et al. (2011) reported a novel method for the growth of CNTs that uses three different iron-containing proteins: hemoglobin, myoglobin, and cytochrome c, to control precisely the size and atomic structure of the CNTs. Most of the literature devoted to the CCVD synthesis of CNTs has investigated different catalysts and carbon sources and focused on the content and characterization of the nanotubes in the final product while neglecting some parameters such as the efficiency and lifetime of the catalyst. On the other hand, despite the extensive research on the formation mechanism of CNTs, some aspects of CNT growth are still poorly understood. A detailed understanding of CNT growth mechanism required to succeed in performing the control of CNT geometrical characteristics, which is one of the most important challenges in the science of nanotubes, is reported (Sarno et al. 2012b). A growth mechanism in particular in terms of the support role has been proposed. It was found that not only the size but also the chemical composition influences the activity of the catalyst nanoparticles.

CVD conducted in the presence of inorganic structure-directed or templated structure has been explored to synthesize a wide variety of new carbon materials. During templated synthesis methods, substrate or precursor materials are included into the template framework in such a way that their final structure reflects the shape and dimension of the templates. Organic templates are commonly used to synthesize inorganic materials. For example, surfactants have been used to template the synthesis of MCM-41 (Asefa et al. 1999) or other porous silicates (Kresge et al. 1992; Sandi et al. 1999) and oxides (Bagshaw et al. 1995). In the templated synthesis method, CNTs are obtained as a print of a directional structure. In this case, addition of metal particles to the system is not necessary. It can lead to the production of aligned CNTs. A major hurdle in applying CVD to template synthesis is the necessity to control the deposition rate. In fact, the pores can become blocked before the chemical vapor traverses the length of the pores. The first example of CNT synthesis by CVD template-based method has been reported by Martin et al. (Hulteen and Martin 1997; Che et al. 1998), using alumina membrane, with and without the addition of transition metal catalysts. The use of microporous zeolites, such as Y, ZSM-5, mordenite, and L, as templates (Kyotani et al. 2003) and, only recently, BEA has been investigated. CNTs were grown by a template-based hydrocarbon CVD in the channels of alumina membrane (AO), in the absence of transition metal catalyst (Ciambelli et al. 2004a,b). The nanotubes have open tips and a well-formed central hole and are very pure. In particular, the template-based CVD allows high control of the growth of aligned arrays of CNTs, which permits to overcome the main limitation concerning the use of CNTs as new electrical interconnections: the difficulties in the control of the microstructural characteristics and the use of expensive electron beam lithography. One of the key points that emerged concerns the improvement of the graphitization of CNTs, synthesized in the absence of transition metal catalyst. Very high temperatures permit to increase the carbon order. On the other hand, a change in the less expensive operating conditions can affect also the tube quality. To improve the graphitization of the tube walls, the effect of changing the operating conditions has been explored (Gommes et al. 2004b; Sarno et al. 2012a). Yield, selectivity, and quality of CNT in terms of diameter (up to very thin CNT); carbon order; length; arrangement (i.e., number of tubes for each channel); and purity, which are critical requisites for several applications, have been optimized. Samples produced by using methane have a well-ordered structure. The role of alumina channel surface during CNT growth has been investigated, and its catalytic activity, thanks to the quantitative online analysis of the product in the outlet reactor, has been proved for the first time (Ciambelli et al. 2011; Sarno et al. 2013). A vertical arrangement that does not suffer from longitudinal diffusive limitations along the reactor axes was chosen for the syntheses, permitting the control of the thickness and concentration homogeneity of the boundary layer established in steady-state conditions at the membrane interface, and forces the flow to pass through the membrane. Furthermore, the reaction zone arrangement was modified to preserve flatness, at the macro-scale and, consequently, micro-scale of the alumina, which is a prerequisite for a better integration in microelectronic devices (Sarno et al. 2013).

13.3 Physical Limitations in a CVD Reactor, Kinetic Studies, and Reactor Design

There is no doubt that the final result of a synthesis of CNTs should normally depend as much on the catalyst and feed gas as on the reactor operating conditions. Most of the literature devoted to CVD synthesis of CNTs deals with the use of different catalysts and carbon sources and overlooks the possible effects of the reactor operating conditions on the product. On the contrary, in the ambit of chemical engineering, reactor operating conditions are considered to be the main factors influencing the chemical reaction, and the detailed features of a given catalyst are secondary. It is obvious that successful development of a large-scale facility for producing MWCNTs can be achieved only by integrating both approaches. Moreover, to design a reactor for large-scale production, it is necessary to know the kinetic parameters of the reaction and, before this, to have evaluated the operating condition that ensures the highest productivity together with the required quality. The catalytic synthesis of CNTs in a CVD reactor is a typical catalysis case study, aggravated by the solid-state evolution of the reaction product. In particular, in a typical CVD horizontal reactor, physical limitation can be present both longitudinally and in the thickness of the catalytic bed (Gommes et al. 2004b). Figure 13.10 sums up schematically the effects of the two limiting transport mechanisms discussed previously: (1) If the reaction rate is not negligibly small compared with the total ethylene feed, a marked longitudinal concentration gradient develops along the catalytic bed. This results in a lower reaction rate than if the reactor was all bathed in a gas of the same hydrocarbon content as the feed. (2) If the reaction rate is not negligibly small with respect to the maximum hydrocarbon flux that could diffuse through the catalytic bed, at a given position along the reactor, there will be a difference in ethylene concentration between the surface and the basis of the catalytic bed. This results in a lower reaction rate than if the concentration all over the thickness of the catalytic bad was the same as on its surface. Therefore, before performing a kinetic study, to obtain the kinetic expression and develop the design of the reactor, experimental study and mathematical modeling can be used together to find the best conditions. From the point of view of the longitudinal concentration profile by Gommes et al. (2004) the Damkohler number (*Da*), which can be interpreted as the ratio of the chemical reaction rate to the ethylene supplied by the feed, has been used to evaluate the extent of the reaction on the respect of the material flow. While the square of Thiele modulus, which can be interpreted as the ratio of the reaction rate per unit surface of the catalyst bed to the diffusive flux of hydrocarbon per unit surface of the catalytic bed, has been used to evaluate diffusive limitation into the catalyst thickness. This study and the successive evaluation of the kinetic expression permit to perform a reactor design. Pilot-scale production reactors using the CCVD method are already running, and companies such as Arkema S.A., Bayer Material-Science AG, and Nanocyl S.A. are already present on the market, with an annual CNT production of over 10,000 kg/year (Anonymous 2006a,b; Thayer 2007; Pirard 2008). Other companies such as Hyperion, Iljin Nanotech, and Shenzhen Nanotech Port are also leading MWCNT producers, while Norman, South-West Nanotechnologies, Carbolex, and Raymor Industries are active in single-walled CNT production (Anonymous 2006b). Bayer and Arkema

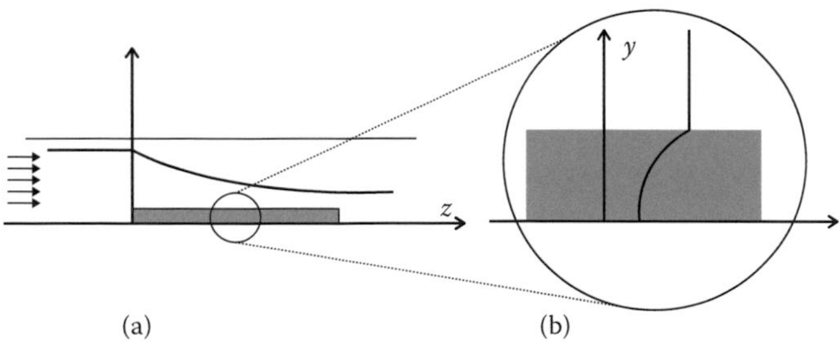

FIGURE 13.10 Schematic representation of the two transport mechanisms: (a) convective flow along the reactor and (b) diffusion through the catalytic bed.

use a fluidized-bed process to produce CNTs (Anonymous 2006b). This technology is easily scalable and allows a high carbon yield, a high space time yield, and a good mixing of the catalyst and of the produced CNTs. However, a fluidized bed also presents several drawbacks, such as a great variation in material density during the reaction, the cutting up of the catalytic support, and the breaking up of CNTs because of great agitation. Nanocyl S.A. produces CNTs in a mobile-bed inclined rotating reactor (Bossuot et al. 2004). This technology was chosen because the kinetics of CNT synthesis by hydrocarbon decomposition is quite slow and because the ratio between the volume of product and of catalyst is very large (larger than 50). Furthermore, the rotation of the reactor allows the rolling of particles inside the reactor. So a good mixing of particles is possible during the reaction. Furthermore, the flow regime does not correspond to a heterogeneous phase between gas and solid, and it tends to be a homogeneous regime because of the best gas–solid contact. The design and the development of a catalytic reactor require an integrated approach, according to chemical engineering methodology, which includes catalyst preparation and characterization, kinetic studies, and product characterization. Chemical reaction engineering consists of analysis of the factors governing the process to establish the reactor equations. The factors are geometric, hydrodynamic, physical, and physicochemical (Villermeaux 1993). While the first factor consists of the description of the geometry of the reactor, the second is the hydrodynamic factor dealing with the flow of each phase and with the way those phases are put into contact and mixed. The hydrodynamic factor is also closely linked with the physical factor relating the phenomena of transport and transfer of mass, heat, and momentum quantity (Gommes et al. 2004a). Finally, the last factor consists of physicochemical factors in relation to thermodynamic data and the kinetics of the reaction. These four factors allow the linking of the results of the reaction to the operating variables, and knowledge of relationships between all the operating factors is necessary to establish the equations of the reactor.

13.4 Properties of CNTs

13.4.1 Mechanical Properties

CNTs are the strongest and stiffest materials yet discovered in terms of tensile strength and elastic modulus respectively. This strength results from the covalent sp^2 bonds formed between the individual carbon atoms, which is lower in the presence of defects and disorder in the sidewalls. The Young modulus value of an SWNT is estimated as high as 1 Tpa to 1.8 Tpa. The modulus of an SWNT depends on the diameter and chirality. For MWNTs, experiments have indicated that only the outer graphitic shell can support stress when the tubes are dispersed in an epoxy matrix, and for SWNT bundles, a significant reduction in the individual modulus for a rope has been demonstrated owing to weak intertube cohesion (Harris 1999). A single perfect nanotube is about 10 to 100 times stronger (Young modulus = 1000 GPa) than steel per unit weight. The tensile strength, or breaking strain, of nanotubes can be up to 63 GPa, around 50 times higher than that of steel. These properties, coupled with the small weight of carbon atoms, give them great potential in applications such as aerospace. It has even been suggested that nanotubes could be used in the space elevator, an Earth-to-space cable first proposed by Arthur C. Clarke. The electronic properties of CNTs are also extraordinary. Especially notable is the fact that nanotubes can be metallic or semiconducting depending on their chirality. Thus, some nanotubes have conductivities higher than that of copper, while others behave more like silicon. There is great interest in the possibility of constructing nanoscale electronic devices from nanotubes, and some progress is being made in this area. However, to construct a useful device, we would need to arrange many thousands of nanotubes in a defined pattern, and we do not yet have the degree of control necessary to achieve this. There are several areas of technology where CNTs are already being used. These include flat-panel displays scanning probe microscopes and sensing devices. The unique properties of CNTs will undoubtedly lead to many more applications (Nardelli et al. 2000; Yu et al. 2000; Dai et al. 2006; Meo and Rossi 2006; Pop et al. 2006; Prabhakar Bandaru 2007).

13.4.2 Electrical Properties

CNTs have very interesting electrical properties. A single graphite sheet is a semimetal, which means that it has properties intermediate between semiconductors (like the silicon in computer chips, where electrons have restricted motion) and metals (like the copper used in wires, where electrons can move freely). When a graphite sheet is rolled into a nanotube, not only do the carbon atoms have to line up around the circumference of the tube, but also the quantum mechanical wave functions of the electrons must match up. In theory, metallic nanotubes can carry an electrical current density 1000 times greater than metals such as copper (Dai et al. 2006). Individual nanotubes, like macroscopic structures, can be characterized by a set of electrical properties—resistance, capacitance, and inductance—which arise from the intrinsic structure of the nanotube and its interaction with other objects. Electrical transport inside CNTs is affected by scattering by defects and by lattice vibrations that lead to resistance, similar to that in bulk materials. However, the one-dimensional (1D) nature of CNTs and their strong covalent bonding drastically affect these processes. Scattering by small angles is not allowed in a 1D material, only forward and backward motion of the carriers. Most importantly, the 1D nature of the CNT leads to a new type of quantized resistance related to its contacts with 3D macroscopic objects such as the metal electrodes. Of course, in addition to this quantum resistance, there are other forms of contact resistance such as that attributable to the presence of Schottky barriers at metal–semiconducting nanotube interfaces and parasitic resistance, which is simply caused by bad contacts. At the other extreme, in long CNTs, or at high bias, many scattering collisions can take place and the so-called diffusive limit of transport that is typical of conventional conductors is reached. In this limit, the carriers have a finite mobility. However, in CNTs, this can be very high, as much as 1000 times higher than in bulk silicon. The intrinsic electronic structure of a CNT also leads to a capacitance that is related to its density-of-states (i.e., how its energy states are distributed in energy) and it is independent of electrostatics. In addition to quantum capacitance (CQ), a CNT incorporated in a structure has an electrostatic capacitance, gate capacitance (CG), that arises from its coupling to surrounding conductors and as such depends on the device geometry and dielectric structure. Finally, CNTs have inductance, which is resistance to any change in the current flowing through them. Again, there is a quantum and a classical contribution. Classical self-inductance depends on the CNT diameter, geometry of the structure, and the magnetic permeability of the medium. In response to an alternating current (ac) signal, a CNT behaves like a transmission line owing to its inductance (Dai et al. 2006; Prabhakar Bandaru 2007).

13.4.3 Thermal Properties

All nanotubes are expected to be very good thermal conductors along the tube, exhibiting a property known as "ballistic conduction," but good insulators laterally to the tube axis. It is predicted that CNTs will be able to transmit up to 6000 $W \cdot m^{-1} \cdot K^{-1}$ at room temperature; compare this to copper, a metal well known for its good thermal conductivity, which transmits 385 $W \cdot m^{-1} \cdot K^{-1}$. The temperature stability of CNTs is estimated to be up to 2800°C in vacuum and about 750°C in air. Thermal expansion of CNTs will be largely isotropic, which is different than conventional graphite fibers, which are strongly anisotropic. This may be beneficial for carbon–carbon composites. It is expected that low-defect CNTs will have very low coefficients of thermal expansion (Pop et al. 2006; Stahl 2000).

13.4.4 Chemical Properties

The chemical reactivity of a CNT is, compared with a graphene sheet, enhanced as a direct result of the curvature of the CNT surface. This curvature causes the mixing of the π- and σ-orbital, which leads to hybridization between the orbitals. The degree of hybridization becomes larger as the diameter of an SWNT gets smaller. Hence, CNT reactivity is directly related to the π-orbital mismatch caused by an increased curvature. Therefore, a distinction must be made between the sidewall and the end caps of a nanotube. For the same reason, a smaller nanotube diameter results in increased reactivity. Covalent chemical modification of either sidewalls or end caps has shown to be possible. For example, the solubility of CNTs in different solvents can be controlled this way. However, covalent attachment of molecular species to fully sp^2-bonded carbon atoms on the nanotube sidewalls proves to be difficult. Therefore, nanotubes can be considered as usually chemically inert (Lordi and Yao 2000).

13.4.5 Optical Properties

The optical properties of SWNT are related to their quasi 1D nature. Theoretical studies have revealed that the optical activity of chiral nanotubes disappears if the nanotubes become larger; therefore, it is expected that other physical properties are influenced by these parameters too. Use of the optical activity might result in optical devices in which CNTs play an important role (Kataura et al. 1999).

13.5 CNT Applications

1. CNTs can be used in carrier for drug delivery (Sebastien et al. 2005).
2. Functionalized CNTs are reported for targeting of Amphotericin B to cells (Barroug and Glimcher 2002).
3. Cisplatin-incorporated oxidized single-wall carbon nanohorns (SWNHs) have shown slow release of cisplatin in aqueous environment. The released cisplatin had been effective in terminating the growth of human lung cancer cells, while the SWNHs alone did not show anticancer activity (Ajima 2005).
4. Anticancer drug polyphosphazene platinum given with nanotubes had enhanced permeability, distribution, and retention in the brain owing to controlled lipophilicity of nanotubes (Pai et al. 2006).
5. An antibiotic, doxorubicin, given with nanotubes is reported for enhanced intracellular penetration. The gelatin CNT mixture (hydrogel) has been used as a potential carrier system for biomedical (Pai et al. 2006).
6. A CNT-based carrier system can offer a successful oral alternative administration of erythropoietin (EPO), which has not been possible so far because of the denaturation of EPO by the gastric environment conditions and enzymes (Pai et al. 2006).
7. They can be used as lubricants or glidants in tablet manufacturing because of the nanosize and sliding nature of graphite layers bound with van der Waals forces (Pai et al. 2006).
8. In genetic engineering, CNTs are used to manipulate genes and atoms in the development of bioimaging genomes, proteomics, and tissue engineering (Pantarotto 2003).
9. For biomedical applications, Bianco et al. have prepared soluble CNTs and have covalently linked biologically active peptides with them. This was demonstrated for viral protein VP1 of foot-and-mouth disease virus showing immunogenicity and eliciting antibody response. In chemotherapy, drug embedded nanotubes attack directly on viral ulcers and kills viruses. No antibodies were produced against the CNT backbone alone, suggesting that the nanotubes do not possess intrinsic immunogenicity. In vitro studies by Kam et al. (2005) showed selective cancer cell killing obtained by hyperthermia as a result of the thermal conductivity of CNT internalized into those cells.
10. For artificial implants, normally, the body shows a rejection reaction for implants with post-administration pain, but miniature sized nanotubes get attached to other proteins and amino acids, avoiding rejection. Also, they can be used as implants in the form of artificial joints without host rejection reaction. Moreover, because of their high tensile strength, CNTs filled with calcium and arranged/grouped in the structure of bone can act as a bone substitute (Deng et al. 2013).
11. As a preservative, CNTs are antioxidant in nature (Pai et al. 2006).

12. As a diagnostic tool, protein-encapsulated or protein/enzyme-filled nanotubes, because of their fluorescence ability in the presence of specific biomolecules, have been tried as implantable biosensors (Deng et al. 2013).
13. As biosensors, CNTs act as sensing materials in pressure, flow, thermal, gas, optical, mass, position, stress, strain, chemical, and biological sensors.

13.6 CNTs as Effective Additive for Lubricants

Nanoparticles as friction modifier additives and in liquid lubricant or in solid media have been studied extensively in the last years. It has been found that nanoparticle inclusion in base oil enhances the extreme-pressure property and load-carrying capacity of the thin film in between the sliding surfaces, with a reduction in the friction coefficient and wear rate.

Among the class of nanocarbon materials, CNTs have received a great attention by tribology researchers in the last decade because of high load-bearing capacity, low surface energy, high chemical stability, and weak intermolecular and strong intramolecular bonding. At the nanometer scale, the friction of CNTs has been investigated using the AFM technique (Falvo et al. 1999; Ohmae et al. 2005). The rolling motion of an individual CNT on graphite and the transition between stiction and sliding motions were observed (Falvo et al. 1999). The authors showed that adhesion hysteresis was the dominant energy loss responsible for the stick-slip dynamics. The paper (Ohmae et al. 2005) introduces the results of friction testing of a gold tip over nanotube forest. This combination gave high friction and low adhesion. At low contact pressure, it was observed that nanotubes slide in the contact, while at high contact pressure, they were flattened and could both slide and roll according to their orientation and bonding to the sliding surface (Saint-Aubin et al. 2009). Regarding CNT dispersions in oil, the key role of the surfactant on dispersion stability has been confirmed in a few papers (Ni and Sinnott 2001; Chen et al. 2005; Saint-Aubin et al. 2009). In one paper (Hong et al. 2010), original inclusion of CNTs in the domain of semisolid lubrication through grease was introduced. The authors achieved a stable and homogeneous lubricant grease based on single-wall CNTS and MWCNTs in polyalphaolefin (PAO) oil produced without using a chemical surfactant, with weight concentration from 10.5% to 20.0%. Thus, in the latter case, the CNTs were used as friction modifiers, whose action is limited to the film-forming ability on rubbing steel surfaces with loading of 0.1–0.5 wt% (Yang et al. 2011), as is usual in CNTs dispersed in oil.

Moreover, to conjugate and to enhance the performances of the CNTs with the well-tested good behavior of nanochalcogenides, the synthesis of hybrid-nanostructures has become an important goal in lubrication (Miura 2004; Shu et al. 2010; Church et al. 2012). An original synthetic strategy to obtain hybrid organic–inorganic oleylamine@MoS2-CNT nanocomposites with different compositions is introduced along with their application as antifriction and antiwear additives for grease lubricants based on lithium and calcium soaps (Altavilla et al. 2013).

The lubrication mechanism of carbon nanoparticles has not been yet completely understood, but their properties as frictional media are definitely based on structural modification. A deepening about the interfacial mechanism introduced by CNTs dispersed in liquid media has been proposed by Chauveau et al. (2012). Through an original experimental approach, the authors achieved the measurement of the film thickness and the frictional properties of CNT dispersions in PAO base oil in elastohydrodynamic lubrication regime. They observed that the CNT clusters are chemically inert and do not adsorb on the considered surfaces. The contact inlet side behaves like a filter for the aggregates, separating the nanotubes that penetrate and travel between the rubbing surfaces. CNTs reduced the frictional losses with their transient passage in the contact area. At lower sliding speed, the CNTs propagate easily within the contact and the frictional behavior of the tribopair is dominated by these additives. At higher speed, the shear induces a reduction in the number of aggregates passing through the contact and the base oil rheology is the key factor on the control of the frictional response of the tribological pair.

13.7 Conclusion

Among the various methods, CVD clearly emerges as the best one for large-scale production of CNTs. With the prospect of gene therapy, cancer treatments, and innovative new answers for life-threatening diseases on the horizon, the science of nanomedicine has become an ever-growing field that has an incredible ability to bypass barriers. The properties and characteristics of CNTs are still being researched heavily and scientists have barely begun to tap the potential of these structures. Single-walled CNTs and MWCNTs have already proven to serve as safer and more effective alternatives to previous drug delivery. Also, little imagined and not expected attention as that received by tribology research in the last decade will contribute to the CNT commercialization.

Acknowledgment

The authors gratefully acknowledge Dr. Claudia Cirillo for helpful discussion and precious support.

References

Afolabi, A. S., A. S. Abdulkareem, S. D. Mhlanga and S. E. Iyuke (2011), Synthesis and purification of bimetallic catalysed carbon nanotubes in a horizontal CVD reactor. *J. Exp. Nanosci.*, 6, 248.

Ajima, K., M. Yudasaka, T. Murakami, A. Maigné, K. Shiba and S. Iijima (2005), Carbon nanohorns as anticancer drug carriers. *Mol. Pharm.*, 2, 475.

Altavilla, C., M. Sarno and P. Ciambelli (2009), Synthesis of ordered layers of monodisperse $CoFe_2O_4$ nanoparticles for catalyzed growth of carbon nanotubes on silicon substrate. *Chem. Mater.*, 21, 4851.

Altavilla, C., M. Sarno, P. Ciambelli, A. Senatore and V. Petrone (2013), New 'chimie douce' approach to the synthesis of hybrid nanosheets of MoS2 on CNT and their anti-friction and anti-wear properties. *Nanotechnology*, 24, 125601.

Anonymous (2006a), Arkema inaugurates carbon nanotube pilot plant; other CNT developments. *Addit. Polym.*, 2006, 6.

Anonymous (2006b), Arkema opens carbon nanotube pilot plant. *Plast. Addit. Compound.*, 8, 11.

Asefa, T., M. J. MacLachlan, N. Coombs and G. A. Ozin (1999), Periodic mesoporous organosilicas with organic groups inside the channel wall. *Nature*, 402, 867.

Atthipalli, G. R. Epur, P. N. Kumta, M. J. Yang, J. K. Lee and J. L. Gray (2011), Nickel catalyst-assisted vertical growth of dense carbon nanotube forests on bulk copper. *J. Phys. Chem. C*, 115, 3534.

Bagshaw, S. A., E. Pruzet and T. J. Pinnavaia (1995), Templating of mesoporous molecular sieves by nonionic polyethylene oxide surfactants. *Science*, 269, 1242.

Barroug, A. and M. Glimcher (2002), Hydroxyapatite crystals as a local delivery system for cisplatin: Adsorption and release of cisplatin in vitro. *J. Orthopaed. Res.*, 20, 274.

Bossuot, C., P. Kreit, J.-P. Pirard (2004), Procede et installation pour la fabrication de nanotubes de carbone. Patent WO 2004/069742, EP1594802B1.

Chauveau, V., D. Mazuyer, F. Dassenoy and J. Cayer-Barrioz (2012), In situ film-forming and friction-reduction mechanisms for carbon-nanotube dispersions in lubrication. *Tribol. Lett.*, 47, 467.

Che, G., B. B. Lakshmi, C. R. Martin, E. R. Fisher and R. R. Ruoff (1998), Chemical vapor deposition based synthesis of carbon nanotubes and nanofibers using a template method. *Chem. Mater.*, 10, 260.

Chen, C. S., X. H. Chen, L. S. Xu, Z. Yang and W. H. Li (2005), Modification of multi-walled carbon nanotubes with fatty acid and their tribological properties as lubricant additive. *Carbon*, 43, 1660.

Chesnokov, V. V., V. I. Zaikovskii, R. A. Buyanov, V. V. Molchanov and L. M. Plyasova (1994), Morphology of carbon from methane on nickel-containing catalysts. *Kinet. Katal.*, 35, 146.

Church, A. H., X. F. Zhang and B. Sirota (2012), Carbon nanotube-based adaptive solid lubricant composites. *Adv. Sci. Lett.*, 5, 1, 188–191.

Ciambelli, P. D. Sannino, M. Sarno and C. Leone (2011), Wide characterisation to compare conventional and highly effective microwave purification and functionalization of multi-wall carbon nanotubes. *Thin Solid Films*, 519, 2121.

Ciambelli, P., D. Sannino, M. Sarno, C. Leone and U. Lafont (2007), Effects of alumina phases and process parameters on the multiwalled carbon nanotubes growth. *Diam. Relat. Mater.*, 16, 1144.

Ciambelli, P., D. Sannino, M. Sarno, A. Fonseca and J. B. Nagy (2004a), Hydrocarbon decomposition in alumina membrane: An effective way to produce carbon nanotubes bundles. *J. Nanosci. Nanotechnol.*, 4, 779.

Ciambelli, P., D. Sannino, M. Sarno and J. B. Nagy (2004b), Characterization of nanocarbons produced by CVD of ethylene in alumina or alumino-silicate matrices. *Adv. Eng. Mater.*, 6, 804.

Ciambelli, P., D. Sannino, M. Sarno, A. Fonseca and J. B. Nagy (2005), Selective formation of carbon nanotubes over Co-modified beta zeolite by CCVD. *Carbon*, 43, 631.

Ciambelli, P., L. Arurault, M. Sarno, S. Fontorbes, C. Leone, L. Datas, D. Sannino, P. Lenormand and S. Le Blond Du Plouy (2011), Controlled growth of CNT in mesoporous AAO through optimized conditions for membrane preparation and CVD operation. *Nanotechnology*, 22, 265613.

Colbert, D. T., J. Zhang, S. M. McClure, P. Nikolaev, Z. Chen, J. H. Hafner, D. W. Owens et al. (1994), Growth and sintering of fullerene nanotubes. *Science*, 266, 1218.

Dai, H., E. W. Wong and C. M. Lieber (1996), Probing electrical transport in nanomaterials: Conductivity of individual carbon nanotubes. *Science*, 272, 523.

Dai, H., A. Javey, E. Pop, D. Mann and Y. Lu (2006), Electrical transport properties and field effect transistors of carbon nanotubes. *Nano*, 1, 1.

Deng, P., Z. Xu and J. Li (2013), Simultaneous determination of ascorbic acid and rutin in pharmaceutical preparations with electrochemical method based on multi-walled carbon nanotubes–chitosan composite film modified electrode. *J. Pharm. Biomed.*, 76, 234.

Di Bartolomeo, A., M. Sarno, F. Giubileo, C. Altavilla, L. Iemmo, S. Piano, F. Bobba et al. (2009), Multiwalled carbon nanotube films as small-sized temperature sensors. *J. Appl. Phys.*, 105, 06451.

Dumpala, S. J. B. Jasinski, G. U. Sumanasekera and M. K. Sunkara (2011), Large area synthesis of conical carbon nanotube arrays on graphite and tungsten foil substrates. *Carbon*, 49, 2725.

Ebbesen, T. W. and P. M. Ajayan (1992), Large-scale synthesis of carbon nanotubes. *Nature*, 358, 220.

Falvo, M. R., R. M. Taylor II, A. Helser, V. Chi, F. P. Brooks Jr., S. Washburn and R. Superfine (1999), Nanometre-scale rolling and sliding of carbon nanotubes. *Nature*, 397, 236.

Fan, S., M. Chapline, N. R. Franklin, T. W. Tombler, A. M. Cassel and H. Dai (1999), Self-oriented regular arrays of carbon nanotubes and their field emission properties. *Science*, 283, 512.

Feng, J.-M., R. Wang, Y.-L. Li, X.-H. Zhong, L. Cui, Q.-J. Guo and F. Hou (2010), One-step fabrication of high quality double-walled carbon nanotube thin films by a chemical vapor deposition process. *Carbon*, 48, 3817–3824.

Flahaut, E., C. Laurent and A. Peigney (2005), Catalytic CVD synthesis of double and triple-walled carbon nanotubes by the control of the catalyst preparation. *Carbon*, 43, 375.

Fotopoulos, N. and J. P. Xanthakis (2010), A molecular level model for the nucleation of a single-wall carbon nanotube cap over a transition metal catalytic particle. *Diam. Relat. Mater.*, 19, 557.

Funaro, M., M. Sarno, P. Ciambelli, C. Altavilla and A. Proto (2013), Real time radiation dosimeters based on vertically aligned multiwall carbon nanotubes and graphene. *Nanotechnology*, 24, 075704.

Giubileo, F. A. Di Bartolomeo, M. Sarno, C. Altavilla, S. Santandrea, P. Ciambelli and A. M. Cucolo (2012), Field emission properties of as-grown multiwalled carbon nanotube films. *Carbon*, 50, 163.

Gommes, C., S. Blacher, C. Bossuot, P. Marchot, J. B. Nagy and J. P. Pirard (2004a), Comparison of different methods for characterizing multi-walled carbon nanotubes. *Carbon*, 42, 1473.

Gommes, C., S. Blacher, C. Bossuot, P. Marchot, J. B. Nagy and J.-P. Pirard (2004b), Influence of the operating conditions on the production rate of multi-walled carbon nanotubes in a CVD reactor. *Carbon*, 42, 1473.

Harris, P. (1999), *Carbon Nanotubes and Related Structures: New Materials for the 21st Century*. Cambridge University Press, Cambridge.

Hatta, N. and K. Murata (1994), Very long graphitic nano-tubules synthesized by plasma-decomposition of benzene. *Chem. Phys. Lett.*, 217, 398.

He, D. L., H. Li, W. L. Li, P. Haghi-Ashtiani, P. Lejay and J. B. Bai (2011), Growth of carbon nanotubes in six orthogonal directions on spherical alumina microparticles. *Carbon*, 49, 2273.

Hong, H., D. Thomas, A. Waynick, W. Yu, P. Smith and W. Roy (2010), Carbon nanotube grease with enhanced thermal and electrical conductivities. *J. Nanopart. Res.*, 12, 529.

Hou, B., R. Xiang, T. Inoue, E. Einarsson, S. Chiashi, J. Shiomi, A. Miyoshi and S. Maruyama (2011), Decomposition of ethanol and dimethyl ether during chemical vapor deposition synthesis of single-walled carbon nanotubes. *Jpn. J. Appl. Phys.*, 50, 4.

Hulteen, J. C. and C. R. Martin (1997), A general template-based method for the preparation of nanomaterials. *J. Mater. Chem.*, 7, 1075.

Iijima, S. (1991), Helical microtubules of graphitic carbon. *Nature*, 354, 56.

Ivanov, V. B., J. Nagy, P. Lambin, A. Lucas, X. B. Zhang, X. F. Zhang, D. Bernaerts, G. Van Tendeloo, S. Amelinckx and J. Van Landuyt (1994), The study of carbon nanotubules produced by catalytic method. *Chem. Phys. Lett.*, 223, 329.

Jiang, Q., L. J. Song, H. Yang, Z. W. He and Y. Zhao (2008), Preparation and characterization on the carbon nanotube chemically modified electrode grown in situ. *Electrochem. Commun.*, 10, 424.

Kam, N. W. S., M. O'Connell, J. A. Wisdom and H. Dai (2005), Carbon nanotubes as multifunctional biological transporters and near-infrared agents for selective cancer cell destruction. *Proc. Natl. Acad. Sci. U.S.A.*, 102, 11600.

Kataura, H. Y. Kumazawa, Y. Maniwa, I. Umezu, S. Suzuki, Y. Ohtsuka and Y. Achiba (1999), Optical properties of single-wall carbon nanotubes. *Synth. Met.*, 103, 2555.

Kim, H.-J. E. Oh, J. Lee, D.-S. Shim and K.-H. Lee (2011), Synthesis of carbon nanotubes with catalytic iron-containing proteins. *Carbon*, 49, 3717.

Kresge, C. T., M. E. Leonowicz, W. J. Roth, J. C. Vartuli and J. S. Beck (1992), Ordered mesoporous molecular sieves synthesized by a liquid-crystal template mechanism. *Nature*, 359, 710.

Kyotani, T., Z. Ma and A. Tomita (2003), Template synthesis of novel porous carbons using various types of zeolites. *Carbon*, 41, 1451.

Li, H., D. L. He, T. H. Li, M. Genestoux and J. B. Bai (2010), Chemical kinetics of catalytic chemical vapor deposition of an acetylene/xylene mixture for improved carbon nanotube production. *Carbon*, 48, 4330.

Li, W. Z., S. S. Xie, L. X. Qian, B. H. Chang, B. S. Zou, W. Y. Zhou, R. A. Zhao and G. Wang (1996), Large-scale synthesis of aligned carbon nanotubes. *Science*, 274, 1701.

Liu, Z. F., L. Y. Jiao, Y. G. Yao, X. J. Xian and J. Zhang (2010), Aligned, ultralong single-walled carbon nanotubes: From synthesis, sorting, to electronic devices. *Adv. Mater.*, 22, 2285.

Lordi, V. and N. Yao (2000), Molecular mechanics of binding in carbon-nanotube-polymer composites. *J. Mater. Res.*, 15, 2770.

Lyu, S. C., B. C. Liu, C. J. Lee, H. K. Kang, C. W. Yang and C. Y. Park (2003), High-quality double-walled carbon nanotubes produced by catalytic decomposition of benzene. *Chem. Mater.*, 15, 3951.

Meo, M. and M. Rossi (2006), Prediction of Young's modulus of single wall carbon nanotubes by molecular-mechanics based finite element modelling. *Compos. Sci. Technol.*, 66, 1597.

Miura, K. (2004), *Encyclopedia of Nanoscience and Nanotechnology*, vol. 9, ed. H. S. Nalwa. American Scientific Publishers, Valencia, CA, p. 947.

Nardelli, M., J.-L. Fattebert, D. Orlikowski, C. Roland, Q. Zhao and J. Bernholc (2000), Mechanical properties, defects and electronic behavior of carbon nanotubes. *Carbon*, 38, 1703.

Narkiewicz, U., M. Podsiadly, R. Jedrzejewski and I. Pelech (2010), Catalytic decomposition of hydrocarbons on cobalt, nickel and iron catalysts to obtain carbon nanomaterials. *Appl. Catal. A-Gen.*, 384, 27.

Ni, B. and S. Sinnott (2001), Tribological properties of carbon nanotube bundles predicted from atomistic simulations. *Surf. Sci.*, 487, 87.

Ohmae, N., J. M. Martin and S. Mori (2005), *Micro and Nanotribology*. ASME Press, New York.

Pai, P., K. Nair, S. Jamade, R. Shah, V. Ekshinge and N. Jadhav (2006), Pharmaceutical applications of carbon tubes and nanohorns. *Curr. Pharm. Res. J.*, 1, 11.

Palizdar, M., R. Ahgababazadeh, A. Mirhabibi, R. Brydson and S. Pilehvari (2011), Investigation of Fe/MgO catalyst support precursors for the chemical vapour deposition growth of carbon nanotubes. *J. Nanosci. Nanotechnol.*, 11, 5345.

Pan, Z. W., S. S. Xie, B. H. Chang, L. F. Sun, W. Y. Zhou and G. Wang (1999), Direct growth of aligned open carbon nanotubes by chemical vapor deposition. *Chem. Phys. Lett.*, 299, 97.

Pantarotto, D., C. Partidos, J. Hoebeke, F. Brown, E. Kramer and J. Briand (2003), Immunization with peptide-functionalized carbon nanotubes enhances virus-specific neutralizing antibody responses. *Chem. Biol.*, 10, 961.

Pirard, J.-P. (2008), Nanotech stewardship. *Chem. Eng. News*, 86, 5.

Pop, E., D. Mann, Q. Wang, K. Goodson and H. Dai (2006), Negative differential conductance and hot phonons in suspended nanotube molecular wires. *Nano Lett.*, 6, 96.

Prabhakar Bandaru, R. (2007), Electrical properties and applications of carbon nanotube structures. *J. Nanosci. Nanotechnol.*, 7, 1.

Ren, Z. F., Z. P. Huang, J. W. Xu, J. H. Wang, P. Bush, M. P. Siegal and P. N. Provencio (1998), Synthesis of large arrays of well-aligned carbon nanotubes on glass. *Science*, 282, 1105.

Rodriguez, N. M. (1993), A review of catalytically grown carbon nanofibers. *J. Mater. Res.*, 8, 3233.

Saint-Aubin, K., P. Poulin, H. Saadaoui, M. Maugey and C. Zakri (2009), Dispersion and film-forming properties of poly(acrylic acid)-stabilized carbon nanotubes. *Langmuir*, 25, 13206.

Sandi, G., K. A. Corrado, R. E. Winans, C. S. Johnson and R. J. Csencsits (1999), Carbons for lithium battery applications prepared using sepiolite as inorganic template. *J. Electrochem. Soc.*, 146, 3644.

Sano, N., S. Ishimaru and H. Tamaon (2010), Synthesis of carbon nanotubes in graphite microchannels in gas-flow and submerged-in-liquid reactors. *Mater. Chem. Phys.*, 122, 474.

Santangelo, S., G. Messina, G. Faggio, M. Lanza, A. Pistone and C. Milone (2010), Calibration of reaction parameters for the improvement of thermal stability and crystalline quality of multi-walled carbon nanotube. *J. Mater. Sci.*, 45, 783.

Sarno, M., D. Sannino, C. Leone and P. Ciambelli (2012a), CNTs tuning and vertical alignment in anodic aluminium oxide membrane. *J. Nat. Gas. Chem.*, 21, 639.

Sarno, M., D. Sannino, C. Leone and P. Ciambelli (2012b), Evaluating the effects of operating conditions on the quantity, quality and catalyzed growth mechanisms of CNTs. *J. Mol. Catal. A-Chem.*, 357, 26.

Sarno, M., A. Tamburrano, L. Arurault, S. Fontorbes, R. Pantani, L. Datas, P. Ciambelli and M. S. Sarto (2013), Electrical conductivity of carbon nanotubes grown inside a mesoporous anodic aluminium oxide membrane. *Carbon*, 55, 10.

Scheibe, B., E. Borowiak-Palen and R. J. Kalenczuk (2010), Oxidation and reduction of multiwalled carbon nanotubes—Preparation and characterization. *Mater. Charact.*, 61, 185.

Sebastien, W. W. Giorgia, P. Monica, B. Cedric, K. Jean-Paul and B. Renato (2005), Targeted delivery of amphotericin b to cells by using functionalized carbon nanotubes. *Angew. Chem.*, 117, 6516.

Shirazi, Y., M. A. Tofighy, T. Mohammadi and A. Pak (2011), Effects of different carbon precursors on synthesis of multiwall carbon nanotubes: Purification and functionalization. *Appl. Surf. Sci.*, 257, 7359.

Shu, L. Y. Feng and X. Yang (2010), Influence of adding carbon nanotubes and graphite to Ag-MoS2 composites on the electrical sliding wear properties. *Acta Metall. Sin.*, 23, 1, 27–34.

Skukla, B., T. Saito, S. Ohmori, M. Koshi, M. Yumura and S. Iijima (2010), Interdependency of gas phase intermediates and chemical vapor deposition growth of single wall carbon nanotubes. *Chem. Mater.*, 22, 6035.

Stahl, H., J. Appenzeller, R. Martel, P. Avouris and B. Lengeler (2000), Intertube coupling in ropes of single-wall carbon nanotubes. *Phys. Rev. Lett.*, 85, 5186.

Thayer, A. M. (2007), Carbon nanotubes by the metric ton. *Chem. Eng. News*, 85, 29.

Tomie, T., S. Inoue, M. Kohno and Y. Matsumura (2010), Prospective growth region for chemical vapor deposition synthesis of carbon nanotube on C–H–O ternary diagram. *Diam. Relat. Mater.*, 19, 1101.

Villermeaux, J. (1993), *Génie de la réaction chimique—Conception et fonctionnement des réacteurs.* Lavoisier, Paris.

Xiang, X., L. Zhang, H. I. Hima, F. Li and D. G. Evans (2009), Co-based catalysts from Co/Fe/Al layered double hydroxides for preparation of carbon nanotubes. *Appl. Clay Sci.*, 42, 405.

Yang, Y., T. Yamabe and B.-S. Kim (2011), Lubricating characteristic of grease composites with CNT additive. *Tribol. Online*, 6, 5, 247–250.

Yong, Z., L. Fang and Z. Zhi-hua (2011), Synthesis of heterostructured helical carbon nanotubes by iron-catalyzed ethanol decomposition. *Micron*, 42, 547.

Yu, M. F., B. S. Files, S. Arepalli and R. S. Ruoff (2000), Tensile loading of ropes of single wall carbon nanotubes and their mechanical properties. *Phys. Rev. Lett.*, 84, 5552.

Zhang, D. S., L. Y. Shi, J. H. Fang, K. Dai and X. K. Li (2006), Preparation and desalination performance of multiwall carbon nanotubes. *Mater. Chem. Phys.*, 97, 415.

Zhang, D. S., L. Y. Shi, J. H. Fang, X. K. Li and K. Dai (2005), Preparation and modification of carbon nanotubes. *Mater. Lett.*, 59, 4044.

14

Carbon Nanotube Fibers

Jandro L. Abot

Cijo Punnakattu Rajan

14.1 Introduction

The excellent electronic and mechanical properties of individual carbon nanotubes (CNTs) have led to the development of one-, two-, or three-dimensional structures composed of CNTs including fibers, films or sheets, and arrays or foams. These structures could be used in microscale and macroscale applications, unlike individual CNTs. CNT fibers or yarns constitute long and continuous arrangements of intertwined or interlocked CNTs that are highly packed, mostly aligned, and parallel to one another (Figure 14.1). CNT fibers contain thousands of CNTs in their cross-sections, and their electrical, thermal, and mechanical properties depend not only on the corresponding properties of the CNTs themselves but also on the relative arrangements among the CNTs within the fiber. These CNT fibers could have much higher specific elastic modulus and specific tensile strength than those of existing commercial carbon and polymeric fibers. Furthermore, CNT fibers exhibit also satisfactory electrical and thermal conductivities, although at least two orders of magnitude lower than those of CNTs themselves. A key feature of CNT fibers is their continuity, which makes them very amenable for a wide range of applications, including electricity transmission, sensing, reinforcement, energy storage, actuation, and many others. Several excellent articles have been published in the past few years reviewing the fabrication and performance of CNT fibers and they serve as a reference in this chapter (Lu et al. 2012; Park and Lee 2012; Miao 2013; Lekawa-Raus et al. 2014a). The fabrication methods of CNT fibers are presented in Section 14.2, including spinning or drawing from CNT solutions, arrays, and aerogels and others. The mechanical response of the CNT fibers is presented in Section 14.3 and discussed in terms of the parameters that affect that response. Sections 14.4 and 14.5 present the electrical and piezoresistive responses,

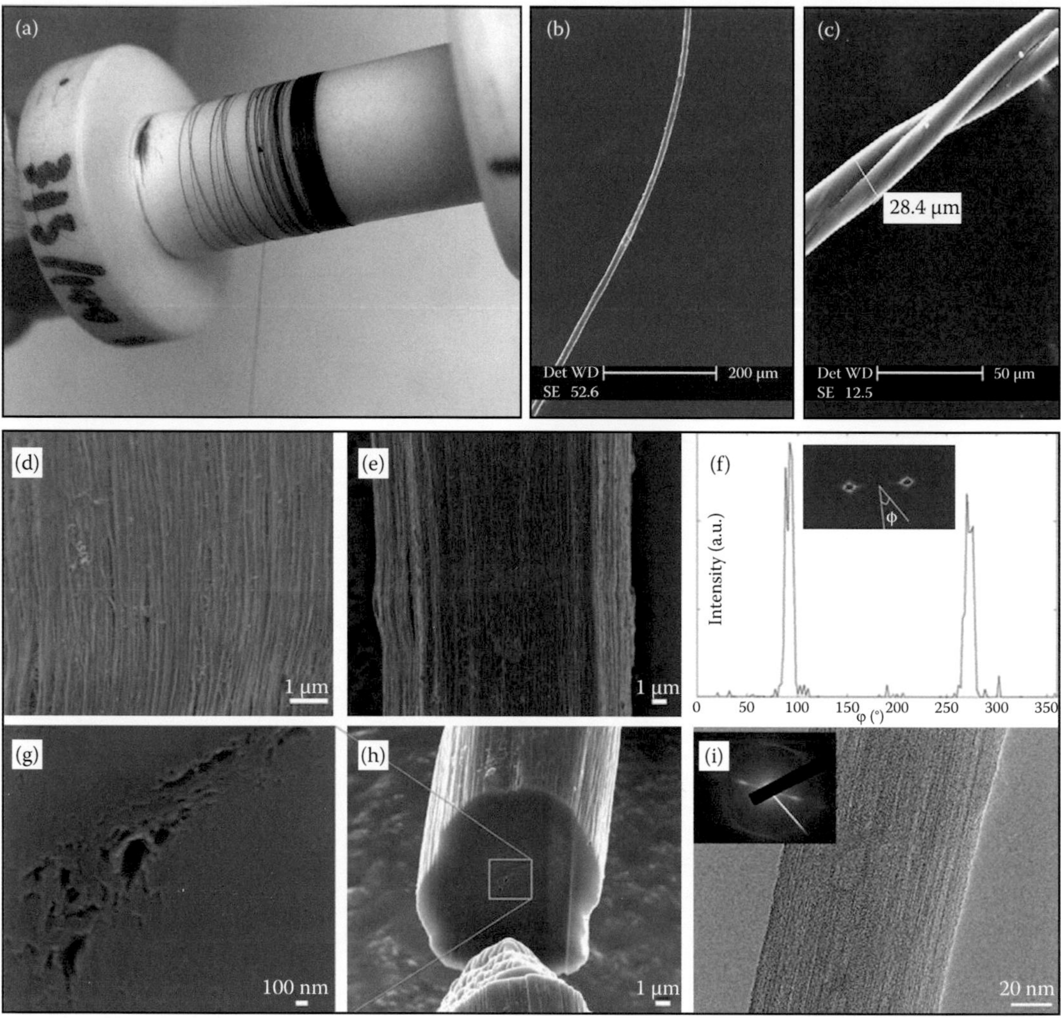

FIGURE 14.1 CNT fiber: (a) unwinding from spool; (b) SEM image of one-thread fiber; (c) SEM image of two-thread fiber; (d, e) high- and low-magnification SEM images showing the typical morphology composed of ~100-nm-thick fibrils aligned along the fiber axis; (f) single-fiber x-ray diffraction (inset) and azimuthal scan showing the high fiber alignment; (g, h) SEM images of cross-section after cutting by focused ion beam showing few hundred-nanometer-sized voids; (i) single-fibril TEM micrograph and electron diffraction (inset) showing near-crystalline packing within a fibril. (a–c, Courtesy of M. J. Schulz and V. Shanov, University of Cincinnati; d–i, From Behabtu, N. et al., *Science*, 339, 182–186, 2013.)

respectively, making special emphasis on the parameters that affect them. Section 14.6 presents the thermal response of CNT fibers. Existing and proposed applications of CNT fibers are presented and considered in Section 14.7.

14.2 Fabrication

The fabrication methods of CNT fibers or yarns that offer potential for realization of the essential properties of CNTs are briefly described next. The primary method to fabricate CNT fibers is through spinning from a CNT forest, which involves two steps (Zhang, Atkinson, and Baughman 2004). First, a CNT forest is grown and then CNT fibers are drawn from it. As the CNTs at the edge of the forest are drawn out, the neighboring CNTs follow the same, thus forming a long, continuous CNT fiber or ribbon. This initial drawing process was modified by adding a twist during nanotube drawing, which yields interlocked CNT fibers (Zhang, Atkinson, and Baughman 2004; Zhang et al. 2007b). The tensile strength values of CNT yarns produced by the forest spinning are still lower than those of high-strength fibers such as carbon or Kevlar. However, the ultimate strains of CNT fibers are larger than those of most commercial fibers because of the nonmonolithic nature of the latter (Jia et al. 2011). The degree of spinnability of CNTs is closely related to the morphology of CNT arrays (Zhang et al. 2009; Huynh and Hawkins 2010). This spinning method is not very productive as the maximum length is limited by the amount of CNTs that can be synthesized on the substrate. A schematic of this fabrication process is shown in Figure 14.2a. An optical image of the twist insertion following the drawing of the nanotubes and a scanning electron microscope (SEM) image of the CNT fiber itself are shown in Figure 14.2b and c, respectively.

Direct spinning from a CNT aerogel is a highly productive method, which involves mechanical drawing of a continuous CNT fiber from a highly porous carbon cluster known as the CNT aerogel (Liu, Ma, and Zhang 2011). A schematic of this process is shown in Figure 14.3. Fibers and ribbons of CNTs are spun directly from the chemical vapor deposition (CVD) synthesis zone of a furnace through the appropriate choice of reactants, control of the reaction conditions, and continuous withdrawal of the product with a rotating spindle used in various geometries (Li, Kinloch, and Windle 2004).

In the wet spinning or solution spinning process, a polymer solution is extruded into a bath that contains a second liquid in which the solvent is soluble but the polymer is not (Zhang et al. 2008a). A schematic of this method is shown in Figure 14.4. In this method, single-wall CNTs (SWCNTs) are homogeneously dispersed in aqueous solutions of sodium dodecyl sulfate (SDS), which can be adsorbed onto the surface of the nanotube bundles stabilizing the nanotubes against van der Waals attractions.

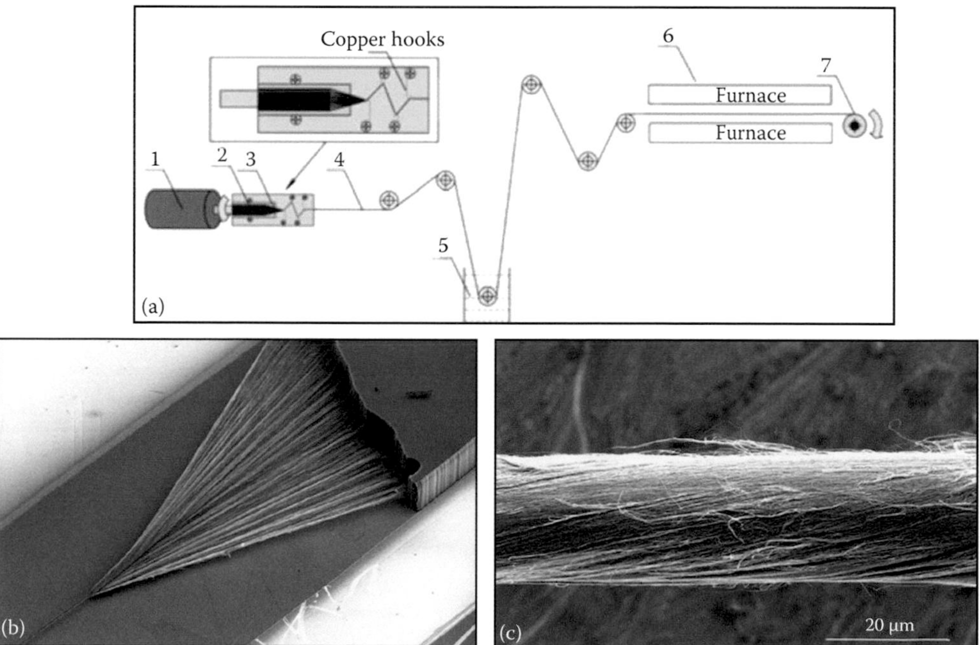

FIGURE 14.2 (a) Schematic illustration of the spinning process: (1) high-speed rotating motor for twisting, (2) CNT array, (3) CNT film, (4) twisting fiber, (5) vessel filled with coating solution, (6) tube furnace, (7) rotating motor for fiber collection. (b) Optical image showing the fiber formation by twist insertion. (c) SEM image of the corresponding spun fiber. (a, From Wei, H. et al., *Nano Res.*, 6, 208–215, 2013; b, c, From Miao, M., *Particuology*, 11, 378–393, 2013.)

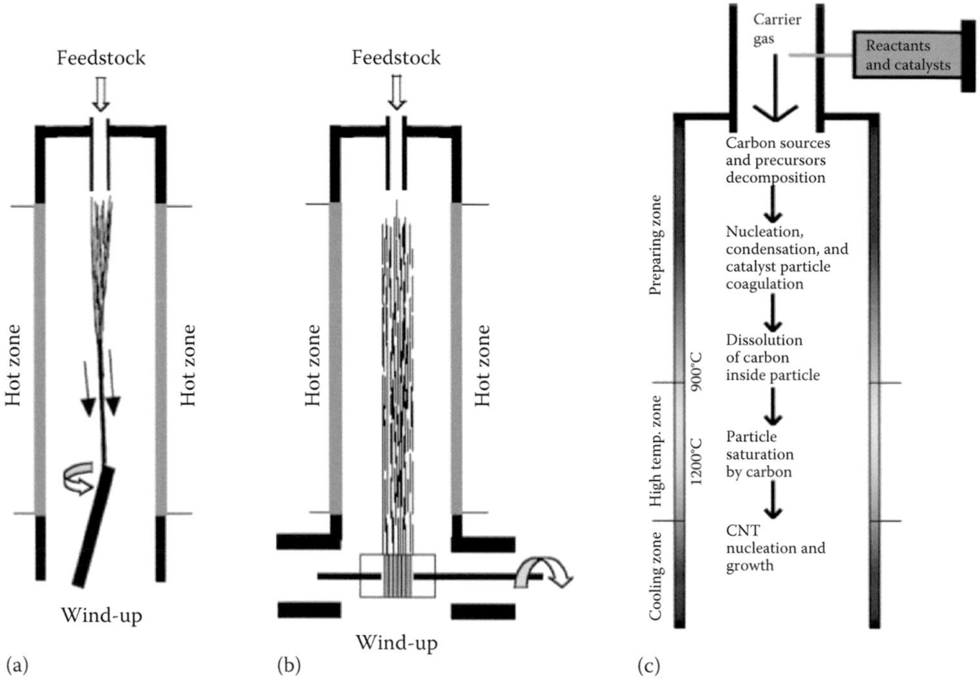

FIGURE 14.3 Schematics of the direct spinning process: (a) drawing of a CNT fiber, (b) drawing of a CNT ribbon, and (c) the reaction steps of CNT synthesis in the vapor phase. (a, b, From Li, Y.L. et al., *Science*, 304, 276–278, 2004; c, From Moisala, A. et al., *J. Phys. Condens. Mat.* 15, S3011–S3015, 2004; a, b, c, From Park, J., Lee, K-H., *Korean J. Chem. Eng.*, 29, 277–287, 2012.)

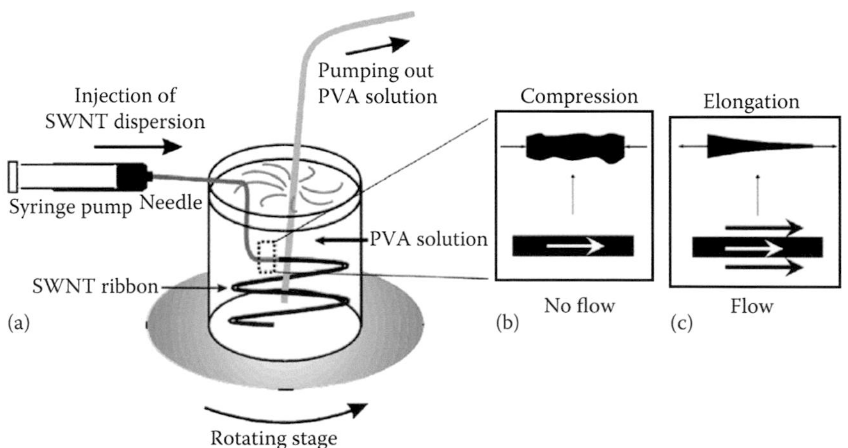

FIGURE 14.4 Schematics of wet spinning process: (a) rotating bath used for coagulating surfactant-dispersed SWNTs into a fiber; (b) a net compressive force acts on the proto-fiber compromising alignment when the coagulation bath is not flowed; (c) a net tensile force results and increases alignment when the coagulant flows along with the extruded fiber. (From Behabtu, N. et al., *Nano Today*, 3, 24–34, 2008.)

This SWCNT solution is injected through a syringe into the coflowing stream of polymer solution containing polyvinylalcohol (PVA). The PVA adsorbs onto the nanotubes and displaces some SDS molecules to form a nanotube ribbon (Lu et al. 2012). Superacids are the only known solvents for SWNTs (Behabtu, Green, and Pasquali 2008). Another method to produce pure multiwall CNT (MWCNT) fibers without using superacid is by using ethylene glycol and a diethyl ether bath (Zhang et al. 2008a).

CNT fibers could also be synthesized by twisting or rolling of CNT films. The SWCNT film is first prepared by a floating catalyst CVD method (Ma et al. 2007, 2009). A CNT strip is then sliced from this thin film and twisted into a CNT fiber. However, the continuous production of CNT fibers is limited to the size of the CNT films. High-quality double-walled CNT (DWCNT) thin films are also fabricated in one step by the catalytic CVD gas-flow reaction process with acetone as the carbon source in an argon flow. The DWCNT film is self-supported and consists of preferentially aligned high-quality DWCNT bundles. This film is then rolled into a fiber to form a multiple membrane structure (Feng et al. 2010). A schematic of this fabrication method and images of the actual fibers are shown in Figure 14.5.

Compaction of a CNT fiber is an efficient post-treatment method done by adding a solvent and then evaporating it. This increases the van der Waals forces between individual nanotubes and the total contact area among them. The capillary force attracts adjacent nanotubes during evaporation, thus resulting in a CNT fiber with increased linear density (Liu et al. 2010). Twisting is a method to increase the strength of a CNT fiber. However, large twisting angles decrease the strength (Liu et al. 2010), so careful optimization is necessary. Introduction of covalent bonds between the nanotubes can also increase the fiber's strength. Applying chemicals and/or heat can achieve this (Park and Lee 2012). More recently,

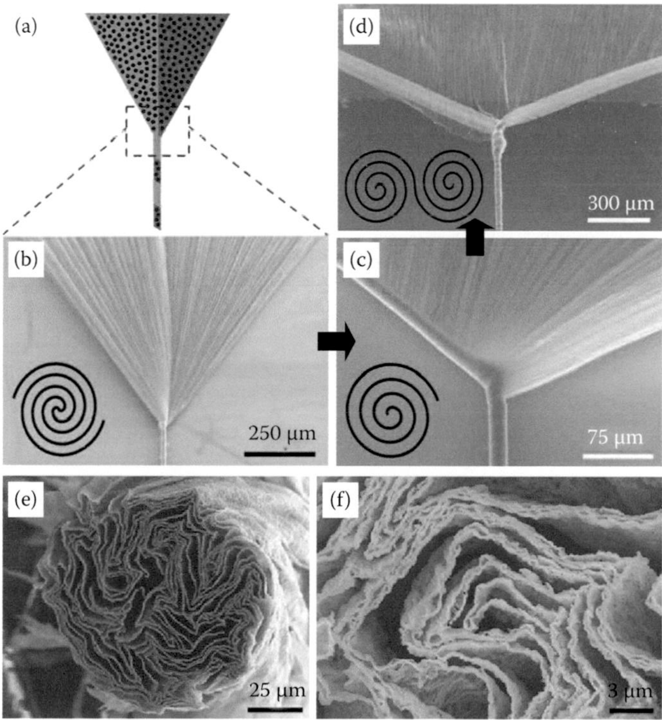

FIGURE 14.5 (a) Illustration of biscrolling by twist insertion in a spinning wedge (black dots represent guest particles). SEM micrographs of (b) Fermat-type twist insertion during spinning from a MWCNT forest (inset illustrates a Fermat scroll); (c, d) guest-free spinning wedges showing an Archimedean scroll and dual Archimedean scrolls, respectively (insets illustrate the Archimedean scrolls). (e, f) SEM images of the cross-sections of biscrolled fibers. (From Lima, M.D. et al., *Science*, 331, 51–55, 2011.)

gamma irradiation of the CNT fiber has been shown to also increase its tensile strength (Miao et al. 2011). The electrical conductivity of pure CNT yarns can be improved by metal doping but significant improvements are yet to be realized (Randeniya et al. 2010).

14.3 Mechanical Response

The mechanical response of CNT fibers determines their potential use as reinforcements and many applications that require maintaining their integrity at relatively high strain or stress levels. The mechanical properties of CNT fibers are affected by parameters and phenomena occurring at various scales. At the nanoscale, the structure of the CNTs, including their length, number of walls, thickness, and defects, plays a significant role. The spatial entanglements of CNTs, their propensity to pack and collapse, and the load transfer mechanisms between them affect the mechanical properties at the microscale. At the level of the CNT fiber, its diameter, degree of twist, and a variety of processing parameters affect the mechanical properties.

14.3.1 Effect of CNT Parameters

The nanotube length is a parameter that plays an important role in the mechanical properties of the CNT fibers. In general, longer nanotubes yield a larger tensile strength (Figure 14.6a) because of a higher effective load transfer that is achieved because of an increase in intertube contact area, circumferential wrapping (Zhang et al. 2007a,b; Fang et al. 2010), and a smaller number of tube ends per unit length (Zhang et al. 2007b; Zhu et al. 2011). Although CNT fibers have been spun from several millimeter-tall

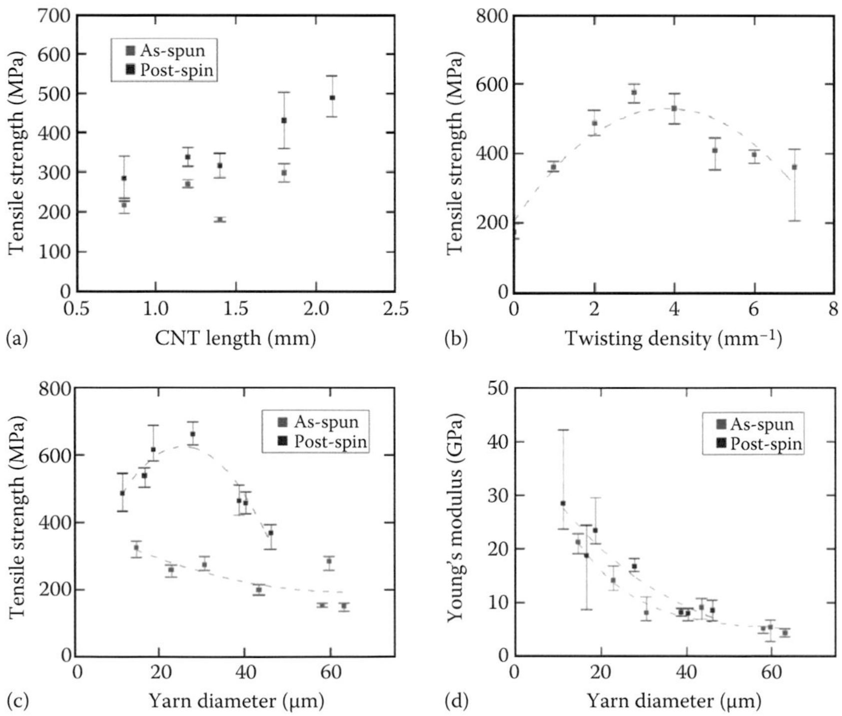

FIGURE 14.6 Tensile strength of CNT fiber: (a) data as a function of CNT length or array height; (b) data and fit as a function of fiber twist; (c) data and fit as a function of fiber diameter. Elastic modulus of CNT fiber: (d) data and fit as a function of fiber diameter. (From Ghemes, A. et al., *Carbon*, 50, 4579–4587, 2012.)

CNT arrays, the tensile strength of these fibers has been reported to be lower than that of fibers spun from shorter CNT array yarns (Jayasinghe et al. 2011). The most spinnable fibers are obtained using CNT arrays that have a height corresponding to the end of the linear height–time growth period of the nanotubes since after that, the array tends to have very thick nanotubes at the bottom, hindering its spinnability (Li et al. 2007; Huynh and Hawkins 2010; Jia et al. 2011).

The diameter and thickness of CNTs and, more importantly, their relative spatial arrangement, which determines the interfacial load transfer between neighboring nanotubes, play also an important role in the mechanical properties of the CNT fiber. It has been shown that the load transfer is maximized in the case of longer, smaller-diameter, and thin nanotubes like single-walled and double-walled ones that are easily collapsible and thus increase the contact area (Elliott et al. 2004; Motta et al. 2005; Jia et al. 2011). This results in an increase in the elastic modulus and the tensile strength of the CNT fibers (Jia et al. 2011).

The morphology of CNT fibers resembles that of a traditional textile yarn, where the individual nanotubes or CNT bundles can be considered analogous to the filaments in a textile yarn. Therefore, the established mechanics of traditional yarns apply to CNT fibers, although the frictional forces between the nanotubes differ somehow. The effect of CNT waviness was investigated, and it was determined that the elastic modulus increases and the ultimate strain decreases with decreasing tube waviness. However, fibers with different CNT waviness have similar tensile strength because the fibers become straight upon external loading (Jia et al. 2011).

14.3.2 Effect of CNT Fiber Parameters

The effect of the CNT fiber diameter has been investigated and determined to play also a role in its mechanical properties. Experimental results show that the tensile strength increases almost linearly with decreasing fiber diameter for the same twist level (surface twist or helix angle) while varying the width of array side wall that is spun and the amount of inserted twist in turns per unit fiber length (Fang et al. 2010). This effect could be attributed to the larger number of joined CNT ends that are weakly connected and to the amount of inserted twist in turns per meter of fiber, which is smaller in larger diameter fibers, resulting in less compacted nanotubes in the fiber (Fang et al. 2010; Liu et al. 2010). Other studies showed that the tensile strength and the elastic modulus increase with fiber diameter by keeping constant the amount of inserted twist in turns per unit length instead of the helix angle (J. Zhao et al. 2010; Ghemes et al. 2012). This is attributed to a smaller compaction after twisting inside a thinner fiber because of a smaller radial pressure.

The degree of twist in the CNT fiber, which is given by the angle between the longitudinal direction of the CNT fiber and that of individual nanotubes, affects the load transfer between nanotubes and, thus, the mechanical properties of the CNT fiber. Increasing the twist angle of the CNT fiber after being initially spun is an effective method to densify them and improve the friction coefficient between the nanotubes and consequently increases the fiber tensile strength according to Equation 14.1. The cosine function in Equation 14.1 would yield a reduction in tensile strength, but the experimental results show that its effect is minimal. The experimental observations show that the tensile strength of the CNT fiber initially increases with increasing twisting angles, followed by a decrease for larger twist angles as seen in Figure 14.6b (Fang et al. 2010; Ghemes et al. 2012). A double peak has been reported in the tensile strength of CNT fibers spun from thin-walled nanotubes at a large twist angle (J. Zhao et al. 2010) when the nanotubes collapse, reducing the bundle cross-sectional area and making them all contribute to take the load (Koziol et al. 2007; Zhang and Li 2010). Other studies reported a decrease in both the elastic modulus and the tensile strength with increasing twist angle in the range of 10°–40°, while the ultimate strain increases (Liu et al. 2010). The response of the CNT fiber is thus similar to that obtained in traditional textile yarns, although the individual nanotubes in a CNT fiber are hollow compared with the solid filaments in a textile yarn.

The tensile strength of the CNT fiber, σ_y, can be expressed by (Zhang, Atkinson, and Baughman 2004):

$$\sigma_y \cong \sigma_f \cos(2\alpha)(1 - k\operatorname{cosec}(\alpha)), \tag{14.1}$$

where σ_f is the tensile strength of the filaments (nanotubes in this case), α is the twist angle of the surface filaments with the yarn axis, and k is a constant given by

$$k \cong \left(\frac{dQ}{\mu}\right)^{1/2} \frac{1}{3L}, \tag{14.2}$$

where d is the filament diameter, Q is the filament migration length, which represents the distance along the yarn over which a filament shifts from the yarn surface to the deep interior and back again, μ is the coefficient of static friction between the filaments, and L is the filament length.

14.3.3 Enhancement of Properties through Post-Processing Treatment

The mechanical performance of as-spun CNT fibers can also be improved through liquid densification, which involves imbibing it with a solvent followed by its evaporation (Liu et al. 2010). The elastic modulus and the tensile strength of the CNT fiber could also be improved by one order of magnitude using polymer impregnation, a process by which the load transfer via weak van der Waals forces between the nanotubes is enhanced by both additional intertube load transfer and by the crystallinity of the impregnated polymer (Liu et al. 2010). Raman spectra shows that there is a significant amount of strain transfer between the individual nanotubes and the CNT fiber that also leads to a decrease in the ultimate strain of the impregnated CNT fiber compared with that of the neat CNT fiber (Ma et al. 2009b).

14.3.4 Measured and Calculated Properties

The experimentally obtained tensile modulus, tensile strength, and toughness of the CNT fibers reported in the literature are summarized in Table 14.1. The elastic modulus ranges from 70 to 350 GPa, and the tensile strength ranges from 0.23 to 8.8 GPa. The large variation in these properties is attributed to the different hierarchical structures and post-treatments of the CNT fibers. Other studies show the elastic modulus ranging from 26 to 57 GPa, the tensile strength ranging from 0.6 to 1.8 GPa, and the ultimate tensile strain ranging from 1.7% to 4.3% (Zu et al. 2012). The specific elastic modulus and tensile strength of the CNT fiber clearly distinguish from the same properties of other materials such as metals and composite materials (Figure 14.7).

The tensile strength of the CNT fibers could also be predicted using micromechanics models and Monte Carlo simulations and assuming a statistical distribution. The results corresponding to a micromechanical model where the CNT fiber consists of twisted and densely packed nanotubes show a decreasing tensile strength with increasing twist angle, number of nanotubes in the cross-section, and gauge length of the CNT fiber (Porwal, Beyerlein, and Phoenix 2007; Beyerlein et al. 2009). The predicted tensile strength ranges between 3 and 7 GPa, much higher than the values obtained experimentally. A similar model consisting of parallel, rigid rods that can slide with respect to each other approximately predicts the tensile strength of a CNT bundle (Vilatela, Elliott, and Windle 2011). It appears that larger diameter nanotubes with fewer walls are ideal because of their higher degree of polygonization or higher propensity to collapse as determined by the elasticity theory of a continuum medium (Lu, Chou, and Kim 2011), atomistic simulations (Elliott et al. 2004), and analysis of transmission electron microscopy images (Motta et al. 2007). The previously described helical and aligned CNT arrangements in the fiber models constitute highly idealized configurations. Both theoretical analyses and atomistic simulations

TABLE 14.1 Mechanical Properties of CNT Fibers

			Mechanical Properties			
Treatment	Linear Density (tex)	Volumetric Density (g cm^{-3})	Tensile Strength (GPa)	Specific Tensile Strength (GPa SG^{-1})	Stiffness (GPa)	Specific Stiffness (GPa SG^{-1})
			Electrospinning			
Spun with PVB, annealed (400 °C)	5.01	0.47	0.0127 ± 0.0014	0.027 ± 0.003	3.48 ± 0.28	7.40 ± 0.59
CNTs refluxed in 6 N HNO_3 for 24 h before spinning	–	–	>0.4	–	–	–
Wet-spun fibers from liquid crystalline phase annealed (280°C, 2 h)	–	–	0.15 ± 0.06	–	69 ± 41	–
Nitrogen-doped CNTs	–	–	0.17 ± 0.07	–	142 ± 70	–
102% H_2SO_4 doped	–	0.87–1.1	0.116 ± 0.01	0.1–1.16	120 ± 10	109–138
Vacuum-annealed at 1100°C for 1 h	–	–	–	–	–	–
Spun using HSO_3 Cl, H_2SO_4	–	–	0.05–0.32	–	120	–
Doped with HSO_3 Cl	–	1.3 ± 0.1	1 ± 0.2 (max 1.3)	0.77	120 ± 50 (max 200)	92.3
Doped with HSO_3 Cl and I_2	–	1.4				
			Other Wet-Spun Fibers			
As-made CNT/PVA	–	1.3–1.5 ± 0.2	0.15	0.11	9–15	6–11
HCl used for coagulation annealed at 1000°C	–	–	–	0.65		12
DNA dispersion, annealed (320°C–350°C, 2 h)	–	–	0.101	–	14.5	–
Hyaluronic acid used as dispersant and 0.3 M HNO_3 as coagulant	–	–	0.11 ± 0.003	–	13 ± 1	–
			Dry Spinning, Dirctly from CVD Reactor			
As-made	–	1.7–2	0.4–1.25	0.2–0.75	34	20
As-made	0.03–0.1	0.4–1	0.5–1.5	0.5–1.5	50–150	50–150
As-made (metallic CNTs)	0.03–0.05	–	–	–	–	–
Oxidized (1 h, 400°C, air), soaked (72 h, 30% H_2O_2 and 24 h, 37% HCl) DI water wash, soaked (24 h, 98% H_2SO_4) DI water wash	–	0.28	0.32	1.14	–	–
As previous line, additionally I_2 doped at 200 °C for 12 h	–	0.33	0.64	1.94	–	–
Soaked with DI water, annealed at 200 °C	–	1.8	–	–	–	–
Doped with $KAuBr_4$	–	1.97	–	–	–	–

(Continued)

TABLE 14.1 (CONTINUED) Mechanical Properties of CNT Fibers

			Mechanical Properties			
Treatment	Linear Density (tex)	Volumetric Density (g cm^{-3})	Tensile Strength (GPa)	Specific Tensile Strength (GPa SG^{-1})	Stiffness (GPa)	Specific Stiffness (GPa SG^{-1})
		Dry Spinning from CNT Array				
As-spun (single fiber)	0.01	–	0.15–0.30	–	4.5	–
As-made	–	0.8	0.55	0.69	20	25
As made	–	0.8	–	–	–	–
As-made	–	–	0.85	–	275	–
As-made, twisted	–	–	1.91	–	300	–
As-made arrays grown by water-assisted CVD	–	0.9	–	–	–	–
As-made, arrays grown by water-assisted CVD	–	–	0.35 ± 0.07	–	25 ± 5	–
As-made	0.35–0.9	0.38–0.84	–	0.65	–	18
As-made shrunk (and twisted)	–	0.38–0.64	0.6–1.1	1.56–1.71	56	87

Source: Lekawa-Raus, A., Patmore, J., Kurzepa, L. et al., *Adv. Funct. Mater.*, 24, 3661–3682, 2014.
Note: SG refers to Specific Gravity.

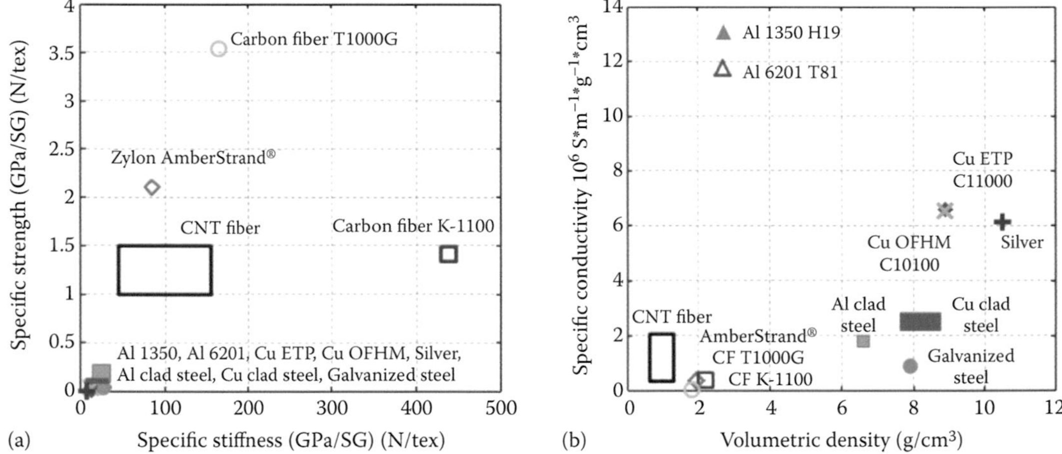

FIGURE 14.7 Comparison of specific mechanical and electrical properties between CNT fibers and those of commonly used metals and fibers in engineering applications such as aluminum 1350 and 6201, copper ETP (electrolytic tough pitch) and OFE (oxygen free electronic), galvanized steel, copper and aluminum clad steels, silver, carbon fiber T1000G, carbon fiber K-1100, and polybenzoxazole (PBO) Zylon fiber clad with conductive metals: (a) specific tensile strength as a function of specific stiffness map; (b) specific electrical conductivity as a function of volumetric density map. (From Lekawa-Raus, A. et al., *Carbon*, 68, 597–609, 2014.)

have found that at large aspect ratios, a CNT is energetically favorable to be self-folded because of the van der Waals interactions between its different parts (Lu and Chou 2011).

The Poisson's ratio of the CNT fibers plays a very important role in the calculation of the true stress, which is needed for an accurate calculation of the tensile strength. Experimental studies have shown that the Poisson's ratio of low-twist CNT fibers is close to 1.0 at 1.2% axial strain, quickly increases to about 7.0–8.0 when the axial strain increases to about 3%, and is largely maintained until the fiber breaks (Miao et al. 2010). High-twist or less porous CNT fibers would tend to exhibit lower Poisson's ratios (Figure 14.8).

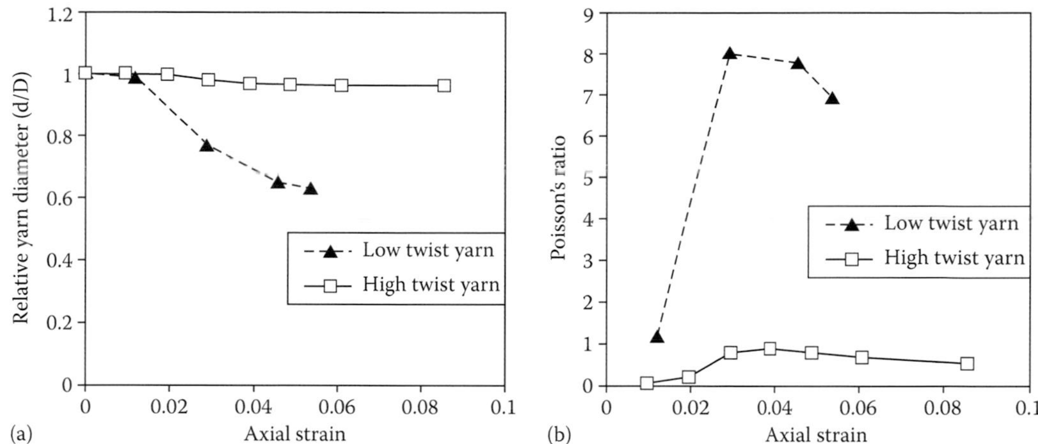

FIGURE 14.8 Experimental data and fit as function of axial strain the CNT fiber: (a) fiber diameter; (b) Poisson's ratio. (From Miao, M. et al., *Carbon*, 48, 2802–2011, 2010.)

The response of the CNT fibers subjected to axial compressive loading is cumbersome to determine due to their inherent difficulties with fiber compressive tests. Most testing methods provide some lateral constraint to the fiber, which clouds its actual pure compressive response. However, studies have been conducted to determine the axial compression of CNT fibers embedded in an epoxy matrix using the single-fiber-composite method (Gao et al. 2010). The compressive stress in the fiber is created through thermal shrinkage, which arises from mismatch of the coefficients of thermal expansion (CTEs) between the polymer matrix and the fiber. The axial compressive stress in the fiber is obtained from Raman spectroscopy analysis, while the compressive strain is obtained by considering the CTEs of the fiber and the matrix. It was determined that the elastic modulus of the CNT fiber is 350 GPa, and it can deform to large strains without permanent deformation or fracture (Gao et al. 2010).

The interfacial shear strength (IFSS) of CNT fibers is also an important mechanical property because it determines their feasibility of being used as reinforcement in composite materials. The IFSS between CNT fibers and an epoxy matrix using a single-fiber fragmentation test yields values in the range of 12–20 MPa (Deng et al. 2011), comparable with those of E-glass/epoxy and carbon/epoxy without fiber surface modification. A more accurate method to determine the IFSS is the microdroplet test, which yields a value of 14.4 MPa using also an epoxy matrix (Zu et al. 2012). SEM imaging reveals that the epoxy infiltrates the surface layer of the CNT fiber and that the shear failure occurs along the interface between CNT bundles. Because of the statistical increase in failure with increase in volume, CNT fibers may not achieve the strength of the constituent nanotubes. However, both the high surface area of nanotubes and the twisting ability of CNT fibers may provide interfacial coupling that mitigates their reduction in strength (De Volder et al. 2013).

14.4 Electrical Response

As in the case of the mechanical response, the electrical response of the CNT fiber does not preserve the superior electrical conductivity of individual nanotubes, which exhibits a ballistic conduction response. This may be attributed to the nonuniformity of individual nanotubes in terms of diameter, length and number of walls and chirality, impurities like amorphous carbon, as well as to microscale effects like nanotube misalignment, entanglements, or low densification. The electrical conductivity of a pure CNT fiber is dominated by the electrical properties of the nanotubes, the intertube contact resistance, and the applied temperature. Coating, doping, and post-processing treatment of the CNT fiber have also been demonstrated as effective techniques to modify their electrical response.

To ensure a reliable comparison of the electrical properties of widely used conducting materials, the specific conductivity could be determined by (Miao 2011; Zhao et al. 2011):

$$\sigma' = \frac{GL}{LD} 10^9, \tag{14.3}$$

where σ' is the specific conductivity (in Siemens per meter over gram per cubic centimeter), G is the conductance (in Siemens), L is the length (in meters), and LD is the linear density or volumetric density multiplied by the cross-sectional area of the CNT fiber (in tex). Crucially, the specific conductivity takes into account the volumetric density, which ponders the weight of an electrical device when selecting a conductive material, particularly in aerospace applications or overhead power lines.

14.4.1 Effect of Electrical Properties of Individual Nanotubes

The electrical properties of individual nanotubes play a significant role in determining the electrical properties of CNT fibers. Nanotubes may be either metallic or semiconducting depending on their atomistic structures. Metallic nanotubes possess better electrical conductivity than semiconducting nanotubes (Bernholc et al. 2002). Thus, CNT fibers spun from pure metallic nanotubes are anticipated to have higher electrical conductivity.

14.4.2 Effect of Contact Resistance between Nanotubes

The highest electrical conductivity of CNT fibers is still at least two orders of magnitude lower than that of individual nanotubes, and this is mostly because of the contact resistance between neighboring nanotubes. The contact resistance between nanotubes in the fiber depends strongly on the contacting area as well as the intertube spaces (Miao 2011; Jakubinek et al. 2012). Increasing the CNT length and increasing CNT alignment are two effective ways of enhancing the intertube conducting areas, which in turn increases the electrical conductivity of CNT fibers (Li et al. 2007). This is likely attributed to the fact that a fiber made of longer nanotubes has fewer end connections and larger contact area between neighboring tubes. Increasing the CNT alignment under stretching also reduces the contact resistance of the CNT fiber accordingly (Badaire et al. 2004). Another way to minimize the contact resistance between nanotubes is to reduce the intertube space by increasing the twist angle, which results in a CNT fiber with lower porosity and higher conductivity, as shown in Figure 14.9. It is worth mentioning that the specific conductivity of CNT fibers has been found to be independent of CNT fiber porosity (Miao 2011).

14.4.3 Effect of Temperature

The temperature dependence of the electrical resistivity of CNT fibers differs from both metals and semiconductors. In general, the resistivity decreases with decreasing temperature down to a crossover temperature that could vary over a very wide range of temperatures (40 K to well above room temperature), below which it will start increasing (Behabtu et al. 2013; Lekawa-Raus et al. 2014a). The models used to fit the experimental data include variable range hopping, fluctuation-assisted tunneling, or weak/strong localization (Li et al. 2007; Alvarenga et al. 2010; Behabtu et al. 2013), although none of them can describe the turnover response from metallic to semiconducting-like behavior. The presence of a higher share of metallic nanotubes in the CNT fiber, along with better purity, crystallinity, alignment, and packing density of the nanotubes as well as doping, will yield a more metal-like response of the resistance–temperature dependence (Zhang et al. 2007b; Alvarenga et al. 2010; Behabtu et al. 2013). Figure 14.10 shows the temperature-dependent conductivity of annealed CNT fibers and isotropic films, as well as acid-doped and iodine-doped CNT fibers. They show two distinctive regimes: At low

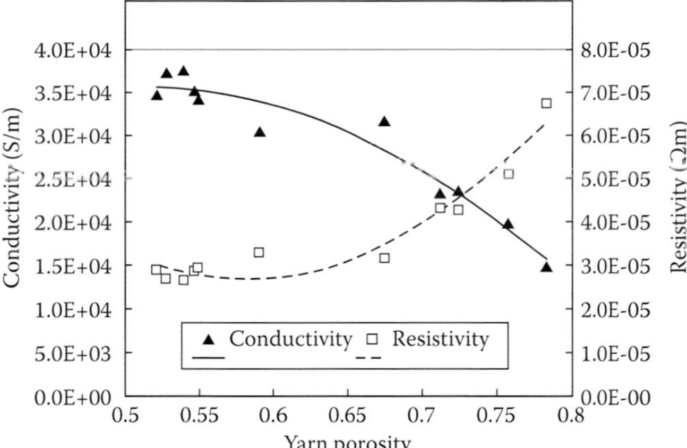

FIGURE 14.9 Effect of yarn porosity on the electrical conductivity and resistivity of CNT fibers: electrical conductivity and resistivity as a function of fiber porosity. (From Miao, M., *Carbon*, 49, 3755–3761, 2011.)

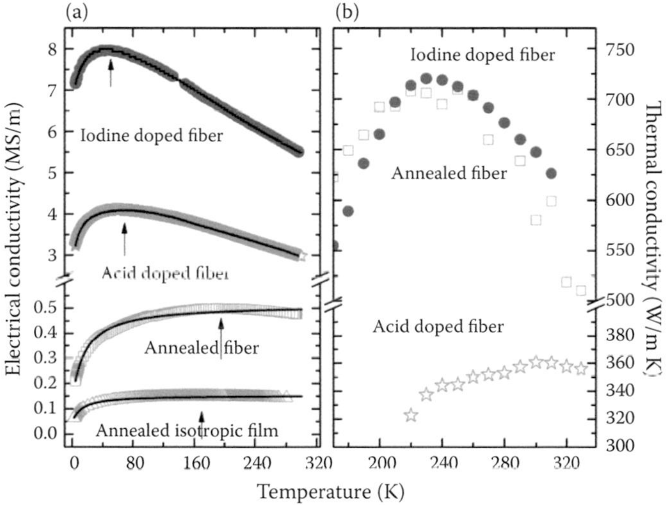

FIGURE 14.10 Temperature dependence of electrical and thermal properties of acid-doped, annealed, and iodine-doped fibers and an annealed random film: (a) electrical conductivity; (b) thermal conductivity. (From Behabtu, N. et al., *Science*, 339, 182–186, 2013.)

temperatures, the conductivity rises with temperature (semiconducting behavior), whereas at high temperatures, the conductivity drops (metallic behavior).

14.4.4 Effect of Coating, Doping, and Post-Processing Treatment

The electrical response of CNT fibers can be further modified through its coating and doping with metallic nanoparticles such as Cu, Au, or Pt. The nanoparticles are doped on the surfaces of CNT fibers using the self-fuelled electrodeposition method. Experimental results show that the electrical conductivity of Cu- or Au-coated CNT fibers at room temperature is about 600 times greater than that of pure CNT fibers (Randeniya et al. 2010). An alternative doping approach is to introduce iodine atoms within the CNT fibers (Zhao et al. 2011; Behabtu et al. 2013). Post-process annealing of the CNT fiber at 500°C in high vacuum of above 10^{-6} torr for several hours should remove most air and non-CNT contaminants of the CNT fiber and thus increase its electrical conductivity.

14.4.5 Measured Electrical Properties

The electrical conductivity values of the CNT fiber reported in the literature vary widely from about 10 to 6.7×10^4 S cm^{-1} depending on the fabrication and treatment conditions, as summarized in Table 14.2 (Lekawa-Raus et al. 2014a). CNT fibers spun with the aid of a polymer will have a conductivity of about 10 S cm^{-1}. Annealing it will increase the conductivity to a maximum value of approximately 200 S cm^{-1}, and the use of hydrochloric acid as coagulant will improve the conductivity to about 140 S cm^{-1}. The electrical conductivity of CNT fibers drawn from CNT forests reaches hundreds of S cm^{-1}, which is higher than that obtained in the case of CNT fibers spun via wet methods, although lower than the thousands of S cm^{-1} in the case of fibers spun directly from a CVD reactor. The latter is, in turn, several times higher than that of most CNT fibers produced using other dry spinning methods and is in the range of the standard fibers spun from superacid liquid crystals. The adsorption of oxygen and air humidity cause an increase in the electrical conductivity of CNT fibers. Up until now, the highest absolute electrical conductivity has been reached in very fine CVD furnace-grown CNT fibers purified with strong acids, 3.0×10^4 S cm^{-1}, or doped with iodine, 5.5×10^4 S cm^{-1}, which are an order of magnitude higher than those of CVD-grown and standard liquid crystalline spun fibers (Behabtu et al. 2013). It has been reported that CNT yarns doped with iodine may reach a specific electrical conductivity of 19.6×10^6 S m^{-1} g^{-1} cm^3, exceeding the specific electrical conductivity of aluminum and copper, which amounts to 14.15×10^6 S m^{-1} g^{-1} cm^3 and 6.52×10^6 S m^{-1} g^{-1} cm^3, respectively (Zhao et al. 2010). However, the values of specific electrical conductivity reported for undoped material are still below that of copper (Alvarenga et al. 2010).

14.5 Piezoresistive Response

The piezoresistive response of CNT fibers is complex and of significant relevance toward their use in sensors for structural health monitoring and other applications. Initial theoretical and modeling studies on the piezoresistive response of assemblies of CNTs with diameters between 0.5 and 4 nm, with each chiral index given equal weight, show that their piezoresistance is strongly dependent on chirality and strain level (Cullinan and Culpepper 2010). For small strains, the resistance is dominated by metallic nanotubes because they have a lower resistance and the change in resistance is small and slightly negative. This is because of the closing of the secondary bandgap. At approximately 0.3% strain, the bandgap in the metallic nanotubes is zero and a new energy gap starts to open, thereby causing the bandgap to increase and the resistance increases between 0.3% and 2% strain. At 2.5% strain, the contributions from the semiconducting nanotubes become significant and the resistance starts to decrease. A few experimental and theoretical studies are already available on the piezoresistive response of CNT fibers. Detailed experiments on laterally unconstrained CNT fibers subjected to uniaxial tension using four-probe measurements showed a definite negative piezoresistivity as observed in Figure 14.11 (Abot, Alosh, and Belay 2014). From a macroscopic perspective, as the CNT fiber is subjected to a tensile stress, the nanotubes that comprise the fiber are brought closer together, reducing the voids between them. As the nanotubes come into closer contact, the electrical resistance of the fiber decreases as the interfacial resistance decreases. However, there are two competing phenomena in this process. As the CNT fiber is loaded, it elongates and its cross-sectional area reduces by an amount proportional to the elongation, the factor of which is determined by the Poisson's ratio of the CNT fiber. As the cross-sectional area decreases, the electrical resistance of the CNT fiber increases. Therefore, as the CNT fiber is loaded, its reduction in cross-sectional area yields a change in resistance that counteracts the reduction of electrical resistance by means of minimizing the interfacial contact resistance of the nanotubes comprising the CNT fiber. As there is a negative piezoresistive response observed in all the tested CNT fiber samples,

TABLE 14.2 Electrical Properties of CNT Fibers

Treatment	Type of CNTs	Length (cm)	Diameter (μm)	Electrical Properties	
				σ (S cm^{-1})	σ' (S m^{-1}/g cm^{-3})
Spun with PVB, annealed (400°C)	MWNT	No limit	–	154 ± 9.6	0.03×10^6
CNTs refluxed in 6 N HNO_3 for 24 h before spinning	SWNT	–	0.2–5	355	–
Wet-spun fibers from liquid crystalline phase annealed (280°C, 2 h)	MWNT	No limit	10–80	80	–
Nitrogen-doped CNTs	MWNT	No limit	10–80	300	–
102% H_2SO_4 doped	SWNT	No limit	Tens of μm	5000	0.5×10^6
Vacuum-annealed 1100°C for 1 h	SWNT	No limit	Tens of μm	500	–
Spun using HSO_3Cl, H_2SO_4	SWNT	No limit	Tens of μm	8333	–
100% H_2SO_4 (24 h, 100°C)	SWNT	No limit	60	4170	–
Annealed at >1000°C	SWNT	No limit	60	382	–
Doped with HSO_3Cl	SWNT	No limit	8–10	29,000 ± 3000	2.2×10^6
Doped with HSO_3Cl and I_2	SWNT	No limit	8–10	50,000 ± 5000	4×10^6
Doped as previous line—annealed at 600°C in Ar and H_2	SWNT	No limit	8–10	4000	–
		Other Wet-Spun Fibers			
As-made CNT/PVA	SWNT	100	2–100	10	0.07×10^4
Spun with PVA, annealed (1000°C in H_2) stretched	SWNT	–	–	200	–
HCl used for coagulation, annealed at 1000°C	SWNT	–	10–50	140	–
DNA dispersion, annealed (320°C–350°C, 2 h)	SWNT	50	20–30	166.7	–
Hyaluronic acid used as dispersant and 0.3 M HNO_3 as coagulant	SWNT	<1000	Tens of μm	537 ± 56	–
		Dry Spinning, Directly from CVD Reactor			
As-made	SWNT	20	50–500	1400	–
As-made	DWNT	No limit	10–200	5000	0.29×10^6
As-made	Mixture	No limit	10–20	3150	0.7×10^6
As-made (metallic NTs)	SWNT	No limit	10–20		2×10^6
Oxidized (1 h, 400°C, air), soaked (72 h, 30% H_2O_2 and 24 h, 37% HCl), DI water wash, soaked (24 h, 98% H_2SO_4), DI water wash	DWNT	–	5.5	20,000	6×10^6
As previous line,	DWNT	–	4.1–44.7	11,200	4×10^6
additionally I_2 doping at	DWNT	–	5.5	67,000	19.6×10^6
200°C for 12 h	DWNT	–	4.1–44.7	23,100	7×10^6

(*Continued*)

TABLE 14.2 (CONTINUED) Electrical Properties of CNT Fibers

Treatment	Type of CNTs	Length (cm)	Diameter (μm)	Electrical Properties σ (S cm^{-1})	σ' (S m^{-1}/g cm^{-3})
Soaked with DI water, annealing at 200°C	–	–	330	2970	0.16 × 10^6
Doped with $KAuBr_4$	–	–	330	13,000	0.66 × 10^6
		Dry Spinning, from CNT Array			
As-made (single yarn)	MWCNT	No limit	2–10	300	–
As-made	MWCNT	No limit	10	300	0.037 × 10^6
As-made	MWCNT	No limit	8.5–10	300	0.037 × 10^6
As-made	MWCNT	No limit	3	595.2	–
Annealed in air 480°C	MWCNT	No limit	3	818.3	–
Oxidized in 5 N HNO_3	MWCNT	No limit	3	969	–
As-made	MWCNT	No limit	4	170	–
As-made, twisted	MWCNT	No limit	3	410	–
As-made, arrays grown by water-assisted CVD	MWCNT	No limit	10–34	400–810	0.04–0.09 × 10^6
As-made, arrays grown by water-assisted CVD	MWCNT	No limit	12 × 5	500	–
As-spun	MWCNT	No limit	12–27	150–370	0.045 × 10^6
As-spun shrunk (and twisted)	MWCNT	No limit	4–34	500–900	0.13–0.14 × 10^6
		Dry Spinning, from CNT Cotton			
Oxidized (350°C, 24 h), immersion (37% HCl, 48 h)	DWCNT	1	5–20	5900	–

Source: Lekawa-Raus, A., Patmore, J., Kurzepa, L. et al., *Adv. Funct. Mater.*, 24, 3661–3682, 2014.

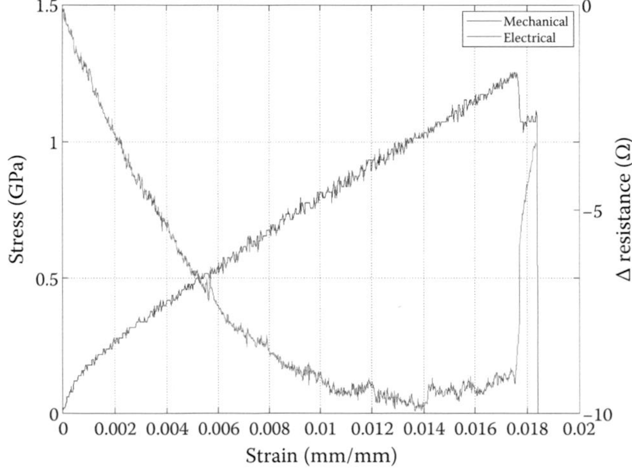

FIGURE 14.11 Coupled mechanical and electrical responses of the CNT fiber: stress as a function of strain (ascending curve) and resistance as a function of strain (descending curve). (From Abot, J. L. et al., *Carbon*, 70, 95–102, 2014.)

it can be concluded that for at least some initial portion of loading, the reduction in the electrical resistance by means of minimizing the interfacial contact resistance of the nanotubes in the CNT fiber is the dominant factor in its piezoresistive response. As the CNT fiber is further loaded, there is a point in which the nanotubes are brought into a minimally spaced distance from each other, which corresponds to a minimum value of the resistance, as the interfacial resistance between the nanotubes can no longer be reduced. The point of inflection (i.e., the point at which the resistance reaches its minimum) precedes the ultimate failure of the CNT fiber. There is a subsequent failure of individual nanotubes that will either further reduce the cross-sectional area of the CNT fiber or introduce electrically entropic pathways in the CNT fiber, both of which will increase the electrical resistance of the CNT fiber. The electrical resistance corresponding to this subresponse exhibited a strong and sharp linear increase that could be attributed to a sizable decrease in the cross-sectional area of the CNT fiber.

More recently, a comprehensive study of the piezoresistive response of CNT fibers spun through the floating catalyst CVD process was performed including the effects of time-dependent loading and humidity (Lekawa-Raus, Koziol, and Windle 2014c). It was shown that the resistance increases with increasing strain and stress. Initially, the resistance increases exponentially, consistent with the opening of bandgaps of metallic nanotubes and the decrease in the contact area of the CNT bundles. Afterward, the resistance increases proportionally to the strain and the effect of change of contact area of CNT bundles dominates the resistance decrease resulting from the straining of semiconducting nanotubes (Figure 14.12). The application of cyclic loading as well as stress relaxation or creep to the CNT fiber leads

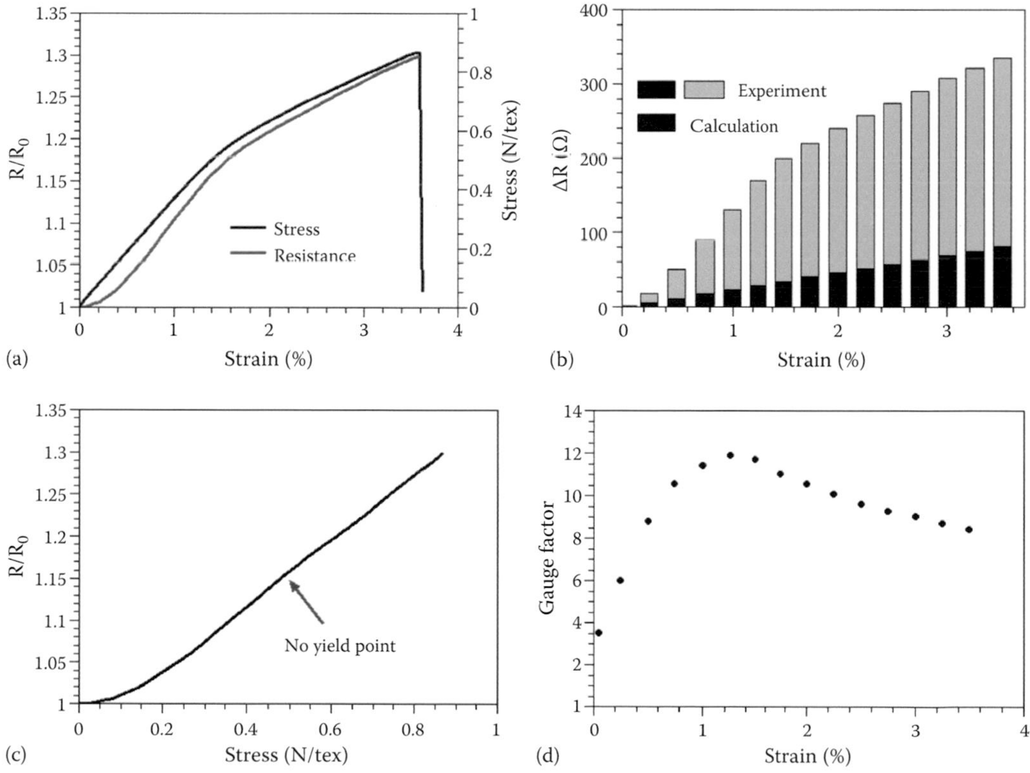

FIGURE 14.12 Piezoresistive response under cyclic loading (10 cycles to 1% at speed of 0.5 mm/min): (a) relative resistance- and stress–strain curves; (b) relative resistance as a function of strain; (c) relative resistance as a function of stress; (d) gauge factor as a function of strain. (From Lekawa-Raus, A. et al., *ACS Nano*, 8, 11214–11224, 2014.)

to an additional decrease of the electrical resistivity effects. The uptake of water by the CNT fiber has also been shown to lead to an additional decrease in the electrical resistivity.

14.6 Thermal Response

As previously mentioned, the thermal properties of CNT fibers are not as superb as those of the individual nanotubes. Most of the reported values on the thermal conductivity of the CNT fiber range from 5 to 60 W m^{-1} K^{-1} for the axial direction (Ericsson et al. 2004; Jakubinek et al. 2012), which are much lower than the axial thermal conductivity of individual nanotubes (Figure 14.13). Recent experimental results show that CNT fibers doped with chlorosulfonic acid during spinning could reach a thermal conductivity of 380 W m^{-1} K^{-1}, and further doping could increase it to 635 W m^{-1} K^{-1} (Figure 14.13), which is already much higher than that of metals used in electrical wiring (Behabtu et al. 2013). A few experimental results are available on other thermal properties of the CNT fiber spun from a forest including the thermal diffusivity, 62 mm^2 s^{-1}, and the heat capacity, 700 ± 50 J kg^{-1} K^{-1} (Aliev et al. 2007). Such low values of axial thermal conductivity and diffusivity in CNT assemblies are mostly attributed to morphology issues including defects in the nanotubes, interconnections of CNT bundles, and misalignment. More

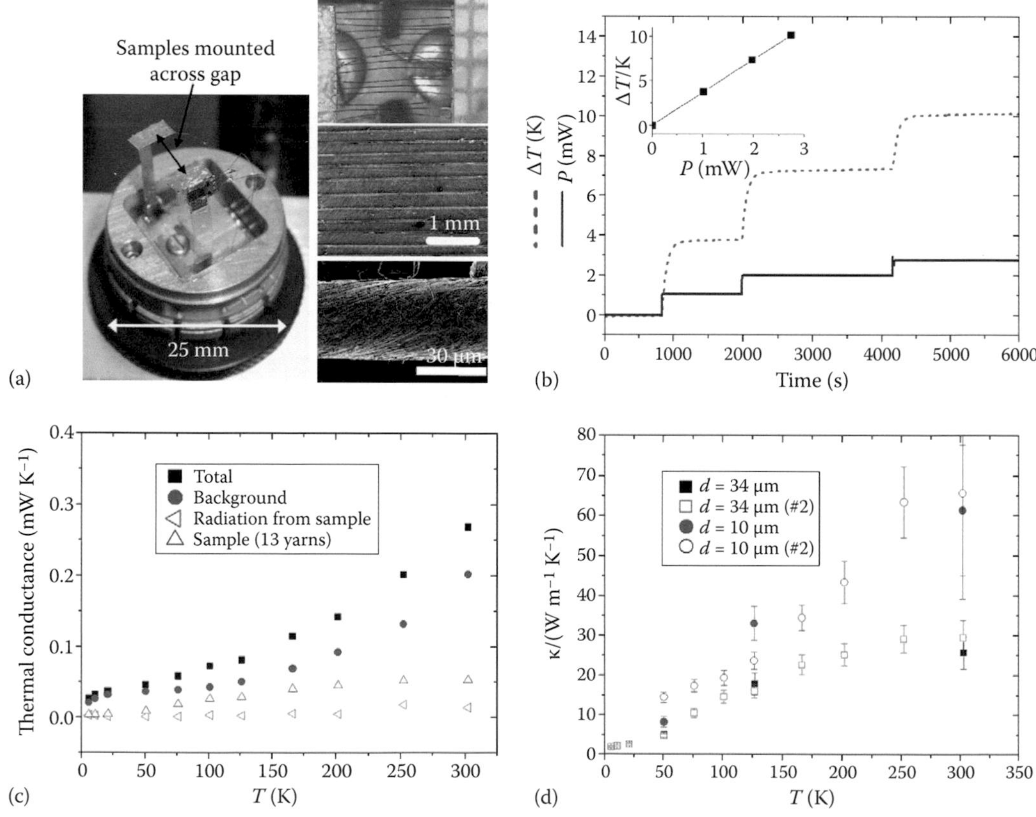

FIGURE 14.13 (a) Experimental setup for parallel thermal conductance measurements and images of 34-μm-diameter CNT fiber segments in parallel arrangement; (b) power and temperature histories for a single conductance measurement; (c) thermal conductance as a function of temperature data of thirteen 34-μm-diameter CNT fibers; (d) thermal conductivity as a function of temperature data of CNT fibers. (From Jakubinek, M.B. et al., *Carbon*, 50, 244–248, 2012.)

theoretical calculations and experimental measurements are still needed to shed light into the actual thermal response of CNT fibers and in particular the intertube thermal transport.

14.7 Applications

The engineering applications of CNT fibers will continue to expand and one day perhaps reach large-scale use when their fabrication techniques and properties could be enhanced. Among the most promising applications are transmission lines, sensors, energy storage devices, biotechnology sensors, artificial muscles, and many others.

14.7.1 Electricity Transmission

CNT fibers could become the next-generation power transmission lines because of the superb electrical conductivity of metallic nanotubes that form the CNT fibers. Electrons move ballistically in the nanotubes and do not scatter as they do in conducting metallic materials such as copper and aluminum. Because scattering increases the thermal resistance of metals, it causes power lines to heat up, expand, and then sag. Figure 14.14 shows the relative resistance-temperature curves of a CNT fiber made of purely armchair SWNTs, a fiber spun directly from a CVD furnace with mixed types of nanotubes and a standard copper wire. Often, the resistivity-temperature dependence around room temperature may be simplified to a linear function and the temperature coefficient of resistivity, α, is found as the gradient of this function, measured in units of K^{-1}. In the case of CNT fibers at room temperature, this coefficient, α, varies between -0.001 and 0.002 K^{-1}, which in absolute terms is approximately half that of copper or aluminum (Zhang, Atkinson, and Baughman 2004; Behabtu et al. 2013). This constitutes a superb characteristic of CNT fibers that may pave the way for their use as transmission lines in the future. Other applications of CNT fibers are presented in Figure 14.15, including a light emitting diode, an Ethernet cable, and a transformer.

14.7.2 Structural Health Monitoring

CNT fibers exhibit piezoresistance characteristics that can be tapped for sensing purposes. Sensors composed of CNT fibers can be integrated in polymeric and composite materials and detect initiating

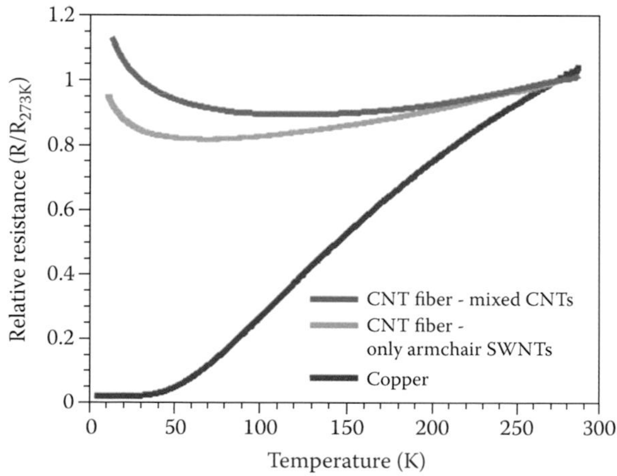

FIGURE 14.14 Temperature dependence of relative electrical resistance of copper and two CNT fibers spun directly from CVD reactor including a standard one and one made purely of metallic SWCNTs (resistance of a given sample at 273 K). (From Lekawa-Raus, A. et al., *Adv. Funct. Mater.*, 24, 3661–3682, 2014.)

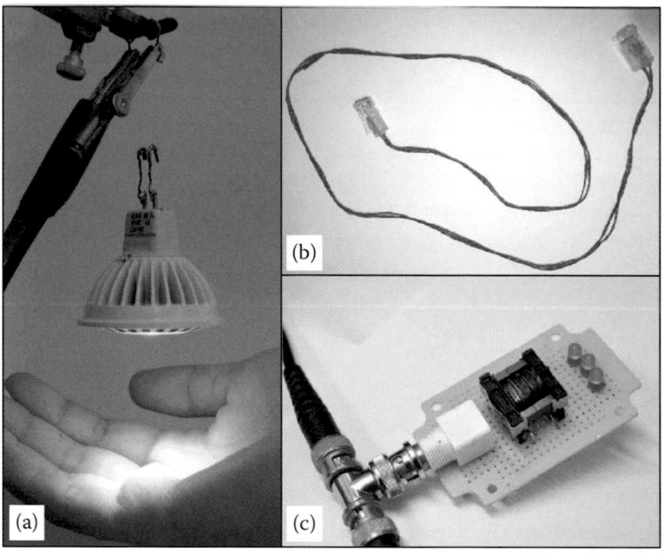

FIGURE 14.15 (a) 46-g light-emitting diode lit and suspended by two 24-mm-thick CNT fibers; (b) Ethernet cable produced using CNT fibers; (c) transformer with copper wiring replaced with CNT fibers. (From Behabtu, N. et al., *Science*, 339, 182–186, 2013; Lekawa-Raus, A. et al., *Adv. Funct. Mater.*, 24, 3661–3682, 2014.)

damage or measure distributed strain. These CNT fiber sensors, for example, can detect damage and delamination through resistance measurements without adding significant weight and without altering the integrity of the host laminated composite material (Abot et al. 2010a,b). It was shown that the CNT fiber sensors capture instantaneously delamination and minor damage as demonstrated by the resistance response as a function of time or displacement, as seen in Figure 14.16. Furthermore, the fiber sensors are shown to be sensitive enough to provide a significant resistance increase output ahead of an impending delamination or minor damage and could thus be used to not only monitor a structure but also to prevent further damage propagation. The experimental results also show the ability of a combination of different CNT fiber sensors to detect the exact location and extent of the delamination throughout the entire loading process. Furthermore, early results indicate the feasibility of CNT fiber sensors of measuring distributed strain inside polymeric or composite materials (Abot et al. 2010a; Zhao et al. 2010). Preliminary modeling results show the potential to assemble strain gauges composed of CNT fiber and the ability to reach gauge factors higher than that of metallic foil strain gauges (Abot et al. 2015).

14.7.3 Energy Storage

The as-spun CNT fiber behaves like a typical electrical double-layer capacitor due to the small charge layer separation within the CNT fiber and to the high nanotube surface area accessible to the electrolyte. The specific capacitance of the as-spun and annealed fibers was determined to be 48 and 100 F g^{-1}, respectively, using a three-electrode electrochemical cell (Kozlov et al. 2005). CNT fibers could also be used to form supercapacitors by coating them with polymeric wires such as PVA and polyaniline (PANI) and further coating them with an electrolyte material such as PVA. Further oxidation of the CNT fiber by gamma irradiation treatment can significantly increase the specific capacitance of the CNT fiber (Su et al. 2014). This improvement can be attributed to the pseudo capacitance characteristics introduced by functional groups developed in the irradiation treatment. Two-thread fiber supercapacitors based on gamma-irradiated CNT fibers and randomly orientated PANI nanowires (IR-CNT@PANI) demonstrated excellent specific capacitance, maintained their capacitance at high current densities, and exhibited increases in tensile strength and electrical conductivity (Figure 14.17). The gram capacitance of the

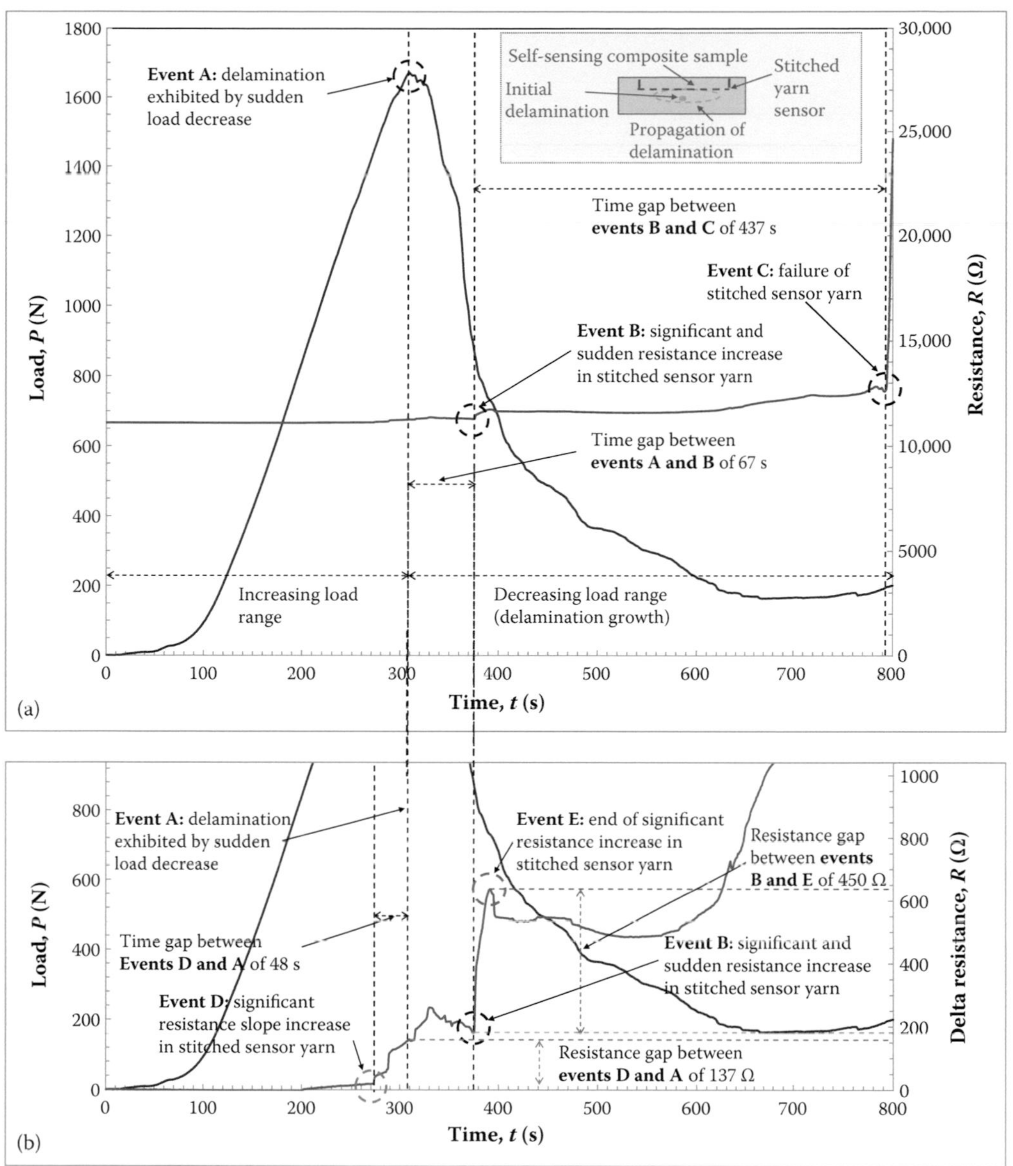

FIGURE 14.16 Detection of mode II-dominated delamination in six-layer glass/epoxy laminated composite sample using a single stitched yarn sensor configuration: (a) load as a function of time (curve starting from 0) and resistance as a function of time; (b) load as a function of time (curve starting from 0) and delta resistance as a function of time. (From Abot, J.L. et al., *J. Multifunc. Compos.* (in press), submitted.)

two-thread fiber supercapacitor was reported to be 78 F g^{-1} or equivalently an area capacitance of 43 mF cm^{-2}. The resulting supercapacitors demonstrate high cyclic charge–discharge stability, high flexibility, and high strength, thus making them suitable for electronic textile applications (Zhang et al. 2014).

Other energy storage devices offering a combination of high energy density and high power density could be built using CNT fiber springs (Hill et al. 2014). When loaded in tension, the CNT fiber stores energy with an energy per unit length of up to 13.4 mJ m^{-1} and an energy density of up to 7720 kJ m^{-3} or 6.7 kJ kg^{-1}. Two electric power supply devices were built, including one that converted stored mechanical energy into electric power using an escapement-based power regulation mechanism and piezoelectric

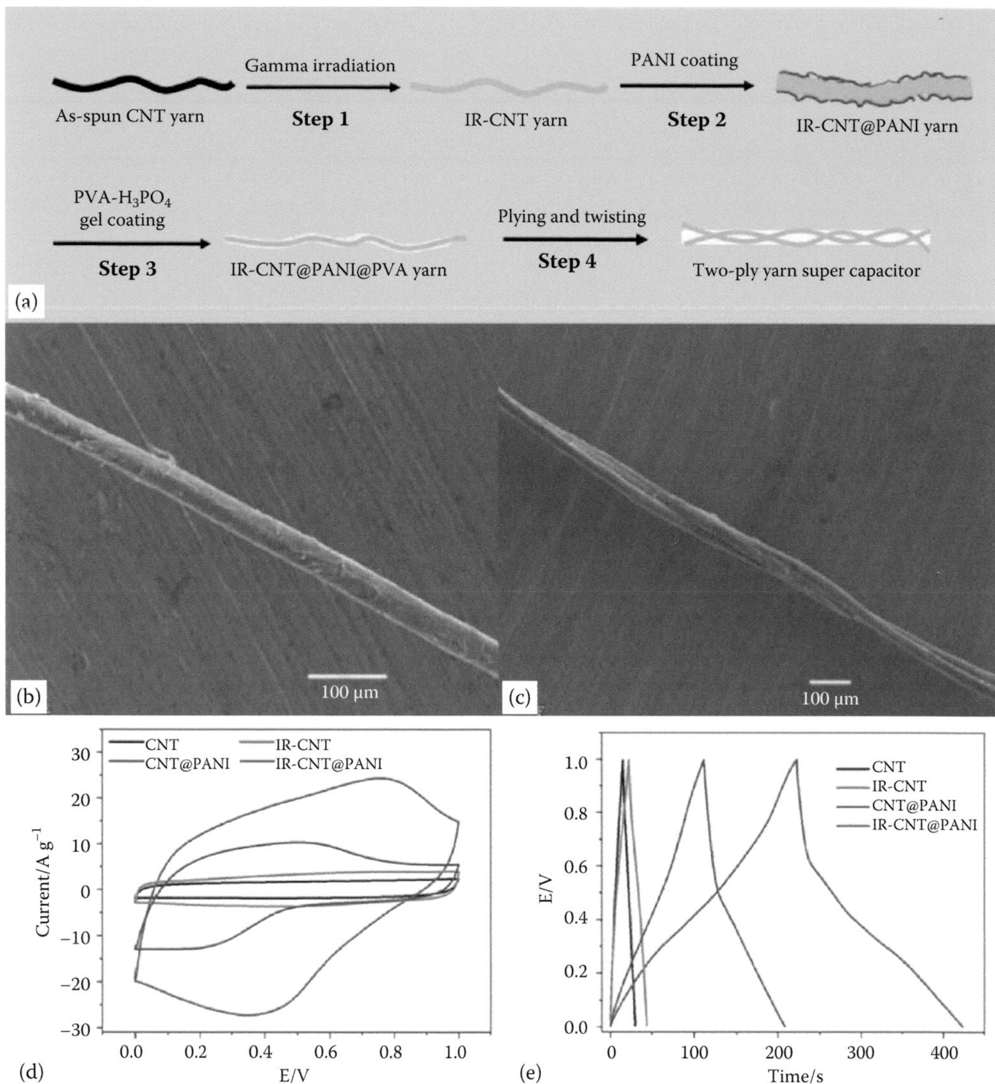

FIGURE 14.17 (a) Schematic illustration of steps to prepare irradiated CNT fiber-polyaniline (IR-CNT@PANI) two-thread fiber supercapacitor; (b) SEM image of IR-CNT@PANI single-thread yarn; (c) SEM image of IR-CNT@PANI two-thread fiber supercapacitor. Electrochemical properties of all four types of two-ply yarn supercapacitors; (d) cyclic voltammograph curves at scan rate 0.5 V s^{-1}; (e) galvanostatic charge/discharge curves at current density 0.2 A g^{-1}. (From Su, F. et al., *ACS Appl. Mater. Interfaces*, 6, 2553–2560, 2014.)

energy conversion, and another one using an electromagnetic conversion system to convert energy stored in the CNT spring to electric output power without regulating the rate of energy release from the spring. Further improvements are expected to increase the energy densities, power densities, and efficiencies of power supplies.

14.7.4 Bioengineering

In addition to their good mechanical and electrical properties, CNT fibers have high specific surface area and good electro-catalytic properties, making them promising for electrochemical devices. CNT fibers resemble electric wires with nanoscale surface topography and porosity, which facilitate molecular-scale interactions with enzymes and other chemicals to efficiently capture and promote electron transfer reactions (Zhu et al. 2012a). Crucially, by forming a continuous structure, CNT fibers may avoid the potential toxicity caused by asbestos-like CNTs when implanted in vivo. Wet-spun CNT fibers using a coagulating polymer like PVA or polyethylenimine were first introduced as microelectrodes for electrochemical sensors to detect nicotinamide adenine dinucleotide, hydrogen peroxide, and dopamine (Wang et al. 2003). It was demonstrated that the CNT fiber microelectrodes could accelerate the redox process of these biomolecules allowing for high sensitivity of low-potential detection. Glucose-sensing electrodes were built by adsorption of a mediator on the surface of a CNT fiber (Zhu et al. 2010, 2012b). Figure 14.18 shows a schematic illustration of a brush-like CNT fiber microelectrode and its working mechanism for sensing glucose. Other microelectrode applications of CNT fibers include deep brain stimulation (Jiang, Li, and Hao 2011) and detection of dopamine in brain slices (Schmidt et al. 2013). CNT fiber-based microelectrodes may become alternatives to carbon fiber microelectrodes for the detection of neurotransmitters because they are more sensitive, exhibit fast electron transfer kinetics, and are more resistant to surface fouling.

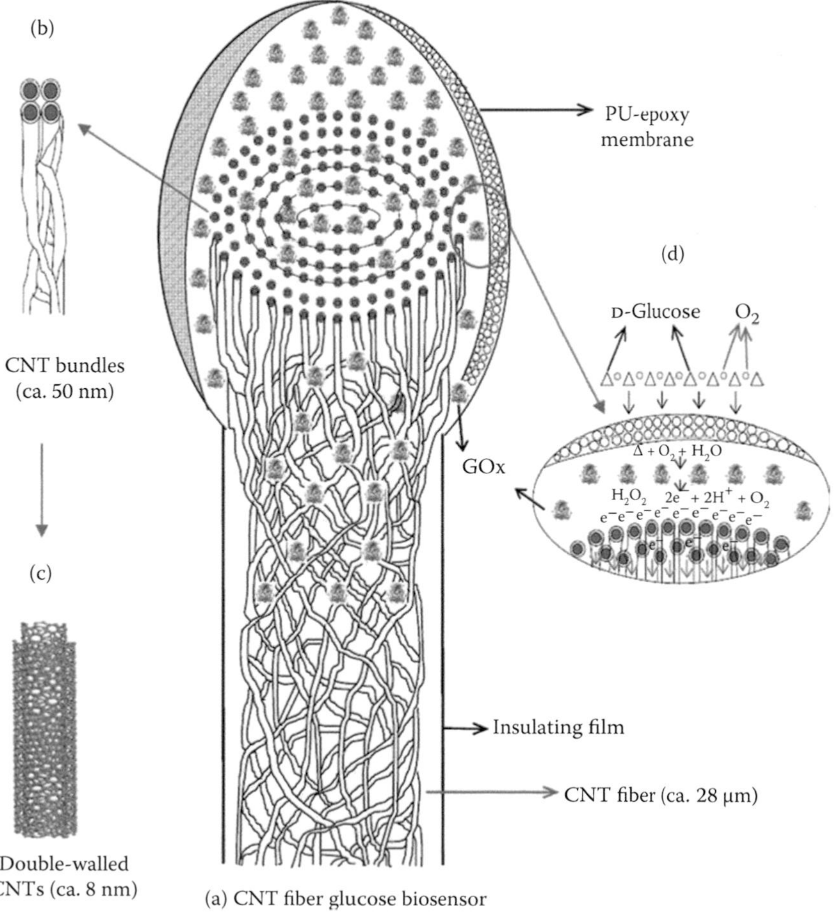

FIGURE 14.18 Schematic illustrations of (a) CNT fiber-based glucose biosensor; (b) CNT bundles; (c) DWCNT; (d) working principle of biosensor. (From Zhu, Z. et al., *Nanotechnology*, 21, 165501, 2010.)

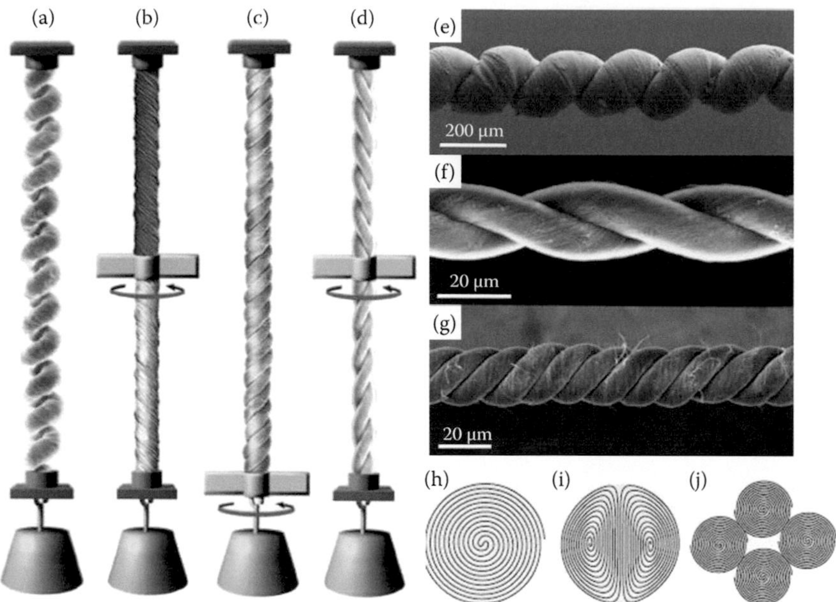

FIGURE 14.19 Muscle configurations and CNT fiber structures for tensile and torsional actuation. Tensile load and paddle positions for (a) a two-end-tethered, fully infiltrated homochiral yarn; (b) a two-end-tethered, bottom-half-infiltrated homochiral yarn; (c) a one-end-tethered, fully infiltrated homochiral yarn; and (d) a two-end-tethered, fully infiltrated heterochiral yarn. SEM micrographs of (e) a fully infiltrated homochiral coiled yarn, (f) a neat two-thread fiber, and (g) a neat four-thread yarn. Illustrations of ideal cross-sections of CNT fibers: (h) Fermat, (i) dual-Archimedean, and (j) infiltrated four-thread Fermat. (From Lima, M.D. et al., *Science*, 338, 928–932.)

14.7.5 Other Applications

Another application of CNT fibers is in artificial muscles that could provide fast, high-force, large-stroke torsional and tensile actuation (Foroughi et al. 2011; Lima et al. 2012). It was demonstrated that torsional motors and contractile muscles with the corresponding sensors could generate more than a million torsional and tensile actuation cycles specifically delivering 11,500 revolutions/minute and a 3% tensile contraction (Figure 14.19). A hydrostatic actuation mechanism explains the simultaneous occurrence of lengthwise contraction and torsional rotation during the CNT fiber volume increase caused by an electrochemical charge injection.

The use of CNT fibers for antennas is also being explored. Measurements of a CNT fiber dipole antenna demonstrate its functionality as a radiofrequency radiator (Schulz, Vesselin, and Yin 2014). Experimental results confirm a 4% downward frequency shift due to kinetic inductance effects and material losses greater than 12 dB at 2.45 GHz, as predicted by the method of moments.

References

Abot, J.L., Kiyono, C.Y., Thomas, G.P. et al., "Strain gauge sensors comprised of carbon nanotube yarn: Parametric numerical analysis of their piezoresistive response," *Smart Mater. Struct.* 24 (2014): 075018.

Abot, J.L., Schulz, M.J., Song, Y. et al., "Novel distributed strain sensing in polymeric materials," *Smart Mater. Struct.* 19 (2010a): 1–19.

Abot, J.L., Song, Y., Medikonda, S. et al., "Delamination detection with carbon nanotube thread in self-sensing composite materials," *Compos. Sci. Technol.* 70 (2010b): 1113–9.

Abot, J.L., Alosh, T. and Belay, K., "Strain dependence of electrical resistance in carbon nanotube yarns," *Carbon* 70 (2014): 95–102.

Abot, J.L., Wynter, K., Mortin, S.P., Borges de Quadros, H., Renner, D.C. and Belay, K. Localized detection of damage in laminated composite materials using carbon nanotube yarn sensors. *J. Multifun. Compos. Special Issue: Novel Sensing Techniques and Approaches in Composite Materials (in press).*

Aliev, A.E., Guthy, C., Zhang, M. et al., "Thermal transport in MWCNT sheets and yarns," *Carbon* 45 (2007): 2880–8.

Alvarenga, J., Jarosz, P.R., Schauerman, C.M. et al., "High conductivity carbon nanotube wires from radial densification and ionic doping," *Appl. Phys. Lett.* 97 (2010): 182106.

Badaire, S., Pichot, V., Zakri, C. et al., "Correlation of properties with preferred orientation in coagulated and stretch-aligned single-wall carbon nanotubes," *J. Appl. Phys.* 96 (2004): 7509–13.

Behabtu, N., Green, M.J. and Pasquali, M., "Carbon nanotube-based neat fibers," *Nano Today* 3 (2008): 24–34.

Behabtu, N., Young, C.C., Tsentalovich, D.E. et al., "Strong, light, multifunctional fibers of carbon nanotubes with ultrahigh conductivity," *Science* 339 (2013): 182–6.

Bernholc, J., Brenner, D., Nardelli, M.B. et al., "Mechanical and electrical properties of nanotubes," *Annu. Rev. Mater. Res.* 32 (2002): 347–75.

Beyerlein, I.J., Porwal, P.K., Zhu, Y.T. et al., "Scale and twist effects on the strength of nanostructured yarns and reinforced composites," *Nanotechnology* 20 (2009): 485702.

Cullinan, M.A. and Culpepper, M.L., "Carbon nanotubes as piezoresistive microelectromechanical sensors: Theory and experiment," *Phys. Rev. B* 82 (2010): 115428.

De Volder, M.F.L., Tawfick, S.H., Baughman, R.H. et al., "Carbon nanotubes: Present and future commercial applications," *Science* 339 (2013): 535–9.

Deng, F., Lu, W., Zhao, H. et al., "The properties of dry-spun carbon nanotube fibers and their interfacial shear strength in an epoxy composite," *Carbon* 49 (2011): 1752–7.

Elliott, J.A., Sandler, J.K.W., Windle, A.H. et al., "Collapse of single-wall carbon nanotubes is diameter dependent," *Phys. Rev. Lett.* 92 (2004): 095501.

Ericsson, L.M., Fan, H., Peng, H. et al., "Macroscopic, neat, single-walled carbon nanotube fibers," *Science* 305 (2004): 1447–50.

Fang, S., Zhang, M., Zakhidov, A.A. et al., "Structure and process-dependent properties of solid-state spun carbon nanotube yarns," *J. Phys. Condens. Matter* 22 (2010): 334221.

Feng, J.-M., Wang, R., Li, Y.-L. et al., "One-step fabrication of high quality double walled carbon nanotube thin films by a chemical vapor deposition process," *Carbon* 48 (2010): 3817–24.

Foroughi, J., Spinks, G.M., Wallace, G.G. et al., "Torsional carbon nanotube artificial muscles," *Science* 334 (2011) 494–7.

Gao, Y., Li, J., Liu, L. et al., "Axial compression of hierarchically structured carbon nanotube fiber embedded in epoxy," *Adv. Funct. Mater.* 20 (2010): 3797–803.

Ghemes, A., Minami, Y., Muramatsu, J. et al., "Fabrication and mechanical properties of carbon nanotube yarns spun from ultra-long multi-walled carbon nanotube arrays," *Carbon* 50 (2012): 4579–87.

Hill, F.A., Havel, T.F., Lashmore, D. et al., "Storing energy and powering small systems with mechanical springs made of carbon nanotube yarn," *Energy* 76 (2014): 318–25.

Huynh, C.P. and Hawkins, S.C., "Understanding the synthesis of directly spinnable carbon nanotube forests," *Carbon* 48 (2010): 1105–15.

Jakubinek, M.B., Johnson, M.B., White, M.A. et al., "Thermal and electrical conductivity of array-spun multi-walled carbon nanotube yarns," *Carbon* 50 (2012): 244–8.

Jayasinghe, C., Chakrabarti, S., Schulz, M.J. et al., "Spinning yarn from long carbon nanotube arrays," *J. Mater. Res.* 26 (2011): 645–51.

Jia, J., Zhao, J., Xu, G. et al., "A comparison of the mechanical properties of fibers spun from different carbon nanotubes," *Carbon* 49 (2011): 1333–9.

Jiang, C., Li, L. and Hao, H., "Carbon nanotube yarns for deep brain stimulation electrode," *IEEE Trans. Neural Syst. Rehabil. Eng.* 19 (2011): 612–6.

Koziol, K., Vilatela, J., Moisala, A. et al., "High performance carbon nanotube fiber," *Science* 318 (2007): 1892–5.

Kozlov, M.E., Capps, R.C., Sampson, W.M. et al., "Spinning solid and hollow polymer-free carbon nanotube fibers," *Adv. Mater.* 17 (2005): 614.

Lekawa-Raus, A., Patmore, J., Kurzepa, L. et al., "Electrical properties of carbon nanotube based fibers and their future use in electrical wiring," *Adv. Funct. Mater.* 24 (2014a): 3661–82.

Lekawa-Raus, A., Kurzepa, L., Peng, X. et al., "Towards the development of carbon nanotube based wires," *Carbon* 68 (2014b): 597–609.

Lekawa-Raus, A., Koziol, K.K.K. and Windle, A.H., "Piezoresistive effect in carbon nanotube fibers," *ACS Nano* 8 (2014c): 11214–24.

Li, Y.L., Kinloch, I.A. and Windle, A.H., "Direct spinning of carbon nanotube fibers from chemical vapor deposition synthesis," *Science* 304 (2004): 276–8.

Li, Q., Li, Y., Zhang, X. et al., "Structure dependent electrical properties of carbon nanotube fibers," *Adv. Mater.* 19 (2007): 3358–63.

Lima, M.D., Fang, S., Lepró, X. et al., "Biscrolling nanotube sheets and functional guests into yarns," *Science* 331 (2011): 51–5.

Lima, M.D., Li, N., de Andrade, M.J. et al., "Electrically, chemically, and photonically powered torsional and tensile actuation of hybrid carbon nanotube yarn muscles," *Science* 338 (2012): 928–32.

Liu, K., Sun, Y., Zhou, R. et al., "Carbon nanotube yarns with high tensile strength made by a twisting and shrinking method," *Nanotechnology* 21 (2010): 045708.

Liu, L., Ma, W. and Zhang, Z., "Macroscopic carbon nanotube assemblies: Preparation, properties, and potential applications," *Small* 7 (2011): 1504–20.

Lu, W. and Chou, T.-W., "Analysis of the entanglements in carbon nanotube fibers using a self-folded nanotube model," *J. Mech. Phys. Solids* 59 (2011): 511–24.

Lu, W., Chou, T.-W. and Kim, B.-S., "Radial deformation and its related energy variations of single-walled carbon nanotubes," *Phys. Rev. B* 83 (2011): 134113.

Lu, W., Zu, M., Byun, J.-H. et al., "State of the art of carbon nanotube fibers: Opportunities and challenges," *Adv. Mater.* 24 (2012): 1805–33.

Ma, W., Song, L., Yang, R. et al., "Directly synthesized strong, highly conducting, transparent single-walled carbon nanotube films," *Nano Lett.* 7 (2007): 2307–11.

Ma, W., Liu, L., Yang, R. et al., "Monitoring a micromechanical process in macroscale carbon nanotube films and fibers," *Adv. Mater.* 21 (2009a): 603–8.

Ma, W., Liu, L., Zhang, Z. et al., "High-strength composite fibers: Realizing true potential of carbon nanotubes in polymer matrix through continuous reticulate architecture and molecular level couplings," *Nano Lett.* 9 (2009b): 2855–61.

Miao, M., "Electrical conductivity of pure carbon nanotube yarns," *Carbon* 49 (2011): 3755–61.

Miao, M., "Yarn spun from carbon nanotube forests: Production, structure, properties and applications," *Particuology* 11 (2013) 378–93.

Miao, M., McDonnell, J., Vuckovic, L. et al., "Poisson's ratio and porosity of carbon nanotube dry-spun yarns," *Carbon* 48 (2010): 2802–11.

Miao, M., Hawkins, S., Cai, J. et al., "Effect of gamma irradiation on the mechanical properties of carbon nanotube yarns," *Carbon* 49 (2011): 4940–7.

Motta, M., Li, Y.L., Kinloch, I. et al., "Mechanical properties of continuously spun fibers of carbon nanotubes," *Nano Lett.* 5 (2005): 1529–33.

Motta, M., Moisala, A., Kinloch, I.A. et al., "High performance fibres from 'dog bone' carbon nanotubes," *Adv. Mater.* 19 (2007): 3721–6.

Park, J. and Lee, K.-H., "Carbon nanotube yarns," *Korean J. Chem. Eng.* 29 (2012): 277–87.

Porwal, P.K., Beyerlein, I.J. and Phoenix, S.L., "Statistical strength of twisted fiber bundles with load sharing controlled by frictional length scales," *J. Mech. Mater. Struct.* 2 (2007): 773–91.

Randeniya, L.K., Bendavid, A., Martin, P.J. et al., "Composite yarns of multiwalled carbon nanotubes with metallic electrical conductivity," *Small* 6 (2010): 1806–11.

Schmidt, A.C., Wang, X., Zhu, Y. et al., "Carbon nanotube yarn electrodes for enhanced detection of neurotransmitter dynamics in live brain tissue," *ACS Nano* 7 (2013): 7864–73.

Schulz, M., Vesselin, S. and Yin, Z. (eds.), *Nanotube Superfiber Materials: Changing Engineering Design* (Oxford: Elsevier, 2014).

Su, F., Miao, M., Niu, H. et al., "Gamma-irradiated carbon nanotube yarn as substrate for high-performance fiber supercapacitors," *ACS Appl. Mater. Interfaces* 6 (2014): 2553–60.

Vilatela, J.J., Elliott, J.A. and Windle, A.H., "A model for the strength of yarn-like carbon nanotube fibers," *ACS Nano* 5 (2011): 1921–7.

Wang, J., Deo, R.P., Poulin, P. et al., "Carbon nanotube fiber microelectrodes," *J. Am. Chem. Soc.* 125 (2003): 14706–7.

Wei, H., Wei, Y., Wu, Y. et al., "High-strength composite yarns derived from oxygen plasma modified super-aligned carbon nanotube arrays," *Nano Res.* 6 (2013): 208–15.

Zhang, X. and Li, Q., "Enhancement of friction between carbon nanotubes: An efficient strategy to strengthen fibers," *ACS Nano* 4 (2010): 312–6.

Zhang, M., Atkinson, K.R. and Baughman, R.H., "Multifunctional carbon nanotube yarns by downsizing an ancient technology," *Science* 306 (2004): 1358–61.

Zhang, X.F., Li, Q.W., Holesinger, T.G. et al., "Ultrastrong, stiff, and lightweight carbon-nanotube fibers," *Adv. Mater.* 19 (2007a): 4198–201.

Zhang, X.F., Li, Q.W., Tu, Y. et al., "Strong carbon-nanotube fibers spun from long carbon-nanotube arrays," *Small* 3 (2007b): 244–8.

Zhang, S., Koziol, K.K.K., Kinloch, I.A. et al., "Macroscopic fibers of well-aligned carbon nanotubes by wet spinning," *Small* 4 (2008a): 1217–22.

Zhang, S., Zhu, L., Minus, M.L. et al., "Solid-state spun fibers and yarns from 1-mm long carbon nanotube forests synthesized by water-assisted chemical vapor deposition," *J. Mater. Sci.* 43 (2008b): 4356–62.

Zhang, Y., Zou, G., Doorn, S.K. et al., "Tailoring the morphology of carbon nanotube arrays: From spinnable forests to undulating foams," *ACS Nano* 3 (2009): 2157–62.

Zhang, D., Miao, M., Niu, H. et al., "Core spun carbon nanotube yarn supercapacitors for wearable electronic textiles," *ACS Nano* 8 (2014): 4571–9.

Zhao, H., Zhang, Y., Bradford, P.D. et al., "Carbon nanotube yarn strain sensors," *Nanotechnology* 21 (2010): 305502.

Zhao, J., Zhang, X., Di, J. et al., "Double-peak mechanical properties of carbon-nanotube fibers," *Small* 6 (2010): 2612–7.

Zhao, Y., Wei, J., Vajtai, R. et al., "Iodine doped carbon nanotube cables exceeding specific electrical conductivity of metals," *Sci. Rep.* 1 (2011): 83.

Zhu, Z., Song, W., Burugapalli, K. et al., "Nano-yarn carbon nanotube fiber based enzymatic glucose biosensor," *Nanotechnology* 21 (2010): 165501.

Zhu, C., Cheng, C., He, Y.H. et al., "A self-entanglement mechanism for continuous pulling of carbon nanotube yarns," *Carbon* 49 (2011): 4996–5001.

Zhu, Z., Garcia-Gancedo, L., Flewitt, A.J. et al., "Design of carbon nanotube fiber microelectrode for glucose biosensing," *J. Chem. Technol. Biotechnol.* 87 (2012a): 256–62.

Zhu, Z., Garcia-Gancedo, L., Flewitt, A.J. et al., "A critical review of glucose biosensors based on carbon nanomaterials: Carbon nanotubes and graphene," *Sensors* 12 (2012b): 5996–6022.

Zu, M., Li, Q., Zhu, Y. et al., "The effective interfacial shear strength of carbon nanotube fibers in an epoxy matrix characterized by a microdroplet test," *Carbon* 50 (2012): 1271–9.

15

Endohedrally Doped Carbon Nanotubes

Hidetsugu Shiozawa

15.1 Introduction

Single-walled carbon nanotubes (SWCNTs) possess unique physical, chemical, and mechanical properties that are defined by their atomic structure (Saito et al. 1998). Since their discovery (Iijima 1991), continuous efforts have been made toward their production in larger yield and higher purity. A variety of SWCNT synthesis techniques established so far include chemical vapor deposition, arc discharge, and laser ablation methods using combinations of various metal catalysts, carbon feedstocks, and carrier gases.

Postsynthesis functionalization of SWCNT properties has been explored via both chemical and physical approaches, including chemical functionalizations of the nanotube's outer surfaces, adsorption, substitutions of carbon with other elements, electrochemical doping, and intercalations of internal and intersticial nanospaces within bundled carbon nanotubes (CNTs) (Tasis et al. 2006; Shiozawa et al. 2008d). Chemical functionalizations played an important role in the development of purification and sorting techniques such as the density gradient centrifugation (Arnold et al. 2006; Ghosh et al. 2010), ion exchange (Tu et al. 2009), and gel chromatography (Tanaka et al. 2009; Liu et al. 2011, 2013). Recent progresses allowed an as-grown mixture of various SWCNTs to be sorted according to their metallicity (semiconducting or metallic) and further to the chiral vector (n,m) that define the nanotube's atomic lattice.

Intercalation allows electron and hole doping of bundled SWCNTs. Reports on the intercalation with alkali metals (Pichler et al. 1999; Liu et al. 2003; Rauf et al. 2004; Kramberger et al. 2009) as an electron donor or with hole dopants such as I_2, Br_2 (Rao et al. 1997; Cambedouzou et al. 2004), and $FeCl_3$ (Liu et al. 2004) revealed a wide variety of property changes. Doping induced polymerization of C_{60} inside SWCNTs was reported (Pichler et al. 2001).

In turn, filling the tubular interior of SWCNTs is an interesting route of functionalization. The method is unique since it allows the SWCNT interiors to be filled with substances possessing different chemical and physical properties (Monthioux et al. 2006) while the intrinsic properties of SWCNT's outer surfaces remain unchanged. Hence, this offers an effective way to tailor the SWCNT's electronic properties in a controlled manner that is insensitive to environmental factors and can be combined with all the other aforementioned functionalization techniques.

In this chapter, I review recent progresses on this topic. First, in Section 15.2, I discuss on the endohedral doping of SWCNTs with fullerene. Some of the pioneering works on the synthesis and studies of fullerene peapods, SWCNTs filled with C_{60} and other fullerene compounds, are shown. In Section 15.3, I review progresses on the endohedral doping with nonfullerenes that attract increasing attention in the field. A large variety of inorganic and organic compounds encapsulated within SWCNTs and their unique structures and properties reported are summarized. In Section 15.4, I show that reactions of molecules within CNTs offer a means to create novel nanostructures and hence allow unique properties to emerge. The electronic and magnetic properties of nanostructures created inside SWCNTs are discussed in detail. Finally, in Section 15.5, growth of other types of one-dimensional (1D) molecular structures demonstrated inside SWCNTs is summarized.

15.2 Endohedral Doping with Fullerenes

15.2.1 Buckminsterfullerene

The well-studied peapods are formed via the insertion of Buckminsterfullerene C_{60} into the hollow tubular spaces inside SWCNTs. These so-called bucky peapods were first observed under a transmission electron microscopy (TEM) (Smith et al. 1998) as minor natural products in an SWCNT material synthesized by pulsed laser vaporization (PLV) (Thess et al. 1996). Later, the synthesis of peapods in high yield was carried out using an SWCNT material exposed to a vapor of C_{60} in vacuo (Hirahara et al. 2000; Smith and Luzzi 2000; Bandow et al. 2001; Kataura et al. 2001). Figure 15.1 is a TEM micrograph of bundled SWCNTs encapsulating bucky fullerenes in high yield (Shiozawa et al. 2006). X-ray diffraction (XRD) and Raman spectroscopy are useful tools and have been repeatedly used for the characterization of peapod materials on a bulk scale (Bandow et al. 2001; Kataura et al. 2001; Abe et al. 2003). The XRD pattern of empty SWCNT consists of several peaks originating from the 2D hexagonal lattice of the bundled tubes (see the top spectrum in Figure 15.2). The diffraction peak intensity is scaled by the form factor that changes as a result of filling the interior of SWCNTs with foreign substances (Abe et al. 2003). For well-filled

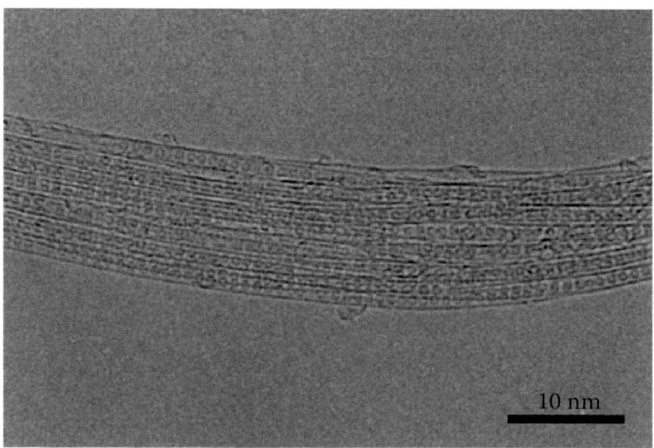

FIGURE 15.1 TEM micrograph of the bundled single-walled nanotubes containing the C_{60}; $(C_{60})_n$@SWCNTs. (From Shiozawa, H. et al., *Phys. Rev. B*, 73, 075406, 2006. Copyright 2006 by the American Physical Society.)

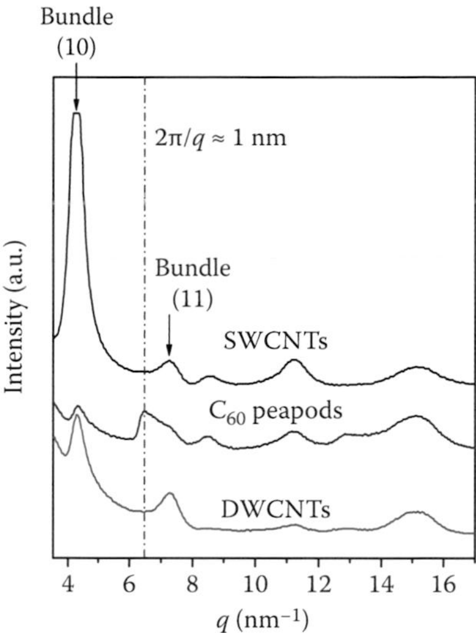

FIGURE 15.2 XRD profiles of bundled SWCNT (top), C_{60} peapods (center), and DWCNT (bottom) taken at room temperature. The peak superimposed by the dash-dotted vertical line is of 1D C_{60} crystals formed inside SWCNTs. (Courtesy of Y. Maniwa, after Abe, M. et al., *Phys. Rev. B*, 68, 041405, 2003.)

bucky peapods, the peak of the two-dimensional hexagonal lattice of bundled SWCNTs, indexed as 10, is reduced in intensity while the new peak corresponding to the 1D chain of fullerene peas emerges, as marked with the dash-dotted line in Figure 15.2. From the 10 peak position evaluated from electron and XRD measurements, the mean distance between the fullerene peas was evaluated to be 0.95–0.97 nm (Liu et al. 2002; Abe et al. 2003; Maniwa et al. 2003), reduced from $d = 1.0$ nm for the C_{60} fcc solid.

Scanning tunneling microscopy (STM) allows mapping of the local electronic states of individual peapods. An STM study unveiled the local electronic structure of the SWCNT host modified by the encapsulation of C_{60} fullerenes (Hornbaker et al. 2002).

Typical Raman spectra of SWCNTs and C_{60} peapods are shown in Figure 15.3. In both data collected at 90 K with a 488 nm laser, the G band of E_{2g} symmetry is observed around 1580 cm^{-1}, while the

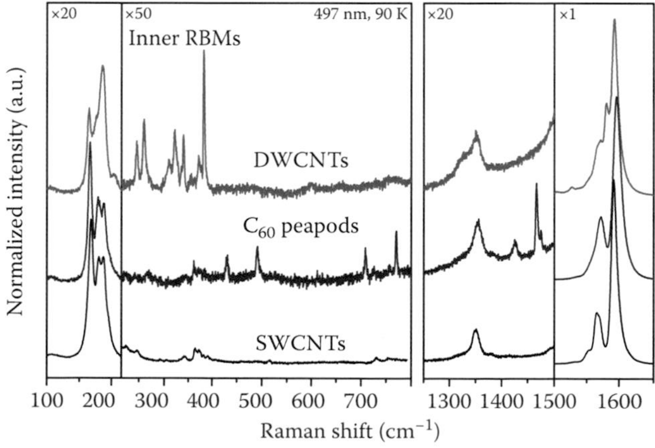

FIGURE 15.3 Raman spectra of SWCNT (bottom), C_{60} peapods (center), and DWCNT (top) measured at 90 K with a laser wavelength of 497 nm. (Courtesy of H. Kuzmany, adapted from Pfeiffer, R., *Dispersion of Raman Lines in Carbon Nanophases* [dissertation]. University of Vienna, 2004.)

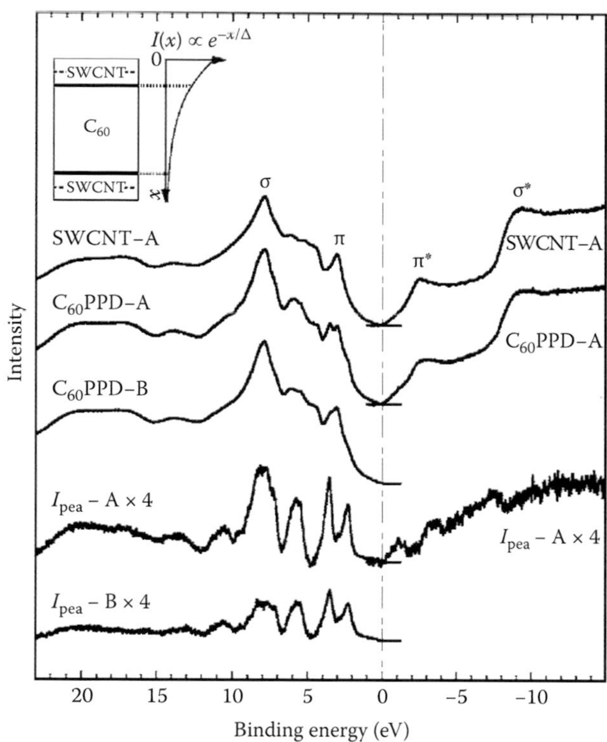

FIGURE 15.4 Photoemission and inverse photoemission spectra of C_{60} peapods and SWCNTs as well as the corresponding C_{60} pea spectra $I_{pea} = I_{peapods} - I_{SWCNT}$ multiplied by 4. (From Shiozawa, H. et al., *Phys. Rev. B*, 73, 075406, 2006. Copyright 2006 by the American Physical Society.)

defect-induced D mode is around 1350 cm^{-1}. The peaks observed in the frequency range of 150–210 cm^{-1} are attributed to the radial breathing modes of SWCNTs (Bandow et al. 2001), whose Raman frequency is scaled inversely by the tube diameter. The small features around 360 cm^{-1} are the overtone of the radial breathing modes (RBM). The spikes observed only in the peapod spectrum are the signal from the peas. The peapod spectrum is basically a weighted sum of the C_{60} spectrum and SWCNT spectrum (Kuzmany et al. 2004).

Details of the molecular electronic structure of bucky peapods on a bulk scale were studied by means of photoemission and inverse photoemission spectroscopy (Shiozawa et al. 2006) as well as electron energy-loss spectroscopy (EELS) and optical absorption spectroscopy (Liu et al. 2002). In Figure 15.4 are the photoemission and inverse photoemission spectra of an SWCNT material before and after filling with bucky fullerenes. In the photoemission (inverse photoemission) spectrum of the peapods, the valence (conduction) band structures originating from the fullerene pea molecular orbitals are observed. The difference spectra for the peas plotted at the bottom are very similar to those of a C_{60} fcc solid. In other words, the peapod spectrum is a weighted sum of the fullerene spectrum and the SWCNT spectrum. No signs of charge transfer or hybridization between the molecular states were observed within the limit of experimental error, which points to only a van der Waals interaction between the SWCNT host and the fullerene guest.

15.2.2 Doping of C_{60} Peapods

The response of C_{60} peapods to doping provides further information about the electronic properties. While only the hosting SWCNTs accept hole doping, electron doping allows both SWCNTs and C_{60} peas to be doped. The pentagonal pinch mode $A_g(2)$ of C_{60} is known to be down-shifted by 6.5 cm^{-1} per elementary charge given to the fullerene cage (Kuzmany et al. 2004). The heavily doped state achieved by the potassium intercalation was assigned to a C_{60}^{-6} phase in which doping-induced polymerization of C_{60} inside SWCNTs was reported (Pichler et al. 2001).

15.2.3 Other Fullerenes

Apparently, metallofullerenes were sought to be good candidates as filling species. Aberration corrected TEM allowed the encapsulated metals in fullerene cages to be resolved (Hirahara et al. 2000; Smith and Luzzi 2000; Suenaga et al. 2000), and their role as a catalyst on bimolecular reactions between adjacent fullerene cages was elucidated under electron-beam irradiation (Koshino et al. 2010). In Figure 15.5, TEM images of the isolated and bundled SWCNTs containing the Gd@C_{82} fullerenes are shown. The undulating electronic bandgap of a SWCNT encapsulating metallofullerenes was probed by scanning tunnelling microscopy (see Figure 15.6).

Later, exohedrally functionalized fullerenes were encapsulated (Koshino et al. 2008) and metals reacting with nanotube walls were observed under electron beam irradiation (Chamberlain et al. 2011).

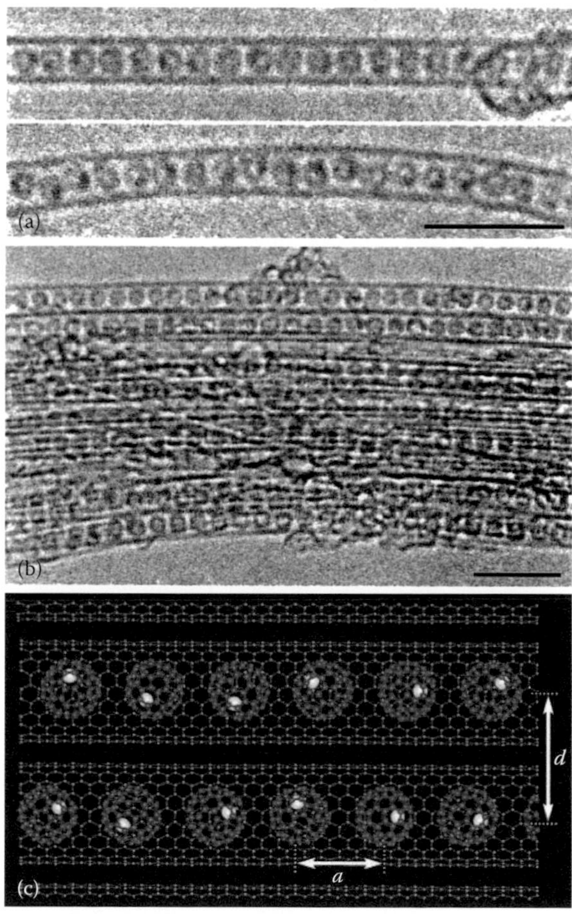

FIGURE 15.5 (a, b) TEM images of the isolated and bundled single-walled nanotubes containing the Gd@C_{82} fullerenes; (Gd@C_{82})$_n$@SWCNTs. Dark spots seen on most of the fullerene cages correspond to the encapsulated Gd atoms that are oriented randomly in respect to the tube axis (bar = 5 nm). (c) A schematic representation of (Gd@ C_{82})$_n$@SWCNTs. (From Hirahara, K. et al., *Phys. Rev. Lett.*, 85, 5384–5387, 2000. Copyright 2000 by the American Physical Society.)

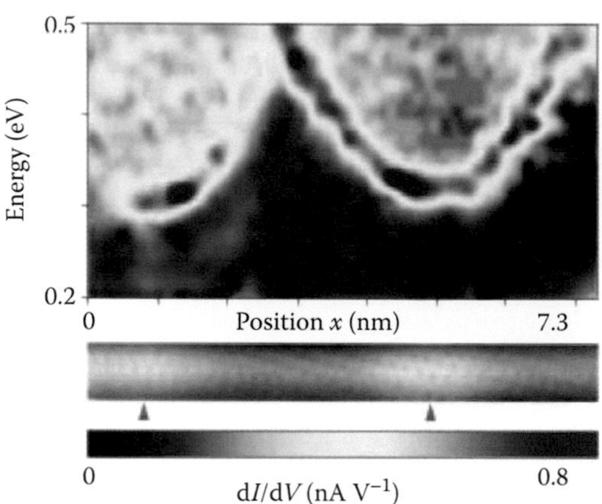

FIGURE 15.6 d*I*/d*V* spectra (top) and topographic image (bottom) of a 7.3-nm-long (11,9) SWCNT encapsulating GdC_{82} endohedral fullerenes at a sample bias voltage of 0.5 V. The sites of inserted fullerenes are indicated by the arrows. (Reprinted by permission from Macmillan Publishers Ltd: Lee, J. et al., *Nature*, 415, 1005–1008, 2002, Copyright 2002.)

It was demonstrated that fullerene derivatives can be bonded to and carry transition metal complexes into SWCNTs (Fan et al. 2011; Maggini et al. 2014).

Peapods can be transformed into double-walled CNTs (DWCNTs) (Bandow et al. 2001; Pfeiffer et al. 2005) by heating in vacuum or electron irradiation, which provided proof of fullerenes encapsulated inside SWCNTs on a bulk scale as observed via Raman spectroscopy. Further details will be discussed later in this review.

15.3 Endohedral Doping with Nonfullerene Compounds

15.3.1 Inorganic Compounds

The CNT interiors can be filled with inorganic substances having different chemical and physical properties (Monthioux et al. 2006) such as metal oxides (Ajayan et al. 1995), metals (Ajayan et al. 1993; Sloan et al. 1998; Jeong et al. 2003; Corio et al. 2004), and binary crystals (Meyer et al. 2000; Sloan et al. 2002a,b).

When the nanotube's diameters are reasonably small, low-dimensional crystal structures that differ from those in the bulk could be formed. For example, SWCNTs can accommodate water molecules that were found to undergo a structural transition to form tube-like solid structures as probed by XRD (Maniwa et al. 2002, 2005, 2007) (Figure 15.7).

Molecular dynamics calculations revealed that inside SWCNTs with diameters ranging from 0.6 to 2.0 nm, oxygen molecules can be frozen to become various 1D structures from a chain to a helical geometry depending on the chiral index (n,m) of SWCNTs (Hanami et al. 2010). It was suggested that the oxygen inside SWCNTs exhibits various features of a quasi-1D spin 1 magnet.

Iodine helical chains encapsulated inside SWCNTs were observed under an electron microscope (Fan et al. 2000). As shown in Figure 15.8, double helices of bonded selenium atoms were made within the narrow cavity inside DWCNTs (Fujimori et al. 2013a), and conducting monatomic sulphur chains as long as 160 nm were found to be stabilized up to c.a. 800 K in the constraining volume of CNTs (SWCNTs and DWCNTs) with its 1D phase assigned by high-resolution TEM and XRD (Fujimori et al. 2013b).

Some binary compounds encapsulated in CNTs exhibit unique 1D structures. Potassium iodide encapsulated in SWCNTs was found to exhibit a high-pressure phase structure (Urita et al. 2011) and twisting mercury iodide crystals were observed in a triple-walled nanotube (Sloan et al. 2003). The

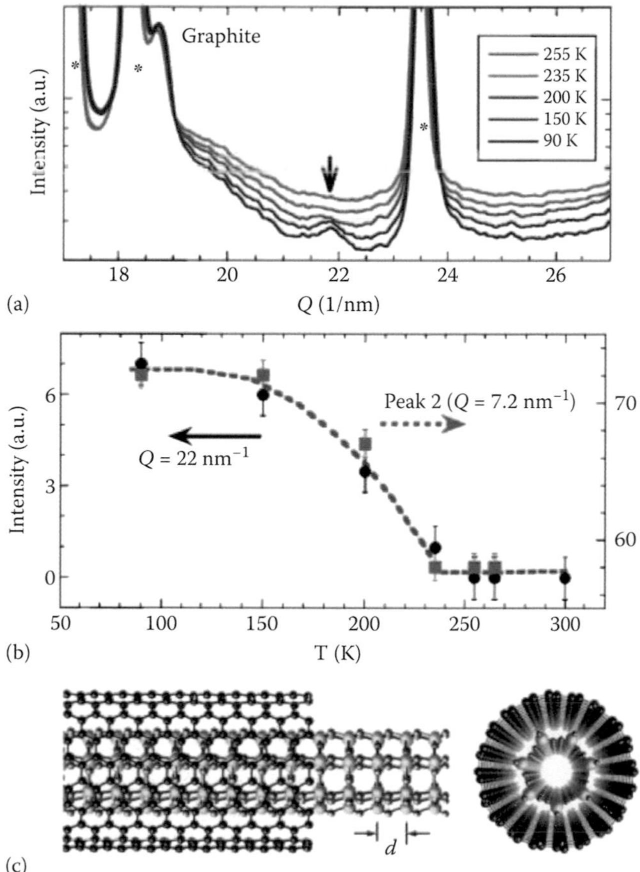

FIGURE 15.7 (a) Temperature dependence of XRD profiles in water-exposed single-walled nanotube bundles. Stars denote peaks due to bulk ice. (b) Temperature dependence of the intensity of the peak around $Q = 22\ nm^{-1}$ shown by the arrow in (a), along with that of peak 2 at around $Q = 7.2\ nm^{-1}$. (c) The proposed structure of the ice nanotube inside a single-walled nanotube. The estimated d-spacing is 0.287 nm at 90 K. (From Maniwa, Y. et al., *J. Phys. Soc. Jpn.*, 71, 2863–2866, 2002.)

smallest phase change material was created via filling of SWCNTs with germanium telluride and their amorphous-to-crystalline phase change properties observed under transmission and scanning electron microscopes (Giusca et al. 2013) (Figure 15.9).

15.3.2 Organic Compounds

Various organic molecules such as metal porphyrin (Kataura et al. 2002), perylene-3,4,9,10-tetracarboxylic dianhydride (Fujita et al. 2005), beta-carotene (Yanagi et al. 2006), or coronene (Okazaki et al. 2011; Talyzin et al. 2011; Botka et al. 2014), and others (Takenobu et al. 2003) were inserted into CNTs. Figure 15.10 shows the TEM image of a SWCNT encapsulating coronenes.

One-dimensional structures such as linear carbon chains (carbyne or polyynes) (Nishide et al. 2006, 2007; Malard et al. 2007; Kuwahara et al. 2011; Moura et al. 2011; Zhao et al. 2011; Shi et al. 2013) and thiophene olygomers (Kalbac et al. 2010; Loi et al. 2010; Alvarez et al. 2011; Gao et al. 2011; Yumura and Yamashita 2014) were encapsulated inside SWCNTs. Raman studies suggested the presence of the charge transfer between the 1D guest and the CNT host (Malard et al. 2007; Moura et al. 2011).

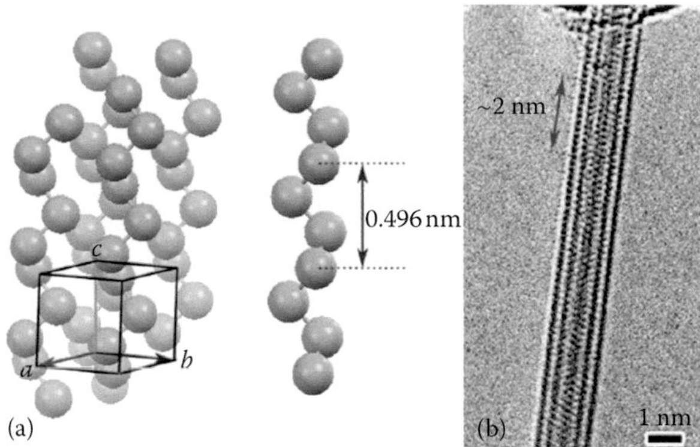

FIGURE 15.8 (a) Crystallographic structure of trigonal Se consisting of four Se chains in a unit cell (left) and the constituent single Se chain (right) with a pitch length of 0.496 nm. (b) HR-TEM image of Se@DWCNT, revealing the Se double-helix structure with a pitch length of ~2 nm. (Reprinted with permission from Fujimori, T. et al., *ACS Nano*, 7, 5607–5613, 2013. Copyright 2013, American Chemical Society.)

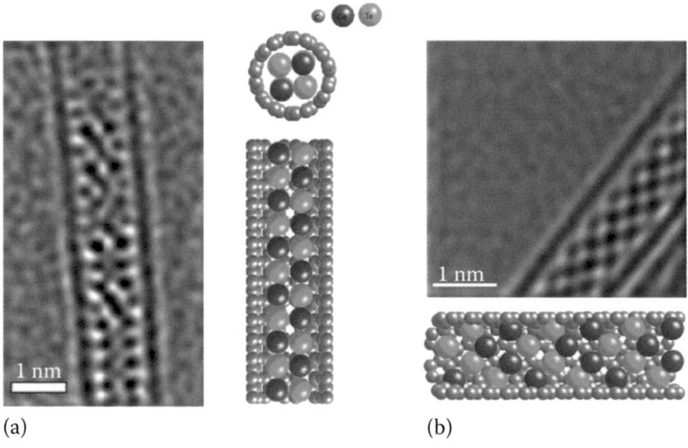

FIGURE 15.9 (a) Aberration-corrected TEM image of GeTe rock salt in a 2 × 2 crystal form within a single-walled nanotube, displaying expanded lattice along and across the tube capillary. Structural model of a 2 × 2 GeTe crystal within a CNT not incorporating lattice distortions. (b) TEM image of a single-walled nanotube incorporating GeTe with rhombohedral arrangement, also showing the corresponding structural model. (Courtesy of Giusca, C.E. et al., *Nano Lett.*, 13, 4020–4027, 2013.)

Organometallic and coordination compounds are promising functional fillers as reported in some of the recent works with metallocenes (Guan et al. 2005; Li et al. 2005, 2006a,b; Shiozawa et al. 2008b; Kharlamova et al. 2015), metal phthalocyanine (Schulte et al. 2007; Alvarez 2015), metal acetylacetonates (Cambre et al. 2009; Shiozawa et al. 2010, 2015), and carbonyl complexes (Chamberlain et al. 2012b).

Metallocenes are some of the most studied materials as filling species. Theoretical studies have shown that filling with ferrocene ($FeCp_2$) or cobaltocene ($CoCp_2$) can induce changes in SWCNT's electrical properties depending on the type of metallocene and the chirality of the tubes (Lu et al. 2004; Garcia-Suarez et al. 2006; Sceats and Green 2006, 2007).

Experimentally, several different metallocenes have been encapsulated into SWCNTs (Li et al. 2005, 2006a,b; Shiozawa et al. 2008b, 2009; Kharlamova et al. 2015). The encapsulated ferrocene ($FeCp_2$) was

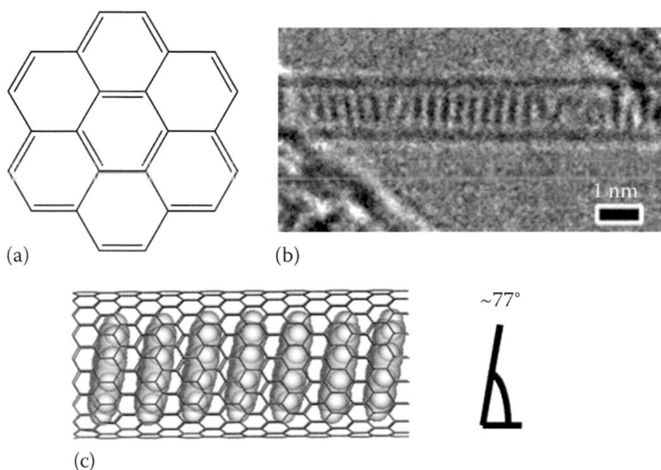

FIGURE 15.10 (a) Molecular structure of coronene. (b) TEM images of SWCNTs encapsulating coronenes. (c) Schematic illustration of SWCNTs encapsulating coronenes. (Courtesy of Okazaki, T. et al., *Angew. Chem. Int. Ed.*, 50, 4853–4857, 2011. Copyright Wiley-VCH Verlag GmbH & Co. KGaA. Reproduced with permission.)

first evidenced by TEM (Li et al. 2005, 2006a; Shiozawa et al. 2008b), backed up by bulk characterizations of the filling factor using Raman and photoemission spectroscopy (Shiozawa et al. 2008b,c; Sauer et al. 2013). From the valence band photoemission data, it appeared that 0.14 electrons per ferrocene or 4.2×10^{-4} electrons per nanotube carbon atom are transferred from the ferrocene guest to the SWCNT host (see Section 15.4.3 for details). The degree of charge transfer is even greater in cerocene-filled SWCNTs for which the electron screening effect was enhanced as observed by means of x-ray absorption spectroscopy (Shiozawa et al. 2009).

Photoluminescence spectroscopy was done recently on SWCNT accommodating ferrocene (Liu et al. 2012; Iizumi et al. 2014). The photoluminescence signal was found to be altered as a result of the intermolecular charge transfer.

15.3.3 Synthesis Process

Before SWCNTs can be filled, they have to be opened. This is usually done by oxidizing them at temperatures in a range from 400°C up to 500°C in air for about (half) an hour (Kataura et al. 2001). After this, filling can be carried out by exposing an oxidized SWCNT material to a vapor or liquid of filling molecules (fullerenes [Kataura et al. 2001], metallocene [Li et al. 2005; Shiozawa et al. 2008b, 2009; Kharlamova et al. 2015], metal phthalocyanine [Schulte et al. 2007], carbonyl complexes [Chamberlain et al. 2012b], and metal acetylacetonates [Shiozawa et al. 2010]) or in aqueous, organic solvent (Shiozawa et al. 2008a; Chamberlain et al. 2011), or supercritical CO_2 (Gimenez-Lopez et al. 2011) solutions.

For the filling using the vapor phase, an oxidized SWCNT material as powder, film, or buckypaper is sealed together with a filling material as powder, crystal, or liquid form into an evacuated glass or quartz ampoule. Then, the ampoule is heated to a temperature at which the filling material sublimates or evaporates and kept for some hours or days. The advantage of this process is its very clean filling conditions while one requires a vacuum setup and glass sealing using a blowtorch. This process is used for filling SWCNT with C_{60}, organometallic, and coordination compounds. Obviously, this process requires filling materials that can sublime or evaporates in vacuum without decomposition.

Filling with inorganic alloys with low eutectic points can be done by wetting an SWCNT material with the molten alloy in a vacuum sealed ampoule (Sloan et al. 2002b).

Another option is to use a solution. An oxidized SWCNT material is dispersed or immersed in a saturated aqueous/organic solvent solution of a filling substance. Sonication can accelerate the filling process. The major advantage of this method is the low reaction temperature, usually slightly above room temperature, although some solvent molecules may also enter the SWCNTs (Simon and Kuzmany 2006). Additionally, the filling yield depends on the solubility of the filling material in the solvent. Examples are C_{60} in toluene, ferrocene in acetone (Shiozawa et al. 2008a), or N@C_{60} in *n*-hexane (Simon et al. 2004).

Finally, SWCNT can be filled using super-critical (sc) CO_2. Here, a mixture of an oxidized SWCNT sample and a filling material is immersed in sc-CO_2 at about 50°C and 150 bar for several days. This process is very clean and takes place at rather low temperatures, although it requires special equipment for handling the sc-CO_2. Examples are fullerenes and functionalized fullerenes (Britz et al. 2004; Khlobystov et al. 2004).

After the filling process, the excess filling material trapped within or on the surface of bundled SWCNTs has to be removed. This can be done by rinsing the sample in a clean solvent or by evaporating the excess material by heating the sample in vacuum.

15.4 Nano-Reactions inside CNTs

The chemical reactions taking place inside SWCNTs allows novel heterogeneous nanostructures to be created. To mention a few, the internal structures can be small-diameter SWCNTs, boron nitride (BN) nanotubes, and metal wires. In this section, the mechanism for the inner CNT formation from different precursors as well as the electronic and magnetic properties of SWCNTs encapsulating metallic clusters are discussed in detail.

15.4.1 Transformation of C_{60} Peapods to DWCNTs

Fullerene peapods can be transformed into DWCNTs by heating (Bandow et al. 2001) (see Figure 15.11) or electron irradiation in vacuum. The coalescence of fullerene peas to the formation of tubular structures was observed by TEM with intensive or controlled electron beam irradiation (Smith et al. 1999; Hernandez et al. 2003; Koshino et al. 2010).

The modified local electronic structure in DWCNTs with respect to their individual inner and outer constituent single-walled nanotubes was observed by scanning tunneling microscopy and spectroscopy (Giusca et al. 2007).

In the XRD, the (10) bundle peak regains its intensity as the 1D fullerene chain peak disappears once the peapods are fully transformed into DWCNTs (see the bottom spectrum in Figure 15.2).

Further evidence of the DWCNT formation on a bulk scale can be obtained from Raman spectroscopy. The top spectrum in Figure 15.3 is the Raman data for a bucky peapod material after being heated at 1250°C in vacuum. It exhibits the sharp Raman lines located at frequencies ranging from 250 to 400 cm^{-1}. Provided that the RBM frequency is inversely scaled by the SWCNT diameter, they can be attributed to the RBM modes of different chirality inner tubes formed inside the parent outer CNTs (Pfeiffer 2004).

One peculiarity of the inner-tube RBMs is their very small line width down to 0.35 cm^{-1} (Pfeiffer et al. 2003). These narrow lines indicate rather long phonon lifetimes and, thus, very clean growth conditions inside the outer nanotubes. Another peculiarity is the fact that there are about 10 times as many RBM lines as there should be for geometrically possible inner tubes (Pfeiffer et al. 2005; Simon et al. 2006b). This can best be seen by comparing the inner-tube RBMs with the RBMs of SWCNTs with similar diameters. Such a comparison is shown in Figure 15.12, in which the left panel depicts the Raman response of the (6,5) and (6,4) tubes in a CoMoCat material and the right panel depicts the corresponding data for a DWCNT material. In the latter, one inner tube type does not only contribute one peak to the spectrum, but a whole cluster of peaks spread over ~30 cm^{-1} (Pfeiffer et al. 2005).

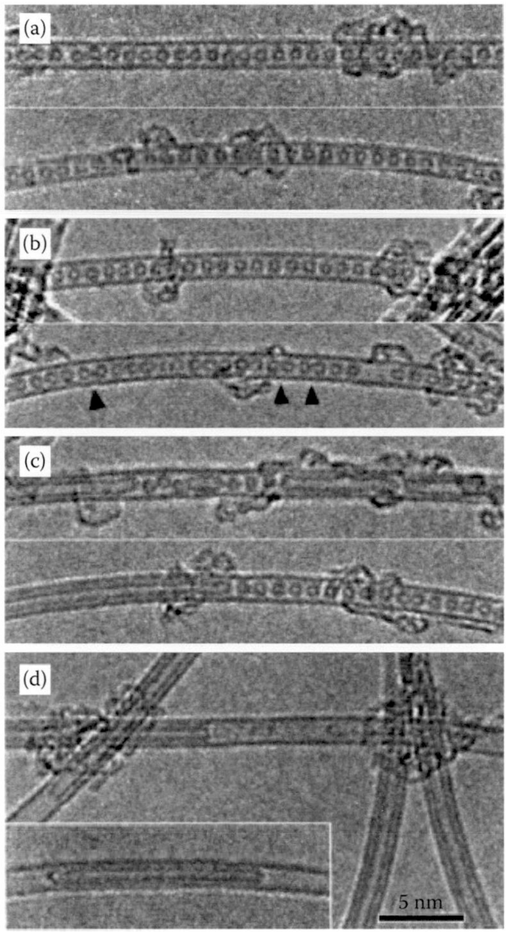

FIGURE 15.11 TEM micrographs of (a) $(C_{60})_n$@SWCNTs, annealed at (b) 800°C, (c) 1000°C, and (d) 1200°C. (From Bandow, S. et al., *Chem. Phys. Lett.*, 337, 48–54, 2001. Reprinted with permission from Elsevier.)

This was explained as a result of RBM frequency shifts for inner tubes encapsulated in different outer tubes. The fact is that one species of (n,m) inner tube can grow in several different outer-tube species. The interaction between inner and outer tubes is larger for those with smaller-diameter distances whose inner-tube RBMs are shifted to higher frequencies with respect to the corresponding freestanding ones (Pfeiffer et al. 2004). Based on this model, each line can be assigned to one inner tube–outer tube pair with defined chiral vectors (Pfeiffer et al. 2006).

15.4.2 Growth of Inner Tubes with Metal Catalysts

Metallocenes and metal acetylacetonates were encapsulated in SWCNTs, transformed into metal particles. It was reported that some transition (Fe, Ni, and Co) and precious (Pt) metals can catalyze the formation of inner CNTs in a confined tubular environment (Shiozawa et al. 2008b, 2010; Kharlamova et al. 2015). In such cases, the encapsulated molecules are also the source of carbon atoms to form the inner tubes (Shiozawa et al. 2008b).

As discussed in the previous section, the bucky peapods can be transformed into DWCNTs by electron irradiation or annealing in vacuum at temperatures above 1000°C (Smith and Luzzi 2000; Bandow

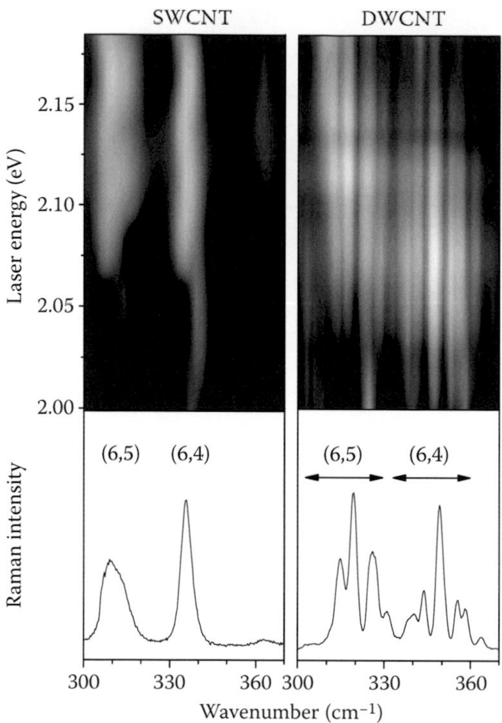

FIGURE 15.12 Energy dispersive Raman contour plot for SWCNT (CoMoCat) and DWCNT samples at 80 K. Lower panels show individual Raman spectra collected with a 2.1 eV (590 nm) laser. (Courtesy of Simon, F. et al., *Phys. Rev. B*, 74, 121411, 2006.)

et al. 2001). This transformation occurs even at temperatures as low as 800°C, but it literally takes weeks (Bandow et al. 2001). In contrast, the inner-tube growth from ferrocene is catalytically triggered and can be processed at temperatures as low as 600°C. With nickelocene as a precursor, the inner-tube growth temperature can be as low as 400°C (Kharlamova et al. 2015).

15.4.2.1 Growth Properties as Probed by Raman Spectroscopy

Raman spectroscopy offers a useful noninvasive way to determine how the inner tubes grow from different precursor molecules. Examples are shown in Figure 15.13. Panel a shows the Raman response from a ferrocene-filled SWCNT material, $FeCp_2$@NT. For the annealed samples, the sharp inner-tube RBM lines are clearly observed.

Comparing the high-resolution Raman data for the (6,4) inner tube produced from the $FeCp_2$ precursor with that from the C_{60} precursor yields several surprising results, as shown in Figure 15.14. First, the line positions are exactly the same in both cases, which means the same assignment to inner–outer tube pairs can be used. Second, the intensity distribution for the various tube pairs is different between the C_{60}-grown DWCNTs and the $FeCp_2$-grown DWCNTs. In the latter case, the (6,4) tubes in the rather large outer tubes are dominating. This feature is independent of the annealing temperature, as compared between 600°C and 1150°C in Figure 15.13a, which means the same growth mechanism is relevant at both temperatures and it is apparently different from the mechanism involving the coalescence of C_{60} peas.

Importantly, the RBM line distribution for the $FeCp_2$-grown inner tubes is also observed for those grown from some other metallocenes, such as $NiCp_2$ and $GdCp_3$. Hence, it seems that the fine structures of the inner-tube RBM are identical among DWCNT materials from different metallocene-filled SWCNTs. Moreover, since DWCNTs from $GdCp_3$ composed of three planar aromatic ligands around a

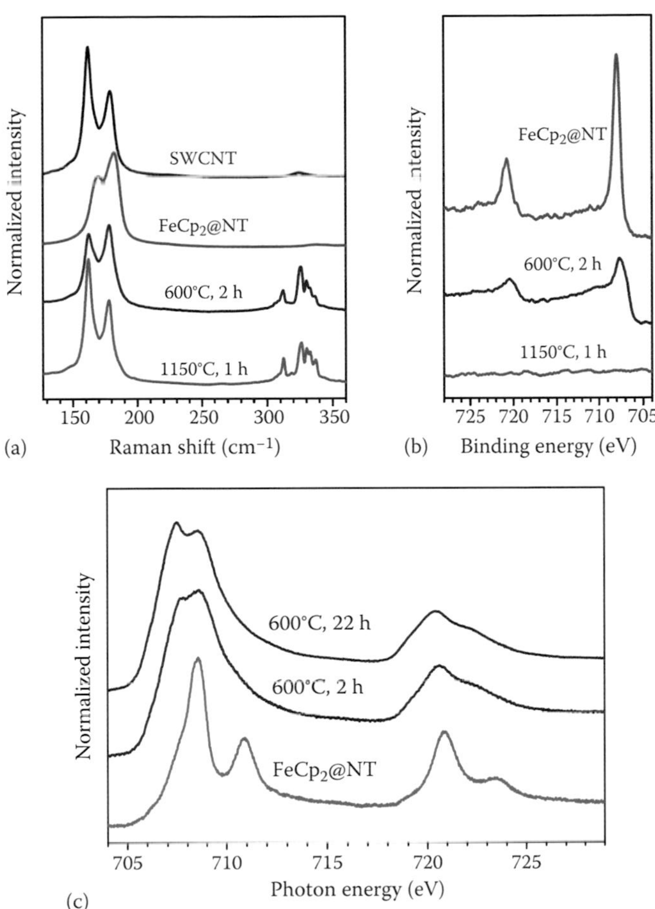

FIGURE 15.13 (a) Raman spectra in the RBM region for pristine SWCNT, $FeCp_2$@NT, and $FeCp_2$@NT annealed at 600°C for 2 h and $FeCp_2$@NT annealed at 1150°C for 1 h; recorded with a photon energy (wavelength) of 1.16 eV (1060 nm) at room temperature. (b) Fe 2*p* photoemission spectra for $FeCp_2$@NT and those annealed at 600°C for 2 h and 1150°C for 1 h. (From Shiozawa, H. et al., *Adv. Mater.*, 20, 1443–1449, 2008. Copyright Wiley-VCH Verlag GmbH & Co. KGaA. Reproduced with permission.) (c) Fe 2*p* absorption spectra for $FeCp_2$@NT and those annealed at 600°C for 2 h and 22°C. (From Shiozawa, H. et al., *Phys. Rev. B*, 77, 153402, 2008. Copyright 2008 by the American Physical Society.)

rare-earth ion shows the same spectral feature in the RBM region as that from $FeCp_2$ composed of two planar aromatic ligands around an iron ion, the growth mechanism seems to be independent of the structural form as well as the number of carbon atoms in the precursor materials.

In Figure 15.14b are the RBM line intensity versus the inner-tube and outer-tube diameter differences of the corresponding DWCNTs, clearly showing the difference between two DWCNT products in tube-pair distribution. For the $FeCp_2$-grown DWCNT, the diameter difference is greater for the smaller-diameter tubes, while the opposite trend is seen in the case for the C_{60}-grown DWCNT. This difference can be attributed to different growth mechanisms. Since in the catalytic process with ferrocene, each inner tube is grown from scratch, the diameter can be well adapted to the outer-tube diameter and an optimum diameter difference can be established. In contrast, in the noncatalytic process with C_{60}, predefined caps of growing inner tubes are available, which apparently have a strong influence on the diameter of eventually grown inner tubes.

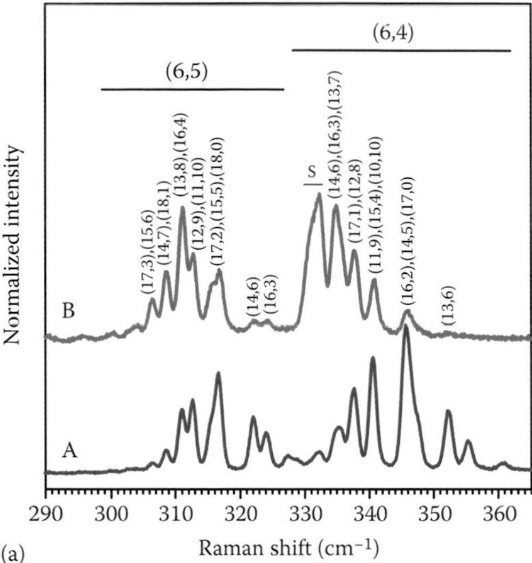

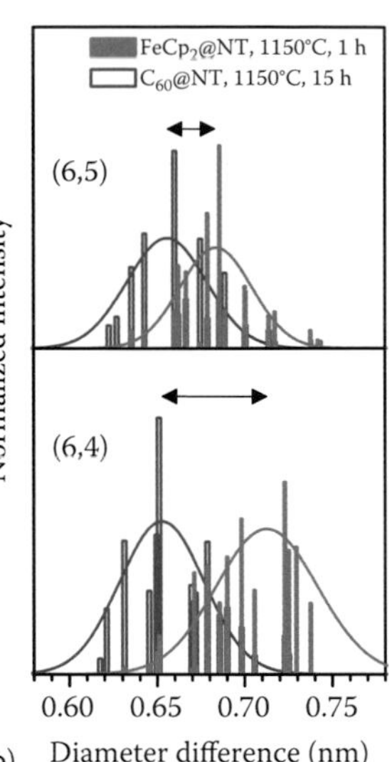

FIGURE 15.14 (a) Raman responses recorded with a 2.1 eV (590 nm) laser for the (6,5) and (6,4) inner tubes grown from C_{60} (spectrum A) and for those from $FeCp_2$ at 1150°C (spectrum B). (b) RBM line intensity versus the inner-tube and outer-tube diameter difference for the DWCNT with (6,5) and (6,4) inner tubes. The solid curves are Gaussian fits of the inner-tube RBM line distributions. (From Shiozawa H. et al., *Adv. Mater.*, 20, 1443–1449, 2008. Copyright Wiley-VCH Verlag GmbH & Co. KGaA. Reproduced with permission.)

Recent studies compared the inner-tube growth properties with different metal catalysts. Pt(II) acetylacetonate, $Pt(C_5H_7O_2)_2$, was introduced into SWCNTs, and it was found that pure Pt clusters act as a catalyst for the inner-tube growth, but require much higher temperatures to be activated as compared to Fe carbide clusters from ferrocene (Shiozawa et al. 2010); see Figure 15.15b, in which the RBM lines are compared for the (6,4) and (6,5) inner tubes grown from ferrocene and those from Pt(II) acetylacetonate at different temperatures. On the contrary, Ni carbide clusters from nickelocene, $NiCp_2$, grow inner tubes at temperatures as low as 400°C (Kharlamova et al. 2015). A general trend found for the tube growth with metal catalysts is that the smaller the tube diameter, the lower the growth temperature (Shiozawa et al. 2010) (see Figures 15.15a and c). At the low temperature limit, DWCNTs of optimal interwall distances are formed. In such DWCNTs, the intertube interaction is too small to be resolved by Raman scattering so that one inner-tube species exhibits one single RBM peak (see the RBM signal for the (7,3) inner tubes grown at lower temperatures in Figure 15.15a or the (6,4) and (6,5) inner tube lines in Figure 15.15b). It means that the inner tubes are well distanced from the outer tubes and the growth process is solely determined by the properties of the catalytic metal. The fact that thinner tubes grow at lower temperatures points to a lower activation energy for the thinner SWCNT growth. This is in line with previous reports on the bulk scale synthesis of SWCNTs where the reaction temperature dependence of the nanotube's diameter and chirality distribution was studied using the pulsed laser vaporization (PLV) method with mixed Fe/Ni, Co/Ni, RhPt, or RhPd catalysts (Bandow et al. 1998; Kataura et al. 2000) and the alcohol chemical vapor deposition method with Fe/Co catalyst (Miyauchi et al. 2004). A similar trend was also observed for the (7,2) and (8,3) inner-tube growths from C_{60} peas, as determined from Raman spectroscopy (Pfeiffer et al. 2007).

15.4.2.2 Characterization of the Interior Chemical Status

The chemical status inside DWCNT products made from filled SWCNTs depends hugely on the annealing temperature, as observed by means of x-ray photoemission and absorption spectroscopy (Shiozawa

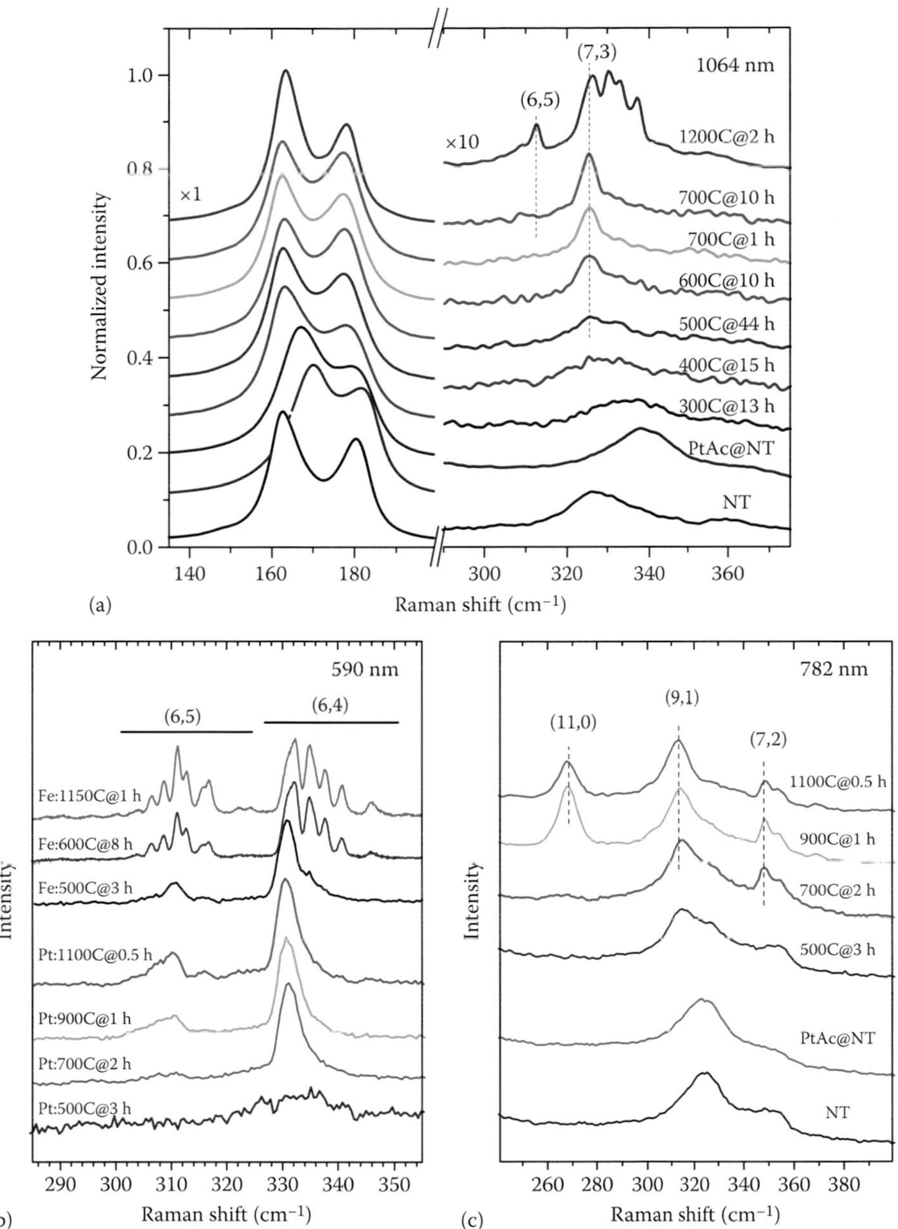

FIGURE 15.15 (a) Raman spectra in the outer-tube and inner-tube RBM regions obtained at a wavelength of 1064 nm for the empty SWCNT (NT) and Pt(II) acetylacetonate-filled SWCNT (PtAc@NT) before and after heating in vacuum at various temperatures and durations. (b) Raman spectra in the inner-tube RBM region obtained at 590 nm for the annealed PtAc@NT samples (lower set of spectra) and for the annealed ferrocene-filled SWCNTs (upper set of spectra). (c) Raman spectra in the inner-tube RBM region collected at 782 nm for the empty SWCNT (NT) and the pristine and annealed PtAc@NT samples. (From Shiozawa, H. et al., *Adv. Mater.*, 22, 3685–3689, 2010. Copyright Wiley-VCH Verlag GmbH & Co. KGaA. Reproduced with permission.)

et al. 2008b,c). The Fe 2*p* absorption spectrum of the $FeCp_2$@NT exhibits the characteristic features corresponding to the molecular orbitals of ferrocene (Figure 15.13c). After heating at 600°C, it shows a set of broad and asymmetric doublets corresponding to various chemical compositions of iron. Drastic changes in the Fe 2*p* edge are direct evidence for the decomposition of encapsulated $FeCp_2$ molecules (Shiozawa et al. 2008b). After the conversion to DWCNTs by annealing at 1150°C, all iron atoms of $FeCp_2$ are removed from the CNTs, as observed in the X-ray photoemission spectrum in Figure 15.13b. In contrast, when $FeCp_2$@NT is transformed into DWCNTs at 600°C, some of the iron atoms stay in the tube material. The latter demonstrates that iron-doped DWCNTs were produced from $FeCp_2$@NT.

Further evidence for the DWCNT growth from $FeCp_2$@NT was obtained from TEM observations. Figure 15.16a shows a TEM image of a $FeCp_2$@NT specimen. The image exhibits ferrocene molecules as dark contracts indicated by arrows distributed inside the SWCNT, which are rarely observed outside SWCNT and not observed in pristine SWCNT samples. Figure 15.16c–f display TEM images of the $FeCp_2$@NT annealed at 600°C for 2 h. Panel e shows SWCNTs containing a capped inner tube. Figure 15.16b was recorded after annealing at 1500°C for 1 h, showing a clean DWCNT. Importantly, nanocrystals were observed inside SWCNTs, such as the one displayed in Figure 15.16f. The crystal lattice is matched to that of an iron carbide Fe_3C (a = 0.4523 nm, b = 0.5089 nm, c = 0.67428 nm, space group is Pbnm) from a structure simulation. It was reported that the encapsulated iron carbide nanocrystals act as a catalyst that absorb carbon atoms from one side and generate an inner tube to another side, as sketched in Figure 15.16g.

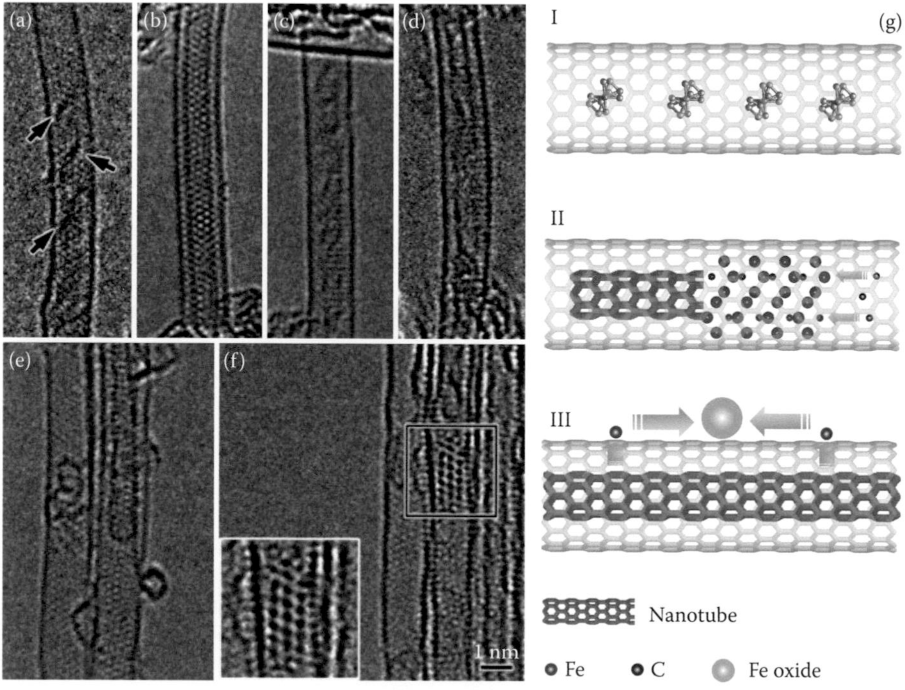

FIGURE 15.16 (a) TEM image of a pristine $FeCp_2$@NT. Arrows indicate dark contrasts. (b) $FeCp_2$@NT annealed at 1150°C for 1 h. (c–f) $FeCp_2$@NT annealed at 600°C for 2 h. The inset of (f) depicts a structure simulation output as dark dots superimposed upon the scaled image of the observed nanocrystal. The lattice of the nanocrystal is well matched to that of an iron carbide Fe_3C (a = 0.4523 nm, b = 0.5089 nm, c = 0.67428 nm, space group is Pbnm). (From Shiozawa, H. et al., *Adv. Mater.*, 20, 1443–1449, 2008. Copyright Wiley-VCH Verlag GmbH & Co. KGaA. Reproduced with permission.) (g) Schematic diagram for the inner-tube growth process. (I) $FeCp_2$ filled SWCNT. (II) Catalytic reaction for the inner-tube growth after decomposition of $FeCp_2$. (III) Precipitation of iron atoms onto the DWCNT surface and subsequent oxidation and aggregation into a nanoparticle. (From Shiozawa, H. et al., *Adv. Mater.*, 20, 1443–1449, 2008. Copyright Wiley-VCH Verlag GmbH & Co. KGaA. Reproduced with permission.)

15.4.3 Charge Transfer between the Filling Guest and SWCNT Host

Electronic interactions between the SWCNT host and molecular guest allow the SWCNT's properties to be altered. The interaction changes further as the encapsulated molecules react with one another as well as the formation of inner tubes and/or metal clusters takes place. The understanding of how the SWCNT's properties change as the interior chemical status changes is the key toward the functionalization of SWCNTs in a controlled manner. The ability to control the SWCNT's doping level via the interconversion of encapsulated molecules makes SWCNTs promising components for electronic devices.

Changes in charge transfer between the SWCNT and the filling material can be analyzed by means of various spectroscopy techniques, such as optical absorption, Raman, photoluminescence, and photoemission spectroscopy.

Photoemission spectroscopy using an ultraviolet radiation source allows a direct observation of the density of valence-band states with a high energy resolution in comparison with x-ray photoemission spectroscopy. This is complimentary to the interband density of states obtained from optical absorption spectroscopy. SWCNTs have spiky features in the density of carbon sp^2 π orbital state, known as the 1D van Hove singularity (VHS) characteristic to their quantized states. The VHS peaks of a SWCNT material containing bundled SWCNTs with diameters around 1.2–1.5 nm can be observed as three prominent peaks in both optical absorption and valence-hand photoemission spectra (Kataura et al. 1999; Ishii et al. 2003). The 1D Tomonaga-Luttinger liquid state of SWCNTs was observed as a power law scaling of the photoemission response in the vicinity of the Fermi level (Ishii et al. 2003).

When SWCNTs are electron (hole) doped, the lowest energy VHS peak in the optical absorption spectrum is reduced in intensity as a result of the occupation (depletion) of the corresponding conduction (valence) VHS state (Takenobu et al. 2003; Liu et al. 2004). In the photoemission spectrum, the VHS peaks shift to higher (lower) binding energies as bundled SWCNTs are electron (hole) doped, for example, by the intercalation with alkali metals (Rauf et al. 2004, 2005; Larciprete et al. 2005). From the energy shifts, one can estimate the number of electrons (holes) transferred to the SWCNTs.

As shown in Figure 15.17, the VHS peaks of $FeCp_2$@NT are located at 0.5, 0.8, and 1.1 eV, which are shifted toward higher binding energies as compared with those for the pristine SWCNT. From a close similarity to alkali metal-doped SWCNT materials (Rauf et al. 2004; Larciprete et al. 2005), this behavior can be attributed to the electron charge transfer from the $FeCp_2$ to the SWCNT. Upon annealing at

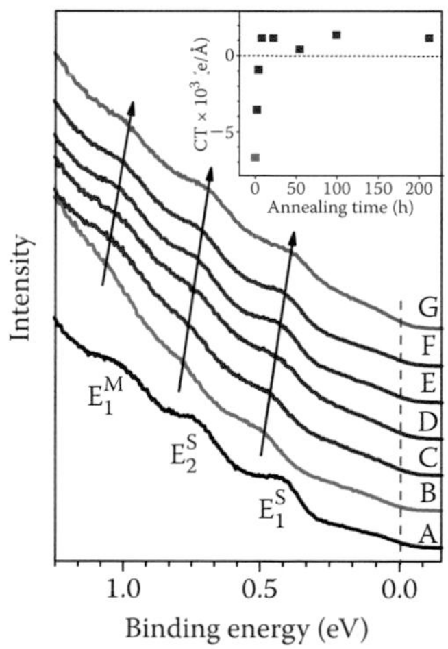

FIGURE 15.17 Valence band photoemission spectra for (A) pristine SWCNT, (B) $FeCp_2$@NT, (C–F) Fe-doped DWCNT or $FeCp_2$@NT annealed at 600°C for 2, 8, 54, and 212 h, respectively, (G) $FeCp_2$@NT annealed at 1150°C for 1 h. Inset: Charge transfer density versus annealing time. (Reprinted with permission from Shiozawa, H. et al., *Phys. Rev. B*, 77, 153402, 2008. Copyright 2008 by the American Physical Society.)

600°C, the VHS peaks are gradually shifted toward lower binding energies and become stable after 22 h. At this stage, the material is transformed into iron-doped DWCNT (Shiozawa et al. 2008b). In turn, after annealing at 1150°C, the iron atoms have left the DWCNT, but the VHS peaks are further shifted toward lower energies. This suggests that there is electron transfer from the outer tube to the inner tube in the empty DWCNT. The direction of the charge transfer is consistent with a theoretical prediction (Zolyomi et al. 2008).

A quantitative analysis of the charge transfer between the filling and the SWCNT was provided both experimentally and theoretically (Shiozawa et al. 2008c). Charge transfer values for $FeCp_2$@NT can be derived from the filling factor and energy shifts of the VHS peaks by comparing them with those for potassium doped SWCNT (Shiozawa et al. 2008b). The inset of Figure 15.17 shows the charge transfer density along the tube axis for $FeCp_2$@NT samples annealed at 600°C for different time intervals. The charge transfer reaches a maximum negative value of −0.0067 e/Å for the pristine sample and drops rapidly at the first few annealing steps. It becomes positive and approaches a constant value around +0.0010 e/Å after 8 h of annealing.

The direction and the values are in very good agreement with the theoretical values. From a detailed comparison to the calculations, a possible covalent bonding between the ferrocene and the nanotube host was suggested, which was also supported by the sp^3 character of carbon atoms observed by C 1*s* photoemission and absorption spectroscopy (Shiozawa et al. 2008c). Recently, doping of SWCNTs was controlled to a larger extent by filling them with nickelocene in larger yield followed by nano-reactions (Kharlamova et al. 2015) (see Figure 15.18). Use of nickelocene allowed the initial reaction temperature to be reduced down to 250°. Synchronous charge transfer among the molecular components (outer and inner tubes, nickel carbide, and nickel clusters) led to bipolar doping of the CNTs to be achieved in a wide range of ±0.0012 e^- per carbon as nickelocene-filled SWCNTs are heated in vacuum at various temperatures from 250°C up to 1200°C. As mentioned before, in DWCNTs, electrons are transferred from the inner tube to the outer tube, which means hole doping of the parent SWCNT. The type of nickel compounds created inside SWCNTs depends on the temperature. From the x-ray photoemission and TEM analysis, it was found that nickel carbide clusters are produced at temperatures up to 400°C, which are then transformed into pure nickel clusters at temperatures up to 600°C. The atomic concentration of nickel to carbon is reduced as the temperature is elevated (Figure 15.18b). Taking the quantity of inner tubes and nickel at each temperature into account, one gets the doping efficiency per nickel atom as plotted in Figure 15.18c. It is observed that the nickel carbide clusters have an electron doping efficiency of ~0.055 electrons per nickel, while with the pure nickel clusters it gets as high as ~0.12 electrons per nickel.

Furthermore, electron doping of the 1D VHS state of SWCNTs was realized via a chemical reaction of an encapsulated organo-cerium compound, $CeCp_3$ (Shiozawa et al. 2009). It was found that the formation of cerium clusters inside the CNTs increases the doping level and greatly enhances the density of conduction electrons. The transition of the cerium encapsulating semiconducting tubes to metallic results in enhanced screening of the photoexcited core-hole potential.

Figure 15.19a shows the x-ray absorption spectra for pristine and annealed $CeCp_3$-filled SWCNT at the C 1*s* → π^* excitation region. The pristine $CeCp_3$@SWCNT shows discernible fine structures corresponding to the unoccupied S_1^*, S_2^*, M_1^*, and S_3^* VHS peaks, similar to those for empty SWCNT materials (Shiozawa et al. 2008c). Upon annealing, S_1^* and S_2^* are suppressed and the two structures emerge near the onset of the π^* peak for the Ce@DWNT. The corresponding C 1*s* photoemission spectra exhibit energy shifts to lower binding energies by up to 0.3 eV (see Figure 15.19b).

Upon C 1*s* → π^* near-edge absorption of a SWCNT, an incoming photon excites one electron from the C 1*s* core level to the unoccupied π^* state, as illustrated in the left panel of Figure 15.19c. The final state of this process includes one hole (h^+) at the C 1*s* and one electron (e^-) at the previously unoccupied state. The energy shifts observed in both x-ray absorption and photoemission data at the C 1*s* edge were explained as results of screening of the core hole potential enhanced by doping-induced transition of the semiconducting SWCNTs to metallic phases.

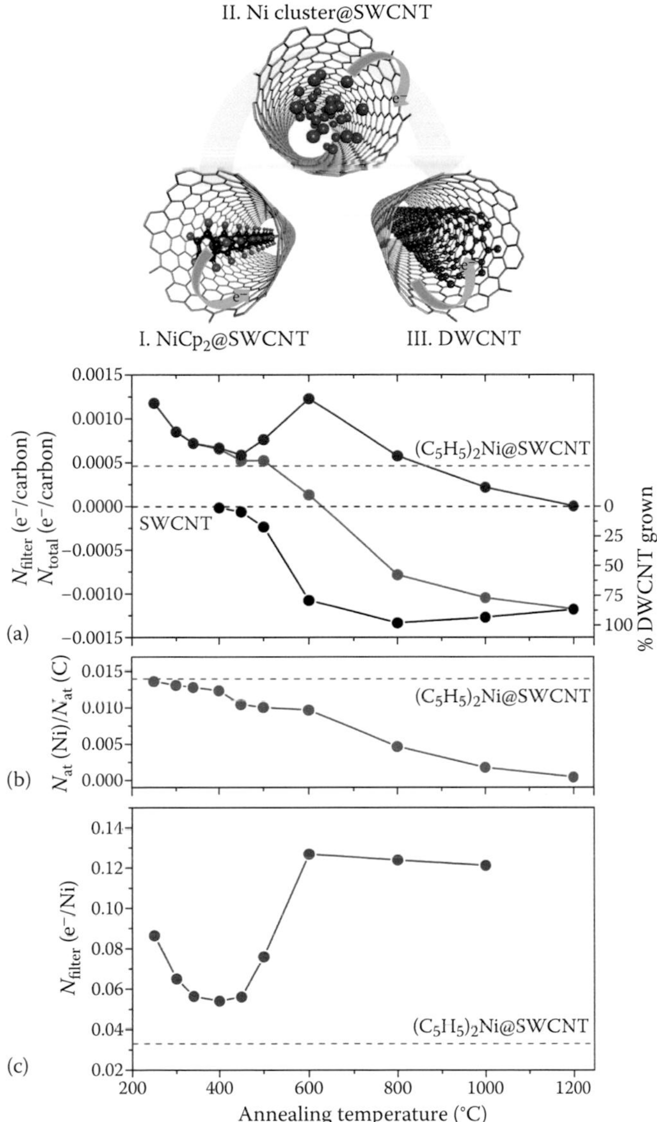

FIGURE 15.18 (Top) Schematic of electron bipolar doping of SWCNTs via the chemical transformation of encapsulated nickelocene. (a) The number of electrons transferred to SWCNTs in total (light gray), the normalized area intensity of the mean-diameter inner tube RBM peak (black), and the number of electrons transferred from encapsulated nickel substances apart from inner tubes (dark gray) plotted against annealing temperature. (b) The nickel-to-carbon atomic ratio. (c) The number of electrons transferred from nickel substances per nickel atom. Dashed horizontal lines show the values for SWCNTs and nickelocene-filled SWCNTs. (From Kharlamova, M.V. et al., *Nanoscale*, 7, 1383–1391, 2015. Published by The Royal Society of Chemistry.)

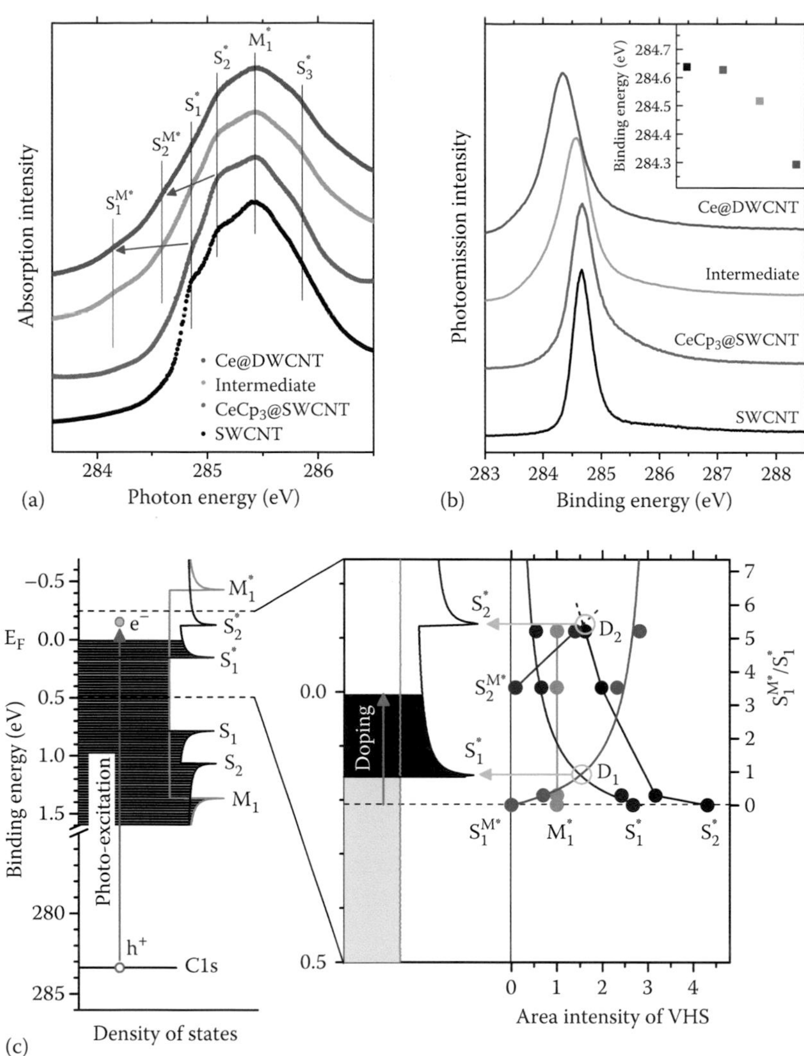

FIGURE 15.19 (a) C 1*s* → π* absorption of cerocene-filled SWCNTs. (b) C 1*s* photoemission. Inset shows C 1*s* main peak positions. (c) (left) Schematic diagram for C 1*s* → π* photoabsorption process. (Right) Normalized area intensity of VHS absorption peaks, S_1^*, S_2^*, M_1^*, S_1^{M*}, and S_2^{M*}, as a function of the intensity ratio of S_1^{M*}/S_1^*. The vertical arrow indicates chemical-reaction-induced doping of the conduction states of tubes. The curves S_1^* and S_1^{M*} are the results of fitting using the equations $f_1(\alpha) = C(1 + \alpha)^{-1}$ and $f_2(\alpha) = C\alpha(1 + \alpha)^{-1}$, respectively, where C = const. and $\alpha = S_1^{M*}/S_1^*$. The open circles, D_1 and D_2, indicate the crossing points between S_n^* and S_n^{M*} (n = 1, 2) and correspond to the doping levels at the S_1^* and S_2^* peaks in the density of states, respectively. (From Shiozawa, H. et al., *Phys. Rev. Lett.*, 102, 046804, 2009. Copyright 2009 by the American Physical Society.)

15.4.3.1 Formation of Metal Clusters

Encapsulated in CNTs, organometallic, coordination compounds and metallofullerenes can be reacted to form metal clusters. Metallocenes and metal acetylacetonates inside SWCNT were transformed into metal particles (Li et al. 2006b; Shiozawa et al. 2008b, 2009, 2010; Briones-Leon et al. 2013; Kharlamova et al. 2015). $(Gd@C_{82})_n$@SWCNTs can be converted into Gd nanowires@DWCNTs via annealing in vacuum (Kitaura et al. 2008). Carbonyl complexes of transition metals inserted in SWCNT were transformed into metal particles by heating or electron beam irradiation (Chamberlain et al. 2012b). As mentioned before, ferrocene in SWCNTs can be transformed into iron carbide clusters by annealing in vacuum (Shiozawa et al. 2008b). This process allows magnetic nanoparticles to be prepared within the interior of SWCNT. The magnetic properties of those encapsulated in pure metallic or semiconducting SWCNT hosts were studied with regard to the effect of cluster size and metalicity of SWCNTs by means of superconducting quantum interference device (SQUID) and x-ray magnetic circular dichroism (XMCD) (Briones-Leon et al. 2013).

15.4.4 Magnetic Properties of Clusters inside CNTs

Magnetic substances can be arranged in the nanotubular interior, which allows novel low-dimensional hybrid magnetic nanostructures to be made (Gimenez-Lopez et al. 2011). $C_{59}N$ magnetic fullerene peapods were made via vacuum annealing of functionalized $C_{59}N$-filled SWCNTs, as probed by electron spin resonance spectroscopy (Simon et al. 2006a).

Furthermore, encapsulated reactions inside SWCNTs provide a unique opportunity for the synthesis of magnetic metal clusters (Briones-Leon et al. 2013; Shiozawa et al. 2015). Ferrocene in SWCNTs can be transformed into iron carbide clusters by annealing in vacuum. This process allows magnetic nanoparticles to be prepared within the interior of SWCNTs. The magnetic properties of those encapsulated in pure metallic or semiconducting SWCNT hosts were studied with regard to the effect of cluster size and metalicity of SWCNT by means of SQUID and XMCD (Briones-Leon et al. 2013). The XMCD spectroscopy allows the magnetic states of minority elements and specific atomic orbitals in compounds and alloys to be probed, as well as the spin and orbital magnetic moments to be evaluated separately. Hence, the XMCD is a powerful tool for the study of metal clusters encapsulated in SWCNTs. It was shown that the Fe 3d spin magnetic moments of iron carbide clusters are reduced as compared with the bulk metallic iron, while the 3d orbital magnetic moments are unchanged. The field dependence data showed no hysteresis and can be fitted to the Langevin function, which indicates that the iron clusters inside SWCNTS are superparamagnetic (see Figure 15.20). Interestingly, irons in metallic SWCNTs have smaller magnetic moments than those in semiconducting SWCNTs. This could indicate that the size effect on the magnetism becomes greater when the magnetic clusters are in contact with metallic SWCNTs, which have a higher density of states at the Fermi level.

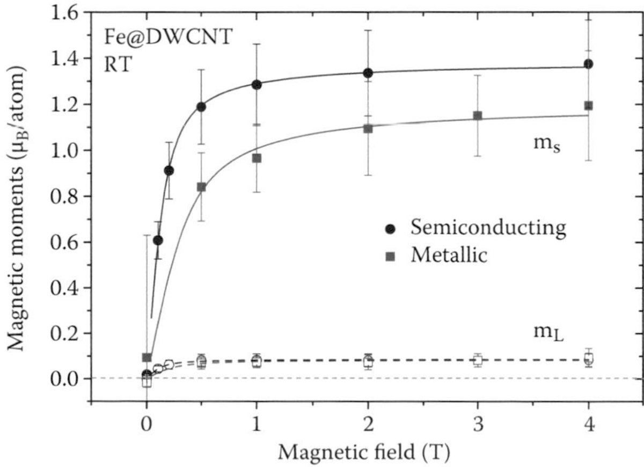

FIGURE 15.20 Spin and orbital magnetic moments of the metallic and semiconducting Fe@DWCNT. The data are fitted with the modified Langevin function. The measurements were done at 300 K. (From Briones-Leon, A. et al., *Phys. Rev. B*, 87, 195435, 2013. Copyright 2013 by the American Physical Society.)

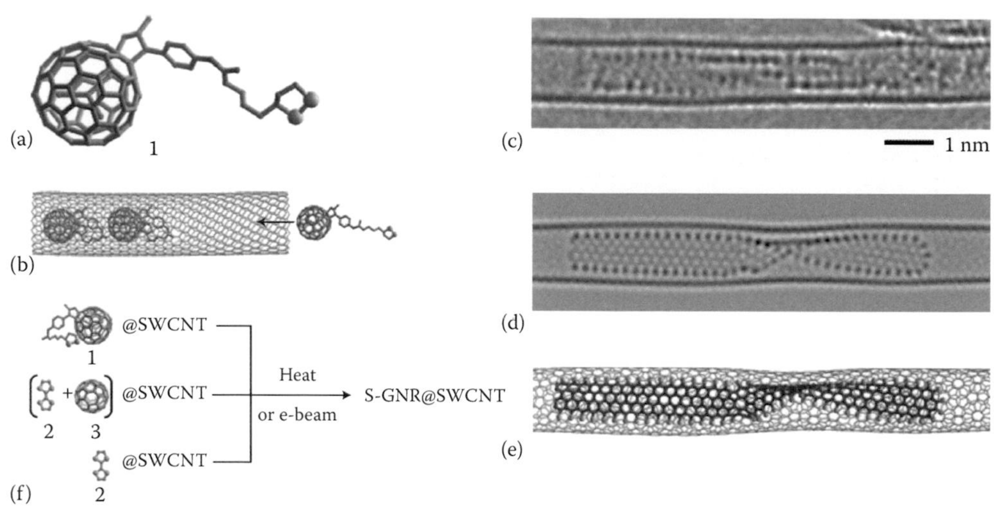

FIGURE 15.21 (a) Functionalized fullerene 1 bearing an organic group with sulphur atoms on their surface. (b) 1 encapsulated into a (14,5)-SWCNT. (c) An experimental TEM image for sulphur atoms terminating the edges of the graphene nanoribbon (GNR) appears as chains of dark atoms. (d) An image simulated from the model. (e) A model of S-GNR@SWCNT. (Reprinted by permission from Macmillan Publishers Ltd: Chuvilin, A. et al., *Nat. Mater.*, 10, 687–692, 2011, Copyright 2013.)

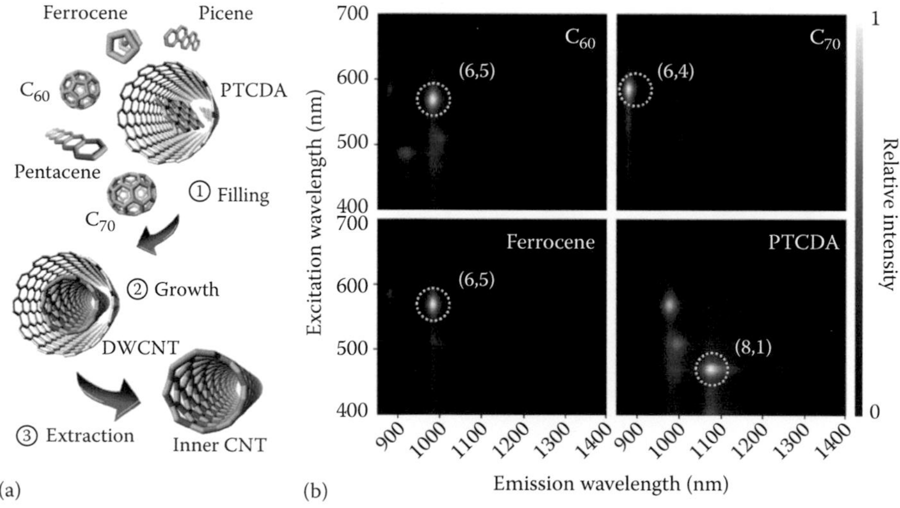

FIGURE 15.22 (a) Schematic diagram of noncatalytic synthesis of single-wall CNTs within a parent tube template. Precursors were first encapsulated into the interior nanospace and subsequently annealed at 1200°C to induce growth. Newly formed inner tubes were then extracted via sonication and further separated from the bulk outer tube templates using the density gradient ultracentrifugation method. (b) Photoluminescence contour maps of the separated inner tubes. (Reprinted by permission from Macmillan Publishers Ltd: Lim, H.E. et al., *Nat. Commun.*, 4, 2548, 2013, Copyright 2013.)

15.5 Growth of 1D Molecular Nanostructures

It was reported that similar reactions can be made with SWCNTs encapsulating organic molecules to the formation of inner CNTs (Fujita et al. 2005; Lim et al. 2013), oligomers of coronene (Okazaki et al. 2011; Botka et al. 2014), polymers, and graphene ribbons from coronene or perylene, tetrathiafulvalene, or functionalized fullerene (Chuvilin et al. 2011, Talyzin et al. 2011; Chamberlain et al. 2012a; Fujihara et al. 2012; Chernov et al. 2013) (see Figures 15.21 and 15.22). Figure 15.22 shows examples of noncatalytic synthesis of single-wall CNTs via twisted graphene nanoribbons within a parent tube template. Linear carbon chains with single–triple bonds (carbyne chains) were formed by heating a DWCNT material at temperatures in a range of 1400°C–1500°C, as evidenced by strong Raman signals observed at frequencies between 1820 and 1880 cm^{-1} (Shi et al. 2013). Heterogeneous structures can also be produced in a similar manner. BN inner nanotubes were synthesized from encapsulated ammonia borane complexes (ABCs) (Nakanishi et al. 2013), as identified from TEM and EELS observations. In detail, ABC-filled carbon tubes were caped with C_{60}, then annealed at 1400°C to produce the BN-CNT double-wall structure with an optical gap of 6 eV.

Acknowledgment

I acknowledge valuable discussions with Prof. Hans Kuzmany.

References

Abe, M., H. Kataura, H. Kira, T. Kodama, S. Suzuki, Y. Achiba, K. Kato et al. Structural transformation from single-wall to double-wall carbon nanotube bundles. *Physical Review B*, 68 (4): 041405, 2003. 10.1103/PhysRevB.68.041405.

Ajayan, P. M., T. W. Ebbesen, T. Ichihashi, S. Iijima, K. Tanigaki, and H. Hiura. Opening carbon nanotubes with oxygen and implications for filling. *Nature*, 362 (6420): 522–525, 1993. 10.1038/362522a0.

Ajayan, P. M., O. Stephan, P. Redlich, and C. Colliex. Carbon nanotubes as removable templates for metal-oxide nanocomposites and nanostructures. *Nature*, 375 (6532): 564–567, 1995. 10.1038/375564a0.

Alvarez, L. E. A. One-dimensional molecular crystal of phthalocyanine confined into single-walled carbon nanotubes. *The Journal of Physical Chemistry C*, 119 (9): 5203–5210, 2015. 10.1021/acs.jpcc.5b00168.

Alvarez, L., Y. Almadori, R. Arenal, R. Babaa, T. Michel, R. Le Parc, J., L. Bantignies et al. Charge transfer evidence between carbon nanotubes and encapsulated conjugated oligomers. *Journal of Physical Chemistry C*, 115 (24): 11898–11905, 2011. 10.1021/jp1121678.

Arnold, M. S., A. A. Green, J. F. Hulvat, S. I. Stupp, and M. C. Hersam. Sorting carbon nanotubes by electronic structure using density differentiation. *Nature Nanotechnology*, 1 (1): 60–65, 2006. 10.1038/nnano.2006.52.

Bandow, S., S. Asaka, Y. Saito, A. M. Rao, L. Grigorian, E. Richter, and P. C. Eklund. Effect of the growth temperature on the diameter distribution and chirality of single-wall carbon nanotubes. *Physical Review Letters*, 80 (17): 3779–3782, 1998. 10.1103/PhysRevLett.80.3779.

Bandow, S., M. Takizawa, K. Hirahara, M. Yudasaka, and S. Iijima. Raman scattering study of double-wall carbon nanotubes derived from the chains of fullerenes in single-wall carbon nanotubes. *Chemical Physics Letters*, 337 (1–3): 48–54, 2001. 10.1016/S0009-2614(01)00192-0.

Botka, B., M. E. Fuestoes, H. M. Tohati, K. Nemeth, G. Klupp, Z. Szekrenyes, D. Kocsis et al. Interactions and chemical transformations of coronene inside and outside carbon nanotubes. *Small*, 10 (7): 1369–1378, 2014. 10.1002/smll.201302613.

Briones-Leon, A., P. Ayala, X. J. Liu, K. Yanagi, E. Weschke, M. Eisterer, H. Jiang, H. Kataura, T. Pichler, and H. Shiozawa. Orbital and spin magnetic moments of transforming one-dimensional iron inside metallic and semiconducting carbon nanotubes. *Physical Review B*, 87 (19): 195435, 2013. 10.1103/PhysRevB.87.195435.

Britz, D. A., A. N. Khlobystov, J. W. Wang, A. S. O'Neil, M. Poliakoff, A. Ardavan, and G. A. D. Briggs. Selective host-guest interaction of single-walled carbon nanotubes with functionalised fullerenes. *Chemical Communications*, (2): 176–177, 2004. 10.1039/b313585c.

Cambedouzou, J., J. L. Sauvajol, A. Rahmani, E. Flahaut, A. Peigney, and C. Laurent. Raman spectroscopy of iodine-doped double-walled carbon nanotubes. *Physical Review B*, 69 (23): 235422, 2004. 10.1103/PhysRevB.69.235422.

Cambre, S., W. Wenseleers, and E. Goovaerts. Endohedral copper(ii)acetylacetonate/single-walled carbon nanotube hybrids characterized by electron paramagnetic resonance. *Journal of Physical Chemistry C*, 113 (31): 13505–13514, 2009. 10.1021/jp903724h.

Chamberlain, T. W., J. C. Meyer, J. Biskupek, J. Leschner, A. Santana, N. A. Besley, E. Bichoutskaia, U. Kaiser, and A. N. Khlobystov. Reactions of the inner surface of carbon nanotubes and nano-protrusion processes imaged at the atomic scale. *Nature Chemistry*, 3 (9): 732–737, 2011. 10.1038/NCHEM.1115.

Chamberlain, T. W., J. Biskupek, G. A. Rance, A. Chuvilin, T. J. Alexander, E. Bichoutskaia, U. Kaiser, and A. N. Khlobystov. Size, structure, and helical twist of graphene nanoribbons controlled by confinement in carbon nanotubes. *ACS Nano*, 6 (5): 3943–3953, 2012a. 10.1021/nn300137j.

Chamberlain, T. W., T. Zoberbier, J. Biskupek, A. Botos, U. Kaiser, and A. N. Khlobystov. Formation of uncapped nanometre-sized metal particles by decomposition of metal carbonyls in carbon nanotubes. *Chemical Science*, 3 (6): 1919–1924, 2012b. 10.1039/c2sc01026g.

Chernov, A. I., P. V. Fedotov, A. V. Talyzin, I. S. Lopez, I. V. Anoshkin, A. G. Nasibulin, E. I. Kauppinen, and E. D. Obraztsova. Optical properties of graphene nanoribbons encapsulated in single-walled carbon nanotubes. *ACS Nano*, 7 (7): 6346–6353, 2013. 10.1021/nn4024152.

Chuvilin, A., E. Bichoutskaia, M. C. Gimenez-Lopez, T. W. Chamberlain, G. A. Rance, N. Kuganathan, J. Biskupek, U. Kaiser, and A. N. Khlobystov. Self-assembly of a sulphur-terminated graphene nanoribbon within a single-walled carbon nanotube. *Nature Materials*, 10 (9): 687–692, 2011. 10.1038/NMAT3082.

Corio, P., A. P. Santos, P. S. Santos, M. L. A. Temperini, V. W. Brar, M. A. Pimenta, and M. S. Dresselhaus. Characterization of single wall carbon nanotubes filled with silver and with chromium compounds. *Chemical Physics Letters*, 383 (5–6): 475–480, 2004. 10.1016/j.cplett.2003.11.061.

Fan, X., E. C. Dickey, P. C. Eklund, K. A. Williams, L. Grigorian, R. Buczko, S. T. Pantelides, and S. J. Pennycook. Atomic arrangement of iodine atoms inside single-walled carbon nanotubes. *Physical Review Letters*, 84 (20): 4621–4624, 2000. 10.1103/PhysRevLett.84.4621.

Fan, J., T. W. Chamberlain, Y. Wang, S. Yang, A. J. Blake, M. Schroeder, and A. N. Khlobystov. Encapsulation of transition metal atoms into carbon nanotubes: A supramolecular approach. *Chemical Communications*, 47 (20): 5696–5698, 2011. 10.1039/c1cc10427f.

Fujihara, M., Y. Miyata, R. Kitaura, Y. Nishimura, C. Camacho, S. Irle, Y. Iizumi, T. Okazaki, and H. Shinohara. Dimerization-initiated preferential formation of coronene-based graphene nanoribbons in carbon nanotubes. *Journal of Physical Chemistry C*, 116 (28): 15141–15145, 2012. 10.1021/jp3037268.

Fujimori, T., R. B. dos Santos, T. Hayashi, M. Endo, K. Kaneko, and D. Tomanek. Formation and properties of selenium double-helices inside double-wall carbon nanotubes: Experiment and theory. *ACS Nano*, 7 (6): 5607–5613, 2013a. 10.1021/nn4019703.

Fujimori, T., A. Morelos-Gomez, Z. Zhu, H. Muramatsu, R. Futamura, K. Urita, M. Terrones et al. Conducting linear chains of sulphur inside carbon nanotubes. *Nature Communications*, 4: 2162, 2013b. 10.1038/ncomms3162.

Fujita, Y., S. Bandow, and S. Iijima. Formation of small-diameter carbon nanotubes from PTCDA arranged inside the single-wall carbon nanotubes. *Chemical Physics Letters*, 413 (4–6): 410–414, 2005. 10.1016/j.cplett.2005.08.033.

Gao, J., P. Blondeau, P. Salice, E. Menna, B. Bartova, C. Hebert, J. Leschner et al. Electronic interactions between "pea" and "pod": The case of oligothiophenes encapsulated in carbon nanotubes. *Small*, 7 (13): 1807–1815, 2011. 10.1002/smll.201100319.

Garcia-Suarez, V. M., J. Ferrer, and C. J. Lambert. Tuning the electrical conductivity of nanotube-encapsulated metallocene wires. *Physical Review Letters*, 96 (10): 106804, 2006. 10.1103/PhysRevLett.96.106804.

Ghosh, S., S. M. Bachilo, and R. B. Weisman. Advanced sorting of single-walled carbon nanotubes by non-linear density-gradient ultracentrifugation. *Nature Nanotechnology*, 5 (6): 443–450, 2010. 10.1038/NNANO.2010.68.

Gimenez-Lopez, M. D. C., F. Moro, A. La Torre, C. J. Gomez-Garcia, P. D. Brown, J. van Slageren, and A. N. Khlobystov. Encapsulation of single-molecule magnets in carbon nanotubes. *Nature Communications*, 2: 407, 2011. 10.1038/ncomms1415.

Giusca, C. E., Y. Tison, V. Stolojan, E. Borowiak-Palen, and S. R. P. Silva. Inner-tube chirality determination for double-walled carbon nanotubes by scanning tunneling microscopy. *Nano Letters*, 7 (5): 1232–1239, 2007. 10.1021/nl070072p.

Giusca, C. E., V. Stolojan, J. Sloan, F. Boerrnert, H. Shiozawa, K. Sader, M. H. Ruemmeli, B. Buechner, and S. R. P. Silva. Confined crystals of the smallest phase-change material. *Nano Letters*, 13 (9): 4020–4027, 2013. 10.1021/nl4010354.

Guan, L. H., Z. J. Shi, M. X. Li, and Z. N. Gu. Ferrocene-filled single-walled carbon nanotubes. *Carbon*, 43 (13): 2780–2785, 2005. 10.1016/j.carbon.2005.05.025.

Hanami, K.-I., T. Umesaki, K. Matsuda, Y. Miyata, H. Kataura, Y. Okabe, and Y. Maniwa. One-dimensional oxygen and helical oxygen nanotubes inside carbon nanotubes. *Journal of the Physical Society of Japan*, 79 (2): 023601, 2010. 10.1143/JPSJ.79.023601.

Hernandez, E., V. Meunier, B. W. Smith, R. Rurali, H. Terrones, M. B. Nardelli, M. Terrones, D. E. Luzzi, and J. C. Charlier. Fullerene coalescence in nanopeapods: A path to novel tubular carbon. *Nano Letters*, 3 (8): 1037–1042, 2003. 10.1021/nl034283f.

Hirahara, K., K. Suenaga, S. Bandow, H. Kato, T. Okazaki, H. Shinohara, and S. Iijima. One-dimensional metallofullerene crystal generated inside single-walled carbon nanotubes. *Physical Review Letters*, 85 (25): 5384–5387, 2000. 10.1103/PhysRevLett.85.5384.

Hornbaker, D. J., S. J. Kahng, S. Misra, B. W. Smith, A. T. Johnson, E. J. Mele, D. E. Luzzi, and A. Yazdani. Mapping the one-dimensional electronic states of nanotube peapod structures. *Science*, 295 (5556): 828–831, 2002. 10.1126/science.1068133.

Iijima, S. Helical microtubules of graphitic carbon. *Nature*, 354: 56–58, 1991. 10.1038/354056a0.

Iizumi, Y., H. Suzuki, M. Tange, and T. Okazaki. Diameter selective electron transfer from encapsulated ferrocenes to single-walled carbon nanotubes. *Nanoscale*, 6 (22): 13910–13914, 2014. 10.1039/c4nr04398g.

Ishii, H., H. Kataura, H. Shiozawa, H. Yoshioka, H. Otsubo, Y. Takayama, T. Miyahara et al. Direct observation of tomonaga-luttinger-liquid state in carbon nanotubes at low temperatures. *Nature*, 426 (6966): 540–544, 2003. 10.1038/nature02074.

Jeong, G. H., A. A. Farajian, R. Hatakeyama, T. Hirata, T. Yaguchi, K. Tohji, H. Mizuseki, and Y. Kawazoe. Cesium encapsulation in single-walled carbon nanotubes via plasma ion irradiation: Application to junction formation and ab initio investigation. *Physical Review B*, 68 (7): 075410, 2003. 10.1103/PhysRevB.68.075410.

Kalbac, M., L. Kavan, S. Gorantla, T. Gemming, and L. Dunsch. Sexithiophene encapsulated in a single-walled carbon nanotube: An in situ raman spectroelectrochemical study of a peapod structure. *Chemistry—A European Journal*, 16 (38): 11753–11759, 2010. 10.1002/chem.201001417.

Kataura, H., Y. Kumazawa, Y. Maniwa, I. Umezu, S. Suzuki, Y. Ohtsuka, and Y. Achiba. Optical properties of single-wall carbon nanotubes. *Synthetic Metals*, 103 (1–3): 2555–2558, 1999. 10.1016/S0379-6779(98)00278-1.

Kataura, H., Y. Kumazawa, Y. Maniwa, Y. Ohtsuka, R. Sen, S. Suzuki, and Y. Achiba. Diameter control of single-walled carbon nanotubes. *Carbon*, 38 (11–12): 1691–1697, 2000. 10.1016/S0008-6223(00)00090-7.

Kataura, H., Y. Maniwa, T. Kodama, K. Kikuchi, K. Hirahara, K. Suenaga, S. Iijima, S. Suzuki, Y. Achiba, and W. Kratschmer. High-yield fullerene encapsulation in single-wall carbon nanotubes. *Synthetic Metals*, 121 (1–3): 1195–1196, 2001. 10.1016/S0379-6779(00)00707-4.

Kataura, H., Y. Maniwa, M. Abe, A. Fujiwara, T. Kodama, K. Kikuchi, H. Imahori, Y. Misaki, S. Suzuki, and Y. Achiba. Optical properties of fullerene and non-fullerene peapods. *Applied Physics A—Materials Science & Processing*, 74 (3): 349–354, 2002. 10.1007/s003390201276.

Kharlamova, M. V., M. Sauer, T. Saito, Y. Sato, K. Suenaga, T. Pichler, and H. Shiozawa. Doping of single-walled carbon nanotubes controlled via chemical transformation of encapsulated nickelocene. *Nanoscale*, 7 (4): 1383–1391, 2015. 10.1039/c4nr05586a.

Khlobystov, A. N., D. A. Britz, J. W. Wang, S. A. O'Neil, M. Poliakoff, and G. A. D. Briggs. Low temperature assembly of fullerene arrays in single-walled carbon nanotubes using supercritical fluids. *Journal of Materials Chemistry*, 14 (19): 2852–2857, 2004. 10.1039/b404167d.

Kitaura, R., N. Imazu, K. Kobayashi, and H. Shinohara. Fabrication of metal nanowires in carbon nanotubes via versatile nano-template reaction. *Nano Letters*, 8 (2): 693–699, 2008. 10.1021/nl073070d.

Koshino, M., N. Solin, T. Tanaka, H. Isobe, and E. Nakamura. Imaging the passage of a single hydrocarbon chain through a nanopore. *Nature Nanotechnology*, 3 (10): 595–597, 2008. 10.1038/nnano.2008.263.

Koshino, M., Y. Niimi, E. Nakamura, H. Kataura, T. Okazaki, K. Suenaga, and S. Iijima. Analysis of the reactivity and selectivity of fullerene dimerization reactions at the atomic level. *Nature Chemistry*, 2 (2): 117–124, 2010. 10.1038/NCHEM.482.

Kramberger, C., H. Rauf, M. Knupfer, H. Shiozawa, D. Batchelor, A. Rubio, H. Kataura, and T. Pichler. Potassium-intercalated single-wall carbon nanotube bundles: Archetypes for semiconductor/metal hybrid systems. *Physical Review B*, 79: 195442, 2009. 10.1103/PhysRevB.79.195442.

Kuwahara, R., Y. Kudo, T. Morisato, and K. Ohno. Encapsulation of carbon chain molecules; in single-walled carbon nanotubes. *Journal of Physical Chemistry A*, 115 (20): 5147–5156, 2011. 10.1021/jp109308w.

Kuzmany, H., R. Pfeiffer, M. Hulman, and C. Kramberger. Raman spectroscopy of fullerenes and fullerene-nanotube composites. *Philosophical Transactions of the Royal Society A—Mathematical Physical and Engineering Sciences*, 362 (1824): 2375–2406, 2004. 10.1098/rsta.2004.1446.

Larciprete, R., L. Petaccia, S. Lizzit, and A. Goldoni. Transition from one-dimensional to three-dimensional behavior induced by lithium doping in single wall carbon nanotubes. *Physical Review B*, 71 (11): 115435, 2005. 10.1103/PhysRevB.71.115435.

Lee, J., H. Kim, S. J. Kahng, G. Kim, Y. W. Son, J. Ihm, H. Kato, Z. W. Wang, T. Okazaki, H. Shinohara, and Y. Kuk. Bandgap modulation of carbon nanotubes by encapsulated metallofullerenes. *Nature*, 415 (6875): 1005–1008, 2002. 10.1038/4151005a.

Li, L. J., A. N. Khlobystov, J. G. Wiltshire, G. A. D. Briggs, and R. J. Nicholas. Diameter-selective encapsulation of metallocenes in single-walled carbon nanotubes. *Nature Materials*, 4 (6): 481–485, 2005. 10.1038/nmat1396.

Li, Y. F., R. Hatakeyama, T. Kaneko, T. Izumida, T. Okada, and T. Kato. Synthesis and electronic properties of ferrocene-filled double-walled carbon nanotubes. *Nanotechnology*, 17 (16): 4143–4147, 2006a. 10.1088/0957-4484/17/16/025.

Li, Y. F., R. Hatakeyama, T. Kaneko, and T. Okada. Nano sized magnetic particles with diameters less than 1 nm encapsulated in single-walled carbon nanotubes. *Japanese Journal of Applied Physics, Part 2*, 45: L428–L431, 2006b. 10.1143/JJAP.45.L428.

Lim, H. E., Y. Miyata, R. Kitaura, Y. Nishimura, Y. Nishimoto, S. Irle, J. H. Warner, H. Kataura, and H. Shinohara. Growth of carbon nanotubes via twisted graphene nanoribbons. *Nature Communications*, 4: 2548, 2013. 10.1038/ncomms3548.

Liu, X., T. Pichler, M. Knupfer, M. S. Golden, J. Fink, H. Kataura, Y. Achiba, K. Hirahara, and S. Iijima. Filling factors, structural, and electronic properties of c-60 molecules in single-wall carbon nanotubes. *Physical Review B*, 65 (4): 045419, 2002. 10.1103/PhysRevB.65.045419.

Liu, X., T. Pichler, M. Knupfer, and J. Fink. Electronic and optical properties of alkali-metal-intercalated single-wall carbon nanotubes. *Physical Review B*, 67 (12): 125403, 2003. 10.1103/PhysRevB.67.125403.

Liu, X., T. Pichler, M. Knupfer, J. Fink, and H. Kataura. Electronic properties of fecl3-intercalated single-wall carbon nanotubes. *Physical Review B*, 70 (20): 205405, 2004. 10.1103/PhysRevB.70.205405.

Liu, H., D. Nishide, T. Tanaka, and H. Kataura. Large-scale single-chirality separation of single-wall carbon nanotubes by simple gel chromatography. *Nature Communications*, 2: 309, 2011. 10.1038/ncomms1313.

Liu, X. J., H. Kuzmany, P. Ayala, M. Calvaresi, F. Zerbetto, and T. Pichler. Selective enhancement of photoluminescence in filled single-walled carbon nanotubes. *Advanced Functional Materials*, 22 (15): 3202–3208, 2012. 10.1002/adfm.201200224.

Liu, H., T. Tanaka, Y. Urabe, and H. Kataura. High-efficiency single-chirality separation of carbon nanotubes using temperature-controlled gel chromatography. *Nano Letters*, 13 (5): 1996–2003, 2013. 10.1021/nl400128m.

Loi, M. A., J. Gao, F. Cordella, P. Blondeau, E. Menna, B. Bartova, C. Hebert, S. Lazar, G. A. Botton, M. Milko, and C. Ambrosch-Draxl. Encapsulation of conjugated oligomers in single-walled carbon nanotubes: Towards nanohybrids for photonic devices. *Advanced Materials*, 22 (14): 1635–1639, 2010. 10.1002/adma.200903527.

Lu, J., S. Nagase, D. P. Yu, H. Q. Ye, R. S. Han, Z. X. Gao, S. Zhang, and L. M. Peng. Amphoteric and controllable doping of carbon nanotubes by encapsulation of organic and organometallic molecules. *Physical Review Letters*, 93 (11): 116804, 2004. 10.1103/PhysRevLett.93.116804.

Maggini, L., M.-E. Fuestoes, T. W. Chamberlain, C. Cebrian, M. Natali, M. Pietraszkiewicz, O. Pietraszkiewicz et al. Fullerene-driven encapsulation of a luminescent eu(iii) complex in carbon nanotubes. *Nanoscale*, 6 (5): 2887–2894, 2014. 10.1039/c3nr05876j.

Malard, L. M., D. Nishide, L. G. Dias, R. B. Capaz, A. P. Gomes, A. Jorio, C. A. Achete et al. Pimenta. Resonance raman study of polyynes encapsulated in single-wall carbon nanotubes. *Physical Review B*, 76 (23): 233412, 2007. 10.1103/PhysRevB.76.233412.

Maniwa, Y., H. Kataura, M. Abe, S. Suzuki, Y. Achiba, H. Kira, and K. Matsuda. Phase transition in confined water inside carbon nanotubes. *Journal of the Physical Society of Japan*, 71 (12): 2863–2866, 2002. 10.1143/JPSJ.71.2863.

Maniwa, Y., H. Kataura, M. Abe, A. Fujiwara, R. Fujiwara, H. Kira, H. Tou et al. Suematsu. C-70 molecular stumbling inside single-walled carbon nanotubes. *Journal of the Physical Society of Japan*, 72 (1): 45–48, 2003. 10.1143/JPSJ.72.45.

Maniwa, Y., H. Kataura, M. Abe, A. Udaka, S. Suzuki, Y. Achiba, H. Kira, K. Matsuda, H. Kadowaki, and Y. Okabe. Ordered water inside carbon nanotubes: Formation of pentagonal to octagonal ice-nanotubes. *Chemical Physics Letters*, 401 (4–6): 534–538, 2005. 10.1016/j.cplett.2004.11.112.

Maniwa, Y., K. Matsuda, H. Kyakuno, S. Ogasawara, T. Hibi, H. Kadowaki, S. Suzuki, Y. Achiba, and H. Kataura. Water-filled single-wall carbon nanotubes as molecular nanovalves. *Nature Materials*, 6 (2): 135–141, 2007. 10.1038/nmat1823.

Meyer, R. R., J. Sloan, R. E. Dunin-Borkowski, A. I. Kirkland, M. C. Novotny, S. R. Bailey, J. L. Hutchison, and M. L. H. Green. Discrete atom imaging of one-dimensional crystals formed within single-walled carbon nanotubes. *Science*, 289 (5483): 1324–1326, 2000. 10.1126/science.289.5483.1324.

Miyauchi, Y. H., S. H. Chiashi, Y. Murakami, Y. Hayashida, and S. Maruyama. Fluorescence spectroscopy of single-walled carbon nanotubes synthesized from alcohol. *Chemical Physics Letters*, 387 (1–3): 198–203, 2004. 10.1016/j.cplett.2004.01.116.

Monthioux, M., E. Flahaut, and J. P. Cleuziou. Hybrid carbon nanotubes: Strategy, progress, and perspectives. *Journal of Materials Research*, 21 (11): 2774–2793, 2006. 10.1557/JMR.2006.0366.

Moura, L. G., C. Fantini, A. Righi, C. Zhao, H. Shinohara, and M. A. Pimenta. Dielectric screening in polyynes encapsulated inside double-wall carbon nanotubes. *Physical Review B*, 83 (24): 245427, 2011. 10.1103/PhysRevB.83.245427.

Nakanishi, R., R. Kitaura, J. H. Warner, Y. Yamamoto, S. Arai, Y. Miyata, and H. Shinohara. Thin single-wall bn-nanotubes formed inside carbon nanotubes. *Scientific Reports*, 3: 1385, 2013. 10.1038/srep01385.

Nishide, D., H. Dohi, T. Wakabayashi, E. Nishibori, S. Aoyagi, M. Ishida, S. Kikuchi et al. Single-wall carbon nanotubes encaging linear chain c10h2 polyyne molecules inside. *Chemical Physics Letters*, 428 (4–6): 356–360, 2006. 10.1016/j.cplett.2006.07.016.

Nishide, D., T. Wakabayashi, T. Sugai, R. Kitaura, H. Kataura, Y. Achiba, and H. Shinohara. Raman spectroscopy of size-selected linear polyyne molecules c2nh2 (n = 4–6) encapsulated in single-wall carbon nanotubes. *Journal of Physical Chemistry C*, 111 (13): 5178–5183, 2007. 10.1021/jp0686442.

Okazaki, T., Y. Iizumi, S. Okubo, H. Kataura, Z. Liu, K. Suenaga, Y. Tahara, M. Yudasaka, S. Okada, and S. Iijima. Coaxially stacked coronene columns inside single-walled carbon nanotubes. *Angewandte Chemie—International Edition*, 50 (21): 4853–4857, 2011. 10.1002/anie.201007832.

Pfeiffer, R. *Dispersion of Raman Lines in Carbon Nanophases*. PhD thesis, University of Vienna, 2004.

Pfeiffer, R., H. Kuzmany, C. Kramberger, C. Schaman, T. Pichler, H. Kataura, Y. Achiba, J. Kurti, and V. Zolyomi. Unusual high degree of unperturbed environment in the interior of single-wall carbon nanotubes. *Physical Review Letters*, 90 (22): 225501, 2003. 10.1103/PhysRevLett.90.225501.

Pfeiffer, R., C. Kramberger, F. Simon, H. Kuzmany, V. N. Popov, and H. Kataura. Interaction between concentric tubes in DWCNTs. *European Physical Journal B*, 42 (3): 345–350, 2004. 10.1140/epjb/e2004-00389-0.

Pfeiffer, R., F. Simon, H. Kuzmany, and V. N. Popov. Fine structure of the radial breathing mode of double-wall carbon nanotubes. *Physical Review B*, 72 (16): 161404, 2005. 10.1103/PhysRevB.72.161404.

Pfeiffer, R., F. Simon, H. Kuzmany, V. N. Popov, V. Zolyomi, and J. Kurti. Tube-tube interaction in double-wall carbon nanotubes. *Physica Status Solidi B—Basic Solid State Physics*, 243 (13): 3268–3272, 2006. 10.1002/pssb.200669176.

Pfeiffer, R., M. Holzweber, H. Peterlik, H. Kuzmany, Z. Liu, K. Suenaga, and H. Kataura. Dynamics of carbon nanotube growth from fullerenes. *Nano Letters*, 7 (8): 2428–2434, 2007. 10.1021/nl071107o.

Pichler, T., M. Sing, M. Knupfer, M. S. Golden, and J. Fink. Potassium intercalated bundles of single-wall carbon nanotubes: Electronic structure and optical properties. *Solid State Communications*, 109 (11): 721–726, 1999. 10.1016/S0038-1098(98)00614-0.

Pichler, T., H. Kuzmany, H. Kataura, and Y. Achiba. Metallic polymers of c-60 inside single-walled carbon nanotubes. *Physical Review Letters*, 87 (26): 267401, 2001. 10.1103/PhysRevLett.87.267401.

Rao, A. M., P. C. Eklund, S. Bandow, A. Thess, and R. E. Smalley. Evidence for charge transfer in doped carbon nanotube bundles from raman scattering. *Nature*, 388 (6639): 257–259, 1997. 10.1038/40827.

Rauf, H., T. Pichler, M. Knupfer, J. Fink, and H. Kataura. Transition from a tomonaga-luttinger liquid to a fermi liquid in potassium-intercalated bundles of single-wall carbon nanotubes. *Physical Review Letters*, 93 (9): 096805, 2004. 10.1103/PhysRevLett.93.096805.

Rauf, H., H. Shiozawa, T. Pichler, M. Knupfer, B. Buchner, and H. Kataura. Influence of the c-60 filling on the nature of the metallic ground state in intercalated peapods. *Physical Review B*, 72 (24): 245411, 2005. 10.1103/PhysRevB.72.245411.

Saito, R., G. Dresselhaus, and M. Dresselhaus. *Physical Properties of Carbon Nanotubes*. Imperial College Press, London, 1998. Available at http://books.google.at/books?id=ZhXjJeiefZsC.

Sauer, M., H. Shiozawa, P. Ayala, G. Ruiz-Soria, X. Liu, A. Chernov, S. Krause, K. Yanagi, H. Kataura, and T. Pichler. Internal charge transfer in metallicity sorted ferrocene filled carbon nanotube hybrids. *Carbon*, 59: 237–245, 2013. 10.1016/j.carbon.2013.03.014.

Sceats, E. L. and J. C. Green. Noncovalent interactions between organometallic metallocene complexes and single-walled carbon nanotubes. *Journal of Chemical Physics*, 125 (15): 154704, 2006. 10.1063/1.2349478.

Sceats, E. L. and J. C. Green. Charge transfer composites of bis(cyclopentadienyl) and bis(benzene) transition metal complexes encapsulated in single-walled carbon nanotubes. *Physical Review B*, 75 (24): 245441, 2007. 10.1103/PhysRevB.75.245441.

Schulte, K., J. C. Swarbrick, N. A. Smith, F. Bondino, E. Magnano, and A. N. Khlobystov. Assembly of cobalt phthalocyanine stacks inside carbon nanotubes. *Advanced Materials*, 19 (20): 3312–3316, 2007. 10.1002/adma.200700188.

Shi, L., P. Rohringer, P. Ayala, T. Saito, and T. Pichler. Carbon nanotubes from enhanced direct injection pyrolytic synthesis as templates for long linear carbon chain formation. *Physica Status Solidi B—Basic Solid State Physics*, 250 (12): 2611–2615, 2013. 10.1002/pssb.201300148.

Shiozawa, H., H. Ishii, H. Kihara, N. Sasaki, S. Nakamura, T. Yoshida, Y. Takayama et al. Photoemission and inverse photoemission study of the electronic structure of c-60 fullerenes encapsulated in single-walled carbon nanotubes. *Physical Review B*, 73: 075406, 2006. 10.1103/PhysRevB.73.075406.

Shiozawa, H., C. E. Giusca, S. R. P. Silva, H. Kataura, and T. Pichler. Capillary filling of single-walled carbon nanotubes with ferrocene in an organic solvent. *Physica Status Solidi B—Basic Solid State Physics*, 245 (10): 1983–1985, 2008a. 10.1002/pssb.200879626.

Shiozawa, H., T. Pichler, A. Grüneis, R. Pfeiffer, H. Kuzmany, Z. Liu, K. Suenaga, and H. Kataura. A catalytic reaction inside a single-walled carbon nanotube. *Advanced Materials*, 20: 1443–1449, 2008b. 10.1002/adma.200701466.

Shiozawa, H., T. Pichler, C. Kramberger, A. Gruneis, M. Knupfer, B. Buchner, V. Zolyomi et al. Fine tuning the charge transfer in carbon nanotubes via the interconversion of encapsulated molecules. *Physical Review B*, 77 (15): 153402, 2008c. 10.1103/PhysRevB.77.153402.

Shiozawa, H., T. Pichler, R. Pfeiffer, and H. Kuzmany. Carbon nanotubes, single-walled: Functionalization by intercalation. In *Nanomaterials: Inorganic and Bioinorganic Perspectives*. Edited by Charles M. Lukehart and Robert A. Scott. John Wiley & Sons, Ltd., Chichester, UK, 2008d.

Shiozawa, H., T. Pichler, C. Kramberger, M. Ruemmeli, D. Batchelor, Z. Liu, K. Suenaga, H. Kataura, and S. R. P. Silva. Screening the missing electron: Nanochemistry in action. *Physical Review Letters*, 102 (4): 046804, 2009. 10.1103/PhysRevLett.102.046804.

Shiozawa, H., C. Kramberger, R. Pfeiffer, H. Kuzmany, T. Pichler, Z. Liu, K. Suenaga, H. Kataura, and S. R. P. Silva. Catalyst and chirality dependent growth of carbon nanotubes determined through nano-test tube chemistry. *Advanced Materials*, 22 (33): 3685–3689, 2010. 10.1002/adma.201001211.

Shiozawa, H., A. Briones-Leon, O. Domanov, G. Zechner, Y. Sato, K. Suenaga, T. Saito et al. Nickel clusters embedded in carbon nanotubes as high performance magnets. *Scientific Reports*, 5: 15033, 2015. 10.1038/srep15033.

Simon, F. and H. Kuzmany. Growth of single wall carbon nanotubes from c-13 isotope labelled organic solvents inside single wall carbon nanotube hosts. *Chemical Physics Letters*, 425 (1–3): 85–88, 2006. 10.1016/j.cplett.2006.04.094.

Simon, F., H. Kuzmany, H. Rauf, T. Pichler, J. Bernardi, H. Peterlik, L. Korecz, F. Fulop, and A. Janossy. Low temperature fullerene encapsulation in single wall carbon nanotubes: Synthesis of n@c-60@swcnt. *Chemical Physics Letters*, 383 (3–4): 362–367, 2004. 10.1016/j.cplett.2003.11.039.

Simon, F., H. Kuzmany, B. Nafradi, T. Feher, L. Forro, F. Fulop, A. Janossy, L. Korecz, A. Rockenbauer, F. Hauke, and A. Hirsch. Magnetic fullerenes inside single-wall carbon nanotubes. *Physical Review Letters*, 97 (13): 136801, 2006a. 10.1103/PhysRevLett.97.136801.

Simon, F., R. Pfeiffer, and H. Kuzmany. Temperature dependence of the optical excitation lifetime and bandgap in chirality assigned semiconducting single-wall carbon nanotubes. *Physical Review B*, 74 (12): 121411, 2006b. 10.1103/PhysRevB.74.121411.

Sloan, J., J. Hammer, M. Zwiefka-Sibley, and M. L. H. Green. The opening and filling of single walled carbon nanotubes (SWTs). *Chemical Communications*, (3): 347–348, 1998. 10.1039/a707632k.

Sloan, J., S. Friedrichs, R. R. Meyer, A. I. Kirkland, J. L. Hutchison, and M. L. H. Green. Structural changes induced in nanocrystals of binary compounds confined within single walled carbon nanotubes: A brief review. *Inorganica Chimica Acta*, 330: 1–12, 2002a. 10.1016/S0020-1693(01)00774-5.

Sloan, J., A. I. Kirkland, J. L. Hutchison, and M. L. H. Green. Integral atomic layer architectures of 1d crystals inserted into single walled carbon nanotubes. *Chemical Communications*, (13): 1319–1332, 2002b. 10.1039/b200537a.

Sloan, J., A. I. Kirkland, J. L. Hutchison, and M. L. H. Green. Aspects of crystal growth within carbon nanotubes. *Comptes Rendus Physique*, 4 (9): 1063–1074, 2003. 10.1016/S1631-0705(03)00102-6.

Smith, B. W. and D. E. Luzzi. Formation mechanism of fullerene peapods and coaxial tubes: A path to large scale synthesis. *Chemical Physics Letters*, 321 (1–2): 169–174, 2000. 10.1016/S0009-2614(00)00307-9.

Smith, B. W., M. Monthioux, and D. E. Luzzi. Encapsulated c-60 in carbon nanotubes. *Nature*, 396 (6709): 323–324, 1998. 10.1038/24521.

Smith, B. W., M. Monthioux, and D. E. Luzzi. Carbon nanotube encapsulated fullerenes: A unique class of hybrid materials. *Chemical Physics Letters*, 315 (1–2): 31–36, 1999. 10.1016/S0009-2614(99)00896-9.

Suenaga, K., T. Tence, C. Mory, C. Colliex, H. Kato, T. Okazaki, H. Shinohara, K. Hirahara, S. Bandow, and S. Iijima. Element-selective single atom imaging. *Science*, 290 (5500): 2280–2282, 2000. 10.1126/science.290.5500.2280.

Takenobu, T., T. Takano, M. Shiraishi, Y. Murakami, M. Ata, H. Kataura, Y. Achiba, and Y. Iwasa. Stable and controlled amphoteric doping by encapsulation of organic molecules inside carbon nanotubes. *Nature Materials*, 2: 683–688, 2003. 10.1038/nmat976.

Talyzin, A. V., I. V. Anoshkin, A. V. Krasheninnikov, R. M. Nieminen, A. G. Nasibulin, H. Jiang, and E. I. Kauppinen. Synthesis of graphene nanoribbons encapsulated in single-walled carbon nanotubes. *Nano Letters*, 11 (10): 4352–4356, 2011. 10.1021/nl2024678.

Tanaka, T., H. Jin, Y. Miyata, S. Fujii, H. Suga, Y. Naitoh, T. Minari, T. Miyadera, K. Tsukagoshi, and H. Kataura. Simple and scalable gel-based separation of metallic and semiconducting carbon nanotubes. *Nano Letters*, 9 (4): 1497–1500, 2009. 10.1021/nl8034866.

Tasis, D., N. Tagmatarchis, A. Bianco, and M. Prato. Chemistry of carbon nanotubes. *Chemical Reviews*, 106 (3): 1105–1136, 2006. 10.1021/cr050569o.

Thess, A., R. Lee, P. Nikolaev, H. J. Dai, P. Petit, J. Robert, C. H. Xu et al. Crystalline ropes of metallic carbon nanotubes. *Science*, 273 (5274): 483–487, 1996. 10.1126/science.273.5274.483.

Tu, X., S. Manohar, A. Jagota, and M. Zheng. DNA sequence motifs for structure-specific recognition and separation of carbon nanotubes. *Nature*, 460 (7252): 250–253, 2009. 10.1038/nature08116.

Urita, K., Y. Shiga, T. Fujimori, T. Iiyama, Y. Hattori, H. Kanoh, T. Ohba et al. Confinement in carbon nanospace-induced production of ki nanocrystals of high-pressure phase. *Journal of the American Chemical Society*, 133 (27): 10344–10347, 2011. 10.1021/ja202565r.

Yanagi, K., Y. Miyata, and H. Kataura. Highly stabilized beta-carotene in carbon nanotubes. *Advanced Materials*, 18 (4): 437–441, 2006. 10.1002/adma.200501839.

Yumura, T. and H. Yamashita. Modulating the electronic properties of multimeric thiophene oligomers by utilizing carbon nanotube confinement. *Journal of Physical Chemistry C*, 118 (10): 5510–5522, 2014. 10.1021/jp5006555.

Zhao, C., R. Kitaura, H. Hara, S. Irle, and H. Shinohara. Growth of linear carbon chains inside thin double-wall carbon nanotubes. *Journal of Physical Chemistry C*, 115 (27): 13166–13170, 2011. 10.1021/jp201647m.

Zolyomi, V., J. Koltai, A. Rusznyak, J. Kuerti, A. Gali, F. Simon, H. Kuzmany, A. Szabados, and P. R. Surjan. Intershell interaction in double walled carbon nanotubes: Charge transfer and orbital mixing. *Physical Review B*, 77 (24): 245403, 2008. 10.1103/PhysRevB.77.245403.

16

Functionalized Carbon Nanotubes

Vivian Machado de Menezes

16.1 Carbon Nanotubes

From the chemical point of view, carbon is one of the most versatile elements in the periodic table. Its ability to form different types of chemical bonds results in a wide variety of allotropes with peculiar properties. In addition to the amorphous configuration, carbon can be found in the form of carbon fibers, diamond, fullerene, graphite (and graphene), or nanotubes (Pierson 1993). Forasmuch as nanotubes present amazing structural, mechanical, electrical, thermal, and chemical properties, they serve as fundamental elements in the world of nanotechnology, being considered strategic materials in current technologies and in the coming decades. Carbon nanotubes have outstanding mechanical and electronic properties owing to the combination of their dimensionality (high length–diameter ratio), structure, and topology and are extremely flexible and rigid compounds that resist to high tensions (Saito et al. 1998).

In 1991, Iijima noted, through electron microscopy, the existence of compounds formed from multiple layers of rolled graphene sheets in cylindrical shape. That was the first report on the observation of multiwalled carbon nanotubes (MWCNTs), so-called because of its tubular morphology with nanometric size. Two years later, the synthesis of nanotubes as a single layer, the single-walled carbon nanotubes (SWCNTs), as illustrated in Figure 16.1b, was divulged. This material can be thought of as a single graphene sheet wrapped cylindrically forming a one-dimensional (1D) structure (Bethune et al. 1993; Iijima and Ichihashi 1993).

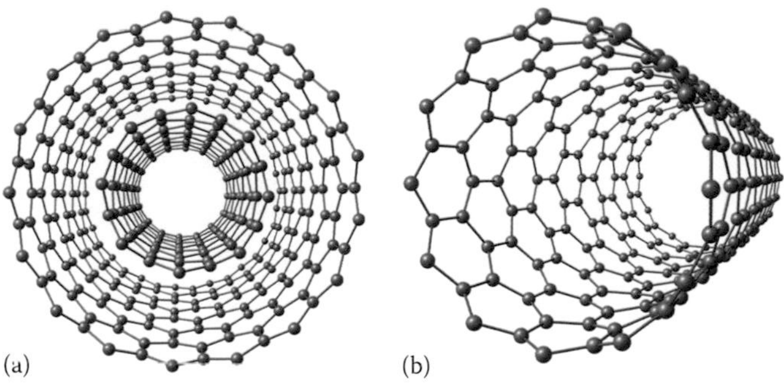

FIGURE 16.1 (a) A double-wall and (b) a single-wall carbon nanotube.

16.1.1 Properties of Carbon Nanotubes

Carbon nanotubes are cylinder-shaped structures (or tubes) formed by carbon atoms that are arranged in a honeycomb lattice. Considering a graphene sheet rolled up in a cylindrical shape, the beginning and the end of the rolling consist in the chiral vector, which connects two crystallographically equivalent sites along the circumference of the nanotube. The chiral vector is defined by the pair of indices (n,m), where n and m are integers and $0 \leq m \leq n$. Each pair (n,m) corresponds to a chirality or symmetry of the nanotube, fully identifying each carbon nanotube. Since nanotubes can be rolled in many ways, there are several possible orientations of the hexagons. Nanotubes with n = m are called armchair and nanotubes with m = 0 are named zigzag: Both types are achiral nanotubes. The other ones (n,m) correspond to chiral nanotubes. Figure 16.2 shows different types of SWCNTs.

The electronic properties of carbon nanotubes are also well described considering the rolling up of a graphene sheet into a cylinder. However, graphene, which is considered a semimetal or a zero-bandgap semiconductor, has electronic states that are sensitive to the periodic boundary conditions. Consequently, SWCNTs can suffer changes in their conducting character, being metallic or semiconductor depending

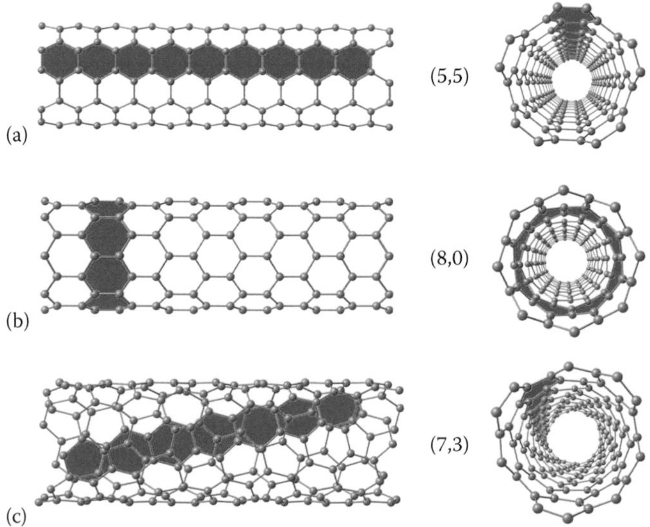

FIGURE 16.2 Examples of three types of SWCNTs identified by the indices (n,m): (a) armchair, (b) zigzag, and (c) chiral.

on their geometrical aspects. Moreover, carbon nanotubes are not planar structures as graphene, which has an sp^2 hybridization. The curvature effect leads to distortions in the orbital directions, resulting in an $sp^{2+\alpha}$ hybridization pattern ($0< \alpha <1$), where α is related to the degree of the surface's curvature (Endo et al. 1996).

Because of the length–diameter relation, nanotubes are considered as having infinite length. Thus, edge effects are neglected and the boundary conditions result in the quantization of the electronic states in the circumferential direction of the nanotube, whereas they are continuous in the axial direction. Hence, the energy bands consist in a set of relations of 1D energy dispersion, which are transversal to the set of 2D graphites (graphene).

All the armchair nanotubes (n,n) are metallic. The ones in which n − m is a multiple of 3 are semimetallic or almost zero-bandgap semiconducting. All others are semiconducting (Saito et al. 1992).

The electronic transport properties of SWCNTs have raised considerable experimental and theoretical interest as a result of the possibility of many applications in nanoelectronics. In the case of perfect nanotubes, the electronic movement is quantized: Researchers have observed a ballistic conduction regime (Javey et al. 2003; White and Todorov 1998) and neglected the inelastic scattering process. Nanotubes behave as quantum wires (Tans et al. 1997) inasmuch as the electrons move through the nanotube elastically with a probability of electronic transmission over the tube equal to 1 (Frank et al. 1998).

In the ballistic transport regime, the conductance is independent of the wire length, since the scattering region is smaller than the electron mean free path. In this scheme, there is no energy dissipation in the conductor, and all energy is dissipated in the electrical contacts that connect the ballistic conductor to macroscopic elements. This occurs even in the presence of impurities or defects on the tube surface, when scattering effects must be taken into account. Nonetheless, the scattering mechanism affects the transmission coefficient, decreasing the conductance, which then ceases to be quantized.

Besides possible uses in nanoelectronics, nanotubes are attractive for chemical and biological applications on behalf of their susceptibility to chemical interactions, which are ascribed to the relationship between diameter/surface and s–p hybridizations. Two main sources of reactivity in SWCNTs are the geometry of the curvature and the overlap of π orbitals. The curvature at the end of the nanotubes with closed ends leads to a modification of the carbon atoms' molecular orbitals and a displacement of the electronic density to outside of the tube, corresponding to an enhanced reactivity of the exterior surface. Thus, closed nanotubes have their ends even more reactive, a region where the carbon atoms present a character next to the sp^3 structure (Loiseau et al. 2006).

Additional parameters such as structural order, vacancies, presence of impurities, among others, also affect reactivity. Furthermore, nanotubes may show molecular adsorption and charge transfer behaviors (Meyyappan 2005), for some elements tend to bind to the carbon atoms (i.e., oxygen, hydrogen, nitrogen), introducing new features that change the chemistry of the carbon surface. An example is its solubility and adsorption behavior, which can suggest new applications of these materials (Vardharajula et al. 2012). The sensitivity of the nanotubes to the presence of other molecules can both compromise measurements of their properties and enable their use as chemical sensors in nanoscale.

Carbon nanotubes are similar in size to many biological species, having important implications in the development of nanomaterials for medical and pharmaceutical areas, including biosensors, drug delivery vehicles, and new biomaterials. However, to successfully use these embedded systems for biomedical devices, knowledge of their biocompatibility with the body and their biological features is extremely relevant. Although there is no full understanding of the effects of nanotubes to the human body, it is known, so far, that pristine nanotubes have a certain toxicity. Nonetheless, they can present good biological compatibility when functionalized (Smart et al. 2006).

The hydrophobicity of the nanotubes introduces an obstacle to their use in biomedical components, since most of the solvents compatible with biological fluids are aqueous. Functionalization has been used to work around the problem of insolubility. This process permits to control and change the original properties of the nanotubes through their interaction with atoms, molecules, or functional groups. On the other hand, the hydrophobicity of nanotubes can allow a stronger interaction of the hydrophobic

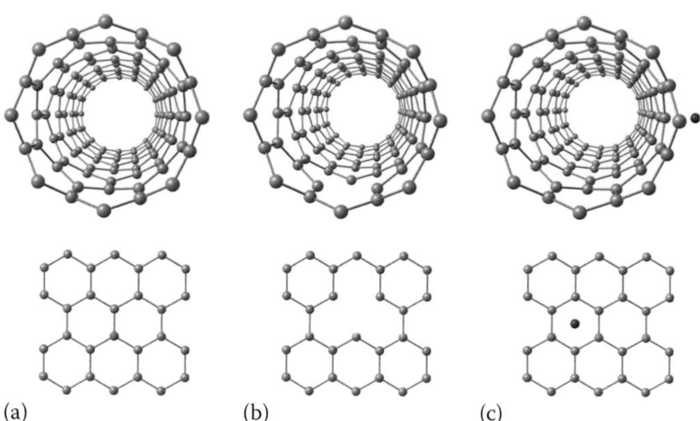

FIGURE 16.3 Schematic functionalized carbon nanotubes by (a) substitutional doping, (b) deformation (vacancy), and (c) adsorption of an atom.

molecules (or molecules with hydrophobic groups) with the tube than with the solvent molecules, which is an interesting property for the use of nanotubes as drug carriers (Vardharajula et al. 2012).

Functionalization of carbon nanotubes is also an important tool to overcome drawbacks from its stability due to the aromaticity and strong bonds between their sp^2 carbon atoms. This technique can be performed through their walls or ends by encapsulation, adsorption of atoms or molecules directly, substitutional doping (Souza Filho and Fagan 2007), deformation (Fagan et al. 2003), and adsorption of chemical functional groups (Veloso et al. 2006) (Figure 16.3). The binder system may also be removed from the nanotube through defunctionalization (Sun et al. 2002a), which is an essential property for the development of nanotube-based devices for drug delivery or reversible sensors of molecules (Tasis et al. 2006).

16.2 Functionalization of Carbon Nanotubes

Any changes in the original structure of a carbon nanotube correspond to a functionalization process. The approaches for the chemical modification of these quasi-1D structures are the covalent attachment of chemical groups, the noncovalent adsorption of functional atoms or molecules, formation of defects on the nanotube surface through vacancies or structural deformations, and the endohedral filling of their inner empty cavity.

16.2.1 Chemical Doping

An alternative to render a nanotube soluble in many solvents is the modification of their sidewalls and ends by organic functionalization. Aqueous solubility can be guaranteed, for instance, by a covalent bond (chemical adsorption) of hydrophilic ligands.

Two main strategies are routinely used for attaching functional groups to the tube. The first one consists of the oxidative treatment using strongly acidic solutions, like concentrated H_2SO_4 or HNO_3, or oxidizing gases such as oxygen (O_2) or carbon dioxide (CO_2), which can break the tubes by covering their ends and their defect points with oxygenated groups as carboxyl (–COOH), carbonyl (–CO–), and hydroxyl (–OH) (Vardharajula et al. 2012; Veloso et al. 2006). The functional groups are then used to incorporate a variety of other groups to solubilize the nanotube. This treatment not only generates various functional groups but also cuts and shortens the nanotubes into smaller pieces (Mehra et al. 2014). The second type of covalent functionalization is based on addition reactions, allowing different functional groups to be attached. Each functionalization strategy can have different effects on

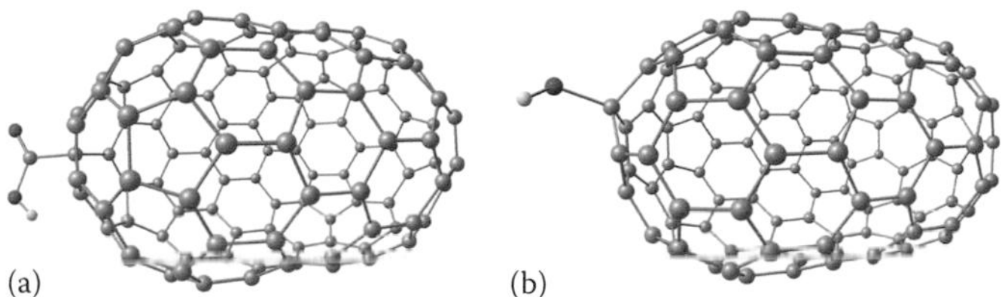

FIGURE 16.4 Functionalized carbon nanotubes with their ends capped by the groups (a) carboxyl and (b) hydroxyl.

the intrinsic properties of the nanotubes (Klumpp et al. 2006). Functionalization of nanotubes with molecules containing amine radicals is also chemically obtained (Tasis et al. 2006). Figure 16.4 shows schematic nanotubes functionalized by carboxyl and hydroxyl, and Figure 16.5 illustrates one of these nanotubes interacting with a molecule with biological interest.

The covalent bond offers the advantage of being more robust than the noncovalent one during manipulation and processing (Bhushan 2004; Klumpp et al. 2006).

Another parameter that profoundly changes the chemical reactivity of carbon nanotube sidewalls is the presence of defects, which are important in the covalent chemistry of the SWCNTs because they can either serve as anchor groups for further functionalization or be created by covalent attachment of further groups. Defects such as vacancies or pentagon–heptagon pairs (Stone-Wales defects) result in locally enhanced chemical reactivity of the graphitic nanostructures (Hirsch 2002; Karousis et al. 2010).

There is also the possibility of introducing some kind of impurity (i.e., atoms of another chemical element) on the structure of the carbon nanotubes to change their properties deliberately under controlled conditions. This process, referred to as substitutional doping, can be conducted through substitution reactions. The introduction of donor/acceptor levels by substitutional doping seems an attractive alternative to control the physical and chemical properties of nanotubes. When small quantities of boron and nitrogen are introduced at a substitutional position of the tube lattice, a p- or n-type extrinsic semiconductor is obtained. Several studies on atom substitution have suggested that doping atoms can introduce impurity states near the Fermi level, affecting the electronic properties of the carbon nanotubes (Carroll 1998; Zhang et al. 2002).

Another selective and controlled route of functionalization of carbon nanotubes is electrochemistry. A constant potential or current is applied to a nanotube electrode immersed in a solution that contains an appropriate reagent. The application of this potential/current creates an extremely reactive species through electron transfer between the nanotube and the reagent. Several organic radical species are likely to react with the reagent or to self-polymerize, resulting in polymer-coated nanotubes (Balasubramanian and Burghard 2005).

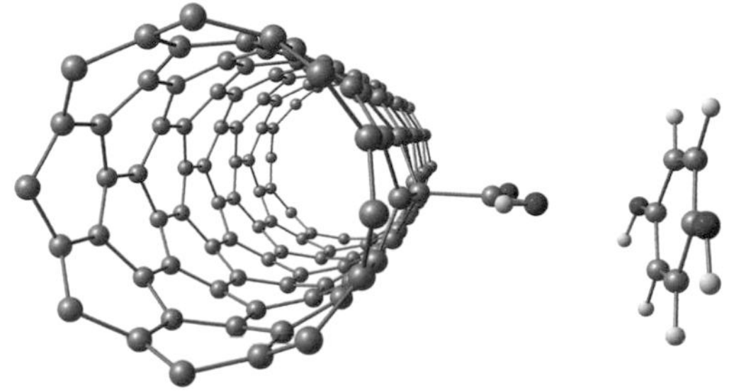

FIGURE 16.5 Carboxylated carbon nanotube interacting with a hydroquinone molecule.

Finally, another way to functionalize carbon nanotubes is by a photochemical approach. In this case, reactive species such as nitrene are generated by photoirradiation during sidewall addition reactions. For that purpose, the nitrene precursor is the photoactivated azido compound (Balasubramanian and Burghard 2005).

16.2.2 Physical Adsorption

Because of the formation of big bundles held strongly together, it is very difficult to disperse carbon nanotubes homogeneously in a solution (Tasis et al. 2006). Noncovalent functionalization (or physical adsorption) is an additional procedure used to solubilize nanotubes, exfoliating the carbon nanotubes' bundles through weak interactions. The nanotube surface can be modified via van der Waals forces and π–π interactions, electrostatic interactions, hydrogen bonding, and by adsorption or wrapping of polynuclear aromatic compounds, surfactants, nucleic acids, peptides, polymers, oligomers, or biomolecules. The hydrophobic π–π interactions are generally considered as responsible for the noncovalent stabilization (Karousis et al. 2010; Klumpp et al. 2006; Mehra et al. 2014; Vardharajula et al. 2012). Figure 16.6 illustrates an aromatic molecule adsorbed on a carbon nanotube by a noncovalent interaction.

The aforementioned process has the advantage of preserving the electronic structure of the aromatic surface of the nanotube forasmuch as the sp^2 structure and the union of carbon atoms in the tube are preserved (Karousis et al. 2010; Klumpp et al. 2006).

Hollow carbon nanotubes can be used to produce quasi-1D nanostructures by filling the inner cavity with selected materials, a case in which the capillarity forces introduce substances into the nanometric systems. From a chemist's perspective, the ability of nanotubes to encapsulate molecules and confine them inside their empty cavity to form nearly 1D arrays is one of their most fascinating properties (Karousis et al. 2010; Ugarte et al. 1999).

The inner cavity of carbon nanotubes offers space for the storage of guest molecules. The filling process involves an opening phase by oxidation and a subsequent immersion of the nanotube into different molten substances. Afterward, the materials introduced by capillarity are transformed into metals or oxides by thermal treatment (Ugarte et al. 1999).

The absorption of fullerenes is a particularly notable example of the endohedral chemistry of nanotubes. This incorporation is achieved at defect points at the ends or on the sidewalls of the tube. The encapsulated fullerenes tend to form chains coupled by van der Waals forces (Hirsch 2002).

There are many aspects to be considered before a complete understanding about the capillarity of carbon nanotubes is achieved, but the dependence of the nanotube's size on capillary forces was already observed. Filling nanotubes represent an outstanding example of manipulation of the nanomaterial's properties (Ugarte et al. 1998, 1999). In Figure 16.7, it is possible to see a carbon nanotube with a fullerene inside.

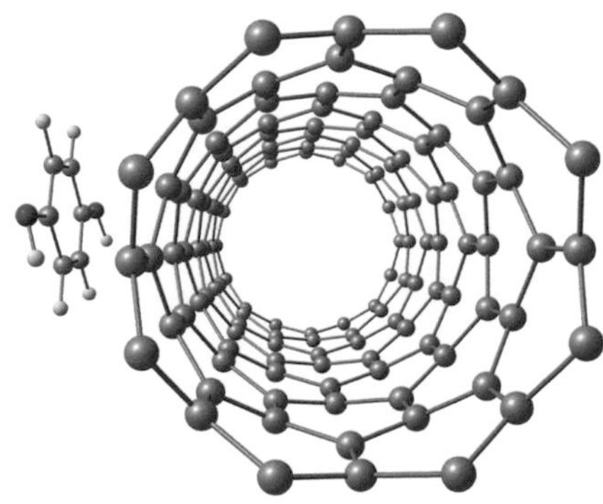

FIGURE 16.6 A hydroquinone molecule adsorbed on a carbon nanotube sidewall.

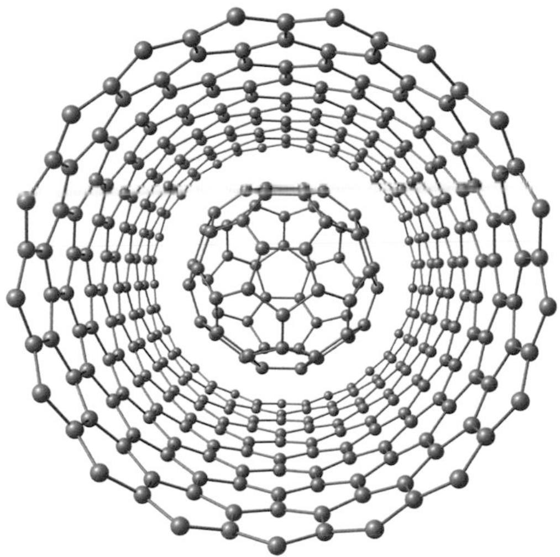

FIGURE 16.7 A carbon nanotube encapsulated with a fullerene molecule.

16.3 Properties of Functionalized Carbon Nanotubes

16.3.1 Electronic and Electronic Transport Properties

Covalent functionalizations commonly induce local impurity states, which turn out to control the electronic and transport properties of the carbon nanotubes. For metallic nanotubes, a monovalent addend bonding induces an impurity state located near the Fermi level. In consequence, it acts as a strong scattering center and the ballistic transport properties can be considerably affected. The conductance decreases with the addend concentration. On the other side, a divalent addend has a weak effect on the conductance near the Fermi level, creating impurity states far away from it. Each functional group exhibits its own position of the impurity state and creates considerable dips in the ballistic conductance of metallic carbon nanotubes (Park et al. 2006). For the case of semiconducting carbon nanotubes, the functionalization process can close the bandgap, enhancing significantly the conductance around the Fermi energy (Ranjan and Seifert 2006).

16.3.2 Capacitance Increasing

Supercapacitors (or electrochemical capacitors or ultracapacitors) are devices used to store and deliver a huge amount of electrical energy in high-power pulses. The most promising supercapacitors are based on two materials regarding the charge storage mechanism: double-layer or redox pseudocapacitive. The electrical charge stored in these materials is typically several orders of magnitude higher than in conventional capacitors. Recently, it has been demonstrated that even higher capacitance values can be achieved in composites obtained by combining carbon nanotubes (a double-layer capacitive material) with a conducting polymer (a redox pseudocapacitive material). These composites are capable of combining the high surface area and electrical conductivity of carbon nanotubes with the redox electrochemistry of the conducting polymers (Hughes 2004). The increased capacitance was well explained by the enhancement of the specific surface area and the abundant pore distributions of lower pore sizes. In conventional supercapacitors, most of the surface area resides in micropores, which are incapable of supporting an electrical double layer (An et al. 2001). Capacitance is strongly dependent on the type of the nanotube and on the presence of the functionalizations (Frackowiak et al. 2002).

16.3.3 Optical Properties

The degree of dispersion in a solution can greatly affect the observed characteristics of the functionalized and solubilized carbon nanotubes, such as their Raman and luminescence properties (Sun et al. 2002b). It was shown that the polymer-bound carbon nanotubes in homogeneous organic and aqueous solutions are luminescent or strongly luminescent. The dependence of the strong excitation wavelength of luminescence indicates a distribution of emitters (Riggs et al. 2000). Strong luminescence at room temperature has been detected in nanotube solutions, with the source of this luminescence being attributed to the presence of extensive conjugated electronic structures and the excitation-energy trapping related to defects in the nanotubes (Banerjee and Stanislau 2002; Sun et al. 2002b). Supposedly, functionalization facilitates the development of intrinsic luminescence, which emanates from the carbon nanotubes by dispersion of tubes and by trapping of the excitation energy on the nanotube surface itself (Riggs et al. 2000). Zhou et al. (2007) also reported a method that induces strong blue luminescence by electrochemical treatment of MWCNTs. In addition, formation of bound trions in functionalized carbon nanotubes may generate a bright photoluminescence peak that is considerably red-shifted from those of the excitons and trapped excitons. This is obtained through chemical doping of controlled sp^3 defect sites in semiconducting SWCNTs. The ability to shift the excitation wavelength from the visible range to the near-infrared can be interesting for potential applications such as photonics, electronics, and bioimaging (Brozena et al. 2014).

16.3.4 Mechanical Properties

Carbon nanotubes possess some of the most remarkable mechanical properties among all known materials, such as very high Young's modulus, high fracture strength and strain, and low mass density. Considering these properties, they are perfect candidates for reinforcements in nanocomposites (Zhang et al. 2008). A recent research indicates that nanotube functionalizations can lead to degradation of some of their properties, depending on the chirality and configuration of the process (Yang et al. 2014). Some functionalizations introduce impurities atoms on the nanotube surface, affecting the mechanical properties. The presence of impurities destroys the perfect atomic structure, which can weaken the fracture strain, tensile strength, and torsion behavior of nanotubes. Functionalized carbon nanotubes can have their critical bulking load reduced (in about 15%). In addition, the elastic modulus decreases gradually with the increasing functionalization (Garg and Sinnott 1998; Yang et al. 2014; Zhang et al. 2008).

16.3.5 Thermal Conductivity

Because of their outstanding properties, one of the most attractive applications of carbon nanotubes is their incorporation into materials to improve the thermal, mechanical, and electrical responses of high-performance composites (Lima et al. 2012; Padgett and Brenner 2004). However, carbon nanotubes functionalized by chemical attachment of some functional groups can have their thermal conductivity reduced by a decrease in the phonon scattering length rather than changes in the vibrational frequencies of the carbon atoms. The system exhibits a drop in thermal conductivity ascribed to chemisorption. The thermal and the electrical properties seem very sensitive to the degree of functionalization and probably also to the chemical nature of the functional group (Lima et al. 2012).

16.3.6 Magnetic Properties

Carbon nanotubes are promising candidates for the production of efficient spin transistors. Ferromagnetic contacts can be used to inject a spin-polarized current inside a nanotube, allowing the observation of the spin-valve effect (Cottet et al. 2006). Spin-valves made of nanotubes contacted to magnetic electrodes may display significant values of magnetoresistance (Cottet et al. 2006; Kirwan et al. 2009). It is also possible to produce the spin-valve effect using magnetically coupled impurities on metallic nanotubes to generate an efficient spin filter. The presence of magnetically coupled substitutional impurities prevents the occurrence of magnetic frustration, enhances the scattering contrast between different spin channels, and substantially amplifies the significance of the magnetoresistance (Kirwan et al. 2009).

16.3.7 Superhydrophobicity

The wettability of a solid surface is a very essential property for practical applications. It depends on the surface energy and the geometrical structure of the solid surface. As a result of their wettability properties, the research related to superhydrophobic materials based on carbon nanotubes has been attractive in the last decades. Recently, superhydrophobic carbon nanotubes have been fabricated using nanotubes functionalized by covalent attachment and noncovalent adsorption or wrapping of long-chain hydrophobic molecules on the surface of the nanotube (Hong and Uhum 2006). Superhydrophobic surfaces with angles of water contact and hysteresis larger than 150° and smaller than 10°, respectively, have been reached (Chakradhar et al. 2014). Electrically conductive superhydrophobic carbon nanotube films have received considerable attention for use in various applications. The surface roughness and surface energy state of the material must be considered to fabricate superhydrophobic films. The control of nanotubes' film texture through functionalization can aid the manipulation of their wettability and lead to superhydrophobicity (Jeong et al. 2014). Electrical conductivity measurements confirmed that the functionalization reaction is mild and reasonably nondestructive to the electronic structure of carbon nanotubes' surfaces. The combination of superhydrophobicity and high electrical conductivity is highly desirable for effective applications (Bayer et al. 2013; Yao et al. 2013).

16.3.8 Biological Properties

Carbon nanotubes' solubility and biocompatibility are the most crucial factors for their successful application in the biomedical area. Functionalization of carbon nanotubes allows obtaining enhanced solubility and reduced toxicity. Functionalization of the nanotube bundles is essential to achieve the nanotube's solubility (and dispersion) (Sun et al. 2002a; Vardharajula et al. 2012). Further properties are required to achieve successful use of carbon nanotubes in biomedical applications, i.e., some specificity for recognition of only one type of target biomolecule and rejection of others, which can be obtained through specific ligands (Shim et al. 2002).

16.4 Applications of Functionalized Carbon Nanotubes

As a consequence of the unique properties of functionalized carbon nanotubes, researchers and companies are considering their use in several fields of knowledge.

Figure 16.8 illustrates the exponential increase in publications regarding the subject "functionalized carbon nanotubes" and "applications" in the last decade. Considering the elevated number of possible uses, the following survey introduces only some applications of carbon nanotubes.

16.4.1 Sensors

As a result of their high structural porosity and high specific surface area, carbon nanotubes are potential materials for gas sensors. Gas sensors based on carbon nanotubes can provide significant advantages over metal oxides and conducting polymer sensors in relation to higher sensitivity, small sizes, fabrication of massive nanosensor arrays, no power loss, among others. Defects (pentagon or heptagon, impurities, Stone-Wales) present on the nanotube sidewalls and end caps facilitate the adsorption of gas molecules, improving the sensitivity and selectivity of the nanotube gas sensor when compared with

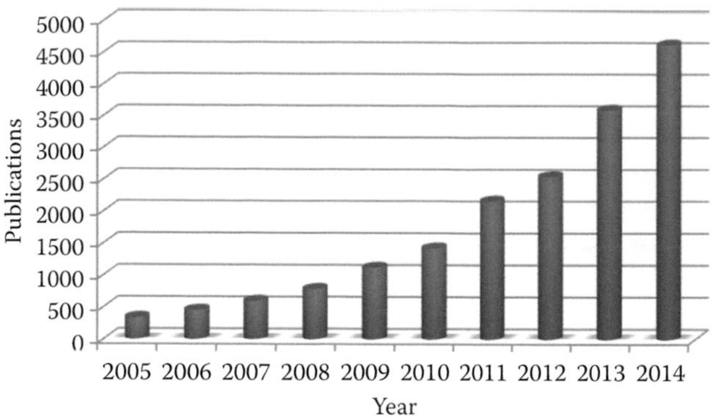

FIGURE 16.8 Articles regarding the applications of functionalized carbon nanotubes. (Available at http://www.sciencedirect.com, accessed on November 9, 2014.)

the pristine nanotube. Changes in the physical properties ascribed to electrical charge transfer upon gas exposure allow the application of carbon nanotubes as chemical nanosensors for novel chemiresistors in environmental monitoring applications of air pollutants or released gases from burning fossil fuels, for example. The electrical resistance of nanotubes decreases when oxidized gas molecules are adsorbed on their surface, whereas the adsorption of reducing gas molecules results in the increase of the nanotubes' electrical resistance. Hence, nanotube-based gas sensors can act based on specific changes in the electrical conductance, for instance. The high thermal stability of nanotubes renders them resistant under different reaction conditions, preserving their basic structure (Mittal and Kumar 2014; Penza et al. 2008).

16.4.2 Biomedical Applications

Functionalized carbon nanotubes are widely applied in the biomedical area; therefore, the understanding of their biocompatibility and improvement of their dispersion in aqueous solvents are mandatory. As a consequence of the recent advances in tissue engineering, many materials are being formulated for applications in different tissues, such as bone and the cardiovascular system, among others. The use of carbon nanotubes for preservation of cells, delivery of growth factors or genes, and as scaffolding matrices has been reported to promote integration with the host tissue (Vardharajula et al. 2012). The fabrication of titanium materials coated with functionalized carbon nanotubes for applications in orthopedics has been proposed as an implantable electronic device, with properties such as electrical conductivity, nontoxicity, and ability to inhibit cell growth and proliferation of connective tissue (Przekora et al. 2014). Because of their excellent electrochemical properties, carbon nanotubes are being preferred in the field of biosensors at the expense of other materials. In addition, the direct covalent attachment of desired biomolecules onto MWCNTs has some advantages, like no leaching as compared with the noncovalent attachment, resulting in enhanced sensing response and stability of the biosensors (Reza et al. 2014). As important future tools for research and biomedical diagnosis, carbon-nanotube-based transistors and electrochemical biosensors are applied for their unique characteristics. As long as carbon nanotubes have a diameter similar to or smaller than that of individual proteins, they presumably can be used as high-performance electrical conduits for interfacing with biological systems (Kim et al. 2007). Carbon nanotubes are constantly studied as nanovectors in targeted and controlled drug delivery as a result of their physicochemical properties. These nanotubes have a great surface area and are able to carry several biomolecules and deliver them to specific targets when functionalized (Mehra et al. 2014). The application of carbon nanotubes as vectors for a molecule allows the tube to control the release and stability of the drug active principle, avoiding some undesirable side effects (Menezes et al. 2012). Furthermore, the linkage of antigenic targets to functionalized nanotubes can enhance the immunopotentiating effects, representing an opportunity to produce an immune response with high specificity to the antigen presented by the carbon nanotubes (Fadel and Fahmy 2014). Finally, the functionalization of carbon nanotubes can greatly expand their potential applications without causing any side effects.

16.4.3 Energy Storage

A challenge in the 21st century is to achieve cleanliness and renewability of energy. Solar energy conversion is one of the most natural and abundant ways to produce alternative energy to carbon fuels and satisfies the required characteristics mentioned previously. Because of that, the conversion, storage, and application of solar energy are attracting considerable attention from the scientific community. Therefore, materials converted by solar energy with efficient full-band light-harvesting, photothermal conversion, and thermal storage characteristics are very promising for developing renewable and clean energy sources (Tang et al. 2014). SWCNTs are interesting platforms for photovoltaic applications. These novel materials, in combination with photoexcitable electron donors, could be helpful for the fabrication of ultrahigh efficient photoelectrochemical cells. SWCNTs are efficient electron acceptors and transporters. Thus, the combination of SWCNTs with donor groups represents an innovative idea to collect solar energy and convert it into useful electricity (Guldi et al. 2005). Studies revealed that polyethylene glycol (PEG)/SiO_2/MWCNT composites, designed and synthesized, demonstrated a wider absorption range for sunlight, highlight-to-heat conversion and energy storage efficiencies, excellent form-stable properties, as well as high thermal conductivities. These systems can improve the usage efficiency of the solar radiation, which can be extensively applied in the areas of energy conversion and storage (Tang et al. 2014). Hydrogen is up-and-coming as a green fuel for transportation applications. It has been accepted as an ideal energy source because its manipulation does not produce air pollution or greenhouse-gas emissions. Currently, the storage and delivery of hydrogen remain a topic of technological importance. Carbon materials have attracted the scientific community's interest as one of the possible materials for hydrogen storage. Extensive theoretical and experimental surveys have been recently conducted to obtain highly efficient hydrogen storage materials. As carbon nanotubes present a unique tubular structure with high specific surface, hollowness, interstitial sites, nanometer scale diameter, porosity, and thermal and chemical stability, they are widely investigated as potential candidates for hydrogen storage materials (Silambarasan et al. 2014; Soleymanabadi and Kakemam 2013).

16.4.4 Electromagnetic Wave Absorption

The problems brought by electromagnetic wave absorption, such as signal interference, back-radiation of microstrip radiators, and the impact of electromagnetic wave radiation on human health, have been assessed by researchers. The application of electromagnetic wave absorbing materials has become a focus of the scientific community with regard to the development of military and commercial applications. Carbon nanotubes are also being used as electromagnetic wave absorbents through certain modifications in mass, wide range, and strong absorption of light. Two special structures of carbon nanotubes containing composites to enhance electromagnetic wave absorption are proposed to date from analyses of conductivity, permittivity, and permeability: a foam and a laminated structure (Ren et al. 2014).

16.4.5 Reinforcement of Materials

The introduction of carbon nanotubes to create reinforced structures is attracting the interest of academics because of their peculiar structures with large aspect ratio, outstanding tensile strength and Young's modulus, and excellent electrical and thermal properties. All these characteristics render carbon nanotubes as the most appropriate candidates for the reinforcement of a variety of materials, as fiber or polymer composites, which have had their mechanical and functional properties of high performance improved (Li et al. 2013; Singhal et al. 2011). Studies demonstrated that the adoption of nanotube reinforcements leads to important improvements, namely, interfacial shear strength, fatigue resistance, interlaminar fracture toughness, glass transition temperature, and electrical conductivity (Li et al. 2013). The potential improvement of the strength of polymers by carbon nanotubes/polymer has encouraged some researchers to investigate the use of carbon nanotubes as reinforcements for metal and ceramic matrices (Singhal et al. 2011).

In summary, this chapter presents a brief review of some routes to achieve the functionalization of carbon nanotubes, as well as some related consequences. Forasmuch as the process of functionalization allows the manipulation of physical and chemical properties of carbon nanotubes, it has a very important role in the research and development of nanotube-based materials and systems to be applied in nanotechnology.

References

An, B. K. H. et al. "Electrochemical properties of high-power supercapacitors using single-walled carbon nanotube electrodes," *Adv. Funct. Mater.* 11 (2001): 387–392.

Balasubramanian, K. and Burghard, M. "Chemically functionalized carbon nanotubes," *Small* 1 (2005): 180–192.

Banerjee, S. and Stanislau, S. W. "Structural characterization, optical properties, and improved solubility of carbon nanotubes functionalized with Wilkinson's catalyst," *J. Am. Chem. Soc.* 124 (2002): 8940–8948.

Bayer, I. S. et al. "Superhydrophobic and electroconductive carbon nanotube-fluorinated acrylic copolymer nanocomposites from emulsions," *Chem. Eng. J.* 221 (2013): 522–530.

Bethune, D. S. et al. "Cobalt-catalysed growth of carbon nanotubes with single atomic-layer walls," *Nature* 363 (1993): 605–607.

Bhushan, B. *Springer Handbook of Nanotechnology* (Berlin: Springer-Verlag, 2004).

Brozena, A. H. et al. "Controlled defects in semiconducting carbon nanotubes promote efficient generation and luminescence of trions," *ACS Nano* 5 (2014): 4239.

Carroll, D. L. "Effects of nanodomain formation on the electronic structure of doped carbon nanotubes," *Phys. Rev. Lett.* 81 (1998): 2332–2335.

Chakradhar, R. P. S. et al. "Stable superhydrophobic coatings using PVDF-MWCNT," *Appl. Surf. Sci.* 301 (2014): 208–215.

Cottet, A. et al. "Nanospintronics with carbon nanotubes," *Semicond. Sci. Technol.* 21 (2006): S78–S95.

Endo, M. et al. *Carbon Nanotubes* (London: Pergamon, 1996).

Fadel, T. R. and Fahmy, T. M. "Immunotherapy applications of carbon nanotubes: From design to safe applications," *Trends Biotechnol.* 32 (2014): 198–209.

Fagan, S. B. et al. "Ab initio study of radial deformation plus vacancy on carbon nanotubes: Energetics and electronic properties," *Nano Lett.* 3 (2003): 289–291.

Frackowiak, E. et al. "Enhanced capacitance of carbon nanotubes through chemical activation," *Chem. Phys. Lett.* 361 (2002): 35–41.

Frank, S. et al. "Carbon nanotube quantum resistors," *Science* 280 (1998): 1744–1746.

Garg, A. and Sinnott, S. B. "Effect of chemical functionalization on the mechanical properties of carbon nanotubes," *Chem. Phys. Lett.* 295 (1998): 273–278.

Guldi, D. N. et al. "Single-wall carbon nanotubes as integrative building block for solar-energy conversion," *Angew. Chem.* 117 (2005): 2051–2054.

Hirsch, A. "Functionalization of single-walled carbon nanotubes," *Angew. Chem. Int.* 41 (2002): 1853–1859.

Hong, Y. C. and Uhum, H. S. "Superhydrophobicity of a material made from multiwalled carbon nanotubes," *App. Phys. Lett.* 88 (2006): 244101-1–244101-3.

Hughes, M. "Carbon nanotube-conducting polymer composites in supercapacitors," in *Dekker Encyclopedia of Nanoscience and Nanotechnology*, eds. Schwarz, J. A. et al. (New York: Marcel Dekker, 2004), 447–459.

Iijima, S. "Helical microtubules of graphitic carbon," *Nature* 354 (1991): 56–58.

Iijima, S. and Ichihashi, T. "Single-shell carbon nanotubes of 1-nm diameter," *Nature* 363 (1993): 603–605.

Javey, A. et al. "Ballistic carbon nanotube field-effect transistors," *Nature* 424 (2003): 654–657.
Jeong, D.-W. et al. "Stable hierarchical superhydrophobic surfaces based on vertically aligned carbon nanotube forests modified with conformal silicone coating," *Carbon* 79 (2014): 442–449.
Karousis, K. et al. "Current progress on the chemical modification of carbon nanotubes," *Chem. Rev.* 110 (2010): 5366–5397.
Kim, B. S. N. et al. "Carbon nanotubes for electronic and electrochemical detection of biomolecules," *Adv. Mater.* 19 (2007): 3214–3228.
Kirwan, D. F. et al. "Enhanced spin-valve effect in magnetically doped carbon nanotubes," *Carbon* 47 (2009): 2528–2555.
Klumpp, C. et al. "Functionalized carbon nanotubes as emerging nanovectors for the delivery of therapeutics," *Biochim. Biophys. Acta* 1758 (2006): 404–412.
Li, M. et al. "Interfacial improvement of carbon fiber/epoxy composites using a simple process for depositing commercially functionalized carbon nanotubes on the fibers," *Carbon* 52 (2013): 109–121.
Lima, A. M. F. et al. "Electrical conductivity and thermal properties of functionalized carbon nanotubes/polyurethane composites," *Polímeros* 22 (2012): 117–124.
Loiseau, A. et al. *Understanding Carbon Nanotubes: From Basics to Applications* (Berlin: Springer, 2006).
Mehra, N. K. et al. "A review of ligand tethered surface engineered carbon nanotubes," *Biomaterials* 35 (2014): 1267–1283.
Menezes, V. M. de et al. "Carbon nanostructures interacting with Vitamins A, B3 and C: Ab initio simulations," *J. Biomed. Nanotechnol.* 8 (2012): 1–5.
Meyyappan, M. *Carbon Nanotubes: Science and Applications* (Boca Raton, FL: CRC Press, 2005).
Mittal, M. and Kumar, A. "Carbon nanotube (CNT) gas sensors for emissions from fossil fuel burning," *Sens. Actuators B* 203 (2014): 349–362.
Padgett, C. W. and Brenner, D. W. "Influence of chemisorption on the thermal conductivity of single-wall carbon nanotubes," *Nano Lett.* 4 (2004): 1051–1053.
Park, H. et al. "Effects of sidewall functionalization on conducting properties of single wall carbon nanotubes," *Nano Lett.* 6 (2006): 916–919.
Penza, M. et al. "Pt- and Pd-nanoclusters functionalized carbon nanotubes networked films for sub-ppm gas sensors," *Sens. Actuators B* 135 (2008): 289–297.
Pierson, H. O. *Handbook of Carbon, Graphite, Diamond and Fullerenes: Properties, Processing and Applications* (New Jersey: Noyes Publications, 1993).
Przekora, A. et al. "Titanium coated with functionalized carbon nanotubes—A promising novel material for biomedical application as an implantable orthopaedic electronic device," *Mater. Sci. Eng. C* 45 (2014): 287–296.
Ranjan, N. and Seifert, G. "Transport properties of functionalized carbon nanotubes: Density-functional Green's function calculations," *Phys. Rev. B* 73 (2006): 153408-1–153408-3.
Ren, F. et al. "Current progress on the modification of carbon nanotubes and their application in electromagnetic wave absorption," *RSC Adv.* 4 (2014): 14419–14431.
Reza, K. K. et al. "Biofunctionalized carbon nanotubes platform for biomedical applications," *Mater. Lett.* 126 (2014): 126–130.
Riggs, J. E. et al. "Strong lumin essence of solubilized carbon nanotubes," *J. Am. Chem. Soc.* 122 (2000): 5879–5880.
Saito, R. et al. "Electronic structure of chiral graphene tubules," *Appl. Phys. Lett.* 60 (1992): 2204–2206.
Saito, R. et al. *Physical Properties of Carbon Nanotubes* (London: Imperial College Press, 1998).
Shim, M. et al. "Functionalization of carbon nanotubes for biocompatibility and biomolecular recognition," *Nano Lett.* 2 (2002): 285–288.
Silambarasan, D. et al. "Reversible hydrogen storage in functionalized single-walled carbon nanotubes," *Physica E* 60 (2014): 75–79.
Singhal, S. K. et al. "Fabrication and characterization of Al-matrix composites reinforced with amino-functionalized carbon nanotubes," *Compos. Sci. Technol.* 72 (2011): 103–111.

Smart, S. K. et al. "The biocompatibility of carbon nanotubes," *Carbon* 44 (2006): 1034–1047.

Soleymanabadi, H. and Kakemam, J. "A DFT study of H2 adsorption on functionalized carbon nanotubes," *Physica E* 54 (2013): 115–117.

Souza Filho, A. G. de and Fagan, S. B. "Funcionalização de nanotubos de carbono," *Quim. Nova* 30 (2007): 1695–1703.

Sun, Y.-P. et al. "Functionalized carbon nanotubes: Properties and applications," *Acc. Chem. Res.* 35 (2002a): 1096–1104.

Sun, Y.-P. et al. "Luminescence anisotropy of functionalized carbon nanotubes in solution," *Chem. Phys. Lett.* 351 (2002b): 349–353.

Tang, B. et al. "A full-band sunlight-driven carbon nanotube/PEG/SiO_2 composites for solar energy storage," *Sol. Energy Mater. Sol. Cells* 123 (2014): 7–12.

Tans, S. T. et al. "Individual single-wall carbon nanotubes as quantum wires," *Nature* 386 (1997): 474–477.

Tasis, D. et al. "Chemistry of carbon nanotubes," *Chem. Rev.* 106 (2006): 1105–1136.

Ugarte, D. et al. "Filling carbon nanotubes," *Appl. Phys. A* 67 (1998): 101–105.

Ugarte, D. et al. "Capillarity in carbon nanotubes," in *The Science and Technology of Carbon Nanotubes*, eds. Tanaka, K. et al. (Kidlington: Elsevier, 1999), 128–142.

Vardharajula, S. et al. "Functionalized carbon nanotubes: Biomedical applications," *Int. J. Nanomedicine* 7 (2012): 5361–5374.

Veloso, M. V. et al. "*Ab initio* study of covalently functionalized carbon nanotubes," *Chem. Phys. Lett.* 430 (2006): 71–74.

White, C. T. and Todorov, T. N. "Carbon nanotubes as long ballistic conductors," *Nature* 393 (1998): 240–242.

Yang, Q.-S. et al. "Modeling the mechanical properties of functionalized carbon nanotubes and their composites: Design at the atomic level," *Adv. Cond. Matter Phys.* 2014 (2014): 482056-1–482056-8.

Yao, H. et al. "Electrically conductive superhydrophobic octadecylmine-functionalized multiwall carbon nanotube films," *Carbon* 53 (2013): 366–373.

Zhang, G. et al. "Effect of substitutional atoms in the tip on field-emission properties of capped carbon nanotubes," *Appl. Phys. Lett.* 80 (2002): 2589–2591.

Zhang, Z. Q. et al. "Mechanical properties of functionalized carbon nanotubes," *Nanotechnology* 19 (2008): 395702-1–395702-6.

Zhou, J. et al. "An electrochemical avenue to blue luminescent nanocrystals from multiwalled carbon nanotubes (MWCNTs)," *J. Am. Chem. Soc.* 129 (2007): 744–745.

17

Carbon Nanotube Networks

Luca Camilli

Abstract

Carbon nanotubes (CNTs) can be thought of as quasi one-dimensional hallow cylinders made up by carbon atoms arranged in a hexagonal lattice. Owing to their outstanding mechanical, electrical, and thermal properties, CNTs have been the subject of intense study over the past three decades. Nowadays, CNTs can be synthesized in large quantities and manipulated with high precision in order to create interesting two- and three-dimensional networks. In this chapter, we are going to review the most fashionable ways of producing these CNT networks, and then we will focus on their applications, notably in the field of optoelectronics, biomedicine, and environment protection.

17.1 Introduction

Among the numerous carbon nanostructures that have been deeply investigated in universities and research centers over the past few decades—e.g., nanodiamonds, fullerenes, nanotubes, graphene, nanofibers, and nanohorns—carbon nanotubes (CNTs) are perhaps the most studied ones after graphene. Indeed, after Iijima's landmark publication in 1991 (Iijima 1991), numerous experiments have been carried out to understand and disclose their remarkable chemical and physical properties, arising from the nanoscale dimension and the quasi-one-dimensional (1D) periodicity (Ajayan 1999; Dresselhaus et al. 2001; Dai 2002).

Few fundamental experiments have been performed on individual CNTs, some aimed at investigating their mechanical properties (Yu et al. 2000a,b), some at probing their electronic and optoelectronic properties (Tans et al. 1998; Lee 2005). However, when it comes to applications, a network—a structure made of many individual CNTs overlapping and touching each other—is far more used than isolated tubes. Indeed, dealing with single CNTs is challenging because of their tiny dimensions, i.e., diameter of 20 nm down to below 1 nm and length ranging from microns up to few millimeters, whereas 2D CNT films, deposited for instance on a substrate, or 3D self-standing CNT solids are much easier to handle.

This chapter focuses on 2D and 3D CNT networks. In particular, we first overview the state-of-the-art methods used for their synthesis and then explore some of their most promising applications, especially in photovoltaic for the case of 2D films and environment protection and biomedicine for the case of 3D networks.

Finally, we conclude the chapter by acknowledging the great deal of research that has been carried out on CNT networks and by speculating on possible future contributions of CNT networks in our everyday life.

17.2 Synthesis of CNT Networks

17.2.1 Synthesis of 2D CNT Networks

CNTs have been largely integrated in electronic devices because of their remarkable electronic properties (Anantram and Léonard 2006; Marulanda 2011). For instance, owing to their good electrical conductivity and optical transparency, thin CNT films can be used as a transparent electrode in organic solar cells, replacing the expensive indium tin oxide layers (Schindler 2012); also, because of their large surface area, they are employed in gas and biomolecule sensing devices (Balasubramanian and Burghard 2006, Wang and Yeow 2009). In addition, CNT films deposited on doped silicon wafers have been reported to generate a remarkable photocurrent upon illumination (Jung et al. 2013). In all these cases and more, the 2D CNT films need to be deposited on a substrate that, depending on the device under exam, could either work only as a rigid support (e.g., the case of sensing devices) or be necessary to build up a working hetero-junction (e.g., the case of CNT-silicon solar cells). The easiest approach to deposit CNTs on a chosen substrate is growing them directly on it. In this scenario, chemical vapor deposition (CVD) is probably the most used synthesis method (Kumar and Ando 2010). Briefly, within such an approach, a thermal decomposition of a hydrocarbon vapor, such as methane, acetylene, and ethylene, is achieved in the presence of a nanostructured metal catalyst deposited on a rigid substrate (see Figure 17.1).

CVD allows for growth of high-quality CNTs on a large scale and with relatively high yield. However, there are two main issues within this approach: (1) a limited range of suitable substrates, since transition metals and some of their alloys are the only catalytic substrates that can be used for the CNTs' growth; (2) a postsynthesis purification step, which is necessary for eliminating catalytic particles inside the grown CNTs (Hou et al. 2008) and which cannot be performed without removing the CNTs from the growth substrate. Besides, within the CVD process, it might be difficult to control both homogeneity and thickness of the synthesized CNT film, which are two key parameters for devices. For all these reasons, the best option to make a 2D CNT film is to (1) synthesize CNTs through CVD, (2) remove

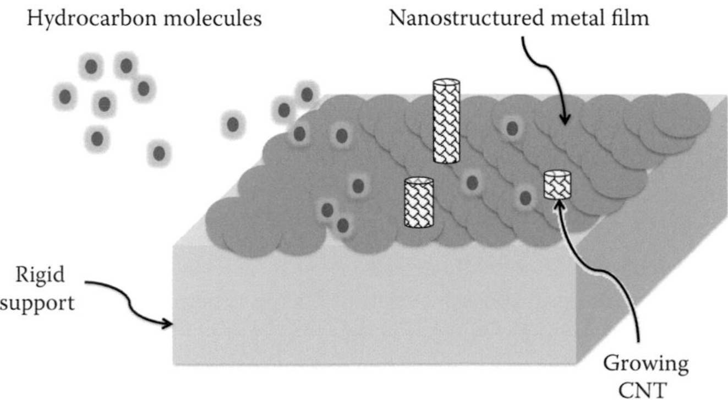

FIGURE 17.1 Schematic representation of a CVD process. At high temperature, hydrocarbon molecules are converted into CNTs by nanostructured metal film deposited on a rigid support, such as alumina or silicon.

CNTs from the growth substrate and chemically purify them, (3) deposit them on the chosen, arbitrary substrate. Regarding the first step, nanostructured 3*d* group transition metals deposited on silicon or alumina support are the generally used growth substrates. When it comes to remove them from the growth substrate, it is sufficient to immerge it in an organic solvent—for instance ethanol—and then apply ultrasounds for a few hours. The resulting dispersion can then undergo several purification steps, like centrifugation, acid etching, and so forth (Hou et al. 2008). Eventually, to deposit the purified CNTs on the chosen substrate, several protocols have been developed. Here, we review the three which are most used, due to their simplicity and scalability.

17.2.1.1 Drop Casting

Drop casting represents perhaps the easiest method allowing for deposition of a CNT film on an arbitrary substrate. The first step consists of realizing a stable dispersion of (purified) CNTs. At this aim, organic solvents, e.g., dimethylformamide, toluene, and chlorinated aromatic solvents, or aqueous solutions with surfactants are generally used, and a prolonged sonication treatment is applied to (1) make the CNT dispersion more stable over time and (2) isolate and unbundle CNT aggregates (Ausman et al. 2000; Rastogia et al. 2008). Once the dispersion is stable and optically almost transparent (see Figure 17.2), the CNTs are ready to be deposited on arbitrary substrates.

According to the drop casting approach, droplets of the dispersion are simply dropped on the substrate. To remove residuals from either the organic solvent or the surfactants, thermal treatment at relatively low temperature (around 300°C) can also be applied. Reducing or increasing the number of dropped droplets modifies the thickness of the resulting CNT film. Also, the concentration of the CNT dispersion affects the overall film thickness.

As a matter of fact, although very simple, this method is no longer largely used. Indeed, it is not possible to finely adjust the homogeneity of the film. The CNTs dispersed in the droplet are located mainly at the edge of the droplet itself because of the surface tension of the liquid. Therefore, as the droplet expands on the substrate, the CNTs are driven farther and farther away from the droplet center, until the CNTs are finally deposited on the substrate when the droplet evaporates. To improve film homogeneity, the substrate can be kept hot, using for instance a simple hot plate, to allow fast evaporation of the droplets. Nevertheless, the film homogeneity is typically poor. This method has been used, for instance, to realize solar cells based on the photoactive heterojunction between CNT film and doped silicon (Le Borgne et al. 2010).

17.2.1.2 Spray Coating

The spray coating method is a simple solution, providing great potential to scaling up for large area coatings. The ability to form thick films is another advantage of this method over, for instance, the drop casting.

FIGURE 17.2 Pictures of a CNT dispersion in water and sodium dodecyl sulfate. After 2 h of ultrasonication, the dispersion is stable and appears transparent.

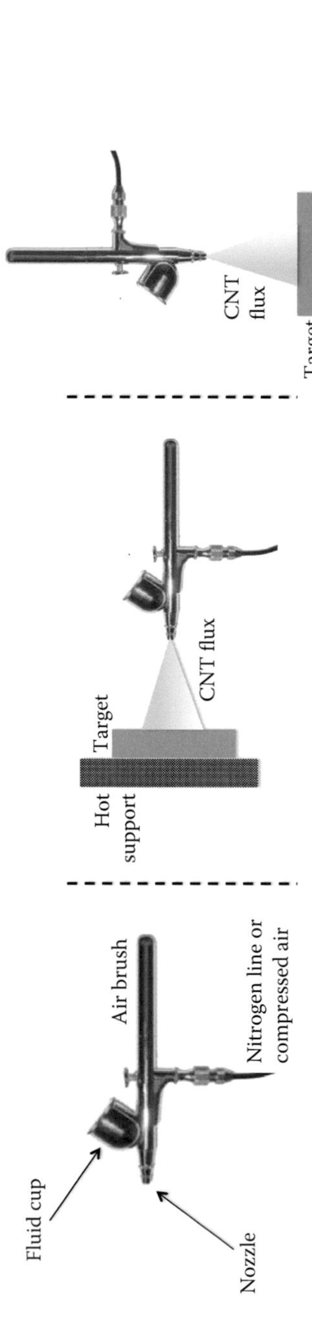

FIGURE 17.3 Left panel: a commercially available airbrush that could be used for spraying CNTs. Center panel: horizontal setup, with the hot support behind the target. Right panel: vertical setup.

Once the CNT's dispersion is stable, it is loaded in a spray tool, as the one displayed in Figure 17.3, and then sprayed on an arbitrary substrate. Several substrates may be used, i.e., glass, silicon wafer, indium tin oxide, and polymers, which makes this technique suitable for a large range of device fabrication (Kaempgen et al. 2005; Camilli et al. 2012; Park et al. 2012).

Film homogeneity depends on the quality of CNT dispersion. If the quality of the starting CNT dispersion is high, then homogeneity is very good. Besides, film thickness can be adjusted either by deposition time or by the concentration of the dispersion; i.e., keeping the deposition time constant, the higher the CNT concentration is, the thicker the resulting film.

Two experimental configurations may be used. The first one is achieved with the target substrate being held horizontally and the spraying tool over it (vertical configuration; see Figure 17.3). However, if the CNT dispersion is not stable, the CNTs end up agglomerating and the gravity force pushes them downward, toward the spray ejector, eventually clogging it. The second configuration is represented by the target and the spraying tool being one in front of each other in an imaginary horizontal line (horizontal configuration; see Figure 17.3). In this last geometry, the disadvantage is represented by the fact that the sprayed solvent might condense once it reaches the substrate surface, thus forming droplets, which move downward because of the gravity force. The result is then a film, which is thicker at the bottom. To avoid such inhomogeneity, the substrate could be mildly heated up so that the solvent evaporates as soon as it reaches the substrate surface, thus leaving on it only the CNTs. A low-boiling-point solvent may be chosen to reduce this issue as well.

A drawback of this simple approach is that the film stability is strongly affected by the used substrate. In particular, if the adhesion to substrate is weak, then the films are likely to peel off eventually. In this case, pretreatments can be applied to the substrate before the CNT dispersion is sprayed. For instance, it has been reported that coating a glass substrate with a layer of self-assembled polyimide or 3-aminopropyl triethoxysilan improves the CNT–substrate interaction (Schindler et al. 2007).

17.2.1.3 Vacuum-Assisted Filtration

The third method used for depositing CNTs on arbitrary substrates is filtration. In particular, vacuum-assisted filtration is largely used since it allows for a fine control over thickness and homogeneity of the resulting CNT film.

Within this approach, a cellulose filter is mounted on top of a porous filter of a vacuum filtration apparatus, like the one reported in Figure 17.4, and the CNT dispersion is loaded in it.

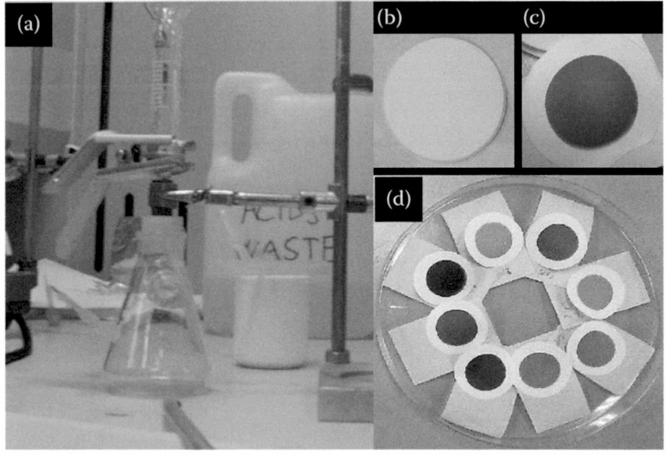

FIGURE 17.4 (a) Example of an experimental setup used for vacuum filtration experiments. Cellulose filter as-received (b) and after vacuum filtration experiment, with a CNT film on top (c). (d) Several CNT films deposited on cellulose filters. Darker means thicker CNT films.

Whereas the solvent passes through the porous filter, the CNTs do not, thus forming a black film on the cellulose filter. After rinsing it for several times using water and other specific solvents able to dissolve the surfactant used in the dispersion, the cellulose filter with the CNT film on top is removed from the apparatus and pressed on the chosen substrate. Then, the cellulose filter is easily removed by acetone and the CNT film eventually sticks to the substrate owing to van der Waals interaction.

The thickness and homogeneity of the CNT film depend on the quality and concentration of the starting CNT dispersion. One of the cons of this approach is the impossibility of depositing CNT on a large area, which is defined by the diameter of the used vacuum apparatus.

17.2.2 Synthesis of 3D CNT Solids

Assembling CNTs into bulk materials (i.e., 3D macrostructures) with desired structures and properties is an important step toward CNTs' practical applications in our everyday life. In this scenario, CNT aerogels (Zou et al. 2010; Sun et al. 2013), buckypapers (Cooper et al. 2003; Endo et al. 2005), long fibers or yarned sheets (Zhang et al. 2004, 2005), and sponges (Camilli et al. 2014) have been successfully fabricated. To form 3D scaffolds of CNTs, like sponges and aerogels, three approaches have been used: (1) a modified version of the CVD method, which enables the direct formation of 3D CNT solids; (2) chemical reactions to covalently cross-link CNTs; and (3) stacking CNTs and other carbonaceous allotropies via π-π interaction.

17.2.2.1 Direct CVD Synthesis of 3D CNT Solids

To obtain 3D self-standing CNT solids, the individual CNTs forming the solid must be all linked together, interacting, to work as a single macrostructure. The key factors leading to the formation of such a structure are a considerable length of the individual CNTs and their contorted, not aligned, geometry. To synthesize CNTs with these particular features, the standard CVD method can be ad hoc modified. The main modifications concern (1) how to continuously grow CNTs with high yield and (2) how to increase their length and contorted geometry at the same time.

The first issue is solved by continuously adding a catalyst to the process. In a standard CVD technique, the catalyst is represented by a thin film of nanostructured 3*d*-group transition metals deposited on a rigid support, such as silicon, silicon dioxide, or alumina. After a few minutes, the catalytic nanostructures become poisoned, since they end up being covered with carbonaceous layer, thus inhibiting further CNT growth. This is the reason why although the carbonaceous precursor is still present in the growth chamber, after a limited period, the growth process ends. In the modified CVD process, instead, the catalyst is continuously supplied in the growth chamber in gas form, along with the carbon precursor (floating catalyst CVD technique). In this kind of approach, ferrocene is one of the most used catalysts. It is generally dispersed in organic solvents, like ethanol, and when injected into the growth chamber, the high temperature immediately vaporizes it (Camilli et al. 2014). The advantages of the floating catalyst technique are that it does not require the stage of catalyst preparation, as in the case of the fixed catalyst (or standard) method; there is always a fresh catalyst in the chamber so that the growth can run as long as the carbon precursor is supplied as well.

The floating catalyst technique provides CNTs with considerable length (Zhong et al. 2010). Regarding how to make them contorted, so that they would touch and interact with each other, this can be achieved by introducing, into CNT lattice, atoms other than carbon, to introduce defects and imperfection in the honeycomb structure. The most used atoms are sulfur, which is also a CNT growth enhancer (Camilli et al. 2014), and boron, which creates bent CNTs with elbow-like features (Hashim et al. 2012).

17.2.2.2 Cross-Linking Individual CNTs

Chemically cross-linking CNTs is an alternative way of realizing 3D CNT frameworks. This approach relies on making organic bridges, which are bonded to CNTs. As an example, Ozden et al. (2014) have recently reported the scalable synthesis of 3D macroscopic solids made of covalently connected CNTs via Suzuki cross-coupling reaction. The Suzuki reaction is one of the most widespread synthetic routes to create carbon–carbon single covalent bonds between organic halides and boronic acid derivatives using Pd-based catalysts (Miyaura and Suzuki 1995). In the case of the work of Ozden and collaborators, commercial CNTs are refluxed in concentrated HNO_3 for 18 h to oxidize CNTs, the majority of which would contain carboxyl groups (–COOH). The carboxyl groups are then converted to the corresponding acid chloride (–COCl) by reacting with $SOCl_2$ at 80°C for 24 h. Next, a palladium-based catalyst [tetrakis(triphenylphosphine) palladium(0), $Pd(PPh_3)_4$)] is added under argon along with a mixture of 1,4-phenyldiboronic acid and anhydrous Cs_2CO_3 in anhydrous toluene. The mixture is eventually heated at 100°C for 5 days. The resulting material is filtered and rinsed to remove excess Cs_2CO_3 and unreacted 1,4-phenyldiboronic acid. After this treatment, the CNTs result in being covalently bonded through a –CO–Ph–CO– group. Eventually, after lyophilization, self-standing 3D solids are formed (Ozden et al. 2014).

17.2.2.3 π–π Stacking of CNTs

All-carbon aerogels with a monolith 3D framework made up of interconnected CNTs and graphene sheets have been recently reported as well. By freeze-drying aqueous solutions of CNTs and giant graphene oxide (GGO) sheets, followed by chemical reduction of GGO into graphene with hydrazine vapor, Gao and collaborators have in fact fabricated all-carbon ultralight aerogels, whose density ranges between 0.2 and 22 mg cm^{-3} (Sun et al. 2013).

The key step in this procedure is represented by the chemical reduction of GGO into graphene, since reduced graphene exhibits a stronger π–π interaction with both CNTs and other reduced graphene, thus providing the backbone of the CNT solid. The resulting all-carbon net shows remarkable properties, such as temperature-invariant elasticity, ultralow density, excellent thermal stability, extremely high absorption capacities for organic liquid, and good electrical conductivity (Sun et al. 2013). All these features could pave the way for the use of such aerogels as elastic and flexible conductors, organic absorbents, and supercapacitors, and in high-performance conductive polymer composites.

17.3 Applications of CNT Networks

17.3.1 2D CNT Films in Photovoltaic

After Lee's paper in 2005 showing that an individual single-walled CNT bridging two metal contacts can indeed form a p–n junction and produce power conversion efficiency under illumination (Lee 2005), a great deal of effort has been dedicated at realizing solar cells using CNTs as a building block (Kislyuk and Dimitriev 2008). Owing to the abundance of C element and the cheapness of CNT synthesis and deposition methods, it is understandable that C-based solar cells are indeed very appetitive.

At the base of the photovoltaic effect in CNTs themselves, there is the creation of electron–hole (e–h) pairs induced by the electronic transition between van Hove singularities upon light absorption, in a wavelength range between the ultraviolet (UV) and infrared (IR) region, as shown in Figure 17.5 (Chen et al. 1998).

So far, the most promising results have been obtained by solar cells with the photoactive junction being made by a CNT film in intimate contact with an n-doped crystalline or amorphous silicon substrate. In this kind of solar cell, the 2D CNT film actually plays a triple role. The first one is to create the active metal–semiconductor Schottky junction (Behnam et al. 2008; Wadhwa et al. 2010) with the silicon substrate. As is well known, if a semiconductor material is illuminated by photons with energy higher than the semiconductor's energy gap, the electrons present in the valence band absorb energy and jump to the conduction band (excited electron), leaving a hole in the valence band (creation of e–h pair, or exciton). If no potential is present within the semiconductor substrate, then electrons and holes will soon recombine (electrons go back to their position in the valence band). To separate the e–h

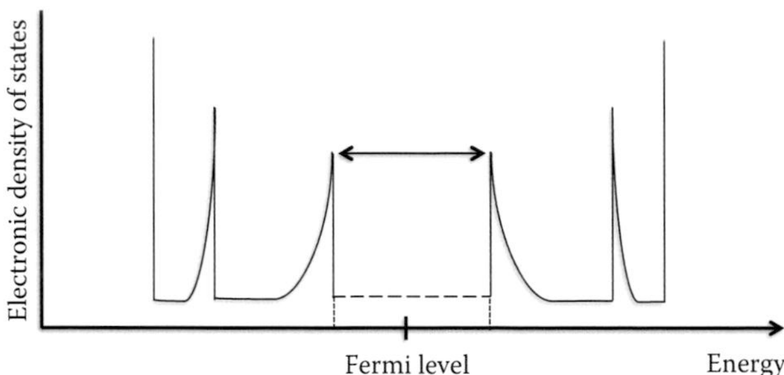

FIGURE 17.5 Typical density of states as a function of energy for single-walled CNTs. The double-arrow connector highlights the electronic transition between an electron in a Van Hove singularity below the Fermi level and a Van Hove singularity above. (—) or (- -) for semiconductive and metallic tubes, respectively.

pair, thus creating electron current or photocurrent (photovoltaic effect), a built-in potential (Galvani potential) needs to be present. This electrical potential, in fact, generates an electromotive force, which drives electrons and holes away owing to their opposite charge, thus avoiding recombination. A junction is necessary to create the Galvanic potential. In the case of silicon, the most used junction is the *p*–*n* junction, obtained by doping a region of the substrate with donor atoms and another, adjacent region with acceptor ones. In this case, the different doping levels of the two regions will create a potential difference (built-in voltage) at the interface, able to split the e–h pair. The doping procedure is, however, quite expensive and time consuming. The advantage of the CNT/Si solar cells over the regular Si ones is therefore to replace the step of the *p*-doping with the formation of a 2D CNT film on the surface of the *n*-Si substrate. This, as discussed in the previous paragraph, can be easily done and scaled for large area depositions. The difference in the Fermi level of the two materials (CNTs on one hand and *n*-Si on the other) creates the photoactive heterojunction (i.e., Schottky junction), hence the built-in potential, necessary for the e–h pair splitting and creation of photocurrent upon illumination.

Extending the Si spectral range toward the near-UV and the near-IR regions is then the second role played by CNT film. In fact, upon illumination, the CNTs absorb (Chen et al. 1998) light and the electrons below the Fermi level are excited to the electronic states in the conduction band. This happens in the case of single-walled and multiwalled CNTs as well (Castrucci et al. 2011; Del Gobbo et al. 2013). Therefore, in the case of CNT-Si solar cells, the collected photocarriers are generated both in the CNT film and in the Si substrate.

As a matter of fact, the creation of photocarriers is far more effective in the Si substrate rather than in the CNT layer, this being a result of the ultrafast recombination of the excitons in CNTs, especially metallic ones (Avouris et al. 2008). Indeed, it has been proved that, keeping the other parameters constant, the thinner the CNT layer is, the better the power conversion efficiency (Le Borgne et al. 2010; Del Gobbo et al. 2011). In this context, the majority of the incident light would pass through the CNT layer without being absorbed, eventually reaching the Si substrate, where it is absorbed. According to this scenario, the CNT film actually acts more as a semitransparent electrode, by collecting and driving holes to the metal electrode, rather than a photoabsorbing layer (Del Gobbo et al. 2011).

17.3.2 3D CNT Networks in Environment Protection

Owing to their porous structure and high surface area, their hydrophobic nature, the lightweight, and the remarkable elastic properties, self-standing 3D CNT networks such as sponges and aerogels have been recently employed to remove pollutants and organic contaminants from water (Gui et al. 2010; Hashim et al. 2012; Camilli and Bahramifar 2014; Camilli et al. 2014).

Human society requires fresh and clean water for its survival, and this water demand has been growing rapidly because of rapid urbanization and the increase in the number of global population. Besides, chemical farming practices (i.e., application of fertilizers, pesticides, and insecticides) and some industrial activities (i.e., chemical and oil industry) have been using and contaminating fresh water, especially with trace amounts of aromatic and chlorinated hydrocarbons. Chlorobenzenes, for instance, are extremely dangerous because of their chemical stability and limited photochemical degradation. Oil spills, during extraction and distribution, could be catastrophic for the environment as well.

In this context, CNTs have been considered as excellent candidates for wastewater cleanup because of their remarkable sorption capabilities toward a wide range of organic chemicals and inorganic agents, even more than conventional systems such as clay and activated carbon (Mauter and Elimelech 2008; Pan and Xing 2008). However, using CNTs in water treatment is not straightforward, since in the form of powders or 2D films, they are hard to be handled, can be dispersed and lost in water, and need to be retrieved through further filtration. However, these issues might be addressed by using self-standing millimeter-size 3D CNT networks, as the ones displayed in Figure 17.6. Such structures, made of long and interconnected nanotubes, are so porous that they are extremely lightweight (below 10 mg cm^{-3}) and can be compressed up to values as high as 75% without showing plastic deformation (Camilli et al. 2014).

Taking advantage of their hydrophobic nature and high surface area, 3D CNT sponges have been efficiently used to uptake amounts of vegetable and exhaust oil from water (see Figure 17.7).

Owing to their ultra lightweight, sponges easily float on water surface, and when they come across contaminated regions, they selectively uptake oil (Gui et al. 2010; Hashim et al. 2012; Camilli et al. 2014). For instance, it has been reported that a 20 mm × 10 mm × 3 mm CNT sponge of 1.5 mg is able to absorb vegetable oil of up to 150 times its initial weight, thus exhibiting extraordinary absorption capacity (Camilli et al. 2014). This superior uptake efficiency is ascribed to (1) the high number of open pores in sponges, which can be filled with oil during absorption, and (2) the high density of defects in the carbon

FIGURE 17.6 Picture of a few self-standing 3D CNT networks.

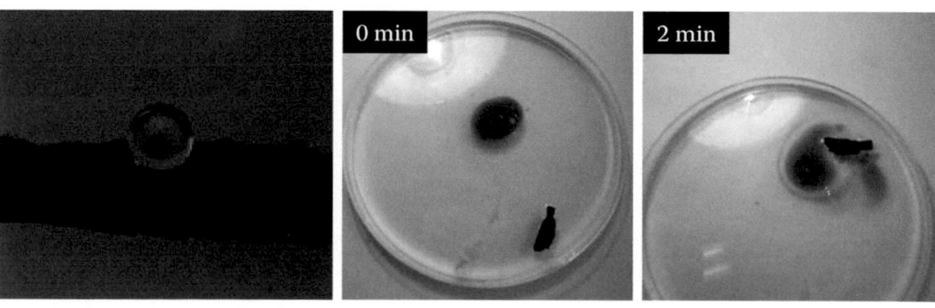

FIGURE 17.7 Left panel: a water droplet on a self-standing CNT network. The highly porous structure and its intrinsic hydrophobicity make the CNT solid a super-hydrophobic material. Central and right panels: a CNT sponge is selectively absorbing exhaust oil from water.

nanostructures making the sponges: It has been reported in fact that rough and contorted surfaces allow for a faster and more efficient absorption of organics with respect to smooth ones (Peng et al. 2003). It is noteworthy here of reminding that the 3D sponges described previously are made through a modified CVD process, as described in Section 17.2.2.1, and therefore, the individual carbon nanostructures forming the sponges exhibit, in fact, a very contorted and defective morphology.

Once sponges become saturated, the absorbed oil can be removed either by mechanical squeezing or by ignition. In the latter case, as soon as all the oil is burnt away, the fire blows out without destroying the sponge, which is hence ready to be reused again (Hashim et al. 2012; Camilli et al. 2014).

In addition, since catalyst residues are present inside the CNTs owing to the CVD approach, sponges could be either dragged by tweezers or driven by a magnet without any direct contact (Hashim et al. 2012; Camilli et al. 2014). In fact, the presence of metallic nanoparticles inside individual CNTs confers magnetic properties to the sponge itself.

The absorption capability of such 3D networks has been demonstrated for a large number of oils, such as gasoline, pump oil, and diesel oil, and also organic solvents, such as chloroform, hexane, ethanol, dimethyl formamide, and dichlorobenzene. If one compares the absorption capacity, defined by the ratio between the final and initial weight after full absorption, of CNT sponges, natural products (cotton, loofah), polymeric sponges (polyurethane and polyester), and activated carbon to different organics, that of the CNT sponges results in being always higher (Gui et al. 2010), thus proving the incredible potential of such 3D porous networks in environment protection.

17.3.3 3D CNT Networks in Biomedicine

Perhaps, thinking of CNTs, the first possible applications that come to mind are in the field of electronics or chemical engineering (i.e., as a filler in composite materials). Nevertheless, in the past decade, CNTs have shown great potential in drug delivery, in imaging of engineering tissues, and in tissue engineering scaffolds.

Concerning their application in drug delivery, CNTs can be functionalized with pharmaceutical agents or biological species that target specific cells or tissues. In this scenario, CNTs act as a cargo vehicle to selectively deliver therapeutic agents. Since functionalized CNTs display low toxicity and are not immunogenic, such systems hold great potential in the field of nanobiotechnology and nanomedicine (Bianco et al. 2005). When it comes to imaging of engineering tissues, CNTs can be functionalized with contrast agents and imaged with standard techniques (Liu et al. 2007; Richard et al. 2008).

In all the previously mentioned examples, CNTs are used in form of powders; no 3D architecture is required for that kind of application. However, 3D CNT networks have been also largely studied in the field of biomedicine. A 3D CNT network, in fact, closely resembles the framework of the extracellular matrix, that is, the collection of extracellular molecules secreted by cells that provide structural and biochemical support to the surrounding cells. The presence of large and open pores, in fact, makes 3D CNT films an optimal scaffold where cells can anchor, settle, and start communicating with other cells (Figure 17.8).

In particular, the importance of the CNT film thickness (i.e., the importance of the 3D nature of CNT films) over cell proliferation and growth has been recently addressed (Lee et al. 2015). In their manuscript, in fact, Lee and coworkers report not only that single-walled CNT films show no cytotoxicity to mesenchymal stem cells but also that the thicker the CNT network, the better in terms of cell growth rate and differentiation. Since the adhesion of proteins to CNTs is quite strong, and CNTs exhibit a large surface area, when the CNT film is filled with growth factor proteins and nutrients, it also promotes cell proliferation and differentiation.

Particularly interesting is taking advantage of the excellent electrical properties of CNTs to make it easier for the cells to "talk" to each other. At this aim, neurons (i.e., neuronal cells) represent the optimum cell target since they transmit information through both electrical and chemical signals. Several reports have indeed shown not only that CNTs anchored on planar substrates can promote

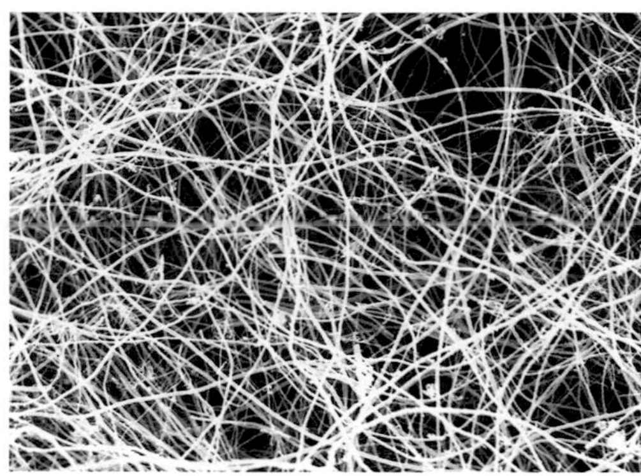

FIGURE 17.8 Scanning electron microscope micrograph of the highly porous structures of a 3D CNT network. The large and open pores make it easier for cells to anchor and communicate with each other.

cell attachment, growth, differentiation, and long-term survival of neurons (Mattson et al. 2000; Hu et al. 2004, 2005) but also that neurons grown on a conductive nanotube network always display more efficient signal transmission (Lovat et al. 2005; Mazzatenta et al. 2007). Although a precise mechanism is still lacking, it is sure that the presence of nanotubes improves the responsiveness of neurons by forming electrical shortcuts between the proximal and distal compartments of the neuron and enhances the overall efficacy signal propagation (Cellot et al. 2009).

The fact that CNT films might affect neuronal information processing pushes researchers into gaining insights into the functioning of hybrid neuronal/nanotubes networks. Indeed, knowing how to engineer the interaction between neurons and CNTs would be important for the design of smart materials, which could trigger specific synaptic reorganization in the neuronal network.

17.4 Future Perspectives

With the rise of graphene taking place (Geim and Nvoselov 2007), the majority of research groups studying CNTs moved to the new, exciting field. The era of CNTs seemed to be at its end. However, as discussed in this chapter, novel applications of 2D and 3D CNT networks are being developed, not only in the field of material science but also in nanomedicine as well as in environment protection. CNT technology, from its synthesis to manipulation, seems to be mature enough to allow CNTs to move from a lab-scale to an industry-scale perspective. Using CNTs as a building block in 2D and 3D networks generated outstanding results, which foresees an intense use of these nanostructures in the near future.

References

Ajayan, P. M., "Nanotubes from carbon," *Chemical Review* 99 (1999): 1787–1799.

Anantram, M. P., Léonard, F., "Physics of carbon nanotube electronic devices," *Reports on Progress in Physics* 69 (2006): 507.

Ausman, K. D., Piner, R., Lourie, O., Ruoff, R. S., Korobov, M., "Organic solvent dispersions of single-walled carbon nanotubes: Toward solutions of pristine nanotubes," *Journal of Physical Chemistry B* 104 (2000): 8911–8915.

Avouris, P., Freitag, M., Perebeinos, V., "Carbon-nanotube photonics and optoelectronics," *Nature Photonics* 2 (2008): 341–350.

Balasubramanian, K., Burghard, M., "Biosensors based on carbon nanotubes," *Analytical and Bioanalytical Chemistry* 385 (2006): 452–468.

Behnam, A., Johnson, J. L., Choi, Y. H. et al., "Experimental characterization of single-walled carbon nanotube film-Si Schottky contacts using metal–semiconductor–metal structures," *Applied Physics Letters* 92 (2008): 243116.

Bianco, A., Kostarelos, K., Prato, M., "Applications of carbon nanotubes in drug delivery," *Current Opinion in Chemical Biology* 9 (2005): 674–679.

Camilli, L., Bahramifar, N., "Carbon nanotubes remove contaminants from water," *Membrane Technology* 3 (2014): 8.

Camilli, L., Scarselli, M., Del Gobbo, S., Castrucci, P., Gautron, E., De Crescenzi, M., "Structural, electronic and photovoltaic characterization of multiwalled carbon nanotubes grown directly on stainless steel," *Beilstein Journal of Nanotechnology* 3 (2012): 360–367.

Camilli, L., Pisani, C., Gautron, E. et al., "A three-dimensional carbon nanotube network for water treatment," *Nanotechnology* 25 (2014): 065701–065707.

Castrucci, P., Scilletta, C., Del Gobbo, S. et al., "Light harvesting with multiwall carbon nanotube/silicon heterojunctions," *Nanotechnology* 22 (2011): 115701.

Cellot, G., Cilia, E., Cipollone, S. et al., "Carbon nanotubes might improve neuronal performance by favouring electrical shortcuts," *Nature Nanotechnology* 4 (2009): 126–133.

Chen, J., Hamon, M. A., Hu, H. et al., "Solution properties of single walled carbon nanotubes," *Science* 282 (1998): 95–98.

Cooper, S. M., Chuang, H. F., Cinke, M., Cruden, B. A., Meyyappan, M., "Gas permeability of a buckypaper membrane," *Nano Letters* 3 (2003): 189–192.

Dai, H., "Carbon nanotubes: Synthesis, integration and properties," *Accounts of Chemical Research* 35 (2002): 1035–1044.

Del Gobbo, S., Castrucci, P., Scarselli, M. et al., "Carbon nanotube semitransparent electrodes for amorphous silicon-based photovoltaic devices," *Applied Physics Letters* 98 (2011): 183113–183115.

Del Gobbo, S., Castrucci, P., Fedele, S., "Silicon spectral response extension through single wall carbon nanotubes in hybrid solar cells," *Journal of Materials Chemistry C*, 1 (2013): 6752–6758.

Dresselhaus, M. S., Dresselhaus, G. and Avouris, P. *Carbon nanotubes* (Springer-Verlag Berlin, Heidelberg, New York, 2001).

Endo, M., Muramatsu, H., Hayashi, T., Kim, Y. A., Terrones, M., Dresselhaus, N. S., "Nanotechnology: 'Buckypaper' from coaxial nanotubes," *Nature* 433 (2005): 476.

Geim, A. K., Nvoselov, K. S., "The rise of graphene," *Nature Materials* 6 (2007): 183–191.

Gui, X., Wei, J., Wang, K. et al., "Carbon nanotube sponges," *Advanced Materials* 22 (2010): 617–621.

Gurvan, M., Tonon, T., Scornet, D., Cock, J. M., Kloareg, B., "The cell wall polysaccharide metabolism of the brown alga Ectocarpus siliculosus. Insights into the evolution of extracellular matrix polysaccharides in Eukaryotes," *New Phytologist* 188 (2010): 82–97.

Hashim, D. P., Narayanan, N. T., Romo-Herrera, J. et al., "Covalently bonded three-dimensional carbon nanotube solids via boron induced nanojunctions," *Scientific Reports* 2 (2012): 363–370.

Hou, P. X., Liu, C., Cheng, H. M., "Purification of carbon nanotubes," *Carbon* 46 (2008): 2003–2025.

Hu, H., Ni, Y., Montana, V., Haddon, R. C., Parpura, V., "Chemically functionalized carbon nanotubes as substrates for neuronal growth," *Nano Letters* 4 (2004): 507–511.

Hu, H., Ni, Y., Mandal, S. K. et al., "Polyethyleneimine functionalized single-walled carbon nanotubes as a substrate for neuronal growth," *Journal of Physical Chemistry B* 109 (2005): 4285–4289.

Iijima, S., "Helical microtubules of graphitic carbon," *Nature* 354 (1991): 56–58.

Jung, Y., Li, X., Rajan, N. K., Taylor, A. D., Reed, M. A., "Record high efficiency single-walled carbon nanotube/silicon p–n junction solar cells," *Nano Letters* 13 (2013): 95–99.

Kaempgen, M., Duesberg, G. S., Roth, S., "Transparent carbon nanotube coatings," *Applied Surface Science* 252 (2005): 425–429.

Kislyuk, V. V., Dimitriev, O. P., "Nanorods and nanotubes for solar cells," *Journal of Nanoscience and Nanotechnology* 8 (2008): 131–148.

Kumar, M., Ando, Y., "Chemical vapor deposition of carbon nanotubes: A review on growth mechanism and mass production," *Journal of Nanoscience and Nanotechnology* 10 (2010): 3739–3758.

Le Borgne, V., Castrucci, P., Del Gobbo, S. et al., "Enhanced photocurrent generation from UV-laser-synthesized-single-wall-carbon-nanotubes/n-silicon hybrid planar devices," *Applied Physics Letters* 97 (2010): 193105.

Lee, J. U., "Photovoltaic effect in ideal carbon nanotube diodes," *Applied Physics Letters* 87 (2005): 073101.

Lee, J. H., Shim, W., Khalid, C. N. et al., "Random networks of single-walled carbon nanotubes promote mesenchymal stem cell's proliferation and differentiation," *ACS Applied Materials and Interfaces* 7 (2015): 1560–1567.

Liu, Z., Cai, W., He, L. et al., "In vivo biodistribution and highly efficient tumor targeting of carbon nanotubes in mice," *Nature Nanotechnology* 2 (2007): 47–52.

Lovat, V., Pantarotto, D., Lagostena, L. et al., "Carbon nanotube substrates boost neuronal electrical signaling," *Nano Letters* 5 (2005): 1107–1110.

Marulanda, J. M. *Carbon nanotubes applications on electron devices* (InTech, 2011).

Mattson, M. P., Haddon, R. C., Rao, A. M., "Molecular functionalization of carbon nanotubes use as substrates for neuronal growth," *Journal of Molecular Neuroscience* 14 (2000): 175–182.

Mauter, M. S., Elimelech, M., "Environmental applications of carbon-based nanomaterials," *Environmental Science & Technology* 42 (2008): 5843–5859.

Mazzatenta, A., Giugliano, M., Campidelli, S. et al., "Interfacing neurons with carbon nanotubes: Electrical signal transfer and synaptic stimulation in cultured brain circuits," *Journal of Neuroscience* 27 (2007): 6931–6936.

Miyaura, N., Suzuki, A., "Palladium-catalyzed cross-coupling reactions of organoboron compounds," *Chemical Reviews* 95 (1995): 2457–2483.

Ozden, S., Narayanan, T. N., Tiwary, C. S. et al., "3D macroporous solids from chemically cross-linked carbon nanotubes," *Small* (2014) doi: 10.1002/smll.201402127.

Pan, B., Xing, B., "Adsorption mechanisms of organic chemicals on carbon nanotubes," *Environmental Science & Technology* 42 (2008): 9005–9013.

Park, C., Kim, S. W., Lee, Y. S., Lee, S. H., Song, K. H., Park, L. S., "Spray coating of carbon nanotube on polyethylene terephthalate film for touch panel application," *Journal of Nanoscience and Nanotechnology* 12 (2012): 5351–5355.

Peng, X., Li, Y., Luan, Z. et al., "Adsorption of 1,2-dichlorobenzene from water to carbon nanotubes," *Chemical Physics Letters* 376 (2003): 154–158.

Rastogia, R., Kaushala, R., Amit, S. K. T., Sharma, A. L., Kaur, I., Bharadwaj, L. M., "Comparative study of carbon nanotube dispersion using surfactants," *Journal of Colloid and Interface Science* 328 (2008): 421–428.

Richard, C., Doan, B., Beloeil, J., Bessodes, M., Toth, E., Scherman, D., "Noncovalent functionalization of carbon nanotubes with amphiphilic Gd3+ chelates: Toward powerful T1 and T2 MRI contrast agents," *Nano Letters* 8 (2008): 232–236.

Schindler, A., "ITO replacements: Carbon nanotubes," in *Handbook of Visual Display Technology*, eds. J. Chen, W. Cranton and M. Fihn (Springer-Verlag, Berlin, Heidelberg, 2012), 795–808.

Schindler, A., Brill, J., Fruehauf, N., Novak, J. P., Yaniv, Z., "Solution-deposited carbon nanotube layers for flexible display applications," *Physica E: Low-Dimensional Systems and Nanostructures* 37 (2007): 119–123.

Sun, H., Xu, Z., Gao, C., "Multifunctional, ultra-flyweight, synergistically assembled carbon aerogels," *Advanced Materials* 25 (2013): 2554–2560.

Tans, S. J., Verschueren, A. R. M., Dekker, C., "Room-temperature transistor based on a single carbon nanotube," *Nature* 393 (1998): 49–52.

Wadhwa, P., Liu, B., McCarthy, M. A., Wu, Z. C., Rinzler, A. G., "Electronic junction control in a nanotube-semiconductor Schottky junction solar cell," *Nano Letters* 10 (2010): 5001–5005.

Wang, Y., Yeow, J. T. W., "A review of carbon nanotubes-based gas sensors," *Journal of Sensors* 2009 (2009): 493904–493929.

Yu, M.-F., Lourie, O., Dyer, M. J., Moloni, K., Kelly, T. F., Ruoff, R. S., "Strength and breaking mechanism of multiwalled carbon nanotubes under tensile load," *Science* 287 (2000a): 637–640.

Yu, M.-F., Kowalewski, T., Ruoff, R., "Investigation of the radial deformability of individual carbon nanotubes under controlled indentation force," *Physical Review Letters* 85 (2000b): 1456–1459.

Zhang, M., Atkinson, K. R., Baughman, R. H., "Multifunctional carbon nanotube yarns by downsizing an ancient technology," *Science* 306 (2004): 1358–1361.

Zhang, M., Fang, S., Zakhidov, A. A., Lee, S. B., Aliev, A. E., Williams, C. D., "Strong, transparent, multifunctional, carbon nanotube sheets," *Science* 309 (2005): 1215–1219.

Zhong, X. H., Li, Y. L., Liu, Y. K. et al., "Continuous multilayered carbon nanotube yarns," *Advanced Materials* 22 (2010): 692–696.

Zou, J. H., Liu, J. H., Karakoti, A. S. et al., "Ultralight multiwalled carbon nanotube aerogel," *ACS Nano* 4 (2010): 7293–7302.

18

Carbon Nanotubes for Sensing Applications

Abha Misra

18.1 Introduction

18.1.1 Carbon and Its Allotropes

Carbon is one of the most common elements owing to its unique electronic configuration that allows it to achieve various hybridizations. The different types of hybridization in fact allow carbon to exist in different structural configurations while retaining the purity of the carbon atom. Such structural configurations are known as allotropes. In general, carbon mostly exists in three well-known allotropic forms, like diamond, graphite, and fullerenes. Diamond is an sp^3 hybridized tetrahedral structure, whereas graphite has sp^2 hybridized planar hexagonal structure stacked together by van der Waals forces. Recently, the invention of monolayer of graphene has revolutionized the field of research because of its unique semimetal characteristics. While zero-dimensional sp^2 hybridized fullerene demonstrated a cage like spherical structure, cylindrical carbon nanotubes (CNT) presented a one-dimensional structure. CNT has generated immense interest in researchers due to its exceptional multifunctional properties applicable in various interdisciplinary fields, which will be discussed in the later sections.

18.1.1.1 Introduction of CNTs

Since the discovery of CNT by Iijima in 1991 (Iijima 1991), an exponential rise has been seen in the interest of researchers from various fields, as reflected in the number of publications. Exceptional properties that make CNT exciting lie in their structural characteristics that make them a supermaterial. The basic structure of CNT is considered as a rolled sheet of graphene in the form of both single-walled CNT (SWCNT) and multiwalled CNTs (MWCNT). The rolling direction as shown in Figure 18.1 mainly controls electronic properties. Unit cell of CNT where two axes of symmetry along which if the sheet of graphene is rolled we obtain armchair (n, n) and zigzag $(n, 0)$ structure, where n is an integer. Any other rolling axis besides these two results in a chiral structure of CNT. These geometries, which depend on the growth process, as discussed in the next section, play an important role in determining the electronic properties of CNT, whether it is semiconducting or metallic (Odom et al. 1998).

18.1.1.2 Synthesis of CNTs

As discussed in the previous section, the electronic properties of nanotubes are greatly affected by the arrangement of carbon atoms in the lattice; hence, it is imperative to have a growth process that ensures

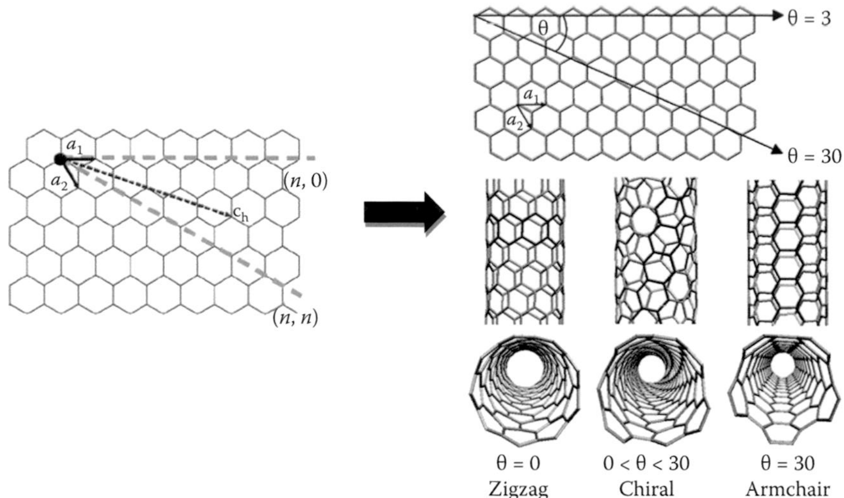

FIGURE 18.1 Atomic arrangement in zigzag, chiral, and armchair CNTs. (From Odom, T.W. et al., *Nature*, 391, 62, 1998.)

high crystallinity and uniformity in CNT. Nanotubes are obtained mainly by three synthesis methods, namely, arc discharge, laser ablation, and thermal chemical vapor deposition (CVD). Among these, although arc discharge and laser ablation provide highly crystalline CNT with minimal defects, CVD is the most popular. This is because the first two methods have a very high operation temperature, ranging around thousands of degree Celsius, and suffer from high initial setup cost. In addition, CNT contains a lot of unwanted residues/by-products. On the contrary, the CVD process can be operated at comparatively lower temperatures (around 500°C–900°C), it has a low initial setup cost, and also the yield of CNT is higher and more uniform as compared with the other two methods. However, none of these methods guarantee to provide complete control over the electronic structure of CNT as mentioned previously.

Thermal CVD process utilizes hydrocarbon (toluene, xylene, CH_4, C_2H_4, etc.) as a source in the presence of metal (ferrocene, Ni, Fe, etc.) nanoparticles as catalyst for CNT growth. Hence, the nanotubes grown using this method are also known as catalytically grown CNT. These chemicals in the solution are first vaporized at 100°C–200°C and then passed into the reaction zone. Chemical reaction takes place at 700°C–800°C (Reddy et al. 2013). CVD reactions cause the synthesis of both SWNT and MWNT nanotubes, depending on the nature of catalyst (details can be found in various literatures (Dai 2002).

18.1.1.3 Properties of CNTs

The cylindrical geometry of CNT as mentioned earlier has significant impact on its electronic properties. As mentioned earlier, depending upon their diameter and helicity, CNTs can be either metallic or semiconducting (Odom et al. 1998). The zigzag configuration of SWCNT was considered as metallic, but later, it was found using scanning tunneling microscopy analysis that it actually contains a narrow bandgap. However, isolated armchair CNTs are truly metallic in nature (Odom et al. 1998). Electron transport in CNT is ballistic along the tube axis and takes place through quantum effects. Hence, CNT is capable of carrying current without scattering and dissipation of heat. The unique band structure of CNT imparts them high current carrying capacity (theoretically, it is capable of conducting high current density of 1000 times higher than that of metals such as copper), high thermal conductivity of 3000 W/m-K, which is greater than diamond. At room temperature, it is predicted that the thermal conductivity of CNT could reach 6000 W/m-K, which is 15 times higher than that of copper. At low-temperature, CNT also demonstrated to exhibit super conductivity.

Covalent bonding between individual carbon atoms puts CNT among one of the strongest and stiffest materials. CNTs have excellent mechanical properties, like high compressibility and recoverability. CNTs exhibit extremely high strength in the axial direction, which can be attributed to the high Young's modulus, which is on the order of 270–950 GPa, and tensile strength of 11–63 GPa (Yu et al. 2000), which is 50 times greater than that of steel.

18.1.2 CNT-Based Sensors

18.1.2.1 Properties Relevant for Sensing

One of the most promising applications of CNT is the capability to sense various inputs generated from interactions based on mechanical-deformation, chemical-molecules, biological, optical-radiation, electrical, etc. CNT utilization in sensors varies from individual SWNTs and MWNTs to CNT networks, composites, and even to bulk (forests or bundles) CNT material. CNT for sensing devices presents several key advantages, including nanoscale devices for large-scale integration and high surface-to-volume ratio for high sensitivity. The high aspect ratio of CNT enables more area of the sensor to be exposed to the incoming analyte. The π-electron conjugation in the CNT structure provides more active sites where the molecular doping by the analytes changes its electrical properties, thus making it easier to detect a number of gases (Meyyappan 2005). CNT curvature also plays an important role in increasing sensitivity for gases by reducing the energy barrier required for a molecule to get adsorbed on the CNT surface (Ruffieux 2002). Importantly, power consumption, which is a major issue, is found lowest for CNT-based devices/sensors. Their high photoabsorption capability (~97%), coupled with high thermal conductivity in metallic CNT and photovoltaic mechanism in case of semiconducting CNT (Barkelid and Zwiller 2014), presents them as a potential candidate for practical photo devices. Decorating aligned quantum dots eventually enhances the external quantum efficiency. Furthermore, chemical doping is used to tune the bandgap of the CNT to generate more electron–hole pairs (exciton) when the photons are incident. Given their extraordinary mechanical strength, coupled with ultra small size, they have proven to be a potential replacement of conventional materials in atomic force microscopy (AFM), which actually acts as a force sensor, and scanning tunneling microscope tips. Electrical and magnetic fields can induce polarization in CNT; hence, these are being used for actuators and resonators.

18.1.2.2 CNT-Based Gas Sensors

Gas sensing has probably been one of the most preferred applications of CNT owing to their unique properties as discussed in the previous section. Thus, it falls into a class of chemical sensors that detects both the organic and inorganic species present in the environment, and the resulting responses are predictable in nature, thus allowing an accurate determination of the type of analyte. In the detection of the presence of such analytes (gases in this case), there exist mainly two mechanisms, namely, physisorption and chemisorption. Physisorption of gas molecules is caused by the van der Waals force that binds the adsorbate onto the surface of the adsorbent physically; therefore, it is also termed as *physical adsorption*. The energy involved in such a weak bond formation is very low and the extent of adsorption increases with the increase in the pressure at which the gas is exposed and normally decreases upon increasing the temperature. The characteristic of such adsorption is a fast response upon gas exposure followed by equally fast desorption in the absence of gas exposure. Thus, the fast adsorption and desorption process is governed by the interaction of gas molecules with low-energy binding sites of the surface of the adsorbent that require minimal energy to bind and detach from the analyte. *Chemisorption*, on the other hand, involves actual charge transfer between the gas molecules and the surface of the CNT. The process of charge transfer is possible only if the activation energy barrier is crossed. Such sites that facilitate chemisorption are called high-energy binding sites that include structural defects, an oxygen functional group adsorbed onto the surface of CNT.

In general, the rate of adsorption of analyte depends on the rate of collision on the sensing material surface, which can be estimated by the kinetic theory of gases. The rate of collision at pressure P can be approximated as $PN_0/(2\pi MRT)^{1/2}$, where N_0 is the Avogadro number, M is the molar mass of the gas, and R is the gas constant. In addition, the rate of collision also depends on the surface area occupied by the single adsorbed particle (S), the number of available sites for adsorption $(1 - \Theta)$, and adhesion coefficient K, which is expressed as $K = K_0 \exp(-E_A/RT)$, where K_0 is called condensation coefficient. On the other hand, rate of desorption depends on the number of particles adsorbed on the surface and the average time of the contact of adsorption particles on the surface ($\tau = \upsilon^{-1}\exp(-E_f/RT)$). Thus, the adsorption rate of certain gases on the surface can be described as the difference between the rate of adsorption and the rate of desorption as follows,

$$\frac{d\Theta}{dt} = K_0 \frac{SPN_0}{\sqrt{2\pi MRT}}(1-\Theta)e^{\frac{E_A}{RT}} - \upsilon\Theta e^{\frac{-E_D}{RT}}$$

where, Θ is the surface coverage, defined as $\Theta = N(t)/N_0$, where $N(t)$ is the concentration of the particles adsorbed on the solid surface at the moment of time t, N_0 is the concentration of total available adsorption sites. E_A and E_D are the activation energies of adsorption and desorption, respectively (Adamson and Gast 1997).

In addition, the adsorption mechanism depends mainly on the reactivity of the analyte and also on the presence of metallic and/or semiconducting nanotubes. For example, for NO_x species, it has been reported that metallic nanotubes contribute to chemisorption while physisorption is caused by semiconducting nanotubes, which led to the conclusion that the metallic nature, which has abundance of conduction electrons, helps in adsorption of gases (Ruiz-Soria et al. 2014). A clear difference in detection is reported (Chang et al. 2001) when CNTs are exposed to nitrogen dioxide (NO_2) and ammonia (NH_3). It is revealed that the change in conductance of CNT also depends on the nature of the exposed gas, whether it is electron donating or accepting. It was shown that NO_2, being electron accepting, depleted electrons on interaction, thus making the CNT more p-type, resulting in increased conductivity. NH_3, being an electron donating gas, caused the recombination of holes and therefore showed a decreased conductance. Here, CNTs are considered as p-type because of oxygen molecules adsorbed on their surface, which, because of its high electronegativity, depletes the electrons by forming a bond and induces holes as majority carriers.

However, since all the gases do not posses opposite polarity as that of NO_2 and NH_3, there arises a bottleneck that CNT alone cannot distinguish between different gases, and it is necessary for practical applications. The practical solution is to functionalize CNT specifically for a particular gas so that we obtain a unique signature in the form of response upon exposure. This method has led to numerous research articles focusing on the selectivity of CNT-based gas sensors. Various methods have evolved to achieve selectivity, like decoration of metal nanoparticles (Kong et al. 2001), metal oxides (Lu et al. 2009), and formation of hybrid structures using a mix of CNT and polymers (Wang and Swager 2011). Again, these modifications have some drawbacks; for instance, metal-oxide-decorated CNTs require high operational temperature of around 150°C–300°C to selectively detect different gases. Polymers, although supporting room-temperature operations, are very sensitive to humidity and temperature, which causes a change in their inherent properties, thus adversely affecting their ability to sense.

To obtain a more precise sensing response of CNT, CNT field effect transistors (CNTFETs) are preferred to the chemiresistive sensors. Kong et al. (2000) reported the first CNTFET-based gas sensor where a backgated transistor was fabricated and exposed to NO_2 and NH_3 gases for observation of sensor response. A general observation was made that the conductivity of CNTFET changes with an increase or decrease in the concentration of the exposed gases. The concentration of NO_2 was varied from 2 to 200 ppm, while for NH_3, it was from 0.1% to 1%, and the corresponding change in the conductance was revealed. For NO_2 gas, conductance increased by an order of 3, whereas for NH_3 gas, it decreased by 100-fold. This behavior can now be explained considering the electron donating/accepting ability of the gases, as discussed earlier.

From Figure 18.2, we can observe that for NH_3, the conductance of CNTFET decreases, while for NO_2, it increases. However, the exact mechanisms of charge transfer and, eventually, change in the current, I_{ds}, were not clear in this work. Later, there was a detailed report revealing the exact mechanism for the adsorption of NO_2 onto the CNTFET (Zhang et al. 2006). It was shown that using different combinations of the contact between the CNT and the metal plays an important role in deciding the sensing behavior of the CNTFET sensor. To elucidate the mechanism further, three different masks were used: one with top gated CNTFET, second with contact covered using polymer coating, and finally, a device that was fully covered from the top, again by polymer coating. Figure 18.3 shows the different masks

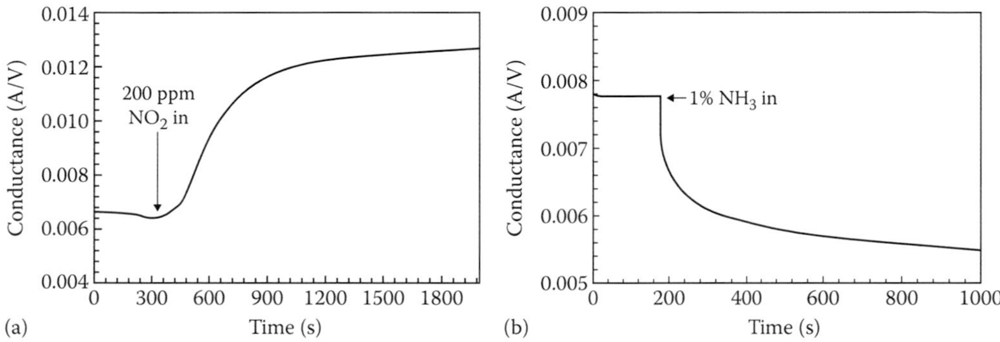

FIGURE 18.2 (a) Conductance is plotted with the time of 200 ppm NO_2 gas exposure. (b) Conductance is plotted with the time of 1% NH_3 gas exposure. (From Kong, J. et al., *Science*, 287, 622, 2000.)

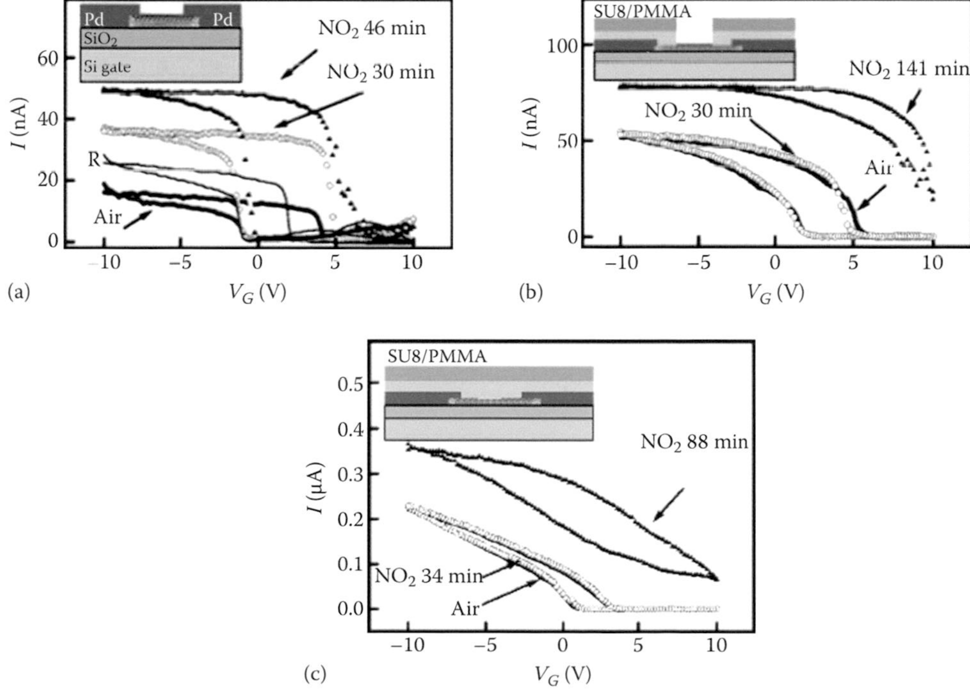

FIGURE 18.3 Electrical transfer characteristics of CNTFET device for NO_2 gas exposure: (a) the uncovered device upon exposure to NO_2, (b) the contact-covered device, (c) the fully covered device. (From Zhang, J. et al., *Appl. Phys. Lett.*, 86, 123112, 2006.)

and their corresponding responses toward NO_2 gas. In Figure 18.3a, the conductance vs. time graph shows a faster response of the sensor when compared with Figure 18.3b and c. This was attributed to the nanotube contact, which acts as a Schottky barrier model, as suggested in previous reports (Heinze et al. 2002), whereas for the partially covered and completely covered device, diffusion-assisted change in conductance was observed.

18.1.2.3 CNT-Based Organic Vapor Sensors

For a given CNT, the adsorption affinity for organic molecules depends on various factors such as surface hydrophobicity, high surface area of the CNT, and electronic polarizability. In addition, modified CNTs add to the variation in shape, size, and morphology along with the impurity (e.g., attached chemical groups, metals etc.). The adsorption of nitrobenzene is much stronger than that of benzene, toluene, and cholorobenzene. Hence, nitroaromatic compounds adsorbed more strongly than the nonpolar aromatic compounds, and aromatic compounds adsorbed more strongly than the aliphatic compounds. Nitroaromatic compounds are highly polar and strong electron acceptors while interacting with adsorbents with high electron polarizability, while other organic compounds are either nonpolar or non-electron acceptors. The general features of organic chemical adsorption on CNTs are heterogeneous adsorption and hysteresis. Heterogeneous adsorption can be explained through the presence of high-energy adsorption sites such as defects in CNTs, functional groups, and sites between the CNT bundles and condensation, such as surface and capillary condensation of liquid analytes. Concentration-dependent thermodynamics and kinetics have been reported that indicate that organic molecules may occupy high-energy adsorption sites first and then the lower-energy sites. A deviation from adsorption to desorption curve presents the hysteresis, which could be related to the strong π–π coupling of CNT surface with the benzene ring containing chemicals or alteration of adsorbent structure or reorganization after adsorption of organic molecule. Small molecules present a rate-limiting process for diffusion into inner sites of CNT that leads to extremely low diffusivity. On the other hand, bigger molecules show a much faster sorption rate because of low adsorption contribution in the inner pores of CNT. Bigger molecules can twist themselves to match with the CNT curvature and thus form stable complexes.

18.1.2.4 CNT-Based Optical Sensors

CNTs have gained wide popularity as an optical sensor and specifically infrared sensor because of their ultrahigh absorption capability, which is around 99.98% (Kaul et al. 2013). The studies were conducted on a vertically aligned MWCNT grown using a plasma-enhanced CVD (PECVD) process. Two different templates (Co/Ti and Co/Ti/NbTiN) were used for the growth of MWCNTs on silicon substrate and their effect on various properties of MWCNT was studied. It was observed that for Co/Ti, the surface looked mostly reflective, whereas for Co/Ti/NbTiN, very dense and dark arrays of MWCNT were obtained. Also, the contact resistance for the latter case was expected to be considerably low, which was attributed to the metal nitride–MWCNT interface. Further, the optical reflection properties were studied and found to be directly proportional to the catalyst thickness, as shown in Figure 18.4. It was revealed that increased absorption might be a result of the minimal back scattering, which again arises from the weak electron coupling in the grown MWCNT.

However, for single CNTs, the mechanism is entirely different when compared with that of the bulk. To understand the mechanism of photocurrent generation in single CNT, a controlled experiment was performed as reported in a previous work (Adamson and Gast 1997). The scanning photocurrent microscopy (SPCM) technique was used to investigate the electronic structure of the suspended metallic and semiconducting CNT. This technique is highly efficient in providing accurate information about the photocurrent generation mechanism in nanostructures. It was observed that for metallic CNT forming p–n junctions, multiple polarity reversals occurred as a result of the change in gate voltage, resulting in the generation of photothermal current along the nanotube axis owing to the difference in the Seebeck coefficients at two ends of the tube. While for semiconducting CNT p–n

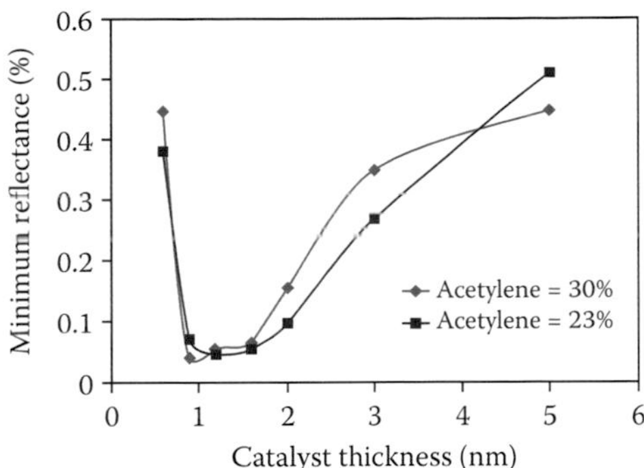

FIGURE 18.4 The dependence of reflectance of vertically aligned CNT ensemble grown using PECVD on catalyst thickness. (From Kaul, A.B. et al., *Small*, 9, 1058, 2013.)

junctions with a uniform doping profile, the Seebeck coefficients will be the same along the nanotube, thus eliminating the possibility of photothermal current. However, in this case, electric field at the junction catalyzed, and the exciton formation thus generated photocurrent as the laser in SPCM was traversed from one end to the other, which was found similar to photovoltaic effect (Figure 18.5). Thus, two different mechanisms were revealed for different types of CNT depending upon the process of carrier generation and recombination.

The excellent optical absorption capabilities of CNT very recently led to their application as terahertz detectors. Terahertz optics is an emerging field that has the potential to revolutionize imagining for biomedical, security, and environmental monitoring. A film of n-type CNT was chemically doped to p-type to achieve a p–n junction. This junction, when illuminated by a terahertz beam, showed extreme sensitivity to polarization of the beam. Similar to the previous work, Seebeck effect was found to be a major phenomenon contributing to the response, which was confirmed further by optothermal and thermoelectric measurements (He et al. 2014). Further, there were attempts to elucidate the mechanism of photocurrent in various morphologies like random assembly of CNTs (Sarker et al. 2009), where it was found that diffusion plays a dominant role in the photocurrent generation, with nanotube–metal contact giving the highest value of photocurrent due to Schottky barrier modulation, which is consistent with previous reports. Decoration of metal oxide on MWCNTs revealed altogether different mechanisms of photodetection. Zhu et al. (2006) demonstrated that combining the three-photon absorption properties of ZnO (Gu et al. 2008) with the saturable absorption properties of MWCNT leads to ultrafast optical detection, which was shown to be applicable for optical switches.

18.1.2.5 CNT-Based Resonators or Resonant Sensors

Resonant sensors are the devices that can be used to detect changes in the mechanical stress or strain in the form of stimulus. This change is measured in terms of variation in the resonant frequency due to external force. CNTs, owing to their high Young's modulus, ultrasmall cross-section, and great tensile strength, are ideal candidates for ultrasensitive resonators. In general, for such measurements, a cantilever configuration or suspended nanotube between two supports is preferred. The first such experiment on MWCNT was reported by Poncharal et al. (1999), where a cantilever configuration was used to study the field-induced vibrations (Figure 18.6). It was observed that even on application of high electric fields, the bending observed was in the elastic region, suggesting high elastic modulus. Furthermore, in this study, mass of a carbon particle that was the order of femtogram [$M = 22 \pm 6$ fg (1 fg = 10^{-15} g)] was measured from the resonance frequency of the cantilever structure of MWCNT.

However, when it comes to the modeling of CNT resonators, classic mechanics concepts do not hold true given their nanoscale dimensions, and hence, the atomistic modeling technique is a preferred

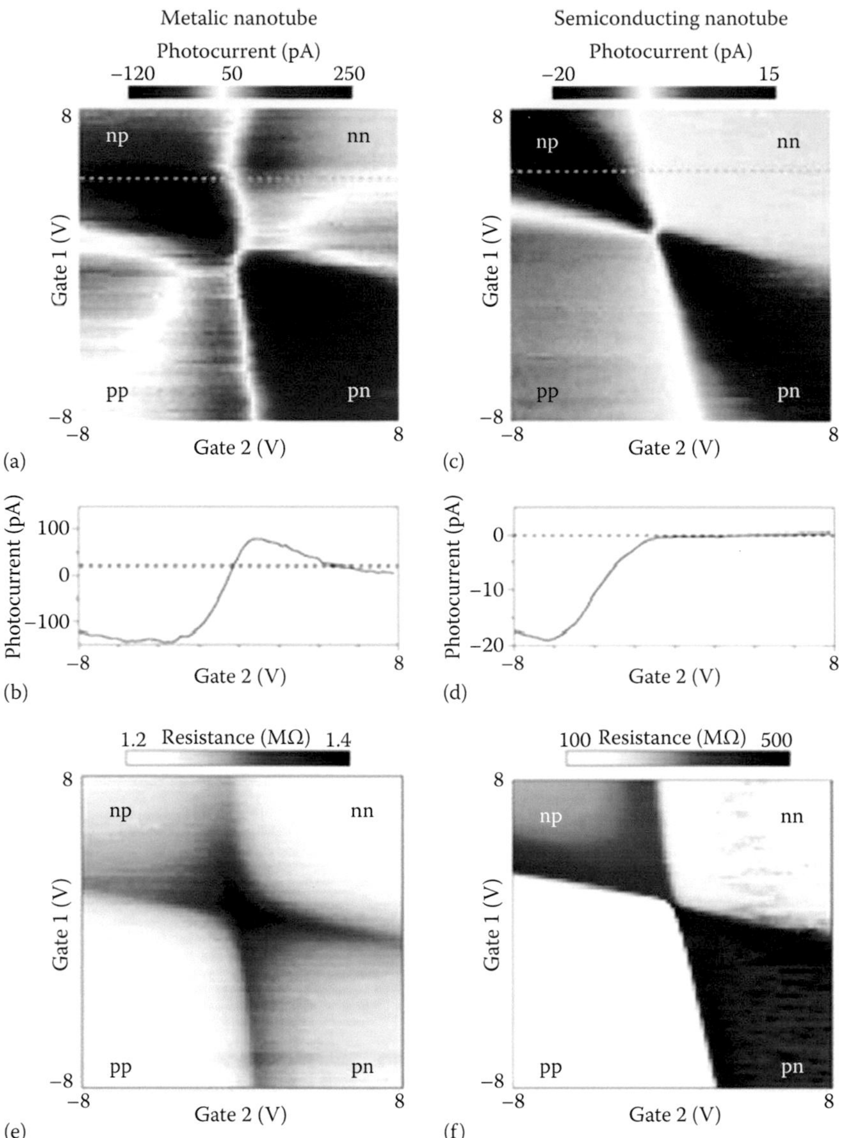

FIGURE 18.5 Variation of photocurrent in metallic and semiconducting CNTs with gate voltage. (a) Photocurrent map of a metallic nanotube, displaying a six-fold pattern characteristic of the photothermal effect. (b) Line profile of the photocurrent as a function of gate voltage. (c) Photocurrent map of a semiconducting nanotube. (d) Line profile of the photocurrent as a function of gate voltage. (e) Resistance as a function of gate voltage for the metallic nanotube. (f) Resistance map for the semiconducting nanotube. (Adamson, A.W., Gast, A.P.: *Physical Chemistry of Surfaces, 6th Edition*. 1997. Copyright Wiley-VCH Verlag GmbH & Co. KGaA. Reproduced with permission.)

method. In the first of such reports (Li and Chou 2003), covalent bonding between carbon–carbon atoms in a SWCNT was simulated as structural beams, and thereafter, three stiffness parameters were evaluated: tensile resistance, flexural rigidity, and torsional stiffness. Assuming CNT as a space-frame-like structure under the condition of free vibration, fundamental frequencies were calculated for both cantilever and bridged configurations. It was found that fundamental frequency decreases for an increase in the nanotube length at boundary conditions, while keeping the aspect ratio constant; a decrease in the diameter increases the fundamental frequency.

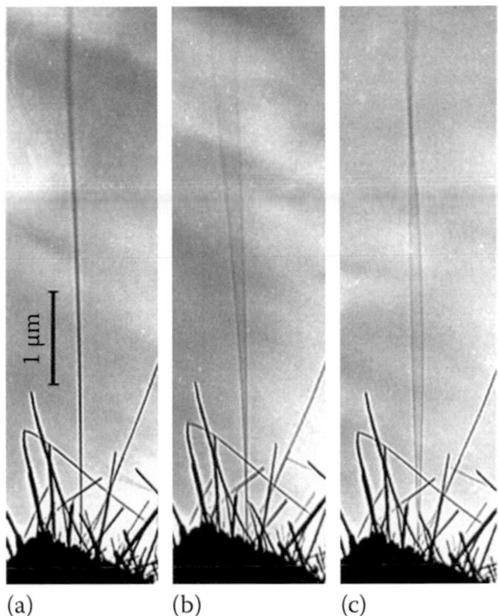

FIGURE 18.6 Transmission electron microscopy image of suspended device. Nanotube responds to resonant alternate applied potentials. (a) In the absence of a potential, the nanotube tip (vibrated slightly because of thermal effects). (b) Resonant excitation of the fundamental mode of vibration (1530 kHz). (c) Resonant excitation of the second harmonic (23.01 MHz). (From Poncharal, P. et al., *Science*, 283, 1513, 1999.)

Since the properties of both MWCNT and SWNT have been discussed, now the question arises which one of them is a better candidate. It was then revealed by Garcia-Sanchez et al. (2007) that the resonant properties of MWCNTs are more reproducible than those of SWNTs. This is because MWCNTs are more rigid structures and hence are highly resistant to deformation. This presents them as a more suitable candidate for various applications such as signal processing.

18.1.2.6 CNT-Based Peizoresistive Sensors

Peizoresistive sensors can detect a change in the electrical resistance on application of strain. CNTs present an interesting aspect in the class of peizoresistive materials owing to their favorable mechanical and electrical properties, as discussed earlier. They have shown the ability to detect the slightest strain induced into them mostly using AFM. This experiment was first demonstrated by Tombler et al. (2000), where bridge configuration, as discussed in the previous section, between two metal electrodes on SiO_2/Si substrate was used (Figure 18.7a). Figure 18.7b shows the AFM image of the SWNT suspended between the metal electrodes.

The suspended SWNT was then forced in the downward direction toward the trench using an AFM tip positioned at the center of the suspended nanotube (Figure 18.7c). During the process, deflection in the AFM tip and the change in resistance of the SWNT were obtained, and since the deflection in the AFM tip can be correlated to the deflection in the SWNT, actual movement by SWNT on application of strain was also obtained for different angles. One important observation made was that even after repeated pushing using the AFM tip, the sample was retracting to its initial position, which is an important factor for a robust strain sensor. Further, it was revealed that conductance decreased as the tube was bent using AFM. When the bending angle was small, the decrease in conductance too was slow, but as the bending increases, the rate of decrease in conductance increased, and for maximum bending angle, the decrease in conductance was observed to be by two orders of magnitude. Simulations were performed to find the exact mechanism for such decrease in conductance. It was revealed that for lower strain or for lower bending angles, the SWNT structure remains in the sp^2 state, but as the bending is increased, the local bonding of the structure just beneath the tip changes from sp^2 to sp^3, while in the regions away from the tip, it remains sp^2. The sp^3 configuration facilitates σ-electron as the major

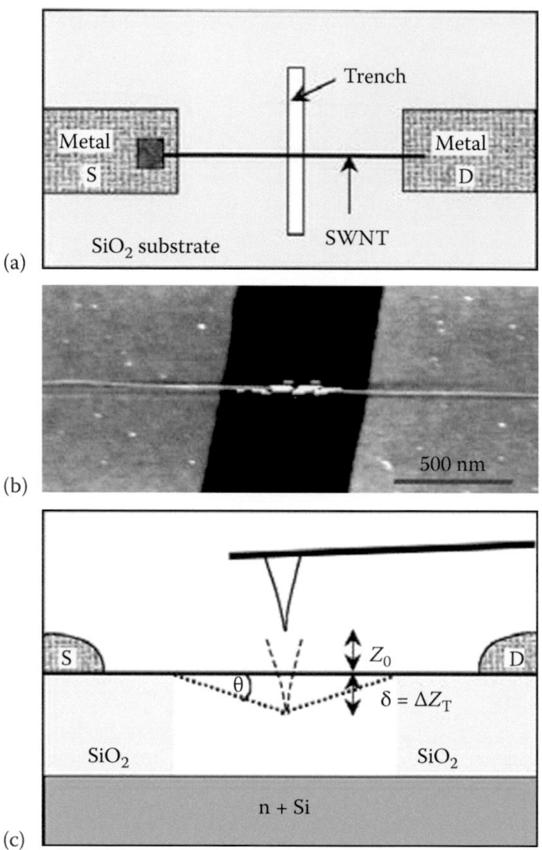

FIGURE 18.7 (a) Top view of the device. The trench is about 500 nm wide and 175 nm deep. (b) AFM image of the cantilever. (c) Side view of the AFM pushing experiment. (From Tombler, T.W. et al., *Nature*, 405, 769, 2000.)

contributor to the density of states instead of π-electrons. Now, since the π-electrons are more free to move (as they are delocalized), their concentration decreases because of the formation of σ-electrons, which eventually causes the conductance of SWNT to decrease.

The previously presented theory was argued by another report by Minot et al. (2003), who suggested that instead of the change in the hybridization state of SWNT, it was the strain-induced bandgap modification that led to a change in the conductance. The strain was induced using an AFM tip and the same acted as a voltage source. The change in the bandgap was explained using the following equation:

$$dE_{gap}/d\sigma = \beta kT,$$

which is derived from the chirality-dependent bandgap of CNTs as suggested in a previous report (Yang and Han 2000). Two processes, tunneling and thermal activation, were proposed to contribute to the change in the conductance.

References

Adamson, A. W. and A. P. Gast, *Physical Chemistry of Surfaces*, 6th Edition, Wiley, New York, 1997.

Barkelid, M. and V. Zwiller, Photocurrent Generation in Semiconducting and Metallic Carbon Nanotubes, *Nat. Photon.*, 8, 47, 2014.

Chang, H., J. D. Lee, S. M. Lee, and Y. H. Lee, Adsorption of NH_3 and NO_2 Molecules on Carbon Nanotubes, *Appl. Phys. Lett.*, 79, 3863, 2001.

Dai, H. Carbon Nanotubes: Synthesis, Integration, and Properties, *Acc. Chem. Res.*, 35, 1035, 2002.

Garcia-Sanchez, D., A. S. Paulo, M. J. Esplandiu, F. Perez-Murano, L. Forró, A. Aguasca, and A. Bachtold, Mechanical Detection of Carbon Nanotube Resonator Vibrations, *Phys. Rev. Lett.*, 99, 085501, 2007.

Gu, B., J. He, W. Ji, and H. Wang, Three-Photon Absorption Saturation in ZnO and ZnS Crystals, *J. Appl. Phys.*, 103, 073105, 2008.

He X. et al. Carbon Nanotube Terahertz Detector, *Nano Lett.*, 14, 3953, 2014.

Heinze, S., J. Tersoff, R. Martel, V. Derycke, J. Appenzeller, and P. Avouris, Carbon Nanotubes as Schottky Barrier Transistors, *Phys. Rev. Lett.*, 89, 106801, 2002.

Iijima, S. Helical Microtubules of Graphitic Carbon, *Nature*, 354, 56, 1991.

Kaul, A. B., J. B. Coles, M. Eastwood, R. O. Green, and P. R. Bandaru, Ultra-High Optical Absorption Efficiency from the Ultraviolet to the Infrared Using Multi-Walled Carbon Nanotube Ensembles, *Small*, 9, 1058, 2013.

Kong, J., M. G. Chapline, and H. Dai, Functionalized Carbon Nanotubes for Molecular Hydrogen Sensors, *Adv. Mater.*, 13, 1384, 2001.

Kong, J., N. R. Franklin, C. Zhou, M. G. Chapline, S. Peng, K. Cho, and H. Dai, Nanotube Molecular Wires as Chemical Sensors, *Science*, 287, 622, 2000.

Li, C. and T. Chou, Single-Walled Carbon Nanotubes as Ultrahigh Frequency Nanomechanical Resonators, *Phys. Rev. B*, 68, 073405, 2003.

Lu, G., L. E. Ocola, and J. Chen, Room-Temperature Gas Sensing Based on Electron Transfer between Discrete Tin Oxide Nanocrystals and Multiwalled Carbon Nanotubes, *Adv. Mater.*, 21, 2487, 2009.

Meyyappan, M. *Carbon Nanotube Science and Applications*, CRC Press, 2005.

Minot, E. D., Y. Yaish, V. Sazonova, J. Park, M. Brink, and P. L. McEuen, Turning Carbon Nanotube Band Gaps with Strain, *Phys. Rev. Lett.* 90, 156401, 2003.

Odom, T. W., J.-L. Huang, P. Kim, and C. M. Lieber, Atomic Structure and Electronic Properties of Single-Walled Carbon Nanotubes, *Nature*, 391, 62, 1998.

Poncharal, P., Z. L. Wang, D. Ugarte, and W. A. de Heer, Electrostatic Deflections and Electromechanical Resonances of Carbon Nanotubes, *Science*, 283, 1513, 1999.

Reddy, S. K., A. Suri, and A. Misra, Influence of Magnetic Field on the Compressive Behavior of Carbon Nanotube with Magnetic Nanoparticles, *Appl. Phys. Lett.*, 102, 241919, 2013.

Ruffieux, P., O. Groning, M. Bielmann, P. Mauron, L. Schlapbach, and P. Groning, Hydrogen Adsorption on sp^2-bonded Carbon: Influence of the Local Curvature, *Phys. Rev. B.*, 66, 245416, 2002.

Ruiz-Soria G. et al., Revealing the Adsorption Mechanisms of Nitroxides on Ultrapure, Metallicity-Sorted Carbon Nanotubes, *ACS Nano*, 8, 1375, 2014.

Sarker, B. K., M. Arif, P. Stokes, and S. I. Khondaker, Diffusion Mediated Photoconduction in Multiwalled Carbon Nanotube Films, *J. Appl. Phys.*, 106, 074307, 2009.

Tombler, T. W., C. Zhou, L. Alexseyev, J. Kong, H. Dai, L. Liu, and C. S. Jayanthi, Nanotubes under Local-Probe Manipulation, *Nature*, 405, 769, 2000.

Wang, F. and T. M. Swager, Diverse Chemiresistors Based upon Covalently Modified Multiwalled Carbon Nanotubes, *J. Am. Chem. Soc.*, 133, 11181, 2011.

Yang, L., and J. Han, Electronic Structure of Deformed Carbon Nanotubes, *Phys. Rev. Lett.*, 85, 154, 2000.

Yu, M.-F., O. Lourie, M. J. Dyer, K. Moloni, T. F. Kelly, and R. S. Ruoff, Strength and Breaking Mechanism of Multiwalled Carbon Nanotubes under Tensile Load, *Science*, 287, 367, 2000.

Zhang, J., A. Boyd, A. Tselev, M. Paranjape, and P. Barbara, Mechanism of NO_2 Detection in Carbon Nanotube Field Effect Transistor Chemical Sensors, *Appl. Phys. Lett.*, 86, 123112, 2006.

Zhu Y. et al. Multiwalled Carbon Nanotubes Beaded with ZnO Nanoparticles for Ultrafast Nonlinear Optical Switching, *Adv. Mater.*, 18, 587, 2006.

19

Biomedical Carbon Nanotubes

Diogo Mata

Maria H. Fernandes

Maria A. Lopes

Rui F. Silva

19.1 Introduction

Carbon nanotubes (CNTs) have been widely investigated in respect to their excellent electrical, thermal, chemical, and mechanical properties. These result from CNT's unique sp^2 structure with quantum confinement in the circumferential direction, which promotes their use as ultimate fillers in various composite materials with diverse technological applications. Pioneering solutions based on CNTs—single walled (SWCNTs) and multiwalled (MWCNTs)—have been widely explored to endorse relevant breakthroughs in biology's and medicine's upmost demands of scaffolding, drug delivering, cell tracking and sensing, and in situ control of cell proliferation and differentiation. Moreover, the outstanding idea that all of these functionalities could be accumulated in unique multifunctional CNT-based smart templates triggered a huge excitement in the tissue engineering field. Their potential inclusion in materials aiming the contact with biological tissues has led to a widespread evaluation of the biological prolife and biosafety of CNTs and functionalized CNTs, reaching the hand of in vitro and in vivo testing. Recent data, attained with highly pure materials, broadly converge to report their biocompatibility, likeminded for clinical applications. Furthermore, the biological activity of living cells, in terms of cell adhesion, viability, and proliferation, seems to be enhanced in a CNT-rich microenvironment, thus supporting the idea to incorporate CNTs in synthetic substrates for tissue engineering applications, namely, those pointing toward replacement, healing, and tissue growth. Among several tissues, CNT applications in muscle-skeletal tissue engineering have caught particular interest, in part because of the relevance of locomotion and mobility for quality of life. The inclusion of either SWCNTs or MWCNTs has been found to improve the mechanical performance of bulk materials, scaffolds, and cements aiming at the regeneration of the bone tissue. CNTs may also provide the opportunity for custom chemical functionalization based on potential application and microenvironment characteristics. Furthermore, CNTs' inclusion also seems to prompt and improve the biological response related to the bone tissue metabolism and regeneration. In vitro, both plain CNT constructs and CNT-containing composites seem to improve the functional activity of osteoblastic and osteoblastic-precursor cells, while in vivo, CNT-containing scaffolds and coatings have a proven record of bone tissue compatibility, permitting the favorable bone tissue ingrowth in their vicinity.

This chapter is a survey in the biomedical applications of CNTs, with emphasis on smart CNT-bioceramic composites applied to bone tissue engineering. The synthesis, structure, and realistic properties of CNTs are covered in the present chapter. Detailed description through the CNT in vitro and in vivo toxicological profiles, CNT-engineered multifunctional biomedical systems, and an overview on smart CNT-bioceramic composites applied to bone tissue engineering is presented. The last topic will be addressed in more detail, stressing the composite's major processing challenges, attractive properties, in vitro toxicological profiles, and efficiency as electron stimuli delivering systems to bone cells while in vitro electrically stimulated. Last, a brief summary will cover an integrated discussion of the major subjects and main conclusions and prospects for future research lines.

19.2 Synthesis of CNTs

Over the past decades, an extensive research work has been carried out to develop cost-effective and high-quality production methods of one-dimensional (1D) carbon nanofilaments such as carbon nanofibers (CNFs) and CNTs. The earlier reported synthesis of CNFs by a chemical vapor deposition (CVD) method dates back to the 1970s, in investigation works carried out by Baker's group (Baker et al. 1972) and Endo's group (Oberlin et al. 1976). The CVD growth of MWCNTs was first reported by Endo et al. (1993), following Endo's previous experience in CNFs, 2 years later to Iijima's discovery using an arc discharge apparatus. Soon after, in 1996, Dai et al. reported the CVD growth of SWCNTs (Dai et al. 1996).

CVD methods had cached superior attention over the arc-discharge and laser ablation methods because of their much higher simplicity and higher growth yield. While the latter two use high vacuum conditions and high temperatures (up to 3000°C) and need a continuous replacement of the carbon source (graphite target), the CVD process operates at mild conditions such as atmospheric pressure and low temperature (below 1200°C), with a continuous flow of a gas carbon source—carbon monoxide (CO) or a hydrocarbon (e.g., C_2H_2, CH_4). Moreover, in CVD methods, the dissociation of the gas carbon source is assisted by transition-metal catalyst nanoparticles (e.g., Co, Ni, Cu, Fe) placed on substrates (e.g., Si, glass, metal foils—catalyst-supported methods) (Terrones and Grobert 1997) or injected directly into the CVD chamber (floating-catalyst methods) (Andrews et al. 1999). The use of catalytic particles in the CVD growth processes of 1D materials has been reported since 1964, with the pioneer studies of Wagner and Ellis (1964) describing the synthesis of Si whiskers via the vapor–liquid–solid mechanism and, later on, with the growth studies of carbon nanofilaments by Baker et al. (1972) and Endo et al. (Oberlin et al. 1976). Thereafter, it has been found that catalytic particles have key roles in the production of carbon nanofilaments (Gorbunov et al. 2002): (1) lowers the nucleation/growth temperature of the carbon filament by promoting the dissolution of the initial elementary amorphous carbon (or methyl radicals) in the particle and subsequent saturation and precipitation of carbon in graphitic form; (2) works as a transporting medium in the diffusion process that starts with the initial amorphous carbon being dissolved in the particle and ends when the final graphitic carbon phase is precipitated; (3) helps to mold the tubular graphitic structure, making the perpendicular precipitation of graphene layers to the particle surface more energetically favorable than the parallel precipitation one. Therefore, the use of the catalyst is meant to facilitate the control of the nanofilament's structure, such as number of graphene layers, diameter, length, and alignment (Anna et al. 2003; Dupuis 2005).

CVD methods can be also classified according to the activation source of the gas carbon precursor: (1) chemical activation—combustion flame (CCVD) (Height et al. 2004); (2) electrical activation—plasma enhanced (PECVD) (Chhowalla et al. 2001); and (3) thermal activation–hot-filament (HFCVD) (Dillon et al. 2003) and hot-wall or thermal (TCVD) (Lee et al. 1999). Despite the simplicity of the CCVD method, involving the oxidization of a gas flow mixture of O_2 + hydrocarbon (carbon source) + organometallic (metallic particles source) + carrier gas, it yields a carbon soot with a high amount of carbon by-products and low graphitized filaments (Height et al. 2004). On the other hand, PECVD becomes a very promising method to grow high-quality nanofilaments at low temperature (below 500°C) on low-melting-point substrates such as polymers and glass, highly desirable for electronic applications (Hofmann et al. 2003). Yet, this method is not recommended for large-scale synthesis because of the low yields (limited by the plasma area) and the great expense of the plasma power source.

In contrast, the HFCVD methods use inexpensive metallic filaments (e.g., tungsten, W) heated up at 2000°C–2500°C to thermally dissociate the carbon gas precursor at mild pressure (5–100 mbar). These filaments might have different 3D arrangements over large areas, allowing the use of growth substrates with a wide range of sizes and geometries. Thus, the HFCVD methods present higher yields and are more versatile and less costly than PECVD ones. Nevertheless, some limitations of the HFCVD method can be pointed-out. An extra step of carburization of the filaments is needed to avoid the consumption of carbon from the gas mixture during the growth step. Besides this, after the carburization of the filaments, they become highly brittle, with high appetency to break under severe conditions, so they limit the range of the growth parameters that might be used. Also, during the growth run, small metallic clusters of the filaments or components near the hot filaments might evaporate and condensate on the substrates and then catalyze the growth of carbon by-products and/or interfere with the as-placed catalyst particles used to nucleate and grow carbon filaments (Gan et al. 2000). Another disadvantage is that the filaments have an upper limit temperature (melting point temperature), so an extra heating source, a heated substrate (600°C–800°C), is needed to increase the gas activation temperature.

The TCVD method consists of a simple furnace running in the temperature range of 600°C–1200°C, sufficiently high to fully activate the gas mixture. The slightly higher temperature used in TCVD, compared with HFCVD, is explained by the former being purely thermally activated. Nonetheless, the absence of filaments has huge advantages, as referred previously. In some cases, the gas mixture flows through a quartz tube housed in the furnace at room pressure (~1000 mbar) and reacts with the growth substrates positioned in a quartz boat, at the hot reaction zone (center of the furnace). This zone is totally clean, containing only the metal catalyst and the substrate, therefore allowing the growth of nanofilaments with high graphitization level and structural control. After this, TCVD methods emerged as the most promising approach for mass production of CNTs as they offer the highest degree of growth control and upscaling potential. Various TCVD-based scalable processes had been developed over the last years, including (1) floating catalyst methods—"HiPCO process" of the Rice University in 1999 (Nikolaev et al. 1999); (2) "CoMoCat process" of the Oklahoma University in 2000 (Kitiyanan et al. 2000); (3) "nanoagglomerate fluidized process" of the Tsinghua University in 2002 (Wang et al. 2002); and (4) catalyst-supported methods—the "supergrowth process" of the National Institute of Advanced Industrial Science and Technology in 2004 (Hata et al. 2004).

Among these several CVD methods, the catalyst-supported TCVD provides superior versatility and yields and the purest CNTs (~99.98%) (Hata et al. 2004; Zhao et al. 2008a). This approach has two important advantages: It allows the growth of highly dense vertically aligned CNTs (VACNTs) or "forests" (~10^{11} CNTs cm^{-2}) with a high structural control (Zhao et al. 2008a), and the catalyst can be patterned on the substrate to obtain any desired 1D, 2D, or 3D shape of VACNTs (Hata et al. 2004).

19.3 CNT Structure and Properties

Carbon is present in all building blocks of life owing to its chemical bonding versatility. Each carbon atom has six electrons that occupy $1s^2$, $2s^2$, and $2p^2$ atomic orbitals. The $2s^2$ and $2p^2$ valence orbitals are filled by four weakly bonded electrons that are free to change position; in crystalline carbon forms, they are as $2s^1$, $2p_x^1$, $2p_y^1$, and $2p_z^1$. When these orbitals are combined, $2s^1 + 2p_x^1$, $2s^1 + 2p_x^1 + 2p_y^1$, or $2s^1 + 2p_x^1 + 2p_y^1 + 2p_z^1$, hybridization states are formed, sp, sp^2, and sp^3, respectively. These states promote covalent sigma (σ) bonding with specific structural arrangements: sp—chain structures; sp^2—planar structures; and sp^3—tetrahedral structures (Dresselhaus and Endo 2001).

Those of the most common sp^2- and sp^3-bonded crystalline structures of carbon (i.e., carbon allotrope phases) are graphite and diamond. According to a recent version of the phase diagram (Bundy et al. 1996), graphite is the most stable form of solid carbon at ambient conditions. The 3D graphite structure concerns multilayers of graphene stacked on top of each other, bonded together by weak non-covalent van der Waals bonding (Figure 19.1). In a graphene layer, each carbon atom is covalently sp^2-σ bonded to three neighbored atoms in a hexagonal (hcc) structure (Figure 19.1).

Beyond the 3D configuration, sp^2-bonded carbon clusters of few atoms were found to be able to yield highly stable nanostructures by simply eliminating high-energy dangling bonds (Dresselhaus and Endo 2001). Following this, carbon clusters had been catching great interest from the scientific community over the last five decades (Mildred 2012). Relevant landmarks in the carbon field were accomplished with the discovery of novel carbon sp^2 allotropes (Figure 19.1) of closed cage nanostructures—fullerenes (0D) (Kroto et al. 1985) and CNTs (1D) (Iijima 1991; Iijima and Ichihashi 1993)—or open nanostructures—cones (0D) (Ge and Sattler 1994), graphene (2D) (Novoselov et al. 2004), and ribbons (2D) (Berger et al. 2006).

These nanostructures are related between them by being formed by a single layer of graphene, the building block of all sp^2-bonded carbon allotropes (Geim and Novoselov 2007). Simply, the graphene layer can be modulated (e.g., buckling, rolling) to promote the bonding of carbon atoms of the extremities of the layer and reach a low-energy state (Geim and Novoselov 2007).

Regarding CNT structures, they are formed by concentrically rolled-up graphene layers. Those having several layers are MWCNTs, and when formed by a single layer, they are denominated SWCNTs (Figure 19.1). Both CNT types were discovered by Iijima in 1991 and 1993, respectively (Iijima 1991; Iijima and Ichihashi 1993). Other nanofilaments, with similar cross-sectional appearance in transmission electron microscopy (TEM), were observed some decades earlier, in the 1970s, by Baker et al. (1972), the CNFs. However, these carbon nanofilaments present structural differences. While CNFs are filaments with graphene layers having different angles to the filament axis, CNTs are filaments with tubular graphene layers parallel to the axis (Martin-Gullon et al. 2006; Rodriguez et al. 1995).

SWCNTs and MWCNTs do not present a simple and universal rolling direction. They reveal a large number of potential directions, i.e., helicities or chiralities, that define the structural arrangement of the tube. To describe such fundamental structural characteristic of the nanotubes, an SWCNT is adopted for simplification. This structure is formed by rolling up a single graphene layer (Figure 19.2) with hexagonal lattice unit vectors ($|a_1| = |a_2| = 0.246$ nm). Considering this lattice, the unit cell of a nanotube can be defined by two vectors, C_h and T (Dresselhaus et al. 2004; Harris 2009).

The C_h is the chiral vector or rolling vector defining the circumference on the surface of the tube connecting two equivalent carbon atoms positions. This vector can be expressed as

$$C_h = na_1 + ma_2, C_h = (n, m), \tag{19.1}$$

where n and m are positive integers ($0 \leq n \leq m$), also known as chiral indices. The tube is designated armchair when $n = m$, zigzag when $m = 0$, and chiral in all other cases. Examples for each type of CNT structure are given in Figure 19.3.

The magnitude of the C_h vector defines one dimension of the unit cell and might be used to determine the diameter of the tube (d_t), as follows,

$$|C_h| = d_t\pi = 0.246\sqrt{n^2 + nm + m^2}. \tag{19.2}$$

The T vector, perpendicular to C_h, is defined in the direction of the axis of the tube.

$$T = t_1a_1 + t_2na_2, T = (t_1, t_2), \tag{19.3}$$

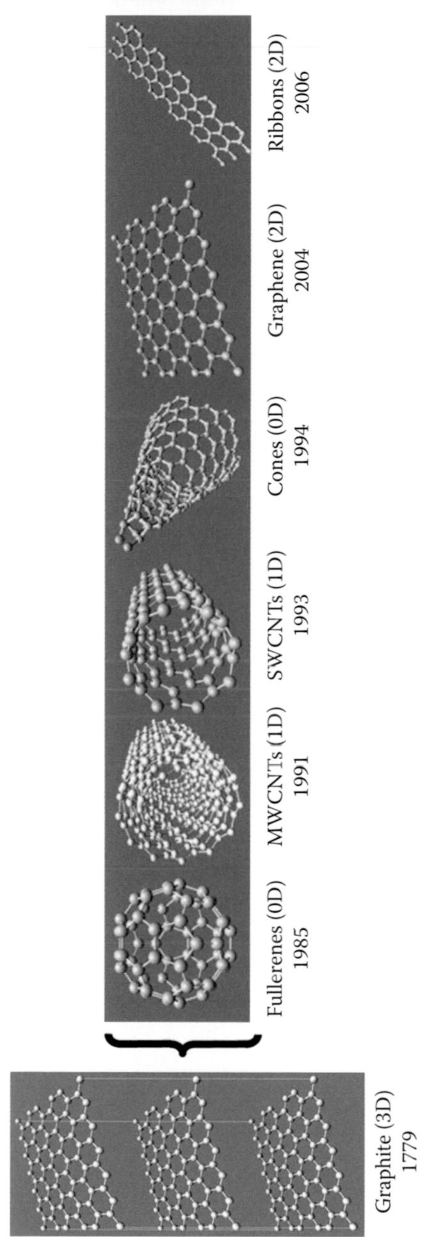

FIGURE 19.1 Carbon sp^2 allotropes (generated using the nanotube modeler freeware—http://www.jcrystal.com/products/wincnt/index.htm).

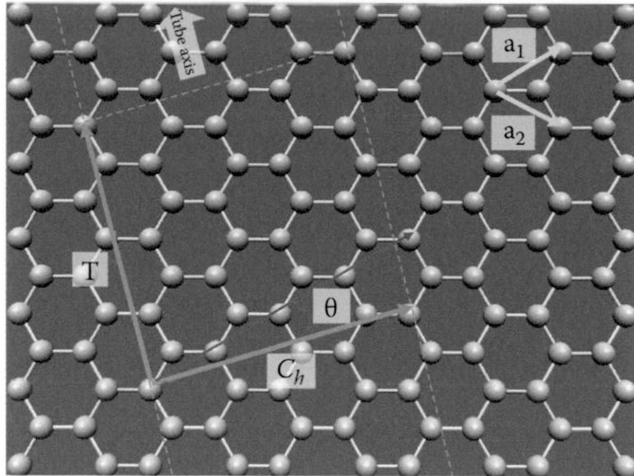

FIGURE 19.2 Unrolled hexagonal lattice of an SWCNT structure (generated using the nanotube modeler freeware—http://www.jcrystal.com/products/wincnt/index.htm).

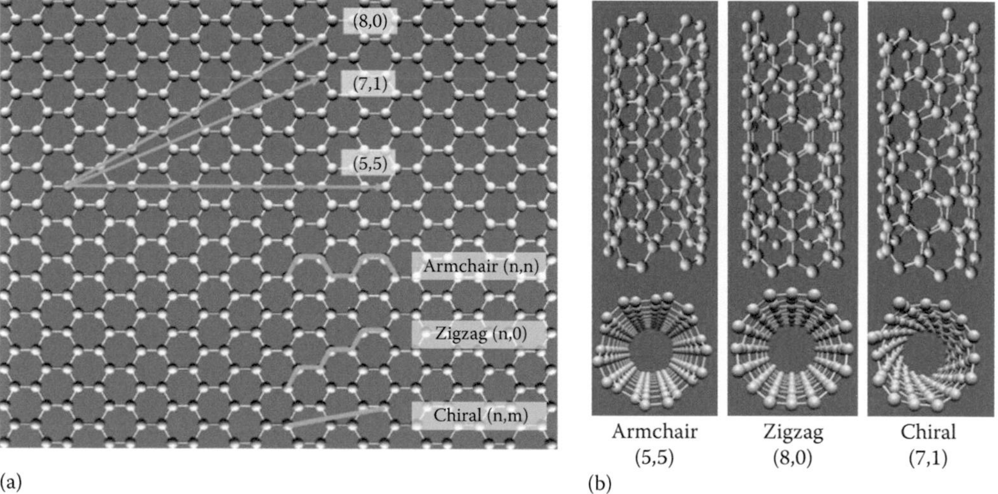

FIGURE 19.3 (a) Definition of chiral vectors on a hexagonal lattice for an armchair, zigzag, and chiral tube. (b) Respective 3D structures of the CNTs of (a) (generated using the nanotube modeler freeware—http://www.jcrystal.com/products/wincnt/index.htm).

where t_1 and t_2 are positive integers that are related with (n, m), as follows,

$$t_1 = \frac{(2m+n)}{d_R} \tag{19.4}$$

and

$$t_2 = -\frac{(2n+m)}{d_R}, \tag{19.5}$$

where d_R is given by,

$$d_R = d, \text{ if } n - m \neq \text{multiple of } 3d \tag{19.6}$$

and

$$d_R = 3d, \text{ if } n - m = \text{multiple of } 3d, \tag{19.7}$$

where d_R is the highest common divisor of (n, m).

The magnitude of the T vector corresponded to the other dimension of the unit cell, i.e., the shorter repeat distance along the axis of the tube, determined as

$$|T| = \frac{\sqrt{3}C_h}{d_R}. \tag{19.8}$$

The number of carbon atoms in the unit cell of a nanotube is $2N$, where N is given by

$$N = \frac{2(n^2 + m^2 + nm)}{d_R}. \tag{19.9}$$

The angle between C_h and the zigzag direction (Figure 19.2) is the chiral angle (θ), calculated as follows:

$$\theta = \tan^{-1}\left(\frac{\sqrt{3}m}{m + 2n}\right). \tag{19.10}$$

A nule chiral angle, $\theta = 0°$, corresponds to a zigzag tube, whereas an armchair tube has a $\theta = 30°$, and a chiral tube to intermediate angles, $0° < \theta < 30°$. Thus, a chiral angle (θ) or a pair of integers (n, m) is commonly used to define the type of the nanotube structure.

Besides the evident rolling structural differences between CNTs and graphene, tubes have some other singularities. CNTs are closed-cage clusters with extreme curvature sections, and thus, higher strain energies are predictable than in graphene (Dresselhaus and Endo 2001). Also, the curvature geometry may lead to some overlapping of the orbitals π ($2p_z$) and σ($2s + 2p_x + 2p_y$) (Saito et al. 1992), resulting in a small amount of sp^3 bonds ($2s + 2p_x + 2p_y + 2p_z$). Moreover, the circumference of nanosized dimensions allowed a very limited number of electron states in the circumference direction. Therefore, a quantum confinement of the waveforms in the circumferential directions is expectable with only a few wave vectors needed to describe the periodic boundary conditions (Saito et al. 1992).

The CNT singularities of curvature and quantum confinement offer them unique properties, different from those of 3D graphite, 2D graphene, and other 1D fibrillar carbon allotropes. Table 19.1 shows some of the main physical properties of 3D graphite (highly oriented pyrolytic graphite [HOPG]), 2D graphene, and several 1D fibrillar carbon allotropes.

Comparing the SWCNTs and MWCNTs, it can be seen that the former has slightly higher mechanical and conductive values. This occurs because SWCNTs have smaller diameters and, therefore, higher curvature confinement than MWCNTs do, so the strain energy of the CNT curvature (Dresselhaus and Endo 2001) and the σ-π rehybridization phenomenon that promotes a higher π orbital delocalization outside the tube (Dresselhaus and Endo 2001; Han 2005) are more pronounced. Besides this, particularly in phonon transport, MWCNTs seem to have lower efficiency than SWCNTs do because of the intercarbon layer scattering phenomenon that ultimately perturbs the magnitude of the thermal conductivity (Dresselhaus and Endo 2001).

TABLE 19.1 Physical Properties of Carbon sp^2 Allotropes

Properties	Graphite (HOPG)	Graphene	Pitch Carbon Fibers	SWCNTs	MWCNTs
			Mechanical		
Young's Modulus (GPa)	28–31 (Pierson)	1000 (Lee et al. 2008)	241–542 (Paris et al. 2002)	1000–1250 (Krishnan et al. 1998; Teo et al. 2004)	1000–1280 (Teo et al. 2004; Wong et al. 1997)
Flexural strength (GPa)	0.08–0.17 (Pierson)	–	2.09–3.04 (Naito et al. 2009)	–	14 (Wong et al. 1997)
Tensile strength (GPa)	0 11 (Pierson)	130 (Kuilla et al. 2010)	1.1–3.21 (Naito et al. 2009; Pierson)	25–135 (Wang et al. 2010)	11–63 (Wang et al. 2010; Yu et al. 2000)
			Electrical and Electronic		
Conductivity ($S\ m^{-1}$)	2×10^5–2.5×10^5 (Pierson)	10^6 (Sruti et al. 2010)	4×10^5–4.5×10^5 (Pierson)	10^6 (Teo et al. 2004)	10^6 (Teo et al. 2004)
Current density ($A\ cm^{-2}$)	–	10^{11}–10^{12} (Chen et al. 2008)	–	10^7–10^9 (Teo et al. 2004)	10^7–10^9 (Teo et al. 2004)
Bandgap (eV)	–	Zero-gap semiconductor (semimetallic) 0 (Avouris 2010)	–	~0.4–0.7 (nonsemiconductive) if $n - m$ is not divisible by 3, 0 (metallic) if $n - m$ is divisible by 3 (Teo et al. 2004)	~0 (metallic) (Teo et al. 2004)
			Thermal and Optical		
Thermal conductivity ($W\ mK^{-1}$)	190–390 (Pierson)	4840–5300 (Balandin et al. 2008)	500–1100 (Arai 1993)	3500–5800 (Balandin et al. 2008; Teo et al. 2004)	>3000 (Balandin et al. 2008; Teo et al. 2004)
Fluorescence emission (nm)	–	NIR range: 900–1500 (oxidized) (Sun et al. 2008) UV-visible range: 400–700 (oxidized) (Sun et al. 2008)	–	NIR range: 800–1600 (pristine, semiconductive) (Cherukuri et al. 2004) UV-visible range: 400–550 (functionalized) (Lacerda et al. 2006)	UV-visible range: 400–550 (functionalized) (Lacerda et al. 2006)

Nonetheless, graphene has key advantages over CNTs. Graphene is a versatile charge carrier with electrons and holes having similar mobility, as high as 200,000 $cm^2\ Vs^{-1}$ and extremely high current densities up to $10^{12}\ A\ cm^{-2}$ (Avouris 2010; Chen et al. 2008). These electron transporting properties make graphene the highest performing nonmetallic conductor at room temperature beyond any 3D and 2D material known (Avouris 2010; Bolotin et al. 2008). However, the conductivity values for graphene are close to or slightly lower than those for CNTs, probably because of the σ-π rehybridization phenomenon referred previously. Also, graphene is a zero-gap semiconductor with an electronic structure much simpler than that of SWCNTs owing to the absence of confined density of states, making the exploration of the electronic properties of graphene greatly attractive than the latter.

Considering optical properties, graphene and SWCNTs show similar fluorescence spectroscopy ranging from ultraviolet (UV) to near-infrared (NIR) regions. The optical properties in CNTs depend on the diameter and symmetry of the tubes, so these emissions become more improbable in MWCNTs (Han 2005). However, some MWCNT fluorescence emission was found, but only in the UV-visible region.

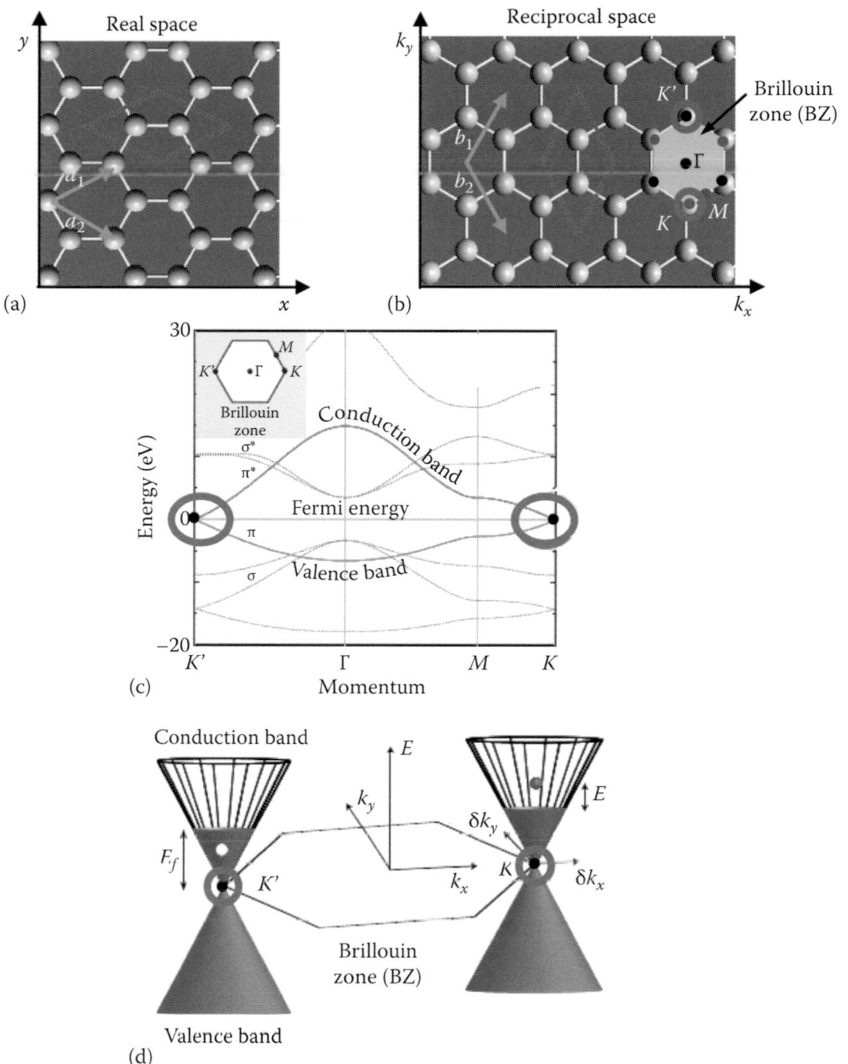

FIGURE 19.4 (a) Real space and (b) reciprocal space of a hexagonal lattice (unit cells highlighted by dark gray dotted line and lattice vectors highlighted by light gray). The BZ is highlighted by a gray hexagonal area (black spots are equivalent Dirac or *K* points, as also the dark gray spots ones) with two symmetric Dirac points labeled *K* and *K*', and two reference positions in the BZ, Γ, and M. (c) 2D electronic band structure (E_{2D}, Equation 19.11) of a graphene layer along the Γ–*K*', Γ–*M*, and Γ–*K* directions. (Reprinted with permission from Beenakker, C.W.J., *Rev. Mod. Phys.*, 80, 1342, 2008. Copyright 2015 by the American Physical Society.) (d) Respective 3D conical band structure of (c) presenting electron (filled circle) and hole (empty circle) excitations. (Reproduced with permission from Geim, A.K., MacDonald, A.H., *Phys. Today*, 60, 36, 2007. Copyright 2015, American Institute of Physics.)

According to Table 19.1, the electronic, electrical, and mechanical properties of CNTs are the most noteworthy and are the ones that have caught more attention caused by the wide range of potential applications (see Section 19.4). To better understand CNT electronic properties, the band structure of the 2D graphene is introduced. The unit cell of the graphene in real space and in reciprocal space is presented (highlighted by the dark gray dotted line in Figure 19.4a and b). These unit cells contain two atoms, the core atoms ($1s^2$). The other four valence electrons form the valence bands, three σ ($2s^1$, $2p_x^1$, and $2p_y^1$) and one π $\left(2p_z^1\right)$. The respective basic vectors of the hexagonal lattice are defined as a_1 and a_2, for the real space,

$$a_1 = \left(\frac{\sqrt{3}}{2}a, \frac{a}{2}\right);\ a_2 = \left(\frac{\sqrt{3}}{2}a, -\frac{a}{2}\right),$$

and b_1 and b_2 for the reciprocal space,

$$b_1 = \left(\frac{2\pi}{\sqrt{3}a}, \frac{2\pi}{a}\right); b_2 = \left(\frac{2\pi}{\sqrt{3}a}, -\frac{2\pi}{a}\right).$$

The reciprocal vectors, b_1 and b_2, delineate the Brillouin zone (BZ; gray area in Figure 19.4b) with six corners, called *K* points or Dirac points. The three corners labeled with a black dot are connected to each other by reciprocal vectors, so they are equivalent. The same occurs for the dark gray dotted corners (Beenakker 2008). Two of these *K* points are represented as *K* and *K*' in Figure 19.4b (highlighted with dark gray circles).

The 2D energy (E_{2D}) dispersion relations for π band in graphene (Saito et al. 1992) are given by

$$E_{2D}(k_x, k_y) = \pm\delta_0 \left\{1 + 4\cos\left(\frac{\sqrt{3}k_x a}{2}\right)\cos\left(\frac{k_y a}{2}\right) + 4\cos^2\left(\frac{k_y a}{2}\right)\right\}^{1/2}, \tag{19.11}$$

where (k_x, K_y) is a wave vector *K* defined in the BZ, δ_0 is the nearest-neighbor transfer integral, and $a = |a_1| = |a_2| = 0.246$ nm.

The allowed *K* vectors in the Γ–*K*', Γ–*M*, and Γ–*K* directions and respective E_{2D} energies are represented in Figure 19.4c. At the Dirac points, the conduction/antibonding band (π*) and the valence/bonding band (π) meet at the Fermi energy (E_f), with a 2D linear/3D conical dependence of the energy dispersion (E_{2D}) (Figure 19.4b–d). Under this condition, the electrons are able to run freely through the π* band, with this band being completely empty of electrons and the π band completely full. This occurs by an electron–hole conversion mechanism (Beenakker 2008). An electron is defined as a filled state with an energy $E > E_f$, and a hole excitation, an empty state with an energy $E < E_f$. For an undoped graphene, when $E_f = 0$ (Figure 19.4c and d), the electron belongs to the π* band and the hole to the π band (Beenakker 2008). Thus, undoped graphene is a zero-gap material (metallic) that behaves as "nonsemiconductor" because of the linear energy dispersion relation (E_{2D}) at the Dirac points (Figure 19.4c).

To determine the band structure of the 2D graphene presented previously, an infinite graphene layer without boundary conditions was assumed (Saito et al. 1992). However, the number of allowed states or *K* waveform vectors in the circumferential direction of a CNT is very limited, which might induce the formation of a gap. This means that only specific slices of the conical bands of graphene at the *K* points in reciprocal space are allowed. As proposed by Dresselhaus et al. (2004), the allowed *K* values in the circumferential direction of a CNT can be represented by parallel lines in the 2D graphene BZ (Figure 19.5).

A semiconducting CNT presents *K* vectors that do not cross a *K* point; otherwise, it is a metallic CNT (Dresselhaus et al. 2004). The spacing of the lines is inversely proportional to the CNT diameter and their angle depends on the nanotube chirality. Thus, another form to assess the electronic behavior of a specific tube contemplates its chiral integers (n, m) (Dresselhaus et al. 2004). The metallic conduction is observed only for

$$n - m = 3q \tag{19.12}$$

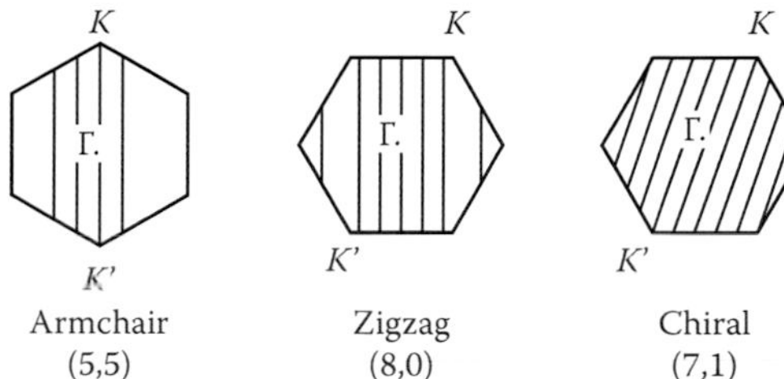

FIGURE 19.5 Allowed *K* vectors (solid lines) in the graphene Brillioun zone for the (5,5), (8,0), and (7,1) tubes. (Reproduced with permission from Dresselhaus, M.S., Dresselhaus, G., and Jorio, A., *Ann. Rev. Mater. Res.*, 34, 247–278. Copyright 2015 by Annual Reviews.)

where $(n, m) = C_h$ and q is an integer. Thus, armchair CNT is metallic, while zigzag and chiral tubes can be metallic or semiconductive.

In relation to MWCNTs, their electronic properties should be similar to those of several independent SWCNTs concentrically rolled up, since the neighboring layers should not interact (Dresselhaus and Endo 2001). Therefore, a more complex band structure is expected comparatively with an SWCNT. The several tubes/layers in the MWCNTs might present different chiralities with different electronic behaviors. For example, if one tube is semiconductive and the other is metallic, the low-energy properties are dictated by the metallic tube, with neglected changes in the density of states around the Fermi level. In this sense, MWCNTs are considered as metallic conductors with an electronic behavior between those of an SWCNT and the 3D graphite (Dresselhaus and Endo 2001).

Regarding CNT electrical properties, electrical conduction appears to be almost ballistic along the tube length (Chico et al. 1996). This is attributed to the restriction of allowed electronic states in CNTs that implies that the electron transport should follow discrete electron states (Chico et al. 1996). The ballistic conduction means that there are no scattering losses and no overheating. Thus, high current densities passing through the CNT's structure are allowed (Zhu et al. 1999) (Table 19.1).

Turning to the mechanical properties (Table 19.1), CNTs present the highest Young's modulus ever measured (Krishnan et al. 1998) but also present extreme flexibility (Falvo and Clary 1997), thus giving proof of their high resilience and toughness.

19.4 Potential Applications of CNTs

The supreme properties of CNTs had been promoting huge enthusiasm over the last decades, guiding the exploration of a wide range of potential applications. Table 19.2 compiles some of these applications, divided in two main categories: properties and scale.

Considering the high aspect ratio of CNTs, they are indicated to work as field emission materials (Bonard et al. 1999) in electron microscopes or in field emission displays/lightening, with some advantages over the common Si and W tips such as the higher current densities and resistance to electromigration (Robertson 2004). Also, the high aspect ratio, high mechanical strength, and flexibility convert them into ideal materials to be used as scanning probes (e.g., atomic force microscopy and scanning tunneling microscopy) (Hafner et al. 2001) and microcatheters (Endo et al. 2004). In addition, CNTs can be applied as electrodes (Britto et al. 1999), in electrochemistry and in electrocatalysis of fuel cells (Li et al. 2002; Serp et al. 2003), as simple electron donors in photovoltaic devices (Pradhan et al. 2006), or as sensor devices (e.g., chemical and physical) (Bekyarova et al. 2004) because of their high surface area, high conductivity, and chemical stability in a wide range of potentials. The high surface area of CNTs has also caught some attention for supercapacitors and lithium-ion battery applications (Leroux et al. 1999). Finally, CNTs have been applied in several nanoelectronic devices (e.g., field effect transistors and interconnects) because of their high chemical stability, high current density (over 1000 that for copper),

TABLE 19.2 CNT Applications

Properties	Limited-Volume Applications	Large-Volume Applications
Mechanical	• Microcatheters (Endo et al. 2004) • Probes (Hafner et al. 2001)	• Structural composites (Robertson 2004)
Electrical and electronic	• Electrodes (Britto et al. 1999) • Sensor devices (Bekyarova et al. 2004; Portney and Ozkan 2006) • Nanoelectronics (Robertson 2007) • Electrostimulated drug-delivery systems (Bianco et al. 2005; LaVan et al. 2003)	• Supercapacitors (Robertson 2004) • Li-ion batteries (Leroux et al. 1999) • "Smart" multifunctional composites (Supronowicz et al. 2002; Zhang and Webster 2009) • Fuel cells (Li et al. 2002) • Field emission displays/lighting (Bonard et al. 1999)
Thermal and optical	• Thermal rectifiers (Chang et al. 2006) • Fluorescent markers for cell targeting (Cherukuri et al. 2004; Welsher et al. 2008)	• Photovoltaic devices (Pradhan et al. 2006) • Transparent conducting films (Wu et al. 2004)

and high thermal conductivity (see Section 19.3, Table 19.1) (Robertson 2007). The latter property allows CNTs to work as thermal rectifiers to prevent overheating in devices (Chang et al. 2006). However, one of the biggest problems for the electronic application of CNTs is that their bandgap depends on the structure's chirality. So far, strict control of the CNT structure is not possible, but some progresses have been made by engineering the metal catalyst particles involved in the CNT growth mechanisms.

Moreover, low-weight structural CNT-based composites have caught huge interest in several areas (e.g., aerospace and automobile). These high-performance composites were possible to be designed because of the combination of a high CNT Young's modulus (see Section 19.3, Table 19.1) and a low density of 2 g cm^{-3} (Robertson 2004). Another use of CNTs in composites is as high-electroconductive fillers (see Section 19.3, Table 19.1) to produce conductive composites at low percolation threshold owing to the CNT high aspect ratio (length to radius ratio up to 10^4). The low electrical percolation threshold and the high flexibility of CNTs make them promising in producing transparent and flexible conducting composite substrates (Wu et al. 2004) for the displays industry, in replacement of brittle indium tin oxide ones.

In addition, the high surface area and the ability to bond chemical groups make CNTs as ultimate fillers in "smart" multifunctional composites with escalating interest for biological applications. These CNT-based composites had triggered particular excitement in tissue engineering (Harrison and Atala 2007; Zhang and Webster 2009), with major attention in nervous and bone tissues (Stevens 2008; Zhang and Webster 2009), owing to the accumulation of several CNT functionalities in a unique scaffold material, such as electrical conductivity for cell sensing (Portney and Ozkan 2006) and control of cell functions under electric stimulation (Supronowicz et al. 2002), fluorescence for cell tracking (Cherukuri et al. 2004; Welsher et al. 2008), and in situ delivery of functional biomolecules, from CNT outer wall or hollow core, under (or not) an electrostimulated release (Bianco et al. 2005; LaVan et al. 2003).

19.5 CNTs for Biomedical Applications

19.5.1 Toxicological Profile of CNTs

There are escalating concerns regarding the hazards of CNTs in the human body because of their apparent similarities to asbestos and other carcinogenic fibers (Kane and Hurt 2008; Zhao et al. 2008b). The toxicity of CNT is still controversial because of contradictory reports, mostly a result of the wide range of variables that dictate the toxicological profile of CNTs. These variables are commonly classified in two main groups: (1) physicochemical characteristics of CNTs (morphology, purity, chemical functionalization, agglomeration state) and (2) the applied experimental protocol (type of cell/tissue, dosage, administration route). Nonetheless, several studies clearly identified two common factors that are responsible for the toxic behavior of pristine CNTs (Firme and Bandaru 2010; Kobayashi et al. 2010; Kostarelos 2008; Pulskamp et al. 2007; Stone and Donaldson 2006; Wick et al. 2007). First, the residual amount of metals used to produce CNTs can act as a catalyst to oxidative stress in cells, leading to toxicity (Firme and Bandaru 2010; Pulskamp et al. 2007). Investigation works had shown that only CNTs that are highly pure, above 99%, are nontoxic (Kagan et al. 2006; Pulskamp et al. 2007). Second, CNTs may present high biopersistency owing to their nonbiodegradability, hydrophobicity, and morphology (Firme and Bandaru 2010; Kobayashi et al. 2010; Kostarelos 2008; Stone and Donaldson 2006; Wick et al. 2007). Large CNT agglomerates >20 μm in diameter result from their feeble dispersion (Wick et al. 2007); long (>10 μm) and thick (>50 nm) individual CNTs (Kobayashi et al. 2010; Kostarelos 2008; Stone and Donaldson 2006) cannot be totally eliminated either through urinary and/or lymphatic mechanisms, inducing inflammation and granuloma lesions (Kane and Hurt 2008; Kostarelos 2008; Stone and Donaldson 2006; Wick et al. 2007). In line with this, the chemical functionalization of CNT becomes essential to restrict the formation of big aggregates, to accelerate the digestion process accomplished by phagocytes, and to increase the CNT mobility in physiological serums. Also, CNT dosages above 10 mg L^{-1} (Hirano et al. 2008; Pulskamp et al. 2007; Simon-Deckers et al. 2008; Wu et al. 2005) or 2 mg kg^{-1} (Lacerda et al. 2008; Liu et al. 2006) substantially increase cellular apoptosis or tissue necrosis. Summarizing, the window of opportunity for CNTs in biology and biomedicine concerns their use at low dosages with short and thin dimensions and in the functionalized form.

19.5.2 CNT-Engineered Multifunctional Biomedical Systems

Three decades ago, CNTs became accountable for cotriggering the nanoscience and nanotechnology revolution worldwide. A couple of years later, the boom in the nanomedicine field occurred in the following sectors: in vivo imaging, in vitro diagnosis, drug delivery, and biomaterials (Mazzola 2003; Wagner et al. 2006). CNTs have caught increasing enthusiasm in nanomedicine because of their low cost and supreme electrical, mechanical, and optical properties (Kostarelos et al. 2009; Liu et al. 2009; Lu et al. 2009). Pioneering solutions based on CNTs have been designed to endorse significant advances in the sectors referred to previously (Kostarelos et al. 2009; Liu et al. 2009; Lu et al. 2009).

CNTs (single-walled and multiwalled) can be easily tracked and monitored in vitro and in vivo by Raman spectroscopy (Liu et al. 2008) and fluorescence spectroscopy (Cherukuri et al. 2006; Lacerda et al. 2006, 2007). Of relevance, both Raman scattering and fluorescence are stable with neither blinking nor photobleaching, even after prolonged exposure to excitation (Heller et al. 2005). In fluorescence spectroscopy, the photoluminescence of CNTs might occur at different emission wavelengths, ranging from the NIR region for pristine SWCNTs (Cherukuri et al. 2006) to UV-visible region for chemical functionalized SWCNTs and MWCNTs (Lacerda et al. 2006, 2007) (see Section 19.3, Table 19.1). For example, in vitro confocal laser scanning microscopy (CLSM) imaging using an argon laser with a wavelength near 488 nm is able to detect and show CNT-NH3+, with the same excitation wavelength (Figure 19.6a), interacting with cells (Figure 19.6b and c) (Lacerda et al. 2007). These optical absorption properties presented by CNTs have been used for cell tracking and laser heating cancer therapy (Kam et al. 2005).

In addition, the broad electronic properties of CNTs have been applied in the fabrication of highly sensitive and specific nanoscale biosensors (Yun et al. 2007) and electromechanical actuators for artificial muscles (Fennimore et al. 2003).

The fast progress in the in vivo toxicological and biodistribution studies of CNTs, alongside a wider knowledge in chemical functionalization processes, opened new bioapplication opportunities in areas such as drug delivery (Kostarelos et al. 2009b). The high surface area and high versatile surface functionalization chemistry of CNTs make them promising delivery vehicles of therapeutic molecules such as glycoproteins and bone morphogenetic proteins (BMP2, BMP4, and BMP7) or antibiotics (Kostarelos et al. 2009; Liu et al. 2009).

Solo CNTs have been applied as scaffolds presenting a general trend for a biocompatible profile. Regarding the response to osteoblastic cells, appropriate cellular adhesion and early proliferation have

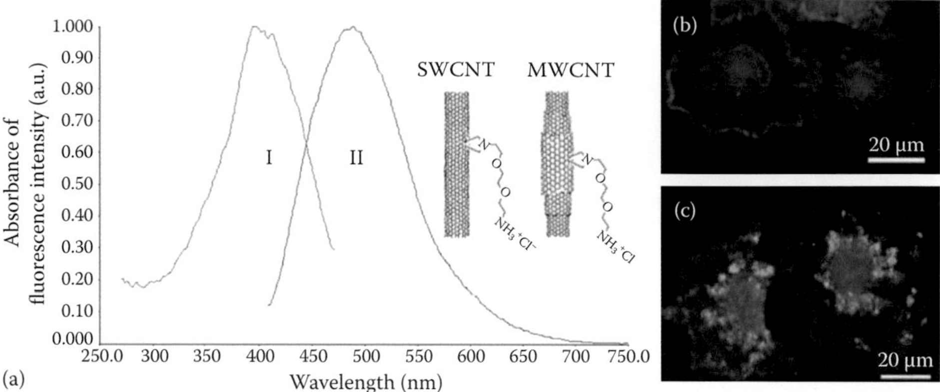

FIGURE 19.6 (a) Excitation (I) and emission (II) spectra of SWCNT-NH_3^+ and MWCNT-NH_3^+. (Lacerda, L., Pastorin, G., Wu, W. et al.: *Adv. Funct. Mater.* 16. 2006. 1839–1846. Copyright Wiley-VCH Verlag GmbH & Co. KGaA. Reproduced with permission.) (b and c) Confocal images of A549 cells (plasma membrane stained at red [shown in dark gray] and nucleus stained at blue [shown in medium gray]) in the (b) absence and (c) presence of SWCNT-NH_3 at green (shown in light gray) ($\lambda_{emission}$ = 510 nm). Scale bar corresponds to 20 μm. (Reprinted by permission from Macmillan Publishers Ltd. *Nature Nanotechnol.* Kostarelos, K. et al., *Nat. Nanotechnol.*, 2, 110, 2007. Copyright 2015.)

been reported over SWCNT films prepared by different techniques and reporting different degrees of purity (Kalbacova et al. 2007), MWCNT sheets (Akasaka et al. 2009), and compact MWCNT constructs (Xiaoming et al. 2009). Concerning functional activity, increased alkaline phosphatase (ALP) activity, osteoblastic gene expression, and calcium deposition were seen in $SaOS_2$ grown over compact MWCNTs (Xiaoming et al. 2009). Moreover, CNTs have risen as an exciting filler to turn ordinary bone grafts into multifunctional composite ones, the third generation of improved bone grafts for bone tissue engineering (Navarro et al. 2008; Spear and Cameron 2008; Stevens and George 2005). CNTs are inexpensive fillers, easily mass produced by CVD methods, with controlled morphologies and high purities. Also, CNTs are simply made of carbon, which, when incorporated in synthetic hydroxyapatite (HA) matrices, make them more chemically similar to those of carbonate apatites of natural bone (Legeros and Legeros 1993). Moreover, CNT characteristics—nonmetallic phases, high aspect ratio (quasi-1D material), and ultimate mechanical strength and electrical conductivity (see Section 19.3, Table 19.1)—make them the best-performing filler to obtain highly tough and conductive biomaterials at low percolation thresholds, without damaging the biological profile of the matrix.

These nonmetallic electroconductive bone grafts have become an exciting solution in clinical electrostimulation practices as they are seen as a breakthrough in the delivery of mechanisms of stimulus to the damaged bone (Guiseppi-Elie 2010; Spear and Cameron 2008).

19.5.3 Smart CNT-Bioceramic Composites for Bone Tissue Engineering

19.5.3.1 Electroconductive CNT Composites to Promote In Situ Electrostimulation of Bone

Bone regenerative medicine has experienced huge progress in the past decades, driven by the great socioeconomic interest in treating skeletal diseases (Stevens 2008). While there have been tremendous improvements in synthetic bone grafts materials, most are still incapable of fully repairing and regenerating severe bone injuries (Navarro et al. 2008). The development of solutions followed the routes of complex tissue engineering, involving cell manipulation, and of simpler strategies such as "smart" bone grafts with new functionalities, able to stimulate specific phenotype expressions of osteoblastic cells (Navarro et al. 2008; Spear and Cameron 2008). One such exciting functionality is to take advantage of the piezoelectric effect of bone. The forces exerted internally on the bone generate electrical signals that are carried to the bone cells, thus helping to regulate their biological functions (Bassett and Becker 1962; Yasuda 1954). These electrical currents found in healthy bone are not expected to exist at the fractured bone site, in case of bone tissue damage or loosening. Thus, it was thought that the use of exogenous electric fields would mimic the mislaid endogenous electric signals and accelerates bone healing (Bassett and Becker 1962).

Electrical stimulation, broadly attained by capacitive coupled electrical field, direct current field, and electromagnetic field, has been used to enhance the bone healing process by activating voltage-gated Ca^{2+} channels in osteoblast cell plasma membranes (Zayzafoon 2006). In vitro, the direct application of electric current has been found to induce cell elongation, modulate cell alignment, and favor the migration and invasion of human mesenchymal stem cells (hMSCs) into injured sites (Griffin et al. 2012). Electrical stimulation also seems to enhance the cell proliferation and the osteogenic differentiation of osteoblastic precursor cells (Woo et al. 2009). These findings seem to substantiate the improved musculoskeletal healing attained with electrical stimulation in clinical trials aiming at different applications of bone repair/regeneration, as reviewed in Behrens et al. (2013) and Novicoff et al. (2008).

These data have inspired the development of conductive synthetic matrices to assist in the bone regeneration process under electric stimulation (Meng et al. 2013; Supronowicz et al. 2002). Ideally, for clinical application, the electric stimuli should be confined within the tissue volume expected to be regenerated. As such, it is expected that conductive bone grafts may be able to increase the electrical conductivity at the fracture bone site, implying the use of lower applied voltages, and to confine exogenous electrical fields on their surface and directly deliver current to cells, boosting the spatial and temporal control of bone tissue regeneration. To further elucidate this behavior under an in vivo scenario, an example is shown considering a capacitive coupling stimulation method with parallel electrodes (Figure 19.7). The 2D distributions of the current density (electric field) lines interacting with a conductive and a dielectric bone graft granule implanted in a homogeneous fracture bone site are illustrated in Figure 19.7c and d (Edmonds 2001; Grimnes and Martinsen 2008).

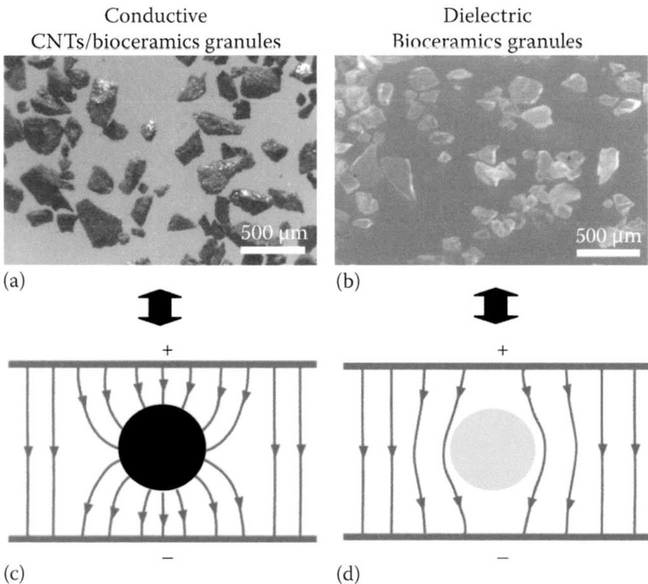

FIGURE 19.7 Optical images of (a) conductive and (b) dielectric CNT/bioceramic granules. Schematic images of the current density (electric field) lines distributions of (c) conductive and (d) dielectric spheres implanted in a homogeneous material.

If a conductor is placed between parallel electrodes, positive charges are attracted in one direction and negative charges in the opposite one, originating a small degree of polarization, but there are no fields within the conductor ($E = 0$). Therefore, all the uncompensated charges must reside in the surface layer, where all the external fields are confined at the surface of the conductor in perpendicular angles (Figure 19.7c). Thus, in this case, cells or tissues in contact with the surface of the conductor are crossed by electrical fields. If the charge is reversed, then the polarity of the conductor will change. This causes a small shift in charge, and an alternating electric current is generated locally at the surface of the conductor. Contrarily, in a dielectric material, the external fields are parallel to its surface (Figure 19.7d), so they are not confined; thus, a much lower efficiency of stimuli delivering at the bone fracture site is observed relatively to the conductor material. This becomes particularly useful since electric/electromagnetic strategies are spatially nondelimited, indiscriminately affecting both targeting and nontargeting anatomical locations. In the latter, the likely activation of voltage-dependent pathway signals in the surrounding tissues may lead to an overloaded concentration of the internal Ca^{2+} concentration in cells, causing functional disorders (e.g., oxidative stress, cytotoxicity) and its premature apoptosis (Orrenius et al. 2003).

19.5.3.2 Processing and Physical Properties of CNT/Glass/HA Bone Grafts

Bone grafts are expected to restore skeletal integrity by offering biological functional and mechanical supports during bone repairing. To ensure these functionalities, bone graft scaffolds should fulfill the following main requisites (Navarro et al. 2008): (1) biodegradable, with noncytotoxic degradation by-products and (2) mechanical properties close to those of bone and to keep its structural integrity and thus to preserve the porous network during the first stages of the new bone formation. To match this biological profile, it becomes mandatory to control the CNT agglomeration state in the consolidated bioceramic composites.

HA powder was planetary comilled with P_2O_5-glass (65 P_2O5, 15CaO, 10CaF_2, 10Na_2O mol%) in 97.5/2.5 proportion in weight, yielding to a final particle diameter size of $D_{0\cdot5} = 1.8 \pm 1.4$ μm. CNT/glass/HA composite powder suspensions were mixed by a two-step process with different volume fractions (0, 0.4, 0.9, 1.8, 3, 4.4, and 7.8 vol%) of CNTs (purified NC7000, Nanocyl, Belgium, in isopropyl alcohol; ≥99.8%, Sigma-Aldrich). This process contemplates a mechanical approach, not used excessively to avoid damage to CNTs: (1) a high-speed shearing for 15 min (IKA T25-Ultra-Turrax, working at 20,500 rpm) followed by (2) a 60-min sonication step (Selecta, working at 60 kHz, 200 W). Afterward, the composite powders were dried, also by a two-step method to avoid phase separation: (1) fast evaporation, by combining heating at 80°C and vacuum under a magnetic shearing until a high viscosity slurry is obtained; (2) slow evaporation, in an oven at 60°C for 24 h. Once dried, the powders were crushed in an agate mortar and sieved to less than 75 μm. CNT/glass/HA composite powders were consolidated by hot-pressing at a fixed pressure of 30 MPa for 60 min at 1100°C, under vacuum conditions (Mata et al. 2014a,b,c).

For the preparation of the composite materials, the hot-pressing temperature of 1100°C made the glass react with the HA, causing its partly conversion into β-tricalcium phosphate (β-TCP) phase, as depicted in the x-ray diffraction patterns in Figure 19.8a and b, respectively, for the HA/glass matrix and the 4.4 vol% CNT/HA/glass composite. The peaks correspond to those reported in the Joint Committee on Powder Diffraction Standards (JCPDS) files for HA (JCPDS 72-1243) and β-TCP (JCPDS 09-0169). Comparing those two plots, it becomes quite clear that there are almost no changes in phase composition when the CNTs are added to the matrix.

The use of Raman spectroscopy in imaging mode permitted clearly identifying the areas in the dense materials where HA, β-TCP, and CNTs are present, as Figure 19.8c and d illustrate (Cuscó et al. 1998). Combining the code colors in Figure 19.8d with the corresponding spectra in Figure 19.8f, it becomes clear that CNTs (addressed by the G' [2D] signature and higher G/D ratio) are present mainly as elongated agglomerates but can be identified throughout the matrix, within the resolution limits of the technique. Thus, from Raman observations only, there are no clearly identifiable zones containing individualized CNTs thereof.

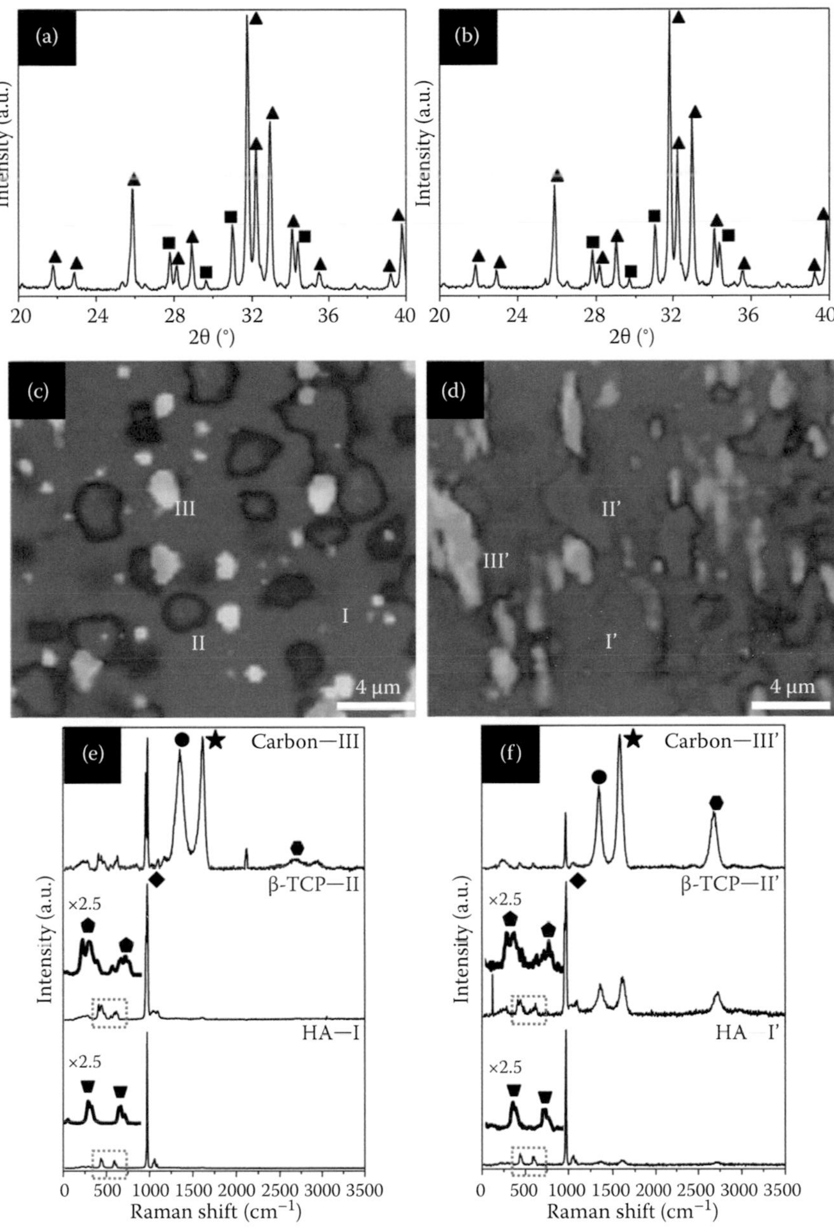

FIGURE 19.8 XRD spectra of hot-pressed (a) glass/HA and (b) CNT (4.4 vol%)/glass/HA compacts (■ β-TCP, ▲ HA). (c and d) Color Raman maps and respective (e and f) spectra of (a) and (b) (carbon and CNTs: ● D-band, ★ G-band, ⬢ G'(2D)-band; β-TCP: ◆ $\nu_1\_PO_3^{2-}$, ⬟ $\nu_2\_PO_3^{2-}$; HA: ◆ $\nu_1\_PO_3^{2-}$, ▼ $\nu_2\_PO_3^{2-}$). (Mata, D. et al., *J. Mater Chem B*, 3, 1836, 2015. Reproduced by permission of The Royal Society of Chemistry.)

High magnified scanning electron microscopy (SEM) micrographs of the microstructure of wet-etched surfaces give further insights on the location of individual CNTs. The top view of microetched polished samples clearly show the cross-sectional profile of CNTs (highlighted with arrows) placed in the grain boundaries (Figure 19.9a and b). To further reveal the 3D position of CNTs, respective 30° tilted macroetched surfaces are also shown (Figure 19.9c and d). The composites have predominantly individual CNTs located at the grain boundaries, corroborating the 2D observations. This points toward

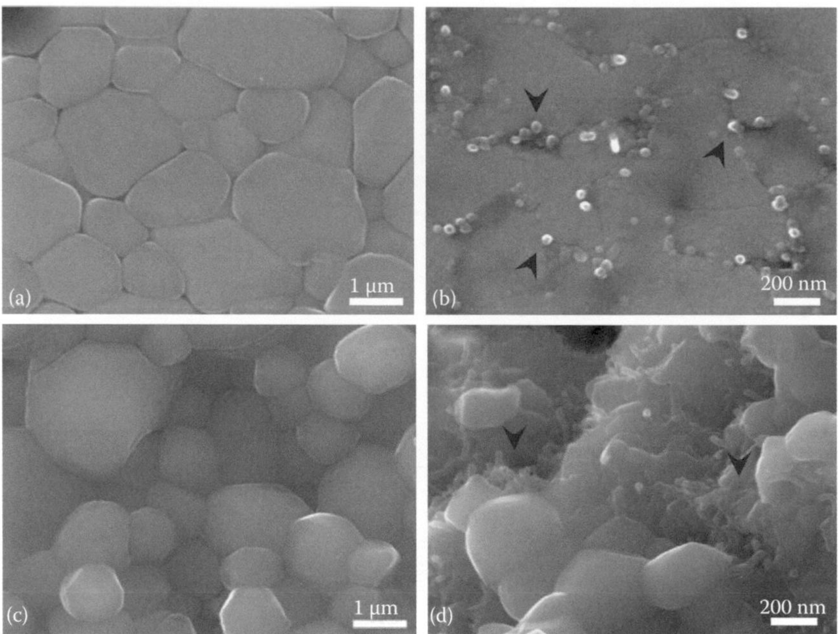

FIGURE 19.9 (a–d) SEM images of microetched and macroetched samples of hot pressed glass/HA and CNT (4.4 vol%)/glass/HA samples showing, respectively, the 2D and 3D intergranular disposition of CNTs. (Reproduced from Mata, D. et al., *Nanotechnology* 25 (2014):145602-8 by permission of IOP Publishing. Copyright 2015.)

the existence of a microstructure with a 3D CNT network formed by microsized agglomerates (3 μm) together with individualized CNTs located at the grain boundaries.

Despite most of the CNTs being located in the agglomerates, TEM analysis revealed that there are also CNT bundles and ropes around individual HA grains, as can be observed in Figure 19.10. The low-magnification TEM image (Figure 19.10a) reveals the CNT agglomerates and the intergrain CNT disposal. CNT bundles of submicron size are easily identified at the HA grain triple points (Figure 19.10b), but more interestingly, at the matrix grain boundaries, there are ropes of CNTs connecting those small bundles. The CNT agglomerates, oriented and ellipsoid shaped, with a main axis having a few micrometers, are thus interconnected at a submicrometric scale, ensuring that a network of contiguity is effectively established throughout the volume of the material.

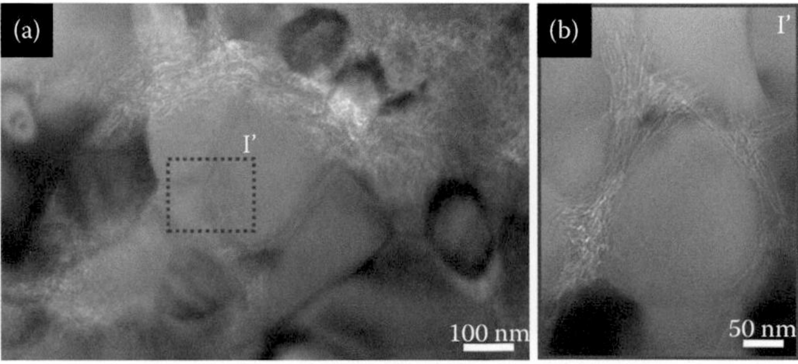

FIGURE 19.10 TEM micrographs of the CNT (4.4 vol%)/glass/HA composite. (a) Low-magnification image of the microstructure showing CNT agglomerates connected by CNT bundles; (b) high-magnification images of CNT bundles. (Reprinted from *Mater. Sci. Eng. C*, 34, Mata, D., Horovistiz, A.L., Branco, I. et al., Carbon nanotube-based bioceramic grafts for electrotherapy of bone, 360–368, Copyright 2015, with permission from Elsevier.)

The increasing amounts of CNT incorporation raise the porosity from 0.2% for the matrix, to about 6% for the 4.4 vol% composite, as is evidenced by the bottom plot of Figure 19.11. The evolution of porosity with CNT content reflects not only the damping effect of the CNTs and the difficulty in transmitting the pressure to the glass/HA matrix for densification but also the increased interactions and entanglement of CNT agglomerates that further prevent matrix contiguity from occurring.

Such a porosity effect is clearly mirrored in the bending strength of the matrix and composites tested (top plot of Figure 19.11) and, to some extent, in the compressive strength (middle graph in Figure 19.11). For fully dense materials, CNT mechanical reinforcing is expected owing to toughening by grain bridging or pull-out, crack deflection, and crack branching (Zhan et al. 2002). This is observed for the compressive strength of the 0.9 vol% CNT composite, but as porosity has an exponentially detrimental effect on the rupture strength of a given material (Liu and Fu 1996), a decrease in the mechanical properties is observed for higher CNT contents. This effect was clearly identified by others for the indentation fracture toughness and bending strength in HA/CNT composites produced under hot pressing, although no information regarding porosity levels was given (Meng et al. 2008). Both the compressive and bending strengths of our materials have values that are well within or above cortical bone specifications for adult, healthy individuals (Evans and Vincentelli 1974; Hench 1991) and are much larger than those reported in other works (Lei et al. 2011; Meng et al. 2008; Zhao and Gao 2004). The porosity of the composite with 4.4 vol% of CNTs is closer to that of the cortical bone (~5.4%–14.2%) (Dong and Guo 2004) than the one of the glass/HA matrix alone (1%).

The foremost condition for obtaining electrically conductive biocomposites is that a 3D CNT network might form, in the present case made of CNT ropes, bundles, and large agglomerates, as was previously established. The existence of electrical conductivity due to the presence of CNTs in ceramic materials at various loading levels has been demonstrated by several authors and is reviewed in some works (Cho et al. 2009).

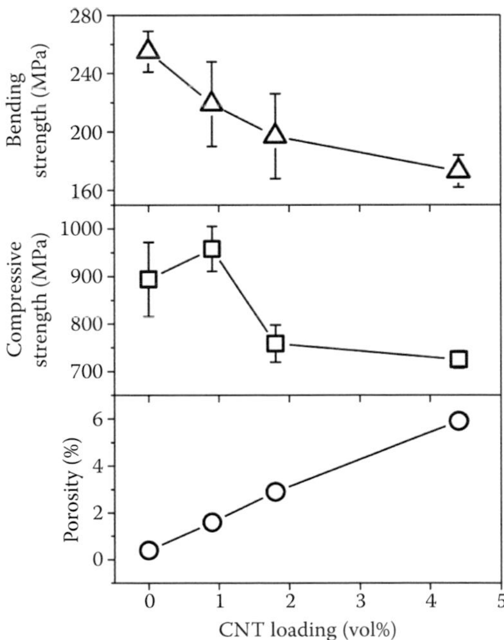

FIGURE 19.11 Porosity and mechanical properties of CNT/glass/HA composites with different CNT loadings. (Reprinted from *Mater. Sci. Eng. C*, 34, Mata, D., Horovistiz, A.L., Branco, I. et al., Carbon nanotube-based bioceramic grafts for electrotherapy of bone, 360–368, Copyright 2015, with permission from Elsevier.)

However, the minimum volume fraction of CNTs that permits this has to be experimentally evaluated since it is a function of the processing conditions. The increase in conductivity with the amount of CNTs in otherwise insulating materials should follow the following scaling law (Stauffer and Aharony 1994):

$$\sigma = k(p - p_c)^t, \tag{9.13}$$

where p is the volume fraction of CNTs, p_c is the percolation threshold given as the critical amount of CNTs that gives rise to an abrupt increase in electrical conductivity, and the exponent t is a measure of the dimensionality of the system. Both the exponent t and the percolation threshold have been attributed different values in various researches, and for ceramic composites, typical t exponent and p_c values are within the 1.3–2 range (low values—two dimensional; and high values close to 2—three dimensional) (González-Julián et al. 2011; Rul et al. 2004; Stauffer and Aharony 1994) and 0.6–5.5 vol% range (González-Julián et al. 2011; Rul et al. 2004), respectively. For CNT/ceramic matrix composites densified by hot-pressing, Rul et al. (2004) determined experimental percolation threshold and exponent values of p_c = 0.64 vol% and t = 1.73, respectively. These parameters were later successfully used by González-Julián et al. (2011) to fit their electrical conductivity data as a function of CNT content in quite a different system prepared under rapid pressure assisted sintering. The scatter of values is usually attributed to differences in the type of CNTs used, their size, and aspect ratio. Moreover, CNT dispersion in the ceramic matrix and temperature-inducing oxidation effects also has relevant contributions (White et al. 2010).

In the present work, a sudden increase of five orders of magnitude is recorded in the plot of the electrical conductivity data as a function of CNT volume fraction, as given in Figure 19.12. The lines fitting the data points and those of the inset graph correspond to the application of the scaling law to the experimental results, yielding a percolation threshold of p_c = 1.5 vol% and an exponent of t = 1.98. The p_c value is close to those obtained in other ceramic/CNT systems (González-Julián et al. 2011) and the exponent value obtained is consistent with electrical conductivity due to a 3D network of CNTs. In general, lower t values mean that the electrical conductivity between CNTs is hindered by insulating barriers that are

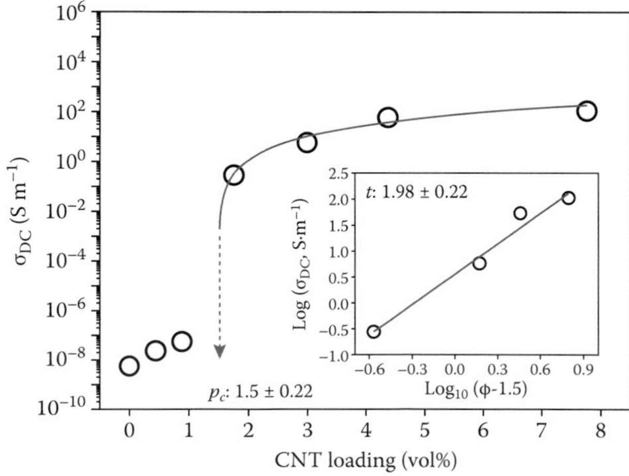

FIGURE 19.12 DC conductivity plot of the composites as a function of the CNT loading: data points fitted by an electrical percolation curve and log–log plot of the conductivity as a function of p–p_c as inset. (Reprinted from *Mater. Sci. Eng. C*, 34, Mata, D., Horovistiz, A.L., Branco, I. et al., Carbon nanotube-based bioceramic grafts for electrotherapy of bone, 360–368, Copyright 2015, with permission from Elsevier.)

surpassed only by thermally induced hopping (Rul et al. 2004) for interparticle distances below 10 nm (Li et al. 2007).

To further investigate the influence of the CNT percolation in electrical and phonon conductivity of the composites under simulated in vivo bone stimulation conditions, four CNT loads around that of the p_c value (1.5 vol%) were selected and closer studied: two below p_c (0 and 0.9 vol%) and two above p_c (1.8 and 4.4 vol%).

Impedance spectroscopy was used to evaluate the AC conductivity of the unreinforced matrix and of the composites at the body temperature (37°C), as shown in Figure 19.13a.

The glass/HA matrix follows a purely capacitive or rectifying-like (Schottky) behavior over the 40–10^5 Hz range of frequencies, as also reported by others for HA (Gittings et al. 2009). The same was observed for the composites below the percolation threshold. On the other hand, composites with nominal CNT loadings above 1.5 vol% have a predominantly resistive electrical or metallic-like behavior, with constant conductivities at all frequencies. That is, the composites behave either as insulating materials, below p_c, or as conducting materials, above p_c. Although a similar trend was found in CNT/ceramic

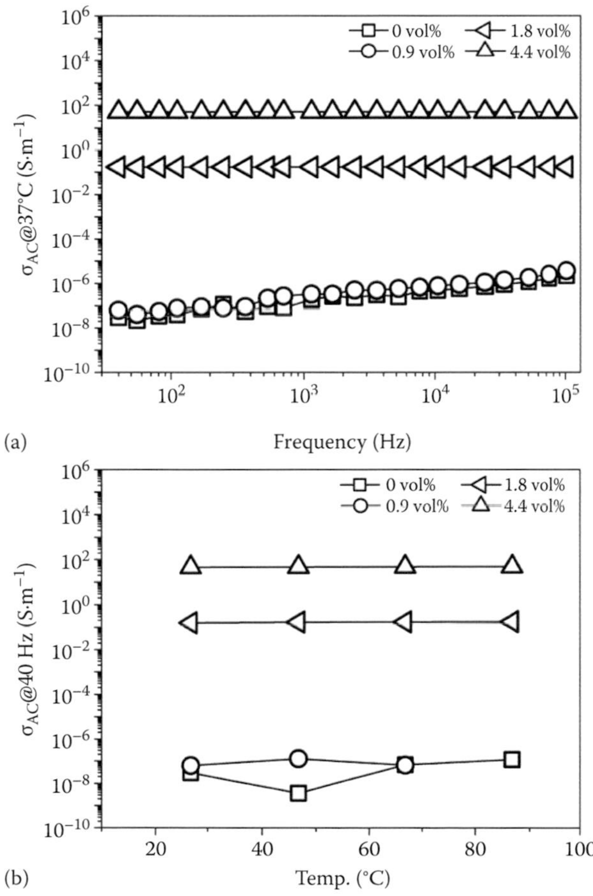

FIGURE 19.13 AC conductivity plot of the glass/HA and CNT/glass/HA composites for 0.9, 1.8, and 4.4 vol% CNT loadings as a function of the (a) signal frequency and (b) temperature. (Reprinted from *Mater. Sci. Eng. C*, 34, Mata, D., Horovistiz, A.L., Branco, I. et al., Carbon nanotube-based bioceramic grafts for electrotherapy of bone, 360–368, Copyright 2015, with permission from Elsevier.)

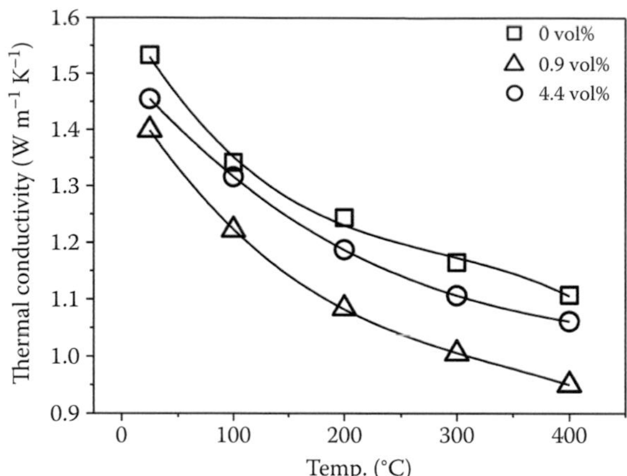

FIGURE 19.14 Thermal conductivity plots of the glass/HA and CNT/glass/HA composites for 0.9 and 4.4 vol% CNT loadings as a function of the temperature. (Reprinted from *Mater. Sci. Eng. C*, 34, Mata, D., Horovistiz, A.L., Branco, I. et al., Carbon nanotube-based bioceramic grafts for electrotherapy of bone, 360–368, Copyright 2015, with permission from Elsevier.)

composites (Ahmad and Pan 2009), the rate-limiting conductivity mechanisms in dielectric/CNT composites may vary as a function of the CNT's content, both below and above the percolation threshold (González-Julián et al. 2011), and are dependent upon temperature. Nevertheless, the electrical conductivity of the materials of the present work does not vary with temperature, above the body temperature, under a frequency typical of bone stimulation (40 Hz) (Figure 19.13b).

Moreover, for the CNT/glass/HA composites above the p_c value, the independence of electrical conductivity on the applied frequency means an increased flexibility on the selection of the best value for the electrical stimulation, since the current values will be kept constant, depending only on the magnitude of the applied voltage.

The CNT percolation also affects the phonon conductivity in the composite, as it is shown in Figure 19.14. Composites with CNT contents below p_c, with an interrupted phonon transportation pathway, have lower thermal conductivities than do those with CNT loads above the percolation limit, where a contiguous CNT pathway exists. The thermal barriers include porosity, which increases not only with CNT content but also with CNT–CNT and CNT–matrix contacts (Miranzo et al. 2012). The balance between these will determine the thermal conductivity, and from the results, it becomes evident that the porosity effect overcomes the increased thermal conductivity that arises with CNT loads. Thus, the matrix is always more thermally conductive than the CNT composites, both below and above the percolation threshold.

However, at low temperatures, in the range of 47°C to 50°C (Figure 19.14), the threshold for bone thermonecrosis (Eriksson and Albrektsson 1984), the matrix, and the CNT composites have approximate values of thermal conductivity, around 1.5 W m^{-1} K^{-1}. This value should be high enough to dissipate minor Joule's heat generation by microcurrents delivered on the composite–cell/tissue interface during the electrical stimulation protocols (Ormsby et al. 2011).

19.5.3.3 In Vitro Biological Characterization of CNT/Glass/HA Bone Grafts under Electrical Stimulation

The MG63 cell line was used to address the potential role of the electroconductivity of the CNT/HA/glass composite on the osteoblastic cell response under electrical stimulation and to select an optimized set of conditions for improved cell behavior (Mata et al. 2015). This cell system provides a homogeneous, phenotypically stable, and proliferative population that shows many phenotypic features of normal osteoblastic cells, including hormonal responsiveness and expression of early- and late-stage osteogenic genes, being widely used as an osteoblast cell model for in vitro research (Czekanska et al. 2012). Cell cultures, performed on standard cell culture coverslips, HA/glass matrix, and CNT/HA/glass composite were submitted to electrical stimuli of 5 μA/15 min, 5 μA/30 min, 15 μA/15 min, or 15 μA/30 min, one time a day for up to five consecutive days, and were assessed 24 h after one, three, and five stimuli for the DNA content, metabolic activity, and gene expression of runt-related transcription factor 2 (Runx2), collagen type I (Col 1), ALP, osteocalcin (OC), and osteoprotegerin (OPG) (Table 19.3).

The DNA content on the cultured coverslip and on the two materials, in nonstimulated conditions, increased throughout the culture time. This parameter reflects the number of cells present on the substrates, being an index of the cell proliferation. Under electrical stimulation, DNA content on the CNT/HA/glass composite was greatly increased after 1 and 3 stimuli (~25% to 60%) and was similar after 5 days of treatments, compared with nonstimulated cultures (Figure 19.15c). The inductive effect was dependent on the stimulus intensity and duration and was higher with 15 μA, 30 min (after one stimulus, ~60%) and 15 μA, 15 min (after three stimuli, ~62%). Cell response over the dielectric surfaces (coverslip and HA/glass matrix) was similar under electrical stimulation (Figure 19.15a and b). Compared with nonstimulated cultures, DNA content was slightly increased, similar, and significantly lower after one, three, and five daily treatments, respectively. Results for the MTT (3-(4,5-dimethylthiazol-2-yl)-2,5-diphenyltetra-zolium reduction (Figure 19.15d–f), a metabolic activity/proliferation assay based on the reduction ability of cell mitochondrial dehydrogenases, showed a pattern similar to that observed for the DNA content, in the conductive and dielectric substrates. However, in both types of surfaces, for the inductive effects, the percentage of increase in the optical density (OD) values reflecting the metabolic activity was always higher than that found for the DNA content. Thus, on the CNT/HA/glass composite, increases of ~40% to 130% were found after one and three stimuli. Hence, under selected stimulation conditions, both the cell proliferation and the metabolic activity were induced, suggesting the presence of a higher number of cells with increased metabolic activity, compared with nonstimulated conditions. However, these cellular parameters decreased after repeated stimuli. This behavior has been previously reported and appears to be conditioned by the stage of cell differentiation (Jansen et al. 2010; Tsai et al. 2009).

Cultures were characterized for the mRNA expression of several osteoblastic genes, 24 h after five daily electrical stimuli, under 15 μA, 15 min (Figure 19.16). Results show that the expression of Runx2 was particularly sensitive to the inductive effects of the electrical stimulation. It was up-regulated on the cultures grown over the CNT/HA/glass composite and also on those over the HA/glass matrix.

Even in the cultured coverslip, which showed low expression of Runx2 in the nonstimulated cultures, there was a trend for an increase under stimulation conditions. This is an interesting finding, as Runx2 is the earliest transcription factor for osteogenic differentiation and, in addition, it activates the expression of multiple late stage osteoblastic genes (Isaacson and Bloebaum 2010). Over the stimulated CNT/HA/Glass composite, there was also increased mRNA expression for ALP and OC, respectively an early

TABLE 19.3 Primers Used on RT-PCR Analysis

Gene	Forward Primer	Reverse Primer
GAPDH	CAGGACCAGGTTCACCAACAAGT	GTGGCAGTGATGGCATGGACTGT
Runx2	CAGTTCCCAAGCATTTCATCC	TCAATATGGTCGCCAAACAG
COL1	TCCGGCTCCTGCTCCTCTTA	ACCAGCAGGACCAGCATCTC
ALP	ACGTGGCTAAGAATGTCATC	CTGGTAGGCGATGTCCTTA
OC	CACTCCTCGCCCTATTG	CCCACAGATTCCTCTTCT
OPG	AAGGAGCTGCAGTAGGTCAA	CTGCTCGAAGGTGAGGTTAG

Source: Mata, D. et al., *J. Mater. Chem. B*, 3, 1835, 2015. Reproduced by permission of The Royal Society of Chemistry.

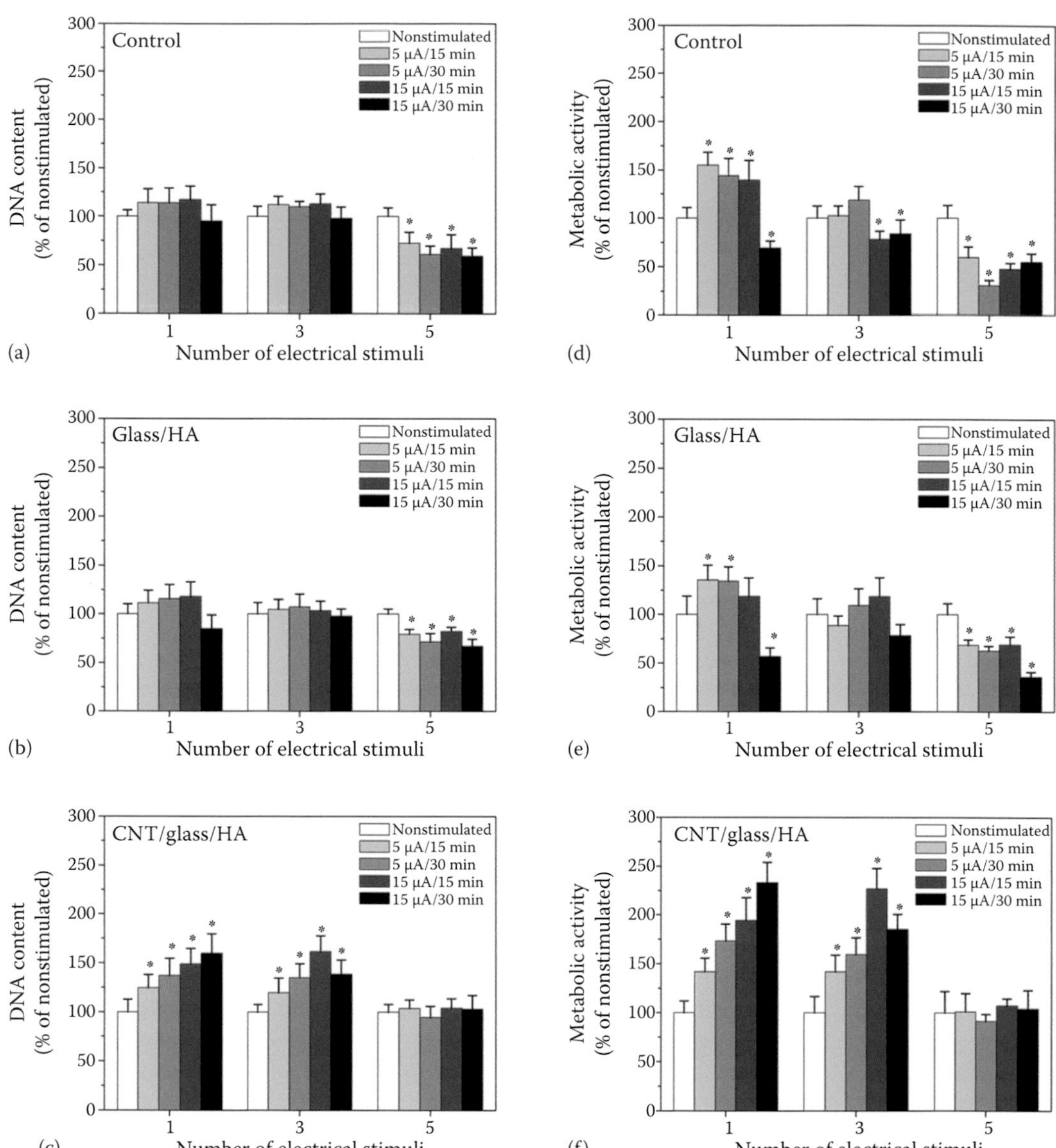

FIGURE 19.15 (a–c) DNA content and (d–f) cell viability/proliferation of osteoblastic cell cultures grown over standard cell culture coverslips, glass/HA matrix, and CNT/Glass/HA composite and characterized 24 h after one, three, and five daily electrical stimuli, under different stimulation conditions. Results are expressed as percentage of variation from nonstimulated cultures. *Significantly different from nonstimulated cultures. (Mata, D. et al., *J. Mater. Chem. B*, 3, 1840, 2015. Reproduced by permission of The Royal Society of Chemistry.)

and a late stage marker in the osteogenic differentiation pathway. As both molecules have a role in the extracellular matrix mineralization, respectively in the initiation of mineral deposition (ALP) and in the regulation of crystal growth (OC) (O'Donnell et al. 1989), this observation points to an enhanced osteogenic differentiation under electrical stimulation.

No effect was noted on the expression of Col 1 and OPG; Col 1 is the most abundant extracellular bone matrix, being considered an early bone differentiation marker, which has a role in osteoblastic differentiation and also in the nucleation site and growth space of HA (Datta et al. 2008). On the other hand,

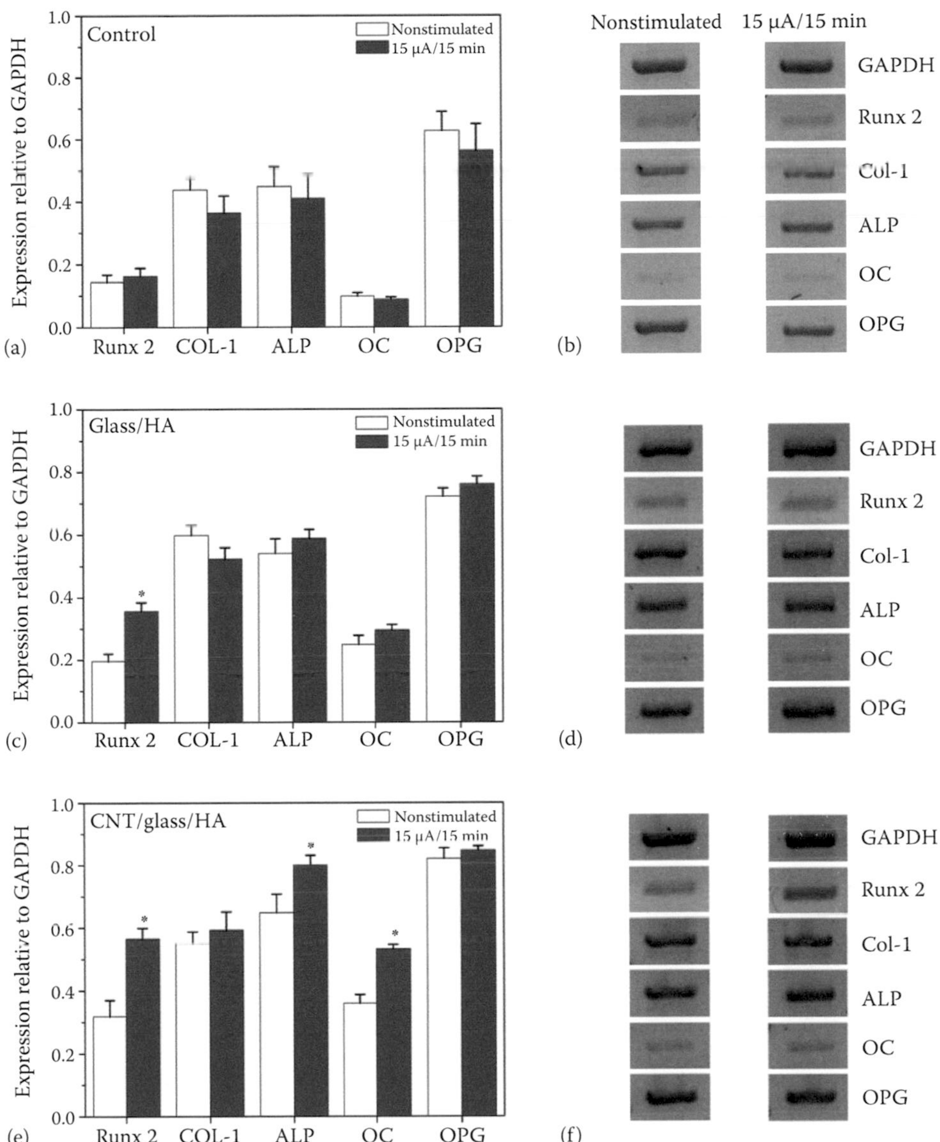

FIGURE 19.16 Expression of osteoblastic-related genes by cell cultures grown over standard cell culture coverslips, glass/HA matrix, and CNT/glass/HA composite and characterized 24 h after five daily electrical stimuli, under 15 μA, 15 min. Results are expressed as percentage of variation from nonstimulated cultures. Densitometric analysis of the reverse transcriptase-polymerase chain reaction (RT-PCR) bands normalized to the corresponding GAPDH value (a, c, e) and representative images of the PCR products in the agarose gel (b, d, f). *Significantly different from nonstimulated cultures. (Mata, D. et al., *J. Mater. Chem. B*, 3, 1841, 2015. Reproduced by permission of The Royal Society of Chemistry.)

OPG is a key molecule in the interplay between osteoblasts and osteoclasts during bone remodeling. It is a decoy receptor that binds to receptor activator of nuclear factor-κB ligand (RANKL), blocking its binding to the receptor activator of nuclear factor-κB (RANK) receptor on osteoclasts, inhibiting osteoclastogenesis (Matsuo and Irie 2008). These results on the gene expression might suggest that electrical stimulation induces the osteogenic differentiation and, eventually, may also enhance the level of matrix mineralization of the collagenous extracellular matrix. In addition, the osteoblast–osteoclast interactions regarding the modulation of osteoclastogenesis via OPG appear not to be affected. Overall, such a profile would be interesting to speed up the initial material stabilization leading to a faster osseointegration process.

On CLSM (Figure 19.17), nonstimulated and stimulated cells (observed 24 h after one and three daily electrical stimuli, 15 μA/15 min) displayed a polygonal/elongated morphology on the HA/glass matrix and, essentially, an elongated/fusiform appearance on the CNT/HA/glass composite. In both surfaces, cells exhibited a well-organized F-actin cytoskeleton, with intense staining at the cell boundaries, prominent nucleus, and ongoing cell division, signs of mechanical integrity and healthy behavior. Over the CNT/HA/Glass composite, a specific cell orientation was visible already at day 1, and a characteristic aligned cell growth was evident at later culture times, compared with the random pattern of cell growth seen over the HA/glass matrix.

Results showed that under an appropriate electrical stimulation protocol, the conductive CNT/HA/glass composite exhibited significantly improved cell proliferation, metabolic activity, and osteoblastic gene expression. The lower inductive effect on cell proliferation but higher expression of relevant osteogenic genes and ALP activity found over the CNT/HA/glass composite after few stimuli suggest an inductive effect on the osteoblastic differentiation, taking into account the established reciprocal relationship between proliferation and differentiation during the development of the osteoblastic phenotype (Datta et al. 2008).

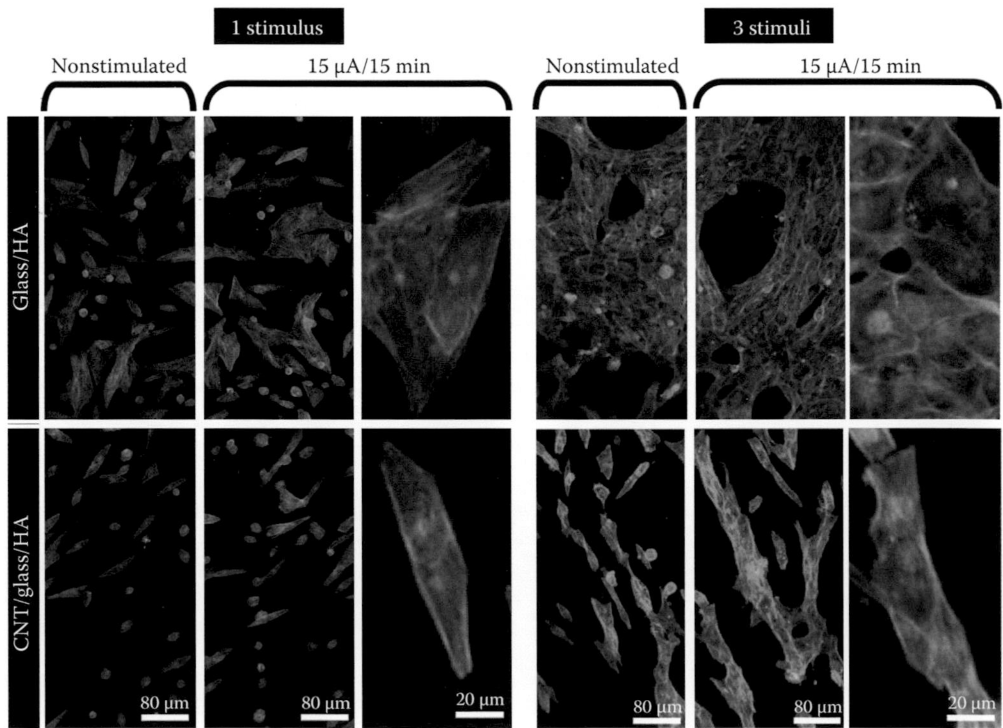

FIGURE 19.17 CLSM appearance of osteoblastic cell cultures grown over glass/HA matrix and CNT/glass/HA composite and observed 24 h after one and three daily electrical stimuli under different stimulation conditions. Cells were stained for F-actin cytoskeleton (green, shown in gray) and nucleus (red, shown in light gray). (Mata, D. et al., *J. Mater. Chem. B*, 3, 1842, 2015. Reproduced by permission of The Royal Society of Chemistry.)

Briefly, cell functional activity was significantly improved on the CNT/HA/glass composite, which is intimately related with its efficiency in the electrical stimuli delivering to cells. According to Section 19.5.3.1, conductive substrates increase the local conductivity of the culture medium and render the confinement of the exogenous electrical fields on the surface of the material. In conductive substrates, the electrical field confinement (higher current density) at perpendicular angles to the surface (Martinsen and Grimnes 2011) imposes osteoblastic cells in contact to be crossed by the electrical fields (Figure 19.18).

In contrast, in a dielectric material, the external fields are parallel to its surface, so they are not confined; thus, a much lower efficiency of stimuli delivering at the sample surface is observed relative to the conductor material.

Regarding the involved mechanisms, electrical stimulation seems to modulate cell behavior essentially by altering the intracellular calcium dynamics (Titushkin et al. 2010). Although being a versatile process and likely to depend on the cell type and microenvironmental conditions, the modulation of the intracellular calcium levels in hMSCs seems to be involved in directing their osteogenic differentiation (Titushkin et al. 2010). Associated mechanisms appear to be related with the clustering and activation of cell-surface receptors (e.g., integrins), interaction with G-protein-coupled receptors (e.g., phospholipase C), adenosine 5' triphosphate release, and activation of ion channels. Regarding the latter, the intracellular calcium can be raised inside osteoblastic cells by Ca^{2+} transport via L-type Ca^{2+} channels (Chesnoy-Marchais and Fritsch 1988). Of interest, it was found that these ion channels preferentially regulate Ca^{2+}-dependent genes and enzymes, such as the Ca^{2+}/calmodulin-dependent protein kinase II (CaMKII) (Peterson et al. 1999), a key pathway toward regulation of osteoblast proliferation/differentiation (Zayzafoon 2006). In the present work, the high current–voltage thresholds of osteoblastic Ca^{2+} ion channels give further evidences that cell response to stimuli appears to be L-type Ca^{2+}/CaMKII mediated, since cell proliferation and metabolic activity were maximized for the high current stimuli conditions (15 μA, 15 min). Moreover, compared with glass/HA, the conductive CNT/glass/HA composite reveals a higher activation of L-type ion channels (i.e., intracellular Ca^{2+} levels) and further larger efficiency of bone cell stimulation. This may be related to the high efficiency of charge transfer voltage at the substrate–cell interface, related to the high density of electric field lines crossing the cell, which boosts opening of voltage-gated channels at multilocations in the cell membrane and raises action potentials' magnitude.

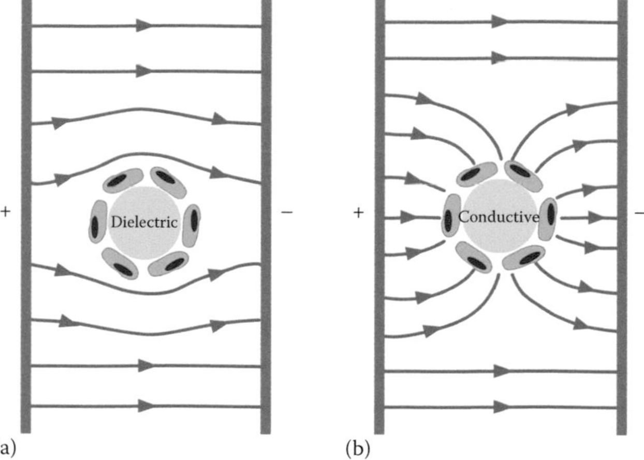

FIGURE 19.18 Top view schematic images of the electric field (current density) line distribution of (a) dielectric and (b) conductive spherical samples interfacing with osteoblastic cells (elliptical purple sketches) immersed in a homogeneous α-modified minimum essential media (α-MEM) culture medium. (Mata, D. et al., *J. Mater. Chem. B*, 3, 1843, 2015. Reproduced by permission of The Royal Society of Chemistry.)

19.6 Summary and Future Perspectives

The findings of the present chapter have shown that pioneering solutions based on CNTs—single-walled and multiwalled—have been widely explored to endorse relevant breakthroughs in tissue engineering's upmost demands of scaffolding, drug delivering, cell tracking and sensing, and in situ control of cell proliferation and differentiation. Moreover, the outstanding idea that all of these functionalities could be accumulated in a unique multifunctional CNT-based smart bone graft was also shown, offering further contributions for the potential clinical use of CNTs in bone tissue engineering, a field of study that has noticed huge progress in the past decades driven by the great socioeconomic interest in treating skeletal diseases. The incorporation of biologically safe CNTs, with high flexibility and electrical conductivity, in a bioactive and mechanically reinforced glass/HA matrix gives rise to unique bone grafts—CNT/glass/HA—mimicking the biofunctionalities of collagen I fibers and apatite-like phase of natural bone tissue, respectively. The CNT/glass/HA composites integrating impurity-free CNT agglomerates with maximized interfacing area, but with respective diameter sizes below the biologically safe threshold, allowed the in situ control of bone cell functions and thus are apt to interface with bone tissue. Also, these composites have a conjugation of suitable biocompatibility and exciting functional properties, high thermal and electrical conductivities that make them attractive to turn the noninvasive bone electrotherapies highly selective, by maximizing the confinement of electrical fields at the bone damaged site. Therefore, the presented results validate the hypothesis that CNT/bioglass/HA bone grafts could be used together with a noninvasive electrical stimulation technique to boost cell growth by an efficient stimuli delivering mechanism. This stimulation efficiency is thought to be related to the high density of electric field lines crossing the cell, which boosts the opening of voltage-gated channels at multilocations in the cell membrane, and raises action potentials' magnitude. Reducing the cost and time of bone treatment can make the electrotherapies of clinical relevance. These novel nonmetallic graft materials offer new possibilities, not only in bone regeneration but also in other excitable tissues (e.g., neuronal).

Despite the enthusiasm for CNT-engineered biomedical products, the risk–benefit balance for CNTs dictates their clinical fate. So far, the CNT potential toxicological risks made some of their applications unrealistic to clinic. CNT toxicity is still a controversial topic: While some studies have shown biocompatibility with cells and tissues, others have demonstrated their toxic effect. Indeed, up to now, studies clearly show that metal impurities, morphology, hydrophobicity, and nonbiodegradability dictate the in vitro and in vivo toxicological profiles of CNTs. Purifying and shortening pristine CNTs just solve part of their toxicological risks. To take a step forward in this clinical crossroad, it is compulsory to modulate the CNT in vivo biocompatibility and biodegradability via chemical functionalization. Chemical charging of CNTs restricts the formation of big agglomerates and increases the CNT mobility in physiological serums and ultimately avoids their accumulation into cell, tissues, and organs. Also, CNT functionalization will provide new opportunities to modulate the interaction between these materials and biological environments, holding a strong promise as novel systems for drug delivery, or just systems with improved biocompatibility and biodegradability.

This chapter further highlights the relevance of considering all contributing factors, such as the discussed toxicological profile when preparing a CNT-based system for use in biology and medicine. On the whole, it is expected in the coming decades that CNTs continue to arouse interest of both fundamental scientists and engineers in biomedical applications due to the unique combination of their properties.

Acknowledgments

This work was developed in the scope of the project CICECO-Aveiro Institute of Materials (Ref. FCT UID/CTM/50011/2013), financed by national funds through the Foundation for Science and Technology (FCT)/Ministry of Education and Science (MEC) and cofinanced by European Regional Development Fund (FEDER) under the PT2020 Partnership Agreement.

References

Ahmad, K., and Pan, W. "Dramatic effect of multiwalled carbon nanotubes on the electrical properties of alumina based ceramic nanocomposites." *Composites Science and Technology* 69 (2009):1016–21.

Akasaka, T., Yokoyama, A., Matsuoka, M. et al. "Adhesion of human osteoblast-like cells (Saos-2) to carbon nanotube sheets." *Bio-Medical Materials and Engineering* 19 (2009):147–53.

Andrews, R., Jacques, D., Rao, A.M. et al. "Continuous production of aligned carbon nanotubes: A step closer to commercial realization." *Chemical Physics Letters* 303 (1999):467–74.

Anna, M., Albert, G.N., and Esko, I.K. "The role of metal nanoparticles in the catalytic production of single-walled carbon nanotubes—A review." *Journal of Physics: Condensed Matter* 15 (2003):S3011.

Avouris, P. "Graphene: Electronic and photonic properties and devices." *Nano Letters* 10 (2010):4285–94.

Baker, R.T.K., Barber, M.A., Harris, P.S., Feates, F.S., and Waite, R.J. "Nucleation and growth of carbon deposits from the nickel catalyzed decomposition of acetylene." *Journal of Catalysis* 26 (1972):51–62.

Bassett, C.A.L., and Becker, R.O. "Generation of electric potentials by bone in response to mechanical stress." *Science* 137 (1962):1063–4.

Beenakker, C.W.J. "Colloquium: Andreev reflection and Klein tunneling in graphene." *Reviews of Modern Physics* 80 (2008):1337–54.

Behrens, S.B., Deren, M.E., and Monchik, K.O. "A review of bone growth stimulation for fracture treatment." *Current Orthopaedic Practice* 24 (2013):84–91.

Bekyarova, E., Davis, M., Burch, T. et al. "Chemically functionalized single-walled carbon nanotubes as ammonia sensors." *The Journal of Physical Chemistry B* 108 (2004):19717–20.

Berger, C., Song, Z., Li, X. et al. "Electronic confinement and coherence in patterned epitaxial graphene." *Science* 312 (2006):1191–6.

Bianco, A., Kostarelos, K., and Prato, M. "Applications of carbon nanotubes in drug delivery." *Current Opinion in Chemical Biology* 9 (2005):674–9.

Bolotin, K.I., Sikes, K.J., Jiang, Z. et al. "Ultrahigh electron mobility in suspended graphene." *Solid State Communications* 146 (2008):351–5.

Bonard, J.M., Salvetat, J.P., Stöckli, T., Forró, L., and Châtelain, A. "Field emission from carbon nanotubes: Perspectives for applications and clues to the emission mechanism." *Applied Physics A* 69 (1999):245–54.

Britto, P.J., Santhanam, K.S.V., Rubio, A., Alonso, J.A., and Ajayan, P.M. "Improved charge transfer at carbon nanotube electrodes." *Advanced Materials* 11 (1999):154–7.

Bundy, F.P., Bassett, W.A., Weathers, M.S. et al. "The pressure–temperature phase and transformation diagram for carbon; updated through 1994." *Carbon* 34 (1996):141–53.

Chang, C.W., Okawa, D., Majumdar, A., and Zettl, A. "Solid-state thermal rectifier." *Science* 314 (2006):1121–4.

Chen, J.-H., Jang, C., Xiao, S., Ishigami, M., and Fuhrer, M.S. "Intrinsic and extrinsic performance limits of graphene devices on SiO_2." *Nature Nanotechnology* 3 (2008):206–9.

Cherukuri, P., Bachilo, S.M., Litovsky, S.H., and Weisman, R.B. "Near-infrared fluorescence microscopy of single-walled carbon nanotubes in phagocytic cells." *Journal of the American Chemical Society* 126 (2004):15638–9.

Cherukuri, P., Gannon, C.J., Leeuw, T.K. et al. "Mammalian pharmacokinetics of carbon nanotubes using intrinsic near-infrared fluorescence." *Proceedings of the National Academy of Sciences of the United States of America* 103 (2006):18882–6.

Chesnoy-Marchais, D., and Fritsch, J. "Voltage-gated sodium and calcium currents in rat osteoblasts." *The Journal of Physiology* 398 (1988):291–311.

Chhowalla, M., Teo, K.B.K., Ducati, C. et al. "Growth process conditions of vertically aligned carbon nanotubes using plasma enhanced chemical vapor deposition." *Journal of Applied Physics* 90 (2001):5308–17.

Chico, L., Benedict, L.X., Louie, S.G., and Cohen, M.L. "Quantum conductance of carbon nanotubes with defects." *Physical Review B* 54 (1996):2600–6.

Cho, J., Boccaccini, A., and Shaffer, M.P. "Ceramic matrix composites containing carbon nanotubes." *Journal of Materials Science* 44 (2009):1934–51.

Cuscó, R., Guitián, F., Aza, S.D., and Artús, L. "Differentiation between hydroxyapatite and β-tricalcium phosphate by means of μ-Raman spectroscopy." *Journal of the European Ceramic Society* 18 (1998):1301–5.

Czekanska, E.M., Stoddart, M.J., Richards, R.G., and Hayes, J.S. "In search of an osteoblast cell model for in vitro research." *European Cells & Materials* 24 (2012):1–17.

Dai, H., Rinzler, A.G., Nikolaev, P. et al. "Single-wall nanotubes produced by metal-catalyzed disproportionation of carbon monoxide." *Chemical Physics Letters* 260 (1996):471–5.

Datta, H.K., Ng, W.F., Walker, J.A., Tuck, S.P., and Varanasi, S.S. "The cell biology of bone metabolism." *Journal of Clinical Pathology* 61 (2008):577–87.

Dillon, A.C., Mahan, A.H., Parilla, P.A. et al. "Continuous hot wire chemical vapor deposition of high-density carbon multiwall nanotubes." *Nano Letters* 3 (2003):1425–9.

Dong, X.N., and Guo, X.E. "The dependence of transversely isotropic elasticity of human femoral cortical bone on porosity." *Journal of Biomechanics* 37 (2004):1281–7.

Dresselhaus, M., and Endo, M. "Relation of carbon nanotubes to other carbon materials." In *Carbon Nanotubes: Synthesis, Structure, Properties and Applications*, edited by Dresselhaus, M., Dresselhaus, G., and Avouris, P. (Germany: Springer, 2001).

Dresselhaus, M.S., Dresselhaus, G., and Jorio, A. "Unusual properties and structure of carbon nanotubes." *Annual Review of Materials Research* 34 (2004):247–78.

Dupuis, A.-C. "The catalyst in the CCVD of carbon nanotubes—A review." *Progress in Materials Science* 50 (2005):929–61.

Edmonds, D. *Electricity and Magnetism in Biological Systems* (USA: Oxford University Press, 2001).

Endo, M., Takeuchi, K., Igarashi, S. et al. "The production and structure of pyrolytic carbon nanotubes (PCNTs)." *Journal of Physics and Chemistry of Solids* 54 (1993):1841–8.

Endo, M., Koyama, S., Matsuda, Y., Hayashi, T., and Kim, Y.-A. "Thrombogenicity and blood coagulation of a microcatheter prepared from carbon nanotube–nylon-based composite." *Nano Letters* 5 (2004):101–5.

Eriksson, R.A., and Albrektsson, T. "The effect of heat on bone regeneration: An experimental study in the rabbit using the bone growth chamber." *Journal of Oral and Maxillofacial Surgery* 42 (1984):705–11.

Evans, F.G., and Vincentelli, R. "Relations of the compressive properties of human cortical bone to histological structure and calcification." *Journal of Biomechanics* 7 (1974):1–10.

Falvo, M.R., and Clary, G.J. "Bending and buckling of carbon nanotubes under large strain. (Cover story)." *Nature* 389 (1997):582.

Fennimore, A.M., Yuzvinsky, T.D., Han, W.-Q. et al. "Rotational actuators based on carbon nanotubes." *Nature* 424 (2003):408–10.

Firme, C.P., III, and Bandaru, P.R. "Toxicity issues in the application of carbon nanotubes to biological systems." *Nanomedicine: Nanotechnology, Biology and Medicine* 6 (2010):245–56.

Gan, B., Ahn, J., Zhang, Q. et al. "Branching carbon nanotubes deposited in HFCVD system." *Diamond and Related Materials* 9 (2000):897–900.

Ge, M., and Sattler, K. "Observation of fullerene cones." *Chemical Physics Letters* 220 (1994):192–6.

Geim, A.K., and MacDonald, A.H. "Graphene: Exploring carbon flatland." *Physics Today* 60 (2007):35–41.

Geim, A.K., and Novoselov, K.S. "The rise of graphene." *Nature Materials* 6 (2007):183–91.

Gittings, J.P., Bowen, C.R., Dent, A.C.E. et al. "Electrical characterization of hydroxyapatite-based bioceramics." *Acta Biomaterialia* 5 (2009):743–54.

González-Julián, J., Iglesias, Y., Caballero, A.C. et al. "Multi-scale electrical response of silicon nitride/multi-walled carbon nanotubes composites." *Composites Science and Technology* 71 (2011):60–6.

Gorbunov, A., Jost, O., Pompe, W., and Graff, A. "Role of the catalyst particle size in the synthesis of single-wall carbon nanotubes." *Applied Surface Science* 197–198 (2002):563–7.

Griffin, M., Iqbal, S., Sebastian, A., Colthurst, J., and Bayat, A. "Electrical stimulation enhances migration and invasion of bone marrow stem cells: Implications for fracture healing." *Journal of Bone & Joint Surgery, British Volume* 94-B (2012):33.

Grimnes, S., and Martinsen, G. *Bioimpedance and Bioelectricity Basics* (UK: Academic Press, 2008).

Guiseppi-Elie, A. "Electroconductive hydrogels: Synthesis, characterization and biomedical applications." *Biomaterials* 31 (2010):2701–16.

Hafner, J.H., Cheung, C.-L., Oosterkamp, T.H., and Lieber, C.M. "High-yield assembly of individual single-walled carbon nanotube tips for scanning probe microscopies." *The Journal of Physical Chemistry B* 105 (2001):743–6.

Han, J. "Structures and properties of carbon nanotubes." In *Carbon Nanotubes Science and Applications*, edited by Meyyappan, M. (USA: CRC Press, 2005).

Harris, P.J.F. *Carbon Nanotube Science: Synthesis, Properties and Applications* (UK: Cambridge University Press, 2009).

Harrison, B.S., and Atala, A. "Carbon nanotube applications for tissue engineering." *Biomaterials* 28 (2007):344–53.

Hata, K., Futaba, D.N., Mizuno, K. et al. "Water-assisted highly efficient synthesis of impurity-free single-walled carbon nanotubes." *Science (New York, N.Y.)* 306 (2004):1362–4.

Height, M.J., Howard, J.B., Tester, J.W., and Vander Sande, J.B. "Flame synthesis of single-walled carbon nanotubes." *Carbon* 42 (2004):2295–307.

Heller, D.A., Baik, S., Eurell, T.E., and Strano, M.S. "Single-walled carbon nanotube spectroscopy in live cells: Towards long-term labels and optical sensors." *Advanced Materials* 17 (2005):2793–9.

Hench, L.L. "Bioceramics: From concept to clinic." *Journal of the American Ceramic Society* 74 (1991):1487–510.

Hirano, S., Kanno, S., and Furuyama, A. "Multi-walled carbon nanotubes injure the plasma membrane of macrophages." *Toxicology and Applied Pharmacology* 232 (2008):244–51.

Hofmann, S., Ducati, C., Kleinsorge, B., and Robertson, J. "Direct growth of aligned carbon nanotube field emitter arrays onto plastic substrates." *Applied Physics Letters* 83 (2003):4661–3.

Iijima, S. "Helical microtubules of graphitic carbon." *Nature* 354 (1991):56–8.

Iijima, S., and Ichihashi, T. "Single-shell carbon nanotubules of 1-nm diameter." *Nature* 363 (1993):603.

Isaacson, B.M., and Bloebaum, R.D. "Bone bioelectricity: What have we learned in the past 160 years?" *Journal of Biomedical Materials Research Part A* 95A (2010):1270–9.

Jansen, J., van der Jagt, O., Punt, B. et al. "Stimulation of osteogenic differentiation in human osteoprogenitor cells by pulsed electromagnetic fields: An in vitro study." *BMC Musculoskeletal Disorders* 11 (2010):188.

Kagan, V.E., Tyurina, Y.Y., Tyurin, V.A. et al. "Direct and indirect effects of single walled carbon nanotubes on RAW 264.7 macrophages: Role of iron." *Toxicology Letters* 165 (2006):88–100.

Kalbacova, M., Kalbac, M., Dunsch, L., and Hempel, U. "Influence of single-walled carbon nanotube films on metabolic activity and adherence of human osteoblasts." *Carbon* 45 (2007):2266–72.

Kam, N.W.S., O'Connell, M., Wisdom, J.A., and Dai, H. "Carbon nanotubes as multifunctional biological transporters and near-infrared agents for selective cancer cell destruction." *Proceedings of the National Academy of Sciences of the United States of America* 102 (2005):11600–5.

Kane, A.B., and Hurt, R.H. "Nanotoxicology: The asbestos analogy revisited." *Nature Nanotechnology* 3 (2008):378–9.

Kitiyanan, B., Alvarez, W.E., Harwell, J.H., and Resasco, D.E. "Controlled production of single-wall carbon nanotubes by catalytic decomposition of CO on bimetallic Co–Mo catalysts." *Chemical Physics Letters* 317 (2000):497–503.

Kobayashi, N., Naya, M., Ema, M. et al. "Biological response and morphological assessment of individually dispersed multi-wall carbon nanotubes in the lung after intratracheal instillation in rats." *Toxicology* 276 (2010):143–53.

Kostarelos, K. "The long and short of carbon nanotube toxicity." *Nature Biotechnology* 26 (2008):774–6.

Kostarelos, K., Bianco, A., and Prato, M. "Promises, facts and challenges for carbon nanotubes in imaging and therapeutics." *Nature Nanotechnology* 4 (2009):627–33.

Kostarelos, K., Lacerda, L., Pastorin, G. et al. "Cellular uptake of functionalized carbon nanotubes is independent of functional group and cell type." *Nature Nanotechnology* 2 (2007):108–13.

Krishnan, A., Dujardin, E., Ebbesen, T.W., Yianilos, P.N., and Treacy, M.M.J. "Young's modulus of single-walled nanotubes." *Physical Review B* 58 (1998):14013–9.

Kroto, H.W., Heath, J.R., O'Brien, S.C., Curl, R.F., and Smalley, R.E. "C 60: Buckminsterfullerene." *Nature* 318 (1985):162–3.

Lacerda, L., Pastorin, G., Wu, W. et al. "Luminescence of functionalized carbon nanotubes as a tool to monitor bundle formation and dissociation in water: The effect of plasmid–DNA complexation." *Advanced Functional Materials* 16 (2006):1839–46.

Lacerda, L., Pastorin, G., Gathercole, D. et al. "Intracellular trafficking of carbon nanotubes by confocal laser scanning microscopy." *Advanced Materials* 19 (2007):1480–4.

Lacerda, L., Ali-Boucetta, H., Herrero, M.A. et al. "Tissue histology and physiology following intravenous administration of different types of functionalized multiwalled carbon nanotubes." *Nanomedicine* 3 (2008):149–61.

LaVan, D.A., McGuire, T., and Langer, R. "Small-scale systems for in vivo drug delivery." *Nature Biotechnology* 21 (2003):1184–91.

Lee, C.J., Kim, D.W., Lee, T.J. et al. "Synthesis of aligned carbon nanotubes using thermal chemical vapor deposition." *Chemical Physics Letters* 312 (1999):461–8.

Legeros, R., and Legeros, J. "Dense hydroxyapatite." In *An Introduction to Bioceramics*, edited by Hench, L., and Wilson, J. (USA: World Scientific, 1993).

Lei, T., Wang, L., Ouyang, C., Li, N.-F., and Zhou, L.-S. "In situ preparation and enhanced mechanical properties of carbon nanotube/hydroxyapatite composites." *International Journal of Applied Ceramic Technology* 8 (2011):532–9.

Leroux, F., Méténier, K., Gautier, S. et al. "Electrochemical insertion of lithium in catalytic multi-walled carbon nanotubes." *Journal of Power Sources* 81–82 (1999):317–22.

Li, W., Liang, C., Qiu, J. et al. "Carbon nanotubes as support for cathode catalyst of a direct methanol fuel cell." *Carbon* 40 (2002):791–4.

Li, C., Thostenson, E.T., and Chou, T.-W. "Dominant role of tunneling resistance in the electrical conductivity of carbon nanotube-based composites." *Applied Physics Letters* 91 (2007):223114-3.

Liu, D.-M., and Fu, C.-T. "Effect of residual porosity and pore structure on the mechanical strength of SiC-Al2O3-Y2O3." *Ceramics International* 22 (1996):229–32.

Liu, Z., Cai, W., He, L. et al. "In vivo biodistribution and highly efficient tumour targeting of carbon nanotubes in mice." *Nature Nanotechnology* 2 (2006):47–52.

Liu, Z., Davis, C., Cai, W. et al. "Circulation and long-term fate of functionalized, biocompatible single-walled carbon nanotubes in mice probed by Raman spectroscopy." *Proceedings of the National Academy of Sciences of the United States of America* 105 (2008):1410–5.

Liu, Z., Tabakman, S., Welsher, K., and Dai, H. "Carbon nanotubes in biology and medicine: In vitro and in vivo detection, imaging and drug delivery." *Nano Res* 2 (2009):85–120.

Lu, F., Gu, L., Meziani, M.J. et al. "Advances in bioapplications of carbon nanotubes." *Advanced Materials* 21 (2009):139–52.

Martin-Gullon, I., Vera, J., Conesa, J.A., González, J.L., and Merino, C. "Differences between carbon nanofibers produced using Fe and Ni catalysts in a floating catalyst reactor." *Carbon* 44 (2006):1572–80.

Martinsen, O.G., and Grimnes, S. *Bioimpedance and Bioelectricity Basics* (UK: Academic Press, 2011).

Mata, D., Horovistiz, A.L., Branco, I. et al. "Carbon nanotube-based bioceramic grafts for electrotherapy of bone." *Materials Science and Engineering: C* 34 (2014a):360–8.

Mata, D., Oliveira, F.J., Ferreira, N.M. et al. "Processing strategies for smart electroconductive carbon nanotube-based bioceramic bone grafts." *Nanotechnology* 25 (2014b):145602.

Mata, D., Oliveira, F.J., Ferro, M. et al. "Multifunctional carbon nanotube/bioceramics modulate the directional growth and activity of osteoblastic cells." *Journal of Biomedical Nanotechnology* 10 (2014c):725–43.

Mata, D., Oliveira, F.J., Neto, M.A. et al. "Smart electroconductive bioactive ceramics to promote in situ electrostimulation of bone." *Journal of Materials Chemistry B* 3 (2015): 1831–45.

Matsuo, K., and Irie, N. "Osteoclast–osteoblast communication." *Archives of Biochemistry and Biophysics* 473 (2008):201–9.

Mazzola, L. "Commercializing nanotechnology." *Nature Biotechnology* 21 (2003):1137.

Meng, Y.H., Tang, C., Tsui, C., and Chen, D. "Fabrication and characterization of needle-like nano-HA and HA/MWNT composites." *Journal of Materials Science: Materials in Medicine* 19 (2008):75–81.

Meng, S., Rouabhia, M., and Zhang, Z. "Electrical stimulation modulates osteoblast proliferation and bone protein production through heparin-bioactivated conductive scaffolds." *Bioelectromagnetics* 34 (2013):189–99.

Mildred, S.D. "Fifty years in studying carbon-based materials." *Physica Scripta* 2012 (2012):014002.

Miranzo, P., García, E., Ramírez, C. et al. "Anisotropic thermal conductivity of silicon nitride ceramics containing carbon nanostructures." *Journal of the European Ceramic Society* 32 (2012):1847–54.

Navarro, M., Michiardi, A., Castano, O., and Planell, J.A. "Biomaterials in orthopaedics." *Journal of the Royal Society Interface* 5 (2008):1137–58.

Nikolaev, P., Bronikowski, M.J., Bradley, R.K. et al. "Gas-phase catalytic growth of single-walled carbon nanotubes from carbon monoxide." *Chemical Physics Letters* 313 (1999):91–7.

Novicoff, W.M., Manaswi, A., Hogan, M.V. et al. "Critical analysis of the evidence for current technologies in bone-healing and repair." *Journal Bone Joint Surgery* 90 (2008):85–91.

Novoselov, K.S., Geim, A.K., Morozov, S.V. et al. "Electric field effect in atomically thin carbon films." *Science* 306 (2004):666–9.

O'Donnell, P.T., Collier, V.L., Mogami, K., and Bernstein, S.I. "Ultrastructural and molecular analyses of homozygous-viable Drosophila melanogaster muscle mutants indicate there is a complex pattern of myosin heavy-chain isoform distribution." *Genes & Development* 3 (1989):1233–46.

Oberlin, A., Endo, M., and Koyama, T. "Filamentous growth of carbon through benzene decomposition." *Journal of Crystal Growth* 32 (1976):335–49.

Ormsby, R., McNally, T., Mitchell, C. et al. "Effect of MWCNT addition on the thermal and rheological properties of polymethyl methacrylate bone cement." *Carbon* 49 (2011):2893–904.

Orrenius, S., Zhivotovsky, B., and Nicotera, P. "Regulation of cell death: The calcium–apoptosis link." *Nature Reviews. Molecular Cell Biology* 4 (2003):552–65.

Peterson, B.Z., DeMaria, C.D., and Yue, D.T. "Calmodulin is the Ca2+ sensor for Ca2+-dependent inactivation of L-type calcium channels." *Neuron* 22 (1999):549–58.

Portney, N., and Ozkan, M. "Nano-oncology: Drug delivery, imaging, and sensing." *Analytical and Bioanalytical Chemistry* 384 (2006):620–30.

Pradhan, B., Batabyal, S.K., and Pal, A.J. "Functionalized carbon nanotubes in donor/acceptor-type photovoltaic devices." *Applied Physics Letters* 88 (2006):093106–3.

Pulskamp, K., Diabaté, S., and Krug, H.F. "Carbon nanotubes show no sign of acute toxicity but induce intracellular reactive oxygen species in dependence on contaminants." *Toxicology Letters* 168 (2007):58–74.

Robertson, J. "Realistic applications of CNTs." *Materials Today* 7 (2004):46–52.

Robertson, J. "Growth of nanotubes for electronics." *Materials Today* 10 (2007):36–43.

Rodriguez, N.M., Chambers, A., and Baker, R.T.K. "Catalytic engineering of carbon nanostructures." *Langmuir* 11 (1995):3862–6.

Rul, S., Lefèvre-Schlick, F., Capria, E., Laurent, C., and Peigney, A. "Percolation of single-walled carbon nanotubes in ceramic matrix nanocomposites." *Acta Materialia* 52 (2004):1061–7.

Saito, R., Fujita, M., Dresselhaus, G., and Dresselhaus, M.S. "Electronic structure of chiral graphene tubules." *Applied Physics Letters* 60 (1992):2204–6.

Serp, P., Corrias, M., and Kalck, P. "Carbon nanotubes and nanofibers in catalysis." *Applied Catalysis A* 253 (2003):337–58.

Simon-Deckers, A., Gouget, B., Mayne-L'Hermite, M. et al. "In vitro investigation of oxide nanoparticle and carbon nanotube toxicity and intracellular accumulation in A549 human pneumocytes." *Toxicology* 253 (2008):137–46.

Spear, R., and Cameron, R. "Carbon nanotubes for orthopaedic implants." *International Journal of Material Forming* 1 (2008):127–33.

Stauffer, D., and Aharony, A. *Introduction to Percolation Theory* (UK: Taylor & Francis, 1994).

Stevens, M.M. "Biomaterials for bone tissue engineering." *Materials Today* 11 (2008):18–25.

Stevens, M.M., and George, J.H. "Exploring and engineering the cell surface interface." *Science* 310 (2005):1135–8.

Stone, V., and Donaldson, K. "Nanotoxicology: Signs of stress." *Nature Nanotechnology* 1 (2006):23–4.

Supronowicz, P.R., Ajayan, P.M., Ullmann, K.R. et al. "Novel current-conducting composite substrates for exposing osteoblasts to alternating current stimulation." *Journal of Biomedical Materials Research. Part A* 59 (2002):499–506.

Terrones, M., and Grobert, N. "Controlled production of aligned-nanotube bundles." *Nature* 388 (1997):52.

Titushkin, I., Sun, S., Shin, J., and Cho, M. "Physicochemical control of adult stem cell differentiation: Shedding light on potential molecular mechanisms." *Journal of Biomedicine & Biotechnology* 2010 (2010):743476.

Tsai, M.-T., Li, W.-J., Tuan, R.S., and Chang, W.H. "Modulation of osteogenesis in human mesenchymal stem cells by specific pulsed electromagnetic field stimulation." *Journal of Orthopaedic Research* 27 (2009):1169–74.

Wagner, R.S., and Ellis, W.C. "Vapor–liquid–solid mechanism of single crystal growth." *Applied Physics Letters* 4 (1964):89–90.

Wagner, V., Dullaart, A., Bock, A.-K., and Zweck, A. "The emerging nanomedicine landscape." *Nature Biotechnology* 24 (2006):1211–7.

Wang, Y., Wei, F., Luo, G., Yu, H., and Gu, G. "The large-scale production of carbon nanotubes in a nano-agglomerate fluidized-bed reactor." *Chemical Physics Letters* 364 (2002):568–72.

Welsher, K., Liu, Z., Daranciang, D., and Dai, H. "Selective probing and imaging of cells with single walled carbon nanotubes as near-infrared fluorescent molecules." *Nano Letters* 8 (2008):586–90.

White, A.A., Kinloch, I.A., Windle, A.H., and Best, S.M. "Optimization of the sintering atmosphere for high-density hydroxyapatite–carbon nanotube composites." *Journal of the Royal Society Interface* 7 (2010):S529–S39.

Wick, P., Manser, P., Limbach, L.K. et al. "The degree and kind of agglomeration affect carbon nanotube cytotoxicity." *Toxicology Letters* 168 (2007):121–31.

Woo, D.G., Shim, M.-S., Park, J.S. et al. "The effect of electrical stimulation on the differentiation of hESCs adhered onto fibronectin-coated gold nanoparticles." *Biomaterials* 30 (2009):5631–8.

Wu, Z., Chen, Z., Du, X. et al. "Transparent, conductive carbon nanotube films." *Science* 305 (2004):1273–6.

Wu, W., Wieckowski, S., Pastorin, G. et al. "Targeted delivery of amphotericin B to cells by using functionalized carbon nanotubes." *Angewandte Chemie International Edition* 44 (2005):6358–62.

Xiaoming, L., Hong, G., Motohiro, U. et al. "Maturation of osteoblast-like SaoS2 induced by carbon nanotubes." *Biomedical Materials* 4 (2009):015005.

Yasuda, I. "On the piezoelectric activity of bone." *Journal of Japanese Orthopedic Surgery Society* 28 (1954):267–71.

Yun, Y., Dong, Z., Shanov, V. et al. "Nanotube electrodes and biosensors." *Nano Today* 2 (2007):30–7.

Zayzafoon, M. "Calcium/calmodulin signaling controls osteoblast growth and differentiation." *Journal of Cellular Biochemistry* 97 (2006):56–70.

Zhan, G.-D., Kuntz, J.D., Wan, J., and Mukherjee, A.K. "Single-wall carbon nanotubes as attractive toughening agents in alumina-based nanocomposites." *Nature Materials* 2 (2002):38–42.

Zhang, L., and Webster, T.J. "Nanotechnology and nanomaterials: Promises for improved tissue regeneration." *Nano Today* 4 (2009):66–80.

Zhao, L., and Gao, L. "Novel in situ synthesis of MWNTs–hydroxyapatite composites." *Carbon* 42 (2004):423–6.

Zhao, B., Futaba, D.N., Yasuda, S. et al. "Exploring advantages of diverse carbon nanotube forests with tailored structures synthesized by supergrowth from engineered catalysts." *ACS Nano* 3 (2008a):108–14.

Zhao, Y., Xing, G., and Chai, Z. "Nanotoxicology: Are carbon nanotubes safe?" *Nature Nanotechnology* 3 (2008b):191–2.

Zhu, W., Bower, C., Zhou, O., Kochanski, G., and Jin, S. "Large current density from carbon nanotube field emitters." *Applied Physics Letters* 75 (1999):873–5.

20

Ultrashort Carbon Nanocapsules for Biomedicine

Stuart J. Corr

Steven A. Curley

Lon J. Wilson

20.1 US-SWNTs as High-Performance MRI Contrast Agents and pH Probes

Carbon-based nanomaterials such as single-walled carbon nanotubes (SWNTs), graphene, C60 buckministerfullerene, and graphene nanoribbons have received considerable attention over the last two decades. For SWNTs, because of their unique physical attributes such as electronic configuration (Strano et al. 2003), high aspect ratio (Dresselhaus, Dresselhaus, and Saito 1995), mechanical properties (Dresselhaus et al. 2004), and ease of functionalization, they have become an attractive candidate in the field of nanomedicine, especially in such applications as drug delivery vectors (Klumpp et al. 2006; Prato, Kostarelos, and Bianco 2008), noninvasive radiofrequency (RF) hyperthermia cancer therapy (Gannon et al. 2007), and contrast agents (CAs) for diagnostic imaging modalities.

SWNTs of length 2–10 μm can easily be chemically "cut" into smaller sections (20–80 nm length) by fluorination followed by pyrolysis at 1000°C under an inert atmosphere (Gu et al. 2002). We have shown these SWNTs, termed *ultrashort SWNTs* (US-SWNTs), to be T_2-weighted magnetic resonance imaging (MRI) CAs owing to their inherent superparamagnetic nature (Ananta et al. 2009), high-performance T_1-weighted MRI CAs when internally loaded with Gd^{3+} ions (Sitharaman et al. 2005), x-ray CAs when internally filled with molecular iodine (Ashcroft et al. 2007), and R-radiotherapeutic agents when internally doped with 211AtCl molecules (Hartman et al. 2007).

These US-SWNTs have also recently been explored in our laboratories as nanocapsules for enhanced chemotherapeutic drug delivery when loaded with complexes such as *cis*-diamminedichloroplatinum(II)

(CDDP, cisplatin). Further, we were able to successfully remotely "trigger" the release of this cargo using an externally applied, noninvasive RF electric field, which paves the way for using this technology in targeted cancer therapy.

Finally, our most recent work illustrated the prominence of using loaded US-SWNTs as agents for magnetically enhanced stem cell retention in applications such as cellular cardiomyoplasty (Tran et al. 2010, 2014). In this technique, the induced magnetic moment of stem cells labeled with gadolinium-loaded US-SWNTs was exploited to allow the cells to be held in place when used alongside a permanent magnet. These stem cells could be held in place long enough for differentiation to occur without damaging or altering the differentiated cell phonotype or ability to reproduce.

Throughout this chapter, we aim to highlight several of the biomedical applications of US-SWNTs as mentioned previously. These include the use of US-SWNTs as a medium for enhanced MRI CA design and pH probes, the structural origin of MRI enhancements, their use as chemotherapeutic remotely triggered delivery vectors, and finally, their applications in stem cell labeling and enhanced retention.

20.1.1 MRI Principles

MRI is one of the most powerful, noninvasive diagnostic imaging modalities in medicine and biomedical research owing to its superior resolution and its ability to provide in-depth anatomical details in the early diagnosis of many diseases. It relies on the bulk magnetic properties arising from the combined nuclear spins of protons from water molecules found in biological tissue. These nuclear spins align and create a net magnetization vector (NMV) when exposed to a large external magnetic field (B_0) in the z-direction (the alignment of the nuclear spins is parallel to B_0). The NMV precesses along the applied net magnetic field vector or the longitudinal plane at a specific Larmor frequency given by $\omega = \gamma B_0$, where γ is the gyromagnetic ratio, which is a characteristic of the spinning proton. When an RF electric field pulse is applied, perpendicular to the B_0 field, the excess protons aligned in the z-direction absorb this energy, changing their direction and spinning into the transverse xy-plane. The absorbed RF energy is released once the pulse is turned off and the spins return to their initial state, by recovery in the longitudinal z-direction and decay in the transverse xy-direction. These relaxation processes are analyzed using coiled wire detectors in the MRI machine and are termed T_1 for longitudinal recovery and T_2 for transverse decay. These times are sensitive to the local environment and water content of various tissues and organs, which in turn gives rise to variations in contrast between areas of an image. This contrast difference can be further increased through the use of CAs that decrease the relaxation times of these processes and result in either brightening (T_1 weighted) or darkening (T_2 weighted) of the MRI signal.

Chemical CAs are widely used to improve the sensitivity, interpretation, and diagnostic ability in MRI procedures. For example, in 2007, up to 45% of performed MRI procedures in the United States integrated the use of CAs. CA design typically includes either paramagnetic or superparamagnetic species, which have T_1-weighted and T_2-weighted imaging properties, respectively. Paramagnetic elements possess a positively small magnetic susceptibility in the range of 10^{-6}–10^{-1} and are slightly attracted to a magnetic field. Magnetic susceptibility (χ) is the ratio of magnetization (M) in response to an applied magnetic field (H). Superparamagnetic materials, on the other hand, have large magnetic moments and susceptibilities but do not retain magnetism in the absence of an applied magnetic field. This ability to "turn on and off" their magnetic nature has led them to be explored for biological applications. Perhaps the most widely used superparamagnetic material for MRI is dextran-coated iron-oxide nanoparticles, sold under the Food and Drug Administration (FDA)-approved Feridex I.V. (Advanced Magnetics). This T_2-weighted CA has found extensive use for the detection of cancerous liver lesions (Wang 2011).

Other clinical MRI CAs will usually try to employ T_1-weighted chemical species in their design as these are favored over T_2-weighted CAs as they produce a brightening effect on the MRI signal. A popular choice is the use of the gadolinium ion Gd^{3+}, which has seven unpaired electrons and a symmetric electronic ground state. Most CAs that use Gd^{3+} are chelating agents, which afford protection to cells from the toxic nature of the Gd^{3+} ion. Such an example of an FDA-approved paramagnetic CA is

Magnevist (Bayer Healthcare Pharmaceuticals, Inc.). Other types of paramagnetic species and nanoparticles have also been investigated, such as manganese oxide nanoparticles (Lee et al. 2014; Omid et al. 2014) and dysprosium oxide nanoparticles (Lee et al. 2014; Rosenberg et al. 2014).

In the first section of this chapter, we will review the use of US-SWNTs and gadolinium-loaded US-SWNTs (termed *gadonanotubes* [GNTs]) as high-performance T_2-weighted and T_1-weighted MRI CAs, respectively.

20.1.2 US-SWNTs: High-Performance T_2-Weighted MRI CAs

We have previously reported that both raw and purified full-length high-pressure carbon monoxide (HiPco) SWNTs are remarkable high-performance T_2-weighted MRI CAs owing to their inherent superparamagnetic nature (Ananta et al. 2009). Furthermore, US-SWNTs demonstrated superiority to their full-length counterparts, with a T_2 relaxation efficiency of 31.7 ms per mg SWNT at 1.41 T and 37°C, compared with 4.6 and 5.1 for raw SWNTs (r-SWNTs) and purified SWNTs (p-SWNTs). These properties make US-SWNTs important candidates for next-generation molecular and cellular MRI agents.

In our previous studies (Ananta et al. 2009), as-obtained r-SWNTs were purified using a liquid bromine (Br_2) technique that removes the residual iron catalyst (~17% per weight) without causing significant structural damage to the tube sidewalls (Mackeyev et al. 2007). This process reduces the iron concentration to ~6%, the remainder of which is due to iron catalyst particles trapped inside the SWNTs. The p-SWNTS are then chemically cut into US-SWNTs using a fluorination process. In brief, p-SWNTs are fluorinated in a custom-made fluorination system that introduces C–F bonds on the outer SWNT surface. Pyrolysis of these fluorinated SWNTs in an inert atmosphere at 1000°C drives off the C–F bonds in the form of volatile COF_2 and CF_4 and leaves behind chemically "cut" US-SWNTs approximately 20–80 nm in length. This procedure is highlighted in Figure 20.1.

MRI parameters, such as T_2 proton relaxation values, were obtained on a bench-top Bruker Minispec mq60 relaxometer (1.41 T, Bruker Optics, Inc.). Phantom experiments were also performed on a 3-T MRI system (General Electric), with images obtained using two-dimensional spin-echo imaging with a repetition time of 500 ms and echo times ranging from 10 to 50 ms in either 5- or 10-ms increments.

The relaxation times (T_1, T_2) for the three SWNT species as well as their relaxivities (r_1, r_2) are shown in Table 20.1. The relaxivities (mM^{-1} s^{-1}) are synonymous with the efficacy of a CA and are a function of the CA concentration (CA, mM) and are calculated using the expression $r_{1,2} = R_{1,2} - R_0/[CA]$, where $R_{1,2}$ and R_0 are the relaxation rate constants (s^{-1}) of the sample and dispersion medium, respectively.

As can be seen in Table 20.1, all three SWNT types show very short T_2 relaxation times. The efficacy of the SWNTs (r_2) is also exceptionally high, especially for the US-SWNTs ($r_2 > 192$ mM^{-1} s^{-1}), when compared with the clinically utilized Ferumoxtran-10 complex ($r_2 = 65$ mM^{-1} s^{-1}). Also, efficacy enhancement is not proportional to residual iron catalyst content, which is counterintuitive to the assumed role of catalyst particles in the SWNT relaxation properties (Al Faraj et al. 2009; Choi et al. 2007). Also, the large r_2/r_1 ratio suggests that US-SWNTs are an ideal candidate for specialized molecular imaging probes as T_2-shortening CAs are expected to have a more pronounced effect on the T_2 relaxation time than on the T_1 relaxation times. The SWNT types were also analyzed in a 3-T MRI scanner, the data of which are shown in Table 20.1. Similar to the 1.41-T studies, all SWNT types showed enhanced T_2 relaxation properties, with the US-tubes demonstrating superior qualities. Although purified and empty

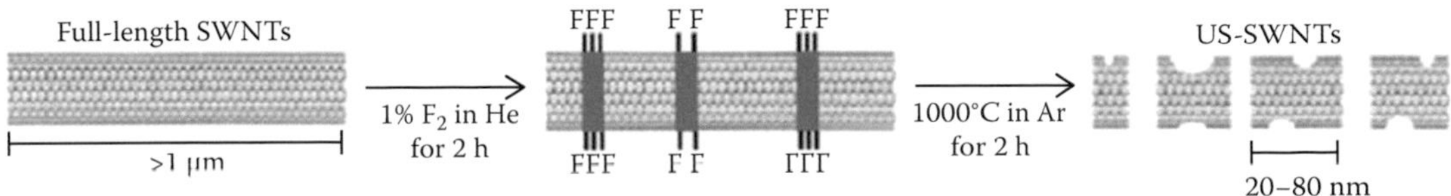

FIGURE 20.1 Synthesis of US-SWNTs from as-purchased full-length SWNTs.

TABLE 20.1 Relaxation Parameters of Aqueous SWNT Solutions at 37°C (Dispersed Using Pluronic F108 Surfactant) Using a Bruker Minispec mq60 (1.41 T) and GE MRI (3 T)

SWNT Material	Fe (mM)	T_2 (ms)	r_2 Relaxivity ($mM^{-1} s^{-1}$)[a]	T_2 (ms mg^{-1})[b]
		Minispec (1.41 T)		
r-SWNT	0.449	34.6	63.6	4.6
p-SWNT	0.167	46.5	126.8	5.1
US-SWNTs	0.075	67.7	192.5	31.7
Ferumoxtran-10			65	
		MRI (3 T)		
r-SWNT	1.11	13.6	65	
p-SWNT	0.4	15.1	166	
US-SWNTs	0.04	94.1	230	

Source: Reprinted with permission from Ananta, J. S., M. L. Matson, A. M. Tang, T. Mandal, S. Lin, K. Wong, S. T. Wong, and L. J. Wilson, Single-walled carbon nanotube materials as T2 weighted MRI contrast agents, *The Journal of Physical Chemistry C* 113 (45):19369–19372. doi:10.1021/jp907891n. Copyright 2009 American Chemical Society.

[a] Based on the Fe concentration.

[b] Based on the SWNT concentration.

US-SWNTs can be used as a T_2-weighted MRI CA, which will darken the contrast for the image, as previously mentioned, it is more desirable to use T_1-weighted MRI CAs, which will increase the brightness of the biological sample.

20.1.3 GNTs: High-Performance MRI CAs and pH Probes

The first reported case of nanoscale loading and confinement of Gd^{3+} ion clusters within US-SWNTs was by Sitharaman et al. in 2005 (Sitharaman et al. 2005). In their seminal work, it was shown that these Gd^{3+}-loaded US-SWNTs, aka GNTs, were linear paramagnetic molecular agents with MRI efficacies 40 to 90 times larger than any currently used clinical Gd^{3+}-based CA.

The synthesis of GNTs is a relatively facile process. As-received SWNTs (manufactured using the arc discharge [Journet et al. 1997] or HiPco technique [Bronikowski et al. 2001], for example) were cut into US-SWNTs via the fluorination process described in the previous section. Loading of Gd^{3+} ion clusters occurs by soaking and sonicating US-SWNTs in high-performance liquid chromatography (HPLC) grade water containing aqueous $GdCl_3$ at normal physiological pH (pH 7). Gd^{3+} ion loading occurs through the sidewall defects or end-of-tube openings created by the fluorination process. It was estimated that each Gd^{3+} ion cluster contains fewer than 10 Gd^{3+} ions. With regard to the crystalline nature of the Gd^{3+} ion clusters, x-ray diffraction analysis did not reveal any Gd^{3+} ion crystal lattice, the absence of which was thought to be attributed to the small cluster size (1 nm × 2–5 nm), low gadolinium content (2.84% [m/m] determined from inductively coupled plasma optical emission spectroscopy [ICP-OES]), and/or the amorphous nature of the hydrated Gd^{3+} ion clusters with their accompanying Cl_2 counter ions.

There, synthesized GNTs were analyzed using the same techniques described in the previous section (i.e., Minispec and MRI), and it was shown that the GNTs significantly reduced the relaxation rates relative to unloaded US-SWNTs or pure solutions (with only the surfactant). In fact, the relaxivity obtained for the GNTs was calculated to be $r_1 \sim 170\ mM^{-1}\ s^{-1}$, which is nearly 40 times greater than any current Gd^{3+} oral-based MRI CA such as $[Gd(DTPA)(H_2O)]^{2-}$ with $r_1 \sim 4\ mM^{-1}\ s^{-1}$.

We have also reported on the ability of GNTs to act as ultrasensitive pH-smart probes for MRI as a clinical tool for cancer diagnosis (Hartman et al. 2008). While the normal physiological conditions of tissue homeostasis operate around pH 7.4, the extracellular microenvironment of cancerous regions is usually less than pH 7.0 and in some instances can fall as low as pH 6.3. This increase in acidity is a result of the elevated rates of glycolysis, which in turn produce higher levels of lactate acid. By developing MRI

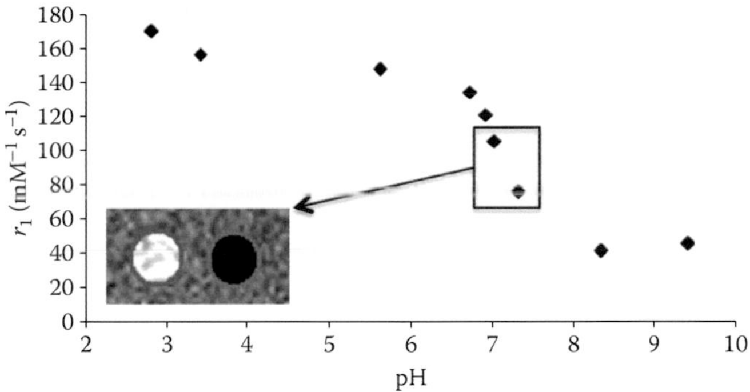

FIGURE 20.2 r_1 relaxivity (per Gd^{3+} ion) as a function of pH for the GNTs at 1.41 T and 37°C (data taken using a Bruker mq60 Minispec). T_1-weighted inversion-recovery scan on two GNT samples at different pH (7.0 inset left and 7.4 inset right) using a 1.5-T Philips magnetic resonance scanner. (Reprinted with permission from Hartman, K. B., S. Laus, R. D. Bolskar, R. Muthupillai, L. Helm, E. Toth, A. E. Merbach, and L. J. Wilson. Gadonanotubes as ultrasensitive pH-smart probes for magnetic resonance imaging, *Nano Letters* 8 (2):415–419. doi:10.1021/nl0720408. Copyright 2008 American Chemical Society.)

TABLE 20.2 Sensitivity (mM^{-1} s^{-1}/pH) of GNTs Compared with Other MRI pH Probe Methods across Various pH Ranges

pH Probe	pH Range	Slope (mM^{-1} s^{-1}/pH)
GNTs	7.0–7.4	98
Pendant arm ligation of Gd^{3+}	6.5–8.5	1.3
Gd^{3+}-DO3A tetrapod system	6.5–8.5	1.4

probes that are pH sensitive, it is thought that the early onset of cancer, as well as small metastatic cancer sites, can be detected using MRI.

In our study (Hartman et al. 2008), the relaxivities (per Gd^{3+} ion) of the GNTs were acquired as a function of pH ranging from 3 to 10 at 1.4 T (37°C). These results are highlighted in Figure 20.2. As can be seen, across the pH range of 8.3–6.7 the relaxivity depicts a threefold increase in pH (40 mM^{-1} s^{-1} to 133 mM^{-1} s^{-1}, respectively). Further, the change in pH across the range 7.4 and 7.0 can be described by the slope 98 mM^{-1} s^{-1}/pH. These results are compared with other MRI pH probes in Table 20.2. These data show that the GNTs outperform other pH-based probe methods, such as a pH-responsive pendant arm that ligates Gd^{3+} ions at high pH and releases them at low pH (Woods et al. 2004), as well as a Gd^{3+}-DO3A tetrapod system (Jebasingh and Alexander 2005). GNTs were also imaged using a T_1-weighted inversion-recovery scan on a 1.5-T Philips magnetic resonance scanner as there can be discrepancies between relaxivity characteristics acquired using nonimaging techniques (such as the Bruker Minispec mq60) and their actual performance in an MRI scanner. As can be seen in the inset of Figure 20.2, there is a marked difference in contrast between the GNT sample at pH 7.4 and 7.0 (inset right and left, respectively), whereby the acquired relaxivities were found to be 200 mM^{-1} s^{-1} and 98 mM^{-1} s^{-1}, respectively. Finally, the aggregation state of the GNTs were shown to not play an important role in relaxivity characteristics, such as those shown for gadofullerene MRI CAs (Sitharaman et al. 2004) (Gd^{3+} internalized into C60 fullerene).

20.1.4 Structural Origin of MRI Enhancement in GNTs

The unpaired electron spins of Gd^{3+} ions interact with the nuclear spins of protons in water molecules to effectively reduce the relaxation time of the NMV back to its initial position. These T_1-weighted longitudinal processes can be described, in part, by the longitudinal relaxivity parameter (r_1), which describes the efficacy with which a CA enhances the relaxation of neighboring water protons after the initial RF pulse has been removed.

There are several important factors that influence this enhancement. These include (1) hydration number (*Q*): the number of fast-exchanging water molecules directly coordinated to the paramagnetic ion; (2) water proton residence lifetime (τ_m) of the direct, ion-coordinated water protons; (3) the rotational correlation time (aka tumbling time) of the water–ion complex (τ_R); and (4) the separation distance (r_{GdH}) between the water protons and the paramagnetic ion.

Much research has recently been done to understand the enhanced relaxivities (r_1) associated with GNTs as current, prevailing theory cannot account for them. This work also extends the results by Ananta et al. (2010), who demonstrated a markedly high r_1 enhancement when GNTs were loaded into hollow pores of porous silicon microparticles. We reported on the use of x-ray absorption spectroscopy to reveal new information regarding the structural origin of the enhancement effects seen in GNTs (Ma et al. 2013a,b). In this work, Gd^{3+} ions were shown to be coordinated by approximately nine oxygen ions, resulting in potentially high hydration numbers (*Q*). As evidenced by the lack of detectable Gd–C interactions and small Gd–O_9 sites, the local environment of the GNT complex was deemed favorable for water protons to be relaxed by more than one Gd^{3+} ion simultaneously. Also, the measured Gd–H bond distances (i.e., r_{GdH}) were found to be much shorter than that of clinically used CAs. This distance was deemed to be a major factor in the enhancement effect as r_1 scales with $1/(r_{GdH})^6$. These results were the first of their kind to clearly demonstrate that the enhancement effects are made up of a plethora of structural contributions.

20.2 US-SWNTs as Chemotherapeutic-Based Delivery Vectors and Stem Cell Nanolabels

20.2.1 Cisplatin-Loaded US-SWNTs for Enhanced Cancer Chemotherapeutic Delivery

One major limitation that needs to be overcome in the field of cancer chemotherapies is the inherent toxic nature of chemotherapeutic drugs to normal, healthy cells. Although there are a multitude of drug platforms available to treat a variety of both metastasized and localized, primary cancers, many of them can induce undesirable side effects such as nausea, vomiting, hair loss, fatigue, infection, and anemia. The efficacy of many drugs is attenuated because of poor solubility and biodistribution, rapid elimination, inability to cross cellular barriers (such as the blood–brain barrier), and the inability to differentiate between healthy and cancerous cells.

A common chemotherapy drug for the treatment of cancers (i.e., testicular, cervical, ovarian, head, neck, and lung cancers) is *cis*-diamminedichloroplatinum(II) (cisplatin, CDDP). Cisplatin (one of the first platinum-based chemotherapy drugs) complexes trigger cellular necrosis by binding to and cross-linking the cellular DNA. However, there are several severe side effects, which limit the use of CDDP in the clinic. These include ototoxicity (hearing loss), nephrotoxicity (kidney damage), neurotoxicity (nerve damage), nausea, and vomiting. Hence, a method to increase CDDP concentrations in the tumor cells while regulating and minimizing the toxicity toward normal cells would be of substantial benefit.

We have previously reported on the synthesis of cisplatin-loaded US-SWNTs (CDDP@US-SWNTs) as a novel nanocapsule for enhanced chemotherapeutic delivery (Guven et al. 2012). The release of CDDP can be controlled by wrapping the US-SWNTs in an amphiphilic surfactant (Pluronic F108). Further, the *in vitro* assessment of the toxicity of CDDP@US-SWNTs to CDDP-resistant breast cancer cell lines was evaluated and shown to be enhanced for CDDP@US-SWNTs when compared with free CDDP. These nanocapsules are thus thought to increase the levels of intercellular CDDP, effectively "dosing" the cancer cells with more CDDP. The synthesis procedure for creating CDDP@US-SWNTs is shown in Figure 20.3.

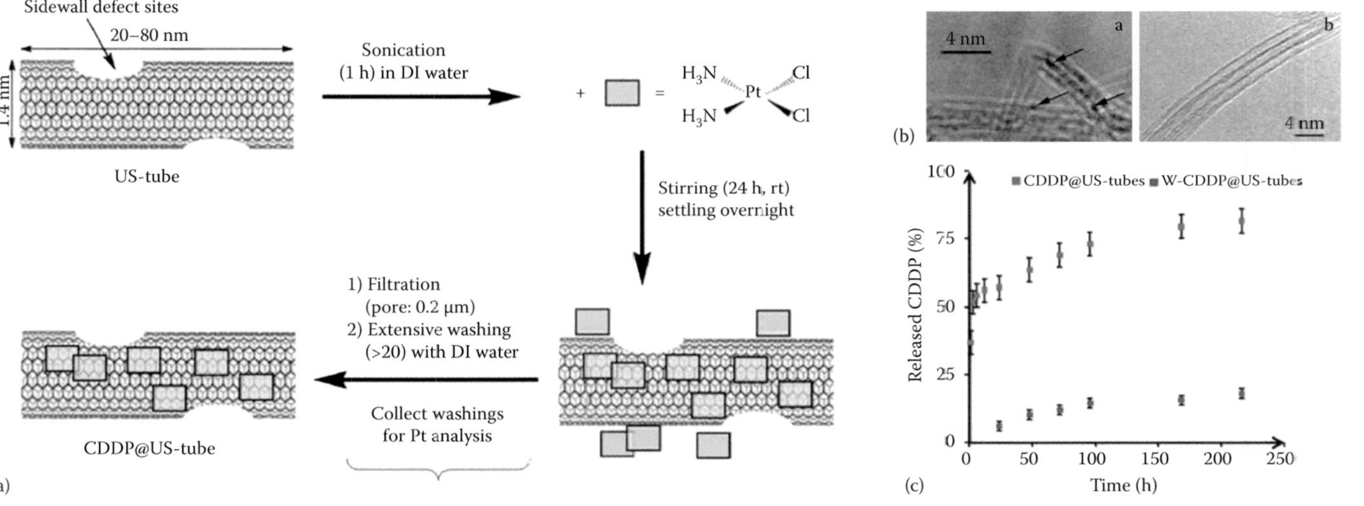

FIGURE 20.3 (a) Synthesis pathway for the creation of cisplatin-loaded ultrashort carbon nanotubes (CDDP@US-SWNTs). (b) TEM image of CDDP@US-SWNTs and empty US-SWNTs (a and b, respectively). (c) Time release study of CDDP@US-SWNTs and pluronic-wrapped W-CDDP@US-SWNTs. Released CDDP was characterized by analyzing platinum (Pt) via ICP-OES. (Reprinted from *Biomaterials*, 33, 5, Adem Guven, Irene A. Rusakova, Michael T. Lewis, and Lon J. Wilson, Cisplatin@US-tube carbon nanocapsules for enhanced chemotherapeutic delivery, 1455–1461, Copyright 2012, with permission from Elsevier.)

As can be seen in Figure 20.3, the protocol for synthesizing CDDP@US-SWNTs consists of sonicating US-SWNTs, which have previously been chemically cut using the fluorination process described in the earlier sections, and then stirring them alongside CDDP overnight at room temperature. Filtration through a 0.2-μm pore size and extensive washing with deionized water removes all residual CDDP from the outer tube surface. Finally, the addition of 0.17% Pluronic F108 surfactant passivates CDDP release. Figure 20.3b shows a transmission electron microscopy (TEM) image of CDDP preferentially loaded into the internal cavities of the US-SWNTs; a clear distinction of contrast can be seen between the loaded and empty US-SWNTs. Also shown in Figure 20.3 is the time-release characteristics of CDDP@US-SWNTs with and without addition of 0.17% Pluronic surfactant (i.e., "wrapped" CDDP-US-SWNTs). Release kinetics of the tubes was characterized by performing ICP-OES for platinum (Pt) analysis in the media (phosphate-buffered saline [PBS] at 37°C) in which the tubes were suspended. As can be seen, the wrapped CDDP@US-SWNTs leaked only 6.1% CDDP in 24 h and 16.9% in a 1-wk period compared with 50% in the first 3 h and 80% after 1 wk for the unwrapped CDDP@US-SWNTs.

The well-established and commonly used MTT (3-(4,5-dimethylthiazol-2-yl)-2,5-diphenyltetrazolium bromide) cell proliferation assay was used to evaluate the concentration-dependent (0–50 μM) cytotoxicity of these wrapped CDDP@US-SWNTs to CDDP-resistant breast cancer cell lines (MCF-7 and MDA-MB-231) at two time points: 24 h and 48 h. The empty US-SWNTs were shown to be relatively nontoxic, maintaining viability profiles >80% across both cell lines and time points. The wrapped CDDP@US-SWNTs, however, generally resulted in a 15%–20% decrease in cell viability for both cell lines and time points, except for the MDA-MB-231 cells at 48 h, where the viabilities were shown to be approximately the same.

The conclusions drawn from this seminal piece of work were that the US-SWNTs apparently assist in the delivery of free CDDP into the cancer cells from the wrapped CDDP@US-SWNTs in the first 24 h. Further, the increase in toxicity for the wrapped CDDP@US-SWNT when compared with free CDDP indicates that these nanocapsules can help overcome CDDP resistance by increasing drug accumulation in resistant cells.

20.2.2 RF-Triggered Drug Release from Cisplatin-Loaded US-SWNTs

Noninvasive RF cancer hyperthermia is currently under development in our laboratories as a novel technique for the treatment of a variety of cancers (Glazer et al. 2010; Raoof et al. 2012, 2014b). It works on the principle of the interactions between RF electric fields (13.56 MHz) and cancerous tissues, whereby a temperature increase will be evident because of the nonzero dielectric properties of tissues, organs, and cancer. Recent work has shown that heating is tumor specific owing to the larger dielectric constants of cancer compared with normal, healthy tissue (Raoof et al. 2013a). It is hypothesized that the long wavelength properties of these radiowaves (~22 m) can allow for full-body penetration, resulting in increased energy deposition into deep-seated tumors. A system schematic is shown in Figure 20.4. It is also hypothesized that the use of targeted, metallic, and semiconducting nanomaterials such as gold nanoparticles, quantum dots, and SWNTs may enhance this heating effect owing to their nanoscale interactions with RF electric fields (Corr et al. 2012; Gannon et al. 2007; Glazer and Curley 2011; Raoof et al. 2014a).

In a bid to expand this field further, we recently reported on remotely triggered cisplatin release from US-SWNTs by RF electric fields (Raoof et al. 2013b). This approach was completely novel in that the use of RF fields induced a temperature-dependent phase change in the surfactant (Pluronic F108) that wrapped around the CDDP@US-SWNTs, resulting in an active and controlled release of the cisplatin cargo. These proof-of-principle experiments can allow for controlled, targeted, and remotely triggered release of cytotoxic payloads, which can ultimately alleviate the harmful side effects induced by toxic cancer drugs such as CDDP. The concept is shown in Figure 20.5.

In these studies, Pluronic-wrapped CDDP@US-SWNTs were prepared as described in the previous section. To evaluate their RF heating properties, wrapped CDDP@US-SWNTs were suspended in a PBS suspension and exposed to the RF heating system with their heating properties recorded using an

FIGURE 20.4 RF hyperthermia system. (a) An amplifier unit generates a 13.56-MHz RF signal of variable power (0–2 kW) defined by the user. (b) This power load drives a capacitively coupled transmitting (Tx) and receiving head (Rx) configuration and creates a uniform electric field of strength ~90 kV/m (dependent on power). The patient is placed between the Tx and Rx heads and is exposed to the noninvasive and nonionizing electric fields.

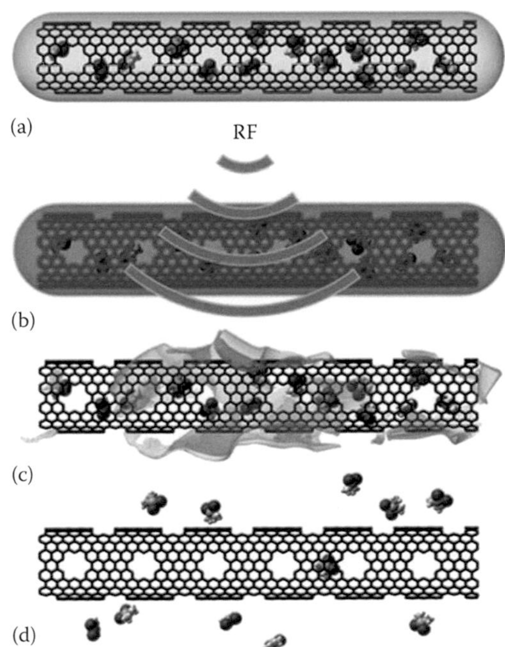

FIGURE 20.5 (a) Pluronic-F108-wrapped CDDP@US-SWNTs. (b) The interaction of RF electric fields with tubes causes a phase change in the Pluronic F108 and starts to unwrap (c) from the CDDP@US-SWNTs, releasing the toxic cisplatin cargo (d). (Reprinted from *Biomaterials*, 34, 7, M. Raoof, B. T. Cisneros, A. Guven, S. Phounsavath, S. J. Corr, L. J. Wilson, and S. A. Curley, Remotely triggered cisplatin release from carbon nanocapsules by radiofrequency fields, 1862–1869, Copyright 2013, with permission from Elsevier.)

infrared camera. No clear distinction was observed between the bulk heating properties of the suspension and the filtrate (i.e., the filtrate that was isolated by filter centrifugation of the tubes to remove them from the suspension). The authors stated that although there was no significant enhancement of the bulk temperature, localized RF-induced heating could not be ruled out. This led them to develop an indirect "thermometer" approach whereby they exploited the selective binding affinity of fluorescent proteins for SWNTs. Using the thermally stable fluorescent protein allophycocyanin (APC), which would freely fluoresce in the presence of wrapped CDDP@US-SWNTs as the surfactant layer prevents binding of APC to the SWNT surface, the authors mixed this protein in with the CDDP@US-SWNT suspension and exposed them to RF. By monitoring the fluorescence of the medium before and after RF exposure (at several time points), they were able to work out the localized temperature of the CDDP@US-SWNTs when exposed to RF electric fields by comparing with a control group (i.e., using a water bath as the heat source). As the melting point of Pluronic F108 is ~57.5°C, fluorescence quenching would be expected to occur around this temperature region, which was in fact evident from their studies. These results suggested that even though the bulk properties of the CDDP@US-SWNT suspension were not significantly enhanced, this did not reflect on what was happening on the localized, nano level, whereby temperature variations above 57.5°C were indirectly observed.

An application of this was to investigate the level of CDDP release as a function of RF exposure (CDDP release was characterized by analyzing platinum using ICP-OES) using water bath experiments as a control. Raoof et al. (2013b) demonstrated that RF electric field exposure approximately doubles the amount of CDDP release from wrapped CDDP@US-SWNTs after only 2.5 or 5 min of RF exposure, compared with equivalent durations in the control water bath experiments. These results were also propagated in *in vitro* experiments whereby liver cancer cells were exposed to RF-exposed CDDP@US-SWNT suspensions through a trans-well membrane system (i.e., the released CDDP would be transported through the trans-well pore so that only free CDDP was available to the cells). This system mimics the limitations of *in vivo* drug delivery systems in that the drug carriers are usually separated from the main tumor sites because of the multiscale barriers to effective transport.

In their work, Raoof et al. (2013b) found that the wrapped CDDP@US-SWNTs reduced the viability of both liver cancer cell lines (Hep3B and HepG2) to 50%, which was a result primarily of the spontaneous release of CDDP (i.e., not RF related). Further, this cytotoxicity was significantly enhanced in both cell lines by 3 min of RF electric field exposure. Although cytotoxicity could further be increased by multiple RF exposures in the Hep3B cell line, this effect was not observed in the HepG2 cell lines. Overall, their results clearly demonstrated the feasibility of using RF electric fields to remotely trigger CDDP release for enhanced and controlled cancer cytotoxicity.

20.2.3 GNTs as Magnetic Nanolabels for Stem Cell Detection

We recently reported on the ability for GNTs to be used as a T_1-weighted intracellular labeling agent for pig bone-marrow-derived mesenchymal stem cells (MSCs). Even without the use of a transfection agent, GNTs were found to deliver up to 10^9 Gd^{3+} ions per cell without negatively affecting cell viability, differentiation potential, proliferation pattern, and phenotype (Tran et al. 2010). These results indicated the high plausibility of using GNTs as an efficient and high-resolution imaging agent for *in vivo* stem cell tracing.

MRI-based tracking agents are thought to be an excellent candidate for stem cell detection owing to the inherently high-spatial-resolution nature of MRI, which is performed in a noninvasive, nonionizing fashion. This has led many laboratories to study the capabilities of MRI for stem cell imaging, with and without the use of nanoparticles. Because of the extremely high T_1-weighted relaxivity of GNTs and tendency to readily translocate through the cellular membrane in the absence of a transfection agent, GNTs are deemed superlative candidates for the intracellular labeling and monitoring of stem cells.

As can be seen in Figure 20.6, the optimal concentration for GNT MSC labeling was found to be 27 μM Gd^{3+}, which delivered ~10^9 Gd^{3+} ions (~0.98 pg) per cell in a period of around 4 h. Longer durations up to

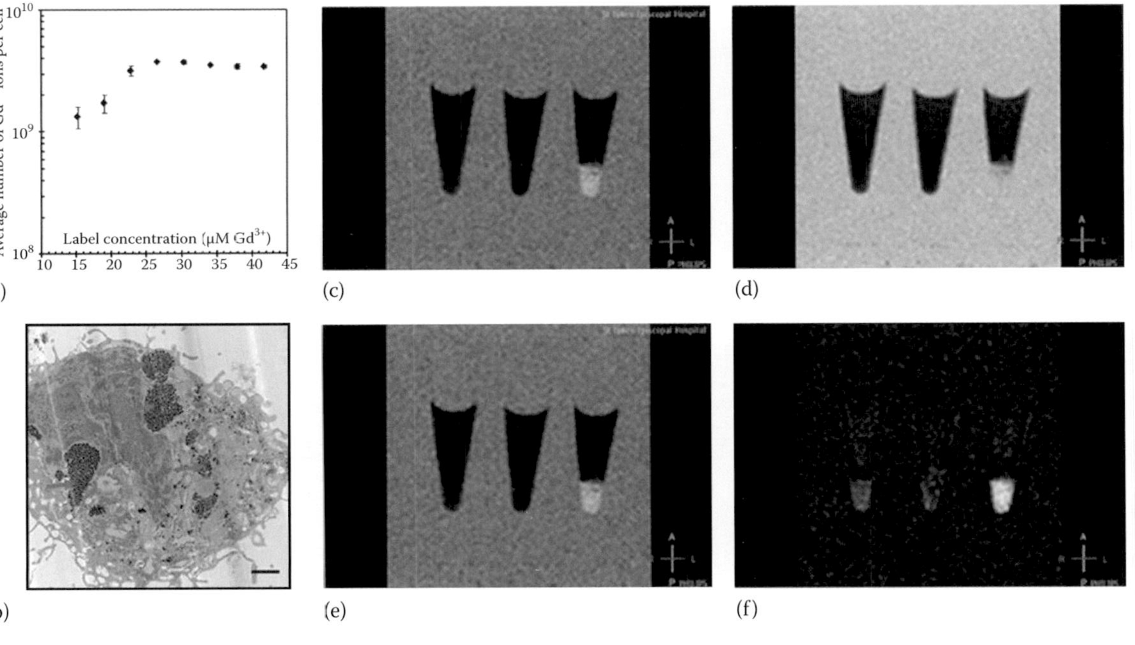

FIGURE 20.6 (a) Gd^{3+} uptake (through GNTs) in MSCs as a function of labeling concentration. (b) TEM images of GNT-labeled MSCs (GNTs were found to predominantly aggregate in the cytoplasm). (c–f) T_1-weighted MRI images (1.5 T at 25°C) of MSCs (left to right: unlabeled, Gd-DTPA-labeled, and GNT labeled) for different T_1 relaxivity times: (c) 150 ms, (d) 300 ms, (e) 500 ms, and (f) 800 ms. (Reprinted from *Biomaterials*, 31, 36, Gadonanotubes as magnetic nanolabels for stem cell detection, 9482–9491, Copyright 2010, with permission from Elsevier.)

24 h were not found to increase intracellular Gd^{3+} ion concentration. At these concentrations, the cells were fully viable. Uptake of the GNT imaging agents was confirmed via TEM image analysis, as shown in Figure 20.6b, and was depicted to predominantly accumulate within the intracellular cytoplasm, suggesting that an active form of endocytosis may be responsible for the uptake mechanism.

The self-renewal properties and proliferation kinetics of the labeled MSCs were also evaluated using colony-forming unit fibroblast and population doubling time assays. Overall, the results indicated that the labeling of MSCs with GNTs did not impair proliferation kinetics or self-renewal activities, although a 20% increase in growth rate was evidenced for the GNT-labeled MSCs across a 7-day period. The ability of MSCs to differentiate into a variety of cell types was also not affected by the GNTs. Results indicated that GNT-labeled cultures were successfully able to differentiate into adipocytes, osteocytes, and chondrocytes, which further suggests that GNT-labeled MSCs may retain their therapeutic potential for applications such as tissue regeneration.

Finally, direct MRI images were taken on Gd-DTPA-labeled, GNT-labeled, and unlabeled MSC cell pellets (10×10^6 MSCs/pellet) at various inversion delay times ($T_i = 150$, 300, 500, and 800 ms), as shown in Figure 20.6c. Overall, there was a substantial difference in contrast observed between the labeled and unlabeled pellets. The extracted T_1 relaxation times were shown to be 1079 ms, 495 ms, and 876 ms for the Gd-DTPA-labeled, GNT-labeled, and unlabeled MSC cell pellets, respectively. Given that there is nearly a twofold reduction in T_1 relaxation times for the GNT-labeled MSCs compared with the unlabeled MSCs, the effectiveness of GNTs to be used as a viable and efficient MRI CA for stem cell detection, even at low loading concentrations, is demonstrated.

20.2.4 GNT Applications in Cellular Cardiomyoplasty

The evaluation of GNTs as a vector to magnetically enhance the retention of transplanted MSCs during cellular cardiomyoplasty was also recently investigated (Tran et al. 2014). In this work, the overall paramagnetic moment of MSCs labeled with GNTs was exploited to enhance their retention when directly injected into cardiac tissue combined with magnets to "hold" them in place. Specifically, epicardial injections of GNT-labeled MSCs were administered around a 1.3 T NdFeB ring magnet sutured onto the left ventricle of female juvenile pigs (Figure 20.7). Compared with the diamagnetic lutetium (Lu) control group, the use of an external magnet enhanced the retention of GNT-labeled MSCs up to three times, with the magnet being tolerated up to 1 wk; an inflammatory response was observed, however, after 48 h. These proof-of-concept studies were invaluable in demonstrating the use of GNTs as a means to enhance cell retention in applications such as cellular cardiomyoplasty.

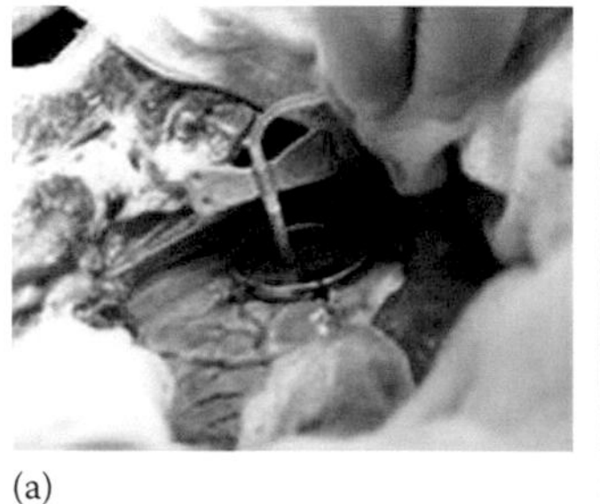
(a)

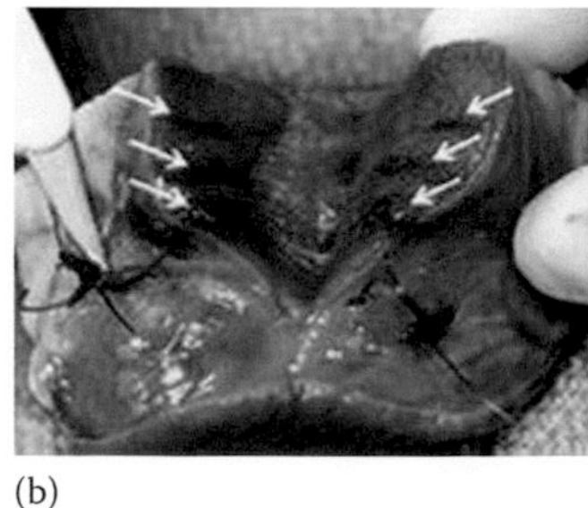
(b)

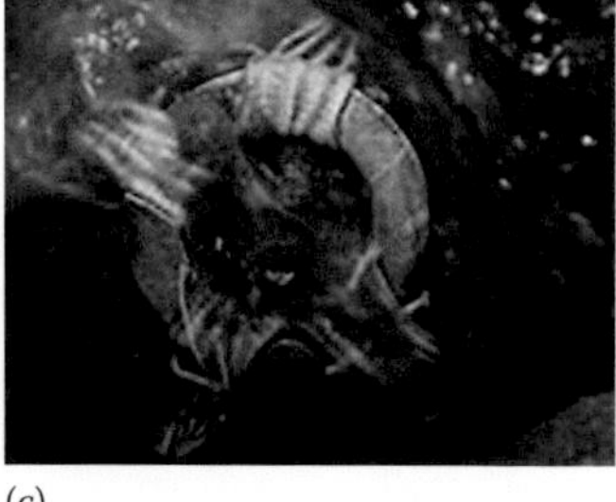
(c)

FIGURE 20.7 *In vivo* retention study: (a) A butterfly needle was used to inject GNT-labeled MSCs around a 1.3 T NdFeB ring magnet sutured into the left ventricular anterior wall. (b) The presence of the GNTs near the injection site (after 24 h) is indicated by the white arrows. (c) An inflammatory response by the heart against the sutured magnet was observed after 48 h. (Reprinted from *Biomaterials*, 35, 2, The use of gadolinium-carbon nanostructures to magnetically enhance stem cell retention for cellular cardiomyoplasty, 720–726, Copyright 2014, with permission from Elsevier.)

The retention, survival, and differentiation potential of transplanted cells into the heart is severely limited by muscle contraction and blood flow at the injections site. The very nature of the highly dynamic and powerful environment of the heart calls for superior technologies that can sufficiently retain the stem cells at the site of interest long enough for stem cell differentiation into cardiomyocytes to take place. Without this enhanced retention ability, the benefits afforded by current stem cell regeneration approaches are nullified.

Previous research has investigated the use of T_2-weighted superparamagnetic iron oxide nanoparticle-labeled MSCs, which are then embedded in grafts, stents, and cardiac tissue and localized near an external magnetic field for enhanced MSC retention. Although this is a viable and plausible method, whereby the MRI properties of the iron oxides nanoparticles can be exploited to monitor MSC retention and performance, the use of iron oxides' nanoparticles requires the use of polycationic transfection agents, many of which have been shown to adversely affect cellular differentiation, which in turn reduces the therapeutic potential of the stem cells. Further, the "brightening" effect of T_1-weighted MRI CAs such as GNTs is preferred to the T_2-weighted "darkening" effect, such as that observed from iron oxide nanoparticles.

The recent results of Tran et al. (2014) clearly illustrate a novel technology toward enhancing the retention of MSCs for cellular cardiomyoplasty. Although the use of a magnet directly sutured onto the heart can be considered a somewhat invasive technique, the use of minimally invasive magnetic catheters may offer an alternating platform for MSC retention.

20.3 Conclusion

US-SWNTs of length 20–80 nm and diameter 1.4 nm can be applied to a variety of biomedical applications. Foremost, empty US-SWNTs have a superparamagnetic nature arising from residual catalyst content from the manufacturing process (e.g., Fe, Y, Ni), which can be exploited in MRI as a T_2-weighted MRI CA, producing a darkening effect. Second, when loaded with a paramagnetic species such as the gadolinium Gd^{3+} ion that forms clusters at side-wall defect sites, they act as a T_1-weighted MRI CA, with relaxivities of ~170 mM^{-1} s^{-1}, which is approximately 40 times larger than currently available clinical CAs. These GNTs can also be effectively utilized as an ultrasensitive MRI pH smart-probe, possessing a large dynamic range in enhanced MRI contrast across a physiologically significant pH range, which can be used as a means to probe the acidic nature of the microenvironment of cancer. The origins of these MRI contrast enhancement effects were shown to be based on many structural origins, such as hydration number and ion–proton distance. In contrast, we also described the use of US-SWNTs as nanocapsules for enhanced delivery vectors of toxic chemotherapeutic drugs such as cisplatin. These nanocapsules could also be remotely triggered using external noninvasive RF electric fields to release their cargo in a controlled fashion. Finally, the use of GNTs as magnetic nanolabels for stem cell detection was discussed, as well as their application for enhanced retention of labeled MSCs for cellular cardiomyoplasty.

Acknowledgments

This work was funded by the NIH (U54CA143837), the NIH M.D. Anderson Cancer Center Support Grants (CA016672), the V Foundation (SAC), The Robert A. Welch Foundation (LJW; C-0627), and an unrestricted research grant from the Kanzius Research Foundation (SAC, Erie, PA, USA).

References

Al Faraj, A., K. Cieslar, G. Lacroix, S. Gaillard, E. Canet-Soulas, and Y. Crémillieux. 2009. "In vivo imaging of carbon nanotube biodistribution using magnetic resonance imaging." *Nano Letters* 9 (3):1023–1027. doi:10.1021/nl8032608.

Ananta, J. S., M. L. Matson, A. M. Tang, T. Mandal, S. Lin, K. Wong, S. T. Wong, and L. J. Wilson. 2009. "Single-walled carbon nanotube materials as T2-weighted MRI contrast agents." *The Journal of Physical Chemistry C* 113 (45):19369–19372. doi:10.1021/jp907891n.

Ananta, J. S., B. Godin, R. Sethi, L. Moriggi, X. Liu, R. E. Serda, R. Krishnamurthy et al. 2010. "Geometrical confinement of gadolinium-based contrast agents in nanoporous particles enhances T1 contrast." *Nature Nanotechnology* 5 (11):815–821. doi:http://www.nature.com/nnano/journal/v5/n11/abs/nnano.2010.203.html-supplementary-information.

Ashcroft, J. M., K. B. Hartman, K. R. Kissell, Y. Mackeyev, S. Pheasant, S. Young, P. A. W. Van der Heide, A. G. Mikos, and L. J. Wilson. 2007. "Single-molecule I2@US-tube nanocapsules: A new x-ray contrast-agent design." *Advanced Materials* 19 (4):573–576. doi:10.1002/adma.200601424.

Bronikowski, M. J., P. A. Willis, D. T. Colbert, K. A. Smith, and R. E. Smalley. 2001. "Gas-phase production of carbon single-walled nanotubes from carbon monoxide via the HiPco process: A parametric study." *Journal of Vacuum Science & Technology A* 19 (4):1800–1805. doi:http://dx.doi.org/10.1116/1.1380721.

Choi, J. H., F. T. Nguyen, P. W. Barone, D. A. Heller, A. E. Moll, D. Patel, S. A. Boppart, and M. S. Strano. 2007. "Multimodal biomedical imaging with asymmetric single-walled carbon nanotube/iron oxide nanoparticle complexes." *Nano Letters* 7 (4):861–867. doi:10.1021/nl062306v.

Corr, S. J., M. Raoof, Y. Mackeyev, S. Phounsavath, M. A. Cheney, B. T. Cisneros, M. Shur et al. "Citrate-capped gold nanoparticle electrophoretic heat production in response to a time-varying radio-frequency electric field." *The Journal of Physical Chemistry C* 116 (45):24380–24389. doi:10.1021/jp309053z.

Dresselhaus, M. S., G. Dresselhaus, and R. Saito. 1995. "Physics of carbon nanotubes." *Carbon* 33 (7):883–891. doi:http://dx.doi.org/10.1016/0008-6223(95)00017-8.

Dresselhaus, M. S., G. Dresselhaus, J. C. Charlier, and E. Hernández. 2004. "Electronic, thermal and mechanical properties of carbon nanotubes." *Philosophical Transactions of the Royal Society of London A: Mathematical, Physical and Engineering Sciences* 362 (1823):2065–2098.

Gannon, C. J., P. Cherukuri, B. I. Yakobson, L. Cognet, J. S. Kanzius, C. Kittrell, R. B. Weisman et al. 2007. "Carbon nanotube-enhanced thermal destruction of cancer cells in a noninvasive radiofrequency field." *Cancer* 110 (12):2654–2665. doi:10.1002/cncr.23155.

Glazer, E. S., and S. A. Curley. 2011. "Non-invasive radiofrequency ablation of malignancies mediated by quantum dots, gold nanoparticles and carbon nanotubes." *Therapeutic Delivery* 2 (10):1325–1330. doi:10.4155/tde.11.102.

Glazer, E. S., K. L. Massey, C. Zhu, and S. A. Curley. 2010. "Pancreatic carcinoma cells are susceptible to noninvasive radio frequency fields after treatment with targeted gold nanoparticles." *Surgery* 148 (2):319–324. doi:http://dx.doi.org/10.1016/j.surg.2010.04.025.

Gu, Z., H. Peng, R. H. Hauge, R. E. Smalley, and J. L. Margrave. 2002. "Cutting single-wall carbon nanotubes through fluorination." *Nano Letters* 2 (9):1009–1013. doi:10.1021/nl025675+.

Guven, A., I. A. Rusakova, M. T. Lewis, and L. J. Wilson. 2012. "Cisplatin@US-tube carbon nanocapsules for enhanced chemotherapeutic delivery." *Biomaterials* 33 (5):1455–1461. doi:http://dx.doi.org/10.1016/j.biomaterials.2011.10.060.

Hartman, K. B., D. K. Hamlin, D. S. Wilbur, and L. J. Wilson. 2007. "211AtCl@US-tube nanocapsules: A new concept in radiotherapeutic-agent design." *Small* 3 (9):1496–1499. doi:10.1002/smll.200700153.

Hartman, K. B., S. Laus, R. D. Bolskar, R. Muthupillai, L. Helm, E. Toth, A. E. Merbach, and L. J. Wilson. 2008. "Gadonanotubes as ultrasensitive pH-smart probes for magnetic resonance imaging." *Nano Letters* 8 (2):415–419. doi:10.1021/nl0720408.

Jebasingh, B., and V. Alexander. 2005. "Synthesis and relaxivity studies of a tetranuclear gadolinium(III) complex of DO3A as a contrast-enhancing agent for MRI." *Inorganic Chemistry* 44 (25):9434–9443. doi:10.1021/ic050743r.

Journet, C., W. K. Maser, P. Bernier, A. Loiseau, M. Lamy de la Chapelle, S. Lefrant, P. Deniard, R. Lee, and J. E. Fischer. 1997. "Large-scale production of single-walled carbon nanotubes by the electric-arc technique." *Nature* 388 (6644):756–758.

Klumpp, C., K. Kostarelos, M. Prato, and A. Bianco. 2006. "Functionalized carbon nanotubes as emerging nanovectors for the delivery of therapeutics." *Biochimica et Biophysica Acta (BBA)—Biomembranes* 1758 (3):404–412. doi:http://dx.doi.org/10.1016/j.bbamem.2005.10.008.

Lee, S. H., B. H. Kim, H. B. Na, and T. Hyeon. 2014. "Paramagnetic inorganic nanoparticles as T1MRI contrast agents." *Wiley Interdisciplinary Reviews: Nanomedicine and Nanobiotechnology* 6 (2):196–209. doi:10.1002/wnan.1243.

Ma, Q., M. Jebb, M. F. Tweedle, and L. J. Wilson. 2013a. "X-ray absorption spectroscopy study of Gd 3+-loaded ultra-short carbon nanotubes." *Journal of Physics: Conference Series* 430 (1):012085.

Ma, Q., M. Jebb, M. F. Tweedle, and L. J. Wilson. 2013b. "The gadonanotubes: Structural origin of their high-performance MRI contrast agent behavior." *Journal of Materials Chemistry B* 1 (42):5791–5797. doi:10.1039/C3TB20870B.

Mackeyev, Y., S. Bachilo, K. B. Hartman, and L. J. Wilson. 2007. "The purification of HiPco SWCNTs with liquid bromine at room temperature." *Carbon* 45 (5):1013–1017. doi:http://dx.doi.org/10.1016/j.carbon.2006.12.026.

Omid, H., M. A. Oghabian, R. Ahmadi, N. Shahbazi, H. R. M. Hosseini, S. Shanehsazzadeh, and R. N. Zangeneh. 2014. "Synthesizing and staining manganese oxide nanoparticles for cytotoxicity and cellular uptake investigation." *Biochimica et Biophysica Acta (BBA)—General Subjects* 1840 (1):428–433. doi:http://dx.doi.org/10.1016/j.bbagen.2013.10.001.

Prato, M., K. Kostarelos, and A. Bianco. 2008. "Functionalized carbon nanotubes in drug design and discovery." *Accounts of Chemical Research* 41 (1):60–68. doi:10.1021/ar700089b.

Raoof, M., S. J. Corr, W. D. Kaluarachchi, K. L. Massey, K. Briggs, C. Zhu, M. A. Cheney, L. J. Wilson, and S. A. Curley. 2012. "Stability of antibody-conjugated gold nanoparticles in the endolysosomal nanoenvironment: Implications for noninvasive radiofrequency-based cancer therapy." *Nanomedicine: Nanotechnology, Biology and Medicine* 8 (7):1096–1105. doi:http://dx.doi.org/10.1016/j.nano.2012.02.001.

Raoof, M., B. T. Cisneros, S. J. Corr, F. Palalon, S. A. Curley, and N. V. Koshkina. 2013a. "Tumor selective hyperthermia induced by short-wave capacitively-coupled RF electric-fields." *PLoS One* 8 (7):e68506. doi:10.1371/journal.pone.0068506.

Raoof, M., B. T. Cisneros, A. Guven, S. Phounsavath, S. J. Corr, L. J. Wilson, and S. A. Curley. 2013b. "Remotely triggered cisplatin release from carbon nanocapsules by radiofrequency fields." *Biomaterials* 34 (7):1862–1869. doi:http://dx.doi.org/10.1016/j.biomaterials.2012.11.033.

Raoof, M., S. J. Corr, C. Zhu, B. T. Cisneros, W. D. Kaluarachchi, S. Phounsavath, L. J. Wilson, and S. A. Curley. 2014a. "Gold nanoparticles and radiofrequency in experimental models for hepatocellular carcinoma." *Nanomedicine: Nanotechnology, Biology and Medicine* 10 (6):1121–1130. doi:http://dx.doi.org/10.1016/j.nano.2014.03.004.

Raoof, M., C. Zhu, B. T. Cisneros, H. Liu, S. J. Corr, L. J. Wilson, and S. A. Curley. 2014b. "Hyperthermia inhibits recombination repair of gemcitabine-stalled replication forks." *Journal of the National Cancer Institute* 106 (8):pii: dju183. doi:10.1093/jnci/dju183.

Rosenberg, J. T., B. T. Cisneros, M. Matson, M. Sokoll, A. Sachi-Kocher, F. C. Bejarano, L. J. Wilson, and S. C. Grant. 2014. "Encapsulated gadolinium and dysprosium ions within ultra-short carbon nanotubes for MR microscopy at 11.75 and 21.1 T." *Contrast Media & Molecular Imaging* 9 (1):92–99. doi:10.1002/cmmi.1542.

Sitharaman, B., R. D. Bolskar, I. Rusakova, and L. J. Wilson. 2004. "Gd@C60[C(COOH)2]10 and Gd@C60(OH)x: Nanoscale aggregation studies of two metallofullerene MRI contrast agents in aqueous solution." *Nano Letters* 4 (12):2373–2378. doi:10.1021/nl0485713.

Sitharaman, B., K. R. Kissell, K. B. Hartman, L. A. Tran, A. Baikalov, I. Rusakova, Y. Sun et al. 2005. "Superparamagnetic gadonanotubes are high-performance MRI contrast agents." *Chemical Communications* (31):3915–3917. doi:10.1039/B504435A.

Strano, M. S., C. A. Dyke, M. L. Usrey, P. W. Barone, M. J. Allen, H. Shan, C. Kittrell, R. H. Hauge, J. M. Tour, and R. E. Smalley. 2003. "Electronic structure control of single-walled carbon nanotube functionalization." *Science* 301 (5639):1519–1522. doi:10.1126/science.1087691.

Tran, L. A., R. Krishnamurthy, R. Muthupillai, M. da Graça Cabreira-Hansen, J. T. Willerson, E. C. Perin, and L. J. Wilson. 2010. "Gadonanotubes as magnetic nanolabels for stem cell detection." *Biomaterials* 31 (36):9482–9491. doi:http://dx.doi.org/10.1016/j.biomaterials.2010.08.034.

Tran, L. A., M. Hernández-Rivera, A. N. Berlin, Y. Zheng, L. Sampaio, C. Bové, M. da Graça Cabreira-Hansen, J. T. Willerson, E. C. Perin, and L. J. Wilson. 2014. "The use of gadolinium–carbon nanostructures to magnetically enhance stem cell retention for cellular cardiomyoplasty." *Biomaterials* 35 (2):720–726. doi:http://dx.doi.org/10.1016/j.biomaterials.2013.10.013.

Wang, Y.-X. J. 2011. "Superparamagnetic iron oxide based MRI contrast agents: Current status of clinical application." *Quantitative Imaging in Medicine and Surgery* 1 (1):35–40.

Woods, M., G. E. Kiefer, S. Bott, A. Castillo-Muzquiz, C. Eshelbrenner, L. Michaudet, K. McMillan et al. 2004. "Synthesis, relaxometric and photophysical properties of a new pH-responsive MRI contrast agent: The effect of other ligating groups on dissociation of a p-nitrophenolic pendant arm." *Journal of the American Chemical Society* 126 (30):9248–9256. doi:10.1021/ja048299z.

IV

Nanodiamonds

21

Detonation Nanodiamonds

Valerie Dolmatov

Diamonds possess a unique combination of high chemical resistance, highest hardness and wear resistance, low thermal expansion coefficient, highest heat conductivity, and wide energy gap. The key distinctive feature of detonation nanodiamonds (DNDs) from other diamonds originates from the ability to have on their surface a large amount of oxygen-containing functional groups predetermining the hydrophilcity of DNDs on graphene layers lying on a diamond core. Only the manufacture of DNDs in large-tonnage amounts can be mastered in a short time, which sets them apart from other highly active forms of carbon.

All the application areas of DNDs are based on the chemical and structural stability of the nanodiamond core (crystallinity), specific structural features, and surface reactivity. It should be noted that there is no universal material "detonation diamond." The properties of DNDs from each manufacturer are different, depending on synthesis conditions and methods used for chemical purification and modification.

21.1 Fundamental Aspects of the Detonation Synthesis of Nanodiamonds

The difference of the DND synthesis from other dynamic schemes consists in that carbon contained in molecules of explosives serves as a raw material for diamonds (Volkov et al. 1990). In detonation of carbon-containing explosives with a negative oxygen balance in a chamber filled with a medium that is inert toward the diamonds being synthesized (N_2, Ar, gaseous detonation products), the diamond-containing blend (DB) is not only composed of carbon black but also has a significant part of the diamond fraction, ~40%–50%. DNDs can be synthesized from various individual and mixed powerful explosives. In detonation of explosives, they build up, on average, temperatures of up to 4000 K and pressures of up to 30 GPa. The many-years' experience of obtaining DNDs shows that a mixture of TNT (2,4,6-trinitrotoluene) with hexogen (TH) is best suitable for synthesis of DNDs among the wide variety of explosives (Dolmatov 2011a). The characteristics of these substances have been sufficiently well studied and their manufacture is mastered on a large scale. Processing of these explosives poses no difficulties, and charges can be produced by compaction and casting into any required shape. The yield of DNDs under industrial conditions may be as high as ~8%–10%.

The maximum yield of DB and DNDs is observed for formulations that contain hexogen and 40 to 70 wt% TNT and have densities of ≥1650 kg/m^3 (Vereschagin 2001).

21.1.1 On the Mechanism of Detonation Synthesis of Nanodiamonds

The concepts of the DND formation mechanism vary between different authors and occasionally do so rather strongly. There is qualitative incompatibility of the strongly nonequilibrium conditions in the "chemical peak" zone with the formation of a stable crystalline phase of diamond. Therefore, this may, in all probability, refer to the process of self-organization of carbon into a condensed phase in conformity with the basic chemical properties of carbon atoms (namely, formation of various kinds of C–C bonds), rather than to the crystallization of nanosize diamond (because of the short existence of the necessary P, T, and carbon concentration). The process of nucleation and growth of DND particles has an exclusively universal chemical nature irrespective of the starting carbon-containing, hydrogen-containing (TNT, hexogen), or hydrogen-free (benzotrifuroxane) explosive, with the universal species for DND formation being a radical-like dimer C_2 (Dolmatov et al. 2013c). The C_2 dimer is the reason for the appearance of a nanodiamond formation center, radical-like molecule of adamantane, which is supposedly the only stable carbon formation in the chemical reaction zone. The process in which a nanodiamond crystallite grows presumably occurs by the diffusion mechanism as a result of the reaction of the addition of the C_2 dimer at free carbon bonds in the adamantane molecule and then at free bonds on the surface of a diamond particle. DNDs cease to grow because of the accumulation of structural defects and termination of transitions of sp^2-hybridized carbon to sp^3-hybridized "diamond" carbon and the exhaustion of carbon radicals owing to their recombination.

21.1.2 New Method for Obtaining Detonation Diamonds

The strongest influence on the yield and quality of the DB and DNDs is exerted by the initial composition of explosives, detonation synthesis medium, and very high gradient of temperature and pressure. This leads to graphitization (or amorphization) of a part of DNDs and to the active attack on the DB by gases aggressive under the detonation conditions (H_2O, CO_2, N_xO_y), with the result that a part of the DB is gasified and the rest is saturated at the surface by oxygen-containing functional groups. Addition of more readily oxidized substances (reducing agents) to the charge shield strongly affects the result of a detonation synthesis (Dolmatov 2011b, Pat. US 7,862,792 B2). Naturally, the oxidizing agents (CO_2, H_2O, O_2, NO_2, N_2O_3) are primarily bound with more readily oxidized reducing agents (in our case, hydrazine, urotropin, urea, ammonia), rather than with the difficultly oxidized carbon and, especially, DND. The DB is contaminated with metal-containing impurities (Fe), which were previously contained in walls of the explosion chamber and means of explosion initiation (Cu, Pb). A new method for synthesis in the presence of reducing agents provides an increase in the yield of DB by a minimum factor of 1.6, and that of DND, by a factor of 1.5–2. In the case of urotropin, the amount of incombustible admixtures in DNDs is anomalously small. The reason is that, under the explosion conditions, urotropin was found to be, among other things, a rather strong complex-forming agent and the incombustible metal-containing impurities are present in the form of intricate complex compounds that are easily dissolved in the course of purification, even in weak nitric acid, rather than in the common form of carbide or oxide (Dolmatov 2008, Pat. US 7,862,792 B2).

21.2 Industrial Synthesis of DNDs

21.2.1 Production of DNDs

The production of DNDs includes the following: detonation synthesis, chemical purification and washing of DNDs to remove the acid, and conditioning of the product. The process of industrial detonation synthesis is performed periodically by manually placing an explosive charge equipped with an electrical detonator in the explosion chamber through the upper hermetically closed hatch. A TH charge is placed in a bag with a solution of a reducing agent and suspended with a string to the hook of the upper hatch of the explosion chamber, with an electrical wire for explosion initiation connected to terminals of a special device on the cover of the explosion chamber. The explosion is initiated from a console situated in a neighboring room, with the compartment door closed. The whole cycle takes 5–7 min. The resulting mass is poured from the receiving vessel into an apparatus with a sieve on its top and a magnetic separator. The squeezed mass is delivered to the finishing purification.

21.2.2 Nitric Acid Purification of DB

The most efficient technology with the highest output capacity for chemical purification of DNDs is the treatment of DB with diluted nitric acid under elevated pressures (up to 100 atm) at high temperatures (230°C–240°C) (Pat. of Russia 2109683). In this way, it is possible to reach a DND purity of ≥99 wt% at a DB residence time in the high-pressure and high-temperature zone as short as 30–40 min. This technology has been successfully implemented by the authors in Russia, China, and Belarus. The process in which nondiamond carbon is oxidized is combined with the processes in which metal-containing impurities are acid-dissolved and converted to water-soluble products. The thermal-oxidative treatment under pressure yields a DND suspension in an aqueous solution of nitric acid. Nearly all impurities are converted either to gaseous products or to a dissolved state and are removed in washing of DNDs with desalinated water. Taking into account that titanium is the only convenient construction material for the reactor, the specifics of its operation restrict the highest possible nitric acid concentrations to 65%–70%. In the general case, the DND purification technology includes six main stages: (1) blend pretreatment (grinding, homogenization, removal of magnetic impurities, magnetic separation, and drying to certain humidity), (2) preparation of DB suspensions in aqueous solutions of nitric acid, (3) thermal-oxidative treatment of blend suspensions in the continuous mode under pressure (this stage is the key procedure in the technology. A special apparatus and procedures for startup, tests, operation, and emergency and planned shutdowns have been developed by the authors.), (4) separation of the products formed in the thermal-oxidative treatment, (5) washing of DNDs with distilled water to remove acids in the multistage counterflow scheme, and (6) obtaining normalized stabilized suspensions of DNDs in distilled water or a dry homogeneous powder. At present, this DND purification technology provides the highest and most stable purification quality parameters, is easily scaled, and is best tested and ecologically safest.

21.2.3 Industrial Chemical Purification of DB

The stages of the thermal-oxidative treatment of the DB under pressure are fully remote controlled. The equipment is enclosed in special-purpose protective compartments. The entire body of information about the process is displayed at centralized control consoles. The parameters are maintained automatically by local control systems supplemented with safety-lock and signalization units by analysis of computer-processed information. The console can receive signals from 128 channels and actuate 40 channels of local control units.

The reactor unit is a cascade of electrically heated flow-through column apparatus with internal coaxial partitions made of high-purity titanium, which enables thermal treatment at temperatures of up to 250°C and pressures of up to 100 atm. The processed volume of the fluid flow is up to 20 L/h. The temperature mode is controlled individually for each reactor.

21.3 Properties of Detonation Diamonds and Diamond-Containing Stock

21.3.1 Determining the Structure of the Particle Core of DNDs by X-ray Diffraction Analysis with Synchrotron Radiation

To determine the DND size, it is necessary to compare the diffraction reflection profiles of the theoretical and experimental x-ray diffraction line (XRD) patterns. Various researchers have determined the DND diameter from experimental XRD patterns to be 30–50 Å and point to the existence of a diamond structure but rarely determine the lattice constants and nearest interatomic distances (Dolmatov 2008). It has been found that the new DNDs synthesized in the presence of reducing agents have the lattice constants of diamond, there are no impurity phases, and all the samples are polydisperse: The minimum, maximum, and average DND sizes correspond to the radii $R \approx$ 10.5–14.8, 43–51, and 27–35 Å, respectively. Experimental XRD patterns of the DNDs demonstrate that the intensity ratios of reflections from different faces frequently disagree with those for ideal diamond.

21.3.2 Particle Structure of Chemically Modified Detonation Nanodiamods

The goal of Kulakova et al. (2009) was to determine the state of carbon in the surface layer of DND particles, both original and treated under various conditions, including those subjected to chemical modification by covalent grafting. The photoemission from the valence band makes it possible to identify the state of carbon atoms in seven monolayers. The spectra of the characteristic electron energy loss (CEEL) of the original and reduced DNDs nearly coincide with the CEEL spectrum of natural diamond, which points to the same state of carbon atoms (i.e., the sp^3 hybridization) in the samples. Consequently, the defective carbon shell is constituted by carbon atoms in the sp^3-hybridization state. The fact that nitrogen is uniformly distributed over the volume of DND particles was confirmed by the experiment on step-by-step oxidation of DNDs (with mass-spectrometric analysis of the released oxidation products, CO_2 and N_2) in which a symbate release of carbon dioxide and nitrogen was recorded. Thus, the model of a particle of chemically modified DNDs includes a diamond core with a diameter of ≤3 nm in which nitrogen impurity atoms and paramagnetic centers (10 spins/particle) are more or less uniformly distributed, a layer with disrupted diamond structure (with a thickness of about 1 nm), and a cover composed of functional groups.

21.3.3 Clusters of DNDs

In Ozerin et al. (2008), the authors demonstrated, based on calculations of the atomic distribution in DNDs, that the factors restricting the growth of the diamond core in DNDs (≤8 nm) are the disrupted long-range order in DNDs and the accumulation of crystal-structure defects in the course of growth of a carbon nanoparticle in postdetonation processes. It follows from a comparison of the atomic distributions in DNDs and defect-free diamond, which have different lattice constants, that the coordination number changes on the nanosize scale in the outer coordination spheres (CSs). Absent atoms (defects) are observed in the CS, which changes the structure and morphology of the DND surface. Thus, the accumulated defects remove the surface layers of carbon from the diamond structure of the core. The crystal size distribution function was found using the modified Fourier analysis of the profile of the XRD line (Ozerin et al. 1986). The XRD patterns at large scattering angles were identical within experimental error for all the powdered DND samples, with observed reflections corresponding to a diamond-type lattice with parameter a = (3.565 ± 0.005) Å. The effective crystallite sizes found for different crystallographic directions on the assumption of a prismatic crystallite shape of the crystallites were used to estimate the volume of a unit DND crystallite (8×10^4 Å^3). As also in the case of large-angle scattering, the small-angle scattering curves of all the DND samples under study were found to be identical upon mutual normalization. For the DND samples under study, the small-angle x-ray scattering pattern corresponds to the scattering on so-called indestructible aggregates (Kruger et al. 2005). Comparison of the substructure particle volume found from the small-angle scattering data with the unit crystallite volume estimated from XRD measurements at large diffraction angles shows that the substructure particle may be constituted by (4×10^5 Å^3)/(8×10^4 Å^3) = 5 crystallites; i.e., it may be a cluster of crystallites. The most natural form of a cluster of this kind is that of a system of five equivalent tetrahedrons. The average value of the excluded volume of the "indestructible" aggregate was 3.5×10^6 Å^3, which corresponds to (3.5×10^6 Å^3)/(4×10^5 Å^3) ≈ 9–10 clusters constituting an aggregate of DNDs.

Thus, DND aggregates produced by detonation synthesis are extended three-dimensional structures constituted by 9–10 clusters, each including four to five crystallites with a diamond-type crystal lattice (Ozerin et al. 2008).

21.3.4 Spectral Studies of DNDs

The strongly developed DND surface and the large C_{surf}/C_{tot} ratio, which predetermine the high concentration of surface functional groups, enable application of infrared (IR) spectroscopy for sufficiently reliably determining the nature of the functional groups (Kulakova et al. 2000). It was found in Dolmatov et al. (2014) that the amount (0.04–9.3 wt%) and type of incombustible impurities have no effect on the most characteristic absorption bands of DND samples in the IR spectrum. As a rule, attempts to dope DND crystallites by exploding an explosive charge in a shell composed of aqueous solutions of boron and phosphorus compounds or an explosive charge with various introduced elements had no effect on the IR spectra of DNDs.

The IR spectra of various new DND particles show a certain shift of all the characteristic bands but at the same time confirm the existence of intracrystalline nitrogen impurity centers and the presence of the following functional groups on the surface of crystallites: NO_2 (825.5, 829, 1361.7, 1365, and 1373 cm^{-1}), CH (636, 652, 1111, and 1134 cm^{-1}), NO_3 (883 and 1385 cm^{-1}), CH_2 (1458, 2854, 2858, 2924, 2928, and 2932 cm^{-1}), OH (1628, 1635, and 3236–3433 cm^{-1}), C=O (1732 and 1736 cm^{-1}), and –C–O–C– (1149 and 1157 cm^{-1}).

The nanodiamond surface was also analyzed using Auger electron spectroscopy, CEEL spectroscopy, and x-ray photoelectron spectroscopy. Specific features of DNDs were manifested in Auger spectra (Kulakova et al. 2000). The Auger spectra of the starting and hydrogen-reduced DNDs coincide but strongly differ both from the Auger spectrum of natural diamond and from the spectrum of graphite. Thus, the upper monolayer of carbon atoms cannot be strictly described either as sp^3 or as sp^2 hybridization state of carbon. For this monolayer, the electronic structure can be qualitatively described as follows: The occupancy of carbon atoms is the same as that in diamond, and the electron energy near the Fermi level (E_F) is the same as that in graphite. The latter follows from the coincidence of peaks in the spectra of modified nanodiamods and graphite.

21.3.4.1 EPR Study of Modified Nanodiamonds

Similarly to natural diamonds, DNDs have paramagnetic properties owing to the presence of paramagnetic centers. It was demonstrated using the electron paramagnetic resonance method and Raman spectroscopy that there are nanodiamonds in the DND samples under study. The aim of Dolmatov 2013b was to identify features of the EPR spectra of DNA modified at the time of synthesis by the introduction of doping elements in the aqueous shell of the explosive charge (compound of boron and phosphorus), or directly into the explosive charge (boron, aluminum, silicon, sulfur, boron compounds, phosphorus, and germanium). It is expected that after the explosion of the charge doping elements can enter into the crystal structure of DNA. The structural perfection of these nanodiamonds depends on technological features of the synthesis procedure. A dependence of the spectroscopic splitting factor on the number of free radicals was observed in the DNDs under study. The most perfect DND structure is obtained with dopant elements (Ge, Si, S) introduced directly into the explosive charge (with its subsequent detonation). A clearly pronounced correlation was observed between the g-factor and the number of paramagnetic centers in DNDs (Dolmatov et al. 2013b). EPR method does not allow to confirm or deny the fact of doping crystallites DNA. Exploring relaktsionnye processes DNA samples, indirectly shown the presence of nanocrystals centers paramagnetic relaxation times of more than 10^{-5} seconds. Discovers facts indicating the removal of stresses arising in the process of detonation synthesis, nanocrystals of diamond, and self-organization of the atoms in the crystal lattice in the process of long-term storage (more than 1 year).

21.3.4.2 XRD Patterns of DNDs

The observed broad symmetric diffraction peaks at angles $2\theta = 43.9°$, 75.3°, and 91.5°, well described by Lorentz profiles, correspond to the (111), (220), and (3111) reflections, respectively, from a diamond-type lattice with parameter $a_0 = 3.565 \pm 0.005$ Å. The average size of DND particles, determined by the Selyakov-Scherrer formula from the half-width of these lines for all the three diffraction peaks, was found to be $L = 45 \pm 5$ Å (Alexenskij et al. 1997). The coincidence of the average sizes of particles for different diffraction peaks indicates that the broadening of the diffraction peaks is mostly a result of the small particle size, rather than to internal stresses. This section presented rather different opinions on the structure of DND particles, which means that, first, it is rather difficult to develop a real concept of the DND structure on the basis of the available instrumental analysis methods, and second, the structure of nanoparticles in general is not obvious. This, however, in no way curtails the ultimate advantages of using DNDs in their application technologies, which are steadily increasing in number.

21.4 Application of DNDs and Diamond-Containing Stock

When introduced into materials, DNDs always serve as a high-power structuring agent and thereby provide a dispersion reinforcement of a composite. The potential demand of the market is huge, but its development is hindered by the lack of DNDs in necessary amounts (~10 ton/year) that could ensure the main thing, the regular supply to any customer.

21.4.1 Electrochemical and Chemical Metal–Nanodiamond Coatings

The electrochemical deposition of metals, including that with nanomaterials, is among the most technologically convenient and controllable processes. Composite electrochemical coatings based on DNDs are an example of the highly effective influence exerted by small inclusions of nanoparticles on the microscopic structure and macroscopic properties of a material.

DND particles have not only an intricate structure but also unlike charges on different faces (Barnard and Sternberg 2007). According to Barnard (2008), the strong aggregation of primary DND particles is caused by the coherent interphase Coulomb interactions caused by the high surface charges on faces of the primary particles. The motion of DND particles suspended in an electrolyte to the cathode is of complex nature. Each DND particle (in an aggregate or in the individual state) has a multilayer solvate shell, which is partly displaced by ions of the electrolyte because of the electrostatic and adsorption forces. Introduction of DNDs into electrochemical coatings substantially raises their microhardness and wear and corrosion resistance, improves their outward appearance, reduces their porosity, makes the friction coefficient noticeably lower, and raises the throwing power of the electrolyte. The main reason for this behavior is that the metal domains in a coating become smaller.

21.4.1.1 Chrome–Nanodiamond Coatings

Chrome plating is among the most frequently employed galvanic processes. Chrome coatings are widely used in the industry. Coatings produced from an electrolyte without additives characteristically have a nonuniform layered (loose) structure. When DNDs are introduced into the electrolyte, the structure of the resulting deposits becomes more uniform and has a higher density (Dolmatov 2011a). DNDs positively affect the coating structure up to concentrations of 15 g/L, but with their concentration raised to 20 g/L, the coating structure approaches that obtained in an electrolyte having no additives. The effect of DB and DNDs on the microstructure of chrome coatings deposited at 50°C and cathode current density of 50 A/dm^2 was studied. It was shown that the structure of chrome coatings produced from electrolytes containing DB and DNDs is microgranulated; the microhardness of chromium coatings increases by 20%–25% and 25%–30%, respectively, as compared with the microhardness of chromium deposited from the standard electrolyte; and the wear resistance of the coatings becomes 2–3 times higher. The following additive concentrations and electrolysis modes can be recommended for deposition of composite coatings with improved functional properties from the standard chrome plating electrolyte: DB, 2.5–5 g/L (DND 10–15 g/L); cathode current density, 30–70 A/dm^2; and optimal electrolyte temperature 45°C–55°C. The results of these studies lead to the following main conclusions. The chrome coating electroplated without DNDs is comparatively smooth but has parts with a large number of irregularly shaped pits with depths of up to 3 μm; the coating is constituted by densely packed crystallites with average transverse sizes exceeding 150 nm and nearly rectangular shape. The electroplated chrome coating with DNDs has no characteristic pits; the average sizes of chrome crystallites are less than 50 nm; the nearly complete absence of DNDs on the surface of the electroplated coating suggests that the nanodiamonds are situated at the substrate and serve as nuclei in the initial growth stage of an electroplated coating, which leads to the formation of nanosize crystallites and provides the uniformity of the coatings.

21.4.1.1.1 Chrome Plating in the Presence of Mixed Nanodiamonds Formed in Detonation and Static Syntheses

Use of a mixed additive of DNDs and ASM powder (static-synthesis diamonds crushed to nanometer dimensions with crystallites up to 100 nm in size) in a chrome plating electrolyte leads to results fundamentally different from those obtained with DNDs and ASM introduced separately.

Properties of ASM:

- Diamond cubic system: a = 0.357 nm
- Pycnometric density: 3.49 g/cm^3
- Particle size: 2–100 nm
- Specific surface area: 56 m^2/g
- Chemical composition, %, mas.: C–99.0; (Ni, Mn, Cr, Fe)–0.5; Si–0.2; B–0.2; (O,H)–0.1
- Fireproof residue, %, mas.: ~0.1
- Start temperature oxidation in air: 723 K
- Start temperature graphitization in a vacuum: 1373 K
- The electrical resistance: $1{\cdot}10^{11}$ Om·m
- Specific magnetic susceptibility: $\chi{\cdot}10^8$, m^3/kg 0.5
- Electrophoretic surface charge mV: –6.53

Table 21.1 shows how the microhardness of a Cr-DND-ASM coating obtained in the wear-resistant chrome-plating mode depends on the current density mixed additive concentration and on the ratio

TABLE 21.1 The Dependence of Microhardness of Wear-Resistant Cr-DND-ASM Coatings on the Cathode Current Density, the Content of Mixed Additives in the Standard Electrolyte, and DND/ASM Ratio (55°C ± 1°C)

Cathode Current Density (A/dm^2)	Content of DND-ASM in the Electrolyte (g/L)							
	0	0.25/0.25	2.5/2	2.5/5	5/5	2.5/10	2.5/15	10/10
	Microhardness of Cr-DND-ASM-Coating (kg/mm^2)							
30	733	758	834	1025	1040	1163	1109	1100
40	921	941	973	1063	–	1191	1152	–
50	762	860	1191	1302	1339	1404	1342	1368
60	818	863	1029	1212	–	1382	1463	–
70	882	–	–	1340	1392	1422	–	1505

TABLE 21.2 Corrosion Resistance of Chrome-Nanodiamond Coatings Depending on the Content of a Nanodiamond Additive in the Standard Electrolyte and Temperature Regime (Hard [100 A/dm^2] and Wear-Resistant [50 A/dm^2] Chrome Plating), and Coatings of Titanium Nitride and Pure Chrome[a]

Content of Nanodiamonds in the Electrolyte (g/L)		Mass Loss (as a Result of Corrosion) (wt% 10^3)
Titanium nitride (no additives)		19.0
Chrome (no additives, hard chrome plating)		6.7
Chrome (no additives, wear-resistant chrome plating)		3.2
Cr-DND hard	30	6.6
Cr-DND wear-resistant	30	2.1
Cr-DND-ASM hard	2.5/5.0	8.2
Cr-DND-ASM wear-resistant	2.5/5.0	5.5
Cr-DND-ASM hard	2.5/10	8.2
Cr-DND-ASM wear-resistant	2/10	6.3
Cr-DND-ASM hard	5/5	0.8
Cr-DND-ASM wear-resistant	5/5	2.1
Cr-DND-ASM hard	10/10	0
Cr-DND-ASM wear-resistant	10/10	1.0

[a] Thickness of all coatings: 3–5 μm, "korrodkot" testing (Vjacheslavov and Shmeleva 1977).

between diamonds of two different types (DND and ASM). It was demonstrated that the microhardness substantially increased (especially at the practically important current densities of 50–60 A/dm^2): by 20%–50% as compared with the use of only DNDs and by 8%–24% as compared with the use of only ASM. In this case, the microhardness of the Cr-DND-ASM coating reaches a large value of 1300–1400 kg/mm^2 even in the wear-resistant chrome-plating mode at low total concentrations of diamonds (4.5–7.5 g/L). Addition of mixed nanodiamonds (DND + ASM) to the chrome plating electrolyte, on average reduced the coating wear compared with pure chrome in 13–18 times; compared with chrome + DND in 2,5–3,5 times; and compared with chrome + ASM in 10–45 times.

It can be seen in Table 21.2 that the chrome coating has a significant corrosion resistance (3–6 times higher), compared with a titanium nitride coating (at the same coating thickness).

Use of pure DNDs in a coating only slightly affects the corrosion resistance, whereas application of mixed nanodiamonds at a total concentration of no less than 10 g/L and a 50:50 ratio makes the corrosion resistance of a coating 2–10 times that for a chrome or chrome-DND coating. No chrome carbide formation was observed in deposition of chrome with DNDs (DS).

21.4.1.2 Nickel–Nanodiamond Coatings

Nickel coatings are used, together with chrome coatings, in machine building, instrument making, jewelry industry, watch making, ship building, and aircraft construction. Studies have been carried out with very small amounts (and, accordingly, concentrations) of best quality DNDs: DND-TAN (ammonia-treated DNDs manufactured by FGUP "SKTB "Technolog," St. Petersburg, Russia) and DND-OS (manufactured by NanoCarbon Research Institute Ltd., Japan) (Dolmatov et al. 2011a). Nickel coatings were deposited with a thickness of 14–16 μm. The microhardness of the coatings was measured with a PMT-3 microhardness meter under a load of 20 g. The nickel-plating electrolyte was chosen from the set of sulfamate electrolytes with the following optimal electrolysis modes: temperature of 39°C–43°C, cathode current densities of 3 and 4 A/dm^2, pH ~3. Both these additives noticeably affect the microhardness and wear resistance of the coatings, beginning at 0.5 g/L, with the wear resistance becoming 2 times higher. The porosity is affected by the DND-OS additive, beginning at the lowest concentrations, being noticeably diminished to nearly zero porosity. The DND-TAN additive also diminishes the porosity of the coatings (Table 21.3).

Table 21.4 presents the reproducible results obtained with DND-TAN used to deposit various metals.

The work on this application has been accomplished at FGUP "SKTB "Technolog" under the support provided by the Ministry of Education and Science of the Russian Federation within the Federal Target Program "R&D along Priority Directions of Development of the Scientific and Technological Complex of Russia for Years 2014–2020 (Agreement on Grants No. 14.579.21.0001, RFMEF157914X0001)."

TABLE 21.3 Influence of Nanodiamond Additives DND-TAN and DND-OS on the Physicochemical Properties of Nickel Coatings

	Weight Loss at Attrition (%)[a]	Microhardness (MPa)		Porosity (Number of Pores/1 cm^2)		Corrosion Rate (gm^2/y)	
	Current Density (A/dm^2)						
Additive (g/L)	3	3	4	3	4	3	4
No additives	6.2	2529.4	2418.6	34	29	2.6	2.1
DND-OS—0.1	Not determined	2464.5	2403.9	23	3	3.25	4.25
DND-OS—0.5	3.2	2835.1	2980.0	0	8	3.5	3.5
DND-TAN—0.1	Not determined	2659.7	2803.9	25	18	2.0	3.375
DND-TAN—0.5	3.2	2969.4	3000.6	10	10	2.25	1.25

[a] In 19 h of attrition, the samples with pure nickel were worn to the base. The samples with a nickel–diamond coating have been tested within 20 h without the base exposure.

TABLE 21.4 Physicochemical Parameters of Metal–Diamond Coatings

Coating	Increase in Wear Resistance *n* (Times Where *n*)[a]	Microhardness (kg/mm^2)	Decrease in Porosity *m* (Times Where *m*)[a]	Increase in Corrosion Resistance *p* (Times Where *p*)[a]
Cr	2–12	To 1200	2–3	3–6
Ni	5–6	To 580	4–5	2–5
Cu	9–10	160	No pores	Does not corrode
Au	1.8–5.5	180	6	Does not corrode
Ag	10–15	120	8	Does not corrode
Zn	–	–	6–8	2–4
Sn	4–5	150	7	1.5
Al	10–13	600–700	3	Does not corrode
Fe	6–8	800	8–10	3–4
Ni + B	6	8000	–	Does not corrode
Sn + Pb	–	31	No pores	5–7
Sn + Sb	2.5	25	5–6	2–3
Ag + Sb	1.5–1.7	150	–	–

[a] Times Where *n* (*m*, *p*) is the number of times specified in this table is greater or less than the known value (without DNA).

21.4.2 Composite Materials Based on Polymeric Matrices Filled with DNDs and DB

It is common knowledge that introduction of filler nanoparticles into a polymer can improve the construction behavior of large-tonnage polymers and impart new functional properties to these materials. This means that, to create materials with unique or improved properties, it is not necessary to develop new chemical industries for obtaining new and, as a rule, expensive polymers.

21.4.2.1 Fluorinated Elastomers

At present, the demand for chemically and thermally resistant polymeric materials is steadily growing. Promising materials in this regard are perfluorinated and polyfluorinated polymers and elastomers having high chemical and thermal stability (Dolmatov 2011a). Two disadvantages of fluoroelastomers should be attributed to their poor mechanical strength and low resistance to abrasive wear. It follows from the data obtained that coatings containing DB (up to 20 wt%) surpass in tribotechnical properties such known materials as the fluorinated polymers of Tedlar (polyvinyl fluoride) and Teflon FEP (hexafluoride-co-tetrafluoroethene) brands, manufactured by Du Pont company (f_{fr} = 0.4 against f_{fr} = 0.01 for the coating developed), providing at the same time a substantially wider working range of temperatures and mechanical loads.

21.4.2.2 Polyurethanes

As a polymeric matrix, Voznjakovskij (2007) used polyurethane of Carbotane brand (manufactured in the United States) and SKU-60md polyurethane synthesized as a model. The uniform distribution of the ultrasmall nanodiamond additives in the polyurethane matrix was provided by modification of their surface with amino groups.

The data in Table 21.5 demonstrate that even the minimum content of the modifying nanodiamond additives provide a substantial reinforcement effect, which is more significant the stronger the deformation. The maximum effect obtained in this set of experiments is observed with the modifying additive taken in an amount of 0.2 wt%. It should be noted that despite the significant rise in the conditional strength (by nearly a factor of 3), the elasticity of the polymer also substantially increased (from 400% to 600%). This manner of reinforcement is characteristic of nanocomposites and is not predicted by the classical reinforcement theory.

21.4.2.3 Various Type of Polymers

Table 21.6 shows the results obtained by determining the wear resistance of polymeric materials with the DB additive. The samples under study were prepared by mixing in an impact mill, followed by molding and sintering. Tests were carried out on an SMT-1 friction machine by the method of intersecting cylinders at a sliding velocity of 0.78 m/s. The wear resistance was found from the area of the wear crater on a sample (S_{wear}). It can be seen in Table 21.6 that the wear resistance of the materials grows with increasing

TABLE 21.5 Physicochemical Properties of Nanocomposites of Polyurethane Carbotane

	Content of DND (wt%)				
Parameters	0	0.025	0.05	0.1	0.2
Stress at 50% elongation (MPa)	2.0	1.6	1.1	2.1	2.9
Stress at 200% elongation (MPa)	4.5	5.5	3.2	5.9	–
Stress at 300% elongation (MPa)	10.0	10.0	5.4	8.9	12.3
Conditional strength, *P* (MPa)	19.5	42.8	28.0	23.9	59.6
Specific elongation at rupture (%)	400	450	410	410	625
Residual (elongation) strain upon rapture (%)	10	10	15	10	29

TABLE 21.6 Wear Resistance of Polymer Materials with DB Additive

	Wear Surface (S_{wear}) (mm^2)			
	Content of DB (wt%)			
Material (Sintering Temperature, K)	0	1.0	1.5	2.0
Polyacrylamide (513)	13.2	12.4	–	10.3
Polymethylmetakrylate (413)	13.5	–	11.3	–
Low-pressure polyethylene (423)	11.9	9.8	9.1	–
Polytetrafluoroethylene (PTFE), sort F-4 (643)[a]	84.2	31.0	–	18.2

[a] Load: 3.74 kg; testing time: 18 min.

TABLE 21.7 Wear Resistance of PTFE, Sort F-4, with Different Additives in Number 5 wt%

Additive	S_{BET} of an Additive (m^2/g)	S_{wear} (mm^2)
DB	344	7.2
DB	468	5.9
DND	282	10.6
Silicon dioxide	400	19.0
Silicon nitride	28	13.7
Cobalt aluminate	10	17.5
Soot P-803	15	19.7
Graphite C-1	10	28.2
Molybdenum sulphide[a]	–	28.0
DB doped with cobalt	387	8.3

[a] $d_r = 10$ μm.

DB content. The most significant result was obtained for F-4. As demonstrated by further studies, the maximum wear resistance of the polymer of F-4 brand is obtained upon addition of DB in an amount of 5% (Pat. of Russia 2005741). Compared with the known dispersed additives, DB occupies the leading position in the influence it exerts (Table 21.7).

21.4.2.4 Conclusions

(1) The polymer of F-4 brand with addition of DB approaches bronze in wear resistance, with the friction coefficient of the composite remaining at the level of the pure polymer (0.2). The combination of the high wear resistance and low friction coefficient makes it possible to regard the given composite as a promising antifriction material, surpassing the known analogs. (2) Use of DNDs leads to an increase in the initial elasticity modulus and tensile strength of nanocomposites. (3) Filling of an elastomeric matrix (polyfluorinated elastomers, polyisoprenes, butadien-styrene and butadien-nitrile [frost-resistant] rubbers, polysiloxanes, polyurethanes) with DNDs leads to the following: 1.5–3.0 times increase in the conditional stress and cohesion strength of rubbers, 1.35–2.0 times increase in the tearing resistance, 50%–70% higher elasticity, 1.3–2.0 times smaller wear resistance, 1.3–5.0 smaller friction coefficient, ~50% higher frost resistance, minimum thermal aging of polyfluorinated rubbers, higher resistance to corrosive media, and better technological convenience and rheological properties. (4) Introduction of DNDs into polymeric matrices (ED-20 epoxy resin, polyamide, fluoroplastic, polycarbonate, polyvinyl alcohol, polystyrene–polybutadiene–polystyrene block copolymer) leads to the following: 1.3–1.7 higher tensile strength, 1.3–2.1 times higher elasticity, up to 4.6 times increase in wear resistance, decrease in the friction coefficient to 0.16 at room temperature and to 0.08°C at 150°C, up to 7.5 times higher limiting load on a polymeric nanocomposite at its preserved working capacity.

21.4.3 DNDs in Oils and Lubricants

DND and DS serve as effective friction modifiers in oil formulations by saturating friction surfaces, filling irregularities, and creating new friction surfaces. At high loads and maximum displacement of the liquid phase from between the friction surfaces, DNDs operate as microscopic rolling bearings, which provides an increase in the limiting load (Table 21.8) without seizures (Dolmatov 2001, 2011a, Pat. of Russia WO 93/01261).

21.4.4 Biological Activity of DNDs

The toxicity of the presently used antioxidant preparations in the chemotherapy of cancer is very high, and a search for new low-toxic preparations with a similar effect is exceedingly topical. It was first demonstrated by Dolmatov and Kostrova (Pat. of Russia 2203068, Dolmatov and Kostrova 2000) that DNDs belong to novel anticancer preparations. DNDs can be regarded as polyfunctional supramolecular structures bearing on their surface such functional groups as $-OH$, $-NH_2$, and $-C(O)NH_2$, which can provide their antiradical activity and the ability to be actively involved in free-radical processes in living cells. The multitude of functional groups on the surface of DND particles can provide both additional generation and deactivation of the excessively generated radicals in metabolic processes. It is the control over free-radical processes under the action of DNDs that can serve as a basis for their anticancer effect.

Present-day studies demonstrate that DNDs are invariably better sustained by cells, compared with other carbon-based nanomaterials, such as carbon nanotubes or fullerenes, which are continued to be studied with respect to their in vitro toxicity or biocompatibility (Grabinski et al. 2007, Pulskamp et al. 2007).

DNDs not only have a large specific surface area and high absorption capacity but also can, remaining biocompatible, penetrate into cells of diverse cell strains (Liu et al. 2007, Schrand 2007). Similarly to nanodiamonds of micrometer dimensions, DNDs do not cause inflammation or such cytotoxic reactions as formation of active oxygen species (AOS), morphological changes, changes in viability, or formation of cytokines in quite a number of cell strains, including those of microphages, fibroblasts, epithelial cells, etc. The effect of the surface chemistry of DNDs purified with strong acids or bases on the viability of cells has been studied by Schrand et al. (2007). At concentrations of up to 100 μg/L, all DNDs

TABLE 21.8 Characteristics of Lubricating Diamond-Containing Compositions and the Effect of Their Use

№ п/п	Characteristic Name (Unit)	Value
1	Content of DND (w.%)	0.01–0.30
2	Concentration of particles in the DND (1 cm^3)	~10^{14}
3	Decrease in temperature in the contact zone (%)	16–20
4	Decrease in friction coefficient (%)	20–30
5	Increase in limiting loads of friction pair (steel-bronze) (MPa)	From 16 to 72 (~4 times)
6	Decrease in wear of mating parts (n)	n = 1.5–3.0
	Internal Combustion Engine	
7	Decrease in running-in time (m times)	m = 10–12
8	Decrease in fuel consumption (%)	3–6
9	Increase in power of an engine (%)	4–8
10	Increase in cylinder compression (%)	10–17
	Lubricoolant	
11	Decrease in cutting moment (%)	20–30
12	Increase in instrument life (p times)	p = 1.5–4.0

were nontoxic for various kinds of cells, including those of neuroblastomas, macrophages, RS-12, and keratinocytes, according to an MTT (3-[4,5-dimethylthiazd-2-yl]-2,5 diphenyl tetrazolium bromide) analysis. It has been shown that internalized DND aggregates are situated in cytoplasma, rather than in the nucleus (Fu et al. 2007, Neugart et al. 2007, Schrand 2007).

Because many body cells are subject to natural death, it is believed that the body would free itself of a small amount of DNDs via natural renewal of cells without formation of AOS or negative side effects.

Dolmatov and Gorbunov first obtained radioactive nanodiamonds for medical purposes (Dolmatov et al. 2013a, Pat. Application 2013115568/05(023053). Radioactive nanodiamonds (R-DND and R-ND-ASM) are powders of light gray to black color, with individual particles having sizes of 2 to 100 nm, as determined by the light-scattering method and XRD analysis. The specific surface area, found using Brunauer–Emmett–Teller isotherms by the method of thermal desorption of argon, is 200 to 450 m^2/g for DNDs and 20 to 60 m^2/g for ND-ASM. An analysis of the IR absorption spectra of R-DND and R-ND-ASM samples revealed absorption bands characteristic of carbonyl (=CO), carboxy (=COOH), and hydroxy (–OH) groups. The radioactive nanodiamonds have a radioactivity with a γ-radiation dose rate of less than 5.0 μR/s (180 μSv/h) and $\gamma + \beta$ radiation dose rate of less than 20 μR/s (720 μSv/h). The radioactivity of DNDs and ND-ASM upon neutron irradiation can be attributed to radionuclides formed from Na, Ca, Ti, Fe, Al, W, V, and Cu impurities. When DNDs are used as an anticancer preparation (Pat. of Russia 2203068), it is impossible to determine in a living organism in real-time mode the places of DND accumulation, efficiency of action upon cancer cells, and relative amounts of DNDs at places of clearance from the organism. Only use of radioactive DNDs makes this possible.

21.4.4.1 Conclusions

(1) No analogs to DNDs in the combined effect on an organism are known. (2) Use of DNDs simultaneously with chemotherapy and radiotherapy may prove promising for curing of malignant tumors to preclude the mutagenicity of preparations, with their therapeutic effect preserved and the appearance of mutations in normal cells and the induction of secondary tumors under the action of antitumor preparations ruled out. (3) DNDs are believed to be promising components in regeneration of bones (Pramatarova et al. 2007).

21.4.5 Polishing

Because of the structural disorder in the surface layers of DND particles and absence of rigid cutting faces, DNDs do not exhibit the conventional abrasive properties but behave as elastic and highly persistent particles that control mechanochemical processes on the submolecular level. DND particles have no deforming effect on a material being processed, and as a result, it retains an ordered and unstrained structure, including its boundary layers. This provides a high quality of the surface; in particular, some electrical parameters approach their theoretical values. DNDs can provide ideal surface parameters of materials being treated, with microscopic irregularities having dimensions on the order of interatomic distances.

For epitaxial, vacuum-evaporation, and lithographic processes, atomically and molecularly smooth and clean surfaces can be only obtained by using solid nanosize particles. Therefore, DNDs with particle size smaller than 1 μm fill the niche unoccupied by static-synthesis diamonds, which is limited in practice to ASM1/0 micropowders.

It has been reliably demonstrated that DND formulations can provide a high planarity (down to 1 μm) and reproducibly form nanometer irregularities with heights of several nanometers to fractions of a nanometer for more than 30 crystals (semiconductors [elementary and compound], insulators [Si_3N_4], oxides [α-Al_2O_3, ZrO_2:Y, α-SiO_2], garnets [GGG, YAG], glasses, etc.) belonging to six crystallographic systems and also for AlN ceramic and CuW, VT6, and other alloys. The assortment of materials and treatment quality steadily grows, which points to the versatility of this technology.

21.4.6 Other Application Technologies of DNDs

21.4.6.1 Organo-Silicate Bioprotective Matrices

The study of Khamona et al. (2012) is concerned with sol–gel synthesis and biological activity of epoxysiloxane nanocomposite coatings formed on the basis of epoxysiloxane sols doped with mild biocides of varied origin with respect to mold fungi being the most frequently found in the air medium of major cities. It is suggested for the first time to use DNDs as an effective and ecologically safe biocide. The authors selected four micromycete species to test the bioresistance of their epoxysiloxane coatings to a number of micromycetes of mold fungi: *Cladosporium cladosporioides*, *Cladosporium sphaerospermum*, *Ulocladium chartarum*, and *Aspergillus niger*. The sol–gel method was used to synthesize organo-inorganic coatings and microcomposite powders on the basis of siloxane sols modified with mild biocides. The concentration of the additive introduced into the sols was varied from 0.05 to 0.25 wt% DND. When the DND concentration in a sol reaches a value of 0.20–0.25 wt%, no indications of growth and spore formation is observed on the surface of protective coatings.

21.4.6.2 Modification of Cements

In Vlasov et al. (2004), Portland cement manufactured by Sukholozhskii cement plant (Russia) was taken for strength tests. To prepare samples for studies, Portland cement clinker was ground and batched with distilled water in an amount of 5–10 wt%. Simultaneously, a biocide additive, dry DND powder manufactured by Federal State Unitary Plant "SKTB" "Technolog," was introduced into the stock in an amount of 1–5 wt% because it is known that DNDs are mild biocides (Kondratieva et al. 2006). Cubes produced from Portland cement with the addition of a varied amount of the DND powder were subjected to unique climatic tests near the Russian Antarctic station Bellingshausen. Samples of these materials were exposed during 2 months and after that were examined to find the extent to which their strength characteristics were preserved. The experimental results suggested that, upon addition of 0.05 to 0.5 and 1.0 to 5.0 wt% dry DND powder, the strength parameters of the cement samples are nearly doubled, but upon further increase in the powder content to more than 5 wt%, they start to decrease. It was also found that, after the samples are exposed to the climatic conditions of West Antarctic for 2 months, their strength characteristics remain unchanged. Thus, addition of DNDs to the Portland cement clinker presumably affects its strength characteristics by changing the phase composition of the cement stone, with the result that its bioresistance is improved.

Also, DNDs and DS have a considerable application potential:

- As crystallization centers in static syntheses of diamond macrocrystals from nondiamond carbon and in CVD technologies;
- As technological lubricants (with DB) for pressure processing of metals (cold sheet-metal pressing, cold and hot forging, cold and hot drawing, sizing, and hot pressing). This reduces the friction coefficient (to 0.035) and lubricant expenditure, rules out metal pickup by tools, and improves the quality of finished articles;
- As a material with heat-insulating properties comparable with those of asbestos, which are improved with increasing density of such a material; and
- As a pigment for paints and varnishes and as an active additive to corrosion-resistant paint coatings.

References

Alexenskij, A.E., M.V. Baidakova, A.Y. Vul' et al. "Phase transition of diamond–graphite in the ultradispersed diamond clusters," *Fizika tverdogo tela* 39, no. 6 (1997): 1125–34.

Barnard, A.S. "Self-assembly in nanodiamond agglutinates," *Journal of Materials Chemistry* 18 (2008): 4038–41.

Barnard, A.S., M. Sternberg. "Crystallinity and surface electrostatics of diamond nanocrystals," *Journal of Materials Chemistry* 17 (2007): 4811–19.

Dolmatov, V.Y. "Detonation synthesis ultradispersed diamonds: Properties and application," *Russian Chemical Reviews* 70, no. 7 (2001): 607–26.

Dolmatov, V.Y. "Biologically active detonation synthesis ultradispersed diamonds," Pat. of Russia 2203068 of 12.04.2001, A61K33/44, publ. April, 2003.

Dolmatov, V. "Modified method for synthesis of detonation nanodiamonds and their real elemental composition," *Russian Journal of Applied Chemistry* 81 (2008): 1747–53.

Dolmatov, V. *Detonation Nanodiamonds. Production, Properties and Application*. Saint-Petersburg: Professional, 2011a.

Dolmatov, V.Y. "Diamond–carbon material and a method for the production thereof," Pat. US 7,862,792 B2 of December, 30, 2005, publ., January, 4, 2011b.

Dolmatov, V.Y., L.N. Kostrova. "Detonation-synthesized nanodiamonds and the possibility to develop a new generation of medicines," *Journal of Superhard Materials* 3 (2000): 79–82.

Dolmatov, V.Y., E.K. Gorbunov. "Synthetic radioactive nanodiamond and production method thereof," Pat. Application 2013115568/05(023053) of prior 01.04.2013, reg. June, 2014.

Dolmatov, V.Y., V.G. Suschev, V.A. Marchukov. "A method of isolation of synthetic ultradispersed diamonds," Pat. of Russia 2109683, 01B 31/06, publ. April 1998.

Dolmatov V.Y., V. Myllymaki, and A. Vehanen. "A possible mechanism of nanodiamonds formation during detonation synthesis," *Journal of Superhard Materials* 35, no.3 (2013c): 143–50.

Dolmatov, V.Y., N.M. Lapchuk, A.N. Panova et al. "Peculiarities of paramagnetism and Raman scattering in detonation nanodiamonds synthesized in the presence of doping elements" (paper presented at the Annual 16th International Conference, Rock Cutting and Metal-Working Tools—Techniques and Technology of Its Making and Using, Morskoe, the Crimea, September, 15–21, 2013b).

Dolmatov, V.Y., E.K. Gorbunov, M.V. Veretennikova et al. "Radioactive nanodiamonds," *Journal of Superhard Materials* 35, no. 4 (2013a): 251–55.

Dolmatov, V.Y., I.I. Kulakova, V. Myllymaki. "IR spectra of detonation nanodiamonds modified during the synthesis," *Journal of Superhard Materials* 36, no. 5 (2014): 344–57.

Fu, C.C., H.Y. Lee, K. Chen et al. "Characterization and application of single fluorescent nanodiamonds as cellular biomarkers," *Proceedings of the National Academy of Sciences of the United States of America* 104 (2007): 727–32.

Grabinski, C., S. Hussain, K. Lafdi et al. "Effect of particle dimension on biocompatibility of carbon nanomaterials," *Carbon* 45 (2007): 2828–35.

Khamona, T.V., O.A. Shilova, D.Y. Vlasov et al. "Bioactive coatings based on nanodiamond-modified epoxy siloxane sols for stone materials," *Inorganic Materials* 48, no. 7 (2012): 702–08.

Kondratieva, I.A., A.A. Gorbushina, A.I. Boikova et al. "The study of biodeteriorations of industrial clinker compositions," *Stroitel'nie Materiali* 7 (2006): 58–60.

Kovalev, V.V., E.A. Petrov. "Antifriction material," Pat. of Russia 2005741 of 03.01.1992, publ. January, 1994.

Kruger, A., F. Kataoka, M. Ozawa et al. "Unusually tight aggregation in detonation nanodiamond: Identification and disintegration," *Carbon* 43, no. 8 (2005): 1722–30.

Kulakova, I.I., V.Y. Dolmatov, T.M. Gubarevich. "Chemical properties of ultradispersed detonation diamonds," *Journal of Superhard Materials* 1 (2000): 42–8.

Kulakova, I.I., V.V. Korol'kov, R.Y. Yakovlev et al. "Structure of particles of chemically modified detonation nanodiamond" (paper presented at the Annual 12th International Conference, Rock Cutting and Metal-Working Tools—Techniques and Technology of Its Making and Using, Morskoe, the Crimea, September, 20–26, 2009).

Liu, K.K., C.L. Cheng, C.C. Chang et al. "Biocompatible and detectable carboxylated nanodiamond on human cell," *Nanotechnology* 18 (2007): 325102–03.

Neugart, F., A. Zappe, F. Jelezko et al. "Dynamics of diamond nanoparticles in solution and cells," *Nano Letters* 7 (2007): 3588–91.

Ozerin, A.N., S.A. Ivanov, S.N. Chvalun et al. "Calculation of function of crystallite distribution function by sizes in the polycrystalline samples by means of Fourier analysis of a profile of X-ray diffraction line," *Zavodskaya laboratoriya* 52 (1986): 20–3.

Ozerin, A.N., T.S. Kurkin, L.A. Ozerina et al. "X-ray diffraction study of the structure of detonation nanodiamonds," *Crystallography Reports* 53 (2008): 80–7.

Pramatarova, L., E. Pecheva, S. Stavrev et al. "Artificial bones through nanodiamonds," *Journal of Optoelectronics and Advanced Materials* 9 (2007): 236–39.

Pulskamp, K., S. Diabate, H.F. Krug. "Carbon nanotubes show no. sign of acute toxicity but induce intracellular reactive oxygen species in dependence on contaminants," *Toxicology Letters* 168 (2007): 58–74.

Schrand, A.M. "Characterization and in vitro biocompatibility of engineered nanomaterials," in: *The School of Engineering*. Dayton, OH: University of Dayton, 2007, 276.

Schrand, A.M., L. Dai, J.J. Schlager et al. "Differential biocompatibility of carbon nanotubes and nanodiamonds," *Diamond and Related Materials* 16 (2007): 2118–23.

Vereschagin, A. *Detonation Nanodiamonds*. Bijsk: AlGTU, 2001.

Vjacheslavov, P.M., N.M. Shmeleva. *Testing Methods for Electrolytic Coatings*. Leningrad: Mashinostroenie, 1977.

Vlasov, D.Y., M.S. Zelenskaya, E.V. Safronova. "Mycobiotics of a stony substrate in urban environment," *Mikologiya i Phytopatologiya* 38, no. 4 (2004): 13–22.

Volkov, K.V., V.V. Danilenko, V.I. Elin. "Synthesis of diamond from carbon of detonation products of explosives," *Fizika goreniya i vzriva* 26, no. 3 (1990): 123–25.

Voznjakovskij, A.P. "Composition materials based on polyurethanes and nanocarbons" (paper presented at the Annual 10th International Conference, Rock Cutting and Metal-Working Tools—Techniques and Technology of Its Making and Using, Morskoe, the Crimea, September 15–21, 2007.

Zakharov, A.A., V.E. Red'kin, A.M. Staver et al. "Lubricating composition," Pat. of Russia WO 93/01261, publ. January, 1993.

22

Surface-Modified Nanodiamonds

Wesley Wei-Wen Hsiao

Hsin-Hung Lin

Feng-Jen Hsieh

Huan-Cheng Chang

22.1 Introduction

Diamonds have been admired throughout time for their beauty, endurance, and rarity as jewels. Interestingly, they are now playing a new role as diagnostic and therapeutic tools in biology and medicine, particularly for nanometer-sized diamonds (Ho 2009). Diamond nanoparticles, or nanodiamonds (NDs) in short, were first synthesized by detonation in the former Soviet Union in the 1960s (Danilenko 2004). Although the discovery was not made public for more than two decades because of the Cold War (Osawa 2008), these particles brought about by explosive destruction have found wide applications in modern science and technology (Mochalin et al. 2012).

As an allotrope (sp^3 hybridization) of carbon, diamond is renowned for its superlative physical properties, including (1) being the hardest known material to date, (2) having the highest thermal conductivity of any bulk material, and (3) possessing the largest refractive index of all dielectric materials (Field 1992). Some of these properties are preserved even for nanoscale diamonds. Scientists have developed a number of methods of making NDs. Aside from detonation, NDs can be produced by crushing or ball milling of micrometer- or millimeter-sized diamond crystallites synthesized by high-pressure–high-temperature (HPHT) and chemical vapor deposition (CVD) methods (Schrand et al. 2009). These HPHT-NDs and CVD-NDs are similar to detonation NDs (DNDs) in lattice structure; however, their sizes and surface properties markedly differ. Specifically, CVD-NDs are grown as thin films on substrates. They have been utilized in industries for surface coating of cutting and dressing tools. DNDs are particulate (average diameter ~5 nm), with a highly uniform size distribution for the primary particles. They form tightly bound agglomerates during synthesis through covalent linkage between disordered (sp^2) carbon atoms on their surfaces. Biologists are applying these DND clusters as drug delivery vehicles. HPHT-NDs, on the other hand, are monocrystalline crystallites with high optical transparency. Although their size distribution is broad, they contain less sp^2 carbon on the surface and thus have less agglomeration, which are more suitable for bioimaging applications.

Composed of only carbon atoms, NDs are intrinsically biocompatible and noncytotoxic. Biologists have become interested in these remarkable biochemical properties and their potential applications in nanomedicine (Xing and Dai 2009). However, for practical use in biology, one must overcome three major challenges that stem from the material's inherent properties: (1) difficulty in direct modification on the inert surface of NDs, (2) low colloidal stability due to aggregation of the ND particles in physiological medium, and (3) nonspecific protein adhesion onto the ND surface. The colloidal stability of unmodified NDs is not static and, in fact, it drops at higher ionic strengths in solution, similar to the behavior of many other nanoparticles. This poses a serious problem for the use of NDs in biological systems since physiological buffers and cultivation media all contain high concentrations of salts. Thus, before biological applications of NDs, researchers must understand their surface properties and the aspects determining their interactions with environments, which in turn influence the colloidal stability of these carbon-based nanomaterials.

In this chapter, we will reexamine three approaches to the surface modification of NDs, including functionalization, encapsulation, and bioconjugation. We will then review the toxicity assessments of these nanomaterials, both in vitro and in vivo, and place particular emphasis on recent advances in the development of surface-modified NDs for applications in biology and nanoscale medicine.

22.2 Surface Modification of NDs

Only in form of a stable colloidal dispersion of NDs is valuable in biological research. Dynamic light scattering is a useful tool to characterize the colloids and their stability. The technique measures the hydrodynamic sizes of the particles in solution, yielding information on the size change before and after surface modification. The zeta potential is another key parameter that describes the stability of a colloid and its tendency toward agglomeration. The colloid is considered stable when the zeta potential is lower than −30 mV or higher than +30 mV. It has been reported that NDs can form colloids with zeta potentials ranging from −40 mV to +40 mV, depending on their surface terminations (Vial et al. 2008; Williams et al. 2010; Kaur et al. 2012). Thus, it is expected that through proper surface modification, the effects of particle agglomeration and nonspecific interaction can be avoided.

22.2.1 Functionalization

NDs, once produced by detonation, HPHT, or CVD methods, are always contaminated with sp^2 carbon atoms on the surface. To effectively remove these contaminants as well as residual chemical compounds, researchers have treated the as-formed ND materials with strong oxidative acids (Huang and Chang 2004), air oxidation (Osswald et al. 2006), or ozone (Shenderova et al. 2011). As a result, these particles are terminated with a variety of functional groups including carboxylic (COOH) and carbonyl (C=O) groups, in addition to different alcohol (tertiary, secondary, primary) and ether groups, etc. Fourier-transform infrared (FTIR) spectroscopy, Raman spectroscopy, x-ray photoelectron spectroscopy, and thermal desorption mass spectrometry are commonly used tools to characterize the phase composition and surface terminations of these NDs. Through FTIR, researchers can differentiate different types of functional groups and adsorbed species on the ND surface. Moreover, they can detect subtle changes in the chemical composition before and after surface modification using these spectroscopic methods. For example, the FTIR spectra of NDs often exhibit absorption bands at 1700–1800 cm^{-1}, which can be associated with carbonyl, ketone, aldehyde, carboxylic acid, ester, and other oxygen-containing groups. The spectra will be much simplified if a homogeneous layer of hydroxyl groups is formed on the surface by borane reduction (Krueger et al. 2006).

High-temperature gas treatments also serve as a way of modifying the ND surface. For instance, heating NDs in NH_3 can lead to the production of a variety of nitrogen-containing functional groups, including NH_2, C≡N, and moieties containing C=N. In contrast, heating NDs in Cl_2 produces acylchlorides and in F_2 forms C–F groups. Treatment in H_2 completely reduces C=O to C–OH and even creates C–H

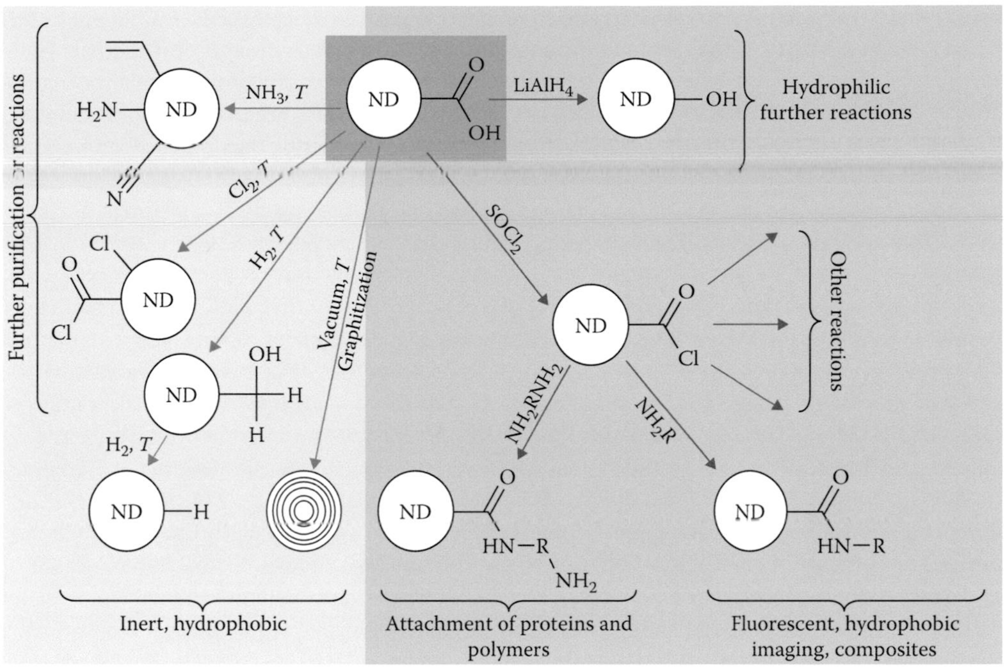

FIGURE 22.1 Overview of the commonly used methods for chemical modification of ND surfaces. (From Mochalin, V.N. et al., *Nat. Nanotechnol.*, 7, 11–23, 2012.)

groups. Hydroxyl groups can be removed at higher temperatures through longer hydrogenation times or via hydrogen plasma treatment. Annealing them in N_2, Ar, or vacuum at high temperatures completely eliminates all these functional groups and converts DNDs into graphitic carbon nano-onions. Krueger and Lang (2012) have presented a full description of all possible techniques for the chemical treatment and functionalization of ND surfaces.

To summarize, both wet chemistry techniques and high-temperature gas treatments have enabled the attachment of an extensive variety of functional groups to the surface of NDs. These chemical alterations give researchers access to further conjugation of NDs with bioactive ligands or biomolecules for biotechnological and biomedical applications. An overview of the commonly used strategies for the chemical modifications of ND surfaces is shown in Figure 22.1 (Mochalin et al. 2012).

22.2.2 Encapsulation

Although alteration of the ND properties can be readily attained by direct surface modification, polymer grafting appears to be a promising alternative. The polymer grafting on ND surface can be prepared by either "grafting to" or "grafting from" approaches. The "grafting to" technique involves a chemical reaction between end-functionalized polymer chains and complementary functional groups on the solid substrate surface, whereas the "grafting from" refers the technique of synthesis of a covalently attached polymer in situ on the solid substrate (Minko 2008). Poly-L-lysine was one of the first polymers covalently grafted on the carboxylated surface of NDs through amide bond formation (Fu et al. 2007). Researchers later used polyelectrolytes (such as polyallylamine) or organosilanes for the surface functionalization with amino groups (Vial et al. 2008). They also found atom-transfer radical-polymerization to be an efficient method for attaching polymer brushes onto the ND surfaces terminated with an initiator (Dahoumane et al. 2009; Zhang et al. 2012a). Coating of NDs with hyperbranched polyglycerol by ring-opening polymerization significantly improved their colloidal stability in water and buffers (Zhao et al. 2011). Lately, copolymers consisting of poly[*N*-(2-hydroxypropyl)methacrylamide] were introduced as a highly biocompatible, protein-resistant coating. These polymeric molecules are hydrophilic and highly flexible, allowing them to capably prevent ND aggregation in biological medium and enabling bioorthogonal attachment of various types of molecules by the so-called click chemistry (Rehor et al. 2014a).

Recently, forward-thinking researchers have introduced NDs with novel silica coatings (Bumb et al. 2013; Prabhakar et al. 2013; von Haartman et al. 2013; Rehor et al. 2014b,c; Slegerova et al. 2015). These coatings provide a platform for subsequent chemical treatment based on silica chemistry. For example, the silica-encapsulated ND surface contains a variety of free silanol groups that allow the conjugation of biomolecules to the encapsulated particles. Figure 22.2 illustrates the concept of coating NDs with methacrylamide copolymers grown from an ultrathin silica shell. The copolymer can bear both fluorescent probes (e.g., Alexa Fluor 488) and targeting peptides (e.g., cyclic Arg-Gly-Asp peptide [cRGD]) via click chemistry (Slegerova et al. 2015). Through the chemical treatment, researchers can selectively attach the molecules of interest and drastically improve the colloidal stability of the particles. In addition, the shell coating can normalize the irregular shape of the original ND particle, as justified in Figure 22.3 (Rehor et al. 2014b).

Researchers have recently also reported that NDs can form complexes with liposomes. Chang and coworkers have shown that HPHT-NDs could be encapsulated within cationic cholesterol-based lipids after surface reduction and silanization (Hui et al. 2010). The encapsulation enhanced the diffusion of the particles in the cytoplasm of living cells by more than one order of magnitude. Ho and coworkers have independently developed self-assembled ND-lipid hybrid particles, which allowed for a potent

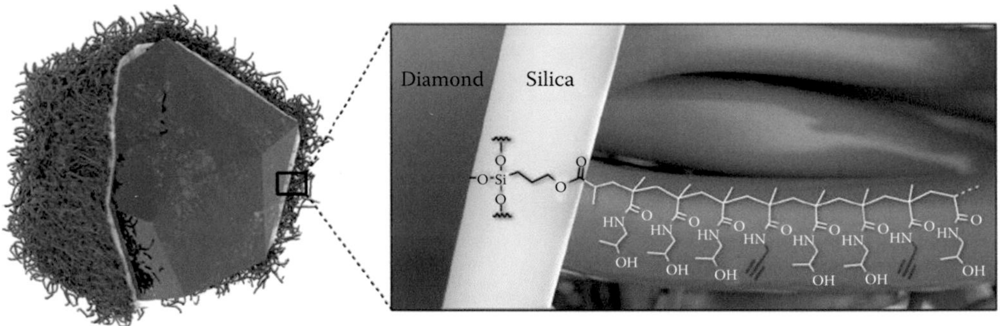

FIGURE 22.2 Illustration of the structure of silica coating on an ND particle for further conjugation with polymeric molecules. (From Slegerova et al., *Nanoscale*, 7, 415–420, 2015.)

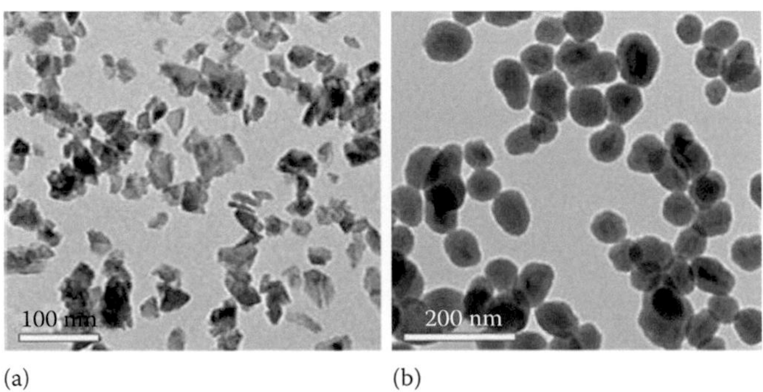

FIGURE 22.3 Transmission electron microscopy images of (a) as-received and (b) silica-coated ND particles. (From Rehor, I. et al., *Small*, 10, 1106–1115, 2014.)

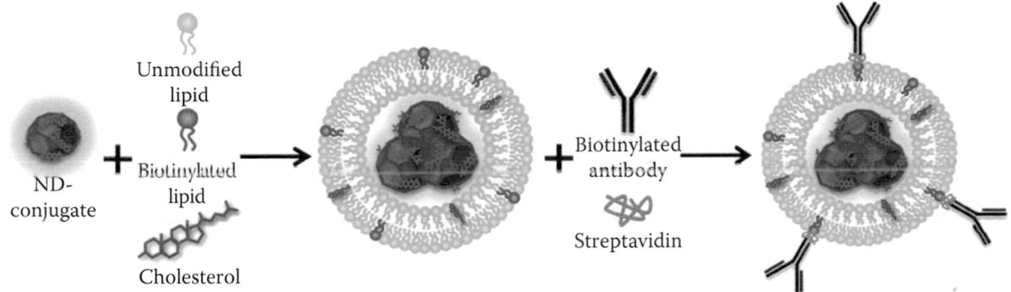

FIGURE 22.4 Encapsulation of NDs in liposome by rehydration of lipid thin films containing cholesterol and biotinylated lipids in concentrated ND solution. (From Moore, L. et al., *Adv. Mater.*, 25, 3532–3541, 2013.)

interaction between the ND surface and small molecules. At the same time, they provided a mechanism for the cell-targeted delivery of imaging or therapeutic payloads, as shown in Figure 22.4 (Moore et al. 2013). The ease of the ND encapsulation in liposomes opens a promising new avenue to conjugate the particles with bioactive ligands or proteins on the lipid layer for specific targeting applications.

22.2.3 Bioconjugation

The surface of NDs can be rich in a variety of functional groups, depending on the methods of diamond synthesis and subsequent chemical treatments. These functional groups can yield diverse behaviors ranging from inertness to high reactivity or even high toxicity in cells and living organisms. Conjugation of NDs with biomolecules can be achieved by either physical adsorption through noncovalent interactions or covalent linkage through the surface functional groups for stable and site-specific bonding. It has been reported that noncovalent coating of the nanoparticles with proteins such as serum albumin can stabilize the ND colloids in phosphate-buffered saline (PBS) over weeks (Tzeng et al. 2011; Lee et al. 2013).

A variety of organic and biological molecules have been conjugated onto NDs either covalently or noncovalently (Figure 22.5). These include chemotherapeutic drugs (Huang et al. 2007; Li et al. 2010; Liu et al. 2010), carbohydrates (Hartmann et al. 2012; Barras et al. 2013), peptides (Huang and Chang 2004; Vial et al. 2008), proteins (Nguyen et al. 2007; Shimkunas et al. 2009; Chang et al. 2013), small interfering RNA (Chen et al. 2010; Alhaddad et al. 2011), and DNA (X.Q. Zhang et al. 2009; Kaur et al. 2012). When attaching biomolecules like proteins to the ND surface, care must be taken to conserve their functionalities. One of the examples is given by Nguyen et al. (2007) in the immobilization of lysozymes onto HPHT-NDs. The researchers detected the hydrolytic activity of the enzymes after adsorption but found that their activity decreased as the surface coverage was lowered. To solve this issue, they blocked the ND surface with supplementary proteins such as cytochrome *c*, thereby creating a more "crowded" environment. This tactic increased the activity of the ND-bound lysozymes from 60% to 70%.

To reduce steric constraints and retain the adsorbate's activity more effectively, researchers have developed methods to insert spacers between the biomolecules of interest and NDs. This step is crucial for enzymes with active sites that may be sterically hindered after attachment to the ND surface. Another function of the spacers is that it helps suppress nonspecific interactions and protein conformational changes caused by strong biomolecule–particle interactions. One of the most commonly used spacers is polyethylene glycol (Zhang et al. 2012a), which has a low toxicity and a low degree of specific interactions with biomolecules. They can be covalently conjugated with carboxylated NDs via the amino groups on their termini or other conjugation methods (Figure 22.6). These spacers are available in various discrete lengths and also provide carboxyl groups on the other termini for carbodiimide crosslinking with bioactive ligands (B. Zhang et al. 2009) or proteins (Chang et al. 2013) for ensuing cell labeling and targeting applications.

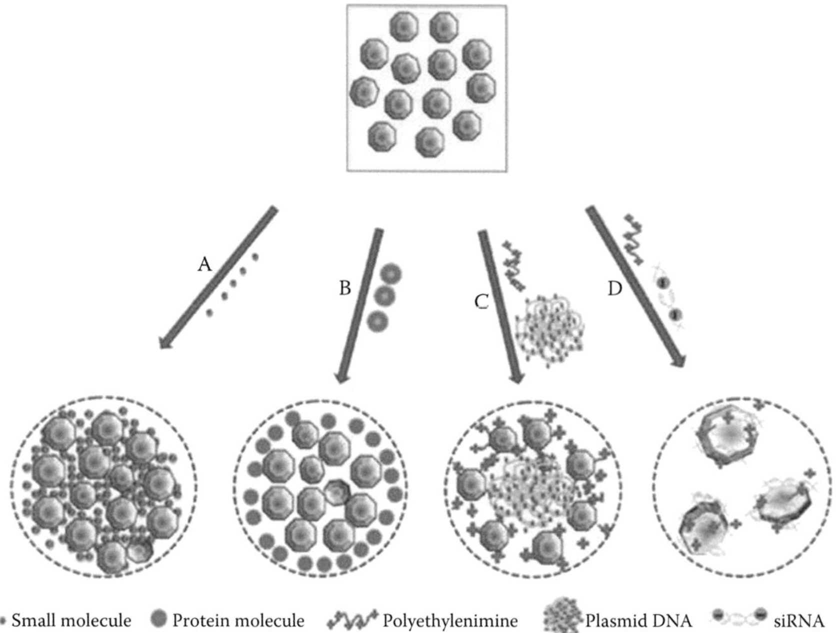

FIGURE 22.5 A schematic diagram representing the bioconjugation of NDs with different types of biomolecules. (From Kaur, R., Badea, I., *Int. J. Nanomed.*, 8, 203–220, 2013.)

Many potent drugs, such as those treating cancer, are bundled with delivery challenges. For instance, some of the drugs are not soluble in polar protic solvents (such as water) but are soluble in polar aprotic solvents that are harmful to the body. Through surface modification in conjunction with drug loading on NDs, researchers can create new delivery methods to solve this problem. The benefits are many, as NDs are nontoxic and biocompatible and have the ability to transport significant amounts of drugs. While some studies have already used ND surfaces to conjugate with drugs via chemical bonding (Liu et al. 2010; Li et al. 2011), a majority of the researches have been focusing on physical adsorption procedure (Chen et al. 2009; Chow et al. 2011).

22.3 Toxicity Assessments of Surface-Modified NDs

Scientists have become increasingly interested in the potential applications of surface-modified NDs in biological research. These applications span from in vitro cell-based systems to in vivo animal models. A large body of studies on the biocompatibility and toxicity of nanoparticles has concluded that carbon-based nanomaterials are likely more biocompatible and less toxic toward cells and organisms as compared with metal- or semiconductor-based nanomaterials (Lewinski et al. 2008). Carbon nanoparticles themselves, however, differ in the degree of toxicity both in vitro and in vivo. When cells were treated with carbon black, carbon nanotube, graphene, and ND, results showed that NDs have the highest biocompatibility and lowest cytotoxicity (Schrand et al. 2007a; Zhang et al. 2012b). The relative toxicity of these nanomaterials in vivo, however, awaits further exploration.

22.3.1 In Vitro Studies

HPHT-NDs and DNDs are two of the most popular NDs used in biological research. Chang and coworkers were the first to examine the biocompatibility of oxidized HPHT NDs in human cell lines (Yu et al. 2005). They employed techniques such as the 3-[4,5-dimethylthiazol-2-yl]-2,5-diphenyltetrazolium

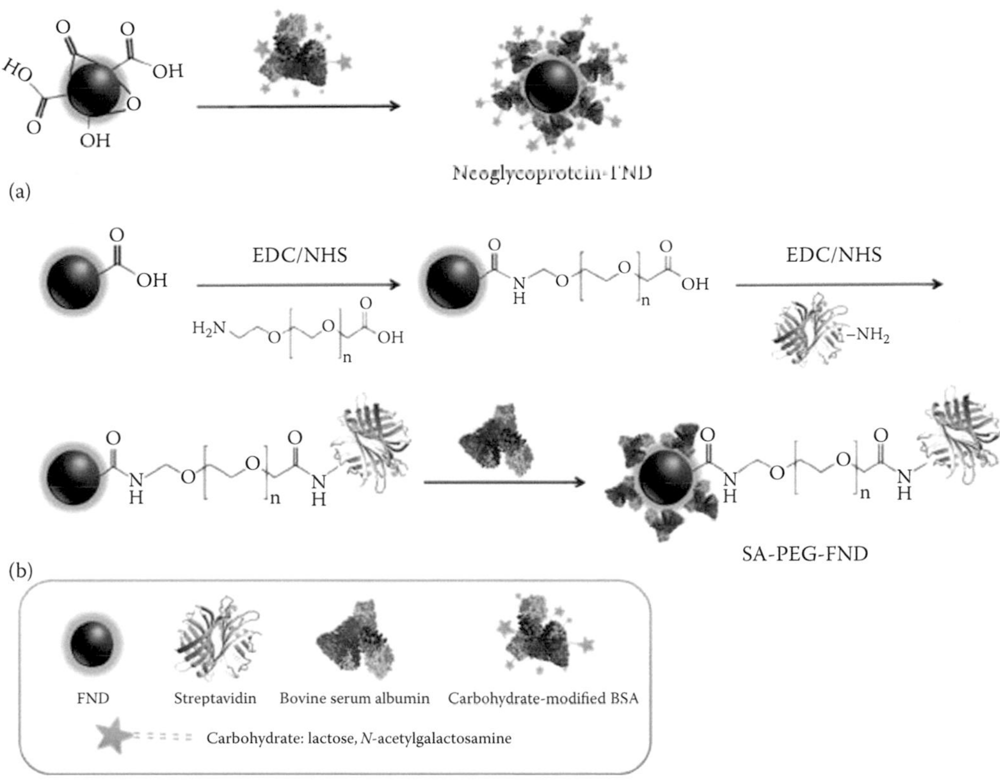

FIGURE 22.6 Work flows of the bioconjugation of HPHT-NDs by (a) physical adsorption and (b) covalent linkage through activation of the surface carboxyl groups. EDC, *N*-(3-dimethylaminopropyl)-*N*'-ethylcarbodiimide hydrochloride; FND, fluorescent ND; NHS, *N*-hydroxysuccinimide; PEG, polyethylene glycol. (From Chang, B.M. et al., *Adv. Func. Mater.*, 23, 5737–5745, 2013.)

bromide (MTT) assay to assess toxicity. The principle of this assay is that it takes the reduction activity of MTT as an indicator of the cell's viability according to its mitochondrial function. Subsequent studies showed that the surface modification with biomolecules such as peptides (Vaijayanthimala et al. 2009) does not alter the biocompatibility of HPHT-NDs. Further studies by the same research group indicated that HPHT-NDs purified by air oxidation and strong oxidative acid treatments have excellent hemocompatibility with negligible hemolytic and thrombogenic activities in human blood. No cytotoxicity effect was detected for human primary endothelial cells treated with oxidized HPHT-NDs of dimensions of 35–500 nm (Li et al. 2013). These conclusions were augmented by a recent study showing that HPHT-NDs can effectively enter human cells but do not induce any significant cytotoxic or genotoxic effects on six cell lines with an exposure dose of up to 250 μg/mL (Paget et al. 2014). Furthermore, these NDs are biocompatible with stem cells and therefore can serve as useful labels of these cells for localization and tracking in vivo (Blaber et al. 2013; Wu et al. 2013).

A number of experiments have provided results supporting the intrinsic biocompatibility of DNDs (Schrand et al. 2007a,b, 2009); however, some studies have refuted this intrinsic biocompatibility, arguing that DNDs can induce toxic responses under certain conditions (Xing et al. 2011; Solarska et al. 2012; Zhang et al. 2012b). Studies that made this argument have found that the toxicity of DNDs varies, depending on the surface chemistry (Marcon et al. 2010; Xing et al. 2011), the type of cell lines used for the assessment (Xing et al. 2011), and the composition of the treatment medium (Li et al. 2010). Concerns about the toxicity of DNDs typically arise as a result of the small size of DNDs and their ability to enter cells and localize in critical organelles. The high biocompatibility and low cytotoxicity of DNDs, however, have been well founded as per the studies of Schrand et al. (2009) on the subject. They failed to observe any disruption of mitochondrial membrane permeability, morphological alterations, or viability changes (using luminescence measurement of adenosine triphosphate [ATP] production) when using 5–100 μg/mL DNDs. Moreover, their DNDs did not induce the generation of reactive oxygen species nor cause oxidative stress, which could have led to membrane dysfunction, protein degradation, or DNA damage. The researchers confirmed the lack of change in the expression of genes that served as indicators of inflammation and protection against apoptosis in macrophages and colorectal cancer cells incubated with DNDs. A most recent study with an array of ND subtypes (both DNDs and HPHT-NDs) showed that the carbon-based nanoparticles are well tolerated by multiple cell types at both functional and gene expression levels (Moore et al. 2014).

22.3.2 In Vivo Studies

In contrast to the in vitro studies with living cells, in vivo biocompatibility evaluations employ animal models to explore the potential impacts of NDs on human health. A variety of animal models have been used to assess the toxicity of NDs in vivo, including *Paramecium caudatum* and *Tetrahymena thermophile* (Lin et al. 2012), *Caenorhabditis elegans* (Mohan et al. 2010), *Xenopus* embryos (Marcon et al. 2010), *Drosophila melanogaster* embryos (Simpson et al. 2014), zebrafish embryos (Mohan et al. 2011; Chang et al. 2012), mice (Yuan et al. 2009, 2010; Zhang et al. 2010; Vaijayanthimala et al. 2012), rats (Vaijayanthimala et al. 2012), and rabbits (Puzyr et al. 2007).

C. elegans is an ideal model organism to assess the toxicity of a nanoparticle. The worm is small (~1 mm in length) and optically transparent. It has also been employed as a benchmark system for ecotoxicological studies owing to its short life cycles, ease of handling, and high sensitivity to various types of stresses. Researchers introduced surface-oxidized HPHT-NDs into the nematodes by either feeding or microinjection and found that both the growth and development of the ND-containing worms were normal (Mohan et al. 2010; Kuo et al. 2013). Toxicity assessments, performed by using longevity and reproductive potential as the physiological indicators, showed that the NDs are nontoxic and do not cause any detectable stress to the worms.

Rabbits are one of the earliest animal models used for this study (Puzyr et al. 2007). Researchers gave a high dose (125 mg) of DNDs to the rabbits through intravenous administration yet did not cause any deaths. Both the red blood cell count and the hemoglobin level of the rabbits remained stable for 15 min after the administration of the NDs. However, after a longer period (48 h after the injection), the levels of biochemical molecules such as total bilirubin, triglyceride, and low-density lipoprotein changed to a statistically significant degree. But 3 months later, the rabbits showed no signs of inflammation, suggesting long-term biocompatibility of the NDs.

In the studies using murine models, it has been found that the intratracheal instillation of DNDs has a negative effect on the lungs, liver, kidneys, and hematological systems of mice (Zhang et al. 2010). These effects did not lead to differences in body weight or abnormal pathologies. Of the organs, the lungs suffered the most severe toxicological effects, becoming inflamed and suffering tissue damage, a result of the high uptake and long retention time of the NDs in the lung tissue. However, the results of Yuan et al. (2010) did not show any pulmonary toxicity in mice after intratracheal instillations of DNDs, in contrast to the result of Zhang et al. (2010), who used a lower dose in the treatment. The contradictions in the results could be associated with the sources of DNDs and how their surfaces are modified, as found in cell studies. For example, researchers observed no in vivo toxicity and no visible side effects (e.g., stress response) after HPHT-NDs were microinjected into rats over a 5-month period. The differences between the control and ND-treated organisms were insignificant (Vaijayanthimala et al. 2012). No significant elevation of the inflammatory cytokine levels of interleukin-1β (IL-1β) and IL-6 was detected in mice after intravenous injection of the HPHT-ND particles either (Li et al. 2013).

To summarize, the literature currently published in the field has shown the high biocompatibility of NDs in various cell lines and some animal models with minimal or no cytotoxicity. Thus, NDs have much potential for biomedical applications. However, other factors, such as the purity, surface chemistry, and dimensions of particles, could play a role in the presence of cytotoxicity. Further problems to be considered include immunogenicity, diffusivity, and metabolism of NDs such as the nonspecific accumulation of the particles in reticuloendothelial systems when administrated in vivo. As scientists are planning to use NDs in nanoscale medicine, it is our responsibility to attest to the safety of the nanomaterials via more animal studies before progressing to human clinical trials.

22.4 Application of Surface-Modified NDs

The modern era of nanoscience and nanotechnology has brought about diverse applications of surface-modified NDs in biology. These applications are emerging in various branches of life sciences, including biotechnology and biomedicine. Notable examples of these applications are the use of NDs as solid-phase extraction (SPE) supports, enzyme immobilization substrates, bactericidal agents, and biomolecule delivery vehicles. As NDs have been harnessed for the delivery of many classes of molecules with a major focus on chemotherapeutic agents, a long-term goal of these activities is to implement the surface-modified NDs for clinical use to facilitate the diagnosis and treatment of various diseases.

22.4.1 Biotechnical Applications

22.4.1.1 Solid Phase Extraction

Selective extraction of biomolecules such as proteins from a complex mixture is a key step in bioanalytical chemistry. Different approaches have been developed to use solid supports such as micrometer-sized agarose beads, which are typically packed in affinity chromatography columns, to capture proteins of interest from crude sample solutions. These beads are then recovered and analyzed with gel electrophoresis or mass spectrometry. However, direct analysis of these surface-bound proteins is often accompanied with reduced mass resolution and accuracy associated with the interference from the micrometer-sized beads in ion formation and extraction.

Given its excellent chemical stability, small size, and ease of purification, NDs have been proposed as an SPE device for bioanalytical applications (Wu et al. 2010). Chang and coworkers demonstrated the utility of acid-washed NDs as a tool to facilitate protein analysis with HPHT-ND particles of ~100 nm in size (Kong et al. 2005a). These particles have a remarkably high affinity for polypeptides and proteins and a large protein loading capacity of up to 100 mg/g. They can extract proteins from complex medium or highly diluted solution in minutes. A combination of ionic interactions, hydrogen bonding, and hydrophobic interactions might be the origin for the high affinity. The affinity is so high that these protein–ND complexes can sustain repeated washing with deionized water without much loss. Moreover, after separation by either centrifugation or filtering, they can be immediately analyzed by gel electrophoresis or mass spectrometry. A platform called "SPEED" (SPE and elution on diamond) has been developed for proteomic analysis by the researchers (Chen et al. 2006). The advantage of this SPEED platform is that it facilitates purification and concentration of intact proteins and their enzymatic digests for ensuing sodium dodecyl sulfate–polyacrylamide gel electrophoresis (PAGE) or matrix-assisted laser desorption/ionization mass spectrometry (MALDI-MS) analysis without previous removal of the ND adsorbent (Figure 22.7). Moreover, one-pot work flow involving the reduction of disulfide bonds, protection of free cysteine residues, and proteolytic digestion of the adsorbed proteins can be directly carried out on the particles.

Through chemical alteration, it is possible to adjust ND's affinity for different types of biomolecules. For example, covalent or noncovalent coating of NDs with polylysine (PL) or polyarginine (PA) creates a positively charged layer on the ND surface. These PL- and PA-coated NDs exhibit a strong affinity for

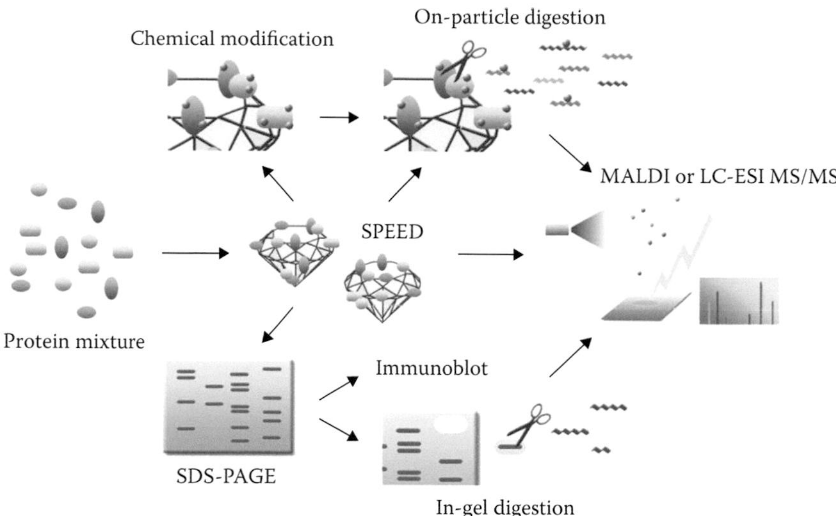

FIGURE 22.7 Scope of typical applications of the SPEED platform to proteome analysis. ESI, electrospray ionization; LC, liquid chromatography; SDS sodium dodecyl sulfate. (From Wu, C.C. et al., *J. Chin. Chem. Soc.*, 57, 582–594, 2010.)

negatively charged species such as DNA (Kong et al. 2005b), phosphopeptides (Chang et al. 2008), and sulfated glycosaminoglycans (Hsieh et al. 2013). They provide a speedy platform for quantitative analysis of these biomolecules by high-affinity purification and MALDI-MS. Likewise, NDs functionalized with boronic acid show specificity for glycopeptides (Xu et al. 2013). Finally, when playing the role of selective extractors of biomolecules from solution, NDs have a clear advantage—their inertness, which prevents them from reacting with the adsorbed analytes before PAGE or MS analysis.

22.4.1.2 Enzyme Immobilization

Immobilization of enzymes on a substrate has been an important process in biotechnology for commercial uses. Many of the disadvantages of enzymes, such as their short-term stability and complications in recovery and recycling, can be eliminated and improved by noncovalent or covalent immobilization of the molecules on HPHT-ND or DND surfaces. This immobilization augments stability, facilitates the repeated use and easy separation from the reaction mixture, as well as improves the catalytic properties of an enzyme (Bornscheuer 2003). In such applications, one benefit of using NDs as opposed to other nanoparticles is their biocompatibility. For example, one study found that the catalytic activity of immobilized trypsin on DND surfaces was not significantly altered (Wei et al. 2010). Both the Michaelis constant and the maximum effective velocity of ND-bound trypsin were within the same order of magnitude as free trypsin. Notably, the immobilized trypsin could catalyze protein cleavage with an even higher efficiency than that of free trypsin. In contrast to free trypsin, the immobilized enzyme did not undergo autolysis. This implies that performing the catalytic reaction with immobilized enzyme is faster and has in effect higher enzyme concentrations than the reaction with trypsin alone. MS analysis showed that 5 min of bovine serum albumin and myoglobin cleavage by immobilized trypsin was comparable with 12 h of the cleavage by free trypsin.

22.4.1.3 Antibacterial Activity

The applications of NDs as bactericidal agents consist of two parts: NDs and ND–protein complexes. An early study reported that NDs noncovalently modified with specific immunoglobulin antibodies can interact with *Salmonella typhimurium* and *Staphylococcus aureus* in a solution (Huang et al. 2004).

Later experiments showed that HPHT-NDs conjugated with lysozymes can bind to *Escherichia coli*, yet nonmodified NDs have no affinity for the bacteria (Perevedentseva et al. 2007). Through a bacterial survival test, the ND–lysozyme complex was demonstrated to have high antibacterial activity. These results highlight the importance that the ND–protein conjugates can serve as an effective probe for the detection and treatment of bacteria or other pathogens (including viruses) in solution.

A recent study found that DNDs, despite their good compatibility with eukaryotic cells, can kill Gram-positive as well as Gram-negative bacteria rapidly and efficiently. Researchers investigated six different types of DNDs pretreated with annealing in H_2 or air at high temperatures (450°C–500°C) and determined how their surface compositions affect the bactericidal activities (Wehling et al. 2014). They attributed the activities to the partially oxidized and negatively charged surfaces, especially those containing acid anhydride groups. As the bactericidal potential of these NDs can be inhibited by proteins such as fetal bovine serum, the extent and duration of their antibacterial action can be properly controlled and modulated for specific applications.

22.4.2 Biomedical Applications

22.4.2.1 Protein Delivery

To be useful as a tool for biomolecule delivery, a nanoparticle carrier should have the characteristics of (1) a sufficient loading capacity in proportion to its weight, (2) strong binding of bioactive ligands or biomolecules to its surface, and (3) a functional mechanism for targeted release. NDs comply with all these requirements. As discussed in the earlier sections, the ND particles (diameter <100 nm) have large specific surface areas and good biocompatibility and can be readily surface-modified with various functional groups. The high loading capacity grants the ND a high concentration of the payload to be delivered yet requires less of the delivery agent itself. With a slow release mechanism, ND allows for the development of novel applications for drug delivery, such as controlled and sustained release delivery. In addition, the ability to protect and retain the inherent therapeutic effects of the attached entities makes ND a suitable carrier.

ND has a role in the therapeutic release of insulin and transforming growth factor-β (TGF-β) serving as the model therapeutics (Shimkunas et al. 2009; Smith et al. 2011). ND-delivered insulin is a potential promoter of wound healing and vascularization in patients with severe burns and other conditions. The ND delivery of insulin is pH-triggered, a fact that allows this method to target bacterial infections accompanying the serious wounds. NDs can also deliver the protein, TGF-β, which is a potential antiscarring agent. The TGF-β–ND complexes can be soundly stored in PBS and triggered release follows their incubation in serum-containing media. Studies have confirmed via enzyme-linked immunosorbent assay that the structure of the adsorbed protein molecules was preserved (Smith et al. 2011).

Taken together, these studies demonstrate that NDs have applications not just in cancer drug delivery but also in wound healing and antiscarring. What is more is that the field of regenerative medicine, which includes bone, cardiovascular, and neuron tissue engineering, might also benefit from the ND-enabled technology as it often relies on controlled growth factor release. The ND's gift of mediating protein elution as a way of minimizing burst release and the recent synthesis of polymeric matrices or planar deposition of NDs could prove to be foundations to its extension toward regenerative medicine applications.

22.4.2.2 Gene Delivery

Researchers have been investigating the combination of gene delivery and nanomedicine because of its potential of combining efficacy and safety into a single platform (Chen et al. 2009; X.Q. Zhang et al. 2009; Zhang et al. 2011). Conventional approaches are often extreme: The efficiency is directly proportional to the toxicity. Naked nucleic acids alone fail as therapeutics because of their inefficient cellular delivery. Physicochemical characteristics such as high negative surface charges and large molecular weights are additional barriers to the efficient cellular internalization of DNAs (Wiethoff and Middaugh 2003). The relatively small size of interfering RNAs (siRNAs) also obstructs cellular delivery, as particles with molecular weight under 50 kDa are susceptible to excretion through glomerular filtration (Gary et al. 2007; Whitehead et al. 2009).

NDs have entered the scene as gene delivery platforms, finding usefulness when combined with a conventional polyethylenimine 800 polymer (PEI800). This polymer, while less efficient than its counterparts, is also less toxic. Compared with PEI800 alone, ND-PEI800 mediates a 70-fold increase in transfection efficacy, with maintained biocompatibility for plasmid DNA. In addition, ND-PEI800s mediates a 400- and 800-fold increase in transfection efficacy when compared to amine- and carboxyl-terminated NDs (X.Q. Zhang et al. 2009). This polymer also effectually silences the expression of green fluorescent protein via siRNA delivery, only with improved efficacy and reduced toxicity than that of Lipofectamine in serum-containing media (Chen et al. 2010). The studies that followed employed ND-PEI complexes in addressing select disease models, such as Ewing's sarcoma, a type of pediatric bone cancer (Alhaddad et al. 2011). The siRNA delivery can inhibit EWS-Fli1 expression, which in turn lowers the capacity for cancer cell proliferation. It is concluded from these studies that NDs have the potential to heighten the transfection ability of polymers while remaining biocompatible with the cell lines.

22.4.2.3 Preclinical Validation of Cancer Therapy

Studies of ND-based drug delivery have been focused on the use of NDs in doxorubicin (DOX) release (Huang et al. 2007; Chow et al. 2011). DOX, while a common form of treatment, is highly toxic and can lead to complications such as myelosuppression and cardiotoxicity. The study that introduced the concept of ND drug delivery showed NDs to potently sequester DOX on their surface, delaying drug release and activity. To assess the safety of NDs, the researchers conducted maintained drug efficacy against multiple cell lines and quantitative real-time polymerase chain reaction studies. The results did not show an increase in the expression of inflammatory cytokines, including IL-6, tumor necrosis factor-α, or inducible nitric oxide synthase (Huang et al. 2007). In addition, DOX bound to the ND surface was significantly more toxic toward DOX-resistant mouse LT2-Myc liver and 4T1 mammary tumor models than was bare DOX (Chow et al. 2011).

As a second benefit, in long-term treatment, DOX bound to ND was superior to bare DOX in preventing tumor growth. NDs both circumvented the premature efflux of DOX from tumor cells and decreased the adverse effects of naked DOX through a significant reduction in myelosuppression and early mortality (Chow et al. 2011). In addition, NDs increased the circulation half-time of DOX from 0.83 to 8.43 h, which further ratified the sequestering behavior of NDs, as anticipated in earlier studies (Huang et al. 2007).

In addition, the ND-enabled delivery of water-insoluble drugs 4-hydroxytamoxifen and purvalanol A, drugs under development as breast cancer and liver cancer therapies, respectively, showed preserved efficacy following an efficient synthesis process (Chen et al. 2009). Chemotherapeutic drugs such as 4-hydroxytamoxifen and purvalanol A have poor water solubility but are soluble in the polar, organic solvents dimethyl sulfoxide (DMSO) and ethanol, respectively. The use of nonaqueous solvents limits the parenteral administration of these formulations in clinically relevant settings. However, when formulated with NDs, the aqueous dispersibility of purvalanol A and 4-hydroxytamoxifen markedly increased. Through the adsorption of the drugs on their surfaces, NDs considerably decreased particle size while improving the zeta potential of these drugs in water, which in turn fosters their dispersibility and potential for cellular uptake. These findings allude to the role that NDs could play in designing injectable formulations of water insoluble drugs.

In summary, through providing high surface loading, better aqueous dispersibility, continued tunable release, and improved retention in chemoresistant cells, the formulation of chemotherapeutic drugs on NDs advances the pharmaceutical properties of the agents. The hydrophilic functional group-enriched surface, significant surface-to-volume ratio, adeptness at forming loose clusters, superior cellular delivery, and excellent biocompatibility of the NDs allow for the aforementioned properties. Research has shown that the ND is a suitable platform on which to assemble chemotherapeutic drugs for surmounting many of the major challenges faced in medicine.

22.5 Conclusions and Perspectives

Over a brief time frame, NDs have emerged as an important and promising nanomaterial for myriad biological applications. Their diversity in structure, size, and surface properties has resulted in their use in a diverse selection of fields, including bioanalysis, drug delivery, cancer therapy, and bioimaging. Compared with other nanoparticles, NDs have higher potential to offer an all-inclusive drug delivery platform—one with superior safety profiles and heightened efficacy. Further studies on their structure, surface chemistry and coatings, improved control of their colloidal properties, as well as the continued development of surface modification and targeting methods promise successful creation of effective ND-based drug delivery systems. It is expected that the promising applications of NDs in biology and nanoscale medicine will continue to attract the attention of researchers in the fields, making a steady flow of progress. But a greater control over the properties of this nanomaterial will only come when we better understand their structure and surface chemistry. Likewise, an increase in our knowledge on this subject will result in improved manufacturing volumes, conceivably to levels that will transcend those of fullerenes, carbon nanotubes, graphenes, and other carbon nanomaterials. As we search for innovated ways to create surface-modified NDs, the resultant buildup in supply will give ways to novel applications. Thus, NDs might one day forge ahead and prove to have more practical uses in research, technology, and medicine.

Acknowledgments

The authors gratefully acknowledge Dr. Petr Cigler at the Institute of Organic Chemistry and Biochemistry, v.v.i. Academy of Sciences of the Czech Republic, for his help and suggestions. They also thank Damon Verial for proofreading the manuscript. This work was supported by the research project (NSC 103-2628-M-001-005-) granted by the National Science Council of Taiwan.

References

Alhaddad, A., Adam, M. P., Botsoa, J. et al., "Nanodiamond as a Vector for siRNA Delivery to Ewing Sarcoma Cells." *Small* 7 (2011): 3087–3095.

Barras, A., Martin, F. A., Bande, O. et al., "Glycan-Functionalized Diamond Nanoparticles as Potent *E. coli* Anti-Adhesives." *Nanoscale* 5 (2013): 2307–2316.

Blaber, S. P., Hill, C. J., Webster, R. A. et al., "Effect of Labeling with Iron Oxide Particles or Nanodiamonds on the Functionality of Adipose-Derived Mesenchymal Stem Cells." *PLoS One* 8 (2013): e52997.

Bornscheuer, U. T., "Immobilizing Enzymes: How to Create More Suitable Biocatalysts." *Angewandte Chemie-International Edition* 42 (2003): 3336–3337.

Bumb, A., Sarkar, S. K., Billington, N., Brechbiel, M. W., and Neuman, K. C., "Silica Encapsulation of Fluorescent Nanodiamonds for Colloidal Stability and Facile Surface Functionalization." *Journal of the American Chemical Society* 135 (2013): 7815–7818.

Chang, C. K., Wu, C. C., Wang, Y. S., and Chang, H. C., "Selective Extraction and Enrichment of Multiphosphorylated Peptides Using Polyarginine-Coated Diamond Nanoparticles." *Analytical Chemistry* 80 (2008): 3791–3797.

Chang, C. C., Zhang, B., Li, C. Y. et al., "Exploring Cytoplasmic Dynamics in Zebrafish Yolk Cells by Single Particle Tracking of Fluorescent Nanodiamonds." *Proceedings of SPIE* 8272 (2012): 827205.

Chang, B. M., Lin, H. H., Su, L. J. et al., "Highly Fluorescent Nanodiamonds Protein-Functionalized for Cell Labeling and Targeting." *Advanced Functional Materials* 23 (2013): 5737–5745.

Chen, W. H., Lee, S. C., Sabu, S. et al., "Solid-Phase Extraction and Elution on Diamond (SPEED): A Fast and General Platform for Proteome Analysis with Mass Spectrometry." *Analytical Chemistry* 78 (2006): 4228–4234.

Chen, M., Pierstorff, E. D., Lam, R. et al., "Nanodiamond-Mediated Delivery of Water-Insoluble Therapeutics." *ACS Nano* 3 (2009): 2016–2022.

Chen, M., Zhang, X. Q., Man, H. B. et al., "Nanodiamond Vectors Functionalized with Polyethylenimine for siRNA Delivery." *Journal of Physical Chemistry Letters* 1 (2010): 3167–3171.

Chow, E. K., Zhang, X. Q., Chen, M. et al., "Nanodiamond Therapeutic Delivery Agents Mediate Enhanced Chemoresistant Tumor Treatment." *Science Translational Medicine* 3 (2011): 73ra21.

Dahoumane, S. A., Nguycn, M. N., Thorel, A. et al., "Protein-Functionalized Hairy Diamond Nanoparticles." *Langmuir* 25 (2009): 9633–9638.

Danilenko, V. V., "On the History of the Discovery of Nanodiamond Synthesis." *Physics of the Solid State* 46 (2004): 595–599.

Field, J. E. (Ed.) *Properties of Natural and Synthetic Diamond*. London: Academic Press (1992).

Fu, C. C., Lee, H. Y., Chen, K. et al., "Characterization and Application of Single Fluorescent Nanodiamonds as Cellular Biomarkers." *Proceedings of the National Academy of Sciences of the United States of America* 104 (2007): 727–732.

Gary, D. J., Puri, N., and Won, Y. Y., "Polymer-Based siRNA Delivery: Perspectives on the Fundamental and Phenomenological Distinctions from Polymer-Based DNA Delivery." *Journal of Controlled Release* 121 (2007): 64–73.

Hartmann, M., Betz, P., Sun, Y. et al., "Saccharide-Modified Nanodiamond Conjugates for the Efficient Detection and Removal of Pathogenic Bacteria." *Chemistry—A European Journal* 18 (2012): 6485–6492.

Ho, D. (Ed.) *Nanodiamonds: Applications in Biology and Nanoscale Medicine*. Morwell: Springer (2009).

Hsieh, C. C., Guo, J. Y., Hung, S. U. et al., "Quantitative Analysis of Oligosaccharides Derived from Sulfated Glycosaminoglycans by Nanodiamond-Based Affinity Purification and Matrix-Assisted Laser Desorption/Ionization Mass Spectrometry." *Analytical Chemistry* 85 (2013): 4342–4349.

Huang, L. C. L. and Chang, H. C., "Adsorption and Immobilization of Cytochrome c on Nanodiamonds." *Langmuir* 20 (2004): 5879–5884.

Huang, T. S., Tzeng, Y., Liu, Y. K. et al., "Immobilization of Antibodies and Bacterial Binding on Nanodiamond and Carbon Nanotubes for Biosensor Applications." *Diamond and Related Materials* 13 (2004): 1098–1102.

Huang, H., Pierstorff, E., Osawa, E., and Ho, D., "Active Nanodiamond Hydrogels for Chemotherapeutic Delivery." *Nano Letters* 7 (2007): 3305–3314.

Hui, Y. Y., Zhang, B., Chang, Y. C. et al., "Two-Photon Fluorescence Correlation Spectroscopy of Lipid-Encapsulated Fluorescent Nanodiamonds in Living Cells." *Optics Express* 18 (2010): 5896–5905.

Kaur, R., and Badea, I., "Nanodiamonds as Novel Nanomaterials for Biomedical Applications: Drug Delivery and Imaging Systems." *International Journal of Nanomedicine* 8 (2013): 203–220.

Kaur, R., Chitanda, J. M., Michel, D. et al., "Lysine-Functionalized Nanodiamonds: Synthesis, Physiochemical Characterization, and Nucleic Acid Binding Studies." *International Journal of Nanomedicine* 7 (2012): 3851–3866.

Kong, X. L., Huang, L. C. L., Hsu, C. M. et al., "High-Affinity Capture of Proteins by Diamond Nanoparticles for Mass Spectrometric Analysis." *Analytical Chemistry* 77 (2005a): 259–265.

Kong, X. L., Huang, L. C. L., Liau, S. C. V., Han, C. C., and Chang, H. C., "Polylysine-Coated Diamond Nanocrystals for MALDI-TOF Mass Analysis of DNA Oligonucleotides." *Analytical Chemistry* 77 (2005b): 4273–4277.

Krueger, A., and Lang, D., "Functionality Is Key: Recent Progress in the Surface Modification of Nanodiamond." *Advanced Functional Materials* 22 (2012): 890–906.

Krueger, A., Liang, Y., Jarre, G., and Stegk, J., "Surface Functionalisation of Detonation Diamond Suitable for Biological Applications." *Journal of Materials Chemistry* 16 (2006): 2322–2328.

Kuo, Y., Hsu, T. Y., Wu, Y. C., and Chang, H. C., "Fluorescent Nanodiamond as a Probe for the Intercellular Transport of Proteins *In Vivo*." *Biomaterials* 34 (2013): 8352–8360.

Lee, J. W., Lee, S., Jang, S. et al., "Preparation of Non-Aggregated Fluorescent Nanodiamonds (FNDs) by Non-Covalent Coating with a Block Copolymer and Proteins for Enhancement of Intracellular Uptake." *Molecular Biosystems* 9 (2013): 1004–1011.

Lewinski, N., Colvin, V., and Drezek, R., "Cytotoxicity of Nanoparticles." *Small* 4 (2008): 26–49.

Li, J., Zhu, Y., Li, W. et al., "Nanodiamonds as Intracellular Transporters of Chemotherapeutic Drug." *Biomaterials* 31 (2010): 8410–8418.

Li, Y. Q., Zhou, X. P., Wang, D. X., Yang, B. S., and Yang, P., "Nanodiamond Mediated Delivery of Chemotherapeutic Drugs." *Journal of Materials Chemistry* 21 (2011): 16406–16412.

Li, H. C., Hsieh, F. J., Chen, C. P. et al., "The Hemocompatibility of Oxidized Diamond Nanocrystals for Biomedical Applications." *Scientific Report* 3 (2013): 3044.

Lin, Y. C., Perevedentseva, E., Tsai, L. W., Wu, K. T., and Cheng, C. L., "Nanodiamond for Intracellular Imaging in the Microorganisms *In Vivo*." *Journal of Biophotonics* 5 (2012): 838–847.

Liu, K. K., Zheng, W. W., Wang, C. C. et al., "Covalent Linkage of Nanodiamond-Paclitaxel for Drug Delivery and Cancer Therapy." *Nanotechnology* 21 (2010): 315106.

Marcon, L., Riquet, F., Vicogne, D. et al., "Cellular and *In Vivo* Toxicity of Functionalized Nanodiamond in *Xenopus* Embryos." *Journal of Materials Chemistry* 20 (2010): 8064–8069.

Minko, S., "Grafting on Solid Surfaces: 'Grafting to' and 'Grafting from' Methods." In Stamm, M. (Ed.) *Polymer Surfaces and Interfaces*. Heidelberg: Springer (2008).

Mochalin, V. N., Shenderova, O., Ho, D., and Gogotsi, Y., "The Properties and Applications of Nanodiamonds." *Nature Nanotechnology* 7 (2012): 11–23.

Mohan, N., Chen, C. S., Hsieh, H. H., Wu, Y. C., and Chang, H. C., "*In Vivo* Imaging and Toxicity Assessments of Fluorescent Nanodiamonds in *Caenorhabditis elegans*." *Nano Letters* 10 (2010): 3692–3699.

Mohan, N., Zhang, B., Chang, C. C. et al., "Fluorescent Nanodiamond—A Novel Nanomaterial for *In Vivo* Applications." *MRS Online Proceedings Library* 1362 (2011): mrss11-1362-qq06-01.

Moore, L., Chow, E. K. H., Osawa, E., Bishop, J. M., and Ho, D., "Diamond–Lipid Hybrids Enhance Chemotherapeutic Tolerance and Mediate Tumor Regression." *Advanced Materials* 25 (2013): 3532–3541.

Moore, L., Grobárová, V., Shen, H. et al., "Comprehensive Interrogation of the Cellular Response to Fluorescent, Detonation and Functionalized Nanodiamonds." *Nanoscale* 6 (2014): 11712–11721.

Nguyen, T. T. B., Chang, H. C., and Wu, V. W. K., "Adsorption and Hydrolytic Activity of Lysozyme on Diamond Nanocrystallites." *Diamond and Related Materials* 16 (2007): 872–876.

Osawa, E., "Monodisperse Single Nanodiamond Particulates." *Pure and Applied Chemistry* 80 (2008): 1365–1379.

Osswald, S., Yushin, G., Mochalin, V., Kucheyev, S. O., and Gogotsi, Y., "Control of sp^2/sp^3 Carbon Ratio and Surface Chemistry of Nanodiamond Powders by Selective Oxidation in Air." *Journal of the American Chemical Society* 128 (2006): 11635–11642.

Paget, V., Sergent, J. A., Grall, R. et al., "Carboxylated Nanodiamonds Are Neither Cytotoxic Nor Genotoxic on Liver, Kidney, Intestine and Lung Human Cell Lines." *Nanotoxicology* 8 (2014): 46–56.

Perevedentseva, E., Cheng, C. Y., Chung, P. H. et al., "The Interaction of the Protein Lysozyme with Bacteria E-Coli Observed Using Nanodiamond Labelling." *Nanotechnology* 18 (2007): 315102.

Prabhakar, N., Nareoja, T., von Haartman, E. et al., "Core-Shell Designs of Photoluminescent Nanodiamonds with Porous Silica Coatings for Bioimaging and Drug Delivery II: Application." *Nanoscale* 5 (2013): 3713–3722.

Puzyr, A. P., Baron, A. V., Purtov, K. V. et al., "Nanodiamonds with Novel Properties: A Biological Study." *Diamond and Related Materials* 16 (2007): 2124–2128.

Rehor, I., Mackova, H., Filippov, S. K. et al., "Fluorescent Nanodiamonds with Bioorthogonally Reactive Protein-Resistant Polymeric Coatings." *Chempluschem* 79 (2014a): 21–24.

Rehor, I., Slegerova, J., Kucka, J. et al., "Fluorescent Nanodiamonds Embedded in Biocompatible Translucent Shells." *Small* 10 (2014b): 1106–1115.

Rehor, I., Lee, K. L., Chen, K. et al., "Plasmonic Nanodiamonds: Targeted Core-Shell Type Nanoparticles for Cancer Cell Thermoablation." *Advanced Healthcare Materials* (2014c): 1–9.

Schrand, A. M., Dai, L., Schlager, J. J., Hussain, S. M., and Osawa, E., "Differential Biocompatibility of Carbon Nanotubes and Nanodiamonds." *Diamond and Related Materials* 16 (2007a): 2118–2123.

Schrand, A. M., Huang, H. J., Carlson, C. et al., "Are Diamond Nanoparticles Cytotoxic?" *Journal of Physical Chemistry B* 111 (2007b): 2–7.

Schrand, A. M., Hens, S. A. C., and Shenderova, O. A., "Nanodiamond Particles: Properties and Perspectives for Bioapplications." *Critical Reviews in Solid State and Materials Sciences* 34 (2009): 18–74.

Shenderova, O., Koscheev, A., Zaripov, N. et al., "Surface Chemistry and Properties of Ozone-Purified Detonation Nanodiamonds." *Journal of Physical Chemistry C* 115 (2011): 9827–9837.

Shimkunas, R. A., Robinson, E., Lam, R. et al., "Nanodiamond–Insulin Complexes as pH-Dependent Protein Delivery Vehicles." *Biomaterials* 30 (2009): 5720–5728.

Simpson, D. A., Thompson, A. J., Kowarsky, M. et al., "*In Vivo* Imaging and Tracking of Individual Nanodiamonds in *Drosophila melanogaster* Embryos." *Biomedical Optics Express* 5 (2014): 1250–1261.

Slegerova, J., Hajek, M., Rehor, I. et al., "Designing the Nanobiointerface of Fluorescent Nanodiamonds: Highly Selective Targeting of Glioma Cancer Cells." *Nanoscale* 7 (2015): 415–420.

Smith, A. H., Robinson, E. M., Zhang, X. Q. et al., "Triggered Release of Therapeutic Antibodies from Nanodiamond Complexes." *Nanoscale* 3 (2011): 2844–2848.

Solarska, K., Gajewska, A., Bartosz, G., and Mitura, K., "Induction of Apoptosis in Human Endothelial Cells by Nanodiamond Particles." *Journal of Nanoscience and Nanotechnology* 12 (2012): 5117–5121.

Tzeng, Y. K., Faklaris, O., Chang, B. M., Kuo, Y., Hsu, J. H., and Chang, H. C., "Superresolution Imaging of Albumin-Conjugated Fluorescent Nanodiamonds in Cells by Stimulated Emission Depletion." *Angewandte Chemie International Edition* 50 (2011): 2262–2265.

Vaijayanthimala, V., Tzeng, Y. K., Chang, H. C., and Li, C. L., "The Biocompatibility of Fluorescent Nanodiamonds and Their Mechanism of Cellular Uptake." *Nanotechnology* 20 (2009): 425103.

Vaijayanthimala, V., Cheng, P. Y., Yeh, S. H. et al., "The Long-Term Stability and Biocompatibility of Fluorescent Nanodiamond as an *In Vivo* Contrast Agent." *Biomaterials* 33 (2012): 7794–7802.

Vial, S., Mansuy, C., Sagan, S. et al., "Peptide-Grafted Nanodiamonds: Preparation, Cytotoxicity and Uptake in Cells." *Chembiochem* 9 (2008): 2113–2119.

von Haartman, E., Jiang, H., Khomich, A. A. et al., "Core-Shell Designs of Photoluminescent Nanodiamonds with Porous Silica Coatings for Bioimaging and Drug Delivery I: Fabrication." *Journal of Materials Chemistry B* 1 (2013): 2358–2366.

Wehling, J., Dringen, R., Zare, R. N., Maas, M., and Rezwan, K., "Bactericidal Activity of Partially Oxidized Nanodiamonds." *ACS Nano* 8 (2014): 6475–6483.

Wei, L. M., Zhang, W., Lu, H. J., and Yang, P. Y., "Immobilization of Enzyme on Detonation Nanodiamond for Highly Efficient Proteolysis." *Talanta* 80 (2010): 1298–1304.

Whitehead, K. A., Langer, R., and Anderson, D. G., "Knocking Down Barriers: Advances in siRNA Delivery." *Nature Reviews Drug Discovery* 8 (2009): 129–138.

Wiethoff, C. M., and Middaugh, C. R., "Barriers to Nonviral Gene Delivery." *Journal of Pharmaceutical Sciences* 92 (2003): 203–217.

Williams, O. A., Hees, J., Dieker, C., Jäger, W., Kirste, L., and Nebel, C. E., "Size-Dependent Reactivity of Diamond Nanoparticles." *ACS Nano* 4 (2010): 4824–4830.

Wu, C. C., Han, C. C., and Chang, H. C., "Applications of Surface-Functionalized Diamond Nanoparticles for Mass-Spectrometry-Based Proteomics." *Journal of the Chinese Chemical Society* 57 (2010): 582–594.

Wu, T. J., Tzeng, Y. K., Chang, W. W. et al., "Tracking the Engraftment and Regenerative Capabilities of Transplanted Lung Stem Cells Using Fluorescent Nanodiamonds." *Nature Nanotechology* 8 (2013): 682–689.

Xing, Y., and Dai, L. M., "Nanodiamonds for Nanomedicine." *Nanomedicine* 4 (2009): 207–218.

Xing, Y., Xiong, W., Zhu, L. et al., "DNA Damage in Embryonic Stem Cells Caused by Nanodiamonds." *ACS Nano* 5 (2011): 2376–2384.

Xu, G. B., Zhang, W., Wei, L. M., Lu, H. J., and Yang, P. Y., "Boronic Acid-Functionalized Detonation Nanodiamond for Specific Enrichment of Glycopeptides in Glycoproteome Analysis." *Analyst* 138 (2013): 1876–1885.

Yu, S. J., Kang, M. W., Chang, H. C., Chen, K. M., and Yu, Y. C., "Bright Fluorescent Nanodiamonds: No Photobleaching and Low Cytotoxicity." *Journal of the American Chemical Society* 127 (2005): 17604–17605.

Yuan, Y., Chen, Y. W., Lui, J. H., Wang, H. F., and Liu, Y. F., "Biodistribution and Fate of Nanodiamonds *In Vivo*." *Diamond and Related Materials* 18 (2009): 95–100.

Yuan, Y., Wang, X., Jia, G. et al., "Pulmonary Toxicity and Translocation of Nanodiamonds in Mice." *Diamond and Related Materials* 19 (2010): 291–299.

Zhang, B., Li, Y., Fang, C. Y. et al., "Receptor-Mediated Cellular Uptake of Folate-Conjugated Fluorescent Nanodiamonds: A Combined Ensemble and Single-Particle Study." *Small* 5 (2009): 2716–2721.

Zhang, X. Q., Chen, M., Lam, R. et al., "Polymer-Functionalized Nanodiamond Platforms as Vehicles for Gene Delivery." *ACS Nano* 3 (2009): 2609–2616.

Zhang, X., Yin, J., Kang, C. et al., "Biodistribution and Toxicity of Nanodiamonds in Mice after Intratracheal Instillation." *Toxicology Letters* 198 (2010): 237–243.

Zhang, P., Yang, J., Li, W. et al., "Cationic Polymer Brush Grafted-Nanodiamond via Atom Transfer Radical Polymerization for Enhanced Gene Delivery and Bioimaging." *Journal of Materials Chemistry* 21 (2011): 7755–7764.

Zhang, X. Y., Fu, C. K., Feng, L. et al., "PEGylation and polyPEGylation of nanodiamond." *Polymer* 53 (2012a): 3178–3184.

Zhang, X. Y., Hu, W. B., Li, J., Tao, L., and Wei, Y., "A Comparative Study of Cellular Uptake and Cytotoxicity of Multi-Walled Carbon Nanotubes, Graphene Oxide, and Nanodiamond." *Toxicology Research* 1 (2012b): 62–68.

Zhao, L., Takimoto, T., Ito, M. et al., "Chromatographic Separation of Highly Soluble Diamond Nanoparticles Prepared by Polyglycerol Grafting." *Angewandte Chemie International Edition* 50 (2011): 1388–1392.

23

Cargo-Delivering Nanodiamonds

Basem Moosa

Niveen M. Khashab

23.1 Introduction

More than 50 years ago, nanodiamonds (NDs) were first discovered in Union of Soviet Socialist Republic (USSR), where they were produced by detonation reaction of carbon-based explosives. They remained essentially unknown to the rest of the world, for many reasons including security and the lack of industrial interests, until the end of the 1980s. A number of important discoveries in the late 1990s led to a wide interest in these nanoparticles (Figure 23.1). By then, mass production of NDs had already started in different countries (Osawa 2010).

By definition, NDs are carbon nanoparticles with a truncated octahedral architecture that are typically about 2 to 8 nm in diameter. They not only exhibit diamond-like characters such as chemical stability and extremely high hardness, stiffness, and strength but also have the advantages of nanomaterials, such as small size, large surface area, and high adsorption capacity. Therefore, NDs have superior physical and chemical properties over conventional materials. The family of diamond materials (Krueger 2008) (Figure 23.2) includes additional members, such as the so-called diamond like carbon (Erdemir and Donnet 2006), sintered diamond phases (Alam 2004), and diamond microparticles and nanoparticles (Palnichenko et al. 1999). Closely related to diamond is a group of organic molecules, the diamondiods (Dahl et al. 2003). Although they are not real diamond materials, they have interesting properties that could eventually lead to new insights into the behavior of nanoscale diamond materials (Fokin et al. 2005).

Today, there is a baffling array of NDs available for research. They have been synthesized by the detonation technique, laser ablation, high-energy ball milling of high-pressure–high-temperature diamond microcrystals, plasma-assisted chemical vapor deposition (CVD), autoclave synthesis from supercritical fluids, chlorination of carbides, ion irradiation of graphite, electron irradiation of carbon onions, and ultrasound cavitation, with the first three of these methods being used commercially (Frenklach et al. 1991; Banhart and Ajayan 1996; Gogotsi et al. 1996; Yang et al. 1998; Daulton et al. 2001; Welz et al. 2003; Galimov et al. 2004; Boudou et al. 2009).

NDs were used in a wide range of applications including catalysis, quantum computing, formation of hard coatings and composites, polishing, seedings of substrates for CVD diamond growth, and biomedical applications. The biological applications of ND includes the use of these nanoparticles in

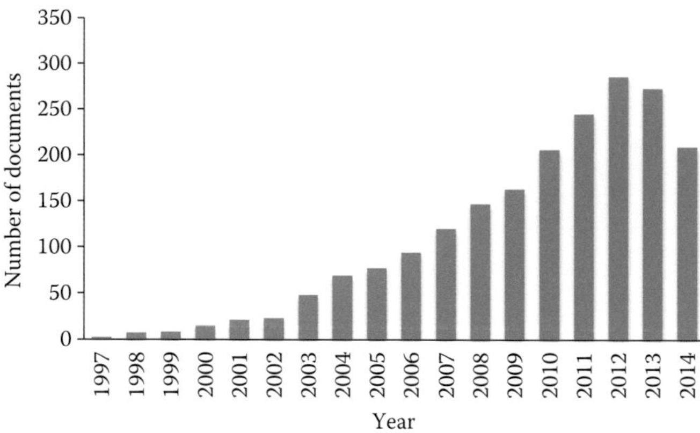

FIGURE 23.1 Number of ND-related articles per year.

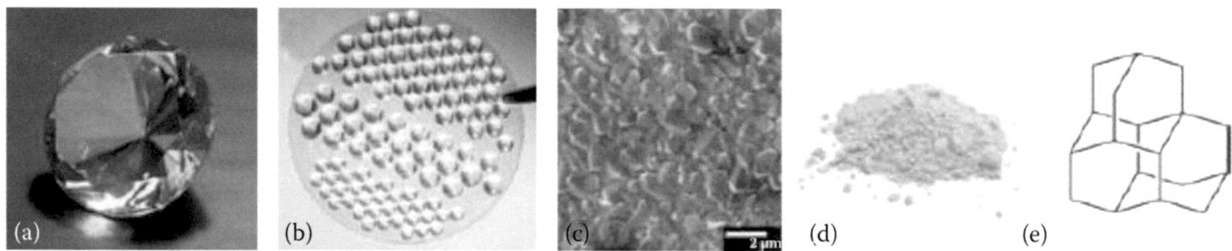

FIGURE 23.2 Different diamond materials: (a) gem quality diamond, (b) single-crystalline diamond film with microlens structures, (c) polycrystalline diamond film, (d) detonation diamond powder, (e) triamantane, a diamondoid molecule. (Krueger, A. *Chem-Eur. J.*, 14, 1382, 2008. Copyright Wiley-VCH Verlag GmbH&Co. KGaA. Reproduced with permission.)

biocompatible composites and implants, targeted and controlled drug delivery, parts of biosensors, as a fluorescent marker for cell imaging, and as a stable solid support for peptide synthesis.

In clinical practice, the controlled delivery and release of drugs are often desirable, as the tailored dosing of any chemotherapeutics is vital toward the reduction of side effects and complications (Langer 1998; Langer and Tirrell 2004; Solarska et al. 2012). The utilization of nanoparticle-based vehicles as multifunctional versatile and biocompatible drug carriers would serve as the ideal technology, as their significant advantages include the ability to target a specific location in the body and the reduction of drug amount used, which results in less side effects (Niemeyer 2001; Allen and Cullis 2004; Niemiec et al. 2011).

Thus, nanoparticle-based chemotherapeutics delivery can resolve the ever lasting problem of random and excessive drug delivery. In recent years, a considerable effort has been devoted to the design and synthesis of novel nanostructural materials with functional biological properties (Caruso 2001; Moghimi et al. 2001; Rao and Cheetham 2001; Gao et al. 2004). Owing to its superior physical properties and biocompatibility, diamond-based nanostructures have merged as promising materials for biomedical application. In this review, a collection of the promising pharmaceutical applications of NDs is presented. We divided our discussion to (1) small molecule and (2) macromolecule delivery by ND platforms.

23.2 Small Molecule Delivery

Dean Ho was among the first scientists who had their fingerprints all over the field of ND and its application in biology and medicine (Huang et al. 2007a,b, 2008; Lam et al. 2008; Pierstorff and Ho 2008; Pierstorff et al. 2008; Chen et al. 2009; Ho 2009; Moore et al. 2013a; Man et al. 2014; Xi et al. 2014). Most recently (Zhang et al. 2011), he presented a multicomponent ND drug delivery system (2–8 nm) that can be used in targeting, imaging, and enhancing therapy of paclitaxel (PTX). This was done through heterofunctionalization of NDs by attaching fluorescently labeled PTX-DNA conjugate and anti-EGFRmAbs onto the ND surface (Figure 23.3).

ND–daunorubicin conjugates were synthesized recently to overcome multidrug chemoresistance in leukemia. Adjusting reaction parameters such as acidity and concentration optimized the loading of daunorubicin onto NDs that shows good cell viability (Man et al. 2014) (Figure 23.4).

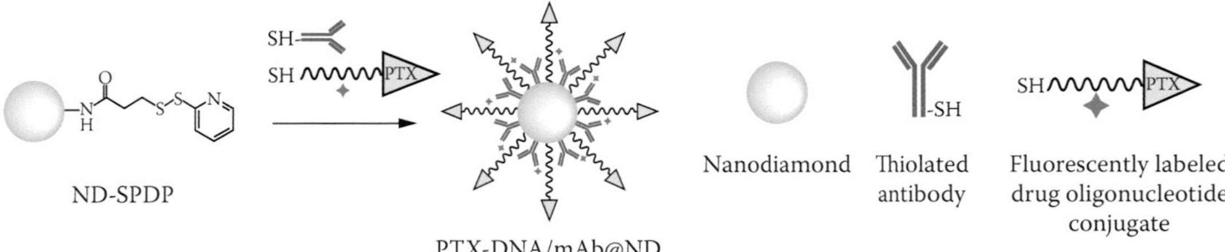

FIGURE 23.3 The synthesis of PTX-DNA/mAb@ND. (Zhang, X.Q., Lam, R., Xu, X.Y. et al.: *Adv. Mater.* 23. 2011. 4770. Copyright Wiley-VCH Verlag GmbH&Co. KGaA. Reproduced with permission.)

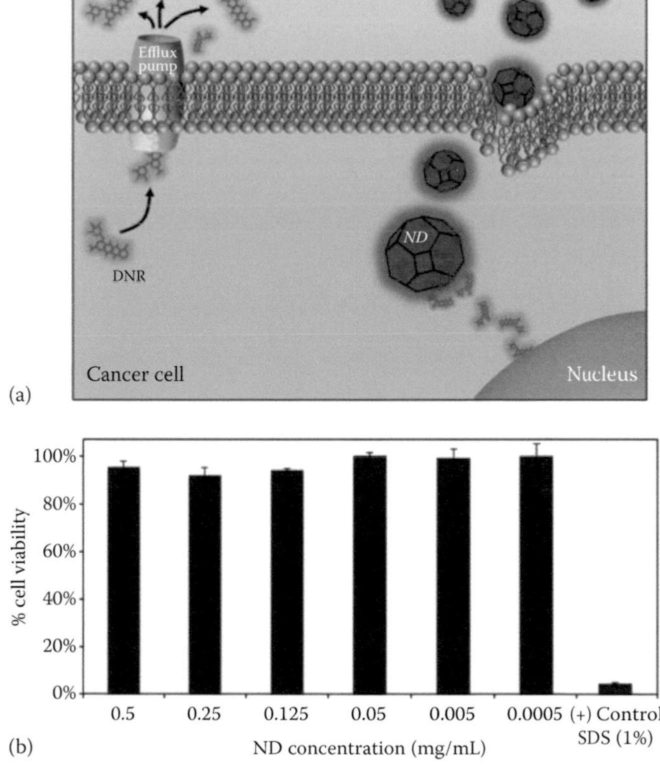

FIGURE 23.4 (a) Response from a resistant cancer cell to treatment from Daunorubicin (DNR) versus nanodiamond conjugated daunorubicin (ND-R). (b) Cell viability after exposure to varying concentrations of NDs, demonstrating the biocompatibility of NDs, even at higher concentrations. (Reprinted from *Nanomed. Nanotechnol. Biol. Med.*, 10, Man, H.B., Kim, H., Kim, H.J. et al., Nanomedicine: Nanotechnology, biology, and medicine, 359. Copyright 2014, with permission from Elsevier.)

Li et al. (2011) also studied the intracellular delivery of doxorubicin hydrochloride (DOX). In this work, doxorubicin was physically adsorbed on red fluorescent NDs (FNDs) of ~140 nm size. To enable FND surface to adsorb positively charged DOX, the authors treated FNDs with the H_2SO_4/HNO_3 mixture and thus introducing negatively charged carboxylic acid functionalities. The saturation concentration of DOX on FND was relatively high, 38.3 ± 2.3 μg/mg^{-1} (3.98 × 10^{16} molecules/mg). Importantly, it was shown by this study that FND-DOX internalized by Hela cells through clathrin-dependent pathway. Confocal red fluorescent microscopy indicated that DOX carrying FNDs were localized mainly in the cytoplasm from where the DOX was later liberated and migrated into the nucleus.

Another approach to overcome the resistance of cancer cell toward chemotherapeutics using ND materials is by synthesizing a novel ND-lipid hybrid particle (NDLP) that is targeted to epidermal growth factor receptor (EGFR) vector (Moore et al. 2013a). NDLPs could be readily self-assembled from a variety of modified NDs and used to specifically deliver imaging or therapeutic molecules to triple negative breast cancer (TNBC) cells (MDA-MB-231) in vitro and in vivo (Figure 23.5).

Guan et al. (2010) successfully loaded NDs with cisplatin (*cis*-dichlorodiamineplatinum (II) [CDDP]), which is a benchmark standard in the field. The loading procedure for CDDP was the same as with DOX, i.e., surface functionalization with carboxylic groups followed by ionic interaction with amino groups of CDDP. To avoid overdosing, the system was designed to have a pH-responsive property. The authors showed that the CDDP–ND composite would deliver a low concentration of CDDP during the circulation period in the blood (pH 7.4) but release a much higher concentration into acidic lysosomes (pH < 6) (Figure 23.6).

Wang et al. (2013) recently used a relatively simpler strategy of PEG-ND preparation with the aim of DOX delivery and slow release. In this work, poly ethylene glycol (PEG) was attached through activated carboxylic groups present on the oxidized ~140-nm-sized FND surface (Figure 23.7).

Li et al. (2010) attached 10-hydroxycamptothecin (HCPT) to NDs through NaOH-aided solubility enhancement of HCPT, which resulted in increased diffusion to ND's interior. This treatment dramatically enhanced the loading capacity of the drug on NDs. Covalent attachment of drugs through both ether and ester linkages on the activated hydroxyl-terminated alkyl-grafted NDs was successfully demonstrated by Zheng et al. (2009) Alkylation of surface hydroxyl groups by 6-(chloro-hexyloxy)-tetrahydropyran with subsequent hydrolysis afforded hydroxy-hexyl-NDs. Subsequent mesylation of this linker made it possible to attach vitamin K_3 analog, which previously demonstrated anticancer property by the same group.

Our group (Yan et al. 2012) combined theoretical and experimental methods to investigate the mechanism of pH-dependent drug release from ND surface. In the endosomal recycling process, the nanoparticle will pass from mildly acidic vesicle to pH ~4.8; thus, it is important to investigate DOX release from NDs at different pH values. Fluorescein-labeled ND (Fc-NDs) released DOX dramatically

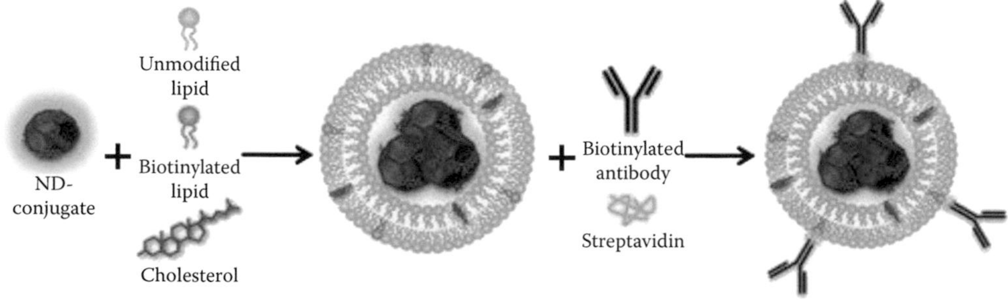

FIGURE 23.5 NDLPs are synthesized by rehydration of lipid thin films containing cholesterol and biotinylated lipids with concentrated ND solutions. Hybrid particles are then targeted using biotinylated antibodies and streptavidin crossbridges. (Moore, L., Chow, E.K., Osawa, E. et al.: *Adv. Mater.* 2013. 25. 3532. Copyright 2013 Wiley-VCH Verlag GmbH&Co. KGaA. Reproduced with permission.)

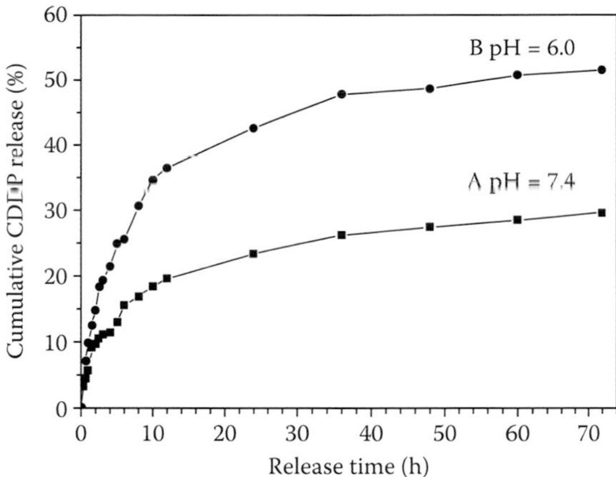

FIGURE 23.6 CDDP release from nanocomposite. (Guan, B., Zou, F., Zhi, J.F.: *Small.* 2010. 6. 1514. Copyright Wiley-VCH Verlag GmbH&Co. KGaA. Reproduced with permission.)

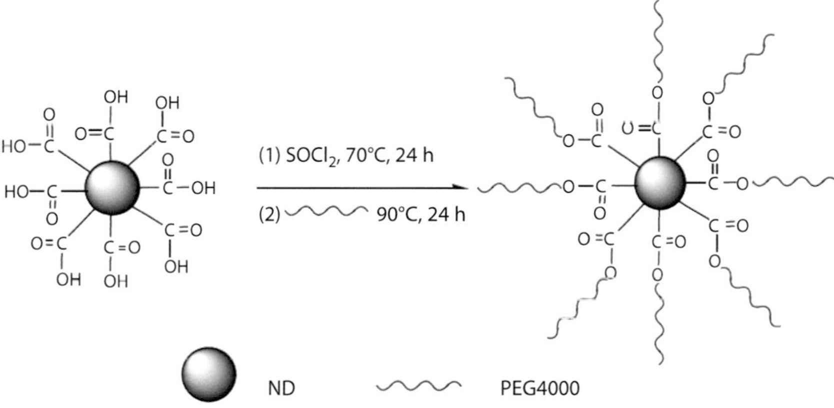

FIGURE 23.7 Synthesis of PEG-FND. (Reprinted from *Diam. Rel. Mat.*, 36, Wang, D.X., Tong, Y.L., Li, Y.Q. et al. PEGylated nanodiamond for chemotherapeutic drug delivery, 26. Copyright 2013, with permission from Elsevier.)

under acidic conditions, while an increase in the DOX loading efficiency (up to 6.4 wt%) was observed under basic conditions (Figure 23.8).

Adnan et al. (2011) demonstrated both theoretically and experimentally the influence of pH on the degree of DOX loading on ND surface, which also supports previous work. They concluded that DOX molecules bind to ND at high pH and needs 10% of free ND surface area for binding to happen. Further theoretical calculations suggest that H^+ weakens the electrostatistic interaction between ND surface carboxyl groups and DOX amino groups, and the interaction energies at pH < 7, pH 7, and pH > 7 are 10.4, 25.0, and 27.0 kcal mol^{-1}, respectively (Figure 23.9).

One recent and important application of ND is the use of ND material itself as antibacterial agents. Depending on their surface composition, NDs kill Gram-positive and -negative bacteria rapidly and efficiently (Wehling et al. 2014). The antibacterial activity came from the presence of partially oxidized and negatively charged surfaces, specifically those containing acid anhydride groups (Figure 23.10).

Fewer reports on in vivo applications of NDs can be found in literature. In vivo evaluation of ND-mediated drug delivery was done by Chow et al. (2011) and was later highlighted and discussed by Ma et al. (2011) Interestingly, the studies showed that the cytotoxic profile of ND was not affected by verapamil, a well-known ATP-binding cassette (ABC) transporter inhibitor, a fact that suggests

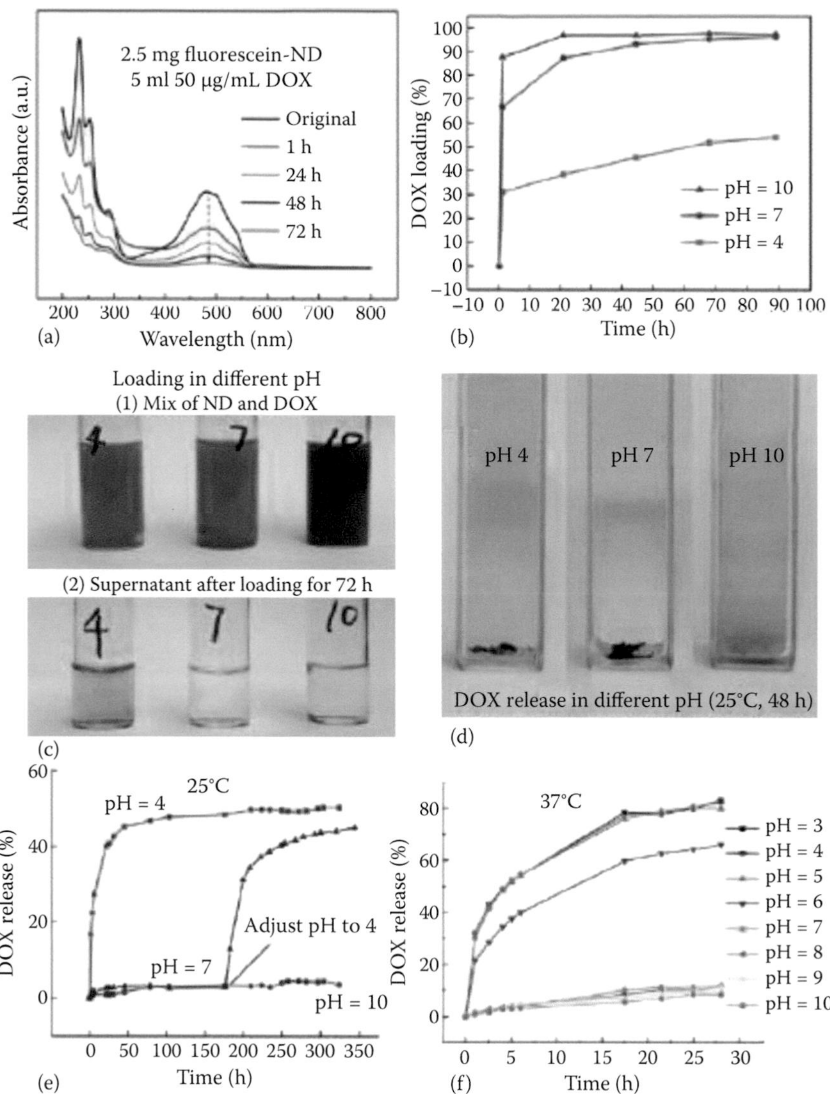

FIGURE 23.8 (a) Loading of DOX onto NDs. The ultraviolet-visible spectra showing the disappearance of free-floating DOX from the solution. (b) The loading of DOX at different pH values. (c) DOX loading and (d) release at different pH values (25°C); DOX desorption from Fc-ND-DOX at different pH values at (e) 25°C and (f) 37°C. (Yan, J.J. et al., *N. J. Chem.*, 36, 1479, 2012. Reprinted by permission of The Royal Society of Chemistry.)

the nonspecific nature of ND-mediated resistance abrogation. The authors mentioned increased drug retention (preventing the "burst release") in tumors and circulation as one of the advantageous reasons underlying the improved in vivo pharmacokinetic profile and overall performance.

23.3 Macromolecule Delivery

The potential of NDs as a targeted macromolecule-delivery vehicle was heavily explored, as NDs have several critical properties such as biocompatibility, the ability to carry a broad range of therapeutics (protein, DNA, and RNA), dispersability in water, and scalability (Grausova et al. 2009; Dhanak et al. 2012; Man and Ho 2012; Mochalin et al. 2012b). In most cases, the physical adsorption of such a charged

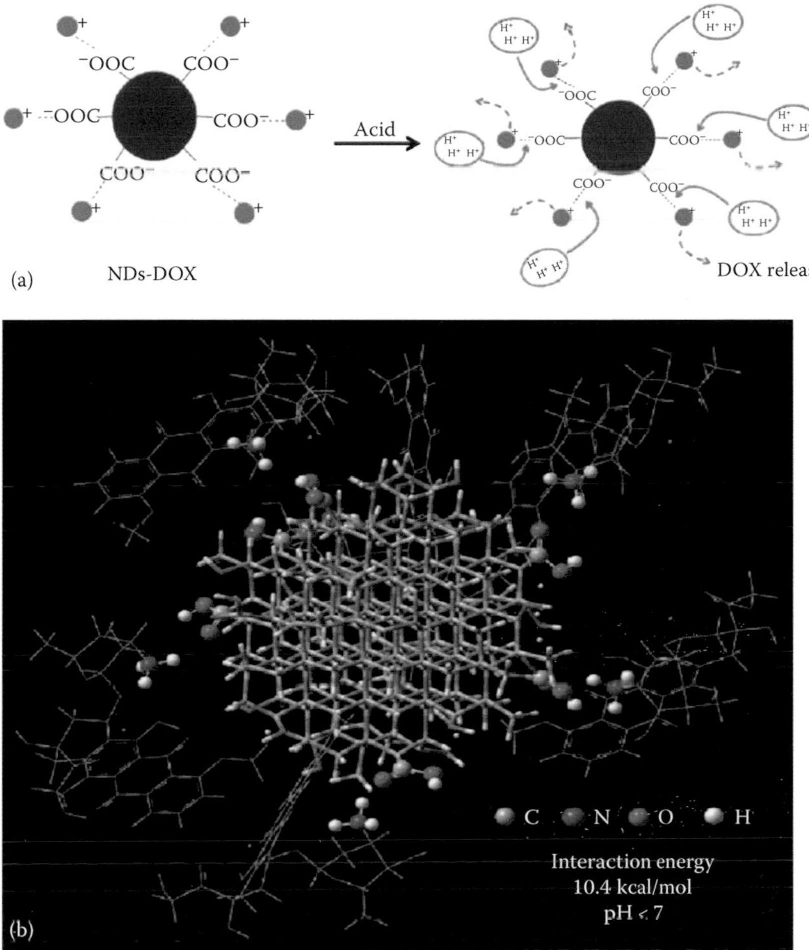

FIGURE 23.9 (a) Proposed scheme for the DOX release from ND-DOX under acidic condition; (b) the computational model and interaction energy of ND carboxyl groups with DOX at pH > 7. (Yan, J.J. et al., *N. J. Chem.*, 36, 1479, 2012. Reprinted by permission of The Royal Society of Chemistry.)

macromolecule was the main reason to use the ND as a vehicle (Huang et al. 2007a, 2008; Yeap et al. 2009).

23.3.1 ND-Mediated Peptide and Protein Delivery

As charged molecules, peptide or protein can be loaded onto the surface of ND, since their surface elements such as anionic end groups (–COO^{-}) and protonated amino groups $\left(-NH_3^+\right)$ provide a favorable condition for charge–charge interaction with ND. Meanwhile, hydrogen bonds may exist between ND and these biomolecules to enhance the adsorption. An example of directly loading proteins on ND is the work done by Ho's group, where the bovine insulin was noncovalently bound to ND via physical adsorption in an aqueous solution (Dhanak et al. 2012). Although insulin presents a slightly negative net charge at neutral pH, the adsorption was still completed, indicating that this loading process may include both electrostatic interaction and H-binding. Meanwhile, insulin could be released from the ND–insulin complex in basic condition (Shimkunas et al. 2009) (Figure 23.11), which can be explained by the change in charge characteristics affected by pH modification. It was concluded that exposure of the ND–insulin

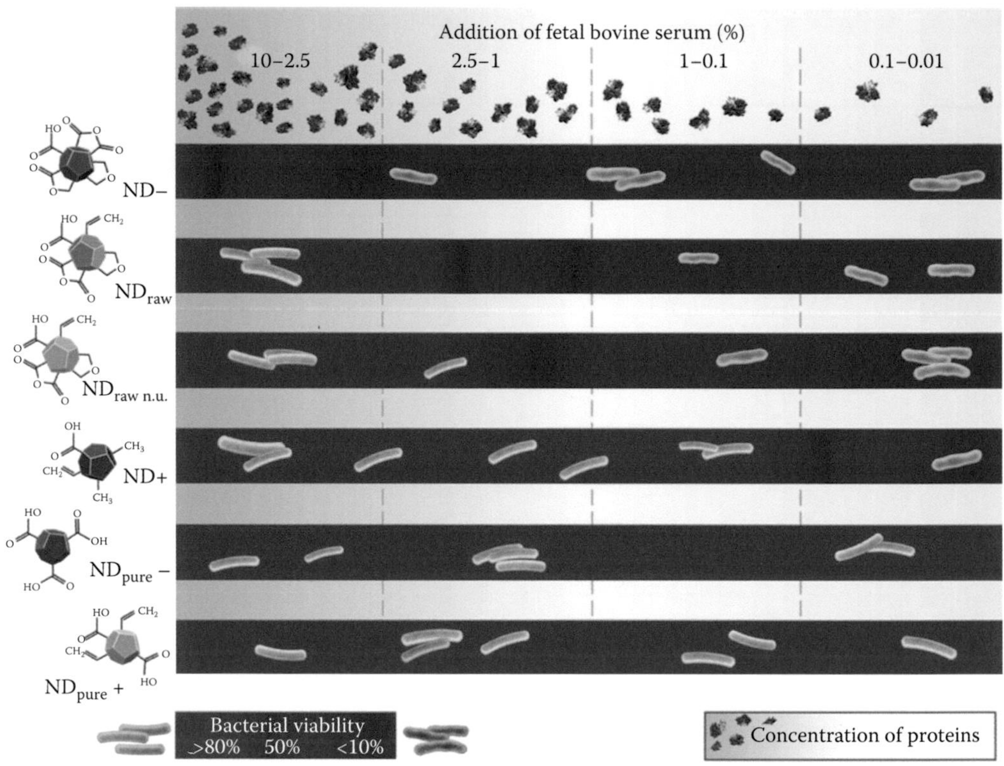

FIGURE 23.10 Antibacterial activity of ND particles toward Gram-positive and -negative bacteria. (Reprinted with permission from Wehling, J. et al., *ACS Nano*, 8, 6475, 2014. Copyright 2014, American Chemical Society.)

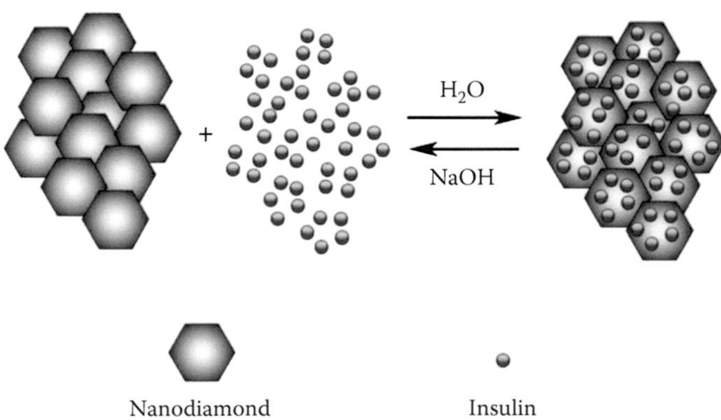

FIGURE 23.11 A schematic illustration showing insulin adsorption to NDs in water and desorption in the presence of NaOH. Insulin noncovalently binds to the ND surface in water by means of electrostatic and other interactions. The shift to an alkaline environment alters the insulin surface charge characteristics, thereby causing release from the ND surface. (Reprinted from *Biomaterials*, 30, Shimkunas, R.A., Robinson, E., Lam, R. et al., 5720. Copyright 2009, with permission from Elsevier.)

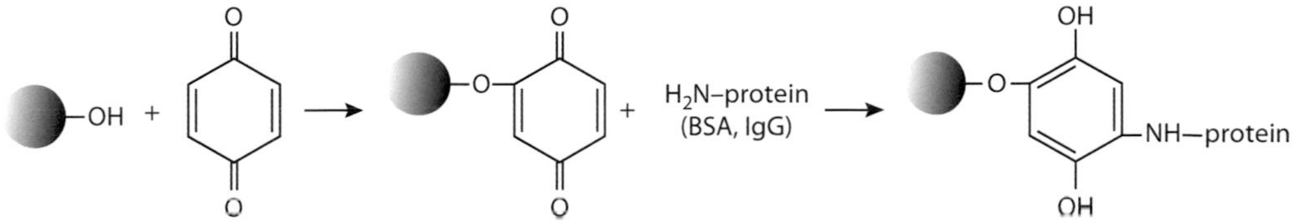

FIGURE 23.12 Schematic representation of protein immobilization on the benzoquinone activated NDs. (With kind permission from Springer Science+Business Media: *Nanoscale Res. Lett.*, 5, 2010, 631, Purtov, K.V., Petunin, A.I., Burov, A.E. et al.)

complex to alkaline environments mediates the interaction between the NDs and insulin, resulting in protein release. Another example is loading transforming growth factor-β antibody (Ab) on ND under dilute saline conditions, which was done in the same group (Smith et al. 2011). While the ND-Ab complex was found to be stable in water, Abs were triggered to release when incubated in serum-containing media, and the preservation of protein activity after release was confirmed via enzyme-linked immunosorbent assay. In an analogous work by Niu's group, another protein, bovine serum albumin (BSA), was also directly loaded on ND (Wang et al. 2011).

With hydroxyl and carboxyl groups on surfaces, NDs can also bind proteins via covalent linkage after certain functionalization treatments. However, unlike physical adsorption, covalent binding limits the release of proteins in physiological conditions. Although enzyme-triggered release might be achieved in the future, current literature reports are focusing on binding specific proteins (such as antibodies) on ND instead of releasing them. Interestingly, covalently linked protein on NDs might not be the main therapeutic drug, but some functional components in the platform to facilitate the delivery of other drugs. With the protein-functionalized surfaces, the ND may load other molecules or target certain cells. In a work by Purtov's group (Purtov et al. 2010), immunoglobulin (IgG) was bonded covalently to NDs after functionalizing NDs with benzoquinone (Figure 23.12) and the resulting ND–IgG complex maintained stability in blood serum. In the same work, a more specific protein, rabbit antimouse (RAM) Ab, was immobilized on ND together with BSA, and the RAM–ND–BSA complex was able to specifically bind the target antigen through Ab–antigen interaction. In a more complicated drug delivery system that was done by Ho's group (Zhang et al. 2011), both Ab and oligonucleotide were covalently bonded to NDs after introducing sulfhydryl groups. While bonded Ab enhanced targeting specificity, oligonucleotide conjugated further with PTX and fluorescein introduced imaging and delivery abilities. Besides antibodies, cell-penetrating peptides, such as TAT (HIV transactivator of transcription protein), were also conjugated on ND to increase the delivery efficiency (Huang et al. 2011; Wang et al. 2013).

Before linking proteins on the surfaces, pristine ND was also functionalized with other macromolecules. In an ongoing work by Yang's group (Wang et al. 2010), *N*,*O*-carboxymethyl chitosan (CMCS) was used to modify the surfaces of ND. This design combined the high surface area of ND and pH-responsive protein release property of CMCS, which might be utilized for protein drug delivery applications in the future.

23.3.2 ND-Mediated Gene Delivery

Combining both efficiency and safety in one gene delivery platform is always a challenge, and conventional approaches are often either highly efficient but more toxic or less toxic but less efficient (Zhang et al. 2009; Purtov et al. 2010). To this end, ND is a promising material as it has no or a very small toxic side effect on biological systems, especially compared with other nanocarbon materials (Zhang et al. 2011). Unlike some positively charged proteins that can be adsorbed directly on ND, negative-charged gene fragments cannot be loaded directly via electrostatic interaction. Therefore, functionalization of ND is necessary to introduce positive-charged surfaces.

Covalent functionalization of ND surface can be achieved by introducing terminal amine, which is promising since the ND surfaces can be easily hydroxylated or carboxylated for further reactions. Several recent literatures have proven the achievability of this functionalization. While Ho's group synthesized ND-NH_2 via covalent attachment of (3-aminopropyl)-trimethoxysilane to hydroxylated ND-OH, (Zhang et al. 2009) (Figure 23.13), Gogotsi's group covalently linked ethylenediamine to carboxylated ND-COOH via amide bonds (Mochalin et al. 2008, 2011), and Badea's group also used amide bonds to bind lysine moieties to ND-COOH surfaces through a three-carbon-length linker (1,3-diaminopropane) (Zhang et al. 2009). With terminated amine on surfaces, ND is expected to bind target nucleic acid via electrostatic interactions. Except for the delivery application, the ND-NH_2 can also be utilized to fabricate epoxy composites, generating other modifications to polymeric systems (Mochalin et al. 2012a). Meanwhile, other cationic groups were also introduced in published reports, including thionine providing S^+ and triethylammonium providing N^+ (Martin et al. 2010). In this work, the authors observed that the ND complex entered HeLa cells, and some reached the cell nuclei, indicating that ND functionalized with more bulky charged groups could still load nucleic acid efficiently.

Ho's group loaded bone morphogenetic proteins (BMP) protein, which is a bone morphogenetic protein, used in bone development and formations, on ND by means of physisorption to form a stable colloid (Figure 23.14). The BMP release was triggered by slightly changing the acidity of the media so it could be used as an injectable composite to promote bone formation (Moore et al. 2013b).

Noncovalent functionalization by attaching positively charged species (especially polymers or macromolecules) on ND is quite feasible. Attaching positively charged polyethyleneimine (PEI) has been

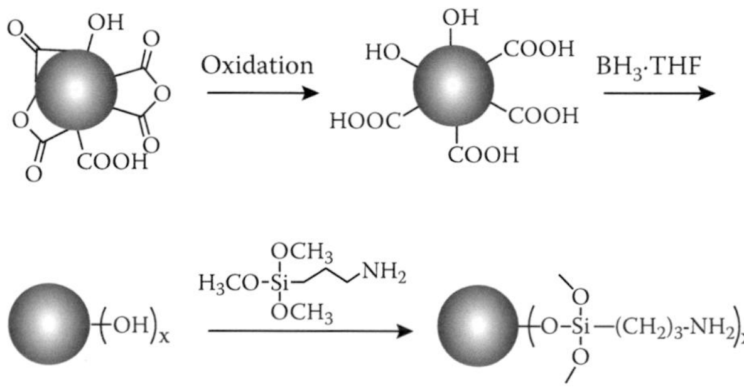

Oxidation

$BH_3 \cdot THF$

$H_3CO\text{-}Si(OCH_3)_2\text{-}(CH_2)_3\text{-}NH_2$

ND$-(OH)_x$ → ND$-(O\text{-}Si\text{-}(CH_2)_3\text{-}NH_2)_x$

FIGURE 23.13 Schematic of ND-NH_2 modification. (Reprinted with permission from Mochalin, V.N. et al., *ACS Nano*, 5, 7494, 2011. Copyright 2011, American Chemical Society.)

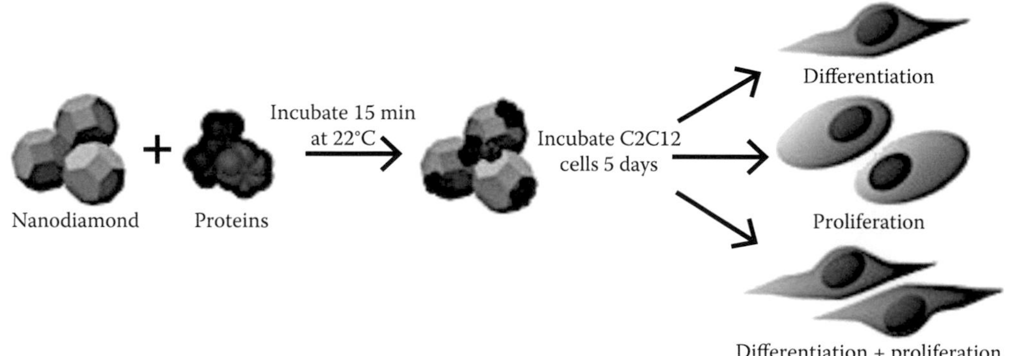

FIGURE 23.14 Recombinant BMP-2, human fibroblast growth factors (FGF)-basic, and BSA proteins were loaded into ND clusters, showing differentiation into osteoblasts and/or proliferation, depending on the protein combination. (Reprinted from Moore, L. et al., *J. Dent. Res.*, 92, 976, 2013. With permission.)

heavily reported in literature (Zhang et al. 2009; Chen et al. 2010; Alhaddad et al. 2011, 2012; Kaur et al. 2012; Kim et al. 2012; Moosa et al. 2014). PEI can be easily adsorbed on ND since oxidized ND surfaces are prone to polar interactions such as H-bonding and electrostatic interactions (Barnard and Sternberg 2007; Krueger 2008). The resulting ND–PEI complex is positively charged on surface and loads both DNA and siRNA (Zhu et al. 2012).

23.4 Conclusion

The use of NDs in biomedical applications is supported by excellent biocompatibility, high specific surface area, chemical stability, and high affinity to biomolecules. The ease of functionalization of NDs opens the door for a wide variety of applications especially in controlled delivery of small molecules (synthetic drugs) and large molecules (protein or gene). The challenge of using NDs effectively in delivery applications is choosing the optimum size and increasing the cargo loading efficiency. New techniques for large-scale modification and incorporation of these promising nanomaterials in industrial composites are needed to move this material forward into real-life practical use.

References

Adnan, A., R. Lam, H. N. Chen, J. Lee, D. J. Schaffer, A. S. Barnard, G. C. Schatz, D. Ho and W. K. Liu, *Mol Pharm* 8 (2), 368 (2011).

Alam, T. M., *Mater Chem Phys* 85 (2–3), 310 (2004).

Alhaddad, A., M.-P. Adam, J. Botsoa, G. Dantelle, S. Perruchas, T. Gacoin, C. Mansuy et al. *Small* 7 (21), 3087 (2011).

Alhaddad, A., C. Durieu, G. Dantelle, E. Le Cam, C. Malvy, F. Treussart and J.-R. Bertrand, *PLoS One* 7 (12), e52207 (2012).

Allen, T. M. and P. R. Cullis, *Science* 303 (5665), 1818 (2004).

Banhart, F. and P. M. Ajayan, *Nature* 382 (6590), 433 (1996).

Barnard, A. S. and M. Sternberg, *J Mater Chem* 17 (45), 4811 (2007).

Boudou, J. P., P. A. Curmi, F. Jelezko, J. Wrachtrup, P. Aubert, M. Sennour, G. Balasubramanian, R. Reuter, A. Thorel and E. Gaffet, *Nanotechnology* 20 (23), 235602 (2009).

Caruso, F., *Adv Mater* 13 (1), 11 (2001).

Chen, M., E. D. Pierstorff, R. Lam, S. Y. Li, H. Huang, E. Osawa and D. Ho, *ACS Nano* 3 (7), 2016 (2009).

Chen, M., X.-Q. Zhang, H. B. Man, R. Lam, E. K. Chow and D. Ho, *J Phys Chem Lett* 1 (21), 3167 (2010).

Chow, E. K., X. Q. Zhang, M. Chen, R. Lam, E. Robinson, H. J. Huang, D. Schaffer, E. Osawa, A. Goga and D. Ho, *Sci Transl Med* 3 (73), 73ra21 (2011).

Dahl, J. E., S. G. Liu and R. M. K. Carlson, *Science* 299 (5603), 96 (2003).

Daulton, T. L., M. A. Kirk, R. S. Lewis and L. E. Rehn, *Nucl Instrum Methods B* 175, 12 (2001).

Dhanak, V. R., Y. V. Butenko, A. C. Brieva, P. R. Coxon, L. Alves and L. Siller, *J Nanosci Nanotechnol* 12 (4), 3084 (2012).

Erdemir, A. and C. Donnet, *J Phys D Appl Phys* 39 (18), R311 (2006).

Fokin, A. A., B. A. Tkachenko, P. A. Gunchenko, D. V. Gusev and P. R. Schreiner, *Chem-Eur J* 11 (23), 7091 (2005).

Frenklach, M., W. Howard, D. Huang, J. Yuan, K. E. Spear and R. Koba, *Appl Phys Lett* 59 (5), 546 (1991).

Galimov, E. M., A. M. Kudin, V. N. Skorobogatskii, V. G. Plotnichenko, O. L. Bondarev, B. G. Zarubin, V. V. Strazdovskii, A. S. Aronin, A. V. Fisenko, I. V. Bykov and A. Y. Barinov, *Dokl Phys* 49 (3), 150 (2004).

Gao, X. H., Y. Y. Cui, R. M. Levenson, L. W. K. Chung and S. M. Nie, *Nat Biotechnol* 22 (8), 969 (2004).

Gogotsi, Y. G., K. G. Nickel, D. Bahloul-Hourlier, T. Merle-Mejean, G. E. Khomenko and K. P. Skjerlie, *J Mater Chem* 6 (4), 595 (1996).

Grausova, L., L. Bacakova, A. Kromka, S. Potocky, M. Vanecek, M. Nesladek and V. Lisa, *J Nanosci Nanotechnol* 9 (6), 3524 (2009).

Guan, B., F. Zou and J. F. Zhi, *Small* 6 (14), 1514 (2010).
Ho, D. A., *ACS Nano* 3 (12), 3825 (2009).
Huang, H., E. Pierstorff, E. Osawa and D. Ho, *2007 7th IEEE Conference on Nanotechnology*, Vols. 1–3, 574 (2007a).
Huang, H., E. Pierstorff, E. Osawa and D. Ho, *Nano Lett* 7 (11), 3305 (2007b).
Huang, H. J., E. Pierstorff, E. Osawa and D. Ho, *ACS Nano* 2 (2), 203 (2008).
Huang, S., J. Shao, L. Gao, Y. Qi and L. Ye, *Appl Surf Sci* 257 (20), 8617 (2011).
Kaur, R., J. M. Chitanda, D. Michel, J. Maley, F. Borondics, P. Yang, R. E. Verrall and I. Badea, *Int J Nanomedicine* 7, 3851 (2012).
Kim, H., H. B. Man, B. Saha, A. M. Kopacz, O.-S. Lee, G. C. Schatz, D. Ho and W. K. Liu, *J Phys Chem Lett* 3 (24), 3791 (2012).
Krueger, A., *Chem-Eur J* 14 (5), 1382 (2008).
Lam, R., M. Chen, E. Pierstorff, H. Huang, E. J. Osawa and D. Ho, *ACS Nano* 2 (10), 2095 (2008).
Langer, R., *Nature* 392 (6679), 5 (1998).
Langer, R. and D. A. Tirrell, *Nature* 428 (6982), 487 (2004).
Li, J., Y. Zhu, W. X. Li, X. Y. Zhang, Y. Peng and Q. Huang, *Biomaterials* 31 (32), 8410 (2010).
Li, Y. Q., X. P. Zhou, D. X. Wang, B. S. Yang and P. Yang, *J Mater Chem* 21 (41), 16406 (2011).
Ma, X. W., Y. L. Zhao and X. J. Liang, *Acta Pharmacol Sin* 32 (5), 543 (2011).
Man H. B. and D. Ho, *Phys Status Solidi A* 209 (9), 1609 (2012).
Man, H. B., H. Kim, H. J. Kim, E. Robinson, W. K. Liu, E. K. Chow and D. Ho, *Nanomedicine* 10 (2), 359–369 (2014).
Martin, R., M. Alvaro, J. R. Herance and H. Garcia, *ACS Nano* 4 (1), 65 (2010).
Mochalin, V. N., J. Giammarco, A. Gurga, J. Detweiler, C. Hobson, Y. Gogotsi, A. Peterson and G. R. Palmese, *Abstr Pap Am Chem S* 236 (2008).
Mochalin, V. N., I. Neitzel, B. J. M. Etzold, A. Peterson, G. Palmese and Y. Gogotsi, *ACS Nano* 5 (9), 7494 (2011).
Mochalin, V. N., I. Neitzel and Y. Gogotsi, PMSE Preprints (2012a).
Mochalin, V. N., O. Shenderova, D. Ho and Y. Gogotsi, *Nat Nanotechnol* 7 (1), 11 (2012b).
Moghimi, S. M., A. C. Hunter and J. C. Murray, *Pharmacol Rev* 53 (2), 283 (2001).
Moore, L., E. K. Chow, E. Osawa, J. M. Bishop and D. Ho, *Adv Mater* 25 (26), 3532 (2013a).
Moore, L., M. Gatica, H. Kim, E. Osawa and D. Ho, *J Dent Res* 92 (11), 976 (2013b).
Moosa, B., K. Fhayli, S. Li, K. Julfakyan, A. Ezzeddine and N. M. Khashab, *J Nanosci Nanotechnol* 14 (1), 332 (2014).
Niemeyer, C. M., *Angew Chem Int Ed* 40 (22), 4128 (2001).
Niemiec, T., M. Szmidt, E. Sawosz, M. Grodzik and K. Mitura, *J Nanosci Nanotechnol* 11 (10), 9072 (2011).
Osawa, E., Nanodiamonds applications in biology and nanoscale medicine. In *Single-Nano Buckydiamond Particles: Synthesis Strategies, Characterization Methodologies and Emerging Applications*, Ho, D., Ed., Springer, New York (2010), Vol. 2010:1–33.
Palnichenko, A. V., A. M. Jonas, J. C. Charlier, A. S. Aronin and J. P. Issi, *Nature* 402 (6758), 162 (1999).
Pierstorff, E. and D. Ho, *Int J Nanomedicine* 3 (4), 425 (2008).
Pierstorff, E., R. Lam and D. Ho, *Nanotechnology* 19 (44), 445104 (2008).
Purtov, K. V., A. I. Petunin, A. E. Burov, A. P. Puzyr and V. S. Bondar, *Nanoscale Res Lett* 5 (3), 631 (2010).
Rao, C. N. R. and A. K. Cheetham, *J Mater Chem* 11 (12), 2887 (2001).
Shimkunas, R. A., E. Robinson, R. Lam, S. Lu, X. Xu, X.-Q. Zhang, H. Huang, E. Osawa and D. Ho, *Biomaterials* 30 (29), 5720 (2009).
Smith, A. H., E. M. Robinson, X.-Q. Zhang, E. K. Chow, Y. Lin, E. Osawa, J. Xi and D. Ho, *Nanoscale* 3 (7), 2844 (2011).
Solarska, K., A. Gajewska, G. Bartosz and K. Mitura, *J Nanosci Nanotechnol* 12 (6), 5117 (2012).
Wang, D. X., Y. L. Tong, Y. Q. Li, Z. M. Tian, R. X. Cao and B. S. Yang, *Diam Relat Mater* 36, 26 (2013).
Wang, H.-D., C. H. Niu, Q. Yang and I. Badea, *Nanotechnology* 22 (14), 145703/1 (2011).

Wang, H.-D., Q. Yang and C. H. Niu, *Diam Relat Mater* 19 (5–6), 441 (2010).
Wehling, J., R. Dringen, R. N. Zare, M. Maas and K. Rezwan, *ACS Nano* 8 (6), 6475 (2014).
Welz, S., Y. Gogotsi and M. J. McNallan, *J Appl Phys* 93 (7), 4207 (2003).
Xi, G., E. Robinson, B. Mainia-Farmell, E. F. Vanin, K. W. Shim, T. Takao, E. V. Allender et al. *Nanomedicine* 10 (2), 381–391 (2014).
Yan, J. J., Y. Guo, A. Altawashi, B. Moosa, S. Lecommandoux and N. M. Khashab, *New J Chem* 36 (7), 1479 (2012).
Yang, G. W., J. B. Wang and Q. X. Liu, *J Phys-Condens Mater* 10 (35), 7923 (1998).
Yeap, W. S., S. M. Chen and K. P. Loh, *Langmuir* 25 (1), 185 (2009).
Zhang, X. Q., M. Chen, R. Lam, X. Y. Xu, E. Osawa and D. Ho, *ACS Nano* 3 (9), 2609 (2009).
Zhang, X. Q., R. Lam, X. Y. Xu, E. K. Chow, H. J. Kim and D. Ho, *Adv Mater* 23 (41), 4770 (2011).
Zheng, W. W., Y. H. Hsieh, Y. C. Chiu, S. J. Cai, C. L. Cheng and C. P. Chen, *J Mater Chem* 19 (44), 8432 (2009).
Zhu, Y., J. Li, W. Li, Y. Zhang, X. Yang, N. Chen, Y. Sun, Y. Zhao, C. Fan and Q. Huang, *Theranostics* 2 (3), 302 (2012).

24

Nanodiamonds in Biomedicine

Saniya Alwani

Ildiko Badea

Abstract

Nanodiamonds (NDs) are the most biocompatible representatives of the carbon nanofamily and are widely researched for diagnostic and therapeutic applications. Unlike other carbon nanomaterials, the surface of NDs is innately reactive and contains diverse combinations of functional groups. Homogenizing the surface through oxidation or reduction allows conjugation of various chemical moieties to target specific actions. To date, NDs are conjugated to a range of inorganic and organic, natural, or synthetic functional groups both through physical adsorption and chemical immobilization. Functionalization of NDs also provides a chemical means to mitigate ND aggregation, which is considered a major challenge toward its biomedical applications. Physical approaches are also broadly utilized to produce well-dispersed stable ND dispersions. To maximize their utilization, the interaction of NDs with biological system at cellular and organism levels is also extensively investigated. High biocompatibility and affinity for a diverse range of biomolecules and therapeutic agents make them attractive for nanomedicine especially as biomarkers and delivery vehicles. This chapter provides a detailed outlook regarding the structure, composition, and properties of NDs. Challenges associated with their utilization and recent approaches to overcome these challenges are elaborated. Diverse array of diagnostic and therapeutic applications of ND-based systems are presented.

24.1 Introduction

Carbon is one of the most abundant elements in the Earth's crust, known to exist in all life forms. Diamond is considered the hardest allotrope of carbon, possessing a tightly packed interpenetrating cubic lattice structure. Extremely high bond energy between two carbon atoms (83 kcal/mol) and the directionality of tetrahedral bonds (Narayan et al., 2011) account for the ultrastable nature of the diamond core.

Diamond nanoparticles (NPs), also called nanodiamonds (NDs) are carbon nanomaterials that exhibit unique biological, thermal, mechanical, and optoelectronic properties. They possess high surface areas and tunable surface structures. Because of their inherent properties, NDs are widely utilized in electronics (Dolmatov, 2001), medicine (Manus et al., 2010), and advanced research technologies (Fedyanina & Nesterenko, 2010; Saini et al., 2010; Wu et al., 2010).

24.1.1 Production of NDs

NDs naturally exist in different levels of the universe. Carbonaceous chondrites, which are a special type of stony meteorite (Alexander, 2013), are the major naturally occurring source of NDs sized 2.5 nm. These NDs also contain hydrogen, nitrogen, and oxygen impurities (Galli, 2010).

NDs are synthetically produced by many different approaches. Each method yields particular sized diamond NPs, which are suitable for specific applications. Based on the primary particle size, NDs are categorized into nanocrystalline particles (1 to ≥150 nm), ultrananocrystalline particles (2 to 10 nm), and diamondoid structures (1 to 2 nm) (Shenderova & McGuire, 2006; Kaur & Badea, 2013).

Ultradispersed diamond NPs (Figure 24.1a) ranging from 2 to 8 nm are produced by detonation and are widely researched for biomedical applications (Mochalin et al., 2011). In this process, NDs are synthesized by detonating a mixture of trinitrotoluene (TNT) and hexogen in a negative oxygen balance to facilitate carbon–carbon interaction (Figure 24.1b). Detonation involves transformation of the explosive material; it requires constant maintenance of optimal thermodynamic conditions. During detonation, pressure and temperature are maintained at levels capable of allowing the appearance of free carbon in the product and, at the same time, inhibiting diamond-to-graphite transition due to explosion (Dolmatov et al., 2004). The process creates soot containing nanosized diamond particles, which are then isolated and purified. To maximize the yield up to 90%, the soot is subjected to high-energy electronic irradiation and heat treatment at 800°C, followed by rapid cooling. Isolation of NDs from the soot is carried out by boiling in acid at high temperature of 3000°C–4000°C and high pressure of 20–30 GPa for 1 to 2 days (Shenderova et al., 2002; Baidakova & Vul, 2007; El-Say, 2011; Kaur & Badea, 2013). Despite various novel approaches for ND synthesis, detonation NDs hold primary focus to generate NDs for nanomedicine research, majorly as therapeutic delivery vectors (Chen et al., 2009; Shimkunas et al., 2009; Kaur et al., 2012).

Novel techniques for generating medical-grade NDs focus primarily on reducing the formation of core aggregates in the synthetic process and producing disaggregated primary diamond NPs (Fang et al., 2013). The core aggregates are formed mainly because of sp^2 graphitic carbon contamination present in the soot resulting from detonation (Kruger et al., 2005). Most of these novel techniques involve conversion of graphite into diamond structures by creating optimal conditions of temperature and pressure through different modes (Sun et al., 2005; Khachatryan et al., 2008; Fang et al., 2013). One high-temperature–high-pressure technique involves decomposition of graphitic carbon nitride to produce uniform-sized, single-phase NDs with no graphite contamination (Fang et al., 2013). The graphitic carbon nitride treated at 800°C–900°C and between 22 and 25 GPa pressure decomposes directly to diamond and nitrogen (Fang et al., 2011). Unlike detonation, this approach is centered upon creating high-temperature and high-pressure conditions during the synthesis phase of NDs. It offers many advantages over detonation. Mainly, it produces pure ND crystals with virtually no graphitic contamination and hence allows their utilization in an as-synthesized form with minimal requirement of postsynthetic purification (Fang et al., 2013).

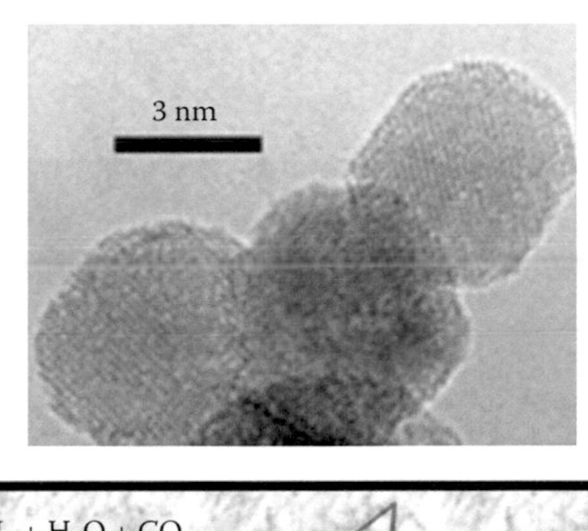

(a)

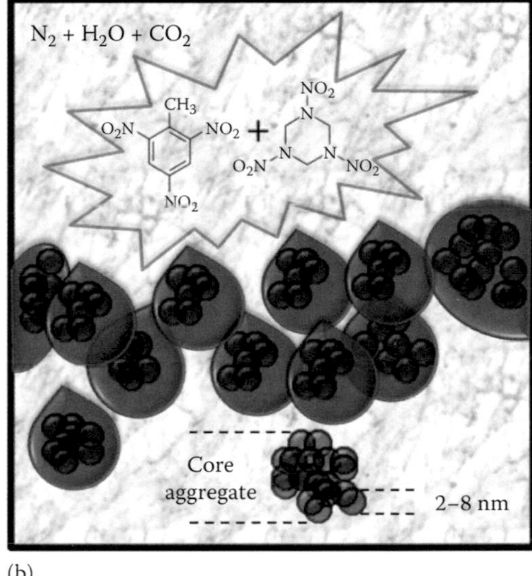

(b)

FIGURE 24.1 (a) High-resolution transmission electron microscopic image of detonation NDs. (Reprinted from Kaur, R., Badea, I., *Int. J. Nanomed.*, 8, 203–220, 2013. Copyright 2013, Dove Medical Press. With permission.) (b) Schematic representation of ND synthesis in a detonation chamber. Explosion of TNT and hexogen in negative oxygen balance created by nitrogen carbon dioxide and water produces soot containing core aggregates of ND particles, which are then isolated and purified.

Ultrasonic cavitation is another approach to synthesize NDs based on the high-pressure–high-temperature technique (Khachatryan et al., 2008). This process involves transition of graphite into diamond in organic liquid at a temperature of 120°C and atmospheric pressure, in the presence of ultrasonic waves. Ultrasonic cavitation acts as super-high-pressure shock waves producing necessary combination of temperature and pressure to facilitate this transformation. However, this process produces very low yield of up to 10% (Khachatryan et al., 2008).

High-energy laser irradiation of graphite is an alternate method of producing NDs (Sun et al., 2005; Hu et al., 2008). This technique involves irradiating the graphite powder with high-energy pulsed laser beam at room temperature and atmospheric pressure. Pulsed laser energy produces high temperature, which heats the raw graphite powder, followed by rapid cooling that generates high pressure. Thus, high temperature and high pressure combination is created to form nanosized diamond particles of 5 nm with fewer impurities as compared with those formed through the detonation process (Sun et al., 2005).

While these alternative techniques offer high-quality NDs, the major source of biomedical applications today remains the detonation NDs, particularly because of high yield and convenience for its application at industrial scale.

24.1.2 Core Structure, Surface Chemistry, and Chemical Composition of NDs

A single detonation ND is composed of a 4- to 5-nm diamond particle having sp^3 hybridized carbon atoms surrounded by an amorphous and graphitic sp^2 hybridized carbon layer, bearing a large number of functional groups on the surface (Kulakova, 2004), as represented in Figure 24.2.

The chemical composition of NDs is well elucidated. They consist of carbon to an extent of 80%–88%, and this carbon predominantly forms the diamond phase. Along with carbon, NDs also contain oxygen (10% or more), hydrogen (0.5%–1.5%), nitrogen (2%–3%), and an incombustible residue (0.5%–8.0%) by weight (Dolmatov, 2001). Recent elemental analysis reports 90% of carbon, 3% of oxygen, and 0.8% of hydrogen contents in ND (Paci et al., 2013). The difference in oxygen and hydrogen contents can be attributed to water, since absolute removal of sorbed water is not possible (Szabo et al., 2006; Paci et al., 2013). A single particle consists of a crystalline diamond core, surrounded by an amorphous shell (coat) composed of functional groups. These groups are covalently bound to ND particles and determine the chemical state of its surface (Spitsyn, 2005). The diamond core consists of sp^3 hybridized carbons where all four carbons are bound via sigma bonds with other carbon atoms and forms a tetrahedral symmetry (Kaur & Badea, 2013), The shell consists of sp^2 hybridized carbons in which a central carbon is attached to three more adjacent carbons via sigma bonds and has a cloud of delocalized pi-electrons. These sp^2 hybridized carbons arrange in two-dimensional planar hexagonal graphite layers (Kaur & Badea, 2013). The functional groups may include hydroxyl, carbonyl, ether, or anhydride. Nitrogen-containing groups like amine, amide, cyano, or nitro groups might also appear; however, evidence suggests that most of the nitrogen, detected by elemental analysis, is accumulated in the core (Kruger et al., 2005). Minute quantities of sulfone, methyl, or methylene groups may also be present on the ND surface (Kulakova, 2004). In addition to the functional groups, sorbed water may also be present in the pores between NDs resulting from aggregation (Kulakova, 2004). Spectroscopic and scattering analysis of detonation NDs reveals a core size distribution that peaks at 5 nm (Ōsawa, 2007; Iakoubovskii et al., 2008). However, clear signals of diamond core are not clearly detectable even after vibrant cleaning because of extensive layering of nondiamond amorphous

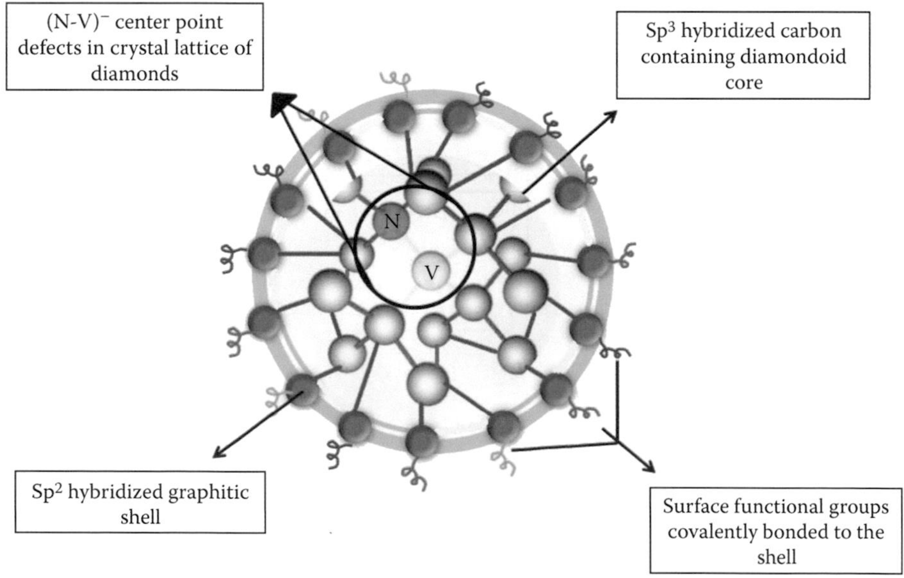

FIGURE 24.2 Structure of a single ND particle exhibiting a layered morphology.

carbon appearing as various functional moieties, like C–C, C=C, C–H, C–O, and C=O (Iakoubovskii et al., 2008). High-temperature–high-pressure synthesis of NDs greatly reduces this graphite contamination and the diamond peaks in spectroscopic analysis are more discernable (Shiryaev et al., 2006; Iakoubovskii et al., 2008).

Detonation NDs are considered to have a faulty and a defective structure, although experimental analysis using high-resolution transmission electron microscopy reveals that the core of an ND particle consists of regularly arranged carbon atoms (Vereschagin et al., 1994). However, a common structural defect in natural and synthetic ND is multiple twinning, in which two individual crystals share same crystal lattice symmetry (Iakoubovskii et al., 2008; Galli, 2010). Twins, once formed, act as preferential sites for impurities and defects (Yan & Vohra, 1999). This process may increase the fragility of NDs (Iakoubovskii et al., 2008). Nitrogen impurity is the most common point defect that may exist in the diamond structure as a result of detonation (Figure 24.2). It consists of a nearest-neighbor pair of nitrogen atom, which substitutes for carbon atom and forms a vacancy in the diamond lattice (Yan & Vohra, 1999). The nitrogen vacancy (NV) centers are responsible for the photoluminescence properties of NDs and are exploited for medical imaging purposes.

Conclusively, NDs appear to have spherical morphology having a well-defined diamond cage in the core. The structure of the surface varies with synthetic and postsynthetic conditions (Galli, 2010). This unique structure of ND provides many advantages over other carbon nanomaterials, which usually feature a chemically inert bare graphitic surface like in the case of graphenes or carbon nanotubes or structure—less mixture of different forms of carbon (amorphous or disordered carbon) with less controllable and less accessible external surface (Zhou et al., 2012).

24.1.3 Key Properties of NDs for Biomedical Applications

NDs have attracted significant attention in numerous applications because of their innate physical and chemical properties. In particular, detonation NDs have the propensity to form a porous cluster structure in solution; therefore, many molecules can be adsorbed on its surface. This phenomenon is exploited for developing novel technologies in nanomedicine, like a smart drug complex providing attractive features such as sustained release and targeting. Moreover, they also allow chemical modification of the surface, which opens avenues for designing smart hi-tech tools.

The following are some fundamental properties of NDs making them attractive for electronic, mechanical, and medicinal research.

24.1.3.1 Toughness

NDs possess many properties of bulk diamond, among which is extraordinary hardness and rigidity (Mochalin et al., 2011), suggesting that NDs are highly resistant to deformation under corrosive environments. This property attracts extensive interest in utilizing NDs for designing biological implants (Grausova et al., 2009). Implantable delivery systems require optimum hardness characteristics to ensure tunable release rates of the incorporated therapeutic molecule (Ho & Lam, 2009). Because of their rigid structure, NDs introduce these properties and can develop biocompatible and mechanically durable implants for localized delivery.

NDs are also investigated for engineering nanocomposites for bone tissue engineering (Q. Zhang et al., 2011). Substitution of the composite with NDs resulted in an increase in Young's modulus (200%) and overall hardness (800%) of the system. It was also shown to have the least negative impact on osteoblastic proliferation even at high concentrations of 100 μg/mL in vitro. In addition, the proliferation of osteoblasts was also maintained upon direct interaction with the scaffolds containing up to 10% of ND complexes. Oestoblastic activity is a critical component of healthy bone mineralization and in reducing bone resorption (Rodan & Martin, 1982; Dvorak-Ewell et al., 2011). This opened new avenues to utilize NDs for making up bone scaffolds (Zhang et al., 2011).

24.1.3.2 Inertness

This property is particularly useful for biomedicinal application of NDs, where the particle must have intrinsic ability to protect its integrity in case of any impact from the surrounding environment. ND shows this chemical inertness to a large extent and is capable of maintaining its inherent structure and properties in the presence of a corrosive biological environment containing acids, bases, enzymes, and many other chemicals. A study conducted by Yuan et al. (2009) revealed that the diameter of NDs remained unchanged even after residing for several days in various organs, highlighting their resistance toward biotransformation and degradation after digestion.

24.1.3.3 Large Surface Area and High Adsorption Potential

Another important feature that makes NDs an interesting tool for biomedicine and engineering is the high ratio of surface area to volume, which allows a large amount of active moieties to be loaded onto its surface. For therapeutic applications, NDs can be functionalized with various chemical groups to allow adsorption and delivery of therapeutic small molecules and biomolecules. Using this property as a tool, many concepts have been generated to utilize NDs as delivery agents for nanomedicine (Chen et al., 2009, 2010; Shimkunas et al., 2009). The large surface area of NDs also facilitates their utilization in specialized areas such as an intestinal adsorbent for removing toxic metabolites of protein and nonprotein origin, like drugs, xenobiotics, radionuclides, metals, and exotoxins out of the body (Puzyr et al., 2007; Gibson et al., 2011).

NDs retain the functional properties of the adsorbed materials and hence can also be used as analytical tools for extraction and purification of proteins from natural origin and recombinant technologies (Puzyr et al., 2007).

24.1.3.4 Photoluminescence

NDs are optically transparent and are capable of producing luminescence from colored centers. These colored centers are point defects in their structure, commonly known as NVs (Figure 24.2). They absorb light at wavelengths of visible, infrared, or ultraviolet spectral region (Kratochvílová et al., 2011) and emit bright fluorescence at 550–800 nm (Chao et al., 2009). Fluorescent NVs in the ND core are formed during synthesis and can be further enhanced by altering the diamond properties through high-energy ion beam irradiations and subsequent thermal annealing (Fu et al., 2007).

In relationship to NDs, fluorescence is often used interchangeably with photoluminescence to define the spectroscopic properties of NDs. Photoluminescence is a broader term encompassing both the fluorescence and phosphorescence properties of spectroscopically active molecules (Smestad, 2002). The photoluminescence of NDs is dependent on the particle size and excitation laser intensity (Chao et al., 2009). For smaller particles (5–50 nm), the spectra show structureless bands in an emission range of 550–700 nm, and for larger particles (100–500 nm), structured emission bands in 580–720 nm range are observed (Chao et al., 2009). Changes in the spectral patterns of NDs are correlated with the differences in structure and features of the differently sized particles, contributing toward photoluminescence. For smaller particles, the surface is well defined, and hence, the natural defects created on the surface are the major contributor toward photoluminescence. In addition, surface carbons in threefold coordinated state (i.e., sp^2 hybridization) and the dangling bonds created with surface carbons of adjacent NDs in the cluster also contribute toward emission. The intrinsic diamond core defects are postulated to account for a structureless region in the spectra (Chao et al., 2009). Unlike small NDs, larger diamond particles do not have a well-defined surface because of aggregation; hence, the vacancy centers in the diamond core and the heterogeneity in the diamond phase resulting during synthesis or postsynthetic treatment are competing origins of photoluminescence (Chao et al., 2009).

Stable photoluminescence, lack of photobleaching, and numerous opportunities of functionalization render NDs an attractive probe for bioimaging. However, some limitations, such as low intensity

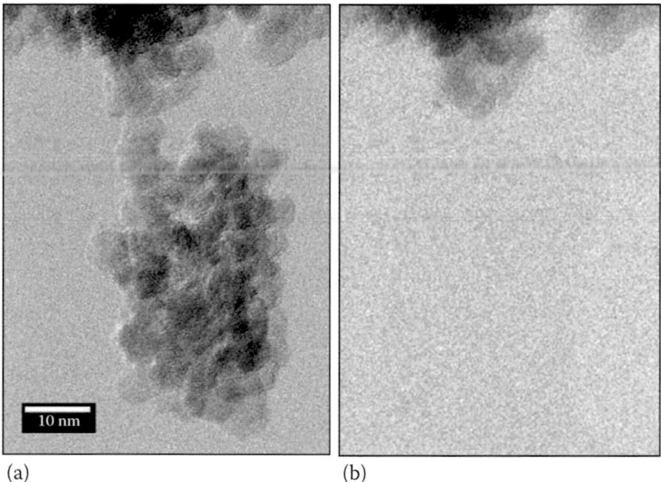

FIGURE 24.3 Bright-field scanning transmission electron micrograph of an ND cluster held in contact by adhesion of a single ND grain (a). When the contact area was irradiated by a 100 times brighter stationary electron beam, creating heat energy, the cluster was cut off within a minute (b). (Reprinted with permission from Iakoubovskii, K., Mitsuishi, K., Furuya, K., High-resolution electron microscopy of detonation nanodiamond. *Nanotechnology*, 19, 155705, 2008. Copyright 2008, IOP Publishing.)

of fluorescence, remain (Faklaris et al., 2010), requiring further optimization before its commercial use in diagnosis.

24.1.3.5 Aggregation

Small (5 to 10 nm) NDs possess a tendency to form aggregates of tens to hundreds of nanometers to minimize their surface free energy (Krueger, 2008). Multiple NDs can adhere together, forming stable core aggregates. Adhesion between two detonation NDs can be strong enough to hold a large cluster of NDs, as depicted in Figure 24.3. Continuous application of mechanical or vibrational energy through methods such as heating or milling can overcome these adhesive forces, leading to disaggregation of ND particles (Iakoubovskii et al., 2008).

Aggregation could be an advantageous property to a certain extent as the size of the aggregate majorly controls the cellular uptake of NPs in the biological system. There is always an optimum particle size that facilitates controlled cellular uptake through a phenomenon called wrapping effect, a process by which the cell membrane encloses the particle for internalization. Very small particle size (5–10 nm) results in minimal receptor–ligand interaction, which is insufficient to stimulate the cell membrane to wrap around the particle (Trono et al., 2011). Aggregation of primary NDs sized 2–8 nm to 20–30-nm particles is favorable for cellular uptake. It could lead to enhanced interactions with receptors on the cell surface (Alkilany & Murphy, 2010), undergoing receptor-mediated endocytosis. Aggregation up to a certain degree will also result in low surface energy, contributing to the increased stability of the system.

However, there is a narrow range where aggregation results in favorable responses. Large aggregates of 100–200 nm might result in formulation challenges and biological incompatibilities (Xing et al., 2011). These issues are discussed in the next chapter.

24.2 Dispersion Stability of NDs—A Challenge for Biomedical Application

The formation and maintenance of well-dispersed NP formulation in biologically relevant media are critical parameters for controlled therapeutic response. The propensity of NDs to aggregate in tight structures of 100 to 200 nm (Kruger et al., 2005) is a major formulation challenge associated with their biological application, such as incompatibilities. Therefore, this issue is of great interest to researchers utilizing addressed NDs as imaging and therapeutic tools.

24.2.1 Theories Related to ND Aggregation

Several theories have been formulated to explain ND aggregation. These theories differ in the mechanisms promoting aggregation and the factors affecting these mechanisms.

24.2.1.1 Aggregation Mediated through Surface Functional Groups

Unlike other carbon nanomaterials, NDs possess a functionalized surface as a result of the detonation process. The functional groups present on detonation NDs facilitate aggregation through van der Waals interactions or hydrogen bonding between adjacent ND crystals (Shenderova et al., 2002; Ho, 2010; El-Say, 2011). The van der Waals forces are inversely proportional to the particle size of NDs. The typical proportionality of attractive forces to ND size is found to be $1/r^6$, where r represents the radius of the particle. For NDs having size less than 10 nm, surface groups may react together, forming covalent bonds (Figure 24.4), and yield aggregated crystal-like ordered structures (Kruger et al., 2006; Chao et al., 2009).

24.2.1.2 Aggregation Mediated through ND Facets

This theory suggests self-assembly through electrostatic interactions between different facets of NDs (Barnard, 2008; Huang et al., 2008a; Chang et al., 2011). Individual ND particles exhibit anisotropic facet variations; i.e., it consists of two distinct facets. These include facets with a net positive charge and facets with either neutral or net negative charge depending upon the degree of graphitization (Barnard & Sternberg, 2007). This polyhedral and multipolar structure of NDs can possibly attract each other and form aggregates. The type of interaction involving neutral facets as primary participants in aggregation of NDs is characterized as *coherent interfacial Coulombic interactions*, while all other alternating configurations resulting from random interactions are termed as *incoherent interfacial Coulombic interactions* (Chang et al., 2011). A variety of different one- or two-dimensional self-assembled networks and suprastructures are possible through these interactions, as depicted in Figure 24.5.

24.2.1.3 Aggregation Mediated through Graphitic Soot

This is the most widely addressed theory of the ND aggregation. It suggests that the major cause of aggregation is the graphitic soot formed as a result of the detonation process around the NDs. There are three sets of major events that occur during the deposition of carbon atoms from the detonation mixture, which explains the mechanisms and the factors affecting the formation of core aggregates (Kruger et al., 2005):

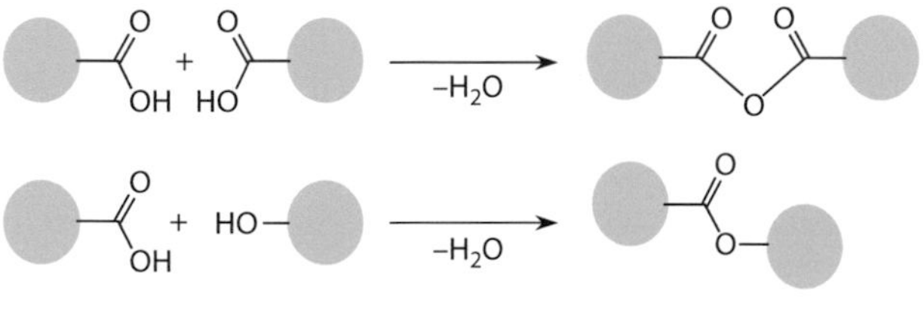

FIGURE 24.4 Example of reactions for interparticle covalent binding in ND agglomerates. (Kruger, A. et al., *J. Mater. Chem.*, 16, 2322–2328, 2006. Reprinted by permission of The Royal Society of Chemistry.)

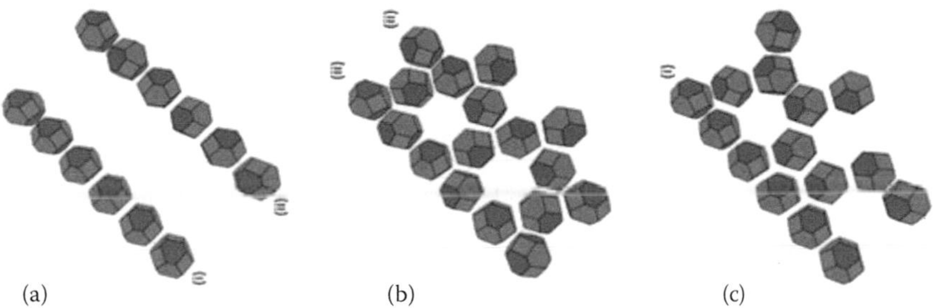

FIGURE 24.5 Schematic illustration of possible networks of NDs: (a) one-dimensional orientation, (b) and (c) two-dimensional orientation based on (111)–(111) and (100)–(111) interaction configurations. (100) represents facets with a net positive charge and (111) represents facets with either neutral or net negative charge. (Barnard, A.S., *J. Mater. Chem.*, 18, 4038–4041, 2008. Reprinted by permission of The Royal Society of Chemistry.)

1. Excess carbon resulting from detonation is subjected to high-temperature and high-pressure conditions inside the chamber, leading to their arrangement as a diamond lattice. This process leads very fast to the growth of several diamond nuclei with uniform morphology.
2. Alteration in the temperature and pressure conditions stops further growth of ND crystals. At this point, the remaining carbons arrange themselves as a stable graphite layer over diamond crystals. Studies report two types of graphite arrangements, including ribbon structures (Kuznetsov et al., 1994) and spherical graphitic shells (Raty et al., 2003). Graphitic ribbons are formed as a result of the carbon redistribution process, which arranges carbons in haphazard and crumpled fashion (turbostratic structure) (Kuznetsov et al., 1994). On the other hand, spherical shells result from diamond-to-graphite phase transition (Raty et al., 2003). Apart from carbon rearrangement as graphitic layers, this stage of synthesis also creates soot embryos that are spiral sub-NPs composed of several hundred sp^2 carbon atoms (Ozawa et al., 2002).
3. The final event that results in formation of a core aggregate involves coagulation of the soot embryos. These embryos continue to generate sphere-shaped but faulty NPs. When they reach an optimum diameter of 10 to 20 nm, they begin to coagulate themselves into aggregate structures. These aggregate structures are commonly called soot areas. They may contain soot embryos along with graphitic shells or graphitic ribbon-like structures. The aggregation of an irregular graphitic shell around the particle forms a core aggregate of NDs (Figure 24.6) (Kruger et al., 2005).

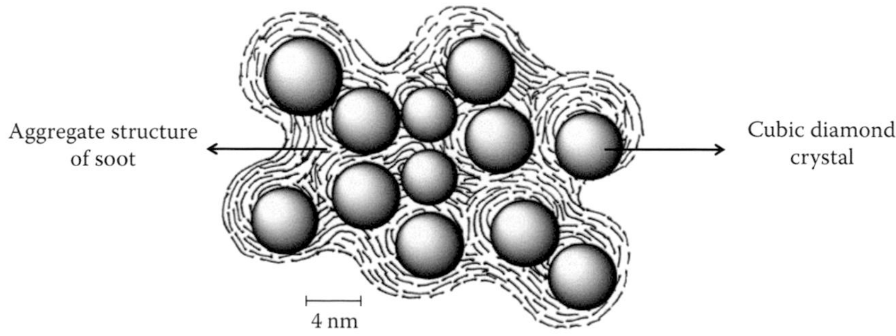

FIGURE 24.6 A simplified model of core aggregate showing NDs and graphitic soot. (Reprinted from *Carbon*, 43, Kruger, A. et al., Unusually tight aggregation in detonation nanodiamond: Identification and disintegration, 1722–1730. Copyright 2005, with permission from Elsevier.)

24.3 Methods to Mitigate Aggregation and Produce Stable ND Dispersion

There are varieties of methods targeting disintegration of the NDs from their core aggregate structures. They are either physical, chemical, or combination methods, selected based on the future applications of the NDs. Physical approaches tend to counter ND aggregation mediated through the removal of the graphitic layer, while chemical methods target the surface through various functionalizations to reduce electrostatic attraction and hydrogen bonding between individual particles.

24.3.1 Physical Methods: Stirred Media Milling and Dry-Media-Assisted Attrition Milling

Physical methods used to deaggregate ND colloidal dispersions are widely employed in biomedical research owing to the ease of procedure. Stirred media milling, also called wet grinding, is a traditional method to disintegrate NDs. It involves preparing ND slurries and adding suitable grinding media (Eidelmana et al., 2005). Conventionally, yttrium stabilized zirconia (YTZ) beads, which are available in a variety of sizes ranging from micrometers to millimeters, are used as the grinding media (Eidelmana et al., 2005). When the ND dispersion containing zirconia beads is subjected to ultrasonication, the beads hit the graphitic soot and on the diamond core. These collisions induce high energy impact and sheer forces (Liang et al., 2009), ultimately causing the aggregation to break. In this process, bath sonication additionally creates shock waves (Liang et al., 2009), adding to the effect of zirconia beads to produce high-intensity impact (Figure 24.7).

The zirconia beads are softer than the ND core; therefore, the impact with the diamond core leads to their erosion during milling (Osawa, 2005; Paci et al., 2013). These eroded zirconia particles are difficult to remove and yield a measurable quantity (0.2%) of contamination in the dispersion (Paci et al., 2013). Certain comprehensive modifications (Figure 24.8) in the parameters of this technique were found to be effective in reducing this contamination. Osawa et al. observed that the milling conditions should be adjusted to mild levels so that the beads hit only the susceptible deformed areas of the core aggregates, the graphitic soot, and not collide with the core of the diamond particles. Reducing the milling agitation, milling duration, and particle size of the beads to 30 μm can reduce the susceptibility of bead destruction. In addition, the broken beads can be removed by centrifugation, adsorption onto activated charcoal, or diffusion through an osmotic membrane (Osawa, 2005). Modification of the dispersion media is also recommended to control contamination. Solvents such as ethanol are preferred instead

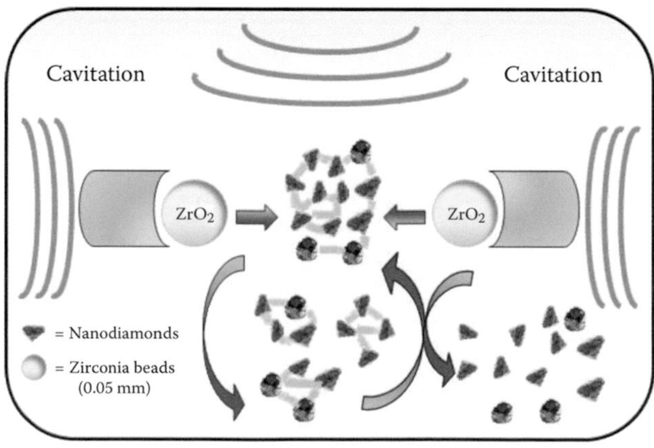

FIGURE 24.7 Bead-assisted deaggregation of NDs under bath sonication.

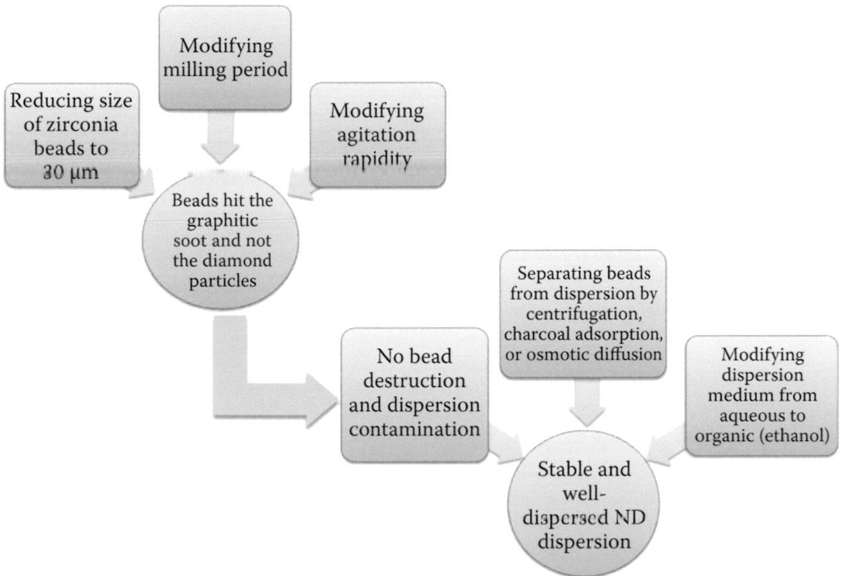

FIGURE 24.8 Parameters that could improve stirred media milling.

of water to avoid reaggregation at the end of the milling process (Osawa, 2005). The aqueous medium shows better grinding properties and produces finer ND particles as compared with organic medium and it is a biocompatible solvent for medical applications. However, it is shown to induce reaggregation, unlike the organic dispersion medium (Knieke et al., 2010). Strategies that could improve stirred media milling are summarized in Figure 24.8.

Dry-media-assisted attrition milling is also a noteworthy approach to formulate stable ND dispersion. In this process, water-soluble nontoxic and noncontaminating crystalline materials like sodium chloride or sucrose are used. These crystals can be later removed by rinsing the NDs with water, hence reducing the susceptibility of contamination, unlike ceramic beads. The process is capable of producing NDs of <10 nm, which is close to the primary particle size (Pentecost et al., 2010).

24.3.2 Chemical Methods: Surface Functionalization of NDs

Unlike other carbon nanomaterials, the surface of NDs is not inert, therefore opening several avenues for functional group modification and conjugation. Conjugation of various organic or inorganic molecules on the surface of NDs can control aggregation and simultaneously impart specific properties for biological applications.

Surface functionalization involves induction of surface homogeneity by modifying a variety of functional groups (hydroxyl, carbonyl, ether, or anhydride) already present on the surface of detonation NDs to facilitate electrostatic repulsion by similar functional groups. NDs are also functionalized to confer chemical affinity to therapeutic molecules. For example, several approaches are employed to synthesize a cationic surface for biomolecular conjugation (X. Zhang et al., 2009; P. Zhang et al., 2011). Recently, ND functionalization has also focused on targeting the therapeutic system to certain cells and organs for enhanced site-specific activity (B. Zhang et al., 2009), reduced toxicity, and limited waste through processes like opsonization (Zhang et al., 2012a). Functionalization of NDs with various chemical moieties to grant specific biomedical application will be discussed later.

Chemical functionalization is a multistep process. First, the surface of the NDs needs to be prepared, and then the actual conjugation of the functional moieties takes place.

24.3.2.1 Pretreatment of NDs before Functionalization

All types of ND functionalization begin with pretreatment to homogenize the surface functionalities. This step converts a heterogeneous surface to contain, as much as possible, a single type of functionality. It ensures a similar conjugation behavior of the entire ND surface during functionalization (Figure 24.9). The pretreatment process can control aggregation mainly by facilitating electrostatic repulsion between similar functional groups on the surface (Shenderova et al., 2006; Petrova et al., 2007; Pichota et al., 2008).

Most often, the pretreatment step involves oxidation (Figure 24.9a). This step removes all impurities like carbon soot and metal ions from the ND surface. Among some complex functional groups involved in oxidation are chromenes (C_9H_8O), pyrones ($C_5H_4O_2$), phenols, and epoxides, which are mainly at the edges of graphite surfaces (Paci et al., 2013). The difference in the nature of the oxidative agents may result in variable functional group distribution on the surface after oxidation. For example, combination of nitric acid and sulfuric acid may result in carboxylate (COO^-)-rich surface (Kaur et al., 2012; Shenderova et al., 2014), while the combination of potassium permanganate and sulfuric acid mainly result in SO_3^- or O^- derivatives of phenol (Paci et al., 2013). Oxidation with strong mineral acids causes conversion of many chemical species into carboxylic functional groups, which imparts hydrophilicity to NDs (Jee & Lee, 2009).

Oxidation of NDs can also be done by various other methods, which include oxidation in air or utilizing other oxidizing agents (Krueger, 2008). Air oxidation is a simple and inexpensive approach for disaggregation of NDs (Petrova et al., 2007; Gaebel et al., 2012). It utilizes either nonprocessed air (Gaebel et al., 2012) or air with enriched oxygen or ozone (Petrova et al., 2007) and does not require any toxic substances or specific catalyst. It is particularly useful for producing NDs with enhanced NVs for biological imaging. It involves a subsequent oxidation step in a tube furnace at a controlled temperature of 600°C for a specific time of 30 min using air as the medium. Depending on the temperature of air oxidation, different materials are removed from the diamond surface. NDs of 8 nm with stable optical characteristics can be obtained by this technique (Gaebel et al., 2012). Air oxidation also allows for selective removal of sp^2-bonded carbon from ND, providing a control over sp^2/sp^3 carbon ratio on the surface (Osswald et al., 2006).

Multistep oxidation, including combinations of more than one technique, is also considered effective to counter aggregation (Shenderova et al., 2006). One of such techniques is the graphitization-oxidation

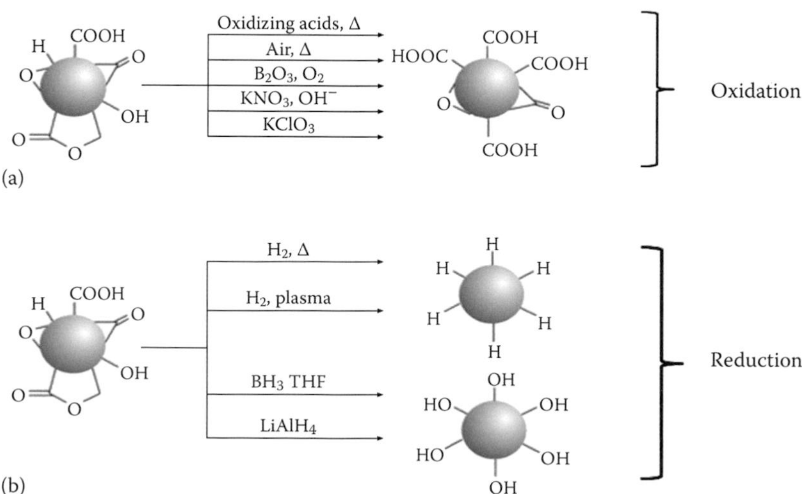

FIGURE 24.9 Modes for pretreatment of ND surface. (Krueger, A., *J. Mater. Chem.*, 18, 1485–1492, 2008. Reprinted by permission of The Royal Society of Chemistry.)

method (Xu & Xue, 2004). It begins with conventional treatment of graphitic soot with strong mineral acids and also involves a subsequent step of regraphitization in which NDs are heat treated in nitrogen at 1000°C for an hour, resulting in complete blackness of ND particulates. Graphitization is then followed by reoxidation, in which the sample is treated in air at 450°C for several hours. The black color of the product starts to fade and eventually disappears, indicating that the thin graphite layer has been completely removed, yielding purified NDs. The NDs are then dispersed in water by ultrasonication. This process produces particles of less than 50 nm in size, which indicates disaggregation of the ND core aggregates (Xu & Xue, 2004).

Another mode of pretreating ND surface is reduction (Figure 24.9b), which is aimed at converting most of the functionalities into hydrogen or hydroxyl groups. Therefore, reduction of NDs can create either positive surface through hydrogenation or negative surface through hydroxylation of ND surfaces (Krueger, 2008).

The choice of the pretreatment method is crucial to control the overall functionalization process. Oxidation or reduction of NDs is selected based on the terminal functional groups required on the surface. For example, carboxylate functionality serves as a typical ligand for covalent conjugation of amine containing chemicals like amino acids or proteins (Liu et al., 2008; Kaur et al., 2012). This conjugation is mediated through the formation of an amide bond (R–CO–NH–R). Therefore, to functionalize NDs with biomolecules, oxidation is the best suited approach for pretreating the surface. On the contrary, certain molecules require hydroxylated surface for conjugation. A typical example is the biotinylation of NDs, before which the surface is reduced through borane-forming hydroxyl groups. The hydroxylated surface is then covalently conjugated to biotin through a silane linker (3-aminopropyl trimethoxysilane) (Krueger et al., 2008).

24.3.2.2 Functionalization Chemistry

Surface funtionalization occurs after the pretreatment step ensures homogeneity of the ND surface. It utilizes three different types of surface chemistries: wet chemistry, gas phase methods, or atmospheric plasma treatments. Different approaches to functionalize NDs may result in variations in the binding energies and electrostatic features of the surface (Datta et al., 2011). Because of the high surface-to-volume ratio of the detonation NDs, extensive functional group conjugation is possible.

Wet chemistry treatment (Figure 24.10, blue shaded scheme) makes use of suitable solvent systems to introduce functional groups of interest. Depending upon the nature of functional groups to be attached on the surface, oxidized carboxylated NDs or reduced hydroxylated NDs can be used (Jee & Lee, 2009). For example, carboxylate functionalized NDs can be reacted with thionyl chloride to form highly reactive acyl chloride functionalities, which can be further attached to amine-containing chemical moieties (Mochalin et al., 2011). The NDs treated by this pathway can be used to conjugate amino acids or proteins on their surface (Mochalin et al., 2011). Wet chemistry treatments maintain the crystal characteristics of diamond core even at the size of 5 nm (Jee & Lee, 2009).

The use of novel concepts has significantly advanced wet chemistry approaches of ND functionalization. A relatively new approach for modifying the surface utilizes a process called atom transfer radical polymerization (Lia et al., 2006; P. Zhang et al., 2011). This process utilizes compounds called radical initiators (benzoyl peroxides, hydroxyethyl-2-bromoisobutyrate, or 2,2,2-trichloroethanol), which are attached covalently to oxidized NDs through esterification. Chemical groups are then introduced in the system, which polymerize and arrange as brush arrays on the surface. This process can create hydrophilic or hydrophobic surface, depending upon the nature of the polymer and hence provide better control over surface reactivity (Lia et al., 2006). The radical polymerization phenomenon is also employed to develop a simple approach for introducing polar and nonpolar functionalities on graphitized ND surface (Chang et al., 2010). It begins with surface graphitization and then introduction of these graphitized NDs in a mixture containing monomers of hydrophilic compounds in aqueous medium and hydrophobic compounds in organic medium. Exposing the reactive NDs to polar and nonpolar groups at the same time resulted in a combined

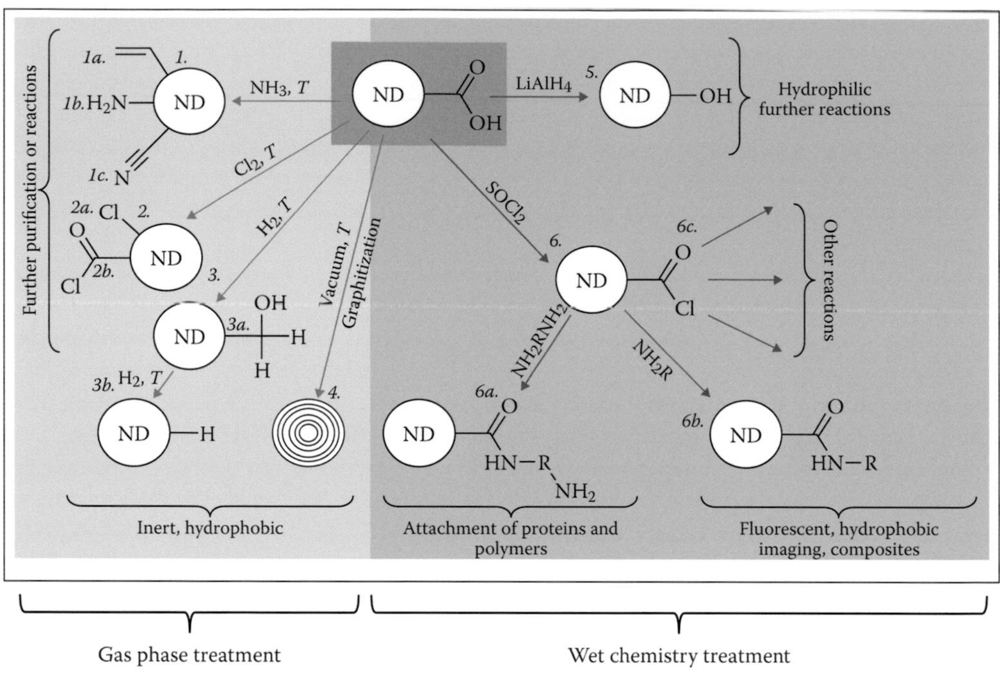

FIGURE 24.10 Wet chemistry and gas phase treatments for ND functionalization. (Reprinted by permission from MacMillan Publishers Ltd. *Nat. Nanotechnol.* Mochalin, V., Shenderova, O., Ho, D. et al., *Nat. Nanotechnol.*, 7, 11–23, 2011. Copyright 2011.)

hydrophilic/hydrophobic surface without undergoing multistep complex procedures (Chang et al., 2010).

Another approach for surface functionalization involves the treatment of NDs in gas or a vapor reactive medium (Figure 24.10, light gray shaded scheme on the left side) (Spitsyn et al., 2005). This approach is efficient to purify a very low dimensional fraction of the powder having narrow particle size distribution. It facilitates greater penetration into the inter particle spaces than do strong acids or salts in liquid phase (Spitsyn et al., 2005). The gas phases include hydrogen, ammonia, carbon tetrachloride, or argon. NDs treated with ammonia yield carbonyl, amine, or cyano groups on the ND surface, while treatment with chlorine results in the formation of chloro-NDs or acyl-chloride functionalized NDs. Acyl chloride, in particular, is extremely reactive since chlorine is readily replaced by the upcoming functional group like alcohol, phenols, or amines (Clark, 2004). Therefore, this mode is particularly important for biomolecular conjugation, where acyl-chloride is reacted with amine containing moieties to facilitate amide bond formation, similar to the natural peptides. High-temperature gas treatments with hydrogen produces alcohol functionalized NDs or hydrogenated NDs. Complete removal of all functional groups on ND surface might result from thermal treatments of NDs in argon, nitrogen or vacuum medium, which gives rise to graphitic carbon nano-onions. Both hydrogenated NDs and carbon nano-onions are inert and hydrophobic in nature (Mochalin et al., 2011).

Gas phase treatments are conducted in quartz reactors, in which NDs are treated with flowing gas or vapors at high temperatures ranging from 400°C to 1100°C and atmospheric pressures for durations ranging from 30 min to 5 h (Spitsyn et al., 2005). Gaseous functionalization can result in hydrophilic or hydrophobic and acidic or basic ND terminations (Spitsyn et al., 2005).

All these functionalization methods lead to the improvement of the dispersibility of the NDs. In addition, the homogeneous surface can serve as a template for binding molecules, discussed in Section 24.4.

24.4 Grafting Biomolecules on NDs

Introducing protonable biomolecules, including amino acids, peptides, and proteins, on the ND surface is a novel approach for functionalization. This is mainly targeted for controlled release of small drug molecules and targeted delivery of biomolecules such as peptides, proteins, RNA, and DNA. Functionalization of NDs with biologically active molecules is achieved through noncovalent adsorption or covalent immobilization at the surface. A first evidence of noncovalent biomolecular functionalization is presented by decorating the surface with poly-L-lysine, a polypeptide that can bind and isolate DNA oligonucleotides during mass spectrometric analysis (Kong et al., 2005). Poly-L-lysine functionalized NDs are also employed to synthesize fluorescent NDs for possible application in human stem cell research (Vaijayanthimala et al., 2009). Recently, a more complex system consisting of polyols like glycerol and polypeptides chained with basic amino acids (glycine, arginine, lysine, and histidine) was also used to functionalize NDs for gene delivery (Zhao et al., 2014). Similarly, NDs were coated by adsorbtion with specific enzymes like lysozyme for antibacterial activities (Perevedentseva et al., 2007). These systems were extensively characterized over time for stability, biological interactions, and efficacy (Perevedentseva et al., 2011; Wang et al., 2011).

Another approach for grafting biomolecules on the surface involves covalent conjugation that permanently immobilizes the molecules of interest on the surface and creates a controlled and relatively stable functionalized system (Krueger et al., 2008). Lysozyme, an antibacterial enzyme, was covalently attached to pretreated carboxylated NDs (Liu & Sun, 2010). Covalent conjugation did not compromise the antibacterial activity of the lysozyme. The fully functional capacity of the system was ensured up to 10 h at room temperature and up to several weeks if stored at 5°C after covalent conjugation of the enzyme (Liu & Sun, 2010). NDs modified through biotinylation were shown to covalently bind coenzymes (vitamins) like biotin (Krueger et al., 2008).

A relatively new concept highlights the use of simpler biomolecules like amino acids through covalent immobilization (Kaur et al., 2012). This approach creates a primary amine-rich ND surface, which facilitates electrostatic interaction with genetic materials (DNA or RNA) to utilize NDs as gene delivery vectors. The concept of grafting amino acids on the NP surface arises from the process of natural DNA wrapping in eukaryotic cells. The nuclei of all eukaryotic cells consist of histone proteins, which are alkaline in nature (Cox et al., 2005; Youngson, 2006). These histone proteins are rich in lysine and arginine amino acids (Klyszejko-Stefanowicz et al., 1989). The DNA wraps around histone proteins, which condense and package the DNA to form nucleosomes (Klyszejko-Stefanowicz et al., 1989). There are many types of interactions between histone proteins and DNA during nucleosome formation (Taverna et al., 2007). Among these interactions, one of major importance is the formation of salt bridges between the side chains of basic amino acids (especially lysine and arginine) and phosphate oxygen of DNA. The salt bridges are a combination of electrostatic interactions and hydrogen bonding between the positively charged amines and the negatively charged phosphate oxygen (Figure 24.11).

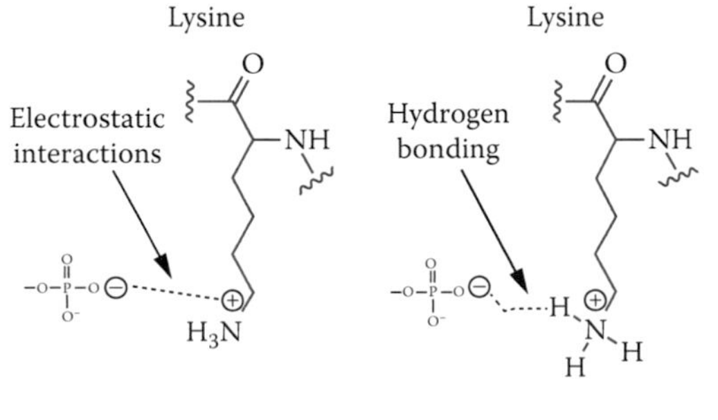

FIGURE 24.11 Electrostatic interaction and hydrogen bonding (salt bridges) between primary amine of lysine residue and phosphate oxygen of DNA.

This natural process is replicated artificially by functionalizing NDs with basic amino acids covalently (Kaur et al., 2012) or noncovalently (Ghosh et al., 2008) to induce similar interactions with anionic nucleic acids forming diamoplexes (complex of diamond NP with DNA or RNA). As an added benefit, the basic amino acids also impart water solubility to the complex for stability in hydrophilic environment of the cell (Taverna et al., 2007).

24.5 Biomedical Applications of NDs

24.5.1 Biodistribution and Cellular Uptake of NDs—Mechanisms, Challenges, and Controlling Factors

The route of administration into the body is one of the factors that influence the design decisions regarding biomedical devices and delivery systems. The functionalization of the NDs is also governed by the route of administration. After systemic administration, NDs, like all other NPs, are distributed in the body through blood circulation. The particle size and the shape, surface functionalities, and the overall charge in dispersing medium play a critical role in defining their distribution in the body (Petros & DeSimone, 2010). NPs less than 5 nm are rapidly cleared, mainly through renal mechanisms and also through extravasations, which promote the discharge of fluids into extravascular spaces and subsequent removal of materials through lymphatics (Owens & Peppas, 2006; Alexis et al., 2008; Wong et al., 2008). As the particle size increases from nanometers to micrometers, new challenges appear, which include mainly NP accumulation in the metabolic organs like liver, spleen, and bone marrow (Ilium, 1982). The metabolism of NDs can be stimulated through a process called opsonization, which is associated with nonspecific protein adsorption on the surface (Owens & Peppas, 2006; Nie, 2010; Narayan et al., 2011; Wei et al., 2012). It causes recognition and uptake of NDs by circulating macrophages, followed by degradation and excretion, a process called mononuclear phagocytosis (Figure 24.12a). The risk of elimination of the NDs through this process is reduced by functionalizing the surface with polyethylene glycol (PEG). Pegylated NDs avoid opsonization and phagocytosis, thus increasing the chance for tissue accumulation (X. Zhang et al., 2011a,b). Another removal pathway that presents a major challenge for ND survival in the body is associated with mechanical filtration through sinusoids in the spleen and Kuppfer cells in the liver. These systems are powerful machinery that removes the NDs through the reticulocyte–endothelial system (Yuan et al., 2009). This process is also stimulated by nonspecific protein adsorption on ND surface.

In addition to general biodistribution, mechanisms and factors involved in cellular uptake is also an extensively studied area to evaluate the biological interactions of NDs. Properties, mainly the particle size, shape, and surface characteristics, influence the cellular uptake processes (Petros & DeSimone, 2010). Differences in particle size define the mechanism involved in its internalization and, in turn, the microenvironment experienced by them intracellularly. Most NPs are internalized in the cells through endocytosis, a process used to internalize fluids and molecules within small vesicles formed by the cell membrane folding (Figures 24.12b–e) (Rothen-Rutishauser et al., 2006; Hillaireau & Couvreur, 2009; Kuhn et al., 2014). Endocytosis of NPs can occur through different mechanisms dictated mainly by the size of the particles. However, not all cell types are equipped with the necessary machinery to allow the entire spectrum of endocytotic processes (Douglas et al., 2008). The major pathway associated with carbon NPs is clathrin-mediated endocytosis (Kam & Dai, 2005; Li et al., 2008; Faklaris et al., 2009). It is a form of receptor-mediated endocytosis, which naturally functions for cellular uptake of proteins such as lipoprotein and transferrin (Schmid, 1997; Conner & Schmid, 2003; Kuhn et al., 2014). It is also involved in cell and serum homeostasis by regulating membrane electrolytic pumps, for example, voltage-gated calcium channels (Conner & Schmid, 2003). This endocytotic process forms coated pits having transmembranous and cytosolic proteins, mainly clarithrin (Figure 24.12d). They detach from the membrane and form a vesicle in the cytoplasm of the cell. The assembly of these vesicles is controlled by a set of proteins called assembly proteins (Conner & Schmid, 2003). Another mechanism elucidated for ND

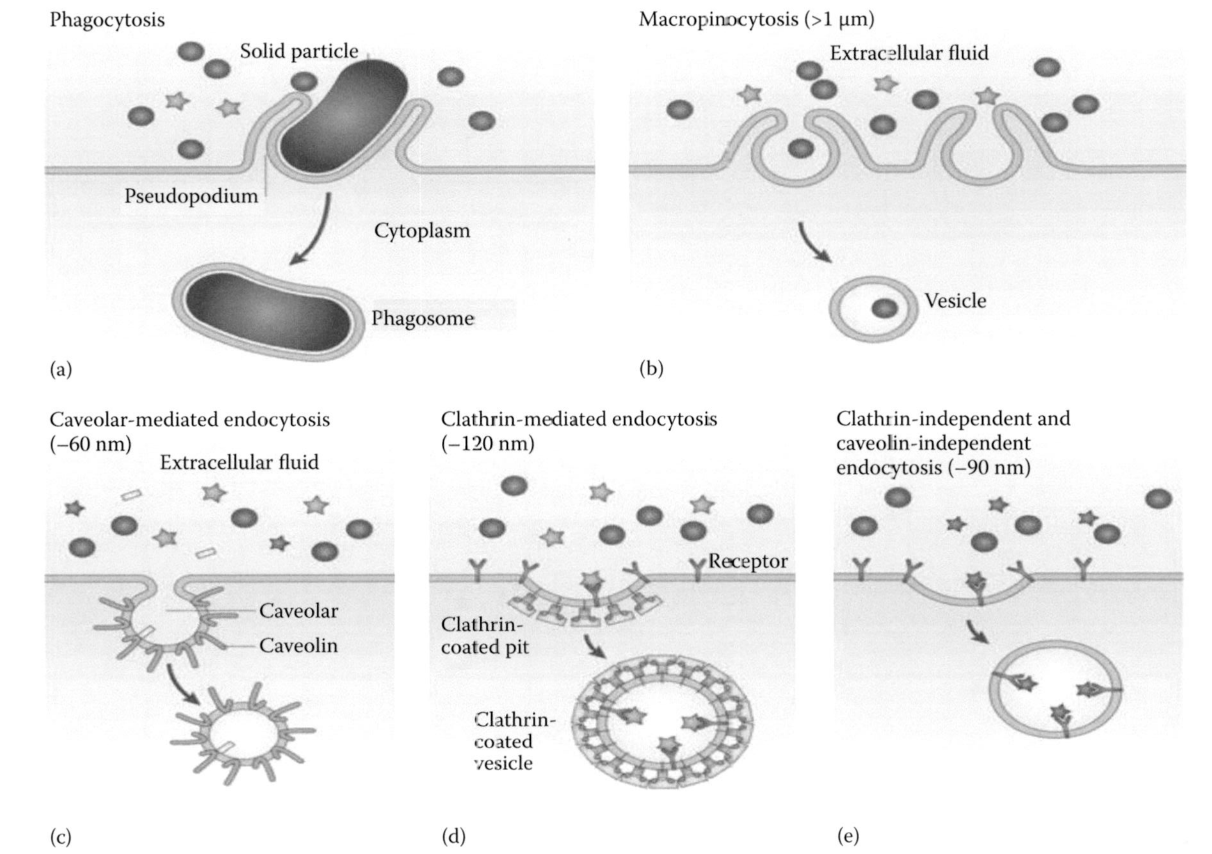

FIGURE 24.12 Uptake of microparticles and NPs. Internalization of large particles is facilitated by phagocytosis (a). Nonspecific internalization of smaller particles (>1 μm) can occur through macropinocytosis (b). Smaller NPs can be internalized through several pathways, including caveolar-mediated endocytosis (c), clathrin-mediated endocytosis (d), and clathrin-independent and caveolin-independent endocytosis (e). NPs are represented by circles (>1 μm), stars (90–120 nm) and rods (about 60 nm). (Reprinted by permission from MacMillan Publishers Ltd. *Nat. Rev. Drug Discovery.* Petros, R., DeSimone, J., *Nat. Rev. Drug Discovery*, 9, 615–627, 2010. Copyright 2010.)

uptake is macropinoccytosis (Liu et al., 2009; Alhaddad et al., 2012), which is responsible for internalizing larger particles of 1 μm (Petros & DeSimone, 2010). It involves actin-driven membrane protrusion to enclose NDs in aggregated forms (Figure 24.12b). Macropinocytosis seems similar to phagocytosis, however; rather than zippering around the engulfed particle, it causes the protrusions to collapse and fuse again to the plasma membrane.

Investigations regarding the uptake of NDs provide evidence for both macropinocytosis (Figure 24.12b) and clathrin-mediated endocytosis (Figure 24.12d); however, the latter is more commonly observed (Faklaris et al., 2009; Vaijayanthimala et al., 2009; Perevedentseva et al., 2013). Clathrin-mediated endocytosis is mediated through intact microfilament architecture, as observed for poly-L-lysine-coated NDs (Vaijayanthimala ct al., 2009). The same mechanism was observed in both healthy and cancerous cells, which indicates that the uptake process is not cell-type specific. However, recent investigation suggests that the uptake is approximately 30% higher in cancerous cells compared with healthy cells at the longest time interval (480 s) (Perevedentseva et al., 2013), presumably owing to structural, genetic, and phenotypic alterations. Liu et al. (2009) showed that slightly larger particles (100 nm) follow a combination of both mechanisms for cellular uptake. This is a result of the aggregation of NDs resulting in nonuniform particle size distribution that may stimulate a combination of pathways, each specific for a particle size range. NDs ranging from 46 to 150 nm were internalized via clarithrin-mediated endocytosis; however, aggregated NDs of 1 μm in size can be internalized via macropinocytosis (Alhaddad et al., 2012).

The kinetics and extent of ND cellular uptake, like all other NPs, can be controlled by two parameters: thermodynamic driving forces for membrane wrapping (determined by particle size) and diffusion kinetics of the receptors responsible for NP interaction (Chithrani & Chan, 2007). Thermodynamic driving forces for wrapping create the required free energy for NP internalization, while the diffusion kinetics of receptors corresponds to the availability of receptors to initiate receptor-mediated endocytosis (Chithrani & Chan, 2007). Research indicates that the optimal particle size for wrapping and receptor interaction is 55 nm (Gao et al., 2005). Smaller NPs do not have enough energy for membrane wrapping and therefore cannot undergo endocytosis. For bigger NPs (>50 nm), more time will be consumed for membrane wrapping and hence will have reduced cellular uptake (Gao et al., 2005).

NDs conjugated to more sophisticated targeting moieties adopt a different pathway for cellular uptake as compared with the NDs without these molecules (B. Zhang et al., 2009). They follow caveolin-dependent endocytosis, which results in the formation of flask-shaped invaginations of the plasma membrane (Figure 24.12c) (Conner & Schmid, 2003). Endothelial cells that lack the machinery for caveolin-mediated endocytosis are unable to bind and uptake serum proteins (Razani et al., 2002), and therefore, this pathway can facilitate the uptake of protein-coated NPs (Wang et al., 2009). Unlike other receptor-mediated endocytotic pathways, caveolin pits can also encapsulate larger amounts and sizes of NPs. A single caveolin-mediated membrane invagination can accommodate three 20-nm, two 40-nm, or one 100-nm polymeric NP (Wang et al., 2009). By analogy, we could hypothesize that ND aggregates may be internalized in the cells through this mode as well. However, limited studies are available to date to support this mechanism for cellular uptake of NDs.

Another interesting phenomenon that affects cellular uptake kinetics is the interaction of NDs with serum proteins. It is extensively researched to build evidence for the suitability for in vivo utilization of ND complexes. NDs have a reactive surface and hence are capable of adsorbing proteins forming a layer called protein corona (Cedervall et al., 2007; Lesniak et al., 2012; Gunawan et al., 2014). Several studies indicate that adsorbed proteins down-regulate the cellular uptake of NPs and also lead to aggregation (Cedervall et al., 2007; Vaijayanthimala et al., 2009; Lesniak et al., 2012; Gunawan et al., 2014). Serum proteins can adsorb rapidly on NP surface, leading to an increase in the overall particle size and alteration of surface charges (Tenzer, 2013). This process can compromise the electrostatic interactions of cationic NPs with cell membrane and hence affects cellular uptake and toxicity profiles. Protein coronas are formed on all NPs irrespective of their surface charges (Alkilany et al., 2009); however the composition of the layer might vary with the positivity or negativity of the surface. In general, adsorption of proteins on NP surface is a dynamic process, such that the identities of adsorbed proteins may change over time but the total amount remains roughly constant (Rahman et al., 2013). There are certain conflicting evidences that suggest that serum proteins, mainly albumin, enhance cellular uptake of carbon nanomaterials (Lu et al., 2014). Therefore, understanding the behavior of NDs in a native biological environment is a prerequisite for their therapeutic evaluation.

After cellular internalization, ND-based delivery systems face many challenges intracellularly. One of such challenges is the entrapment of NDs in endosomes after internalization (Solarska-Ściuk et al., 2014). The endosome is an acidic organelle owing to the presence of digestive enzymes, and therefore, endosomal entrapment especially affects gene transfection and delivery of acid-labile small molecules (Kumar et al., 2003; Khalil et al., 2006). Disruption of the endosomal membrane through the proton sponge effect is one way to protect NDs from endosomal degradation (Petit et al., 2013). It is mediated by chemical groups with high buffering capacity and propensity to swell upon protonation. Protonation causes influx of H^+ ions, causing the membrane to rupture (Varkouhi et al., 2011). These groups may include pH-sensitive molecules like histidine, which can be covalently attached to the ND surface (Badea et al., 2014) to reduce endosomal trapping. Another study also suggests that diverting the uptake process in favor of macropinocytosis also increases the potential of gene (siRNA) transfection (Alhaddad et al., 2012). This diversion is observed to depend on the size and shape of NDs. In aggregated NDs, weak electrostatic interactions may change the spherical shape of the particle. This may result in variations in the surface area acquired during membrane adhesion, which may ultimately change the mechanism of uptake (Alhaddad et al., 2012). Overall, size, surface properties, interactions with biomolecules, serum proteins, cell membranes, and cellular organelles are all factors that contribute to the biological effect and toxicity of the NDs.

24.5.2 Biocompatibility and Biological Affinity

A biocompatible nature is a major prerequisite for all therapeutic and diagnostic aids used in biomedicine. NDs certainly provide well-established evidence of limited toxicity at cellular and organ levels, demonstrating the highest biocompatibility among the members of the nanocarbon family (Schrand et al., 2007).

When the toxicity of ND was compared with other carbon nanomaterials, including carbon black, multiwalled carbon nanotubes, and single-walled carbon nanotubes, the degree of cytotoxicity occurred in the following order: single-walled nanotube > multiwalled nanotube > carbon black > ND (Schrand et al., 2007). NDs do not induce any mitochondrial membrane damage and preserve high cell viability (Schrand et al., 2007). Unlike other carbon nanomaterials, they do not fragment DNA (Huang et al., 2007) or alter the gene and protein expression profiles of living cells at a wide range of concentrations and particle sizes (Liu et al., 2007). Regardless of the terminal functional moieties, NDs show no overexpression of genes like interleukin-6, tumor necrosis factor-α, and inducible nitric oxide synthase and are hence expected to show absence of inflammatory responses upon administration (Huang et al., 2007).

These properties are particularly advantageous to develop ND-based delivery vectors. Serum proteins are also shown to play a significant role in influencing the biocompatibility of NDs (Li et al., 2010). In vitro studies reveal that the cell culture medium containing serum proteins enhances the biocompatibility of oxidized carboxylated NDs even at high doses. The adsorbed serum proteins are not only protective but are also shown to increase hydrophilicity and stability of ND dispersion (Li et al., 2010). This builds an evidence for the safety of NDs in protein-rich biological systems.

The in vivo biocompatibility of NDs is also addressed. The potentially most affected organ system is the respiratory system, since powdered detonation NDs may cause environmental pollution (Zhu et al., 2012). NDs are considered nontoxic to the respiratory system as it did not show any oxidative damage to lung tissues upon long-term residence (Yuan et al., 2010). They are shown to be excreted from the body mainly through lymphatic tissue or directly through cough (Yuan et al., 2010). NDs also maintain their biocompatibility after prolonged exposures through oral (Schrand et al., 2009) or subcutaneous (Puzyr et al., 2007) routes.

24.5.3 Application of NDs in Bioimaging and Research Technologies

Artificial induction and enhancement of fluorescing centers have gained much attention over the years to utilize NDs as biological probes (Y. Chang et al., 2008; Ho, 2009).

24.5.3.1 Fluorescence Imaging

Internal structural defects and impurities cause NDs to emit visible fluorescence, which show some advantages compared with the conventional dyes used for fluorescence imaging. Under the same excitation conditions, a single 35-nm-sized diamond NP is significantly brighter than commonly used dye molecules such as Alexa Fluor 546 (Fu et al., 2007). In addition to higher fluorescence, NDs also resist photobleaching and fluorescence blinking for a longer duration of 5 min, unlike Alexa Fluor, which photobleaches within 10 s (Fu et al., 2007). Photobleaching is one of the major deficiencies of imaging probes, where the fluorophore undergoes chemical degradation because of the generation of reactive oxygen species in biological system as a by-product of florescence excitation (Johnson, 2010). Fluorescence blinking is the phenomenon showing intermittent light and dark bands upon continuous illumination, common in hard nanomaterials like quantum dots (Ko et al., 2011). Fluorescence blinking presents a problem for single-molecule spectroscopy (Ko et al., 2011) owing to lack of image consistency. In addition to the previously mentioned problems, autofluorescence of cellular structures like mitochondria and lysosomes is also a concern in the design of imaging probes (Davis et al., 2010). It usually arises in the emission wavelength range common to many fluorescent dyes (i.e., 510–560 nm) and therefore creates intense background signals, making the detection of the probe increasingly challenging. However, NDs, when excited with 532 nm wavelength light, produce an emission between 650 and 720 nm, thereby reducing the interference of cellular autofluorescence (Fu et al., 2007).

Various approaches have been utilized to induce or enhance the fluorescence centers in NDs. One of such techniques involves irradiation of NDs with helium ion at 40 keV, followed by thermal annealing at 800°C to create vacancies (Y. Chang et al., 2008). Hydrogen ions at 3 MeV are also used in place of helium to induce NV centers in ND core (Wee et al., 2009). Another way to create NVs is to directly incorporate nitrogen atoms as native nitrogen ^{14}N (Meijer et al., 2005), its isotope ^{15}N (Rabeau et al., 2006), or as cyanide (CN^-) ions (Spinicelli et al., 2011) directly in the core. To preserve fluorescence from the NVs, the NDs are encapsulated by bulky groups like phenols, which reduce the nonradioactive decay pathways of colored centers (Bray et al., 2015). Fluorescence can also be produced through surface conjugation. For example, NDs can be functionalized with a hydrophobic molecule octadecylamine, producing bright blue fluorescence, and can be useful for imaging hydrophobic components of the body (Mochalin & Gogotsi, 2009).

Opportunities to induce and enhance the florescence centers in NDs, along with its innate biocompatibility, at all levels has attracted their use as cellular biomarkers (Fu et al., 2007) for tracking natural and therapeutic processes (Liu et al., 2009; Wu et al., 2013; Hsu et al., 2014). Fluorescence of NDs can be clearly detected in the treated cells and allows detailed localization of NDs and the conjugated therapeutic moiety (Figure 24.13).

Fluorescent NDs are also employed for targeted bioimaging to study ligand–receptor binding kinetics. NDs bioconjugated with transferrin were observed through laser scanning fluorescence microscopy to evidence overexpression of these receptors on cancerous cells (Wenga et al., 2009). Here, the NDs act both as a carrier for transferrin and as an imaging tool; hence, no additional imaging probe was required to study these interactions.

Fluorescent NDs are also considered as candidates of interest for long-term cell tracking and sorting using flow cytometry (Hui et al., 2010).

More recently, theranostic applications of NDs have been examined, which enable their utilization for multiple purposes like delivering therapeutic molecules, imaging biological processes, and tracking treatment approaches at the same time. This is particularly done by artificial induction of enhanced fluorescing centers and simultaneously inducing specific surface functionalizations to confer affinity for

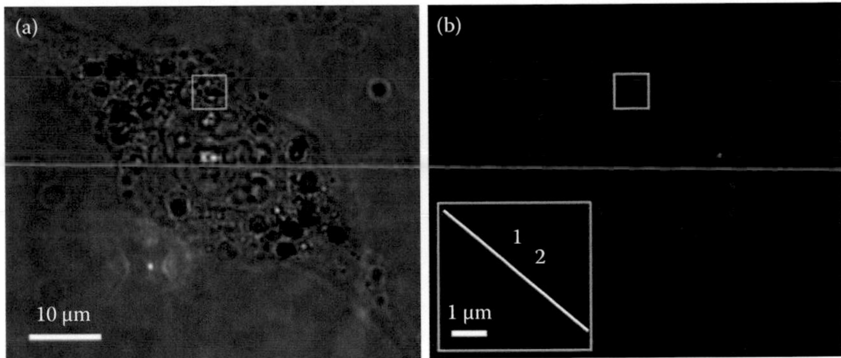

FIGURE 24.13 (a) Bright-field image of a HeLa cell after uptake of 35-nm fluorescent NDs. Most of the internalized NDs are seen to distribute in the cytoplasm. (b) Epifluorescence image of a single HeLa cell after the ND uptake. An enlarged view of the fluorescent spots (denoted by "1" and "2") with diffraction-limited sizes (FWHM ~500 nm) is shown in the inset. The separation between these two particles is ~1 μm. (Fu, C.-C. et al., *PNAS*, 104, 727–732, 2007. Copyright 2007, National Academy of Sciences, U.S.A.)

therapeutic drugs and biomolecules (Zhang et al., 2011). An example of such an approach is the conjugation of ND with an anticancer drug, cisplatin, through both physical and chemical conjugation. Here, NDs acted as a delivery vehicle for cisplatin and also as a probe for cellular imaging (Huynh et al., 2013).

In biomedical analyses, the NV centers in NDs can also be used as a light source for imaging optical microscopic imaging (Hui et al., 2010). It provides a fluorescence alternative to scattering imaging of cells and can ensure less photo damage (Hui et al., 2010).

24.5.3.2 Combined Fluorescence and Magnetic Imaging

Magnetism is a property of NDs that can impart further advantages for bioimaging. Magnetic NDs can serve as contrast agents in magnetic resonance imaging (MRI) (I. Chang et al., 2008), a commonly used diagnostic tool. NDs can provide a safer alternative to traditional gadolinium, which is a free solubilized ion and is considered highly toxic especially in renally impaired patients (Grobner, 2006). Because of their biocompatibility, NDs can alleviate this health concern and serve as MRI reporting marker to elucidate metabolic fate and distribution of nanoparticulate delivery systems (I. Chang et al., 2008).

An alternative to complete replacement of traditional contrast agents, NDs can also be conjugated with these contrast agents for better resolution and localization (Manus et al., 2010). Gadolinium/ND complexes are useful for long-term cellular examination and evaluation of disease progression. NDs tend to attract water molecules to the surface of gadolinium and increase the relaxivity of the contrast agent, resulting in higher resolution imaging compared with traditional gadolinium tracers (Manus et al., 2010).

Recent investigations focus on the modulation of NDs by magnetic resonance and serve to develop a highly proficient technique called deterministic emitter switch microscopy, which can enable super-resolution imaging particularly useful for nanomedicinal research (Chen et al., 2013).

24.5.3.3 Scattering Imaging

Elastically scattered light from a single 55-nm diamond is 300-fold brighter than that of a single cell organelle of the same size. As compared with the metal-based scattering labels like gold and silver, ND also produces a stable image with minimum photo damage (Smith et al., 2007). Therefore, NDs are investigated to improve optical bioimaging (Colpin et al., 2006; Smith et al., 2007). NDs also produce a sharp Raman signal and an isolated signature peak at 1332 cm^{-1} and therefore can be detected in the cells through Raman mapping (Michealson & Hoffman, 2012; Pope et al., 2014). A nanocomplex prepared by conjugating lyzosyme to NDs is shown to provide useful information regarding the interaction of the enzyme with bacterial cells through Raman spectroscopic mapping (Perevedentseva et al., 2007). In this complex, NDs not only act as carrier for the enzyme but also serve as a nanoprobe to track the biomolecular changes resulting from enzyme delivery (Perevedentseva et al., 2007).

While these techniques are employed mainly on laboratory scale for in vitro processes, there is growing evidence that these technologies might be expanded into clinical applications in medical imaging and diagnosis.

24.5.4 Applications of NDs in Therapeutics

Because of continuous advancements in the knowledge of disease conditions at molecular levels, many traditional therapies are now under constant optimizations to treat these diseases at their most basic levels. These novel approaches include targeted and controlled drug delivery, gene therapy, and personalized medicine. Targeted drug and gene therapy appears to be the most widely researched approaches especially for various chronic illnesses like cancers. They require efficient and biocompatible carriers. Because of their innate biocompatibility and extensive functionalization opportunities, NDs can be utilized to protect and deliver new or conventional therapeutic moieties to specific targets in the physiological system, rendering ND technologies highly desirable for biomolecular and drug delivery applications (Figure 24.14).

24.5.4.1 NDs for Small Molecule Delivery

Many studies have evidenced the use of NDs as effective drug delivery vehicles because of their diverse surface characteristics. Opportunities to modify the surface of NDs according to functional groups of the drug attract their utilization for drug delivery to specific targets. High surface loading, physicochemical stability of the complexes in physiological system, and opportunities to target ND complexes toward specific cells encourage the administration of minimal drug quantities to hit the optimum

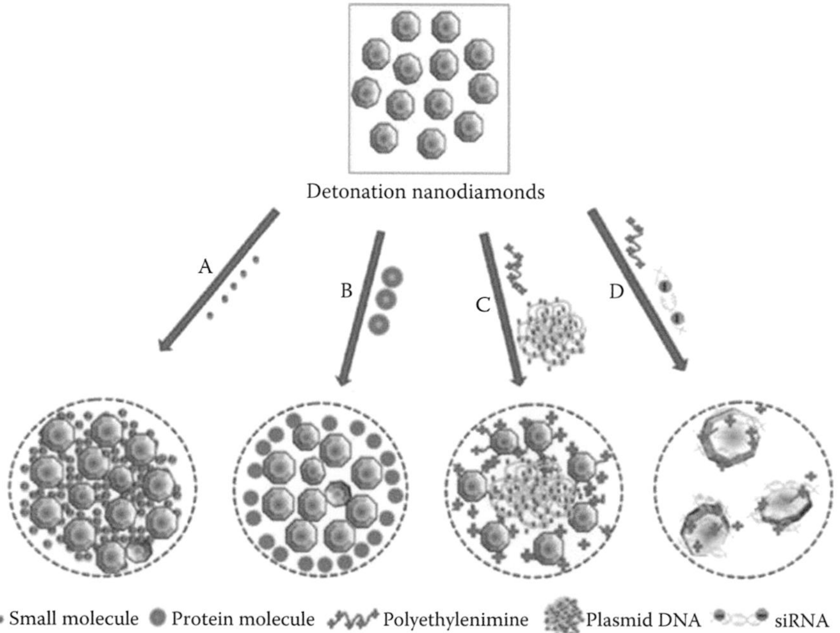

FIGURE 24.14 A schematic diagram representing the binding of detonation NDs with (A) small molecules, (B) proteins, (C) plasmid DNA, and (D) siRNA. (Reprinted from Kaur, R., Badea, I., *Int. J. Nanomed.*, 8, 203–220, 2013. Copyright 2013, Dove Medical Press. With permission.)

efficacy standards. These properties prevent the toxicity and adverse effects associated with the drug. NDs are investigated for both localized and systemic delivery of drugs.

NDs can serve as a potential carrier for systemic drug delivery because of their nanosize, diverse surface area, and several possibilities of surface functionalizations as described earlier (Ho & Lam, 2009). NDs, like other nanosized carriers, can overcome the physiological barriers, reduce systemic exposure times, and allow delivery of potent drugs like chemotherapeutic agent to targeted tissues. An apoptosis-inducing anticancer drug, doxorubicin hydrochloride (DOX), was conjugated successfully to NDs (Xiao et al., 2013; Li et al., 2014). This complex provided slow and sustained dissociation of the drug from the complex, with maximum release after 16 h of administration in vitro. This builds evidence that by synthesizing controlled drug delivery ND systems, the dosing frequency could be reduced. ND–DOX complexes accumulated in the tumor after systemic administration, and subsequently, the efficacy of DOX increased to 4 times. Accumulation in kidneys and associated nephrotoxicity were reduced compared with free DOX (Li et al., 2014). However, it was postulated that NDs alone also possess some anticancer properties, which could add to the effect of the drug and contribute to prolonged life span in tumor-induced mice (Li et al., 2014). Elaborative investigations are required to study this effect more clearly. Release of potent drugs from ND complexes can be tightly controlled by producing pH-dependent adsorption and desorption (Broz et al., 2006). It can be achieved by attaching acid phosphatase, an enzyme responsible for switching the drug on ND surface with other substrates in acidic medium (Broz et al., 2006). This study provides a proof of concept for designing nanocarriers for target-specific actions.

ND complexes also overcome drug efflux processes in chemoresistant tumors (Chow et al., 2011). The cancerous cells that were treated with ND–DOX complexes retained higher amount of the drug (0.26 μg/mL) compared with free DOX (0.02 μg/mL) after 4 h of efflux. The overall intensity of DOX action is slightly reduced after complexation; however, these complexes are still able to elicit significant cell death for tumor suppression (Chow et al., 2011).

Polymers like PEG adsorbed on ND surface can increase the drug loading capacity of NDs (Zhang et al., 2012). The PEGylated ND complex allowed for a slower drug release, producing a peak of drug concentration after 20 h, which remained constant up to 80 h. This sustained release of drug highlights the potential to design controlled delivery systems particularly for potent drugs. The cell death mechanisms of chemotherapeutic drugs like DOX are preserved after ND conjugation (Zhang et al., 2012).

Another example of utilizing NDs for cancer chemotherapy involves their conjugation with active plant alkaloids like paclitaxel (Liu et al., 2010). This complex is generated by covalent immobilization of the alkaloid on ND after a series of surface treatments. This complexation mediated a dose-dependent cellular uptake of paclitaxel and promoted cell death through mitotic arrest and apoptosis (Liu et al., 2010). The activity of paclitaxel conjugated to NDs produced similar activity in vitro at 50 μg/mL concentration when compared to 50 nM of paclitaxel alone after 24 h. The ND–paclitaxel complex also significantly reduced the tumor size at an average of 25 mm^3 (Liu et al., 2010).

NDs are also employed to reduce challenges associated with hydrophobic drug delivery. Administration of drugs with low aqueous solubility always presents a problem to leading to reduced bioavailability (Savjani et al., 2012). Low aqueous solubility also creates formulation challenges to prepare these drugs for systemic delivery. Water insoluble drugs can be conjugated with NDs for passage within the hydrophilic body environment to reach its targets. A variety of hydrophobic drugs like purvalanolA, 4-hydroxytamoxifen, and dexamethasone are investigated for delivery as ND complexes (Chen et al., 2009). The particle size of these drugs greatly reduced from micrometers to nanometers after complexation with NDs (for example, purvalanol changes from 340 μm to 556 nm), indicating greater dispersibility in aqueous medium. The effective surface charge also changed to positive values, presumably because the water molecules have higher affinity to hydrate the charged form of the drugs resulting from complexation compared with the neutral forms (Chen et al., 2009).

Multicomponent ND-based systems were also designed for systemic delivery by conjugating on the surface diagnostic, therapeutic, and targeting chemical species (Zhang et al., 2011). A versatile ND construct was synthesized by conjugating NDs with monoclonal antibodies to induce receptor-mediated endocytosis in tumor cells and with chemotherapeutic drugs like paclitaxel to induce tumor cell death. The drug was attached to the NDs through fluorescently labeled oligonucleotides for intracellular tracking of conjugated therapeutics (Zhang et al., 2011).

In addition to systemic delivery, NDs are also frequently used to design topical drug delivery systems. An effective topical drug delivery system must possess a structure that is innately biocompatible and inert, allows adjustable release rates, and is highly resistant to deformation. Design of multilayered ND films presents as a successful approach to localize combinations of molecules in a single delivery system (Huang et al., 2008b). ND films are synthesized using layer-by-layer deposition, in which, initially, poly-L-lysine is deposited on a plain negatively charged quartz surface. This layer creates a positive surface charge which is then interacted electrostatically with NDs to form a secondary layer. This assembly creates interlayer cavities for drug loading and imparts high stability to the system through electrostatic interactions between NDs and the peptide. A multilayered ND film provides an advantage of high drug loading capacity and controlled release of the drug and therefore is investigated for topical administration of an anti-inflammatory glucocorticoid, dexamethasone (Huang et al., 2008b). The multilayered film continuously eluted the drug in a sustained fashion over 24 h. The film and inter-layer cavities can be constructed with full control; thus, further reduction in drug elution rate can be achieved through incorporating the drug into the inner layers (Huang et al., 2008b). These films can also be used to simultaneously administer both cationic and anionic drugs without using multiple delivery systems. However, limited studies are available regarding the synthesis of multidrug delivery ND systems.

ND microfilms are also synthesized as implants using polymers like parylene C for unidirectional, slow, and continuous release of chemotherapeutic drugs for prolonged periods (Lam et al., 2008). This system was used to investigate the release of an anticancer drug, DOX. The rate of drug elution was at least 3 times higher over the first day for uncovered ND–DOX complexes as compared with the complexes sandwiched between the polymer films. The film created an additional elution control layer, which resulted in constant drug release and prevention of an initial burst release (Lam et al., 2008). The incorporation of such potent drugs into this form can reduce the adverse effects associated with a sudden rise in drug levels immediately after administration.

24.5.4.2 NDs for Protein Delivery

NDs are also investigated for the delivery of proteins, particularly hormones or enzymes. An example of such an application is the development of ND–insulin complexes designed for local delivery of insulin as a growth hormone to heal wounds (Shimkunas et al., 2009). This system specifically elutes insulin in alkaline medium of wounds associated with bacterial infections, thus increasing the local concentration of the therapeutic protein. During the investigation, NaOH was used as an alkaline medium, while water was used for comparison as a nonalkaline medium. The amount of insulin released by the first day was more than 20 times higher in the alkaline medium compared with water. This pattern also continued similarly on the fifth day, with 46% insulin released in alkaline medium in comparison with only 2% drug released in the water. This change in the release of insulin from ND complexes can be correlated with the change occurring in the charge distribution due to pH variations. Insulin in the aqueous medium possesses a net negative charge, which becomes much stronger upon increasing the alkalinity of the medium. Electrostatic interactions between NDs and insulin are disturbed at this instance, causing desorption of the protein. The amounts of the insulin desorbed are directly proportional to the increase in the alkalinity of the medium from pH 9 to pH 11.5. Most importantly, the functional capacity of the protein was not compromised as a result of this pH switching and desorption.

As an analytical tool, cytochrome C was investigated as a potential biomarker to evaluate conformational changes occurring to proteins in solution and on the surface (Kamatari et al., 1996; Cheng et al., 2003). Conjugated to NDs, cytochrome C could generate an optimized probe for fluorescence and optical microscopy (Huang & Chang, 2004). In addition to proteins, fluorescent peptides such as thiolated peptides are also conjugated to ND surface through connecting groups like maleimide (H_2C_2 (CO) $_2$NH) functionality for fluorescence bioimaging (Vial et al., 2008).

NDs are also employed as a matrix for matrix-assisted laser desorption ionization mass spectrometry to specifically capture and analyze glycoproteins (Yeap et al., 2008). This opens new avenues to utilize NDs for proteomic research.

24.5.4.3 NDs for the Delivery of Genetic Materials

NDs have attracted great attention as nonviral vectors for gene therapy. For this application, NDs must possess a cationic surface to bind and carry genetic materials, which are anionic in nature (Zhang et al., 2011). Cationic NDs are synthesized through a variety of methods ranging from noncovalent functionalization with polymers like polyethyleneimine (PEI) to covalent conjugation of derived organic moieties like triethylammonium groups (Martín et al., 2010). NDs adsorbed with low molecular weight (LMW) PEIs produce higher transfection efficiency and reduced toxicity in comparison with polymer used alone (P. Zhang et al., 2009). ND–PEI complex showed a protein expression at four orders of magnitude higher compared with naked DNA. Cell viability remained more than 80% after administration of ND–PEI complexes at a concentration of 90 μg/mL, thus reducing the innate cytotoxicity of cationic polymers.

Hydroxylated NDs functionalized with small moieties such as triethylammonium groups also bind and deliver plasmid DNA in the nucleus of cancerous cells, as evidenced qualitatively by the induction of fluorescence. The cells treated with free plasmid DNA did not show any fluorescence, while the cells treated with the plasmids bound to the functionalized NDs showed optimum fluorescence as well as protein expression (Martín et al., 2010).

Carboxylated NDs functionalized covalently with 2-bromoisobutyric acid to polymerize methacrylates on the surface function as gene delivery vectors and as biolabels to track intracellular processes (Zhang et al., 2011). These functional groups gather on the surface to form a cationic brush-like border, which condenses DNA into stable NPs.

Cationic brushed NDs efficiently transfect the plasmid DNA and produce a gene expression of 10E6-10E8 RLU/mg of proteins. Different types of methacrylate borders were coated on NDs, all of them producing several folds higher transfection efficiency as compared with naked DNA and DNA complexed to PEI.

Functionalizing NDs with protonable biomolecules like amino acids (Kaur et al., 2012) is also a potential approach for gene delivery. Covalent immobilization of basic amino acids directly on the oxidized ND surface can form stable diamoplexes with DNA and RNA (Kaur et al., 2012) and thus are promising as vectors for gene therapy. Peptides like poly-L-lysine can also be coated on NDs to create a positively charged surface (Kong et al., 2005; Vaijayanthimala et al., 2009). Such functionalization can also facilitate electrostatic binding with genetic materials; however, limited investigations are done to utilize peptide-modified NDs for gene delivery.

24.6 Conclusion

Innate biocompatibility at micro and macro levels with biological systems, new methods to overcome aggregation challenges, and novel approaches for specific surface funcionalization have greatly introduced NDs as an encouraging platform for numerous biomedicinal applications. Continual evaluations are still required to understand the surface changes occurring under various biological conditions with special focus on newly derived synthetic approaches. Also, more research is needed to further elaborate on the cellular interaction of NDs, localization in suborganelle levels, biodistribution after in vivo delivery, and consequences of long-term residence in biological systems. This information will encourage the development of more sophisticated nano delivery systems with better control and accurate predictions that will improve modern medicine.

References

Alexander, C. M., 2013. *Chondrite.* Available at: http://www.britannica.com/EBchecked/topic/114270/chondrite (accessed 19/01/2015).

Alexis, F., Pridgen, E., Molnar, L. & Farokhzad, O., 2008. Factors affecting the clearance and biodistribution of polymeric nanoparticles. *Molecular Pharmaceutics*, 5(4), pp. 505–515.

Alhaddad, A. et al., 2012. Influence of the internalization pathway on the efficacy of siRNA delivery by cationic fluorescent nanodiamonds in the Ewing sarcoma cell model. *PLoS One*, 7(12), p. e52207.

Alkilany, A. M. & Murphy, C. J., 2010. Toxicity and cellular uptake of gold nano particles: What we have learned so far? *Journal of Nanoparticle Research*, 12(7), pp. 2313–2333.

Alkilany, A. et al., 2009. Cellular uptake and cytotoxicity of gold nanorods: Molecular origin of cytotoxicity and surface effects. *Small*, 5(6), pp. 701–708.

Badea, I. et al., 2014. *Functionalized Nanodiamonds as Delivery Platforms for Nucleic Acids.* Saskatchewan, Canada, Patent No. 20140314850.

Baidakova, M. & Vul, A., 2007. New prospects and frontiers of nanodiamond clusters. *Journal of Physics D: Applied Physics*, 40(20), pp. 6300–6311.

Barnard, A. S., 2008. Self-assembly in nanodiamond agglutinates. *Journal of Material Chemistry*, 18, pp. 4038–4041.

Barnard, A. S. & Sternberg, M., 2007. Crystallinity and surface electrostatics of diamond nanocrystals. *Journal of Material Chemistry*, 17, pp. 4811–4819.

Bray, K. et al., 2015. Enhanced photoluminescence from single nitrogen-vacancy defects in nanodiamonds coated with phenol–ionic complexes. *Nanoscale*, 7(11), pp. 4869–4874.

Broz, P. et al., 2006. Toward intelligent nanosize bioreactors: A pH-switchable, channel equipped, functional polymer nanocontainer. *Nano Letters*, 6(10), pp. 2349–2353.

Cedervall, T. et al., 2007. Detailed identification of plasma proteins adsorbed on copolymer nanoparticles. *Angewandte Chemie International Edition*, 46(30), pp. 5754–5756.

Chang, I. P., Hwang, K. C. & Chiang, C.-S., 2008. Preparation of fluorescent magnetic nanodiamonds and cellular imaging. *Journal of American Chemical Society*, 130(46), pp. 15476–15481.

Chang, Y. et al., 2008. Mass production and dynamic imaging of fluorescent nanodiamonds. *Nature Nanotechnology*, 3(5), pp. 284–288.

Chang, I. et al., 2010. Facile surface functionalization of nanodiamonds. *Langmuir*, 26(5), pp. 3685–3869.

Chang, L.-Y., Ōsawa, E. & Barnard, A. S., 2011. Confirmation of the electrostatic self-assembly of nanodiamonds. *Nanoscale*, 3, pp. 958–962.

Chao, J. et al., 2009. Chapter 9: Protein–nanodiamond complexes for cellular surgery. In: D. Ho, ed. *Nanodiamonds: Applications in Biology and Nanoscale Medicine.* Springer Science & Business Media, Heidelberg, Germany, pp. 189–224.

Chen, M. et al., 2009. Nanodiamond-mediated delivery of water-insoluble therapeutics. *ACS Nano*, 3(7), pp. 2016–2022.

Chen, M. et al., 2010. Nanodiamond vectors functionalized with polyethylenimine for siRNA delivery. *The Journal of Physical Chemistry Letters*, 1(21), pp. 3167–3171.

Chen, E., Gaathon, O., Trusheim, M. & Englund, D., 2013. Wide-field multispectral super-resolution imaging using spin-dependent fluorescence in nanodiamonds. *Nano Letters*, 13(5), pp. 2073–2077.

Cheng, Y.-Y., Lin, S. H., Chang, H.-C. & Su, M.-C., 2003. Probing adsorption, orientation and conformational changes of cytochrome c on fused silica surfaces with the Soret band. *Journal of Physical Chemistry A*, 107(49), pp. 10687–10694.

Chithrani, B. D. & Chan, W. C. W., 2007. Elucidating the mechanism of cellular uptake and removal of protein-coated gold nanoparticles of different sizes and shapes. *Nano Letters*, 7(6), pp. 1542–1550.

Chow, E. et al., 2011. Nanodiamond therapeutic delivery agents mediate enhanced chemoresistant tumor treatment. *Science Translational Medicine*, 3(73), p. 73ra21.

Clark, J., 2004. *Introducing Acyl Chloride (Acid Chlorides)*. Available at: http://www.chemguide.co.uk/organicprops/acylchlorides/background.html (accessed 01/25/2015).

Colpin, Y., Swan, A., Zvyagin, A. & Plakhotnik, T., 2006. Imaging and sizing of diamond nanoparticles. *Optical Letters*, 31(5), pp. 625–627.

Conner, S. D. & Schmid, S. L., 2003. Regulated portals of entry into the cell. *Nature*, 422, pp. 37–44.

Cox, M., Nelson, D. R. & Lehninger, A. L., 2005. *Lehninger Principles of Biochemistry*. San Francisco. W.H. Freeman.

Datta, A., Kirca, M., Fu, Y. & To, A., 2011. Surface structure and properties of functionalized nanodiamonds: A first-principles study. *Nanotechnology*, 22(6), p. 065706.

Davis, R. W. et al., 2010. Accurate detection of low levels of fluorescence emission in autofluorescent background: Francisella-infected macrophage cells. *Microscopic Microanalysis*, 16(4), pp. 478–487.

Dolmatov, V. Y., 2001. Detonation synthesis ultradispersed diamonds: Properties and applications. *Russian Chemical Reviews*, 70(7), pp. 607–626.

Dolmatov, V. Y., Veretennikova, M. V., Marchukov, V. A. & Sushchev, V. G., 2004. Currently available methods of industrial nanodiamond synthesis. *Physics of Solid State*, 46(4), pp. 611–615.

Douglas, K., Piccirillo, C. & Tabrizian, M., 2008. Cell line-dependent internalization pathways and intracellular trafficking determine transfection efficiency of nanoparticle vectors. *European Journal of Pharmaceutics and Biopharmaceutics*, 68(3), pp. 676–687.

Dvorak-Ewell, M. et al., 2011. Osteoblast extracellular Ca^{2+}-sensing receptor regulates bone development, mineralization, and turnover. *Journal of Bone and Mineral Research*, 26(12), pp. 2935–2947.

Eidelmana, E. et al., 2005. A stable suspension of single ultrananocrystalline diamond particles. *Diamond and Related Materials*, 14(11–12), pp. 1765–1769.

El-Say, K. M., 2011. Nanodiamonds as drug delivery system: Applications and prospective. *Journal of Applied Pharmaceutical Science*, 1(6), pp. 29–39.

Faklaris, O. et al., 2009. Photoluminescent diamond nanoparticles for cell labelling: Study of the uptake mechanism in mammalian cells. *ACS Nano*, 3(12), pp. 3955–3962.

Faklaris, O. et al., 2010. Photoluminescent nanodiamonds: Comparison of the photoluminescence saturation properties of the NV color center and a cyanine dye at the single emitter level, and study of the color center concentration under different preparation conditions. *Diamond and Related Materials*, 19(7–9), pp. 988–995.

Fang, L., Ohfuji, H., Toru, S. & Tetsuo, I., 2011. Experimental study on the stability of graphitic C3N4 under high pressure and high temperature. *Diamond and Related Materials*, 20(5–6), pp. 819–825.

Fang, L., Ohfuji, H. & Irifune, T., 2013. A novel technique for the synthesis of nanodiamond powder. *Journal of Nanomaterials*, 2013.

Fedyanina, O. N. & Nesterenko, P., 2010. Regularities of chromatographic retention of phenols on microdispersed sintered detonation nanodiamond in aqueous—Organic solvents. *Russian Journal of Physical Chemistry A*, 84(3), pp. 476–480.

Fu, C.-C. et al., 2007. Characterization and application of single fluorescent nano diamonds as cellular biomarkers. *Proceedings of the National Academy of Sciences of the United States of America*, 104(3), pp. 727–732.

Gaebel, T. et al., 2012. Size reduction of nanodiamonds via air oxidation. *Diamond and Related Materials*, 2, pp. 28–32.

Galli, G., 2010. Structure, stability and electronic properties of nanodiamonds. In: L. Colombo & A. Fasolino, eds. *Computer-Based Modeling of Novel Carbon Systems and Their Properties, Carbon Materials: Chemistry and Physics 3*. Springer, Heidelberg, Germany, pp. 37–56.

Gao, H., Shi, W. & Freund, L. B., 2005. Mechanics of receptor-mediated endocytosis. *Proceedings of the National Academy of Sciences of the United States of America*, 102(27), pp. 9469–9474.

Ghosh, P. et al., 2008. Efficient gene delivery vectors by tuning the surface charge density of amino acid-functionalized gold nanoparticles. *ACS Nano*, 2(11), pp. 2213–2218.

Gibson, N., Luo, T., Brenner, D. & Shenderova, O., 2011. Immobilization of mycotoxins on modified nanodiamond substrates. *Biointerphases*, 6(4), pp. 210–217.

Grausova, L. et al., 2009. Nanodiamond as promising material for bone tissue engineering. *Journal of Nanoscience and Nanotechnology*, 9(6), pp. 3524–3534.

Grobner, T., 2006. Gadolinium—A specific trigger for the development of nephrogenic fibrosing dermopathy and nephrogenic systemic fibrosis? *Oxford Journals, Medicine, Nephrology Dialysis Transplantation*, 21(4), pp. 1104–1108.

Gunawan, C., Lim, M., Marquisb, C. P. & Amal, R., 2014. Nanoparticle–protein corona complexes govern the biological fates and functions of nanoparticles. *Journal of Materials Chemistry B*, 2, pp. 2060–2083.

Hillaireau, H. & Couvreur, P., 2009. Nanocarriers' entry into the cell: Relevance to drug delivery. *Cellular and Molecular Life Sciences*, 66(17), pp. 2873–2896.

Ho, D., 2009. Beyond the sparkle: The impact of nanodiamonds as biolabeling and therapeutic agents. *ACS Nano*, 3(12), pp. 3825–3829.

Ho, D., 2010. *Nanodiamonds: Applications in Biology and Nanoscale Medicine*, 2010 edition. Springer.

Ho, D. & Lam, R., 2009. Nanodiamonds as vehicles for systemic and localized drug delivery. *Informa Healthcare*, 6(9), pp. 883–895.

Hsu, T.-C. et al., 2014. Labeling of neuronal differentiation and neuron cells with biocompatible fluorescent nanodiamonds. *Scientific Reports*, 4, p. 5004.

Huang, L. & Chang, H., 2004. Adsorption and immobilization of cytochrome c on nanodiamonds. *Langmuir*, 20(14), pp. 5879–5884.

Huang, H., Pierstorff, E., Osawa, E. & Ho, D., 2007. Active nanodiamond hydrogels for chemotherapeutic delivery. *Nano Letters*, 7(11), pp. 3305–3314.

Huang, H. et al., 2008a. Large-scale self-assembly of dispersed nanodiamonds. *Journal of Material Chemistry*, 18, pp. 1347–1352.

Huang, H., Pierstorff, E., Osawa, E. & Ho, D., 2008b. Protein-mediated assembly of nanodiamond hydrogels into a biocompatible and biofunctional multilayer nanofilm. *ACS Nano*, 2(2), pp. 203–212.

Hui, Y. Y., Cheng, C.-L. & Chang, H.-C., 2010. Nanodiamonds for optical bioimaging. *Journal of Physics D: Applied Physics*, 43(37), p. 374021.

Hu, S., Sun, J., Tian, F. & Jiang, L., 2008. The formation of multiply twinning structure and photoluminescence of well-dispersed nanodiamonds produced by pulsed-laser irradiation. *Diamond and Related Materials*, 17(2), pp. 142–146.

Huynh, V. T. et al., 2013. Nanodiamonds with surface grafted polymer chains as vehicles for cell imaging and cisplatin delivery: Enhancement of cell toxicity by POEGMEMA coating. *ACS Macro Letters*, 2(3), pp. 246–250.

Iakoubovskii, K., Mitsuishi, K. & Furuya, K., 2008. High-resolution electron microscopy of detonation nanodiamond. *Nanotechnology*, 19, p. 155705.

Ilium, L. et al., 1982. Blood clearance and organ deposition of intravenously administered colloidal particles. The effects of particle size, nature and shape. *International Journal of Pharmaceutics*, 12(2–3), pp. 135–146.

Jee, A.-Y. & Lee, M., 2009. Surface functionalization and physicochemical characterization of diamond nanoparticles. *Journal of Current Applied Physics*, 9(2), pp. e144–e147.

Johnson, L., 2010. Chapter 17: Practical considerations in selection and applications of fluorescent probes. In: J. Pawley, ed. *Handbook of Biological Confocal Microscopy*. Springer Science & Business Media, Heidelberg, Germany, p. 362.

Kam, N. & Dai, H., 2005. Carbon nanotubes as intracellular protein transporters: Generality and biological functionality. *Journal of American Chemical Society*, 127(16), pp. 6021–6026.

Kamatari, Y. O., Konno, T., Kataoka, M. & Akasaka, K., 1996. The methanol-induced globular and expanded denatured states of cytochrome c: A study by CD fluorescence, NMR and small-angle X-ray scattering. *Journal of Molecular Biology*, 259(3), pp. 512–523.

Kaur, R. & Badea, I., 2013. Nanodiamonds as novel nanomaterials for biomedical applications: Drug delivery and imaging systems. *International Journal of Nanomedicine*, 8, pp. 203–220.

Kaur, R. et al., 2012. Lysine functionalized nano diamonds: Synthesis, physiochemical characterization and nucleic acid binding studies. *International Journal of Nanomedicine*, 7, pp. 3851–3866.

Khachatryan, A. et al., 2008. Graphite-to-diamond transformation induced by ultrasound cavitation. *Diamond and Related Materials*, 17, pp. 931–936.

Khalil, I., Kogure, K., Akita, H. & Harashima, H., 2006. Uptake pathways and subsequent intracellular trafficking in nonviral gene delivery. *Pharmacological Reviews*, 58(1), pp. 32–45.

Klyszejko-Stefanowicz, L., Krajewska, W. M. & Lipinska, A., 1989. Chapter 2: Histone occurrence, isolation, characterization and biosynthesis. In: G. S. Stein & J. L. Stein, eds. *Histones and Other Basic Nuclear Proteins*. CRC Press, Boca Raton, FL, pp. 18–55.

Knieke, C. et al., 2010. Nanoparticle production in stirred media mills: Opportunities and limitations. *Chemical Engineering & Technology Special Issue: Grinding and Milling*, 33(9), pp. 1401–1411.

Ko, H.-C., Yuan, C.-T. & Tang, J., 2011. Probing and controlling fluorescence blinking of single semiconductor nanoparticles. *Nano Reviews*, 2, p. 5895.

Kong, X. et al., 2005. Polylysine-coated diamond nanocrystals for MALDI-TOF mass analysis of DNA oligonucleotides. *Analytical Chemistry*, 77(13), pp. 4273–4277.

Kratochvílová, I. et al., 2011. Tuning of nanodiamond particles' optical properties by structural defects and surface modifications: DFT modelling. *Journal of Materials Chemistry*, 21, pp. 18248–18255.

Krueger, A., 2008. The structure and reactivity of nanoscale diamond. *Journal of Material Chemistry*, 18, pp. 1485–1492.

Krueger, A. et al., 2008. Biotinylated nanodiamond: Simple and efficient functionalization of detonation diamond. *Langmuir*, 24(8), pp. 4200–4204.

Kruger, A. et al., 2005. Unusually tight aggregation in detonation nanodiamond: Identification and disintegration. *Carbon*, 43(8), pp. 1722–1730.

Kruger, A., Liang, Y., Jarre, G. & Stegk, J., 2006. Surface functionalisation of detonation diamond suitable for biological application. *Journal of Materials Chemistry*, 16, pp. 2322–2328.

Kuhn, D. et al., 2014. Different endocytotic uptake mechanisms for nanoparticles in epithelial cells and macrophages. *Beilstein Journal of Nanotechnology*, 5, pp. 1625–1636.

Kulakova, I. I., 2004. Modification of surface and the physicochemical properties of nanodiamonds surface chemistry of nanodiamonds. *Physics of the Solid State*, 46(4), pp. 636–643.

Kumar, V. et al., 2003. Single histidine residue in head-group region is sufficient to impart remarkable gene transfection properties to cationic lipids: Evidence for histidine-mediated membrane fusion at acidic pH. *Gene Therapy*, 10(15), pp. 1206–1215.

Kuznetsov, V. et al., 1994. Effect of explosion conditions on the structure of detonation soots: Ultradisperse diamond and onion carbon. *Carbon*, 32(5), pp. 873–882.

Lam, R. et al., 2008. Nanodiamond-embedded microfilm devices for localized chemotherapeutic elution. *ACS Nano*, 2(10), pp. 2095–2102.

Lesniak, A. et al., 2012. Effects of the presence or absence of a protein corona on silica nanoparticle uptake and impact on cells. *ACS Nano*, 6(7), pp. 5845–5857.

Li, W. et al., 2008. The translocation of fullerenic nanoparticles into lysosome via the pathway of clathrin-mediated endocytosis. *Nanotechnology*, 19(14), p. 145102.

Li, J. et al., 2010. Nanodiamonds as intracellular transporters of chemotherapeutic drug. *Biomaterials*, 31(32), pp. 8410–8418.

Li, Y. et al., 2014. In vivo enhancement of anticancer therapy using bare or chemotherapeutic drug-bearing nanodiamond particles. *International Journal of Nanomedicine*, 9, pp. 1065–1082.

Lia, L., Davidson, J. & Lukehart, C. M., 2006. Surface functionalization of nanodiamond particles via atom transfer radical polymerization. *Carbon*, 44(11), pp. 2308–2315.

Liang, Y., Ozawa, M. & Krueger, A., 2009. General procedure to functionalize agglomerating nanoparticles demonstrated on nanodiamond. *ACS Nano*, 3(8), pp. 2288–2296.

Liu, Y. & Sun, K., 2010. Protein functionalized nanodiamond arrays. *Nanoscale Research Letters*, 5(6), pp. 1045–1050.

Liu, K.-K., Cheng, C.-L., Chang, C.-C. & Chao, J.-I., 2007. Biocompatible and detectable carboxylated nanodiamond on human cell. *Nanotechnology*, 18(32), p. 325102.

Liu, K. et al., 2008. Alpha-bungarotoxin binding to target cell in a developing visual system by carboxylated nanodiamond. *Nanotechnology*, 19(20), p. 205102.

Liu, K., Wang, C., Cheng, C. & Chao, J., 2009. Endocytic carboxylated nanodiamond for the labeling and tracking of cell division and differentiation in cancer and stem cells. *Biomaterials*, 30(26), pp. 4249–4259.

Liu, K. et al., 2010. Covalent linkage of nanodiamond-paclitaxel for drug delivery and cancer therapy. *Nanotechnology*, 21(31), p. 315106.

Lu, N., Li, J., Tian, R. & Peng, Y., 2014. Binding of human serum albumin to single-walled carbon nanotubes activated neutrophils to increase production of hypochlorous acid, the oxidant capable of degrading nanotubes. *Chemical Research in Toxicology*, 27(6), pp. 1070–1077.

Manus, L. et al., 2010. Gd(III)-nanodiamond conjugates for MRI contrast enhancement. *Nano Letters*, 10(2), pp. 484–489.

Martín, R., Alvaro, M., Herance, J. & García, H., 2010. Fenton-treated functionalized diamond nanoparticles as gene delivery system. *ACS Nano*, 4(1), pp. 65–74.

Meijer, J. et al., 2005. Generation of single colour centers by focussed nitrogen implantation. *Applied Physics Letters*, 87, p. 261909.

Michealson, S. & Hoffman, A., 2012. Chapter 8: Bonding and concentration of hydrogen. In: O. A. Shenderova & D. M. Gruen, eds. *Ultrananocrystalline Diamond: Synthesis, Properties and Applications*. William Andrew, p. 254.

Mochalin, V. & Gogotsi, Y., 2009. Wet chemistry route to hydrophobic blue fluorescent nanodiamond. *Journal of American Chemical Society*, 131(13), pp. 4594–4595.

Mochalin, V., Shenderova, O., Ho, D. & Gogotsi, Y., 2011. The properties and applications of nanodiamonds. *Nature Nanotechnology*, 7(1), pp. 11–23.

Narayan, R. J., Boehm, R. D. & Sumant, A. V., 2011. Medical applications of diamond particle and surfaces. *Materials Today*, 14(4), pp. 154–163.

Nie, S., 2010. Understanding and overcoming major barriers in cancer nanomedicine. *Nanomedicine (London)*, 5(4), p. 5230528.

Osawa, E., 2005. Disintegration and purification of crude aggregates of detonation nanodiamonds. In: D. M. Gruen, O. A. Shenderova & A. Y. Vul, eds. *Synthesis, Properties and Applications of Ultrananocrystalline Diamond: Proceedings of the NATO ARW on Synthesis, Properties and Applications of Ultrananocrystalline Diamond*. Springer Science & Business Media, Heidelberg, Germany, pp. 231–240.

Ōsawa, E., 2007. Recent progress and perspectives in single-digit nanodiamond. *Diamond and Related Materials*, 16(12), pp. 2018–2022.

Osswald, S. et al., 2006. Control of sp2/sp3 carbon ratio and surface chemistry of nanodiamond powders by selective oxidation in air. *Journal of American Chemical Society*, 128(35), pp. 11635–11642.

Owens, D. & Peppas, N., 2006. Opsonization, biodistribution, and pharmacokinetics of polymeric nanoparticles. *International Journal of Pharmaceutics*, 307(1), pp. 93–102.

Ozawa, M., Goto, H., Kusunoki, M. & Osawa, E., 2002. Continuously growing spiral carbon nanoparticles as the intermediates in the formation of fullerenes and nanoonions. *The Journal of Physical Chemistry B*, 106(29), pp. 7135–7138.

Paci, J. T. et al., 2013. Understanding the surfaces of nanodiamonds. *The Journal of Physical Chemistry C*, 117(33), pp. 17256–17267.

Pentecost, A. et al., 2010. Deaggregation of nanodiamond powders using salt and sugar assisted milling. *ACS Applied Materials and Interfaces*, 2(11), pp. 3289–3294.

Perevedentseva, E. et al., 2007. The interaction of the protein lysozyme with bacteria E. coli observed using nanodiamond labelling. *Nanotechnology*, 18(31), p. 315102.

Perevedentseva, E., Cai, P., Chiu, Y. & Cheng, C., 2011. Characterizing protein activities on the lysozyme and nanodiamond complex prepared for bio applications. *Langmuir*, 27(3), pp. 1085–1091.

Perevedentseva, E. et al., 2013. Nanodiamond internalization in cells and the cell uptake mechanism. *Journal of Nanoparticle Research*, 15, p. 1834.

Petit, T. et al., 2013. Surface transfer doping can mediate both colloidal stability and self-assembly of nanodiamonds. *Nanoscale*, 5(19), pp. 8958–8962.

Petros, R. & DeSimone, J., 2010. Strategies in the design of nanoparticles for therapeutic applications. *Nature Reviews. Drug Discovery*, 9(8), pp. 615–627.

Petrova, I. et al., 2007. Detonation nanodiamonds simultaneously purified and modified by gas treatment. *Diamond and Related Materials*, 16(12), pp. 2098–2103.

Pichota, V. et al., 2008. An efficient purification method for detonation nanodiamonds. *Diamond and Related Materials*, 17(1), pp. 13–22.

Pope, I. et al., 2014. Coherent anti-stokes Raman scattering microscopy of single nanodiamonds. *Nature Nanotechnology*, 9(11), pp. 940–946.

Puzyr, A. et al., 2007. Nanodiamonds with novel properties: A biological study. *Diamond and Related Materials*, 16(12), pp. 2124–2128.

Rabeau, J. R. et al., 2006. Implantation of labelled single nitrogen vacancy centers in diamond using N 15. *Applied Physics Letters*, 88, p. 023113.

Rahman, M. et al., 2013. Nanoparticle and protein corona. *Protein–Nanoparticle Interactions, Springer Series in Biophysics*, 15, pp. 21–44.

Raty, J. et al., 2003. Quantum confinement and fullerenelike surface reconstructions in nanodiamonds. *Physical Review Letters*, 90(3), p. 037401.

Razani, B., Woodman, S. & Lisanti, M., 2002. Caveolae: From cell biology to animal physiology. *Pharmacology Reviews*, 54(3), pp. 431–467.

Rodan, G. & Martin, T., 1982. Role of osteoblasts in hormonal control of bone resorption—A hypothesis. *Calcified Tissue International*, 34(3), p. 311.

Rothen-Rutishauser, B., Schurch, S. & Gehr, P., 2006. Chapter 7: Interaction of particles with membranes. In: K. Donaldson & P. Borm, eds. *Particle Toxicology.* CRC Press, Boca Raton, FL, p. 140.

Saini, G. et al., 2010. Core-shell diamond as a support for solid-phase extraction and high-performance liquid chromatography. *Analytical Chemistry*, 82(11), pp. 4448–4456.

Savjani, K. T., Gajjar, A. K. & Savjani, J. K., 2012. Drug solubility: Importance and enhancement techniques. *ISRN Pharmaceutics*, 2012, p. 195727.

Schmid, S., 1997. Clathrin-coated vesicle formation and protein sorting: An integrated process. *Annual Review of Biochemistry*, 66, pp. 511–548.

Schrand, A. M. et al., 2007. Differential biocompatibility of carbon nanotubes and nanodiamonds. *Diamond and Related Materials*, 16(12), pp. 2118–2123.

Schrand, A. M., Hens, S. A. C. & Shenderova, O. A., 2009. Nanodiamond particles: Properties and perspectives for bioapplications. *Critical Reviews in Solid State and Materials Sciences*, 34(1–2), pp. 18–74.

Shenderova, O. & McGuire, G., 2006. Types of nanocrystalline diamond. In: O. Shenderova & D. Gruen, eds. *Ultrananocrystalline Diamond: Synthesis Properties and Applications.* New York: William Andrew, pp. 79–114.

Shenderova, O., Zhirnov, V. & Brenner, D., 2002. Carbon nanostructures. *Critical Reviews in Solid State and Materials Science*, 27, pp. 227–356.

Shenderova, O. et al., 2006. Modification of detonation nanodiamonds by heat treatment in air. *Diamond and Related Materials*, 15(11–12), pp. 1799–1803.

Shenderova, O. et al., 2014. Carbon-dot-decorated nanodiamonds. *Particle & Particle Systems Characterization*, 31(5), pp. 580–590.

Shimkunas, R. et al., 2009. Nanodiamond–insulin complexes as pH-dependent protein delivery vehicles. *Biomaterials*, 30(29), pp. 5720–5728.

Shiryaev, A. A., Iakoubovskii, K., Grambole, D. & Dubrovinskaia, N., 2006. Spectroscopic study of defects and inclusions in bulk poly- and nanocrystalline diamond aggregates. *Journal of Physics: Condensed Matter*, 18(40), p. L493.

Smestad, G. P., 2002. Chapter 2: Absorbing solar energy. In: *Optoelectronics of Solar Cells.* SPIE Press, p. 32.

Smith, B. R., Neibert, M., Plakhotnik, T. & Zvyagin, A. V., 2007. Transfection and imaging of diamond nanocrystals as scattering optical labels. *Journal of Luminescence*, 127(1), pp. 260–263.

Solarska-Ściuk, K. et al., 2014. Intracellular transport of nanodiamond particles in human endothelial and epithelial cells. *Chemico-Biological Interactions*, 219, pp. 90–100.

Spinicelli, P. et al., 2011. Engineered arrays of nitrogen-vacancy color centers in diamond based on implantation of CN^- molecules through nanoapertures. *New Journal of Physics*, 13, p. 025014.

Spitsyn, B. et al., 2005. Purification and functionalization of nanodiamond. *Synthesis, Properties and Applications of Ultrananocrystalline Diamond NATO Science Series*, 192, pp. 241–252.

Sun, J. et al., 2005. Preparation of nanodiamonds by laser irradiation of graphite. *Chinese Optic Letters*, 3(5), pp. 287–288.

Szabo, T. et al., 2006. Evolution of surface functional groups in a series of progressively oxidized graphite oxides. *Chemistry of Materials*, 18(11), pp. 2740–2749.

Taverna, S. et al., 2007. How chromatin-binding modules interpret histone modifications: Lessons from professional pocket pickers. *Natural Structural Molecular Biology*, 14(11), pp. 1025–1040.

Tenzer, S. et al., 2013. Rapid formation of plasma protein corona critically affects nanoparticle pathophysiology. *Nature Nanotechnology*, 8, pp. 772–781.

Trono, J. et al., 2011. Size, concentration and incubation time dependence of gold nanoparticle uptake into pancreas cancer cells and its future application to X-ray drug delivery system. *Journal of Radiation Research*, 52(1), pp. 103–109.

Vaijayanthimala, V., Tzeng, Y., Chang, H. & Li, C., 2009. The biocompatibility of fluorescent nanodiamonds and their mechanism of cellular uptake. *Nanotechnology*, 20(42), p. 425103.

Varkouhi, A., Scholte, M., Storm, G. & Haisma, H., 2011. Endosomal escape pathways for delivery of biologicals. *Journal of Controlled Release*, 151(3), pp. 220–228.

Vereschagin, A., Sakovich, G., Komarov, V. & Petrov, E., 1994. Properties of ultrafine diamond clusters from detonation synthesis. *Diamond and Related Materials*, 3(1–2), pp. 160–162.

Vial, S. et al., 2008. Peptide-grafted nanodiamonds: Preparation, cytotoxicity and uptake in cells. *Chembiochem*, 9(13), pp. 2113–2119.

Wang, Z., Tiruppathi, C., Minshall, R. & Malik, A., 2009. Size and dynamics of caveolae studied using nanoparticles in living endothelial cells. *ACS Nano*, 3(12), pp. 4110–4116.

Wang, H., Niu, C., Yang, Q. & Badea, I., 2011. Study on protein conformation and adsorption behaviors in nanodiamond particle–protein complexes. *Nanotechnology*, 22(14), p. 145703.

Wee, T.-L. et al., 2009. Preparation and characterization of green fluorescent nanodiamonds for biological applications. *Diamond and Related Materials*, 18(2–3), pp. 567–573.

Wei, Q. et al., 2012. Biodistribution of co-exposure to multi-walled carbon nanotubes and nanodiamonds in mice. *Nanoscale Research Letters*, 7(1), p. 473.

Wenga, M.-F., Chiang, S. Y., Wang, N. S. & Niu, H., 2009. Fluorescent nanodiamonds for specifically targeted bioimaging: Application to the interaction of transferrin with transferrin receptor. *Diamond and Related Materials*, 18(2–3), pp. 587–591.

Wong, J. et al., 2008. Suspensions for intravenous (IV) injection: A review of development, preclinical and clinical aspects. *Advanced Drug Delivery Reviews*, 60(8), pp. 939–954.

Wu, C. C., Han, C. C. & Chang, H. C., 2010. Applications of surface-functionalized diamond nanoparticles for mass-spectrometry-based proteomics. *Journal of the Chinese Chemical Society*, 57, pp. 583–594.

Wu, T. et al., 2013. Tracking the engraftment and regenerative capabilities of transplanted lung stem cells using fluorescent nanodiamonds. *Nature Nanotechnology*, 8(9), pp. 682–689.

Xiao, J. et al., 2013. Nanodiamonds-mediated doxorubicin nuclear delivery to inhibit lung metastasis of breast cancer. *Biomaterials*, 34(37), pp. 9648–9656.

Xing, Y. et al., 2011. DNA damage in embryonic stem cells caused by nanodiamonds. *ACS Nano*, 5(3), pp. 2376–2384.

Xu, K. & Xue, Q., 2004. A new method for deaggregation of nanodiamonds from explosive detonation: Graphitization—Oxidation method. *Journal of Physics of Solid State*, 46(4), pp. 649–650.

Yan, C.-S. & Vohra, Y. K., 1999. Multiple twinning and nitrogen defect center in chemical vapor deposited homoepitaxial diamond. *Diamond and Related Materials*, 8(11), pp. 2022–2031.

Yeap, W., Tan, Y. & Loh, K., 2008. Using detonation nanodiamond for the specific capture of glycoproteins. *Analytical Chemistry*, 80(12), pp. 4659–4665.

Youngson, R. M., 2006. *Collins-Dictionary of Human Biology.* Glasgow: HarperCollins.

Yuan, Y. et al., 2009. Biodistribution and fate of nanodiamonds in vivo. *Diamond and Related Materials*, 18(1), pp. 95–100.

Yuan, Y. et al., 2010. Pulmonary toxicity and translocation of nanodiamonds in mice. *Diamond and Related Materials*, 19(4), pp. 291–299.

Zhang, B. et al., 2009. Receptor-mediated cellular uptake of folate-conjugated fluorescent nanodiamonds: A combined ensemble and single-particle study. *Small*, 5(23), pp. 2716–2721.

Zhang, X.-Q. et al., 2009. Polymer-functionalized nanodiamond platforms as vehicles for gene delivery. *ACS Nano*, 3(9), pp. 2609–2616.

Zhang, P. et al., 2011. Cationic polymer brush grafted-nanodiamond via atom transfer radical polymerization for enhanced gene delivery and bioimaging. *Journal of Material Chemistry*, 21, pp. 7755–7764.

Zhang, Q. et al., 2011. Fluorescent PLLA-nanodiamond composites for bone tissue engineering. *Biomaterials*, 32(1), pp. 87–94.

Zhang, X. et al., 2011. Multimodal nanodiamond drug delivery carriers for selective targeting, imaging, and enhanced chemotherapeutic efficacy. *Advanced Materials*, 23(41), pp. 4770–4775.

Zhang, X. et al., 2012a. PEGylation and polyPEGylation of nanodiamond. *Polymer*, 53(15), p. 3178.

Zhang, X. et al., 2012b. PolyPEGylated nanodiamond for intracellular delivery of a chemotherapeutic drug. *Polymer Chemistry*, 3(10), pp. 2716–2719.

Zhao, L. et al., 2014. Polyglycerol-functionalized nanodiamond as a platform for gene delivery: Derivatization, characterization, and hybridization with DNA. *Beilstein Journal of Organic Chemistry*, 10, pp. 707–713.

Zhou, G., Lelkes, P. I., Gogotsi, Y. & Mochalin, V., 2012. *Functionalized Nanodiamond Reinforced Biopolymers.* United States, Patent No. US 13/498,436.

Zhu, Y. et al., 2012. The biocompatibility of nanodiamonds and their applications in drug delivery systems. *Theranostics*, 2(3), pp. 302–312.

Index

Note: Page numbers followed by 'f' and 't' denote figures and tables respectively.

B

D

E

F

G

H

I

N

O

P

Q

R

S

T

U

V

W

X

Y

Z